# Spectrochemical Analysis

JAMES D. INGLE, JR.
*Oregon State University*

STANLEY R. CROUCH
*Michigan State University*

Prentice Hall, Upper Saddle River, New Jersey 07458

*Library of Congress Cataloging-in-Publication Data*

INGLE, JAMES D., (date)
    Spectrochemical analysis.

    Bibliography: p.
    Includes index.
    1. Spectrum analysis.  I. Crouch, Stanley R.
II. Title.
QD95.I48 1988     543′.0858     87-29168
 **ISBN**  0-13-826876-2

Editorial/production supervision
    and interior design: KATHLEEN M. LAFFERTY
Cover design: EDSAL ENTERPRISES
Manufacturing buyer: PAULA MASSENARO

ISBN 0-13-826876-2

90000

9 780138 268763

©1988 by Prentice-Hall, Inc.
Upper Saddle River, New Jersey 07458

Printed in the United States of America

10

ISBN  0-13-826876-2

Prentice-Hall International (UK) Limited, *London*
Prentice-Hall of Australia Pty. Limited, *Sydney*
Prentice-Hall of Canada, Inc., *Toronto*
Prentice-Hall Hispanoamericana, S. A., *Mexico*
Prentice-Hall of India Private Limited, *New Delhi*
Prentice-Hall of Japan, Inc., *Tokyo*
Prentice-Hall Asia Pte. Ltd., *Singapore*
Editora Prentice-Hall do Brasil, Ltda., *Rio de Janeiro*

*Dedicated to our parents*

*Miriam and Jimmie Ingle*
*Mildred and Ned Crouch*

# Contents

# 8 FLAME AND PLASMA ATOMIC EMISSION SPECTROMETRY    225

# 9 ARC AND SPARK EMISSION SPECTROMETRY    257

# 10 ATOMIC ABSORPTION SPECTROPHOTOMETRY    273

# 11 ATOMIC FLUORESCENCE SPECTROMETRY    307

# 12 INTRODUCTION TO MOLECULAR SPECTROSCOPY    325

# 13 ULTRAVIOLET AND VISIBLE MOLECULAR ABSORPTION SPECTROPHOTOMETRY    352

# 14 INFRARED SPECTROMETRY    404

# 15 MOLECULAR LUMINESCENCE SPECTROMETRY    438

# 16 MOLECULAR SCATTERING METHODS   494

# 17 SPECTROCHEMICAL TECHNIQUES ON THE HORIZON   525

# A STATISTICAL CONCEPTS   543

# B PROPERTIES OF OPTICAL MATERIALS   550

# Preface

Spectrochemical methods of analysis are among the most widely used of all analytical methods. Trace metal determinations are most frequently carried out by atomic emission or atomic absorption spectroscopy. Ultraviolet and visible molecular absorption and fluorescence methods are extensively employed for the determination of metallic and nonmetallic elements, for inorganic ions, and for organic species. Spectrochemical instrumentation is used not only for direct analysis, but also as a tool to monitor chemical reactions or the effluents from chromatographs and automated flow analyzers. Because of the widespread use of optical spectrochemical methods, it is increasingly important that chemists and other scientists have a firm understanding of their principles. Thus one goal of this book is to provide a thorough treatment of the fundamental principles, terminology, methodology, and instrumentation common to analytical optical methods. A second goal is to discuss specific spectrochemical analysis techniques in terms of their implementation and characteristics. Where appropriate, representative examples of practical applications of the techniques are given.

*Spectrochemical Analysis* is intended as a textbook for graduate and advanced undergraduate students, but should also prove useful as a reference source for practitioners who wish to broaden or update their knowledge of spectrochemical methods. The level is intended to bridge the gap between the survey coverage of spectrochemical methods in undergraduate general instrumental analysis textbooks and the comprehensive coverage in monographs dealing with only one or two spectrometric techniques. Because basic principles and terminology are initially reviewed, the book can be used by students and others who have not had the benefit of a strong survey course in instrumental analysis.

The field of spectrochemical analysis is very broad. The radiation/matter interactions that are the basis of many of the methods are often learned in the context of quantum and statistical mechanics. The instrumentation utilized combines optics, mechanics, electronics, and signal processing principles. The data obtained are properly assessed through the application of statistics, information theory, and increasingly, computer and systems science. It is important to realize, however, that there is a great deal of chemistry in spectrochemical methods. Chemical reactions and interactions are often employed to improve the detection limits and to increase the selectivity of methods. Similarly, chemical and physical interactions are often the cause of those interferences that are broadly called "matrix effects." Thus this book also emphasizes chemistry and the ways in which an understanding of chemistry can aid in ob-

**xiii**

taining high-quality spectrochemical data. Indeed, the title of the book was chosen to emphasize the important interplay of spectroscopy and chemistry in this field.

The breadth of the field of spectrochemical analysis makes it necessary to restrict the coverage to "optical" spectrochemical methods to allow in-depth coverage in a reasonable amount of space. The optical spectrochemical techniques covered involve radiation/matter interactions in the UV-visible-IR regions of the electromagnetic spectrum, which can be used primarily for determination of the concentration of species in mixtures. Thus techniques such as NMR, microwave spectrometry, ESR, refractometry, and Mossbauer spectrometry, which are used primarily for identification and determination of structures and properties of pure compounds, as well as qualitative and quantitative uses of mass spectrometry, are not covered. It is expected that at the graduate level these topics would be covered in detail in a spectral interpretation course or a molecular spectroscopy course taught by organic and physical chemists, respectively. Radiochemical and X-ray techniques are not covered in this textbook because they do not conveniently fit into the common framework of optical methods due to significant differences in methodology, instrumentation, and origins of analytical signals.

Throughout the book we have attempted to stress principles and concepts by presenting them first before the details of a given topic are introduced. This approach has the advantage that it allows a topic to unfold and expand based upon a strong fundamental framework. It also allows and, indeed, encourages considerable flexibility in the use of the book. Instructors of broad, survey-type courses can pursue a given topic to the depth desired without getting "bogged down" in too much detail or without the feeling that fundamental principles and concepts are being skipped if the coverage of a topic is shortened. Sufficient depth is given in each topic, however, for the book to be useful in more specialized courses that might stress one or more of the major topics presented, such as atomic spectroscopy or molecular spectroscopy. A second important aspect of our approach is an emphasis on the quantitative, mathematical relationships that relate the signals measured and the signal-to-noise ratios obtained in spectrochemical methods to physical, chemical, and instrumental factors. This approach gives the user a quantitative feel for the way in which an instrumental variable, for example, will affect the quality of the data obtained and a knowledge of those variables that are the most effective in improving or optimizing the measurement.

The book is divided into four major sections. The first six chapters cover the fundamental aspects that are common to all optical spectrochemical methods. These include the nature of spectrochemical information and measurements, the methodology of spectrochemical analysis, the instrumentation used in optical spectrometers, and the factors that determine signal-to-noise ratio ($S/N$) in spectrometry. Some instructors may choose to delay the introduction of certain topics in Chapters 3 to 6 until the general concepts and equations are modified to apply to the specific spectrometric techniques discussed in later chapters. Chapter 7 introduces the general principles and fundamental equations for atomic spectrometric techniques, while Chapters 8 to 11 cover specific atomic spectroscopic methods, including flame emission, arc, spark, and plasma emission, atomic absorption, and atomic fluorescence spectrometry. Chapter 12 serves to introduce the basis of molecular spectroscopy. Specific molecular spectroscopic methods, including UV-visible absorption, IR absorption, molecular fluorescence, phosphorescence, and chemiluminescence and molecular scattering techniques, are presented in Chapters 13 to 16. Chapter 17 deals with those spectrochemical methods, both atomic and molecular, that are becoming or should become standard "tools of the trade," such as photoacoustic, thermal lensing, and laser ionization techniques. Six appendices contain additional information about statistical concepts, optical materials, optical filters, photomultiplier tubes, sample preparation methods, and the relationships between the different quantities that are used to express the strength of transitions.

The atomic and molecular spectroscopy sections were written to be used independently of each other. Thus some instructors may prefer to introduce atomic spectroscopy prior to molecular spectroscopy because atomic transitions are simple transitions between purely electronic states. On the other hand, a good case can be made for the opposite order because many students are more familiar with molecular absorption spectroscopy and the quantitative treatment of this topic is somewhat easier than the treatment of atomic absorption, which requires a discussion of line profiles and atomization. We feel that the flexibility gained by making these sections independent more than offsets the small amount of redundancy required.

Although the book has been written for a full-semester course in optical spectrochemical methods, it can be readily abridged for a one-quarter course. This can be accomplished by shortening the coverage of topics that should be review material (parts of Chapters 1 to 4, 7, and 12) and by skipping some sections in later chapters on specific techniques or even complete chapters (e.g., Chapters 9, 16, or 17)

We have attempted to make the figures as informative as possible by writing explanatory captions.

These extended captions provide a useful second level of explanation that expands on or complements the textual information. Each chapter ends with a number of illustrative questions and problems and a selected list of references.

A course in modern spectrochemical analysis should include a parallel laboratory. Because of the specialized equipment available at most universities and because many universities do not have graduate laboratories, we have not attempted to include laboratory experiments in this text.

## Acknowledgments

We would like to acknowledge the assistance and support of many people who were involved in the publication of this work. The production staff at Prentice Hall, most of whom we never knew, helped us keep on schedule. Especially we wish to thank Elizabeth Perry, Nancy Forsyth, Curtis Yehnert, and Dan Joraanstad, who each served as the chemistry editor during various stages of the book. Extra special thanks go to the production editor Kathleen Lafferty for her professional and understanding guidance during the final stages of publication. Before we became totally addicted to our word processors, Patty Ramus, Sherree Bittner, Brandy Schuyler, Jenny Harber, Valerie Borst, Dawn Penrose, and Debbie Wuethrich helped in typing earlier versions of chapters. Many students including Max Hineman, Pat Wiegand, Kim Ratanathanawongs, Peter Wentzell, Tom Doherty, Helen Archontaki, Mark Phillips, Jeff Fahey, Cecilia Yappert, and Paul Kraus assisted in developing materials, reading and commenting on the text, and checking the answers to problems. Lucy Ingle, Sara Ingle, and Jeff Louch contributed by checking the accuracy of equations during proofing of the galleys.

Our colleagues contributed in many direct and indirect ways. Many colleagues including Ed Piepmeier, Mike Schuyler, Chris Enke, Joe Nibler, Jim Holler, Jim O'Reilly, Gary Hieftje, Gary Horlick, and Ray Barnes served as responsive sounding boards as we developed ideas for the book. Others including James Winefordner, Alexander Scheeline, Eric Salin, Adrian Wade, Gil Haugen, Gary Christian, Timothy Nieman, Earl Wehry, and Scott Goode reviewed in detail all or sections of the manuscript and offered many useful comments that were incorporated (mostly) into the manuscript. We are especially grateful to our teacher, Howard Malmstadt, who provided us with the foundation for many of the concepts in the book.

Our families and graduate students exhibited thoughtful patience over extended periods of time when we seemed to be invisible. Many of our students also suffered through courses based on early versions of the manuscript and gleefully identified numerous errors and unintelligible discussions. Finally, we apologize to all those who helped that we did not acknowledge specifically. We know there were many and greatly appreciate your assistance.

*James D. Ingle, Jr.*
*Stanley R. Crouch*

# CHAPTER 1

# Spectrochemical Information

We live in an era in which the analysis of chemicals and materials is of extreme importance. Analytical methods are used to help monitor the status of human health, the ingredients in the foods we eat and the water we drink, and the quality of the environment. Analysis is also an important step in industrial processes and in the development of safe and valuable industrial products. Spectrochemical analysis is one of the major tools of analytical chemistry. The applications of spectrochemical methods range from the determination of extremely low levels of noxious materials in the environment to the monitoring of major components, such as sodium and potassium in biological fluids. These methods thus cover a broad scope of materials, a variety of matrices, and a wide range of concentrations.

"Spectrochemical" is a compound word that comes from *spectrum* and *chemical*. A **spectrum** is a display of the intensity of radiation emitted, absorbed, or scattered by a sample versus a quantity related to photon energy, such as wavelength or frequency. The term *spectrochemical* implies that a spectrum or some aspect of a spectrum is used to determine chemical species and to investigate the interaction of chemical species with electromagnetic radiation. Spectrochemical methods can involve a direct optical measurement of the photons emitted or transmitted or an indirect measurement of

a quantity related to the result of photon absorption. As examples of the latter, the number of ions or a quantity related to the kinetic energy produced by absorption can be monitored and plotted as a function of wavelength to obtain a spectrum.

This chapter describes the information that can be obtained from spectrochemical studies and the nature of spectrochemical analysis. The use of spectral data to express analytical information is examined, and several criteria for evaluating spectrochemical techniques are introduced. The general concepts and basic definitions introduced here are used throughout this book as we explore the field of spectrochemical analysis.

## 1-1 RADIATION/MATTER INTERACTIONS

**Spectroscopy** is the science that deals with the interactions of electromagnetic radiation with matter. Several types of interactions are possible. Many of these involve transitions between specific energy states of chemical species and are observed by monitoring the absorption or emission of electromagnetic radiation. In these types of interactions it is useful to consider electromagnetic radiation as being composed of discrete packets of energy which we call **photons**. Electromagnetic radiation also has a wave character, and we can relate the energy of a photon to its wavelength and frequency by

$$E = h\nu = \frac{hc}{\lambda} \qquad (1\text{-}1)$$

where $E$ is the energy in joules (J), $\nu$ is the frequency (Hz or $s^{-1}$), $h$ is Planck's constant ($6.63 \times 10^{-34}$ J s), $c$ is the speed of light ($3.00 \times 10^8$ m $s^{-1}$ in a vacuum), and $\lambda$ is the wavelength (m). **Spectrometry** is a more restrictive term than *spectroscopy* and denotes the quantitative measurement of the intensity of electromagnetic radiation at one or more wavelengths with a photoelectric detector.

There are several types of radiation/matter interactions, such as reflection, refraction, diffraction, and some types of scattering that do not involve transitions between energy states, but rather cause changes in the optical properties of the radiation (e.g., direction and polarization). These interactions are often related to the bulk properties of the sample rather than to specific chemical species. Several analytical techniques are based on these bulk interactions.

Spectrochemical analysis, in general, deals with electromagnetic radiation of an enormous range of frequencies, from the audio frequencies ($<20$ kHz) to gamma rays ($>10^{19}$ Hz). **Optical spectrochemical analysis**, the primary topic of this book, covers a more restrictive range, the near ultraviolet (UV), the visible, and the infrared (IR) regions, where instrumental requirements are similar, and the materials used for dispersing, focusing, and directing the radiation are conventional optical materials (glass, quartz, or alkali halide crystals). Optical spectrochemical techniques are often divided into atomic spectroscopic techniques and mo-

lecular spectroscopic techniques. **Atomic spectroscopy** deals with spectrochemical phenomena involving free atomic species that are usually in the vapor state, whereas **molecular spectroscopy** deals with optical measurements of molecular species in the vapor state, in solution, or in the solid state.

Photon energies in the optical regions are typically expressed in units of joules, ergs (1 erg $= 10^{-7}$ J), or electron volts (1 eV $= 1.6 \times 10^{-19}$ J). The wavelength of radiation in this region is usually expressed in nanometers (1 nm $= 10^{-9}$ m), angstroms (1 Å $= 10^{-10}$ m), or micrometers (1 $\mu$m $= 10^{-6}$ m). One electron volt of photon energy corresponds to radiation with a wavelength of 1240 nm.

Because the wavelength is inversely proportional to energy, the wavenumber is often used, particularly in the IR region. The wavenumber, $\bar{\nu}$, is the number of cycles per unit length (usually cm) and thus the reciprocal of the wavelength, $\bar{\nu} = 1/\lambda$. The wavenumber is usually expressed in $cm^{-1}$. It is directly proportional to the photon energy,

$$\bar{\nu} = \frac{1}{\lambda} = \frac{\nu}{c} = \frac{E}{hc} \qquad (1\text{-}2)$$

The energy or wavelength of the photon determines the type of transition or interaction that occurs, as shown in Table 1-1.

## 1-2 NATURE OF SPECTROCHEMICAL ANALYSIS

Before getting into the details of spectrochemical methods and procedures, it is useful to consider some general aspects which describe the kinds of determinations that can be performed and the types of samples that can be

**TABLE 1-1**
Regions of electromagnetic spectrum

| Designation | Wavelength range, $\lambda$ | Frequency range, $\nu$ (Hz) | Wavenumber or energy range | Transition |
|---|---|---|---|---|
| $\gamma$-Ray | $<0.05$ Å | $>6 \times 10^{19}$ | $>2.5 \times 10^5$ eV | Nuclear |
| X-Ray | 0.05–100 Å | $3.0 \times 10^{16}$–$6.0 \times 10^{19}$ | $124$–$2.5 \times 10^5$ eV | $K$- and $L$- shell electron |
| Far (vacuum) UV | 10–180 nm | $1.7 \times 10^{15}$–$3.0 \times 10^{16}$ | 7–124 eV | Middle shell electrons |
| Near UV | 180–350 nm | $8.6 \times 10^{14}$–$1.7 \times 10^{15}$ | 3.6–7 eV | Valence electrons |
| Visible | 350–770 nm | $3.9 \times 10^{14}$–$8.6 \times 10^{14}$ | 1.6–3.6 eV | Valence electrons |
| Near IR | 770–2500 nm | $1.2 \times 10^{14}$–$3.9 \times 10^{14}$ | 12,900–4000 $cm^{-1}$ | Molecular vibrations |
| Middle or fundamental IR | 2.5–50 $\mu$m | $6.0 \times 10^{12}$–$1.2 \times 10^{14}$ | 4000–200 $cm^{-1}$ | Molecular vibrations |
| Far IR | 50–1000 $\mu$m | $3.0 \times 10^{11}$–$6.0 \times 10^{12}$ | 200–10 $cm^{-1}$ | Molecular rotations |
| Microwave | 1–300 mm | $1.0 \times 10^9$–$3.0 \times 10^{11}$ | | Molecular rotations |
| Radio waves | $>300$ mm | $<1 \times 10^9$ | | Electron and nuclear spin |

analyzed by spectrochemical methods. In the process, we shall define several important terms and concepts which are used throughout this book.

## Types of Analyses

In spectrochemical methods, spectroscopy is used for the identification of chemical species (*qualitative analyses*) and for the determination of the amount of a particular species (*quantitative analyses*). At first thought, the different goals of qualitative and quantitative analyses would seem to dictate quite different approaches. However, to ascertain the presence of a species with some degree of certainty, we must have at least a rudimentary knowledge of the quantity required. For example, to identify the presence of mercury in river water, we must determine that the mercury concentration exceeds a certain threshold, decision level. Thus qualitative analysis can be considered as merely a low-resolution type of quantitative analysis, often involving a simple binary, yes or no determination of the sought-for species. Although quantitative methods are stressed in this book, it is useful to keep in mind that similar considerations of uncertainty in the results, limits of detectability, upper concentration ranges, and error in the procedures also apply to qualitative methods.

The constituents determined in a spectrochemical analysis can cover a broad concentration range. In some cases spectrochemical methods are used to determine **major constituents**. These are considered here to be species present in the relative weight range 1 to 100%. **Minor constituents** are species present in the range 0.01 to 1%, while **trace constituents** are those present in amounts lower than 0.01% (100 $\mu g\ g^{-1}$).

The size of the sample is also used to classify the type of analysis performed. A **macro analysis** is one carried out on a sample weighing more than 0.1 g. A **semimicro analysis** (sometimes called a *meso analysis*) utilizes a sample size in the range 0.01 to 0.1 g, while a **micro analysis** employs a sample size in the range $10^{-4}$ to $10^{-2}$ g. If the sample size is lower than $10^{-4}$ g, the term **ultramicro** is sometimes employed. The term **ultra-trace analysis** is considered to be the determination of a trace constituent in an ultramicro sample.

## Samples

The nomenclature for dealing with the samples used in spectrochemical analyses is often confusing and contradictory. Here we shall use the term **initial sample** to mean a portion or subset of the bulk material or population about which analytical information is desired.

For example, a liter of water (the initial sample) is obtained from a lake (the bulk material) in order to determine the mercury content, or a few grams of soil (initial sample) are acquired from a field (bulk material) in order to determine the concentration of a pesticide. The **analytical sample** indicates that portion of the initial sample which is presented to the instrument for spectrochemical analyses. In some cases the initial sample is treated prior to the analyses. Such treatment may include grinding, heating, dissolution, preconcentration, dilution, separation, addition of a buffer, adjustment of the pH, or many other steps. In all cases the analytical sample must be representative of the concentration of the sought-for species in the bulk material. The sampling and sample preparation operations, which are the entire set of procedures done to acquire and to prepare the analytical sample from the bulk material, are thus crucial steps in an entire analysis. Errors made in sampling or sample preparation are carried through the entire process and lead directly to errors in the final result (see Section A-2 in Appendix A).

The species to be determined in the analytical sample is designated the **analyte**. There may be several analytes in a given sample, and their concentrations can range from the trace level to the major constituent level. The term **matrix** refers to the collection of all the constituents in the sample. The **analytical matrix** refers specifically to the matrix of the analytical sample which may differ from that of the initial sample due to the substances added or removed in the various sample treatment stages. The matrix as defined here includes the analyte as well as all the other constituents, which are called **concomitants**. In trace analyses the analyte is present in such small amounts that it is convenient to speak of the analyte and the matrix separately. Thus for trace analysis we sometimes think of the matrix as being composed of the concomitants. In major constituent determinations, however, the analyte is a major part of the matrix and this division is no longer helpful. For example, in determining iron in steel, the analyte (iron) is present in such large amounts that it determines to a large extent the bulk properties of the sample. Here the matrix is no longer composed of just concomitants.

In some analyses, information about specific chemical species is of interest. **Chemical speciation** is concerned with determining the concentration of specific chemical forms of the analyte (e.g., the amount of metal in a particular oxidation state or the amount of a drug bound to protein). The nature of the sample matrix and the effect of the matrix on a determination depend upon the chemical interactions among matrix components and between matrix components and the analyte. Thus the chemical form(s) of the analyte and of the matrix components is critical.

## Spectrochemical Phenomena

To obtain spectroscopic information about chemical samples, the species to be determined is usually stimulated in some way by the application of energy in the form of heat, electrical energy, radiation, particles, or a chemical reaction. Several spectroscopic phenomena depend on transitions between energy states of particular chemical species. Prior to the application of external energy, the analyte is often in its lowest-energy or **ground state**. The applied energy then causes the analyte species to be momentarily in a higher-energy or **excited state**. Spectrochemical information is provided by measuring the electromagnetic radiation emitted by the species as it returns to the ground state from an excited level or by measuring the amount of electromagnetic radiation absorbed in the excitation process. Other spectroscopic techniques depend on the changes in the optical properties of electromagnetic radiation that occur when it interacts with the sample or analyte or on photon-induced changes in chemical form (e.g., ionization or photochemical reactions).

Many of the primary spectroscopic phenomena that can occur are shown in Figure 1-1, where external electromagnetic radiation impinges upon a collection of analyte species (atoms, molecules, or ions) in a sample. In the absence of external radiation (i.e., the sample is in a light-tight box or the external source is turned off), the analyte can be stimulated by collisional processes, or by electrical or chemical energy, and can emit photons when excited species return to lower-energy states. The term **emission** is a broad term that does not imply a particular excitation process but only the process of a photon being emitted. However, **emission spectroscopy** usually refers to spectral information that results from nonradiational activation processes. The emission that results from species excited by chemical reactions is usually termed **chemiluminescence**. Note here that chemiluminescence is a form of emission in which the excitation process is specifically said to be chemical. The simple, two-level energy diagram of Figure 1-2a shows the transitions that occur in emission techniques.

When the sample is exposed to an external source of electromagnetic radiation, many additional spectrochemical phenomena are possible (see Figure 1-1). **Reflection** and **scattering** are optical phenomena resulting in a change in direction of the incident photon. Reflection and elastic light scattering do not involve a change in the frequency of the incident photon; however, there are spectroscopic techniques such as Raman spectroscopy that involve inelastic scattering in which the change in the energy of the scattered photon is related to molecular energy levels.

**Absorption** of the incident photons by the analyte promotes the analyte to an excited state; this results in a reduction in the intensity of the electromagnetic radiation transmitted by the sample. The absorption pro-

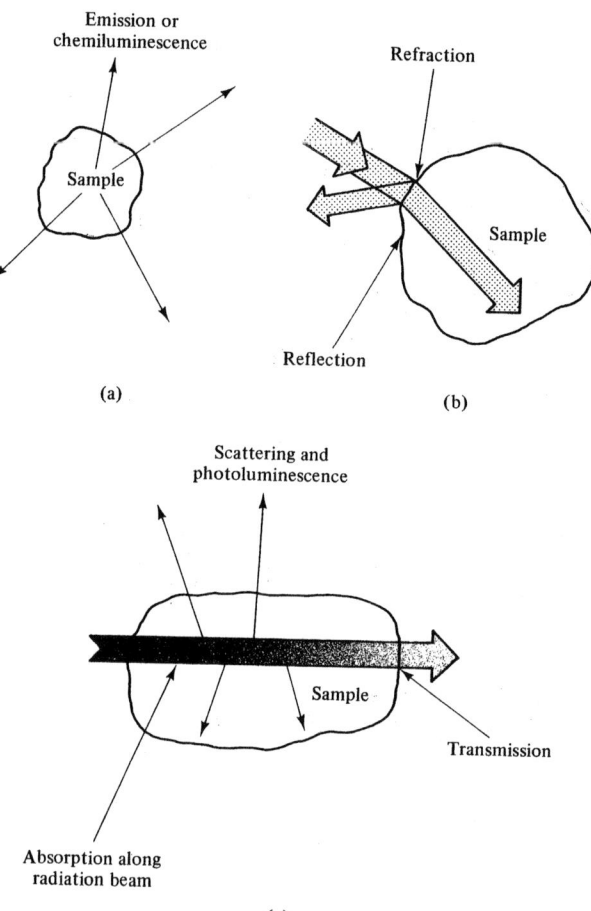

**FIGURE 1-1**   Some types of optical interactions. In the absence of an external radiation source (a), the analyte can be excited by collisional processes or chemical reactions and the resulting emission or chemiluminescence measured. In (b) and (c), a radiation beam from an external source is directed into the sample. As shown in (b), interactions such as reflection and refraction cause a change in direction of the beam at the sample interface. Within the sample (c), radiation can be scattered or absorbed by the analyte and the decreased intensity of the transmitted beam can be measured. Absorption of photons produces analyte species in excited states or ions in some cases. Deactivation of excited species can proceed by emission of photons (photoluminescence) or release of kinetic energy. The heat released in the latter case alters properties of the sample which can be measured.

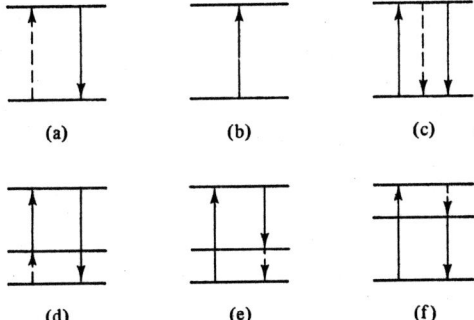

**FIGURE 1-2** Common types of optical transitions. In (a) the basis of emission or chemiluminescence techniques is illustrated in which the analyte is excited by a thermal, a chemical or some other nonradiative process (dashed line and upward-pointing arrow). The analyte can return to the lower-energy state by emission of a photon (solid line and arrow indicate a radiative transition). In (b) the analyte is excited by absorption of a photon and the resulting reduction in intensity of the photon signal is measured in absorption techniques. In (c) the emission of a photon following radiative excitation, termed photoluminescence, is measured. The dashed line and arrow indicate that the excited state can also lose its energy by a nonradiative pathway. As shown in (d) and (e), excitation or deactivation can involve a combination of radiative and nonradiative transitions in which the wavelength of the emitted photon can be less (Stokes transition) or greater (anti-Stokes transition) than that of the excitation photon. The latter is also true in (f), where the species undergoes a nonradiative deactivation to a lower excited level before photon emission occurs.

cess is illustrated by Figure 1-2b. The species excited by absorption of photons can lose the excess energy by radiational or nonradiational processes; the latter leads to an increase in the kinetic energy of the sample. Radiational deactivation processes result in emission of photons. The emission of photons from excited states produced by radiational activation (absorption) is called **photoluminescence** and is illustrated in Figure 1-2c. *Fluorescence* and *phosphorescence* are particular types of photoluminescence. In photoluminescence, the frequency of the emitted photon may be the same as the frequency of the incident photon, or it may be different, as illustrated in Figure 1-2d to f. The term **luminescence** refers to emission from cool bodies or to emission from hot bodies that is not due to thermal excitation. Thus in this text chemiluminescence of molecules in solution is considered together with molecular fluorescence and phosphorescence as being molecular luminescence.

## Analysis of Real Samples

The analyses of actual samples is complicated by the fact that the analyte is present as part of a sample matrix. Concomitant species in the matrix can undergo the same spectrochemical process (absorption, luminescence, scattering, etc.) as the analyte, or they can affect the ability of analyte species to undergo the desired process. In some cases concomitants can affect our ability to observe or measure the optical interaction of the analyte. All these effects due to the concomitants can give rise to interference effects which are often termed **matrix effects**. Concomitant species in the analytical matrix can come from the initial sample itself, from reagents and solvents used to treat the sample prior to the analysis, or from contamination during the sample acquisition, storage, and preparation steps. Concomitants can also interfere by processes not involving optical interactions, such as by chemical reaction with the analyte species. The matrix can change in other ways, too. For example, cigarettes in storage prior to the determination of nicotine can lose water by dehydration; this leads to nicotine values higher than in the original sample.

In many spectrochemical methods we desire to obtain selective information about specific chemical species. Selectivity in emission, absorption, and luminescence methods arises because the spectral signals from the analyte occur at certain frequencies (wavelengths). Thus the optical information concerning the analyte can often be distinguished from information from concomitants by using instrumentation that allows monitoring of specific wavelengths and/or excitation of the sample by photons of specific wavelengths. As is discussed in later chapters, selectivity can be enhanced by using another optical property, such as polarization, in conjunction with intensity and wavelength information or by employing selective chemical reactions involving the analyte.

In qualitative analysis the frequencies (wavelengths) of optical transitions are used to identify the presence of a chemical species. Quantitative determinations are based on the intensity of the electromagnetic radiation at a given frequency and its relationship to the analyte concentration. We will follow the accepted convention that samples are *analyzed*, but that concentrations or species are *determined*. Thus we can properly speak of the analysis of paint for lead or the determination of lead in paint, but it is incorrect to speak of the analysis of lead in paint.

In nearly all spectrochemical techniques a **blank** measurement is required. The **ideal blank** contains all the sample constituents except the analyte. In practice the blank is treated as identically to the sample as possible. For example, if the analytical sample is a solution, the blank is a solution composed of the same solvent as the sample plus any reagents used to prepare the sample. The instrumental response from the blank is then subtracted from that of the sample in order to compensate for the effects of concomitants. An ideal blank can eliminate some types of interference effects due to concomitants but cannot compensate for concomitant species that affect the production and measurement of the analyte response. It is difficult to prepare an ideal blank because the concomitants and their concentrations are not usually known. A more desirable approach is to arrange measurement conditions to minimize the response due to concomitants.

Actual analyses are further complicated because the sample is almost always confined during measurement by a container, except for a few in situ measurements. In molecular spectroscopy, the container is typically a glass, quartz, or salt cell. Optical interactions with the container walls can give rise to additional interference effects. In atomic spectroscopy the container is typically a flame, a plasma, or a heated chamber. The hot gases produced can emit or absorb radiation which can also be a potential source of interference. In some cases the sample matrix can alter the interference effects of the container. Since there are many possible interferences in a spectrochemical procedure, a major part of this book deals with their identification, their effects, and their minimization (i.e., ways to compensate for them).

## 1-3 EXPRESSIONS OF ANALYTICAL INFORMATION

Several different types of information are required in order to develop, apply, and optimize an analytical technique. This section considers how to express spectrochemical information in a convenient manner so that analyte concentration data can readily be extracted or the dependence of the results on chemical, physical, or instrumental variables can easily be summarized.

### Calibration Data

The desired result of a spectrochemical analysis is the concentration of the analyte. It is almost never obtained directly as the result of an absolute measurement of an optical signal, but is obtained indirectly through calibration, subtraction of blanks, comparison with standards, and other procedures.

The **total spectrochemical signal** is defined as the unmodified readout signal obtained from measurement of a sample or standard. The **blank** or **reference** signal is defined as the readout signal obtained from measurement of the blank. It includes the **background signal** due to optical signals from the sample container and the concomitants in the blank. The **analytical signal** is extracted from the total spectrochemical signal. Ideally, it is directly related to the analyte concentration. In emission spectrometry, for instance, the total emission signal contains emission components from the analyte, the concomitants, and other sources (e.g., the sample container). The analytical signal is obtained by subtracting the blank emission signal from the total emission signal. In some automated instruments the readout signal may be the analytical signal if the instrument carries out the appropriate modifications internally.

The analytical signal is related to the analyte concentration and other variables by the **calibration function**, $f$, defined by

$$S = f(c_a, \lambda, x_i) \tag{1-3}$$

where $S$ is the analytical signal, $c_a$ is the analyte concentration, $\lambda$ is the analysis wavelength, and $x_i$ is the magnitude of the $i$th experimental variable (e.g., temperature, time). Often the calibration function is considered only a function of $c_a$.

The dependence of the calibration function on a specific variable is often given a particular name. For example, a plot of the analytical signal versus analyte concentration, with all the other variables in the calibration function held constant, is called the **calibration curve**, the **analytical curve**, or the **working curve**. Typically, the calibration curve is obtained from measuring the analytical signals for a series of **standards** (analyte solutions of known concentrations) of different analyte concentrations. The calibration curve in conjunction with the measurement of the analytical signal from a sample is then used to determine the analyte concentration in the sample.

The **analytical function**, $g$, is the inverse of the calibration function:

$$c_a = g(S) \tag{1-4}$$

For example, if $S = 10^{-4}c_a$, $c_a = 10^4S$. The analytical function provides a direct method for calculating the analyte concentration from the measured analytical signal for a given set of experimental conditions. In modern computerized instruments, the calibration function is determined by the instrument from measurments on standards. The analytical function is then calculated and used to provide the analyte concentration directly from the measured analytical signal. Such internal calculations can eliminate operator errors and the tedium associated with using calibration data.

## Atomic and Molecular Spectra

A **spectrum** is a plot of the analytical signal versus wavelength, frequency, or wavenumber with all other variables held constant. The peaks (lines or bands) are characterized by their shape, height (intensity), width, and position (wavelength). Usually, the width is expressed as the **half-width** ($\Delta\lambda$), which is the width in wavelength units at half the net peak height. The half-width is also called the **full width at half maximum** (FWHM). A spectrum is an essential summary of spectral information of any spectrometric technique because it indicates the wavelength to use for quantitative analysis in order to obtain the maximum analytical signal. The analyte spectrum along with spectra of concomitants allows the wavelength of analysis to be chosen both to maximize the analytical signal and to discriminate against background signals.

The spectrum of the analyte is the basis of qualitative analysis since the wavelengths and relative intensities of peaks in a spectrum are characteristic of the analyte. Qualitative analysis through spectral characteristics is favorable for species which provide "rich" spectra (spectra with numerous resolvable peaks), so that these can be "fingerprints" for the analyte(s).

The spectra of atoms in the ultraviolet, visible, and near-infrared regions arise from purely electronic transitions of outer-shell (valence) electrons, as discussed in detail in Section 7-3. Because the quantized energy levels are relatively far apart, atomic spectra are narrow **line spectra**. A transition to or from the ground electronic level is a **resonance transition**, and the resulting spectral line is a **resonance line**. Atomic spectra are often quite simple because many of the possible transitions are **forbidden transitions**. The **allowed transitions** can be predicted from quantum mechanics, and **selection rules** tell us the requirements for an allowed transition. Atomic spectral lines have a finite width (typically, much less than 1 Å) even though the transitions are between two distinct energy levels because of line broadening due to lifetime effects, the Doppler effect, and collisions. Line broadening is discussed in detail in Section 7-4.

Because molecules have quantized vibrational levels and rotational levels in addition to electronic levels, as shown in Figure 1-3, their spectra are necessarily more complicated than the spectra of isolated atoms. There are three distinct types of spectra we can observe for molecules; these are considered in more detail in Chapter 12. **Molecular electronic spectra** are observed in the UV, visible, and near-IR regions of the spectrum

and are due to transitions between a vibrational and rotational level in one electronic state to a vibrational and rotational level in another electronic state. Thus, many transitions between two electronic states are possible and such molecular spectra are called **band spectra**. In the gas phase, we can observe the vibrational and rotational structure. In the condensed phases, much of the structure of molecular spectra is blurred because of frequent molecular collisions and level perturbations due to near-neighbor interactions. As a result, molecular electronic spectra of liquids and solutions often consist of one or more broad, featureless bands (typically, 10 to 100 nm wide), where each band is an envelope of the multitude of possible transitions between vibrational-rotational levels in two electronic states. **Vibration-rotational spectra** involve transitions from the rotational levels of one vibrational level to the rotational levels of another vibrational level of the same electronic state and are observed most often in the IR region of the spectrum. The rotational structure of infrared spectra is usually lost in condensed phases. **Rotational spectra** involve transitions from one rotational level to another rotational level of the same vibrational level of the same electronic state. Purely rotational spectra are normally observed in the microwave region of the spectrum. Raman spectroscopy also gives information

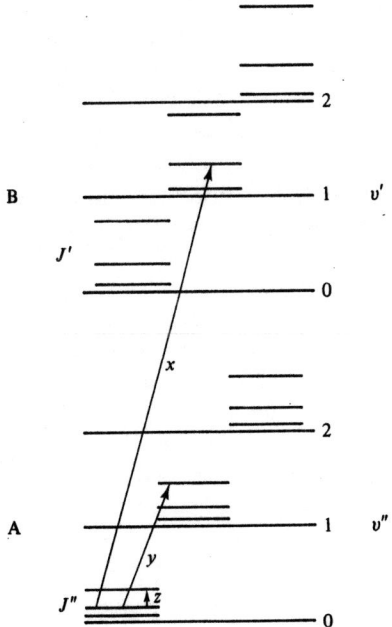

**FIGURE 1-3** Vibrational and rotational levels of molecular electronic states A and B. The vibrational levels are given by $v''$ and $v'$, while the rotational levels are given by $J''$ and $J'$. The three arrows show the three types of transitions possible: electronic ($x$), vibration-rotation ($y$), and pure rotation ($z$).

about vibrational and rotational levels through measurement of the inelastic scattering of UV-visible radiation.

## Optimization of the Response Function

The influence of the other variables (in equation 1-3) on the calibration function or the analytical signal is also important. The dependence of $S$ on all possible variables cannot realistically be determined; in practice, only those variables deemed to be critical or which can be conveniently measured are considered. The other variables of interest depend upon the technique and particular application and can include instrumental variables, such as excitation light source intensity, volume element of total sample in the instrument that is actually probed, and slit width, as well as physical and chemical variables such as temperature, pH, ionic strength, reagent concentrations, and concentrations of concomitant species. Results of studies of the influence of experimental variables are used to choose optimum operating conditions as discussed below.

Measurement conditions are optimized by maximizing (or minimizing) a **response function**, which is an optimization criterion tailored to a specific application. Often the response function is the analytical signal, and conditions are adjusted to maximize its magnitude. Since the variable values may affect the concomitant and sample container spectral signals differently than the analyte spectral signal, another useful response function is the **signal-to-background ratio** ($S/B$). Here $S$ is the analytical signal from the analyte and $B$ is the sum of background spectral signals from other species or the sample container. Changing the value of a variable to double $S$ can make it more difficult to measure $S$ if the background signal increases an order of magnitude.

The foregoing two response functions are not necessarily adequate because they do not take into account the precision with which the analytical signal can be measured. Increasing the magnitude of $S$ or $S/B$ through changing the value of an experimental variable can degrade the quality of the measurement if the relative uncertainty in the magnitude of $S$ is increased. Thus it is often best to optimize the precision of a measurement by minimizing the **relative standard deviation** (RSD) for measuring the analytical signal, where RSD $= \sigma_S/S$ and $\sigma_S$ is the **standard deviation** in measuring the analytical signal. The standard deviation is a quantitative measure of the uncertainty in a quantity due to random variations in the factors which affect the quantity from one measurement to the next, and is discussed in detail in Appendix A.

Sometimes the precision is optimized by maximizing a quantity called the **signal-to-noise ratio** ($S/N$),

where $S$ is the analytical signal and $N$ is the **root-mean-square** (rms) noise in the analytical signal. If noise is what causes the uncertainty in the signal, the signal-to-noise ratio is the reciprocal of the relative standard deviation ($S/N = [\text{RSD}]^{-1}$).

The magnitude of $S$ or $S/B$ is often simpler to measure, and frequently yields the same optimal experimental conditions as are obtained by optimizing the $S/N$. Maximizing the analytical signal may also maximize the $S/N$ or minimize the RSD if $\sigma_S$ is independent of the signal magnitude. The $S/N$ may be optimized by maximizing the $S/B$ for cases in which $\sigma_S$ is determined by noise in the background signal.

Knowledge of the calibration function allows quantitative information about precision to be obtained. Application of propagation of uncertainty mathematics (see Appendix A) to equation 1-3 yields

$$\sigma_S = \left[ \left( \frac{\partial f}{\partial c_a} \right)^2 \sigma_{c_a}^2 + \left( \frac{\partial f}{\partial \lambda} \right)^2 \sigma_{\lambda}^2 + \left( \frac{\partial f}{\partial x_i} \right)^2 \sigma_{x_i}^2 \right]^{1/2} \tag{1-5}$$

From equation 1-5 the effect of variations in a given variable (magnitude of $\sigma$ for the $i$th variable) can be determined. Thus, in general, the magnitude of a variable can be adjusted so that the product of the slope of the calibration function with respect to the variable ($\partial f/\partial x_i$) and the uncertainty in the variable $\sigma_{x_i}$ are small. This is often accomplished by adjusting the magnitude of the variable to a value where the analytical signal is independent of the variable ($\partial f/\partial x_i = 0$). In general, lowering the background signal without decreasing $S$ (maximization of $S/B$) will improve precision if fluctuations in the background signal determine $\sigma_S$ since the absolute standard deviation in the background signal usually increases with the magnitude of the background signal.

The dependence of the calibration function on a variable is graphically expressed as a plot of $S$ vs. the variable. It is becoming more common with sophisticated instrumentation (e.g., graphic capabilities) to collect and display multidimensional data. For example, in molecular fluorescence the analytical signal depends on both the excitation and emission wavelength, so that a three-dimensional plot can be made with the $x$ and $y$ axes representing the two wavelengths and the $z$ axis representing the magnitude of the analytical signal. Despite the difficulty we have in constructing such plots, they can be extremely informative in choosing favorable operating conditions.

The single factor at a time approach to optimization described above does not always yield the optimum conditions if the variables interact with each other (i.e., the optimum value for one variable depends on the magnitude of another variable). In the **simplex optimization** approach several experimental variables are changed simultaneously, and a systematic search for the optimum response is made. The simplex is a geometric figure that is moved across the response surface by a prescribed set of rules until it reaches the optimum response or fails. Simplex optimization is simple enough to be done with the aid of a hand calculator and is well suited for computer-controlled systems where response measurements, calculations, and variable adjustments can all be made very quickly in real time. Use of this and other mathematical optimization techniques (e.g., steepest descents) is increasing.

## 1-4 EVALUATION CRITERIA IN SPECTROCHEMICAL TECHNIQUES

### Practical Considerations

Many factors must be considered in choosing a specific spectrometric technique for a given application. First, there are the practical considerations of cost, sample size required, simplicity, portability, and robustness. Cost considerations must take into account not only the initial costs of the instrument, but also the special facilities (e.g., power) required to install the instrument, the maintenance costs, and the operating cost (e.g., gases, reagents, technician salary). The amount of sample required is critical in situations, such as clinical applications, where sample sizes are limited. The simplest instrument and procedure which fulfill the requirements of the method are often preferred to reduce the probability of error and downtime. Simplicity also reduces the operator skill required. Portability and robustness are important where an instrument must repeatedly be moved between sampling sites with minimal setup time or where conditions are harsh, as on an oceanographic research vessel.

Speed of analysis is important since it affects the cost per analysis and the sample throughout, which may be critical in routine analysis situations involving large numbers of samples. The speed of an analysis is affected not only by the actual measurement time for the sample, but also by the time necessary to obtain the calibration curve and the time needed to prepare the analytical sample. Often sample preparation (e.g., dissolution of biological and geochemical samples) is the most time-consuming step and the step most demanding of technical skill. Techniques which are more prone to interference effects often require additional cleanup or matrix modification steps to remove or suppress interferences before the actual measurement step. Thus, overall, where speed of analysis is paramount, techniques requiring minimal sample preparation are preferred.

### Automation and Multiple-Species Capability

Automation of spectrometric instruments has increased dramatically due to the introduction of microcomputers. Automation, the performance of tasks without operator assistance, can free the operator of tedious tasks and increase precision due to more reproducible performance of steps formerly requiring operator skill. Numerous tasks can be automated, including the selection of preprogrammed instrumental variable values for a given analyte, measurement of spectral signals, construction of calibration curves, and presentation of analytical and statistical information. Further steps which may be automated include sample preparation, cleanup, and introduction. Since some techniques are more suitable for automation than others and because only certain automated instruments are commercially available, the selection of a specific spectrometric technique may depend on the degree of automation desired. Automation is particularly needed for situations involving unattended analysis of samples at night or at remote locations. The new generations of microprocessor-controlled instruments have introduced error checking and instrument diagnostic features not previously available. We can expect to see more instruments with built-in intelligence to assist the operator in optimization, instrumental operation, and complex data interpretation. These have been termed "expert" systems.

Multiple-species analysis is more convenient with automated instrumentation. Multiple-species analysis can be carried out by some instruments which measure each analyte in a sequential manner or by others which measure all analytes simultaneously. Simultaneous analysis can be implemented by measurement of $n$ analytes in each of $n$ identical, but separate, analytical samples, where all samples are processed separately but simultaneously in $n$ parallel channels or by simultaneous measurement of $n$ spectral signals from a given analytical sample. Simultaneous analysis is faster than sequential analysis and requires less sample for the situation where $n$ spectral signals are measured from one analytical sample.

### Interferences and Selectivity

The accuracy of all spectrometric techniques can be degraded by interferences. An **interferent** is a substance present in the analytical sample which affects the magnitude of spectral signal measured for the analyte. Thus the choice of a spectrochemical technique depends on the expected interferences and their concentrations in the sample to be analyzed. Once a technique is chosen, the preparation of the sample for analysis may still have to be carried out in a manner that reduces interference effects to an acceptable level. The **selectivity** of a technique is related to its relative freedom from interference effects. Many analytical techniques are quite selective, but few are truly **specific**, a term which implies complete freedom from interferences.

## Figures of Merit

The characteristics of a spectrochemical technique for a given analyte are indicated by several figures of merit, such as accuracy, precision, sensitivity, detection limit, and dynamic range. The **accuracy** indicates how close the measured analyte concentration is to the true analyte concentration in the sample and is normally expressed as the relative percent error (i.e., a 1% error indicates that the measured concentration is within 1% of the true analyte concentration). The accuracy depends on the analyte concentration, the precision, and interference effects.

The **precision**, usually expressed as a percent RSD, indicates the reproducibility of repetitive measurements of equivalent analyte solutions. Averaging of repetitive measurements can be used to improve precision, and thus the accuracy, if random errors are limiting rather than systematic errors due to interferents or other factors. An inherently high-precision technique can be faster since repetitive measurements are not required.

The precision depends upon the analyte concentration and can depend upon the sample matrix and interferences. As indicated in Section 1-3, instrumental, physical, and chemical variables are often selected to optimize precision. The analyte concentration in samples is often adjusted by dilution or preconcentration to be in the range yielding best precision. Thus a plot of precision (RSD) vs. analyte concentration is useful.

The concepts of sensitivity, detection limit, and dynamic range are illustrated in Figure 1-4. **Sensitivity** may have several meanings, as discussed in greater detail in Section 6-4. It usually indicates the response of the instrument to changes in analyte concentration and is expressed as the slope of the calibration curve or the change in analytical signal per unit change in analyte concentration. This definition is the primary definition of sensitivity used throughout this book. Sometimes *sensitivity* refers to the ability to distinguish concentration differences and thus takes into account the precision of measurements.

The **detection limit** (DL) is typically defined as the analyte concentration yielding an analytical signal equal

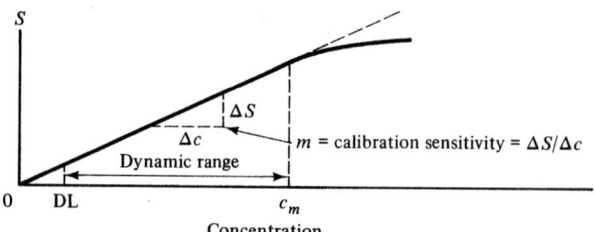

**FIGURE 1-4** Calibration curve illustrating the concepts of sensitivity, detection limit, and dynamic range. The calibration sensitivity ($m$) is the slope of the calibration curve at a particular concentration or the derivative of the calibration function at a particular concentration $[\partial S/\partial c)_c]$. The detection limit (DL) indicates the lowest analyte concentration that can be measured with a specified degree of certainty and is often defined by $DL = 2s_{bk}/m$ or $DL = 3s_{bk}/m$, where $s_{bk}$ is the standard deviation of the blank measurement. The dynamic range is defined by DL and $c_m$, where $c_m$ is the maximum concentration that can be measured before the calibration curve deviates a specific amount from the extrapolated linear portion of the plot.

to two or three times the standard deviation of a blank measurement (33 to 50% RSD). It will be discussed in more detail in Section 6-4, but is indicative of the lowest analyte concentration that can be reported as being detected with a specified degree of certainty. The detection limit is a critical figure of merit since a spectrometric technique cannot be used without preconcentration steps if the analyte concentration in the analytical sample is below the detection limit. In fact, the analyte concentration should normally be higher than 10 times the detection limit to obtain reasonable precision (5% RSD or less). The choice between two spectrochemical techniques both with detection limits well below the analyte concentrations to be measured is based on other criteria. Often variables are optimized for the best detection limit at the expense of other criteria, such as precision at higher concentrations. **Detectability** is a term that denotes the ability to provide low detection limits.

**Dynamic range** can be specified as the concentration range or analytical signal range over which the analytical curve is linear or the calibration slope is constant. It is usually defined at the lower end by the detection limit and at the upper end by an analyte concentration where the analytical signal deviates a specific relative amount (e.g., 5%) from the extrapolated linear portion of the curve or where the slope deviates a specific relative amount from the slope in the linear portion. Nonlinearity can be inherent in the technique or due to the matrix of the standards, nonideal instrumental performance, instrumental distortion, or incorrect utilization of the instrument.

A linear calibration curve is usually preferred because it is easier to detect an abnormality and because it is easier to work with mathematically (i.e., fewer points are needed to establish the calibration curve and a linear, straight-line, least-squares curve-fitting model can be employed). A large dynamic range is preferred because a wide range of analyte concentrations can be used without sample dilution. In some samples, the analyte concentration varies only over a small range, so that a large dynamic range is not required. An example is Na in human serum, which varies typically from 0.135 to 0.148 M.

A nonlinear calibration curve can be used as long as enough standards are measured to establish the calibration function. Many computerized instruments incorporate least-squares software for fitting data.

## Overview

The choice of a specific spectrometric technique and instrument for a given analyte in a particular sample depends on the relative importance of the evaluation criteria and figures of merit to the specific situation or range of situations expected. A systematic listing of the requirements and restrictions can aid in this choice. This would include factors such as the budget, the type of personnel, the number of samples to be analyzed per day, the amount of initial sample available, the required turnaround time from receipt of sample to reporting of the analytical results, where the analysis is to be performed (e.g., on- or off-line), the range of concentrations of the analyte and concomitants in the sample matrix, and the accuracy and precision required.

Similarly, the optimization of experimental variables and sample preparation steps will depend upon many of the factors noted above. Trade-offs must often be made. For example, speed of analysis might be increased with some sacrifice of precision or accuracy.

## PROBLEMS

**1-1.** Express 4000 Å in nm, as a frequency, as a wavenumber (cm$^{-1}$), and as an energy in joules, ergs, and electron volts.

**1-2.** Emission and luminescence techniques both involve the measurement of the emission of photons from excited species. What makes these techniques different?

**1-3.** Why are real blanks not ideal blanks?

**1-4.** What is the difference between *S/B* and *S/N*?

**1-5.** In a photoluminescence spectrometry, the photoluminescence signal is usually proportional to the excitation source intensity. If the lamp intensity is proportional to the fourth power of the lamp temperature, to what degree must the RSD of the lamp temperature be controlled to ensure that the RSD of the photoluminescence signal due to lamp temperature fluctuations is 1% or less?

**1-6.** Why does high precision not necessarily ensure high accuracy?

**1-7.** A certain optical transition occurs in the visible region of the spectrum at 530 nm. Find the energy of the transition in J and in eV. What is the wavenumber of the transition?

**1-8.** A 0.08-g sample is analyzed for a constituent present at approximately 5 mg g$^{-1}$. Classify the analysis in terms of constituent amount and sample size.

**1-9.** A certain determination follows a calibration function $S = 1.5 \times 10^4 c$, where $S$ is the analytical signal and $c$ is the molar concentration. Find the analytical function for this determination.

**1-10.** You have a "30-cm ruler," which is in fact 28 cm long. Briefly discuss (a) measurement precision, and (b) its accuracy.

**1-11.** The RSD of a measurement limited by noise is 0.50%. What is the *S/N*?

**1-12.** Calculate the detection limit for a determination if the blank standard deviation is 0.010 V and the slope of the calibration curve is 2.0 V μM$^{-1}$.

## REFERENCES

1. IUPAC, *Compendium of Analytical Nomenclature*, Pergamon Press, Oxford, 1978. This compendium contains many definitions of terms and figures of merit.

The following monographs or series include chapters on optical spectrochemical methods of analysis and serve as excellent general reference sources.

2. C. N. Reilley, R. W. Murray, and F. W. McLafferty, eds., *Advances in Analytical Chemistry and Instrumentation*, Wiley-Interscience, New York, 1960–1973.

3. P. J. Elving, I. M. Kolthoff, et al., eds., *Treatise on Analytical Chemistry*, Wiley-Interscience, New York, 1959–present.

4. G. H. Morrison, ed., *Trace Analysis: Physical Methods*, Wiley-Interscience, New York, 1965.

5. P. J. Elving, I. M. Kolthoff, and J. D. Winefordner, eds., *Chemical Analysis*, Wiley-Interscience, New York, 1941–present.

6. C. L. Wilson, D. W. Wilson, and G. Svehla, eds., *Comprehensive Analytical Chemistry*, Elsevier, Amsterdam, 1959–present.

7. R. Belcher, L. Gordon, and H. Freiser, eds., *Analytical Chemistry*, Pergamon Press, Oxford, 1961–present.

8. T. Kuwana, ed., *Physical Methods in Modern Chemical Analysis*, Academic Press, New York, 1978–present.

9. D. Glick, ed., *Methods of Biochemical Analysis*, Wiley-Interscience, New York, 1954–present.

10. J. D. Winefordner, ed., *Trace Analysis: Spectroscopic Methods for Elements*, Wiley-Interscience, New York, 1976.

11. J. D. Winefordner, ed., *Spectrochemical Methods of Analysis*, Wiley-Interscience, New York, 1971.

12. G. D. Christian and J. B. Callis, eds., *Trace Analysis: Spectroscopic Methods for Molecules*, Wiley-Interscience, New York, 1986.

# CHAPTER 2

# Spectrochemical Measurements

When we discuss the quantitative determination of a particular species by spectrochemical methods, we are considering the measurement of its concentration using some form of spectroscopy. Before going further it is important to be clear about what the measurement process is and how it applies to spectrochemical procedures. A **measurement** can be defined, in general, as the determination of a particular characteristic of a sample in terms of a number of standard units of that characteristic. A chemical concentration measurement thus consists of determining how many standard concentration units (molarity, $\mu g\ mL^{-1}$, etc.) of a particular species are present in the sample. Reference to a standard is implicit in this definition of measurement and it is an important part of a spectrochemical determination. Spectrochemical measurement systems differ from others, such as electrochemical systems, in that during the measurement process the chemical information is carried, or **encoded**, as an optical signal.

This chapter first considers the complete spectrochemical measurement process and identifies the general components of a spectrochemical analysis system. The various ways in which intensities of optical signals are expressed in the radiometric and photometric systems of units are explored so that we can be quantitative about how concentration information is encoded. Spec-

trochemical methods are classified according to the types of spectral information produced, and the principles of absorption, emission, luminescence, and scattering are introduced. It is shown that optical information can be selected by its wavelength, its time or temporal behavior, its spatial position, or by several combinations of selection criteria. The final section describes how optical information representing chemical information can be converted into a number (measured) that is related to concentration.

## 2-1 COMPLETE SPECTROCHEMICAL MEASUREMENT

Spectrochemical determination of the concentration of the analyte in a sample involves many steps, including acquisition of the initial sample, sample preparation or treatment to produce the analytical sample, presentation of the analytical sample to the instrument, measurement of the optical signals, establishment of the calibration function with standards and calculations, interpretation, and feedback. The spectrochemical measurement process can be conceptualized as shown in Figure 2-1. The initial sample, obtained from the bulk material, is often treated prior to its presentation

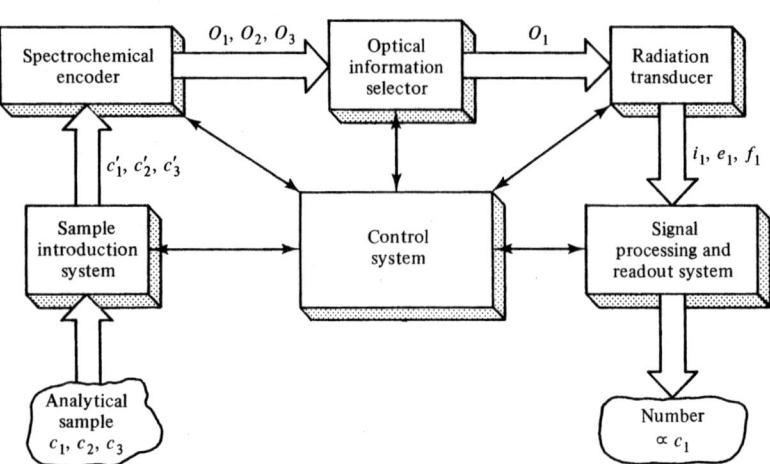

**FIGURE 2-1** Spectrochemical measurement process. Chemical concentration information (e.g., the concentration of species 1, $c_1$) is converted to a number by this process. A sample introduction system presents the sample to the encoding system, which converts the concentrations $c_1'$, $c_2'$, $c_3'$ into optical signals $O_1$, $O_2$, $O_3$. The information selection systems selects the desired optical signal $O_1$ for presentation to the radiation transducer. This device converts the optical signal into an electrical signal (current $i$, voltage $e$, frequency $f$, etc.) that is processed and read out as a number. The control system supervises and coordinates the operations performed. The unidirectional arrows indicate the flow of information, while the bidirectional arrows indicate control signals.

to the encoder by dissolution, preconcentration, solvent extraction, complexation, dialysis, chromatography, or some combination of these and other steps to produce the analytical sample. Thus the concentrations presented to the encoding system might be altered in a known manner from those present in the initial sample or even from those present in the analytical sample. In some automated systems the sample pretreatment steps are an integral part of the sample introduction system and are carried out without manual manipulation. In most cases the sample pretreatment steps are carried out manually. In any case the sample introduction system must present the analytical sample to the spectrochemical encoder in a form that is appropriate to the particular encoder. In molecular absorption spectroscopy the sample presentation step is often the manual or automatic introduction of the sample solution into a cuvette secured by a holder. In some instruments the sample introduction system carries out a complex sequence of operations. In atomic spectroscopy, for example, the sample must be introduced continuously or discretely into an **atomizer**, which converts the analytical sample into free atoms in the vapor phase, a complicated conversion process.

The spectrochemical encoding system produces optical information that represents (encodes) the chemical concentrations. Most often the chemical information is encoded simply as a spectrum, that is as intensities at particular optical frequencies (wavelengths). Many spectrochemical sources, however, also produce time-dependent or position-dependent spectral information, which may or may not be used in the selection process.

The optical information selector sorts out the desired optical signal from the many signals produced in the encoding process. Quite often the selection is made solely on the basis of optical frequency (wavelength) so that the information selector is a wavelength selector. Discrimination against background signals from the concomitants and sample container can also be made on the basis of time and position behavior. In the single-channel instrument depicted in Figure 2-1, the information selector isolates optical information characteristic of species 1, for example, for presentation to the transducer. In multichannel systems, information concerning several species is sorted into different information channels for simultaneous or sequential presentation to the transducer system.

The **radiation transducer**, or **photodetector**, converts the optical signal into a corresponding electrical signal that can readily be processed by modern electronic systems. Many photodetectors produce analog outputs so that the magnitude of an electrical current, voltage, or charge carries the sought-for information. However, some transducers produce output data in one of the time domains (frequency, phase, or pulse width) or directly in the digital domain, inherently a numerical representation. The processing electronics convert the transducer output into a form (*data domain*) that is appropriate for the readout system. This can involve the operations of current-to-voltage conversion, linear amplification, logarithmic amplification, ratioing, analog-to-digital conversion, and so on. The readout system displays the data so that a numerical value can be obtained. Typical readout devices include moving-coil meters, strip-chart recorders, and digital displays as in digital voltmeters, printers, or CRT terminals.

In order for the number obtained as a result of spectrochemical measurement to be related to the concentration of the desired species, it is apparent that the system must be calibrated (standardized) in some manner. If the conversion factors or transfer functions of all the functional elements in Figure 2-1 were known exactly, were stable, and were known functions of concentration, the measurement system could be calibrated at the factory to read directly in concentration units and then checked only occasionally. However, in many spectrochemical instruments in use today, conversion factors change with time and matrix and may not be known exactly. Hence it is usually a requirement that standards of known concentration be introduced regularly for calibration purposes. These standards are taken through the entire process, if possible, to evaluate the total system response. The calibration results are often expressed as a working curve or as the analytical function as described in Chapter 1. In computer-controlled systems these calibration data can be manipulated and stored so that the processing system can produce concentration data directly.

The control system in Figure 2-1 is an integral part of the measurement process. In the past a human operator has been the most common controller. However, we are observing more and more instrument functions being placed under the control of microcomputers. This frees the human operator for the tasks of experimental design, data interpretation, quality assessment, and often, sample preparation.

The final results of a spectrochemical analysis are often part of a larger feedback control system, in which the analytical data are used to indicate the status of a process or the quality of a product. Thus if the information obtained is outside the limits set for a process or product, corrective action must be taken so as to reduce the error signal. Even human health care can be viewed as a large feedback control system in which analytical data are an integral part. Just as in an industrial process, the data are used to monitor the status of the patient and to determine the type, intensity, and timing of any corrective actions that are needed. Spectrochemical measurements can thus be an extremely important part of such control loops.

## 2-2 EXPRESSIONS OF OPTICAL INTENSITY

In spectrochemical systems the chemical information is almost always encoded as the magnitude of optical signals at particular wavelengths. There are two systems that describe optical intensities, the radiometric system

and the photometric system. The radiometric system is employed almost exclusively in modern spectroscopy and will be used throughout this book. The photometric system, based on the response of the human eye, is still used to describe photodetectors and several other optical components. The photometric system is thus introduced here and compared to the radiometric system.

### Radiometric System

The radiometric system of units is based on the actual radiant energy emitted by a source or striking a receiver (e.g., optical transducer) and is preferred in the International System of Units (SI). The basic quantity in this system is the radiant energy $Q$ in joules (J).

*Basic Definitions.* In the radiometric system there are general quantities used to describe radiation, quantities used to describe sources, and quantities that deal with the receiver. Table 2-1 lists and defines the basic radiometric quantities, symbols, and units. The terms *radiant intensity*, *emittance*, *emissivity*, and *radiance* refer specifically to radiation from a source and the volumes, areas, and solid angles are those of the source. *Irradiance* and *exposure* are used to describe the receiver and its area. In some cases more than one symbol is given for a quantity in Table 2-1. The first symbol is that accepted by the International Union of Pure and Applied Chemistry (IUPAC) and used in this book. The symbols in parentheses are sometimes used and are included here for completeness. The units given for radiometric quantities are practical units rather than strict SI units, which use meters instead of centimeters. Most sources and receivers have dimensions on the order of centimeters and practical units are thus easier to visualize.

The quantities given in Table 2-1 are in general functions of spectral position (wavelength, wavenumber, frequency, etc.) in that they are usually employed to represent the magnitude of the quantity over some spectral interval. Often they represent the cumulative magnitude of the quantity over the wavelength interval from 0 to $\lambda$. If the modifier "total" is employed, as in total radiance, it implies the radiance over the wavelength interval from 0 to $\infty$.

It is often useful to consider these radiometric quantities within small spectral intervals. For convenience, radiometric quantities per unit spectral interval evaluated at a particular spectral position are called **spectral quantities** and given a subscript $\lambda$, if wavelength is used, $\bar{\nu}$ if wavenumber is used, and $\nu$ if frequency is used. The **spectral radiance** $B_\lambda$ is the radiance per unit

**TABLE 2-1**
Radiometric system

| Quantity | Symbol(s) | Description | Defining equation[a] | Unit(s) |
|---|---|---|---|---|
| **General** | | | | |
| Radiant energy | $Q$ | Energy in the form of radiation | | J (ergs) |
| (Radiant) energy density | $U$ | Radiant energy per unit volume | $U = \dfrac{\partial Q}{\partial V}$ | J cm$^{-3}$ |
| Radiant flux or radiant power | $\Phi(P)$ | Rate of transfer of radiant energy | $\Phi = \dfrac{\partial Q}{\partial t}$ | W |
| **Source** | | | | |
| Radiant intensity | $I$ | Radiant power per unit solid angle from a point source | $I = \dfrac{\partial \Phi}{\partial \Omega}$ | W sr$^{-1}$ |
| (Radiant) emittance or (radiant) exitance | $M$ | Radiant power per unit area | $M = \dfrac{\partial \Phi}{\partial A}$ | W cm$^{-2}$ |
| (Radiant) emissivity | $J$ | Radiant power per unit solid angle per unit volume | $J = \dfrac{\partial^2 \Phi}{\partial \Omega\, \partial V}$ | W sr$^{-1}$ cm$^{-3}$ |
| Radiance | $B(L)$ | Radiant power per unit solid angle per unit projected area | $B = \dfrac{\partial^2 \Phi}{\partial \Omega\, \partial A_p} = \dfrac{\partial^2 \Phi}{\partial \Omega\, \partial A \cos \theta}$ | W sr$^{-1}$ cm$^{-2}$ |
| **Receiver** | | | | |
| Irradiance | $E$ | Radiant power per unit area | $E = \dfrac{\partial \Phi}{\partial A}$ | W cm$^{-2}$ |
| (Radiant) exposure | $H$ | Integrated irradiance | $H = \displaystyle\int_0^t E\, dt$ | J cm$^{-2}$ |

[a] Since the radiant flux $\Phi$ may vary in space and direction, the solid angle $\Omega$, the volume $V$, and the area $A$ in the defining equations must be small enough to give meaningful local values. If not, more complicated relationships apply (see *Optical Radiation Measurements*, Vol. 1, *Radiometry*, F. Grum and R. J. Becherer, Academic Press, New York, NY, 1979). The radiance, irradiance, and emittance are related to the radiant energy density by $B = Uc/4\pi$, $E = Uc$, and $M = Uc$, where $c$ is the speed of light. For a Lambertian surface (constant $B$ independent of viewing direction), $B = M/\pi$.

wavelength interval (i.e., per nm) and is given by

$$B_\lambda = \frac{\partial B(\lambda)}{\partial \lambda} \qquad (2\text{-}1)$$

where $B(\lambda)$ is the cumulative radiance. In analogous fashion $B_{\bar{v}}$ is the radiance per unit wavenumber and $B_v$ is the radiance per unit frequency. Since $\bar{v} = 1/\lambda$ and $d\bar{v} = -d\lambda/\lambda^2$, the following equation relates $B_{\bar{v}}$ to $B_\lambda$:

$$B_{\bar{v}} = \frac{\partial B(\bar{v})}{\partial \bar{v}} = B_\lambda \lambda^2 \qquad (2\text{-}2)$$

Here the minus sign, which results because changes in $\bar{v}$ are opposite to changes in $\lambda$, is omitted because we are interested in the magnitude of $B_{\bar{v}}$. Similarly, since $v = c/\lambda$ and $dv = -cd\lambda/\lambda^2$,

$$B_v = \frac{B_\lambda \lambda^2}{c} \qquad (2\text{-}3)$$

where $c$ is the speed of light. Other spectral quantities, such as spectral radiant power ($\Phi_\lambda$, $\Phi_v$, or $\Phi_{\bar{v}}$), spectral radiant intensity ($I_\lambda$, $I_v$, or $I_{\bar{v}}$) and spectral irradiance ($E_\lambda$, $E_v$, or $E_{\bar{v}}$) are defined in a similar manner.

The total radiance $B$ from a source is related to its spectral radiance by

$$B = \int_0^\infty B_\lambda\, d\lambda \qquad (2\text{-}4)$$

Often we are interested in the radiance $B_{\Delta\lambda}$ in a wavelength interval $\Delta\lambda = \lambda_2 - \lambda_1$. In this case

$$B_{\Delta\lambda} = \int_{\lambda_1}^{\lambda_2} B_\lambda\, d\lambda \qquad (2\text{-}5)$$

and $B_{\Delta\lambda}$ equals the cumulative radiance $B(\lambda_2)$ to $\lambda_2$ if $\lambda_1 = 0$. If $B_\lambda$ is constant throughout the interval $\Delta\lambda$,

$$B_{\Delta\lambda} = \frac{\partial B(\lambda)}{\partial \lambda} \Delta\lambda = B_\lambda\, \Delta\lambda \qquad (2\text{-}6)$$

Sources that emit narrow spectral lines (typical half-widths $\ll 1$ Å) are usually characterized by reporting the radiance $B$ of each line which is the integrated spectral radiance over the total width of the line. A broadband source is normally characterized by its spectral radiance $B_\lambda$ because only part of its emitted spectral range is selected or observed as determined by a wavelength selector.

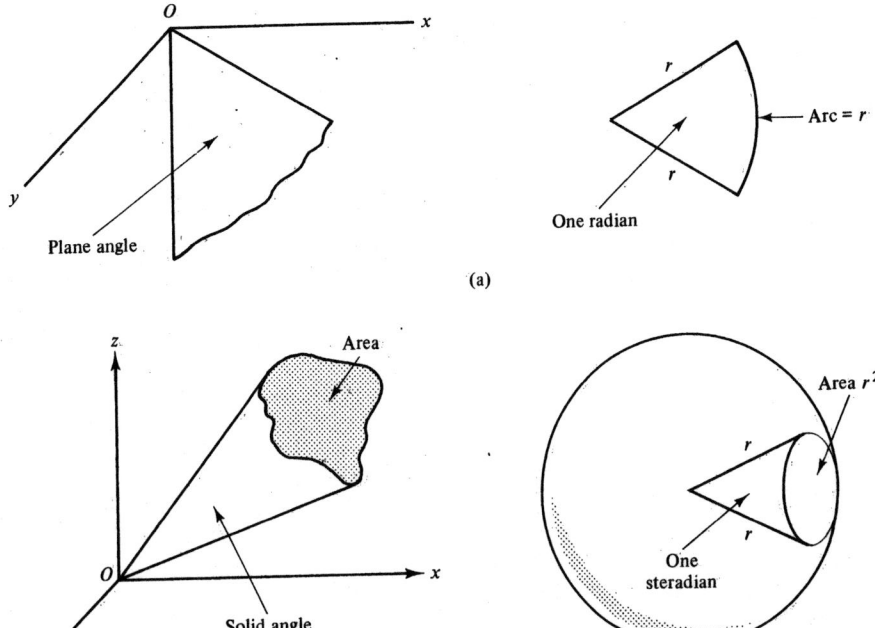

(a)

(b)

**FIGURE 2-2** Plane and solid angles. In (a) a plane angle and one radian of angle are illustrated. One radian is the angle at the center of a circle that intercepts an arc equal in length to the radius. In (b) the solid angle is defined by the cone generated by a line that passes through the vertex $O$ and a point moved along the periphery of the surface. One steradian is the solid angle at the center of a sphere of radius $r$ that subtends an area of $r^2$ units on the surface.

Occasionally, one will encounter radiometric terms with units of photons per second or Einstein's (Avogadro's number of photons) per second instead of watts. Because the energy of a photon is just $h\nu$, these spectral quantities can be directly converted to watt units if the frequency is known. A subscript $p$ is often used to denote units of photons s$^{-1}$ so that $B_{\lambda,p}$ is the spectral radiance in photons s$^{-1}$ cm$^{-2}$ sr$^{-1}$ nm$^{-1}$ and $B_{\lambda,p} = B_\lambda/h\nu$.

*Geometric Factors.* To deal with several of the radiometric quantities it is important to define the geometric factors of solid angle and projected area that are used. Figure 2-2 illustrates the plane angle, the solid angle, and the units used to express these terms. The unit of solid angle, the **steradian** (sr), is defined in a manner similar to that of the unit of angle, the **radian**. Since the area of a sphere of radius $r$ is $4\pi r^2$, a sphere subtends $4\pi r^2/r^2 = 4\pi = 12.57$ sr. The concept of projected area that is used in the definition of radiance is illustrated in Figure 2-3.

*Uses of Radiometric Terms.* The radiometric quantities defined in Table 2-1 are used throughout this book, but a few examples are given here for illustrative purposes. The specific quantity that is used depends on the situation. In most spectroscopic situations one is eventually interested in the radiant power that is incident on a receptor (e.g., an optical transducer). Consider, for example, a point source with dimensions that are small compared to the distance ($d$) from the source to the receptor of projected area $A_p$. The source could

be characterized by the total radiant power $\Phi$ that it emits in all directions. In this case, however, it is more useful to use the radiant power per unit solid angle (the radiant intensity), which is given by

$$I = \frac{\Phi}{4\pi} \qquad (2\text{-}7)$$

if the source emits equally in all directions (i.e., is iso-

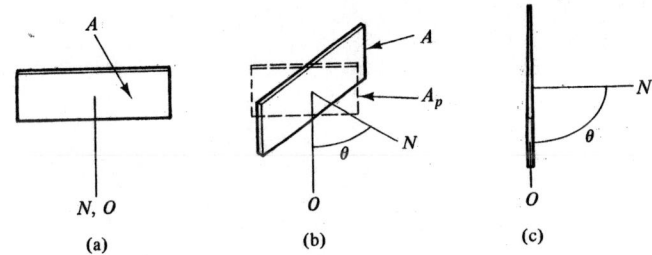

(a)          (b)          (c)

**FIGURE 2-3** Projected area. The projected area is the effective area of a two-dimensional surface as seen from a point, $O$. If the observer looks along the normal ($N$) to the surface (a), the effective area is the same as the true area. As the angle $\theta$ between the line of observation and the normal increases the projected area decreases (b) until at 90° (c) the effective area is zero. The projected area $A_p$ is then $A_p = A \cos\theta$. The radiance of a source is the radiant power per unit solid angle per unit projected area, $B = (\partial\Phi/\partial\Omega) \times (\partial\Phi/\partial A_p) = \partial^2\Phi/(\partial\Phi\partial A \cos\theta)$. The solid angle $\Omega$ viewed from $O$ is $\Omega = A_p/d^2 = A \cos\theta/d^2$, where $d$ is the distance from $O$ to the surface.

tropic). The radiant power incident on the receptor $\Phi_i$ is the source intensity times the solid angle viewed, or

$$\Phi_i = I\Omega = \frac{IA_p}{d^2} \qquad (2\text{-}8)$$

The irradiance $E$ incident on area $A$ of the receptor is

$$E = \frac{I\Omega}{A} = \frac{IA_p}{Ad^2} \qquad (2\text{-}9)$$

Note that equations 2-8 and 2-9 both predict the familiar result that the intensity incident on the receiver is inversely proportional to the square of the distance from the source.

   In the example shown in Figure 2-4, the source has a significant area and it is usually best to use the source radiance. Also in this figure, only a fraction $A_1'/A_1$ of the total source projected area is viewed by the receptor. This could occur, for example, because an aperture between the source and receptor limits the area viewed. If the source emits radiant power $\Phi$ equally in all directions, the source radiance $B$ is

$$B = \frac{\Phi}{4\pi A_1} \qquad (2\text{-}10)$$

The radiant power $\Phi_i$ incident on the area $A_2$ of the receptor is the source radiance times the solid angle viewed times the area viewed, or

$$\Phi_i = B\Omega A_1' = B\frac{A_2}{d^2}A_1' \qquad (2\text{-}11)$$

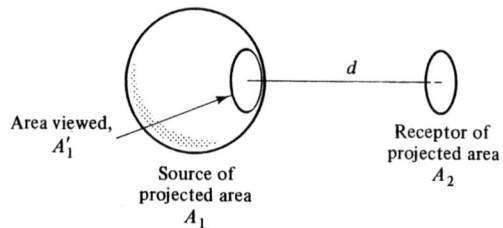

**FIGURE 2-4**  Extended source and receptor separated by distance $d$. An aperture, for example, limits the source area viewed to $A_1'$. The solid angle $\Omega$ viewed by the receptor is $A_2/d^2$.

## Photometric System

The photometric system is a relative system based on the apparent intensity of a source as viewed by the average bright-adapted human eye. The quantities in the photometric system have meaning only in the visible region (350 to 770 nm) of the spectrum. The basic unit of the photometric system is the **lumen** (lm). It is defined in terms of the standard candle, which is called a **candela**. The luminous intensity from a candela is $\frac{1}{60}$ of the luminous intensity of a blackbody radiator (see Chapter 3) of 1 cm$^2$ area at the solidification temperature of platinum (2042 K). A source of 1 candela emits 1 lumen per steradian.

   Photometric quantities exist which correspond to each of the radiometric quantities as shown in Table 2-2. The same symbols are used for photometric quantities as for radiometric quantities. To avoid confusion when both systems are employed or conversions are being made, radiometric quantities should carry the subscript $e$ for energy, while photometric quantities should carry the subscript $v$ for visible.

   Conversions between the two systems are made with the aid of the following equation, which relates

**TABLE 2-2**
Photometric and radiometric quantities

| Radiometric quantity and units | Photometric quantity and units | Definition of photometric quantity |
| --- | --- | --- |
| Radiant energy (J) | Luminous energy (lm s) | Portion of radiant energy in visible region |
| Radiant power or flux (W) | Luminous power or flux (lm) | Luminous energy per unit time |
| Radiant intensity (W sr$^{-1}$) | Luminous intensity (lm sr$^{-1}$) | Luminous power per unit solid angle |
| Radiant emittance (W cm$^{-2}$) | Luminous emittance (lm cm$^{-2}$) | Luminous power per unit source area |
| Irradiance (W cm$^{-2}$) | Illuminance (lm cm$^{-2}$) | Luminous power per unit area incident on a surface |
| Radiance (W sr$^{-1}$ cm$^{-2}$) | Luminance (brightness) (lm sr$^{-1}$ cm$^{-2}$) | Luminous power per unit solid angle per unit projected area |

the spectral luminance to the spectral radiance:

$$B_{v,\lambda} = 680 V(\lambda) B_{e,\lambda} \qquad (2\text{-}12)$$

Here $V(\lambda)$ is called the **spectral luminous efficiency**. It is a dimensionless factor that depends on wavelength and varies in the range 0 to 1 over the visible region (see W. G. Driscoll and W. Vaughan, eds., *Handbook of Optics*, McGraw-Hill, New York, 1978, pp. 1–4). At 556 nm a light-adapted human eye is at its peak efficiency. At this wavelength there are 680 lm $W^{-1}$. At 510 and 610 nm, the spectral luminous efficiency has dropped to 0.5. Total photometric quantities are obtained by integration of the appropriate spectral quantity as shown for luminance in the following equation:

$$B_v = \int_0^\infty B_{v,\lambda}\, d\lambda = 680 \int_{350}^{770} V(\lambda) B_{e,\lambda}\, d\lambda \qquad (2\text{-}13)$$

Equations similar to 2-12 and 2-13 can be used for interconversions of all corresponding quantities (e.g., $\Phi_{v,\lambda}$ to $\Phi_{e,\lambda}$). In this book quantities without the subscripts $v$ and $e$ are always taken as radiometric quantities.

## 2-3 SPECTROCHEMICAL METHODS

The spectral signals that are measured in spectrochemical methods arise from a variety of different physical and chemical phenomena. In Table 2-3 the four major classes of spectrochemical methods as based on the phenomena responsible are summarized. Also included in this table are indirect methods in which the measurement of a nonoptical quantity is used to obtain information about optical phenomena. Methods can also be based on reflection, refraction, diffraction, polarization, and dispersion, but we concern ourselves here primarily with the four major classes. The basic relationships between the quantities measured and concentration are developed in this section and used throughout the book.

### Emission Spectroscopy

In emission spectrochemical methods the sought-for information is the radiation emitted by the analyte due to nonradiational excitation. We further restrict emission spectroscopy in this book so that only techniques that involve excitation by collisional (thermal) processes are considered emission spectrochemical methods. For example, sodium atoms are excited in a flame by collisional processes and emit characteristic radiation. The excitation can also be partly electrical as in a direct-current (dc) arc or high-voltage spark. The excitation step in emission methods produces a statistical distribution of excited states when conditions of thermal equilibrium are maintained. The frequency of the emitted radiation corresponds to the discrete energy differences between levels, as shown in Figure 2-5. When thermal equilibrium is maintained, the number of atoms per $cm^3$ in level $i$, $n_i$, is related to the total number of atoms per $cm^3$, $n_t$, by the Boltzmann distribution

$$n_i = \frac{n_t g_i e^{-E_i/kT}}{Z(T)} \qquad (2\text{-}14)$$

where $g_i$ is the statistical weight of state $i$ (see Chapter 7), $E_i$ is the excitation energy (joules) relative to the ground state, $k$ is Boltzmann's constant, $T$ is the absolute temperature, and $Z$ is the partition function $[Z(T) = \Sigma_{i=0}^\infty g_i e^{-E_i/kT}]$. Here the summation is taken over all the states, including the ground states ($i = 0$).

Sodium and other alkali metals are examples of elements that have excited levels relatively close to the

**TABLE 2-3**
Classification of spectrochemical methods

| Class | Quantity measured | Examples |
|---|---|---|
| Emission | Radiant power of emission, $\Phi_E$ | Flame emission, dc arc emission, spark emission, ICP and DCP emission |
| Absorption | Absorbance or ratio of radiant power transmitted to that incident, $A = -\log(\Phi/\Phi_0)$ | UV-visible molecular absorption, infrared absorption, atomic absorption |
| Luminescence | Radiant power of luminescence, $\Phi_L$ | Molecular fluorescence and phosphorescence, atomic fluorescence, chemiluminescence |
| Scattering | Radiant power scattered, $\Phi_{sc}$ | Turbidimetry, nephelometry, Raman scattering |
| Indirect | Refractive index change, acoustic waves, ion current | Thermal lensing, photoacoustic, photoionization |

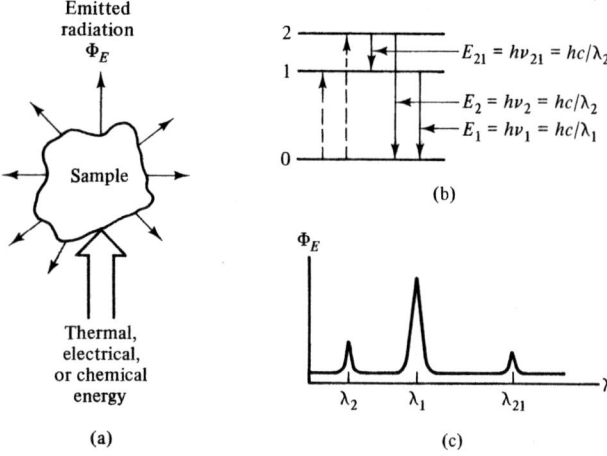

(a)

(b)

(c)

**FIGURE 2-5** Emission and chemiluminescence (bioluminescence) methods. In (a) the addition of thermal, electrical or chemical energy causes nonradiational excitation of the analyte and emission of radiation in all directions (isotropic emission). The energy changes that occur during excitation (dashed lines) or emission (solid lines) are shown in (b). The energies of states 1 and 2 are usually relative to the ground level and often abbreviated $E_1$ and $E_2$, respectively. A typical spectrum is shown in (c).

ground level in energy. The resonance lines of such easily excited atoms occur in the visible and near-IR region of the spectrum. Because the excitation energies of alkali metals are relatively low, their emission spectra are readily observed in media such as flames. Many other elements do not have such low-energy excited levels and more energetic sources such as plasmas are required to produce a significant emission signal. The resonance lines of the more-difficult-to-excite elements often occur in the UV region of the spectrum. Their spectra are often easier to observe by absorption and fluorescence than by emission.

The radiant power of emission $\Phi_E$ from state $j$ to state $i$ is given by the population density of excited atoms $n_j$, times the probability $A_{ji}$ $(\text{s}^{-1})$ that an excited atom will undergo the transition, times the energy per emitted photon $h\nu_{ji}$, times the volume element observed $V$ $(\text{cm}^3)$. Or

$$\Phi_E = A_{ji} h\nu_{ji} n_j V \qquad (2\text{-}15)$$

The Einstein transition probability $A_{ji}$ is discussed further in Chapters 4 and 7 and in Appendix F.

Equation 2-15 shows that the radiant power of emission is proportional to the excited-state population density and thus to the analyte concentration through equation 2-14. Even if the emission source is not in

thermal equilibrium, equation 2-15 still applies and $\Phi_E$ will be proportional to analyte concentration if $n_j$ is proportional to analyte concentration. Although most emission techniques are atomic emission techniques, equations similar to 2-14 and 2-15 apply to molecular emission. In any case, the basis of emission methods is to measure $\Phi_E$ and to establish the exact relationship to the analyte concentration by standards as described in Chapter 1.

## Absorption Spectroscopy

In absorption methods the sought-for information is the magnitude of the radiant power from an external source that is absorbed by the analyte species. Here absorption of radiation is accompanied by excitation of the analyte. For absorption to occur, the frequency of the incident radiation must correspond to the energy difference between the two states involved in the transition as shown in Figure 2-6. For many conditions the absorption of radiation follows Beer's law (see Chapter 3 for derivation):

$$A = -\log T = -\log \frac{\Phi}{\Phi_0} = abc \qquad (2\text{-}16)$$

where $A$ is called the **absorbance**, $T$ is the **transmittance**, $a$ is the **absorptivity**, $b$ is the pathlength of absorption, and $c$ is the concentration of the absorbing species. If $b$ is expressed in cm, $a$ has the units of $(\text{conc})^{-1}\,\text{cm}^{-1}$. As shown in equation 2-16, absorption methods involve measuring the ratio of two radiant powers, calculating the absorbance, and relating the absorbance to concentration.

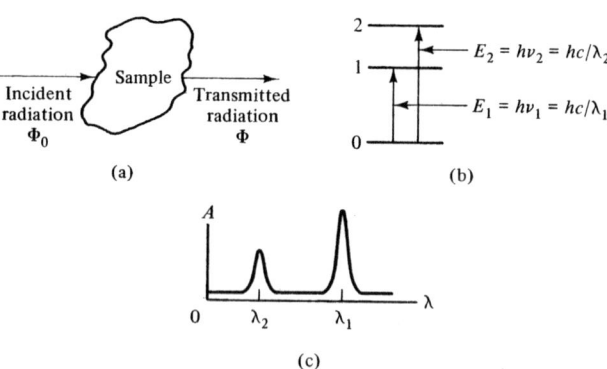

(a)

(b)

(c)

**FIGURE 2-6** Absorption methods. Radiation of incident power $\Phi_0$ can be absorbed by the analyte producing a beam of diminished transmitted power $\Phi$ (a) if the frequency of the incident beam, $\nu_1$ or $\nu_2$ corresponds to energy difference, $E_1$ or $E_2$ (b). The spectrum is shown in (c).

## Luminescence Spectroscopy

Luminescence is radiation emitted from relatively cool bodies. There are several classes of luminescence spectrochemical methods. In **chemiluminescence** and **bioluminescence**, excited analyte species are produced by chemical reactions, and the resulting emission is measured. **Electroluminescence** results from the movement of electrons in a sample and may be caused by an electrical discharge, by recombination of ions and electrons at an electrode, and by interactions of materials with accelerated electrons as in a cathode ray tube. **Triboluminescence** results from the mechanical separation of charges followed by a discharge (e.g., broken crystals of sugar). **Thermoluminescence** is the enhancement of other types of luminescence by the addition of heat. Chemiluminescence and bioluminescence are employed in analytical procedures. The excitation/emission transitions for these are also illustrated in Figure 2-5.

**Photoluminescence** methods utilize an external radiation source for excitation (as in absorption methods), but the sought-for information is the radiation emitted by the sample as shown in Figure 2-7. Molecular fluorescence and atomic fluorescence are examples of photoluminescence spectrochemical methods. If a monochromatic incident beam contains photons of an energy corresponding to the energy differences necessary for absorption, a portion of the incident radiant power $\Phi_0$ is absorbed so that the transmitted radiant power $\Phi$ is

less than the incident radiant power. Under many conditions the radiant power luminesced (for all wavelengths) $\Phi_L$ is proportional to the absorbed radiant power $(\Phi_0 - \Phi)$.

$$\Phi_L = k(\Phi_0 - \Phi) \qquad (2\text{-}17)$$

The constant $k$ is dependent on the species and its environment and is related to the efficiency with which the excited atom or molecule returns to the ground state by emission of a photon. The transmitted radiant power is related to the analyte concentration by Beer's law:

$$\Phi = \Phi_0 10^{-abc} \qquad (2\text{-}18)$$

If equation 2-18 is substituted into equation 2-17, the result is

$$\Phi_L = k\Phi_0(1 - 10^{-abc}) \qquad (2\text{-}19)$$

Equation 2-19 can be expanded in a Taylor series to yield

$$\Phi_L = k\Phi_0\left[2.303abc - \frac{(2.303abc)^2}{2!} + \cdots\right] \qquad (2\text{-}20)$$

When the term $abc$ is less than 0.01, higher-order terms in the expansion contribute less than 1% to $\Phi_L$, and under these conditions,

$$\Phi_L = 2.303k\Phi_0 abc = k'\Phi_0 c \qquad (2\text{-}21)$$

For low absorbances ($abc < 0.01$), the luminescence radiant power is directly proportional to the analyte concentration and to the radiant power incident on the sample $\Phi_0$. Luminescence methods then involve measurements of $\Phi_L$ to obtain the analyte concentration $c$. These methods are discussed further in Chapters 11 and 15.

## Scattering Methods

In addition to being absorbed by the sample, radiation from an external source can be scattered; the intensity, frequency, and angular distribution of scattered radiation can be used in spectrochemical methods. Molecular scattering methods are the subject of Chapter 16. Particles smaller than the wavelength of the incident radiation can scatter that radiation elastically without a change in its energy. Small-particle scattering is called **Rayleigh scattering**; it typically occurs with atoms or molecules. Rayleigh-scattered radiation occurs in all directions from the scattering particle.

Scattering from larger particles with dimensions

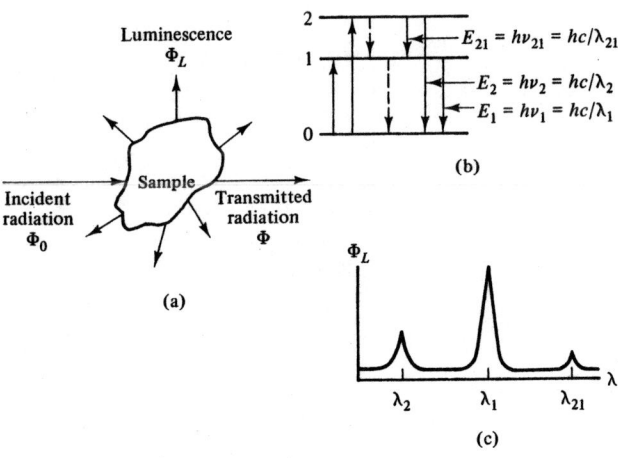

**FIGURE 2-7** Photoluminescence methods. Absorption of incident radiation from an external source (a) causes excitation of the analyte to state 1 or state 2 (b). Excited species can dissipate the excess energy by emission of a photon [luminescence (L)] or by radiationless processes (dashed lines) in (b). Emission is isotropic (a), and the frequencies emitted correspond to the energy differences between levels (c).

on the order of the wavelength of the incident radiation is often called **Debye scattering**. Here the scattered radiation is of the same frequency as the incident radiation, but the angular distribution of the scattered radiation, unlike Rayleigh scattering, is not uniform. Scattering from still larger particles is often termed **Mie scattering**. Large-particle scattering (Debye or Mie) can be used to determine particle sizes and is important in *turbidimetry* and *nephelometry* where suspended particles are the scatterers.

**Brillouin** and **Raman scattering** are forms of inelastic scattering which involve a change in the frequency of the incident radiation. Brillouin scattering results from the reflection of radiant energy waves by thermal sound waves, whereas Raman scattering involves the gain or loss of a vibrational quantum of energy by molecules. These methods are discussed further in Chapter 16.

The equations describing the dependence of the observed scattering signal on experimental parameters are complex, as shown in Chapter 16. However, as for photoluminescence techniques, the scattering signal is proportional to the incident radiant power.

## 2-4 SELECTION OF OPTICAL INFORMATION

Sorting out all the optical information that might be produced in the encoding step is a major step in a spectrochemical measurement. It is a rarity that all the information produced is useful or desirable. In analytical procedures the selection step allows us to separate the analyte optical signal from a majority of the potential interfering optical signals. The selection process is not perfect, however, and we must be aware of the limitations of the instruments and components used. The vast majority of analytical techniques select the desired information based only on its wavelength. As we shall see, there are many additional criteria which can be employed to improve selectivity. The use of time discrimination, position discrimination, polarization, multidimensional information, and chemical selectivity are but a few of the possibilities.

### Wavelength Selection

Wavelength selection in spectrochemical instruments can be based on absorption or interference filters, spatial dispersion of wavelengths, or interferometry. Wavelength selectors which disperse the spectral components of the optical signal spatially are the most common, and some of the major configurations are shown in Figure 2-8. The entrance slit defines the area of the source of radiation that is viewed. The dispersive element can be a prism which spatially separates wavelengths by refraction or a grating which disperses light based on diffraction. The image transfer system produces an image of the entrance slit on the focal plane.

As shown in Figure 2-8, the name of the dispersive wavelength selector depends on the arrangement of apertures or slits in the focal plane where the spectrum is

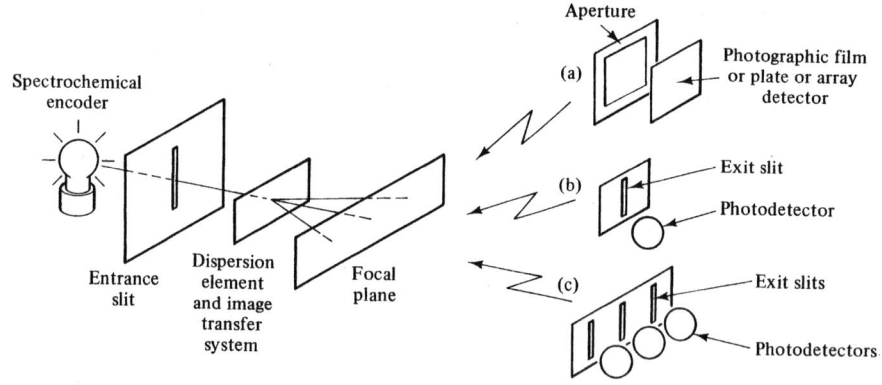

**FIGURE 2-8** Instrumentation for spatial dispersion and detection of optical signals. Some of the radiation from the spectrochemical encoder enters the entrance slit and strikes the dispersion element. The dispersion element and image transfer system cause each wavelength to strike a different position in the focal plane where different photodetector configurations can be used. If a photographic plate or array detector is placed in the focal plane as in (a), the device is called a spectrograph. If one exit slit is used in the focal plane to define the range of wavelengths to be passed to the photodetector as in (b), the dispersion device is called a monochromator. If multiple exit slits with a photodetector detector for each slit are employed as in (c), the dispersion device is called a polychromator.

dispersed as well as the type of detection used. In a **spectrograph**, a large aperture in the focal plane allows a wide range of wavelengths to strike a spatially sensitive detector such as a photographic plate. In recent years, solid-state video-type detectors have become available and are often employed in spectrographs in place of film. These detectors are actually an array of a large number of closely spaced miniature photoelectric detectors. They have the advantage that the spectrum can be obtained immediately without the time required for film development, for obtaining the density of the lines recorded, and so on. A **spectroscope** is a device that allows a visual observation of the spectrum. It is a spectrograph that uses a viewing screen for observing the spectrum in the focal plane.

In a **monochromator**, an exit slit about the same size as the entrance slit is used to isolate a small band of wavelengths from all the wavelengths that strike the focal plane. One wavelength band at a time is isolated and different wavelength bands can be selected sequentially by rotating the dispersion element to bring the new band into the proper orientation so that it will pass through the exit slit. If the focal plane contains multiple exit slits so that several wavelength bands can be isolated simultaneously, the wavelength selector is called a **polychromator**.

A **spectrometer** is a spectrochemical instrument which employs a monochromator or a polychromator in conjunction with photoelectric detection of the isolated wavelength band(s). The photodetector is placed just outside the exit slit. If a polychromator is employed with a separate photodetector for each exit slit, the instrument is often called a **direct-reading spectrometer**. Some spectrometers use optical components to sweep the spectrum quite rapidly across a single exit slit. These **rapid-scanning spectrometers** can obtain a spectrum in a few milliseconds.

A **spectrophotometer** is an instrument similar to a spectrometer except that it allows the ratio of the radiant powers of two beams to be obtained, a requirement for absorption spectroscopy. A **photometer** is a spectrochemical instrument which uses a filter for wavelength selection in conjunction with photoelectric detection.

**Interferometers** are nondispersive devices in which the constructive and destructive interference of light waves can be used to obtain spectral information. Several important interferometer types are discussed in Section 3-7.

Many additional optical components, such as mirrors and lenses, are used in the image transfer system to focus, collimate, and direct the radiation. Their details are presented in Chapter 3.

## Other Selection Criteria

Since many of the optical phenomena employed for spectrochemical analyses are time dependent, or can be made so, it is no surprise that time discrimination techniques are often employed for improving selectivity. The time dependence of the optical signal from the analyte can be used to distinguish the analyte signal from time-independent or steady-state background signals. If background signals are also time dependent, measurements can be made at a time interval which maximizes the $S/B$ if the background signals have a different time dependence than the analytical signal. Time discrimination is usually used along with the wavelength selection techniques discussed previously.

There are many additional ways to distinguish the desired optical signal from interfering signals. With many atomic emission sources, the intensity of the analyte and background emission vary with the spatial region of the source viewed. The $S/B$ can be significantly improved if the analyte emission can be monitored from a region that is relatively free from background emission, by adjustment of viewing position. In some spectrochemical methods, the selectivity for the analyte can be improved by using chemical reactions of the analyte with selective reagents. The change in the optical signal as a result of the reaction is measured and related to the analyte concentration. The degree of polarization of the optical signal is yet another criterion that can aid in selecting the analyte signal.

The incorporation of computers into spectrochemical instruments has made possible the collection and display of data as a function of more than one variable at a time. In many cases the information from multidimensional instruments can be used for optimization purposes, including improvement of the selectivity for the analyte. A great deal of research is under way that will help us treat the vast quantities of data that such multidimensional instruments can produce. The development of computer software to handle and reduce this information will remain a challenging task for many years to come.

## 2-5 MEASUREMENT OF OPTICAL SIGNALS

All spectrochemical techniques that operate in the UV-visible and IR regions of the spectrum employ similar instrumental components, as is evident from the general block diagram of Figure 2-1. The major instrumental differences between emission, photoluminescence, and absorption techniques occur in the arrangement and type of sample introduction system, encoding system,

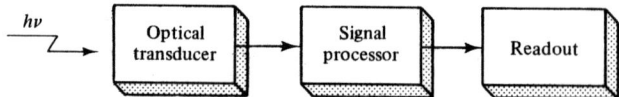

**FIGURE 2-9** Radiant power monitor. The radiant power monitor provides a numerical readout that is related to the radiant power (number of photons per second or watts) impingent on the transducer.

and information selection system. All techniques depend upon the measurement of radiant power. The **radiant power monitor** or optical transducer-signal processing-readout system is shown in block diagram form in Figure 2-9. The specific transducers and signal processing devices used in various regions of the spectrum in specific spectrochemical techniques are described in Chapter 4. In this section we explore how the analytical signal is extracted from the readout data in spectrochemical methods.

## Analytical Signal

As discussed in Chapter 1, the analytical signal is rarely obtained directly as a result of one spectrochemical measurement. Because of the presence of background and other extraneous signals, the analytical signal must be extracted from the raw readout data, although this process may be invisible to the operator in an automated instrument. Here we will define the various kinds of signals produced in spectrochemical methods to learn what is necessary in order to obtain the analytical signal.

The analytical signal for emission and chemiluminescence techniques is defined as the signal to be displayed by the readout device due only to analyte emission. It is given the symbol $E_E$, and we presume that $E_E$ is directly related to the radiant power of emission $\Phi_E$. Similarly, the analytical signal in photoluminescence techniques, $E_L$, is the measured signal due only to radiationally produced emission of the analyte. In the case of absorption methods, the analytical signal

is the absorbance $A$ due only to absorption of radiation by the analyte species.

Because of the presence of extraneous signals, such as signals from concomitants, the sample cell, and room light, at least two measurements are required to obtain the analytical signal. The background or extraneous signal that registers on the readout device is due to two primary sources. The first source is the **dark signal** $E_d$ of the radiant power monitor, which is the signal present when no radiation is impingent on the transducer. The second source is the **background signal**, $E_B$ due to background radiation that strikes the transducer. The background radiation is composed of radiation from all sources other than the desired optical phenomenon from the analyte.

The data domains of the different signals in a spectrochemical instrument prior to the signal processing system are dependent on the observation point. For example, in Figure 2-9, prior to the transducer the data are present as an optical radiant power in watts. The transducer can convert this optical signal to an electrical current, voltage, or charge. Normally, the output of the signal processing system to be displayed on the readout device is an electrical voltage. Hence, in general in this book, analyte and background signals will be written as voltages $E$. However, we will often need to look back at the magnitude and form of the signal prior to this point in the instrument in order to relate it to phenomena being measured, and in some cases the signals may be expressed in the frequency domain.

## Emission and Chemiluminescence Spectrometry

The basic instrumental configuration for wavelength-resolved emission spectrochemical methods is shown in Figure 2-10. The excitation source acting upon the sample in the sample container (sample cell) is the spectrochemical encoder of Figure 2-1. The emission that results from excitation of the analyte species by a flame, a plasma, or a chemical reaction encodes the concen-

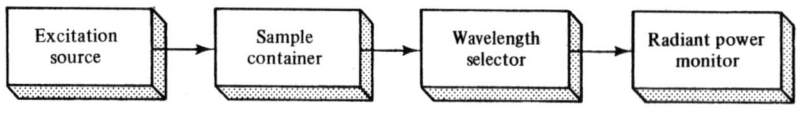

**FIGURE 2-10** Instrumentation for emission spectrochemical methods. The excitation source provides the external energy necessary to excite the analyte species. For example, the excitation source could be a flame, a plasma, a high-voltage spark discharge, or a chemical reaction. The sample container holds the sample. The wavelength selector passes a selected wavelength band emitted by the sample to the radiant power monitor.

tration of the analyte as the radiant power of emission $\Phi_E$. In some spectrochemical methods the excitation source and sample container are an integral unit, as in the nebulizer-burner used in flame emission and the reaction cell used in chemiluminescence.

When the analytical sample is present in the sample cell, a total or composite signal $E_{tE}$ is obtained. This total signal is the sum of analytical signal $E_E$, the dark signal $E_d$, and the background emission signal $E_{bE}$. To extract the analytical signal, a second measurement is required to obtain the sum of the dark signal and the background emission signal. This second measurement is normally made by replacing the analytical sample with a blank that is ideally identical to the analytical sample except that the analyte is missing. Thus the analytical signal is obtained as follows:

$$E_E = E_{tE} - E_{bk} \qquad (2-22)$$

where $E_{bk}$ is the blank signal given by $E_d + E_{bE}$. If desired, the dark signal can be obtained separately by blocking all radiation from reaching the radiant power monitor. The background emission signal could then be obtained from $E_{bk} - E_d$. In many instruments the blank solution is used to adjust the readout device to read zero by suppression of the blank signal. This establishment of the zero position is still, however, a measurement of the blank signal.

## Photoluminescence Spectrometry

The instrumentation used in photoluminescence methods is illustrated in Figure 2-11. Here an external source of electromagnetic radiation excites the analyte (photoexcitation). The analyte concentration is optically encoded as the luminescent radiant power $\Phi_L$, which is measured with the radiant power monitor. The emission wavelength selector that views the luminescence of the sample is typically placed to collect radiation at 90° with respect to the excitation axis. Other geometries, such as front surface and 180°, are used in special situations (see Chapters 11 and 15). In some cases only one of the wavelength selectors is necessary.

When the analytical sample is placed in the sample cell, a composite total luminescence signal $E_{tL}$ is obtained:

$$E_{tL} = E_L + E_E + E_{bk} \qquad (2-23)$$

where $E_L$ is the analyte luminescence signal, $E_E$ is the analyte thermal emission signal, and $E_{bk}$ is the blank signal. The blank signal is composed of the dark signal $E_d$, the background emission signal $E_{bE}$, a scattering signal $E_{sc}$ due to scattered source radiation, and a background luminescence signal $E_{bL}$:

$$E_{bk} = E_d + E_{bE} + E_{sc} + E_{bL} \qquad (2-24)$$

Analyte and background emission in the UV-visible region are usually significant only in atomic spectroscopy.

The analyte luminescence signal $E_L$ can be obtained with two measurements only if the analyte emission signal $E_E$ is small compared to $E_L$, which is often the case. If $E_E$ is significant, subtraction of the blank signal gives a measured analyte luminescence signal $E_L'$ that differs from $E_L$ as shown in the equation

$$E_L' = E_{tL} - E_{bk} = E_L + E_E \qquad (2-25)$$

To obtain the true analyte luminescence signal $E_L$ when $E_E$ is significant, the excitation source must be turned off. Then the two measurements indicated in equation 2-22 are made to obtain $E_E$. Subtraction of $E_E$ from $E_L'$ gives the true analyte luminescence signal. In some cases it is possible to eliminate the measured contribution from analyte emission optically or electronically. For example, if the excitation source is modulated and alternating-current (ac) amplification is used, the ac luminescence signal can be distinguished from the dc emission signal. Often the blank measurement is used to set the zero position of the readout device.

## Absorption Spectrometry

A typical absorption spectrometer is illustrated in Figure 2-12. The absorption spectrometer is essentially identical to the luminescence spectrometer of Figure

**FIGURE 2-11** Instrumentation for photoluminescence spectrometry. Specific wavelengths from an external radiation source are isolated by the excitation wavelength selector to excite the analyte in the sample cell. The emission wavelength selector selects the wavelength band where analyte luminescence is concentrated and passes it to the radiant power monitor.

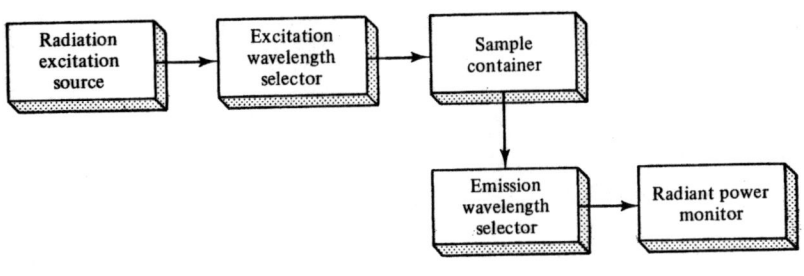

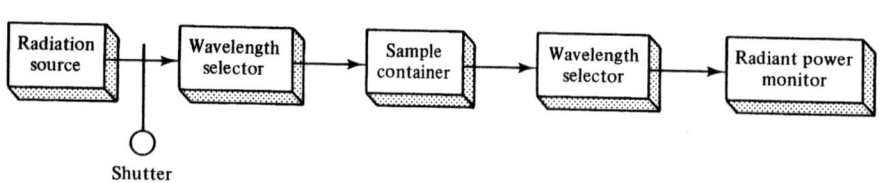

Shutter

**FIGURE 2-12** Absorption spectrometer. A narrow spectral band from the radiation source is passed through the sample. The transmitted radiant power is measured by the radiant power monitor. Replacement of the analytical sample by a reference provides a measure of the reference radiant power. The ratio of the radiant power transmitted by the sample to that transmitted by the reference is used to calculate the absorbance $A$ of the sample.

2-11 except that the source, sample cell, and transducer are all on the same optical axis. This permits the measurement of the transmitted radiant power. The shutter allows the user to block the radiation source in order to obtain the dark signal. Usually, only one wavelength selector is required.

Absorption measurements can be made in two ways: The transmittance $T$ can be displayed by the readout device and the absorbance $A$ calculated manually from $A = -\log T$; or the logarithmic conversion can be done electronically or with computer software and the absorbance $A$ displayed by the readout device. Both of these readout schemes are used, although direct absorbance readout is becoming the more common.

*Transmittance Readout.* The ideal or true transmittance $T$ is the ratio of radiant power passed by the analyte to the radiant power passed by an ideal blank. It could be obtained by (1) measuring the signal $E_s$ that results from the source radiant power passing through the analytical sample; (2) measuring the signal $E_r$ that results from the source radiant power passing through the ideal blank or reference solution; and (3) obtaining the transmittance as in

$$T = \frac{E_s}{E_r} \qquad (2\text{-}26)$$

In practice, however, the presence of other signals (dark signal, background emission) necessitates a third measurement. The measured transmittance $T'$ is defined by the equation

$$T' = \frac{E_{st} - E_{0t}}{E_{rt} - E_{0t}} \qquad (2\text{-}27)$$

where $E_{st}$ is the total sample signal obtained with the source shutter open and the analytical sample in the sample container, $E_{0t}$ is the zero percent transmittance (0% $T$) signal obtained with the shutter closed and the blank in the sample container, and $E_{rt}$ is the 100%

transmittance (100% $T$) signal obtained with the shutter open and the blank (reference) in the sample container.

The 0% $T$ signal $E_{0t}$ is composed of any background emission $E_{bE}$ and dark current $E_d$ as shown by the equation

$$E_{0t} = E_{bE} + E_d \qquad (2\text{-}28)$$

When the blank is in the sample container and the shutter open, the measured total reference signal $E_{rt}$, called the 100% $T$ signal, is composed of the reference transmission signal $E_r$, the 0% $T$ signal, and any background luminescence $E_{bL}$:

$$E_{rt} = E_r + E_{0t} + E_{bL} \qquad (2\text{-}29)$$

When the analytical sample is in the sample container and the shutter is open, the measured signal is $E_{st}$, the total sample signal. This signal is given by

$$E_{st} = E_s + E_E + E_L + E_{bL} + E_{0t} \qquad (2\text{-}30)$$

where $E_s$ is the sample transmission signal, $E_E$ is the analyte emission signal, and $E_L$ is the analyte luminescence signal.

From equations 2-27 to 2-30, the measured transmittance $T'$ is found to be

$$T' = \frac{E_s + E_L + E_E + E_{bL}}{E_r + E_{bL}} \qquad (2\text{-}31)$$

The measured transmittance $T'$ is equal to the true transmittance $T$ (equation 2-26) only when $E_r \gg E_{bL}$ and $E_s \gg E_L + E_E + E_{bL}$. If one or more of the signals $E_L$, $E_E$, or $E_{bL}$ is significant, several additional measurements must be made to correct equation 2-31 to obtain the true transmittance, or additional optical information selection techniques must be used to discriminate against the extraneous signals. As in photoluminescence techniques, analyte and background emission are usually significant only for atomic spectrometry. The measured transmittance $T'$ is then used to calculate an absorbance $A' = -\log T'$ which is an approximation to the true absorbance $A = -\log T$.

### Direct Absorbance Readout.

As part of the signal processing electronics, many modern absorption spectrometers have provision for obtaining and displaying absorbance directly. The true absorbance $A$ is given by

$$A = \frac{E_A}{k'} \qquad (2\text{-}32)$$

where $E_A$ is the voltage proportional to the analyte absorbance and $k'$ is a logarithmic conversion factor in volts per absorbance unit.

The voltage $E_A$ and hence $A$ are found from two measurements: First a reference logarithmic voltage or zero absorbance voltage $E_{lr}$ is obtained with the shutter open and the blank in the sample container; then a sample logarithmic voltage $E_{ls}$ is obtained with the shutter open and the analytical sample in the sample container. The voltage $E_A$ is then given by

$$E_A = E_{ls} - E_{lr} = -k' \log \frac{E_s}{E_r} = -k' \log T \qquad (2\text{-}33)$$

The voltages $E_{ls}$ and $E_{lr}$ are logarithmically related to $E_s$ and $E_r$ as follows:

$$E_{ls} = -k' \log \frac{E_s}{k''} \quad \text{and} \quad E_{lr} = -k' \log \frac{E_r}{k''}$$

$$(2\text{-}34)$$

where $k''$ is a constant reference voltage. Often $E_{lr}$ is set to zero on the readout device so that $E_{ls}$ is read out directly as $E_A$.

Note that in the two-step absorbance measurement scheme, a measurement is not made with the light-source shutter closed (0% $T$) since $A$ would be infinity. Thus $(E_d + E_{bE})$ must be negligible compared to $E_s$ and $E_r$ or electronically or optically set to zero by other means. Also, $E_E + E_{bL} + E_L$ must be negligible so that $E_s = E_{st}$ and $E_r = E_{rt}$; otherwise, the measured absorbance $A'$ only approximates the true absorbance $A$.

We will make use of these equations in later chapters as we explore the factors influencing the accuracy and precision of spectrochemical methods.

## PROBLEMS

**2-1.** Consider the spectrometer below, which can be used for emission, luminescence, or absorption measurements. Note that the radiant power monitor can be placed in two positions in which it is equidistant from the sample, and that shutters 1 and 2 control the radiant power incident upon the sample and radiant power monitor, respectively. The following measurements were made with either an analyte solution (*a*) or an ideal blank (*b*), where C = closed shutter and O = open shutter.

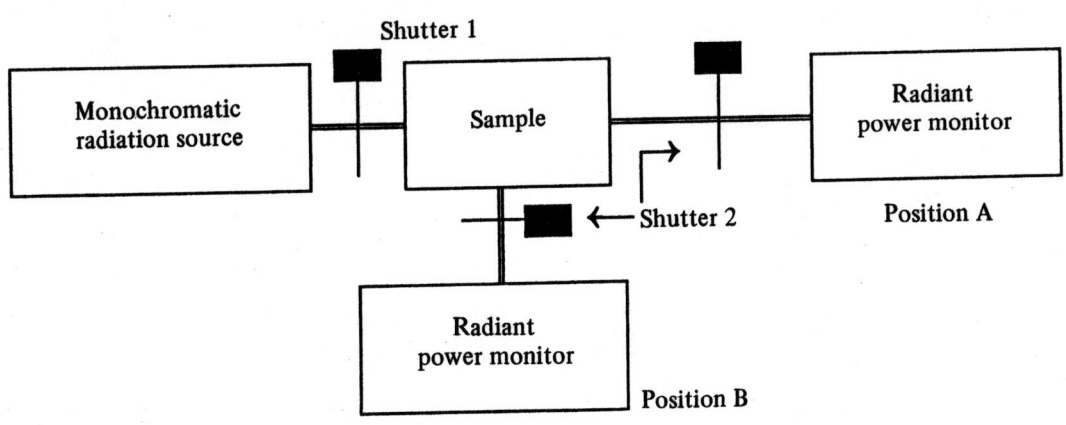

| Position of radiant monitor | Shutter 1 | Shutter 2 | Sample | Signal (V) |
|---|---|---|---|---|
| A | C | O | $a$ | 0.15 |
| A | C | O | $b$ | 0.10 |
| A | C | C | $b$ | 0.07 |
| A | O | O | $b$ | 3.20 |
| A | O | O | $a$ | 1.50 |
| B | C | O | $a$ | 0.15 |
| B | O | O | $a$ | 0.40 |
| B | O | O | $b$ | 0.12 |

In this experiment, scattering and chemiluminescence are insignificant. Calculate or indicate the following signals: $E_d$, $E_{bE}$, $E_{bL}$, $E_E$, $E'_L$, $E_L$, $E_{0t}$, $E_{rt}$, $E_{st}$, $E_s$, $E_r$, $T$, $A$, $T'$, $A'$.

**2-2.** A point source emits 25.13 W. Express the intensity as a radiant intensity.

**2-3.** An extended source is spherical in shape with a radius of 2.00 cm and emits 12.57 W from 399.5 to 400.5 nm. Determine the spectral radiant emittance, the spectral radiant intensity, and the spectral radiance at 400 nm.

**2-4.** A beam of 632.8-nm photons from a He-Ne laser strikes a detector area of 5.0 mm². The laser has a flux of $3.18 \times 10^{15}$ photons s$^{-1}$.

(a) What is the laser radiant power in watts?
(b) What is the laser irradiance at the detector?
(c) Express the answer for part (b) in terms of photons s$^{-1}$ rather than watts.

**2-5.** Why are electronic, vibrational, and rotational spectra observed in different wavelength regions?

**2-6.** Calculate the absorbances corresponding to transmittances of 0.100, 0.0316, and 0.0100.

**2-7.** An extended source emits 2.00 W cm$^{-2}$ sr$^{-1}$ nm$^{-1}$ at 300 nm. Calculate the spectral radiant power impingent on a 1.00-cm² receptor that is 2.00 m away in W nm$^{-1}$ and photons s$^{-1}$ nm$^{-1}$ if a 1.00-mm-diameter aperture is placed at the source. What is the incident radiant power in watts over 10.0 nm centered at 300 nm if the source radiance is constant over this region?

**2-8.** Consider a collection of atoms in thermal equilibrium at 3000°C. What fraction of the atoms are in the first excited state if the energy difference between the ground and the first excited state corresponds to 400 nm? Assume that the statistical weights of the ground and first excited states are the same and that higher excited states are not significantly populated.

**2-9.** The atomic density of an excited state is $4.00 \times 10^{10}$ atoms cm$^{-3}$. Calculate the radiant power of emission from a volume element of 1.00 mm³ of this excited state in photons s$^{-1}$ if the Einstein transition coefficient is $1.00 \times 10^9$ s$^{-1}$.

**2-10.** What is the difference between a monochromator and a polychromator?

**2-11.** An ordinary 3.0-V flashlight bulb draws roughly 0.25 A of current and converts about 1.0% of the electrical power into light ($\lambda \approx 550$ nm). If the flashlight beam initially has a cross-sectional area of 10 cm²:
(a) How many photons are emitted per second?
(b) How many photons are found in a cubic meter of the beam?
(c) What is the radiant emittance (W cm$^{-2}$) of the beam as it leaves the flashlight?

**2-12.** A 1.0-mW laser has a beam diameter of 2 mm. Assume that the beam has negligible divergence and calculate the radiant energy density (J cm$^{-3}$) of the laser beam.

**2-13.** The 4p level of the Na atom is 3.75 eV above the 3s ground level. What wavelength of radiation (in nm) would be required to excite Na from the ground level to the 4p level? What frequency is required? What is the energy of this transition in joules?

**2-14.** For the copper atom, there are resonance lines at 324.7 and 327.4 nm. What is the energy difference in eV between the two excited states involved in these two transitions?

**2-15.** A sample is illuminated with 1.0 μW of radiation and transmits 90% of the incident radiation. Calculate the maximum value of the luminescence radiant power that could be observed.

**2-16.** The irradiance on a receptor 50 cm from a source with a projected area of 0.010 cm² is 2.0 μW cm$^{-2}$. Calculate the radiance of the source.

# REFERENCES

1. F. Grum and R. J. Becherer, *Optical Radiation Measurements*, vol. 1, *Radiometry*, Academic Press, New York, 1979.

2. W. G. Driscoll and W. Vaughan, eds., *Handbook of Optics*, McGraw-Hill, New York, 1978. A very useful handbook with a good deal of information about optical principles and optical materials. Contains a chapter on "Radiometry and Photometry" (J. F. Snell) that is quite practical.

3. F. E. Nicodemus, ed., *Radiometry: Optical Resource Letter*, American Institute of Physics, New York, 1970.

4. E. L. Dereniak and D. G. Crowe, *Optical Radiation Detectors*, Wiley, New York, 1984. Includes a chapter on radiometry.

5. H. T. Betz and G. L. Johnson, "Spectroradiometric Principles," in *Analytical Emission Spectroscopy*, vol. I, pt. II, E. L. Grove, ed., Marcel Dekker, New York, 1971,

pp. 323–381. Presents a good introductory discussion of the principles of radiometry.

6. *Electro-Optics Handbook*, Technical Series EOH-11, RCA Corp., Lancaster, Pa., 1974. Has a good discussion of the radiometric and the photometric systems.

7. J. Wilson and J. F. B. Hawkes, *Optoelectronics: An Introduction*, Prentice-Hall, Englewood Cliffs, N.J., 1983. Includes a brief discussion of the radiometric system.

8. IUPAC, *Compendium of Analytical Nomenclature*, Pergamon Press, Oxford, 1978.

# CHAPTER 3

# Optical Components
# of Spectrometers

We have seen in Chapter 2 that there are four or five basic components that are present in optical spectrometers. In addition to excitation sources for emission, external radiation sources for luminescence and absorption, sample containers, wavelength selection devices, and radiant power monitors, optical spectrometers contain a number of optical elements for collimating, focusing, and directing electromagnetic radiation. In this chapter we consider the optical components and systems used in spectrochemical instruments. Radiation sources and radiant power monitors are described in Chapter 4, while the remaining components of spectrochemical instruments are discussed in later chapters dealing with specific methods.

Prior to considering optical elements in detail, it is necessary to review several important optical relationships. Thus this chapter begins with a discussion of the conservation law, and the laws of reflection, refraction, and absorption. Then the phenomena of interference and diffraction are explored. The nature and production of polarized light are discussed in some detail. Devices that can amplitude modulate a radiant energy beam are then discussed. These modulators can be mechanical, electro-optical, acousto-optical, or magneto-optical devices. The mirrors and lenses that are used for imaging purposes in spectrochemical instru-

ments are introduced, and the nonidealities which must be considered in their use are described. Filters, prisms, and gratings are commonly employed for wavelength selection purposes. These components are first considered as optical elements and then they are combined with lenses and mirrors to form complete wavelength selection systems. The chapter ends with a discussion of Fabry–Perot, Michelson, and other interferometers, devices that are used in nondispersive wavelength selection systems.

## 3-1 BASIC OPTICAL RELATIONSHIPS

There are several basic laws in optics from which many other important relationships are derived. The conservation law, the laws of reflection and refraction, and the absorption law are used throughout this chapter in our discussion of optical components and spectrometric systems.

### The Conservation Law

A basic principle of wave motion states that when a wave strikes a boundary between two media, a portion of the wave is reflected, a portion is absorbed, and a portion is transmitted into the new medium. This is often known as the *conservation law*. Consider radiant energy of wavelength $\lambda$ to be incident on a boundary between two media. The fraction of the incident radiant energy lost by absorption at the interface or surface is known as the **spectral absorptance**, $\alpha(\lambda)$. The fraction reflected at the interface or surface is the **spectral reflectance**, $\rho(\lambda)$, while the fraction transmitted into the entered medium is the **spectral transmittance**, $T(\lambda)$. The quantitative statement of the conservation law is then

$$\alpha(\lambda) + \rho(\lambda) + T(\lambda) = 1 \qquad (3\text{-}1)$$

The conservation law has several practical consequences in spectrochemical measurements. In the absence of significant reflection at an interface, the spectral absorptance is given by $\alpha(\lambda) = 1 - T(\lambda)$. In the absence of absorption, the spectral reflectance is $\rho(\lambda) = 1 - T(\lambda)$, which will be used in discussing reflection and refraction at interfaces. In the absence of any transmission into the second medium, the conservation law tells us that $\alpha(\lambda) = 1 - \rho(\lambda)$. We will use this last relationship in discussing blackbody radiation in Section 4-1. If there is no transmission into the second medium and no absorption, $\rho(\lambda)$ is equal to unity, and we have total internal reflection. In all cases it is important to bear in mind that the complete conservation law is that given in equation 3-1. The other versions are special cases in which assumptions have been made. Throughout this book we will usually use $T$ for transmittance instead of $T(\lambda)$ and understand that $T$ is wavelength dependent. In the optics literature, $\tau$ is often used as the symbol for transmittance (sometimes called the *transmission factor*).

### The Laws of Reflection and Refraction

Maxwell showed that the velocity of all electromagnetic waves in free space ($c$, the speed of light) could be expressed theoretically as $c = 1/\sqrt{\varepsilon_0\mu_0}$, where $\varepsilon_0$ is the **permittivity of free space** ($\varepsilon_0 = 8.854 \times 10^{-12}$ $C^2$ $N^{-1}$ $m^{-2}$) and $\mu_0$ is the **permeability of free space** [$\mu_0 = 4\pi \times 10^{-7}$ N $s^2$ $C^{-2}$; N is newtons (kg m $s^{-2}$) and C is coulombs (A s)]. Thus $c \approx 3 \times 10^8$ m $s^{-1}$. The currently accepted value is $c = 2.99792458 \times 10^8$ m $s^{-1}$.

When radiation propagates through a medium of **electric permittivity** $\varepsilon$ and **permeability** $\mu$, its velocity is $v = 1/\sqrt{\varepsilon\mu}$. The ratio of the speed of an electromagnetic wave in vacuum to that in matter is called the **absolute index of refraction**, $\eta$, which is defined as

$$\eta = \frac{c}{v} = \sqrt{\frac{\varepsilon\mu}{\varepsilon_0\mu_0}} \qquad (3\text{-}2)$$

The permittivity of a medium is related to that of vacuum (free space) by $\varepsilon = K_d\varepsilon_0$, where $K_d$ is the dimensionless **dielectric constant** or relative permittivity of the medium ($K_d \geq 1$). The permeability of a material is related to that of vacuum by $\mu = K_m\mu_0$, where $K_m$ is the dimensionless **relative permeability** ($K_m \geq 1$). Thus the index of refraction can also be written as

$$\eta = \sqrt{K_d K_m} \qquad (3\text{-}3)$$

The refractive index of a medium varies somewhat with the wavelength of the incident radiation, a phenomenon known as **dispersion**. Media that are colorless and transparent have characteristic frequencies of oscillation (the natural frequencies of the atomic and molecular oscillators) in the ultraviolet region, where they become opaque. Glasses, for example, may have characteristic oscillations at wavelengths near 100 nm. For these materials, the refractive index increases as the radiation frequency approaches a natural oscillation frequency. Hence, for materials such as glasses, $\eta$ increases with increasing frequency. This behavior is called **normal dispersion**. If the refractive index of a material is given without reporting the wavelength at which it was measured, $\eta$ is normally assumed to be specified for the Na D line at 589 nm. For glass and visible wavelengths $\eta$ is typically about 1.5, and we will use this value for rough calculations (see Appendix B for optical properties of materials).

The frequency of an electromagnetic wave is determined by the radiation source and is independent of the medium. Thus in a medium of refractive index $\eta > 1$, a wave of frequency $\nu$ undergoes a reduction in wavelength compared to that in vacuum. The wavelength in the medium is $\lambda = v/\nu = c/\eta\nu$. When a wave goes from a medium of refractive index $\eta_1$ into a medium of refractive index $\eta_2$, the ratio of the wavelengths in the two media is given by $\lambda_2/\lambda_1 = \eta_1/\eta_2$. The wavelength is thus smaller in the medium of higher refractive index. For the sodium D line with a wavelength of 589.0 nm in vacuum, the wavelength in glass is $\lambda_{glass} = 589.0$ nm $\times 1.0/1.5 = 393$ nm. The frequency of the light, however, remains $5.09 \times 10^{14}$ Hz. The velocity of the light in glass is $v = c/\eta_{glass} = 3.0 \times 10^8$ m $s^{-1}/1.5 = 2.0 \times 10^8$ m $s^{-1}$.

Let us now consider a monochromatic plane wave that is incident on a smooth interface separating two media of refractive indices $\eta_1$ and $\eta_2$ as shown in Figure 3-1a. We shall assume that both media are nonabsorbing so that $\rho(\lambda) = 1 - T(\lambda)$. The relationships between the angle of incidence $\theta_1$, the angle of reflection $\theta_3$, and the angle of refraction $\theta_2$ are readily derived from trigonometry and the velocities in the two media.

The time required for a wavelet at point $B$ in

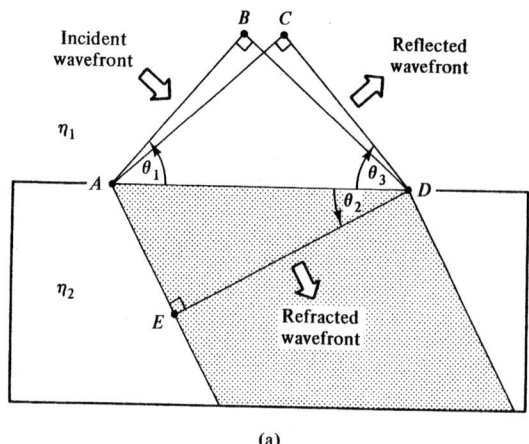

(a)

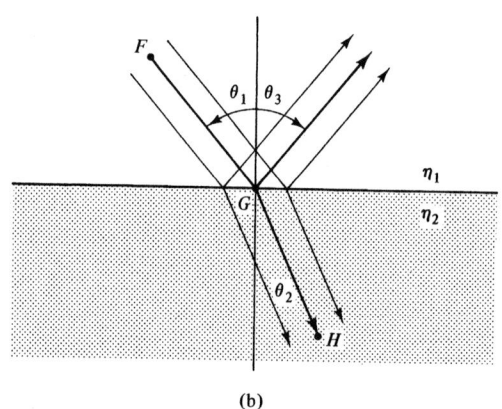

(b)

**FIGURE 3-1**  Wave (a) and ray (b) representation of reflection and refraction. In (a), the incident wavefront makes an angle $\theta_1$ with the surface, the reflected wavefront makes an angle $\theta_3$, and the transmitted wavefront makes an angle $\theta_2$. From trigonometry, $\sin \theta_1 = BD/AD$, $\sin \theta_3 = AC/AD$, and $\sin \theta_2 = AE/AD$. Thus $\sin \theta_1/BD = \sin \theta_3/AC = \sin \theta_2/AE = 1/AD$. The rays shown in (b) are perpendicular to the corresponding wavefronts in (a). In the ray representation, angles $\theta_1$, $\theta_2$, and $\theta_3$ have the same meaning as in the wave representation, but are measured from the normal to the surface.

Figure 3-1a to proceed to point $D$ is $t = BD/v_1$, where $v_1$ is the velocity in the incident medium. Because the velocity of the reflected wave is also $v_1$, a wavelet at point $A$ covers distance $AC$ in time $t$. Thus $AC = BD$, and

$$\theta_1 = \theta_3 \qquad (3\text{-}4)$$

Equation 3-4 is the **law of specular reflection**, and it states that the angle of incidence equals the angle of reflection.

Figure 3-1a shows the transmitted beam in a me-

dium of refractive index $\eta_2$ different from that of the incident beam. It travels distance $AE = v_2 t$ in time $t$. Thus $\sin \theta_1/(v_1 t) = \sin \theta_2/(v_2 t)$, or $\sin \theta_1/\sin \theta_2 = v_1/v_2$. From equation 3-2 we can express the velocity ratio as $v_1/v_2 = \eta_2/\eta_1$. Hence

$$\eta_1 \sin \theta_1 = \eta_2 \sin \theta_2 \qquad (3\text{-}5)$$

Equation 3-5 is known as **Snell's law of refraction**.

Figure 3-1a shows the wavefront representation of reflection and refraction, while the ray representation is shown in Figure 3-1b. A ray follows a trajectory that is orthogonal to that of the wavefront. The ray diagram shows more clearly the bending that occurs upon refraction. Snell's law (Equation 3-5) tells us that a ray bends toward the normal ($\theta_2 < \theta_1$) when it enters a medium of higher refractive index ($\eta_2 > \eta_1$) and away from the normal when it enters a medium of lower refractive index.

Consider the ray in Figure 3-1b advancing from point $F$ to point $H$. The transit time from $F$ to $H$ is given by

$$t = \frac{FG}{v_1} + \frac{GH}{v_2}$$

where $FG$ is the distance traveled in the medium of refractive index $\eta_1$ and $GH$ is the distance traveled in the medium of refractive index $\eta_2$. In general the transit time of a ray in various media will be given by

$$t = \sum_{i=1}^{m} \frac{x_i}{v_i} \qquad (3\text{-}6)$$

where $x_i$ is the distance and $v_i$ is the velocity in medium $i$. Since $\eta_i = c/v_i$, equation 3-6 can be written

$$t = \frac{\sum x_i \eta_i}{c} = \frac{\text{OPL}}{c} \qquad (3\text{-}7)$$

where the summation in equation 3-7 is called the **optical pathlength** (OPL), or the distance a photon would travel in time $t$ in a vacuum. An important principle, called **Fermat's principle** states that light, in going from one point to another, traverses the route having the shortest optical pathlength. Snell's law can also be derived from this general principle.

We have been considering reflection from a smooth surface, or more precisely a surface with irregularities small compared to the wavelength of the radiation. Reflection from such surfaces is **specular reflection**. If the surface is rough, however, with irregularities comparable to the wavelength, **diffuse reflection** occurs, and the reflected light travels in all directions. The law of reflection still holds, however, for any region of the surface small enough to be smooth, but because of different orientations of various regions, the total reflected beam is not uniform in direction.

*Reflection Losses at Interfaces.*   There are many cases in spectrochemical instruments where radiant energy must be transmitted across one or more interfaces separating dielectric media of different refractive indices. For example, there are two air–glass interfaces to traverse when electromagnetic radiation is transmitted through a lens. In a spectrophotometer cuvette filled with solution there are two air–glass interfaces and two glass–solution interfaces to traverse. If it is assumed that no absorption occurs in these situations, the fraction of the incident radiation transmitted is just $1 - \rho(\lambda)$ by the conservation law. Hence, any radiation that is reflected at the interface will cause a loss in the radiant power transmitted into the new medium.

If the incident beam is monochromatic and normal to the interface, the reflectance $\rho(\lambda)$ is given by

$$\rho(\lambda) = \left(\frac{\eta_2 - \eta_1}{\eta_2 + \eta_1}\right)^2 \qquad (3\text{-}8)$$

This equation is known as the **Fresnel equation**. Note that the larger the difference in refractive indices, the larger the reflectance. Also note that when $\eta_2 = \eta_1$, there is no reflection. For an air–glass interface, where $\eta_{air} = 1$ and $\eta_{glass} = 1.5$, approximately 4% of the light incident perpendicular to the interface is reflected and 96% is transmitted unless an antireflection coating is applied (see Appendix B). Some complicated lens systems have 10 or 20 such air–glass interfaces and the reflective losses can become quite large. (Appendix B discusses such losses in more detail.)

If unpolarized radiation strikes the interface at an angle other than 90°, the reflectance varies with the angle of incidence according to Fresnel's complete equation:

$$\rho(\lambda) = \frac{1}{2}\left[\frac{\sin^2(\theta_1 - \theta_2)}{\sin^2(\theta_1 + \theta_2)} + \frac{\tan^2(\theta_1 - \theta_2)}{\tan^2(\theta_1 + \theta_2)}\right] \qquad (3\text{-}9)$$

Here $\theta_1$ is the angle of incidence and $\theta_2$ the angle of refraction measured with respect to the normal to the surface. Figure 3-2 (curve a) illustrates how reflectance changes with the angle of incidence for a beam traveling from air into glass. Note that the reflectance changes only slightly up to an angle of 60° and then increases rapidly until it is 100% at grazing incidence (90°).

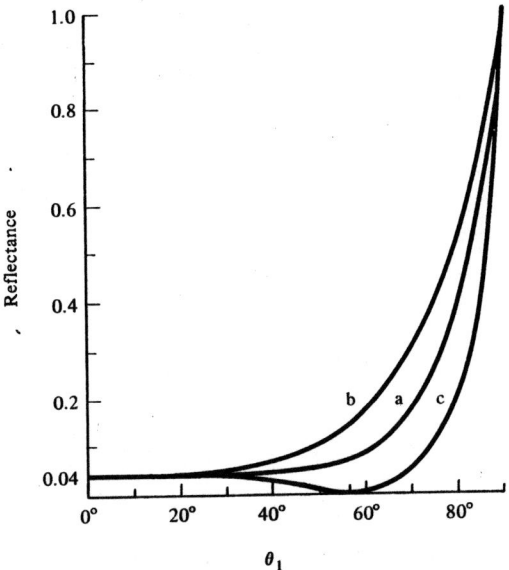

**FIGURE 3-2**   Reflectance vs. angle of incidence for a monochromatic beam traveling from air ($\eta_1 = 1$) into glass ($\eta_2 = 1.5$) as calculated from the Fresnel equation (3-9). For a given angle of incidence $\theta_1$, the angle of refraction $\theta_2$ is calculated from Snell's law (equation 3-5). Curves a, b, and c are for unpolarized, perpendicular-polarized, and parallel-polarized light, respectively. Note that the reflectance of the perpendicular component always exceeds that of the parallel component. Note also that at a particular angle, known as Brewster's angle, the reflectance of the parallel component becomes zero. For a beam traveling from air into glass, this angle is 58°40′.

*Total Internal Reflection.*   Consider the transmission and reflection of radiation where the source is in a medium of refractive index $\eta_1$ that is larger than the refractive index $\eta_2$ of the transmission medium. Equation 3-9 tells us that the transmitted radiant flux $[T(\lambda) = 1 - \rho(\lambda)]$ is a maximum when the incident beam is normal to the surface ($\theta_1 = 0°$). As shown in Figure 3-3, increasing the angle of incidence causes the radiant power of the reflected beam to increase, while the radiant power of the refracted beam grows weaker. At the **critical angle** $\theta_1 = \theta_c$, the transmitted beam is exactly parallel to the boundary ($\theta_2 = 90°$), and the transmitted intensity goes practically to zero. Since $\sin\theta_1 = (\eta_2/\eta_1)\sin\theta_2$, at $\theta_2 = 90°$, $\theta_c = \sin^{-1}(\eta_2/\eta_1)$. For incident angles exceeding $\theta_c$, all of the incident radiant flux is reflected into the more dense medium (see Figure 3-3d) in the phenomenon known as **total internal reflection**.

For an air–glass interface, the critical angle is $\theta_c \approx 42°$. Thus light originating in the glass will be completely reflected at an air–glass interface if the angle of incidence exceeds 42°. Furthermore, little loss of intensity occurs in

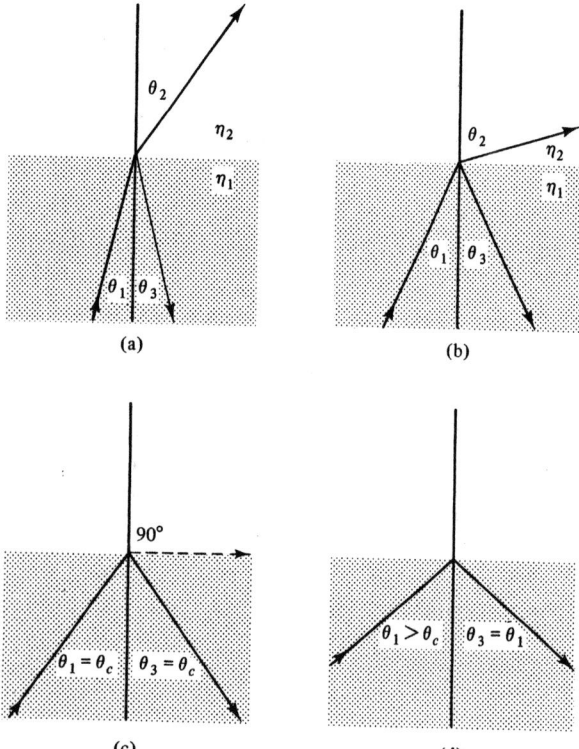

(a)  (b)

(c)  (d)

**FIGURE 3-3** Total internal reflection. The incident beam is in a medium of refractive index $\eta_1$ greater than that of the transmission medium. Hence the refracted beams bend away from the normal (a). As the angle of incidence $\theta_1$ becomes larger (b), the reflected beam grows stronger and the refracted beam grows weaker. At angles of incidence equal to (c) or exceeding (d) the critical angle $\theta_c$, the transmitted beam intensity goes to zero.

total internal reflection. Many reflex cameras, binoculars, and telescopes use totally reflecting prisms for this reason. Some refractometers, which measure refractive index, are based on total internal reflection. This phenomenon is also the basis for fiber optics transmission of light, as discussed in Section 3-3.

## The Absorption Law

Another process that can affect radiation is absorption by the medium. In the absorption process atoms or molecules in the medium are excited, and the energy absorbed can be dissipated as thermal energy, radiant energy (e.g., luminescence), or chemical energy (e.g., photochemical reactions). The amount of radiation absorbed depends on the total number of absorbers that interact with the beam. We should not be surprised, then, that the amount of radiation absorbed depends on the thickness of the medium and the concentration of the absorbing species.

Several assumptions must be made to derive the absorption law. These basic assumptions are listed in Table 3-1.

Let us consider an absorbing medium with a uniform concentration of absorbers. The radiant flux incident on a thin slice of absorbing material is $\Phi$, while that emerging is $\Phi - d\Phi$. When the incident flux is increased, the number of photons removed from the beam is increased in direct proportion; that is, $d\Phi$ is proportional to $\Phi$. Similarly, since the number of absorbers that interact with the beam is directly proportional to the thickness of the slice $db$,

$$d\Phi = -k\Phi\, db \qquad (3\text{-}10)$$

where the proportionality constant $k$ is called the **absorption coefficient** and the minus sign indicates attenuation of the beam with increasing thickness $db$. To obtain the absorption in a container of finite thickness $b$, equation 3-10 is rearranged and integrated from zero thickness, where the incident flux is $\Phi_0$ to thickness $b$, where the transmitted flux is $\Phi$. Thus

$$\int_{\Phi_0}^{\Phi} \frac{d\Phi}{\Phi} = -k \int_0^b db$$

or

$$\ln \frac{\Phi}{\Phi_0} = -kb$$

and

$$\Phi = \Phi_0 e^{-kb} \qquad (3\text{-}11)$$

Equation 3-11 shows that the flux (radiant power) decreases exponentially with increasing distance through an absorber.

The absorption coefficient $k$ has units of $cm^{-1}$. Its magnitude depends on the wavelength of the inci-

**TABLE 3-1**
Assumptions of the absorption law

1. The incident radiation is monochromatic.
2. The absorbers (molecules, atoms, ions, etc.) act independently of each other.
3. The incident radiation consists of parallel rays, perpendicular to the surface of the absorbing medium.
4. The pathlength traversed is uniform over the cross section of the beam. (All rays traverse an equal distance of the absorbing medium.)
5. The absorbing medium is homogeneous and does not scatter the radiation.
6. The incident flux is not large enough to cause saturation effects. (Lasers can cause such effects, as discussed in Chapters 11 and 15.)

dent radiation, the nature of the absorber, and the concentration of the absorber. In atomic spectroscopy, the absorption law is most often used in the form shown in equation 3-11, where $k$ is understood to be concentration dependent. In molecular spectroscopy, however, it is common to state explicitly the concentration dependence of the absorption coefficient by expressing $k$ as $k = k'c$, where $k'$ is a new absorption coefficient independent of concentration. Thus equation 3-11 becomes $\Phi = \Phi_0 e^{-k'bc}$. Most often in absorption spectroscopy we measure the transmittance $T = \Phi/\Phi_0$ or the absorbance $A = -\log T$. Thus

$$T = \frac{\Phi}{\Phi_0} = e^{-kb} = e^{-k'bc} = 10^{-abc} \qquad (3\text{-}12)$$

and

$$A = -\log T = 0.434kb = 0.434k'bc = abc \qquad (3\text{-}13)$$

In the last form of equations 3-12 and 3-13, the term $a$ is called the **absorptivity** of the absorbing species ($a = 0.434k' = 0.434k/c$). The absorptivity $a$ has the units of L g$^{-1}$ cm$^{-1}$ and is used when $b$ is in cm and $c$ is in g L$^{-1}$. When $c$ is expressed in mol L$^{-1}$, and $b$ in cm, the proportionality constant is called the **molar absorptivity** $\varepsilon$, which has the units L mol$^{-1}$ cm$^{-1}$. Thus the absorption law is often expressed as

$$A = \varepsilon bc \qquad (3\text{-}14)$$

For strongly absorbing molecules in the UV-visible region, $\varepsilon$ can be as high as $10^4$ or $10^5$ L mol$^{-1}$ cm$^{-1}$. Equations 3-13 and 3-14 are known as the Beer–Lambert law or more commonly as Beer's law. Equation 3-12 is the exponential form of Beer's law, which is also written in terms of the absorptance $\alpha$ (also called the *absorption factor*), where $\alpha = 1 - T = 1 - e^{-kb} = 1 - 10^{-A} = 1 - 10^{-\varepsilon bc}$. Note that $\alpha$ is used to represent both absorption at an interface or surface and absorption within a medium. Usually, we will use $\alpha$ to mean absorption within a medium.

For some applications, the absorption properties of a species are given in terms of the **absorption** or **transition cross section** $\sigma$ in cm$^2$. Thus Beer's law can also be stated $A = 0.434\,\sigma bn$, where $n$ is the concentration (density) of the species in atoms or molecules per cm$^3$ (see Appendix F). In some cases quantities $k$, $a$, $\varepsilon$, $\alpha$, and $\sigma$ are written as $k(\lambda)$, $\alpha(\lambda)$, and so on, to stress the wavelength dependence. Also, it is common to use $l$ or $L$ as the symbol for pathlength in atomic spectroscopy [e.g., $\alpha(\lambda) = 1 - e^{-k(\lambda)l}$].

The absorption characteristics of optical materials used for windows, lenses, prisms, and sample cells must be carefully considered in spectrochemical applications. The absorption coefficient should be small enough that attenuation by the material is slight. The absorption coefficients of common optical materials are plotted vs. wavelength in Appendix B. In the UV region quartz (natural silica) or synthetic silica is used. Glass is used in the region 320 nm to 2 μm, while halide salts are employed in the infrared and vacuum UV region.

## 3-2 INTERFERENCE, DIFFRACTION, AND POLARIZATION OF ELECTROMAGNETIC WAVES

The phenomena of interference, diffraction, and polarization all deal with what occurs when two or more electromagnetic waves overlap at some point in space. The general concept of the superposition of waves is considered first and the production of overlapping waves through interference and diffraction is then discussed. The polarization of electromagnetic radiation is described next, and the phenomena of optical rotatory dispersion and circular dichroism are introduced. This section ends with a brief consideration of methods to produce polarized radiation.

### Superposition of Waves

An electromagnetic wave is an example of a **transverse wave** in that directions of the vibrating electric and magnetic fields are perpendicular to the direction of propagation. Waves that have disturbances along the direction of propagation, like sound waves, are, by contrast, **longitudinal waves**. A transverse wave is said to be **plane polarized** (linearly polarized) if a fixed plane, the plane of vibration, contains both the vibrating electric field vector and the propagation vector.

For a plane-polarized, monochromatic wave, we can describe the variation of the electric field by

$$\mathbf{E} = \mathbf{E}_m \sin\left[(\omega t) - (kx + \phi_0)\right] \qquad (3\text{-}15)$$

where $\mathbf{E}$ is the instantaneous value of the electric field (V m$^{-1}$ or N C$^{-1}$), $\mathbf{E}_m$ is the maximum amplitude, $\omega$ is the angular frequency ($\omega = 2\pi\nu$), $k$ is the free-space propagation number ($k = 2\pi/\lambda$), $x$ is the distance, and $\phi_0$ is the initial phase or epoch angle. We can write equation 3-15 as $\mathbf{E} = \mathbf{E}_m \sin(\omega t + \phi)$, where $\phi = -(kx + \phi_0)$. The irradiance $E$ of the wave is proportional to the square of the time-average amplitude $\bar{\mathbf{E}}$ of the field. In a vacuum

$$E = c\varepsilon_0\bar{\mathbf{E}}^2 = \frac{c\varepsilon_0\mathbf{E}_m^2}{2} = 1.33 \times 10^{-3}\,\mathbf{E}_m^2 \qquad (3\text{-}16)$$

where the evaluated constant has the units of $F\ s^{-1}$ ($C^2\ N^{-1}\ m^{-1}\ s^{-1}$). In other media, $c\varepsilon_0$ in equation 3-16 is replaced by $v\varepsilon$.

Now let us consider two plane-polarized waves of identical frequency that overlap in space. If the waves are not large-amplitude waves, such as those from high-powered lasers, the **principle of superposition** holds, and the resulting electromagnetic disturbance at a particular point can be considered as the algebraic sum of the constituent disturbances. If we describe the waves by $\mathbf{E}_1 = \mathbf{E}_{m1} \sin(\omega t + \phi_1)$ and $\mathbf{E}_2 = \mathbf{E}_{m2} \sin(\omega t + \phi_2)$, the combination wave can be written

$$\begin{aligned} \mathbf{E} &= \mathbf{E}_1 + \mathbf{E}_2 \\ &= \mathbf{E}_{m1} \sin(\omega t + \phi_1) + \mathbf{E}_{m2} \sin(\omega t + \phi_2) \end{aligned} \tag{3-17}$$

With trigonometry it can be shown that the composite wave is of the form

$$\mathbf{E} = \mathbf{E}_m \sin(\omega t + \phi) \tag{3-18}$$

Thus the composite wave is of the same frequency as the constituent waves, but of different amplitude and phase. The amplitude $\mathbf{E}_m$ is given by

$$\mathbf{E}_m^2 = \mathbf{E}_{m1}^2 + \mathbf{E}_{m2}^2 + 2\mathbf{E}_{m1}\mathbf{E}_{m2} \cos(\phi_2 - \phi_1) \tag{3-19}$$

and the phase $\phi$ is given by

$$\tan\phi = \frac{\mathbf{E}_{m1} \sin\phi_1 + \mathbf{E}_{m2} \sin\phi_2}{\mathbf{E}_{m1} \cos\phi_1 + \mathbf{E}_{m2} \cos\phi_2} \tag{3-20}$$

Equation 3-19 shows that the square of the amplitude of the composite wave is given by the sum of the squares of the amplitudes of the individual waves plus an interference term, $2\mathbf{E}_{m1}\mathbf{E}_{m2} \cos(\phi_2 - \phi_1)$. Since the irradiance of the resulting wave depends on the square of the amplitude (Equation 3-16), the irradiance also depends on this interference term. In fact, if we define $\delta = \phi_2 - \phi_1$, constructive interference (maximum irradiance) will occur when $\delta = 0, \pm 2\pi, \pm 4\pi$, and so on. Or we can write the condition for constructive interference as $\delta = \pm 2m\pi$, where $m$ is 0, 1, 2, and so on. Similarly, destructive interference (minimum irradiance) occurs when $\delta = \pm\pi, \pm 3\pi$, and so on. The resultant electric field vector $\mathbf{E}$ for in- and out-of-phase conditions is shown in Figure 3-4.

The phase difference $\delta$ can arise from a difference in pathlength or a difference in initial phase:

$$\begin{aligned} \delta &= (kx_1 + \phi_{01}) - (kx_2 + \phi_{02}) \\ &= \frac{2\pi(x_1 - x_2)}{\lambda} + (\phi_{01} - \phi_{02}) \end{aligned}$$

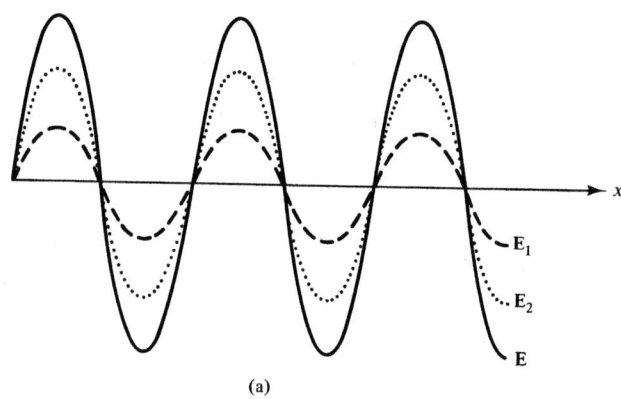

(a)

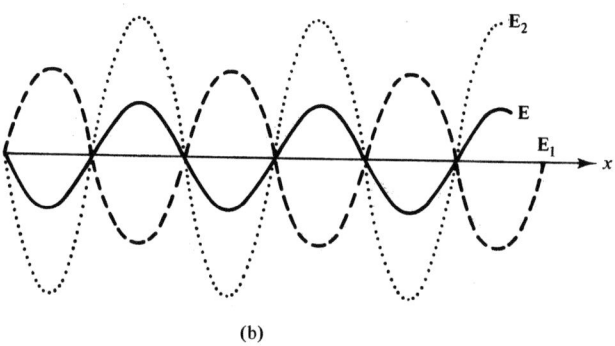

(b)

**FIGURE 3-4** Superposition of sinusoidal waves. In (a) the two constituent waves $\mathbf{E}_1$ and $\mathbf{E}_2$ are in phase and add to produce a larger-amplitude resultant. In (b) the two waves are 180° out of phase and add to produce a smaller-amplitude resultant. Note in both cases that the combination wave is of the same frequency as the constituent waves.

where $x_1$ and $x_2$ are the distances from the sources of the two waves to the point of observation and $\lambda$ is the wavelength in the medium. If the two waves are initially in phase at the sources ($\phi_{01} = \phi_{02}$), or come from the same source by two different routes, we can write

$$\delta = \frac{2\pi(x_1 - x_2)}{\lambda}$$

The refractive index of the medium is $\eta = \lambda_0/\lambda$, where $\lambda_0$ is the wavelength in vacuum. Thus

$$\delta = \frac{2\pi\eta(x_1 - x_2)}{\lambda_0} = \frac{2\pi\,\Delta(\mathrm{OPL})}{\lambda_0} \tag{3-21}$$

where $\Delta(\mathrm{OPL})$ is the difference in optical pathlength of the two waves. Constructive interference occurs when $\delta = m2\pi$ or

$$\Delta(\mathrm{OPL}) = \frac{\delta\lambda_0}{2\pi} = m\lambda_0 \tag{3-22}$$

where $m$ is $0$, $\pm 1$, $\pm 2$, and so on. Thus, constructive interference occurs when the optical pathlength difference is an integral multiple of the wavelength. Similarly, destructive interference occurs when $\Delta(\text{OPL}) = (2m + 1)\lambda_0/2$. We will use this important result in describing a variety of important optical effects.

### Interference

We may define **optical interference** as the interaction of light waves to yield a resultant irradiance that is different from the sum of the component values, as predicted by equation 3-19. Two waves for which the initial phase difference, $\phi_{01} - \phi_{02}$, is zero or constant for a long time are said to be **coherent**. Coherence is thus a condition for achieving interference. Waves emitted by separate sources are incoherent with respect to one another. For this reason we normally observe interference phenomena when light reaching the point of interference comes from a single source by two different paths. Another condition is that the interfering waves must have very nearly the same frequency composition. This again is readily achieved if the radiation comes from a single source. Finally, although we have considered the sources to be polarized, normal unpolarized light also produces interference for reasons that are discussed later in this section.

Interference of light waves can be produced in several ways. In wavefront-splitting devices, portions of the primary wavefront are used to emit secondary waves that produce the interference pattern. The classic Young two-slit interference experiment is an example. In amplitude-splitting devices, the primary wave itself is divided into two segments which travel different routes before recombining and interfering. An example of this type is the Michelson interferometer discussed in Section 3-7. Interference effects are also produced by thin films and by plates of various thicknesses. The Fabry–Perot interference filter and Fabry–Perot interferometer are examples of multiple-beam interference devices; both are discussed in Section 3-7.

### Diffraction

**Diffraction** is the deviation of light from rectilinear propagation when it encounters an obstacle. The distinction between diffraction and interference is somewhat arbitrary since diffraction can also result in the superposition of waves. By custom, interference is usually considered to be the superposition of only a few waves, whereas diffraction is considered to be that of a large number of waves. Consider the single slit of width $W$ shown in Figure 3-5 illuminated by a source at a long distance so that the rays reaching the slit are

parallel. The resulting diffraction pattern observed at a long distance or with a lens is termed **Fraunhofer diffraction**. Rays from the slit that reach $P_0$ (Figure 3-5a) all have identical optical pathlengths. Since they are in phase in the plane of the slit, they are still in phase at $P_0$ and give rise to a central maximum of irradiance. Now consider two rays that reach $P_1$ on the

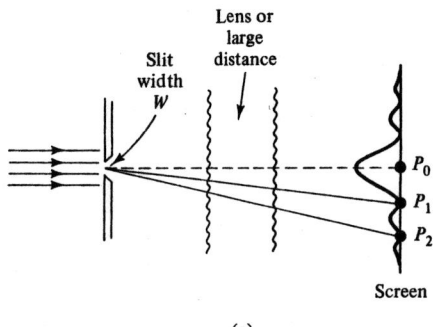

(a)

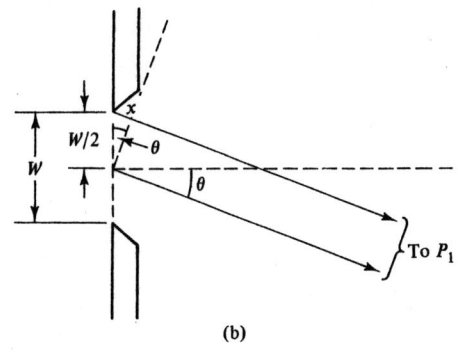

(b)

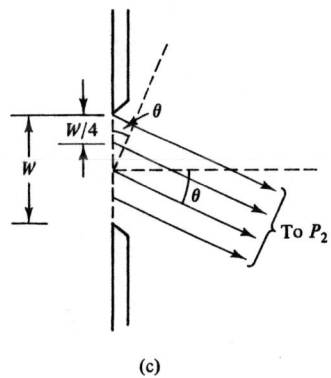

(c)

**FIGURE 3-5** Fraunhofer diffraction by a single slit. In (a) parallel source rays illuminate the slit where diffraction occurs. Observation of the light at a large distance, or focused by a lens, yields a diffraction pattern. The diffraction minimum at point $P_1$ is seen to arise in (b) from the superposition of rays diffracted at angle $\theta$ that are a distance $W/2$ apart. The second minimum at $P_2$ from rays diffracted at angle $\theta'$ is seen to arise in (c) from the rays that are $W/4$ apart.

screen (Figure 3-5b). The ray originating at the top of the slit must travel a distance $x = (W/2) \sin \theta$ farther than the ray from the center of the slit. According to our earlier discussion, we will get destructive interference when this pathlength difference is $\lambda/2$. Thus a minimum irradiance occurs when $(W/2) \sin \theta = \lambda/2$ or $W \sin \theta = \lambda$.

We can locate another of the minima by dividing the slit into four parts as shown in Figure 3-5c. Now any two adjacent rays will destructively interfere when $(W/4) \sin \theta = \lambda/2$ or $W \sin \theta = 2\lambda$. By extension, the general formula for diffraction minima is

$$W \sin \theta = m\lambda \qquad m = \pm 1, \pm 2, \pm 3, \ldots \qquad (3\text{-}23)$$

Now consider the location of the first minimum in the pattern as the slit width $W$ is varied. According to equation 3-23 with $m = 1$, $\sin \theta = \lambda/W$. As $W$ decreases, $\theta$ increases, and the beam spreads. If we consider angle $\theta$ to be small, then $\theta \approx \lambda/W$. The angular half-width (width at half maximum intensity) of the beam in this case is said to be **diffraction limited**. The width cannot be reduced to zero unless $W$ is infinite or $\lambda$ is zero. The total beam half-width $W'$ after the beam

travels a distance $b$ is approximately its original width $W$ plus $b$ times the angular width $\lambda/W$. For large enough $b$, we can neglect the original width and $W' \approx b\lambda/W$. A plot of the irradiance of a diffraction-limited beam upon striking a screen is shown in Figure 3-6.

Now consider multiple slits as shown in Figure 3-7. The difference in OPL between adjacent rays is $x = d \sin \theta$, where $d$ is the distance between slits. Maxima in irradiance occur when this distance is an integral multiple of the wavelength or

$$d \sin \theta = m\lambda \qquad m = 0, \pm 1, \pm 2, \ldots \qquad (3\text{-}24)$$

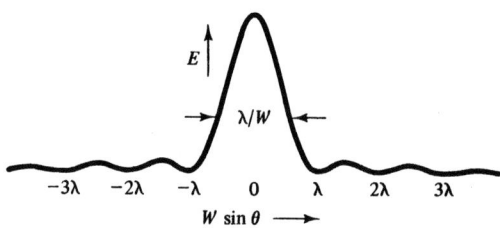

**FIGURE 3-6** Fraunhofer diffraction at a single slit. The angular half-width is $\approx \lambda/W$. The irradiance of the second maximum is about 4.4% of that of the principal or central maximum.

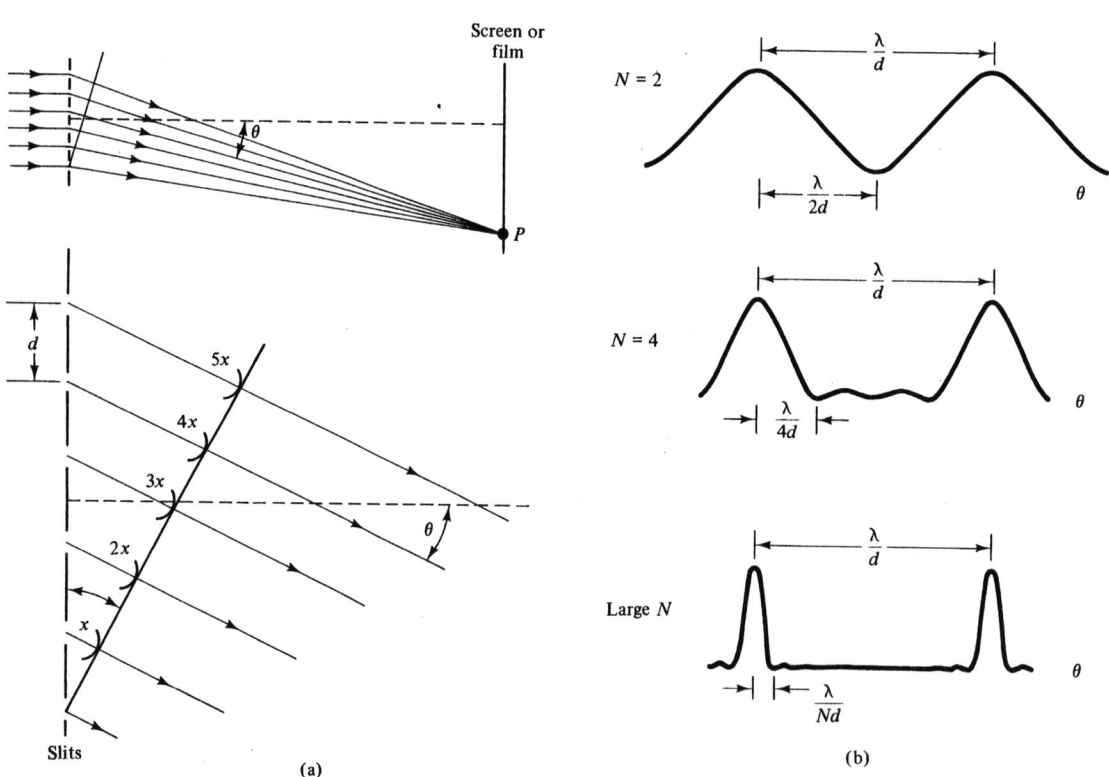

**FIGURE 3-7** Multiple-slit diffraction from six slits separated by distance $d$. In (a) light rays reaching one point $P$ on a screen are shown. If the path difference $x = d \sin \theta$ is an integral multiple of a wavelength, the waves are in phase at $P$. In (b) the irradiance pattern at the screen is shown for $N$ (number of slits) = 2, 4, and large. For large $N$ each principal maximum has the shape of a single slit pattern.

As the number of slits increases, each diffraction maximum grows in irradiance and becomes narrower. The narrowing occurs because with multiple slits only a slight change in angle away from that corresponding to a maximum leads to almost total destructive interference. Diffraction gratings can be considered to consist of many closely spaced slits. Such gratings are considered in detail in Section 3-5.

## Polarization of Light

We have seen that electromagnetic radiation is a transverse wave and have discussed the nature of plane-polarized or linearly polarized light. We will now consider what occurs when we have two plane-polarized light waves of identical frequency moving through the same region of space in the same direction. If the electric field vectors of the two waves are aligned with each other, they will simply combine to give a resultant wave that is also linearly polarized. The amplitude and phase of the resultant depends, of course, on the amplitude and phase of the two superimposing waves. If, however, the electric field vectors of the waves are mutually perpendicular, the resultant wave may or may not be linearly polarized. If the waves are of equal amplitude and orthogonal, superposition can lead to plane-polarized, elliptically polarized, and circularly polarized radiation; which of these is obtained depends on the phase difference between the two waves, as discussed below.

*Linear Polarization.* Let us consider in more detail two waves with mutually perpendicular electric field vectors. If the two waves have a phase difference that is zero or an integral multiple of $\pm 2\pi$, they are in phase, and the resultant wave is linearly polarized, as shown in Figure 3-8a. We can equally well reverse the process and resolve any plane-polarized wave into two mutually perpendicular plane-polarized components. These two orthogonal components are often denoted the parallel ($\parallel$) or $\pi$ component and the perpendicular ($\perp$) or $\sigma$ component.

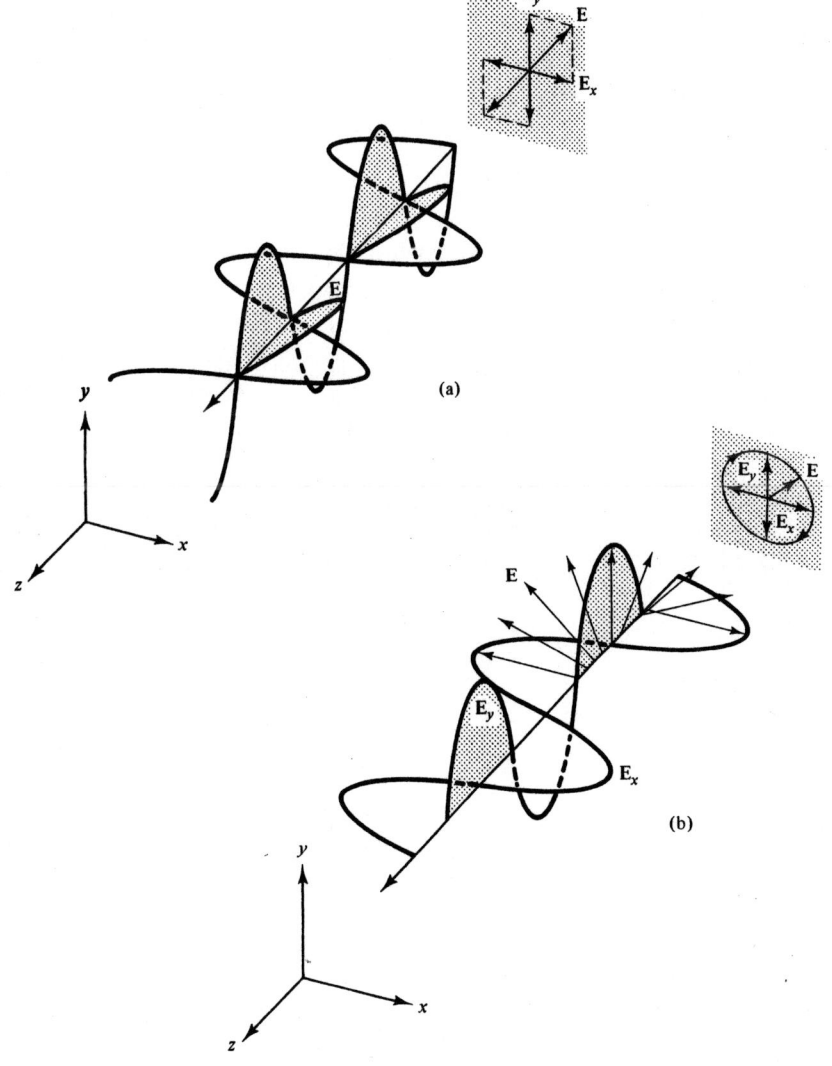

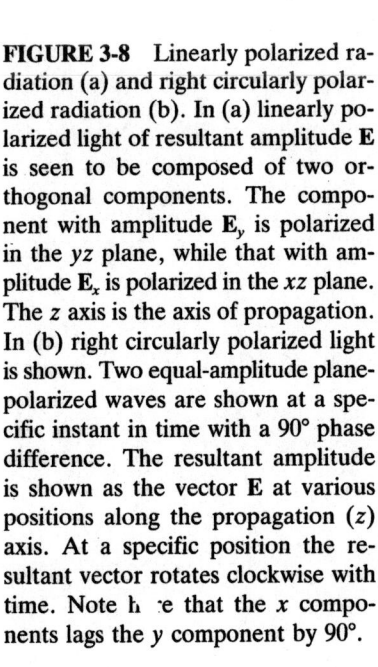

**FIGURE 3-8** Linearly polarized radiation (a) and right circularly polarized radiation (b). In (a) linearly polarized light of resultant amplitude **E** is seen to be composed of two orthogonal components. The component with amplitude $\mathbf{E}_y$ is polarized in the $yz$ plane, while that with amplitude $\mathbf{E}_x$ is polarized in the $xz$ plane. The $z$ axis is the axis of propagation. In (b) right circularly polarized light is shown. Two equal-amplitude plane-polarized waves are shown at a specific instant in time with a 90° phase difference. The resultant amplitude is shown as the vector **E** at various positions along the propagation ($z$) axis. At a specific position the resultant vector rotates clockwise with time. Note h e that the $x$ components lags the $y$ component by 90°.

If the two orthogonal waves have a phase difference that is an odd-integer multiple of $\pm\pi$, they are 180° out of phase, but again combine to give a resultant that is linearly polarized. In this case the plane of vibration is rotated from that of the in-phase case (not necessarily by 90°).

### Circular Polarization.

An interesting case arises if the two orthogonal waves are 90° out of phase. As shown in Figure 3-8b, such a case gives rise to circularly polarized light because the resultant traces out a circle. Note in Figure 3-8b that the resultant vector (when the $x$ component lags the $y$ component by 90°) rotates clockwise. The resultant wave is said, therefore, to be **right circularly polarized**. If, on the other hand, the $x$ component were to lead the $y$ component by 90°, the resultant electric field vector would rotate counterclockwise. The resultant in this case is said to be **left circularly polarized**.

If we were to add two equal-amplitude, in-phase, but oppositely polarized circular waves together, the resultant would be a plane-polarized wave. Hence a plane-polarized wave can be considered to consist of equal-amplitude left and right circularly polarized waves. The two oppositely polarized components are often called the $l$ and $d$ components, where $d$ (right-handed or dextrorotatory) refers to clockwise rotation and $l$ (left-handed or levorotatory) refers to counterclockwise rotation from the point of view of the observer looking toward the propagation vector.

### Elliptical Polarization.

If the two superimposing, plane-polarized waves have a phase difference between 0° (linear polarization) and 90° (circular polarization), the resultant traces out an ellipse and the radiation is said to be elliptically polarized. In actuality, linearly and circularly polarized light are simply special cases of elliptically polarized light where the phase difference is 0° or 90°. The various polarization configurations for different phase differences are illustrated in Figure 3-9.

### Normal Light.

Light from common sources (filament lamps, the sun, arc lamps) is emitted by nearly independent radiators (atoms and molecules). Each radiator produces a polarized wave train for a short time ($\approx 10^{-8}$ s). The light propagating in a given direction consists of many such wave trains whose planes of vibration are randomly oriented around the direction of propagation. As a result of the random superposition of independent polarized wave trains, no single resultant state of polarization is observable, and natural light is referred to as **unpolarized light**. Unpolarized light can be described as two orthogonal plane-polarized waves of equal amplitude with a phase difference between them that varies randomly in time.

### Optical Rotatory Dispersion.

For some substances, the characteristics of optical phenomena such as absorption, refraction, and reflection depend on the polarization of the incident radiation. Substances that rotate the plane of vibration of plane-polarized radiation are termed **optically active**. These include anisotropic crystals and liquids or solutes in solution that can exist as enantiomers (e.g., chiral molecules). In the latter case, optical activity is observed if one of the enantiomers is in excess.

The rotation of plane polarized light by an optically active substance can be viewed as being due to the different propagation rates of the $d$ and $l$ components, as shown in Figure 3-10a and b. The propagation velocities differ because of the different refractive indicies for the $d$ and $l$ components, $\eta_d$ and $\eta_l$, respectively.

The rotation in degrees ($\alpha$), also termed the **optical rotatory power**, is given by

$$\alpha = \frac{180b}{\lambda}(\eta_l - \eta_d) \qquad (3\text{-}25)$$

where $b$ is the pathlength and $\lambda$ is the wavelength of the incident radiation. The difference in refractive index for the two polarization components ($\eta_l - \eta_d$) is denoted the **circular birefringence**. It is often only on the order of $10^{-6}$. However, with a 10-cm pathlength, op-

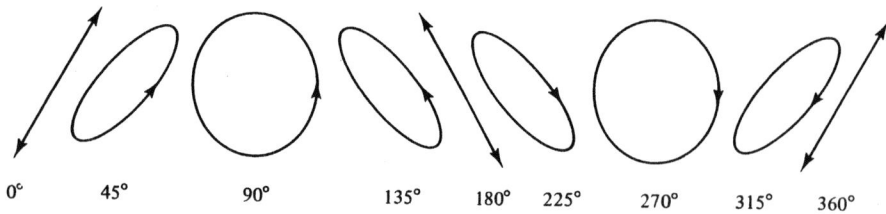

0°    45°    90°    135°    180°    225°    270°    315°    360°

**FIGURE 3-9** Polarization configurations for various phase differences for equal amplitude waves. The $x$ component leads the $y$ component by the phase difference shown. Note that phase differences 0° and 360° give rise to linearly polarized light, while differences of 90° and 270° give rise to circularly polarized light. Other phase differences lead to elliptically polarized radiation.

**FIGURE 3-10** Principles of optical rotatory dispersion and circular dichroism measurements. Plane-polarized radiation is represented from the perspective of an end-on view at a given time as the sum of two circularly polarized components of equal magnitude that are rotating in opposite directions before (a) and after (b) passing through a transparent optically active medium. For (b) it is assumed that the refractive index of the medium is greater for the $l$ component. The rotation of the $l$ vector is seen to be delayed relative to the $d$ component because its propagation rate through the medium is slower. The plane of vibration of the linearly polarized radiation is rotated by angle $\alpha$, which is equal to $(\alpha_d - \alpha_l)/2$. In (c), an end view of the electric vector is shown at a point after the plane-polarized light in (a) passes through an absorbing optically active medium exhibiting circular dichroism but no optical rotation. For the case shown, the length of the electric vector for the $d$ component is less than that for the $l$ component due to greater absorption of this component ($\varepsilon_d > \varepsilon_l$). Thus the resultant (**E**) is due to the vector sum of two circularly polarization components of unequal magnitude that are rotating in opposite directions. In time, the resultant traces out an ellipse as shown in (d) and the radiation has become elliptically polarized. The length of the major and minor axes are equal to the sum and difference, respectively, of the $d$ and $l$ electric vectors. The ellipticity is the angle $\theta$, whose tangent is the ratio of the major to minor axis lengths. This angle is defined as the angle between the major axis and a line drawn through the intersection of the tangents drawn parallel to the major and minor axes. From the drawing $\tan \theta = a/b$. Usually, CD is accompanied by optical rotation [i.e., a combination of the effects shown in (b) and (c)]. Thus in (e), the $d$ component is absorbed more strongly but retarded less than the $l$ component. The major axis of the elliptically polarized light is rotated angle $\alpha$ relative to the original plane of the incident linearly polarized radiation.

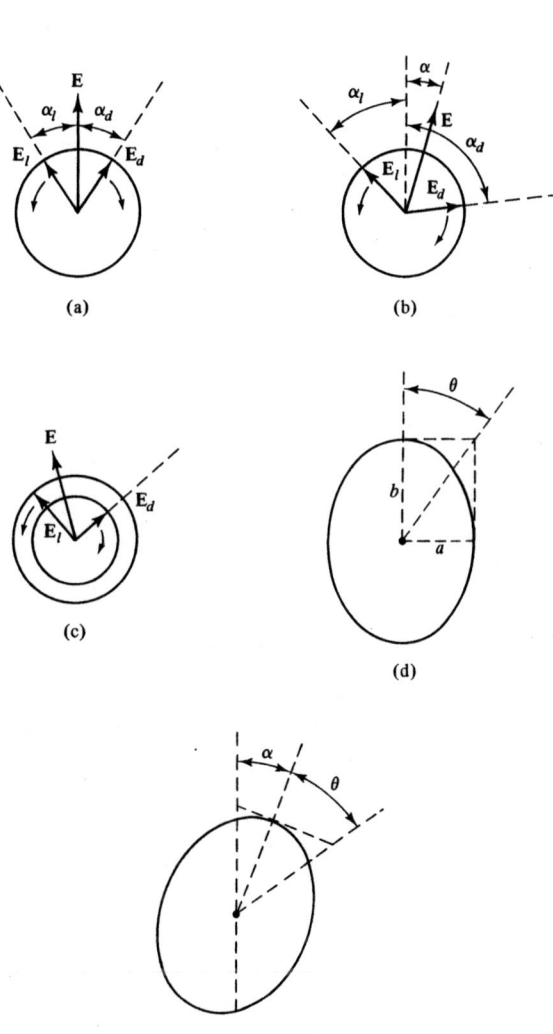

tical rotations of tens to hundreds of degrees are often observed. For a $d$-rotatory compound, the rotation and circular birefringence are positive; these quantities are negative for a $l$-rotatory compound. The rotatory power depends not only on the compound, wavelength, and pathlength, but also on the temperature, and for solutions of an optically active solute, the solvent and the analyte concentration.

The rotation can be normalized to a particular pathlength and concentration as follows:

$$[\alpha] = \frac{\alpha}{bc} \qquad (3\text{-}26)$$

Normally, $b$ is given in decimeters. The quantity $[\alpha]$ is termed the **specific rotation** if $c$ is in g mL$^{-1}$ and has units of degrees per gram per milliliter per decimeter. If the concentration is expressed in molar units, $[\alpha]$ is called the molar rotation with units of degrees per mole per liter per decimeter. For pure liquids, $c$ in equation 3-26 is replaced by the density ($\rho$).

In **polarimetry**, the rotation is measured at one wavelength. The temperature and wavelength are commonly specified by adding a superscript and subscript to $\alpha$. Standard wavelengths include 589 nm (the Na D line) and 546.1 nm (the green Hg line). In **spectropolarimetry**, the dependence of the rotation on wavelength is measured and is termed the **optical rotatory dispersion** (ORD).

*Circular Dichroism.*   **Circular dichroism** (CD) depends on the difference in molar absorptivities for the *d* and *l* components, $\varepsilon_d$ and $\varepsilon_l$, respectively, by optically active materials. If an absorbing sample of an optically active compound is illuminated with plane-polarized radiation, the differential absorption results in one of the circularly polarized components being absorbed more strongly than the other component (see Figure 3-10c). Thus the transmitted radiation is elliptically polarized. The eccentricity of the ellipse can be characterized by a quantity termed the **ellipticity** ($\theta$) in degrees. It is a measure of the magnitude of CD and is related to the length of the major and minor axes of the ellipse as shown in Figure 3-10d. The ellipticity cannot be directly measured as the optical rotation. Rather the absorbance with incident radiation that is circularly polarized in the *d* direction ($A_d$) and the absorbance with *l* circularly polarized radiation ($A_l$) are separately measured. The ellipticity is calculated from these absorbances and the equation

$$\theta = 33 (A_l - A_d) \tag{3-27}$$

The **molar ellipticity** ([$\theta$]) is given by

$$[\theta] = \frac{\theta}{bc} = 3300 (\varepsilon_l - \varepsilon_d) \tag{3-28}$$

where the difference in molar absorptivities is denoted the circular dichroism. The units of [$\theta$] are often in degrees centimeter squared per decimole and the factor of 3300 applies if *b* is expressed in decimeters and *c* is in molarity units. A circular dichroism curve or spectrum is a plot of the molar ellipticity versus wavelength.

In most cases, circular dichroism and optical rotation occur simultaneously as illustrated in Figure 3-10e. The angle of rotation ($\alpha$) in degrees is taken as the angle between the plane of polarization of the incident beam and the major axis of the emerging elliptical beam. The rapid change in the CD and ORD near the maximum of an absorption band is termed the **Cotton effect**. Usually, the sign of the optical rotation changes near the maxima of absorption bands. The ellipticity can also vary in sign in the vicinity of absorption bands. Thus a CD curve can exhibit both positive- and negative-going bands.

*Production of Linearly Polarized Light.*   A device that produces polarized light from unpolarized light in known as a **polarizer**. A **linear polarizer** produces plane-polarized light. Polarizers are based on one of four phenomena: absorption (linear dichroism), reflection, scattering, or linear birefringence (double refraction). The familiar Polaroid sheet contains dichroic material that absorbs the component of unpolarized light perpendicular to the polarizing direction. The transmitted light is plane polarized. The transmitted intensity through a second polarizing sheet positioned as shown

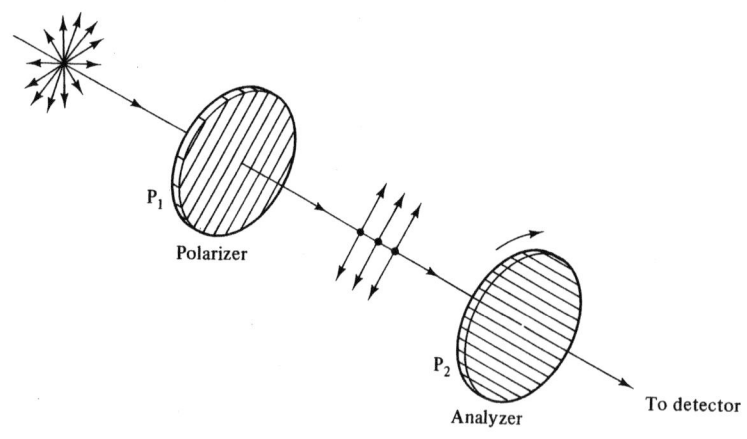

**FIGURE 3-11** Polarizer and analyzer. The most common linear polarizer is the familiar Polaroid sheet. The H-sheet is a molecular analog of a wire grid. It is made from a sheet of polyvinyl alcohol which is heated and stretched in a particular direction. This causes the long hydrocarbon molecules to align. The sheet is then dipped in an ink solution containing iodine. The transmission axis of the polarizer is perpendicular to the direction the sheet was stretched. Many other forms of Polaroid have been developed. A second sheet can be used as an analyzer as shown. Transmittance is at a minimum when the polarizers are crossed. If $\theta$ is the angle between the polarizing directions of $P_1$ and $P_2$, the transmitted irradiance *E* is $E = E_m \cos^2 \theta$, where $E_m$ is the maximum value of the transmitted irradiance (Malus's law).

in Figure 3-11 will be a minimum when the polarizing directions of the two sheets are perpendicular to each other (crossed polarizers) and a maximum when they are parallel.

Light can also be polarized by reflection as illustrated in Figure 3-12. The two orthogonal polarization components have different reflectances. In equation 3-9, the first term represents the reflectance of the perpendicular ($\sigma$) component ($\rho_\sigma = \sin^2(\theta_1 - \theta_2)/[2\sin^2(\theta_1 + \theta_2)]$). The second term represents that of the parallel ($\pi$) component ($\rho_\pi = \tan^2(\theta_1 - \theta_2)/[2\tan^2(\theta_1 + \theta_2)]$). Hence, if the intensities of the two components are equal, $\rho = \rho_\sigma + \rho_\pi$. Curves b and c of Figure 3-2 illustrate how the $\pi$ and $\sigma$ reflectances behave for different angles of incidence. At a certain angle of incidence, known as **Brewster's angle** $\theta_p$, the reflectance of the $\pi$ component becomes zero. From the expression for $\rho_\pi$ above, this occurs when $\theta_1 + \theta_2 = 90°$. From Snell's law the necessary angle occurs when $\tan\theta_1 = \eta_2/\eta_1$. At Brewster's angle, $\theta_p = \tan^{-1}(\eta_2/\eta_1)$, the reflectance for the $\pi$ component is zero so that the reflected beam is plane polarized with its plane of vibration at right angles to the plane of incidence (Figure 3-12a). A plane-polarized incident beam with its plane of vibration in the plane of incidence will undergo no reflections at Brewster's angle (Figure 3-12b). At angles other than Brewster's angle, the reflectance for the $\sigma$ component is greater than the $\pi$ component (see Figure 3-2, curve c). This is also true for reflection at metallic surfaces.

Polarized light is also produced by scattering. The sky appears blue in daytime because short-wavelength light is most readily scattered. Sunlight scattered at right angles is fully polarized; all sunlight is partially polarized. We shall examine the scattering of polarized light and the polarization of light by scattering in more detail in Chapter 16.

Optically anisotropic crystals have optical properties that are not the same in all directions. Polarized, monochromatic rays travel through the crystal with velocities that depend on the state of polarization and the direction of propagation. An exception to this occurs along a particular axis known as the **optic axis** where monochromatic rays travel with the same velocity regardless of the direction of polarization. The optic axis is a property of the crystal and is determined by the crystal structure. Birefringent crystals, such as calcite, show different indices of refraction for the orthogonal polarization states except for rays that propagate along the optic axis.

If unpolarized light is sent into a crystal of calcite at an angle to the optic axis, two beams emerge which have their planes of vibration at right angles to each other. One beam, called the **ordinary ray**, obeys Snell's

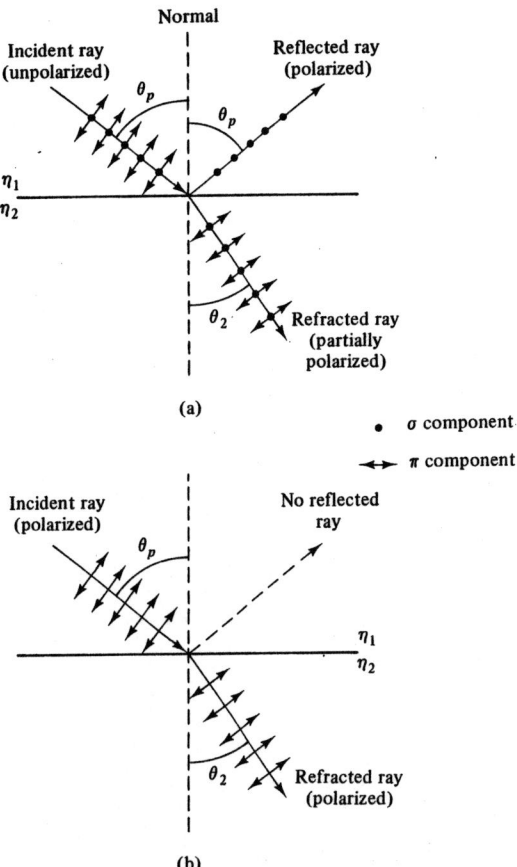

**FIGURE 3-12** Polarization by reflection. In (a) the incident unpolarized beam consists of a component perpendicular to the plane of the propagation vector that is perpendicular to the interface surface ($\sigma$-component) and a component lying in the plane of incidence ($\pi$ component). At a certain angle of incidence $\theta_p$, known as Brewster's angle, the reflected light is completely polarized (only the $\sigma$ component is reflected; the transmitted light is partially polarized). In (b) a polarized incident beam is fully transmitted at Brewster's angle. Since the indices of refraction depend upon wavelength, Brewster's angle is also wavelength dependent.

law at the crystal surface and propagates as though the crystal had a refractive index which is independent of direction (the ordinary refractive index $\eta_o$). The second beam, called the **extraordinary ray**, does not obey Snell's law; it propagates with a velocity determined by the extraordinary refractive index $\eta_e$, which varies with the direction of propagation. Advantage can be taken of the two indices of refraction to separate the two beams; thus birefringent crystals can be used as polarizers. Several polarizing prisms based on this principle are discussed in Section 3-5, and values of $\eta_o$ and $\eta_e$ are tabulated in Appendix B for several important crystals.

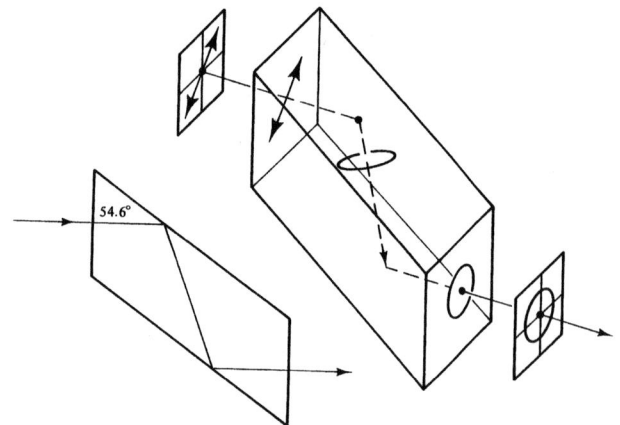

**FIGURE 3-13** Fresnel rhomb. Internal reflection introduces a phase difference between the two perpendicular components of a linearly polarized wave. At an incidence angle of 54.6°, the phase shift is 45°. In the Fresnel rhomb, two such reflections are used to produce a 90° phase shift, which yields circularly polarized light.

*Production of Circularly Polarized Light.*
Circularly and elliptically polarized light can be produced by retardation of one polarization state in a retarder plate. If a birefringent crystal is cut so that its optic axis is parallel to its front and back surfaces, an incident monochromatic beam normal to the axis will propagate through the crystal as an extraordinary and

ordinary ray. Since the two rays move through the crystal at different velocities, a phase difference is introduced. The resultant wave is a superposition of the extraordinary and ordinary waves. If the thickness of the crystal is such that a 90° phase difference results, the plate is known as a **quarter-wave plate**; a half-wave plate introduces a 180° phase difference. Circularly polarized light is most often produced from normal, unpolarized light by a series combination of a linear polarizer and a quarter-wave plate. A half-wave plate can change right circularly polarized light to left, and vice versa.

Rhomboids (rhombs) are also used as retarders and have the advantage of being essentially achromatic. Their use is based on the different phase shifts that occur for the parallel and perpendicular components during internal reflection. The phase difference depends on the angle of incidence and is essentially independent of wavelength. The Fresnel rhomb, shown in Figure 3-13, is commonly used to produce circularly polarized light.

## 3-3 MODULATORS

Several types of devices are used to amplitude modulate a radiation source (see Section 4-5). Modulation is based on mechanical interruption of a light beam or on elec-

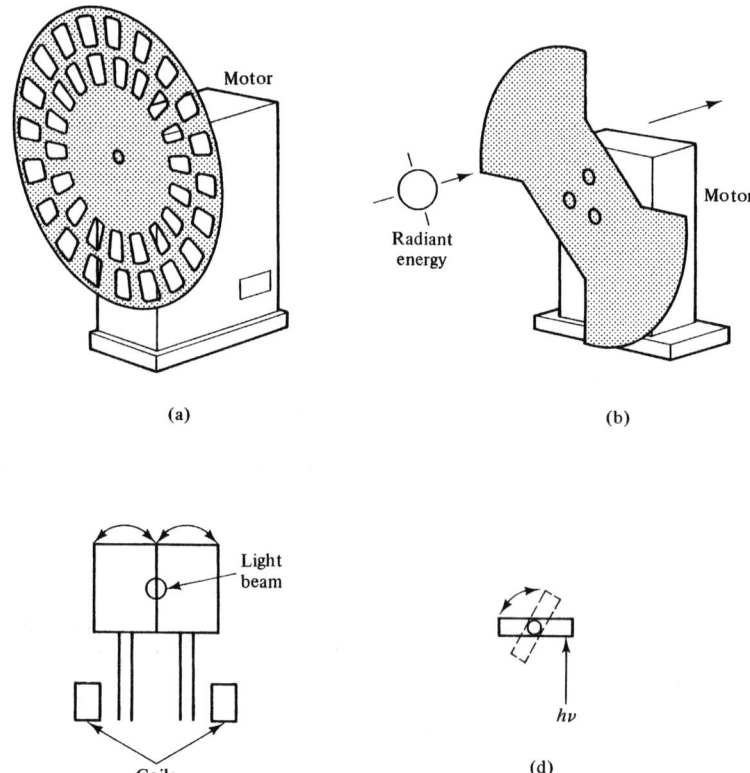

**FIGURE 3-14** Types of mechanical modulators. In (a) a rotating wheel or disk chopper (often called a toothed wheel chopper) is illustrated. A high-quality motor is used to control the rotation rate of the wheel. The periodic interruption of the light beam can also be controlled by rotating vanes as in (b). In the oscillating or tuning fork design of (c), vanes are attached to the tines of a tuning fork. The tines are driven by electromagnets. The resonant frequency is determined by the size, mass, and spring tension of the tines. Oscillating choppers as shown in (d) are based on a periodic rotational oscillation of a vane over a controlled arc (2 to 30°). These are also electromagnetically controlled resonant devices.

tro-optic, magneto-optic, or acousto-optic phenomena. In this section, modulators based upon all of these phenomena are described, including mechanical choppers; Kerr, Pockels, and Bragg cells; and the Debye–Sears modulator.

## Mechanical Choppers

Mechanical modulators or choppers provide a controlled periodic physical blocking of a radiation beam. The most common mechanical modulators are based on a rotating disk or wheel with apertures or vanes as shown in Figure 3-14a and b. The chopping frequency is determined by the number of apertures and the rotation rate of the shaft of the motor. The maximum modulation frequency is generally in the range 1 to 10 kHz. Vibrating or tuning fork modulators are based on periodic lateral movement of one or two vanes in and out of the light path as illustrated in Figure 3-14c. Because these are resonant devices, a given chopper is designed to operate at one frequency (10 to 6000 Hz). The modulation waveform depends on the size, shape, and configuration of the vanes and can be sinusoidal. Other types of resonant choppers include oscillating choppers based on torsion rod and taut band designs (Figure 3-14d), which rely on rotation of a vane in and out of the light path.

Choppers can be used for specialized functions by mounting mirrors, refractor plates, gratings, or filters to the vanes. A chopper with mirrored vanes is commonly used as a chopping beam splitter, as discussed in Section 3-4. A mirrored oscillating chopper can be used to sweep a beam across a target or aperture at high speeds (e.g., up to $10^5$ degrees s$^{-1}$).

In some applications, it is only necessary to block or unblock a radiation beam at certain times in an experiment (e.g., to measure the dark signal). Beam blocking can be accomplished by pulling a vane in an out of the light path by attaching the vane to a shaft whose two rest positions are controlled by a solenoid. Electromechanically controlled camera-type iris shutters are also available.

## Electro-Optic and Magneto-Optic Modulators

Most electro-optic (E-O) modulators are based on the Kerr effect or the Pockels effect. Both of the effects are concerned with the linear birefringence induced in certain materials that are subjected to a strong electrical field. The application of an electric field alters the refractive index and propagation velocity for the ordinary and extraordinary rays. If the medium is illuminated with plane-polarized radiation, the induced phase shift between the two components results in elliptically polarized radiation.

Because the birefringence and retardation are related to the applied electric field, these E-O effects provide a means to construct an electrically controlled variable retarder or wave plate. The **Kerr effect** is a quadratic electro-optic effect because the induced birefringence is proportional to the square of the electric field or voltage applied. By contrast, the birefringence induced by the **Pockels effect** is directly proportional to the electric field or applied voltage and is a linear electro-optic effect. Crystals which are linearly electro-optic are also piezoelectric; application of an electric field produces a change in direction, and application of pressure induces a voltage change.

Electro-optic modulators based on the Kerr and Pockels effects are shown in Figure 3-15a and b. Note that the electro-optic "cell" is sandwiched between two crossed linear polarizers. When no electric field is applied to the cell, no radiation is passed. If a voltage is applied that induces a 180° phase shift, the plane of polarization is rotated 90° and maximum transmission is obtained. By applying the appropriate sinusoidally modulated voltage, a sinusoidally modulated radiation beam is produced. Pockels cells can provide sinusoidal modulation at frequencies as high as 10 to 100 MHz. If the applied voltage is pulsed, rise times from 1 to 30 ns are achieved. Kerr cells can be modulated up to about 10 MHz. Pockels cells are usually preferred because they require 1 to 3 kV for operation, which is 5 to 10 times less than that required for Kerr cells.

Circular birefringence induced by a magnetic field is termed the **Faraday effect**. This magneto-optic effect can be used to construct a Faraday modulator, as shown in Figure 3-15c. The overall design is similar to an E-O modulator except that a magnetic rather than electric field is applied. In this case, the transmitted radiation is always linearly polarized and the rotation of the plane of polarization is proportional to the applied magnetic field.

## Acousto-Optic Modulators

The **acousto-optic** (A-O) **effect** is concerned with changes in optical properties that occur when light passes through a transparent medium into which an acoustic wave is launched. These induced changes can affect the deflection, polarization, phase, frequency, or amplitude of the incident beam. The pressure waves cause a periodic change in the density and refractive index of the medium. More generally, the change in refractive index caused by pressure (mechanical stress or acoustic waves) in crystals is termed the **photoelastic effect** or the **piezo-optic effect**.

Many A-O modulators are based on the Bragg cell design, as illustrated in Figure 3-16a. The moving stress maxima and minima act like multiple reflection

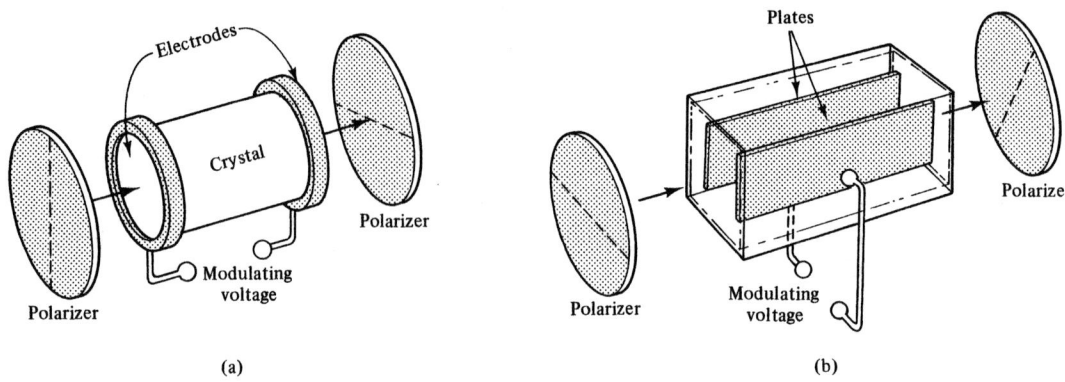

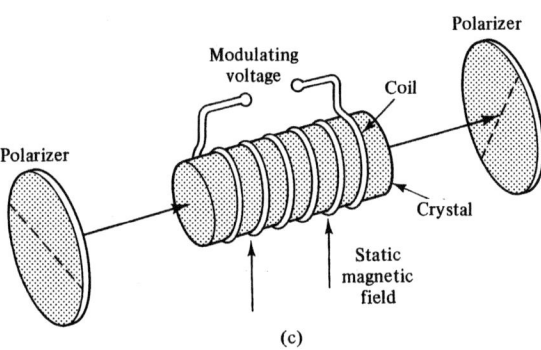

**FIGURE 3-15** Electro-optic and magneto-optic modulators. In all cases shown, a cell which provides an electrically controlled rotation of the plane of vibration of linearly polarized radiation is sandwiched between two crossed linear polarizers. For the Pockels cell shown in (a), a longitudinal electric field is applied to the crystal (commonly KDP, ADP, or deuterated forms of these materials). The electrodes are made of a transparent material (e.g., SnO, InO) or are thin metal films or grids. Transverse electric fields are also employed. For a Kerr cell as in (b), a transverse modulated electric field is applied to a polar liquid such as nitrobenzene or carbon disulfide. Transparent crystals such as barium or strontium titanate or lanthanum-modified lead zirconate titanate (PLZT) ferroelectric ceramics are also used. As shown in (c), a transverse dc magnetic field is applied in a Faraday cell to saturate the magnetization of the medium. The electromagnetic coil provides a longitudinal ac magnetic field to modulate the plane of rotation of the incident radiation. Common media include the transparent ferromagnetic materials.

layers where the separation between layers is the acoustic wavelength in the medium ($\lambda_s$). The incident radiation beam is aligned to intercept the pressure waves at a particular angle called the Bragg angle for which a large fraction of the output beam is periodically diffracted into the first order at the acoustic frequency. The Bragg angle ($\theta_B$) is given by

$$\theta_B = \sin^{-1} \frac{\lambda}{2\lambda_s}$$

If the optical pathlength in the acoustic medium is sufficiently thin [$b < (\lambda_s)^2/4\lambda$], the incident radiation

beam can be parallel to the acoustic waves. The medium becomes a moving transmission grating with "refractive index grooves" with a groove spacing equal to the acoustic wavelength in the medium. This Debye–Sears configuration is commonly used for A-O modulators with liquid media as shown in Figure 3-16b. A portion of the incident beam is periodically diffracted to other angles as regions of low and high refractive index pass across the incident beam.

In both the Bragg and Debye–Sears configurations, the acoustic wave is usually generated with a piezoelectric transducer that is in contact with the acoustic-wave supporting material. The transducer's

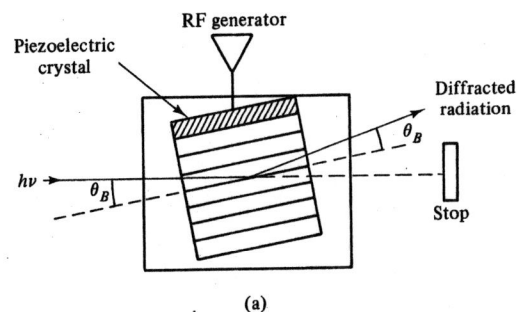

(a)

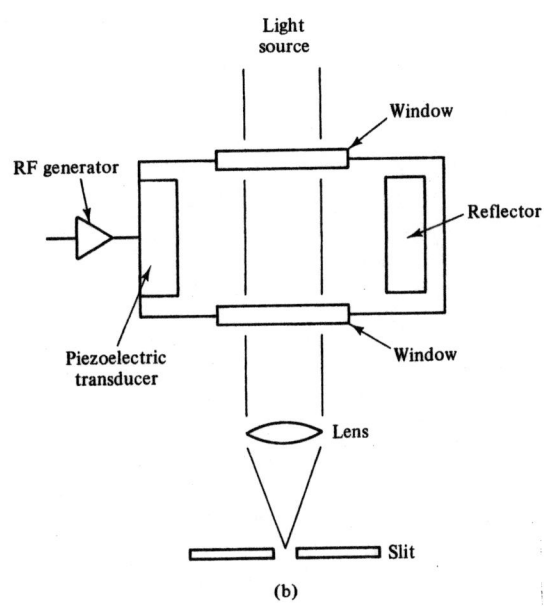

(b)

**FIGURE 3-16** Acousto-optic modulators. In the Bragg cell configuration shown in (a), a piezoelectric transducer in contact with a suitable crystal is driven by radio-frequency generator at typically 1 to 6 W. The orientation of the incident beam and crystal are adjusted so that the incident beam intercepts the acoustic waves at specific angle called the Bragg angle. As the acoustic waves pass through the path of the light beam, the direction of the exiting beam oscillates between two angles (no deflection or zero order and diffraction into the first order). Normally, the zero-order beam is blocked. The Debye-Sears modulator shown in (b) consists of tank filled with an ethanol–water or methanol–water mixture. A piezoelectric crystal and reflector on opposite ends of the tank are used to set up a standing acoustic wave. When the standing wave is at its null point, the incident beam passes straight through the medium and the slit. When the standing wave is at its maximum, up to 50% of the light is diffracted and not passed by the slit.

resonant frequency is determined by the piezoelectric constant of the material used (usually lithium niobate). A given A-O modulator can only be used at a few set frequencies (the fundamental and harmonic frequencies). The bandwidth is controlled by the acoustic velocity in the medium. For Bragg cells, the efficiency with which the incident beam is diffracted into the first order is controlled by the radio-frequency (RF) power applied and the sixth power of the refractive index. Although glass or quartz can be used for crystal media. $TeO_2$ is often the preferred material because of its high refractive index ($\eta = 2.2$). Sinusoidal modulation in the range 3 to 100 MHz is common, with rise times of 4 to 30 ns.

Some A-O modulators use mechanical stress to induce birefringence in isotropic materials. The difference in retardation of the ordinary and extraordinary rays is proportional to the stress applied. Thus an acoustically driven crystal can be used as a variable wave plate, where the stress is applied perpendicular to the light beam. The crystal can be placed between two cross polarizers to construct a modulator similar in design to E-O modulators (Figure 3-15).

In the discussion above we have stressed primarily the application of modulators as an external component used in conjunction with a continuous source. Many of these E-O and A-O devices are also used inside laser cavities for $Q$-switching and mode locking (see Section 4-3). In addition, some of these A-O devices are used as electrically controlled polarizers (i.e., to rotate the plane of linearly polarized light or to produce elliptically or circularly polarized light). In some applications it is necessary to determine the difference in the optical signal observed for different states of polarization (i.e., parallel vs. perpendicular components or $d$ vs. $l$ components). Here electrically controlled polarizers are used as polarization modulators.

## 3-4 IMAGING AND BEAM DIRECTING OPTICS

As discussed in Chapter 2, an optical spectrometer contains an image transfer system to produce an image or set of images of the entrance aperture on the focal plane. In addition, imaging optics are frequently used to collect and focus radiation from an external source onto the sample container and radiation from the sample container onto a filter or the entrance slit of a monochromator. Lenses and mirrors are the most frequently employed optical elements for image formation and transfer. Ordinarily, these elements are large compared to the wavelengths of visible radiation so that we can discuss their principal effects without considering interference or diffraction phenomena. These wave phenomena do, however, play a significant role in limiting the resolution and the sharpness of images formed with simple lenses and mirrors. This section considers the principles of image formation by mirrors and lenses, the irradiance of the images formed, and the aberrations that can limit image quality. The section concludes with a brief discussion of beam splitters, which are used for directing beams, and fiber optics, which are finding increasing applications in spectrochemical methods.

## Mirrors

Mirror systems are extensively used in the UV-visible and IR regions of the spectrum because of the ease with which reflecting systems can be designed to perform over a broad frequency range. In the past, mirrors were frequently made by coating glass with silver, which has high reflectance in the visible and IR. Glass was used as the substrate because of its rigidity, ease of fabrication, and low coefficient of thermal expansion. Today, vacuum-evaporated coatings of aluminum or other metals on highly polished substrates are used in high-quality reflective devices. Often, protective coatings of $MgF_2$ or $SiO_2$ are applied over the aluminum to prevent oxidation. Typical reflectance curves for mirror surfaces are shown in Appendix B. Mirrors made for nontechnical purposes have their reflective coatings behind glass for protection. Mirrors designed for scientific applications, however, are usually **front surface mirrors** with their reflective surfaces in front of the substrate. For use with lasers, which require extremely high reflectivity surfaces (99% or larger), mirrors are often formed from dielectric films (see Appendix B). Mirrors can be planar, spherical, or aspherical. We consider here spherical and aspherical mirrors prior to discussing the familiar plane mirror.

*Spherical Mirrors.*   The spherical mirror is easy to construct and suitable for many applications. Consider a point object O illuminating a spherical mirror as shown in Figure 3-17. This mirror surface is said to be **concave** since viewed along the direction of the incident light the surface curves inward. For the two rays shown a **real image** is formed at point I. The image is said to be real if radiant energy actually passes through the image point. In a **virtual image** the light appears to

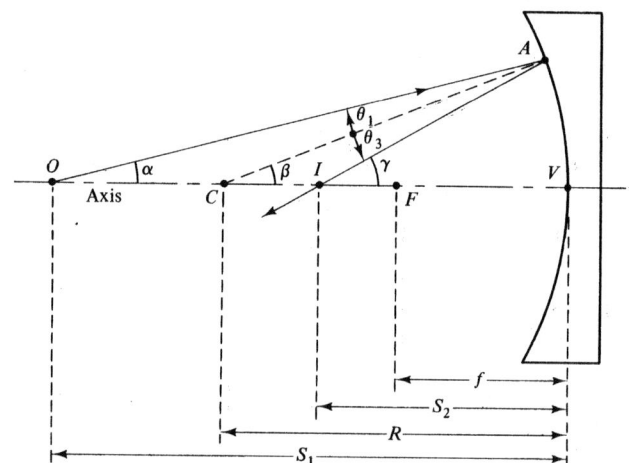

**FIGURE 3-17**   Concave spherical mirror with radius of curvature R and center of curvature C. Object O is located at distance $S_1$ from the mirror vertex V. Image I is located at distance $S_2$. A line through O and C is the optical axis. Distance R is taken as positive if the center of curvature lies to the right of the vertex and negative if it lies to the left as shown. From the external angle theorem, $\beta = \alpha + \theta_1$ and $\gamma = \beta + \theta_3$. From the law of reflection $\theta_1 = \theta_3 = \theta$. Thus $\alpha + \gamma = 2\beta$. If we make the paraxial approximation, angles $\alpha$, $\beta$, and $\gamma$ are (in radians), $\alpha \approx AV/S_1$, $\gamma \approx AV/S_2$, and $\beta = AV/-R$. Thus $1/S_1 + 1/S_2 = -2/R$.

derive from the image point, although in actuality it does not pass through the point. Corresponding object and image points are called **conjugate points**.

The derivation in the caption of Figure 3-17, as well as several other derivations in this chapter, uses the external angle theorem and the paraxial approximation. The **external angle theorem** states that the exterior angle of a triangle is equal to the sum of the opposite interior angles. The **paraxial approximation** assumes that all rays striking the optical element make small angles with respect to the optical axis. As a result we can assume that all distances along the optical axis are near the center of curvature and that the sine or tangent of the angle is approximately equal to the angle in radians. As shown by the derivation in Figure 3-17, the **mirror formula** is

$$\frac{1}{S_1} + \frac{1}{S_2} = \frac{-2}{R} \qquad (3\text{-}29)$$

In the sign convention most commonly adopted for mirrors and lenses, the image distance $S_2$ is positive if the image is real, and negative if the image is virtual. (All object distances considered here will be positive values.) The radius of curvature R is negative if the center of curvature lies to the left of the vertex (concave mirror) and positive if the center lies to the right of the vertex (convex mirror). With this convention equation 3-29 applies equally well to convex as well as concave mirrors.

If the object is sufficiently far from the mirror to ensure parallel light ($S_1 \to \infty$, $1/S_1 \to 0$), the image is formed at a distance $-R/2$ from the vertex $V$, called the **focal length** $f$ of the mirror. Thus

$$f = \frac{-R}{2} \qquad (3\text{-}30)$$

Here $f$ is positive for concave mirrors ($R < 0$) and negative for convex mirrors ($R > 0$). The mirror formula can then be written

$$\frac{1}{S_1} + \frac{1}{S_2} = \frac{1}{f} \qquad (3\text{-}31)$$

Figure 3-18 shows the action of several spherical mirrors. Note that in the case of the convex mirror (Figure 3-18c), the image is virtual ($S_2 < 0$, $R > 0$,

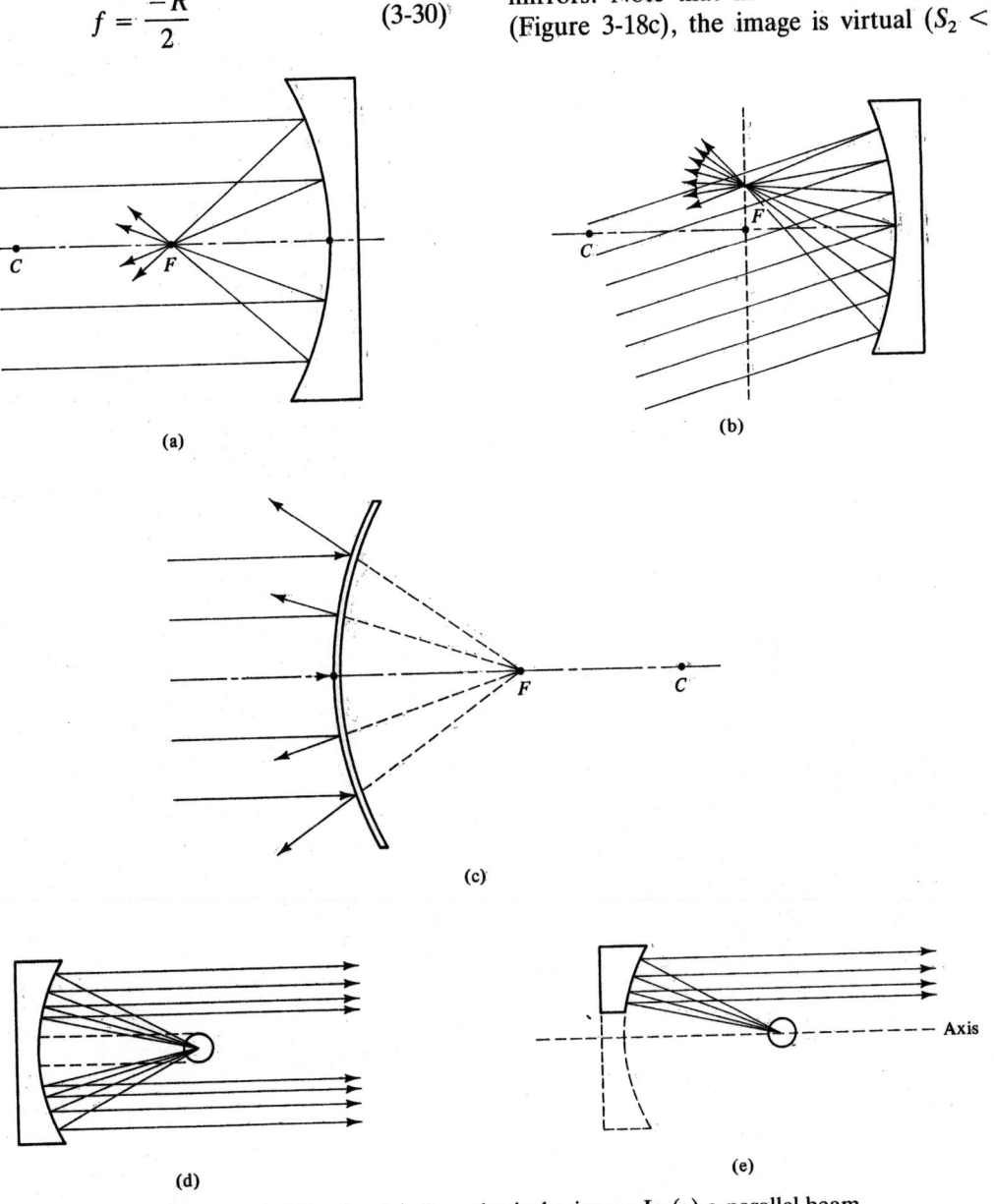

(a)

(b)

(c)

(d)

(e)

**FIGURE 3-18** Focal points for spherical mirrors. In (a) a parallel beam of light strikes a concave mirror and produces a real image at the focal point. In (b) a parallel beam makes an angle with the mirror axis and produces a focus in the focal plane (dashed line). In (c) a convex mirror is shown; the focal point is virtual and the focal length is negative. Placing a source at the focal point of a concave mirror (d) produces a collimated beam. The region of the mirror directly behind the source is masked by the source. In (e) an off-axis mirror is used to avoid having the source in the light path.

$f < 0$) as it is constructed from the intersection of projected rays.

The mirror formula indicates that moving the point source $O$ from infinity toward the mirror or decreasing $S_1$, increases $S_2$, and the image moves away from the focal point toward the center of curvature. If the object is at the center of curvature, the image must also be at that point. As the object approaches the focal point, the image approaches infinity. Thus a point source of light placed at the focal point produces a parallel or **collimated beam** as shown in Figure 3-18d and e.

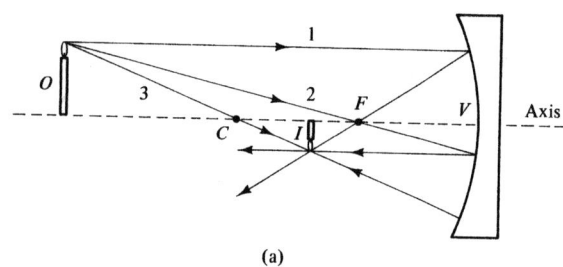

(a)

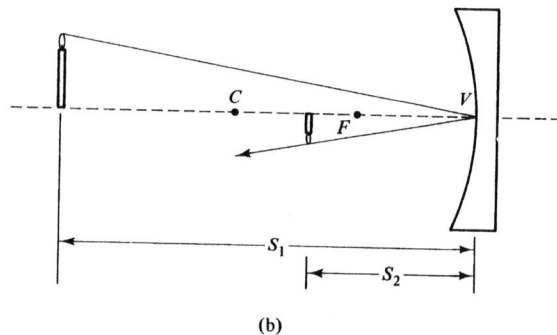

(b)

**FIGURE 3-19** Extended objects imaged by spherical mirrors. In (a) the image is located graphically. Ray 1 strikes the mirror parallel to its axis and reflects through the focal point. Ray 2 passes through the focal point and emerges parallel to the axis. Ray 3 passes through the center of curvature and reflects back on itself. In (b) a ray from the candle tip strikes the vertex and reflects at an equal angle on the other side of the optical axis to illustrate the lateral magnification $m = -S_2/S_1$.

When an extended object (object size significant compared to $S_1$) is used, we can locate the image by the method shown in Figure 3-19a. The **lateral magnification** $m = -i/o$, where $i$ and $o$ are the height of the image and object, respectively, and the minus sign indicates the image is inverted. The two similar right triangles, shown in Figure 3-19, must have proportional sides. Thus $i/o = S_2/S_1$ and $m = -S_2/S_1$.

Table 3-2 summarizes the types of images that can be obtained with concave mirrors. Note that when the object is located closer to the mirror than the focal point, the reflected rays are divergent, and the image is a virtual, erect (noninverted) image.

*Aspherical Mirrors.* Aspherical mirrors are used in place of spherical mirrors when image quality is critical. Although aspherical mirrors are more difficult to produce, and hence more expensive, certain types of aberrations are reduced, as discussed later in this section. **Parabolic mirrors** are the most popular type of aspherical mirror and are widely used in flashlights, automobile headlights, radio-telescope antennas, as well as spectroscopic instruments. Hyperbolic and elliptical mirrors have two foci and are often used as secondary mirrors to change the effective focal length of an optical system.

*Plane Mirrors.* We are all familiar with the image formed when we stand in front of an ordinary plane mirror. The image is virtual in that it appears to originate from behind the mirror. A plane mirror can be considered as a spherical mirror with an infinite radius of curvature. In the mirror formula (equation 3-31), $f = \infty$, and $1/f = 0$. Thus $1/S_1 = -1/S_2$ and $S_1 = -S_2$. The image in a plane mirror is virtual ($S_2 < 0$) and appears to originate from a distance behind the mirror equal to the distance of the object from the mirror. The magnification is $m = -S_2/S_1 = +1$. The plane mirror gives an **erect image** of unity magnification.

If the object is an extended object instead of a point, each point at a particular distance from the mirror is imaged at that same distance behind the mirror. The

**TABLE 3-2**

Images formed by spherical concave mirrors and thin convex lenses

| Object location | Image | | | |
|---|---|---|---|---|
| | Type | Location | Orientation | Size |
| $\infty > S_1 > 2f$ | Real | $f < S_2 < 2f$ | Inverted | $\lvert m \rvert < 1$ |
| $S_1 = 2f$ | Real | $S_2 = 2f$ | Inverted | $\lvert m \rvert = 1$ |
| $f < S_1 < 2f$ | Real | $\infty > S_2 > 2f$ | Inverted | $\lvert m \rvert > 1$ |
| $S_1 = f$ | | $\pm \infty$ | | |
| $S_1 < f$ | Virtual | $\lvert S_2 \rvert > S_1$ | Erect | $\lvert m \rvert > 1$ |

entire image can be constructed by a point-by-point analysis using the law of reflection. As we know, the mirror image of a left hand is a right hand. Thus the plane mirror reverses left and right. The primary use of plane mirrors in spectroscopy is for beam directing.

## Lenses

A lens is an optical element with two or more refracting surfaces. These are most frequently spherical segments made from glass or quartz for visible and UV applications. We restrict ourselves here to **thin lenses** where the thickness of the lens material is small compared to the object and image distances. Several examples of lens types are shown in Figure 3-20. If the lens has a larger index of refraction than the surrounding material, a convex lens tends to make incident wavefronts converge as they traverse the lens, while concave lenses cause divergence.

We consider first refraction at a single spherical surface before proceeding to simple lenses and lens apertures.

*Refraction at a Spherical Surface.* Consider a point source of light illuminating the spherical surface shown in Figure 3-21. If all the angles are small (paraxial approximation), the derivation in the figure caption gives

$$\frac{\eta_1}{S_1} + \frac{\eta_2}{S_2} = \frac{\eta_2 - \eta_1}{R} \tag{3-32}$$

*Thin Lens.* Let us now consider a simple lens with index of refraction $\eta_2 = \eta$ surrounded by air ($\eta_1 = 1.00$), as shown in Figure 3-22. If the first surface

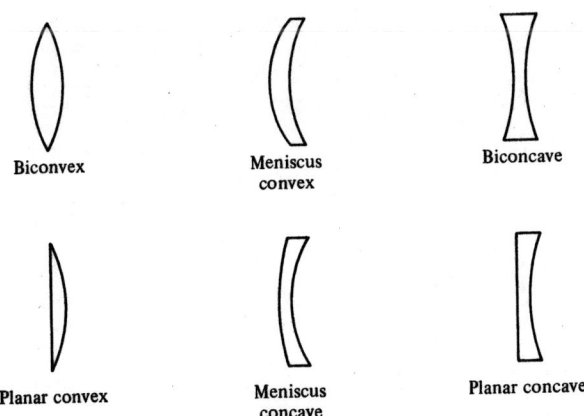

**FIGURE 3-20** Several types of lenses. Lenses which are thicker at the center than at the ends are called convex, converging, or positive lenses. Lenses that are thinner in the center than at the ends are concave, diverging, or negative lenses.

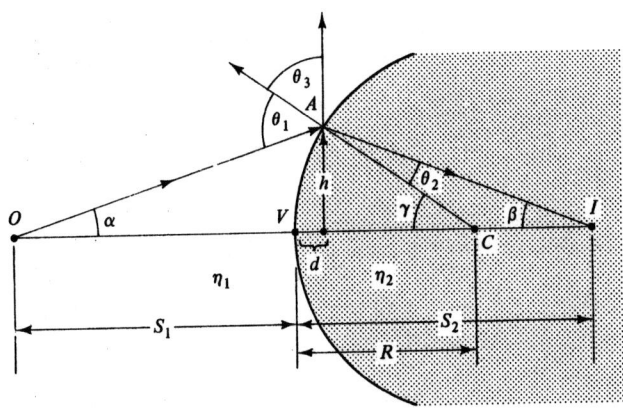

**FIGURE 3-21** Refraction at a spherical surface of radius $R$ and center of curvature $C$. A ray from point source $O$ in a medium of refractive index $\eta_1$ strikes the surface at $A$. If $\eta_2 > \eta_1$, the light entering the medium of refractive index $\eta_2$ is bent toward the normal. If $\theta_1$ and $\theta_2$ are small (paraxial approximation), $\sin \theta \approx \theta$, and Snell's law is $\eta_1\theta_1 = \eta_2\theta_2$. Thus $\theta_2 = \eta_1\theta_1/\eta_2$. Since $\theta_1 = \alpha + \gamma$ (external angle), $\theta_2 = [\eta_1(\alpha + \gamma)]/\eta_2$. Since $\gamma = \beta + \theta_2$ (external angle), $\gamma = \beta + [\eta_1(\alpha + \gamma)]/\eta_2$ and $\eta_1\alpha + \eta_2\beta = (\eta_2 - \eta_1)\gamma$. If angles $\alpha$, $\beta$, and $\gamma$ are small (paraxial rays) and distance $d$ is small compared to object distance $S_1$, image distance $S_2$, and $R$, $\tan \alpha \approx \alpha \approx h/S_1$, $\tan \beta \approx \beta \approx h/S_2$, and $\tan \gamma \approx \gamma \approx h/R$. Thus $\eta_1/S_1 + \eta_2/S_2 = (\eta_2 - \eta_1)/R$.

has radius of curvature $R_1$ ($R_1 > 0$), equation 3-32 yields

$$\frac{1}{S_1} + \frac{\eta}{S_2'} = \frac{\eta - 1}{R_1} \tag{3-33}$$

where $S_2'$ is the image distance from the first surface. For clarity in the derivation that follows, object $O$ in Figure 3-22 has been moved close to the lens such that distance $S_1$ is quite small. If the distance $S_1$ is small, the paraxial rays from $O$ form a virtual image of $O$ at distance $S_2'$ ($S_2' < 0$). The distance $S_2'$ is now the object distance for the second surface. Since we have assumed the lens to be thin, there is no need to modify this distance for the lens thickness. If we use $-S_2'$ as the object distance for the second surface, the image will be formed at distance $S_2$. Or

$$\frac{\eta}{S_2'} + \frac{1}{S_2} = \frac{1 - \eta}{R_2} \tag{3-34}$$

Adding equations 3-33 and 3-34 yields the **lensmakers' formula:**

$$\frac{1}{S_1} + \frac{1}{S_2} = (\eta - 1)\left(\frac{1}{R_1} - \frac{1}{R_2}\right) \tag{3-35}$$

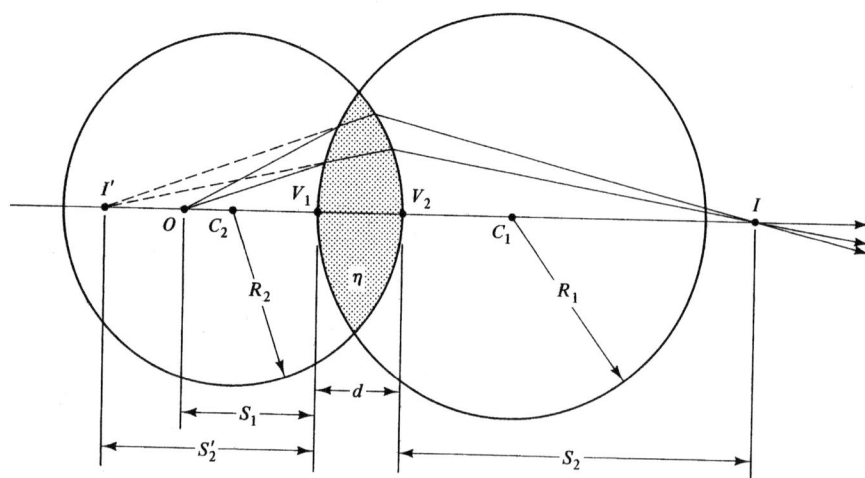

**FIGURE 3-22** Spherical lens. An object $O$ at distance $S_1$ from the lens forms an image $I$ at distance $S_2$. The object rays strike the first surface and forms a virtual image $I'$ at distance $S_2'$, which is now the object distance for the second surface. Lens thickness $d$ is assumed to be negligibly small.

Note that for the biconvex lens of Figure 3-22, $R_1 > 0$ and $R_2 < 0$. Here the sign convention, compatible with that of mirrors, is that $R$ is positive if the center of curvature of the surface is to the right of the vertex and negative if to the left. The focal length $f$ is related to the index of refraction of the lens and its radii of curvature by

$$\frac{1}{f} = (\eta - 1)\left(\frac{1}{R_1} - \frac{1}{R_2}\right) \qquad (3\text{-}36)$$

This follows from equation 3-35 since $f$ is the image distance if the object is at infinity or $f = S_1$ if $S_2 = \infty$. Thus the lens formula, equation 3-35, can also be expressed as

$$\frac{1}{S_1} + \frac{1}{S_2} = \frac{1}{f} \qquad (3\text{-}37)$$

As an example, let us calculate the focal length of a planoconvex lens of radius of curvature 60 mm and index of refraction 1.5. Let us assume that light is incident on the planar surface so that $R_1 = \infty$ and $R_2 = 60$ mm. From equation 3-36 we have $1/f = (1.5 - 1)(0 - 1/-60)$, from which we find that $f =$

120 mm. If the object is placed at 300 mm from the lens, the image position can be found from equation 3-37 as $1/S_2 = 1/120 - 1/300$, from which $S_2 = 200$ mm. If $S_1 = 200$ mm, $S_2 = 300$ mm. For this lens the points 200 mm and 300 mm represent conjugate points. If $S_1 = 240$ mm $(2f)$, $S_2 = 240$ mm. If $S_1 = \infty$, $S_2 = f$. Thus a parallel incident beam of light will form an image at the focal point. Similarly, a point object placed at the focal point will produce a parallel or collimated beam.

The image of an extended object can be located graphically by the method illustrated in Figure 3-23. The lateral magnification $m$ is the ratio of the image height to the object height. Just as with mirrors, $m$ is given by $m = -S_2/S_1$. The types, locations, orientations, and sizes of the images obtained with thin convex lenses are the same as those for concave mirrors and are shown in Table 3-2.

In addition to the simple lenses presented here, cylindrical lenses are also used. These are used to convert a point image into a line image or to change the height of an image without changing its width. Cylindrical surfaces are difficult to produce and they are available only in planar-cylindrical shapes (planar con-

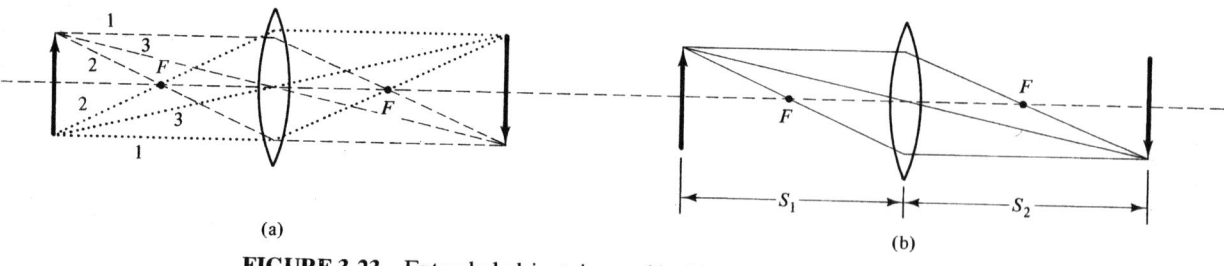

(a)                              (b)

**FIGURE 3-23** Extended objects imaged by biconvex lenses. In (a) graphical location of the image is shown. Ray 1 is parallel to the optical axis and passes through the focal point. Ray 2 passes through the focal point and emerges parallel to axis. Ray 3 is undeviated. In (b) rays from the top of the arrow illustrate the lateral magnification $m = -S_2/S_1$.

vex or planar concave). Aspherical lenses are also available to reduce certain aberrations.

Reflective losses with lenses can be substantial unless they are coated. For example, a typical glass lens can suffer ≈ 8% reflection loss. By coating the lens with a thin layer of a material of index of refraction intermediate to that of air and glass, reflective losses can be greatly reduced. With a lens such **antireflective coatings**, designed for normal incidence, can reduce reflective losses to ≈ 3.5% from ≈ 8%. Antireflective coatings are discussed in Appendix B.

*Lens Stops.*    Any element that determines the amount of radiant energy reaching the image is called an **aperture stop**. In the simple lenses discussed thus far, the size of the lens itself limits the amount of light accepted. In other cases a separate fixed opening (ap-

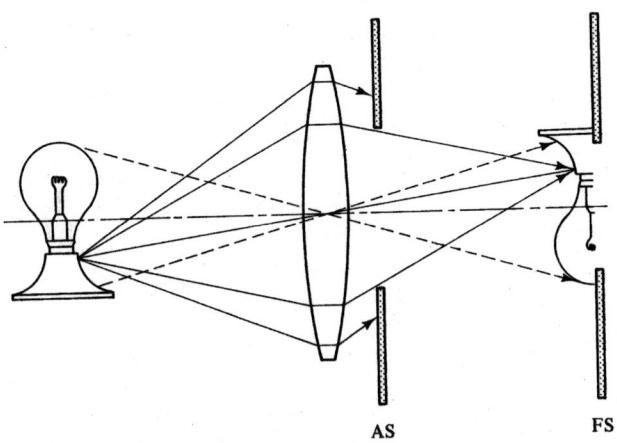

**FIGURE 3-24** Lens with aperture and field stops. The aperture stop (AS) controls the number of rays reaching the image from each object point (solid angle viewed). The field stop (FS) controls the field of view (area of the image viewed). Opening the aperture stop increases the amount of light reaching each image point, while opening the field stop allows more of the object to be imaged.

erture) or an adjustable diaphragm is used to restrict the rays that form the image. This has the effect of limiting nonparaxial rays and improving image quality. A lens with a rear aperture stop is shown in Figure 3-24. In many optical systems an additional element, called a **field stop**, limits the size of the image, as shown in Figure 3-24. In a camera, the field stop is the edge of the film itself.

The image of the aperture stop formed by elements between the object and the aperture stop is called the **entrance pupil**. It is the smallest opening that can be seen at the object point. The image of the aperture stop formed by elements between it and the image is called the **exit pupil**. It is the smallest opening that can be observed at the image point. The entrance pupil controls the cone of light entering the optical system, while the exit pupil controls that leaving it.

Figure 3-25 illustrates how two lenses can be arranged in such a way that a larger cone of light comes from on-axis object points than from those that are off-axis. This is a process known as **vignetting**.

## Image Irradiance

Thus far we have considered only the types of images formed by lenses and mirrors. Also of extreme importance is the irradiance of the resulting image, which depends on the source radiance, the source distance, the characteristics of the optical element itself, and the presence or absence of various apertures. The radiant power $\Phi$ incident on the optical element (lens or mirror) is equal to the source radiance $B_s$ times the solid angle $\Omega$ subtended by the element, times the source area viewed, $A_s$.

$$\Phi = B_s \Omega A_s = B_s \frac{A}{S_1^2} A_s \qquad (3\text{-}38)$$

Here $A$ is the effective or limiting area of the optical element as determined either by the element itself or

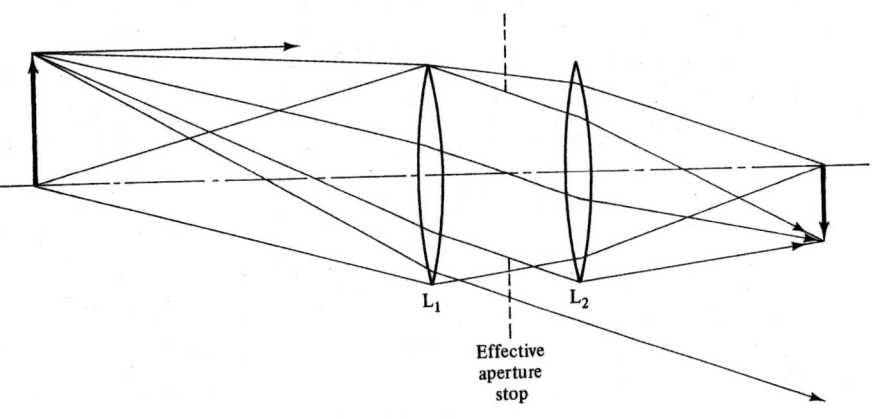

**FIGURE 3-25** Vignetting. The effective aperture stop for on-axis rays is lens $L_1$, whereas for off-axis rays it is much smaller, as shown by the dashed lines. This results in the image becoming gradually weaker in intensity at its extremities.

by an aperture stop, and $S_1$ is its distance from the source. The irradiance of the image $E_i$ in the focal plane at $S_2$ is just the total radiant power of the image divided by its area, $A_i$. If we assume there are no reflection or absorption losses in the optical system, $E_i$ is just

$$E_i = \frac{B_s A}{S_1^2} \frac{A_s}{A_i} \qquad (3\text{-}39)$$

The effective area of the optical element is just $A = \pi(D/2)^2$, where $D$ is the effective or limiting diameter. The ratio $A_s/A_i$ in equation 3-39 is just $1/m^2$, where $m$ is the magnification, if there is no field stop. Thus equation 3-39 becomes

$$E_i = \frac{B_s \pi (D/S_1)^2}{4m^2} \qquad (3\text{-}40)$$

The irradiance of the image is seen to vary with $(D/S_1)^2$. The ratio $D/S_1$ is often called the **relative aperture** of the optical system. The reciprocal of the relative aperture is called the **F-number** or **F/n**. This is

$$\text{F/n} = \frac{S_1}{D} \qquad (3\text{-}41)$$

where F/n is understood to be a single symbol. An F-number of 2 is thus written F/2. If equation 3-41 is substituted into equation 3-40, the result is

$$E_i = \frac{B_s \pi}{4m^2} \frac{1}{(\text{F/n})^2} \qquad (3\text{-}42)$$

and the effective solid angle of the imaging system is $A/S_1^2 = \pi(D/2)^2/S_1^2$. Or

$$\Omega = \frac{\pi}{4} \frac{1}{(\text{F/n})^2} \qquad (3\text{-}43)$$

The image irradiance and the effective solid angle are seen to be inversely proportional to the square of the F-number of collection. Thus the F/n is often referred to as the **optical speed** of the system. An F/1.4 system, for example, is said to be twice as fast as an F/2 system. It is important to note that the magnification $m$ and the F-number are not independent quantities. From equations 3-42 and 3-43, $E_i = B_s \Omega/m^2$. For a given system (fixed $D$ and $f$), if $S_1$ is increased to decrease $m$, $\Omega$ is also decreased. Hence irradiance cannot be increased just by decreasing magnification. Unity magnification usually provides the best compromise between throughput and focal length.

If the extended source is far away ($S_1 \gg f$) from the optical element, the image is formed in the focal plane at a distance $f$ from the optical element ($S_2 \approx f$). With $S_2 = f$ and the relationship $m = -S_2/S_1$, equation 3-40 becomes

$$E_i = \frac{B_s \pi}{4} \left(\frac{D}{f}\right)^2 \qquad (3\text{-}44)$$

The F-number of the optical element $(\text{F/n})_o$ is $f/D$, so

$$E_i = \frac{B_s \pi/4}{(\text{F/n})_o^2} \qquad (3\text{-}45)$$

Notice we distinguish between the F-number of the optical system (determined by $D$ and $S_1$) and the F-number of the optical component (determined by $D$ and $f$). They are equal only if $S_1 = f$. Equation 3-45 shows that the irradiance of the image is also inversely proportional to the F-number or optical speed of the optical element since $(\text{F/n}) = (\text{F/n})_o(S_1/f)$. In photography, the F-number of a camera refers to $(\text{F/n})_o$, which is varied with an aperture stop. The exposure time is doubled if the F-number is increased by $\sqrt{2}$ to keep the irradiance constant for proper film exposure. The F-number of the optical element indicates the solid angle collected (according to equation 3-43) if the object is at the focal point or the solid angle between the optical element and image if $S_1 \gg f$.

In some cases the limiting aperture may not be circular but square or some other shape. By convention, the value of limiting aperture diameter used for F/n calculations is the diameter of a circle with the same area $A$ as the limiting aperture [i.e., $D = (4A/\pi)^{1/2}$].

The total radiant power of the image $\Phi$, is the product of the image irradiance and area or $E_i A_i$. If a field stop restricts the area of the image viewed, then $\Phi = E_i A_{\text{FS}}$, where $A_{\text{FS}}$ is the area of the field stop.

In multicomponent systems, such as used in spectrometers, the image of one component becomes the object for a successive optical component. In this case, the irradiance of the final image and the F/n of the system are usually determined by some limiting optical component.

To determine the irradiance of the image more accurately, reflection and absorption losses must be included and the T-number is often used. The T-number is defined as T-number $= (\text{F/n})/T^{1/2}$, where $T$ is the optical efficiency of the system (the product of the transmittances of all lenses and windows and of the reflectances of all mirrors). Since $T < 1$, the T-number has a lower limit of the F-number; it is wavelength dependent.

## Optical Aberrations

Deviations that optical elements suffer from the ideal or first-order behavior described above are called **optical aberrations**. They fall into two major categories: chromatic aberrations and monochromatic aberrations.

*Chromatic Aberrations.* As mentioned in Section 3-2, the refractive index of a given material is a function of the wavelength of the incident radiation. Because lenses depend on refraction for their imaging properties, they suffer from chromatic aberrations. Front surface mirrors, which depend only on reflection, do not. The equation for the focal length of a thin lens (equation 3-36) shows that the focal length is also a function of wavelength because $\eta$ is a function of wavelength. Consider, for example, fused silica, which has a refractive index of 1.469618 at $\lambda = 404.7$ nm and 1.458404 at $\lambda = 589.0$ nm. Since $1/f \propto (\eta - 1)$, the focal length of a fused silica lens will be 2.4% longer at 589.0 nm than at 404.7 nm. Thus in general $f$ decreases with decreasing wavelength. Chromatic aberrations can be a particular problem if an optical system is aligned and optimized using visible light and the system is later used in the ultraviolet region. Dramatic effects on the size and quality of images and on the throughput can result.

Chromatic aberrations along the optical axis are termed **axial chromatic aberrations**, while in the vertical direction they are termed **lateral chromatic aberrations**. As a result of aberrations, a lens illuminated with white light fills a volume of space with a continuum of overlapping images that vary in size and color.

Chromatic aberrations can be compensated at two wavelengths by using a combination of a positive and a negative lens, called an **achromatic doublet**, as illustrated in Figure 3-26. The most commonly encountered achromat is the cemented Fraunhofer lens. Compensation at two separated wavelengths also reduces chromatic aberrations at intermediate wavelengths. Unfortunately, achromatic lenses are expensive and impractical in the ultraviolet region of the spectrum. One can envision, however, computer-controlled positioners that would adjust the position of a lens dynamically to compensate for chromatic aberrations.

*Monochromatic Aberrations.* Both lenses and mirrors suffer from aberrations even when the incident light is monochromatic. **Spherical aberrations** are a result of deviations from the paraxial approximation. For converging elements, off-axis rays are focused closer to the element than paraxial rays, as demonstrated in Figure 3-27a for a lens. One method to reduce spherical aberrations is to put an aperture stop in front of or

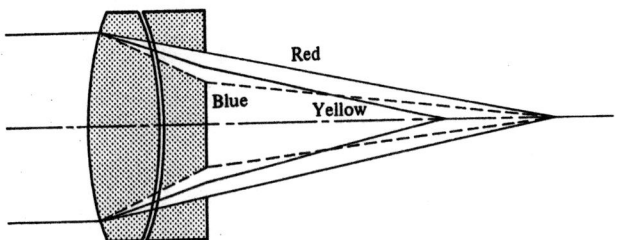

**FIGURE 3-26** Achromatic doublet. The cemented Fraunhofer lens is formed from a crown glass biconvex lens cemented to a flint glass concave-planar lens. The overall shape is roughly convex-planar. The total focal length $f_t$ for two thin lenses in contact is $1/f_t = 1/f_1 + 1/f_2$. Compensation at one wavelength in the red and another in the blue, for example, is achieved by selecting glasses of appropriate refractive indices and choosing their radii of curvature to give the same $f_t$ at the two wavelengths.

behind the lens so that off-axis rays are blocked. This, however, greatly reduces the irradiance of the image. For a planoconvex lens, deviations are also reduced if the lens is used as illustrated in Figure 3-27b. Combination lenses such as achromatic doublets also reduce spherical aberrations. Image quality is improved through reduction of spherical aberrations by use of aspherical optical components. Concave parabolic, ellipsoidal, and hyperboloidal mirrors form perfect images for pairs of conjugate axial points.

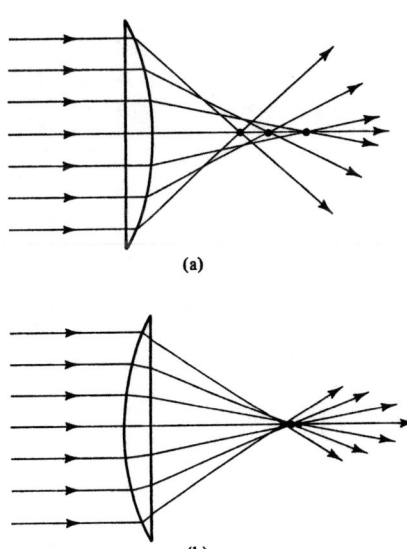

**FIGURE 3-27** Spherical aberrations for a lens. In (a) parallel rays are seen to produce a circle in the focal plane, called the circle of least confusion, due to spherical aberrations. Deviations are reduced if the incident rays make nearly the same angle with the surface as the exiting rays as shown in (b).

**Coma** in lenses and mirrors is readily observed by anyone trying to focus sunlight. A slight tilt of the lens so that the parallel rays from the sun strike the lens at an angle causes the spot to develop a cometlike tail characteristic of coma. Coma arises from differences in the optical pathlengths of various rays to the focal plane in an off-axis system. Coma is of little consequence when the image point is on the optical axis. Coma results in a difference in magnification for rays striking the extremities (marginal rays) than for rays that pass through or are reflected from the center of the element (principal ray). Positive and negative coma are illustrated for a lens in Figure 3-28.

Coma is a strong function of lens shape and object distance. For focusing parallel rays a convex-planar lens gives nearly zero coma. Coma can also be reduced, as can be seen from Figure 3-28, by blocking marginal rays with an aperture stop, but the position of the stop is critical. A second optical element can also compensate for coma if conditions are arranged such that short-pathlength rays to the first focus are long-pathlength rays to the second focus. The symmetric mirror pair shown in Figure 3-29 achieves excellent compensation for coma.

**Astigmatism** arises when the object point is considerably off-axis and results in two distinct focal lengths

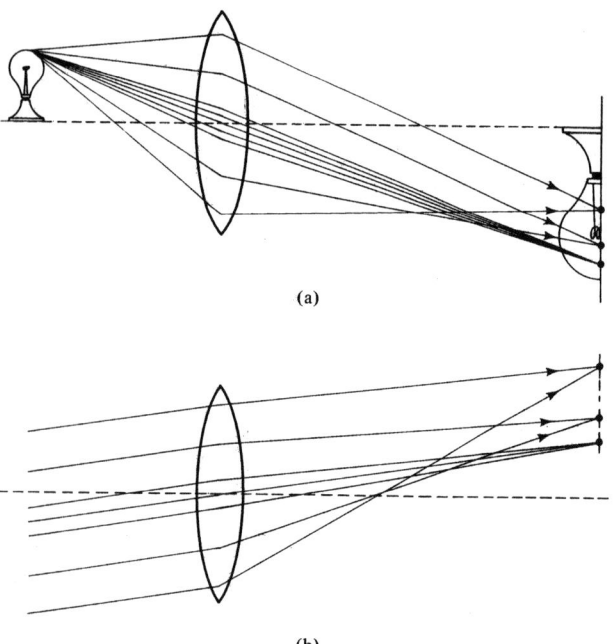

**FIGURE 3-28** Coma. In negative coma (a) marginal rays arrive at the image plane closer to the optical axis than do rays near the principal ray (ray through the lens center). The least magnification is associated with the marginal rays. The opposite is seen in (b), where positive coma is illustrated.

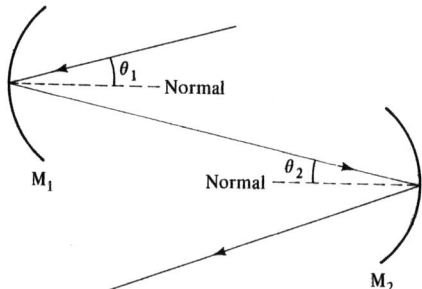

**FIGURE 3-29** Compensation for coma by use of a symmetric mirror pair. If the angles are arranged such that $\sin\theta_2/\sin\theta_1 = S_2^2\cos^3\theta_2/S_1^2\cos^3\theta_1$, the net coma is zero. For a unity magnification system where $S_2 = S_1$ and $f_2 = f_1$, this occurs when $\theta_2 = \theta_1$.

for rays in two different planes. Consider the lens shown in Figure 3-30a. Rays in the tangential plane are focused in the sagittal plane at the tangential focus, while those in the sagittal plane are focused in the tangential plane at the sagittal focus. Figure 3-30b shows an off-axis object point being imaged astigmatically. The separation of the two images is a quantitative measure of the astigmatism. A system in which the separation is zero is said to be **stigmatic**. If the principal ray makes an angle $\theta$ with the optical axis, an optical element with paraxial focal length $f$ has a sagittal focus $f_s$ and tangential focal length $f_t$ given by $f_s = f/\cos\theta$ and $f_t = f\cos\theta$.

An object in a plane normal to the optical axis will give rise to a planar image only for paraxial rays. With finite apertures the image plane is curved as a result of the aberration known as **field curvature**. In an optical system with both astigmatism and field curvature, two parabolic image surfaces are formed, the tangential and sagittal surfaces.

Compound lenses and multiple-mirror systems are also used to reduce astigmatism and field curvature. Because of their freedom from chromatic aberrations, mirrors are often preferred in spectroscopic systems. Examples of aberrations in monochromators are discussed in Section 3-6.

## Beam Splitters

There are many applications in spectroscopy which call for one beam to be split into two beams; the double-beam spectrometer (see Section 4-6) and the two-beam interferometer (see Section 3-7) are but two examples. Devices that accomplish this task are called **beam splitters**. They are available in many different forms for various applications.

One simple beam splitter is based on a partially silvered mirror as shown in Figure 3-31a. They are usu-

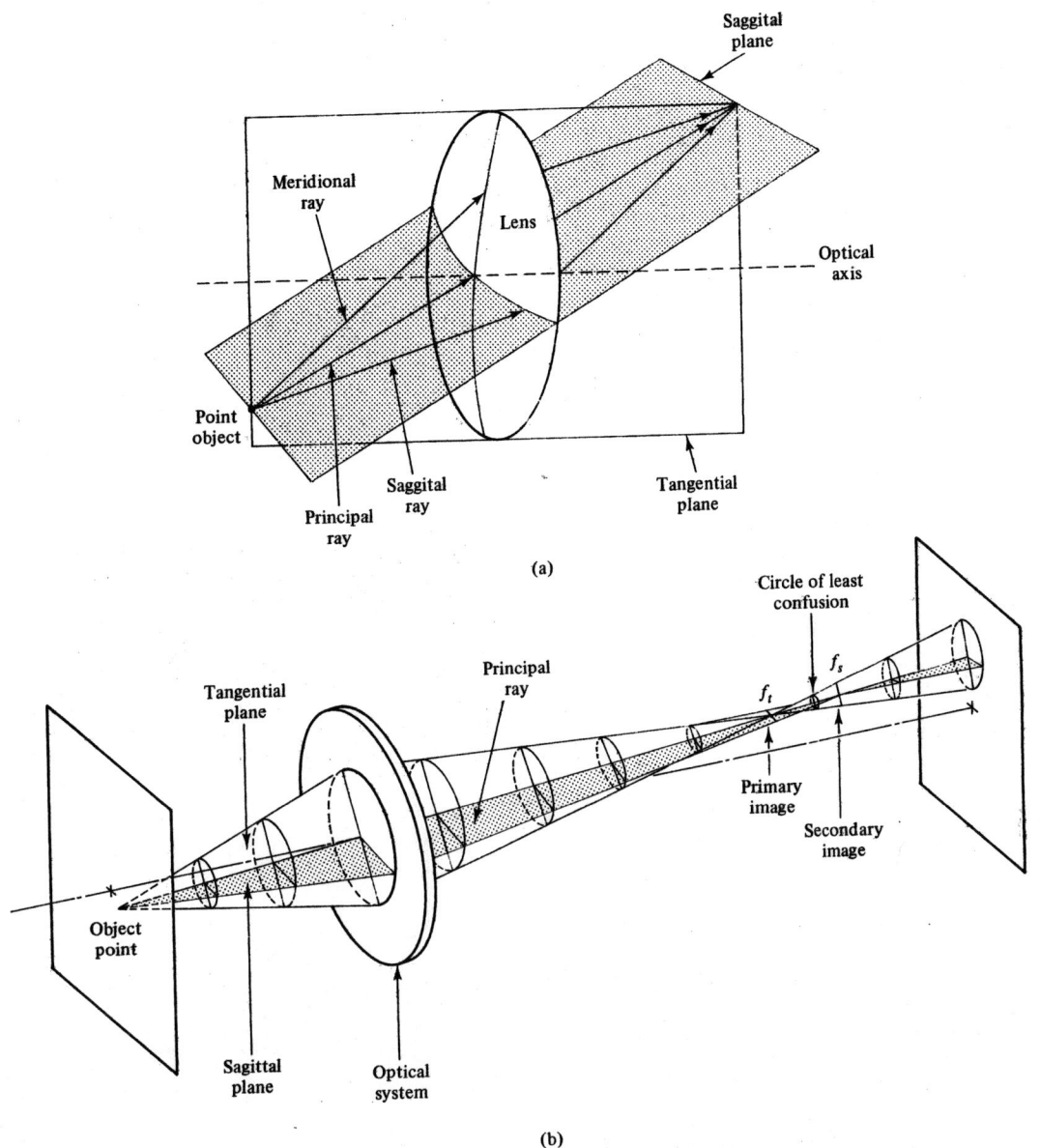

**FIGURE 3-30** Astigmatism. Astigmatism results when rays in two different planes, the tangential and the sagittal planes, are brought to different focal points. The tangential plane (a) contains the principal ray (ray through the aperture center) and the optical axis. The sagittal plane is the plane normal to the tangential plane that contains the principal ray. An off-axis object point (b) is imaged as a line at the tangential focus, becomes a circle beyond this point, and deforms into a line again at the sagittal focus. The circle of least confusion is located halfway between the two foci.

ally front-surface coated. Different coating thicknesses are used to produce various ratios of transmitted to reflected beam intensities. For example, the typical half-silvered mirror might transmit 30% of the incident radiation and reflect 30%. The remaining 40% is lost through absorption. Metallic coatings are useful throughout the UV and visible regions and into the near-infrared region (> 1 μm). Dielectric coatings are also used to make beam splitters of higher efficiency and a wide range of beam intensity ratios. The efficiencies and beam splitting ratios are often wavelength dependent. The second surface of mirror-type beam splitters can be antireflection coated (see Appendix B) to minimize the multiple reflections that can produce unwanted "ghost" beams. Beam splitters are often used in reverse to produce beam combiners.

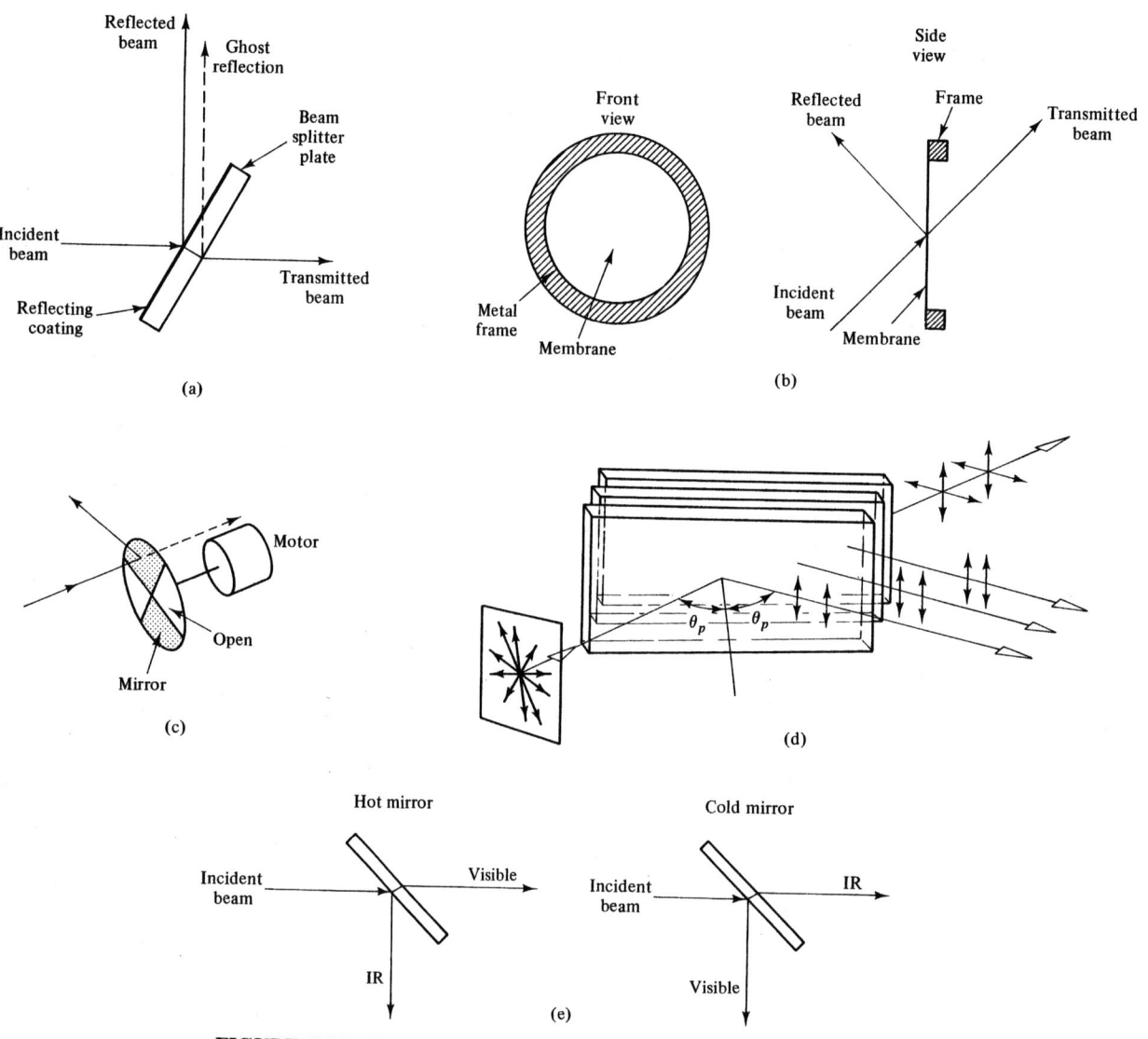

**FIGURE 3-31** Beam splitters. In (a) a partially silvered mirror beam splitter is shown. This type of mirror is partially transparent because the metallic coating is too thin to make it opaque. With such a mirror, one can both look through it and see a reflection simultaneously. In (b) a pellicle beam splitter is shown. These are made from very thin nitrocellulose membranes stretched over a metallic frame. A chopper/beam splitter is shown in (c). Here the chopper wheel consists of alternating transmitting and reflecting segments. The transmitting segments are just open slots, while mirrored surfaces form the reflecting segments. The reflected and transmitted beams are temporally separated as shown. A "pile-of-plates" polarizing beam splitter is shown in (d). Here an unpolarized beam is separated into a perpendicular polarized beam and a partially polarized beam via multiple reflections. The angle of incidence is arranged to be Brewster's angle $\theta_p$. In (e) dichroic mirrors are illustrated. Such mirrors are available to transmit visible radiation (hot mirrors) or to transmit infrared radiation (cold mirrors).

A pellicle beam splitter is made of a durable elastic membrane stretched over an optically flat metal frame and bonded to the edges as illustrated in Figure 3-31b. Typically, the membrane is made from optical-grade nitrocellulose and has a thickness of only 7 μm. An uncoated pellicle beam splitter reflects 8% and transmits 92% of the incident radiation throughout the visible and near-infrared regions. Coatings are used to vary the ratio of the reflected to transmitted beam intensities. Standard coatings provide ratios of (33/67%, 40/40%, and 50/50%). Pellicles are of high efficiency and they do not produce multiple beams.

The beam splitters described above all produce beams that are spatially separated, but temporally overlapping. In some applications it is desirable to split a single beam into two spatially separated beams at different times. The double-beam in-time spectrometer described in Section 4-6 is an example. For such applications, the chopper/beam splitter illustrated in Figure 3-31c is often used. The chopper wheel is usually made by cutting open slots in a circular mirror.

Many other types of beam splitters are used. Several of these separate an unpolarized beam into its two mutually perpendicular components. We have already encountered such a polarizing beam splitter in Section 3-2, where an optically anisotropic crystal, such as calcite, was shown to produce an ordinary and an extraordinary ray. Polarizing prisms, used as beam splitters, are described in Section 3-5. Another polarizing beam splitter is often referred to as the "pile-of-plates" polarizer. This type of beam splitter is based upon polarization by reflection, as described in the preceding section. The pile-of-plates polarizer is illustrated in Figure 3-31d. This type of beam splitter can be made with glass plates for use in the visible region, with quartz plates for the ultraviolet region or with alkali halide plates for use in the infrared region. A crude beam splitter of this type can be made from 10 to 12 microscope slides.

Another type of beam splitter is a wavelength-selective beam splitter or **dichroic mirror**. Such mirrors are made from multilayer, nonabsorbing films. (See Section 3-5 for a description of interference filters based on thin films.) A hot mirror reflects infrared radiation while transmitting visible radiation as illustrated in Figure 3-31e. A cold mirror reflects visible radiation and transmits infrared radiation. There are many uses of dichroic mirrors. Many sources used in the visible also produce intense infrared radiation which can cause heating problems. A dichroic mirror can be used to direct the unwanted infrared radiation away from the sample. In movie projectors, dichroic mirrors are often used to send the infrared radiation to the back of the projector away from the film. Like other beam splitters, dichroic mirrors can be used in reverse as beam combiners. For example, the fundamental infrared beam from a Nd:YAG laser (1.06 μm) is often combined with visible harmonics or with a visible dye laser beam by using a dichroic mirror.

Another type of beam splitter is based on bifurcated fiber optics bundles as described below.

## Fiber Optics

Fiber optics from the communications industry are being used in and with spectrometers to transfer light between various points. The construction of a fiber optic is shown in Figure 3-32. A light ray entering the core at angle $\theta_0$ will be totally internally reflected if $\sin \theta_1' \geq \eta_2/\eta_1$. Since $\sin \theta_1' = \cos \theta_1$ and $\cos^2 \theta_1 = 1 - \sin^2 \theta_1$, the condition for total internal reflection is $1 - \sin^2 \theta_1 \geq (\eta_2/\eta_1)^2$ or $\eta_1 \sin \theta_1 \leq (\eta_1^2 - \eta_2^2)^{1/2}$. The entrance angle $\theta_0$ is related to $\theta_1$ by Snell's law ($\eta_0 \sin \theta_0 = \eta_1 \sin \theta_1$), so total internal reflection occurs when $\eta_0 \sin \theta_0 \leq (\eta_1^2 - \eta_2^2)^{1/2}$. There is thus a maximum value of $\theta_0$ for which total internal reflection occurs given by $\eta_0 \sin \theta_0 = (\eta_1^2 - \eta_2^2)^{1/2}$. Rays that are incident at larger angles are only partially reflected at the core–clad interface and soon pass out of the fiber. The **numerical aperture** NA is defined as $\eta_0 \sin \theta_0$, or

$$NA = \eta_0 \sin \theta_0 = (\eta_1^2 - \eta_2^2)^{1/2} \qquad (3\text{-}46)$$

The NA represents the cone of light accepted by the fiber. The F/n of a fiber optic is related to the NA by F/n = $1/(2 \tan \theta_0)$. Core diameters are typically 50 to 600 μm and the refractive index of the clad is typically

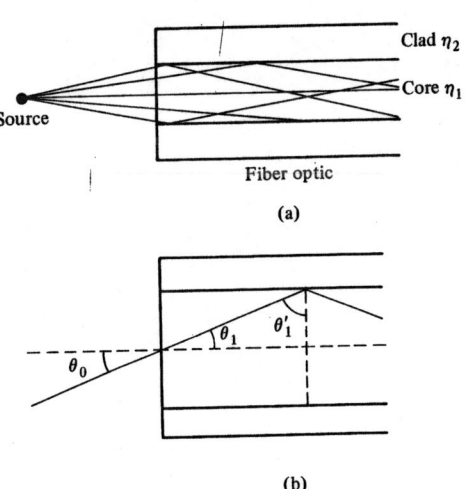

**FIGURE 3-32** Fiber optic. The fiber optic in (a) consists of a core of refractive index $\eta_1$ and a clad of refractive index $\eta_2$, where $\eta_2 < \eta_1$. Normally, additional jackets surround the clad to provide physical strength. In (b) the various angles are defined.

0.02 to 0.2 less than the core. This gives a value for the NA in range 0.1 to 0.5.

Fiber optics are experiencing greater use in spectroscopy for several reasons. Because they are mechanically flexible, light can be transmitted over curved paths. Thus fiber optics can replace several mirrors in directing light between two points in a spectrometer (e.g., from the source to the wavelength selector or from the wavelength selector to the detector). Fiber optics can transmit light over long distances (e.g., 500 m), which allows remote monitoring in hazardous environments since the more delicate components (detectors, monochromators, signal processing electronics) can be far from the monitoring site. Over long distances, the absorption and scattering losses in the fiber can be significant.

A single fiber optic cannot transmit an image because the rays from different parts of the object are scrambled by multiple internal reflections. Bundles of fine glass, quartz, or plastic fibers can, however, be used for image transmission if the fibers are small enough in diameter that each fiber transmits rays from a small area of the object. Some typical uses of fiber optic bundles are illustrated in Figure 3-33.

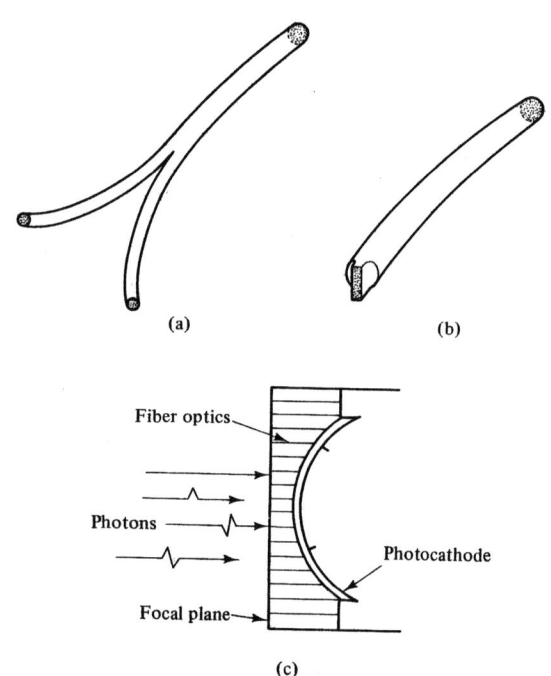

**FIGURE 3-33**   Some applications of fiber optic bundles. In (a) a bifurcated, randomized bundle is shown. This can be used as a beam splitter or to send light down one path and have it return along the other. In (b) one end of the bundle is circular, while the other is rectangular to match a spectrometer slit, for example. In (c) a fiber optic face plate is shown. This allows a flat image to be transferred to a curved surface such as the active surface of a detector.

## 3-5   FILTERS, PRISMS, AND GRATINGS

In the majority of spectrochemical techniques, the optical signals produced by the encoding process are sorted by means of wavelength so that the information concerning one component or a set of components can be isolated and measured. Wavelength selection is achieved in simple, low-resolution systems by filtering the radiation (attenuating all but the desired wavelengths) and in higher-resolution systems by dispersing it (spreading out the wavelengths spatially) into a spectrum. This section considers the optical components used in wavelength selection. Dispersive spectrometer systems are discussed in Section 3-6, while systems based on interferometers are discussed in Section 3-7.

### Filters

Filters are used to pass a band of wavelengths (bandpass filters) or to block wavelengths longer or shorter than some desired value (cutoff filters). The most widely used filters in spectroscopy are based on absorption or on interference.

*Absorption Filters.*   Filters based on absorption by colored glasses, crystals, sintered materials, solutions, and thin films are widely used. Filters made from these materials are simple to use, inexpensive, and relatively insensitive to the angle of incidence of the incoming radiation.

Bandpass filters are characterized by plots of their spectral transmittance $[T(\lambda)]$ vs. wavelength, as shown in Figure 3-34a. The characteristics of interest to the user are shown in the figure. The transmission characteristics of several colored glass filters are shown in Appendix C. Typical values of the full width at half maximum (FWHM) range from 30 to 250 nm. Bandpass filters made with an organic dye impregnated on gelatin and sandwiched between glass plates (Wratten filters) are also commercially available. Liquid filters are often made from colored chemical solutions contained in glass cells.

Cutoff filters contain substances that absorb all the radiation at wavelengths shorter than a given cutoff (short-wavelength cutoff) or all the radiation at long wavelengths (long-wavelength cutoff). The transmittance of a short-wavelength cutoff filter is illustrated in Figure 3-34b. Cutoff filters can be colored glasses, the Wratten type, organic dyes in polymeric hosts, or thin films.

*Interference Filters.*   Interference filters are constructed so that the rays from most wavelengths that strike the filter suffer destructive interference, while only rays within a small wavelength band experience

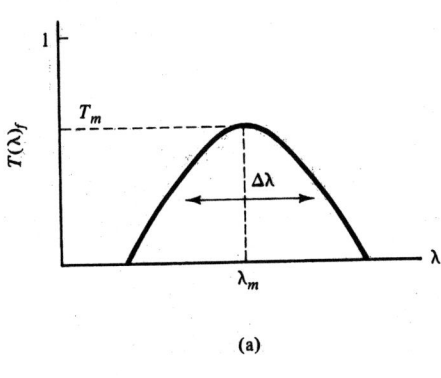

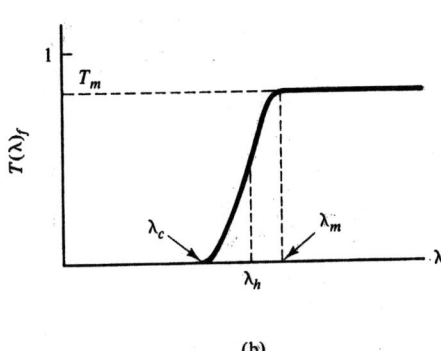

**FIGURE 3-34** Transmittance of bandpass (a) and short-wavelength cutoff (b) filters. With the bandpass filter (a) the wavelength of maximum transmittance $\lambda_m$, the maximum transmittance $T_m$ and the full width at half maximum height $\Delta\lambda$ or FWHM are illustrated. For the cutoff filter (b) the cutoff wavelength $\lambda_c$ is the wavelength where $T(\lambda)_f = 10$, 1 or 0.1%. Wavelength $\lambda_m$ is the wavelength at which the transmittance reaches 90 or 99% of its maximum value $T_m$. Wavelength $\lambda_h$ is the wavelength at which $T(\lambda)_f = 0.5\, T_m$.

constructive interference and are passed. Two types are widely used: the single-layer or Fabry–Perot type and the multilayer dielectric type.

The Fabry–Perot type of interference filter is illustrated in Figure 3-35. The thin dielectric is a material of low refractive index such as quartz, $CaF_2$, $MgF_2$, $ZnS$, $ThF_4$, or sapphire. Waves that reach points $g$ and $e$ in phase undergo constructive interference at the lens focal point. As shown in the figure caption, the condition for constructive interference is

$$2d(\eta^2 - \sin^2\theta)^{1/2} = m\lambda \qquad (3\text{-}47)$$

In many uses of interference filters, the incident beam is normal to the plane of the filter, instead of skewed as in Figure 3-35. In these cases $\theta = 0°$, so $\sin\theta = 0$. The central wavelength passed by the filter $\lambda_m$ can be written

$$\lambda_m = \frac{2d\eta}{m} \qquad (3\text{-}48)$$

where $m$ is called the **order**. For example, for a filter with a $CaF_2$ dielectric with $\eta = 1.35$ and $d = 1.85 \times 10^{-5}$ cm, $\lambda_m = 500$ nm/m. In the first order, $\lambda_m = 500$ nm, in the second order 250 nm, and so on. In reality the rays of one wavelength striking an interference filter have a random phase relationship. However, for $\lambda_m = 2d\eta/m$, all the rays that pass through the filter have the same phase relationship as when they entered. For

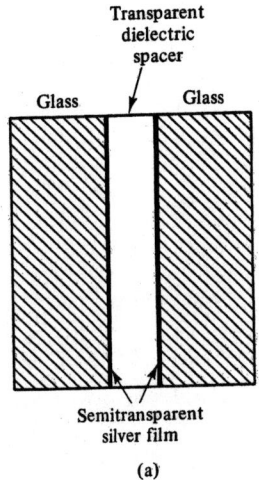

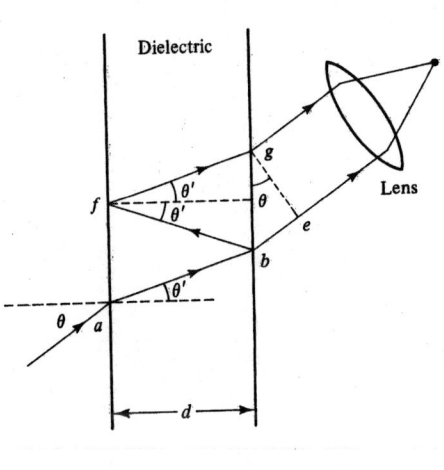

**FIGURE 3-35** Fabry–Perot interference filter: (a) cross section; (b) exploded view showing spacing $d$ between reflective surfaces and angles. The lens in b is equidistant from points $g$ and $e$. Ray 1 travels the path $a{\to}b{\to}e$, while ray 2 travels the path $a{\to}b{\to}f{\to}g$. The two waves will be in phase (constructive interference) at points $g$ and $e$ if $\Delta(OPL)$ is an integral multiple of the wavelength. Because of the law of reflection $ab = bf = fg$. Thus the condition for constructive interference is $2ab\eta - be = m\lambda$. Since $\cos\theta' = d/ab$, $\tan\theta' = \frac{1}{2}\, bg/d$, $\sin\theta = be/bg$, and $\sin\theta' = \sin\theta/\eta$, this can also be expressed as $2d\eta/\cos\theta' - 2d\eta \sin\theta' \tan\theta' = 2d\eta \cos\theta' = m\lambda$. Since $\cos\theta' = [1 - (\sin\theta/\eta)^2]^{1/2}$, the condition for constructive interference in terms of the angle of incidence is $2d(\eta^2 - \sin^2\theta)^{1/2} = m\lambda$.

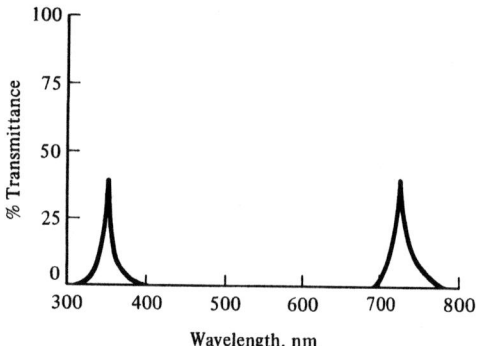

**FIGURE 3-36** Transmission of a typical Fabry–Perot interference filter. The first-order band has $\lambda_m \approx 720$ nm. Note that the second-order band ($\lambda_m \approx 360$ nm) is narrower, with only slightly lower peak transmittance. The free spectral range is 360 nm for the first order.

wavelengths much different from $\lambda_m$, the original phase relationship is destroyed, resulting in destructive interference.

There are several ways to obtain filters with a variable central wavelength. Equation 3-47 shows that for a given filter with fixed $d$ and $\eta$ values, the central wavelength can be changed by changing the angle of incidence of the filter. The new peak wavelength $\lambda$ is related to the central wavelength at normal incidence $\lambda_m$ by $\lambda = \lambda_m(1 - \sin^2 \theta/\eta^2)^{1/2}$. Angle tuning of interference filters is frequently employed when they serve as wavelength selection devices in dye laser cavities. Filters are commercially available in which the $d$ spacing varies along the length of the filter. Such **wedge filters** are also available in circular styles in which $d$ varies with the filter rotation.

The transmission curve of a typical Fabry–Perot interference filter is shown in Figure 3-36 and additional information is given in Appendix C. Higher-order bands

are narrower than first-order bands, and such filters are often used in the second or third order. An additional broadband filter or cutoff filter is frequently employed to isolate the order of interest; this can be an integral part of the filter, in which case the filter is said to be "blocked." The range of wavelengths for which no overlap of adjacent orders occurs is called the **free spectral range**.

Multilayer (or multicavity) filters are made with alternating layers of high- and low-refractive-index dielectrics; they can be considered to be two to six Fabry–Perot cavities cemented together. The multilayer interference filter can achieve quite narrow FWHM values (<1 nm) with high peak transmittances (>50%). Commercial multicavity filters are produced with in-

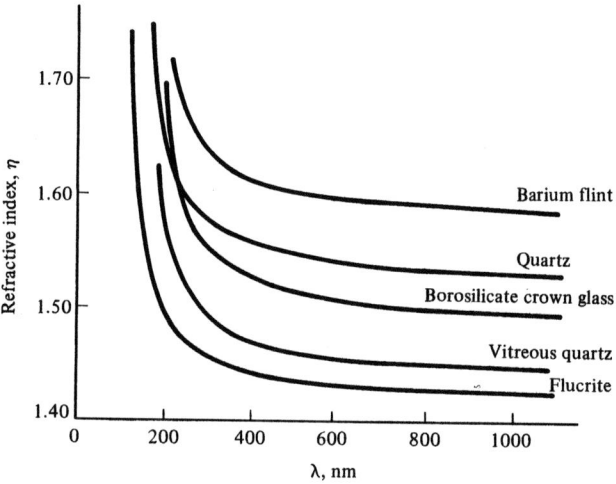

(a)

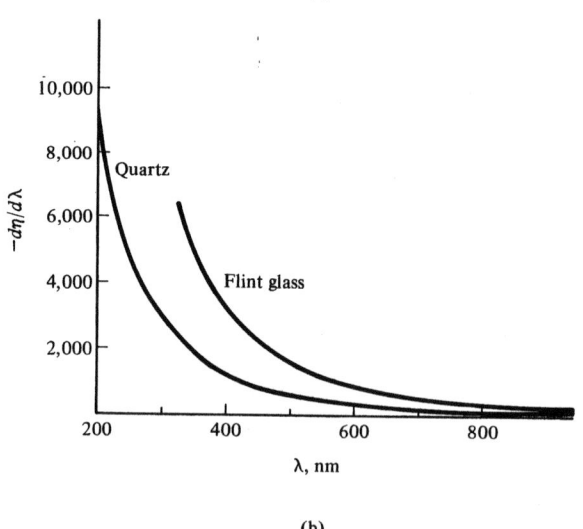

(b)

**FIGURE 3-38** Properties of dispersive prisms: (a) refractive index vs. wavelength for several prism materials; (b) dispersion vs. wavelength for two prism materials.

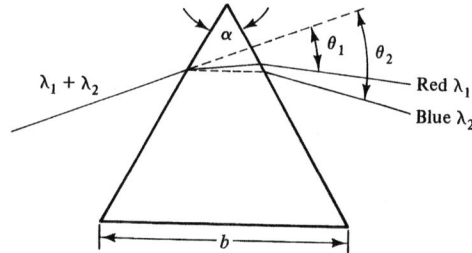

**FIGURE 3-37** Dispersion of light of two wavelengths by a prism of refractive index $\eta$, apex $\alpha$, and baselength $b$. Collimated rays of wavelength $\lambda_1$ (red) and $\lambda_2$ (blue) are refracted upon entering the prism material and upon exiting it according to Snell's law. Normal prism materials show higher refractive indices at shorter wavelengths. Hence blue light of wavelength $\lambda_2$ is more highly refracted than red light ($\lambda_1$).

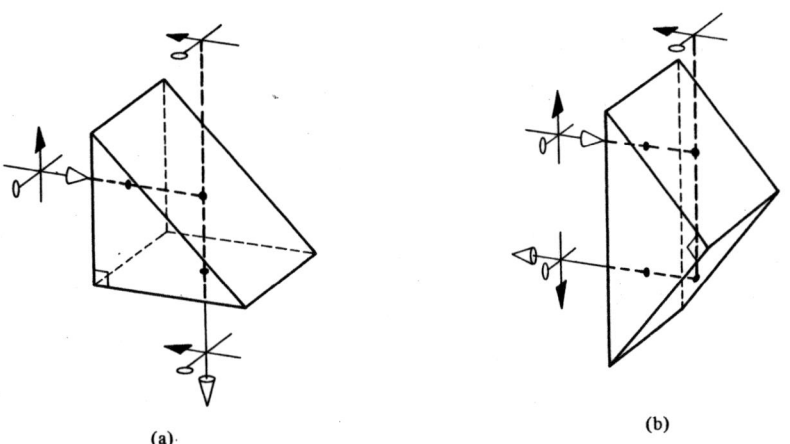

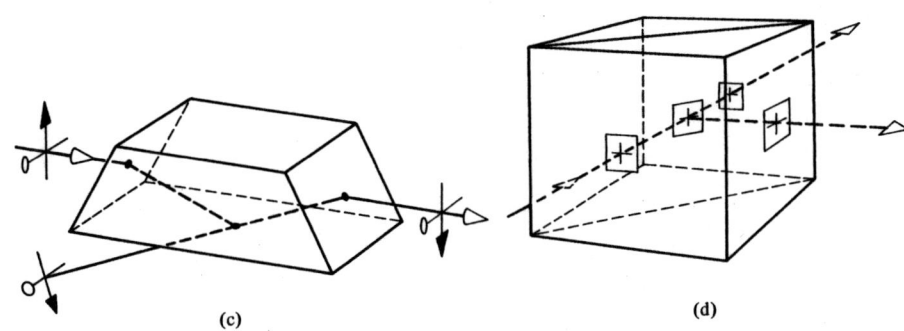

**FIGURE 3-39** Some reflecting prisms. The right-angle prism in (a) produces a 90° deviation and interchanges the top and bottom of the image; the Porro prism in (b) causes a 180° deviation of the image; the Dove prism in (c) reverses the top and bottom of the image. The beam-splitting cube in (d) is made from two 90° prisms with a partially reflective coating at the interface. It nearly eliminates the secondary beams found with ordinary plate beam splitters. Multiple beams are produced, but they are quite faint.

tegral blocking filters to isolate the desired band. Single and multicavity filters are available with peak transmittances in the UV to the IR. Bandwidths (FWHM values) in the range 1 to 100 nm are readily available (see Appendix C).

Dichroic filters (mirrors) are used to separate IR and visible radiation. These either transmit the visible and reflect IR radiation (hot mirrors), or vice versa (cold mirrors). Dichroic mirrors were discussed in Section 3-4.

## Prisms

The prism serves a variety of purposes in optical systems. In spectroscopy prisms are used to change the direction of a beam, to split an incident beam into two outgoing beams, and to produce polarized light, as well as to disperse the incoming beam into a spectrum.

*Dispersing Prisms.* Dispersion occurs in a prism primarily because of the wavelength dependence of the refractive index of the prism material. A typical arrangement for dispersion with a prism is shown in Figure 3-37. The angle $\theta$ that a monochromatic refracted ray makes with the undeviated incident beam is called the **deviation**. The variation of the deviation with wavelength $d\theta/d\lambda$ is called the **angular dispersion**, $D_a$; it is

composed of two factors:

$$D_a = \frac{d\theta}{d\lambda} = \frac{d\eta}{d\lambda}\frac{d\theta}{d\eta} \qquad (3\text{-}49)$$

Here $d\eta/d\lambda$ is often called the dispersion of the prism material; it describes the refractive index variation with wavelength. The factor $d\theta/d\eta$ is a geometrical factor that depends on the prism shape and size ($\alpha$ and $b$) and the incidence angle. At a constant angle of incidence, the factor $d\theta/d\eta$ varies only slightly with wavelength. Thus the variation of angular dispersion with wavelength is primarily a result of the refractive index variation with wavelength as shown in Figure 3-38. Note that the dispersion is high in the ultraviolet region, but decreases dramatically in the visible and near-IR regions. Typical values of $D_a$ are in the range $10^{-6}$ to $10^{-3}$ rad nm$^{-1}$.

*Reflecting Prisms.* Reflecting prisms are designed to change the direction of propagation of a beam, the orientation of the beam, or both. At least one internal reflection takes place within the prism and there is no resulting dispersion. Several examples of reflecting prisms, including a cube beam splitter, are shown in Figure 3-39.

*Polarizing Prisms.* Polarizing prisms are made of birefringent materials. They depend on the different indices of refraction for the ordinary and extraordinary rays to separate these rays. With some polarizing prisms only a single polarized beam emerges; in others, called **polarizing beam splitters**, both rays emerge. The earliest polarizing prism (1828) was the Nicol prism, but it has long since been replaced by more efficient devices; two of these are shown in Figure 3-40. If the two prism sections in the Glan–Foucault prism are cemented together and the interface angle changed, the device is called a **Glan–Thompson polarizer**.

### Diffraction Gratings

A diffraction grating is a plane or concave plate that is ruled with closely spaced grooves. The grating acts like a multislit source when collimated radiation strikes it. Different wavelengths are diffracted and constructively interfere at different angles. Gratings are either designed for transmission of the incident radiation or for reflection. Modern spectrometers invariably use **reflection gratings**.

Spectroscopic gratings are in actuality replicas of master gratings. Contemporary master gratings are ruled by a diamond ruling tool operated by a ruling engine. The tool rules thin aluminum films that are evaporated onto optically flat glass. Machine-ruled gratings are extremely difficult to manufacture and, as a consequence, few master gratings are produced. Replica gratings are produced by vacuum deposition of aluminum on a ruled master and subsequent coating of the Al layer by an epoxy-type material. After polymerization the replica is separated from the master. Today, replica gratings are produced that are superior to the master gratings of only 15 to 20 years ago.

Holographic gratings have also become practical in recent years. These are made by coating a flat glass

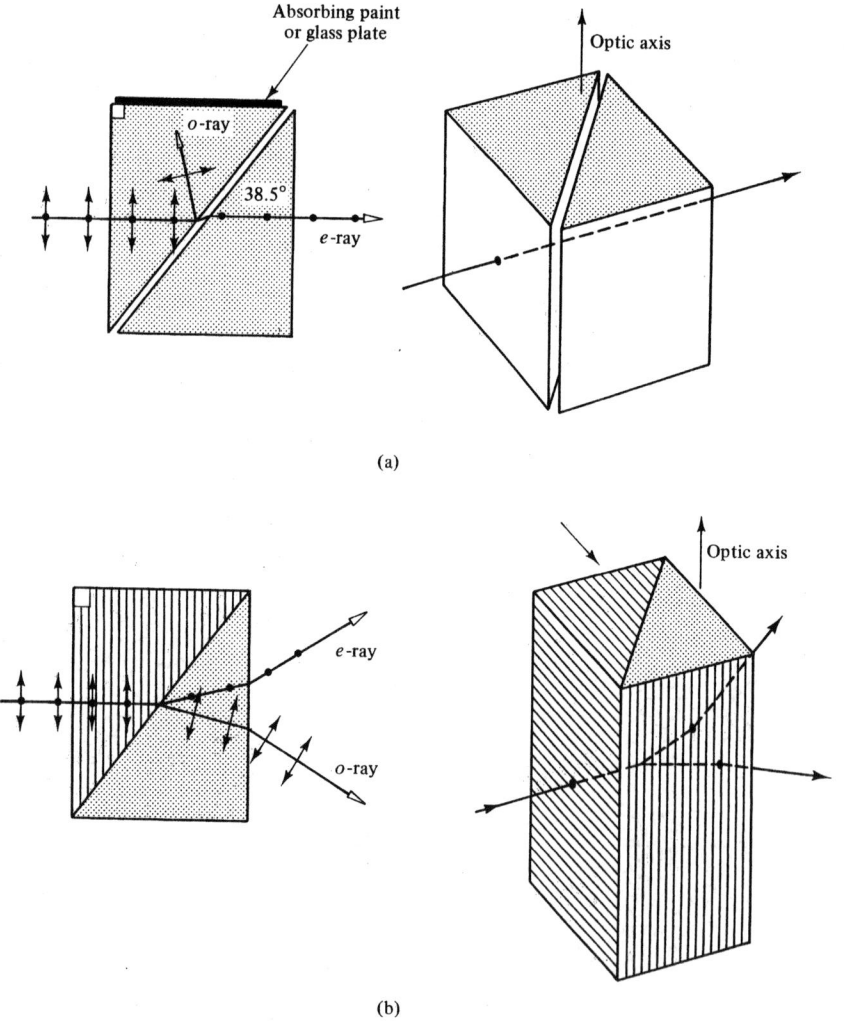

(a)

(b)

**FIGURE 3-40** Some polarizing prisms. The Glan–Foucault prism (a) is constructed of two calcite sections separated by air. The dots indicate that the optic axes for both sections are perpendicular to the top face. The incoming ray is unpolarized and is shown as two orthogonal components. Both components traverse the first section undeviated. At the calcite–air interface, the angle of incidence $\theta$ for the ordinary ray exceeds the critical angle, $\sin (\theta_c)_0 = 1/\eta_0$, where $\eta_0$ is the ordinary refractive index. Thus the ordinary ray is totally internal reflected and absorbed by the layer of paint. This same angle is less than the critical angle for the transmitted extraordinary ray, $\sin (\theta_c)_e = 1/\eta_e$, where $\eta_e$ is the extraordinary refractive index. The Wollaston prism (b) is also made of calcite sections. Note that the prism is formed so that the optic axis for one section is perpendicular to that of the other section. At the interface between sections the two polarization components experience different refractive indices and are thus refracted at different angles. Hence the extraordinary and ordinary rays diverge. The Wollaston prism thus acts as a polarizing beam splitter and both rays are transmitted. Prisms can be purchased with deviation angles between the separated beams ranging from 15 to 45°.

plate with a photosensitive material. Lasers are then employed to project an interference pattern on the plate. The photosensitive layer is developed to produce grooves which are etched into the glass for permanence. Aluminum is then vacuum deposited to make the grating reflective. Holographic gratings are slightly less efficient than ruled gratings. However, their groove pattern is essentially perfect. They eliminate false lines (ghosts) and reduce scattered light produced by errors in ruled gratings.

*Grating Formula.* The surface of a typical ruled grating is shown in Figure 3-41, where the various symbols used are also defined. Note that the angle of incidence and that of diffraction are measured with respect to the grating normal, not the groove normal. The angle of incidence is always taken as a positive angle. The diffraction angle is taken to be positive if the diffracted ray is on the same side of the grating normal as the incident ray, as shown in Figure 3-41a, and negative if it is on the opposite side. A more detailed view is shown in Figure 3-41b, where two parallel monochromatic rays strike adjacent grooves (*A* and *B*) and are diffracted at the same angle β. As we discussed earlier in this chapter, constructive interference of the two exiting rays occurs if the difference in their optical pathlengths is an integral multiple of the wavelength λ. The difference in optical pathlengths is $AC + AD$. Since sin α = $AC/d$ and sin β = $AD/d$, the condition for constructive interference is given by the grating formula:

$$d(\sin \alpha + \sin \beta) = m\lambda \qquad (3\text{-}50)$$

Here *m* is the order of diffraction and assumes the values 0, ±1, ±2, ±3, and so on. The order, like the diffraction angle, is taken as positive for diffraction on the same side of the grating normal as the incident ray and negative on the opposite side. The groove spacing *d* is determined by the groove density in grooves per mm. In the UV-visible region typical groove densities are 300 to 2400 grooves mm$^{-1}$. The groove spacing must be greater than λ/2 for the grating equation to apply because the maximum value of (sin α + sin β) = 2.

Consider polychromatic light striking a grating. Within a given order of diffraction (other than the zero order) the various wavelengths are spatially dispersed because each wavelength undergoes constructive interference at a different diffraction angle. For example, consider light composed of two wavelengths, 500 nm and 600 nm, striking a grating at an incident angle of 10°. If the grating has 1200 grooves per mm, the *d* spacing is $d = 10^6$ nm mm$^{-1}$/1200 grooves mm$^{-1}$ = 833.3 nm/groove. In the first order, the 600-nm radia-

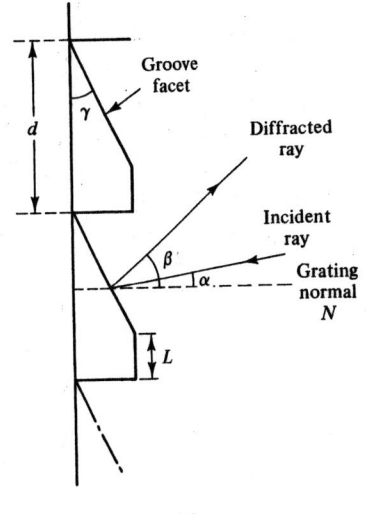

(a)

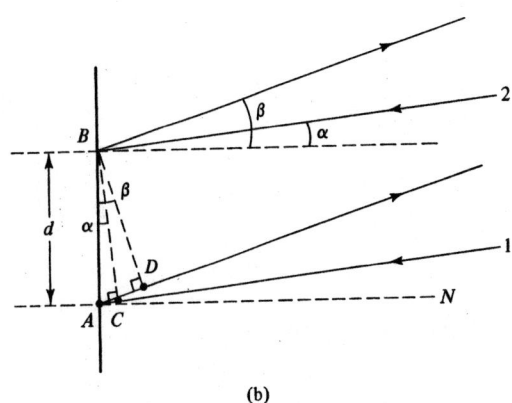

(b)

**FIGURE 3-41** Diffraction from a blazed reflection grating. In (a) the incident ray makes an angle α with the grating normal *N*, while the diffracted ray makes an angle β. The spacing between adjacent grooves is *d*, and the blaze angle is γ. The distance *L* is called the land. In (b) the grooves are removed for clarity, and two monochromatic rays are shown undergoing diffraction at an angle β. Ray 1 travels a distance *AC* farther than ray 2 in reaching the grating surface and a distance *AD* farther than ray 2 on exiting the grating.

tion is diffracted at an angle

$$\sin \beta = \frac{m\lambda}{d} - \sin \alpha = \frac{1 \times 600 \, \text{nm}}{833.3 \, \text{nm}} - \sin 10° = 0.546$$

$$\beta = 33.1°$$

For 500-nm radiation the corresponding first-order diffraction angle is β = 25.2°.

For a given *d* spacing and angle of incidence, α, the grating formula predicts that many wavelengths are observed at a given angle of diffraction β, a phenomenon known as overlapping orders. With the grating discussed in the example above, the second order of 300 nm also occurs at a diffraction angle of 33.1°, as does the third order of 200 nm. In fact, wavelengths $\lambda$, $\lambda/2$, $\lambda/3$, . . . , $\lambda/m$ all appear at the same angle in different orders. The free spectral range $\Delta\lambda_f$ represents the range of wavelengths from the source for which no overlap of adjacent orders occurs, as discussed previously for interference filters. If two lines of wavelengths $\lambda$ (order $m$) and $\lambda - \Delta\lambda_f$ (order $m + 1$) just coincide, the grating formula gives

$$d(\sin \alpha + \sin \beta) = m\lambda = (m + 1)(\lambda - \Delta\lambda_f)$$

from which

$$\Delta\lambda_f = \frac{\lambda}{m + 1} \qquad (3\text{-}51)$$

Thus, for $\lambda = 600$ nm in the first order, the free spectral range is 600 nm/2 or 300 nm.

For common gratings blazed in the first order (see below), $\Delta\lambda_f$ is typically hundreds of nanometers in the UV-visible region. Thus, overlapping orders are a problem only if the source covers a broad spectral range and the detector is responsive at more than one of the wavelengths isolated. Overlapping orders are often separated by limiting the source bandwidth with a broadband filter (order sorter) or with a predisperser.

*Blazed Gratings.* The grating facets in Figure 3-41 are shown tilted at an angle $\gamma$ with respect to the flat surface of the grating. Such a blazed grating is often called an **echellete**. The reason for blazing is to concentrate the radiation in a preferred direction. Note that for $m = 0$, $\sin \alpha = -\sin \beta$ or $\alpha = -\beta$. Thus in the zero order all wavelengths are diffracted at the same angle. If the grating were not blazed, the normal to the grating surface and the normal to the groove would be equal. Specular reflection (angle of incidence = angle of reflection) would then correspond to the zero order. Since most of the incident light undergoes specular reflection, it follows that most of the irradiance is wasted in a unblazed grating. In a blazed grating, however, Figure 3-42a shows that the specularly reflected radiation corresponds to a particular nonzero order determined by the blaze angle. Figure 3-42b shows the particular case of the incident radiation being normal to the grating surface. For $\alpha = 0°$, $\beta$ (0 order) also equals $0°$. The angle of specular reflection now equals twice the blaze angle. The first-order wavelength corresponding to the specularly reflected ray is called the **blaze wavelength** $\lambda_b$, where $\lambda_b = d \sin \beta = d \sin 2\gamma$. If

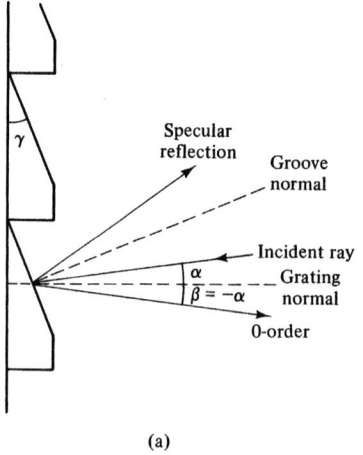

(a)

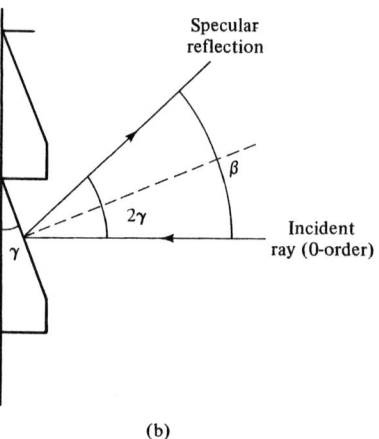

(b)

**FIGURE 3-42** Blazed gratings. By tilting the groove facet an angle $\gamma$ as in (a), the zero order no longer corresponds to the specularly reflected ray. In (b) the incident ray is normal to the grating surface and reflection occurs at $2\gamma$.

$\gamma = 8.73°$ for a grating of 1200 grooves mm$^{-1}$, $\lambda_b = 833.3$ nm $\times$ 0.300 = 250 nm. Thus the blaze wavelength is chosen by varying the blaze angle. The actual relationship between $\lambda_b$ and $\gamma$ depends on how the grating is used, but by convention either $\alpha$ is assumed to be zero as above or $\alpha$ is assumed equal to $\beta$. It is apparent that a grating blazed for $\lambda$ in the first order is also blazed for $\lambda/2$ in the second order, $\lambda/3$ in the third order, and so on.

The **grating efficiency** is the ratio of the intensity diffracted at wavelength $\lambda$ for a given order (usually the first order) to the specular reflectance from a polished blank coated with the same material. The efficiency is highest at $\lambda_b$ and drops to about one-half of the maximum value at $2/3 \lambda_b$ and $3/2 \lambda_b$.

*Dispersion.* The angular dispersion $D_a$ of a grating can be obtained by differentiating the grating for-

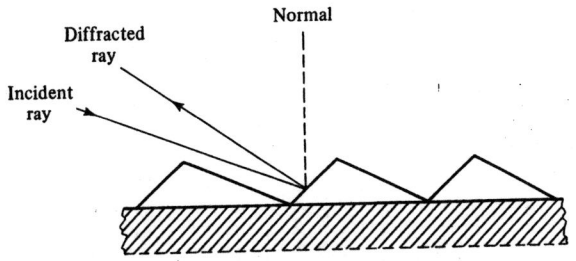

**FIGURE 3-43** Echelle grating. Echelles are relatively coarse gratings (typically 100 grooves mm$^{-1}$) with large blaze angles. The steep side of the groove is employed so as to achieve large diffraction angles. They are typically used in very high orders, as discussed in Section 3-6.

mula (equation 3-50) with respect to wavelength. For a constant angle of incidence $\alpha$, $D_a$ in rad nm$^{-1}$ is given by

$$D_a = \frac{d\beta}{d\lambda} = \frac{|m|}{d\cos\beta} = \frac{\sin\alpha + \sin\beta}{\lambda\cos\beta} \qquad (3\text{-}52)$$

If the grating is operated with nearly normal incidence and small angles of diffraction, $\cos\beta$ is only weakly dependent on $\lambda$. Thus $D_a$ is nearly independent of wavelength. For example, for $\alpha = 3.73°$, $d = 833.3$ nm, and $m = \pm 1$, $\cos\beta$ varies from 0.985 to 0.845 over the range 200 to 500 nm. The angular dispersion for the same grating would vary from $1.22 \times 10^{-3}$ rad nm$^{-1}$ at 200 nm to $1.42 \times 10^{-3}$ rad nm$^{-1}$ at 500 nm.

*Echelle Gratings.* The **echelle grating,** illustrated in Figure 3-43, is a coarsely ruled grating that is used in high-dispersion echelle monochromators, which utilize high diffraction orders and large angles of diffraction. Note in equation 3-52 that $D_a$ is directly proportional to $m$ and to $1/\cos\beta$.

## 3-6 DISPERSIVE WAVELENGTH SELECTION SYSTEMS

As discussed in Section 2-2, complete optical instruments for wavelength selection fall into several categories. This section describes systems based on dispersion. The monochromator, which isolates a single wavelength band at a time, is first discussed and then polychromators and spectrographs are presented. Section 3-7 discusses Fabry–Perot, Michelson, and other interferometers that are finding increasing applications in spectrochemical measurements.

## Monochromators

Monochromators are the most widely used dispersive instruments. They isolate a small wavelength band from a polychromatic source. Monochromators consist of a dispersive element (prism or grating) and an image transfer system (entrance slit, mirrors or lenses, and exit slit). Within the monochromator an image of the rectangular, or sometimes curved, entrance slit is transferred to the exit slit after dispersion of the wavelength components of the incident radiation.

In this section grating monochromators are emphasized because there are few prism systems being made at this time. However, the general principles presented apply equally to both types of monochromators. A grating monochromator based on the Czerny–Turner configuration shown in Figure 3-44 is used as the model for the following discussion because it is a common design; many other configurations are available.

As we will see in more detail later, the slits play an important role in determining the resolution and throughput of a monochromator. A slit is formed by two slit jaws. The slit width is formed by the separation of the two opposing edges of the slit jaws. Slit widths are typically in the range of a few micrometers to several

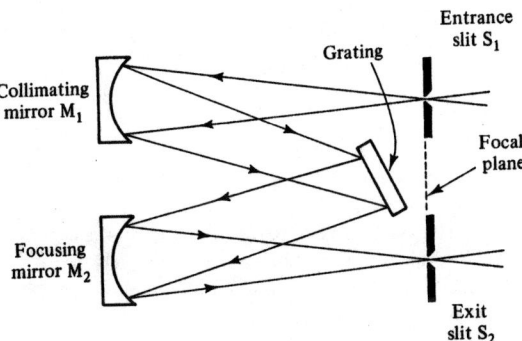

**FIGURE 3-44** Czerny–Turner plane grating monochromator. Incident radiation passes through the entrance slit $S_1$ and strikes the parabolic, collimating mirror $M_1$. The entrance slit, at the focal point of $M_1$, acts as a point source for $M_1$ which produces parallel radiation for the dispersion element, a grating in this case. The grating spatially disperses the spectral components of the incident radiation. Collimated rays of diffracted radiation strike the parabolic focusing or camera mirror $M_2$. The dispersed radiation is focused in the focal plane producing entrance slit images in that plane. Because the parallel rays of a given wavelength are incident on the focusing element at a specific angle, each wavelength is focused to a slit image at a different center position on the focal plane. The exit slit $S_2$ placed in the focal plane isolates a particular wavelength interval.

millimeters. Some monochromators have fixed, un-changeable slit widths, while others have adjustable slits. Slit heights usually vary from 1 to 10 mm.

In most monochromators the exit slit has the same dimensions as the entrance slit so that it can be fully illuminated by the entrance slit image. Many mono-chromators allow both slits to be adjusted in width simultaneously.

We discuss first how different wavelength bands are brought to the exit slit before discussing mono-chromator characteristics.

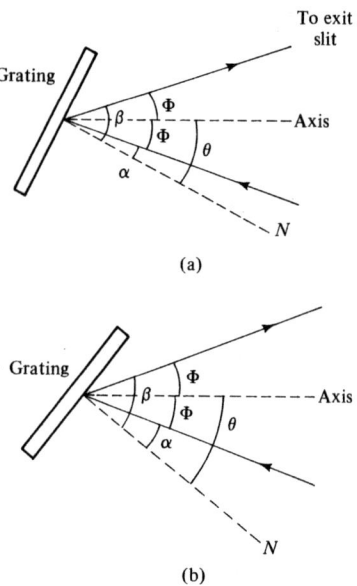

(a)

(b)

**FIGURE 3-45** Typical grating arrangement in a Czerny–Turner monochromator. Angle $\theta$ is the grating rotation angle measured from the grating normal $N$ to the optical axis. Angle $\Phi$ is the exit or takeoff angle and is the angle that the dif-fracted ray which passes through the exit slit makes with respect to the optical axis. In (b) a larger grating angle is shown than in (a). Note that the angle $\Phi$ that the incident and diffracted rays make with respect to the optical axis is fixed.

*Wavelength Selection.* To change the wave-length selected by the monochromator, the dispersive element is rotated to bring a different wavelength band through the exit slit. This is shown more clearly in Fig-ure 3-45 for a Czerny–Turner grating monochromator. Here angle $\Phi$, the takeoff angle, is fixed and the grating angle $\theta$ is changed to isolate a different wavelength. Note that changing $\theta$ changes both the incident angle $\alpha$ and the diffraction angle $\beta$, as can be seen by com-paring Figure 3-45a to Figure 3-45b.

From Figure 3-45 we can see that the grating ro-tation angle $\theta$ is given by $\theta = \alpha + \Phi$. The takeoff angle $\Phi$ is given by $\Phi = \beta - \theta$. Thus the incident angle and diffraction angle can be written in terms of $\theta$ and $\Phi$ as $\alpha = \theta - \Phi$ and $\beta = \theta + \Phi$. If we substitute these relationships into the grating formula (equation 3-50) and use the trigonometric identity $1/2[\sin(\alpha + \beta) + \sin(\alpha - \beta)] = \sin\alpha\cos\beta$, the result is

$$d[\sin(\theta - \Phi) + \sin(\theta + \Phi)] = 2d\sin\theta\cos\Phi \qquad (3\text{-}53)$$
$$= m\lambda$$

Equation 3-53 expresses the grating formula in terms of the experimental variables $\theta$ and $\Phi$. We can take an example to show the consistency of the two versions of the grating formula. Let us assume that in the monochromator of Figure 3-44, $d = 833.3$ nm, $\Phi$ is fixed at 6.71° and the grating is operated in the first order. Table 3-3 gives the grating rotation angle, the angular dispersion $D_a$ calculated from equation 3-52, the linear and reciprocal linear dispersion as discussed below, and the values of $\alpha$ and $\beta$ required to isolate 300-, 400-, and 500-nm radiation.

One common mechanism for mechanically vary-ing the grating angle $\theta$, called a sine bar drive, is illus-trated in Figure 3-46. Since the wavelength is directly proportional to $\sin\theta$ (see equation 3-53), distance $x$ is directly proportional to $\lambda$. A wavelength counter geared to the leadscrew will read out directly in wavelength. The drive is connected to a constant-speed motor, typ-ically a stepping motor, for scanning at a constant rate.

**TABLE 3-3**

Angles and dispersion characteristics of a Czerny–Turner monochromator in the first order[a]

|  | 300 nm | 400 nm | 500 nm |
|---|---|---|---|
| $\theta$ | 10.44° | 13.98° | 17.58° |
| $\alpha$ | 3.73° | 7.27° | 10.87° |
| $\beta$ | 17.15° | 20.69° | 24.29° |
| $D_a$ | $1.26 \times 10^{-3}$ | $1.28 \times 10^{-3}$ | $1.32 \times 10^{-3}$ |
| $R_d$ | 3.17 | 3.12 | 3.03 |

[a]Takeoff angle $\Phi = 6.71°$; groove density $= 1200$ mm$^{-1}$; $f = 250$ mm; $D_a$ in rad nm$^{-1}$; $R_d$ in nm mm$^{-1}$.

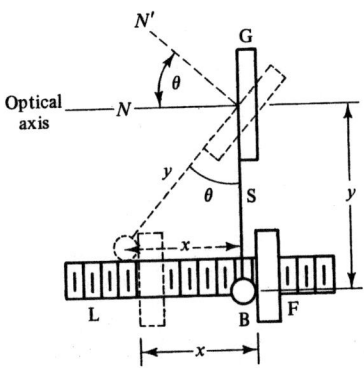

**FIGURE 3-46** Linear wavelength drive mechanism. A precision leadscrew L is used to move a contact flat F. The grating is attached to sine bar S of length $y$. Movement of the flat by distance $x$ pushes contact ball B and rotates the grating through angle $\theta$. (The optical axis is parallel to the leadscrew.) Note that $\sin\theta$ equals $x/y$, the hypotenuse length $y$ is fixed, and $x$ equals $np$, where $n$ is the number of turns of the leadscrew and $p$ is the pitch. If these terms are substituted into the grating formula 3-53, solving for $\lambda$ gives $\lambda = 2dnp\cos\Phi/my$. Thus the wavelength is directly proportional to the number of leadscrew turns $n$. It is also possible to construct a linear wavenumber or cosecant drive. However, modern monochromators usually use a linear wavelength drive and calculate wavenumber with a microprocessor.

*Dispersive Characteristics.* We have already shown how angular dispersion $D_a$ can be calculated for a prism (equation 3-49) and a grating (equation 3-52). The angular dispersion is the angular separation $d\beta$ ($d\theta$ for a prism) corresponding to the wavelength separation $d\lambda$. It is a property of the grating or prism itself. For a monochromator, the dispersion in the focal plane is of more interest. The **linear dispersion**, $D_l = dx/d\lambda$, tells how far apart in distance two wavelengths are separated in the focal plane. Figure 3-47 illustrates how $D_l$ is determined for a dispersion element illuminated by collimated radiation of wavelengths $\lambda_1$ and $\lambda_2$. For the configuration shown, the linear dispersion is given by $D_l = fD_a$, where $f$ is the focal length of the focusing element. The linear dispersion $D_l$ is conveniently expressed in units of mm nm$^{-1}$ in the UV-visible region of the spectrum. Most often it is the **reciprocal linear dispersion** $R_d$ that is specified for a spectrometer. The reciprocal linear dispersion represents the number of wavelength intervals (e.g., nm) contained in each interval of distance (e.g., mm) along the focal plane. For the configuration shown in Figure 3-47,

$$R_d = D_l^{-1} = (fD_a)^{-1} = \frac{d\lambda}{dx} \qquad (3\text{-}54)$$

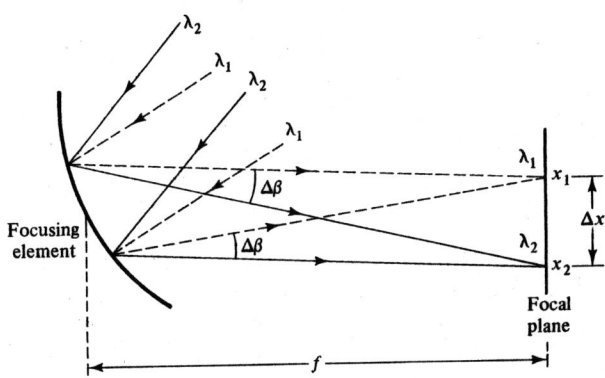

**FIGURE 3-47** Relationship between angular and linear dispersion. Rays of $\lambda_1$ from the dispersion element are parallel to each other and are focused at point $x_1$ by the focusing element. Parallel rays of $\lambda_2$ strike the focusing element at a different angle as determined by the dispersion element and are focused at $x_2$. For a grating monochromator, $\Delta\beta$ is the difference in diffraction angles of $\lambda_1$ and $\lambda_2$ with respect to the grating normal. For a prism monochromator, $\Delta\beta = \Delta\theta$, where $\Delta\theta$ is the difference in refraction angles for $\lambda_1$ and $\lambda_2$ with respect to a ray undeviated by the prism. Since $\tan(\beta_2 - \beta_1) = \tan\Delta\beta = \Delta x/f$, for small angles $\Delta\beta = \Delta x/f$ and $d\beta \approx dx/f$. Thus $dx/d\lambda = f(d\beta/d\lambda)$.

For a prism monochromator the reciprocal linear dispersion is usually quite wavelength dependent because of the variation of $D_a$ with wavelength. Some prism monochromator configurations, like the Pellin–Broca, provide dispersion that is wavelength independent.

The relatively constant dispersion properties of a grating monochromator are illustrated by the monochromator considered in Table 3-3. For 300 nm, $D_a$ from equation 3-52 is $1.26 \times 10^{-3}$ rad nm$^{-1}$. Radiation of 310 nm would be separated from that of 300 nm by $\Delta\beta \approx 1.26 \times 10^{-2}$ rad. When the grating is rotated to isolate 400 and then 500 nm radiation, similar calculations show $D_a = 1.28 \times 10^{-3}$ and $1.32 \times 10^{-3}$ rad nm$^{-1}$, only a small difference. For the grating monochromator of Table 3-3 and a collimating focal length $f$ of 250 mm, the linear dispersion $D_l = fD_a = 250$ mm $\times D_a$. The reciprocal linear dispersion $R_d$, for a grating monochromator is found by combining equations 3-52 and 3-54. Values calculated from the equations below are shown in Table 3-3 for the 250-mm-focal-length monochromator described there.

$$R_d = (D_a f)^{-1} = d\,\frac{\cos\beta}{f|m|} \qquad (3\text{-}55)$$

$$R_d = \frac{\lambda\cos\beta}{f(\sin\alpha + \sin\beta)} \qquad (3\text{-}56)$$

It should be noted from Table 3-3 that although the reciprocal linear dispersion does not change drastically with wavelength as it does for a prism instrument, it *does* change with wavelength.

*Solid Angle and F-Number.* In a monochromator, the size of the dispersion element and the angles used play a key role in determining the solid angle, the F-number, the throughput, and the resolution. In the following discussion of these characteristics we define the physical area and width of the dispersion element as $A_D$ and $W_D$, respectively. We will assume the usual case where the limiting aperture is determined by the projected area of the dispersion element, the most expensive optical element in the monochromator. The effective or projected area $A_p$, or the incident beam cross-sectional area, is given by $A_p = A_D \cos \theta$, where $\theta$ is the angle of incidence with respect to the normal to the limiting aperture. The projected limiting aperture diameter is $D_p$. Usually, the limiting aperture is rectangular, so that by convention, the value of $D_p$ used is the diameter of a circle with an area equal to $A_p$ ($D_p = [4A_p/\pi]^{1/2}$). The F-number is given by

$$\text{F/n} = \frac{f}{D_p} \qquad (3\text{-}57)$$

The solid angle is given by

$$\Omega = \frac{A_p}{f^2} = \frac{\pi/4}{(\text{F/n})^2} \qquad (3\text{-}58)$$

For a grating monochromator with a square grating of width $W_G$, $A_D = W_G^2$, $A_p = W_G^2 \cos \alpha$, and $D_p = 2W_G(\cos \alpha/\pi)^{1/2}$.

In symmetrical configurations, the collimating and focusing element focal lengths are the same. For nonsymmetrical configurations, the collimating element focal length determines the F/n and $\Omega$, while the focusing element focal length determines $R_d$. Typical focal lengths are in the range 0.25 to 2 m; values of F/n usually range from 3 to 20.

*Spectral Bandpass and Slit Function.* The **spectral bandpass** $s$ (nm) is the half-width of the wavelength distribution passed by the exit slit. Except at small slit widths, where aberrations and diffraction effects must be considered, the spectral bandpass is controlled by the monochromator dispersion and the slit width. Under these conditions the spectral bandpass is equal to the **geometric spectral bandpass** $s_g$, given by

$$s_g = R_d W \qquad (3\text{-}59)$$

where $W$ is the slit width. For a monochromator with $R_d = 1$ nm mm$^{-1}$, a slit width of 100 μm would give a spectral bandpass of

$$s_g = (1 \text{ nm mm}^{-1}) \times 100 \text{ μm} \times (10^{-3} \text{ mm μm}^{-1})$$
$$= 0.1 \text{ nm}$$

Many monochromators have equal entrance and exit slit widths. The slits determine the spectral profile of the output observed at the exit slit. If we illuminate the entrance slit with monochromatic light, for example, the image transfer system produces a monochromatic image of the entrance slit in the plane of the exit slit (focal plane). As we rotate the dispersion element, the entrance slit image is swept across the exit slit as illustrated in Figure 3-48. The resulting convolution of the entrance and exit slit images is a triangular shaped function $t(\lambda)$ called the **slit function**. If the slits were of unequal width, $t(\lambda)$ would have a trapezoidal rather than triangular shape.

Mathematically the slit function can be described as

$$t(\lambda) = \{1 - |(\lambda - \lambda_0)/s_g|\} \qquad \text{for} \quad \lambda_0 - s_g \le \lambda \le \lambda_0 + s_g$$

$$t(\lambda) = 0 \qquad \text{elsewhere}$$

$$(3\text{-}60)$$

where $\lambda_0$ is the wavelength setting of the monochromator and $\lambda$ is the incident wavelength. Note that the slit function is normalized to unity to indicate maximum transmission at $\lambda_0$. Also, note that the base of the triangle is $2s_g$ in length. Thus for a monochromatic source, $t(\lambda)$ indicates the fraction of the image on the focal plane at a given wavelength that is passed by the exit slit. In other words, the slit function indicates the shape of a narrow spectral line recorded on a spectrometer during a spectral scan.

For a broadband (continuum) source, the focal plane contains a distribution of overlapping slit images. If the monochromator is fixed at a single wavelength $\lambda_0$, the five images shown in Figure 3-48a could represent the focal plane images from five different wavelengths out of the source spectrum. Each wavelength is passed to a degree that depends on the overlap of its image with the exit slit. Thus for a broadband source, $t(\lambda)$ represents the fractional transmission vs. wavelength of the continuum spectrum impingent on the focal plane. From equation 3-60 the image of wavelength $\lambda = \lambda_0 \pm s_g/2$ would overlap 50% with the exit slit and thus be transmitted 50% compared to the image with $\lambda = \lambda_0$.

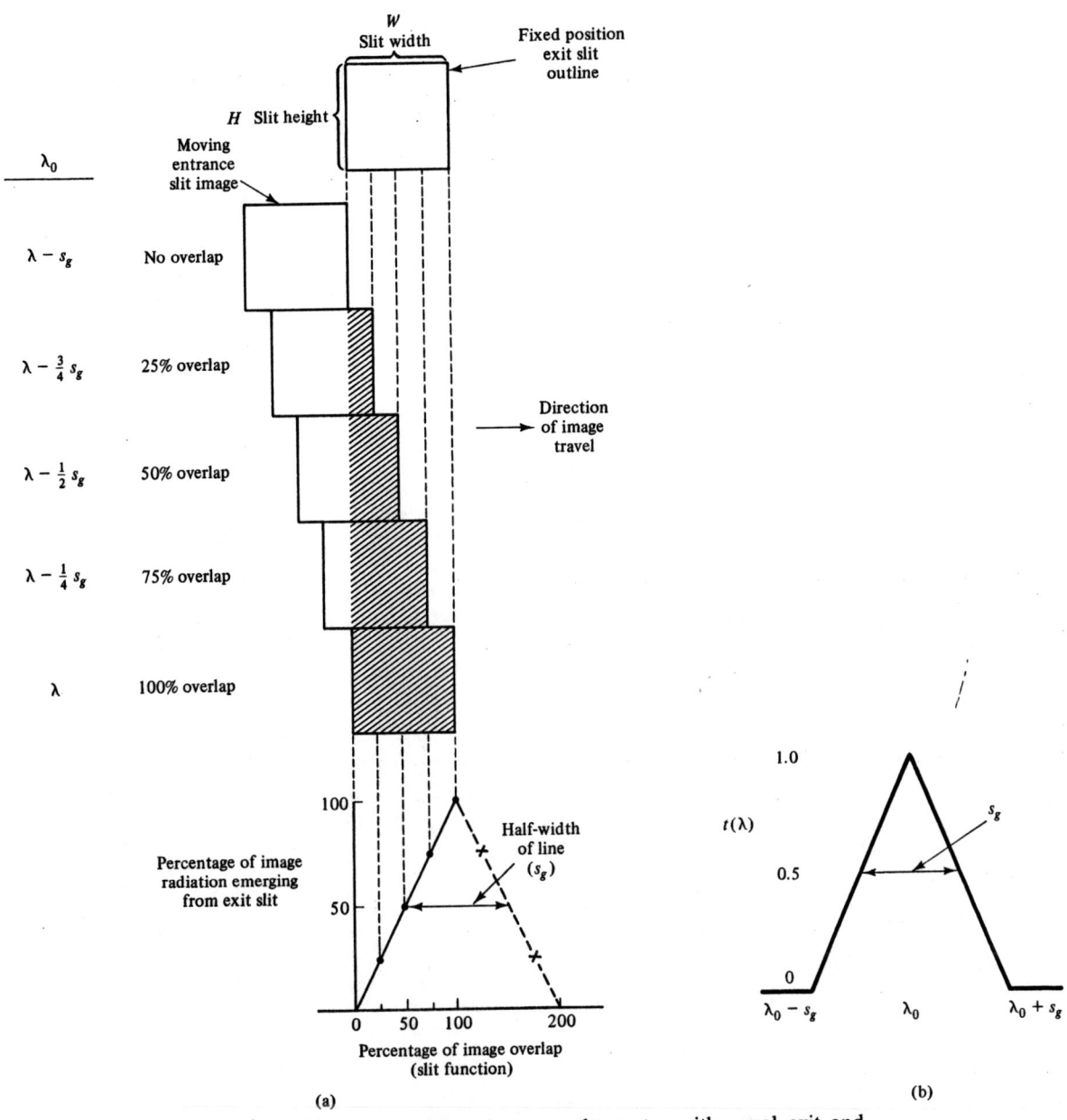

**FIGURE 3-48** Slit function of a monochromator with equal exit and entrance slits. In (a) the slit function is shown as the convolution of the entrance and exit slit images. In (b) the slit function is shown as a function of wavelength.

*Resolution.* The **resolution** of a monochromator is closely related to its dispersion. While the dispersion determines how far apart two wavelengths are separated linearly or angularly, the resolution determines whether the separation can be distinguished. In many cases the resolution is determined by the monochromator spectral bandpass. If the slit width $W$ is large enough that we can neglect aberrations and diffraction ($s = s_g$), a scan of two closely spaced monochromatic lines of peak wavelengths $\lambda_1$ and $\lambda_2$ would appear as idealized in Figure 3-49. Clearly, the two lines will be just separated (baseline resolution) if $\lambda_2 - \lambda_1 = 2s$. Thus the slit-width-limited resolution $\Delta\lambda_s$ is given by

$$\Delta\lambda_s = 2s_g = 2R_d W \qquad (3\text{-}61)$$

If the monochromator wavelength control is adjusted so that $\lambda_0 = \lambda_1$ and the slit width has been adjusted so that $W = \Delta\lambda/R_d$, the image of $\lambda_1$ will be completely passed, while that of $\lambda_2$ will be just at one side of the exit slit.

Equation 3-61 does not apply at small slit widths

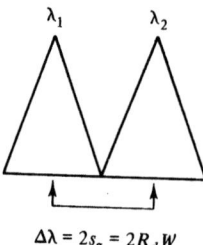

$$\Delta\lambda = 2s_g = 2R_d W$$

**FIGURE 3-49** Baseline resolution for two spectral lines. The centers of the entrance slit width images of the two wavelengths are separated by $2W$ (or $2R_d W$ in wavelength), so that the image of one wavelength passes the exit slit before the image of the other wavelength overlaps the exit slit.

since it predicts that $\Delta\lambda_s \to 0$ as $W \to 0$. At small slit widths, the width of the image of a given wavelength at the focal plane is no longer $W$; instead, it is larger than $W$ because the spectral bandpass is controlled by aberrations and diffraction as well as dispersion. The total spectral bandpass $s$ is

$$s \approx s_g + s_a + s_d \tag{3-62}$$

where $s_a$ is the aberration-limited bandpass in nm and $s_d$ is the diffraction-limited bandpass in nm.

In the case of diffraction-limited resolution, the image of a monochromatic line in the focal plane is no longer a slit image, but rather, a diffraction pattern as described in Section 3-2, where $W$ is now $W'_D$, the width of the beam emerging from the dispersion element ($W'_D$ is proportional to $W_D$, the width of the dispersion element). Figure 3-6 indicates that the angular half-width of the central maximum of this diffraction pattern is $\lambda/W'_D$. This can be converted to a physical width from $D_l = f D_a$ or $W_d = \lambda f/W'_D$, where $W_d$ is the diffraction-limited slit width (i.e., the slit width equal to the half-width of the central maximum). The diffraction-limited spectral bandpass (the half-width of the central maximum in nanometers) is $R_d W_d$ and

$$s_d = \frac{R_d \lambda f}{W'_D} = \frac{\lambda}{D_a W'_D} \tag{3-63}$$

Note that $s_d$ is independent of the slit width and dependent only on $\lambda$, $D_a$, and $W'_D$. The diffraction-limited spectral bandpass cannot be decreased by increasing the focal length because the half-width of the central maximum is proportional to $f$.

As the slit width $W$ is decreased, the bandpass $s$ reaches a minimum value $s_{min}$ determined by $s_a$ and $s_d$, $s_{min} \approx s_d + s_a$, while at large slit widths $s \approx s_g$. For any slit width, the spectral bandpass can be experimentally

determined by scanning a line from a line source (natural half-width $\ll s$) and calculating the half-width in nm.

If one considers two wavelengths $\lambda_1$ and $\lambda_2$ incident on a monochromator, the intensity distribution in the focal plane is the superposition of two diffraction patterns. The **Rayleigh criterion** for resolution states that the two wavelengths are resolved if the central maximum of one line falls on the minimum of the other, as shown in Figure 3-50. Since the half-width of the central maximum and the distance from the first maximum to the first minimum are approximately the same, the diffraction-limited resolution $\Delta\lambda_d$ is the same as $s_d$, so from equation 3-63,

$$\Delta\lambda_d = \frac{R_d \bar{\lambda} f}{W'_D} = \frac{\bar{\lambda}}{D_a W'_D} \tag{3-64}$$

where $\bar{\lambda}$ is the average wavelength of the two lines that are just resolved.

The **resolving power** is another way to express the ability to distinguish two wavelengths. The experimental resolving power of a monochromator is $R_{exp} = \bar{\lambda}/s_{min}$, whereas the theoretical resolving power $R_{th}$ (typically, $10^4$ to $10^6$) given for diffraction-limited resolution is $R_{th} = \bar{\lambda}/\Delta\lambda_d$. If the monochromator is free of aberrations, the experimental resolving power at $s_{min}$ is equal to the theoretical value. Normally, monochromators are operated at slit widths considerably larger than the diffraction-limited width. If $\Delta\lambda_d$ from equation 3-64 is sub-

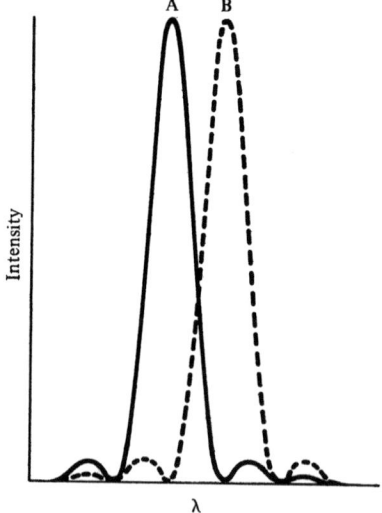

**FIGURE 3-50** Rayleigh criterion for resolution of two spectral lines. The lines are resolved if the central maximum of line A falls on the first diffraction minimum of line B. For normal slit widths $W \gg W_d$, so the slit image is much larger than the central maximum.

stituted into the foregoing definition of $R_{th}$, the result is

$$R_{th} = W'_D D_a \qquad (3\text{-}65)$$

Thus only the size of dispersion element and the angular dispersion determine the resolving power. A larger $W'_D$ results in a narrower central maximum and a larger $D_a$ increases the distance between the central maxima of two wavelengths. Although doubling the focal length doubles the distance between the central maxima of two different wavelengths in the focal plane, it does not change $R_{th}$ since the width of the maxima is also doubled.

For a grating monochromator the angular dispersion is given by equation 3-52, so that the theoretical resolving power can also be expressed as

$$R_{th} = \frac{W'_D |m|}{d \cos \beta} \qquad (3\text{-}66)$$

For a grating instrument, $W_D = W_G$ and the width of the beam emerging from the dispersion element is $W'_D = W_G \cos \beta$. Thus

$$R_{th} = \frac{W_G |m|}{d} = |m| N \qquad (3\text{-}67)$$

where $N = W_G / d$, the total number of grooves used. Thus the resolving power improves directly with the order (increase in $D_a$), the number of grooves used, and the groove density (increase in $D_a$). This expression is strictly valid only for a plane grating, but also holds for a concave grating if the width is smaller than a certain optimum value.

Since the diffraction order is given by $m = [d(\sin \alpha + \sin \beta)]/\lambda$, the theoretical resolving power can also be written

$$R_{th} = \frac{d(\sin \alpha + \sin \beta)N}{\lambda} = \frac{W_G(\sin \alpha + \sin \beta)}{\lambda} \qquad (3\text{-}68)$$

Thus $R_{th}$ increases with increasing $\alpha$ and $\beta$ because $D_a$ is larger. Since the maximum value of $\sin \alpha + \sin \beta$ is 2, the maximum possible theoretical resolving power for a grating of used width $W_G$ is $(R_{th})_{max} = 2W_G/\lambda$.

*Throughput Factors.* The radiant power passing out of the exit slit of a monochromator depends on a number of factors. The output spectral radiant power $(\Phi_\lambda)_o$ can be expressed as the product of the source spectral radiance $B_\lambda$ and the monochromator throughput factor $Y(\lambda)$.

$$(\Phi_\lambda)_o = B_\lambda Y(\lambda) \qquad (3\text{-}69)$$

The monochromator throughput factor is the product of the source area viewed, the solid angle collected by the monochromator, the transmission factor of the monochromator optics, and the slit function. The source area viewed is normally determined by the product of the slit width $W$ and the slit height $H$. Thus $Y(\lambda)$ can be written

$$Y(\lambda) = WH\Omega T_{op} t(\lambda) \qquad (3\text{-}70)$$

and is seen to be the product of a non-wavelength-dependent geometric factor ($WH\Omega$) and a wavelength-dependent factor $[T_{op} t(\lambda)]$. Here $Y(\lambda)$ has the units cm$^2$ sr if $W$ and $H$ are expressed in centimeters, and the solid angle $\Omega$ is expressed in steradians. The optics transmission factor, $T_{op}$, takes into account absorption, reflection, and scattering losses of optical components and is dimensionless, as is the slit function $t(\lambda)$. In the optics literature, the product of the area and solid angle is often called the optical extent, etendue, or optical conductance [it also includes the square of the refractive index of the media between the source (object) and receiver if different from unity].

The total radiant power $\Phi_o$ passed by the monochromator is given by

$$\Phi_o = \int_0^\infty (\Phi_\lambda)_o \, d\lambda = WH\Omega \int_0^\infty B_\lambda T_{op} t(\lambda) \, d\lambda \qquad (3\text{-}71)$$

Two interesting cases can be considered. If the source is monochromatic and the monochromator is set to the appropriate wavelength [$\lambda = \lambda_0$, $t(\lambda) = 1$], then $T_{op}$ is constant over the source width and $\int_0^\infty B_\lambda \, d\lambda = B$, where $B$ is the line radiance. In this case

$$\Phi_o = WH\Omega B T_{op} \qquad (3\text{-}72)$$

The radiant power throughput is proportional to the area viewed ($WH$), the solid angle [$\Omega = (\pi/4)/(F/n)^2$], the line radiance, and the optics transmission factor at wavelength $\lambda_0$. For a broadband source (continuum source), with $B_\lambda$ and $T_{op}$ constant over $\lambda_0 \pm s$, the following equation results:

$$\Phi_o = B_\lambda WH\Omega T_{op} \int_0^\infty t(\lambda) \, d\lambda \qquad (3\text{-}73)$$

The integral of the slit function $t(\lambda)$ is just $s_g$, the geometric spectral bandpass. Thus

$$\Phi_0 = B_\lambda WH\Omega T_{op}s_g = B_\lambda W^2 H\Omega T_{op}R_d \qquad (3\text{-}74)$$

Note that the radiant power passed by the monochromator for a continuum source is proportional to the slit width squared, whereas for a monochromatic source it is proportional to the first power of the slit width. This difference arises because as we open the slits for a continuum source, we not only increase the source area viewed but also pass more of the source spectral distribution. For a monochromatic source, the source line width is much smaller than the monochromator slit width. Opening the slits only increases the source area viewed by the monochromator in this case.

In the equations above we assume that the entrance and exit slits are of equal size, that the light image on the entrance slit is of uniform irradiance and larger than $W$ and $H$ for all $W$ and $H$, and that the limiting aperture of the monochromator is filled. This is usually accomplished by using a lens or mirror to focus an image of the source on the slit with the cone of radiation entering matching the monochromator F/n. Often 1:1 imaging is used, where the optical element is two focal lengths from the source and entrance slit. In this case, the F/n of the lens or mirror should be one-half the F/n of the monochromator. When external optics are used, $T_{op}$ in the expressions above is replaced by $T_{op}T_i$, where $T_i$ is the efficiency factor ($0 < T_i < 1$) for the external imaging optics or $T_{op}$ is redefined to include the losses of the imaging optics. If other than 1:1 imaging is employed, the expressions above are divided by $m^2$, where $m$ is the magnification; the solid angle of the imaging optics is used as $\Omega$ in this case.

If a filter is used in place of a monochromator, similar expressions apply and $Y(\lambda) = A_s\Omega T_i T(\lambda)_f$. Here $A_s\Omega$ is the non-wavelength-dependent geometric factor, where $A_s$ is the source area viewed and $\Omega$ is the solid angle of the source viewed. The area $A_s$ is determined by the actual projected area of the source or an aperture in front of the source and $\Omega$ is determined by the lens or mirror (area and object distance) used to focus or collimate the source radiation on the filter. The product $T_i T(\lambda)_f$ is the wavelength-dependent factor, where $T(\lambda)_f$ is the spectral transmittance of the filter and $T_i$ is the optical efficiency of optical components used in conjunction with the filter.

*Stray Radiation.* Stray radiation or stray light in a monochromator is considered to be any radiation passed that is outside the interval $\lambda_0 \pm s$, where $\lambda_0$ is the wavelength setting and $s$ is the spectral bandpass. In some cases, the range specified above is broadened to include a larger wavelength range or the stray radiation is specified according to the relative amount of radiation passed $x$ nm from a specified laser line. Stray radiation is usually specified as the percentage of the total radiation passing the exit slit at a given wavelength over a specified wavelength range. Thus

$$\% \text{ stray radiation} = \frac{\Phi_{SR}}{\Phi_0} \times 100 \qquad (3\text{-}75)$$

where $\Phi_{SR}$ is the stray radiant power and $\Phi_0$ is the source radiant power passed by the monochromator over $\lambda_0 \pm s$. Stray radiation is very difficult to measure accurately because it depends on the wavelength, the spectral bandpass, and the type of source used.

There are many possible causes of stray radiation. For example, room light can leak into the monochromator or radiation can be reflected internally from walls, optics and baffles. Scattering from dust particles, and fluorescence from optical materials are also sources of stray radiation. An overlapping grating order, grating imperfections, and diffraction from slit edges can also produce stray radiation. With gratings, some stray radiation appears at all wavelengths because there is not complete destructive interference between diffraction orders. This results in the secondary maxima seen in Figure 3-7.

Stray radiation can be reduced in several ways. Good instrument design, including appropriate baffles to intercept various reflected and scattered rays, is one obvious way. Holographic gratings provide much lower stray radiation factors than do ruled gratings. Stray radiation figures can vary from 0.1% to $10^{-5}$%. Typical values are $10^{-3}$% for ruled gratings and $10^{-4}$% for holographic gratings. Broad absorption filters are often employed in front of the entrance slit to narrow the source bandwidth. These can also serve as order-sorting filters. Double monochromators or two monochromators in tandem can achieve extremely low stray radiation figures. These are often employed in situations such as laser Raman spectroscopy where a weak spectral line (Raman scattering) must be measured near a very strong line (Rayleigh scattered line). Note that if a single monochromator produces 0.1% stray radiation, two such monochromators in tandem would provide 0.0001% stray radiation. The slit function of a double monochromator is the convolution of two triangular functions (equal ganged slits).

*Optical Aberrations.* Let us briefly consider optical aberrations with the Czerny–Turner monochromator of Figure 3-44. Parabolic mirrors are used for the collimating and focusing mirrors to eliminate spherical aberrations. Chromatic aberrations are, of course, absent because the image transfer system uses mirrors and not lenses. The remaining problems are principally coma and astigmatism. The mirrors $M_1$ and $M_2$ are of equal focal lengths. If we neglect the grating, temporarily, by considering it to be a reflecting surface, the collimating and focusing mirrors are similar to the symmetric mirror pair of Figure 3-29, which compensates for coma. Addition of the grating and its rotation complicates the situation somewhat. Usually, the grating is placed at one-half of the total distance between the mirrors. Coma is a minimum when the angles of incidence and diffraction are nearly equal. Asymmetric Czerny–Turner mounts, where the centers of curvature of the two mirrors do not coincide, can result in less coma than symmetric mounts.

Astigmatism causes point objects to be imaged as lines. Although the entrance slit is a line object, astigmatism causes blurring of the image and can result in a loss of wavelength resolution (broadening). Since astigmatic images are spread over a larger area than are stigmatic images, astigmatism can cause a loss in throughput and resolution at small $W$. Also spatial structure in the image can be lost in an astigmatic system. To minimize broadening of the slit image horizontally, the entrance and exit slits are located at tangential foci. If the entrance slit is at $f_t$, the grating is illuminated by parallel rays. Thus the entrance slit is a tangential object for the first mirror. Similarly, the tangential object for the second mirror is the tangential image from the first mirror, and the exit slit is the tangential focus for the second mirror. Although this minimizes horizontal broadening, spatial resolution still suffers because a point on the entrance slit is imaged to a vertical line in the focal plane. If point-to-point image transfer is desired, the external optics prior to the monochromator can be arranged to compensate for astigmatism in the monochromator. In some cases, curved slits are employed to intercept curved images.

*Monochromator Types.* Although the Czerny–Turner monochromator is currently a very popular grating design, many other types are currently available and in use. Figure 3-51 shows several of the other popular grating types. The Ebert type is very similar to the Czerny–Turner except that one large mirror serves as both collimator and camera mirror. The Czerny–Turner system has the advantage that focusing the two mirrors is usually somewhat easier than focusing the single mirror of the Ebert monochromator. Also, two smaller mirrors are less expensive than one large mirror.

Even though gratings have essentially replaced prisms in all recent monochromator designs, there are still many older monochromators around with prisms as dispersive elements. Also, prisms are still frequently used as predispersers in double monochromators or as

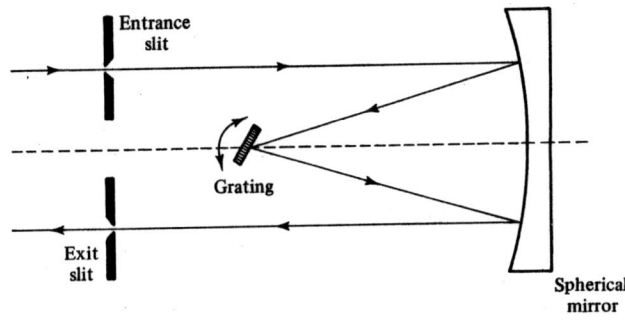

(a)

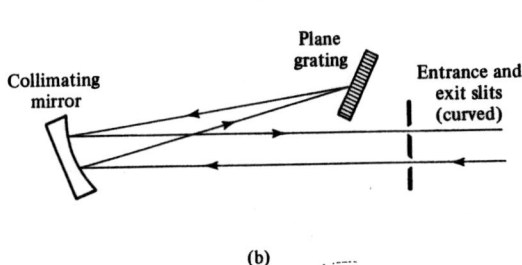

(b)

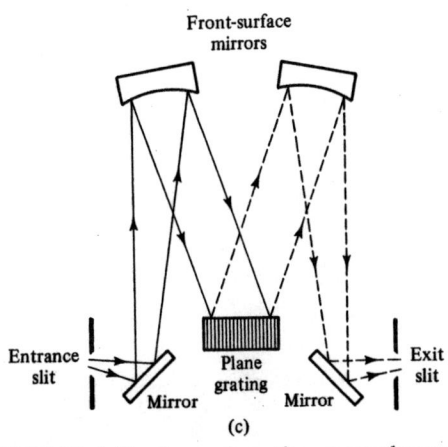

(c)

**FIGURE 3-51** Several grating monochromator designs. The Fastie–Ebert monochromator (a) has entrance and exit slits symmetrically placed on opposite sides of the grating. A single mirror serves as collimator and focusing mirror. Because of the large difference between the angles of incidence and diffraction, the mirror must be much larger than that of the Littrow monochromator (b). This latter monochromator is usually more compact than the Fastie–Ebert design. The Czerny–Turner design in (c) differs from that shown in Figure 3-44 in that it uses two additional folding mirrors, which allows in-line slits to be employed.

cross-dispersers in echelle grating spectrometers (see the next section). The Littrow mounting shown in Figure 3-52a was a widely used prism arrangement. Because each ray traverses the prism twice, it features high dispersion in a compact size. The double pass also avoids

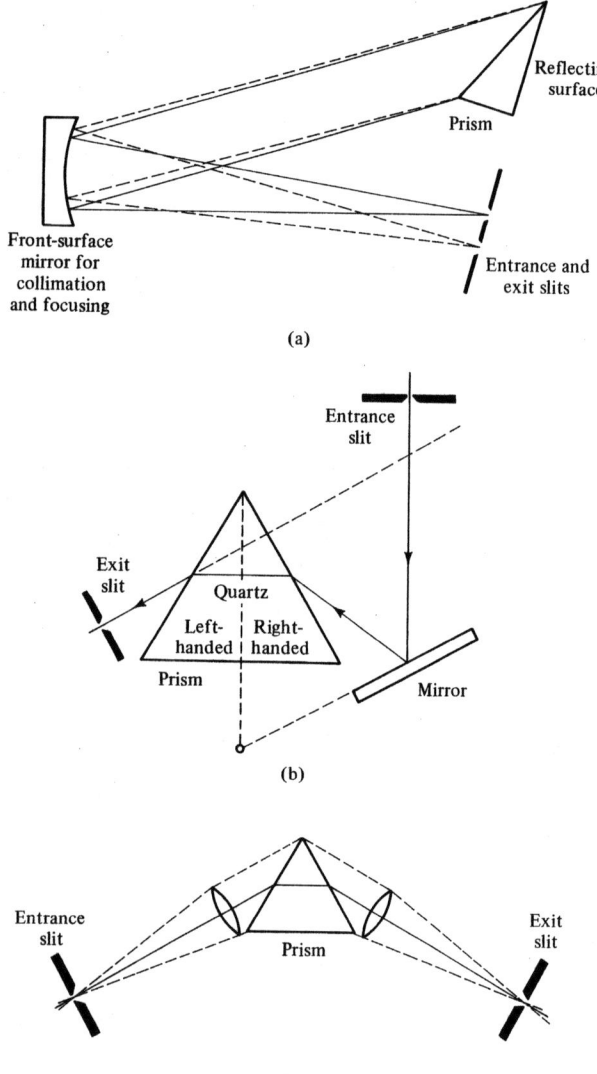

**FIGURE 3-52**   Some prism monochromator designs: (a) Littrow mount; (b) Cornu prism in Wadsworth mount; (c) Bunsen design.

double refraction in an anisotropic material such as quartz. In the Wadsworth mounting shown in Figure 3-52b, a Cornu prism was often employed. Since only a single pass is made through the prism, birefringence is a problem unless the prism is made of right-handed and left-handed quartz pieces cemented together. The circular birefringence in the first half of the prism is canceled by the second half in this case. The Bunsen design used in early prism monochromators is shown in Figure 3-52c.

In the past, prism monochromators were known for high dispersion in the UV and for stray radiation figures lower than grating instruments. The widespread use of holographic gratings in modern monochromators has almost eliminated the stray radiation advantage of prism instruments.

*Echelle Grating Monochromator.* If the expressions for linear dispersion (equation 3-55) and resolving power (equation 3-65) are examined carefully, several methods to improve dispersion and resolution become apparent. Conventional methods to improve linear dispersion include purchasing a finer-ruled grating (smaller $d$ for increased $D_a$) and increasing the focal length of the spectrometer ($D_l = fD_a$). Resolving power was traditionally improved by purchasing a grating with smaller $d$ and by using a larger grating (larger $W_G$ and $N$). Unfortunately, there are physical and monetary limitations to these approaches. Another way to increase dispersion and resolving power is to use large diffraction angles (large $\beta$ for increased $D_a$) and high orders (large $m$ for increased $D_a$). The echelle grating is made specifically to take advantage of this mode of operation.

In an Echelle monochromator, such as the one illustrated in Figure 3-53, diffraction orders of greater than 80 are typically used. Consequently, the free spectral range (recall equation 3-51) is much lower than for a conventional monochromator. Usually, a prism or low-dispersion grating is used as an order sorter or cross-disperser, as shown in Figure 3-53. The echelle grating can given excellent resolution and dispersion in a relatively compact spectrometer. One commercial echelle monochromator provides an average reciprocal linear dispersion of 0.12 nm mm$^{-1}$ and an average resolution of 0.003 nm in an instrument with a 0.75 m focal length. For comparison, a conventional 0.75-m grating monochromator with 2400 grooves mm$^{-1}$ gives a reciprocal linear dispersion of 0.54 nm mm$^{-1}$ and a resolution of $\approx$0.03 nm in the first order.

## Polychromators and Spectrographs

A polychromator system contains multiple exit slits, so that many wavelength bands can be isolated simultaneously. As discussed in Chapters 2 and 4, a separate detector is placed behind each exit slit. A typical configuration using a *Rowland circle* mounting is shown in Figure 3-54. The Rowland circle uses a concave grating so that the grating acts as the dispersion, collimating, and focusing element. The grating, entrance slit, and exit slits are all located on the circumference of a circle; the diameter of the circle is equal to the radius of curvature of the concave grating. The entire circumference of the circle is the focal plane for various wavelengths and various grating orders.

The Paschen–Runge mounting was also widely used in spectrographs with photographic detection. A large portion of the circle is useful for photographing the spectrum. Thus the major advantage of Paschen–Runge mounting is its extraordinarily large wavelength

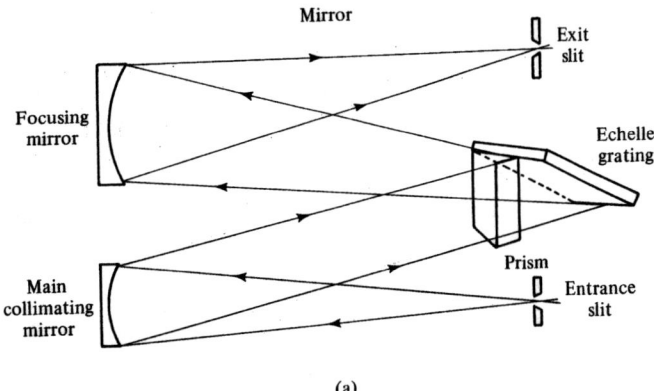

(a)

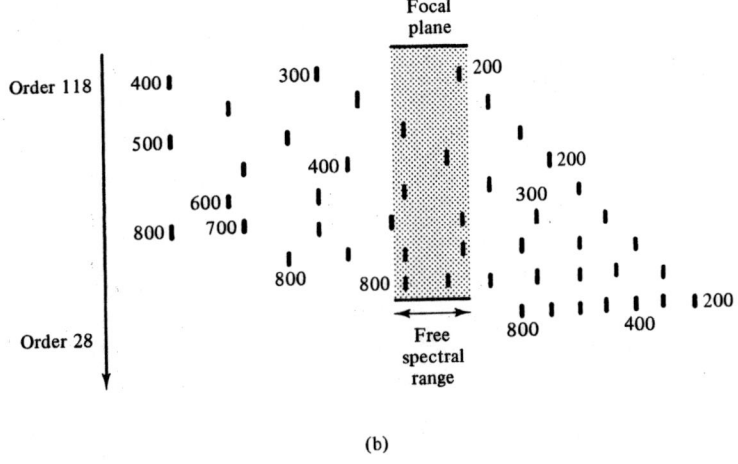

(b)

**FIGURE 3-53** Echelle monochromator. The echelle grating is mounted in a modified Czerny–Turner configuration (a). A prism prior to the grating serves as an order sorter. It disperses the spectrum perpendicular to the diffraction direction, which gives rise to a two-dimensional spectrum as seen in (b). By using the central wavelength region of several different orders, efficiency can be very high at many wavelengths instead of only near the blaze wavelength. Scanning can be accomplished by rotating the grating or by moving the detector along the focal plane.

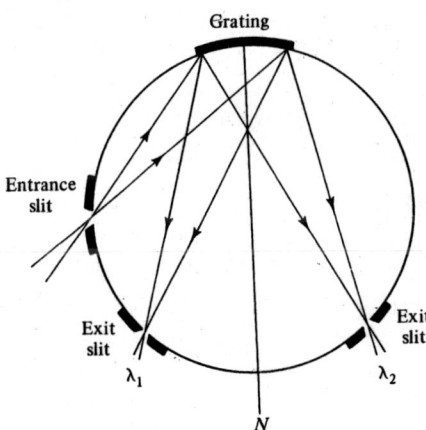

**FIGURE 3-54** Polychromator of Rowland circle type with Paschen–Runge mounting. In a Paschen–Runge mounting, the slits and grating are mounted permanently on the circle.

coverage. Some instruments use long strips of 35 mm film to allow simultaneous recording of wavelengths over the entire UV-visible region. Because of the large diffraction angle employed, dispersion in this configuration is often more wavelength dependent than that described for the Czerny–Turner monochromator. The throughput can be quite different for different diffraction angles, as can coma and astigmatism.

Many modern concave grating spectrographs use the *Eagle* mounting shown in Figure 3-55a. Its principal advantage is its smaller size compared to the *Paschen–Runge* mounting. To accommodate wide wavelength coverage the grating must be rotated, the camera or plate holder rotated or tilted, and the distance between the grating and plate holder must be changed.

The *Wadsworth* mounting shown in Figure 3-55b also utilizes a concave grating, but does not locate the optical elements on a Rowland circle. Instead, a collimator mirror is used to supply the grating with parallel light. The distance from the grating to the focal plane is about half that of an equivalent Rowland circle instrument. This reduces the linear dispersion by an equivalent amount. However, the solid angle of collection is about four times that of an equivalent Rowland circle instrument. The resulting increase in throughput more than makes up for the reflective losses introduced by the collimator mirror.

Plane grating and prism spectrographs of designs similar to the monochromators shown in Figure 3-51 are also widely used. In fact, several monochromators can be readily adapted for spectrographic operation by

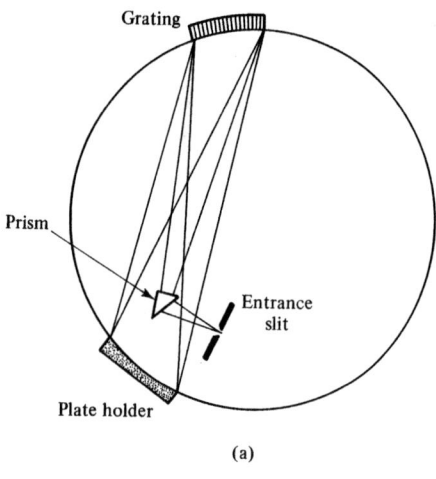

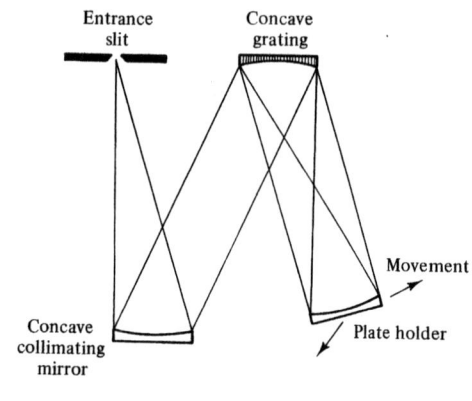

**FIGURE 3-55** Two concave grating mounting configurations: (a) Eagle mounting; (b) Wadsworth mounting. Note that the Eagle mounting is a Rowland circle type, but the Wadsworth is not.

insertion of a mirror into the optical path to intercept the beam prior to the exit slit and direct it to a photographic plate. With Ebert and Czerny–Turner mountings, the wavelength coverage is significantly lower than that of concave grating spectrographs of similar dispersion. The spectral images produced in the focal planes of plane grating instruments are also slightly curved.

Because the solid-state array detectors described in Chapter 4 are not capable of being bent to fit a curved focal plane, a few spectrographs are being manufactured specifically for flat detectors. These systems use holographic gratings and special image transfer optics to minimize curvature of the focal plane.

Echelle gratings are also widely used in polychromators and spectrographs. A commercial echelle grating polychromator utilizes the optical system of Figure 3-53 with an array of 20 exit slits and photomultiplier tubes for simultaneous measurements at 20 wavelengths. The same instrument has provisions for pho-

tographing the spectrum by inserting a mirror into the optical path to redirect the two-dimensional spectrum to a focal plane that has a camera. Echelle spectrographs are also used with array detectors.

## 3-7 NONDISPERSIVE SYSTEMS

Interferometers are widely used in spectroscopy. The Fabry–Perot interferometer is capable of extremely high resolution; it can be used to examine fine structure details of narrow spectral lines. The Michelson interferometer is a multiplex spectrometer in which radiation at all wavelengths is observed simultaneously. It is widely used in the IR and far-IR regions because of its signal-to-noise advantage over dispersive spectrometers. Several other interferometer types are discussed, and Fourier transform spectroscopic methods are introduced.

### Fabry–Perot Interferometer

The **Fabry–Perot interferometer** is conceptually identical to the Fabry–Perot interference filter discussed in Section 3-5. Instead of a thin dielectric film, of thickness on the order of the wavelength of interest, the interferometer uses two plane-parallel, highly reflecting plates separated by a much larger air gap, as shown in Figure 3-56a. The air gap varies in length from several millimeters to several centimeters. If the distance $d$ can be adjusted mechanically by translating one of the mirrors, the instrument is technically a Fabry–Perot interferometer. When the mirror distance is fixed and adjusted for parallelism by a spacer (usually invar or quartz), the system is called a **Fabry–Perot etalon**. The formation of the Fabry–Perot interference pattern is described in Figure 3-56.

The difference in optical pathlength between adjacent rays contributing to an interference spot is $\Delta(\mathrm{OPL}) = 2d\eta \cos \theta$, as given previously in Figure 3-35, where now $\theta$, the angle of incidence on the plates, equals $\theta'$, the angle of refraction in the air gap. Since for an air gap $\eta = 1.00$, $\Delta(\mathrm{OPL}) = 2d \cos \theta$. Constructive interference occurs when $2d \cos \theta = m\lambda$. Expressed in terms of the phase difference $\delta = 2\pi \Delta(\mathrm{OPL})/\lambda$, the condition for maxima in the fringe pattern is

$$\delta_m = 2\pi m \qquad m = 0, 1, 2 \qquad (3\text{-}76)$$

where $m$ is the order.

The fractional radiant power $\Phi/\Phi_m$ transmitted through the etalon can be written as a function of $\delta$. The following function is called the **Airy function**.

$$\frac{\Phi}{\Phi_m} = \frac{1}{1 + C_F \sin^2(\delta/2)} \qquad (3\text{-}77)$$

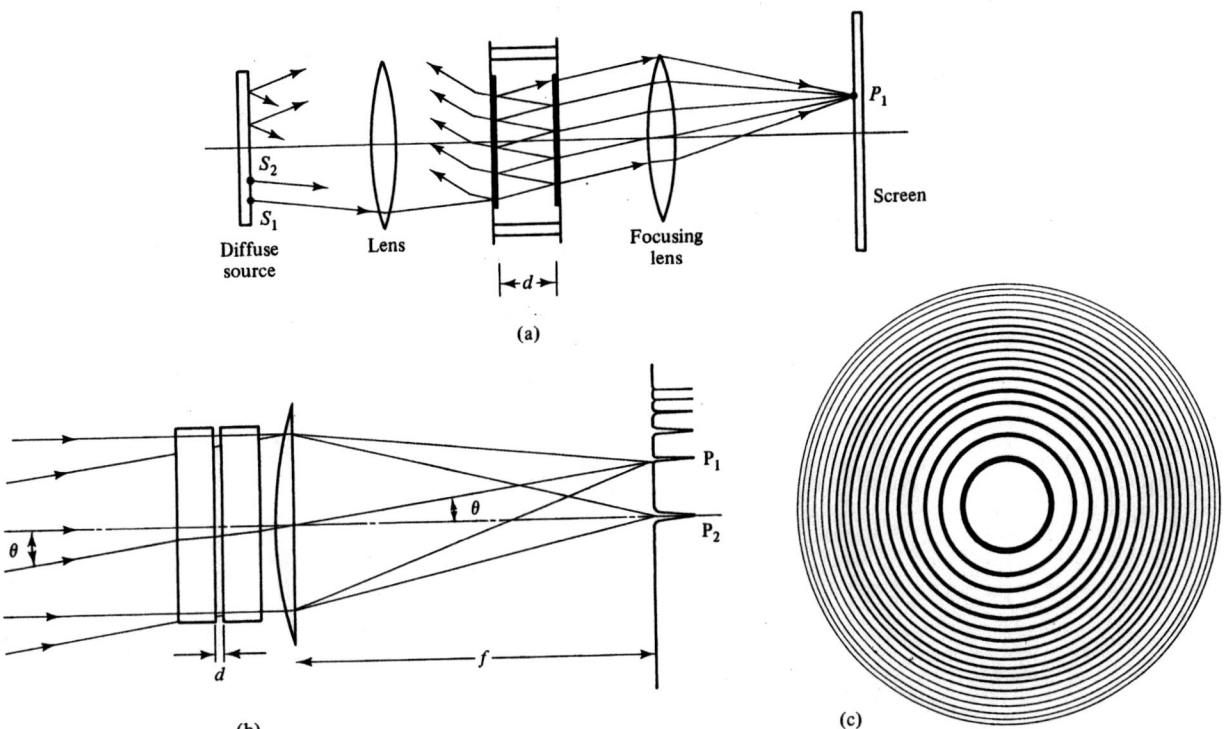

**FIGURE 3-56** Fabry–Perot etalon and interference fringes. In (a) a single ray from a diffuse, monochromatic source is traced through the etalon. The ray undergoes multiple reflections in the etalon. Transmitted rays are focused by a lens onto a screen at point $P_1$ where they interfere to form a bright or dark spot. All rays parallel to that from $S_1$ and in the same vertical plane (e.g., from $S_2$) are focused at point $P_1$ as seen more clearly in (b). Rays from the same plane at a different angle of incidence interfere at a different point $P_2$. All rays incident at a given angle, but from various planes, give rise to a single circular fringe of uniform brightness. The pattern of fringes in (c) results from a diffuse source illuminating the etalon at various angles and in various planes. The central spot can be bright or dark depending on the distance $d$.

The **coefficient of finesse**, $C_F$, is defined by $C_F = 4\rho/(1 - \rho)^2$, where $\rho$ is the reflectance of the mirror coatings (typically, $>0.9$). Here $\Phi_m$ is the maximum radiant power for a given fringe.

A plot of the relative radiant power vs. $\delta$ is shown in Figure 3-57a for a monochromatic source. Note that peaks in the relative transmittance occur at multiples of the phase difference. A measure of the sharpness of each fringe is the full width at half maximum in phase units $(\text{FWHM})_\delta$, which can readily be shown to be

$$(\text{FWHM})_\delta = \frac{4}{\sqrt{C_F}} = \frac{2(1 - \rho)}{\sqrt{\rho}} \qquad (3\text{-}78)$$

Thus the larger the reflectance, the sharper the fringes. Absolute transmittances are typically 1% or less.

A plot of relative transmittance vs. wavelength is

shown in Figure 3-57b. For $\theta = 0°$, interference maxima occur when $\lambda = 2d/m$. Just as with a diffraction grating, we can define the free spectral range $\Delta\lambda_f$ as the range of wavelengths for which no overlap of adjacent orders occurs, or $\Delta\lambda_f = \lambda/(m + 1) \approx \lambda/m$, because $m$ is very large. Thus

$$\Delta\lambda_f = \frac{\lambda}{m} = \frac{\lambda^2}{2d} = \frac{2d}{m^2}$$

The theoretical resolving power $R_{\text{th}}$ is given by $\bar{\lambda}/(\Delta\lambda)_{\text{min}}$, where $(\Delta\lambda)_{\text{min}}$ is the minimum resolvable wavelength difference. Usually, $(\Delta\lambda)_{\text{min}}$ is taken as the FWHM of a transmission peak. Equation 3-78 expresses the FWHM in terms of phase difference $\delta$. From $\delta = 4\pi d/\lambda$, a change in $\delta$ is related to a change in $\lambda$ by $\Delta\delta = 4\pi\Delta\lambda/\lambda^2$.

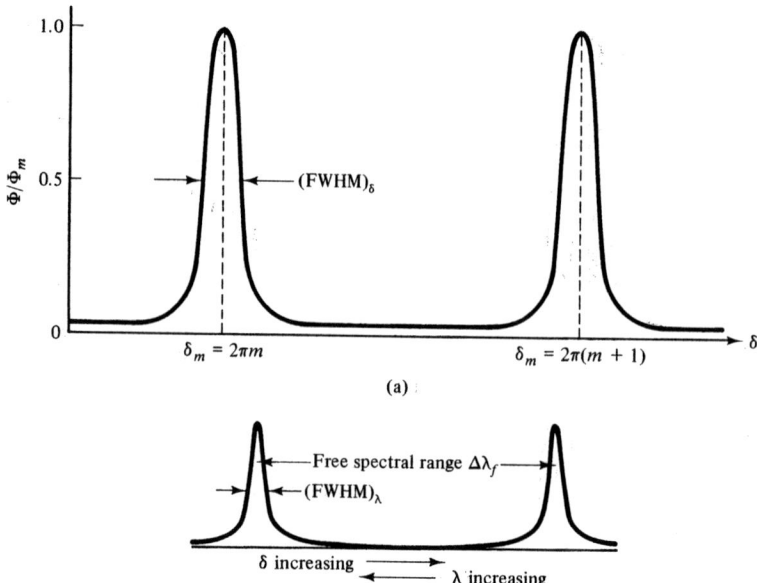

**FIGURE 3-57**  Fabry–Perot fringes for $\rho = 0.9$, $C_F = 360$. In (a) the relative transmittance is plotted vs. $\delta$ and in (b) vs. $\lambda$.

The FWHM in wavelength units $(FWHM)_\lambda$ is thus

$$(FWHM)_\lambda = \frac{(FWHM)_\delta}{4\pi d\lambda^{-2}} = \frac{4/\sqrt{C_F}}{\delta\lambda^{-1}}$$

Since $\delta$ near an interference maximum is $2\pi m$, we can write

$$(FWHM)_\lambda = \frac{2/\sqrt{C_F}}{\pi m\lambda^{-1}}$$

The theoretical resolving power is then

$$R_{th} = \frac{\bar{\lambda}}{(\Delta\lambda)_{min}} = \pi\sqrt{C_F}\frac{m}{2} \qquad (3\text{-}79)$$

The quantity $\pi\sqrt{C_F}/2$ in equation 3-79 is called the **finesse** $F$ of the etalon. It is a measure of the number of resolvable elements in one free spectral range given by the ratio of the separation between adjacent maxima $\Delta\lambda_f$ to the FWHM. Thus

$$F = \frac{\Delta\lambda_f}{(FWHM)_\lambda} = \frac{\pi\sqrt{C_F}}{2} = \frac{\pi\sqrt{\rho}}{1-\rho} \qquad (3\text{-}80)$$

The theoretical resolving power is then written in terms of the finesse as

$$R_{th} = mF = \frac{2dF}{\lambda} \qquad (3\text{-}81)$$

For arbitrary angles of incidence, $R_{th} = 2d \cos\theta \, F/\lambda$.

In the visible region of the spectrum $\bar{F}$ for most Fabry–Perot interferometers is $\approx 30$. The limitation is most often not the reflectances of mirrors as implied by equation 3-80, but the deviation of the mirrors from parallelism. Curved mirror systems with thin dielectric coatings have achieved $F$ values of greater than 1000. Today the **confocal interferometer** is more popular than the plane mirror type. In confocal designs concave spherical reflectors are used separated by a distance nearly equal to their radii of curvature. The focal points of the mirrors are nearly coincident on the optical axis. The confocal design is much easier to align and use than the plane-parallel type.

For use as a spectrometer the Fabry–Perot can achieve remarkable resolution. For $\lambda = 500$ nm, $d = 10$ mm, $F = 30$, equation 3-81 gives $R_{th} \approx 1.2 \times 10^6$ and $(FWHM)_\lambda \approx 4 \times 10^{-4}$ nm. Note, however, that $m = 4 \times 10^4$ and the free spectral range would be $\Delta\lambda_f \approx 0.0125$ nm. Thus spectrometric applications of Fabry–Perot interferometers require order sorting. Often, in fact, a grating monochromator is used as an order sorter.

For spectrometric applications a pinhole can be used to allow radiation to be transmitted to a detector or order sorter. Wavelength scanning is accomplished by varying the $d$ spacing or the refractive index of the gap. Piezoelectric mirror mounts are often used to move one of the mirrors by applying a voltage to the piezoelectric device. Alternatively, by changing the gas pressure within the interferometer, the refractive index of the gap can be changed. Although we have assumed an air gap ($\eta = 1.00$) for simplicity, the exact condition for constructive interference is $\Delta(OPL) = 2d\eta \cos\theta$. For many gases, $\eta$ is proportional to pressure over a considerable range. Hence pressure-scanned Fabry–Perot interferometers (constant $d$) are also common.

Because of their complexity and small $\Delta\lambda_f$ values, Fabry–Perot systems are employed in applications requiring extremely high resolution. Thus they are used to measure the widths of atomic spectral lines, the widths of laser lines and for other demanding applications. Fabry–Perot interferometers can also be used to determine refractive indexes of gases.

## Michelson Interferometer

The Michelson interferometer is an example of a two-beam interferometer, unlike the Fabry–Perot, which uses multibeam interference. Unlike a grating spectrometer, which disperses the spectrum spatially, the Michelson interferometer observes all optical frequencies from the source simultaneously. It is thus an example of a nondispersive or multiplex spectrometer.

A typical Michelson interferometer is shown in Figure 3-58. We will first assume that the source is monochromatic. When the movable mirror is adjusted such that the optical pathlength in the source → beam splitter → movable mirror → detector arm equals that in the source → fixed mirror → beam splitter → detector

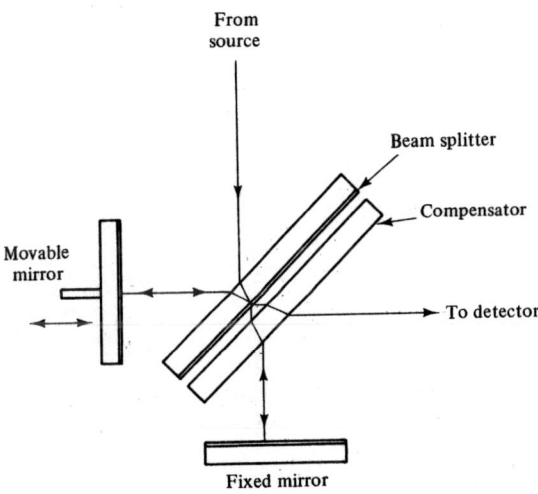

**FIGURE 3-58** Michelson interferometer. The interferometer consists of two plane mirrors, one fixed and one movable, at right angles to each other. A beam splitter reflects $\approx \frac{1}{2}$ of the source intensity to the movable mirror and transmits $\approx \frac{1}{2}$ to the fixed mirror. After reflection the beams are recombined by the beam splitter and sent to a detector. The compensator is used to equalize the optical path in both arms. A displacement $x$ of the movable mirror causes a pathlength difference $2x$ and a phase difference $\delta = 4\pi x/\lambda$ to be introduced between the two beams.

arm, the two beams will be in phase at the detector and thus constructively interfere (bright field at the detector). When the movable mirror is adjusted $\lambda/4$ away from this position, the two beams are 90° out of phase at the mirrors and 180° out of phase when they reach the detector (dark field at the detector). If the mirror is continuously moved at a constant rate, the intensity at the detector alternates sinusoidally between light and dark for each $\lambda/4$ movement, as shown in Figure 3-59a.

The ac component of the detector signal $S(x)$ as a function of mirror displacement is related to the source spectrum $\Phi_\nu$, $\Phi_{\bar\nu}$, or $\Phi_\lambda$ by

$$S(x) = K\Phi_\nu \cos\left(\frac{4\pi x\nu}{c}\right) = K\Phi_{\bar\nu}\cos(4\pi x\bar\nu)$$
$$= K\Phi_\lambda \cos\left(\frac{4\pi x}{\lambda}\right) \quad (3\text{-}82)$$

where $K$ is the constant that includes detector response and geometrical factors, $x$ is the mirror displacement, $\nu$ is the frequency of the source, $\bar\nu$ is the wavenumber of the source, $\lambda$ is the wavelength, and $c$ is the speed of light. The signal $S(x)$ measured vs. displacement $x$ is called the **interferogram**.

If the mirror is moved at a constant rate ($r = dx/dt$), the detector signal oscillates with a frequency $f = 2r\nu/c = 2r\bar\nu$, since the path difference changes at twice the rate of mirror movement. Thus if the monochromatic source is an infrared source with $\bar\nu = 1000$ cm$^{-1}$ ($\nu = 3 \times 10^{13}$ Hz) and the mirror is displaced at a rate of 1 mm s$^{-1}$, the resulting oscillation occurs at 200 Hz. We can see immediately that the Michelson interferometer uniquely encodes a very high frequency optical signal as a low-frequency oscillation. In the mid-IR region of the spectrum, the modulation frequency (oscillation frequency) is conveniently in the audio region of the spectrum, where the signal can be amplified by normal methods. In the far-IR region the modulation frequencies are too low (<1 Hz) for conventional methods and a mechanical chopper is used for upward frequency translation.

If the source is polychromatic, each input frequency can be considered to produce a separate cosine oscillation, and the resulting interferogram is the summation of all the cosine oscillations caused by all the optical frequencies in the source. For example, consider a broadband source containing many optical frequencies. When the movable mirror is placed such that the optical paths in the two arms are identical, all the waves are in phase, and the detector sees a bright field. As the mirror is moved away from the zero position the waves rapidly damp to a steady average value, as shown in Figure 3-59b. The ac signal is the integral over all

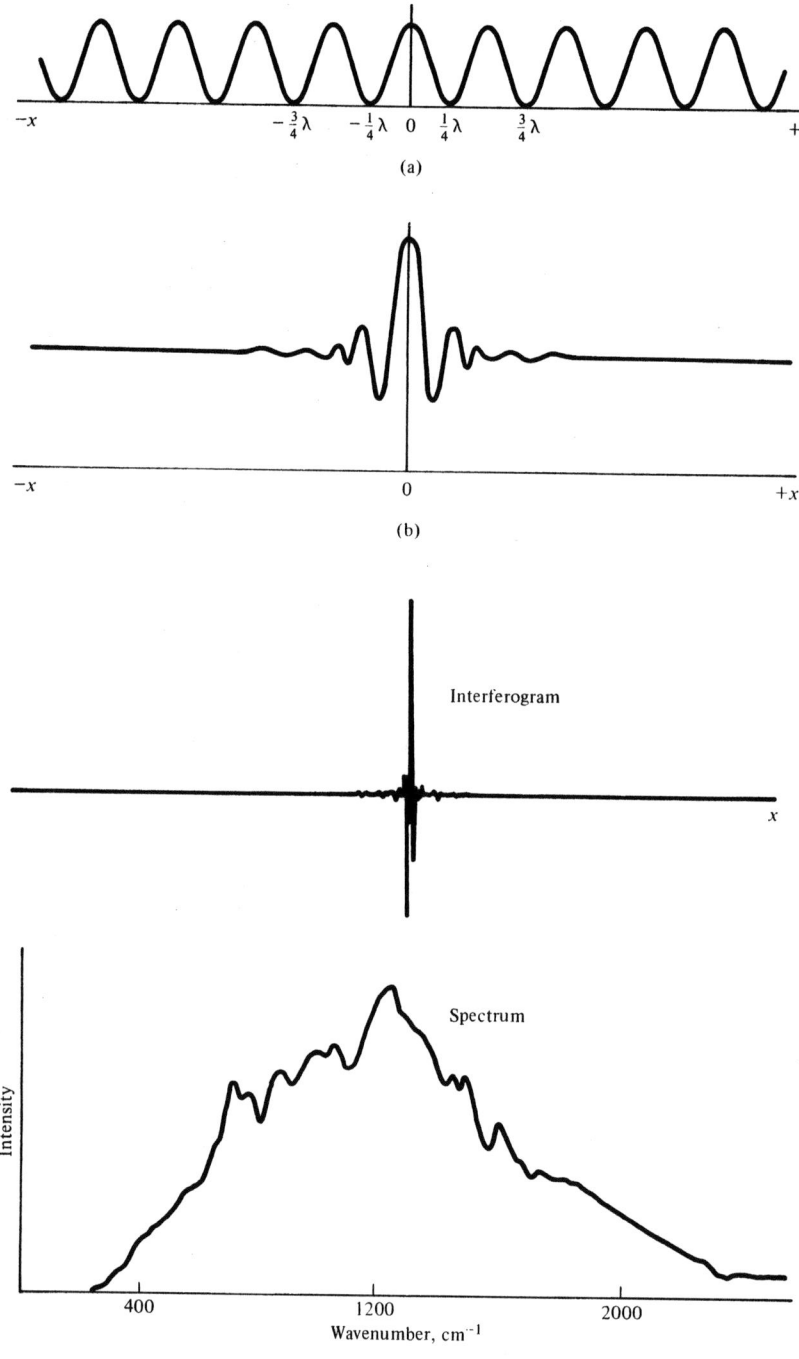

(a)

(b)

Interferogram

Spectrum

(c)

**FIGURE 3-59** Detector output vs. mirror distance in Michelson interferometer for (a) monochromatic source and (b) broadband source. Zero distance refers to equal optical pathlengths in both arms. The interferogram and spectrum from a Fourier transform IR spectrometer are shown in (c).

frequencies:

$$S(x) = \int_{-\infty}^{+\infty} \Phi_{\bar{\nu}} \cos (4\pi x \bar{\nu}) \, d\bar{\nu} \qquad (3\text{-}83)$$

Mathematically, equation 3-83 is a Fourier integral, whose Fourier transform is given by

$$\Phi_{\bar{\nu}} = \int_{-\infty}^{+\infty} S(x) \cos (4\pi x \bar{\nu}) \, dx \qquad (3\text{-}84)$$

Thus by measuring the interferogram, $S(x)$ vs. $x$, and calculating its Fourier transform, the spectrum $\Phi_{\bar{\nu}}$ vs. $\bar{\nu}$ is obtained. The interferogram and its spectrum are called **Fourier transform pairs**. Figure 3-59c shows a typical interferogram of a broadband source and the spectrum that results from transformation. Because the Fourier transform is a fairly extensive calculation, Fourier spectrometers include a computer system for calculating the spectrum.

The limits of integration in equation 3-84 indicate

that an infinitely long mirror travel is required to obtain the true spectrum from an interferogram. In practice the mirror travel is restricted to some finite maximum value, $(\Delta x)_m$, which limits the resolution attainable. The theoretical wavenumber resolution $(\Delta \bar{\nu})_{th}$ is, in fact, reciprocally related to the maximum mirror travel.

$$(\Delta \bar{\nu})_{th} = \frac{1}{(\Delta x)_m} \qquad (3\text{-}85)$$

A spectrometer with a 1-cm travel gives a resolution of $1$ cm$^{-1}$. For $\lambda = 500$ nm, the resolution in nanometers from $\Delta \lambda = \Delta \bar{\nu} \lambda^2$ is $2.5 \times 10^{-2}$ nm. Resolution values of $0.1$ cm$^{-1}$ or less are practical. Fourier transform spectrometers for use in the IR region are described in Chapter 14.

## Other Interferometers

There are several other interferometer types that are useful in spectrochemical applications. The **Mach–Zender** interferometer consists of two beam splitters and two totally reflecting mirrors, as illustrated in Figure 3-60a. Radiation from the source is split into two beams which travel separate paths before being recombined before the detector. A difference between the optical paths can be introduced by slightly tilting one of the beam splitters. Usually, the sample to be tested is place in one arm of the interferometer. This alters the path-length difference and changes the fringe pattern. The Mach–Zender interferometer is a single-pass system; it has only half the sensitivity of the Michelson interferometer, a two-pass system. This type of interferometer is rather difficult to align since the paths of the two beams are different. However, it finds many applications because of this. The Mach–Zender interferometer has been used to measure refractive index variations in gases, to test optical components for flatness, and to measure thickness of components. It is also used to observe density variations in flowing gas streams. The **interference microscope** is usually based on a Mach–Zender interferometer. This device allows transparent objects to be visualized provided that their index of refraction is different from that of their surroundings. The phase difference resulting from the refractive index difference in the two arms generates interference and allows visualization.

Another useful interferometer type is the **common-path** or **Sagnac interferometer** shown in Figure 3-60b. This interferometer features two nearly identical, but opposite paths taken by the two beams before they are recombined. A slight tilt of one of the mirrors introduces a small pathlength difference, which results in an interference pattern at the detector. Because the two

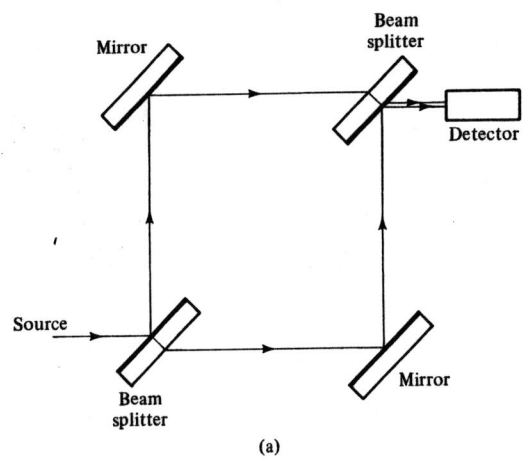

(a)

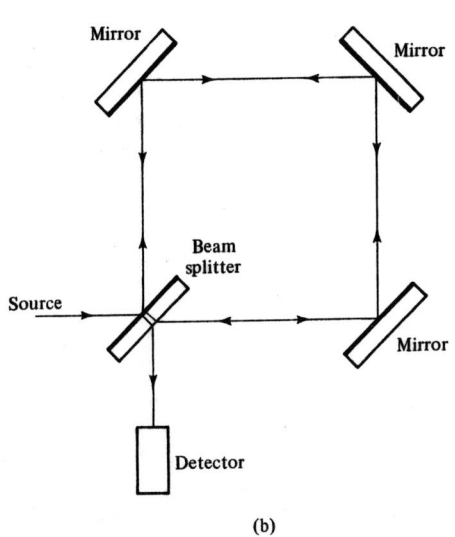

(b)

**FIGURE 3-60** Mach–Zender and common-path interferometer. The Mach–Zender interferometer in (a) is a single-pass interferometer. The beams in the two arms travel different paths. Because of this a test sample can be placed in one arm and the changes in the interference pattern measured. The common-path or Sagnac interferometer (b) is very simple. Here the two beams travel common paths in opposite directions. With a multichannel detector or a movable single-channel detector, it can be used like a Michelson interferometer to produce an interferogram of a polychromatic source that can be treated by Fourier transform methods to yield a spectrum. The common-path interferometer has no moving parts and is readily aligned.

beams travel the same paths and are in fact superimposed, the common-path interferometer is very easy to align and quite stable. In contrast to the Michelson interferometer, it has no moving parts. This device has been used in Fourier transform IR and UV-visible absorption spectrometry. Here the detector must be moved or it can be a multichannel (array) detector. The sample can be placed between the source and the beam splitter or between the beam splitter and the detector.

## Advantages of Fourier Transform Methods

Fourier transform spectroscopic methods have two major advantages over conventional, dispersive spectrometers. First there is a throughput advantage, called **Jacquinot's advantage**, because no entrance slits are present. Radiation must, however, be collimated in a Michelson or common-path interferometer system. Thus a circular aperture and a collimating element are normally used. The effective solid angle is often, however, much larger than that of a conventional spectrometer. Jacquinot's advantage can be a factor of 10 to more than 200. A second advantage, called the **multiplex advantage** (Fellget's advantage), results because the signals from all resolution elements are observed all the time in a multiplex spectrometer. In a conventional scanning spec-

trometer, each resolution element is observed for only a fraction of the scanning time. As discussed in Chapter 5, the multiplex advantage results in a signal-to-noise advantage for multiplex spectrometers when detectors are limited by detector noise. As a result, Fourier transform techniques find their major applications in spectral regions where noisy detectors (see Chapter 4) must be used, such as the mid- and far-IR regions. Because of the increased throughput they are also used in measurements of weak sources. Although most applications have been in mid- and far-IR absorption spectroscopy, Fourier transform spectrometers have also been used in UV-visible absorption, in IR emission, in Raman scattering and in rapid-scanning measurements. Commercial Fourier transform spectrometers are available based on both the Michelson and the common-path interferometer. Inexpensive computers and rapid Fourier transform algorithms have made Fourier transform spectrometers extremely competitive with dispersive systems in the mid-IR and clearly superior in the far-IR region, as discussed in Chapter 14.

## PROBLEMS

**3-1.** Consider a light beam of 579-nm unpolarized radiation in air that is incident a sapphire crystal at an angle of 50.0° with respect to the normal. Calculate the angle of refraction, the reflectance for each polarization component, and the total reflectance. See Appendix B for refractive index data.

**3-2.** Consider an F/2 convex lens that is 5.0 cm in diameter. A light source is placed 10 cm from the lens. Where is the image focused, and what is the magnification? What are the F/n of the system and the solid angle collected?

**3-3.** A monochromator has the following specifications: reciprocal linear dispersion, 2.5 nm mm$^{-1}$; focal length, 0.30 m; F-number, 6.0; grating size, 44.33 × 44.33 mm; groove density, 1200 grooves per mm. Calculate the following at 400 nm assuming the first order is used.
(a) The angular and linear dispersion.
(b) The projected limiting aperture diameter and projected limiting area.
(c) The slit width needed to obtain a 5-nm geometric spectral bandpass.
(d) The angle of incidence and angle of diffraction.
(e) The diffraction-limited slit width, diffraction-limited spectral bandpass, and diffraction-limited resolving power.
(f) The radiant power in watts passed by the monochromator if the special radiance of the continuum source is $2.00 \times 10^{-4}$ W cm$^{-2}$ sr$^{-1}$ nm$^{-1}$ with $H = 5.0$ mm, $W = 0.50$ mm, and $T_{op} = 0.40$.

**3-4** Calculate the velocity of light in $H_2O$ at 579 nm.

**3-5.** The peak absorption coefficient for a molecule with a number density of $1.0 \times 10^{16}$ cm$^{-3}$ is 1.00 cm$^{-1}$. Calculate the absorption cross section and the molar absorptivity.

**3-6.** Collimated laser radiation in 488 nm is impingent on a 100-μm-wide slit. A diffraction pattern is seen on a screen that is 1.0 m behind the slit. What is the half-width (FWHM) of the central maximum?

**3-7.** Calculate the numerical aperture and F/n for a fiber optic with core and cladding refractive indices of 1.50 and 1.48, respectively.

**3-8.** Consider a Fabry–Perot interferometer with a mirror spacing of 5.0 mm and mirror reflectances of 0.95. For $\lambda = 500$ nm and normal incidence, calculate the coefficient of finesse, the order, the FWHM in nanometers, the finesse, the theoretical resolving power, and the free spectral range.

**3-9.** What performance characteristics of a monochromator are affected when only the grating groove density is changed?

**3-10.** A ray in air is incident on a block of crown glass ($\eta_{glass} = 1.52$) at a 30° angle from the normal to the glass surface. At what angle relative to the normal will the ray be transmitted through the glass?

**3-11.** Locate and describe the image of a safety pin 100 cm away from a convex spherical mirror having a radius of curvature of 80 cm.

**3-12.** A thin biconvex lens ($\eta = 1.5$) has radii of curvature of $R_1 = 40$ cm and $R_2 = 20$ cm.

(a) What is the focal length of the lens?

(b) Locate and describe the image of an object 40 cm away from the lens.

**3-13.** A point source of radiant intensity 1.0 W sr$^{-1}$ is placed at the focal point of a 1.0-cm-diameter F/6.0 lens. Calculate the following values:

(a) The focal length of the lens.

(b) The solid angle collected by the lens.

(c) The radiant emittance of the collimated beam (ignore transmission losses).

**3-14.** A Fabry–Perot interference filter has a $d$ spacing of $2.04 \times 10^{-5}$ cm and a dielectric with $\eta = 1.35$.

(a) What central wavelength is passed in the first order?

(b) What wavelength is passed in the second order?

(c) If the filter oscillates such that the angle of incidence varies from $-10$ to $10°$, over what wavelength range is the first-order wavelength scanned?

**3-15.** A grating has a groove density of 1500 grooves per mm. If the incident beam strikes the grating at an angle of $20.0°$;

(a) What diffraction angles will the first order of 300, 400, 500, 600, and 700 nm appear?

(b) What wavelength in the second order overlaps the 600 nm first-order beam?

(c) What is the free spectral range for the first order at 600 nm?

**3-16.** (a) If the grating in problem 3-15 is blazed at an angle of $15.0°$, what is the blaze wavelength?

(b) At what angle would the grating in part (a) need to be blazed to obtain a 450-nm blaze wavelength?

**3-17.** (a) The grating in problem 3-15 is operated at 500 nm in the first order. What is the angular dispersion of the grating?

(b) What is the angular dispersion at 300 nm in the first order?

**3-18.** A thin biconvex lens of refractive index 1.5 and diameter of 2.5 mm has radii of curvature of $R_1 = 1$ cm and $R_2 = 0.5$ cm.

(a) Find the focal point of the lens.

(b) If the object is 2 cm away from the lens, where is the image? What is the magnification?

(c) If the object is a radiation source with a radiance of $2.0 \times 10^{-3}$ W cm$^{-2}$ sr$^{-1}$ and a projected area of 0.10 cm$^2$, what is the irradiance of the image? (Ignore transmission losses.)

(d) Where should the object be placed so that rays emerging from the lens are parallel?

**3-19.** Why are mirrors preferred over lenses for imaging in many spectroscopic instruments that must cover multiple wavelengths? If lenses must be used, how can their imaging properties be idealized at least for two wavelengths? What are the disadvantages of this approach?

**3-20.** A grating has a groove density of 2000 grooves per mm. The incident beam strikes the grating at an angle of $20.0°$.

(a) What wavelength will appear in the first order at a diffraction angle of $30.0°$?

(b) At what diffraction angles will 300, 400, 500, and 600 nm appear?

(c) Find the diffraction angle for the second order of 280 nm. What first-order wavelength is overlapped by the second order of 280 nm?

**3-21.** For the grating in problem 3-20:

(a) What blaze angle would give a blaze wavelength of 300 nm?

(b) The grating is used in a Czerny–Turner monochromator with a takeoff angle of $10.0°$. Find the grating rotation angle $\theta$, the angles of incidence and diffraction $\alpha$ and $\beta$, and the angular dispersion $D_a$ for 300-, 400-, and 500-nm radiation.

(c) What percentage change in angular dispersion occurs over the range 300 to 500 nm?

**3-22.** The monochromator in problem 3-21b has a focal length $f$ of 0.30 m.

(a) Find the linear dispersion $D_l$ and the reciprocal linear dispersion $R_d$ at 300, 400, and 500 nm in the first order.

(b) Find the geometric spectral bandpass and the slit-width-limited resolution at 300, 400, and 500 nm for slit widths of 100, 200, and 500 μm.

(c) What radiant powers in watts are passed by the monochromator for the slit widths in part (b) if a continuum source is incident with a spectral radiance of $1.50 \times 10^{-1}$ W cm$^{-2}$ sr$^{-1}$ nm$^{-1}$ when the monochromator is set to 500 nm; the slit height $H = 5.0$ mm, $T_{op} = 0.43$, and the monochromator F/n = 7.0?

(d) Describe in words why the radiant power throughput of a monochromator shows a different dependence on slit width for a continuum source than for a line source.

(e) What are the theoretical resolving power and the diffraction-limited spectral bandpass for the first order of 500 nm if the grating width is 5.0 cm?

**3-23.** A Michelson interferometer has a mirror driven at 1.5 cm s$^{-1}$. What frequency would the interferogram show if the source radiation were at **(a)** 400 nm; **(b)** 800 nm; **(c)** 10 μm?

**3-24.** What distance must the mirror be driven in a Michelson interferometer to separate **(a)** infrared radiation at 10.15 and 10.18 μm; **(b)** visible radiation at 725 and 730 nm?

**3-25.** Characterize the shape of the slit function, and hence the shape of a scan of a line source, if the exit slit is twice as wide as the entrance slit, and vise versa.

**3-26.** At 589 nm, the focal length of a lens is 5.0 cm and $\eta = 1.544$. At 200 nm, the refractive index is 1.651. Calculate the focal length at this wavelength. If the lens was used at 589 nm for 1 : 1 imaging, where is the image focused at 200 nm?

**3-27.** At 400 nm, a solution of an optical active compound has a circular birefringence of $4.0 \times 10^{-6}$ and a circular dichroism of 10 L mol$^{-1}$ cm$^{-1}$. Calculate the optical rotatory power and the molar ellipticity in a 10-cm cell.

**3-28.** The angular separation of maxima for a multiple-slit mask illuminated with 400-nm radiation is 0.10 rad. Calculate the slit spacing.

**3-29.** Calculate the maximum amplitude of the electric field associated with a beam of irradiance 1 mW cm$^{-2}$.

**3-30.** A line source is scanned with a monochromator at 5.0 nm min$^{-1}$. On a recorder, a triangular peak is observed with a half-width of 1.0 cm with a chart speed of 2.0 cm min$^{-1}$. Calculate the spectral bandpass of the monochromator.

# REFERENCES

The following books, chapters, or reviews treat the general principles and applications of basic optics.

1. E. Hecht and A. Zajac, *Optics*, Addison-Wesley, Reading, Mass., 1974. An excellent modern optics textbook. The illustrations are superb and the discussion is both quantitative and qualitative. Highly recommended.

2. M. Young, *Optics and Lasers*, Springer-Verlag, New York, 1984. A fairly mathematical treatment, but clearly done.

3. W. J. Smith, "Image Formation: Geometrical and Physical Optics," in *Handbook of Optics*, W. G. Driscoll and W. Vaughan, eds., McGraw-Hill, New York, 1978. At an elementary level, but a good review.

4. J. A. Meyer-Arendt, *Introduction to Classical and Modern Optics*, 2nd ed., Prentice-Hall, Englewood Cliffs, N.J., 1984.

5. F. A. Jenkins and H. E. White, *Fundamentals of Optics*, 4th ed., McGraw-Hill, New York, 1976.

6. M. Born and E. Wolf, *Principles of Optics: Electromagnetic Theory of Propagation, Interference and Diffraction of Light*, 6th ed., Pergamon Press, Elmsford, N.Y., 1980.

The following are devoted to more specialized topics in optics.

7. E. Hartfield and B. J. Thompson, "Optical Modulators," in *Handbook of Optics*, W. G. Driscoll and W. Vaughan, eds., McGraw-Hill, New York, 1978.

8. M. Gottlieb, C. L. M. Ireland, and J. M. Ley, *Electro-Optic and Acousto-Optic Scanning and Deflection*, Marcel Dekker, New York, 1983.

9. *Diffraction Grating Handbook*, Bausch and Lomb, Rochester, N.Y., 1970.

10. J. Wilson and J. F. B. Hawkes, *Optoelectronics: An Introduction*, Prentice-Hall, Englewood Cliffs, N.J., 1983. Includes a discussion on optics.

11. *Electro-Optics Handbook*, Technical Series EOH-11, RCA Corp., Lancaster, Pa., 1974. Includes a concise discussion of basic optics and optical components.

The following references are devoted to the optics of spectroscopic instruments.

12. E. L. Grove, ed., *Analytical Emission Spectroscopy*, vol. 1, pt. II, Marcel Dekker, New York, 1971. Contains excellent chapters on "Prism Systems, Spectrographs, and Spectrometers" (H. W. Faust), "Gratings and Grating Instruments" (R. M. Barnes and R. F. Jarrell) and "Spectroradiometric Principles" (H. T. Betz and G. L. Johnson).

13. R. C. Elser, "Optical Instrumentation," in *Trace Analysis; Spectroscopic Methods for Elements*, chap. V, J. D. Winefordner, ed., Wiley-Interscience, New York, 1976.

14. J. F. James and R. S. Sternberg, *The Design of Optical Spectrometers*, Chapman & Hall, London, 1969.

15. R. A. Sawyer, *Experimental Spectroscopy*, Dover, New York, 1963. Includes a good discussion of spectrographs and spectroscopic equipment.

16. E. J. Meehan, "Spectroscopic Apparatus and Measurements," in *Treatise on Analytical Chemistry*, 2nd ed., pt. 1, vol. 7, chap. 3, I. M. Kolthoff and P. J. Elving, eds., Wiley-Interscience, New York, 1981.

17. P. R. Griffiths, ed., *Transform Techniques, in Chemistry*, Plenum Press, New York, 1978. Contains a discussion of interferometers and their design and operation.

18. P. R. Griffiths and J. A. deHaseth, *Fourier Transform Infrared Spectroscopy*, Wiley, New York, 1986. Discusses interferometers and their use in FTIR techniques.

19. J. E. Stewart, *Infrared Spectroscopy: Experimental Methods and Techniques*, Marcel Dekker, New York, 1970. Has a good discussion of conventional dispersive spectrometer designs as well as interferometer systems.

20. G. A. Vanasse, ed., *Spectrometric Techniques*, vols. I–IV, Academic Press, New York, 1977–1985. The first two volumes, in particular, discuss interferometers and their applications in FTIR methods.

The references that follow provide additional details on specialized subjects.

21. R. Bracewell, *The Fourier Transform and Its Applications*, McGraw-Hill, New York, 1965. A classic text on the use of Fourier transforms.

22. E. Charney, *The Molecular Basis of Optical Activity: Optical Rotatory Dispersion and Circular Dichroism*, Wiley, New York, 1979.

In addition to the above, various catalogs and guides are available from optics and spectrometer manufacturers. Many of these provide a wealth of fundamental and practical information.

# CHAPTER 4

# Optical Sources, Transducers, and Measurement Systems

In addition to the optical components presented in Chapter 3, spectrochemical methods require sources and detectors of radiant power. Radiation sources and detectors, along with electronic and computer signal processing systems, are the major topics of this chapter. Because practical radiation sources for absorption, luminescence, and scattering measurements are often compared to blackbody radiators, the laws of blackbody radiation are introduced here first. The Einstein probability coefficients for emission and absorption are then explained. Our discussion of specific radiation sources first deals with conventional line and continuum sources and their characteristics. Then the laser is introduced, and the principles and characteristics of several types of lasers are considered. Radiation transducers (detectors) convert the radiant power of an incident beam into an electrical signal that can be processed by electronic instrumentation. The operating principles and the characteristics of the most widely used thermal and photon detectors are considered here, and the array detectors that are making rapid spectral measurements possible are introduced.

After the optical signal of interest is converted into an electrical signal, it can be manipulated and processed conveniently by analog and/or digital circuits and microcomputers. A brief review of several of the electronic signal processing and readout systems most widely used in spectrochemical methods is presented. The principles of modulation, photon counting, lock-in amplification, and several other signal processing methods are described. The chapter ends by considering complete single-channel, multichannel, and multiplex optical spectrometers and the expressions which govern their use in making emission, absorption, and luminescence measurements.

## 4-1 BLACKBODY RADIATION

We are all familiar with the fact that a heated surface emits electromagnetic radiation. The energy emitted increases, and the spectral distribution of the radiation shifts toward shorter wavelengths as the temperature of the surface increases. This is readily noted by a shift in the color of the surface from red to blue with increasing temperature. A **blackbody** absorbs all radiant energy incident upon it regardless of wavelength. None of the radiant energy is transmitted and none is reflected. Thus a blackbody has unity spectral absorptance $[\alpha(\lambda) = 1]$ at all wavelengths. Because a blackbody is in thermal equilibrium with its environment, it must emit as much radiation as it absorbs. Being a perfect absorber implies that a blackbody is also the most efficient thermal radiator possible. A blackbody, in fact, radiates more total power and more power per unit wavelength interval than any other thermal radiating source at a particular temperature. Although no material is an ideal blackbody, one can approximate a blackbody in the laboratory by a hollow, insulated enclosure or an oven with a small exit hole in one wall. Radiation that enters the hole has little opportunity to escape by reflections so that the enclosure acts nearly as a perfect absorber. If the oven is heated, it can serve as a source and emit radiation through the hole. In thermal equilibrium, absorption balances emission except for the small amount of energy that escapes through the hole.

## Planck's Law

With an ideal blackbody, absorption exactly balances emission. Since absorption occurs at the same rate as emission, the blackbody must contain an equilibrium density of radiation. We will call the spectral energy density of a blackbody radiator $U_\nu^b$ (J cm$^{-3}$ Hz$^{-1}$) or $U_\lambda^b$ (J cm$^{-3}$ nm$^{-1}$), where the superscript $b$ signifies a blackbody.

Classical radiation theory does not explain the spectral distribution of blackbody radiation. Wien produced an expression that fits the observed spectral distribution fairly well at short wavelengths, but it failed at long wavelengths. Rayleigh and Jeans were able to match the observed data, but only in the very long wavelength region. Planck's success in 1900 in fitting the spectral distribution of blackbody radiation proved to be a turning point in the history of physics, for it eventually led to the quantum description of radiation and matter. Planck's model assumed that the walls of the oven were in thermal equilibrium with the radiation field inside. He assumed, furthermore, that the atoms inside behaved as oscillators absorbing and emitting radiant energy. All oscillator frequencies were possible in Planck's model, and thus a continuum distribution was predicted in the spectrum. Planck's final assumption was that each atomic oscillator could absorb or emit only discrete amounts of energy directly related to its oscillation frequency. Each energy value had to be an integer multiple of the basic quantum of energy $h\nu$. Statistical arguments then led directly to **Planck's radiation law**:

$$U_\nu^b = \frac{8\pi h\nu^3}{c^3} \frac{1}{e^{h\nu/kT} - 1} \tag{4-1}$$

where $h$ is Planck's constant and $k$ is Boltzmann's con-

stant. Planck's law was found to be in extremely good agreement with the observed data.

The spectral radiance $B_\nu$ is related to the spectral energy density by $B_\nu = U_\nu c/(4\pi)$, and the spectral radiance in wavelength units is $B_\lambda = B_\nu c/\lambda^2$. Thus two useful alternative forms of Planck's law are

$$B_\nu^b = \frac{2h\nu^3}{c^2} \frac{1}{e^{h\nu/kT} - 1} \tag{4-2}$$

and

$$B_\lambda^b = \frac{2hc^2}{\lambda^5} \frac{1}{e^{hc/\lambda kT} - 1} = \frac{c_1 \lambda^{-5}}{e^{c_2/\lambda T} - 1} \tag{4-3}$$

where $c_1 = 2hc^2 = 1.190 \times 10^{16}$ W nm$^4$ cm$^{-2}$ sr$^{-1}$ and $c_2 = hc/k = 1.438 \times 10^7$ nm K. A plot of $B_\lambda^b$ vs. wavelength is shown in Figure 4-1 for several temperatures. Note from equations 4-1 to 4-3 that the spectral

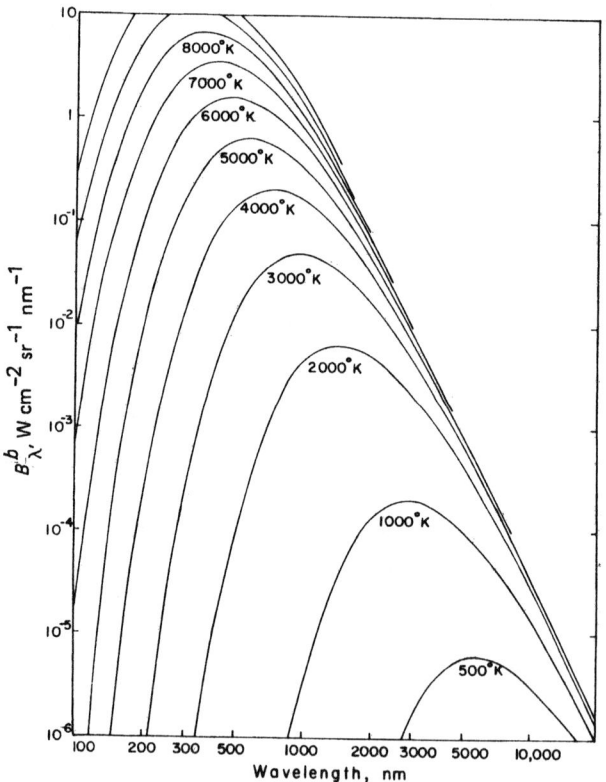

**FIGURE 4-1** Spectral radiance of a blackbody as a function of wavelength for several temperatures. Note the wavelength shift to the blue and the increase in spectral radiance as the temperature increases. (Adapted with permission from J. D. Winefordner, S. G. Schulman, and T. C. O'Haver, *Luminescence Spectrometry in Analytical Chemistry*, Wiley-Interscience, New York, 1972. Copyright © John Wiley & Sons, Inc., 1972).

radiance or energy density at a given frequency or wavelength depends only on temperature.

If equation 4-3 is differentiated with respect to wavelength and the derivative set equal to zero, the wavelength of maximum radiance $\lambda_m$ can be found as a function of temperature:

$$\lambda_m = \frac{c_2}{4.9651T} = \frac{2.897 \times 10^6}{T} \qquad (4\text{-}4)$$

Here $\lambda_m$ is in nanometers and $T$ is in kelvins. At 5000 K, for example, $\lambda_m = 579.6$ nm. At 7000 K, $\lambda_m = 414$ nm. The relationship shown in equation 4-4 is often called **Wien's displacement law**.

The total emittance of a blackbody is found by integrating equation 4-3:

$$M^b = \pi B^b = \pi \int_0^\infty B_\lambda^b \, d\lambda = \sigma T^4 \qquad (4\text{-}5)$$

where $\sigma$ is the Stefan–Boltzmann constant. The value of $\sigma$ is $5.6697 \times 10^{-12}$ W cm$^{-2}$ K$^{-4}$.

## Approximate Blackbody Expressions

Two common approximations to Planck's radiation law are often applied. At long wavelengths where $hc/(\lambda kT) \ll 1$, $(e^{hc/\lambda kT} - 1) \approx hc/(\lambda kT)$. Equation 4-3 then reduces to

$$B_\lambda^b = \frac{2hc^2}{\lambda^5} \frac{\lambda kT}{hc} = \frac{2ckT}{\lambda^4} \qquad (4\text{-}6)$$

This is called the **Rayleigh–Jeans law**; it is correct to within 1% if $\lambda T > 7.2 \times 10^8$ nm K. The Rayleigh–Jeans law obviously fails at short wavelengths because it predicts the spectral radiance will continue to increase as the wavelength gets shorter and shorter. The failure of the Rayleigh–Jeans law is known as the "ultraviolet catastrophe."

At short wavelengths where $hc/(\lambda kT) \gg 1$, $(e^{hc/\lambda kT} - 1) \approx e^{hc/\lambda kT}$, and equation 4-3 becomes

$$B_\lambda^b = \frac{2hc^2}{\lambda^5} e^{-hc/\lambda kT} \qquad (4\text{-}7)$$

Equation 4-7 is known as **Wien's law**. It is correct to 1% if $\lambda T < 3.1 \times 10^6$ nm K.

## Einstein Coefficients

In 1917, Einstein introduced three probability coefficients, $A_{ji}$, $B_{ji}$, and $B_{ij}$, for transitions between two levels in an atomic system (see also Appendix F). For the

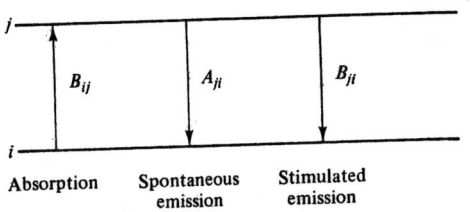

**FIGURE 4-2** Radiative processes in a two-level atomic system. The transition probability for absorption is $B_{ij}$, that for spontaneous emission is $A_{ji}$, while that for stimulated emission is $B_{ji}$.

simple two-level system of Figure 4-2, an atom initially in level $i$ can interact with a radiation field of frequency $v_{ij}$, absorbing the required energy and undergoing a transition to level $j$. The rate of absorption per unit volume (cm$^{-3}$ s$^{-1}$) is proportional to the energy density of the field at the appropriate frequency $U_v$ (J cm$^{-3}$ Hz$^{-1}$) and the population density (atoms per cm$^3$) of state $i$, $n_i$. Or

$$\frac{-dn_i}{dt} = B_{ij} U_v n_i \qquad (4\text{-}8)$$

where $B_{ij}$ is the **Einstein coefficient for absorption** (cm$^3$ J$^{-1}$ s$^{-1}$ Hz). It expresses the number of absorption transitions per second per unit energy density. We assume that $U_v$ is constant over the spectral absorption profile. The units of $B_{ij}$ depend on the units in which $U_v$ are expressed. The product of $B_{ij}$ and $U_v$ is the rate constant for absorption in s$^{-1}$.

An atom in level $j$ can decay to level $i$ by emitting a photon of energy $hv_{ij}$. If the transition is spontaneous, that is, not induced by the radiation field, the emission is **spontaneous emission**. The rate of spontaneous decay per cm$^3$ (s$^{-1}$ cm$^{-3}$) is given by

$$-\left(\frac{dn_j}{dt}\right)_{sp} = A_{ji} n_j \qquad (4\text{-}9)$$

The constant $A_{ji}$ is the **Einstein transition probability** or **Einstein coefficient for spontaneous emission** in s$^{-1}$ and $n_j$ is the population density of state $j$. The reciprocal of the transition probability for spontaneous emission is the **spontaneous emission lifetime** or **radiative lifetime** $\tau_{sp}$ (also called $\tau_r$) and is a characteristic quantity for a given transition. Of course an atom can also decay or be excited by nonradiative processes, but we shall concern ourselves here only with radiative processes.

An atom in level $j$ can also interact with a radiation field of frequency $v_{ij}$, and the transition from level $j$ to level $i$ can be stimulated. This **stimulated emission** or **induced emission** is of the same frequency and phase as the radiation field. The rate of stimulated transitions per unit volume is

$$-\left(\frac{dn_j}{dt}\right)_{st} = B_{ji}U_\nu n_j \qquad (4\text{-}10)$$

where $B_{ji}$ is the **Einstein coefficient for stimulated emission** in $cm^3\ J^{-1}\ s^{-1}\ Hz$. The stimulated emission coefficient $B_{ji}$ is related to the absorption coefficient $B_{ij}$ by $B_{ji}g_j = B_{ij}g_i$, where $g_j$ and $g_i$ are the statistical weights (degeneracies) of the two levels. The statistical weight is related to the orientations of the atom relative to a magnetic field, as discussed in Chapter 7.

Let us consider for the moment a blackbody radiator with two levels $i$ and $j$. For the perfect blackbody the rate of absorption must be exactly balanced at equilibrium by stimulated and spontaneous emission. Hence we can write $B_{ij}U_\nu n_i = A_{ji}n_j + B_{ji}U_\nu n_j$. If we solve for the equilibrium spectral energy density at frequency $\nu_{ij}$ and use $B_{ji} = B_{ij}(g_i/g_j)$, we obtain

$$U_\nu = \frac{A_{ji}n_j}{B_{ij}[n_i - (g_i/g_j)n_j]} \qquad (4\text{-}11)$$

Since the system is at thermal equilibrium, the populations of levels $j$ and $i$ must be related by the Boltzmann distribution $n_j = n_i(g_j/g_i)e^{-h\nu_{ij}/kT}$ (see Section 2-3). If this relationship is substituted into equation 4-11 and we allow a continuous distribution of frequencies $\nu$, the spectral energy density is

$$U_\nu = \frac{A_{ji}}{B_{ij}} \frac{g_j/g_i}{e^{h\nu/kT} - 1} \qquad (4\text{-}12)$$

Equation 4-12 is identical in form to Planck's radiation law (equation 4-1) and a comparison gives the relationship between the Einstein $A$ and $B$ coefficients as

$$A_{ji} = \frac{8\pi h\nu^3 g_i B_{ij}}{g_j c^3} \qquad (4\text{-}13)$$

If stimulated emission is neglected in the steady-state balancing of absorption and emission, it is readily seen that $U_\nu = (A_{ji}/B_{ij})(g_j/g_i)e^{-h\nu/kT}$, and $B_\lambda = [c^2 A_{ji}g_j/(4\pi\lambda^2 B_{ij}g_i)]e^{-h\nu/kT} = (2hc^2/\lambda^5)e^{-hc/\lambda kT}$, which is identical to Wien's law (equation 4-7). Hence the use of Wien's law is equivalent to neglecting stimulated emission.

Einstein coefficients can be related to other measures of transition strength, such as oscillator strength and absorption coefficients. These relationships are discussed in Appendix F.

## Applications to Spectroscopy

The blackbody equations are used to characterize and compare real radiation sources. They are generally used to describe broad spectral distributions from continuum sources or the continuum distribution from sources such as high-pressure arc lamps, which emit a characteristic line spectrum that is superimposed on a broad background continuum as described in the next section. Finally, Planck's law can be applied to describe the emission and self-absorption of atoms in flames under thermal equilibrium conditions (see Chapters 7 and 8).

To describe real sources, often called **gray bodies**, equation 4-3 must be modified to correct for nonidealities. The resulting equation for the spectral radiance of a gray body is

$$B_\lambda = \varepsilon(\lambda)T_w(\lambda)B_\lambda^b \qquad (4\text{-}14)$$

where $\varepsilon(\lambda)$ is the spectral emissivity of the source, $T_w(\lambda)$ is the transmission factor of the lamp window, and $B_\lambda^b$ is the blackbody spectral radiance.

The spectral emissivity $\varepsilon(\lambda)$ is the ratio of the spectral radiance of the real source to that of a blackbody at a specific wavelength and temperature. Values of $\varepsilon(\lambda)$ vary between 0 and 1. In comparison to a blackbody, a gray body does not absorb all the radiation incident upon it. Therefore, $\alpha(\lambda) < 1$ for a gray body, and for a source in thermal equilibrium, $\alpha(\lambda) = \varepsilon(\lambda)$. Thus the spectral emissivity corrects for imperfect absorption by the real source. If a real source is enclosed in an envelope such as glass or quartz, the factor $T_w(\lambda)$ accounts for reflective losses and absorption at the window surface. The transmission factor is wavelength dependent and varies between 0 and 1.

The third modification is the adjustment of the temperature used in equation 4-3 to calculate $B_\lambda^b$. For an ideal blackbody, $T$ is the true temperature of the radiator. In the case of a real source, $T$ is an adjustable parameter that is varied until the wavelength of maximum spectral radiance for the blackbody matches the observed wavelength of maximum spectral radiance. The adjusted temperature is known as the **color temperature**.

To calculate the spectral distribution of a real source from equation 4-14, the color temperature, the spectral emissivity, and the window transmission factor must be known. All of these fitting parameters are wavelength dependent, so that the fit is usually done at a single wavelength or over a small wavelength interval. For some sources, such as tungsten lamps, tables of $\varepsilon(\lambda)$ vs. $\lambda$, $T_w(\lambda)$ vs. $\lambda$, and the color temperature are available. For a tungsten lamp the color temperature is typically 3000 K and $\varepsilon = 0.4$. For a new source, the procedure

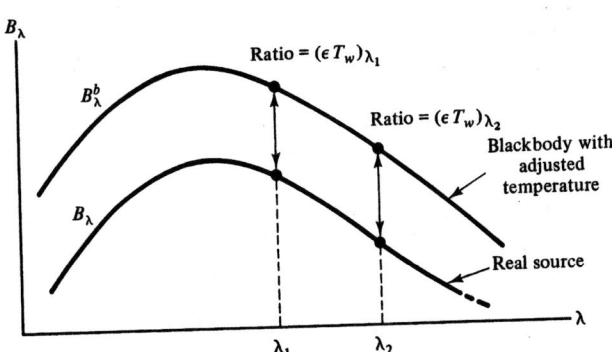

**FIGURE 4-3** Procedure for obtaining spectral distribution of real sources. First the lamp window transmission factor is calculated or measured at specific wavelengths or over the interval of interest. Then the temperature of the blackbody is adjusted so that the maximum spectral radiance occurs at the same wavelength as the observed maximum for the real source. The spectral emissivity $[\epsilon(\lambda) = B_\lambda/(B_\lambda^b \, T_w(\lambda)]$ is calculated for each desired wavelength as shown for $\lambda_1$ and $\lambda_2$. The spectral radiance is then given by equation 4-14.

outlined in Figure 4-3 can be used over a small wavelength interval to obtain the correction factors.

## 4-2 CONVENTIONAL RADIATION SOURCES

Absorption, photoluminescence, and scattering techniques make use of external radiation sources in the spectrochemical encoding process (recall Section 2-6). The source spectral characteristics, wavelength range, radiant flux, directionality, and stability in time and space are all significant parameters in determining the outcome of a spectrochemical procedure. This section considers conventional radiation sources for spectrochemical methods. After first classifying sources and discussing their most important characteristics, the specific sources that are useful in atomic and molecular spectroscopy are presented. Standard sources that are

used in calibrating radiometric and photometric instruments for absolute measurements are also described. Because of their unique operating principles and characteristics, laser sources are treated separately in Section 4-3.

### Characteristics of Sources

Sources for use in spectrochemical methods can be distinguished by the types of spectra they produce. Sources such as blackbodies and incandescent lamps emit radiation over a relatively broad spectral range. The spectral distribution of the radiation produced is said to be a continuum distribution, and such sources are called **continuum sources**. At the other extreme, there are sources, such as hollow cathode discharge lamps, that emit narrow spectral lines, and these are classified as **line sources**. Here the profile (full width at half maximum) and total radiance of a given spectral line are critical. Still other sources produce narrow spectral lines superimposed on a spectral continuum, and these are called **continuum plus line sources**. While the claim is often made that lasers produce "monochromatic" radiation, many lasers, particularly tunable dye lasers, emit a spectral continuum over a relatively narrow spectral interval (i.e., 1 Å). We shall refer to these as **quasi-continuum sources**.

Sources can be further subdivided according to their temporal characteristics. Incandescent lamps, for example, emit radiation continuously with respect to time. Here we use *continuous* to refer to the temporal behavior of a source and *continuum* to refer to its spectral distribution. Other sources emit radiation discontinuously in time and are referred to as pulsed or intermittent sources. Still other subdivisions are possible according to size (point source vs. extended source), wavelength (UV source, visible source, IR source), and spatial behavior (coherent vs. noncoherent).

The characteristics most often considered when selecting a source for a spectroscopic application are listed in Table 4-1.

**TABLE 4-1**
Some important source characteristics

| Characteristic | Examples |
|---|---|
| Spectral output | Continuum, line, continuum plus line |
| Wavelength region | UV, visible, IR, microwave |
| Temporal behavior | Continuous, pulsed, sine wave, coherence |
| Radiance or spectral radiance | |
| Stability | Long term and short term, warm-up time |
| Lifetime | Operating life and shelf life |
| Area of emission | Point source, extended source |
| Spatial behavior | Coherence |

In addition, the power supply requirements, the physical size of the source and its enclosure, as well as the purchase price and the availability of replacements, must be considered. Many of these characteristics are further discussed when specific sources are considered below.

## Continuum Sources

Continuum sources ideally emit a spectral continuum over a broad wavelength region. They are extensively used in molecular spectroscopy and employed in several applications in atomic spectroscopy. The most common of these are incandescent lamps and some arc discharge lamps. Table 4-2 lists the common types of continuum sources that are employed in spectrochemical methods; their construction is illustrated in Figure 4-4.

*Incandescent Lamps.* **Incandescent lamps** are made of resistive materials that are heated electrically. The spectral distribution of these lamps is described by the gray body and blackbody equations (equations 4-3 and 4-14).

The Nernst glower (Figure 4-4a) is a semiconductor material that must be preheated to achieve conduction. Because of a negative temperature coefficient of resistance, it must be used with a ballast resistor in the heating circuit to prevent burnout. The Nernst glower is a useful and inexpensive IR source. It is rather fragile, however, and its lifetime depends on the operating temperature and the care taken in handling it.

The globar is another common source that is more rugged than the Nernst glower. Current through the

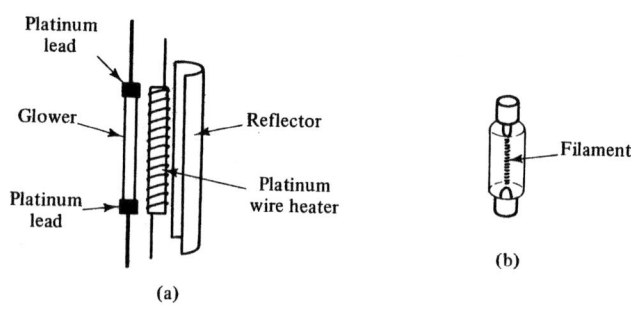

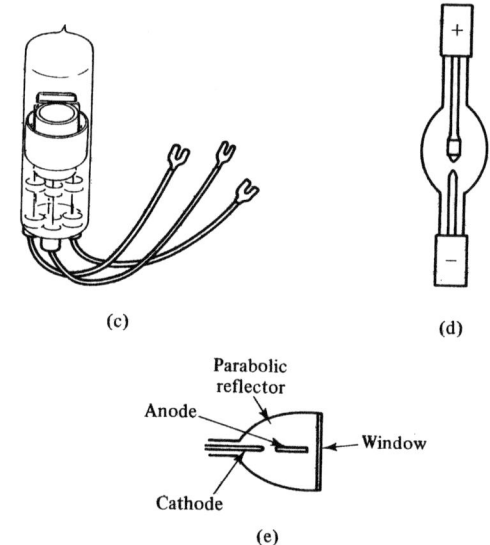

**FIGURE 4-4** Typical continuum sources: (a) Nernst glower; (b) tungsten filament lamp; (c) $D_2$ lamp; (d) conventional Xe arc lamp; (e) EIMAC-type Xe arc lamp with parabolic reflector. The reader should consult Table 4-2 for further details.

**TABLE 4-2**
Common continuum sources

| Type | Radiating material | Window or envelope material | Wavelength range | Approximate[a] spectral radiance ($W\ cm^{-2}\ nm^{-1}\ sr^{-1}$) |
|---|---|---|---|---|
| Nernst glower | Rod of zirconia, yttria, or thoria at 1200–2000 K | None | 0.4–20 μm | $10^{-4}$ |
| Globar | Rod of silicon carbide at 1300–1500 K | None | 1–40 μm | $10^{-4}$ |
| Tungsten | Tungsten filament at 2000–3000 K | Glass | 320–2500 nm | $10^{-2}$ |
| Quartz-iodine ($T \le 3600$ K) | Tungsten filament | Quartz | 200–3000 nm | $5 \times 10^{-2}$ |
| Hydrogen or deuterium | Arc discharge in a few torr of $H_2$ or $D_2$ | Quartz | 180–370 nm | $5 \times 10^{-3}$ |
| Xenon arc | Arc discharge in >10 atm Xe | Quartz | 200–1000 nm | $10^{-1}$ |

[a]Values are rough approximations at specific wavelengths: for Nernst glower and globar, $\lambda = 10$ μm; for tungsten, $\lambda = 500$ nm; for quartz-iodine with iodine scavenger, $\lambda = 400$ nm; for $H_2$, $\lambda = 250$ nm; for Xe arc, $\lambda = 500$ nm (75-W lamp).

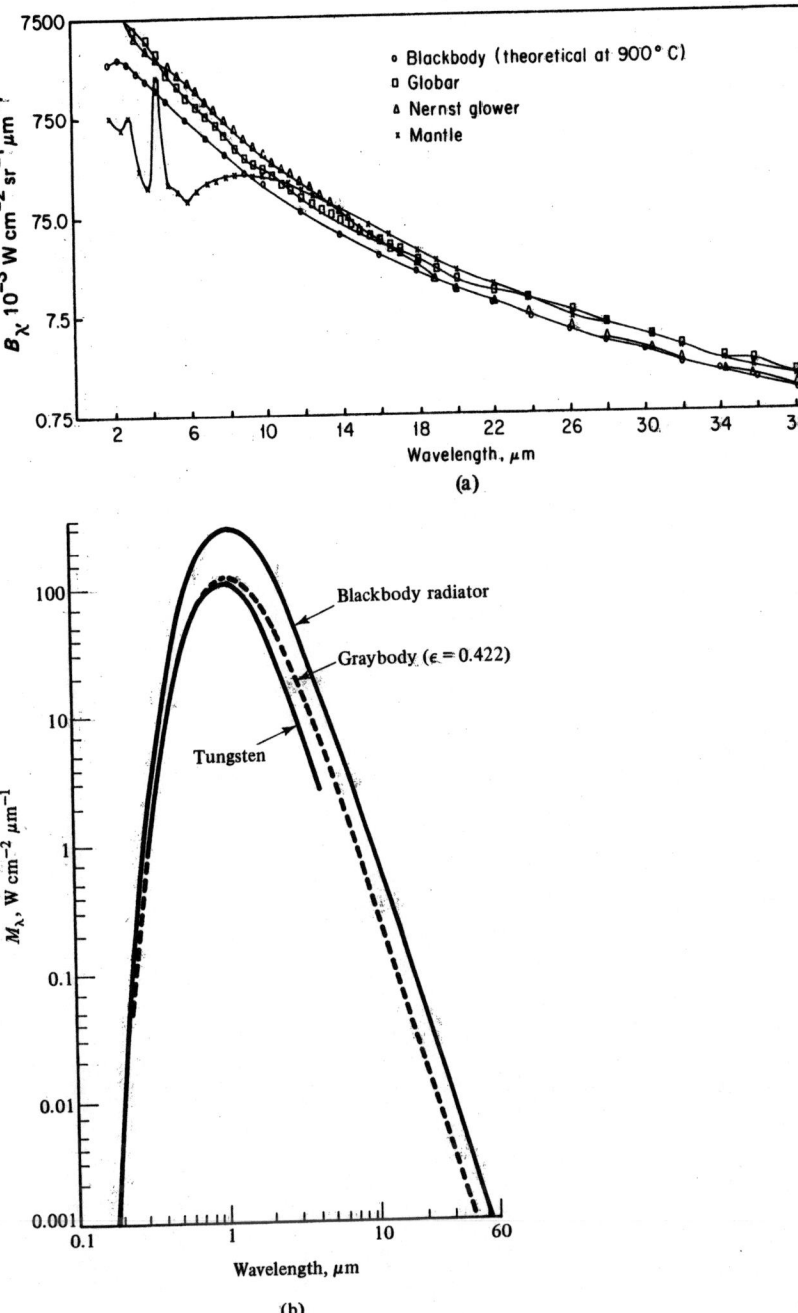

(a)

(b)

**FIGURE 4-5** Spectral distributions of some continuum sources: (a) spectral radiances of Nernst glower, globar, and 900°C blackbody vs. wavelength; (b) spectral emittance of tungsten filament lamp vs. wavelength; (c) spectral irradiance of $D_2$ lamp vs. wavelength measured at 50 cm from source. [(a) Adapted with permission from W. Y. Ramsey and J. C. Alishouse, *A Comparison of Infrared Sources, Infrared Physics,* vol. 8, copyright Pergamon Press, Elmsford, N.Y., 1968, p. 151.]

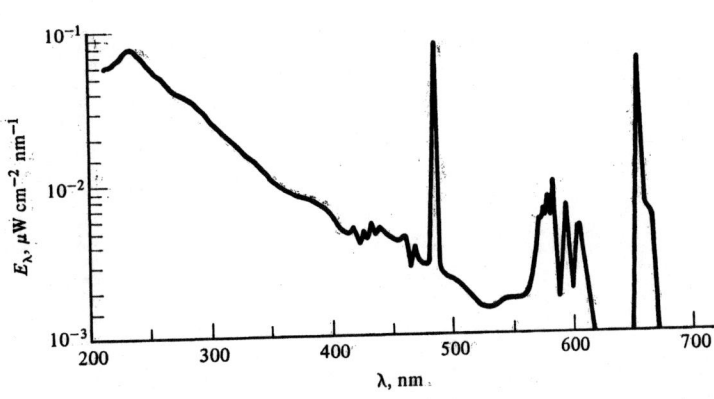

(c)

globar causes the rod to heat and emit radiation at temperatures exceeding 1000°C. The power consumption is normally higher than that of the Nernst glower. Water cooling is needed to cool the metallic electrodes attached to the rod. The spectral distributions of the Nernst glower, the globar, and a 900°C blackbody are shown in Figure 4-5a. The globar has slightly better emissivity at short wavelengths ($\lambda < 6$ μm) than does the Nernst glower. It is less convenient to use and more expensive, however, because of the necessity for water cooling.

Tungsten-filament lamps (Figure 4-4b) are normally used in the visible and near-IR regions. Because of their widespread applications, these lamps are available in a variety of sizes and shapes. Ribbon filament rather than wire filament types are excellent for spectroscopy because the image of the ribbon can be oriented in the plane of a spectrometer slit. The transmission characteristics of the glass envelope limit the wavelength range to 320 to 2400 nm, as shown in Table 4-2 (see also Figure 4-5b). Quartz-halogen lamps have operating lifetimes that are considerably longer than normal tungsten lamps. They can be operated at higher temperatures to provide higher intensities than tungsten lamps. The added iodine reacts with gaseous tungsten near the quartz wall to form $WI_2$ which diffuses to the hot filament and redeposits tungsten. This regeneration process is responsible for the longer lifetime. Because of their higher operating temperatures, quartz-halogen lamps tend to be less stable than normal tungsten lamps unless various feedback stabilization techniques are utilized to control them.

*Arc Lamps.* In an arc lamp an electrical discharge is sustained through a gas or metal vapor enclosed in a sealed glass or quartz envelope. The discharge is initiated and maintained by dc or ac current through the gap between two electrodes (cathode and anode) in the enclosure. Ionization of the vapor within the tube is necessary to achieve conduction. Some discharges operate with a **heated cathode**, and the electron emission from the cathode causes ionization of the gas. With **cold cathode** discharges, ions are formed by cosmic rays, radioactivity, UV radiation, or another external source. To sustain ionization, a potential difference that exceeds the breakdown voltage of the gas is applied. A self-sustaining discharge results when positive ions, accelerated by the field, bombard the cathode to produce secondary electrons. These electrons then collide with gaseous molecules or atoms to maintain ionization in the tube. Since this type of discharge gives a negative dynamic resistance, a ballast resistance is necessary to prevent burnout.

With all arc lamps, starting the arc can be a problem. In some cases an igniter circuit is used to provide a short high-voltage (10 to 20 kV) start pulse to initiate ionization. Some lamps provide a third electrode for ignition. The pressure of the gas in an arc lamp can vary from a few torr to more than 100 atm; the pressure determines the type of discharge and the type of spectral output obtained.

Hydrogen and deuterium arc lamps (Figure 4-4c) provide strong spectral continua in the UV spectral region. As can be seen in Table 4-2, these lamps are useful to wavelengths as short as 185 nm. This short wavelength limit is set by the transmittance characteristics of the quartz envelopes. At wavelengths longer than about 370 nm, the spectra (see Figure 4-5c) from hydrogen and deuterium are no longer continua. Deuterium lamps provide higher intensity than do hydrogen discharges. Spectrophotometric instruments that operate in the UV and visible regions often require two sources: a hydrogen or deuterium lamp is used for the UV region, 200 to 350 nm, and a tungsten-filament lamp is used for the visible region, 350 to 800 nm.

Molecular fluorescence and phosphorescence spectrometry require high-intensity, continuum sources of UV-visible radiation. The xenon arc lamp shown in Figure 4-4d is the most common source for molecular fluorescence spectrometry. It is available in a variety of sizes and shapes. The wattages range from 35 to 10,000. The EIMAC lamp (Varian, Inc.) shown in Figure 4-4e is a xenon arc lamp with the electrodes at the focal point of a parabolic reflector. The beam emerging from this lamp is nearly collimated, which has advantages for certain types of spectroscopy.

In a high-pressure lamp such as a Xe arc lamp, the discharge is constricted to a small volume between the electrodes (typical 2 to 10 mm separation). These lamps are typically operated with a dc power supply of 20 to 30 V and 2 to 20A. Pulsed xenon lamps and power supplies are also available; these provide 1 to 10 μs pulses. Figure 4-6 shows that the spectrum from a Xe arc lamp is effectively a continuum with little fine structure.

## Continuum Plus Line Sources

Other high-pressure arc lamps produce intense line spectra superimposed on an intense spectral continuum. The high pressure and the arc restriction considerably broaden the atomic lines. The most common of these lamps are mercury and xenon-mercury arc lamps. Figure 4-6 compares the spectrum from a xenon arc to that from a mercury arc lamp. The high-pressure mercury arc contains many broad lines from atomic mercury superimposed on a continuum. The spectral radiance is high, exceeding 1 W cm$^{-2}$ nm$^{-1}$ sr$^{-1}$ for the 100-W

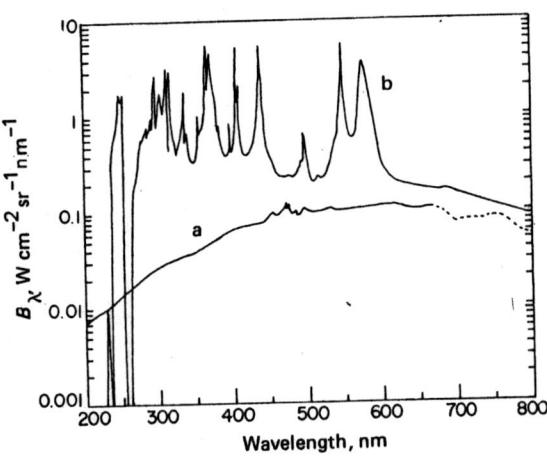

**FIGURE 4-6** Spectral distribution of high intensity arc lamps: curve a, 75-W xenon arc lamp; curve b, 100-W mercury arc lamp. Strong self-absorption (see Chapter 7) causes the lack of radiance at 254 nm. (Adapted with permission from J. D. Winefordner, S. G. Schulman, and T. C. O'Haver, *Luminescence Spectrometry in Analytical Chemistry*, Wiley-Interscience, New York, 1972. Copyright © John Wiley & Sons, Inc., 1972.)

lamp shown in Figure 4-6. Because of the presence of atomic lines, this source is not very useful for molecular absorption spectroscopy, where it is desirable to have a flat spectral distribution; however, it is an excellent source for photoluminescence spectroscopy, especially if a Hg line corresponds to the wavelength of excitation. Mercury arc lamps are available that operate from alternating current or direct current and in wattages up to 2500.

Mercury-xenon arc lamps can provide very high spectral radiances, particularly at wavelengths where mercury emits. They are available in wattages up to 5000. These sources have proven particularly useful in molecular fluorescence spectroscopy, where high intensity is desirable. The presence of the Xe stabilizes the arc compared to an ordinary Hg lamp.

## Line Sources

The line sources used in many atomic spectroscopic applications include low-pressure arc lamps, hollow cathode discharge tubes, electrodeless discharge lamps, and thermal gradient lamps.

*Low-Pressure Arc Lamps.* When an arc discharge is ignited between electrodes in a low-pressure metal vapor atmosphere (<10 mm Hg), a diffuse arc is produced with sharp atomic lines. Metal vapor discharge lamps with volatile elements such as Hg, Cd, Zn, Ga, In, Th and the alkali metals are available commercially and used in some forms of spectroscopy.

The most widely used low-pressure arc lamps are mercury arcs. In a tube filled with mercury vapor at reduced pressure, the dominant line is the 2537-Å line, and these sources are often selected for their ultraviolet emission characteristics. Germicidal lamps are of the

hot-cathode variety and operate at relatively low voltages (105 to 150 V ac). They differ from conventional fluorescent lamps in that they are made to transmit ultraviolet radiation. In a normal fluorescent lamp the wall coating absorbs ultraviolet radiation and reemits visible light.

Mercury Pen-Ray or Sterilamp types are cold-cathode types that start and operate at higher voltages (200 to 410 V ac operating voltage) than the hot-cathode variety. These small pencil-shaped lamps are available commercially; most can be obtained with an appropriate transformer and ballast for starting and sustaining the discharge. The small sizes of these lamps and the sharpness of the atomic lines made them excellent sources for wavelength calibration of spectroscopic equipment, particularly in the ultraviolet region.

*Hollow Cathode Discharge Tubes.* The hollow cathode lamp (HCL) has become the dominant atomic spectral line source. Hollow cathode lamps have undergone extensive commercial development for atomic absorption spectroscopy; consequently, they are available for the majority of the elements in the periodic table. Hollow cathode tubes can be purchased as single-element lamps or as multiple-element lamps. A typical hollow cathode design is illustrated in Figure 4-7. Application of a potential difference across the tube causes ionization of the filler gas. The positive ions strike the interior of the cathode and volatilize the cathode material by cathodic sputtering. A cloud of atomic vapor forms around the cathode. The atoms in the cloud are excited by collisions with rare gas atoms and electrons and emit characteristic lines of the element. Hollow cathode lamps also produce lines that arise from the filler gas and from impurities. Optical windows of Pyrex and quartz are used in commercial lamps.

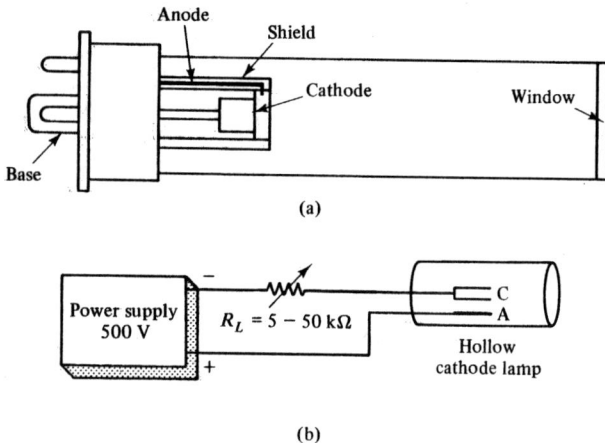

(a)

(b)

**FIGURE 4-7** Shielded hollow cathode lamp (a) and power supply circuit (b). The cathode in (a) is a cuplike hollow cylinder made of or lined with the element or an alloy of the element to be excited. A simple wire serves as the anode. The envelope is filled with neon or argon at a pressure of a few torr. Commercial lamps are available for most elements. Typically, a few hundred volts is applied between the cathode and anode (b).

Hollow cathode lamps can produce extremely narrow atomic lines. When operated properly, the atomic line widths are only 0.01 to 0.02 Å (FWHM). These line widths are narrower than absorption line widths in flames, as discussed in Chapter 7. A typical spectrum from a hollow cathode lamp (HCL) is shown in Figure 4-8.

Unlike most sources, provision is usually made to vary the lamp current in order to vary the line radiance. The operating currents of HCLs should not exceed the maximum current rating specified by the manufacturer. These maximum values vary from element to element

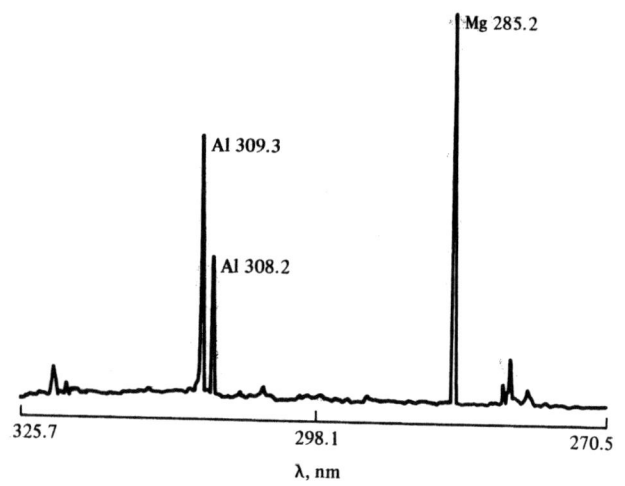

**FIGURE 4-8** Portion of spectrum from a dual-element hollow cathode lamp.

but are usually in the range 10 to 50 mA. High currents can cause appreciable line broadening and self-reversal (see Section 7-5). Operation at high currents can significantly reduce the lamp lifetime. Manufacturers of atomic absorption spectrometers often recommend operating hollow cathodes at currents substantially lower than the maximum values. High intensity or boosted output hollow cathode lamps are also available for many elements. The intensities of these lamps can be severalfold higher than normal HCLs.

For use with synchronous detection systems, HCLs can be electrically modulated or a mechanical chopper can be used. Lock-in detection systems (see Section 4-5) require 50%-duty-cycle modulation, as is commonly used in atomic absorption spectrometry. (The duty cycle is the fraction of the time the lamp is on.) With electrical modulation schemes the hollow cathode current is often not turned completely off, to enhance lamp stability. For atomic fluorescence spectroscopy dc operation of HCLs does not produce sufficient intensity. High intensities can be obtained, without exceeding the maximum current ratings, by pulsing the hollow cathode in a low-duty-cycle mode. Here the current during the lamp on-time can be high, but because of a long off-time, the average current is maintained below the maximum rating. Boxcar integration schemes are often used with pulsed HCLs (Section 4-5).

*Electrodeless Discharge Lamps.* The electrodeless discharge lamp (EDL) is a very intense atomic line source, often producing line radiances 20 to 50 times greater than those of a hollow cathode lamp. Because of this EDLs have found use in atomic fluorescence spectroscopy and in atomic absorption for elements such as As, Se, Te, and Hg, where HCL intensities are relatively low.

The EDL is basically a sealed quartz tube, as shown in Figure 4-9, that contains a few torr of an inert gas and a small amount of the metal (or metal salt) whose spectrum is desired. The lamp is placed in an intense RF or microwave field that is directed on the lamp by an antenna or a waveguide cavity. A Tesla coil is used to start the discharge by ionizing some of the inert gas atoms. The electrons produced are accelerated by the high-frequency component of the field and acquire sufficient energy to ionize other atoms and maintain a plasma. (A plasma is an ionized gas.) The heat produced vaporizes the metal or metal compound, which is excited by collisions to produce the spectrum. Because the gas pressure and temperature are low, broadening effects are small, and narrow spectral lines result.

The microwave or RF generator employed with EDLs is typically a 50- to 1000-W supply operating at 2450 MHz. Although lamps are available commercially

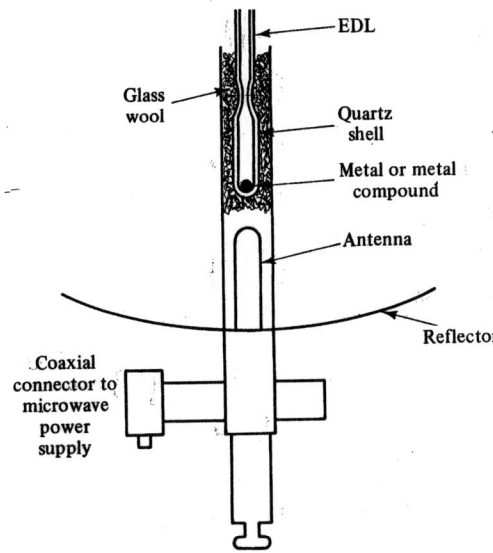

**FIGURE 4-9** Electrodeless discharge lamp. The sealed tube is placed in an intense radio-frequency (RF) or microwave field.

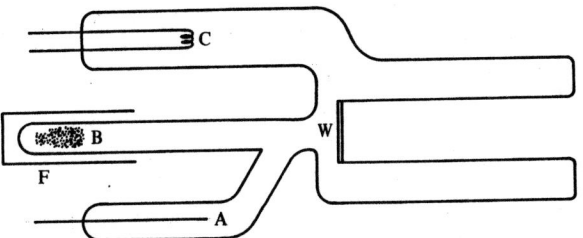

**FIGURE 4-10** Commercial thermal gradient lamp. The lamp is made of glass with a silica viewing window (W) and contains a few torr of argon filler gas. The element to be excited is placed in bulb B, which is heated by furnace F. A discharge occurs between cathode filament C and anode wire A. The discharge ($\approx 0.5$ A) through the atomic vapor provides intense resonance lines of the element.

for some elements, EDLs can be readily produced in the laboratory. Unfortunately, there is still a great deal of art in the preparation of stable, reliable lamps for general-purpose use. Thermostating the lamp often improves stability. Some EDLs have been operated in a pulsed mode.

The use of EDLs in atomic fluorescence spectrometry has declined in recent years due to the availability of lasers. These increasingly important radiation sources are discussed in detail in Section 4-3.

*Thermal Gradient Lamp.* The thermal gradient lamp (TGL) is a relatively new atomic line source that can be constructed for elements of high volatility (As, Cd, P, S, Se, Te, Zn). A commercial TGL is shown in Figure 4-10. Intensities of TGLs are as high as those of EDLs and the lines produced can be narrower. The TGL has a much shorter warm-up time than an EDL of the same element. Because of these factors TGLs could replace EDLs in atomic absorption determinations of As, Se, and Te, and they could become popular atomic fluorescence sources for some elements.

## Miscellaneous Sources

Light-emitting diodes (LEDs) have been used as sources in fixed-wavelength, molecular absorption spectrometers. They are *pn*-junction devices that when forward biased produce radiant energy. Gallium arsenide ($\lambda_m \approx 900$ nm), gallium phosphide ($\lambda_m \approx 650$ nm), and silicon carbide ($\lambda_m \approx 580$ nm) are commonly used. Mixtures of these compounds can be used to shift the wavelength maximum anywhere in the region 540 to 900 nm. The

spectral distribution of an LED is a continuum over a narrow wavelength range (FWHM typically 30 to 50 nm). They are usually operated at low voltages (e.g., 5 V) with currents in the range 1 to 260 mA. Their major advantage is their small size.

Fluorescent lamps are not often used as spectrometric sources, but they are commonly used to initiate photochemical reactions. Fluorescent lamps are discharge devices that produce light by the emission from luminescent powders excited by a low-pressure mercury arc. The spectral distribution of the light is a function of the chemical composition of the phosphor used.

## Standard Sources

Standard sources are used to calibrate monochomators, detectors, and secondary sources for absolute radiometric or photometric measurements. The primary standard photometric source is a blackbody constructed from a fused-thoria ($ThO_2$) crucible that is nearly filled with pure platinum. A small cylinder of pure fused thoria is placed in the center of the crucible. The bottom of the cylinder is packed with powdered fused thoria. The crucible has a small hole in the center from which the light emerges through a funnel-shaped sheath. The crucible is embedded in powdered fused thoria and is heated by enclosing it in an induction furnace. The temperature of the crucible is first raised to well above the melting point of platinum. Then it cools slowly until the solidification temperature (2042 K) is reached. The luminance of the hole at the solidification temperature of platinum is given the value of 60 lm sr$^{-1}$ cm$^{-2}$.

The National Bureau of Standards (NBS) has established working standards of spectral radiance and spectral irradiance based on calibrations relative to blackbody sources. The standards of spectral radiance cover the wavelength region of 250 nm to 2.5 μm. The sources are commercial GE type 30 A/T24/7 tungsten ribbon filament lamps centered 8 to 10 cm behind a fused silica window that is 3 cm in diameter. Values of spectral radiance for these lamps are given as a function of wavelength. They are intended for use with auxiliary optics of specified arrangement. Since these standards are expensive they are used infrequently to calibrate other working standards. When operated properly, the uncertainties in spectral radiance are less than 10%.

Tungsten-filament, quartz-iodine lamps are issued as standards of spectral irradiance for the wavelength region 250 nm to 2.6 μm. They are commercial GE type 6.6 A/T4Q/1CL-200-W lamps. Values of spectral irradiance vs. wavelength are supplied. The irradiance standards require no auxiliary optics. Again these standards are expensive and should be used to calibrate other, less expensive, working standards. The uncertainties in the spectral irradiance standards are also less than 10%.

The primary standard of length is the wavelength in vacuum of the radiation of the orange line of $^{86}$Kr unperturbed by external influences. The accepted wavelength is $\lambda = 6057.80210_5$ Å. The emission is produced in a hot-cathode discharge tube filled with high-purity $^{86}$Kr at a temperature of 64 K.

Despite the existence of working standards for absolute photometry and radiometry, an accuracy of 5% is difficult to achieve and usually obtained only by experts. Accuracies approaching 1% are very rare and realized only in a few types of measurements within a very few laboratories. Fortunately, there are not very many spectrochemical methods that require even an estimate of absolute intensities because relative intensities are used in most procedures.

## 4-3 LASER SOURCES

In spectroscopy laser sources are making it possible to obtain information that was difficult or impossible to obtain with conventional sources. This section presents the basic principles of lasers and describes the types of lasers that are useful in spectrochemical methods. Some of the unique characteristics of laser sources are presented along with the principles of tunable lasers. Applications of lasers to various types of spectrochemical methods are described in later chapters that deal with specific techniques.

### Principles of Lasers

Let us consider a collection of atoms or molecules interacting with an electromagnetic wave that is propagating along the *z* axis. If the wave is of the appropriate

frequency to cause stimulated transitions in the system, each stimulated emission generates a photon while each absorption removes a photon. The change in radiant flux $d\Phi$ in the distance $dz$ due to absorption is given by $d\Phi = -\Phi n_i \sigma dz$, where *s* is the transition cross section in cm² (see Section 3–1 and Appendix F). The change in flux due to stimulated emission in distance $dz$ is $d\Phi = \sigma n_j \Phi \, dz$. Thus we can write the total change in radiant flux in distance $dz$ as

$$d\Phi = \sigma \Phi (n_j - n_i) \, dz \qquad (4\text{-}15)$$

Equation 4-15 shows that there is a net gain in flux when $n_j > n_i$ and a net loss in flux with $n_j < n_i$. When the system is at thermodynamic equilibrium, $(n_i)_{eq} > (n_j)_{eq}$, and the material is a net absorber. However, if the upper-level population can be made to exceed that of the lower level ($n_j > n_i$), the system behaves like an amplifier at frequency $v_{ij}$. Under these conditions, the atomic or molecular system is called an **active medium**, and it has undergone **population inversion**. In the IR-to-visible frequency region, such an amplifier is called a *laser* for *l*ight *a*mplification by *s*timulated *e*mission of *r*adiation.

As in an electronic circuit, we can make an oscillator from an amplifier by introducing positive feedback. This can be done with a laser by placing the active medium in a resonant cavity (Fabry–Perot cavity) made from two mirrors, as shown in Figure 4-11. When the gain of the active medium equals the losses in the system, oscillation begins. If we assume that the major

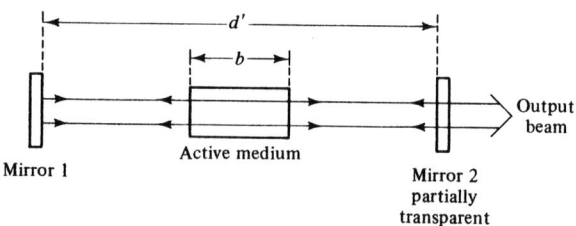

**FIGURE 4-11** Laser cavity. In the laser the active medium is positioned between two mirrors. The length of the cavity is distance $d'$, while the optical pathlength is $d$. Here $d$ is given by $d = \eta b + \eta'(d' - b)$, where $\eta$ is the refractive index of the active medium and $\eta'$ is that of the remainder of the cavity. The electromagnetic wave travels back and forth between the mirrors being amplified at each pass through the medium. A partially transparent mirror allows part of the beam to pass out of the cavity as the output. Although plane parallel mirrors are shown, spherical mirrors are normally used to minimize diffraction losses and improve beam quality.

losses occur because the mirrors are imperfect reflectors, the condition for oscillation is given by

$$\rho_1 \rho_2 G^2 = 1 \qquad (4\text{-}16)$$

where $\rho_1$ and $\rho_2$ are the reflectances of mirrors 1 and 2, respectively, and $G$ is the gain (per pass) of the medium. Here the square of the gain appears because the beam must pass through the active medium twice to make a complete round trip in the cavity. The gain of the medium is related to the transition cross section and the population difference by

$$G = e^{\sigma(n_j - n_i)b} \qquad (4\text{-}17)$$

where $b$ is the length of the active medium. The threshold population inversion can be obtained by substituting $G$ from equation 4-17 into equation 4-16 and solving for $n_j - n_i$. The result is

$$(n_j - n_i)_{th} = \frac{\ln (1/\rho_1\rho_2)}{2\sigma b} \qquad (4\text{-}18)$$

The population inversion can be achieved by several different schemes. In a two-level system, it is not possible to achieve inversion since stimulated emission and absorption must be balanced. Hence systems with three or more levels are used as shown in Figure 4-12. The excitation process, called **pumping**, can be by optical methods with other sources, by electrical methods using an electrical discharge, by chemical reactions, or by a rapid adiabatic expansion. Once inversion is achieved, oscillation can be initiated by spontaneous emission processes. All lasers, of course, have inefficiencies, so that the output energy is significantly less than the pump energy. Even the most efficient lasers may have overall efficiencies of 20% or less. However, as discussed below, the characteristics of laser radiation are remarkably different from those of conventional sources.

## Laser Types

There are many different types of lasers available and several different operational modes.

*Operational Modes.* The four major operational modes are continuous wave, pulsed, $Q$-switched, and mode-locked. In a **continous-wave (CW) laser** the output beam is continuous with respect to time. To achieve CW operation, the upper level of the lasing transition must be longer lived than the lower level. If the upper level were to have a shorter lifetime than the lower level, the laser would be **self-terminating**. Self-

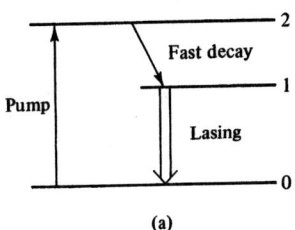

(a)

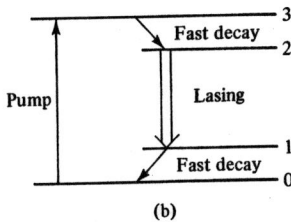

(b)

**FIGURE 4-12** Three- and four-level lasing schemes. In the three-level scheme (a), atoms are excited (pumped) in an appropriate manner to the upper level (2), from which they rapidly decay to an intermediate level (1). If the intermediate level is long-lived compared to the upper level, the population of level 1 can grow relative to level 0 until inversion is achieved. In the four-level scheme (b), the atoms are excited to level 3 from which they rapidly decay to level 2. Population inversion occurs between levels 2 and 1. It is easier to achieve the inversion in the four-level system.

terminating lasers can be operated only as **pulsed lasers**. The maximum useful pump pulse width is on the order of the lifetime of the lasing transition. In a pulsed laser, the output appears as a repetitive pulse train or the pulsed laser can be operated in a single-shot mode. The radiant power output of a pulsed laser, during the pulse, can be very high (kW to > MW) even though the average power is low. Output characteristics of pulsed lasers can be expressed as output energy per pulse, peak output power, or average power. For a laser that produces 5-ns pulses at a repetition rate of 50 pulses $s^{-1}$ with an energy of 10 mJ per pulse, the peak output power is $10 \times 10^{-3}$ J/$(5 \times 10^{-9}$ s) = 2 MW. The average power is $10 \times 10^{-3}$ J/pulse $\times$ 50 pulses/s = 0.5 W. The radiant power outputs of CW lasers are from milliwatts to watts.

With CW and pulsed lasers the population inversion cannot exceed the threshold value to any large extent because lasing occurs when the threshold is reached. In a **Q-switched laser**, often called a **giant pulse laser**, the quality factor ($Q$) of the cavity is spoiled and oscillation is prevented until the population inversion has grown well beyond the threshold value. The cavity is suddenly switched on, and a powerful giant pulse is obtained. Although there are several ways to accomplish $Q$-switching, one simple method employs a **saturable absorber** in the cavity. The saturable absorber is often an organic dye in solution. Until the radiation in the cavity builds up to a high intensity, the dye absorbs the light and prevents laser action. The cavity thus has high losses, and the population inversion must become very large to overcome these losses. When the radiance in the cavity reaches a critical level, the absorber becomes bleached (saturated) and transparent to the light. If the population inversion is high at this time, a very intense laser pulse is produced. Laser output powers on the order of hundreds of megawatts per pulse can be achieved. The $Q$-switched laser can produce a single pulse or a repetitive pulse train.

Before we can understand the fourth operational mode, it is necessary to consider the mode structure of a normal laser beam. In a resonant cavity such as that shown in Figure 4-11, standing waves occur when there is an integral number $n$ of half-wavelengths spanning the optical pathlength $d$ between the mirrors. Consecutive longitudinal (axial) modes, each with a distinctive frequency $v_n = nc/2d$, are separated by frequency differences $\Delta v_n = c/2d$. These longitudinal modes correspond to standing waves set up along the $z$-axis. In addition transverse modes, called TEM for transverse electric and magnetic, can be sustained as well. The transverse modes are given the symbol $\text{TEM}_{ml}$, where $m$ and $l$ are the integer number of transverse nodal lines in the $x$ and $y$ direction across the beam. The light patterns for several modes (observed at the mirrors)

are shown in Figure 4-13. The lowest order $\text{TEM}_{00}$ mode is the most widely used because the flux is Gaussian over the cross section of the beam. (Emittance decreases from the center of the beam cross section according to the Gaussian equation; see Appendix A.) A complete mode specification is of the form $\text{TEM}_{mln}$, where $n$ is the axial mode number. For each $\text{TEM}_{ml}$ mode there are many longitudinal modes each separated by a frequency $\Delta v_n = c/2d$. For a gas laser with a 1-m cavity, $\Delta v_n = 150\,\text{MHz}$. This frequency difference is much smaller than the line width of the atomic transition; hence many modes will oscillate simultaneously.

In **mode locking** the phases of the oscillating modes are forced to be correlated or locked to each other rather than random. This results in ultrashort pulses (picoseconds or less) of very high peak power. Often the phase locking is accomplished by placing a saturable absorber or an acousto-optic modulator (see Section 3-3) in the cavity. If $n'$ modes oscillate in the presence of a saturable absorber, the cavity loss contains a term that oscillates at $\Delta v'$, the beat frequency of adjacent modes. When the absorber bleaches, pulses are produced in which the phases of the modes are locked. The duration of each pulse $t_p$ is $t_p \approx 1/\Delta v' \approx 2d/[c(2n' + 1)]$. The temporal behavior of such a $Q$-switched, mode-locked laser is shown in Figure 4-14.

*Solid-State Lasers.* The first operating laser was a pulsed laser which had as its active medium a ruby crystal. A simplified energy-level diagram of the active

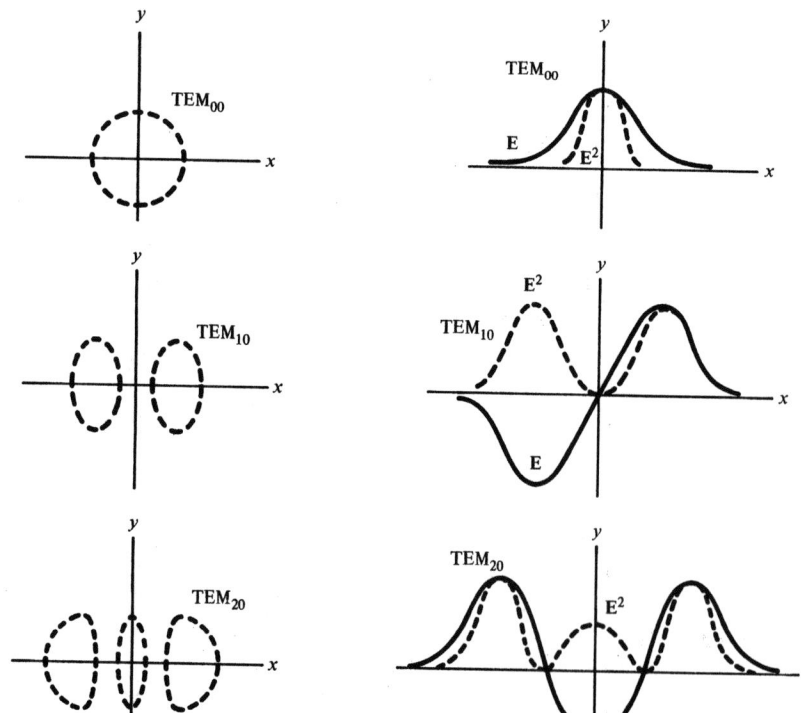

**FIGURE 4-13** Dot, field **E**, and irradiance $\mathbf{E}^2$ patterns for several modes. Note that for the $\text{TEM}_{00}$ mode there are no nodes in either the $x$ or the $y$ directions ($m = 0, l = 0$). For the $\text{TEM}_{10}$ mode there is a single node in the $x$ direction, while for the $\text{TEM}_{20}$ mode there are two nodes in the $x$ direction. The total number of "dots" in a mode pattern is $(m + 1)(l + 1)$. Note also that the $\text{Tem}_{00}$ mode gives a Gaussian irradiance distribution.

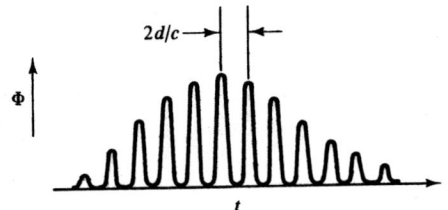

**FIGURE 4-14** Output radiant power vs. time for a *Q*-switched, mode-locked laser. Note that the output consists of short pulses separated by a time difference of $2d/c$, which is one round-trip transit time. The pulse width (FWHM), $t_p$, is approximately $1/\Delta\nu$, where $\Delta\nu$ is the laser line width. For a fairly broadband laser, such as a Nd:YAG laser, where $\Delta\nu \approx 3 \times 10^{12}$ Hz ($\Delta\lambda \approx 100$ Å), $t_p$ is approximately 0.3 ps. Even for a narrow line width laser, such as a He-Ne laser with $\Delta\nu \approx 1500$ MHz ($\Delta\lambda = 2 \times 10^{-2}$ Å), $t_p$ is only 0.6 ns. The peak power is approximately $n'$ times the average power, where $n'$ is the number of locked modes.

$Cr^{3+}$ ion in a ruby laser is shown in Figure 4-15. The ruby rod is surrounded by a flashlamp which provides broadband, pulsed optical pumping. The ends of the rod are silvered to form a resonant cavity. The ruby laser emission occurs at 694.3 nm. The laser pulse is $\approx 0.5$ ms in duration and has a line width of $\approx 0.01$ nm. Typical peak output power is 500 kW, but *Q*-switching has been used to produce peak powers exceeding 150 MW per pulse.

In addition to the ruby laser there are many other solid-state lasers. Of these $Nd^{3+}$ in a glass host or in a yttrium–aluminum–garnet (YAG) host is a particularly

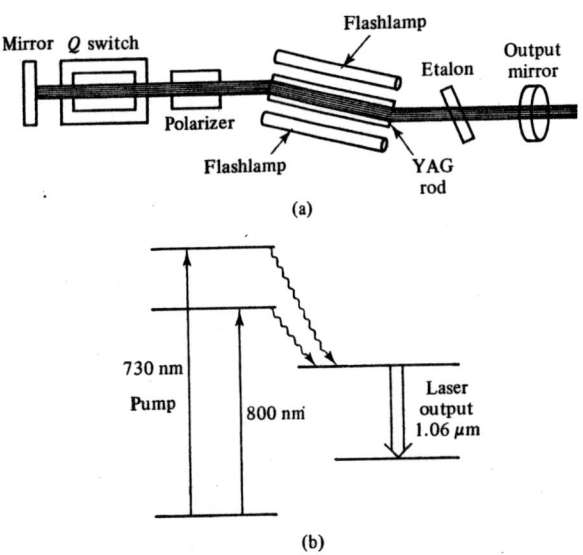

**FIGURE 4-16** Construction (a) and energy levels (b) of a Nd:YAG laser. The pump transitions are in the red region of the spectrum, while the laser output is in the near IR (1.06 μm).

important laser material. The active material is shaped as a rod and is optically pumped. The Nd:YAG laser has been operated in both pulsed and CW modes. The pulsed Nd:YAG laser has a pulse duration on the order of nanoseconds and very high output power at 1.06 μm. It has become a popular laser for pumping tunable dye lasers, as discussed later in this section. The lasing scheme is illustrated in Figure 4-16. Mode-locked Nd:glass lasers can produce pulses as short as 1 ps and peak output powers as high as $10^{13}$ W.

*Gas Lasers.* The helium-neon (He-Ne) laser shown schematically in Figure 4-17 is the most common of all lasers. It is a CW laser pumped by an electrical discharge. The lasing transitions occur between the Ne energy levels, while He is added to increase the pumping efficiency. The helium atoms are ionized to maintain the discharge. Metastable levels in He atoms transfer energy efficiently to Ne. The dominant laser transitions are at 632.8 nm, 1.15 μm, and 3.39 μm. The outputs of many He-Ne lasers as well as other noble gas lasers are linearly polarized as described in Figure 4-17. Simple He-Ne lasers made for use as alignment aids are available without Brewster angle windows and thus with unpolarized outputs. The He-Ne laser is simple and inexpensive. The output power is limited, however, to the range 0.5 to 50 mW because at low currents few metastables form, whereas at high currents the He metastables ionize.

All of the noble gases have been made to lase. The argon ion laser can oscillate at several frequencies, with the most intense outputs being at 488.8 nm in the blue and 514.6 nm in the green. The output power is

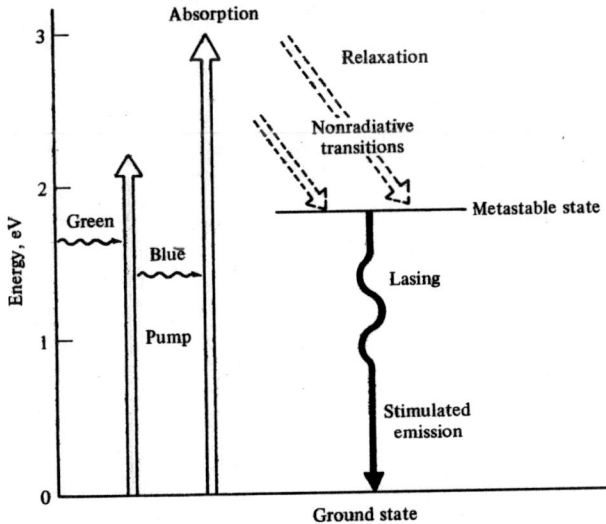

**FIGURE 4-15** Energy levels in a ruby laser. The atomic system consists of absorption bands, a metastable (long-lived) level, and the ground state. The ruby system thus behaves as a three-level laser system.

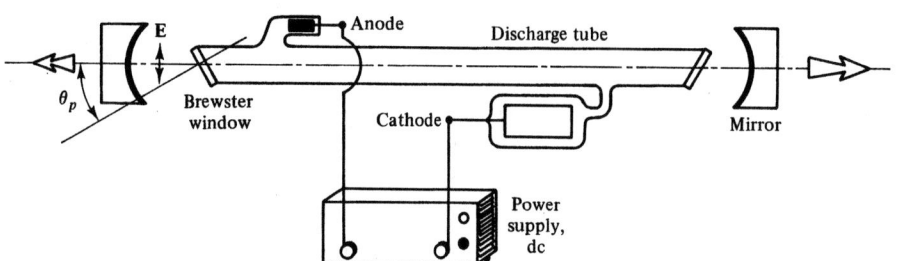

**FIGURE 4-17** Polarized output He-Ne laser. The discharge tube is terminated by end windows tilted at Brewster's angle (see Section 3-1). Thus no reflections occur at the windows for light polarized with the electric field vector in the plane of the figure. The orthogonal component is partially reflected on each pass and quickly becomes an insignificant part of the output beam.

usually several watts CW, but $Ar^+$ lasers have obtained powers as high as 150 W CW. Argon ion and He-Ne lasers are the most reliable lasers at this time.

The $CO_2$ laser emits in the infrared region at 10.6 μm. It is, at present, the most powerful CW laser available; output powers of several hundred kilowatts have been obtained. The $CO_2$ laser can be pumped by collisions with electrons, or by resonant energy transfer from added $N_2$ molecules. The laser output occurs on a series of $CO_2$ rotational lines within a vibrational transition.

The $N_2$ laser utilizes a vibronic transition of the nitrogen molecule at 337.1 nm. It is a pulsed laser that is self-terminating and requires a very short current pulse for pumping. The typical arrangement is the Blumlein circuit, shown in Figure 4-18. The $N_2$ laser is a common pump laser for tunable dye lasers as discussed later.

Several metal vapors (Zn, Hg, Sn, Pb, Cu, Cd) have been made to lase in the visible region, but at present only the Cd vapor laser is in common use. The copper vapor laser is now commercially available. The He-Cd laser emits at 325.0 and 441.6 nm. It utilizes transitions of cadmium excited by collisions with metastable helium.

Excimer lasers are among the newest gas lasers; the rare-gas halide excimer laser was discovered in 1975. Laboratory-type lasers are pumped by electrical discharges. A gaseous mixture such as Ar, $F_2$, and He when subjected to the discharge creates ArF excimers (an excimer is an excited dimer or trimer). A unique property of rare gas halide excimers is that they are dissociated in the ground state. This makes it easy to maintain a population inversion because the ground state of the excimer rapidly dissociates.

Some of the output characteristics of excimer lasers are summarized in Table 4-3. Typically, pulse durations are on the order of 10 to 30 ns with repetition rates as high as 500 Hz.

The excimer laser is inherently tunable over a narrow wavelength range (few nm). Tuning has been accomplished by inserting a prism inside the resonant cavity. The excimer laser is an important ultraviolet radiation source, especially for photochemical studies. Various wavelengths in the UV are available from different media, as shown in Table 4-3. The excimer laser is becoming especially useful for pumping tunable dye lasers. Because of its output wavelength at 308 nm and its long gas lifetime ($>10^7$ shots per gas filling), the XeCl laser is the most suitable excimer laser for pumping dye lasers. The excimer laser has relatively poor beam quality.

*Dye Lasers.* The first liquid laser was successfully operated in 1963, while the first dye laser was introduced in 1966. Since then a great many fluorescent

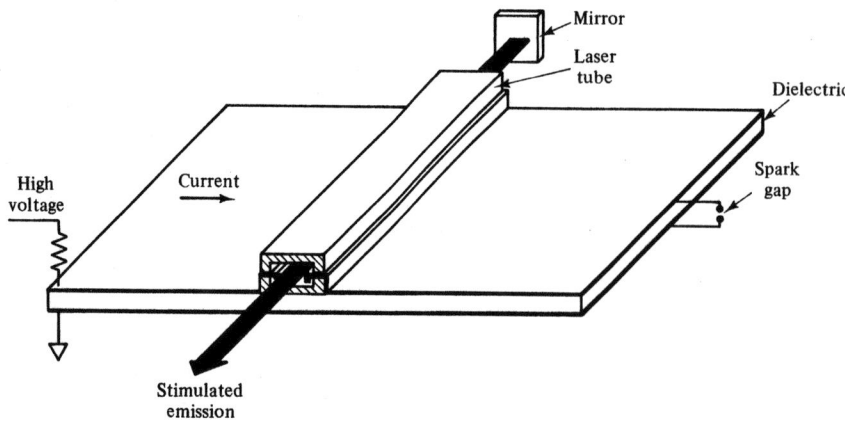

**FIGURE 4-18** Nitrogen laser schematic illustrating a Blumlein pumping circuit. A high current discharge ($\approx 10^5$ A) passes briefly ($\approx 1$ to 5 ns) through the laser tube containing $N_2$ gas. This produces an inverted population. The lasing begins from spontaneous emission, and the inverted population depletes rapidly. Because of the high gain only one mirror opposite the output is needed. Peak output powers as high as megawatts can be obtained. The pulse duration is a few nanoseconds with repetition rates from 1 to 60 Hz.

**TABLE 4-3**

Typical excimer laser characteristics

| Laser medium | Wavelength (nm) | Output energy per pulse (J) | Average output power (W) |
|---|---|---|---|
| ArF | 193 | 0.2–0.3 | 10 |
| KrCl | 222 | 0.03 | 1 |
| KrF | 248 | 0.3–0.4 | 18 |
| XeCl | 308 | 0.08–0.2 | 8 |
| XeF | 351 | 0.08–0.15 | 7 |

dyes (rhodamines, coumarins, fluoresceins, etc.) have been made to lase at frequencies from the IR into the UV region. Energy-level schemes of fluorescent molecules are discussed in Section 12-6. Because of frequent collisions in solution, the spectra of such molecules are fairly broad. Lasing typically occurs between the first excited singlet state and the ground state. The oscillation occurs over a continuous range of wavelengths often on the order of 40 to 50 nm. This broadband oscillation makes the dye laser highly suitable for tuning as discussed below.

Dye lasers are pumped optically by a flashlamp or another laser. Pulsed operation occurs with flashlamp pumping or by pumping with a pulsed laser ($N_2$, Nd:YAG or excimer). Some typical characteristics of pulsed dye lasers are shown in Table 4-4. Because of their UV outputs, the $N_2$ laser and the XeCl excimer laser are suitable for pumping dye lasers directly. A typical dye laser pumped by an $N_2$ laser is illustrated in Figure 4-19. The cavity is formed by a dispersive device (grating) and an output reflector (output coupler). The wavelength selector restricts oscillation to a narrow spectral interval; tuning is accomplished by turning the grating. Wavelength tuning in dye lasers can also be accomplished with prisms, interference filters, and Fabry–Perot etalons. Often more than one dispersive element is used to decrease the bandwidth of the output beam. With the Nd:YAG laser, the fundamental frequency (1.06 μm) is unsuitable for pumping most dyes. Thus the frequency must be increased by doubling, tripling, or even quadrupling with nonlinear optical techniques (see later in this section). These conversion processes can be performed with efficiencies up to 30%; however, they add a great deal of complexity to the laser system.

Continuous-wave operation is also readily available for many dyes pumped with noble gas ion lasers.

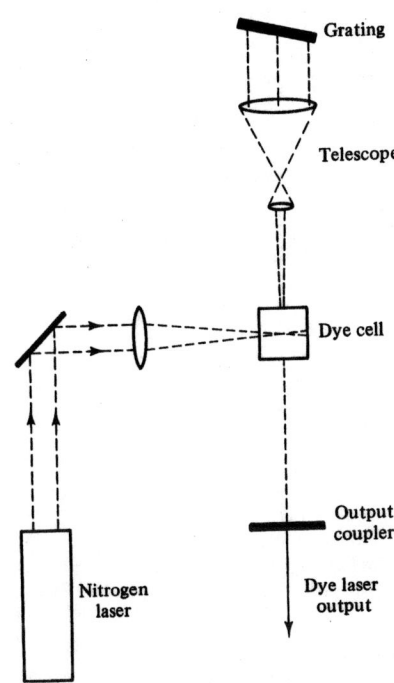

**FIGURE 4-19** Dye laser pumped by $N_2$ laser. The beam expanding telescope increases the number of grooves of the grating hit by the beam and decreases the power density so as to avoid damage to the grating. Tuning is accomplished by rotating the grating.

**TABLE 4-4**

Typical pulsed dye laser characteristics

| Pump source | Range of tunability (nm) | Peak power (kW) | Pulse duration (ns) | Repetition rate (Hz) |
|---|---|---|---|---|
| Flashlamp | 220–960 | 100–500 | 250–750 | 1–10 |
| $N_2$ | 400–970 | 1–100 | 1–8 | 1–100 |
| Nd:YAG | 195–500 | 100–10,000 | 5–10 | 1–30 |
| Excimer | 217–970 | 100–1000 | 10–20 | 1–500 |
| CW[a] | 400–1000 | 0.1–5 | 0.015 | $10^3$–$4 \times 10^6$ |

[a]Synchronously pumped, mode-locked, cavity dumped.

The output powers of CW dye lasers can vary from milliwatts to watts; they are tunable throughout the visible region. Figure 4-20 shows two CW dye laser designs. The simple standing-wave cavity in Figure 4-20a is useful for multimode operation or for low-power single-mode operation. The ring-cavity design in Figure 4-20b can produce higher powers, but it is more complex. Tunable picosecond pulses at repetition rates greater than 1 MHz can be obtained with **synchronously pumped dye lasers** (see Table 4-4). In such a laser a mode-locked argon ion laser is used to pump a dye laser. The cavity of the dye laser is extended so that the intermode spacing of the dye laser is an integral multiple of the argon laser mode locker frequency. With partially transmitting mirrors, the output is a continuous train of picosecond pulses with peak powers on the order of 100 W.

The synchronously pumped dye laser can be modified to achieve wider pulse separations and an order of magnitude more energy per pulse by a technique known as **cavity dumping**. Here the output coupler of the dye laser is replaced by an acoustic-optic deflector. The deflector behaves like a mirror that lies parallel to the laser beam inside the cavity. Whenever the mirror

is switched to the "up" position, the internal energy circulating in the cavity is deflected out of the cavity. Normally, only a small fraction of the total energy available in the cavity is coupled to the output by the output reflector. Pulse separation can be selected by choosing the appropriate cavity dumping rate. Typically, the pulse separation can be varied from less than 1 μs to greater than 1 ms, providing variable repetition rates. Peak powers can be as high as kilowatts.

*Semiconductor Lasers.* In the semiconductor or diode laser, population inversion occurs between the conduction band and the valence band of a *pn*-junction diode. Stimulated transitions of the electrons from the conduction band to the valence band are responsible for laser action, and stimulated emission results from electron–hole recombinations. Since the frequency emitted is directly related to the bandgap energy, various semiconductor compositions can be used to give different wavelengths. Also, since the lasing occurs between bands, these lasers can be tuned over small intervals. Typical materials are GaAs, which lases at 0.84 μm, and lead salt diodes (PbSnTe), which lase in the

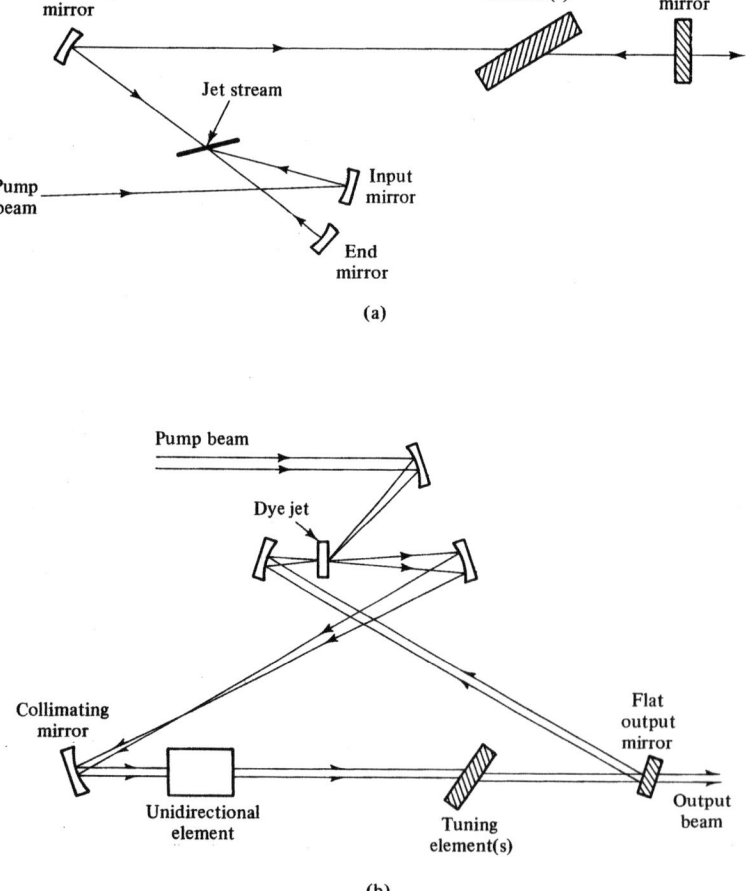

**FIGURE 4-20** Standing-wave cavity (a) and ring-cavity (b) CW dye lasers. In both designs, an argon or krypton ion laser provides the pump beam. The dye solution flows through a nozzle to form a free-flowing jet. The pump beam is tightly focused onto the dye stream. The three-mirror folded cavity in (a) is between the output mirror and the end mirror. The standing-wave cavity design is used for simple, low-power lasers. The ring-cavity design is more efficient for operation at high powers. In both designs, tuning elements such as birefringent filters or interference wedges are used in the cavity to achieve bandwidths as narrow as 0.03 nm. Etalons can be used in conjunction with these to provide even narrower bandwidths. The undirectional device in (b) is usually a Faraday rotator to prevent oscillation in the direction opposite to the arrows.

mid-IR region. Infrared lasers can achieve very narrow line widths ($10^{-6}$ cm$^{-1}$) but must be operated at low temperatures (10 to 20 K). For a given composition tuning can be achieved by changing the diode current, which heats the diode. This causes refractive index changes that alter the effective cavity length. Typical tuning ranges are 20 to 50 cm$^{-1}$.

*Chemical Lasers.* With chemical lasers no optical or electrical pumping is necessary. An exoergic chemical reaction produces molecules in excited states. If population inversion is achieved, lasing can occur on specific transitions. The hydrogen fluoride laser is the best known example; lasing here occurs on vibrational transitions. An electrical discharge is used to produce hydrogen atoms and fluorine atoms. The chemical reaction ($H_2 + F \rightarrow HF^* + H$) produces excited HF. Continuous powers of 475 W have been achieved at 3 $\mu$m. Many other chemical systems have been utilized to produce pulsed or CW lasers.

## Laser Characteristics

Laser radiation has significantly different properties from radiation emitted by conventional sources. Laser radiation is highly directional, spectrally pure, coherent, and of very high radiance.

The directionality of lasers is a direct result of the resonant cavity. Only waves that propagate normal to the mirrors oscillate. Under optimum conditions, this gives rise to a radiant energy beam with an angular divergence $\theta$ (radians) approaching the diffraction limit ($\theta = 2\lambda/\pi w_0$, where $w_0$ is the beam waist or minimum spot size). As a consequence laser radiation can be accurately transmitted over large distances (e.g., to a spot on the moon and back). In the laboratory, the directionality of lasers simplifies the alignment of optical materials and enables samples in remote locations to be probed.

Laser radiation is of high spectral purity. Lasers are often called **quasi-monochromatic sources** or **quasi-continuum sources**, depending on one's point of view. Spectral purity is a result of the resonant interaction of the medium with the pump source, the cavity gain, which enhances the central frequency, and the resonant cavity, which allows oscillation of characteristic frequencies only. In the visible region it is fairly routine to achieve linewidths on the order of 0.01 to 0.1 Å.

Laser radiation is also classified as coherent radiation. Radiation is said to be **temporally coherent**, if for a given point in space, there is always a constant phase difference between the amplitude of the wave at two successive instances in time. The degree of temporal coherence is obviously related to the monochromaticity

(bandwidth) of the laser. A laser with a bandwidth $\Delta\nu$ has a coherence time $\tau_c = 1/\Delta\nu$. Radiation is said to be **spatially coherent** if there is a constant time-independent phase difference for the amplitude at two different points.

The final aspect of laser radiation that makes it different from conventional radiation is its high irradiance. Typically, lasers can achieve irradiances that are 4 to 10 orders of magnitude larger than those from conventional sources. This is a direct consequence of the power and directionality of the laser. For example, a pulsed dye laser might produce a peak output power of 10 kW in a beam of 40-mm$^2$ cross-sectional area. Even if a conventional source were to produce an equivalent output power, it could not be collected and focused into such a narrow beam without many losses.

## Nonlinear Optical Effects

The monochromaticity and high irradiance of laser radiation have led to many new discoveries. Among these are optical effects which depend nonlinearly on the amplitude of the electric field of the laser beam. One of these, optical harmonic generation, is extremely important in the generation of new laser frequencies and is discussed here. An entire new branch of laser spectroscopy involves the observation and application of nonlinear radiation/matter interactions. Many of these techniques are discussed in subsequent chapters.

When an electromagnetic wave propagates through a dielectric medium, the electromagnetic field exerts forces on the outer, valence electrons of the material. Normally, these forces are small, and the resulting electric polarization $P$ (C m$^{-2}$) is directly proportional to the field. Thus we can write

$$P = \varepsilon_0 \chi E \qquad (4\text{-}19)$$

where $\chi$ is a dimensionless constant called the **electric susceptibility**. At extremely high irradiance, we might expect this relationship to break down, particularly if the field becomes comparable to the binding energy of the electrons. At field strengths lower than those that cause such dielectric breakdown of the material, we can describe the polarization by

$$P = \varepsilon_0(\chi_1 E + \chi_2 E^2 + \chi_3 E^3 + \cdots) \qquad (4\text{-}20)$$

where $\chi_n$ is called the $n$th-order susceptibility. If the electric field of the incident wave is of frequency $\omega$ and oscillates according to $E = E_m \sin \omega t$, the resulting polarization is of the form

$$P = \varepsilon_0 \chi_1 E_m \sin \omega t + \varepsilon_0 \chi_2 E_m^2 \sin^2 \omega t$$

$$+ \varepsilon_0 \chi_3 E_m^3 \sin^3 \omega t + \cdots$$

which can be rewritten as

$$P = \varepsilon_0\chi_1 E_m \sin \omega t + \frac{\varepsilon_0\chi_2 E_m^2}{2}(1 - \cos 2\omega t)$$
$$+ \frac{\varepsilon_0\chi_3 E_m^3}{4}(3 \sin \omega t - \sin 3\omega t) + \cdots$$

$$(4\text{-}21)$$

The first term in equation 4-21 is just the normal linear term than predominates at low fields. The polarization that results from this linear term oscillates at the incident frequency $\omega$, and the light that propagates through the medium corresponds to the usual refracted wave. The second term makes a significant contribution only at high irradiances. It contains a constant term, the dc rectification term, plus a $\cos 2\omega t$ term. The light that propagates because of this term is of frequency $2\omega$ with characteristics of monochromaticity and directionality similar to the incident laser beam. The process of generating the light at $2\omega$ is called **second-harmonic generation**. The third term in equation 4-21 contains a component at $3\omega$, and this generates the third harmonic.

A material that is isotropic or has a center of symmetry can produce only odd harmonics. An anisotropic material can, however, produce even and odd harmonics. Among the most important materials used in second-harmonic generation are the piezoelectric crystals potassium dihydrogen phosphate (KDP) and ammonium dihydrogen phosphate (ADP). Centrosymmetric materials like calcite have been used to generate third harmonics.

Harmonic generation is frequently used with dye laser systems. The output of a Nd:YAG laser (1.06 $\mu$m) is frequently doubled or tripled in frequency in order to produce radiation suitable for pumping organic dyes. The outputs of dye lasers are often frequency doubled to produce tunable UV radiation. Efficient harmonic generation must negate the effects of dispersion (variation of $\eta$ with $\lambda$) by a technique known as phase matching.

If two laser beams of frequencies $\omega_1$ and $\omega_2$ are incident on a nonlinear medium, sum ($\omega_1 + \omega_2$) and difference ($\omega_1 - \omega_2$) frequencies can be generated. A device called an **optical parametric oscillator** uses such nonlinear mixing to produce tunable coherent radiation.

Another technique used to shift the frequency of a laser is the **Raman shifter**, discussed in more detail in Chapter 16. Raman shifters use stimulated Raman oscillation in gases such as hydrogen, deuterium, and methane to provide shifts of several thousand wavenumbers. A pump laser beam is focused into a gas cell, where various nonlinear processes occur to generate new shifted frequencies. A prism is usually used to disperse the various wavelengths spatially and allow isolation of the desired beam.

## Summary

Perhaps no invention except the transistor has stimulated the imagination of scientists and the public more than that of the laser. For spectroscopic applications, lasers provide coherent, narrow-bandwidth, high-intensity radiation. Dye lasers have made such radiation tunable over the visible region and parts of the ultraviolet and infrared regions. Despite the great advances that have been made with laser sources, many difficulties must be overcome before they are routinely used in many analytical applications. Only gas lasers are reliable enough for routine uses. The high-powered Nd:YAG and excimer lasers are still somewhat difficult to align and use. Many lasers have "down periods" that approach 50%, which is much too large for routine applications. Users of such lasers must be optical and electronic experts, familiar with elaborate alignment procedures and methods to extract signals from noisy environments. The future should see many advances in laser technology, especially in the development of inexpensive, reliable, and automated systems.

## 4-4  OPTICAL TRANSDUCERS

The radiant power monitor introduced in Section 2-6 consists of a radiation transducer, signal processing electronics and a readout system. The purpose of the radiation transducer is to convert radiant power into an electrical signal or to another physical quantity (e.g., heat or resistance) that can readily be converted to an electrical signal. Regardless of the specific mechanism involved, the characteristics of the transducer (sensitivity, linearity, dynamic range, signal-to-noise ratio, etc.) play an important role in determining the accuracy and precision attainable in spectrochemical methods.

Optical transducers fall into three major categories: thermal detectors, photon detectors, and multichannel detectors. Thermal detectors sense the change in temperature that is produced by the absorption of incident radiation. The temperature change is converted into an electrical signal by methods that depend on the specific transducer. Thermal detectors have a nearly uniform spectral response that is determined by the absorption characteristics of their coatings and window materials. They are thus highly useful for direct radiometric measurements as well as for spectroscopy. Photon detectors, on the other hand, respond to incident photon arrival rates rather than to photon energies. The spectral response of these transducers varies with wavelength. A major advantage of photon detectors is their rapid response time, generally submicrosecond, compared to thermal detectors, which have response times that are slower than milliseconds. Photon detectors can also detect lower radiant powers than thermal detectors in many cases. Multichannel detectors are photographic emulsions, arrays of multiple semiconductor detectors, or arrays of thermal detectors. The elements in array detectors are arranged linearly or in a two-dimensional grid.

The general characteristics of transducers are presented first in order to define many of the terms that are later used in describing specific detectors. Thermal detectors, photon detectors, and multichannel detectors are then discussed.

## Transducer Characteristics

Radiation transducers vary widely in their sensitivity, linearity, spectral response, response speed, electrical output domain, and noise figures. To evaluate a transducer for a particular application, the characteristics discussed below are commonly used.

*Sensitivity and Responsivity.* There are several ways to describe the sensitivity of a detector. The **responsivity** $R(\lambda)$ is the ratio of the rms signal output $X$ (voltage, current, charge) to the rms incident radiant power $\Phi$ evaluated at a particular wavelength and incident power. The **sensitivity** $Q(\lambda)$ is the slope of a plot of electrical output $X$ vs. incident radiant power $\Phi$:

$$Q(\lambda) = \frac{dX}{d\Phi} \qquad (4\text{-}22)$$

For photon detectors and for many thermal detectors, $Q(\lambda)$ and $R(\lambda)$ are wavelength dependent and specified at a particular wavelength. A plot of $Q(\lambda)$ vs. $\lambda$ or $R(\lambda)$ vs. $\lambda$ is called the **spectral response** of the transducer.

The overall functional relationship between the output quantity $X$ and the input quantity $\Phi$ is known as the **transfer function**. It can be expressed by an equation or by means of a plot. The sensitivity is the slope of the transfer function, whereas the responsivity is its magnitude at a given incident power [$R(\lambda) = X/\Phi$]. If $Q(\lambda)$ is constant and independent of $\Phi$, the detector is said to exhibit *linearity*. If a transducer shows linearity and its transfer function passes through the origin, the sensitivity is equal to the responsivity. In many cases these terms are thus used interchangeably. Usually, transducers exhibit linearity over a limited range of incident radiant power. The total range, expressed in powers of 10, for which the transfer function is linear is called the **linear dynamic range**, while the **dynamic range** usually refers to the total range of incident radiant power over which the transducer is responsive.

The sensitivity of a transducer is often not only a function of wavelength and incident radiant power, but may depend on such variables as temperature, bias voltage, and component values. Hence, with some transducers, it might be necessary to keep several variables constant during an experiment to ensure constant sensitivity. The constancy of $Q(\lambda)$ or $R(\lambda)$ with time is known as the **stability** of the transducer. Stability can be expressed as short term (hours) or long term (days or weeks). Often, the stability depends highly on maintaining constant the variables upon which $Q(\lambda)$ or $R(\lambda)$ depend. A long-term change in $Q(\lambda)$ at a constant $\Phi$ is referred to as **degradation**. Some transducers exhibit **hysteresis** in that the responsivity at a particular power changes if the incident power is increased and then brought back to its original value.

*Response Speed.* Transducers vary widely in their ability to detect rapid changes in incident radiant power. Quantitatively, the response time is evaluated in terms of the **time constant** $\tau = 1/(2\pi f_c)$, where $f_c$ is the frequency at which $R(\lambda)$ has fallen to 0.707 of its maximum value (3-dB point) when a sinusoidal input of frequency $f_c$ is incident on the transducer. The **rise time** is the time for the output to rise from 10% to 90% of its final value when an instantaneous (step function) increase in radiant power is incident on the transducer.

*Electrical Output Characteristics.* Transducers also differ in their electrical output domains and their output impedances. The same transducer can, in fact, produce outputs in different data domains. The photomultiplier tube, discussed in detail below, can produce an output charge (an analog domain), an output current (analog), an output pulse rate (a time domain), or a specific number of output pulses (digital domain).

The output impedance of a detector often determines the type of measurement system to be used. Impedances can range from megaohms (phototubes) to a few ohms (thermocouples).

*Dark Signal Characteristics.* The electrical output of a transducer in the absence of radiation is known as its **dark signal**. Because of their different response mechanisms, radiation transducers vary widely in their dark signal characteristics. Photoemissive detectors, which depend on photoelectrons being emitted from photosensitive materials, exhibit dark signals due to thermal emission of electrons. Although dark signals can in principle be subtracted from the total signal in the presence of radiation, noise and unidirectional drifts in the dark signal can become major sources of error.

### Noise Characteristics.

Although we shall delay a detailed discussion of noise and signal-to-noise ratio until Chapter 5, it is useful to define several terms here related specifically to noise characteristics of transducers. The **noise equivalent power** (NEP or $\Phi_n$) is the rms radiant power in watts of a sinusoidally modulated input incident on the detector that gives rise to an rms signal equal to the rms dark noise in a 1-Hz bandwidth. (See Chapter 5 for bandwidth definitions.) The rms dark noise $\sigma_d$ is thus given by $\sigma_d = R(\lambda)\Phi_n$. Since it is a function of wavelength, the NEP is usually specified at a particular wavelength. The modulation frequency, electrical bandwidth, and detector area should also be specified.

The **detectivity** $D$ ($W^{-1}$) is a measure of minimum detectability and is defined as $D = 1/\Phi_n$. As with the NEP, the wavelength, modulation frequency, bandwidth, and detector area should be specified. The **D star** ($D^*$), in cm $Hz^{1/2}$ $W^{-1}$, is a normalization of detectivity to take into account the area and electrical bandwidth dependence. It is related to $D$ by $D^* = DA^{1/2}(\Delta f)^{1/2}$, where $A$ is the detector area in cm$^2$ and $\Delta f$ is the noise equivalent bandwidth in Hz.

### Thermal Detectors

The thermal detectors discussed here are listed in Table 4-5 along with several useful characteristics. Thermal detectors are widely used in the IR region of the spectrum.

### Pneumatic.

A pneumatic detector is based on a thin blackened membrane placed in a gas-filled, airtight chamber. As radiation strikes the detector, the gas is heated and expands against another membrane. The displacement of the membrane is detected in some types. In others the capacitance of the membrane serves as a measure of displacement. A commercial pneumatic cell is the **Golay detector**, which is widely used in infrared spectrometers. The Golay cell is sensitive (see $D^*$ in Table 4-5), but tends to be fragile.

### Thermocouple.

In the thermocouple a thin blackened strip or flake is connected thermally to the junction of two dissimilar metals. Radiation absorbed by the strip causes the junction to increase in temperature and a change in thermoelectric voltage is produced. The thermocouple detector has uniform spectral response in the region 1 to 40 $\mu$m, reasonable sensitivity, and excellent linearity. Thermocouples require no external bias, have high stability, and have low output impedance (5 to 15 $\Omega$). Since their output voltages are often on the order of microvolts, they require large amplification factors. With modern amplifiers, performance is limited by thermal noise (see Chapter 5). Multiple-junction thermocouples, called **thermopiles** are also used.

### Thermistor Bolometer.

The thermistor is made from an intrinsic semiconductor. As the temperature increases the number of valence-band electrons promoted to the conduction band increases, which increases the conductivity and decreases the resistance. A thin blackened tip allows the absorption of radiation, which heats the thermistor. The thermistor is normally placed in a bridge circuit with a reference thermistor

---

**TABLE 4-5**
Thermal detector characteristics

| Type | $D^*$ (cm $Hz^{1/2}$ $W^{-1}$) | $R(\lambda)^a$ | Linear range[b] | Spectral range ($\mu$m) | Time constant (ms) | Output |
|---|---|---|---|---|---|---|
| Pneumatic | $2 \times 10^9$ | Not applicable | $10^{-8}$–$10^{-6}$ W (1%) | 0.8–1000 | 2–30 | Displacement or capacitance |
| Thermocouple | $10^9$ | 5–25 V $W^{-1}$ | $6 \times 10^{-10}$–$6 \times 10^{-8}$ W (0.1%) | 0.8–40 | 10–20 | Voltage |
| Thermistor bolometer | $1.1 \times 10^9 \sqrt{\tau}$ | $\sim 10^3$ V $W^{-1}$ | $10^{-6}$–$10^{-1}$ W cm$^{-2}$ (5%) | 0.8–40 | 1–20 | Resistance change |
| Pyroelectric | $3 \times 10^8$ | $10$–$10^4$ V $W^{-1}$ | $10^{-6}$–$10^{-1}$ W cm$^{-2}$ (5%) | 0.3–1000 | See footnote c | Current |

[a]Voltage responsivity for thermistor assumes constant current of 10 mA; voltage responsivity for pyroelectric detector assumes load resistance of 10 M$\Omega$ (10 V $W^{-1}$) to $10^4$ M$\Omega$ ($10^4$ V $W^{-1}$).
[b]Percentages refer to maximum deviations from linearity in the range shown.
[c]Electrical $\tau$ depends on load resistance; thermal $\tau$ determines low-frequency response.

that is not irradiated. The resistance can be measured by a null-comparison technique, or the out-of-balance voltage of the bridge can be monitored. Bolometers are rugged and exhibit moderate sensitivity and a wide linear range. The thermistor spectral response normally peaks in the near-IR region. Thermistor detectability is limited by thermal noise at frequencies above 20 Hz and by $1/f$ noise (see Chapter 5) at lower frequencies.

*Pyroelectric Detector.* A pyroelectric detector is typically made from triglycine sulfate (TGS). When placed in an electrical field, a surface charge results from alignment of electric dipoles. When a pulse of incident radiation heats the TGS, a change in surface charge results (pyroelectric effect), which is related to the incident radiant power. The output current is proportional to the rate of temperature change of the material $dT/dt$; the detector does not respond to constant radiant energy levels. The pyroelectric detector is fast (<1 ms response time) because only charge-reorientation limits the response speed for modulated inputs. For wavelengths below 2 μm, however, the TGS must be blackened, which can slow the response. The spectral response of a blackened detector is fairly flat over the region 1 to 36 μm. Since the pyroelectric effect is an integrated volume effect, output signals from pyroelectric detectors are not a function of the spatial or temporal distribution of the input radiation. This makes them highly useful for colorimetry, radiometry, and in laser power meters. Linear arrays of pyroelectric detectors are also available.

## Photon Detectors

The photon detectors to be considered here are listed in Table 4-6, where various characteristics are compared. Photon detectors can be broadly classified as photoemissive devices (photomultipliers and photo-tubes), *pn*-junction devices (photodiodes, phototransistors), photoconductive cells, and photovoltaic cells. The operating principles of each of these detectors are discussed below.

*Vacuum Phototubes.* A vacuum phototube (PT), sometimes called a vacuum photodiode, consists of two electrodes sealed in an evacuated glass or silica envelope. A phototube, its circuit symbol, and a typical bias network are shown in Figure 4-21. The photosensitive cathode can be made from a number of photoemissive materials (e.g., $Cs_3Sb$, alkali metal oxides, AgOCs). Typical bias voltages are 75 to 125 V.

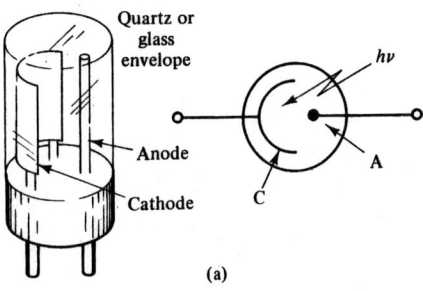

(a)

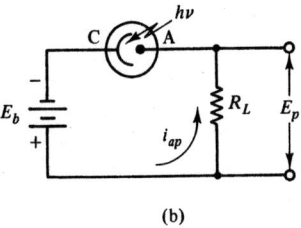

(b)

**FIGURE 4-21** Circuit symbol (a) and bias network (b) for vacuum phototube. The photocathode C consists of a photosensitive material. It is biased negative with respect to the anode A. Irradiation of the cathode causes photoelectrons to be emitted and attracted to the anode. The anodic current $i_{ap}$ is in the direction shown; it produces a voltage drop $E_p$ across load resistor $R_L$.

## TABLE 4-6
Typical photon detector characteristics[a]

| Type | $D^*$ (cm $Hz^{1/2}$ $W^{-1}$) | $R(\lambda)$[b] | Linear range (decades) | Spectral range (nm)[c] | Rise time (ns) | Output |
|---|---|---|---|---|---|---|
| Photomultiplier tube | $10^{12}-10^{17}$ | $10-10^5$ A $W^{-1}$ | 6 | 110–1000 | 1–10 | Current, charge |
| Vacuum phototube | $10^8-10^{10}$ | $10^{-3}-10^{-1}$ A $W^{-1}$ | 5 | 200–1000 | 1–10 | Current |
| Si photodiode | $10^{10}-10^{12}$ | 0.05–0.5 A $W^{-1}$ | 5–7 | 250–1100 | 1–10 | Current |
| Photoconductive cell | $10^9-10^{12}$ | $10^4-10^6$ V $W^{-1}$ | 5 | 750–6000 | $50-10^6$ | Resistance change |
| Photovoltaic cell | $10^8-10^{11}$ | $100-10^6$ V $W^{-1}$ | 3 | 400–5000 | 1000 | Current or voltage |

SOURCE: W. G. Driscoll and W. Vaughan, eds., *Handbook of Optics*, McGraw-Hill, New York, 1978.
[a]Since many different types of detectors are available in each class, the values given represent ranges for several different types.
[b]For the photomultiplier, the value of $R(\lambda)$ includes the internal gain; for the photoconductive cell voltage, responsivity assumes a constant current of 1 to 10 mA.
[c]Extended IR-responsive photoconductors are available (see the text).

The photocathode is characterized by a threshold wavelength $\lambda_t$; at longer wavelengths, photons are ineffective in causing photoemission. The threshold wavelength is related to the photocathode work function $E_c$ by $\lambda_t = hc/E_c = 1240/E_c$. Here $\lambda_t$ is in nanometers when $E_c$ is in electron volts. The work function represents the energy which must be given to an electron in the Fermi level of a metal to raise it to the potential energy of the metal–vacuum interface. Because some electrons occupy states higher than the Fermi level at ordinary temperatures, there is no abrupt cutoff at $\lambda > \lambda_t$. Typical work functions for pure metals are 2 to 5 eV, but semiconductor materials can have values substantially lower. Indeed, extended long wavelength response is obtained with photosensitive materials which show negative electron affinity.

Only a certain fraction of the photons with greater than threshold energy yield photoelectrons with sufficient kinetic energy to escape the photocathode. This fraction is called the **quantum efficiency** $K(\lambda)$ and is the ratio of the number of photoelectrons ejected to the number of incident photons. Typically, $K(\lambda)$ varies from 0 to 0.5. The rate $r_{cp}$ (s⁻¹) at which photoelectrons are emitted from the photocathode for monochromatic radiation of flux $\Phi_p$ (photons s⁻¹) is

$$r_{cp} = \Phi_p K(\lambda) \qquad (4\text{-}23)$$

This rate can also be expressed as a cathodic current $i_{cp}$ by multiplying $r_{cp}$ by the electron charge, $e = 1.6 \times 10^{-19}$ C.

$$i_{cp} = er_{cp} = eK(\lambda)\Phi_p \qquad (4\text{-}24)$$

The efficiency with which photons are converted to photoelectrons is $R(\lambda)$, the **radiant cathodic responsivity** in A W⁻¹. It is related to the quantum efficiency by $R(\lambda) = K(\lambda)e/h\nu = 8.06 \times 10^{-4} K(\lambda) \times \lambda$, where $\lambda$ is in nanometers. Typical values of $R(\lambda)$ are given in Table 4-6 for vacuum phototubes.

A fraction of the photoelectrons emitted are collected at the anode. This fraction $\eta$, called the **collection efficiency**, depends on the bias voltage $E_b$ and approaches unity if $E_b$ is sufficiently high. The arrival rate of photoelectrons at the anode $r_{ap}$ and the anodic photocurrent $i_{ap}$ are thus

$$r_{ap} = \eta r_{cp} \qquad (4\text{-}25)$$

and

$$i_{ap} = \eta i_{cp} \qquad (4\text{-}26)$$

Complete expressions for the anodic pulse rate and anode current in the case of monochromatic incident radiation are then found from substituting equation 4-23 into 4-25 and 4-24 into 4-26. This yields

$$r_{ap} = \eta \Phi_p K(\lambda) \qquad (4\text{-}27)$$

and

$$i_{ap} = \eta \Phi R(\lambda) \qquad (4\text{-}28)$$

where the radiant power in watts is $\Phi = \Phi_p \times h\nu$. If $K(\lambda)$ and $R(\lambda)$ are not constant over the wavelength range incident on the photocathode, these expressions become integrals over this range. For typical phototubes, the anodic current becomes independent of bias voltage ($\eta \rightarrow 1$) above about 60 V. For most phototubes the anode current must be kept lower than $10^{-4}$ to $10^{-5}$ A for linearity to be obtained.

In the absence of radiation, a small anodic **dark current** $i_{ad}$ is obtained, where $i_{ad} = \eta i_{cd}$ and $i_{cd}$ is the cathodic dark current. At moderate bias voltages dark current can be due primarily to thermal emission of electrons at the photocathode. When voltages are low, ohmic leakage can dominate. At low temperatures luminescence and radioactivity can be the primary source of dark current. When high voltages are applied, ionization of residual gas and field emission can be additional sources. Typical values of $i_{ad}$ for vacuum phototubes are in the range $10^{-12}$ to $10^{-14}$ A.

*Photomultiplier Tubes.* The photomultiplier tube (PMT), like the phototube, contains a photosensitive cathode and a collection anode. However, the cathode and anode are separated by several electrodes, called **dynodes**, that provide electron multiplication or gain. Figure 4-22 shows a typical PMT. The cathode is biased negative by 400 to 2500 V with respect to the anode. A photoelectron ejected by the photocathode strikes the first dynode and releases two to five secondary electrons. Each secondary electron is accelerated by the field between the first and second dynode and strikes the next dynode with sufficient energy to release another two to five electrons. Since each dynode down the chain is biased $\approx 100$ V more positive than the preceding dynode, this multiplication process continues until the anode is reached. The result is a large charge packet of a few nanoseconds' duration at the anode for each photoelectron collected by the first dynode. The gain $m$ (the average number of electrons per anode pulse) depends highly on the power supply voltage, but is often $10^4$ to $10^7$. For example, if $m$ were $10^6$, each anode pulse would contain an average charge of $1.6 \times 10^{-19} \times 10^6$ C $= 1.6 \times 10^{-13}$ C. If the pulse were 5 ns in duration, this would result in an average current of 32 μA during the pulse.

It is most common to measure the average current

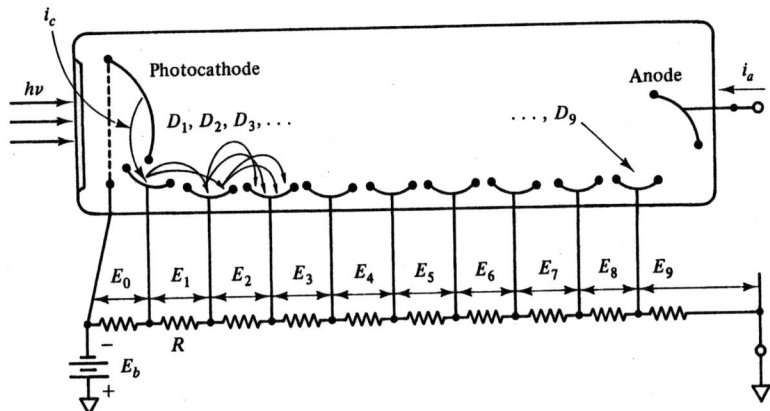

**FIGURE 4-22** Photomultiplier tube. A series of dynodes (typically, 5 to 11) between cathode and anode provide internal gain. The dynodes are made of a secondary emission material (MgO, GaP), which emits two to five electrons when struck by an electron of sufficient energy.

that results from the arrival of many anodic pulses. Here electronics with a time constant much longer than the pulse duration are used to obtain the average photoanodic current $i_{ap}$, which is given by

$$i_{ap} = m\eta i_{cp}^{*} = mi_{cp} = m\eta \int_0^\infty \Phi_\lambda R(\lambda)\,d\lambda \qquad (4\text{-}29)$$

where $i_{cp}^{*}$ is the photocathodic current (A) and $i_{cp} = \eta i_{cp}^{*}$ is the effective photocathodic current (A) or the photocathodic current that reaches the first dynode and causes secondary emission. In the case of monochromatic radiation, or if $R(\lambda)$ is constant over the incident wavelength range, equation 4-29 reduces to

$$i_{ap} = m\eta\Phi R(\lambda) \qquad (4\text{-}30)$$

Equation 4-30 can also be expressed in terms of quantum efficiency as

$$i_{ap} = \frac{me\eta\Phi K(\lambda)}{h\nu} \qquad (4\text{-}31)$$

Since the photomultiplier output is directly dependent on $R(\lambda)$, or $K(\lambda)$, which are functions only of the photocathode material, it is common for manufacturers to supply plots of one or both quantities vs. wavelength. Figure 4-23 shows the cathodic responsivity of several photoemitters vs. $\lambda$. Historically, a photocathode and window material were given an S-response designation (S-1, S-2, etc.), but many of these are now obsolete. Photomultiplier manufacturer's may also supply responsivity curves for the photocathode material only (no window), which is more directly related to quantum efficiency.

In a second mode of operation, the number of anode pulses per unit time or per unit event is counted. This technique is called **photon counting** even though charge packets, and not photons, are actually counted. In this mode, the rate of arrival of anodic pulses for an incident spectral radiant power $\Phi_{p,\lambda}$ (photons $s^{-1}$ nm$^{-1}$) is

$$r_{ap} = \eta r_{cp} = \eta \int_0^\infty \Phi_{p,\lambda} K(\lambda)\,d\lambda \qquad (4\text{-}32)$$

where $\eta$ is the fraction of photoelectrons from the photocathode collected by the first dynode giving rise to secondary emission. For monochromatic radiation or if $K(\lambda)$ is constant over the range of wavelengths incident on the photocathode, equation 4-32 reduces to equation 4-27 for the phototube. Note, however, that each anode pulse contains $m$ ($10^4$ to $10^7$) electrons in the case of the PMT, whereas each pulse contains only a single electron in the case of the phototube.

Because of the high gains of photomultiplier tubes, they are very useful for low-light-level detection. However, they must be operated with very stable power supplies to keep the total gain $m$ constant. If all dynodes are identical $m = \delta^k$, where $\delta$ is the gain of each stage (2-5) and $k$ is the number of dynodes (5-11). Since the gain of each stage $\delta$ is related to the supply voltage, $E_b$, small changes in the supply voltage can result in large overall gain changes. Typically, for 0.1% gain stability, $E_b$ must be regulated to 0.01%. The gain also varies with temperature. Since fluctuations in electric or magnetic fields from other equipment can affect trajectories of electrons in the PMT, and thus the gain, proper shielding around the tube should be used. Since $R(\lambda)$ varies across the photocathode surface, vibrations must be minimized to keep $K(\lambda)$ constant. For many PMTs, a log-log plot of $m$ vs. $E_b$ is linear over several decades of overall gain.

The rise time of photomultipliers depends mostly on the spread in the transit time of electrons during the multiplication process. Typically, this spread is about 10 ns, but some specially designed electron optics can reduce the transit time spread to below 1 ns. Photomultiplier tubes exhibit responses linear to a few percent as long as the anode currents do not exceed max-

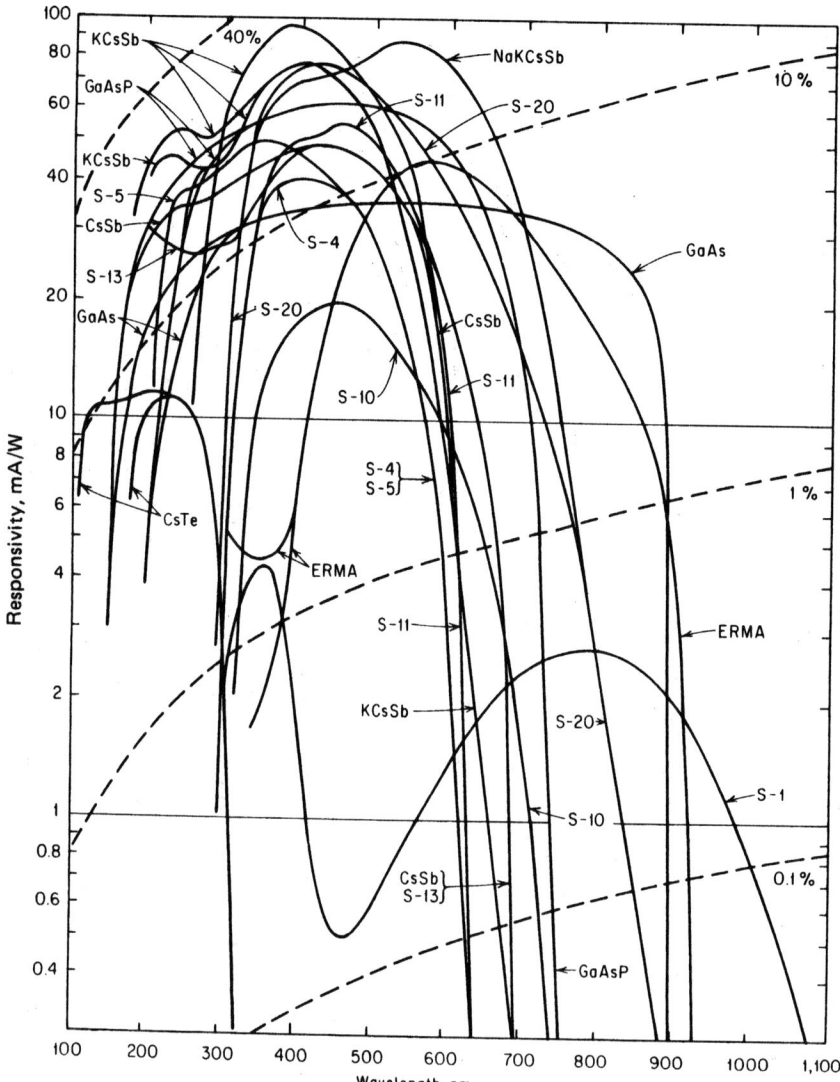

**FIGURE 4-23** Cathodic responsivity curves for various photoemitters and window materials. Dashed lines indicate photocathode quantum efficiency. S-1 = AgOCs (lime or borosilicate, crown-glass window); S-4 = $Cs_3Sb$ (lime or borosilicate, opaque photocathode); S-5 = $Cs_3Sb$ (UV transmitting glass window); S-8 = $Cs_3Bi$ (lime or borosilicate window); S-10 = AgBiOCs (lime or borosilicate); S-11 = $Cs_3Sb$ (lime or borosilicate semi-transparent photocathode); S-13 = $Cs_3Sb$ (fused silica, semitransparent); S-19 = $Cs_3Sb$ (fused silica, opaque); S-20 = $Na_2KCsSb$ (lime or borosilicate); ERMA = extended red multialkali. (From RCA Electronic Components, chart PIT 701B, courtesy of RCA New Products Division, Lancaster, Penn.)

imum ratings (typically, $<1$ μA). The limit of linearity is caused by space-charge effects in the last dynodes. Since the electrons emitted by the dynodes are drawn from the dynode current (current through the bias resistors), the dynode current should be $\approx 100$ times $i_{ap}$ to prevent bias voltage changes and nonlinearity. Operating a PMT at light levels such that the anode current exceeds the maximum rating can cause **fatigue**, a loss in sensitivity as a result of reduced secondary emission. Such sensitivity changes are thought to result from erosion of the dynode surfaces. Tube fatigue varies widely with PMT type and among tubes of the same type. Some PMTs exhibit **hysteresis** when exposed briefly to high light levels. Hysteresis may result in overshoot or undershoot by a few percent when levels are changed. Hysteresis may be caused by capacitance effects within the tube.

Dark current in PMTs arises from effects similar to those in phototubes. In addition, thermal emission

from the dynodes can be a source of dark current in PMTs. Since the cathodic dark current is multiplied by the full gain of the tube, thermal emission from the photocathode (or early dynodes) is often a major component of the total dark current. Typically, dark currents in PMTs are in the range $10^{-7}$ to $10^{-11}$ A. For very low light levels, the dark current can be as large as the photocurrent, and noise in the dark current can limit precision. The thermal dark current $i_{th}$ emitted by the photocathode is given by $i_{th} = CAT^2 e^{-E_c/kT}$, where $C$ is a constant, $A$ is the surface area of the photocathode, $T$ is the photocathode temperature, and $E_c$ is the work function. If the dark current primarily results from thermal emission, then $i_{ad} = m i_{th}$ and the cathodic dark current $i_{cd} = i_{th}$. In this case, cooling the PMT can reduce the dark current. Typically, temperatures of $-60$ to $0°C$ are used. If external radioactivity or cosmic rays are a significant cause of dark current, shielding can be used. Since the anodic dark current is proportional to the gain, except at very low or very high bias voltages, there is often an optimum bias voltage which gives the lowest relative dark current noise.

The limiting source of noise in PMTs (see Chapter 5) depends on the level of illumination. For low levels, the limiting noise is usually traceable to dark current shot noise. At higher illumination levels signal-carried shot or flicker noise can far exceed dark current noise. Therefore, $D^*$ and NEP values, which are calculated by assuming dark current noise is the limiting noise source, must be used with caution. Considerable error in predicting detection capability from $D^*$ values can result if noise in the signal is ignored. Many photomultiplier manufacturers use the term **equivalent noise input** (ENI) instead of noise equivalent power. It is very similar except that a square-wave-modulated blackbody source is used, and the peak-to-peak value is obtained instead of the rms value of a sinusoidally modulated source. Specification sheets for PMTs are discussed in Appendix D.

*Image Dissector Tube.* An image dissector tube (IDT) is a photomultiplier tube which allows spatial resolution of the radiation striking the tube face (photocathode). The photocathode is separated from the dynodes by a plate containing a small aperture. An electromagnetic deflection coil focuses the photoelectrons so that only those produced by a specific region of the photocathode pass through the aperture to the dynodes where amplification occurs. By varying the current applied to the coil, photoelectrons produced by different photocathode regions can be passed through the aperture. Thus an image of the radiation striking the photocathode is converted into an electrical signal. If a portion of the spectrum is dispersed across the photocathode, the IDT can scan the spectrum electromagnetically. Typically, the wavelength range impinging on the photocathode can be scanned in less than 1 µs.

*Photodiodes and Phototransistors.* In a photodiode, absorption of electromagnetic radiation by a *pn*-junction diode causes promotion of electrons from the valence band to the conduction band and thus the formation of electron–hole pairs in the depletion region, as illustrated in Figure 4-24. If the rate of light-induced charge carrier production greatly exceeds that due to thermal processes, the limiting current under reverse bias is directly proportional to the incident radiant power. The photodiode thus acts as a current source when operated in the reverse-biased mode. In some cases photodiodes are operated with no external bias ($E_b = 0$). In either case the $iR$ drop across a load resistor ($R_L$) is measured or a current-to-voltage con-

verter (see the next section) is used to produce a voltage proportional to the incident radiant power.

The spectral response of most photodiodes reaches a maximum in the near-IR region (0.85 to 1.0 µm). They are thus very useful for UV-visible and near-IR detection. These devices often show excellent linearity over six to seven decades of incident radiant power. Responsivities are typically much lower than those of photomultiplier tubes (see Table 4-6) because of the internal gain of the PMT. However, the simplicity, excellent linearity, and very small sizes of photodiodes make them attractive for applications where light levels are relatively high. They have practically replaced vacuum phototubes in all but a few applications. Photodiodes are extremely fast transducers. A special type, known as a **pin junction**, in which the *p*- and *n*-type semiconductor materials are separated by an insulating layer *i*, has a subnanosecond response time. The pho-

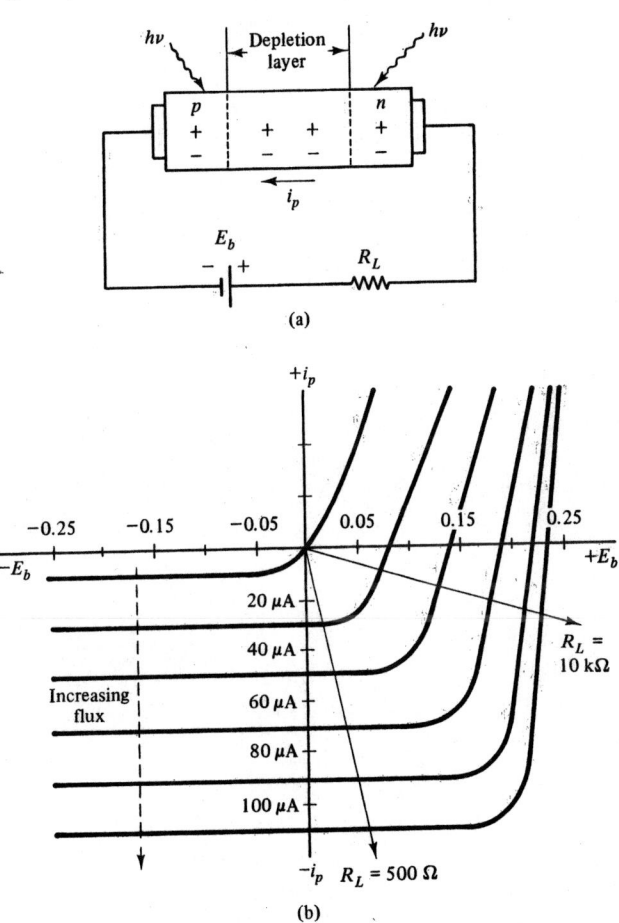

**FIGURE 4-24** Photodiode: (a) pictorial representation showing electron–hole pair production in a reverse-biased photodiode; (b) current–voltage characteristics. For the best linearity and response speed the photodiode is operated under reverse bias where the photocurrent $i_p$ is proportional to the photon flux.

todiode is thus often used to detect rapid high-intensity pulses, such as from a pulsed laser. Because photodiode outputs are usually in the microampere range, photodiode–amplifier combinations have been produced in integrated circuit form to preserve the high response speed.

The **avalanche photodiode** is used where both fast response and high sensitivity are needed. Avalanche photodiodes are operated in the reverse breakdown region of the *pn* junction and provide internal gain (1 to 1000). Very careful control of bias voltage is required to achieve stable current gains. Response times depend on gain, and typical gain–bandwidth product values are 50 to 80 GHz.

Phototransistors also have an internal current gain of up to a few hundred. They are most often used as opto-electronic switches, but find only a few spectroscopic applications because of a limited linear dynamic range and a susceptibility to temperature variations.

*Photoconductive Cells.* The photoconductive cell is made of a semiconductor material such as CdS, PbS, PbSe, InAs, InSb, He-Cd-Te, or Pb-Sn-Te. It acts like a light-dependent resistor, which decreases in resistance when photons are absorbed. Incident photons release electron–hole pairs and increase conductivity. Typically, the detector is put in series with a voltage source and load resistor, and the voltage drop across the load resistor is measured. The PbS cell is still the most sensitive uncooled detector in the near-IR region of the spectrum 1.3 to 3 μm. Photoconductive cells of CdS are very popular as detectors for photographic light meters because of their small size and low cost. Some of the newer photoconductors (GeAu, GeHg, GeCu, GeCd, GeZn) can have extended infrared response (>40 μm) if cooled. Cooling is necessary to avoid thermal excitation of electrons into the conduction band. The GeZn and GeCu detectors have $D^*$ values that exceed those of thermal detectors in the IR region although their spectral response is not flat.

*Photovoltaic Cell.* The photovoltaic cell or barrier-layer cell converts radiant energy into electrical energy. A typical construction is shown in Figure 4-25. In the open-circuit mode, no bias is required and a high-impedance voltage measurement circuit is used to measure the potential difference. In this mode a limited linear dynamic range is achieved. The short-circuit mode, in which a low-resistance current-measuring device is used, often showns better linearity.

Photovoltaic cells made of Se are useful in the spectral range 300 to 700 nm. They often find use in camera exposure meters and simple colorimeters. Sil-

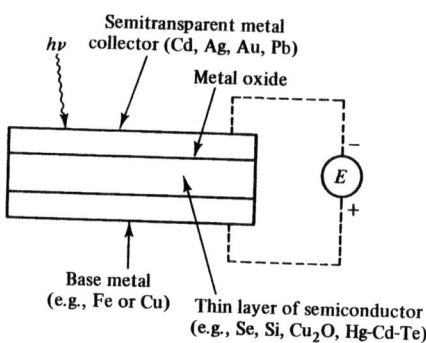

**FIGURE 4-25** Photovoltaic or barrier-layer cell. Radiant power incident on the semiconductor produces electron–hole pairs and a resulting charge separation across the semiconductor–metal interface.

icon photovoltaic cells, known as **solar cells**, can provide relatively large currents and are used as power sources or solar batteries.

## Multichannel Detectors

Multichannel detectors when placed in the focal plane of a spectrograph can provide simultaneous detection of the dispersed radiation. The photographic emulsion was, of course, the original multichannel optical transducer. In recent years, however, video-type multichannel detectors have been shown to have unique capabilities. The discussion here will be limited to photodiode arrays, vidicon tubes, and charge-coupled devices, although many other multichannel devices exist. Since these devices often act as photoelectric replacements for the photographic plate, the characteristics of photographic emulsions are discussed first.

*Photographic Detectors.* Photographic films or plates are emulsions that contain silver halide crystals. Incident photons eventually produce stable clusters of silver atoms within the crystal. Internal amplification is provided in the development process by an electron donor which reduces the remaining silver ions to silver atoms within the exposed crystals. A complexing agent is used to remove the unexposed silver ions. Various emulsions can be obtained with responses from the UV to the near-IR region.

The photographic detection process is an integrating process in that the output (density of silver) is a result of the cumulative effect of all the radiation incident during the exposure time. Photographic detection can be quite sensitive. In the visible, only 10 to 100 photons are required to produce a developable grain.

For quantitative purposes the density of the exposed areas must be obtained. A device called a **den-**

sitometer or microphotometer uses a source, slit, and photoelectric detector to measure the transmittance of the film in the area selected by a positioning mechanism.

The major drawback of the photographic emulsion is, of course, the time required for development and densitometry. The video-type detectors below enormously reduce the time required to obtain a spectrum. Linearity of emulsions is also fairly limited (one to two decades). At low intensities exposure times must be long enough to overcome the inertia of the emulsion, while saturation can readily occur at high intensities.

The application of photographic detectors in arc and spark emission spectroscopy is described in Section 9-3.

*Photodiode Arrays.* Arrays of silicon photodiodes, available in integrated-circuit form, have become extremely useful spectroscopic detectors. They typically contain 256, 512, 1024, or 2048 elements arranged in a linear manner with a center-to-center spacing of 25.4 $\mu$m (0.001 in.) or 50 $\mu$m. They can be obtained with detector-element heights ranging up to 2.5 mm (0.1 in.). The photodiodes are operated in a charge-storage mode, as illustrated in Figure 4-26a.

The integrated-circuit package also contains the necessary scanning circuitry for readout of the array, as shown in Figure 4-26b. Each diode in the array is sequentially interrogated (10 to 25 $\mu$s per diode) after all the diodes have integrated the incident radiation for a specified time period. The integration time is changed by varying the time between the start of each scan. Once a scan has begun, the complete array must be read out. The arrays can be completely scanned in a few milliseconds. The peak value of each signal pulse is sampled and digitized, as shown in Figure 4-26c. Thermal charge generation causes a dark signal to be present. Hence, to take advantage of variable integration times, it is necessary to cool the array to reduce the dark signal. Thermoelectric (Peltier) devices are often used for cooling purposes.

Although diode arrays have no internal gain, they show dynamic ranges of two to four decades. The dynamic range is limited on the lower end by the noise associated with read out of a given diode and on the upper end by the saturation level. The latter effect is determined by the maximum number of electron-hole pairs that can be generated, and typically the saturation charge is 1 to 10 pC. The use of variable integration times can make the effective dynamic range quite large. Diode arrays are also available with front-end micro-channel plate image intensifiers, as illustrated in Figure 4-27. This gives them gain and allows lower light levels to be detected by raising the signal level above the noise associated with readout of the signal.

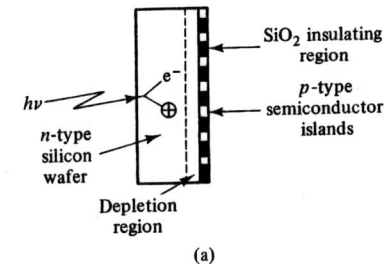

(a)

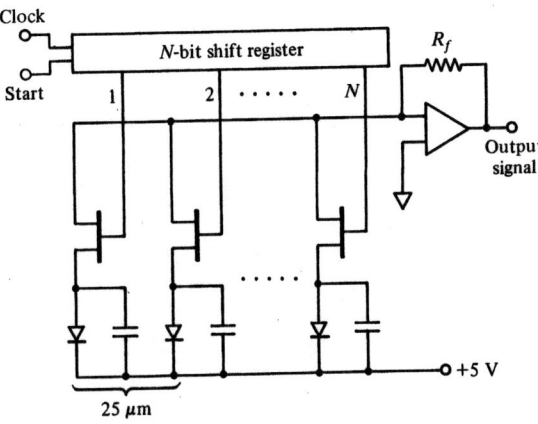

(b)

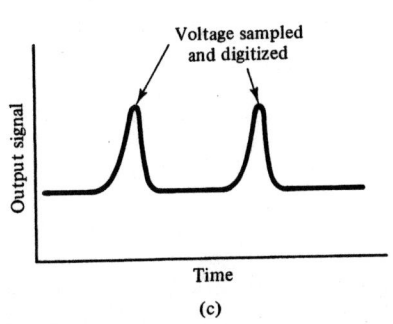

(c)

**FIGURE 4-26** Linear photodiode array: (a) charge-storage mode; (b) readout circuitry; (c) output signals. Each diode is initially reverse biased so that in the *n*-type semiconductor there are minority carriers (holes). The biasing is turned off and photons strike the *n*-type semiconductor for a controlled integration time. Electron–hole pairs are generated and electrons move to the depletion region (a), where they recombine with holes. After the integration period, the charge needed to reestablish reverse bias on the diode is a measure of the integrated light intensity plus dark signal. Diode elements (*p*-islands) are connected to solid-state switches (b) which are sequentially opened to measure the recharging pulse. The recharging pulses appear on the video line, where they are converted to voltage pulses and further amplified. Fast data acquisition circuiting is used to sample and digitize the peak value (c) of each pulse, which is stored in computer memory for later display and manipulation.

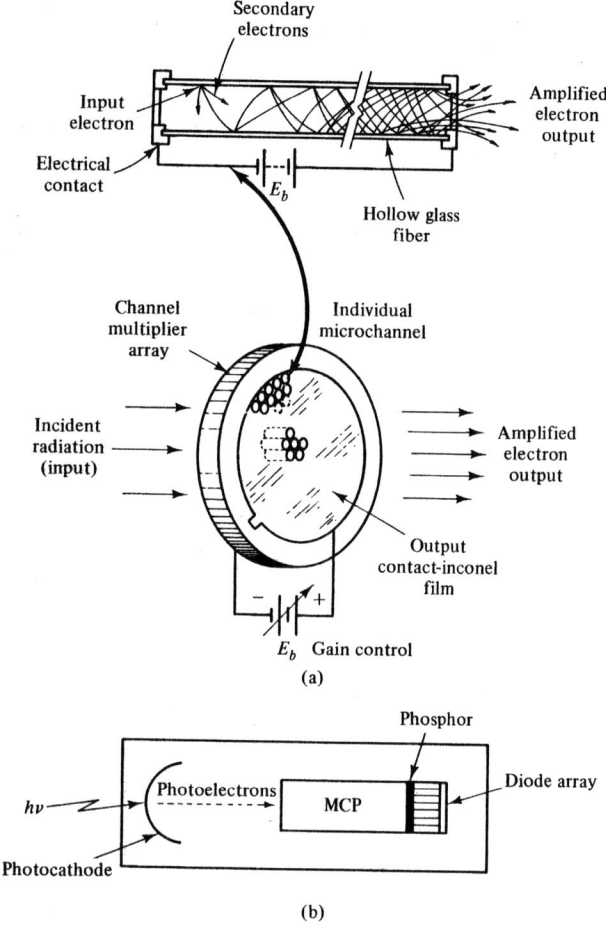

**FIGURE 4-27** Microchannel-plate-intensified diode array. The microchannel plate in (a) consists of a large number of closely packed hollow glass fibers (10 to 40 μm in diameter) with a resistive coating for secondary emission on the inner surface. Each fiber acts as an individual electron multiplier with a gain of $10^3$ or $10^4$. In (b) a photocathode produces photoelectrons which strike the microchannel plate. The amplified electron packets, each with spatial integrity, strike a phosphor, which produces visible radiation detected by the array.

*Vidicon Tube.* The silicon vidicon uses detector elements that are also reverse-biased photodiodes operated in the charge-storage mode. However, scanning electron beam readout techniques are used. Radiation strikes a silicon target where the charge stored on the *pn*-junction capacitance is discharged by generation of electron-hole pairs. The target is then scanned by an electron beam with magnetic deflection coils, and the charge needed to restore each detector to its initial state is measured.

In contrast to the photodiode arrays, which in spectroscopy have been linear arrays, the silicon vidicon is a two-dimensional array containing about $10^5$ resolv-

ing elements, called **pixels**. This two-dimensional imaging capability can be an advantage in some cases. The targets can also in principle be randomly accessed in contrast to the diode array, which must be read out sequentially. Random-access readout has been used to reduce the time required for readout by interrogating only selected elements.

Silicon vidicons suffer from two problems that are greatly reduced in diode arrays. **Blooming** occurs when a strong signal spreads to adjacent sensor elements. In some cases blooming can cause the entire detector to saturate. In spectroscopy where it is often necessary to measure weak signals in the presence of strong signals, blooming can seriously limit the use of the integrating capability and decrease resolution. Another problem is that of **lag**, which is a result of incomplete erasures of the image during a readout cycle. This leads to a carryover of the image from one integration time to another. Lag is also highly undesirable in spectroscopy, particularly in time-resolution applications. Fortunately, lag does not appear to be a significant problem with diode arrays.

Like diode arrays, silicon vidicons have no internal gain. Their dynamic range is similar (two-three decades). To provide gain for the vidicon, the **silicon-intensified target** (SIT) is commonly used. In the SIT, a photocathode is placed in front of the silicon target and the photoelectrons are accelerated by a large 7 to 8 kV potential difference. Here, typically, 1000 electron-hole pairs are produced per incident electron. The silicon vidicon target is operated as an electron detector rather than a photon detector. Gains of several hundred can be obtained and the linear dynamic range can be extended to about four decades. Intensified SIT devices are also available. These have an image intensifier in front of the SIT tube and are called ISITs. Lag and blooming remain as problems for these intensified devices.

*Charge-Coupled and Charge-Injection Devices.* Like photodiode arrays, charge-coupled devices (CCDs) and charge-injection devices (CIDs) are solid-state sensors and are constructed with integrated-circuit technology. With both devices, the charges generated by photons are collected and stored in metal-oxide-semiconductor (MOS) capacitors. The pixel arrangement for the 512 × 320 CCD array made by RCA is illustrated in Figure 4-28a; some CCD arrays are as large as 2000 × 2000 ($4 × 10^6$ pixels). Each pixel consists of a thin conducting electrode and a thin insulating oxide layer on top of a *p* type silicon substrate, as shown in Figure 4-28b.

The MOS capacitors are initially reversed biased by a positive voltage applied to the metallic electrode;

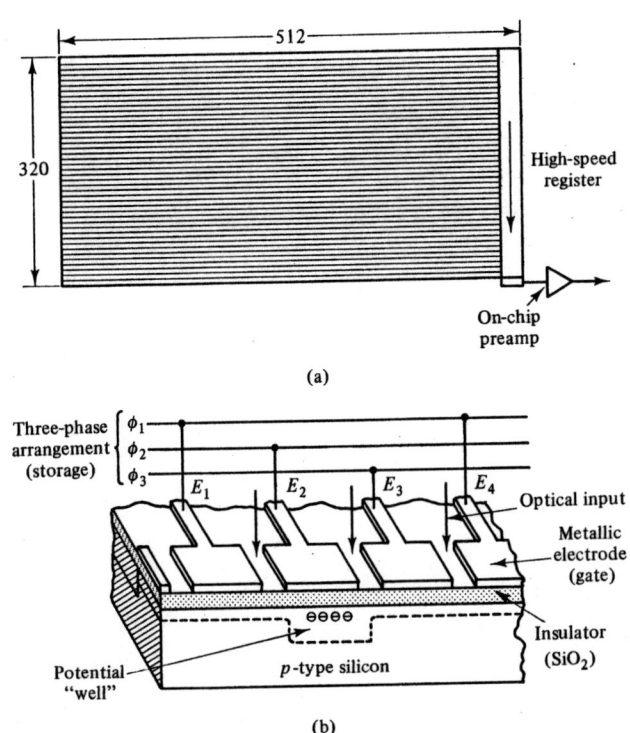

**FIGURE 4-28** Pixel layout (a) and simplified structure (b) for a charge-coupled device array. The array shown is 512 pixels wide and 320 pixels high.

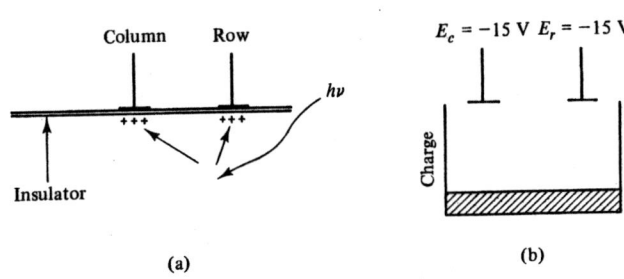

**FIGURE 4-29** Operation of charge injection device. Positive charges produced by photons striking the array are stored below negatively biased capacitor plates (a), as depicted in the well diagram (b). The amount of charge stored is proportional to the light intensity and the integration time. Each pixel is associated with a column and a row capacitor which is normally connected to $-15$ V. When a pixel is selected for readout, both the column and row capacitors are clamped to zero volts. This results in an injection of charge into the substrate and an output signal. If only one of the two capacitors is clamped to zero volts, the charge merely moves under the capacitor with the more negative bias.

this creates a depletion region in the silicon below the electrode (a potential well). Photons striking the array give rise to electron–hole pairs, and the electrons can be stored temporarily in the wells. Typically, each well can store $10^5$ to $10^6$ electrons. The amount of charge accumulated is a linear function of the incident intensity and the integration time. A three-phase clocking arrangement is used to shift the charge horizontally to a high-speed shift register and then down to the readout preamplifier (see Figure 4-28a). The result is a serial row-by-row scan of the charge stored. The CCD array suffers from "blooming" at high light levels since an overilluminated pixel can spill its charge over into a neighboring pixel. This limits the dynamic range at the upper end. For this reason, CCD devices are most often used for detection of weak sources. Commercial CCDs are made for television camera applications. Hence many do not respond in the UV region of the spectrum.

The CID sensor is a two-dimensional array of discrete pixels made from pairs of MOS capacitors. Photons which strike the bulk silicon generate positive charges that can be stored in potential wells beneath negatively biased capacitor plates, as shown in Figure 4-29. Nondestructive readout can also be accomplished. The latter mode can be used to effectively eliminate blooming and improve the *S/N*. Some CID arrays are now available with good response in the UV region as well as in the visible region.

At present, neither the CCD array nor the CID array is as well developed for spectroscopic work as the photodiode array. Both are quite promising detectors and should be especially valuable where advantage can be taken of their two-dimensional nature. A crossed-dispersion echelle grating spectrometer is one example where two-dimensional optical information is produced.

## 4-5 SIGNAL PROCESSING AND READOUT SYSTEMS

In most spectrochemical instruments the output of the radiation transducer is subjected to some signal processing prior to display on the readout device. In some cases the signal processing may be a simple adjustment of signal levels or a bandwidth reduction to reduce noise. In other cases (e.g., Fourier transform spectrometry), extensive calculations must be done to convert the transducer output signal into the quantities presented to the user. It is increasingly common for spectrochemical instruments to contain microcomputers that play an integral role in instrument control, data processing, and display. Such systems are making the new generation of spectrochemical instruments considerably more powerful than earlier systems.

Signal processing and readout systems are extremely important in spectrochemical instruments. The specific type of signal processing used depends highly on the signal form, its level, and the type of environment (i.e., noise sources and the expected noise amplitude). Also, the type of signal processing is dependent on the specific instrumentation used and the format of the data to be presented. The signal processing system can perform various operations, including data conversions (current to voltage, analog to digital, etc.), amplification or voltage division, mathematical operations (e.g., integration, logarithmic conversion, and differentiation), and *S/N* enhancement. Even with a computer, some signal treatment by hardware is almost always necessary prior to acquisition by the computer. Readout systems vary widely from instrument to instrument, but some form of digital readout and, increasingly, graphical display are commonly included.

We will consider in this section many of the standard ways in which signals are conditioned, modified, and processed prior to readout. After presenting some general considerations, we consider the principles of modulation since many signal conditioning methods are discussed. Analog signal processing schemes, including *S/N* enhancement techniques, are described next followed by a consideration of computer data acquisition systems. Digital data processing methods are then briefly introduced. A detailed discussion of analog-to-digital conversion methods and computer data processing techniques is beyond the scope of this book. The reader should consult the references for this chapter for additional information on these methods. The section concludes with a discussion of the common types of readout and display systems.

## General Considerations

Electrical signals can be grouped into three data domain classes: analog, time, and digital. The outputs of most of the optical transducers are in one of the analog domains. The major exception to this is the photomultiplier tube when it is used in a photon counting system. In this case, if the desired quantity is the total number of accumulated counts over some boundary condition (i.e., counts per spark discharge or counts per laser pulse), the photomultiplier output is directly digital. If the desired quantity is the count rate, the photomultiplier output is in one of the time domains (frequency domain). Thus the PMT output can be in any of the three electrical domain classes (analog, time, digital).

In the past completely analog signal processing and readout systems were employed, as illustrated by the general block diagram of Figure 4-30a. Increasingly, the outputs of analog transducers are being converted to digital form as early in the measurement sequence as feasible, as illustrated in Figure 4-30b. Early conversion to the digital domain allows the signal to be treated with highly accurate digital methods or with software in a microcomputer. Even with "digital" systems, some analog signal conditioning is normally done to amplify, filter, or otherwise alter the signal (e.g., convert current to voltage) to make it suitable for analog-to-digital conversion. In many cases it is desirable to shift the signal away from dc, which is a particular noisy region of the frequency spectrum (see Section 5-1). With analog systems, this is accomplished by modulation (see below) and ac signal processing. With digital systems, modulation/demodulation or gated detection schemes are used prior to conversion to digital format.

For photon counting systems, the scheme illustrated in Figure 4-31 is used. The readout is the number of pulses occurring during a preselected counting interval. Because of the high probability of pulse overlap at high count rates, photon counting is used in situations where irradiances are quite low (e.g., Raman scattering and some luminescence methods). Photon counting systems are discussed in more detail in Chapter 5.

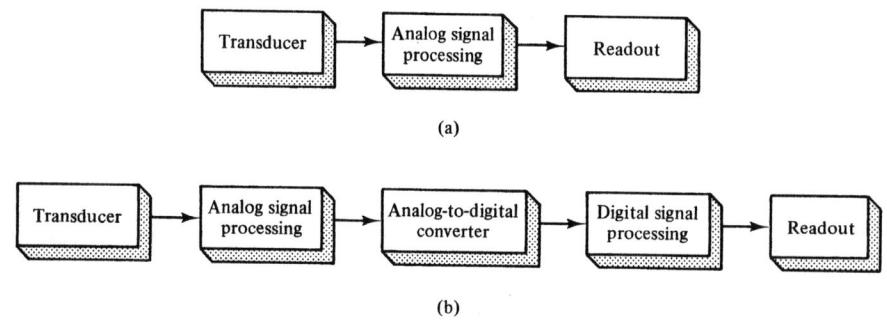

(a)

(b)

**FIGURE 4-30** Typical measurement scheme for analog transducers. After signal conditioning, the signal can be processed and read out in the analog domain (a) or converted into the digital domain (b). In modern systems the trend is toward signal processing in the digital domain with digital hardware or computer software techniques. The processed signal is then read out on an appropriate digital readout device. For simplicity in later discussions, we will consider all the functional blocks between the transducer and the readout to be part of the overall signal processing system.

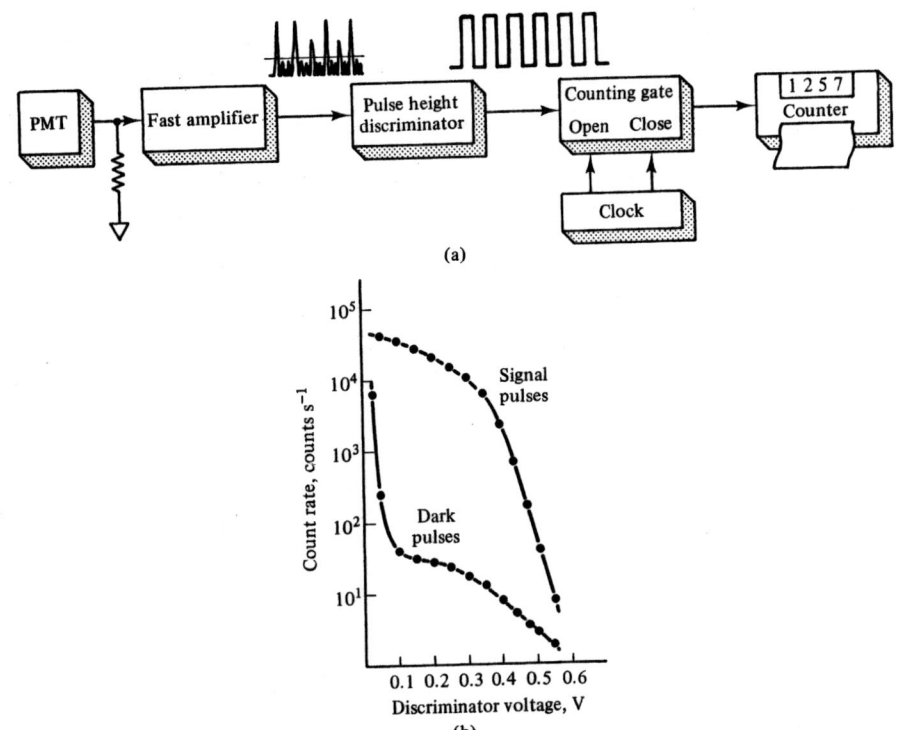

(a)

(b)

**FIGURE 4-31** Photon counting system. In the PMT (a), each photoelectron is converted to a packet of electrons and results in a photoanodic pulse. A fast amplifier increases the amplitude of the individual photoanodic pulses. The discriminator outputs a pulse for every input pulse above a selected threshold level, which allows discrimination against noise pulses and some dark current pulses as shown by the pulse height distribution (b). The average fraction of anodic pulses passed by the discriminator is $f_d$, the discriminator coefficient. A clock opens the counting gate for a selected time interval which allows the pulses to be counted during this interval.

## Modulation Principles

**Modulation** is defined as the alteration of some property of a carrier wave by a signal so that the carrier wave encodes the signal information. Often, the amplitude of the carrier wave is altered in response to the signal amplitude (amplitude modulation). It can also be accomplished by altering the frequency of the carrier wave (frequency modulation), the pulse width, or the phase, but these are not as common in spectrochemical instruments. In spectrometric systems the main purpose of modulation is to move the signal information to a region of the frequency spectrum where it is less subject to noise and more distinguishable from noise than at 0 Hz (dc). Modulation is also used for time discrimination (e.g., lifetime measurements) and for obtaining frequency and phase response information. Section 5-6 contains further information on the use of modulation methods for *S/N* enhancement.

In amplitude modulation, the carrier is typically a square, pulse, or sinusoidal waveform. Pulse waveforms are characterized by the duty cycle or percent of time the signal is in the high-amplitude state. A square wave is just a special case of a pulse waveform where the duty cycle is 50%. In any case, the desired analytical information is transferred by modulation from being the magnitude of a dc signal to being encoded as the amplitude of a carrier waveform. Once the signal information is encoded in this manner, normal signal conditioning and processing techniques (amplification,

bandwidth reduction, etc.) can be applied while discriminating against background signals, other steady-state signals, and low-frequency noise in those signals. To be useful, there must be a means of recovering the signal amplitude information from the modulated carrier, a process called **demodulation**.

The general principles of amplitude modulation are illustrated in Figure 4-32. Typically, demodulation of the filtered carrier waveform is based on correlation techniques. By demodulation, the signal is converted to a dc signal proportional to the ac amplitude of the signal waveform. This is accomplished by electronic or software multiplication or gating of the modulated carrier waveform by a reference waveform of the same frequency which is in phase with the signal (such as the signal used to control the modulation). Signals and random noise at frequencies or phases different from the reference waveform produce an average output of zero.

In spectrochemical instruments there are many different schemes for modulation. It is usually best to modulate as early in the measurement process as feasible so as to achieve the maximum freedom from noise. Thus many techniques modulate at the radiation source itself, which is called **source modulation**. This can be accomplished by inserting a modulator (e.g., a mechanical chopper or an electro-optic modulator as described in Section 3-3) between the excitation or emission source and the detector as shown in Figure 4-33 or by electronic modulation of the excitation source power supply. In absorption, photoluminescence, or

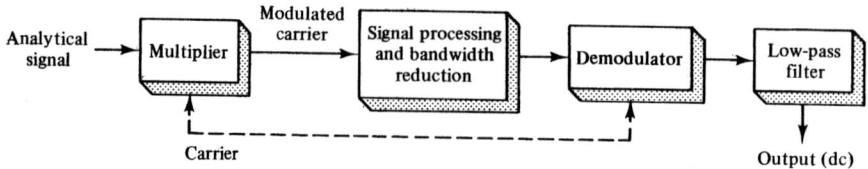

**FIGURE 4-32**   General principles of modulation. The multiplication step produces a carrier wave that varies in amplitude in accordance with variations of the analytical signal amplitude. Any modulator can be considered to be a multiplier. For example, a mechanical chopper multiplies the signal by 1 on one half-cycle and by 0 on the other. The carrier frequency is normally much higher than the frequency components in the signal. Once the signal information is shifted to a higher-frequency region of the spectrum, the signal can be increased in power (amplified) or otherwise treated while bandwidth reduction techniques are employed to discriminate against low-frequency noise and noise at discrete frequencies. The modulation process must be reversible. A demodulator is thus used to recover the signal information (separate it from the carrier). Low-pass filtering then removes the carrier and produces a dc output.

scattering methods, the sample container would be inserted after the chopper so that the source radiation would pass through the sample on alternate half-cycles. Source modulation in absorption or photoluminescence spectroscopy also allows discrimination against any emission by the sample in that the resulting ac output of the transducer is referenced to the dc emission level, which is present during both half-cycles. In emission spectroscopy, the analyte and background emission are both carried by the modulation waveform, and modulation allows discrimination against the dc detector dark signal.

In some cases the source is pulsed on and off in a low-duty-cycle mode, as compared to the 50% duty cycle depicted in Figure 4-33. Although the methods for recovering such signals are quite different, as discussed below, signal information is still moved away from the dc region, and it can be amplified and recovered in the presence of noise.

In another form of modulation, called **sample modulation**, the sample is alternately presented and removed from the excitation light beam or the excitation emission source. In molecular absorption or photoluminescence spectroscopy this can be accomplished by alternately positioning the sample cell and the reference cell in the light path. In atomic emission, fluorescence, and absorption spectrometry, the sample introduction system can alternately present sample and blank to the atomizer. Sample modulation allows automatic referencing of sample information to that of the blank, which can automatically provide correction for blank interferences if the blank is ideal. Unfortuately, most sample modulation techniques operate at only a few hertz, which is often not far enough removed from the noisy dc region to achieve much improvement in *S/N*.

**Wavelength modulation** is the repetitive variation of the wavelength range observed by the detector. This can be accomplished by repetitively scanning a monochromator back and forth across a fixed-wavelength range. Normally, it is possible to achieve only a low

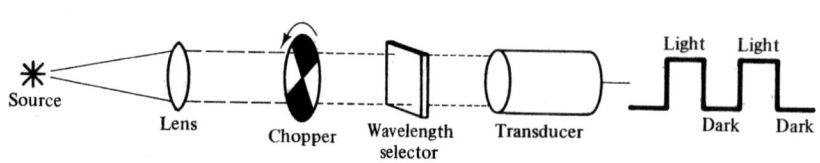

**FIGURE 4-33**   Mechanical chopping of source radiation. The rotating chopper is alternately transparent and opaque resulting in light and dark periods at the transducer. The output is a square wave related to the rotation frequency and number of apertures in the chopper.

modulation frequency with this approach, and the refractor plate approach shown in Figure 4-34a to c is more commonly employed. Wavelength modulation is used primarily in atomic spectrometry, where the variation in wavelength is often very small. Here it is possible to make measurements on and off an atomic line to obtain alternately the analyte and the background signals. Note that wavelength modulation is similar to sample modulation in that essentially a sample–blank measurement is made, although no actual "blank" is measured. Wavelength modulation can also be implemented by oscillation of the spectrometer entrance slit, which effectively changes the angle of incidence to the grating. Oscillation of an interference filter or the splitting of degenerate lines from a source with a modulated magnetic field is also employed. In molecular spectrometry, wavelength modulation can be used to obtain a derivative, as shown in Figure 4-34d. Here the wavelength interval varied is small compared to the width of the molecular band.

Many other characteristics can be used for modulation. Several of these are described in later chapters.

## Signal Conditioning

Usually, the first step in the overall signal processing scheme is to condition the transducer output signal for the circuitry that follows. This often involves data domain conversion and dc or ac amplification (or in some cases, voltage division). After conditioning, the analog signal can be converted to the digital domain or further processed in the analog domain. Some typical signal conditioning elements are given in Figure 4-35 along with their input–output relationships (transfer functions).

*Current-to-Voltage Conversion.* The outputs from several transducers (phototubes, photodiodes, photomultipliers) are currents and must be converted to voltages prior to analog-to-digital conversion or read

**FIGURE 4-34** Wavelength modulation. In (a) a quartz refractor plate oscillates at frequency $f$ near the exit slit (or entrance slit) of a monochromator. Refraction in the plate causes a small wavelength range $\Delta\lambda$ as shown in (b) to be scanned rapidly across the exit slit at $f_m = 2f$. If this range encompasses the analytical band or line, the analytical signal is carried at $f_m$ as shown in (c). If the background signal is constant over $\Delta\lambda$, the background signal is not carried. For atomic absorption and fluorescence measurements, a continuum rather than line excitation source must be employed so that background absorption or fluorescence signals are present over $\Delta\lambda$. In molecular spectrometry, wavelength modulation can be used to take the derivative of a spectrum. As shown in (d), the amplitude of the ac signal during a scan is proportional to the derivative of the band intensity with respect to wavelength if $\Delta\lambda$ is much smaller than the band half-width. Normally, a lock-in amplifier is tuned to the frequency of the refractor plate oscillation to measure the ac signal amplitude. The second derivative can be obtained by tuning the lock-in amplifier to the second harmonic (twice the modulation frequency).

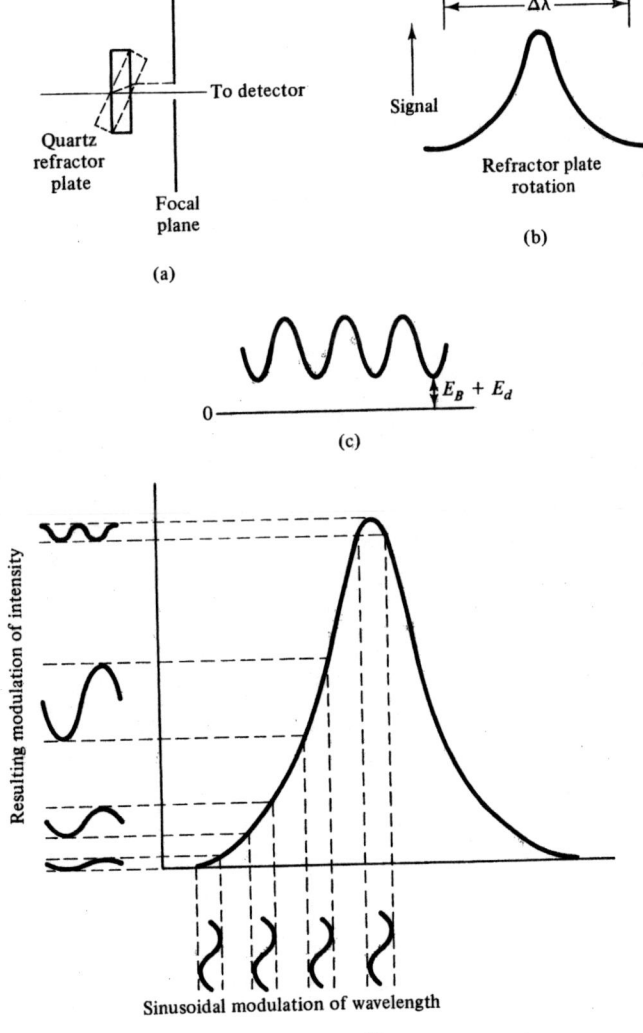

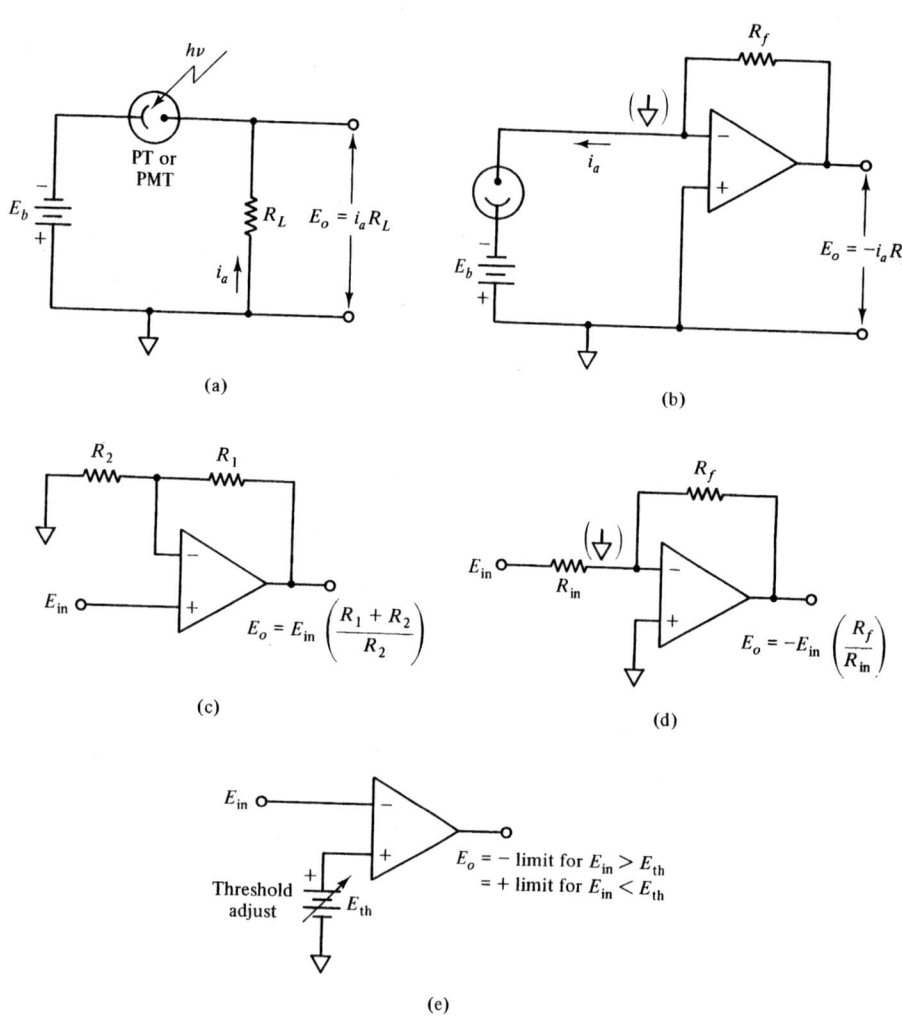

**FIGURE 4-35** Signal conditioning circuits. Common signal conditioners include current-to voltage converters [(a) and (b)], the voltage follower with gain (c), the inverting voltage amplifier (d), and the discriminator (e). Current-to-voltage conversion is used with phototubes (PTs), photomultiplier tubes (PMTs), and photodiodes. The amplifiers shown are used with many transducers. The inverting amplifier (d) can be configured as a resistance-to-voltage converter. The discriminator (e) is used with photon counting systems. The output/input relationships for each circuit are given.

out. Current-to-voltage conversion can be accomplished with a simple load resistor as shown in Figure 4-35a or with an operational amplifier (op amp) current follower as shown in Figure 4-35b. In either case the output voltage $E_o$ is given by

$$E_o = i_o R_L \qquad (4\text{-}33)$$

where $i_o$ is the transducer output current (photoanodic current $i_a$ for a PMT or phototube).

In the case of the op amp current-to-voltage converter, $R_L$ is the feedback resistance $R_f$, and the output voltage is inverted (i.e., $E_o = -i_o R_f$). In the load resistor case, the voltage across the transducer $E_{tr}$ is given by $E_{tr} = E_b - i_o R_L$. At high light levels, if $i_o R_L$ is signficant compared to $E_b$, the bias voltage of the transducer becomes dependent on the light intensity. With the operational amplifier circuit, on the other hand, the anode of the transducer is maintained at virtual common potential by the negative feedback. Thus $E_{tr}$ is maintained at $E_b$ independent of the light level. The operational amplifier technique also reduces stray ca-

pacitance and provides a low output impedance for connecting voltmeters, analog-to-digital converters (ADCs), and so on. With the load resistor technique, the voltage monitoring device is connected directly across the load resistor, which can be as large as $10^{10}$ $\Omega$.

Commercial current-to-voltage converters are available with switch-selectable feedback resistances. Many can measure currents in the picoampere range. These require excellent shielding and grounding and the use of very high quality amplifiers. Some systems have an independently variable time constant so that the response time can be adjusted; they may even have calibrated offset currents for nulling out dark current.

*DC Amplification.* Outputs from voltage-output or current-output transducers after current-to-voltage conversion are often amplified to 0.1 to 10 V prior to conversion to a digital signal or connection to a readout device. For millivolt-level signals, the voltage-follower with gain shown in Figure 4-35c can provide a high input impedance and gains of up to $10^3$ without inversion. Computer data acquisition systems, as discussed later, often provide a programmable-gain amplifier prior to the ADC. Such amplifiers are often voltage followers with remotely selectable gains. An inverting amplifier, such as that shown in Figure 4-35d, can also be used, but the input impedance is only $R_{in}$.

Some transducers, such as thermocouples, produce microvolt-level signals. Special components, such as instrumentation amplifiers and isolation devices, are used to amplify these tiny signals. There are some cases in which a voltage divider must be used to attenuate a signal prior to analog-to-digital conversion or readout. Simple resistive dividers made from precision resistors are typically used.

The outputs of resistive transducers (thermal detectors) can be conditioned in several ways. The resistance can be converted to voltage with the inverting amplifier configuration of Figure 4-35d. Here a fixed input voltage and a fixed input resistor are used, and the transducer is placed in the feedback loop in place of resistor $R_f$. The amplifier output is then a constant times the transducer resistance. If the transducer is substituted for the input resistor, instead, an output voltage proportional to its conductance ($1/R$) is obtained. In many cases, the resistive transducer is placed in one arm of a Wheatstone bridge. For small changes in resistance, the off-balance bridge voltage is proportional to the transducer resistance.

*AC Amplification.* The current-to-voltage converter of Figure 4-35b and the inverting amplifier of Figure 4-35d can be made unresponsive to dc signals by placing a capacitor in the input circuit before the op amp input. The capacitor blocks any dc component of the signal and allows only the varying signal to be amplified.

A capacitively coupled current-to-voltage converter is often used with a modulated source and PMT transducer. The voltage signal at the modulation frequency can be selectively amplified by means of a tuned amplier as discussed later.

*Photon Counting Systems.* For photon counting systems the PMT output is typically converted to voltage with a small load resistor connected to common (e.g., < 100 Ω). The signal conditioning circuitry is usually a fast video voltage amplifier and a discriminator (comparator) such as that shown in Figure 4-35e. Because of the internal gain of the PMT, only modest amplification is needed (10 to 100). However, to avoid errors due to pulse overlap, the rise time of the amplifier should be fast and it should have extended high-frequency response (>100 MHz for modest irradiances).

The discriminator not only allows discrimination against small pulses, but it also shapes the signal prior to the counting circuitry. Again fast response is necessary to avoid pulse pile-up (deadtime) effects. In the discriminator of Figure 4-35e, the user adjusts the threshold voltage level. If the input voltage exceeds this value, the comparator output voltage goes to its negative limit. If the input voltage is below the threshold, the comparator is at its positive limit. Some photon counting systems have a window discriminator, which allows only pulses with amplitudes between a lower and an upper threshold to be counted. These are made from two comparators and a logic circuit.

Commercial photon counters with high-quality components are available from several instrument companies. These can be general-purpose counting units or made specifically for a spectroscopic instrument such as a Raman spectrometer.

### Analog Signal Processing

Although the trend in recent years has been to do as much of the processing as possible in the digital domain, analog signal processing is still widely used. Even with computer systems, some analog processing may be done for reasons of speed or economy. The types of processing considered here are mathematical operations, filtering, ratioing, and *S/N* enhancement. Several of the most common analog processing elements are shown in Fgiure 4-36.

*Mathematical Operations.* Many different mathematical operations can be performed with operational amplifier circuits. The configuration shown in Figure 4-36a is used to perform voltage or current integration. When integration is to begin, the switch is opened and capacitor *C* begins charging. At the end of the integration period the switch is closed to discharge the capacitor. The output voltage just before the switch is closed is given by the expression in Figure 4-36a. Integration is used to measure peak area or to improve the *S/N*.

Differentiation is used to accentuate spectral details during a scan or to measure the time rate of change of signals in reaction-rate measurements. It can be performed by the circuit shown in Figure 4-36b. Here the output voltage is proportional to the time derivative of the input voltage. This simple circuit is usually highly susceptible to noise, which often has a higher rate of change with time than the signal being differentiated. Hence the frequency response of the differentiator is often intentionally degraded by putting a small capacitor in the feedback loop and a resistor in the input circuit. This effectively makes the differentiator an integrator at high frequencies.

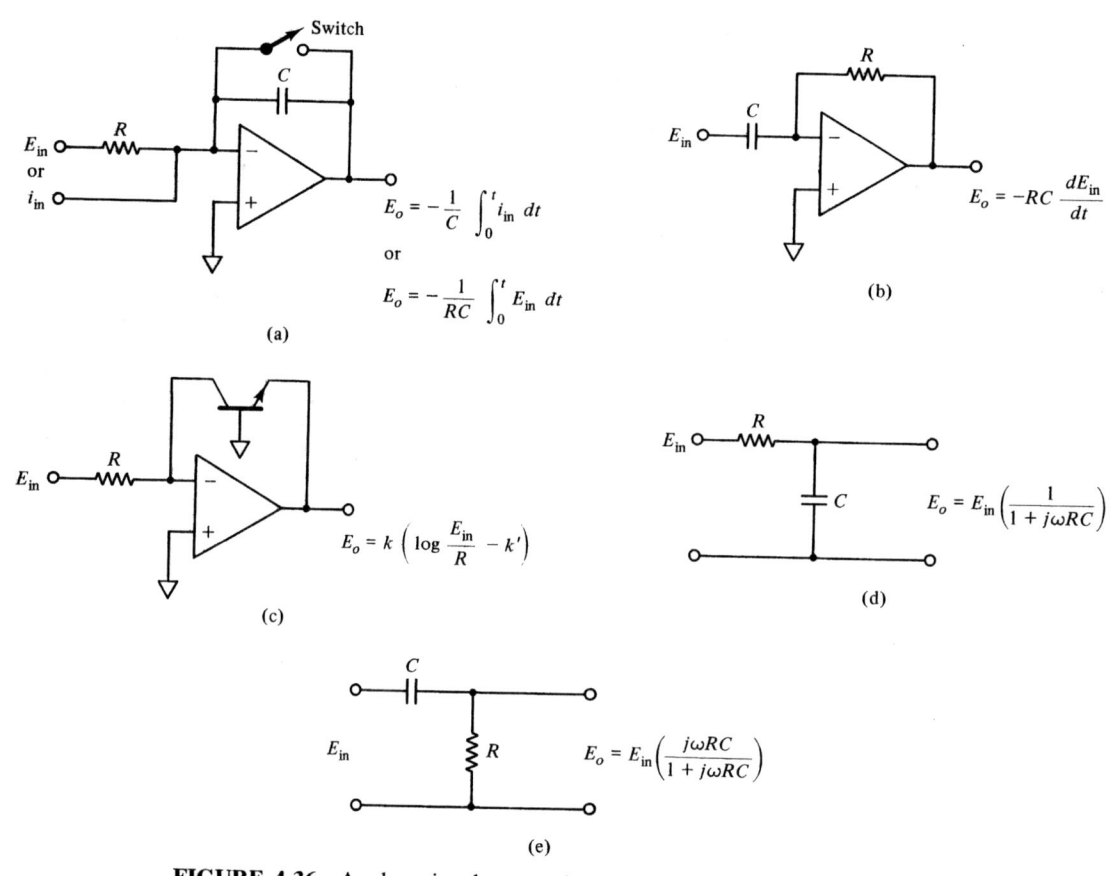

**FIGURE 4-36**   Analog signal processing circuits. Typical analog signal processors include the integrator (a), the differentiator (b), the logarithmic amplifier (c), the low-pass filter (d), and the high-pass filter (e). The input/output relationship for each circuit is given. In the filter transfer functions (d) and (e), $j$ is an operator $\sqrt{-1}$ that indicates a phase shift of 90°.

The circuit of Figure 4-36c is a logarithmic voltage amplifier. It can also be a logarithmic current amplifier if the current input is attached directly to the amplifier inverting input (as in Figure 4-36a). Unfortunately logarithmic elements (transistors and diodes) are quite temperature dependent, and the constants shown in the transfer function are strong functions of temperature. Commercial logarithmic amplifier modules are often temperature compensated. These are usually logarithmic ratio amplifiers. Such amplifiers are frequently used to convert transmittances to absorbances in spectrophotometric instruments (see Section 2-6).

There are many situations in which an output voltage proportional to the difference $(A - B)$ or the ratio $(A/B)$ of two input signals $A$ and $B$ is desired (e.g., dual-beam or dual-channel methods). Although these functions can be implemented in the analog domain, it is increasingly common to use a computer to calculate them.

*Filters.*   Filters are widely used in spectrometric systems. The low-pass filter shown in Figure 4-36d and the high-pass filter shown in Figure 4-36e are first-order passive filters. They provide a decrease in gain of 20 dB per decade change in frequency at frequencies lower or higher than their cutoff frequencies $[f_c = 1/(2\pi RC)]$, respectively. Active filters are based on operational amplifiers; they can avoid loading effects and provide gain. Active filters with orders as high as 6 (rolloff rate = 120 dB/decade) are practical. Commonly, a feedback capacitor $(C_f)$ is used in parallel with the feedback resistor in the op amp current-to voltage converter (Figure 4-35b) to form an active low-pass filter with time constant $\tau = R_f C_f$. The use of filters is discussed in more detail in Section 5-2.

For ac signals, bandpass filters, which transmit a narrow band of frequencies and reject lower or higher frequencies, are readily obtained or constructed. Integrated-circuit (IC) active filters can provide high-pass, low-pass, and bandpass outputs simultaneously. In addition, they can be used to make notch filters which specifically reject a particular frequency band, such as 60 Hz. Bandpass filters can have gain, in which case they are often referred to as **tuned amplifiers**. Such amplifiers are available commercially and are often present in instruments that operate on ac waveforms. One example discussed below is the lock-in amplifier.

*Lock-In Amplifier.* There are many schemes to process and demodulate signals that have been impressed on carrier waves. The **lock-in amplifier** (also called a *synchronous detector*, a *heterodyne detector*, or a *phase-sensitive detector*) is a synchronous demodulation device that can provide recovery of signals literally buried in noise. It is useful for carrier waves that are 50% duty-cycle waves. **Synchronous demodulation** involves multiplication of the modulated carrier (see Figure 4-32) by a reference waveform of identical frequency and of related phase. A block diagram of a lock-in amplifier is shown in Figure 4-37. In the case of the mechanical chopper of Figure 4-33, the reference waveform could be the chopper-drive signal, or another source could be directed through the same chopper to a second detector. In any event, the multiplier has a maximum output for components of the signal of the same frequency as the reference and phase related to it. Since noise is present in a broad frequency range and has random phase (see Chapter 5), a large improvement in *S/N* can be achieved. The low-pass filter removes the component at twice the reference frequency and leaves only the dc component.

The selective amplifiers shown in Figure 4-37 provide gain to obtain signal levels appropriate for the multiplier and subsequent circuits. They also provide discrimination against low-frequency noise and noise at discrete frequencies. Typically, a high-pass filter, a broadband amplifier (bandwidth ≈ 100 Hz), a tuned amplifier (bandwidth a few hertz), or a bandpass filter is used. With 50% duty-cycle modulation techniques (sinusoidal or square carrier wave), the modulated carrier wave becomes sinusoidal after this stage, which also discriminates against harmonics. The sinusoidal carrier wave can then be rectified and filtered to recover the dc waveform.

The synchronous demodulation approach is normally preferred over a narrow bandpass tuned filter followed by simple rectification (asynchronous demodulation) because it is difficult to construct a narrow bandpass filter whose central frequency does not drift. With synchronous demodulation it is not absolutely necessary to remove the dc component of the signal since it averages to zero after the multiplier. Many lock-in amplifiers function properly with slight changes in the carrier frequency because they can track the signal.

The lock-in technique is extremely powerful and is often used in ac measurements. The method allows signals to be amplified in a region of the spectrum that is relatively free from noise and recovered without introducing additional noise. Lock-in amplifiers cannot, however, be used with transients, signals with low duty cycles, or signals with very high repetition rates.

*Boxcar Integrator.* The boxcar integrator is a versatile instrument for measuring repetitive signals, particularly those with short pulse durations and low

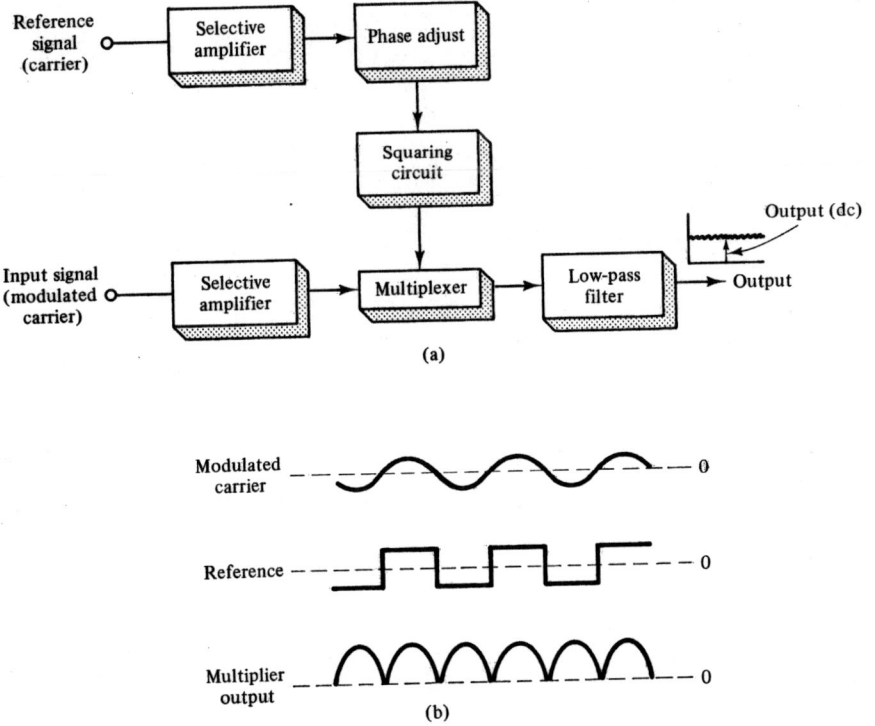

**FIGURE 4-37** Lock-in amplifier block diagram (a) and waveforms (b). The modulated carrier, suitably amplified, is multiplied by a phase-adjusted reference signal derived from the carrier. The multiplier output is then low-pass filtered (a). Only components of the signal that are at the same frequency as the reference and phase-related to it are detected; other signals and noise yield an average output of zero. The low-pass filter removes the carrier and produces a dc output. Demodulation with a square-wave reference signal can be seen in (b) to be equivalent to multiplication by +1 during the positive half-cycle and by −1 during the negative half-cycle. This produces full-wave rectification of the modulated carrier.

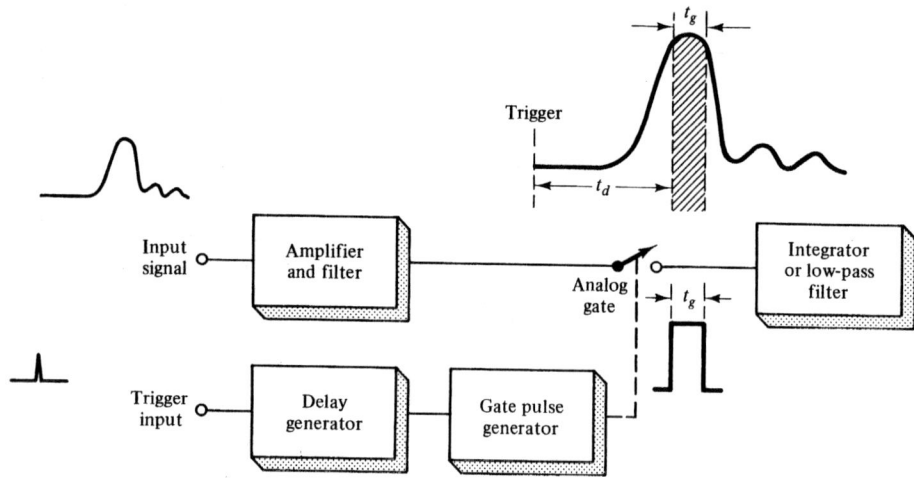

**FIGURE 4-38** Boxcar integrator. The amplified input signal is connected to an integrator for a gate time $t_g$ determined by a gate pulse generator. The gate pulse occurs at a known time delay $t_d$ relative to the trigger signal derived from the start of the repetitive input signal. The same time slice can be averaged for multiple repetitions or the delay generator can slowly scan the gate pulse across the input waveform. Linear integration or exponential averaging can be used.

duty cycles. A block diagram of a boxcar integrator is shown in Figure 4-38.

The boxcar integrator allows the recovery of signals that are time related to a trigger signal and the rejection of those that are not. A typical application might be the measurement of the fluorescence produced by pulsed dye laser excitation. The pumping source for the dye laser could provide the trigger signal, and a PMT would be used to measure the fluorescence. The PMT output, after current-to-voltage conversion, would be the boxcar integrator input. The system could be set up to integrate the PMT output for the entire laser pulse. Averaging over multiple laser firings would improve the *S/N*. Alternatively, a small time slice could be selected and the boxcar time window scanned to produce a time-resolved plot of the fluorescence emission. Gate pulse widths from 100 ps to 10 ms are available on commercial boxcars.

With boxcar integrators a high-pass filter is typically used to remove dc signals and low-frequency noise before they reach the integrator. In some situations it is advantageous to use the dual-channel boxcar approach illustrated in Figure 4-39. Here, one can subtract background or even ratio the signal of interest to a reference signal.

The signal processing techniques provided by lock-in amplifiers and boxcar integrators can also be implemented by digital hardware or by software integration techniques. A synchronous photon counting system can be implemented by using an up/down counter synchronized to a modulated waveform. Here the counter counts up during the on-cycle of the modulation waveform and down during the off-cycle for background subtraction. At this time the hardware approaches are much faster than software-based techniques. A commercial computer-controlled boxcar integrator is available in which analog circuitry is used for high-speed operations and a computer is used to implement several sophisticated

data processing routines. The computer also controls the scanning time and the number of samples averaged.

*Multichannel Averaging.* A multichannel averager acquires a large number of evenly spaced signal readings across a complete waveform on each signal repetition instead of one reading per repetition as in the boxcar integrator. A fast analog-to-digital converter and digital memory are required to digitize and store each reading. On subsequent signal repetitions, the previously stored value for the appropriate channel (memory location) is summed with the new value in a fast

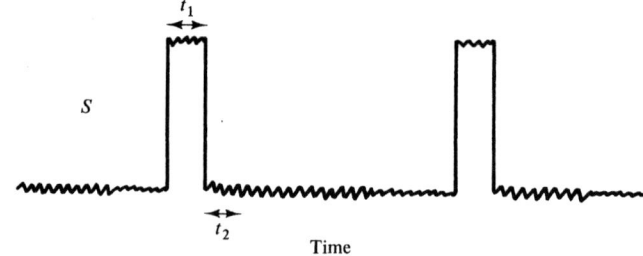

**FIGURE 4-39** Dual-channel boxcar approach. A boxcar integrator can be constructed with two channels (two gates and integrators). The analytical signal information is integrated and stored in one channel during time $t_1$. During time $t_2$ the background is integrated. The difference (integral during time $t_1$ minus integral during $t_2$) is then calculated; this provides correction for dc background signals and discriminates against low-frequency noise. Many dual-channel boxcar integrators also provide a ratioing function so that the ratio of the channel 1 signal to the channel 2 signal is calculated. This is often used with pulsed lasers to reduce the effect of pulse-to-pulse variations in the laser output. Here one photodetector would monitor the laser output itself, while a second would monitor the laser-stimulated process (e.g., fluorescence or transmitted light).

adder, and the new sum is stored. After $n$ repetitions the entire acquired waveform has been averaged over the $n$ repetitions. Multichannel averagers are normally used for acquiring signals that are of lower frequencies than those acquired by boxcar integrators. If averaging is not required, a **transient recorder** can be used to digitize and store a complete single trace of a waveform. One commercial transient digitizer uses a high-speed (5 ns) ADC to digitize the analog information. A computer can then manipulate the stored digital data. Averaging of multiple repetitions can then be done by the computer.

## Computer Data Acquisition

Increasingly microcomputers are being used to acquire and manipulate spectrometric data. A few years ago it was a major task to interface a computer to an experiment. However, today, complete data acquisition systems and analog input–output (I/O) boards are commercially available for almost all popular microcomputers.

A typical data acquisition system is shown in Figure 4-40. The input multiplexer allows 16 analog signals to be selected and converted. The multiplexer address register selects the channel to be converted. The access can be random or sequential. The selected analog input signal is connected to a programmable gain instrumentation amplifier. A sample-and-hold circuit acquires the analog signal at a particular instant in time and holds it constant for the successive approximations ADC. The digital data can then be loaded onto the computer bus through a bus interface (tristate buffers).

Analog I/O boards are also available for many microcomputers. These include a data acquisition system and also one or more output digital-to-analog converters (DACs). Some have on-board memory and even real-time clocks. Prices for these boards can range from a few hundred dollars to over $1000, depending on the options selected. Several firms are now selling complete software packages for laboratory uses. These laboratory notebook packages can be used for many tasks, including data acquisition, manipulation, and graphical display.

## Digital Signal Processing

A substantial amount of signal processing is now being done in the digital domain with computer systems. Although the details are beyond the scope of this book, a few of the common manipulations are briefly considered here.

Many of the mathematical operations discussed above can be done more accurately with computers than

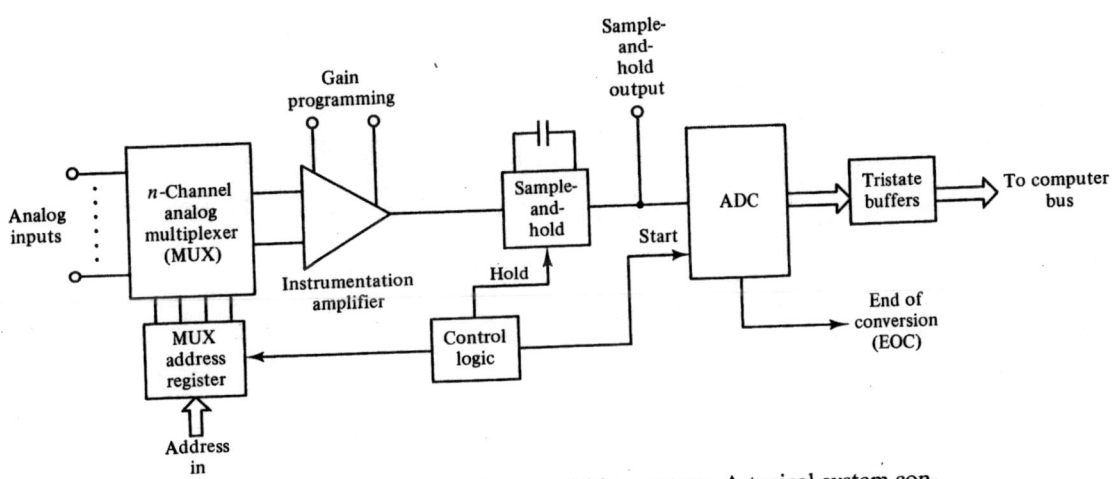

**FIGURE 4-40** Computer data acquisition system. A typical system contains a multiplexer, a programmable gain amplifier, a sample-and-hold circuit, an ADC, and a sequencer. Systems are commercially available for most microcomputers. A wide range of ADC types and conversion speeds are available. Integrating ADCs are relatively slow (ms conversion times), but average the signal during the conversion time. A sample-and-hold circuit is not needed with these converters. Successive approximations converters are relatively fast (microsecond conversion times), but require an absolutely steady voltage during conversion. The fastest converters are flash ADCs developed for digital TV. These convert continuously and have throughput rates in the hundreds of MHz range. They are relatively expensive and of lower resolution than other ADC types.

with analog circuits. Thus such operations as integration, differentiation, ratioing, and logarithmic conversion can readily be performed in the computer. Where high speed is needed, there is still an advantage to using analog circuits for these operations. Although the computations can be quite rapid with a computer, it takes some time for acquiring, manipulating, and storing the data.

Signal-to-noise enhancement techniques are being increasingly implemented with the aid of computers. Digital filtering methods allow the application of filtering functions that are impossible or difficult to implement with analog circuits (see Section 5-6). Algorithms are readily available for common filtering and data smoothing operations. Many commercial spectrochemical instruments with built-in computers contain programs for carrying out a variety of filtering and smoothing functions.

Many computer-based systems have software for taking the Fourier transform or inverse transform of acquired data. Such techniques allow one to manipulate the data in the frequency domain for frequency-domain filtering (see Section 5-6) or to operate on interferometric data.

With computers data can also be readily fit to theoretical or empirical models through nonlinear regression methods. For example, linear and nonlinear calibration curves can be calculated and displayed quickly with computer-based systems. Similarly, data can be fit to standard functions such as exponentials, polynomials, and so on. Again we are increasingly seeing such capabilities being incorporated into commercial instruments.

### Readout and Display Systems

The readout and display system provides a visual representation of the output of the experiment. In the past analog readout and display systems were most common. Most of these are voltage-actuated devices. Computer outputs can be displayed digitally or they can be converted to analog signals by DACs and displayed on analog output devices. Often, an analog display is used for a fast observation of the acquired signals, but the quantitative information is obtained from the digital data.

The output devices used in spectroscopic instruments are analog voltmeters (older instruments), digital voltmeters (panel meters), oscilloscopes, strip-chart recorders, printers, monitors, or terminals. Because of their improved accuracy, freedom from bias, and lower chance of reading error, various digital displays are invariably used in modern instruments unless a continuous plot of output signal vs. time is desired. Even many scanning spectrometers, where recorders are used to display the spectrum, provide digital readouts for quantitative purposes.

For the measurement of voltages, readout devices should have high input impedance to prevent loading. Other characteristics to consider when choosing a readout device for a given application include linearity, resolution (smallest difference in signal that can be resolved), range, response time, stability, and drift.

Monitors used for display with computers can be either monochrome or color. Increasingly, graphics displays are being used to enhance visualization and to provide various plots and correlations that are easy to inerpret and manipulate. High-resolution color graphics display are especially effective in this regard.

## 4-6 OPTICAL SPECTROMETERS

The complete optical spectrometer consists of the optical components discussed in Chapter 3, the radiation source, transducer, and electronic instrumentation discussed in this chapter, and various control systems, sample introduction systems, and sample containers that are specific for each technique. Optical spectrometers can be broadly broken into three categories: single-channel, multichannel, and multiplexing spectrometers. This section considers these spectrometer types and presents the readout expressions for quantitative applications of emission, absorption, and photoluminescence spectrometers.

### Single-Channel Spectrometers

Single-channel spectrometers are considered here to be spectrometers in which information at a single wavelength setting is obtained or a spectrum is recorded sequentially (one wavelength at a time). They are divided into fixed-wavelength systems, conventional scanning systems, and rapid-scanning systems. Single-channel spectrometers can also contain more than one optical beam, as in a double-beam spectrophotometer.

*Fixed-Wavelength Spectrometers.* A spectrometer can be classified as a fixed-wavelength device if it has no scanning mechanism and is intended for operation at a single wavelength even though that wavelength might be alterable. Thus a filter photometer, an atomic absorption spectrometer with manual wavelength adjustment, and a manual colorimeter are all classified as fixed-wavelength spectrometers. These systems are useful only for quantitative analysis after spectral information has been obtained and wavelengths chosen from spectral scans made on a separate instrument.

*Conventional Scanning Spectrometers.* Here a scanning mechanism is present, usually a motor, to turn a grating or prism in a monochromator at conventional scanning speeds (0.01 to 10 nm s$^{-1}$). This sweeps the dispersed spectrum across a fixed exit slit, as described in Chapter 3. Scanning spectrometers can, of course, be operated at fixed wavelengths, which makes them useful for quantitative measurements as well as for spectral scanning. A scanning spectrometer is more complex and costly than a fixed-wavelength system, but considerably more versatile. The conventional spectrometer obtains a complete spectrum in minutes; it can only be used on static systems or on systems in which the spectral properties change very slowly in time.

*Rapid-Scanning Spectrometers.* Some single-channel spectrometers are capable of obtaining a spectrum rapidly. One approach is to oscillate an optical component in the monochromator (grating or mirror) very rapidly. A fast transducer, such as a photomultiplier tube, is required. The mechanical rapid-scanning spectrometer can achieve complete spectral scans in seconds or even milliseconds in some cases. These spectrometers suffer from mechanical complexity and the difficulty in maintaining optical alignment. Mechanical rapid-scanning systems have practically been replaced by multichannel systems.

Another approach to sequential, multiwavelength spectrometry is known as the slew-scan approach. Here the information at certain fixed wavelengths across a broad spectral range might be of interest. The **slew-scan monochromator** is programmed such that it rapidly scans (slews) through the undesired wavelengths and then stops at or slowly scans across the desired wavelength region. The major advantage is that a monochromator is used, for simplicity and low cost, and little time is wasted in spectral regions that are uninteresting.

The image dissector tube described in Section 4-4 is another approach to rapidly scanning a spectrum. The IDT has the advantage of speed in that scanning is accomplished electromagnetically rather than mechanically. The IDT can also scan in two dimensions, which makes it suitable for use with echelle spectrometers. Like any PMT, the IDT has high sensitivity and a wide linear dynamic range. Unlike solid-state array detectors, the IDT does not integrate the incident intensity, which can be a disadvantage in some situations. However, the major drawback of the IDT at present is its high initial cost ($\approx$ \$10,000), which has severely limited its use in spectrochemical instrumentations.

*Double-Beam Spectrometers.* Double-beam systems are used in absorption and photoluminescence measurements. Here the excitation beam is split into sample and reference beams. As shown in Figure 4-41, the beams can pass alternately (double beam in-time) or simultaneously (double beam in-space) through the sample and reference cells. In contrast to the dual-channel spectrometers discussed below, the same wavelength band is viewed by two detectors (double beam in-space) or by a single detector (double beam in-time). In double-beam systems the sample beam always passes through the sample container, while the reference beam may or may not pass through the reference container and hence the blank.

In many molecular double-beam spectrometers both the sample and the blank signals are acquired to allow on-line compensation for blank signals. For absorption spectrometers, where the ratio of the sample signal to the reference signal is calculated, the double-beam system compensates for source intensity drifts. In addition, during scanning, the double-beam system compensates for the source spectral distribution and changes in the monochromator efficiency with wavelength.

In some molecular and all atomic spectrometers, the reference beam does not pass through the reference container, but is monitored directly by the reference detector, often without wavelength selection. Here the double-beam system only provides compensation for source intensity fluctuations.

## Multichannel Spectrometers

Multichannel spectrometers increase sample throughput and reduce sample consumption where many species per sample must be measured. The multichannel spectrometer measures the spectral information at multiple wavelengths simultaneously. It is still a dispersive system in that the spectral information is dispersed spatially by a grating or prism. However, the multichannel spectrometer is either a polychromator with multiple exit slits and detectors or a spectrograph with a multichannel detector (see Chapter 2).

*Direct Readers.* The direct-reading spectrometer has been commercially available for emission spectrometry for many years. Typically, an array of slits and photomultipliers are used, one for each wavelength to be detected. A typical system using a Rowland circle configuration is discussed in Chapter 3, Figure 3-54. Usually, high-dispersion polychromators with focal lengths of 0.75 m or greater used in these systems to make it easier to position the slits and detectors. Instruments with as many as 57 exit slits and detectors have been described.

The direct-reading spectrometer is widely used where the same group of wavelengths must be moni-

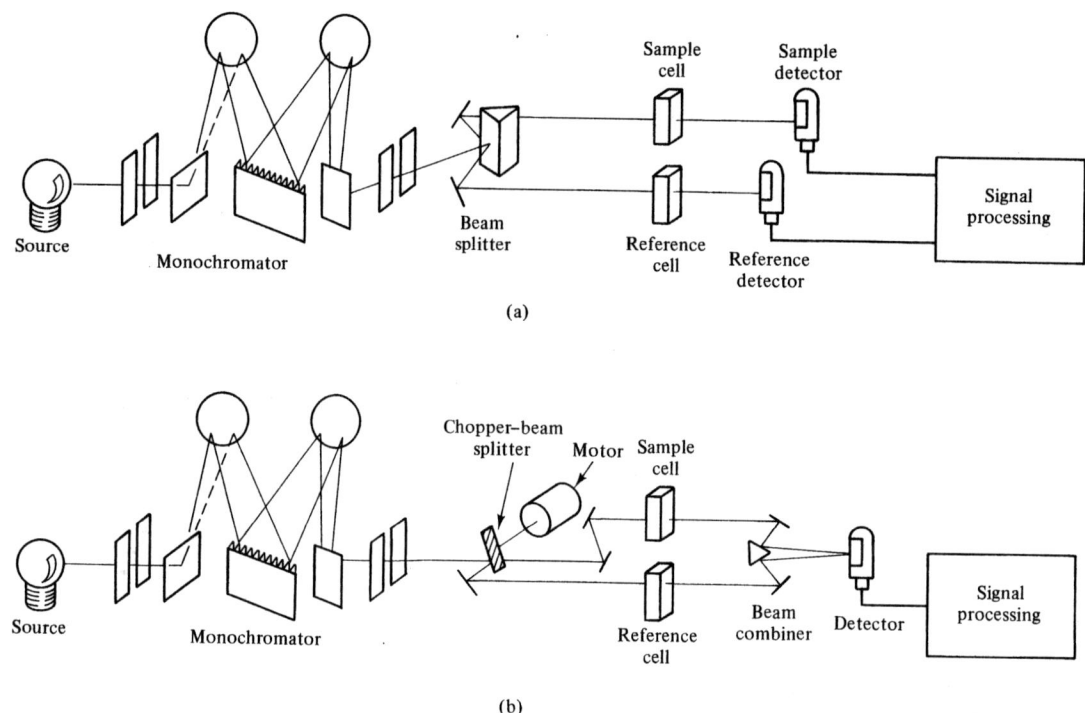

(a)

(b)

**FIGURE 4-41** Double-beam spectrometers. In the double-beam in-space configuration (a), the excitation beam is split and two separate detectors are used. In the double-beam in-time configuration (b), the two beams are recombined and alternately strike a single detector. In either case the signal processor calculates the ratio of the sample signal magnitude to that of the reference signal ($T$) for absorption measurements or the difference between the sample signal strength and the blank signal strength for photoluminescence measurements. Wavelength selection can occur before or after the beams pass through the cells.

tored routinely, as in multielement emission spectroscopy. This type of system is expensive and is not very flexible because alignment of the exit slits is a tedious and time-consuming task. As a result, wavelengths are changed only infrequently. Correction for background is also difficult with a direct reader since fixed wavelengths are used. Oscillating slit arrangements and vibrating quartz plates in front of the exit slits as discussed in the preceding section have been used to scan a small spectral region around the fixed wavelength and thus allow background correction.

*Spectrographs with Multichannel Detectors.* An alternative approach to the direct reader is to remove the exit slits and to obtain all the spectral information simultaneously at the focal plane. A true multichannel detector is then required. The photographic plate has been used with spectrographs for many years. Recently, however, it is being replaced by the video-type multichannel detectors discussed in Section 4-4. Because of their small sizes and rapid response speeds, diode arrays and vidicons are used in small, medium-resolution spectrographs as well as in high-res-

olution systems. They allow spectra to be obtained in situations where the concentrations of the monitored species are changing in time. One commercial emission spectroscopy system uses an echelle grating and a two-dimensional vidicon detector to give high-resolution, multiple-wavelength coverage. Several commercial molecular absorption spectrophotometers use diode arrays; most can provide a complete UV-visible absorption spectrum in seconds. Such spectrophotometers are now becoming quite popular and could make conventional scanning UV-visible instruments obsolete in the near future.

*Dual-Channel Spectrometers.* A dual-channel spectrometer is a special case of a multichannel spectrometer where two wavelengths are monitored. It can be implemented with a direct-reader approach, but is often constructed by using two filters and detectors or two monochromators and detectors. Such spectrometers are used primarily for background or blank correction. One channel is used to monitor the analytical wavelength (analyte plus background signal) and the other channel monitors a wavelength or wavelength range where only the background signal is present (or significant). If the background signal at this second wavelength is equal to the background signal at the analytical wavelength, the difference in the signals from the two channels represents the corrected analytical signal. Note that channels in direct readers can be dedicated to background correction and that with multichannel detectors signals from appropriate wavelengths can be selected to obtain the required information. Scanning a dual-channel spectrometer can be used to obtain derivative information.

### Multiplexing Systems

Multiplexing systems can be classified in general as time-division multiplexing or frequency-division multiplexing systems. **Time-division multiplexing** allows the timesharing of a single optical path or single detector by multiple optical signals. The double-beam-in-time spectrometer is an example of time-division multiplexed spectrometer for two optical beams. Time-division multiplexing has also been used in atomic absorption and fluorescence spectrometry.

In the **frequency-division multiplex** approach all wavelengths strike a single-channel detector simultaneously; the spectral information at each wavelength is encoded in such a manner that it can be recovered using mathematical transform techniques. In Fourier transform spectroscopy the spectral information is encoded with a Michelson interferometer. Since the radiation is not dispersed, the throughput can be very high. Also, the detection of all wavelengths simultaneously leads to a S/N advantage in cases where detector noise is limiting. Fourier transform methods are most widely used in the IR spectral region because detectors are often detector noise limited; in the UV and visible regions the S/N is almost always limited by other sources of noise.

In **Hadamard transform spectroscopy** a conventional dispersive spectrometer is used and a movable mask that contains a series of slits is placed in the focal plane. The positions of the transparent and opaque parts of the mask correspond to the elements of a Hadamard matrix. Computer solution to the linear equations that result from intensity measurements for different mask positions produces a conventional spectrum. Some also have a slit mask for the entrance slit. Like Fourier transform methods, Hadamard transform techniques have been found most useful in the IR region of the spectrum, where detector noise determines the S/N.

### Quantitative Readout Expressions

We are now prepared to expand the general readout expressions presented in Chapter 2 for emission, photoluminescence, and absorption spectrometry to take into account the instrumental parameters discussed in this chapter. We consider here a fixed-wavelength spectrometer of throughput factor $Y(\lambda)$ and analog signal processing with voltage readout. A portion of the radiation emitted or transmitted by the chemical system and wavelength selector is assumed to be incident on a PMT transducer of radiant cathodic responsivity $R(\lambda)$ and gain $m$. For simplicity, here and in later chapters we will assume that the PMT collection efficiency is 1. The gain of the signal modifier in V A$^{-1}$ is $G$. Note that if a current-to-voltage converter is the only signal processing element used, $G = R_f$, the feedback resistance.

*Emission and Chemiluminescence Spectrometers.* The analytical emission signal $E_E$ is obtained by subtraction of a blank signal $E_{bk}$ from the total measured signal $E_{tE}$ ($E_E = E_{tE} - E_{bk}$). The signal $E_E$ is related to instrumental parameters by equations 4-33 (where $i_o$ is the appropriate PMT anodic current or product of the photocathodic current and the PMT gain), 4-30, 3-69, and 3-71:

$$E_E = mGi_E = mG \int_0^\infty B_{\lambda E} Y(\lambda) R(\lambda)\, d\lambda \qquad (4\text{-}34)$$

where $B_{\lambda E}$ is the spectral radiance due to analyte emission and $i_E$ is the analyte emission photocathodic current. Here and elsewhere, for simplicity, we drop the subscript $c$ for cathodic unless there is possible confusion between cathodic and anodic current. The blank signal is composed of a dark signal $E_d$ and a background emission signal $E_{bE}$. The dark signal is given by

$$E_d = mGi_d \qquad (4\text{-}35)$$

while the background emission signal is

$$E_{bE} = mGi_{bE} = mG \int_0^\infty B_{\lambda bE} Y(\lambda) R(\lambda)\, d\lambda \qquad (4\text{-}36)$$

where $B_{\lambda bE}$ is the spectral radiance of the background emission, $i_{bE}$ is the background emission photocathodic current, and $i_d$ is the cathodic dark current. For quantitative analysis $B_{\lambda E}$ can be related to the analyte concentration as is shown in later chapters.

*Photoluminescence Spectrometers.* In photoluminescence the analyte luminescence signal $E_L$ is obtained after blank subtraction from the total luminescence signal $E_{tL}$ ($E_L = E_{tL} - E_{bk}$) if analyte emission is negligible. The photoluminescence signal is related to the spectral radiance of analyte photoluminescence $B_{\lambda L}$ by

$$E_L = mGi_L = mG \int_0^\infty B_{\lambda L} Y(\lambda) R(\lambda)\, d\lambda \qquad (4\text{-}37)$$

where $i_L$ is the photocathodic current due to analyte luminescence. The blank signal $E_{bk}$ is composed of a dark signal, given by equation 4-35, a background emission signal, a scattering signal, and a background luminescence signal. These latter signals are all of the form given by equation 4-37 with their respective spectral radiances substituted for $B_{\lambda L}$. Again $B_{\lambda L}$ is related to the analyte concentration as described in later chapters. Note that for emission and photoluminescence measurements (see equations 4-34, 4-36, and 4-37), the observed signal is the desired spectral radiance distorted by the spectrometer response function [e.g., $Y(\lambda) R(\lambda)$]. Therefore, in a spectral scan, the shape and intensities of the spectral bands or lines obtained are dependent on the instrumental response function.

*Absorption Spectrometers.* In absorption spectrometry the transmittance $T$ is given by $T = E_s/E_r$, where $E_s$ is the signal with the sample in place and $E_r$ is that with the reference in place. If we assume that background luminescence, analyte emission, and analyte luminescence are absent, the transmittance is

$$T = \frac{E_{st} - E_{0t}}{E_{rt} - E_{0t}} = \frac{E_s}{E_r} \qquad (4\text{-}38)$$

Here $E_s$ is obtained by subtracting the 0% $T$ signal $E_{0t}$

from the total sample signal $E_{st}$, and $E_r$ is obtained by subtracting $E_{0t}$ from the 100% $T$ signal $E_{rt}$. With the foregoing assumptions, $E_{0t}$ is due to dark current and background emission given by equations 4-35 and 4-36, respectively. The signals $E_s$ and $E_r$ are given by

$$E_s = mGi_s = mG \int_0^\infty B_\lambda Y(\lambda) R(\lambda) T_s\, d\lambda$$

and

$$E_r = mGi_r = mG \int_0^\infty B_\lambda Y(\lambda) R(\lambda) T_r\, d\lambda \qquad (4\text{-}39)$$

where $B_\lambda$ is the source spectral radiance, $T_s$ is the sample transmission factor, $T_r$ is the reference transmission factor, and $i_s$ and $i_r$ are the reference and sample photocathodic currents. The value of $T_s$ is related to the analyte concentration as detailed in later chapters. Often experimental conditions are arranged such that the analyte absorptivity does not vary over the wavelength range incident on the sample. In this case, the instrumental response function does not appear in a transmittance or absorbance measurement because a ratio is measured.

For direct absorbance measurements the voltage proportional to the analyte absorbance $E_A$ is related to the signals $E_s$ and $E_r$ by equations 2-33 and 2-34. In the case of modulated systems, the radiances and cathodic currents in the expressions above are the average values. For photon counting systems the signals are in terms of $n$, the number of counts in time $t$.

# PROBLEMS

**4-1.** Usually when one buys a radiation source, the manufacturer will supply spectral radiance data so that one can judge if the intensity is sufficient for a given application. For tungsten lamps, the spectral radiance can be estimated from blackbody parameters.
   (a) Calculate the spectral radiance of a tungsten lamp at 500 nm with a color temperature of 2700 K, $\varepsilon_\lambda = 0.40$, and $T_w = 0.92$ in W sr$^{-1}$ cm$^{-2}$ nm$^{-1}$ and photons s$^{-1}$ cm$^{-2}$ sr$^{-1}$ nm$^{-1}$.
   (b) Apply propagation of error mathematics to the blackbody equation to determine how the uncertainty in temperature relates to an uncertainty in radiance. Evaluate the standard deviation and RSD in radiance at 500 nm caused by a 5°C temperature standard deviation under the conditions in part (a).

**4-2.** Consider a ruby laser made with a 10-cm-long ruby rod and mirrors of reflectivity 0.99 and 0.74. Ruby is

$Al_2O_3$ doped with Cr. Assume that the Cr concentration is $1.58 \times 10^{19}$ cm$^{-3}$ and the transition cross section is $1.27 \times 10^{-20}$ cm$^2$ at the lasing wavelength of 694 nm.
   (a) Calculate the transmittance of the ruby rod in the absence of the pump.
   (b) Calculate the threshold population inversion and the number density of the upper energy level of the lasing transition at threshold.

**4-3.** The anode of a 1P28 photomultiplier is connected to a current-to-voltage converter with a $1.0 \times 10^7$-$\Omega$ feedback resistor. The PMT voltage is adjusted to 800 V and the collection efficiency is 1.00. Assume that $1.0 \times 10^{-11}$ W at 500 nm is incident on the photocathode. Calculate or indicate the following values (the specification sheet in Appendix D is required).
   (a) Current gain ($m$).
   (b) Photocathodic current ($i_{cp}$).

(c) Photoanodic current ($i_{ap}$).

(d) Photocathodic pulse rate ($r_{cp}$).

(e) Anodic dark current ($i_{ad}$).

(f) Cathodic dark current ($i_{cd}$).

(g) Photoanodic voltage ($E_p$).

(h) Anodic dark current voltage ($E_d$).

**4-4.** How does one decide between a lock-in amplifier and a boxcar integrator as a signal processor for modulated signals?

**4-5.** When does one choose a multichannel spectrometer instead of a single-channel spectrometer?

**4-6.** Contrast the characteristics and uses of photon detectors versus thermal detectors.

**4-7.** A vacuum phototube has a radiant cathodic responsivity of 0.08 A W$^{-1}$ at 400 nm.
(a) Find the quantum efficiency at 400 nm.
(b) If the incident photon flux at 400 nm is $2.75 \times 10^5$ photons s$^{-1}$, find the anodic pulse rate and the photoanodic current for a collection efficiency of 0.90.

**4-8.** A photomultiplier tube has a radiant cathodic responsivity of 8.0 mA W$^{-1}$ at 365 nm. It is operated under conditions that give a gain $m$ of $1.5 \times 10^5$ and a collection efficiency $\eta$ of 0.88. If $5.8 \times 10^{-11}$ A of anodic current can just be detected, what is the minimum detectable photon flux in photons s$^{-1}$? If photon counting were used instead of dc current signal processing, what is the count rate corresponding to the foregoing photon flux? Note that a lower light level is normally detectable with photon counting.

**4-9.** A PMT has a current gain of $4.5 \times 10^5$ and a quantum efficiency of 0.36 at 450 nm. The PMT output is connected to a current-to-voltage converter with a feedback resistor of 10 MΩ. A monochromatic source with a flux $\Phi_p$ of $2.0 \times 10^4$ photons s$^{-1}$ is incident on the PMT. What is the output voltage of the $I$–$V$ converter? Assume $\eta = 1$.

**4-10.** Compare and contrast the microchannel plate intensified photodiode array with a photomultiplier tube for the detection of low radiation levels. Give the major advantages and disadvantages of both detectors.

**4-11.** Calculate the spectral energy density for a 2500-K blackbody at 600 nm in J cm$^{-3}$ nm$^{-1}$ and photons cm$^{-3}$ nm$^{-1}$.

**4-12.** How do the total radiance and wavelength of maximum emission of blackbodies at 2000 K and 3000 K compare?

**4-13.** For an atomic transition from the first excited state of degeneracy 2 to a nondegenerate ground state, the Einstein coefficient for spontaneous emission is $1.0 \times 10^9$ s$^{-1}$ and the wavelength is 300 nm. Calculate the following values.
(a) The Einstein $B$ coefficients for stimulated absorption and emission.
(b) The absorption oscillator strength (refer to Appendix F).

**4-14.** Calculate the longitudual mode separation for a 0.25-m laser in Hz and nm for a lasing wavelength of 488 nm.

**4-15.** The $D^*$ value for a detector with an area of 0.10 cm$^2$ is $1.0 \times 10^9$ cm Hz$^{1/2}$ W$^{-1}$. Calculate the noise equivalent power with a 1-Hz noise equivalent bandpass.

**4-16.** Consider a linear diode array with diodes that have a saturation charge of a 3.6 pC and a responsivity at 500 nm of 12 pA μW$^{-1}$ cm$^2$. Calculate the exposure time with an incident irradiance of 1.0 μW cm$^{-2}$ that produces saturation. How many electron–holes pairs are produced at saturation?

**4-17.** In photon counting, nonlinearity occurs at high photon fluxes because of pulse overlap effects (i.e., the probability of two photons generating photoelectrons within the response time or dead time of the detector or signal processor becomes significant). For some photon counting systems denoted paralyzable systems, the observed count rate $r'$ is related to the true count rate $r$ by $r' = r(1 - r\tau)$ if $r\tau < 0.13$, where $\tau$ is the system deadtime in seconds. Calculate the true count rate for which the observed count rate is 1% low for a photon counting system with a dead time of 2 ns.

**4-18.** Nonlinear effects are only observed with laser radiation. Calculate the electric field and irradiance necessary for the polarization due to second-order effects to be as large at that due to first-order effects. Assume the second-order electric susceptibility for the material irradiated to be $10^{-3}$ of the first-order susceptibility.

## REFERENCES

The following are general references on optical sources and/or detectors including optical measurements.

1. F. Grum and R. J. Becherer, *Optical Radiation Measurements*; vol. 1, *Radiometry*, Academic Press, New York, 1979. Has a good discussion of sources of radiation and detection.

2. E. L. Dereniak and D. G. Growe, *Optical Radiation Detectors*, Wiley, New York, 1984. Good discussion of all optical detectors. Includes photovoltaic, photocon-ductive, photoemissive, thermal, bolometers, pyroelectric, and charge-transfer devices (CCDs and CIDs).

3. G. J. Zissis and A. J. Larocca, "Optical Radiators and Sources," in *Handbook of Optics*, W. G. Driscoll and W. Vaughan, eds., McGraw-Hill, New York, 1978. A practical description of various optical sources.

4. S. F. Jacobs, "Nonimaging Detectors," in *Handbook of Optics*, W. G. Driscoll and W. Vaughan, eds., McGraw-Hill, New York, 1978.

5. H. T. Betz and G. L. Johnson, "Spectroradiometric Principles," in *Analytical Emission Spectroscopy*, vol. I, pt. II, E. L. Grove, ed., Marcel Dekker, New York, 1971, pp. 323–381.

6. R. C. Elser, "Optical Instrumentation," in *Trace Analysis; Spectroscopic Methods for Elements*, chap. V, J. D. Winefordner, ed., Wiley-Interscience, New York, 1976. A good discussion of optical measurements, including transducer, signal modifier, and readout systems.

7. R. W. Engstrom, *RCA Photomultiplier Handbook*, PMT-62, RCA Corp., Solid State Division, Lancaster, Pa., 1980. A classic reference book for users of photomultiplier tubes. Contains a wealth of useful information.

8. *Electro-Optics Handbook*, Technical Series EOH-11, RCA Corp., Lancaster, Pa., 1974.

9. G. F. Kirkbright and M. Sargent, *Atomic Absorption and Fluorescence Spectroscopy*, Academic Press, London, 1974. Has an excellent chapter on spectral light sources, including hollow cathode lamps, electrodeless discharge lamps, and others.

The following references are devoted to the principles and application of lasers.

10. M. D. Levenson, *Introduction to Nonlinear Laser Spectroscopy*, Academic Press, New York, 1982.

11. N. Omenetto, *Analytical Laser Spectroscopy*, Wiley, New York, 1979. An excellent book dealing with laser principles and the applications of lasers in analytical chemistry.

12. G. M. Hieftje, ed., *New Applications of Lasers to Chemistry*, American Chemical Society, Washington, D.C., 1978. Contains excellent chapters on laser applications written by experts in the field.

13. F. E. Lytle, "Laser Fundamentals," in *Lasers in Chemical Analysis*, G. M. Hieftje, J. C. Travis, and F. E. Lytle, eds., Humana Press, Clifton, N.J., 1981. A good introduction to laser principles.

14. M. J. Wirth, "Tunable Laser Systems," in *Lasers in Chemical Analysis*, G. M. Hieftje, J. C. Travis, and F. E. Lytle, eds., Humana Press, Clifton, N.J., 1981. A practical discussion of the state of the art as of 1981 in tunable lasers.

15. J. M. Harris, "Pulsed Laser Systems," in *Lasers in Chemical Analysis*, G. M. Hieftje, J. C. Travis, and F. E. Lytle, eds., Humana Press, Clifton, N.J., 1981. A good discussion of pulsed lasers.

16. E. H. Piepmeier, ed., *Analytical Applications of Lasers*, Wiley-Interscience, New York, 1986.

17. A. Yariv, *Optical Electronics*, 3rd ed., Holt, Rinehart and Winston, New York, 1985. An excellent laser book. Also discusses beam propagation in fibers, modulation, atomic systems, detectors, and many other topics.

Listed below are more specialized references providing additional details on specific topics covered in this chapter.

18. J. C. Wright, "Nonlinear Optics," in *Lasers in Chemical Analysis*, G. M. Hieftje, J. C. Travis, and F. E. Lytle, eds., Humana Press, Clifton, N.J., 1981. Includes a good discussion of nonlinear polarization effects.

19. H. V. Malmstadt, C. G. Enke, and S. R. Crouch, *Electronics and Instrumentation for Scientists*, Benjamin-Cummings, Menlo Park, Calif., 1981. A modern text that discusses instrumentation and measurement principles.

20. R. J. Higgins, *Electronics with Digital and Analog Integrated Circuits*, Prentice-Hall, Englewood Cliffs, N.J., 1983. A good discussion of modern electronics with emphasis on integrated circuits.

21. Y. Talmi, ed., *Image Devices in Spectroscopy*, ACS Symposium Series 102, American Chemical Society, Washington, D.C., 1982. Papers were presented in a symposium on image detectors.

22. Y. Talmi, ed., *Multichannel Image Detectors*, vol. 2, ACS Symposium Series 236, American Chemical Society, Washington, D.C., 1983. A more recent book based on a 1982 symposium by leaders in the field.

23. D. G. Jones, "Photodiode Array Detectors in UV-VIS Spectroscopy: I and II," *Anal. Chem.*, 57, 1057A, 1207A (1985). Recent reviews on array detectors.

24. G. Horlick and E. G. Codding, "Photodiode Arrays for Spectrochemical Measurements," in *Contemporary Topics in Analytical and Clinical Chemistry*, D. M. Hercules, G. M. Hieftje, L. R. Snyder, and M. A. Evenson, eds., vol. 1, chap. 4, Plenum Press, New York, 1977.

# CHAPTER 5

# Signal-to-Noise
# Ratio Considerations

The precision of instrumental measurements, including spectrochemical measurements, is often limited by noise in the measured signals. Noise is defined here as unwanted fluctuations in the desired signal which obscure its measurement. As an analogy, consider the problem of listening to one person in a crowded room with a multitude of conversations. The noise in this case is the audible noise from all the other conversations which if too loud can obscure the signal, the voice of the person you are trying to hear.

Because noise affects the quality and precision of spectrochemical measurements, the optimization of experimental variables is often based on reducing the relative amount of noise or increasing the signal-to-noise ratio ($S/N$). Noise can be introduced by the sample presentation system, the optical information encoder, the radiation transducer, and the signal processing and readout system. Random errors in sample acquisition and sample preparation can also limit the precision of a technique. Irreproducibility due to these sources is normally not considered noise, although it is sometimes denoted **chemical noise**.

In this chapter the characteristics of noise and the effect of signal processing on the noise observed are first discussed. The various sources of noise and their dependencies on experimental variables are presented next. Complete $S/N$ expressions are developed for emission, luminescence, and absorption techniques. The use of these expressions for optimizing precision is emphasized. Finally, $S/N$ enhancement techniques are reviewed.

## 5-1 CHARACTERISTICS OF SIGNALS AND NOISE

A typical recorder tracing of a dc signal from a spectrometer is shown in Figure 5-1. The resolution of the recorder is sufficient to observe the fluctuations or noise in the signal. The signal is characterized by its mean value, which can be estimated by drawing a line through the center of the fluctuations as shown. Alternatively, $n$ independent measurements of $E_i$ can be made during the observation period, and the arithmetic mean calculated from

$$\overline{E} = \frac{\sum_{i=1}^{n} E_i}{n} \tag{5-1}$$

135

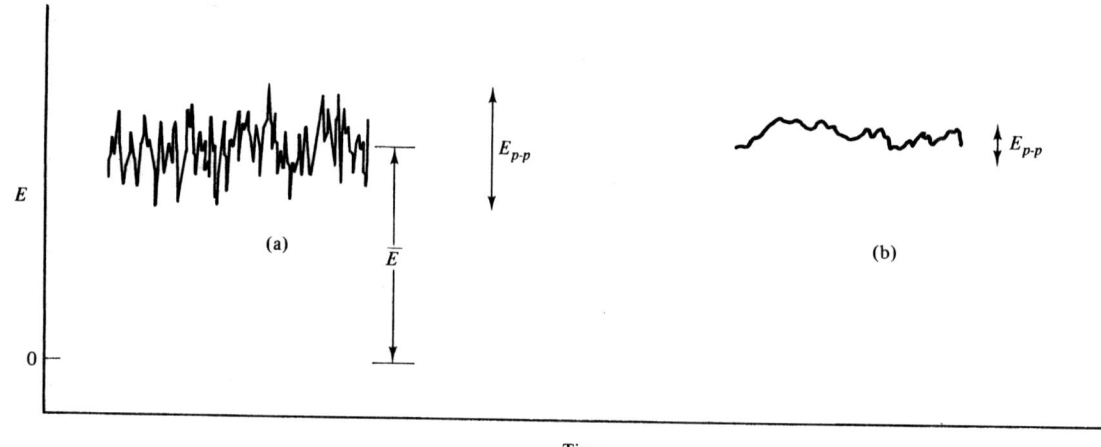

Time

**FIGURE 5-1**   Recorder tracing of dc voltage with noise. The mean value of the signal is $\overline{E}$. The magnitude of the noise is the root-mean-square (rms) noise, $s_E$, the average power content of the noise. The rms noise is approximately one-fifth of the peak-to-peak noise ($E_{p\text{-}p}$). In (a) the *RC* time constant $\tau$ is 0.1 s, while in (b), $\tau$ is 1.0 s. The larger time constant is seen to reduce the higher-frequency noise components and thus to reduce the total amount of noise observed.

## Noise Magnitude

Noise is characterized by its magnitude, frequency, phase, and character. Noise magnitudes are typically expressed as root-mean-square (rms) or as peak-to-peak (p-p) values. The rms noise is the standard deviation of the signal ($s_E$) derived from $n$ measurements (see Appendix A), or

$$\text{rms noise} = s_E = \left[ \frac{\sum_{i=1}^{n} (E_i - \overline{E})^2}{n - 1} \right]^{1/2} \quad (5\text{-}2)$$

The noise magnitude is expressed as an rms value because the instantaneous voltage deviations from the mean ($e_i = E_i - \overline{E}$) average to zero over a large period of time. Thus it is necessary to square the instantaneous voltage deviations and then take the square root as in equation 5-2 to obtain a quantity that is always greater than zero. A fluctuating voltage of 1 V rms produces the same heating (power dissipation) in a resistor as a dc voltage of the same magnitude. The mean square noise $s_E^2$ is the square of the rms value and is thus the variance in the signal due to noise.

The peak-to-peak noise ($E_{p\text{-}p}$) can be extracted from a recorder tracing as shown in Figure 5-1. If $n$ discrete measurements of the signal are made, $E_{p\text{-}p}$ is the range or the difference between the maximum and minimum values of discrete signal measurements. For a normal distribution (see Appendix A), 99.7% of the instantaneous voltage deviations from the mean are within $\pm 2.5$ standard deviations from the mean. Thus the rms

noise is approximately $\frac{1}{5}E_{p\text{-}p}$ for normally distributed noise. This relationship is approximate because in a given observation time, the maximum or minimum instantaneous voltages observed could be less or more than $2.5s_E$ from the mean voltage.

## Noise Types

Noise from a given source may or may not be random. The sign and magnitude of the instantaneous voltage deviation from the mean value are unpredictable for **random noise**. We can imagine random noise to be the result of summing together the outputs of an infinite number of sine-wave generators at some observation point with their magnitudes, frequencies, and phases instantaneously and independently changed. For **nonrandom** noise, the sign and magnitude of the fluctuation can be correlated in time with some event. We can imagine nonrandom noise to be a time-varying signal from a signal generator where the magnitude and time behavior are controlled in a known fashion.

Noise can also be classified as fundamental or nonfundamental. **Fundamental noise** arises from the particle nature of light and matter and can never be totally eliminated. **Nonfundamental noise**, often called **excess noise**, is due to imperfect components and instrumentation; theoretically, it can be eliminated completely. Random noise can be fundamental or nonfundamental noise, but nonrandom noise is never fundamental noise.

The noise observed in a signal at any instant is due to the summation of the fluctuations caused by a large number of random and nonrandom noise sources

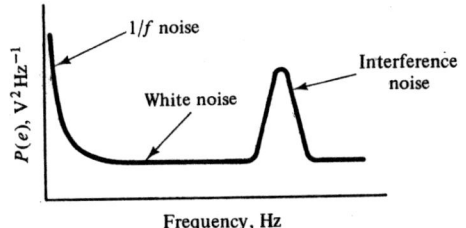

**FIGURE 5-2** Noise power spectrum (NPS). The NPS spectrum is a plot of the mean-square noise per unit frequency interval, $P(e)$ in $V^2\ Hz^{-1}$, vs. the frequency in Hz. The three types of noise obvious in the NPS are white noise [$P(e)$ is independent of f], $1/f$ noise [$P(e)$ increases as $f$ decreases], and interference noise [$P(e)$ is frequency dependent and finite at specific frequencies].

(although the noise from one or two noise sources may dominate). The magnitude and frequency of the noise can be more fully described by a **noise power spectrum** (NPS), as shown in Figure 5-2.

Noise from different sources can be classified according to the shape of the NPS over a specified frequency range. The NPS in Figure 5-2 is seen to be due to a combination of the three basic types of noise: white noise, $1/f$ noise, and interference noise. For **white** or **Gaussian noise**, the magnitude of the noise power $P(e)$ is independent of frequency. White noise is random noise and is usually fundamental noise, although it can arise because of nonfundamental causes.

**Pink** or **1/f noise** is a random noise with a frequency-dependent NPS; it exhibits larger magnitudes at low frequencies. That is, the average amplitude of lower-frequency fluctuations is greater than the amplitude of higher-frequency fluctuations. It is called $1/f$ noise because $P(e)$ is proportional to $1/f^n$, where usually $0 \le n < 1$. Pink noise is due to nonfundamental causes; it often includes slow drifts (low-frequency fluctuations).

**Interference** or **environmental noise** is a frequency-dependent, nonfundamental noise. It often appears at discrete frequencies due to pickup from and parasitic coupling to other signal sources. Thus the amplitude, frequency, and phase are predictable. The most common interference noise in the United States is 60-Hz noise from ac power lines; the NPS has spikes at 60 Hz and its harmonics (e.g., 120, 180 Hz). Interference noise can also occur at frequencies generated by oscillators in instruments and computers or by other timing signals. This type of correlated, nonrandom noise can often be observed readily on a oscilloscope.

Some interference noise, often called **impulse noise**, is correlated noise that is random in time. Examples are noise spikes generated by turning instruments on and off, spikes on the ac power line, and radio-fre-

quency (RF) noise from spark gaps in lasers. The magnitude and time of the fluctuations cannot be easily predicted, but the noise is still correlated to a specific event. By contrast, white and pink noise are uncorrelated noise since the total noise at any given instant is the sum of the noise at all frequencies with random or independent amplitudes and phases.

Noise can also be classified by the dependence of its magnitude on other variables. **Flicker noise** is another name for nonfundamental or excess noise for which the rms noise in the signal is directly proportional to the magnitude of the signal. Sometimes the terms **proportional noise**, **multiplicative noise**, or **fluctuation noise** are used in place of flicker noise. Flicker noise often has $1/f$ character in its NPS, but it can be white in certain frequency regions.

## Signal Characteristics and the Signal-to-Noise Ratio

Electrical signals from the transducer in a spectrochemical instrument almost always vary with time even though the signal may normally be considered to be a static or dc signal. Time-varying signals are also characterized by their frequency spectra, plots of amplitude density ($V\ Hz^{-1}$) vs. frequency (Hz) or power density ($V^2\ Hz^{-1}$) vs. frequency. Knowing the frequency composition of a signal is important if electronic operations are to be performed without loss of information. Also, because the frequency composition of signals and noise sources are, or can be made to be different, advantage can be taken to enhance the signal selectively in the presence of noise.

Most electrical signals in spectrochemical instruments are **bandlimited signals** in that their amplitudes are zero except in a particular frequency range. Limited bandwidth signals can result from intentionally filtering a broadband signal, from modulation, or from bandwidth limitations of transducers, amplifiers, or other system components. We consider here two types of bandlimited signals: dc signals and ac signals.

*Direct-Current Signals.* A **direct-current** (dc) **signal** is one in which the current is unidirectional and the magnitude is "constant" over the period of measurement. Many spectrochemical instruments produce dc electrical outputs. However, no signal can be constant indefinitely; the frequency spectrum of an ostensibly dc signal always has a finite bandwidth. The signal frequency components at frequencies higher than dc may arise from actual radiant power changes due to a varying analyte concentration. Thus the frequency response of the signal processing system must extend from zero to a finite low frequency to process a dc signal and to follow the changes of interest faithfully. High-frequency fluctuations due to undesirable noise can be attenuated by integration or by a low-pass filter, as described in Section 5-2.

*Alternating-Current Signals.* An **alternating-current** (ac) **signal** has significant power density at a nonzero frequency. In spectrochemical systems ac signals can arise because of the nature of the experiment (e.g., type of sample presentation), or modulation can be used to convert a dc signal to an ac signal. In the latter case, shifting the signal power away from dc allows amplification and signal processing to be done in a region of the spectrum that is free from $1/f$ noise and drift. After appropriate operations have been performed, the signal can be converted back to dc by demodulation techniques. Two common types of ac signals and their power spectra are shown in Figure 5-3.

*The Signal-to-Noise Ratio.* The signal-to-noise ratio ($S/N$) is the inverse of the relative noise and is given by

$$\frac{S}{N} = \frac{\overline{E}}{s_E} \qquad (5-3)$$

where $S$ is the mean value of the signal ($\overline{E}$) and $N$ is the rms noise ($s_E$). Note that the relative standard deviation (RSD) of the signal is $s_E/\overline{E}$, the inverse of the $S/N$. Thus the RSD of a measurement equals $(S/N)^{-1}$ if noise limits the measurement precision. A $S/N$ of 1000 corresponds to a relatively high precision measurement

since the RSD equals 0.1%. Although the noise and signal in equation 5-3 are expressed in terms of voltages, they can be expressed in any appropriate signal units or for any observation point within an instrument.

## 5-2 FREQUENCY CHARACTERISTICS OF SIGNAL PROCESSING AND READOUT SYSTEMS

As we have seen, the signal information in a spectrochemical instrument appears, or can be made to appear, in a relatively narrow frequency interval, while noise is present over a broad range of frequencies. Thus the most common way to enhance the signal strength relative to the noise, and thus to improve the $S/N$, is to restrict the frequency response of the electrical measurement system (signal processing and readout system). In fact, the primary function of the signal processor is often to provide such $S/N$ enhancement (see Section 4-5).

### Amplitude Transfer Function

The frequency response of an instrument is controlled by electrical filtering circuits or by time-averaging (integration) circuitry. Even though each specific subsystem (transducer, amplifier, filter, readout device) in an instrument has a frequency response, the overall instrument response is usually limited by a specific circuit. The instrument frequency response is described by a **transfer function** $H(j\omega)$, defined by $H(j\omega) = e_o/e_{\text{in}}$,

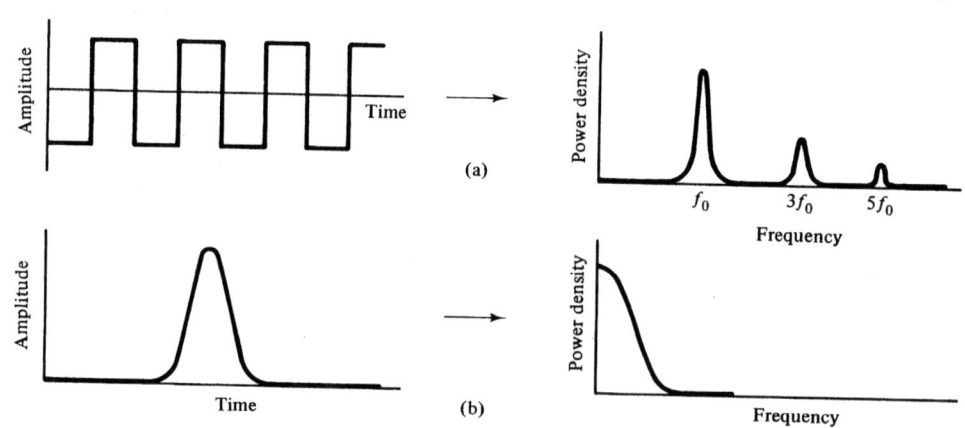

**FIGURE 5-3** Some ac signals and their power spectra. Most of the signal information in the chopped signal (a) is contained at the fundamental frequency $f_0$. Note that chopping produces odd harmonics, as in a square wave. To attenuate noise at other frequencies, a bandpass filter or a tuned amplifier can be used to pass or amplify a narrow frequency interval centered at $f_0$. The peak-shaped signal (b) could result from a spectral scan. Note that its power spectrum is similar to that of a dc signal, but the signal power may extend to relatively high frequencies.

where $e_o$ and $e_{in}$ are the instantaneous output and input voltages, respectively. The transfer function depends on frequency ($\omega = 2\pi f$); the $j$ operator ($\sqrt{-1}$) is a reminder that there may be a phase difference between $e_o$ and $e_{in}$. Note that we use $f$ to represent the electrical frequency, usually in the audio range, while $\nu$ denotes the optical frequency. In our discussion, we will consider only the **amplitude transfer function** $|H(f)|$ and not concern ourselves with phase shifts. It must be kept in mind, however, that phase shifts are introduced by most reactive circuits used for filtering and narrow-band amplification.

The amplitude transfer function is defined by

$$|H(f)| = \left| \frac{E_o}{E_{in}} \right| \qquad (5\text{-}4)$$

where $E_{in}$ is the amplitude of the input signal (or noise) and $E_o$ is the amplitude of the output signal (or noise). A plot of $|H(f)|$ vs. frequency is the frequency response of the system. The amplitude transfer functions for several common signal processing circuits were defined in Figures 4-35 and 4-36. For simple passive filters (filters without gain), the transfer function is unity at the frequency of maximum response and falls off rapidly at higher or lower frequencies. In a spectrochemical instrument, adjustments are made so that $|H(f)| \approx 1$ for the frequency at which the analytical signal is encoded and zero or near zero for frequencies removed from the encoding frequency. Thus noise at frequencies removed from the signal frequency is significantly attenuated.

The amplitude transfer function, as well as the phase shift, can be calculated as a function of frequency for specific filtering networks. It can also be determined experimentally by using a sine-wave signal generator as input to the frequency-limiting circuitry. The input and output voltages are measured sequentially for different signal frequencies to generate a plot of $|H(f)|$ or $|E_o/E_{in}|$ vs. $f$. Similarly, the phase shift between output and input can be measured as a function of $f$.

Three typical amplitude transfer functions are plotted in Figure 5-4. The amplitude transfer function illustrated by curve a is typical of instruments that use a single-stage $RC$ low-pass filter for measuring dc signals. If the signal is ac, or has been converted to ac by modulation, a bandpass filter can be employed. The response of a bandpass filter with a central frequency of 100 Hz is shown in curve b of Figure 5-4. The frequency of maximum response is adjusted to the signal frequency while the width of $|H(f)|$ is adjusted to minimize the range of noise frequencies passed. Curve c in Figure 5-4 shows the frequency response of an integrator with a 1-s integration time.

## System Bandpass

The amplitude transfer function determines the range of frequencies passed by the frequency-limiting components; this range can be expressed as a signal frequency bandpass or a noise equivalent bandpass. The **signal frequency bandpass** is the range of frequencies for which the attenuation is less than 3 decibels (dB) ($|H(f)| > 0.707$) [dB $= 20 \log (E_o/E_{in})$]. The **noise equivalent bandpass** is defined by

$$\Delta f = \int_0^\infty |H(f)|^2 \, df \qquad (5\text{-}5)$$

and represents the frequency range of electrical power passed by an ideal filter with a rectangular amplitude transfer function [i.e., $|H(f)|$ equals 1 or 0], as shown by rectangles d, e, and f in Figure 5-4. The signal frequency bandpass and the noise equivalent bandpass are compared in Table 5-1 for a low-pass filter of time constant $\tau$ ($\tau = RC$), an integrator of integration time $t$, and a bandpass filter with low and high cutoff frequencies (3 dB points) of $f_1$ and $f_2$.

**FIGURE 5-4** Frequency response of filters and integrators. Here curve a corresponds to a low-pass filter with 1-s time constant, curve b to a bandpass centered at 100 Hz, and curve c to an integrator with a 1-s integration time. The equivalent noise bandpass for the low-pass filter, bandpass filter, and integrator are shown by curves d, e, and f, respectively. The area of each rectangle is equivalent to the area under the corresponding curve of the square of the amplitude transfer function.

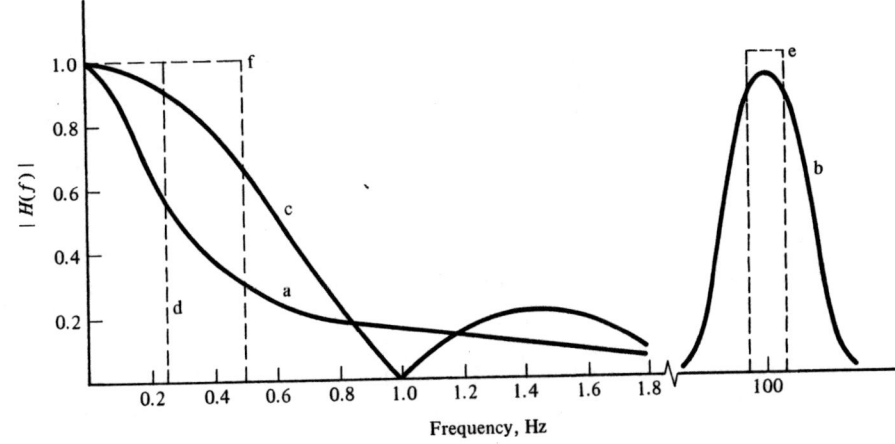

**TABLE 5-1**

Bandpass characteristics of frequency-limiting circuits[a]

| Circuit | $|H(f)|$ | Lower cutoff frequency, $f_1$ (3 dB point) | Upper cutoff frequency, $f_2$ (3 dB point) | Signal frequency bandpass, $f_2 - f_1$ (Hz) | Noise equivalent bandpass, $\Delta f$ (Hz) |
|---|---|---|---|---|---|
| Low-pass filter | $[1 + (2\pi f\tau)^2]^{-1/2}$ | 0 | $(2\pi\tau)^{-1}$ | $(2\pi\tau)^{-1}$ | $(4\tau)^{-1}$ |
| Integrator | $\dfrac{\sin \pi ft}{\pi ft}$ | 0 | $\dfrac{0.433}{t}$ | $\dfrac{0.433}{t}$ | $(2t)^{-1}$ |
| Bandpass filter | — | $f_1$ | $f_2$ | $\dfrac{f_m}{Q}$ | $\dfrac{\pi f_m}{2Q}$ |

[a]$\tau$, Time constant ($RC$) of low-pass filter (s); $t$, integration time of dc integrator (s); $Q$, quality factor for bandpass filter $= f_m/(f_2 - f_1)$; $f_m$, central frequency of bandpass filter (Hz).

To illustrate the differences in the signal bandpass and the noise equivalent bandpass, consider a low-pass filter with a $\tau$ value of 1 s. The signal bandpass (see Table 5-1) is $1/(2\pi\tau) = 0.16$ Hz, while the noise equivalent bandpass is $1/(4\tau) = 0.25$ Hz. At frequencies higher than the 3-dB point (0.16 Hz), the attenuation of the low-pass filter increases with frequency until it asymptotically approaches 20 dB/decade.

**Noise in Readout Signals**

Limiting the bandpass of the signal processing system can drastically attenuate noise, especially if it is white noise [constant power density $P(e)$ at all frequencies]. The mean-square noise (variance) in the readout signal is given by

$$s_E^2 = \int_0^\infty P(e)|H(f)|^2 \, df \qquad (5\text{-}6)$$

where $P(e)$ is the noise density in $V^2 \, Hz^{-1}$. If the NPS is flat or due only to white noise,

$$s_E^2 = P(e) \int_0^\infty |H(f)|^2 \, df = P(e) \, \Delta f \qquad (5\text{-}7)$$

Equations 5-6 and 5-7 indicate that the amount of noise observed decreases as the frequency bandpass of the instrument is reduced. In the case of white noise, $s_E^2$ is proportional to $\Delta f$, or the rms noise ($s_E$) is proportional to $\Delta f^{1/2}$. For $1/f$ noise, decreasing $\Delta f$ by a factor of 2 reduces the rms noise by less than a factor of $2^{1/2}$ because $P(e)$ is greater at lower frequencies. The dependence of the magnitude of the $1/f$ noise observed on $\Delta f$ is controlled by the shape of its NPS. In some cases, there may be no significant improvement or even an increase in the noise observed when $\Delta f$ is decreased.

The equations above indicate that $\Delta f$ should be made as small as possible to reduce the observed rms noise. However, there are practical limitations. For low-pass filters, a small $\Delta f$ is obtained at the expense of a large time constant. The time constant affects the response time of the instrument. The **response time** is the time it takes the output signal to reach some percentage (e.g., 99.7 or 98%) of its equilibrium or final value after a step signal is applied to the input. For a simple low-pass $RC$ filter, the response time is often defined as $5\tau$ ($5RC$), the time for 99.7% of full response.

For dc signals, the time constant of a simple low-pass filter is chosen as a compromise between noise filtering and response time. Thus for a $\tau$ of 1 s, one must wait 5 s after the analytical sample is introduced to the spectrochemical information encoder before taking a signal reading. Time constants larger than 10 s are rarely used because it would take 50 s or more to obtain one sample reading. Also, reduction of $\Delta f$ below about 0.02 Hz rarely provides any $S/N$ additional advantage, due to the dominance of $1/f$ noise and drift.

For single-event, transient optical signals (a peak-shaped, ramp, or exponential time dependence), $\tau$ is usually chosen to be as large as possible to reduce $\Delta f$, but still small enough to prevent distortion or attenuation of the signal information. Generally, $\tau$ should be no greater than one-tenth of the time duration over which the transient signal is measured. In the case of recording spectra, the time constant should be about one-tenth of the time it takes to scan through the narrowest peak. Here one is making a compromise between the time to scan a spectrum (e.g., the scan rate in nm s$^{-1}$) and $\Delta f$.

With an $RC$ filtering system, noise is effectively averaged over a few time constants. Higher-frequency noise components are highly attenuated because the period of their fluctuations is much less than the response time of the $RC$ network (see Figure 5-1). Similarly, with an integration system, the contributions of higher-frequency fluctuations with periods much less than the integration time tend to average to nearly zero because many such fluctuations occur during the integration period. If the noise is at a fixed frequency, such as 60-Hz noise from the ac power lines, choosing the integration time to be an integral multiple of the noise period can provide nearly perfect noise rejection since an integral number of noise cycles are averaged. Many integrating analog-to-digital converters use this principle to provide very high rejection of 60-Hz noise.

When integrators are used for noise reduction in dc signals, the integration period is chosen to be large to reduce $\Delta f$, but small enough for time convenience. Integration times much greater than 1 min are not usually used for the same reasons that very large time constants are not employed.

Integration systems can also be used for transient signals, but the integration time is often chosen to be small enough that the signal does not change significantly over $t$. If the signal amplitude does change over $t$, the time integral of the signal is recorded; this can be used to find, for example, the area of a peak-shaped signal.

An integration system can provide higher throughput since there is not an $RC$ response time. An $RC$ circuit with a $\tau$ value of 1.0 s and an integration system with an integration time $t$ of 0.5 s have identical noise bandwidths ($\Delta f = 0.5$ Hz). However, signal data can be obtained in 1 s with the integrator, but the $RC$ filter circuit requires 2.5 s to respond fully.

A simple $RC$ filter is a first-order filter. Higher-order low-pass filters with two or more $RC$ networks are available. Compared to first-order filters, higher-order filters, particularly active filters based on operational amplifiers, can provide steeper roll-off at higher frequencies and equivalent $\Delta f$ with a faster response time. This means that greater noise reduction can be achieved (i.e., smaller $\Delta f$ values) without waiting as long for the filter response and with less distortion and attentuation of transient signals.

Finally, it should be noted that with computer data acquisition filtering can be carried out with software rather than hardware, as discussed in Section 5-6.

## 5-3  NOISE SOURCES

In this section, the sources of noise in a spectrometric instrument are discussed. This allows us to subdivide the total noise observed in a signal into its individual components. Knowledge of the dominant noise source(s) allows one to optimize experimental variables systematically or to modify the instrumental design for improved $S/N$. The discussion here is based on a typical spectrometer with a photomultiplier tube (PMT) as the radiant power monitor (see Section 2-4). The modifi-

cations for photodiodes, phototubes, and other types of transducers are discussed later in this chapter. Two common signal processor–readout systems, analog and photon counting, are considered. In the analog signal processing system, the signal processor contains circuitry for converting the transducer anodic current into a voltage (possibly with additional stages of amplification) before display on a voltage readout device. The electrical bandwidth is controlled by an analog filter (or integrator). The photon counting signal processor system consists of a fast amplifier, discriminator, and pulse counter; the integration or counting time can be varied (see Section 4-5).

Let us establish at the outset several general principles. First, in each type of signal (dark, background, and analytical), there can be a fundamental and a nonfundamental noise component. Additionally, the signal processing system generates noise (both fundamental and nonfundamental) in the absence of any signals from the optical transducer. Second, the progression of signals through an instrument can be considered to be a series of sequential transformations where both the signal and noise experience the same transformation. Moreover, additional noise or signals can be injected during each transformation step. The transformation steps are illustrated in Figure 5-5.

### Quantum, Secondary Emission, and Shot Noise

First, consider the flux of background and signal photons (from the optical information encoder and selector system) that strikes the photocathode of the optical transducer. If we could count the number of photons impingent per unit time, we could establish a mean value and a standard deviation in the radiant power expressed in photons per second or watts. In many cases, the photons arrive randomly (the spacing between photons is random) because the emission of photons from the excitation source or from the analyte is random. Under these conditions, the Poisson probability distribution function applies,

$$P(n) = \frac{(rt)^n e^{-rt}}{n!} \tag{5-8}$$

where $P(n)$ is the probability of counting $n$ events in time $t$ and $r$ is the mean rate of events. For the Poisson distribution, the mean value of $n$ ($\bar{n}$) is $rt$ and the standard deviation in $n$ ($\sigma_n$) is $(rt)^{1/2}$.

For spectrochemical measurements, we equate $r$ to $\Phi_p$, the mean photon rate or radiant power in photons per second. Thus $\bar{n}$ is the mean number of photons

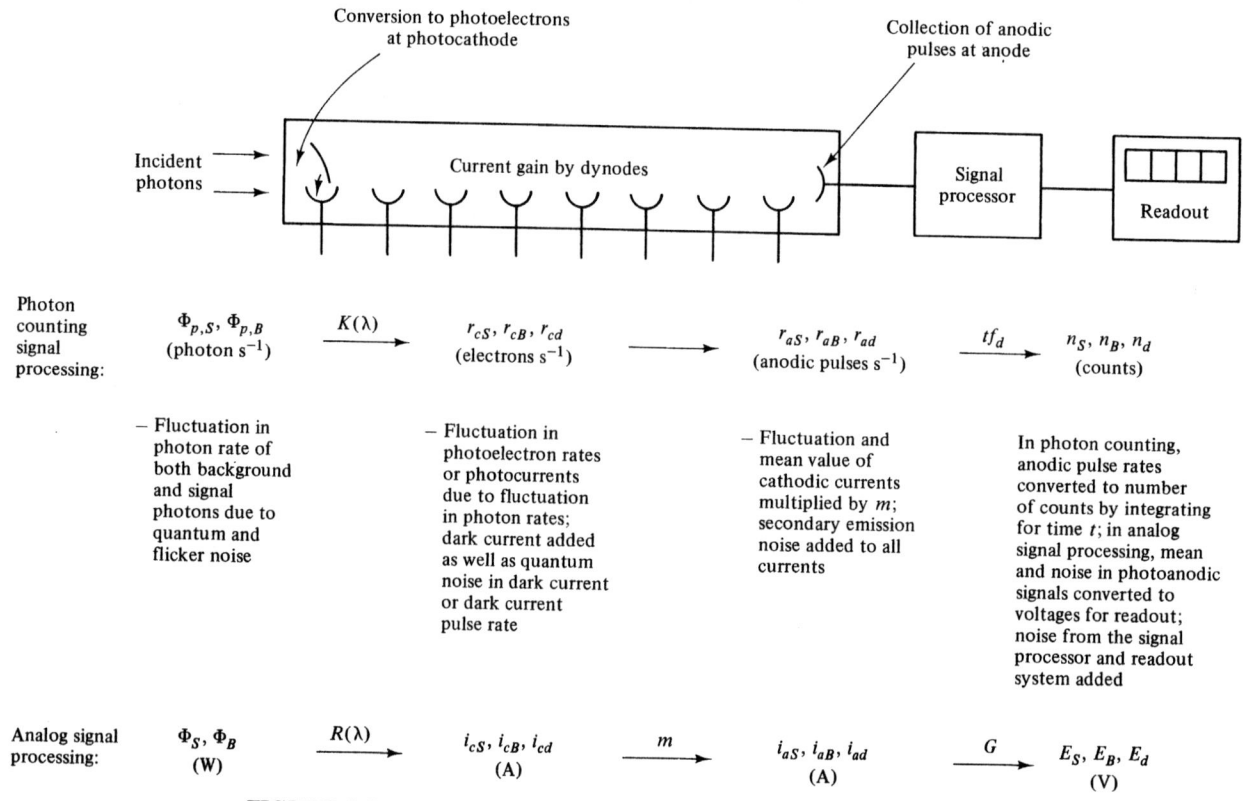

**FIGURE 5-5** Transformation of signals and noise in a typical spectrochemical measurement. In photon counting signal processing, the signal is expressed in terms of a rate ($\Phi_p$ or $r$) or a number of events (pulses) counted in time $t$. In analog signal processing, the signal is expressed in terms of a radiant power $\Phi$, a current $i$, or a voltage $E$. The subscripts $S$, $B$, and $d$ refer to analytical signal, background signal, and dark signal, respectively, and the subscripts $c$ and $a$ refer to the photocathode and anode, respectively.

counted in integration time $t$ and is given by

$$n = \Phi_p t \qquad (5\text{-}9)$$

Here and in further discussions, we will drop the bar and understand that the mean value of $n$ or other signal quantities is implied. The standard deviation or rms noise in the number of photons counted in time $t$ ($\sigma_n$) is found from

$$(\sigma_n)_q = (\Phi_p t)^{1/2} = n^{1/2} \qquad (5\text{-}10)$$

The subscript $q$ denotes **quantum noise**, the fundamental noise due to the random emission of photons from a source. If the background photon flux is negligible, the *S/N* is given by

$$\frac{S}{N} = \frac{n}{(\sigma_n)_q} = n^{1/2} \qquad (5\text{-}11)$$

Equation 5-11 indicates that the *S/N* increases as the square root of $n$ and can be improved by increasing $\Phi_p$ or $t$. The equations above apply equally to signal and background photons. However, the signal in equation 5-11 is normally taken to be only due to signal photons.

The quantum noise in the original photon flux is seen in the signals at the photocathode, the anode, and the readout device because all conversions are random or multiply the signal and noise by linear factors. Quantum noise is often called Schottky noise or shot noise. **Schottky noise** is observed whenever a current passes through or is generated at an interface. The signal and quantum noise equations that apply at the transformation points for both analog and photon counting signal processing are summarized in Table 5-2. In this discussion $n$, $i$, and $E$ are the symbols for signals expressed in terms of the number of events or pulses counted, the current, or the readout voltage, respectively. These symbols are also used as subscripts with $\sigma$ to denote the noise in a particular type of signal. Subscripts $c$, $a$, and $r$ are used to designate the signal at the photocathode, the anode, or the readout device (only for photon counting), respectively. For simplicity, these subscripts are not used with $\sigma$; the application of the symbol to the noise at the photocathode, the anode, or the readout device is determined from the context.

Let us first consider the signals and noise at the photocathode. Equations 1 to 4 in Table 5-2 apply to the analytical signal and the background signal, where $r_c$ is either the signal or background photoelectron pulse rate. The $S/N$ due to quantum noise in the analytical signal is $n_c/\sigma_n = n_c^{1/2}$ or $i_c/\sigma_i = (i_c t/e)^{1/2}$. Notice that this $S/N$ is less than the $S/N$ calculated in the original photon beam ($n^{1/2}$) because some information has been lost at the photocathode due to the quantum efficiency being less than 1. However, the characteristic dependence of the $S/N$ on the square root of the signal and the square root of the integration time under quantum-noise-limited conditions is still obvious.

The photocathode is a transformation point. The cathodic dark signal and dark signal quantum noise are injected at this point and can be calculated from equations 1 to 4 if the cathodic dark pulse rate is known.

Equations 5 to 8 of Table 5-2 apply to the analytical, background, and dark anodic signals or the noise in these signals. We will use the term **shot noise** (subscript

$s$) to denote specifically noise in the anodic current or readout voltage signals that is due both to quantum noise (subscript $q$) and PMT multiplication noise (subscript $m$). **PMT multiplication** or **secondary emission noise** accounts for random fluctuations in the gain due to the random nature of secondary emission of electrons at the dynodes. This is a multiple Poisson process so that the expression for the PMT multiplication noise in the anodic current has the form of the quantum noise equation. Thus equation 8 of Table 5-2 is a summation of quantum noise $[(\sigma_i)_q = (mei_a/t)^{1/2}]$ and multiplication noise $[(\sigma_i)_m = (\alpha mei_a/t)^{1/2}]$, where $\alpha^{1/2}$ is just some fraction of the quantum noise in the anodic current. The factor $\alpha$ is often between 0.1 and 0.5 for a good-quality PMT and can be estimated to be $(\delta - 1)^{-1}$ where $\delta$ is the gain per dynode stage (see Section 4-4). This formula for $\alpha$ does not correctly account for the PMT multiplication noise if secondary emission noise at the dynodes is not uniform or if the statistics are not Poisson (e.g., exponential).

Equations 9 to 12 of Table 5-2 are the readout equations for the mean signal and noise. For photon counting, equations 9 and 10 can be used to calculate the mean ($n_S$, $n_B$, and $n_d$) and shot noise $[(\sigma_S)_s, (\sigma_B)_s,$

## TABLE 5-2
Signal and noise expressions for quantum or shot noise at transformation points[a]

| Transformation point | Photon counting signal processing | | | | Analog signal processing | | | |
| --- | --- | --- | --- | --- | --- | --- | --- | --- |
| | Signal | | Noise | | Signal | | Noise | |
| Photocathode[b] | $n_c = K(\lambda)\Phi_p = r_c t$ $= K(\lambda)n$ | (1) | $(\sigma_n)_q = n_c^{1/2}$ $= [K(\lambda)n]^{1/2}$ | (2) | $i_c = \dfrac{e}{t}n_c = er_c$ $= R(\lambda)\Phi$ | (3) | $(\sigma_i)_q = \dfrac{e}{t}n^{1/2}$ $= \left(\dfrac{ei_c}{t}\right)^{1/2}$ | (4) |
| Anode[c] | $n_a = n_c$ | (5) | $(\sigma_n)_q = n_a^{1/2}$ | (6) | $i_a = mi_c = mR(\lambda)\Phi$ | (7) | $(\sigma_i)_s$ $= m(1+\alpha)^{1/2}\left(\dfrac{ei_c}{t}\right)^{1/2}$ $= \left[mei_a\dfrac{(1+\alpha)}{t}\right]^{1/2}$ | (8) |
| Readout[d] | $n_r = f_d n_a$ | (9) | $(\sigma_n)_q = (n_r)^{1/2}$ $= (f_d n_a)^{1/2}$ | (10) | $E = Gi_a = mGR(\lambda)\Phi$ | (11) | $(\sigma_E)_s = G(\sigma_i)_s$ $= \left[\dfrac{GmeE(1+\alpha)}{t}\right]^{1/2}$ | (12) |

[a] $n_c$, mean number of photocathodic electrons counted in time $t$ (dimensionless); $(\sigma_n)_q$, SD in $n_c$, $n_a$, or $n_r$ due to quantum noise, dimensionless; $r_c$, photocathodic electron pulse rate (s$^{-1}$); $i_c$, mean cathodic current (A); $(\sigma_i)_q$, SD in $i_c$ or $i_a$ due to quantum noise when integrated for time $t$ (A); $n_a$, mean number of photoanodic pulses counted in time $t$ (dimensionless); $i_a$, mean anodic current (A); $(\sigma_i)_s$, SD in $i_a$ due to shot noise when integrated for time $t$ (A); $n_r$, readout counts in time $t$ (dimensionless); $f_d$, discriminator level (dimensionless); $E$, readout voltage (V); $G$, gain factor (V A$^{-1}$); $(\sigma_E)_s$, SD in $E$ due to shot noise when integrated for time $t$ (V); $\alpha$, secondary emission factor (dimensionless).

[b] At the photocathode, the photon flux is converted with quantum efficiency $K(\lambda)$ into a photoelectron pulse rate in photoelectrons s$^{-1}$ or a photocathodic current in A. Conversion of $n$ or $\sigma_n$ to the analog current domain is seen to involve multiplication by the factor $(e/t)$. Division of $n$ or $\sigma_n$ by $t$ converts counts to counts per second and multiplication by $e$ ($1.6 \times 10^{-19}$ A) converts counts per second into amperes. Alternatively, $i_c$ can be expressed as $\Phi R(\lambda)$ or the product of the incident radiant power ($\Phi$) in W and the cathodic responsivity [$R(\lambda)$] in A W$^{-1}$.

[c] With a PMT, the photoelectrons receive an average gain of $m$, or the average number of electrons in an anodic pulse packet is $m$. The gain process does not change the pulse rate so the cathodic pulse rate ($r_c$) and the anodic pulse rate ($r_a$) are the same (assuming a collection efficiency of 1), as are $n_c$ and $n_a$. The conversion factor from cathodic to anodic current for the signal is the PMT gain $m$, while for the noise is $m(1+\alpha)^{1/2}$, where $\alpha$ takes into account the PMT multiplication noise.

[d] For photon counting, a fraction of the anodic pulses reach the readout counter as determined by the discriminator level $f_d$. With analog signal processing, the anodic currents are converted to voltages according to the gain factor $G$ of the signal processing system in V A$^{-1}$.

**TABLE 5-3**

Signal and noise expressions for quantum and shot noise in analytical, background, and dark signals

| Type of signal | Photon counting signal processing | | Analog signal processing | |
|---|---|---|---|---|
| | Mean | Rms noise | Mean | Rms noise |
| Analytical | $n_s = f_d K(\lambda) r_s t$ | $(\sigma_S)_s = n_S^{1/2}$ | $E_S = mGR(\lambda)\Phi_S$ | $(\sigma_S)_s = (mGKE_S)^{1/2}$ |
| Background | $n_B = f_d K(\lambda) r_B t$ | $(\sigma_B)_s = n_B^{1/2}$ | $E_B = mGR(\lambda)\Phi_B$ | $(\sigma_B)_s = (mGKE_B)^{1/2}$ |
| Dark | $n_d = f_d r_{cd} t$ | $(\sigma_d)_s = n_d^{1/2}$ | $E_d = mGi_{cd}$ | $(\sigma_d)_s = (mGKE_d)^{1/2}$ |

and $(\sigma_d)_s$], in the analytical, background, and dark signals in counts, respectively, as shown in Table 5-3. The shot noise in this case is only due to quantum noise because variations in the gain from pulse to pulse change the pulse heights but not the pulse rate.

For analog signal processing, it is convenient to write equation 12 of Table 5-2 in terms of $\Delta f$ so that it can be employed with analog signal processing using filters or integrators. Thus if $\Delta f = (2t)^{-1}$ is used for an integration system,

$$(\sigma_E)_s = [2e\,\Delta f(1 + \alpha)mGE]^{1/2} \qquad (5\text{-}12)$$

where $\Delta f = (4\tau)^{-1}$ when a simple low-pass filter restricts the noise equivalent bandpass. Equation 5-12 can be further simplified by defining a bandwidth constant $K$ equal to $2e\Delta f(1 + \alpha)$; thus

$$(\sigma_E)_s = (mGKE)^{1/2} \qquad (5\text{-}13)$$

The analytical, background, and dark signal mean voltages and standard deviations due to shot noise (see Table 5-3) are thus derived from equations 11 and 12 of Table 5-2.

The equations for the dark signal and noise shown in Table 5-3 assume that all the dark current is due to thermal emission from the photocathode. In reality $n_d$ can be greater due to other sources of dark current. For these, a different discriminator coefficient might be required in the photon counting equations due to differences in the pulse height distribution for these sources of dark current.

If the transducer used is a vacuum phototube or a semiconductor photodiode, the equations in Tables 5-2 and 5-3 must be modified to account for the unity gain ($m = 1$) and the lack of secondary emission noise ($\alpha = 0$). Phototransistors and avalanche photodiodes have gain and the equations for shot noise are similar to those in Table 5-3. With multichannel array detectors (vidicons and photodiode arrays), $m = 1$, $\alpha = 0$, and the quantity related to photoelectron rate is the number of electron–hole pairs generated per unit time. Although these devices integrate the flux over a certain integration time, their outputs can be related to an average input flux, and equations similar to those in Table 5-3 apply. Intensified arrays (SIT tubes, intensified diode arrays) have gain and thus multiplication noise. Although the noise observed with thermal detectors has shot noise components, they are almost invariably limited by thermal noise, which is independent of illumination level, as discussed later in this section.

### Flicker Noise

The fluctuations or SD in the signal and background photon fluxes at the optical transducer can be larger than predicted by quantum or shot noise. This in turn causes the noise in the readout signal or at various transformation points in the system to be greater than predicted by the quantum or shot noise equations. The excess noise present is denoted flicker noise because it is a nonfundamental noise in which the rms noise is proportional to the magnitude of the photon signal.

Thus for analog and photon counting systems we can define the rms flicker noise in the analytical and background readout signals as shown in Table 5-4. The signal flicker factor $\xi$ and the background flicker factor (sometimes $\zeta$ in the literature) are just the RSDs in the analytical and background photon fluxes, due to excess fluctuations. Thus the flicker noise in the signal at any

**TABLE 5-4**

Analytical and background flicker noise

| Type of signal | Photon counting | Analog |
|---|---|---|
| Analytical | $(\sigma_S)_f = \xi n_S$ | $(\sigma_S)_f = \xi E_S$ |
| Background | $(\sigma_B)_f = \chi n_B$ | $(\sigma_B)_f = \chi E_B$ |

point in the instrument (e.g., photocathode, photo-anode) is the product of the mean signal at that point and the flicker factor.

A flicker factor must be evaluated experimentally for a particular situation since it depends on the noise equivalent bandpass and the magnitude of other experimental variables. Different symbols are used for the analytical signal and background flicker factors since they can be independent of each other. However, in some cases, they can be equivalent if the background and signal flicker noise have the same origin. The flicker noise in either the analytical or background signal is often composed of two or more components (i.e., separate flicker factors for each noise component) of independent origins.

The direct proportionality between the magnitude of the absolute flicker noise and the signal level is not derived from fundamental principles as is the square-root relationship between noise and signal level in the case of shot noise. Rather, it is empirically observed that in many cases, fluctuations in the variables which affect the analytical or background signal radiance are such that the relative fluctuation in the radiance is independent of the absolute radiant power. Consider viewing a light source through a filter whose transmittance varies randomly in time with a 1% RSD. The RSD in the transmitted photon flux will be 1% if the RSD in the photon flux incident on the filter due to quantum noise is much less than 1%. If the incident photon flux is doubled, the absolute fluctuation in the transmitted flux will double, but the relative fluctuation will still be 1%.

Specific sources of flicker noise are described in some detail in discussion of specific techniques in later chapters. Here only a few general sources of flicker noise will be identified. In absorption and photoluminescence spectrometry, fluctuations in the readout signals are often caused by flicker noise in the excitation source radiance. **Source flicker noise** is caused by variations in experimental variables that control the source radiance, such as electrical power (not 60-Hz ripple) or temperature; it can also be caused by variations that affect the source radiant power viewed, such as vibrations. Thus a source can be characterized by a **source flicker factor**, denoted $\xi_1$.

Source flicker can cause both analytical and background flicker noise in a photoluminescence experiment because the analyte fluorescence, background (non-analyte) fluorescence, and scattering signals are all proportional to the excitation source radiant power and suffer an equivalent fluctuation due to source flicker. In this case the flicker noise in the analytical and the background signals due to source flicker noise are totally correlated. Flicker noise in the background emission signal is uncorrelated with the foregoing source flicker noises, so that a different flicker factor must be used.

In all spectrometric measurements and, in particular, absorption measurements, the fluctuations in the transmission characteristics of the sample container (independent of the analyte) can be significant. This flicker noise is denoted **transmission** or **convection flicker noise** and is characterized by a transmission flicker factor, denoted $\xi_2$. Here one can see that the absolute fluctuations in a light beam transmitted through a sample container with fluctuating transmission characteristics are proportional to incident radiant power, but that the relative fluctuations are independent of the incident radiant power.

Particularly in atomic spectrometry, the analyte population in the probed volume element in the sample container can vary with time. This **analyte flicker noise** is often caused by fluctuations in the sample presentation system. In emission and chemiluminescence, flicker noise can also be caused by fluctuations in the excitation conditions.

The flicker noise equations (Table 5-4) apply to any radiation transducer if the appropriate output quantity is used and multiplied by the flicker factor.

### Other Noise Sources

The equations for dark current noise in Table 5-3 are incomplete if there are significant contributions from sources other than thermal emission from the photocathode. Some additional sources are thermal emission from the dynodes, cold field emission, ohmic leakage, and radioactivity. Often the shot noise contribution from thermal emission from the dynodes is small because the dynode thermal dark current is small and receives a gain less than $m$. In analog signal processing systems, the noise in the dark current voltage due to nonthermal sources is denoted **excess dark current noise** with the symbol $(\sigma_d)_{ex}$. As the PMT bias voltage increases, the relative contribution of ohmic leakage to the total dark current decreases, but that due to cold field emission increases.

The noise observed in the absence of analytical, background, and dark current signals is **amplifier-readout noise**, $\sigma_{ar}$. It is typically measured with the bias voltage of the PMT or PT turned off. This noise is due to fundamental and nonfundamental noise generated in the signal processing and readout circuitry ($\sigma_a$) and quantization noise ($\sigma_q$). Flicker noise in electronic circuitry can be caused by the random appearance of impurity centers in components. **Quantization noise** is due to the finite resolution of any readout device and is given by

$$\sigma_q = \frac{q}{12^{1/2}} = 0.29q \qquad (5\text{-}14)$$

where $q$ is the quantization level or readout resolution. The quantization uncertainty acts like a white noise source if the rms electrical noise is equal to or greater than one-half of the quantization level. If no noise is obvious in the readout signal, measurements are limited by readout resolution and $\sigma_q = q/2$.

In photon counting, the discriminator level is usually set above the noise level in the pulse amplifier circuitry so that $\sigma_{ar} = \sigma_q$, where $q$ equals 1 count. With analog signal processing, $\sigma_{ar}$ is equal to or greater than $\sigma_q$, where $q$ is determined by the voltage resolution corresponding to the least significant digit of a digital readout or the minimum signal difference that can be distinguished for an analog readout. Often, $\sigma_{ar} > \sigma_q$ because of noise generated in the readout circuitry or prior signal processing circuitry.

The fundamental noise component of $\sigma_{ar}$ arises from the random thermal motion of electrons (Brownian motion) in resistors and is called **Johnson noise** or thermal noise. The rms Johnson noise, $\sigma_J$, generated in a resistor is given by the Nyquist formula,

$$\sigma_J = (4kTR\,\Delta f)^{1/2} \qquad (5\text{-}15)$$

where $k$ is Boltzmann's constant ($1.38 \times 10^{-23}$ J K$^{-1}$) and $T$ is the absolute temperature. Thus noise power $P(e)$ for Johnson noise equals $4kTR$ and has units of (J K$^{-1}$)(K)($\Omega$)(s$^{-1}$) or V$^2$ s$^{-1}$. Note that Johnson noise in a resistive element is independent of current and at 25°C, $\sigma_J = (1.6 \times 10^{-20}R\,\Delta f)^{1/2}$.

Because $\sigma_J$ is proportion to $R^{1/2}$, the largest resistor in a circuit may be the primary cause of the Johnson noise. Thus if a standard op amp current-to-voltage converter is used, the Johnson noise from the feedback resistor can be calculated to give a minimum estimate of $\sigma_{ar}$. Thermal noise is almost always the major source of noise in thermal detectors and can be very important for photon detectors that have no internal gain (phototubes, photodiodes, etc.).

Other sources of noise such as interference noise (e.g., 60-Hz pickup) can be important in certain situations. Fluctuations in any variable affecting the signal can cause an additional noise which is often reflected in the magnitude of flicker noise. For example, variations in the PMT bias voltage or in magnetic fields can cause the gain to fluctuate more than predicted by secondary emission noise.

## 5-4 SIGNAL-TO-NOISE EXPRESSIONS FOR EMISSION AND LUMINESCENCE MEASUREMENTS

The total noise observed in a measurement can arise from several sources, although one or two noise sources often dominate. If two independent noise sources contribute to the total rms noise ($\sigma_t$), the contribution from each source, $\sigma_1$ and $\sigma_2$, adds quadratically (additive variances):

$$\sigma_t = (\sigma_1^2 + \sigma_2^2)^{1/2} \qquad (5\text{-}16)$$

If the two noise sources are nonindependent or correlated, then

$$\sigma_t = (\sigma_1^2 + \sigma_2^2 + 2C\sigma_1\sigma_2)^{1/2} \qquad (5\text{-}17)$$

where $C$ is the correlation coefficient. The correlation coefficient varies between $+1$ (totally correlated noise sources) and $-1$ (anticorrelated noise sources) and equals zero for totally uncorrelated noise, in which case equation 5-17 reduces to equation 5-16. Note that if $\sigma_1$ and $\sigma_2$ are equal, $\sigma_t$ for totally correlated noise ($C = 1$) is larger by a factor of $2^{1/2}$ than it would be for totally uncorrelated noise sources $C = 0$). Equation 5-17 is often written in terms of the covariance $\sigma_{12}$, which equals $C\sigma_1\sigma_2$.

In most cases we will assume that all noise sources are totally uncorrelated. This causes a maximum error of $n^{1/2}$ in estimating the total noise, where $n$ is the number of significant noise sources. The primary exception to this assumption will be the summation of certain types of flicker noise where the different types of flicker noise are known to originate from a common source, such as light source flicker.

### General Expressions

In luminescence and emission measurements, the analytical signal $E_S$ ($E_L$ or $E_E$) is extracted from the total signal $E_t$ measured for a standard or sample by subtraction of the blank signal $E_{bk}$. The total signal equals $E_S + E_{bk}$, where $E_{bk}$ includes contributions from the background signal, the dark signal, and offsets from the signal processor system. The rms noise in the total signal, $\sigma_t$, is therefore due to the sum of the noise contributions from the analytical and the blank signals, $\sigma_S$ and $\sigma_{bk}$, respectively, or

$$\sigma_t = (\sigma_S^2 + \sigma_{bk}^2)^{1/2} \qquad (5\text{-}18)$$

The noise in the blank and analytical signals can

be subdivided into their individual components. The blank noise is due to noise in the background signal (shot and flicker), dark current noise (shot and excess), and amplifier-readout noise or

$$\sigma_{bk} = (\sigma_B^2 + \sigma_{dt}^2)^{1/2} \qquad (5\text{-}19)$$

$$\sigma_B = [(\sigma_B)_s^2 + (\sigma_B)_f^2]^{1/2} \qquad (5\text{-}20)$$

where $\sigma_B$ is the background signal rms noise, $(\sigma_B)_s$ is the background signal shot rms noise, $(\sigma_B)_f$ is the background signal flicker rms noise, and $\sigma_{dt}$ is the total dark current rms noise.

For emission and chemiluminescence techniques, the background signal and noise ($E_B$ and $\sigma_B$) are due to background (concomitant and sample container) emission and/or blank chemiluminescence, respectively. In photoluminescence measurements, the background signal and noise are due to background emission as well as scattering and background luminescence.

The total dark current noise is observed in measuring the dark current signal and includes the actual noise in the dark current ($\sigma_d$) plus amplifier-readout noise, so

$$\sigma_{dt} = (\sigma_d^2 + \sigma_{ar}^2)^{1/2} \qquad (5\text{-}21)$$

$$\sigma_d = [(\sigma_d)_s^2 + (\sigma_d)_{ex}^2]^{1/2} \qquad (5\text{-}22)$$

where $(\sigma_d)_s$ is the dark current rms shot noise and $(\sigma_d)_{ex}$ is the dark current excess rms noise.

The rms noise in the analytical signal ($\sigma_S$) is due to shot and flicker noise, so that

$$\sigma_S = [(\sigma_S)_s^2 + (\sigma_S)_f^2]^{1/2} \qquad (5\text{-}23)$$

where $(\sigma_S)_s$ is the analytical signal shot noise and $(\sigma_S)_f$ is the analytical signal flicker noise. In photoluminescence measurements, one must also add the contributions from analyte emission shot and flicker noise to equation 5-23 if these noise sources are significant.

Combining equations 5-18 to 5-21 yields for the total noise

$$\sigma_t = [\sigma_S^2 + \sigma_B^2 + \sigma_{dt}^2]^{1/2} \qquad (5\text{-}24)$$

$$\sigma_t = [(\sigma_S)_s^2 + (\sigma_S)_f^2 + (\sigma_B)_s^2 + (\sigma_B)_f^2 + \sigma_{dt}^2]^{1/2} \qquad (5\text{-}25)$$

These equations illustrate that the noise arises from three primary sources: the noise in the analytical signal, the noise in the background signal, and noise from the dark current and amplifier-readout system. The noise terms in equation 5-25 can be further expanded with

the equations in Table 5-3 and 5-4 to yield

$$\sigma_t = [mGK(E_S + E_B + E_d) + (\xi E_S)^2 + (\chi E_B)^2 + (\sigma_d)_{ex}^2 + \sigma_{ar}^2]^{1/2} \qquad (5\text{-}26)$$

or in terms of cathodic currents,

$$\sigma_t = (mG)\left\{ K(i_S + i_B + i_d) + (\xi i_S)^2 + (\chi i_B)^2 + \left[\frac{(\sigma_d)_{ex}}{mG}\right]^2 + \left(\frac{\sigma_{ar}}{mG}\right)^2 \right\}^{1/2} \qquad (5\text{-}27)$$

For simplicity here and elsewhere in this chapter, one background flicker factor ($\chi$) is used, and all currents are understood to be cathodic currents (the subscript $c$ is dropped). However, as mentioned previously, different flicker factors may apply to the different components of the background signal.

The *S/N* is the analytical signal divided by the noise in the total signal or

$$\frac{S}{N} = \frac{E_S}{\sigma_t} = \frac{E_S}{(\sigma_S^2 + \sigma_{bk}^2)^{1/2}} = \frac{E_S}{(\sigma_S^2 + \sigma_B^2 + \sigma_{dt}^2)^{1/2}} \qquad (5\text{-}28)$$

$$\frac{S}{N} = \frac{E_S}{[(\sigma_S)_s^2 + (\sigma_S)_f^2 + (\sigma_B)_s^2 + (\sigma_B)_f^2 + \sigma_{dt}^2]^{1/2}} \qquad (5\text{-}29)$$

$$\frac{S}{N} = \frac{E_S}{[mGK(E_S + E_B + E_d) + (\xi E_S)^2 + (\chi E_B)^2 + (\sigma_d)_{ex}^2 + \sigma_{ar}^2]^{1/2}} \qquad (5\text{-}30)$$

$$\frac{S}{N} = \frac{i_S}{\{K(i_S + i_B + i_d) + (\xi i_S)^2 + (\chi i_B)^2 + [(\sigma_d)_{ex}/mG]^2 + (\sigma_{ar}/mG)^2\}^{1/2}} \qquad (5\text{-}31)$$

Equation 5-31 is a useful form because for a given photodetector, the photocathodic currents are directly proportional to the analytical and background signal photon fluxes. Equation 5-31 indicates that variation of the gain factors ($m$ and $G$) will in general have a minor effect on the *S/N* except in some cases where excess dark current or amplifier-readout noise is significant. Variation of $m$ has a slight effect on the shot noise contribution since the secondary emission factor $\alpha$ has a slight dependence on $m$. Also, the collection efficiency of the first dynode in a PMT can vary slightly with $m$.

Sometimes it is useful to express the *S/N* for measuring the analytical signal by including the noise contribution from the blank measurement. Thus the noise from the blank measurement must be counted twice (once for the measurement of the total signal and once for the blank measurement). In this case the $S/N$ is given by

$$\frac{S}{N} = \frac{E_S}{(\sigma_t^2 + \sigma_{bk}^2)^{1/2}} = \frac{E_S}{(\sigma_S^2 + 2\sigma_{bk}^2)^{1/2}}$$

We will use the simpler definition of the $S/N$ ($E_S/\sigma_t$); all the foregoing $S/N$ expressions can be converted to this alternative form by multiplying all noise variances due to the blank by 2. The ramifications of equation 5-31 can be better understood by considering several limiting cases.

### Blank-Noise-Limited $S/N$ Expressions

If the analyte concentration (and hence the analytical signal) is small ($E_S << E_B + E_d$ or $E_t \approx E_{bk}$) such that shot and flicker noise in the analytical signal are negligible, then $\sigma_t = \sigma_{bk}$, and equation 5-31 reduces to

$$\frac{S}{N} = \frac{i_S}{\{K(i_B + i_d) + (\chi i_B)^2 + [(\sigma_d)_{ex}/mG]^2 + (\sigma_{ar}/mG)^2\}^{1/2}}$$

$$(5\text{-}32)$$

Note that the $S/N$ is directly proportional to the analytical signal and inversely proportional to the blank noise and will be so until the analytical signal is increased (for a constant blank noise) to the point that analytical signal shot or flicker noise becomes significant. Here, and for the rest of this chapter, it will be understood that changing the analytical or background signals means changing the analytical and background signal photon fluxes and hence the photocathodic currents rather than changing the gain factors $m$ or $G$, which alters only the photoanodic currents or readout voltage signals.

*Shot-Noise-Limited Case.* If background flicker noise, excess dark current noise, and amplifier-readout noise are negligible, equation 5-32 reduces to

$$\frac{S}{N} = \frac{i_S}{[K(i_B + i_d)]^{1/2}}$$

$$(5\text{-}33)$$

This represents the best $S/N$ obtainable for given values of $i_S$, $i_B$, and $i_d$ since fundamental noise sources are limiting. Under these conditions the $S/N$ is proportional to $K^{-1/2}$ or $\Delta f^{-1/2}$ until $\Delta f$ is reduced to the point that the dark current or background signal shot noise becomes negligible compared to the nonwhite components of the dark current excess or amplifier-readout noise. If dark current shot noise is the limiting noise source for a given analyte concentration, the $S/N$ can be im-

proved by adjusting experimental conditions to increase the analytical signal or to reduce the dark current. The latter can be accomplished by cooling the PMT or by selecting a PMT with lower dark current but the same cathodic responsivity.

If background signal shot noise is limiting, the $S/N$ is proportional to $i_S/i_B^{1/2}$. The $S/N$ can be improved by adjusting experimental variables to increase the foregoing ratio.

*Nonfundamental Noise-Limited Case.* If dark current and background signal shot noise are negligible, the $S/N$ is given by

$$\frac{S}{N} = \frac{i_S}{\{(\chi i_B)^2 + [(\sigma_d)_{ex}/mG]^2 + (\sigma_{ar}/mG)^2\}^{1/2}}$$

$$(5\text{-}34)$$

The dependence of the $S/N$ on noise bandwidth must be experimentally determined, but in general the improvement is less than for shot noise (i.e, $S/N$ improves as $\Delta f^{-x}$, where $x < \frac{1}{2}$). If amplifier-readout or dark current excess noise is limiting, the $S/N$ can be improved by increasing $i_S$, by selecting higher-quality PMT and amplifier-readout components, or by employing photon counting, as discussed later. Also, use of higher PMT gain reduces the relative noise contribution from amplifier noise. When background flicker noise is dominant [$S/N = i_S/(\chi i_B)$], the $S/N$ can be improved by increasing the $S/B$ ($i_S/i_B$), or by reducing $\chi$.

*Background-Signal-Noise-Limited Case.* In many emission and luminescence measurements with PMT detection, the background signal is significant ($E_B > E_d$) such that background signal noise is the dominant part of the blank noise ($\sigma_{bk} \approx \sigma_B$). Here equation 5-32 reduces to

$$\frac{S}{N} = \frac{i_S}{[Ki_B + (\chi i_B)^2]^{1/2}}$$

$$(5\text{-}35)$$

where background flicker noise will dominate if $i_B$ or $\chi$ is large. Signal-to-noise ratio enhancement is achieved by adjusting experimental variables to improve the the signal-to-background ratio ($S/N \propto i_S/i_B^{1/2}$ if background shot noise is dominant) or by reducing $\Delta f$, which is most effective if background shot noise is limiting.

*Detector-Noise-Limited Case.* If background and amplifier-readout noise are negligible so that $\sigma_{bk} = \sigma_d$, equation 5-32 reduces to

$$\frac{S}{N} = \frac{i_S}{\{Ki_d + [(\sigma_d)_{ex}/mG]^2\}^{1/2}}$$

$$(5\text{-}36)$$

The $S/N$ can be improved by increasing $i_S$ or by reducing the dark current signal or noise as noted previously. In a more general sense, any detector (photodiode, thermal detector) can be characterized by its detector noise $\sigma_d$. One measures the total rms noise in the detector output $\sigma_{dt}$ when the detector is not exposed to radiation and subtracts the noise contribution of the signal processing electronics $[\sigma_d = (\sigma_{dt}^2 - \sigma_{ar}^2)^{1/2}]$. In infrared spectroscopy, the detector noise from thermal detectors is often dominant at all analytical signal levels.

Detector noise characteristics can also be expressed as a noise equivalent power (NEP or $\Phi_n$) or detectivity ($D$), as discussed in Section 4-4. The NEP is a useful figure of merit because it indicates the minimum radiant power that can be detected. If the radiant power incident on the detector equals the NEP, the $S/N$ is unity. The detectivity $D$, the inverse of the NEP, is higher for detectors with less noise and represents the theoretical $S/N$ for a photon flux of 1 W on the detector, if only detector noise is significant. The $D^*$ value is just the detectivity normalized to a noise equivalent bandpass of 1 Hz and an area of 1 cm$^2$. Often detector noise is proportional to area, so that $D^*$ allows different types of detectors with different areas to be compared. Typical values of $D^*$ are found in Tables 4-5 and 4-6.

For NEP calculations, the detector and signal processor gain factors must be taken into account since the NEP is usually defined as $\sigma_d/R(\lambda)$, where $\sigma_d$ is the rms dark noise before gain. For a PMT limited by shot noise in the thermal dark current, the detector (dark current) noise in the readout voltage equals $(mGKE_d)^{1/2}$. Thus $\sigma_d$ referenced to the photocathode equals $(Ki_d)^{1/2}$, NEP $= (Ki_d)^{1/2}/R(\lambda)$, $D = R(\lambda)/(Ki_d)^{1/2}$, and $D^* = R(\lambda)A^{1/2}/[2e(1 + \alpha)i_d]^{1/2}$. If excess dark current noise is important, a plot of $D$ vs. the PMT bias voltage can be used to choose the optimum bias voltage (maximum $D$).

## Signal-Shot-Noise-Limited Expressions

Eventually, as $i_S$ is increased by changing experimental conditions or by increasing the analyte concentration until $i_S > i_B$ and $\sigma_S > \sigma_{bk}$, signal shot noise or the $Ki_S$ component in equation 5-31 dominates, and we have

$$\frac{S}{N} = \left(\frac{i_S}{K}\right)^{1/2} \tag{5-37}$$

Here we see that the $S/N$ is proportional to the square root of the analytical signal and also proportional to $K^{-1/2}$ or $\Delta f^{-1/2}$. This holds true until the noise bandwidth is reduced to the point that either signal or background flicker noise, excess dark current noise, or amplifier-readout noise becomes dominant.

## Signal-Flicker-Noise-Limited Expressions

As $i_S$ is further increased, the analytical signal flicker noise increases more rapidly than signal shot noise because it is proportional to signal. Eventually, the analyte signal flicker noise limit is reached, so that equation 5-31 reduces to

$$\frac{S}{N} = \xi^{-1} \tag{5-38}$$

Here the $S/N$ can be improved by reducing the noise bandwidth or by adjusting experimental conditions to reduce the magnitude of the signal flicker factor. The dependence of $(\sigma_S)_f$ or $\xi$ on $\Delta f$ must be determined experimentally; in general, the improvement is less than for shot noise ($S/N$ improves as $\Delta f^{-x}$ where $x < \frac{1}{2}$). Often, $\xi$ has only a minor dependence on experimental variables and significant improvements can be achieved only through substantial redesign of the spectrometer.

## Dependence of *S/N* on Analytical Signal

The dependence of the $S/N$ on the analytical signal (expressed as a photocathodic current) under different conditions is shown in Figure 5-6. The curves also represent plots of $S/N$ versus analyte concentration in the linear portion of the calibration curve and are calculated from the equation

$$\frac{S}{N} = \frac{i_S}{[Ki_S + (\xi i_S)^2 + \sigma_{bk}^2]^{1/2}} \tag{5-39}$$

for different values of $\xi$ and $\sigma_{bk}$ (normalized to the photocathode) with $\Delta f = 1$ Hz and $\alpha = 0.3$. Curve a (dashed line) represents the dependence of $S/N$ on $i_S$ if only signal shot noise is present. Thus curve a indicates the highest $S/N$ obtainable for a given photocathodic current at the noise equivalent bandpass chosen; the log-log slope is $\frac{1}{2}$ because of the square root dependence of signal shot noise on signal. The range of photocathodic currents plotted ($10^{-17}$ to $10^{-8}$ A) represents the typical range of signal levels expected in most emission and luminescence measurements.

In curves b to e, blank and/or signal flicker noise is significant or dominant compared to signal shot noise at certain signal levels. The presence of signal flicker noise or blank noise reduces the $S/N$ to a value below that achievable if only signal shot noise is present. Curve b illustrates that increasing the analyte concentration from the detection limit to higher concentrations gives

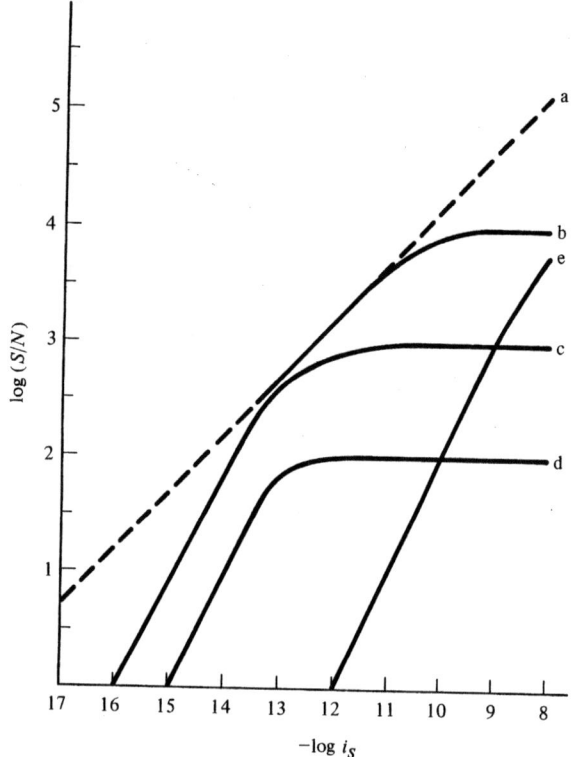

**FIGURE 5-6** Dependence of the *S/N* on signal photocathodic current. Curve a represents the signal shot noise limit with a log-log slope of $\frac{1}{2}$. Curves b ($\xi = 10^{-4}$, $\sigma_{bk} = 1 \times 10^{-16}$ A), c ($\xi = 10^{-3}$, $\sigma_{bk} = 1 \times 10^{-16}$ A), d ($\xi = 10^{-2}$, $\sigma_{bk} = 1 \times 10^{-15}$ A), and e ($\xi = 10^{-4}$, $\sigma_{bk} = 1 \times 10^{-12}$ A) illustrate the effect of increasing the signal flicker factor or the blank noise (referenced to the photocathode). Signal flicker noise becomes dominant at higher signal levels because it increases in proportion to the signal while signal shot noise increases only with the square root of the signal. As the signal flicker factor increases, flicker noise is seen to become significant at lower signal levels and the *S/N* reaches a plateau (log-log slope of 0) at a lower value (*S/N* = $\xi^{-1}$). Where blank noise is dominant, the log-log slope is unity because the noise is independent of the signal. As the signal level rises, the signal shot or signal flicker noise increases while the blank noise is constant. The greater the blank noise, the more severe the reduction of the *S/N* at low signal levels and the higher the signal level before the signal shot or signal flicker noise limit is reached.

rise to three regions in which the dependence of the *S/N* upon the analytical signal or analyte concentration is different. The *S/N* is first proportional to the signal (blank noise dominant), then proportional to the square root of the signal (signal shot noise dominant), and finally independent of the signal (signal flicker noise dominant). If the magnitude of the blank or signal flicker

noise is higher (curves c and d), the signal shot noise limit is not observed because signal flicker noise is greater than signal shot noise when the signal reaches a level high enough that signal-carried noise is significant. Note in curve e (relatively high blank noise) that neither the signal flicker nor shot noise limit is observed. This behavior is also observed if readout resolution is insufficient. Thus, if the gain factors (*m* or *G*) are reduced by a factor of 10 at higher analyte concentrations to avoid overranging the readout device, the relative contribution of flicker noise in the analytical or background signals is reduced by a factor of 10 compared to the readout noise or resolution. In some cases, the *S/N* may not be totally independent of the signal level under signal-flicker-noise-limiting conditions if the signal flicker factor is dependent on the analyte concentration.

In later chapters that deal with specific emission or photoluminescence techniques, the *S/N* expressions discussed are explored further and noise sources explicitly identified. These expressions are also used to discuss theoretical limits of detection and specific optimization methods.

## 5-5 SIGNAL-TO-NOISE EXPRESSIONS FOR ABSORPTION MEASUREMENTS

The same basic *S/N* considerations discussed for emission and luminescence measurements apply to absorption measurements. However, the *S/N* equations have a different form because the measured photon flux from the excitation source is highest for the blank or reference measurement and decreases as the transmittance decreases (increasing analyte concentration or analytical signal *A*). Therefore, we expect the limiting noise sources to progress from signal flicker noise, to signal shot noise, and finally to signal-independent noise as the analyte concentration increases. This is just the opposite of the progression observed for emission and luminescence measurements.

### General Equations

With transmittance (*T*) readout, the absorbance is calculated from $A = -\log T = -0.43 \ln T$. Applying propagation of uncertainty mathematics to the latter form shows how the SD in absorbance $\sigma_A$ is related to the SD in transmittance $\sigma_T$ by

$$\sigma_A = \frac{0.43 \sigma_T}{T} \qquad (5\text{-}40)$$

Here it is assumed that $\sigma_T \ll T$. The RSD in absorbance is thus given by

$$\frac{\sigma_A}{A} = \frac{0.43\sigma_T}{TA} = \frac{-\sigma_T}{T \ln T} \qquad (5\text{-}41)$$

and the $S/N$ of an absorbance measurement is $A/\sigma_A$. If the absorbing system follows Beer's law ($A \propto c$), the RSD of an absorbance measurement is equal to the relative concentration uncertainty ($\sigma_A/A = \sigma_c/c$).

Actually, the transmittance is obtained from measuring the total reference signal ($E_{rt}$), the total sample or standard signal ($E_{st}$), and the 0% $T$ signal ($E_{0t}$) and applying the equation

$$T = \frac{E_{st} - E_{0t}}{E_{rt} - E_{0t}} \qquad (5\text{-}42)$$

Thus the uncertainty in the transmittance is determined by the uncertainty in the three measured signal components. Note in this expression that only $E_{st}$ varies with $T$. Thus the dependence of $\sigma_T$ on $T$ is due primarily to the dependence of the uncertainty in measuring $E_{st}$ on $T$. If it is assumed there is no uncertainty in measuring $E_{rt}$ and $E_{0t}$, application of propagation of error mathematics to equation 5-42 yields

$$\sigma_T = \frac{\sigma_{st}}{E_r} \qquad (5\text{-}43)$$

Substitution of equation 5-43 into equations 5-40 and 5-41 yields

$$\sigma_A = \frac{0.43\sigma_{st}}{TE_r} \qquad (5\text{-}44)$$

$$\frac{\sigma_A}{A} = \frac{-\sigma_{st}}{TE_r \ln T} \qquad (5\text{-}45)$$

where $\sigma_{st}$ is the standard deviation for measuring $E_{st}$. Actually, $\sigma_T$ is slightly larger because of the uncertainty

in measuring $E_{rt}$ and $E_{0t}$, but the approximation is good to within a factor of $2^{1/2}$.

The noise in $E_{st}$ is due to signal flicker noise, signal shot noise, and 0% $T$ noise, as shown by

$$\sigma_{st} = [\sigma_{0t}^2 + (\sigma_s)_s^2 + (\sigma_s)_f^2]^{1/2} \qquad (5\text{-}46)$$

$$\sigma_{st} = [\sigma_{0t}^2 + mGKTE_r + (\xi_s TE_r)^2]^{1/2} \qquad (5\text{-}47)$$

where $\sigma_{0t}$ is the 0% $T$ noise or rms noise in $E_{0t}$, $(\sigma_s)_s$ is the rms signal shot noise in the sample signal $E_s$ $[(\sigma_s)_s = (mGKE_s)^{1/2} = (mGKTE_r)^{1/2} = mG(Ki_s)^{1/2} = mG(KTi_r)^{1/2}]$, $(\sigma_s)_f$ is the rms signal flicker noise in $E_s$ $[(\sigma_s)_f = \xi_s E_s = \xi_s TE_r = mG\xi_s Ti_r]$, $\xi_s$ is the sample signal flicker factor, $i_r$ is the reference photocathodic current ($i_r = E_r/mG$), and $i_s$ is the sample photocathodic current ($i_s = E_s/mG = i_r T$). Substitution of equation 5-47 into equations 5-43 to 5-45 yields

$$\sigma_T = \frac{[KmGE_r T + (\xi_s TE_r)^2 + \sigma_{0t}^2]^{1/2}}{E_r} \qquad (5\text{-}48)$$

$$\sigma_A = \left(\frac{0.43}{E_r}\right)\left[KmGE_r T^{-1} + (\xi_s E_r)^2 + \left(\frac{\sigma_{0t}}{T}\right)^2\right]^{1/2} \qquad (5\text{-}49)$$

$$\sigma_A/A = (-E_r \ln T)^{-1}\left[KmGE_r T^{-1} + (\xi_s E_r)^2 + \left(\frac{\sigma_{0t}}{T}\right)^2\right]^{1/2} \qquad (5\text{-}50)$$

Table 5-5 shows the dependence of $\sigma_T$ and $\sigma_A$ on $A$ or $T$ for the three limiting cases. Note from the table and equation 5-48 that $\sigma_T$ is independent of $T$ for 0% $T$ noise-limiting conditions, but it decreases with decreasing $T$ if signal shot or flicker noise is limiting. With the latter types of noise, $\sigma_T$ and $\sigma_{st}$ are proportional to the square root and first power, respectively, of the sample signal ($E_s$ or $i_s$) and the transmittance. The data in the table also show that the contribution of signal flicker noise to $\sigma_A$ is constant, the contribution

**TABLE 5-5**
Dependence of the standard deviation in transmittance and absorbance on absorbance for limiting cases[a]

| A | T | 0% T noise limited | | Signal shot noise limited | | Signal flicker noise limited | |
|---|---|---|---|---|---|---|---|
| | | $\sigma_T$ | $\sigma_A$ | $\sigma_T$ | $\sigma_A$ | $\sigma_T$ | $\sigma_A$ |
| 0 | 1 | 0.001 | 0.00043 | 0.001 | 0.00043 | 0.001 | 0.00043 |
| 1 | 0.1 | 0.001 | 0.0043 | 0.00033 | 0.0014 | 0.0001 | 0.00043 |
| 2 | 0.01 | 0.001 | 0.043 | 0.0001 | 0.0043 | 0.00001 | 0.00043 |
| 3 | 0.001 | 0.001 | 0.43 | 0.000033 | 0.014 | 0.000001 | 0.00043 |

[a] In all cases, $\sigma_T = 0.001$ or $\sigma_A = 0.43\sigma_T/T = 0.00043$ at $A = 0$ or $T = 1$.

of signal shot noise to $\sigma_A$ increases as $T^{-1/2}$, and the contribution of 0% $T$ noise to $\sigma_A$ increases as $T^{-1}$. This occurs because $\sigma_A$ is proportional to $\sigma_{st}/T$ (equations 5-44 and 5-49), due to the logarithmic relationship between $A$ and $T$.

The dependence of $\sigma_A/A$ shown in Figure 5-7 (plots of equation 5-50) is seen to be more complex because $\sigma_A$ is constant or increasing and $A$ is increasing as $T$ decreases. In all cases $\sigma_A/A$ initially decreases from its value at $T = 1$ ($A = 0$) because $A$ increases faster than $\sigma_A$ increases. In the case of the 0% $T$ noise limit (curve a), $\sigma_A$ increases faster than $A$ increases above $A \approx 0.4$. For the signal-shot-noise-limiting case (curve b), the absorbance at which $\sigma_A$ increases faster than $A$ ($\sigma_A/A$ increases) is larger ($A \approx 0.9$), the precision is better at a given absorbance, and $\sigma_A/A$ increases at high absorbances at a slower rate than for the 0% $T$ noise-limiting case. Finally, for the signal-flicker-noise-limiting case (curve c), $\sigma_A/A$ continually improves because $\sigma_A$ is constant.

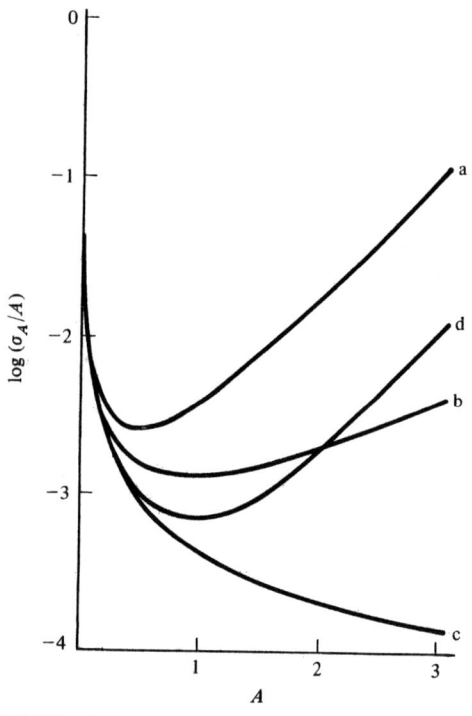

**FIGURE 5-7** Dependence of $\sigma_A/A$ on $\ddot{A}$. In all cases, $\sigma_T = 0.001$ or $\sigma_A = 0.00043$ at $A = 0$. Curves, a, b, and c represent 0% $T$ noise, signal shot noise, and signal flicker noise limited conditions, respectively. In curve d, all three types of noise are considered. At $A = 0$, the contributions from signal flicker noise, signal shot noise, and 0% $T$ noise to the total noise are 90, 9, and 1%, respectively. At absorbances above 1, signal flicker noise becomes insignificant; above an absorbance of 2, signal shot noise becomes insignificant.

In a real instrument a combination of the different limiting noise sources determines the dependence of $\sigma_A/A$ on $A$, as shown in curve d of Figure 5-7. As $T$ decreases ($A$ increases), $E_s$ decreases, so that signal flicker noise becomes insignificant compared to the signal shot noise. Eventually, the signal shot noise is insignificant compared to the 0% $T$ noise. Thus the $\sigma_A/A$ curve is between curves a and c in Figure 5-7 (for a given value of $\sigma_{st}$ at $T = 1$).

If we divide equation 5-50 by $mG$ to express the reference signal in terms of the photocathodic current,

$$\frac{\sigma_A}{A} = (-i_r \ln T)^{-1} \left[ Ki_r T^{-1} + (\xi_s i_r)^2 + \left(\frac{\sigma_{0t}}{mGT}\right)^2 \right]^{1/2}$$

(5-51)

Thus we see that $\sigma_A/A$ for a given $K$ is strongly dependent on the magnitudes of $i_r$, $\xi_s$, and $\sigma_{0t}$. We also note that $m$, $G$, or the radiant power incident on the sample container is usually adjusted so that $E_r = E_{fs}$, where $E_{fs}$ is the full-scale setting of the readout device (100% $T$ voltage). Thus $mG = E_{fs}/i_r$. Substitution of this relationship into equation 5-51 yields

$$\frac{\sigma_A}{A} = (-i_r \ln T)^{-1} \left[ Ki_r T^{-1} + (\xi_s i_r)^2 + \left(\frac{\sigma_{0t} i_r}{TE_{fs}}\right)^2 \right]^{1/2}$$

(5-52)

For high values of $i_r$, the relative magnitudes of $\xi_1$ and $\sigma_{0t}/TE_{fs}$ determine whether signal flicker noise or 0% $T$ noise dominates.

## 0% *T* Noise-Limited Expressions

More detailed information about optimization is obtained by examination of individual limiting cases. For absorption measurements that are limited by 0% $T$ noise or noise independent of the excitation source, equations 5-50 and 5-52 reduce to

$$\frac{\sigma_A}{A} = \frac{\sigma_{0t}}{-E_r T \ln T} = \frac{\sigma_{0t}}{-E_{fs} T \ln T}$$

(5-53)

If equation 5-53 is differentiated with respect to $T$ and set equal to zero, the optimum %$T$ (minimum $\sigma_A/A$) is 36.8 ($A = 0.434$).

The 0% $T$ noise can be due to amplifier-readout noise, dark current noise, or background emission noise, which are all still present when the excitation radiation is blocked. Thus

$$\sigma_{0t} = (\sigma_d^2 + \sigma_{ar}^2 + \sigma_{bE}^2)^{1/2}$$

(5-54)

Expansion of the noise terms yields

$$\sigma_{0t} = [(\sigma_{bE})_s^2 + (\sigma_{bE})_f^2 + (\sigma_d)_s^2 + \sigma_{ar}^2 + (\sigma_d)_{ex}^2]^{1/2}$$

(5-55)

where $(\sigma_{bE})_s = (mGKE_{bE})^{1/2} = mG(Ki_{bE})^{1/2}$, $(\sigma_{bE})_f = \chi E_{bE} = \chi mGi_{bE}$, and $(\sigma_d)_s = (mGKE_d)^{1/2} = mG(Ki_d)^{1/2}$. Thus, for a given $T$ under 0% $T$ noise-limiting conditions, the precision can be improved by reducing the relative 0% $T$ noise or $\sigma_{0t}/E_{fs}$ through reduction of $\sigma_{0t}$ for a given $i_r$, or by increasing $i_r$ for a given $\sigma_{0t}$. A higher value of $i_r$ allows reduction of $G$, which reduces the relative contribution of background emission shot and flicker noise, detector or dark current noise, and possibly amplifier noise if it is dependent on $G$. Similarly, if $i_r$ is increased, $m$ can be reduced, which reduces the contribution from background emission noise and dark current shot noise. When 0% $T$ noise is predominantly readout or quantization noise, equation 5-53 can be written as

$$\frac{\sigma_A}{A} = \frac{\sigma_r}{-E_{fs}T \ln T}$$

Precision can only be improved by using a higher-resolution readout device or scale expansion. In infrared absorption spectrometry, the relative high levels of detector noise leads to the dominance of 0% $T$ noise at all absorbances.

### Signal-Shot-Noise-Limited Expressions

For the case of signal shot noise dominance, equation 5-52 reduces to

$$\frac{\sigma_A}{A} = \frac{[mGK/(TE_r)]^{1/2}}{-\ln T} = \frac{(K/Ti_r)^{1/2}}{-\ln T}$$

(5-56)

Differentiation reveals that the optimum precision is achieved at 13.5% $T$ ($A = 0.87$). When measurements are signal shot noise limited, precision can be improved by increasing the lamp reference signal $i_r$ ($\sigma_A/A \propto i_r^{-1/2}$) or by reducing the noise bandwidth constant $K$ ($\sigma_A/A \propto K^{1/2}$) until the point is reached that signal flicker noise or 0% $T$ noise becomes dominant. The reference photocathodic current can be increased by increasing the radiance of the source or by using a larger monochromator slit width.

### Signal-Flicker-Noise-Limited Expressions

For measurements that are signal flicker noise limited, equation 5-52 reduces to

$$\frac{\sigma_A}{A} = \frac{\xi_s}{-\ln T}$$

(5-57)

in which case $\sigma_A/A$ continually improves with increasing $A$ (if $\xi_s$ is independent of $T$) until the point is reached that signal shot or 0% $T$ noise become limiting. The sample flicker factor is given by

$$\xi_s = (\xi_1^2 + \xi_2^2)^{1/2}$$

(5-58)

Here $\xi_1$ is the **source flicker factor** or the RSD of the source spectral radiance over the measurement bandwidth due to flicker noise. The flicker factor $\xi_2$ is the **sample container transmission flicker factor** or the RSD of the transmission of the sample container due to fluctuating transmission properties independent of the analyte. (It is sometimes called the *convection* or *cell positioning flicker factor*.) Thus $\sigma_A/A$ can be improved at a given $T$ by reducing $\xi_1$ through use of a more stable light source or double-beam compensation, or by reducing $\xi_2$, through better control of the factors influencing it.

### Other Noise Sources

Other noise sources and limiting cases can be important in certain situations in absorption measurements; terms to account for analyte or background luminescence noise must be considered in some rare cases. For atomic absorption spectrophotometry (see Section 10-3), analyte emission noise can be significant, and an additional signal flicker noise, called analyte absorption flicker noise, is usually important. It is due to fluctuations in the analyte absorptivity, in the viewed analyte population, or in the sample container pathlength.

### Direct Absorbance Readout

For instruments with direct absorbance readout, the *S/N* expressions differ from those for transmittance readout only in the readout noise term, since the readout occurs after logarithmic conversion rather than before conversion. Thus $\sigma_A = \sigma_{ls}/k'$, where $\sigma_{ls}$ is the rms noise in the sample logarithmic voltage $E_{ls}$ and $k'$ is the logarithmic conversion factor (absorbance units per volt) see Section 2-5). Therefore,

$$\frac{\sigma_A}{A} = \left\{ (-E_r \ln T)^{-1} \left[ (\sigma_s)_s^2 T^{-1} + (\sigma_s)_f^2 + \left(\frac{\sigma_{0t}'}{T}\right)^2 \right] + \left(\frac{\sigma_{ar}'}{Ak'}\right)^2 \right\}^{1/2}$$

(5-59)

where $\sigma_{0t}'$ is the 0% $T$ noise at the input of the logarithmic circuitry, and $\sigma_{ar}'$ is the rms noise due to the logarithmic conversion circuitry or the readout device. The shot and flicker noise and $E_r$ are also referenced to the input of the logarithmic circuitry. If $\sigma_{ar}'$ is the limiting noise source, equation 5-59 simplifies to

$$\frac{\sigma_A}{A} = \frac{\sigma'_{ar}}{Ak'} \qquad (5\text{-}60)$$

Note that under these conditions, $\sigma_A/A$ continues to improve as $A$ is increased as for the signal-flicker-limited case since the standard deviation in absorbance is independent of $A$. When the logarithmic conversion is performed by computer software, the $S/N$ expressions for transmittance readout still apply. If round-off errors in the software are negligible, the precision can be limited by the resolution of the analog-to-digital converter used to measure $E_s$, and $\sigma_{0t}$ in equations 5-52 and 5-53 is determined by the resolution of the ADC.

In Chapters 10, 13, and 14, the $S/N$ expressions for absorption are considered further as we discuss detection limits, optimization methods, and the precision of specific absorption techniques.

## 5-6  SIGNAL-TO-NOISE RATIO ENHANCEMENT TECHNIQUES

If precision is limited by electrical noise, the relative amount of noise can be decreased ($S/N$ enhancement) to the point that other factors limit the precision. It is crucial to identify the $S/N$ expression applicable to a given type of spectrometric measurement and to the specific instrument and experimental conditions used; this allows one to pinpoint the limiting noise sources.

As discussed in Sections 5-4 and 5-5, the strategy for $S/N$ optimization depends on the limiting noise source(s). Thus identification of these source(s) is the first step in carrying out a systematic $S/N$ optimization. Measurement of the noise in the readout signal under a few sets of conditions can quickly identify its source. For example, consider emission or luminescence measurements. Measurement of the rms noise or SD in the total signal ($\sigma_t$) for an analyte solution and then a blank ($\sigma_{bk}$) indicates if $\sigma_t$ is due to blank noise ($\sigma_t \approx \sigma_{bk}$) or signal-carried shot or flicker noise ($\sigma_t \gg \sigma_{bk}$). The blank-noise-limiting case occurs for low analyte concentrations where the analytical signal is small. Note that if the blank does not contain concomitants which contribute significantly to the blank component of the total signal and noise, the blank noise measurement does not reflect the true blank noise in the total signal. If the blank signal noise is dominant, blank noise can be compared to the total dark current noise ($\sigma_{dt}$), measured with all radiation to the detector blocked, to ascertain if background signal noise is dominant ($\sigma_B = \sigma_{bk} \gg \sigma_{dt}$) or if dark current or amplifier-readout noise is limiting ($\sigma_{bk} \approx \sigma_{dt}$).

If signal carried noise is limiting ($\sigma_t \gg \sigma_{bk}$), one must determine which component (signal shot or signal flicker noise) dominates by comparing the noise in the analytical signal ($\sigma_S$) to the signal shot noise calculated from the appropriate form of the Schottky equation. Alternatively, the dependence of $\sigma_S$ on the analytical signal in the analyte concentration range of interest indicates if signal shot noise ($\sigma_S \propto i_S^{1/2}$) or signal flicker noise ($\sigma_S \propto i_S$) dominates. Similar techniques can be applied to absorption measurements where $\sigma_{st}$ is first compared to $\sigma_{0t}$ (light source blocked) to determine if lamp signal carried noise or noise independent of the lamp signal (0% $T$ noise) dominates.

Once the limiting noise sources are identified, experimental conditions can be changed to optimize the $S/N$. It must be remembered that one is often optimizing the $S/N$ for a particular analyte concentration range; the $S/N$ may be decreased for a different analyte concentration range. Also, conditions that optimize precision may have adverse effects on linearity or on interferences.

### Frequency-Domain Filtering

We have already discussed the techniques of low-pass filtering and integration to enhance the $S/N$ for dc signals. These techniques have two major limitations: the noise bandwidth extends to 0 Hz (dc), which is the region of greatest $1/f$ noise, and measurements are limited to dc or slowly changing signals. Because filtering a signal can be considered as a process of weighting the data (convolution), filtering can be readily carried out by digital techniques in hardware or in software. Two general schemes are employed: time-domain filtering (see later in this section) and frequency-domain filtering.

In the frequency-domain approach, the data are digitized at a high-enough rate to follow the signal changes. The Fourier transform of the signal is taken to convert the information from the time to frequency domain, and then multiplied by the desired frequency response $|H(f)|$ of the filter. An important theorem of Fourier transformation states that multiplication in the frequency domain is equivalent to convolution in the time domain. The inverse Fourier transform of the product then yields the filtered signal in the time domain. Essentially, any frequency response desired can be programmed. Filters with no phase shift, higher-order filters, sharp cutoff filters, and unique bandpass filters are all readily implemented. Digital filtering in the frequency domain is analogous to smoothing in the time domain (see later in this section). The frequency-domain approach is more versatile and allows visualization of the filtering operation.

## Adjustment of Analytical and Background Signal Levels

We have also discussed adjustment of such experimental conditions as wavelength, spectral bandpass, excitation source radiance, and the volume element viewed to increase the analytical signal, reduce the background signal, or improve the $S/B$ ($E_L/E_B$, $E_E/E_B$, or $E_r/E_B$). If measurements are limited by signal shot noise, the $S/N$ varies with the square root of the analytical signal ($\sigma_A/A$ is proportional to $i_r^{-1/2}$ in absorption measurements), as long as background noise or signal flicker noise does not become significant. Once the signal flicker noise limit is reached for a specific instrument, the $S/N$ can only be increased by reducing the signal flicker factor through instrumental modifications.

If measurements are limited by background noise (or 0% $T$ noise limited in absorption measurements), the $S/N$ is proportional to the analytical signal ($\sigma_A/A$ is proportional to $i_r^{-1}$ in absorption measurements) for a constant background signal until the point is reached that signal-carried noise becomes significant. When background noise dominates, lowering the background signal not only reduces interference effects due to background signals uncompensated by the blank measurement, but also reduces the blank noise (or 0% $T$ noise in absorption measurements); an exception must be made for background-flicker-noise-limited conditions, where changing the value of a variable increases the background flicker factor to a greater extent than the background signal is decreased. Since adjustment of experimental variables often affects both the analytical and background signals, care must be taken that the $S/B$ (or $E_r/E_{0t}$ in absorption measurements) is increased. If background shot noise is limiting, the ratio $S/B^{1/2}$ must be increased to improve the $S/N$.

Techniques that use polarization, chemical selectivity, or time resolution can also provide enhancement of the $S/B$ and $S/N$. These are discussed in Chapter 6.

### Photon Counting

Photon counting signal processing can provide a $S/N$ advantage over analog signal processing in certain situations. If signal and noise terms in equation 5-31 are expressed in terms of the readout signal counts, the $S/N$ expression for emission or luminescence measurements becomes

$$\frac{S}{N} = \frac{n_S}{[n_S + n_B + n_d + (\xi n_S)^2 + (\chi n_B)^2]^{1/2}} \quad (5\text{-}61)$$

where $n_S$, $n_B$, and $n_d$ are the number of analytical, back-

ground, and dark signal counts, respectively, accumulated in the measurement time.

The $S/N$ ($\sigma_A/A$) expression for absorption measurements in terms of photon counting is a modification of equation 5-51 or

$$\frac{\sigma_A}{A} = (n_r \ln T)^{-1}\left[ n_r T^{-1} + (n_r \xi_s)^2 \right.$$
$$\left. + \frac{n_{bE}}{T^2} + \left(\frac{n_{bE}\chi}{T}\right)^2 + \frac{n_d}{T^2} \right]^{1/2} \quad (5\text{-}62)$$

where $n_r$ and $n_{bE}$ are the number of counts due to the lamp reference signal and background emission, respectively.

The $S/N$ obtained with photon counting signal processing [$(S/N)_{pc}$] can be compared to the $S/N$ obtained with analog signal processing [$(S/N)_a$] by dividing equation 5-61 or 5-62 by equation 5-31 or 5-51, respectively; the ratio must be evaluated under equivalent experimental conditions, including integration time. When measurements are limited by analytical or background signal flicker noise, the ratio is unity, and there is no $S/N$ advantage with photon counting signal processing. This occurs because the noise in the original signal or background photon fluxes dominates, and the detector or type of signal processing does not change the relative noise in the signal.

When measurements are limited by signal shot noise,

$$\frac{(S/N)_{pc}}{(S/N)_a} = \frac{n_S^{1/2}}{(i_S/K)^{1/2}} = [f_d(1 + \alpha)]^{1/2} \quad (5\text{-}63)$$

Thus, if the discriminator coefficient $f_d$ is unity, the $S/N$ advantage of photon counting is $(1 + \alpha)^{1/2}$; this is typically 5 to 25%. This advantage results because photon counting is not susceptible to multiplication noise; all photoanodic pulses above the discriminator level are weighted equally. Note that if $f_d$ is adjusted to be less than $(1 + \alpha)^{-1}$, the $S/N$ for photon counting is lower than for analog signal processing because the effective analytical signal is significantly reduced. The conclusions above apply equally to absorption measurements limited by signal shot noise or to emission, luminescence, or absorption measurements limited by background shot noise or photocathode thermal emission dark current noise.

Photon counting is particularly advantageous in low-light-level situations where both the analytical and background signals are small and where analog signal processing is limited by amplifier-readout or dark current noise. The $S/N$ advantage is slight if photocathodic thermal emission shot noise is dominant as discussed above, but can be significant if noise from thermal emission at the dynodes or excess dark current noise is significant. With photon counting, amplifier-readout noise is essentially zero, and the signal and noise due to sources of dark current other than photocathode thermal emission are significantly reduced. The *S/N* advantage of photon counting is particularly significant when long integration times are required to accumulate enough signal to obtain a reasonable *S/N*. Photon counting is less susceptible to 1/*f* noise and to slow drifts in dc offset levels, in amplifier gains, and in photomultiplier gains.

### Modulation Techniques

Amplitude modulation techniques translate the signal information to a higher-frequency region by encoding it as the amplitude of a carrier waveform. This process can move dc and low-frequency signal information out of the region of high background 1/*f* noise into a region of lower noise, as shown in Figure 5-8. In spectrometric systems, the techniques of source modulation, sample modulation, or wavelength modulation are used to translate the signal information to higher frequencies, as discussed in Section 4-5.

*S/N Advantages.* It is useful to define the terms *additive* and *multiplicative noise* as an aid to understanding the potential *S/N* advantage provided by modulation techniques. **Additive noise** is independent of the analytical signal and includes detector noise, amplifier noise, and background signal noise (0% *T* in absorption measurements). **Multiplicative noise** is carried by the analytical signal. Although signal shot noise is analytical signal-carried noise, multiplicative noise usually refers only to analytical signal flicker noise that is directly proportional to the signal.

Signals and noise are carried by the modulation waveform if they are present only during the on-cycle of the modulation waveform. Thus the noise and the signal carried affect the amplitude or appear as sidebands of the modulation waveform. Multiplicative noise is always carried by the modulation waveform because it is present only during the on-cycle of modulation when the analyte is probed or viewed. Background signals and noise are not carried if they are present during the on and off cycles of the modulation waveform. They contribute to the magnitude of the baseline but not to the amplitude of the ac waveform. Additive noise may or may not be carried by the modulation waveform. Dark current and amplifier noise are always additive noises that are not carried by the modulation waveform. The type of modulation in a given spectrometer technique determines which types of background signals and additive background noise are carried by the modulation waveform.

Modulation improves the *S/N* when low-frequency additive noise that is not carried by the modulation waveform is limiting, and the analytical signal is moved to a frequency region where the additive noise in the interval Δ*f* is less (i.e., the additive noise has 1/*f* character). No *S/N* improvement results when multiplicative noise or carried additive noise dominates because the noise originally at 0 Hz is moved to $f_m$.

In emission experiments with source modulation (e.g., a mechanical chopper), both the background and analyte emission signals are translated to the modulation frequency; the background and analyte emission flicker noise are carried. No *S/N* advantage results in this case unless 1/*f* noise in the dark signal or amplifier noise is limiting. As discussed in later chapters, sample and wavelength modulation techniques can discriminate against 1/*f* background emission flicker noise. With these modulation techniques, some types of background emission signals are present during the on and off cycles of modulation, and hence the 1/*f* noise at 0 Hz in these background emission signals is not moved to $f_m$.

In photoluminescence experiments with modulation of the excitation source, neither the analyte nor the background emission signal is carried by the modulation waveform. Thus 1/*f* noise in these signals is additive and is discriminated against. This applies equally

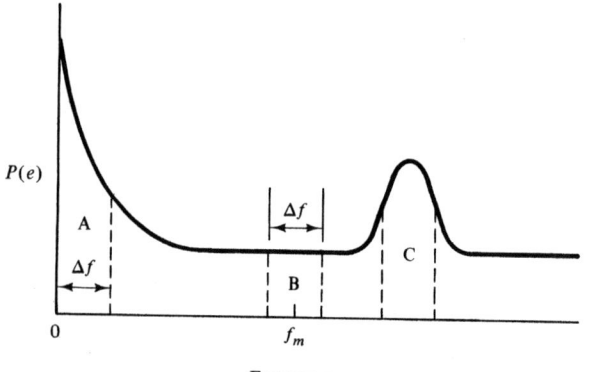

**FIGURE 5-8** Advantage of modulation. Modulation is used to encode the signal information at the modulation frequency $f_m$. The background noise observed is less at frequency $f_m$ (region B) than at dc frequencies (region A) for the same noise equivalent bandpass (Δ*f*). The rms noise is proportional to the square root of the area under the NPS defined by Δ*f*. To discriminate against 1/*f* noise, the signal processor is adjusted to respond only to the signal encoded at $f_m$ and the noise in the bandwidth centered around $f_m$. It is important to choose $f_m$ to be high enough that the 1/*f* noise has fallen off and to be at a frequency where interference noise is negligible ($f_m$ should not be in region C).

to absorption measurements with source amplitude modulation. However, analyte luminescence, background luminescence, and any scattering signals are carried (present only during on-time when radiation is impingent on the sample container) and no discrimination against flicker noise in these signals (source flicker noise and other causes) results. Sample and wavelength modulation techniques can be used to discriminate against $1/f$ noise in background luminescence and scattering signals, but not against analyte fluorescence signal flicker noise. In some cases, double modulation (source plus sample modulation or source plus wavelength modulation) provides more $S/N$ enhancement than only one form of modulation.

The $S/N$ expressions previously presented for dc measurements apply to modulated systems with some modification. First, the signal magnitudes (e.g., $i_S$, $i_r$, $i_B$) used in $S/N$ expressions for modulated systems must correspond to the average value of the signal. For square-wave modulation (50% duty cycle), the average signal magnitude for all modulated signals is one-half of the peak magnitude of the ac waveform (this factor can be slightly less if a bandpass filter is used to filter out all frequencies except the fundamental). More generally, for any pulse-shaped waveform, the average magnitude is the product of the peak amplitude and duty cycle. Second, background flicker noise, amplifier noise, and dark current excess noise must be evaluated at the modulation frequency. The magnitude of these types of noise or the background flicker factor may be less at the modulation frequency than at or low frequencies, but not necessarily zero, as they can have a white noise component.

Modulation can decrease the $S/N$ in certain situations. Often, the dc signal is converted to a square-wave signal with a device like a mechanical chopper in front of a continuous light source or an emission source. Here the number of signal photons observed per unit time is reduced by 50% compared to the dc case. Thus the $S/N$ for a given $\Delta f$ is decreased by a factor of $2^{1/2}$ when signal shot noise dominates and by a factor of 2 when white noise in the blank signal (or 0% $T$ noise in absorption measurements) dominates. If a bandpass filter is used to filter out all frequencies except the fundamental before demodulation, the factors above are slightly higher.

Modulation discriminates against nonwhite additive noise that is not carried because the amplitude of the ac waveform is the difference between the signal levels of the on and off cycles of the modulation waveform. Thus a drift or low-frequency fluctuation in the dc level of the ac waveform does not affect the modulation amplitude as long as it is much longer in duration than the period of modulation. This is illustrated in

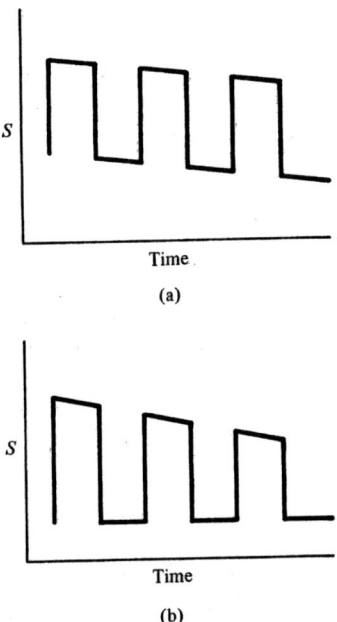

**FIGURE 5-9** Effect of additive and multiplicative noise with modulation techniques. In (a), a low-frequency additive noise component is represented by a drift in the dc level or background baseline. Note that the peak magnitude of the ac waveform is constant over the three periods of the modulation waveform but that the absolute magnitude of signal plus background varies. If the signal processor responds only to the peak amplitude, the low-frequency additive noise will be discriminated against. In (b), a low-frequency multiplicative noise component is represented by a drift in the peak amplitude of the ac waveform. This noise would affect dc and ac measurements in an equivalent way.

Figure 5-9a. However, a fluctuation in the analytical signal (i.e., a multiplicative noise) or an additive $1/f$ background noise that is carried, or present only during the on-cycle of the modulation period, is not compensated. Here the peak amplitude varies with time even if there is no noise or drift in the dc level. This is illustrated in Figure 5-9b. White noise in any signal is not discriminated against because the average magnitude of the fluctuations over the one period of the modulation waveform is the same at any frequency. In other words, modulation techniques do not eliminate noise in background signals; they can just change the center of the frequency range over which the noise is observed. Note also that any noise (including $1/f$ noise) that is added after demodulation or gating is not rejected except by normal filtering or integration techniques.

*Signal Processing and Demodulation.* Once the desired signal has been moved to a higher frequency, it can be amplified, filtered, integrated, or averaged. These operations can be performed without the $1/f$ noise and drift problems from additive noise which is not carried by the modulation waveform because a high-pass or a bandpass filter is used to remove the low-frequency components before demodulation. The resulting bandlimited waveform is demodulated and then low-pass filtered to convert the amplitude of the ac waveform into a proportional dc signal. Low-pass filtering or another $S/N$ enhancement technique can now be used more effectively because $1/f$ noise components have been removed to leave only white additive noise.

With square-wave or sinusoidal modulation, a lock-in amplifier is usually employed, as discussed in Section 4-5. The carrier waveform is multiplied by a reference waveform of the same frequency. This continually reinforces the signal amplitude information; noise at frequencies or phases different from the reference waveform produces an average output of zero. The synchronous demodulation step produces an ac output at twice the reference frequency with the ac amplitude proportional to the signal information. A low-pass filter removes the ac component and determines the overall noise equivalent bandpass. Even though the lock-in amplifier output averages zero for noncarried background or dark signals, the actual dc signal level ($i_B$ or $i_d$) must be used to calculate the background or dark current shot noise in $S/N$ expressions.

For pulse modulation waveforms (duty cycle less than 50%), such as with a pulsed laser, a boxcar integrator is used. As discussed in Section 4-5, a boxcar integrator is a gated integrator. The signal is switched to an integrator during an aperture time controlled by a reference signal. Thus the aperture window can be synchronized to pass signal information only during the actual signal pulse. The signal need not be periodic, as with a lock-in amplifier, but must be repetitive and synchronized to a reference waveform. Typically a high-pass filter is used to remove noncarried $1/f$ additive noise and signals prior to the integrator. As with a lock-in amplifier, a low-pass filter at the output can be used to determine $\Delta f$. Alternatively, the signal information from each pulse can be summed. If $n$ pulses of integration time $t$ are summed, the total averaging time is $nt$. Thus, as a first approximation $\Delta f = 1/(2nt)$. It can be shown that actually, $\Delta f$ is equal to $0.886/(nt)$.

An additional $S/N$ advantage can result from using a high-intensity pulsed source in absorption, and particularly, photoluminescence experiments. Consider the difference between a dc measurement with a continuous light source and a measurement with a pulsed light source of 100 times greater peak intensity and a 1% duty cycle. The signal shot noise in both experiments is the same since the number of photons observed per unit time is the same. However, the dc background emission signal and noise are observed for 1/100 of the time in the pulsed experiment. This means that 1/100 of the background emission signal photons or thermal dark current electrons are observed per unit time compared to the dc experiment. Thus the absolute background emission noise and dark current shot noise are reduced by a factor of 10 ($100^{1/2}$) and the background signal flicker noise (that is carried) by a factor of 100. Also, in some cases, the aperture time can be delayed to measure the analytical signal after a background signal has decayed to a smaller magnitude.

Although modulation techniques usually employ a sinusoidal, square wave, or pulse waveform, any periodic waveform or even nonperiodic waveforms can be used. The **cross-correlation function** $[C_{1,2}(\tau)]$ is defined as

$$C_{1,2}(\tau) = \lim_{T \to \infty} \frac{1}{2T} \int_{-T}^{T} E_1(t)E_2(t - \tau)\, dt \qquad (5\text{-}64)$$

where $E_1(t)$ and $E_2(t)$ are two functions describing the dependence of voltage on time and $\tau$ is the delay time. If $E_1(t) = E_2(t)$, then $C_{1,2}(\tau)$ is the **autocorrelation function** $[C_{1,1}(\tau)]$. Thus correlation techniques depend on the relationship between a signal and a delayed version of itself (*autocorrelation*) or of another signal (*cross correlation*). A plot of $C_{1,2}(\tau)$ vs. $\tau$ is called a **correlogram**. Correlation techniques are less susceptible to impulse noise since the effect of noise spikes will show up at only a certain $\tau$ in a correlogram.

Autocorrelation is useful when there is no reference signal or means of synchronizing a sampling system to a signal. Only at $\tau = 0$ will random noise be correlated in an autocorrelogram. At $\tau$ different from zero, random noise decays to zero, but the desired signal information is enhanced at certain values of $\tau$ or scanned across the waveform.

A lock-in amplifier is a single point correlator where $E_1$ is the measured optical signal and $E_2$ is the reference modulation signal with $\tau$ adjusted to zero. Similarly, a boxcar integrator is a single-point correlator, but in this case, $\tau$ can be adjusted to any desired value.

## Double-Beam and Dual-Channel Techniques

Flicker noise and drift in signals can be reduced by double-beam or dual-channel optical arrangements (see Section 4-6), which monitor both the analytical signal and another signal correlated to it. For double-beam instrumentation in absorption and photoluminescence experiments, the source radiation is split into an analytical beam which probes the sample, and a reference beam, which is viewed directly or after passing through the blank. These two beams are detected simultaneously with two optical transducers (double beam in-space) or alternatively with a single optical transducer (double beam in-time). Ratioing the sample and reference signals compensates for drift and low-frequency flicker noise due to the source since both signals track the source intensity. This compensates for source flicker noise in absorption measurements and analyte luminescence flicker noise, background luminescence flicker noise, and scattering flicker noise due to source flicker in luminescence measurements.

Often in double-beam in-time spectrometers, the amplitudes of both the analytical and reference signals are measured relative to a baseline established by blocking the source radiation. This source modulation provides discrimination against nonwhite noise that is independent of the source. The chopping frequency is chosen to be in a frequency region where $1/f$ additive noise is insignificant.

Particularly in emission and photoluminescence measurements, the dual-channel approach can be used to compensate for $1/f$ noise in background signals. Here the reference channel is adjusted to a wavelength where the analytical signal is zero but the background signal is the same as for the analytical channel. The signal processor continuously calculates the difference between the analytical and reference channel signals. This compensates for flicker noise if the background flicker noise in both channels is correlated. Note that wavelength modulation performs this operation in time.

None of the double-beam or dual-channel approaches compensate for any type of shot noise since different photons are viewed by two detectors or by one detector during two different time periods. Thus when shot noise limits measurement precision, the dual-channel or double-beam approach can result in a reduced $S/N$ since the shot noise in the reference signal adds to the total noise.

## Time-Domain Filtering

**Time-domain filtering**, **signal averaging**, or **ensemble averaging** is accomplished by summing $n$ repetitive measurements of a signal. It can be employed for repetitive signals and is essentially an instrumental superposition of a number of signal traces accomplished by recording and storing $n$ signal traces in exactly the same way (i.e., same period of time in the signal waveform). Thus, if a repetitive waveform is coherent, the signal will increase in direct proportion to $n$, but the random noise will sum as $n^{1/2}$ to provide an increase in the $S/N$ proportional to $n^{1/2}$. This is the principle used with boxcar integrators and signal averagers.

Ensemble averaging is often used to process the signal after frequency-domain filtering with an electrical filter in dc (low-pass filter) or ac (lock-in amplifier) systems. Alternatively, $n$ measurements of the readout signal from an integrating system (box-car integrator, photon counting system) can be summed. In a broader sense, the $S/N$ or precision can be improved by making repetitive measurements of spectra, of total and blank signal pairs, or of sample and standard pairs or even by repetitions of the complete measurement scheme (samples, standards, and establishment of the calibration function).

The $S/N$ does not always improve with $n^{1/2}$. The improvement depends on the limiting types of noise and their nature, the $S/N$ enhancement techniques used prior to ensemble averaging, and the way measurements are conducted. If white noise (e.g., shot noise in the analytical, background, or dark signals) is limiting, the $S/N$ improves with $n^{1/2}$. If nonwhite noise such as $1/f$ noise is limiting, the $S/N$ may improve as $n$ is increased with the maximum improvement limited by the $n^{1/2}$ rule. In some cases, the $S/N$ can even decrease with repetitive measurements.

A related question is whether it is better to make one measurement (or spectral scan) for time $T$ or to average $n$ measurements (or spectral scans) of time $t$ where $T = nt$. The $S/N$ is equivalent in these two cases if white noise is limiting. If nonwhite noise such as $1/f$ noise (drift) is limiting, the use of $n$ repetitive measurements in time $T$ generally provides higher $S/N$ up to a factor of $n^{1/2}$ compared to one measurement over time $T$.

Again, the $S/N$ enhancement observed if nonwhite noise is limiting depends on the nature of the noise and the procedure used. For example, consider a case where there is low-frequency noise or drift only in the blank signal. For this situation, it is better to make 10 paired 1-s measurements (difference between sequential 1-s integrations of total and blank signal) than one paired 10-s measurement. Note that this is effectively manual sample modulation and that pairing blank and sample readings discriminates against any drift or low-frequency noise in the blank signal with a period much greater than the time between blank and sample measurements. Similarly, making many sample and standard measurements in time $T$ can provide higher precision than one sample and standard measurement in time $T$ if multiplicative noise or flicker noise or drift in the analytical signal is limiting.

Digital time-domain filtering, as digital frequency-domain filtering discussed previously, can also be used to smooth a signal spectrum or transient signal. Here the signal data are digitized at a high enough rate to follow a time-dependent signal faithfully with a relatively large $\Delta f$ and then stored in computer memory. Software is then used to take a weighted average of a selected number of signal data points over a chosen time period, which provides software control of $\Delta f$. Such a filter window can be sequentially moved over the data points selected to smooth out the data, such as in the case of a spectrum. The algorithm developed by Savitsky and Golay is now widely used in digital filtering. Other digital filtering algorithms are gaining popularity in certain applications. The availability of low-cost microcomputers has resulted in a rapid increase in the use of digital filtering techniques.

## Multichannel and Multiplex Systems

Multichannel spectrometers [direct readers or systems based on multichannel detectors (diode arrays, vidicons)] and multiplex [interferometers and Hadamard spectrometers (see Section 4-6)] can provide a multiplex advantage by allowing many analytical wavelengths to be monitored simultaneously. Contrast a spectrum taken by scanning with a conventional monochromator-PMT configuration to a spectrum obtained with a spectrograph and multichannel detector where the wavelength coverage and resolution are equivalent. If the spectrum is taken over time $T$ for both cases, the number of photons observed per spectral resolution element is $n$ times greater for the multichannel detector, where $n$ is the number of spectral resolution elements. If measurements with both systems are limited by amplifier white noise, dark signal white noise, signal shot noise, or background shot noise of equivalent magnitude, the multichannel detector provides a $n^{1/2}$ $S/N$ advantage. No advantage results if both detectors are limited by signal or background flicker noise.

A precise comparison is more complex since in the discussion above it is assumed that the same amount of light from the sample container strikes the PMT photocathode or a given array element; that the detector responsivities, gain, and multiplication noise are equivalent; and that the amplifier-readout and dark signal noise are equivalent. The $S/N$ advantage provided by multichannel detectors for spectrum acquisition is usually less than predicted above because of the small detector element area and larger relative amplifier-readout and dark signal noise of multichannel detectors.

Readout noise from multichannel detectors is often independent of the integration time (it arises only during the readout process). Under these readout-noise-limiting conditions, the $S/N$ is directly proportional to the integration time until other noise sources become significant. Here it is better to allow signal levels to approach detector saturation levels than to perform $n$ readouts in the same time (i.e., a $n^{1/2}$ $S/N$ advantage results).

The $S/N$ considerations for multiplex systems are somewhat different because many or all wavelengths are simultaneously striking a single detector. For simplicity consider the comparison of a spectrum taken with a conventional single-channel scanning spectrometer to the spectrum taken with a multiplex spectrometer with equivalent wavelength coverage, number of resolution elements $n$, and measurement time $T$. As a first approximation, the multiplex spectrometer provides a $n^{1/2}$ $S/N$ advantage if detector or amplifier noise is limiting because the number of photons observed per spectral element in time $T$ is $n$ times greater and the noise is only $n^{1/2}$ times greater. This is why the multiplex approach is commonly used in infrared spectrometers limited by detector noise.

If shot or flicker noise is limiting, the $S/N$ comparison depends on the nature and complexity of the spectrum. Note that shot or flicker noise in the multiplex spectrometer can arise from signal or background photons in the resolution element of interest or from photons in other resolution elements. If the spectrum consists of only one strong line in one resolution element, a $n^{1/2}$ $S/N$ advantage is realized with the multiplex spectrometer if it is signal or background shot noise limited; no $S/N$ advantage is realized if it is signal or background flicker noise limited.

When a significant number of photons is observed from other resolution elements (e.g., a complex or continuum spectrum), the $S/N$ advantage decreases or a $S/N$ disadvantage (the **multiplex disadvantage**) results. For example, consider a spectrum with 10 equally intense lines each occupying one of 100 possible spectral resolution elements. Compared to the scanning spectrometer case, the number of signal photons observed in time $T$ in the spectral element of interest increases by 100, but the total number of photons observed increases by 1000. This yields a $S/N$ advantage of $100/(1000)^{1/2}$ or $10^{1/2}$ rather than 10 under shot-noise-limiting conditions. For the case of 100 equally intense lines (a continuum) or a weak line of interest and 10 strong lines (10 times as intense), no $S/N$ advantage results with the multiplex spectrometer. If flicker noise is limiting in the foregoing situation of 10 equally intense lines, the signal increases by 100, but the noise increases by $10 \times 100 = 1000$, yielding a factor of 10 decrease in $S/N$. For a continuum spread across all resolution elements (or a weak line of interest plus 10 other lines of 10 times the intensity), the $S/N$ disadvantage is a factor of 100. For the flicker noise case above, it is assumed the flicker noise for all resolution elements is totally dependent or correlated. When the flicker noise in each channel is totally independent, the $S/N$ disadvantages above are reduced to $10^{1/2}$ and 10.

Again a precise comparison is complex and must take into account the throughput of the device and differences in detector responsivity vs. wavelength and in detector noise. With a Michelson interferometer the solid angle of collection, and hence the throughput, are much larger than for a monochromator with an entrance slit. However, the *S/N* advantage is slightly reduced (factor of $2^{1/2}$ to $8^{1/2}$) due to such factors as light loss at the beam splitter. For a Hadamard transform spectrometer, only a fraction of all spectral elements are viewed simultaneously due to the design of the slit mask(s).

The discussion above indicates the multichannel and multiplex spectrometers provide a significant *S/N* advantage in certain situations, particularly when measurements are limited by detector noise. In the case of a line spectrum consisting of a few lines of interest, the *S/N* with a single-channel spectrometer can be improved by slew scanning between the analytical lines so that most of the measurement time is spent at the analytical wavelengths of interest.

## PROBLEMS

**5-1.** A PMT, op amp *I–V* converter combination was used to monitor the emission of 10 ng mL$^{-1}$ Na in a flame. The PMT current gain was $1.0 \times 10^6$, the feedback resistor was $1.0 \times 10^7 \Omega$, the secondary emission factor was 0.30, the analyte emission flicker factor ($\xi_E$) was $0.50 \times 10^{-2}$, the background emission flicker factor ($\chi$) was $2.0 \times 10^{-2}$, the time constant was 0.50 s, the amplifier-readout noise was $1.0 \times 10^{-6}$ V, the analyte emission signal was 1.0 V, the background emission signal was 0.10 V, and the dark current voltage was 0.0010 V. Assume that the excess dark current noise is negligible.

(a) Calculate the following. Here *S* is always taken as the analytical signal and all noises are rms noises.
  (1) $\Delta f$, *K*, and *mGK*.
  (2) Dark current shot noise.
  (3) Analytical signal shot noise.
  (4) Background signal shot noise.
  (5) Analytical signal flicker noise.
  (6) Background signal flicker noise.
  (7) Total noise and *S/N* for measurement of the analytical signal.
(b) What are the dominant noise sources for this measurement?
(c) Determine the effect of $\Delta f$, *m*, $\alpha$, $\xi_E$, and $\chi$ on the results in part (a).

**5-2.** In molecular absorption, the relative precision ($\sigma_A/A$) depends on the absorbance, the photocathodic current employed, the signal flicker factors, and the 0% *T* noise. The reference photocathodic current can vary from about $1.0 \times 10^{-12}$ to $1.0 \times 10^{-6}$ A depending on the wavelength of analysis, the radiance of the lamp, the slit width employed, and the F/n and optical efficiency of the monochromator. For this problem assume that $\xi_s = 2.0 \times 10^{-4}$, $\sigma_{0r} = 4.0 \times 10^{-5}$ V, 100% *T* or full scale on the readout device ($E_{fs}$) is 1.00 V, and a 0.50-s integration time, $\alpha = 0.30$, and $i_r = 1.0 \times 10^{-11}$ A.
(a) Calculate $\Delta f$, *K*, and *mGK*.
(b) Calculate $\sigma_A/A$ at *A* = 1 and 2 and $\sigma_A$ at *A* = 0.
(c) Indicate the limiting noise sources at each absorbance.

(d) Determine the effect of $i_r$ and $E_{fs}$ on the results in parts (b) and (c).

**5-3.** In this problem, the *S/N* characteristics of a PMT, *I–V* converter configuration are compared to a PT, *I–V* converter configuration. Assume that the background signal noise is negligible, as well as amplifier noise, readout noise, and excess dark current noise. Thus the only noise sources to consider are signal and dark current shot noise, signal flicker noise, and Johnson noise. Assume that the Johnson noise is due solely to the feedback resistor of the op amp *I–V* converter and that the resistance is $1.0 \times 10^6 \Omega$ (25°C). Assume a 0.25-s time constant, a PMT gain of $1.0 \times 10^6$, a cathodic dark current of $1.0 \times 10^{-15}$ A, a collection efficiency of 1, a signal flicker factor of $1.0 \times 10^{-3}$, and $\alpha = 0.30$.

(a) Write a *S/N* expression for each configuration referenced to the output of the op amp.
(b) Calculate the photocathodic current at which the *S/N* = 1 for each configuration. What is the limiting noise at this point?
(c) At what photocathodic current do measurements become signal flicker noise limited (i.e., two-thirds of the total noise) for both configurations?
(d) Under what conditions would you use each configuration?

**5-4.** Calculate the rms Johnson noise for a $10^8$-$\Omega$ resistor at 25°C with a 1-Hz noise equivalent bandwidth.

**5-5.** What specific types of noise becoming limiting with large time constants or long integration times?

**5-6.** Show that the noise equivalent bandpass for an integrator equals $(2t)^{-1}$.

**5-7.** Why is signal flicker noise often limiting at high concentrations in emission and luminescence measurements but at low concentrations in absorption measurements?

**5-8.** If the standard deviations in the blank and sample measurements are quite different, what are the possible limiting noise sources?

**5-9.** When are modulation techniques advantageous?

**5-10.** In an emission experiment the analyte concentration is increased by a factor of 10. Identify the possible limiting noise sources for the following conditions.
   **(a)** The noise in the total signal increases by a factor of 10.
   **(b)** The noise in the total signal increases by a factor of $10^{1/2}$.
   **(c)** The noise in the total signal remains constant.

**5-11.** Why is it imprecise to say that measurements are limited by shot noise?

**5-12.** A photomultiplier tube is operated under illumination conditions that produce a photoanodic current of $6.5 \times 10^{-8}$ A. Repeated measurements of the photoanodic current give a standard deviation of $1.61 \times 10^{-13}$ A. When the illumination level is reduced by a factor of 10, the standard deviation in the photoanodic current is $5.1 \times 10^{-14}$ A. Is the limiting source of noise in the measurement source flicker noise, photocurrent shot noise, or dark current shot noise?

**5-13.** What types of noise must be limiting in a photoluminescence measurement for source modulation to improve the $S/N$?

**5-14.** The rms noise in a signal is measured to be 2.0 mV. The resolution of the readout device is 1 mV. Does quantization noise make a significant contribution to the total noise?

**5-15.** In an absorption measurement, the RSD in the absorbance continually improves as the absorbance is increased from 1 to 2. What type of noise is limiting?

**5-16.** If signal shot noise is limiting, what instrumental variable can be changed to improve the $S/N$?

**5-17.** A spectrum is obtained with a multichannel spectrometer with 500 pixels and with a scanning spectrometer with equivalent wavelength resolution. With the former, an integration time of 10 s is employed, while the scan time with the latter is 100 s. Determine the $S/N$ advantage of the multichannel spectrometer when each of the following types of noise is limiting: signal shot noise, background shot noise, signal flicker noise.

**5-18.** In an atomic emission experiment, the analyte emission signal, background emission signal, and $S/N$ are measured before and after the slit width is doubled. The increase in slit width causes the analytical signal to double and the background signal to quadruple. For each of the cases below, indicate what types of noise are limiting.
   **(a)** The $S/N$ does not change.
   **(b)** The $S/N$ improves by a factor of $2^{1/2}$.
   **(c)** The $S/N$ improves by a factor of 2.
   **(d)** The $S/N$ decreases by a factor of 2.

**5-19.** A series of noise measurements are made in a photoluminescence experiment. From the information given in each step, indicate the types of noise that are not limiting in the measurement of the sample signal.
   **(a)** The noise in the blank signal and total sample signal are approximately equal.
   **(b)** The PMT shutter is closed and the noise decreases by a factor of 10.
   **(c)** The noise in the blank signal increases by a factor of 2 when the excitation radiant power is doubled.

**5-20.** For a photon counting measurement system and a thermal detector radiant power monitor, determine the detector dark noise, the noise equivalent power, and the minimum detectable radiant power ($S/N = 2$) in watts and photons per second at 600 nm with a noise equivalent bandpass of 1 Hz. For the photon counting system, assume that the photocathode quantum efficiency is 10% and that the dark current pulse rate is $100 \ s^{-1}$. For the thermal detector, assume a typical detectivity of $1.0 \times 10^9 \ W^{-1}$ limited by detector thermal noise and a responsivity of $10 \ V \ W^{-1}$. What incident radiant power would be necessary for the signal shot noise to be equal to the thermal detector noise with the thermal detector?

# REFERENCES

1. W. J. McCarthy, "The Signal-to-Noise Ratio in Spectrochemical Analysis," appendix in *Spectrochemical Methods of Analysis*, J. D. Winefordner, ed., *Advances in Analytical Chemistry and Instrumentation Series*, vol. 9, Wiley-Interscience, New York, 1971.

2. C. Th. J. Alkemade, G. D. Boutilier, et al., "A Review and Tutorial Discussion of Noise and Signal-to-Noise Ratios in Analytical Spectrometry," *Spectrochim. Acta*, *33B*, 383 (1978); *35B*, 261 (1980).

3. G. M. Hieftje, "Signal-to-Noise Enhancement through Instrumental Techniques," *Anal. Chem.*, *44*, 81A (May 1972); 69A (June 1972).

4. R. King, *Electrical Noise*, Chapman & Hall, London, 1966.

5. K. D. C. MacDonald, *Noise and Fluctuations, An Introduction*, Wiley, New York, 1962.

6. A. Van der Ziel, *Noise in Measurements*, Wiley, New York, 1976.

7. W. R. Bennett, *Electrical Noise*, McGraw-Hill, New York, 1960.

8. M. S. Epstein and J. D. Winefordner, "Summary of the Usefulness of Signal-to-Noise Ratio Treatment in Analytical Spectrometry, *Prog. Anal. At. Spectrosc.*, *7*(1) (1984). This comprehensive article also contains a very complete reference list.

9. J. D. Ingle, Jr., and S. R. Crouch, "A Critical Comparison of Photon Counting and Direct Current Meas-

urement Techniques for Quantitative Spectrochemical Methods," *Anal. Chem.*, *44*, 785 (1972).

10.  J. D. Ingle, Jr., and S. R. Crouch, "Evaluation of Precision of Quantitative Molecular Absorption Spectrometric Measurements Using Signal-to-Noise Ratio Theory," *Anal. Chem.*, *45*, 333 (1973).

11.  C. G. Enke and T. A. Nieman, "Signal-to-Noise Ratio Enhancement by Least-Squares Polynomial Smoothing," *Anal. Chem.*, *48*, 705A (1976).

12.  J. J. Cetorelli, W. J. McCarthy, and J. D. Winefordner, "The Selection of Optimum Conditions for Spectrochemical Methods: IV. Sensitivity of Absorption, Fluorescence, and Phosphorescence Spectrometry in the Condensed Phase," *J. Chem. Educ.*, *45*, 98 (1968). Parts I to III of this series in the same journal are also classics.

# CHAPTER 6

# Methodology in Spectrochemical Analysis

This chapter considers in detail the principles on which practical analytical methods are based. Instruments are most often calibrated with external standards. Thus the chapter begins with a discussion of external standard calibration. Systematic errors are next considered, and the various types of interferences encountered in analytical procedures are carefully defined. The random errors that can limit precision in spectrochemical methods and statistical methods to assess data quality are described. There then follows a detailed discussion of the meaning of two important figures of merit for analytical procedures, the sensitivity and the detection limit. The chapter continues by describing various techniques for minimizing random and systematic errors; it includes the principles of such important techniques as separations, internal standard methods, standard addition methods, optical encoding methods, and instrumental correction techniques. Finally, the concepts and instrumentation for automated measurements are presented.

## 6-1 EXTERNAL STANDARD CALIBRATION

In most spectrometric methods, the spectrometer is calibrated by determining the analytical signal $S$ for a series of **external standards** of known analyte concentration

$c_s$. It will be understood that determining $S$ implies that $S$ is derived from a total signal measurement and a blank measurement (or reference and sample measurements in absorption measurements). A calibration curve ($S$ vs. $c_s$) is established from the results by plotting the data on graph paper or by fitting them to a suitable mathematical function by use of least-squares techniques. Next, $S$ is obtained for the analytical sample. Then the unknown analyte concentration in the analytical sample $c_x$ can be determined from the calibration curve. The concentration $c_x$ can then be related back to the analyte concentration in the original bulk sample by appropriate sample preparation factors (e.g., amount of dilution).

The ideal standard is identical to the sample except that for the latter, the analyte concentration ($c_a$) is unknown. However, real external standards are usually made from purified chemicals containing a known fraction of the analyte. Thus the external standard does not contain most of the matrix components or concomitants in the original bulk sample. Usually, an attempt is made to add all reagents used in sample preparation (or reagents used for selective analytical reactions in solution techniques) to the external standards such that the concentration of these added species is the same in the analytical sample and the external standards. In the case of solutions, the same solvent is used for analytical samples and external standards (e.g., aqueous external standards are used for analysis of water samples).

In some cases one can use a real sample as an external standard for other samples if the analyte concentration in the calibration sample has previously been determined. The National Bureau of Standards (NBS) and other organizations provide many types of real samples (e.g., geological, biological) in which the concentration of many species is certified. Thus one can choose as a standard a calibration sample with a matrix similar to the analytical sample. The calibration sample can be used to check the suitability of the technique and any external standards prepared from purified chemicals.

To determine the analytical signal, a blank measurement is required. The ideal blank is identical to the sample except that the analyte concentration is zero. Real blanks are usually solvent or reagent blanks. The same solvent in which the sample is dissolved is used as a **solvent blank**. A **reagent blank** contains the solvent plus all reagents used in the sample preparation procedure at the same concentrations as in the analytical sample.

The basic assumption of the external standard procedure is that sample and standard of the same analyte concentration ($c_a = c_s = c_x$) will yield the same analytical signal. Thus, for an accurate determination, the calibration function established from the standards must apply equally to the sample. There will be an error in the determination ($c_x$ determined is incorrect) if this basic assumption is not true.

Several factors can be responsible for the breakdown of the basic assumption. Matrix effects, due to the presence of species in the analytical sample but not in the standards or blank, can alter the sample analytical signal relative to that from a standard of the same analyte concentration. Thus significant differences between an ideal and real standard or an ideal and real blank manifest themselves in the sample presentation or measurement steps. Differences in the magnitudes of experimental variables (instrumental, physical, or chemical) at the times of sample, standard, and blank measurements, random errors, or noise can alter the analytical signals determined and the validity of the calibration function established. Even if the basic assumption is correct, there can be error in the determination if there is loss or gain of analyte (contamination) in the sample acquisition or preparation steps.

In a few cases, calibration with standards is not employed and absolute measurements are made. This implies that the instrument is precalibrated and that the calibration function, and hence the calibration curve, are known so that the analytical signal determined for a sample is converted directly to concentration. Absolute methods can be in error for the same reasons as relative methods based on calibration standards except there are no errors involving preparation and measurement of standards. However, the error due to the difference between the true calibration function and the precalibrated calibration function must be considered.

## 6-2 SYSTEMATIC ERRORS IN SPECTROCHEMICAL METHODS

A systematic error (see Appendix A) in a determination is an error in which the analyte concentration determined in a given sample is either too low or too high compared to the true analyte concentration regardless of the number of repetitive measurements made on the sample. Systematic errors can be classified as matrix errors, calibration errors, sample acquisition and preparation errors, or measurement errors. Any classification is somewhat arbitrary since certain types of errors have characteristics of two or more classes. A systematic error results in an error in determination only if the magnitude of the error is different for samples and standards.

### Matrix Errors

Matrix errors arise because of differences between the matrix of the analytical sample and the matrix of the external standards or blanks used for calibration. The analytical sample matrix can be different because of the matrix of the original bulk sample or because of species added or removed during sample preparation. If a technique has perfect specificity, the difference between ideal and real standards or blanks does not cause an error.

Often matrix errors are called *interference errors* since they are due to the presence of concomitants or interferents in the analytical sample. An **interferent** is a species or material (e.g., particle) in the sample, standard, or blank which affects the mean value of the analytical signal determined. Interference effects are often quantitatively described by indicating the error in determination caused by the specified concentration of the concomitant at a particular analyte concentration under defined experimental conditions. For example, the interference effect of vitamin $B_2$ on the determination of vitamin $B_1$ might be stated as follows: The presence of $10^{-6}$ M vitamin $B_2$ causes the determined concentration of $10^{-8}$ M vitamin $B_1$ to be high by 1% (a 1% positive error). Matrix errors or interferences can be subdivided into blank and analyte interferences as discussed below.

*Blank Interferences.* A **blank interference** (sometimes called an *additive interference*) is due to a species or material which produces an uncompensated signal independent of the analyte concentration. A blank interference causes an error in determination if the blank component of the total signal measured for the sample is different from the blank component of the total signal measured for the standards. This is shown in Figure 6-1.

Usually, a blank interference is caused by a con-

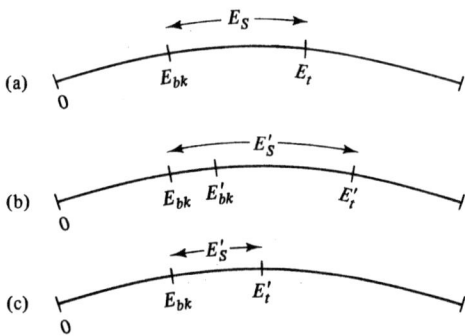

**FIGURE 6-1** Systematic errors caused by blank and analyte interference. The total, blank, and analytical signals are shown for a given analyte concentration with (a) no interferences, (b) a blank interference, and (c) an analyte interference. In (a) subtraction of the blank signal $E'_{bk}$ from the total signal $E_t$ yields the proper analytical signal $E_S$. For (b) the total signal $E'_t$ is larger due to a larger blank signal component $E_{bk}$. If the blank in (a) is used, the analytical signal $E'_S$ determined is too high and a blank interference results. In (c), $E'_S$ is too low and hence the magnitude of $E'_t$ is too low because an analyte interference has suppressed the analytical signal.

comitant in the sample that is not present in the standards, although it could be caused by improper preparation of the standards such that they contain species not in the analytical sample. If the blank interferent is absent from the blank, but equivalent in the standard and sample, the analytical signal extracted from the total signal for the standard and sample is incorrect, but no systematic error in determination results. If the species responsible for a blank interference is present in all the standards at the same concentration, the calibration curve exhibits a nonzero intercept.

Many blank interferences in spectrochemical methods are spectral interferences. A **spectral interference** is caused by an overlap within the monitored wavelength range (e.g., the spectral bandpass) of the emission, luminescence, scattering, or absorption spectral profile of the analyte with one or more such profiles due to concomitants in the sample. An interference effect can also result if stray radiation effects cause the background radiation from concomitants to appear in the monitored wavelength range. A blank spectral interference causes the analytical signal determined to be too high or too low if the blank measurement does not compensate for background spectral signals or for absorption by the concomitants in the sample. Specific types of spectral interferences are summarized in Table 6-1.

Some blank interferences are not spectral interferences. A **nonspectral interference** is caused by a species or substance which itself does not necessarily emit or absorb radiation over the wavelength range monitored, but which affects the spectral signals from other species. A nonspectral blank interference effect occurs if a concomitant depresses or enhances the background spectral signal from the sample container or another concomitant in the sample.

**TABLE 6-1**

Types of blank spectral interferences

| Technique | Potential uncompensated signals |
| --- | --- |
| Emission or chemiluminescence | Background emission from concomitants only in the sample; absorption or scattering of concomitant or sample container emission by concomitants only in the sample |
| Photoluminescence | Background emission, photoluminescence, and scattering from concomitants only in the sample; absorption or scattering of background emission, photoluminescence, or scattering signals from the concomitants or the sample container or of excitation source radiation by concomitants only in the sample |
| Absorption | Background absorption or scattering from concomitants only in the sample; background emission and luminescence from concomitants only in the sample; absorption of emission and luminescence radiation from concomitants and from the sample container by concomitants only in the sample |

*Analyte Interferences.* **Analyte interferences** (sometimes denoted *multiplicative interferences*) are caused by species or substances which alter the magnitude of the analytical signal component of the total signal as shown in Figure 6-1. Even if the blank signal is negligible, an analyte interference affects the measured analytical signal; it causes an error in determination if the analyte interferent is in the sample, but not the standards. If a given concentration of the analyte interferent is added to all standard solutions, it affects the calibration function or, more specifically, it changes the slope or shape of the calibration curve. Analyte interferences occur in both the sample presentation and measurement steps.

An analyte interferent can exert its effect in many ways. In many techniques, the analyte in the sample is converted by the spectrochemical encoder or prior sample preparation steps into a specific active chemical form that is spectrometrically probed in the sample container. In solution molecular spectrometric techniques, this is often done by adding an analytical reagent to react with the analyte to form a product that absorbs or luminesces. In atomic spectrometric techniques, the analyte is converted into an atomic vapor. Concomitants can cause an analyte interference if they alter the production or concentration of the active form of the analyte by changing the efficiency with which the analyte is converted to its active or probed form. For techniques in which the analyte is continuously introduced to the sample container, a concomitant can affect the rate of delivery and thus the amount of analyte in the sample container per unit time.

An analyte interferent can also alter the ability of the active form of the analyte to produce its analytical signal. For example, a concomitant can change the efficiency with which a given analyte or its active form absorbs, emits or luminesces. Concomitants can also alter the efficiency with which the optical signal from the active form of the analyte is transferred to the optical transducer by changing the optical properties of the sample or the sample container.

The majority of analyte interferences are nonspectral interferences. Analyte nonspectral interferences can be further subdivided into two classes, nonspecific and specific interferences. **Nonspecific interferences** are caused by species or substances which influence the analytical signal for a number of similar analytes in a manner fairly independent of analyte type. They are often called **physical interferences**. For example, substances at high concentration can change the bulk properties of the sample (e.g., ionic strength or density in solutions or electron densities in atomization cells) and thereby alter the slope or shape of the calibration curve. **Specific nonspectral interferences** (some-times called **chemical interferences**) are more analyte specific and can be related to chemical reactions between the analyte and a concomitant (e.g., complexation, precipitation). They usually affect the efficiency with which the analyte is converted to its active form.

Some analyte interferences are spectral interferences. For example, concomitants can absorb photons emitted by the analyte in emission techniques. In photoluminescence measurements, they can absorb photons from the excitation and/or emission beam.

The classification above is somewhat arbitrary since a given concomitant can cause spectral and nonspectral interference effects or even blank and analyte interference effects. Consider the determination of analyte A, where it is converted with analytical reagent R into an active form AR, which is photoluminescent.

$$A + R \rightleftharpoons AR \qquad (6\text{-}1)$$

A concomitant X could also react with R or A as follows:

$$X + R \rightleftharpoons XR \qquad (6\text{-}2)$$

$$A + X \rightleftharpoons AX \qquad (6\text{-}3)$$

Species X causes a blank spectral interference if X or XR photoluminesces at the same wavelength at which AR is monitored. An analyte spectral interference effect results if X, AX, or XR absorbs some of the luminescence photons emitted by AR or if AX photoluminesces. Species X could also cause an analyte nonspectral interference if it reduces the amount of R available to react with A (equation 6-2) or prevents all or part of A from reacting with R (equation 6-3).

As discussed in Chapter 2, analyte emission can contribute to the analytical signal measured in photoluminescence. Similarly, analyte emission and photoluminescence can affect the magnitude of the measured absorbance in absorption measurements. These extra spectral signals do not really cause an interference effect as they arise from the analyte, but they can affect the magnitude of the analytical signal and the slope and shape of the calibration curve. Both spectral and nonspectral analyte interferences can alter the primary analytical signal measured as well as additional signals arising from the analyte.

### Calibration Errors

Systematic errors can be caused by incorrectly determining the calibration curve (function) if errors are made in preparing the standards, in presenting the standards to the instrument, in measuring the standards, or in fitting the calibration curve. The determination of the analyte in the sample can be no more accurate than the accuracy with which the analyte concentrations in the standards are known. The accuracy of the standards is determined by the accuracy of the gravimetric and volumetric techniques and equipment employed. For the highest accuracy work, temperature and pressure effects on mass and volume measurements must be considered. The purity and stability of materials used to prepare the standards must be confirmed.

The chemical form of the standards should be identical to that of the analyte since the state of oxidation, complexation, or isomerization of the analyte could influence the analytical signal. Often, the chemical speciation of the analyte must be considered. Thus one must be careful to designate exactly what is meant by the "analyte." Consider the determination of chromium. In one case, total chromium might be of interest, while in another determination, the amount present as Cr(III) or Cr(VI) might be important. If Cr(III) is to be determined, is it the total Cr(III) or just the Cr(III) that is complexed with a specific ligand that is of interest? Such chemical speciation information is very important because the chemical form of an element affects its biological activity, transport, and environmental impact.

Once the standards are prepared, care must be taken, especially for trace analysis, to ensure that the analyte concentration in the standard is the same at the time of measurement as when it was prepared. During storage, the analyte concentration can decrease due to decomposition, volatilization, or adsorption onto container walls. The analyte concentration can be higher than expected due to contamination. Contamination arises from desorption of the analyte from surfaces of the apparatus used to prepare or transfer the standards to the sample container or of the sample container itself. Dust particles and reagent impurities are additional sources of contamination.

Systematic errors can also result from bias or improper fitting of the calibration curve to the data as can occur when using a straight-line least-squares program on nonlinear data. Enough standards must be measured to ensure that the calibration curve is well established since usually the analytical signal for the sample is found by interpolation between the analytical signals from the standards.

### Sample Acquisition, Preparation, and Measurement Errors

The accuracy of a determination is limited by the integrity of the analytical sample presented for measurement. Systematic errors can result if the analyte con-

centration in the analytical sample is not representative of the analyte concentration in the bulk sample. This can occur if the original sample acquired is not representative of the bulk sample or if some of the analyte is lost during sample acquisition, storage, or preparation. Also, the analyte concentration determined can be too high, due to contamination in the steps noted above. The accuracy of volumetric and gravimetric operations must be considered carefully as mentioned for standards.

Systematic errors can arise during sample presentation and measurement of samples, standards, and blanks due to numerous factors, including "cockpit error" (i.e., blunders). Drifts in critical experiment variables (e.g., $T$, pH) that affect the analytical or blank signals cause errors if the magnitude of these variables is significantly different between sample, standard, and blank measurements. If drifts occur in the blank signal, standard and blank measurements or sample and blank measurements should be paired and made close to one another in time. If drifts in the analytical signal are significant, sample measurements should be bracketed by standard measurements.

## 6-3 RANDOM ERRORS IN SPECTROCHEMICAL MEASUREMENTS

It is important to consider the precision of measurements since random errors also affect accuracy. Random errors are caused by fluctuations in the magnitudes of experimental variables or by the particle nature of light and matter; they result in the measured values being distributed around the mean value. The relative uncertainty due to random error decreases with the number of measurements averaged since positive and negative deviations tend to cancel. Some knowledge of statistics is required to assess the influence of random error, and Appendix A should be consulted for review purposes.

### Determination of Standard Deviation in Concentration

The magnitude of the random error is obtained by making $n$ repetitive determinations of the analytical signal for a given sample or standard. From these measurements, the mean $\overline{S}$ and standard deviation $s_S$ of the analytical signal are calculated. The magnitude of $s_S$ in a given technique can be related, with propagation of uncertainty mathematics, to the magnitude of the standard deviation (SD) for each of the measurements used to extract the analytical signal as shown in Table 6-2. With these equations, one can estimate the SD for de-

**TABLE 6-2**
Relationships with measured component signals and the determined analytical signal

| Technique | Component signals measured | Means and standard deviation of component signals | Calculation of mean analytical signal | Calculation of standard deviation for determining analytical signal |
|---|---|---|---|---|
| Emission or luminescence | $E_t, E_{bk}$ | $\overline{E}_t, \overline{E}_{bk}$ | $\overline{E}_s = \overline{E}_t - \overline{E}_{bk}$ | $s_S = (s_t^2 + s_{bk}^2)^{1/2}$ |
| Absorption with transmittance readout | $E_{rt}, E_{st}, E_{0t}$ | $\overline{E}_{rt}, \overline{E}_{st}, \overline{E}_{0t}$ $s_{rt}, s_{st}, s_{0t}$ | $\overline{T} = \dfrac{\overline{E}_{st} - \overline{E}_{0t}}{\overline{E}_{rt} - \overline{E}_{0t}}$ | $s_T = \dfrac{\overline{T}}{\overline{E}_r}\left(\dfrac{s_{st}^2}{\overline{T}} + s_{rt}^2\right)^{1/2}$ [a] $s_T = \dfrac{\overline{T}}{\overline{E}_r}\left\{\left(\dfrac{s_{st}}{\overline{T}}\right)^2 + s_{rt}^2 + \left[\dfrac{(1 - \overline{T})s_{0t}}{\overline{T}}\right]^2\right\}^{1/2}$ [b] $s_T = \dfrac{\overline{T}}{\overline{E}_r}\left[\left(\dfrac{s_{st}}{\overline{T}}\right)^2 + s_{rt}^2 + s_{0t}^2(1 + \overline{T}^{-2})\right]^{1/2}$ [c] |
| Absorption with direct absorbance readout | $E_{ls}, E_{lr}$ | $\overline{E}_{ls}, \overline{E}_{lr}$ $s_{ls}, s_{lr}$ | $\overline{A} = \dfrac{\overline{E}_A}{k'}$ $= \dfrac{\overline{E}_{ls} - \overline{E}_{lr}}{k'}$ | $s_A = \dfrac{(s_{ls}^2 + s_{lr}^2)^{1/2}}{k'}$ [d] |

[a] No 0% $T$ measurement, $s_A = (0.43) s_T/T$.

[b] One 0% $T$ measurement, $s_A$ as in footnote a.

[c] Two 0% $T$ measurements (one for total sample and one for total reference measurement); $s_A$ as in footnote a.

[d] $k'$ is logarithmic conversion factor.

termining the analytical signal from repetitive measurements of each of the signal components (e.g., $n$ measurements of the blank signal followed by $n$ measurements of the total signal) to evaluate if measurement of a particular signal component limits the precision. However, this procedure can yield too high an estimate of $s_S$ if significant drift occurs during repetitive measurements of a given signal component. Thus one usually makes $n$ repetitive determinations of the analytical signal (all component signals measured in each repetition), since this reflects the manner in which measurements are normally conducted and minimizes drifts in the values of variables which affect all signal component measurements.

The SD for determining the analytical signal can be used to determine the uncertainty or SD in the analytical concentration. Application of propagation of uncertainty mathematics to the calibration function $[S = f(c_a);$ see Chapter 1] yields $s_S = (\partial f/\partial c)s_c$ or

$$s_c = \frac{s_S}{(\partial f/\partial c)_c} \tag{6-4}$$

where $(\partial f/\partial c)_c$ is the slope of the calibration curve at the analyte concentration of interest and $c$ is understood to be the analyte concentration $c_a$.

**Statistical Statements**

If the statistical distribution of the data is known, one can make quantitative statements about the random uncertainty of the results (see Appendix A). For normally distributed data, the interval defined by the equation

$$\mu_c = \overline{c} \pm \frac{t_\alpha s_c}{n^{1/2}} \tag{6-5}$$

encloses the true concentration $\mu_c$ with a confidence level of $(1 - 2\alpha)$. Where $s_c$ is the SD for an individual measurement of the analyte concentration calculated from the SD of the measured analytical signal for the sample (equation 6-4) and $t_\alpha$ is the Student $t$ statistic (see appendix A). If more than 30 measurements are made to determine the SD, the $z$ statistic can be substituted for the $t$ statistic. The confidence level chosen depends on the situation, but is typically in the range 90 to 99%. Often data are reported as the mean $\pm$ SD/$n^{1/2}$, which corresponds to a confidence level of only 68% for $n \geq 30$ or less if $n < 30$ (50% for $n = 5$).

It must be stressed that the interval defined by equation 6-5 encloses the true value with only the confidence level chosen and only if random errors described by the normal distribution are dominant. A systematic error could cause an error much larger than $t_\alpha s_c/n^{1/2}$.

In some cases, one may question whether the concentration determined is significantly different from an accepted value ($\mu_c$) (such as for analysis of an NBS certified material). Here the experimental value of $t$ (or $z$ if $n \geq 30$) is calculated from the equation

$$t = \frac{\overline{c} - \mu_c}{s_c/n^{1/2}} \tag{6-6}$$

assuming normally distributed data. The largest value of $t_\alpha$ that is still less than $t$ is found in a probability table (see Table A-2). From the value of $\alpha$ corresponding to this $t_\alpha$, one can say that the difference between the experimental mean and the accepted mean is not totally due to random error with a confidence level greater than $(1 - \alpha)$. If $(1 - \alpha)$ is large (typically $\geq 90\%$), one suspects a significant systematic error, although there is still a probability $\alpha$ that the difference is due to random error, and in any case, the observed difference can be due to both random and systematic error. More complex tests are available to account for the uncertainty in the "accepted" value.

### Other Considerations

The SD in the analytical signal can arise from imprecision in all steps that affect its magnitude as follows:

$$s_S = [(s_S)_a^2 + (s_S)_t^2 + (s_S)_p^2 + (s_S)_m^2]^{1/2} \qquad (6\text{-}7)$$

where $(s_S)_a$ is the sample acquisition SD due to inhomogeneity of the starting materials, $(s_S)_t$ is the sample treatment SD, $(s_S)_p$ is the sample placement SD, and $(s_S)_m$ is the sample measurement SD. Often, the SD in the analytical signal, used to calculate the SD in the concentration or to make statistical statements, is the SD calculated from repetitive paired measurements on a given analytical sample and blank. This procedure only takes into account imprecision in the sample placement and measurement steps and thus often underestimates the SD of the total analysis procedure. Comparison of the SD calculated from repetitive measurements under different conditions allows one to determine what step or steps limit the precision of a total analytical scheme. For example, if discrete sampling is employed, the SD from repetitive measurements on one portion of an analytical sample can be compared to the SD from repetitive measurements of equivalent samples (e.g., different 1-mL aliquots from 100 mL of a given analytical sample solution) to determine if sample placement imprecision is critical. These SDs can be compared to the SD obtained from measurements of $n$ separate samples from the same bulk sample each carried through the complete sample treatment, placement, and measurement scheme to determine if sample acquisition or treatment imprecision is critical.

The factors affecting measurement precision and sometimes sample placement precision can be calculated and understood with $S/N$ theory and measurements as discussed in Chapter 5. Random errors in

sample acquisition or preparation are more difficult to assess; they result from inhomogeneity of the bulk sample and random variations in analyte concentration due to contamination or losses. Random errors due to volumetric and gravimetric operations in these steps can be predicted more quantitatively.

Equation 6-4 does not account for random error in the calibration function due to imprecision in the preparation, presentation, and measurement of the standards and the blank. The uncertainty in the calibration function can be evaluated from the measured SDs of the analytical signals for the standards. For example, with a linear calibration curve, the SD in the slope and intercept can be evaluated. From this information, the SD of the analytical concentration in the sample can be calculated; this reflects both the SD in determining the sample analytical signal and the SD in the calibration function. Calibration curve uncertainty is illustrated pictorially in Figure 6-2.

Experimentally, the SD of the complete analytical procedure can be determined from taking $n$ samples through the entire analytical scheme, including for each sample the preparation of standards and a blank and the establishment of a calibration curve. This SD can be compared to the SD obtained with only one calibration curve determination to evaluate if calibration imprecision is critical.

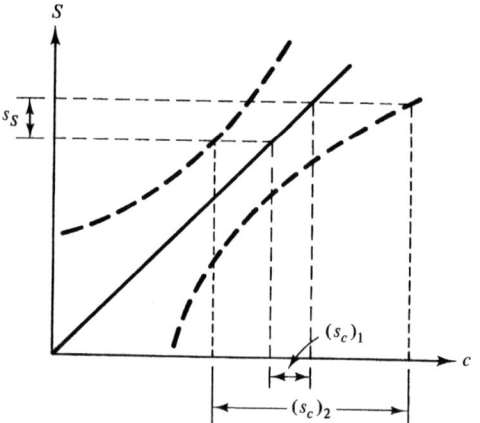

**FIGURE 6-2** Effect of calibration curve uncertainty. The uncertainty in the analyte concentration $[(s_c)_1]$ is usually evaluated from the SD in the determined analytical signal ($s_s$) for the sample and the calibration function (light dashed lines). The heavy dashed curved lines are uncertainty limits for a certain level of confidence for the calibration curve. When the additional uncertainty in the calibration curve is considered, the actual uncertainty in the determined analyte concentration $(s_c)_2$ is seen to be greater.

## 6-4 SENSITIVITY AND DETECTION LIMIT

Several figures of merit for spectrochemical procedures were defined in Chapter 1. In this section, the meanings of the terms *sensitivity* and *detection limit* are examined in detail since they relate directly to the accuracy and precision.

### Sensitivity

The word *sensitivity* is used commonly in analytical chemistry and other areas. We have all seen such statements as: "technique *x* is sensitive for Mg," "technique *x* is more sensitive than technique *y*," "changing variable *x* increases the sensitivity," "for low concentrations, the most sensitive range setting must be used," or "the technique is sensitive to temperature changes." Unfortunately, the term *sensitivity* is often used indiscriminately and often confused with the detection limit.

The three basic definitions of sensitivity as applied to analytical spectrometry are given in Table 6-3. The **calibration sensitivity** is seen to be merely the slope of the calibration curve evaluated in the concentration range of interest; this is the definition of sensitivity accepted by IUPAC. For a linear calibration curve, the sensitivity is independent of concentration and specified by a single value. The calibration sensitivity is based on the general engineering definition of a transfer function as the change in the output quantity per unit change in the input quantity. For example, the sensitivity of a detector is defined in this manner in Chapter 4. Thus in analytical spectrometry, a large value of the calibration sensitivity means that a large change in analytical signal is seen for a small change in concentration. This same definition is used when we speak of the "sensitivity control" on many spectrometers. Here a higher-sensitivity setting produces a greater change in the analytical signal or readout signal for a given change in the analyte concentration. Also, statements about the sensitivity of a given technique to an experimental variable (e.g., temperature) mean that the partial derivative of the analytical or blank signal with respect to the variable is large. The use of the AA sensitivity should be discouraged, but it is mentioned in Table 6-3 for completeness.

The **analytical sensitivity** $\gamma$ is defined as $m/s_S$ or the ratio of the calibration sensitivity to the SD in the analytical signal at a given analyte concentration. It has units of reciprocal concentration (or mass). From equation 6-4, $s_c = s_S/m$, and hence $\gamma = 1/s_c$, or the reciprocal of the SD in concentration. The analytical sensitivity takes into account measurement precision. It is rarely reported because $\gamma$ varies with concentration since $s_S$ or $s_c$ often varies with concentration.

Any of these uses of the term *sensitivity* can be employed if the definition is clearly stated. Confusion arises when the magnitude of the sensitivity is used to make quantitative statements and comparisons. If the word *sensitivity* is used to convey the ability to detect a concentration difference, only the analytical sensitivity should be employed. The ability to detect a concentration difference depends upon the ability to detect a readout signal difference which is limited by readout resolution, noise, or other random errors involved in determining the analytical signal.

### TABLE 6-3
Basic definitions of sensitivity

| Name | Symbol[a] | Descriptive definition | Mathematical definition[b] |
|---|---|---|---|
| Calibration[c] sensitivity | $m$ | Slope of analytical curve | $\left(\dfrac{\partial S}{\partial c}\right)_c$ |
| Atomic absorption sensitivity[d] | $m_A$ | Concentration or mass which yields 1% absorption or $T = 99\%$ or $A = 0.004365$ | $\left(\dfrac{0.0044}{m}\right)_{T=99\%}$ |
| Analytical sensitivity[e] | $\gamma$ | Slope of analytical curve divided by SD of analytical signal measurement | $\dfrac{m}{s_S}$ |

[a]The names and symbols are not universally accepted and are used here for identification.

[b]If the analyte amount is expressed in units of mass rather than concentration, the symbol $c$ is replaced by $g$.

[c]For emission and luminescence measurements, $m$ has units of signal units per concentration unit (e.g., V $(\mu g/mL)^{-1}$ while for absorbance measurements, $m$ has reciprocal concentrations units since $S$ is in dimensionless absorbance units.

[d]The atomic absorption (AA) sensitivity is defined specifically for AA spectrometry and is seen to be the concentration (or mass) which causes a 1% absorption ($\alpha = 0.01$) and hence a transmittance $T$ of 99% ($T = 1 - \alpha$) or $A = 0.004365$. It is inversely proportional to the calibration slope at low concentrations and is specified at one concentration. For a given AA spectrophotometer, the sensitivity for element $x$ is said to be better than for element $y$, if its AA sensitivity is smaller because a given concentration of $x$ produces a larger absorbance than the same concentration of $y$. The term **characteristic concentration** is preferred for $m_A$.

[e]Units are always (concentration units)$^{-1}$.

A calibration sensitivity of 1 V $(\mu g/mL)^{-1}$ does not in any way indicate the concentration differences that can be detected. However, if the SD in the analytical signal at 1 $\mu g\ mL^{-1}$ is 0.01 V, then $\gamma = 1/0.01 = 100\ mL\ \mu g^{-1}$, and concentration differences of 0.01 $\mu g\ mL^{-1}$ $(\gamma^{-1})$ can be detected with a reasonable confidence (64% for $n = 1$ from the $z$ statistic).

The calibration sensitivity should not, in general, be used to compare two techniques or even two similar spectrometers. First, the units can vary between techniques (absorption versus emission or luminescence techniques). Second for emission or luminescence spectrometry, the calibration sensitivity varies with many instrumental factors, including amplification (PMT or electronic). Increasing the electronic gain by a factor of 10 increases $m$ by 10; however, in many cases, $s_S$ also increases by the same amount so that the analytical sensitivity or ability to detect a concentration difference does not really change.

One must even be careful about quantitative comparisons of the calibration slopes obtained for two analytes on a particular spectrometer since the difference between the slopes may be related to the instrumental conditions used for each analyte. If the calibration slopes in emission or luminescence experiments were expressed in photons $s^{-1}$ (or watts) per concentration unit, they would have a more fundamental meaning. Even in this case, a larger value of $m$ does not necessarily mean that smaller concentration differences can be detected since the SD in determining the analytical signal may be considerably larger for the analyte with a higher $m$ due to differences in wavelength (e.g., higher background noise).

For absorption measurements, $m$ is more characteristic of the inherent spectroscopic properties of the analyte since instrumental gain factors cancel when the sample and reference signals are ratioed. If $m$ is greater for species $x$ than $y$ for a given absorption spectrometer, species $x$ is inherently a stronger absorbing species. However, $\gamma$ could still be larger for species $y$ if the magnitude of the SD is lower for $y$ than for $x$.

The differences between calibration and analytical sensitivity are summarized in Figure 6-3. Other advantages of the analytical sensitivity are that it is fairly independent of instrumental amplification factors, it can be used to compare different techniques or different instruments for a given analyte (i.e., the units are the same), and it is independent of mathematical transformations of the signal such as scaling or logarithmic transformation. The analytical sensitivity is also related to the RSD or $S/N$ since $\gamma = (m/s_S) = (m/S)(S/s_S)$ and $S/s_S$ equals the $S/N$ or $(RSD)^{-1}$.

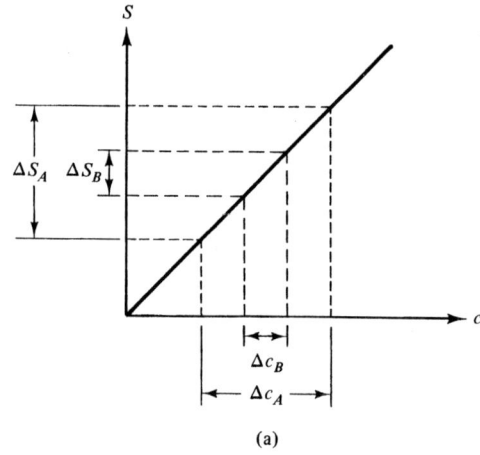

(a)

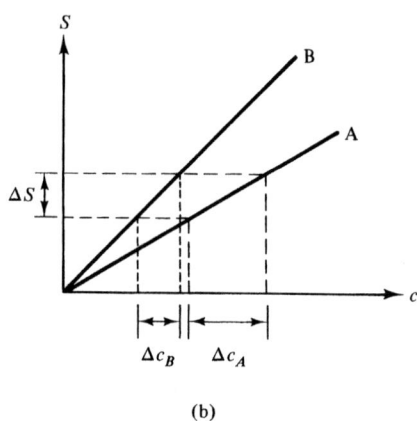

(b)

**FIGURE 6-3** Calibration and analytical sensitivity. Analytes A and B are determined by one technique. In (a), the calibration sensitivity for both analytes is the same, but the analytical sensitivity for analyte B is larger or $\Delta c$ is smaller because the analytical signal for B is measured with more certainty. In (b), the uncertainty in measuring the analytical signal is the same for both analytes, but the analytical sensitivity for B is larger (smaller $\Delta c$) because the calibration sensitivity for B is larger than that for A.

## Detection Limit

*Basic Definitions.* The **detection limit** (DL), **limit of detection** (LOD), or **limiting detectable concentration** (LDC) is defined as the smallest concentration that can be reported as being present in a sample with a specified level of confidence. The DL is a useful figure of merit for comparing two or more techniques for determining one analyte, or for determining a number of elements by one technique. The meaning of the DL is often interpreted incorrectly and confused with sensitivity. To add to the confusion, many definitions and forms of definitions exist.

The word definition of the DL is made more concrete by defining the DL in mathematical terms with the aid of statistics. This allows levels of confidence to be stated if desired. Definitions not based on statistics are invalid.

Most commonly, the DL is defined operationally as the analyte concentration yielding an analytical signal equal to some confidence factor $k$ times the SD of the blank measurement ($s_{bk}$) or the concentration where $S = ks_{bk}$. Alternatively, it can be defined as the analyte concentration where $S/N = k$, which is equivalent to the first definition if $N = s_{bk}$, so $S/N = ks_{bk}/s_{bk} = k$. The DL can also be directly calculated from

$$DL = \frac{ks_{bk}}{m} \qquad (6\text{-}8)$$

Here a linear calibration curve near the DL with calibration slope $m$ is assumed, so $S = mc$ or $c = S/m$ and $S = ks_{bk}$. The factor $k$ is most often chosen to be 2 or 3. These definitions are illustrated graphically in Figure 6-4.

From these definitions, the analytical signal at the DL is about equal to the variability of the blank measurement. Thus we are near the point where the ana-

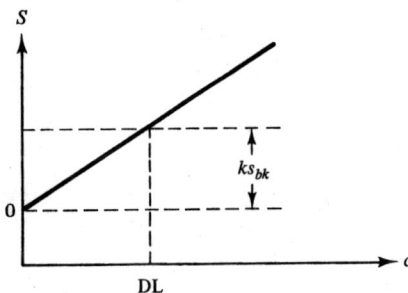

**FIGURE 6-4** Graphical interpretation of detection limit definitions. A linear calibration with slope $m$ is shown. The detection limit (DL) is seen to be the concentration where analytical signal $S$ is some confidence factor $k$ times the SD of the blank measurement ($s_{bk}$). The DL is calculated from $ks_{bk}/m$. From this formula or the graph it can be seen that a lower DL is obtained if the uncertainty in the blank measurement is decreased for a constant $m$. Similarly, the DL can be improved for a given $s_{bk}$ if the calibration sensitivity is increased since a smaller analyte concentration can produce the same analytical signal. The larger the magnitude of $k$ chosen, the smaller the probability that the analytical signal observed is due to the variability of the blank measurement.

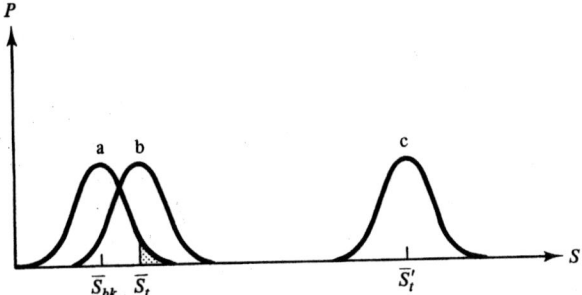

**FIGURE 6-5** Probability distribution interpretation of the detection limit. Curves a, b, and c represent the probability distributions obtained from repetitive measurements on a blank, a sample containing the analyte at a concentration near the DL, and a sample with an analyte concentration well above the DL, respectively. The respective means for these are $\overline{S}_{bk}$, $\overline{S}_t$, and $\overline{S}'_t$. In the case of the low analyte concentration, observe that there is a certain probability (shaded area under curve a) that a blank measurement could yield a signal equal to or greater than the mean of the signal from the sample ($\overline{S}_t$). If curve b represents the distribution for a sample with an analyte concentration exactly equal to the calculated DL, then $\overline{S}_t - \overline{S}_{bk} = ks_{bk}$. One chooses $k$ so that the probability that $\overline{S}_t - \overline{S}_{bk}$ could be totally due to random error is reasonably small. Note for this case, that 50% of the time the measured analytical signal for the sample will be above the defined detectable signal level ($ks_{bk}$) and 50% of the time below it. For the sample with the higher analyte concentration, there is nearly zero probability that $\overline{S}'_t - \overline{S}_{bk}$ could be totally due to random error in the blank measurement.

lytical signal observed could be due to random error. That is, we could measure two blank signals and treat one as the total signal and obtain by difference an apparent analytical signal which is really just due to the variability in the blank measurement. The larger the value of $k$ we choose in defining the DL, the more certain we are that the calculated analytical signal is due to the presence of the analyte and not just random error. This is illustrated in terms of probability distributions in Figure 6-5.

*Calculation and Use of the Detection Limit.* The DL is usually evaluated from equation 6-8 and experimental values for $m$ and $s_{bk}$. Measurements on a series of standards are used to construct a calibration curve to determine $m$, and repetitive blank measurements are made to determine $s_{bk}$. Often, $s_{bk}$ is estimated from one-fifth of the peak-to-peak noise in the blank signal (see Section 5-1). It is prudent to measure a series of standards of decreasing concentration down to concentrations below the DL to confirm that concentrations near the DL can indeed be detected and that the slope of the calibration curve near the DL is linear and known. Often, contamination problems can limit the ability to prepare solutions below a certain concentration and cause the calibration slope to approach zero at low concentrations.

The DL is a useful figure of merit because it defines the lower end of the dynamic range of an analytical technique and instrument. If several analytical techniques are available, one would not normally choose the technique which had a DL higher than the suspected analyte concentration in the sample, although sample preconcentration could be employed. One would prefer not to use a technique where the analyte concentration in the sample is at or slightly above the DL. By definition the RSD at the DL is 50% ($k = 2$). Thus measurements are normally made at concentrations at least a factor of 10 above the DL.

The DL reported by an instrument manufacturer or a researcher may not apply to real samples. Usually, such DLs are obtained with a completely optimized instrument and pure standards. The presence of concomitants in a real sample can affect the slope of the calibration curve or the blank SD. In real samples, it is often the sample-to-sample variation of the uncompensated blank component of the total signal (or of $E_s$ in absorption measurements) that limits detectability rather than the SD obtained from repetitive measurements of one reagent blank solution.

How the blank SD is evaluated is also important. Blank measurements should be made in a way that accounts for sample placement imprecision. It is more realistic to make repetitive measurements on $n$ separately prepared blank solutions than $n$ measurements on one blank to account for variability in blank preparation and contamination.

*Statistical Significance of Detection Limit.* The DL should be defined as the concentration that yields a mean analytical signal which we are confident is not solely due to fluctuations in the blank measurement. The magnitude of the confidence factor $k$ is chosen to correspond to the desired confidence level.

If the blank measurements are normally distributed, the $t$ or $z$ statistic can be used. The $z$ statistic requires sufficient blank measurements ($n \geq 30$) that the SD measured is a good estimate of the population SD ($s_{bk} = \sigma_{bk}$). For one measurement of the mean ($n = 1$), $z$ can be expressed as $z = (S - \mu_S)/s_S$. To

prove that there is analyte in the sample, the null hypothesis that the analyte concentration is zero is tested. Thus $\mu_S = 0$, $s_S = s_{bk}$, and $z = S/s_{bk}$ or $S = zs_{bk}$. Since operationally the DL is defined as the concentration where $S = ks_{bk}$, the confidence factor $k$ corresponds to the $z$ statistic. If an analytical signal $S$ equal to $ks_{bk}$ is obtained, the analyte is detected ($S$ is significantly different from zero) with a $(1 - \alpha)$ significance level, where $\alpha$ is the probability associated with $z = k$.

The most common values of $k$ are 1, 2, and 3; these correspond to levels of confidence (see Table A-2) of 84.1, 97.7, and 99.9%. Note that a one-tailed test is used here since negative concentrations are physically impossible. Thus the alternative hypothesis ($S > 0$) is the basis for the common statement that defining the DL as the concentration yielding an analytical signal equal to twice the SD of the blank corresponds to a confidence level of 97.7%. Unfortunately, this is not correct because it does not take into account that two measurements (i.e., total and blank) are required to obtain the analytical signal. Thus the SD for determining the analytical signal is really $s_S = (s_t^2 + s_{bk}^2)^{1/2}$, not $s_t$. So for zero analyte concentration, $s_t = s_{bk}$ and $s_S = \sqrt{2}\, s_{bk}$. By definition, $S = ks_{bk}$ at the DL and $z = S/s_S = S/\sqrt{2}\, s_{bk}$. Hence $k = \sqrt{2}\, z$ and a confidence factor of $k$ corresponds to the level of confidence associated with the value of $z = k/\sqrt{2}$. For $k = 1$, 2, and 3, the corresponding confidence levels are 76.1, 92.1, or 98.3%, respectively. A 97.7% confidence level applies only if the DL is defined as the concentration where $S = 2\sqrt{2}\, s_{bk}$.

If $s_{bk}$ is determined with less than 30 measurements, the confidence level for $k$ must be determined from the $t$ statistic [$t = (S_S - \mu_S)/s_S = S/\sqrt{2}\, s_{bk}$ for $\mu_S = 0$, $s_S = \sqrt{2}\, s_{bk}$, and $n = 1$]. In an analogous fashion it can be shown that a confidence factor of $k$ corresponds to the level of confidence associated with $t = k/\sqrt{2}$ rather than $t = k$. Thus for $k = 2$, $t = 2/\sqrt{2} = 1.41$, and for $n = 8$, the confidence level is 90% rather than the 92.1% predicted by the $z$ statistic.

Even though many blank measurements are commonly used to estimate $s_{bk}$, the DL is usually calculated as if only one paired measurement ($n = 1$) of the sample and blank is made to determine if the analyte is detected. If $n$ paired measurements of a sample are made, the DL is reduced by $n^{1/2}$ assuming that random error limits the blank measurement precision.

In some cases drift can be significant during the time interval required to make a number of blank measurements; as a result the estimate of $s_{bk}$ is too high. Here it is preferable to obtain the SD of the difference of $n$ paired blank measurements where the time separation between paired blank measurements is the same as between normal paired sample/blank measurements. In this case, $s_{bk}$

equals the measured SD divided by $\sqrt{2}$ and it reflects the SD of making a paired measurement of a sample with zero analyte concentration.

Once a DL is established, statements can be made about measured samples. If for a given sample, $S \geq ks_{bk}$, we can say that the analyte *is detected*, report the analyte concentration determined, and even specify a confidence interval for the true analyte concentration. If $S = ks_{bk}$, we can say that the analyte is detected with a $(1 - \alpha)$ level of confidence calculated from the $z$ or $t$ statistic as discussed above. This does not say that the measured concentration is exact, only that there is an $\alpha$ probability that the analytical signal measured is due totally to random error in the blank measurement.

If $S < ks_{bk}$, one should not report zero analyte concentration, but only state that the analyte was *not detected*. Even if the analyte concentration in the sample is exactly equal to the DL, it is detected only 50% of the time. Thus we cannot even state that the analyte concentration is at least below the DL since a sample with an analyte concentration equal to twice the DL could yield an analytical signal less than $ks_{bk}$ due to random error.

Clearly, when a DL is reported, the definition used should be stated along with information about the calibration slope, $s_{bk}$, and $n$. This allows interconversion between different definitions. Care should be used in reporting a confidence level.

The detection limit and sensitivity are different even in units (except the AA sensitivity), and sensitivity applies over the whole calibration curve. If one wishes to state that small concentrations can be detected with a given technique, the term "low DL" or "high detectability" is preferable to "high sensitivity." A large calibration sensitivity $m$ yields a low DL only if $s_{bk}$ is small. The analytical sensitivity $\gamma$ and DL are related at zero analyte concentration. Here $\gamma = m/s_S = m/s_{bk}$ and DL $= ks_{bk}/m = k/\gamma$. Hence a large $\gamma$ means a low DL, because $\gamma$ represents the ability to detect the difference between zero and a finite analyte concentration. Note that the DL defines the lower limit of the linear dynamic range over which the calibration sensitivity is constant.

*S/N Considerations.* When the blank SD is limited by noise in the blank measurement, $S/N$ theory can be used to predict the DL. For emission and luminescence measurements $E_S = \sigma_S = 0$, so $\sigma_t = \sigma_{bk} = s_{bk}$ for the blank. Thus equations 5-24, 5-26, and 5-27 reduce to

$$\sigma_{bk} = (\sigma_B^2 + \sigma_{dt}^2)^{1/2} \tag{6-9}$$

$$\sigma_{bk} = [mGK(E_B + E_d) + (\chi E_B)^2 + (\sigma_d)_{ex}^2 + \sigma_{ar}^2]^{1/2} \tag{6-10}$$

$$\sigma_{bk} =$$
$$mG\left[ K(i_B + i_d) + (\chi i_B)^2 + \left(\frac{(\sigma_d)_{ex}}{mG}\right)^2 + \left(\frac{\sigma_{ar}}{mG}\right)^2 \right]^{1/2} \tag{6-11}$$

For absorption measurements of the blank, $\sigma_A = \sigma_{bk} = s_{bk}$ and $T = 1$. Under these conditions, equations 5-50 and 5-52 reduce to the equations

$$\sigma_{bk} = \frac{0.43}{E_r}[KmGE_r + (\xi_s E_r)^2 + \sigma_{0t}^2]^{1/2} \tag{6-12}$$

$$\sigma_{bk} = \frac{0.43}{i_r}\left[ Ki_r + (\xi_s i_r)^2 + \left(\frac{\sigma_{0}i_r}{E_{fs}}\right)^2 \right]^{1/2} \tag{6-13}$$

when multiplied by $A$, where $\sigma_{bk}$ is in absorbance units (A.U.). Note that $m$ in equations 6-10 to 6-12 is the PMT gain, not the calibration slope. Thus with equation 6-8 and the equations above, we can predict the DL, determine the limiting noise terms, and evaluate which factors to change to minimize $\sigma_{bk}$ and to improve the DL, as discussed in Chapter 5. We will underestimate $s_{bk}$ with equations 6-9 to 6-13 when factors other than noise limit $s_{bk}$ or when the blank used does not account for all background signals ($E_B$) in a real sample.

*Other Definitions.* The conventional definition of the detection limit presented above is an a posteriori definition because a decision about the analyte being detected or not detected is based on the measured analytical signal for the sample and a previously established blank standard deviation. The signal level $ks_{bk}$ is sometimes denoted the **critical level** or **decision limit**.

Some authors have defined an a priori quantity termed the **limit of identification** or the **limit of guarantee for purity**. It represents the minimum concentration that will always be detected with a specified level of confidence. This figure of merit is defined as the analyte concentration which yields an analytical signal equal to DL $+ k's_t$, where $k'$ is another confidence factor. Often it is assumed that the probability distribution is constant in the vicinity of the DL (i.e., $s_t = s_{bk}$). Using this assumption and taking $k = k'$, the limit of identification is $2ks_{bk}/m$ or twice the detection limit. Hence, if $k = 2$ corresponds to 92% confidence, a sample containing an analyte concentration equal to twice the DL will yield a signal that indicates the analyte is detected (i.e., $\geq ks_{bk}$) 92% of the time.

Another related quantity is the **limit of determination** or **limit of quantitation**. It is the lowest analyte concentration which can be determined with a specified RSD such as 5 or 10%. Typically, it is the analyte concentration which produces an analytical signal equal to a factor of 10 to 20 times the blank standard deviation.

More complex methods have been proposed to calculate the detection limit. These methods take into account the uncertainty in the slope and intercept of the calibration curve.

## 6-5   TECHNIQUES FOR MINIMIZATION OF SYSTEMATIC AND RANDOM ERRORS

The ideal spectrometric technique is totally selective in that the presence of concomitants does not affect the measured analytical signal (i.e., no blank or analyte interferences), and simple external standards can be employed for calibration. This is, of course, rarely the case, especially for complex samples in which the analyte concentration is orders of magnitude lower than the concentration of many species in the analytical sample matrix. In this section, many of the sample treatment and instrumental techniques used to minimize or correct for systematic errors caused by the presence of interferents are reviewed. Some of these techniques can also reduce random error and improve the $S/N$ if they reduce the magnitude of background signals (i.e., blank interferences) and thus background noise.

In an ideal instrument, the experimental variables which affect the magnitude of the analytical and blank signals are controlled so that they do not cause significant errors. However, fluctuations and drifts in the magnitude of experimental variables do occur and cause random and systematic errors.

The methods discussed in this section primarily minimize or correct for errors involved in the sample measurement step. It must be stressed that errors in calibration, sample acquisition, and sample preparation can limit the overall accuracy and precision of an analytical procedure. It is assumed here that relevant experimental variables (wavelength, slit width, reagent concentrations) have been adjusted to minimize the effect of interferences and to optimize the $S/B$ or $S/N$.

### Separations

Various separation techniques, including filtration, dialysis, solvent extraction, volatilization, ion exchange, precipitation, and chromatography, can be used for sample cleanup (see Appendix E). The cleanup can involve separation of the analyte from the major constituents in the sample matrix or removal of the primary interferences from the sample matrix, leaving the analyte behind in the analytical sample. Although separations are extremely valuable, they increase the time required for sample preparation and the probability of

analyte loss or gain (contamination). Still, in many situations, the only way to eliminate an interferent is to remove it, and such procedures are actually gaining popularity with the development of automated instruments which incorporate separation technology.

Many separation techniques also preconcentrate the analyte by transferring the analyte in the original sample to a smaller volume. This is necessary if the analyte concentration is below the DL of a technique or is only slightly above the DL where measurement precision is poor.

Spectrometric detectors are now widely used in liquid chromatography, in particular with high-performance liquid chromatography, and to some extent in gas chromatography. The most common spectrometric HPLC detectors are based on refraction, UV-visible molecular absorption, and molecular fluorescence.

There is growing need for detectors with greater selectivity that compensate for imperfect chromatography. Often, it is difficult to separate completely all components in a complex sample; two or more components can coelute or have approximately the same retention time. In such cases it is advantageous to use a detector that responds to only one of the unresolved components in a chromatographic band. Refractive index detectors are good universal detectors but are quite unselective. Molecular absorption detectors provide moderate selectivity and are discussed in more detail in Section 13-5. Molecular fluorescence detection yields even greater selectivity, as detailed in Section 15-3.

Other spectrometric phenomena on which HPLC detection can be based include IR absorption (see Chapter 14), circular dichroism, optical activity, photoionization, the photoacoustic effect, and thermal lensing. (See Chapter 17 for a discussion of the last three techniques.) In addition to detectors based on molecular properties, it is becoming increasingly common to use atomic emission, absorption, or fluorescence element-specific detection. With this type of detection, the detector responds only to compounds that contain the element of interest. This makes it possible to separate and detect different chemical forms of the same metal (e.g., methyl mercury and ethyl mercury).

### Saturation, Buffer, and Masking Methods

Certain analyte interferences depress or enhance the analytical signal by a fixed fraction once the interferent concentration is above a certain level. In the **saturation method**, the interferent is added to all samples, standards, and blanks to the same high concentration level such that the interference effect is independent of the original concentration of interferent in the sample (i.e., the analyte in the sample and standards experiences the same interference effect). The fractional change in the analytical signal due to the presence of the interferent must be independent of the analyte concentration. If the analytical signal is significantly depressed, the detection limit can be degraded.

In some cases, the presence of another species can suppress the effect of a blank or analyte interference. In the **buffer method**, this third species, often called the **buffer** or **matrix modifier**, is added to all samples, standards, and blanks in sufficient amounts that the analytical signal is independent of the interferent concentration, as shown in Figure 6-6. A pH buffer is used in many wet-chemical procedures to keep the pH of a reaction mixture constant (within limits) regardless of the initial sample pH. Sometimes the buffer is a **masking agent** which selectively reacts with the interferent to form a complex which does not contribute a blank signal or alter the analyte signal. Some buffers are added to ensure that bulk properties such as ionic strength are constant.

In many solution techniques, the buffer is part of the reagent solution that is mixed with the analytical sample. Care must be taken in buffer methods that the buffer is not contaminated with a significant amount of analyte or other interferents and that the buffer reagent does not itself cause an interference effect.

### Dilution, Matrix Match, and Parametric Methods

Simply diluting the sample can sometimes minimize blank or analyte matrix interference if the interferent produces no significant interference effect below a certain

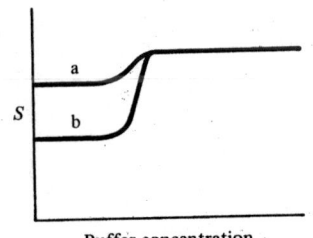

Buffer concentration

**FIGURE 6-6**  The buffer method. In this method, a buffer, matrix modifier, or masking agent is added at a sufficient concentration to all samples, standards, and the blank. Here the signal from the analyte is depressed without the buffer. Curves a and b represent the dependence of the analytical signal on buffer concentration for two solutions with a low and high interferent concentration and the same analyte concentration. Once the buffer is added to a critical concentration level, the analytical signal is seen to be independent of the buffer and interferent concentrations.

concentration level. The degree of dilution is limited by the ratio of the original analyte concentration to the analyte detection limit.

In the **matrix match**, **simulation**, or **synthetic** duplication method, an attempt is made to duplicate the matrix of the analytical sample in the standard and blank solutions. This can compensate for certain blank or analyte interferences if the interference effect is the same for samples, standards, and blank. For example, in the analysis of seawater for trace species, standards can be prepared in synthetic seawater which contains all the major components of seawater (e.g., $Na^+$, $K^+$, $Cl^-$, $Ca^{2+}$, $Mg^{2+}$, $CO_3^{2-}$). The matrix match method can also be used for industrial products where the concentrations of major components are known and controlled. Sometimes the matrix solution can be prepared from a real sample where the analyte is removed from the original sample matrix; known amounts of the analyte are then added to make standards.

The **parametric method** is a special case of the matrix match method; it is usually employed if there is only one major blank or analyte interference. The interferent is added to all standards and blanks at the same concentration level as it is in the analytical sample. A series of calibration curves is prepared with different concentrations of the interferent in the standards; the appropriate calibration curve that matches the interferent concentration in the sample is then used for the determination. Of course, this means that the interferent concentration must be known or accurately determined.

In either the matrix match or parametric methods, care must be taken that the substances added are not contaminated with analyte. Also, the added interferents may reduce precision or increase the detection limit, due to a decrease in the calibration sensitivity or an increase in the background signal and noise.

### Internal Blank and Standard Methods

Preparation of a suitable blank to compensate for blank interferences is often very difficult. In some cases it is possible to make a blank out of the sample; this blank is called an **internal blank**. The internal blank can be made by adding a substance (the suppressor) to the sample; the added substance should totally suppress the analyte signal without affecting the blank signal from other species in the sample. Such specificity is often difficult to achieve; thus the use of an internal blank is limited. For example, in the determination of urinary estrogen by molecular fluorescence, nitrate can be added to quench the fluorescence of the analyte estrogen; supposedly nitrate does not quench the background fluorescence from other species in urine. The suppressor can also be a substance that selectively reacts with the analyte to form a complex which does not produce a spectral signal.

In the **internal standard method**, a known concentration ($c_r$) of a reference species is added to all samples, standards, and blanks. The analytical signals from the analyte ($S_a$) and the reference species ($S_r$) are measured simultaneously with a dual-channel spectrometer with appropriate blank measurements. Alternatively, $S_a$ and $S_r$ are measured rapidly in succession with a single-channel spectrometer. A calibration curve is prepared in which the ratio of the analyte to reference analytical signal ($S_a/S_r$) is plotted vs. the analyte concentration in the standards. The $S_a/S_r$ value for the sample is then used to determine the analyte concentration in the sample.

The internal standard method can compensate for several types of random or systematic errors. If the reference species is chosen to have chemical and spectroscopic properties similar to those of the analyte, the analytical signals from both signals change proportionally when analyte interferences or fluctuations in experimental conditions occur.

If the reference species is added to the initial sample before sample treatment, the internal standard method can potentially compensate for systematic or random errors due to loss of analyte during sample preparation. For example, in sample cleanup steps such as solvent extraction, the efficiency of extraction is often less than 100% and variable; a reference species with similar characteristics can partially compensate for these losses. In the sample presentation and measurement steps, the internal standard method can partially compensate for drifts or random nonfundamental fluctuations in experimental conditions which cause systematic error or random error, respectively, as shown in Figure 6-7.

Analyte interference effects can also be reduced in a similar manner by this technique. If an interferent in the sample depresses the analytical signals from both the analyte and reference species by 50% relative to standards, the ratio $S_a/S_r$ in the sample and standard of equivalent analyte and reference species concentrations is identical.

The internal standard method is limited by the availability of a suitable reference species. The reference species must be similar to the analyte and accurately determined. It should not suffer from its own unique interferences. The analyte and reference concentrations must be adjusted so that both are in the linear ranges of their respective calibration curves. Care must be taken that there is no analyte contamination in the materials used to prepare the reference and that the amount of the reference species in the original sam-

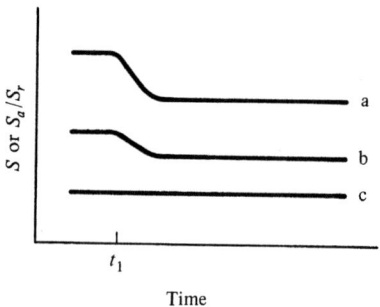

**FIGURE 6-7** Compensation with the internal standard method. The internal standard method compensates for fluctuations in experimental variables that affect the magnitude of both the analytical signal from the analyte ($S_a$) or the reference species ($S_r$). At time $t_1$, both $S_a$ (curve a) and $S_r$ (curve b) decrease due to a change in the magnitude of an experimental variable. However, the ratio of the signals (curve c or $S_a/S_r$) remains constant if the two signals are affected by the same proportionate amount.

ple is insignificant compared to the amount added. In general, the internal standard method provides no compensation for blank interferences so that a good blank must be available. The analytical signals from the analyte and reference species must be totally distinguishable such that neither species interferes with the measurement of the other. Internal standard methods can degrade precision, because the overall precision depends on the reproducibility of measuring both the analyte and reference species.

## Method of Standard Additions

The **method of standard additions** (known additions) is used when it is difficult to duplicate the sample matrix to compensate for analyte interferences. Thus it is used to compensate for nonspectral interferences and certain types of spectral interferences (e.g., nonanalyte absorption in emission and luminescence spectrometry) which enhance or depress the analytical signal by a fixed fraction independent of analyte concentration. The basic idea is to make a "standard" out of the sample.

The procedure for carrying out the single standard addition method is as follows: (1) the analytical signal for the sample $S_x$ is obtained from a total signal and blank signal measurement; (2) a small volume $V_s$ of a concentrated standard solution of known analyte concentration $c_s$ is added to a relatively large volume $V_x$ of another aliquot of the analytical sample; (3) the analytical signal for the standard addition solution $S_{x+s}$ is obtained from a total signal and blank signal measurement.

If the analytical signal is proportional to the analyte concentration in the sample matrix, the equations that apply before and after the standard addition are

$$S_x = mc_x \qquad (6\text{-}14)$$

$$S_{x+s} = mc_{x+s} = \frac{m(c_x V_x + c_s V_s)}{V_x + V_s} \qquad (6\text{-}15)$$

where $m$ is the slope of the calibration curve in the sample matrix and $c_{x+s}$ is the analyte concentration in the standard addition solution. If we solve equation 6-14 for $m$ and substitute this result into equation 6-15 and solve for $c_x$, we obtain

$$c_x = \frac{S_x V_s c_s}{S_{x+s}(V_x + V_s) - S_x V_x} \qquad (6\text{-}16)$$

Usually, the volume added is such that $V_s \ll V_x$ (e.g., $V_s = 100\ \mu L$ and $V_x = 100\ mL$), so that equation 6-16 reduces to

$$c_x = \frac{S_x V_s c_s}{(S_{x+s} - S_x)V_x} \qquad (6\text{-}17)$$

Note that the analyte concentration due to the added standard after dilution in the sample matrix, the *effective analyte standard concentration* $(c_s')$, is $c_s V_s/V_x$. Also, $m$ equals $(S_{x+s} - S_x)V_x/V_s c_s$ or $(S_{x+s} - S_x)/c_s'$. The condition that $V_s \ll V_x$ not only simplifies the equations but also ensures that the sample matrix is not significantly changed by dilution with the added standard; thus interferent concentrations are the same in the analytical sample and the standard addition solution.

In some cases where the amount of sample is limited, the sample and the same volume of sample plus standard addition are diluted with an appropriate solvent to the same final volume $V_f$, where $V_f > V_x + V_s$. It can be shown that equation 6-17 also applies to this situation without the assumption that $V_s \ll V_x$. In either case, the volume and concentration of standard added is typically adjusted so that the addition increases the analyte concentration by a factor of 0.5 to 2.

In the **multiple standard addition procedure**, several standard addition solutions are prepared by adding different volumes of the standard analyte solution to several different sample aliquots of the same volume. The measured analytical signal is plotted vs. the effective analyte standard concentration as shown in Figure 6-8. The equation that applies is a simplification of equation 6-15 if $V_s \ll V_x$ or

$$S_{x+s} = mc_x + mc_s' \qquad (6\text{-}18)$$

$$c_x = \frac{S_{x+s} - mc_s'}{m} = \frac{S_{x+s}}{m} - c_s' \qquad (6\text{-}19)$$

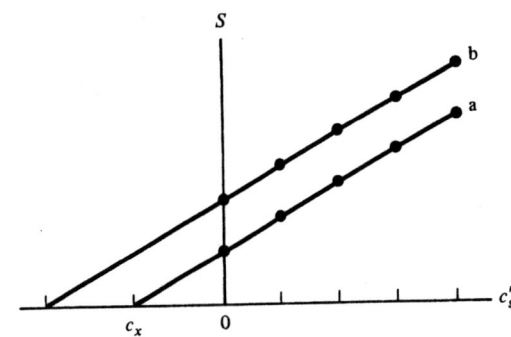

**FIGURE 6-8** Multiple standard additions procedure. A calibration plot is prepared by plotting the analytical signal versus the effective analyte standard concentration due to addition of a standard to the sample, including $c_s' = 0$ (the original analytical sample). The absolute value of the intercept is $c_x$, the analyte concentration in the sample. This intercept represents the amount of analyte which would have to be removed from the sample solution to make the analytical signal zero. Curve a is obtained with a proper blank measurement, while in curve b the blank measurement does not compensate for a blank interference, so all analytical signals are too high by a fixed amount, and the value of $c_x$ determined is too high.

Also $c_x$ can be found graphically from the absolute value of the x-axis intercept as shown in Figure 6-8.

The standard addition procedure is a powerful technique that is often used improperly due to a failure to understand the assumptions involved. First, there must be a good blank measurement (no blank interferences) so that there is no contribution to the measured analytical signals from other species. The error caused by a blank signal is shown in Figure 6-8. Second, the analytical signal must be proportional to the analyte concentration; the calibration curve for the analyte in the sample matrix must be linear. The multiple addition procedure provides a check on this assumption. Third, analyte interferences must increase or decrease the analytical signal from the original analyte in the sample and the analyte added in the addition by the same constant fraction independent of analyte concentration. If an interferent complexes a fixed amount of analyte in the sample solution such that it does not produce an analytical signal, the standard addition procedure will not give accurate results.

A variation on the standard addition procedure is the standard subtraction procedure. Here a reagent is added to mask selectively a certain known amount of analyte so that it does not produce an analytical signal. Similar equations and assumptions apply.

The standard addition procedure also provides a check on recovery and the presence of analyte interferences. The slope of a standard additions plot can be compared to the slope of a conventional calibration curve with external standards. If the slopes are different, the recovery of the analyte is incomplete or analyte interferences in the sample matrix affect the slope.

## Methods Based on Optical Encoding

As discussed in Section 2-4, wavelength selection is the most common method to separate the analyte optical signal from other signals. In addition to wavelength, however, time, position, and polarization can be used to increase selectivity in the optical domain. The use of these selection criteria can increase the $S/B$ and thus reduce systematic errors due to blank interferences; they can also reduce random errors due to a reduction in the background signal and noise.

*Time Discrimination.* Time can be used as a selection criterion if the analytical and background signals have inherently different time behavior or can be forced to do so. In some techniques, the analytical signal is necessarily time dependent due to the nature of the sample presentation and spectrochemical encoder systems; these produce a time-varying population of the form of the analyte that is spectrometrically probed. For example, in several atomic spectrochemical techniques, a fixed amount of sample is heated and vaporized to produce an atomic population which first increases and then decreases as the analyte atoms dissipate from the volume element viewed. Because the time dependence of the atomic population or the optical signal varies with the species, the measurement time can be selected to maximize the signal from the analyte compared to the optical signal from other species. This occurs in arc and spark emission techniques where both the atomic population and excitation conditions (e.g., temperature) vary with time and in electrothermal atomic absorption techniques where the atomic and molecular population of different species are time dependent. Often the discrimination can be enhanced by adding matrix modifiers.

The inherent time dependence of many optical phenomena can be used to gain selectivity if proper measurement conditions are employed. Usually, this is done by controlling the time dependence of the radiation that is incident on the sample. For example, in photoluminescence techniques, the luminescence signal observed is usually not time dependent because the source radiant power incident on the sample is held at a constant level. If a pulsed source or a source-shutter arrangement is employed, the decay of the luminescence signal after the excitation radiation is turned off can be monitored (see Sections 12-6 and 15-6). Such

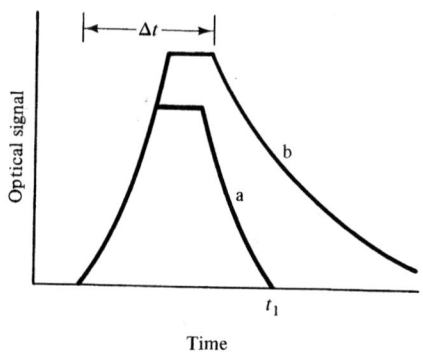

**FIGURE 6-9**   Discrimination based on lifetimes. Excitation source radiation is impingent on sample only during time $\Delta t$. Curve a represent the background optical luminescence signal from concomitants or other optical phenomena (e.g., scattering) while curve b represents the desired optical luminescence signal from the analyte. If the measurement time is delayed until time $t_1$, the background optical signal will decay to an insignificant level.

systems can be used to measure lifetimes of excited states and to improve selectivity. Selectivity is achieved against other species which undergo the same optical phenomenon if the signals from these species decay at significantly different rates than that of the analyte, as illustrated in Figure 6-9.

Selectivity can also be achieved against other optical phenomena which interfere with the optical phenomenon measured. For example, Raman or Rayleigh scattering and fluorescence are phenomena which are very often difficult to distinguish. However, Raman or Rayleigh scattering are extremely rapid phenomena (subpicosecond), whereas fluorescence is relatively long-lived in comparison. The selectivity of fluorescence measurements can be improved by delaying the observation until the scattering signal dies away (see Section 15-6). Similarly, the selectivity of Raman measurements can be improved by making the measurements extremely early in time before a substantial fluorescence signal has developed (see Section 16-2). Phosphorescence can be distinguished from fluorescence because phosphorescence is usually long lived (see Section 15-4).

As discussed in Sections 4-5 and 5-6, modulation techniques can be used to discriminate against certain steady-state signals if the analytical signal can be encoded at some frequency through source, wavelength, or sample modulation. For example, source modulation is commonly used in atomic absorption and fluorescence spectrometry to discriminate against analyte and background emission. Note here that the background signal levels are not reduced, but their average contribution to the measured total signal is zero. The noise from the background signal at the modulation frequency is still present, although $1/f$ additive noise that is not carried by the modulation waveform is discriminated against as discussed in Section 5-6. Wavelength modulation is often used in atomic spectrometry to discriminate against continuum background signals (see Figure 4-34) and in molecular fluorescence spectrometry to minimize spectral interference from concomitants (see Section 15-1).

There are many other examples in which time dependence is or could be exploited to increase selectivity, and we will certainly see more use being made of time-discrimination techniques in the future.

*Spatial Discrimination.* Spatial position is an important selection criterion, particularly in atomic emission, absorption, and fluorescence measurements. In these techniques, the number density of the atomic or molecular species of interest varies substantially with the volume element viewed or probed. Analytical flames, inductively coupled plasmas, dc arcs, high-voltage sparks, and electrothermal cells are examples of atomization sources that produce position-dependent information. If the analyte signal can be monitored in a region of the sample container that is relatively free from background, significant improvements in $S/B$'s can be achieved. Many commercial atomic spectrometer systems allow the user to change the viewing region as well as the region irradiated by the external source in atomic absorption and fluorescence measurements so as to optimize $S/B$'s. The complete spatial profiles of analyte and background signals are of great assistance in optimization. In atomic absorption and atomic fluorescence spectrometry, optimization of the region that is irradiated by the external source and detected can also be of benefit.

*Polarization Discrimination.* The interaction of polarized radiation with optically active substances is the basis of polarimetry, optical rotatory dispersion, and circular dichroism (see Section 3-2). The direction of polarization of Raman scattering can give information about the symmetry of molecular vibrations (see Section 16-2). Depolarization of polarized fluorescence radiation by molecular rotation is used to study the characteristics of macromolecules (see Section 12-6).

Differences in the interaction of polarized radiation with the analyte and concomitants provide the means to enhance selectivity in absorption and photoluminescence techniques. For example, as discussed in Section 10-3, the differences in the intensities of transmitted radiation polarized parallel to and perpendicular to a magnetic field surrounding an atomic vapor can allow discrimination against background absorption in Zeeman atomic absorption spectrophotometry. Differences in the polarization of fluorescence and scattering signals can sometimes be used to attenuate scattering signals in fluorescence measurements (see Section 15-6). Molecules of different molecular weight can be distinguished based on the degree to which their fluorescence radiation is depolarized (see Section 15-6).

### Chemical Selectivity

Primarily in molecular absorption and luminescence spectrometry, the selectivity for the analyte is often enhanced by using chemical reactions of the analyte with selective reagents. Often, these reactions convert the analyte into a form more suitable for spectrochemical measurement. For example, in molecular absorption spectrophotometry many analyte species do not significantly absorb radiation in the UV or visible regions of the spectrum; these must be converted into absorbing species by chemical means. Most methods based on a selective analytical or derivatization reactions are *equilibrium-based methods* since spectrometric measurements are made after the analytical reaction has come to equilibrium. The magnitude of the measured steady-state signal (e.g., absorbance) from the reaction product is related to the analyte concentration.

Chemical selectivity is determined by the availability of selective analytical reagents. Other species can cause a blank interference by reacting with the analytical reagent to form reaction products that produce an optical signal at the analytical wavelength. Also, analyte interferences can occur if the magnitude of the analytical signal from the new chemical form of the analyte is affected or the efficiency or rate of production of the reaction product is altered by the presence of concomitants. A very high degree of chemical selectivity is possible with enzymatic reactions (see Section 13-5) or antibody-antigen reactions as used in immunoassays (see Section 15-3).

When analytical reactions are used, the blank measurement is usually based on a mixture of the analytical reagent and the solvent. Sometimes an internal blank measurement is made on the sample mixture minus one of the necessary analytical reagents. This internal blank does not compensate for concomitants that react with the analytical reagent to produce interfering spectral signals.

In spectrophotometric *reaction-rate* (kinetics-based) *methods*, the rate of change of absorbance with time is the analytical signal measured. This represents a case where both chemical and time selectivity are employed. Concomitants that contribute only a steady-state background absorbance or that react with the analytical reagent at a significantly different rate than the analyte do not interfere.

In *spectrophotometric titrations*, the equivalence point in a titration is indicated by a rapid change in the solution absorbance. Concomitants which absorb radiation at the analytical wavelength, but which do not react with the titrant, do not interfere with the detection of the endpoint.

Derivatization reactions, reaction-rate methods, and spectrometric titrations are covered in detail in Section 13-5 because these techniques are most commonly used with molecular absorption spectrophotometry. These same techniques are also used to increase selectivity in molecular fluorescence spectrometry (see Section 15-3). In chemiluminescence spectrometry, a chemical reaction is mandatory to produce the luminescent species (see Section 15-5).

Selective chemical reactions are normally not used with atomic spectrometric techniques because the atomization process converts all chemical forms of the element of interest into an atomic vapor. However, selective chemical reactions are sometimes used for sample introduction. For example, some elements can be chemically converted to volatile products which are swept out of the sample solution and directed into the atomizer (see Section 10-1).

## Instrumental Correction Methods

Fluctuations and drifts in the magnitude of experimental variables that affect the analytical signal can cause random and systematic errors. These types of errors can be minimized by better controlling the critical variables or by correcting the measured analytical signal for changes in the magnitude of the variable that occur during the measurement process. Instrumental correction techniques for changes in experimental conditions can be based on either feedback systems (e.g., servo systems) or mathematical schemes. In either case, it is necessary to obtain a reference signal that is related to the magnitude of the variable of interest.

In feedback correction schemes, the reference signal is used to control the magnitude of the experimental variable. The reference signal is continually compared to a preset signal level and any error signal detected generates a response that minimizes the error signal and thus maintains the variable value at its preset value. The most common example is optical feedback correction in absorption and photoluminescence spectrome-

ters. The reference signal is provided by an optical transducer that monitors the light source intensity. If the light source intensity suddenly decreases 1%, the optical feedback circuit increases the lamp power supply current or voltage to increase the intensity to its initial value.

With mathematical correction schemes, the reference signal is not used to maintain the variable magnitude at a constant level. Instead, the reference signal provides a means to track changes in the magnitude of the variable. The measured analytical signal is corrected for the changes in the variable magnitude that occur in accordance with the previously established mathematical relationship between the analytical signal and the variable magnitude. As for feedback correction, the most common example of mathematical correction involves compensating for changes in the intensity of the excitation source in absorption and luminescence spectrometers. As discussed in Sections 4-6 and 5-6, the reference beam in double-beam spectrometers provides a signal proportional to the light source intensity. The correction in this case involves electronic or software ratioing of the analytical signal to the reference signal and provides compensation for source flicker noise and drift. Note that we use the terms *correction* and *compensation* interchangeably even though the term *correction* sometimes refers only to feedback correction schemes.

Instrumental correction techniques can be extended to any variable that has a significant effect on the magnitude of the analytical signal (e.g., pH or temperature in solution measurements). Note that the internal standard method is a type of mathematical correction technique. The signal from the reference species is used to correct for changes in experimental conditions, or even analyte interferences, that affect the analyte and reference analytical signals in an equivalent manner. Microcomputer-based instrumentation enables mathematical correction schemes to be more widely used because acquisition and correction can be implemented on-line.

Instrumental correction techniques are also used to compensate for spectral interference from concomitants. In cases where the analytical signal contains optical signal contributions from concomitants uncompensated by the blank measurement (i.e., a blank interference), a measurement under different conditions (e.g., at a different wavelength, with a different source, or with a different polarization state) can yield the background signal, which, in turn, is subtracted from the measured analytical signal. For example, in atomic spectrometry it is common to correct for continuum background signals. Here the analytical signal contains optical contributions from both the analyte line and from the continuum produced by concomitants. The background signal at the analytical wavelength can be estimated by measuring the signal at a wavelength slightly removed from the analytical line and then subtracted from the analytical signal. This can be accomplished with dual-channel (dual-wavelength) spectrometers or spectrometers fitted with refractor plates (see Figure 4-34). Several other techniques are used to estimate the continuum background absorption signal in atomic absorption spectrophotometry (see Section 10-3). Measurements at two or more wavelengths are also used in molecular absorption and photoluminescence spectrometry to compensate for absorption or luminescence by concomitants (see Sections 13-1 and 15-1). Note also that the modulation techniques discussed earlier in this section often are used as instrumental techniques for background signal correction.

In some mathematical correction techniques, signal information at different wavelengths is used to estimate a background signal correction term. For example, a multichannel atomic spectrometer can measure analytical signals (atomic line intensities) from many elements. If $10 \ \mu g \ mL^{-1}$ of element X is known to produce (due to spectral overlap) an apparent signal in the channel for element Y equivalent to the signal from $0.1 \ \mu g \ mL^{-1}$ of Y, the concentration of X can be measured and used to subtract an appropriate signal from the measured signal in channel Y.

Different types of spectral signals can also be used for mathematical correction. For example, it is possible to correct for the attenuation of photoluminescence signals by concomitants that absorb a significant fraction of the excitation or emission radiation if the absorbance at the excitation and emission wavelengths is measured (see Sections 15-1 and 15-2). In this case the correction scheme compensates for analyte rather than blank spectral interferences.

## 6-6 AUTOMATED SPECTROCHEMICAL MEASUREMENTS

In automated instruments, often called *automated analyzers*, a spectrometer or another detection system is one component in a larger system which has provision for automating sample placement or delivery steps, and often sample treatment steps. At a minimum, an automated system usually includes a pump and an automated sampler (e.g., a sample turntable). The samples, standards, and blanks are first loaded into the separate containers on the sampler. By rotation or other means, each container is sequentially positioned under a sampling tube, which is then lowered into the test solution. Each test solution is pumped through tubing into the spectrochemical encoder in a sequential manner. Often, the sampling tube dips into a container containing rinse solution between samples to minimize contamination between samples. Many automated spectrometric systems include additional hardware for adding reagents, mixing, or performing sample cleanup.

Automated systems are widely used in clinical and water analysis laboratories where a few species must be determined in a large number of similar samples. In such applications, the high sample throughput rate of automated systems is critical. Such systems require less skilled operators and often can be operated unattended. Operator blunders are minimized and precision can be improved because all operations are strictly timed and performed by mechanical or electrical devices.

In this section we emphasize automated spectrometric systems based on molecular absorption or molecular fluorescence detection as these molecular detection schemes are most commonly used. Usually, it is necessary to convert the analyte to an appropriate chemical form for measurement. Thus the sample must be mixed with reagents before the reaction mixture is sent to a sample cell. Automated systems using atomic spectrometric detection are briefly covered at the end of this section. The basic characteristics of batch and continuous automated analyzers are now reviewed.

For **batch** or **discrete analyzers**, the sample is maintained as a separate entity in a separate vessel throughout the analysis sequence. Most batch analyzers are sequential analyzers. Pumps, valves, or automatic syringes are configured and controlled to dispense one sample plus appropriate reagents and diluents into a sample cell. After mixing, an appropriate delay time, and possibly heating to speed slow analytical reactions, the molecular spectrometric measurement is made. If one sample cell is employed, it is evacuated and rinsed between samples. Alternatively, different sample cells for each sample are transported sequentially to the spectrometer sample compartment. In this case it is often possible to obtain disposable sample cells with prepackaged reagents; it is necessary only to add the sample.

A unique version of a batch analyzer is the **centrifugal analyzer**; it is a parallel analyzer which allows many samples to be analyzed simultaneously. The operation of the instrument is based on a plastic rotor which typically has 17 dual compartments arranged concentrically (see Figure 6-10a). Each sample, standard, or blank is dispensed into one of the compartments

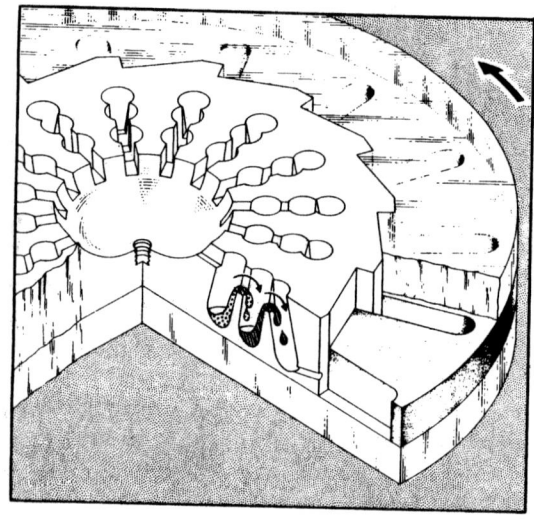

**(a)**

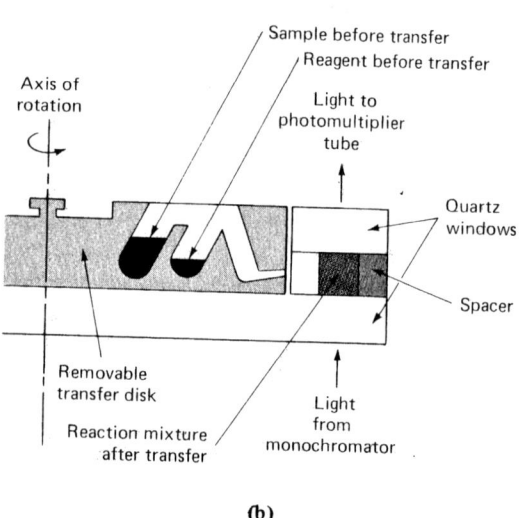

**(b)**

**FIGURE 6-10** Centrifugal analyzer. In (a) the rotor assembly is illustrated. During rotation the sample and reagents are mixed and transferred toward the circumference of the rotor. A more detailed view of a portion of the rotor is shown in (b). The barrier between the reagent and sample compartments prevents mixing until the rotor is rotated. [(a) Adapted with permission from C. D. Scott and C. A. Burtis, *Anal. Chem.*, *45*, 327A (1973), copyright 1973 American Chemical Society; (b) adapted from R. L. Coleman, W. D. Shults, M. T. Kelly and J. A. Dean, *Am. Lab.*, *3* (7), 26 (1971), with permission of publisher.]

while a common reagent mixture is added to the other compartment as illustrated in Figure 6-10b. Once the rotor is started and reaches a critical rotation rate, each sample–reagent pair is mixed simultaneously by centrifugal action and transferred to another compartment at the circumference of the rotor. For the configuration shown, this compartment has windows and serves as the sample cell for molecular absorption measurements. The absorption signal for each sample cell is measured every time the cell passes over the incident light beam. This operation provides virtually simultaneous measurement of the absorption signals of many samples and standards for any desired length of time; it is particularly well suited for kinetics-based measurements of a large number of samples. The centrifugal analyzer can also be configured for molecular luminescence measurements.

In **continuous analyzers**, the sample is a segment, or a group of segments, in a flowing stream; samples are sequentially inserted and carried to a flow cell in a spectrometer. Various operations, such as reagent addition, dilution, incubation, mixing, and separation steps, are carried out between the point of injection and the flow cell while the sample is flowing. Continuous flow analyzers are often mechanically less complex than batch analyzers.

Continuous flow analyzers based on segmented and unsegmented designs are illustrated in Figures 6-11 and 6-12, respectively. In **segmented analyzers**, a given sample is divided into discrete segments by gas bubbles, whereas no bubbles are used with **unsegmented analyzers**. Unsegmented flow analysis is often termed **flow injection analysis** (FIA).

Segmented flow analyzers are more established and commercial systems are available from several companies. As shown in Figure 6-11a, a proportioning (peristaltic) pump is used to pump the sample, reagents, diluents, and air through plastic tubing and other modules and finally to the detector module. The airflow causes a given sample to be divided into many segments, as illustrated in Figure 6-11b. A given sample is aspirated into the system for 30 to 90 s. With a rinse cycle of 10 to 30 s, the sample throughput rate is 30 to 90 samples per hour. Bubbles are introduced at a rate of 30 to 90 bubbles per minute. The reagents and diluents are mixed with the sample plugs at tees. Commonly, helix-shaped coils between the mixing tees and the flow cell promote mixing and provide a delay time for reactions to proceed to the desired degree of completion; they are often placed in water baths to increase the rates of slower reactions. The air bubbles minimize sample dispersion and cross contamination between

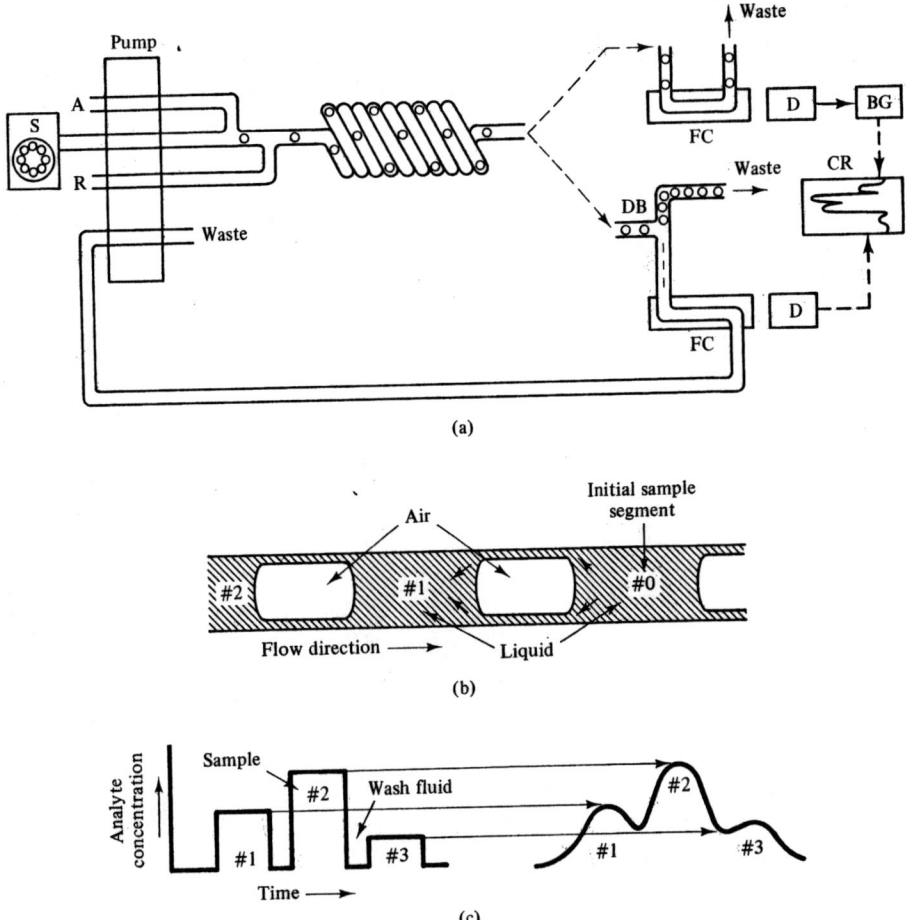

**FIGURE 6-11** Segmented continuous flow analyzers. In (a) a generalized diagram of a segmented flow analyzer is shown. Samples are aspirated from sample cups in the sampler (S), pumped into the manifold, and mixed with the reagent (R). Air (A) is also injected to segment the sample with bubbles as shown in more detail in (b). The bubbles are usually removed by a debubbler (DB) before the stream is sent to the flow cell (FC). If the bubbles are not removed, a bubble gate (BG) signals the detector (D) when no bubbles are in the sample cell. In (c) the analyte concentration profiles at the sampler and at the flow cell are shown. Dispersion causes some merging of the information from different samples. Normally, data from the plateau region (the central segments of the total sample plug injected) are used in quantitative measurements.

samples and enhance mixing between the sample and reagents.

Other modules can be added to the flow stream to cleanup the sample by dialysis, solvent extraction, or filtration. Normally, a debubbler removes the bubbles before the flow stream is sent to the flow cell. Alternatively, a bubble gate is used to trigger the acquisition of spectrometric data only between bubbles.

In unsegmented continuous flow analyzers (Figure 6-12a), a sample loop valve is often used to inject a precise volume of sample (typically 10 to 100 μL) into the flow stream. Peristaltic or positive-displacement pumps force the carrier stream through the valve and to the flow cell. The sample plug disperses into the carrier stream as it flows through tubing to the flow cell as illustrated in Figure 6-12b. If the carrier stream is a reagent mixture, the analyte and reagents react to yield a peak-shaped reaction product profile. In some cases, the sample and reagent are injected with separate sample loop valves into two carrier streams. With proper timing, the two plugs merge at a tee and are mixed as the combined plug is carried to the flow cell. In either

case, the amount of sample dispersion is controlled by the flow rate, the sample size, and the length and diameter of the tubing. It is possible to stop the flow when the sample plug reaches the flow cell to make reaction-rate measurements. Solvent extraction or dialysis modules can be incorporated. Typical sample throughput rates range from 60 to 300 samples per hour.

Automated atomic spectrometric systems are also available. Because sample treatment is not often required, simple automated systems consist of an automatic sampler and pump that allow sequential delivery of samples and standards to the atomizer. In more sophisticated systems, there is provision for automated standard addition or addition of matrix modifiers to reduce interference effects. Flow injection analysis systems are well suited for high-throughput, automated atomic spectrometric measurements.

Microcomputers have greatly enhanced the power and ease of use of all automated analyzers. They allow convenient and precise control of all operations and automated data manipulation and reporting.

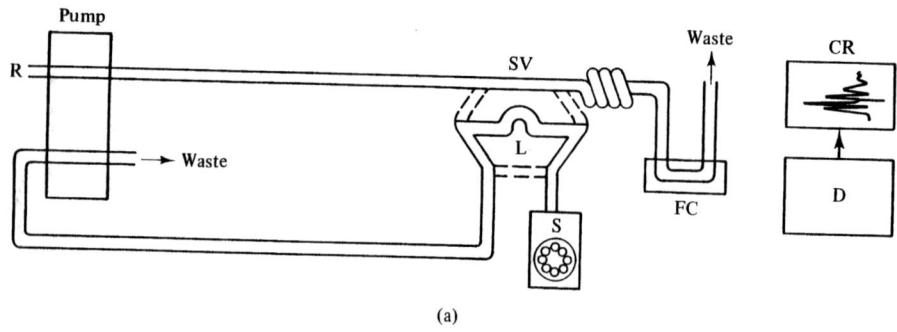

(a)

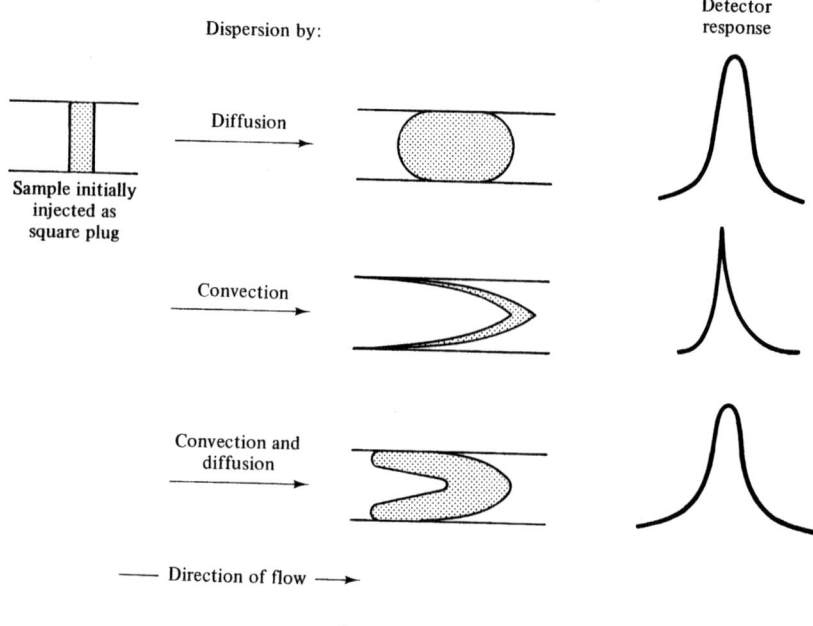

(b)

**FIGURE 6-12** Unsegmented continuous flow analyzer. In (a) a generalized diagram of an unsegmented system is illustrated. The sample from the sampler (S) is first loaded into the sample loop (L) of a sample loop valve (SV). A sample loop valve is a multiport valve which has two positions. In the load position, the ports connected by the solid lines are connected, which allows the sample loop to be filled with sample by aspiration with the pump. The valve is then switched to the inject position, which connects the ports in the configuration pathway shown with the dashed lines. This operation inserts the sample loop into the flowing reagent (R) stream. As the sample plug proceeds toward the flow cell (FC) and the detector (D), it is dispersed by convection and radial diffusion, as shown in (b). The resulting concentration profile is dependent on the degree of dispersion.

## PROBLEMS

**6-1.** Copper was determined in a river water sample by flame atomic absorption spectroscopy with the method of standard additions. For the standard addition, 1.00 mL of a 1000-$\mu$g mL$^{-1}$ Cu standard was added to 1.00 L of sample. The following data were obtained:

Reagent blank: $A = 0.050$
Sample: $A = 0.550$
Sample + std. addition: $A = 1.050$

**(a)** Calculate the concentration of Cu in the sample.
**(b)** Further studies revealed that the reagent blank was not adequate and that the actual blank absorbance in the sample was 0.200. What is the magnitude of the error caused by using an improper blank?

**6-2.** The following atomic emission measurements were made for Na. All measurements were made on standards or the unknown with an appropriate reagent blank ($c_a = 0$) and all sample measurements are paired with a blank measurement.

| $c_a$ (ng mL$^{-1}$) | $E_{tE}$ (V) | $E_{bk}$ (V) |
|---|---|---|
| 0.0 | −0.03 | −0.02 |
| 0.0 | −0.01 | 0.00 |
| 0.0 | −0.01 | −0.01 |
| 0.0 | 0.01 | 0.00 |
| 0.0 | 0.03 | 0.02 |
| 10.0 | 1.04 | 0.02 |
| 10.0 | 1.05 | 0.05 |
| 10.0 | 1.03 | 0.04 |
| Unknown | 0.54 | 0.03 |
| Unknown | 0.52 | 0.05 |
| Unknown | 0.54 | 0.06 |

Calculate the following values.
**(a)** The mean analyte emission signal for 0.0 and 10.0 ng$^{-1}$ mL Na and the unknown.
**(b)** The standard deviation ($s_S$) for the measurement of $E_E$ for 0 and 10 ng mL$^{-1}$ Na and the unknown. Notice that there is an apparent drift in the blank signal, so be sure to use the paired measurement scheme for all calculations.
**(c)** The calibration sensitivity ($m$) for $0 - 10$ ng mL$^{-1}$ (assume linearity).
**(d)** The analytical sensitivity ($\gamma$) at 10 ng mL$^{-1}$.
**(e)** The detection limit based on a confidence factor ($k$) of 3. To what level of confidence (within 1%) does this correspond?
**(f)** The range of voltages in which the population mean for 10 ng mL$^{-1}$ Na is contained with a 98% level of confidence.
**(g)** The concentration of Na in the unknown has been determined accurately by another method to be 5.20 ng mL$^{-1}$. Do the data indicate that there might be some systematic error or interference in this determination by flame emission? Assume that there is no uncertainty in the calibration curve.

In general, watch the significant figures and the difference between a one- and a two-tailed test. Be sure to use the paired measurement scheme for all calculations. Note that $s_S$ for $c_a = 0$ equals $\sqrt{2}\, s_{bk}$.

**6-3.** The following atomic absorption measurements were made for Zn. All measurements were made on standards with an appropriate reagent blank ($c_a = 0$) and all sample measurements are paired with a reference measurement. The signals due to dark current, background and analyte emission, and background and analyte fluorescence are negligible.

| $c_a$ (ng mL$^{-1}$) | $E_r$ (V) | $E_s$ (V) |
|---|---|---|
| 0 | 1.010 | 1.000 |
| 0 | 0.990 | 1.010 |
| 0 | 1.000 | 0.990 |
| 0 | 1.010 | 1.021 |
| 10.0 | 1.000 | 0.798 |
| 10.0 | 1.020 | 0.804 |
| 10.0 | 0.980 | 0.792 |
| Unknown | 1.000 | 0.897 |
| Unknown | 0.990 | 0.900 |
| Unknown | 1.013 | 0.911 |

Calculate the following values.
**(a)** The mean absorbance for 0 and 10 ng mL$^{-1}$ Zn and the unknown.
**(b)** The standard deviation for the measurement of $A$ ($s_A$) for 0 and 10 ng mL$^{-1}$ and the unknown.
**(c)** The calibration sensitivity ($m$) for $0 - 10$ ng mL$^{-1}$ (assume linearity).
**(d)** The analytical sensitivity ($\gamma$) at 10 ng mL$^{-1}$.

**(e)** The detection limit for 97.5% confidence based on the $z$ statistic and the $t$ statistic. Note that $s_A$ for $c_a = 0$ equals $\sqrt{2}\, s_{bk}$. For the $z$-statistic calculation, assume that $s_A = \sigma_A$.
**(f)** The characteristic concentration (atomic absorption sensitivity).
**(g)** The range of Zn concentration in which the true concentration of Zn in the unknown is contained with a 95% level of confidence. Assume no error in the calibration curve.

In general, watch the significant figures and the difference between a one- and a two-tailed test. Be sure to use the paired measurement scheme for all calculations.

**6-4.** Can a single concomitant cause both spectral and nonspectral interference?

**6-5.** Will a spectral interferent in the standards or the blank result in a positive or negative intercept for a calibration curve? Discuss the possibilities.

**6-6.** Strontium was determined in seawater by flame emission spectrometry. The standard addition method was used. The following table indicates the volume of a 3.00 g L$^{-1}$ Sr solution added to 1000 mL of the sample solution and the blank-corrected emission signal recorded.

| Volume of Sr added (mL) | Analytical signal (mV) |
|---|---|
| 0 | 18 |
| 1 | 24 |
| 2 | 30 |
| 3 | 36 |
| 4 | 42 |
| 5 | 48 |

Calculate the concentration of Sr in the unknown by the graphical method and by the single-addition technique for 2 mL.

**6-7.** In absorption measurements, what methods can be used to correct for the effect of analyte luminescence on the measured absorbance signal?

**6-8.** When do noise calculations or measurements give a poor estimate of the detection limit?

**6-9.** A photoluminescence measurement is limited by background luminescence shot noise. Calculate the detection limit for a confidence factor of 2 based on the following information: signal processor gain, $1.0 \times 10^6$ V A$^{-1}$; PMT gain, $1.0 \times 10^6$; noise equivalent bandpass, 1.0 Hz; secondary emission factor, 0.30; calibration curve slope, 1.0 mV nM$^{-1}$; $E_B$, 0.10 V.

**6-10.** What advantages do automated instruments provide?

**6-11.** What techniques can be used to compensate for nonfundamental variations in the excitation source intensity that occur during a series of measurements?

**6-12.** What is indicated when the slope of a calibration curve prepared from external standards and the slope of a standard addition plot differ significantly?

# REFERENCES

Figures of merit such as the detection limit and sensitivity are discussed in the following references.

1. G. L. Long and J. D. Winefordner, "Limit of Detection," *Anal. Chem.*, 55, 712A (1983).

2. A. L. Wilson, "Performance-Characteristics of Analytical Methods: I–IV," *Talanta, 17,* 21 (1970); *17,* 31 (1970); *20,* 725 (1973); *21,* 1109 (1974).

3. H. Kaiser, "Quantitation in Elemental Analysis," *Anal. Chem.*, 42, 24A (1970); 42, 26A (1970).

4. P. W. J. M. Boumans, "A Tutorial Review of Some Elementary Concepts in the Statistical Evaluation of Trace Element Measurements," *Spectrochim. Acta, 33B,* 625 (1978).

5. R. K. Skogerboe, "Concepts of the Definitions of the Terms Sensitivity and Detection Limit," *Spectrosc. Lett.,* 3, 215 (1970).

6. G. H. Morrison, "General Aspects of Trace Analytical Methods: I. Methods of Calibration in Trace Analysis," *Pure Appl. Chem.,* 41, 397 (1975).

7. J. D. Ingle, Jr., "Sensitivity and Limit of Detection in Quantitative Spectrometric Methods," *J. Chem. Educ.,* 51, 100 (1974).

8. J. D. Winefordner, "Quantitation in Analysis," appendix A in *Trace Analysis: Spectroscopic Methods for Elements,* J. D. Winefordner, ed., Wiley-Interscience, New York, 1976.

9. K. Eckschlager and V. Stepanek, *Information Theory as Applied to Chemical Analysis,* Wiley-Interscience, New York, 1979.

Analytical methodology, errors, and data analysis are considered in the following references.

10. R. Klein, Jr., and C. Hach, "Standard Additions—Uses and Limitations in Spectrophotometric Analysis," *Am. Lab.,* 9 (7), 21 (1977).

11. M. A. Sharaf, D. L. Illman, and B. R. Kowalski, *Chemometrics,* Wiley-Interscience, New York, 1986.

12. H. A. Laitinen and W. E. Harris, *Chemical Analysis,* 2nd ed., chaps. 26 and 27, McGraw-Hill, New York, 1975.

13. T. C. O'Haver, "Analytical Considerations," chap. II in *Trace Analysis: Spectroscopic Methods for Elements,* J. D. Winefordner, ed., Wiley-Interscience, New York, 1976.

14. M. Zief and J. W. Mitchell, *Contamination Control in Trace Element Analysis,* Wiley-Interscience, New York, 1976.

15. L. Larsen, N. A. Hartmann and J. J. Wagner, "Estimating Precision for the Method of Standard Additions," *Anal. Chem.*, 45, 1511 (1973).

16. E. B. Sandell, "Errors in Chemical Analysis," chap. 2 in *Treatise on Analytical Chemistry,* pt. I, vol. 1, p. 19, Wiley-Interscience, New York, 1959.

17. B. Kratochvil and J. K. Taylor, "Sampling for Chemical Analysis," *Anal. Chem.*, *53,* 925A (1981).

18. I. M. Kolthoff and P. J. Elving, eds., *Treatise on Analytical Chemistry,* 2nd ed., pt. I, vol. 1, sec. B, Wiley-Interscience, New York, 1978. Chapters 3, 4, 5, and 6 are excellent reviews of analytical methodology, sources of error, and sampling.

Automated analysis is considered in the following references.

19. M. Margoshes and D. A. Burns, "Automation: Instrumentation for Analysis Systems," in *Treatise on Analytical Chemistry,* 2nd ed., pt. I, vol. 4, P. J. Elving and I. M. Kolthoff, eds., Wiley, New York, 1984.

20. W. B. Furman, *Continuous Flow Analysis—Theory and Practice,* Marcel Dekker, New York, 1976.

21. D. Betteridge, "Flow Injection Analysis," *Anal. Chem.,* 50, 832A (1978).

22. L. Snyder, J. Levine, R. Stoy and A. Conetta, "Automated Chemical Analysis: Update on Continuous-Flow Analysis," *Anal. Chem.,* 48, 942A (1976).

23. C. B. Ranger, "Flow Injection Analysis," *Anal. Chem.,* 53, 20A (1981).

24. K. K. Stewart, "Flow Injection Analysis," *Anal. Chem.,* 55, 931A (1983).

25. J. Růžička, "Flow Injection Analysis," *Anal. Chem.,* 55, 1040A (1983).

26. C. D. Scott and C. A. Burtis, "A Miniature Fast Analyzer System," *Anal. Chem.,* 45, 327A (1973).

27. J. Růžička and E. H. H. Hansen, *Flow Injection Analysis,* Wiley, New York, 1981.

28. *Anal. Chim. Acta, 179* (1986). A special issue devoted to flow analysis.

29. *Flow Injection Analysis Bibliography,* Tecator, Höganäs, Sweden, 1985.

# CHAPTER 7

# Introduction to Atomic Spectroscopy

In previous chapters we have considered the aspects of spectrochemical analyses that are generally applicable to all methods. In the remaining chapters of this book, specific spectrochemical methods are described. We follow here the conventional classification of such methods into atomic spectrochemical methods and molecular spectrochemical methods. The specific atomic techniques to be discussed are flame and plasma atomic emission (Chapter 8), arc and spark atomic emission (Chapter 9), atomic absorption (Chapter 10), and atomic fluorescence (Chapter 11) spectrometry. Although these methods involve different excitation and radiative processes, they have much in common. For example, all depend on having the analyte present in the form of a free vapor of neutral atoms (or occasionally ions). A sample presentation system, consisting of a sample introduction system and the atomization device, converts the analyte into the appropriate form for spectrochemical encoding (recall Figure 2-1). The sample introduction step is extremely important in atomic spectrochemical methods and is often the step that limits the amount of analyte converted into the appropriate form for subsequent observation. The most widely employed sample introduction methods are treated here along with the processes that lead to the formation of atomic vapors since these are common to atomic spectrochemical methods.

Many of the interferences observed in emission, absorption, and fluorescence have similar origins that can be traced to inefficiences in the sample introduction or atomization steps. Because of these similarities, it is useful to consider the common interferences together. Similarly, all atomic spectrochemical methods deal with similar electronic transitions of the analyte atoms. The electronic states of atoms and the term symbols that are used to describe these states are reviewed here. The spectral line profiles of analyte atoms play an important role in determining the shapes of calibration curves, the selectivities of the techniques, and the spectral line intensities. The line intensities are a major factor in determining the limits of detection of the methods. Thus this chapter concludes with a discussion of spectral line profiles and line intensities. The general principles and concepts presented here are intended to lay an important foundation for the specific atomic spectrochemical methods that follow.

## 7-1 SAMPLE INTRODUCTION AND ATOMIZATION

We consider first an overview of the sample introduction and atomization steps that must occur in all atomic spectrochemical methods. Later in this section, specific sample introduction methods and the processes that lead to free atoms are considered in some detail.

## Overview

In all atomic spectroscopic methods, the analyte must be converted into the appropriate chemical form to emit or absorb radiation. Almost always this involves converting the analyte into free atoms, although occasionally spectroscopic transitions of ions are used. Samples for atomic spectrochemical analysis may be in the form of liquids, solids, or gases. Most commonly, sample preparation steps produce an analytical sample that is a solution. Thus the sample presentation system has a complex task to perform in order to convert analyte species in solution into vapor-phase free atoms. This usually entails the application of heat to break up molecules into their component atoms. The general routes for introducing solution samples into flame and plasma atomization devices are summarized in Figure 7-1.

*Atomization Devices.* The sample container in which the spectroscopic measurements are made is usually a hot gas or an enclosed furnace. Flames, plasmas, electrical discharges (arcs and sparks), and electrically heated furnaces (electrothermal devices) are commonly used. Flames are formed by combustion of an *oxidant* and a *fuel*, whereas plasmas are partially ionized gases maintained either by an electrical discharge or by coupling to a microwave or RF field. An arc is a continuous electrical discharge between conducting electrodes, while a spark is an intermittent discharge.

The process of forming free atoms by applying heat to a sample is known as **atomization,** and devices that carry out the atomization process are called **atomizers.** These devices can be continuous or pulsed (noncontinuous) atomizers. With *continuous atomizers* such as flames or plasmas, the atomization conditions (e.g., temperature) are constant with time. With a *noncontinuous atomizer* these conditions vary with time. Usually, electrothermal atomizers such as furnaces are used in a noncontinuous mode. The electrical power supplied for heating is varied so that atomization occurs when the temperature reaches a critical value. Spark discharges are also noncontinuous atomizers where conditions can change very rapidly (microseconds).

*Sample Introduction.* The nature of the atomic population and hence the signals obtained in atomic spectrochemical methods depend on the type of atomizer employed and often the method of sample introduction. With continuous sample introduction the sample is constantly introduced in the form of droplets, a dry aerosol, or a vapor. A device called a **nebulizer**

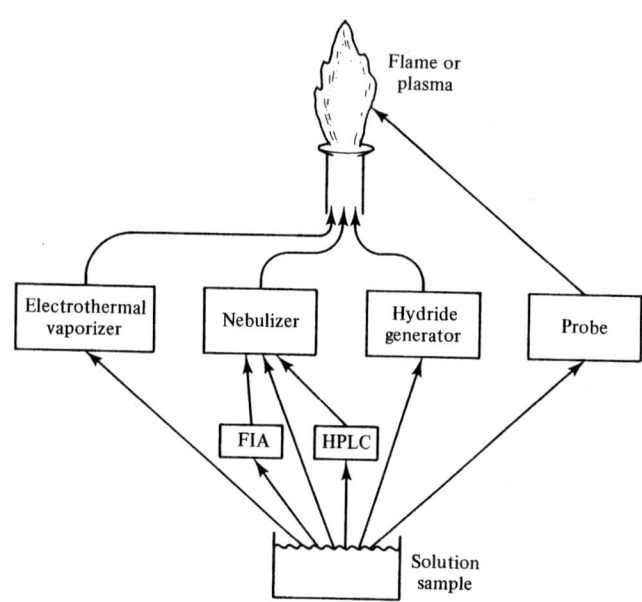

**FIGURE 7-1** Plasma and flame sample introduction schemes for solutions. Solution samples are most commonly introduced into flames or plasmas directly via a continuous nebulizer. In some cases discrete samples are introduced as vapors with an electrothermal vaporizer. Samples can also be introduced as solution plugs with flow injection methods followed by nebulization, or the flame or plasma emission system can be used as a detector for liquid chromatography. If the analyte forms a volatile hydride, it can be introduced as a vapor after chemical reactions form the hydride. Finally, discrete samples can be introduced by direct insertion of a probe into the flame or plasma.

is used to convert the solution sample into a fine spray of droplets. Usually, continuous sample introduction is used only with continuous atomizers, in which case a steady-state atomic population is produced. Samples can also be introduced in fixed or discrete amounts to continuous atomizers, in which case a transient atomic population is produced. Discontinuous sampling is almost exclusively used with noncontinuous atomizers to produce a transient atomic vapor cloud.

Discrete samples can be introduced into atomization devices in numerous ways. With electrothermal atomizers, a syringe is often used to transfer an aliquot of sample to the atomizer. A transient signal is obtained because the atomization conditions vary with time and because the fixed amount of sample is completely consumed during the measurement period. In direct insertion or probe techniques, the sample is placed on a probe (e.g., a carbon rod) and mechanically moved into the atomization region. The atomic vapor cloud produced is a transient because of the limited amount of sample. In flow injection techniques, a plug of the analytical sample solution in a carrier stream is introduced into the atomizer as a mist with the aid of a nebulizer. Dry aerosols and vapors can also be introduced as plugs of material. With vapor introduction (e.g., hydride techniques), the volatile analyte species is often stripped from the analytical solution and carried by a gas to the atomizer. This stripping step can be preceded by a specific chemical reaction that converts the analyte into a volatile form. It is now common to use gas and liquid chromatography to introduce samples into atomizers. Nebulizers are often used for liquid chromatographic eluates. Chromatographic introduction is a type of discontinuous sample introduction since the atomizer receives a time-varying concentration of analyte as components elute from the column.

With arcs and sparks, the sample introduction and atomization processes are more difficult to separate. In many arc and spark determinations, solid samples are employed. These are often shaped into the form of an electrode, and the discharge struck between the sample electrode and a second electrode. Alternatively, the sample electrode can be in the form of a cup into which powdered samples are packed, a porous cup, or a rotating disk for solutions. In any case the atomic signals produced are transient because the discharge conditions vary as the discrete amount of sample is atomized.

In some cases one atomizer can be used as a sample introduction system for a second atomizer. This can be particularly advantageous in atomic emission, where the atomization device is usually called upon to excite the analyte as well as atomize it. Separating the sampling/atomization step from the excitation step can allow optimization of the energy input for each step. As an example of this approach, samples can be introduced into flame or plasma excitation sources by means of electrothermal atomizers. Here the first atomizer converts the analyte into an atomic vapor, while the flame or plasma must now only supply the energy needed to excite the sample.

In **laser microprobe techniques,** a laser beam is directed onto a small portion of a solid sample. The sample is vaporized and atomized by radiative heating. Either the plume of sample formed can be directly probed by the encoding system or the vapor produced can be swept into a second atomization cell for observation of emission, absorption, or fluorescence.

Detailed descriptions of many atomizers and sample introduction systems are presented in Chapters 8 to 11, which deal with specific methods. To provide an understanding of the atomization process, general characteristics of continuous and noncontinuous sample introduction/atomization systems are discussed below. In both systems, the analytical sample is assumed to be a solution, which is the most common sample form.

*Processes Occurring during Atomization.* Let us consider the processes that must take place in order to transform a solution sample into an atomic vapor with continuous sample introduction into a continuous atomizer. As shown in Figure 7-2, the sample introduction system disperses the sample into the high-temperature environment of a flame or plasma, usually as a fine spray or mist. This process is called **nebulization.** Heat from the flame or plasma evaporates the solvent and volatilizes the dry aerosol that remains as Figure 7-2 illustrates. Once free atoms are formed, they can be excited by collisions to produce characteristic spectral lines or an external radiation source can be used to obtain atomic absorption or fluorescence information.

The processes that occur with discrete sample in-

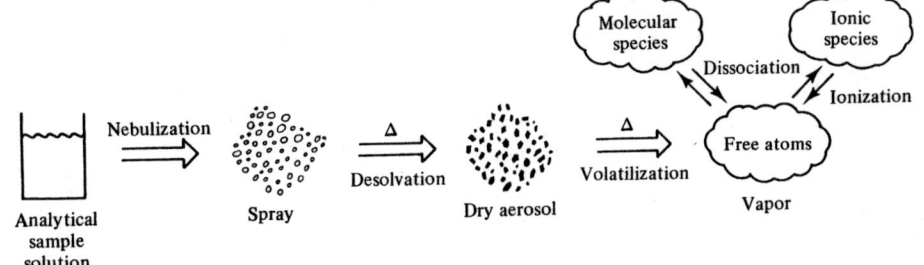

**FIGURE 7-2** Overview of process leading to atomic vapor with continuous sample introduction into a continuous atomizer. The analytical solution is first converted into a spray or mist of fine droplets. The flame or plasma heats the spray and causes the solvent to vaporize leaving dry aerosol particles. Further heating volatilizes the particles and produces atomic, molecular and ionic species containing the analyte. These species are often in thermodynamic equilibrium at least in confined (localized) spatial regions.

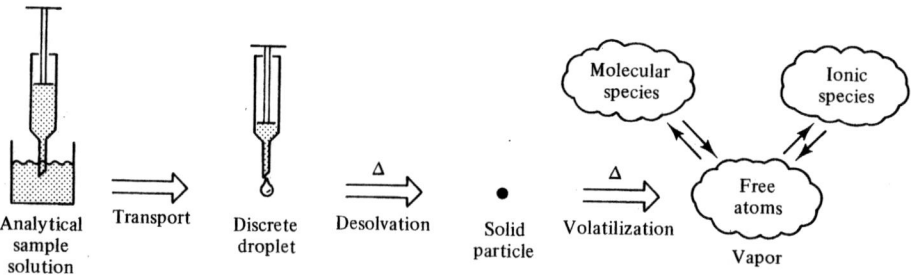

**FIGURE 7-3** Formation of atomic vapor with discrete sample introduction. A discrete sample is transported to the atomization device. The atomizer heats the sample to the boiling point of the solvent and a solid particle remains. The solid particle is then heated to a much higher temperature to produce vapor-phase species. With electrothermal atomizers the heating stages are separated in time. The volatilization step is often carried out in two stages, one of moderate temperature to drive off organic material (ashing or charing), and one of high temperature to vaporize the analyte material (atomization).

troduction into a continuous or noncontinuous atomizer are illustrated in Figure 7-3. With continuous sample introduction, a spray is created, while with discrete sample introduction, a fixed amount of sample is transported to the atomizer. With electrothermal atomizers, the various heating stages (desolvation and vaporization) are often separated in time by a temperature program, while with continuous atomizers these steps occur sequentially under the influence of heat from the atomization device. With arc and spark discharges and solution samples, these same processes occur, although it is difficult to separate the stages. Solid samples introduce additional complexities as discussed in Chapter 9.

### Nebulizers

The formation of free atoms in flames and plasmas depends critically on the properties of the sample transport-nebulizer-atomizer system. The nebulization step and the remaining steps shown in Figure 7-2 are all interactive, and the details of the processes that occur are very complex. The type of nebulizer used influences directly the efficiency of the nebulization, desolvation, and volatilization steps. We consider here the most common types of nebulizers employed in atomic spectrometry and leave it for later chapters to discuss more specialized devices.

*Pneumatic Nebulizers.* The most commonly used nebulization devices are pneumatic devices in which a jet of compressed gas (the *nebulization gas*) aspirates and nebulizes the solution. Two common types are shown in Figure 7-4. The transport of solution to the nebulizer tip is known as **aspiration**. With the concentric nebulizer (Figure 7-4a) the nebulization gas flows through an opening that concentrically surrounds the capillary tube, causing a reduced pressure at the tip and thus suction of the sample solution from the container (Bernoulli effect). In the angular or cross-flow nebulizer (Figure 7-4b), a flow of compressed gas over the sample capillary at right angles produces the same Bernoulli effect and aspirates the sample. In most cases the flow of solution is laminar, and the aspiration rate is proportional to the pressure drop along the capillary and to the fourth power of the capillary diameter; it is inversely proportional to the viscosity of the solution. In some cases a pump (peristaltic or syringe pump) can be used to transport the solution at a rate that is independent of the compressed gas pressure.

The solution drawn up the capillary tube encounters the high velocity of the nebulizing gas, which causes the formation of droplets of various sizes. Typically, pneumatic nebulizers produce droplet diameters in the range 1 to 50 μm. In most modern nebulizers, the mist produced enters a **nebulization chamber**, where most of the droplets with diameters larger than a certain *cutoff diameter* are removed from the stream and drained away. Devices that remove large droplets from the stream are often called **aerosol modifiers**. These can be impact beads, as shown in Figure 7-4, spray chamber surfaces, or impaction surfaces. Impact beads may also produce small droplets from larger drops by causing them to shatter. Only the smaller droplets are transported to the atomization source. For plasmas, the nebulization gas (e.g., Ar) containing the fine sample mist is directed into the plasma (usually formed from the same gas). For flame systems, the nebulization gas is usually the oxidant, and the fuel is brought into the nebulization chamber through a separate port. The mixed oxidant, fuel, and sample mist are then carried to the burner

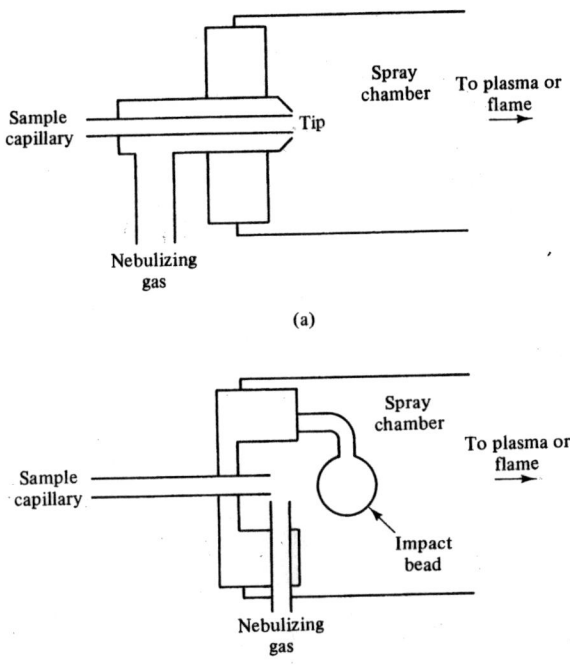

(a)

(b)

**FIGURE 7-4** Pneumatic nebulizers. In (a) a concentric tube nebulizer is shown that is used with a spray chamber. The nebulizing gas flows through an orifice that surrounds the sample-containing capillary concentrically. In the angular or crossed-flow nebulizer (b), the nebulizing gas flows over the sample capillary at right angles and causes aspiration and nebulization of the sample solution. Both nebulizers are examples of pneumatic nebulizers in which a jet of compressed gas aspirates and nebulizes the solution. A spray chamber separates large droplets from small droplets. The latter are carried into the plasma or flame, while the former are drained away. The nebulizer can be made of glass or metal.

head, which supports the flame. In this case, the nebulization-chamber-burner system is called a chamber or **premixed burner system**.

With the pneumatic nebulizers shown in Figure 7-4, aspiration rates are typically in the range 1 to 7 mL min$^{-1}$. The average diameter of the droplets formed by the nebulizer is about 20 μm. Aerosol modifiers in the spray chamber reduce the diameters of the droplets delivered to the atomizer to the range 1 to 20 μm and the average diameter to about 9 to 10 μm. Smaller droplets with a narrower size distribution can be produced with some organic solvents of lower surface tension and viscosity than water.

In flame emission spectrometry, a concentric nebulizer is sometimes used without a spray chamber. A third concentric orifice transports fuel around the nebulizer tip so that the nebulizer is an integral part of

the burner. The flame is supported immediately above the nebulizer tip. These pneumatic, concentric nebulizer-burners are called **total consumption nebulizer-burners** because all the aspirated sample is directed into the flame without droplet size selection. They are discussed in more detail in Chapter 8.

*Frit Nebulizers.* A major disadvantage of conventional pneumatic nebulizers is the wide range of droplet diameters they produce. When only the small droplets are delivered to the atomization cell, the overall transport efficiency is reduced because of discrimination against the larger droplets. Some interference effects may also be related to the process of droplet size discrimination within the spray chamber. The glass frit nebulizer shown in Figure 7-5 produces a much finer aerosol than the conventional pneumatic nebulizer. It is not as susceptible to clogging as a conventional nebulizer and has thus found some use with samples of a high salt or particulate content. A mean droplet diameter of less than 1 μm has been found for this nebulizer; transport efficiencies on the order of 90% have been obtained. To achieve this efficiency, the total sample flow rate must be quite low (0.1 mL min$^{-1}$) so that the overall mass transport rate is not large. However, this nebulizer may be advantageous where sample volumes are limited and where long nebulization times must be used (e.g., for some multielement determinations).

*Ultrasonic Nebulizers.* The ultrasonic nebulizer has often been suggested as a replacement for the

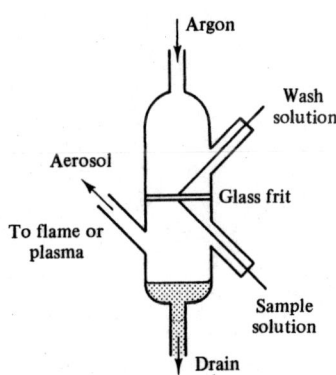

**FIGURE 7-5** Glass frit nebulizer. The sample solution flows over the surface of a fritted glass disk, while nebulizing gas is passed through the many small holes in the disk. A frit with a diameter of 200 mm and a pore size of 4 to 8 μm produces a dense aerosol at nebulizing gas flow rates between 0.2 and 1 L min$^{-1}$. To reduce memory effects, a wash solution is applied to the frit between samples. The drain removes the excess wash solution.

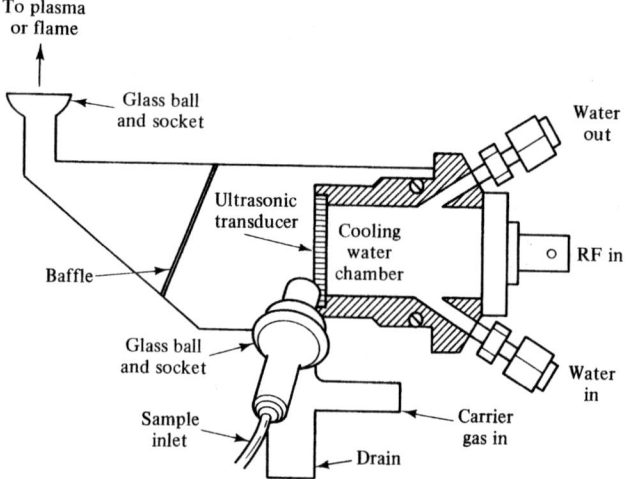

**FIGURE 7-6** Ultrasonic nebulizer. Solution is fed onto the surface of the piezoelectric crystal by gravity flow or by a pump. Vibrations of the crystal cause the solution to break into small droplets, which are transported by the carrier gas to the flame or plasma.

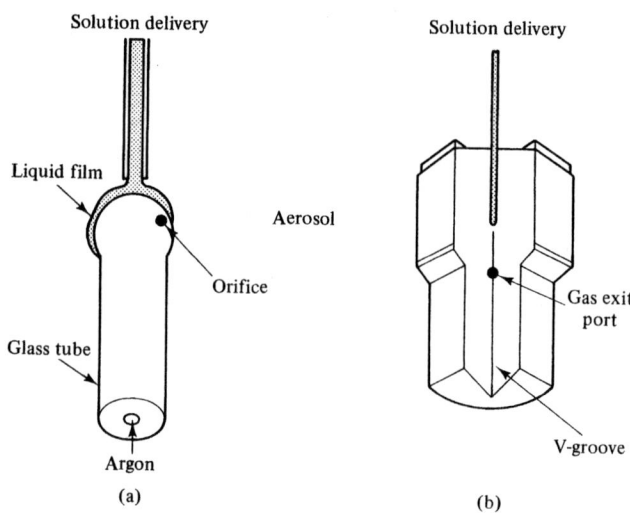

**FIGURE 7-7** Babington-type nebulizers. In the normal Babington nebulizer (a), the sample is fed by gravity or a pump and forms a film over the outside surface of a sphere or rounded tip with a small orifice for the nebulizing gas. The gas blows a hole through the film to produce the aerosol. In the V-groove nebulizer (b) the sample solution passes down a V-groove with a small hole in the center for the nebulizing gas. Both nebulizers can be made of glass, metal or plastic materials such as Teflon.

conventional pneumatic nebulizer. With these devices a piezoelectric crystal (a ceramic crystal transducer) is vibrated at ultrasonic frequencies (20 kHz to 5 MHz). Ultrasonic vibrations are coupled to the sample solution by a coupling liquid, by a velocity transformer, or by directly flowing the solution onto the vibrating surface. The vibrations cause the solution to break up into small droplets which are transported by the carrier gas through the nebulization chamber to the flame or plasma. In the vertical, direct-coupled design shown in Figure 7-6, the sample is fed onto the surface of the vibrating crystal, where nebulization occurs. Such nebulizers can produce dense aerosols that are more homogeneous in droplet size than those produced by pneumatic nebulizers. Ultrasonic nebulizers have the advantage that the nebulizer parameters (frequency of vibration, power applied to the transducer) are independent of any flame or plasma gas flow rates so that separate optimizations can be made. The solution flow rates must, however, be fairly low with these nebulizers (0.1 to 1.0 mL min$^{-1}$). The droplet diameters produced depend on the frequency of the ultrasonic energy as well as the surface tension and viscosity of the liquid. High frequencies favor the formation of small droplets. With a frequency of 1 MHz and a solution flow rate of 0.3 mL min$^{-1}$, droplets with diameters of 1.5 to 2.5 μm have been produced with up to 30% efficiency. The major limitation of ultrasonic nebulizers is their poor efficiency with viscous solutions and with solutions that have high particulate content.

*High Solids Nebulizers.* The capillary tubes used

in pneumatic nebulizers have typical inside diameters of 0.1 mm and are thus subject to clogging when samples with a high particulate content are aspirated. The nebulizer designs shown in Figure 7-7 have been found to be quite tolerant to such solutions. Both nebulizers are variations on the Babington design, which was originally developed for spraying paint. In both designs the solution is delivered through a tube of much larger inside diameter than those used with pneumatic nebulizers. Thus these nebulizers are essentially free from clogging, and they are excellent nebulizers when high particulates cannot be avoided. The Babington-type devices are, however, less efficient than pneumatic nebulizers in producing small droplets with normal samples. Because of this the processes of desolvation, volatilization, and atomization may be less efficient than with conventional nebulizers and some interference effects may be enhanced.

*Isolated Droplet Generator.* In order to study closely the processes of desolvation, vaporization, and atomization, solution samples have been introduced into flames as individual, isolated droplets. Figure 7-8 shows a system for producing isolated droplets. Droplet diameters depend on the size of the capillary used and are typically in the range 10 to 200 μm. By charging some of the droplets produced and electrostatically de-

**FIGURE 7-8** Isolated droplet generator. An alternating voltage is applied to a piezoelectric (bimorph) transducer. The resulting vibrations cause an attached capillary, through which the sample is pumped, to vibrate, and result in a pressure wave along the stream emerging from the capillary. A stream of equally spaced and uniformly sized droplets is produced at a frequency of 50 to 200 kHz. To decrease the droplet introduction rate, some can be charged by applying a high-voltage charging pulse at the desired time and deflected out of the main stream by electrostatic deflection plates. [Reprinted with permission from G. M. Hieftje and H. V. Malmstadt, *Anal. Chem.*, 40, 1860 (1968). Copyright © American Chemical Society, 1968.]

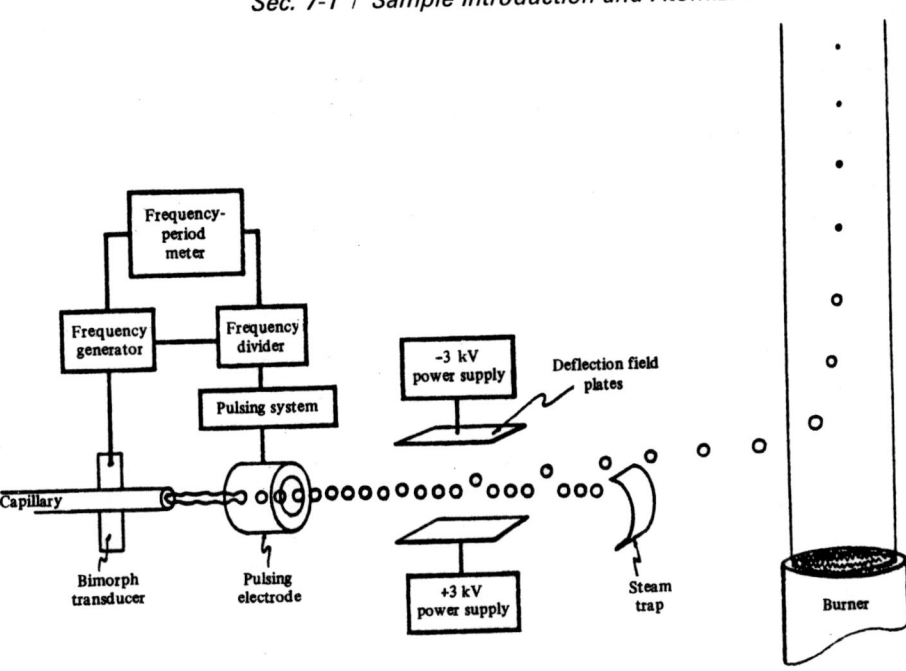

flecting them, the introduction rate can be varied from 1 Hz to $2 \times 10^5$ Hz. Observation of the size of the droplets injected into a flame allows the desolvation and volatilization processes to be studied, while measurement of the emission in a spatially isolated region of the atomizer allows investigation of the atomization process.

### Free-Atom Formation after Nebulization

Once the sample solution has been nebulized and transported to the atomization device, a variety of complex processes begins. We consider here the desolvation of the droplets, the vaporization of the solid or molten particles, the dissociation of molecular species, the ionization of analyte atoms, and the influence of these processes on the shapes of calibration curves.

*Desolvation.* Desolvation of the droplets is the first step that must occur after nebulization. Desolvation leaves a dry aerosol of molten or solid particles and often begins in the nebulization chamber. The efficiency of the desolvation step is determined by several experimental variables, such as the atomizer temperature, the trajectories, diameters and residence times of the droplets in the atomization cell, the nature of the solvent, and the design of the nebulizer. High-temperature atomizers increase desolvation efficiency, as do

relatively low gas flow rates, which increase residence times. Organic solvents evaporate more rapidly than does water, and with flame atomizers, the desolvation can be accelerated by the heat of combustion of the vapor. Desolvation efficiencies can be quite high for pneumatic nebulizers with spray chambers that reject larger droplets. With the total consumption nebulizers, once popular for flame emission, all of the droplets may not reside in the high-temperature environment long enough for complete solvent evaporation, particularly with water as the solvent. When flames are used, water also tends to lower the flame temperature.

*Volatilization.* The solid or molten particle remaining after desolvation must be vaporized to obtain free atoms. Incomplete volatilization leads to a loss of analyte free atoms and thus a reduction in the analytical signal. It may also cause nonlinearity in analytical curves, continuous background emission from incandescent particles, and light scattering in the case of fluorescence measurements. The volatilization efficiency depends on a number of factors, including the atomizer temperature, the composition of the analytical sample (nature and concentration of the analyte, solvent, and concomitants), the size distribution of the dry aerosol, the trajectories and residence times of the particles, and the type of nebulizer. For many metals, oxides are less volatile than the metal itself or salts of the metal. The bond dissociation energies of the compounds that are formed with the analyte are important in determining volatilization efficiencies. With flames, high temperatures and a reducing environment tend to increase the volatilization efficiency and reduce the formation of refractory oxides. The volatilization rate increases as the size of the droplets introduced into the atomizer decreases.

*Dissociation and Ionization.* In the vapor phase the analyte can exist as free atoms, molecules, or ions. The formation of molecular species and ions reduces the concentration of free atoms and thus degrades the detection limit. Molecular species, such as CaO, CaOH, KCl, and LiOH, can be formed by reactions of analyte atoms with gaseous flame constituents or with volatilized species from the analytical sample. Ionic species, such as $K^+$, $Na^+$, $CaOH^+$, and $Ca^+$, are formed by the loss of an electron by an atom or molecule.

In many cases we consider molecules to be in equilibrium with free atoms, at least in localized regions of the atomizer (flame or plasma). If chemical equilibrium exists, the law of mass action describes the degree of dissociation of molecular species, and we can write a dissociation constant for a molecular species (MX) into its components (MX $\rightleftharpoons$ M + X) as

$$K_d = \frac{n_M n_X}{n_{MX}} \qquad (7\text{-}1)$$

where $n$ is the number density (number per $cm^3$) and the subscript indicates the species. From statistical mechanics, the dissociation constant for a diatomic molecule is a function of temperature and the type of reactants according to

$$\log K_d = 20.274 + \tfrac{3}{2} \log \frac{M_M M_X}{M_{MX}}$$
$$+ \log \frac{Z_M Z_X}{Z_{MX}} + \tfrac{3}{2} (\log T) - 5040 \frac{E_d}{T} \qquad (7\text{-}2)$$

where $M_i$ is the molecular or atomic weight of species $i$, $Z_i$ is the partition function of species $i$ (see Section 7-3), $E_d$ is the dissociation energy in eV, and $T$ is temperature in K. The final term in equation 7-2 describes most of the temperature dependence. Small dissociation energies and high temperatures lead to large values of $K_d$ and thus high degrees of dissociation. This results because such dissociation reactions require energy to produce free atomic species (reactions are endoergic). As an example, equation 7-2 predicts that $K_d$ for BaOH ($E_d = 4.7$ eV) at 2200 K is $2.5 \times 10^{12}$ $cm^{-3}$, while that for CaOH ($E_d = 4.3$ eV) is $2.5 \times 10^{13}$ $cm^{-3}$.

The ionization of metal atoms can also be considered an equilibrium process in many flames and plasmas, at least in localized regions. Such ionization ($M \rightleftharpoons M^+ + e^-$) can be described by the **Saha equation** for the ionization constant $K_i$:

$$K_i = \frac{n_{M^+} n_e}{n_M} \qquad (7\text{-}3)$$

where $n_e$ is the number density of free electrons. Again ionization constants can be calculated from statistical mechanics and are of the form

$$\log K_i = 15.684 + \log \frac{Z_{M^+}}{Z_M} + \tfrac{3}{2}(\log T) - \frac{5040 E_{ion}}{T} \qquad (7\text{-}4)$$

where the $Z_i$'s are partition functions for the ion and metal (see Section 7-3) and $E_{ion}$ is the ionization energy in eV. Small values of $E_{ion}$ and high temperatures favor the formation of ions. For example, if all electrons come from the ionization reaction, at moderate concentrations, $K$ ($E_{ion} = 4.34$ eV) is about 50% ionized at 2500 K, whereas Na ($E_{ion} = 5.14$ eV) is only 7% ionized at the same temperature. Usually, only the first ionization needs to be considered in most flames and in the observation region of most plasmas.

The fraction atomized, also called the *free atom fraction*, $\beta_a$, is a measurable quantity that describes the loss of analyte atoms due to compound formation and ionization. It can be expressed as

$$\beta_a = \frac{n_M}{n_T} \qquad (7\text{-}5)$$

where $n_M$ is the number density of free metal atoms and $n_T$ is the total number density of all metal-containing species (e.g., M, MO, MOH, MX, $M^+$).

*Ground-State Atom Density.* We will now obtain an expression for the analyte ground-state, free atom density (atoms per $cm^3$) in terms of the analyte solution concentration $c$ and the efficiencies of the various processes that must occur. The number of analyte atoms that are aspirated per second is given by $10^{-3}NFc$, where $N$ is Avogadro's number (atoms per mol), $F$ is the solution flow rate ($cm^3$ $s^{-1}$), $c$ is the analyte concentration (mol $L^{-1}$), and $10^{-3}$ is a conversion factor (L $cm^{-3}$). Only a fraction of the aspirated atoms pass through the observation zone per second as free atoms. This fraction is known as the overall atomization efficiency, $\varepsilon_a$. It is a product of the nebulization efficiency $\varepsilon_n$, the local desolvation efficiency $\beta_s$, the local volatilization efficiency $\beta_v$, and the local free atom fraction $\beta_a$. Thus

$$\varepsilon_a = \varepsilon_n \beta_s \beta_v \beta_a \qquad (7\text{-}6)$$

The number of free atoms per second entering the atomizer and passing through the observation zone is then $10^{-3}NF\varepsilon_a c$. To calculate the number of free atoms per cm³ of atomizer vapor, we must divide this quantity by the volume of the atomizer vapor per second. For flames, this is given by the flow rate of the unburned gases out of the burner head opening, $Q$ (L s⁻¹), multiplied by a factor $e_f$, which accounts for the expansion of the gases upon combustion. For a flame, the total gas flow rate $Q$ is given by

$$Q = Q_o + Q_f + Q_s \qquad (7\text{-}7)$$

where $Q_o$ is the oxidant flow rate, $Q_f$ is the fuel flow rate, and $Q_s$ is the flow rate of any inert support gases (all flow rates are in L s⁻¹). The expansion factor for gases initially at room temperature (298 K) is found from the ideal gas law to be

$$e_f = \frac{n(T) \times T}{n(298) \times 298} \qquad (7\text{-}8)$$

where $n(T)$ is the number of moles of burned gases, $T$ is the flame temperature in K, and $n(298)$ is the number of moles of unburned gases. Normally $n(T)$ and $n(298)$ are estimated from flame stoichiometry (see Table 8-3) and the fuel and oxidant flow rates. The contribution to $Q$ from vaporized solvent can be shown to be negligible if $\varepsilon_n F < 1$ mL min⁻¹. For a plasma the total flow rate $Q$ is just $Q_s$, the plasma support gas flow rate, and the expansion factor $e_f$ is $T/298$.

The number of free atoms per cm³ of atomizer gases is then $10^{-3}NF\varepsilon_a c/(10^3 Q e_f)$, where the factor of $10^3$ in the denominator is a conversion factor from L to cm³. The fraction of analyte free atoms existing as ground-state atoms is $g_0/Z(T)$, where $g_0$ is the statistical weight of the ground electronic state and $Z(T)$ is the internal partition function $[Z(T) = \Sigma\, g_i e^{-E_i/kT}]$ (see Section 7-3). Thus, after substituting the numerical value for $N$, the final equation for the number density of ground-state analyte atoms in the observation zone $n_0$ is

$$n_0 = \frac{6 \times 10^{17} F \varepsilon_a g_0 c}{Q e_f Z(T)} = kc \qquad (7\text{-}9)$$

where $k$ is the overall nebulization-atomization factor given by $6 \times 10^{17} F \varepsilon_a g_0/[Q e_f Z(T)]$. The number density can be related to the partial pressure $p$ by assuming

that the ideal gas law holds. From the latter, $p/RT$ is the number of moles per unit volume. The number density $n_0$ is then $n_0 = pN/RT = 7.34 \times 10^{21}\, p/T$.

*Calibration Curves.* Later in this chapter theoretical relationships are developed that express optical quantities (radiance of emission, fluorescence radiance, absorbance, etc.) in terms of the ground-state number density $n_0$. In many cases we assume that such expressions also give the relationship of the appropriate optical quantity to the solution concentration $c$. Thus we desire that $n_0$ be linearly related to $c$. From equation 7-9 this linear relationship can occur only if factors such as $F$, $Z(T)$, $\varepsilon_a$, $Q$, and $e_f$ are constant and independent of analyte concentration. This implies indirectly that atomizer temperature, solution viscosity, surface tension, solvent composition, droplet trajectories, residence times, observation heights, and many other variables remain constant and independent of concentration. The variation of these factors with analyte concentration can be one cause of nonlinearity. At high analyte concentrations, nebulization efficiency may decrease because of solution viscosity and surface tension changes; this causes reduced atomization efficiency (decrease in $\varepsilon_n$ and thus $\varepsilon_a$ with increasing $c$) and downward curvature of the analytical curve. At higher analyte concentrations, negative deviations can also occur because $\beta_s$ and $\beta_v$ decrease as the solids content of the droplets or the size of the dry aerosol particles increases.

The free atom concentration can also vary with the analyte concentration when the analyte is involved in equilibria with other species. For example, consider the atomization of an easily ionized element. If the only metal-containing species in the atomizer are free metal atoms M and singly charged ions M⁺, the mass balance condition for the total concentration of metal $n_T$ is $n_T = n_M + n_{M^+}$. The degree of ionization $\alpha_i$ is defined as

$$\alpha_i = \frac{n_{M^+}}{n_T} = \frac{n_{M^+}}{n_M + n_{M^+}} \qquad (7\text{-}10)$$

If the only ions in the atomizer are M⁺ ions, electroneutrality requires that $n_{M^+} = n_e$. With these relationships, equation 7-3 can be written

$$K_i = \frac{(n_{M^+})^2}{n_M} = \frac{\alpha_i^2 n_T}{1 - \alpha_i} \qquad (7\text{-}11)$$

If $n_T \ll K_i$, the fraction ionized $\alpha_i$ is nearly unity and $n_{M^+} \approx n_T$. Solving equation 7-11 for $n_M$ under these conditions shows that $n_M \approx (n_T)^2/K_i$. If $n_T \gg K_i$, the fraction ionized $\alpha_i$ is small, so that $n_M \approx n_T$. These relationships predict that $n_M$ is proportional to $c^2$ at low concentrations and to $c$ at high concentrations. Thus a nonlinearity is predicted at low values of $c$.

In most atomizers, $n_e$ is determined by the ionization of several species, not just the analyte. If curvature due to ionization is a problem for a particular analyte, a large concentration of a more easily ionized element can be added to all samples and standards to suppress ionization of the analyte. We see from equation 7-3 that if $n_e \gg n_T$, the ratio $n_{M^+}/n_M$ and the fraction ionized $\alpha_i$ are constant and $n_M \propto n_T \propto c$. Furthermore, if $n_e \gg K_i$, $\alpha_i \approx 0$.

Similar nonlinearities arise due to dissociation equilibria (equation 7-2) for small values of $K_d$. As for ionization, the fraction dissociated is larger at small concentrations compared to large concentrations. Since, in this case, the dissociated species M is measured, a negative deviation in the calibration curve can result. Many other factors affect linearity in specific techniques, as we shall see in later chapters, but the causes noted above are common for all atomic spectrochemical methods. We will see in the next section that many interferences in atomic spectrochemical methods occur because the conversion of the analyte into free atoms is affected by a concomitant that changes the overall nebulization-atomization efficiency for a given amount of analyte.

### Free-Atom Formation with Discrete Sample Introduction

With discrete sample introduction into a furnace, a droplet of solution is deposited directly in the atomizer (usually a furnace). Hence there is no nebulization term in the expression for the overall atomization efficiency ($\varepsilon_a = \beta_s \beta_v \beta_a$). These factors are usually time dependent because of the discrete sample and the fact that the furnace is usually operated in a noncontinuous mode. In most cases the furnace is heated to the boiling point of the solvent for a time long enough that complete desolvation occurs ($\beta_s = 1$). With these differences in mind, the expression for the instantaneous ground-state free atom density $n_0(t)$ is similar to that given in equation 7-9 except that $F$ is replaced by $V$, the volume of solution delivered to the furnace. Note that $\varepsilon_a$, $e_f$, $Z(T)$, and thus $n_0(t)$ are all time dependent because the atomizer temperature varies with time and the amount of sample is limited.

In many cases we expect $n_0(t)$ to be linearly related to the analyte concentration [$n_0(t) = kc$]. For this to be true, $\varepsilon_a$, $Q$, $e_f$, and $Z(T)$ must be constant and independent of analyte concentration. This implies that measurements must be made at constant temperature, time, flow rate of support gas, observation height, atomization efficiency, and so on. Just as was true for continuous sample introduction, there are many factors that can lead to nonlinearities and many substances that cause interference in the conversion of analyte to ground-state free atoms.

## 7-2 INTERFERENCES IN ATOMIC SPECTROSCOPY

Many of the interferences caused by concomitants are quite similar in all atomic spectrometric techniques. However, certain techniques may show freedom from some types of interference or exhibit increased susceptibility to others. In this section some of the interferences that were introduced in Chapter 6 are discussed in the context of atomic emission, absorption, and fluorescence spectroscopy. Interference effects that are specific to the medium used (flame, plasma, electrothermal atomizer, arc, etc.) are discussed in later chapters dealing with specific techniques.

### Blank Interferences

As discussed in Chapter 6, a blank or additive interference produces an uncompensated signal independent of the analyte concentration. In atomic spectroscopic methods these are usually spectral in nature, although nonspectral blank interferences are possible.

*Atomic Emission Interferences.* Spectral interferences, sometimes called *cross-spectral interferences*, are most troublesome in emission spectroscopy because the selectivity is normally determined by a wavelength selection device (filter, monochromator, polychromator, etc.). Here any concomitant that emits radiation within the bandpass of the wavelength selection device or radiation that appears in that bandpass due to stray radiation causes such a spectral interference. A quick look at any fairly complete set of wavelength tables, such as the MIT tables, will convince the reader that almost every sensitive emission line has several potential interfering lines very near it. Whether a spectral interference is observed or not depends on the composition of the sample and the technique used to excite emission. Because arcs, sparks, and plasmas have higher excitation energies than flames, spectral interferences are potentially more troublesome with these excitation sources.

Atomic line interferences in emission methods can, in principle, be reduced by improving the spectral resolution of the wavelength selection device since true spectral line overlaps are rare. Thus high excitation energy emission sources normally use (and need) much higher resolution wavelength selection devices than flame sources, which often use moderate resolution monochromators or even filters. Spectral interferences are, however, common in flame emission methods. Since the practicing analyst rarely has the option of switching to a higher-resolution spectrometer when spectral interferences occur, the choices usually involve decreasing the spectral bandpass (if possible), changing analyte spectral lines, changing source excitation conditions, or trying to compensate or correct for the effect (see Section 6-5).

Because of their dependence on excitation conditions and spectral resolution, spectral interferences in emission methods are highly instrument dependent. For example, in flame emission with filter photometers, an interference of the Mn triplet (403.1, 403.3, and 403.4 nm) is found on the K doublet (404.4, 404.7 nm). Since the Mn interfering lines are 1 nm or more from the nearest K line, a moderate-resolution instrument can eliminate this interference. On the other hand, the interference of Ga (403.298 nm) on Mn (403.307 nm) is much more difficult to overcome. This latter interference has been eliminated by using an echelle grating spectrometer operating in the twelfth order. Another troublesome interference is that of Na (285.28 nm) on Mg (285.21 nm) in flame emission. This interference is rarely noted, however, in plasma or arc emission methods with higher-resolution spectrometers.

Molecular band emission can occur with flame excitation when the sample contains combustible materials or constituents that give rise to molecular species. For example, high concentrations of Ca in the sample produce CaOH band emission, which can be a blank interference if it appears at the analyte wavelength. In contrast to atomic line interferences, the narrow atomic analyte line is superimposed on the broad emission band from the molecular species, and thus band emission interferences cannot be eliminated by improving the spectral resolution of the spectrometer. In flame emission the normal background emission from the flame itself is not usually an interference since the blank measurement can compensate for unchanging background. However, if a concomitant alters the background emission, it can cause a nonspectral blank interference.

*Atomic Absorption Interferences.* Atomic line interferences are rarely observed in atomic absorption spectroscopy with narrow line sources since the source width effectively determines the spectral resolution. As discussed in Chapter 4, hollow cathode lamps produce atomic lines with FWHM values on the order of 0.001 nm. In flame AA, for example, there is no interference of Na (285.28 nm) on the Mg 285.21-nm line because of this high selectivity. Line sources in AA, however, emit more than just the desired analyte line; lines from

metallic impurities and the filler gas are also emitted. Such lines can cause spectral interferences if they are absorbed and not separated from the analyte line by the wavelength selection device. Multielement line sources are also employed in AA and these can increase the risk of spectral interferences. Continuum source AA, like flame emission, depends highly on the spectral resolution of the wavelength selection device for selectivity. If the spectral bandpass is large, the probability of absorption by other atomic species is enhanced.

Background emission is normally not a problem in AA because modulation and ac detection are used to discriminate against it. However, absorption by molecular species and scattering of the source radiation by nonvolatile salt particles or oxides can occur and can give rise to a blank interference (see Section 10-3). Absorption by molecular species is particularly a problem with electrothermal atomizers and in relatively cool flames. Scattering is also said to cause an interference in cool flames. At wavelengths shorter than 220 nm, the flame itself becomes absorbing. While this is normally taken care of by the blank measurement, any change in the background absorption of the flame by a concomitant in the sample causes a blank interference.

*Atomic Fluorescence Interferences.* Flame atomic fluorescence (AF) with line source excitation is potentially as free from spectral interferences as line source AA. Again continuum sources enhance the risk that concomitants will produce fluorescence within the bandpass of the wavelength selection device.

In AF, normal background emission from the atomization system is usually compensated by modulation techniques. As in AA, alteration of the background by concomitants in the sample can cause a nonspectral blank interference. Resonance AF is particularly susceptible to interference from scattering of the source radiation by droplets and nonvolatilized salt particles. Since resonance AF is measured at the source wavelength, it is often difficult to distinguish such scattering from true fluorescence. Nonresonance AF methods do not suffer from this interference since the fluorescence signal is measured at a wavelength different from the source excitation wavelength.

### Analyte Interferences

Analyte or multiplicative interferences alter the magnitude of the analytical signal itself. In atomic spectroscopy they are invariably nonspectral interferences because of the narrowness of the lines involved.

*Nonspecific Interferences.* Nonspecific interferences, often called *physical interferences,* are fairly independent of analyte type. They can also have nearly the same effect in emission, absorption, and fluorescence with the same atomizer type. With nebulizers, nonspecific interferences can affect the aspiration, nebulization and desolvation processes. Concomitants that affect the sample solution viscosity, surface tension, and density can alter the solution flow rate and the nebulization efficiency. A 1% decrease in viscosity, for example, increases the flow rate about 1%, and a 1% increase in temperature causes about a 2% decrease in viscosity. Sample constituents that change the evaporation rate of the solvent can affect the desolvation efficiency. Similarly, sample components can alter the flame or plasma shape, and this can influence the local nebulization and desolvation efficiency for the analyte. Combustible constituents of the sample can influence atomizer temperature and thus indirectly affect analyte atomization through changing desolvation efficiencies.

With discrete sample introduction, nonspecific interferences affecting sample transport are not nearly as troublesome since sample transport is normally 100% efficient. With electrothermal atomizers, concomitants can, however, affect desolvation rates, and, unless the atomizer temperature program is adjusted to ensure complete desolvation, this can be an interference. With arc excitation of solid samples, the matrix can exhibit a large influence on the rate of volatilization of the analyte, as discussed in Chapter 9.

*Specific Interferences.* Specific interferences, often called *chemical interferences,* are more analyte specific in that the magnitude of the effect depends strongly on the analyte. These interferences can occur in the conversion of the solid or molten particle that remains after desolvation into free, neutral, ground-state analyte atoms. Such interferences occur in emission, absorption, or fluorescence measurements. In emission, specific interferences can alter the fraction of analyte atoms excited; in fluorescence, specific interferences can alter the quantum yield of fluorescence.

Concomitants in the sample can influence the volatilization of the particles containing the analyte and these **solute volatilization interferences** can be analyte specific. For example, in some flames the presence of phosphate in the sample can alter the atomic concentration of calcium in the flame. The interference is attributed to the formation of relatively involatile complexes between the metal and phosphate. Since other elements, such as strontium and magnesium, differ in their tendency to form phosphate complexes and the complexes vary in volatility, the effect is analyte specific. Addition of phosphoric acid to a sample containing Ca causes a depression of the Ca emission, absorption, or fluorescence that is highly dependent on atomizer temperature, residence time of the particles, and the

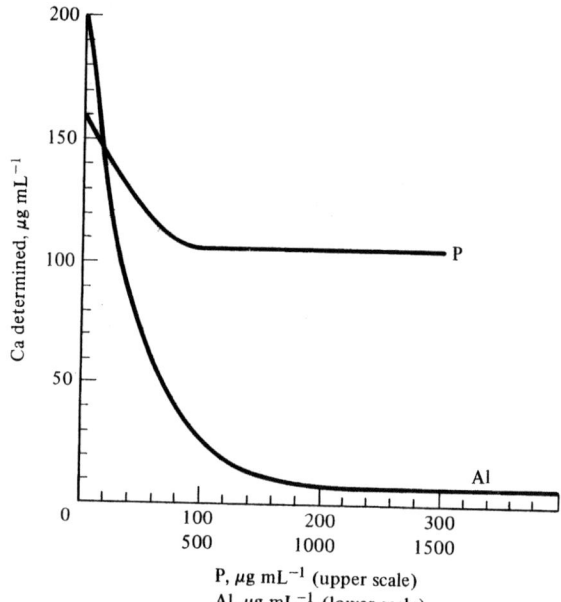

**FIGURE 7-9** Influence of P and Al on calcium flame emission. The Ca concentration determined is proportional to the Ca emission intensity. The actual concentrations were 100 μg mL⁻¹ (P) and 200 μg mL⁻¹ (Al). The phosphate interference on Ca is attributed to the formation of involatile complexes, while the Al interference has been interpreted as the formation of refractory Al-Ca-O species that are undissociated. Both effects can be reduced by adding a complexing agent like EDTA, which prevents Ca from forming the nonvolatile species. Since the Ca-EDTA complex is quite volatile, the interference can be greatly reduced. The phosphate interference can also be reduced by adding a releasing agent such as La, which ties up all the phosphate when added in excess. (Adapted with permission from R. Mavrodineanu and H. Boiteux, *Flame Spectroscopy,* Wiley, New York, 1965. Copyright © John Wiley & Sons, Inc., 1965.)

viewing position in the atomizer. The interference is most pronounced in relatively cool flames. Figure 7-9 illustrates the effect of adding phosphoric acid or aluminum to a solution containing a constant amount of calcium. It can be seen that the depressing effects of both elements eventually reach a plateau.

Enhancement effects can also be observed in which the volatilization of analyte-containing species is increased by the presence of a concomitant. For example, Al normally forms involatile oxides in the solid phase. The particular salt present in the solution can greatly influence the volatility of the particles formed. Nitrate salts are less volatile than chlorides. Hydrofluoric acid in the sample can greatly enhance the amount of aluminum that volatilizes by forming aluminum fluoride. With electrothermal atomizers, the nature of the salt particle that is formed upon desolvation can greatly influence the volatilization of the analyte and thus its atomization efficiency. Because of their complexity and the difficulty in identifying what species are actually formed, solute volatilization effects are not always well understood. Solute volatilization interferences are less important in plasmas because of the high temperatures.

Another type of interference occurs with hot flames (nitrous oxide–acetylene) that burn on burners with a slot orifice. Involatile salts and certain acids have been found to enhance emission and absorption when the slot is aligned parallel to the optical axis, but not when it is aligned perpendicular. These effects have been attributed to a change in the lateral diffusion of analyte species in the gas phase and are called **lateral diffusion interferences** (see Figure 10-22).

The specific interferences discussed above involve the condensed phase as well as the gas phase in the case of the lateral diffusion effect. There are also specific interferences that occur in the vapor phase only. Concomitants in the sample can influence the degree of dissociation of the analyte and cause analyte-specific **dissociation interferences**. For example, the presence of HCl in the sample can affect the dissociation of NaCl and KCl to a different extent because the dissociation energies of the two compounds are different. These interferences are often more pronounced at high concentrations in total consumption nebulizer/burners. Analytes that form oxides or hydroxides in flames can be particularly susceptible to dissociation interferences. Concomitants in the sample that alter flame temperature can vary the degree of dissociation of such compounds. Certain concomitants, such as Cr and Sn, at high concentrations càn catalyze the recombination of H and OH radicals and thus change the degree of dissociation of hydroxides that contain the analyte. In AA the tendency has been to use very hot flames to minimize such dissociation interferences. Again for plasmas, the high temperatures reduce the susceptibility to dissociation interferences.

Another gas-phase interference is an **ionization interference** which occurs when a concomitant alters the degree of ionization of the analyte. The presence of an easily ionized element, such as K, can affect the degree of ionization of a less easily ionized element such as Ca by increasing the electron concentration. This is particularly a problem in flames, where electron densities are usually fairly low. In plasmas, electron densities from the ionization of the plasma support gas are much higher, and small changes caused by small amounts of easily ionized elements have little effect. In flames the degree of ionization of the analyte can also be altered when concomitants change the flame temperature. Ionization interferences are in general worse in hot flames and at low analyte concentrations. Because high tem-

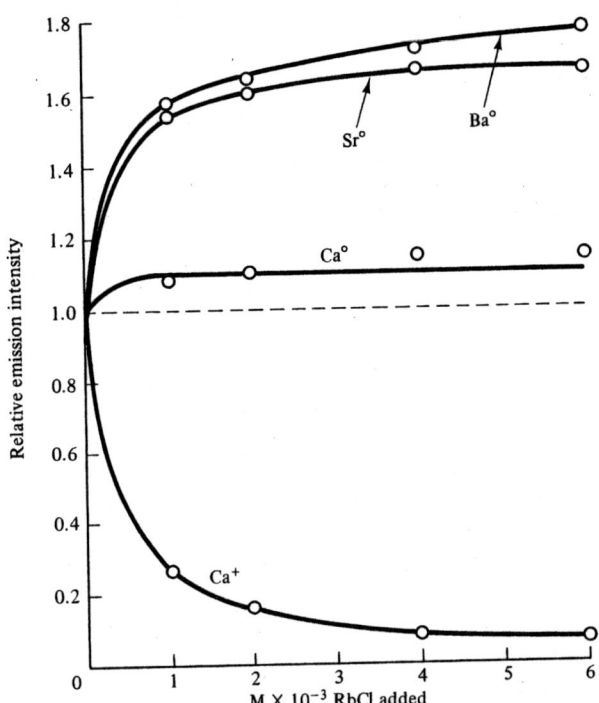

**FIGURE 7-10** Variation of alkaline earth emission in a flame with added RbCl: The lines used were the Ba atomic line at 554 nm, the Sr atomic line at 461 nm, the Ca atomic line at 423 nm, and the Ca ionic line at 393 nm. The alkaline earths were all present at the 20 µg mL$^{-1}$ level. Note that the ionic emission of Ca is suppressed by the addition of Rb, while the atomic emission is enhanced. [Reprinted with permission from W. H. Foster and D. N. Hume, *Anal. Chem.*, *31*, 2033 (1959). Copyright © American Chemical Society, 1959.]

peratures increase volatilization efficiency and compound dissociation, the modern tendency is to use high-temperature media and to add an ionization suppressant to the sample. The suppressant is a more easily ionized element than the analyte, for example K, Na, Li, Cs, or Rb. When added to the samples in a constant high concentration, the ionization of the suppressant produces electrons that shift the ionization equilibrium of the analyte in the direction of neutral analyte atoms. Figure 7-10 shows the effect of RbCl on the flame emission of Ba, Sr, and Ca. Note that a plateau is reached in the case of Sr and Ca, where the analyte is essentially converted entirely to neutral species. Barium has the lowest ionization energy of this group and the greatest tendency to form molecular species. Hence the Ba emission does not completely reach a constant level with added Rb.

Atomic emission methods are also subject to **excitation interferences**, which arise when a concomitant alters the atomizer temperature and thus the fraction of analyte atoms excited. Since the temperature dependence of emission is a function of excitation energy, these interferences can be considered to be analyte specific. Although excitation interferences occur only in emission, it should not be implied that changes in atomizer temperature do not also influence absorption and fluorescence. Atomizer temperature also affects desolvation, solute volatilization, dissociation of compounds, and analyte ionization, and thus changes in temperature influence all three techniques.

In atomic fluorescence spectroscopy, concomitants in the sample can alter atomizer composition and thus change the fluorescence quantum yield. Fluorescence efficiencies are usually higher in inert-gas atmospheres. The presence of oxygen or nitrogen in the atomization gas can decrease the fluorescence yield through quenching. If such an effect is produced by sample constituents, a quenching interference results. It is unlikely that direct quenching by collisions with concomitant atoms or molecules occurs with most atomizers because of the low concentrations compared to the atomizer gases.

## 7-3 ELECTRONIC STATES OF ATOMS

The general characteristics of atomic spectra were introduced briefly in Chapter 1. Here we examine in more detail the electronic configurations of atoms and categorize spectral transitions according to spectroscopic term symbols, shorthand descriptions of electronic states. The statistical weights of states are discussed, and the concept of the partition function is introduced.

### Quantum Numbers

The electronic configuration of the hydrogen atom can be specified by four quantum numbers: $n$, the *principal quantum number*; $l$, the *orbital angular momentum quantum number*; $m_l$, the *orbital magnetic quantum number*; and $m_s$, the *spin magnetic quantum number*. The description of each of these quantum numbers and their allowed values and symbols are given in Table 7-1. We are all familiar with the simple atomic orbital description of the electronic configurations of polyelectronic atoms. Here the electrons are assigned to hydrogen-like atomic orbitals using the quantum numbers of Table 7-1. For many purposes the state of an atom can be classified by the quantum numbers $n$ and $l$ for each electron in the atom, because these are the two quantum numbers that largely determine the energy of the atom. The value of $l$ determines the shape of the orbital. An orbital with $l = 0$ is spherically symmetric, has no orbital angular momentum, and for historical reasons, is designated as an $s$ orbital. Orbitals with $l = 1$ are called $p$ orbitals, those with $l = 2$ are $d$ orbitals, and so on. There are three $p$ orbitals corresponding to the allowed values of $m_l$, five $d$ orbitals, and so on.

**TABLE 7-1**

Summary of quantum numbers for individual electrons

| Quantum number | Description | Allowed values | Symbols |
|---|---|---|---|
| $n$ | Principal quantum number; determines energy (size of elliptical orbit) | 1, 2, 3, 4, . . . | 1 = $K$ shell<br>2 = $L$ shell<br>3 = $M$ shell<br>etc. |
| $l$ | Orbital (azimuthal) angular momentum quantum number; determines magnitude of orbital angular momentum (shape of orbital) | 0, 1, 2, . . . , $n - 1$ | 0 = $s$ orbital<br>1 = $p$<br>2 = $d$<br>3 = $f$<br>etc. |
| $m_l$ | Orbital magnetic quantum number; describes orientation of angular momentum vector | $l, l - 1, . . . , 0, . . . , -l$ | $s$<br>$p_x, p_z, p_y$<br>$d_{x^2-y^2}, d_{xz},$<br>$d_{z^2}, d_{yz}, d_{xy}$ |
| $s$ | Electron spin quantum number; determines magnitude of spin angular momentum | $+\frac{1}{2}$ | |
| $m_s$ | Spin magnetic quantum number; describes orientation of spin angular momentum vector | $+\frac{1}{2}, -\frac{1}{2}$ | $\uparrow, \downarrow$ |

A given assignment of $n$ and $l$ values to all the electrons in an atom produces an **electronic configuration**. The ground-state configuration of an atom can often be deduced by knowing the number of electrons in the atom (equal to its atomic number $Z$), the maximum number of electrons in a given orbital type (2 for $s$ orbitals, 6 for $p$ orbitals, 10 for $d$ orbitals), and the order of filling orbitals. This **Aufbau order** is $1s$, $2s$, $2p$, $3s$, $3p$, $4s$, $3d$, $4p$, $5s$, $4d$, $5p$, $6s$, $4f$, $5d$, $6p$, $7s$, and so on. Thus, for carbon ($Z = 6$), the ground-state configuration is designated $1s^2\,2s^2\,2p^2$, where the superscript indicates the number of electrons in the orbital. For sodium ($Z = 11$), the ground-state configuration is $1s^2\,2s^2\,2p^6\,3s^1$.

The **Pauli exclusion principle** states that no two electrons in the same atom can have the same four quantum numbers. Thus, for He ($Z = 2$ with configuration $1s^2$), the two electrons must have $m_s$ values of $+\frac{1}{2}$ and $-\frac{1}{2}$ respectively (opposite spins).

## Coupling Schemes

For many-electron atoms, the hydrogen quantum numbers can be thought of as describing the individual electrons, but they are not "good" quantum numbers for the entire atom. In the language of quantum mechanics, good quantum numbers are associated with operators (squared angular momentum and $z$-component of angular momentum) that commute with the total atom Hamiltonian. For polyelectronic atoms, three resultant quantum numbers, $L$, $S$, and $J$, are good quantum numbers for the entire atom. For light atoms ($Z < 30$), with weak spin-orbit interactions, a resultant orbital angular momentum specified by quantum number $L$ is produced by coupling the orbital angular momenta of each electron, a process of vector addition. The allowed values of $L$ and the symbols used to abbreviate them are given in Table 7-2. For a single electron the resultant $L$ is equal to $l$ for that electron.

For more than two electrons, the overall coupling can be broken down into a chain of pairwise couplings of the individual $l_i$. Closed shells possess no net orbital motion, and thus the electrons in these shells do not need to be considered when the resultant is computed.

The individual spin moments couple to give a *resultant spin quantum number $S$* as shown in Table 7-2. Two electrons thus give $S = 1$ and $S = 0$ since $s_1 = s_2 = \frac{1}{2}$. States with $S = 0$ are called singlets since there is only a single value for $M_S$, the *resultant spin magnetic quantum number*. States with $S = 1$ are doublets, those with $S = 2$ are triplets, and so on. The value of $2S + 1$, which corresponds to the number of possible orientations of the resultant spin angular momentum

vector, is called the **multiplicity** of the state. The Pauli exclusion principle requires that $S = 0$ for a completely filled subshell.

For atoms with weak spin-orbit interactions, the resultant orbital and spin angular momenta couple to produce a total angular momentum denoted by quantum number $J$, sometimes called the *inner quantum number*, for the entire atom. This Russell–Saunders or $LS$ coupling scheme leads to the allowed values of $J$ in Table 7-2.

When spin-orbit interactions are large, $LS$ coupling no longer describes the electron interaction and the $jj$ coupling scheme is employed. Here a resultant $j$ for each electron is computed according to the allowed values $j = l + s,\ l + s - 1,\ \dots\ |l - s|$. The individual $j$'s then combine to give a $J$ value for the entire atom.

These vector models provide a convenient physical picture and are useful in calculations. However, neither the $LS$ nor the $jj$ coupling scheme is a completely adequate description. Instead these are two limits with intermediate cases possible. The two limiting cases blend smoothly into each other as a function of atomic number. In fact, even a heavy atom, such as lead, does not exhibit pure $jj$ coupling. We shall assume for the remainder of this discussion that we can use the $LS$ coupling scheme.

No matter which coupling scheme applies, the internal energy, the $J$ value and the parity are well-defined quantities for a given state. According to quantum mechanics, the **parity** is *odd* if the atomic wave function changes sign upon inversion of the electron coordinates and *even* if it does not. Configurations with $\Sigma l_i$ an even number lead to states with even parity, and those with this sum an odd number lead to states of odd parity. The electrons in closed subshells contribute to the sum by an even number; therefore, the parity depends only on the valence electrons.

## Term Symbols

A multiplet of closely spaced states with the same $L$ and $S$ values, but different $J$ values, is called a **spectroscopic term**. A complete term symbol is a designation for a particular electronic state and is written $n^{2S+1}\{L\}_J$, where $n$ is the principal quantum number for the valence electron(s), $\{L\}$ is the symbol ($S$, $P$, $D$, $F$, etc.) for the quantum number $L$, $2S + 1$ is the multiplicity, and $J$ is the total angular momentum quantum number. Thus the complete term symbol for the ground state of sodium is $3^2S_{1/2}$. Sometimes the entire electronic configuration is used instead of $n$ in the term symbol and often the value of $n$ is omitted. Thus the ground state of sodium can be written $1s^2\,2s^2\,2p^6\,3s^2\,S_{1/2}$ or just

**TABLE 7-2**

Quantum numbers for many-electron atoms with LS coupling

| Quantum number | Description | Allowed values | Symbols |
|---|---|---|---|
| $L$ | Resultant orbital angular momentum quantum number; determines magnitude of orbital angular momentum | $L = l_1 + l_2, l_1 + l_2 - 1, \ldots, \lvert l_1 - l_2 \rvert$ for two electrons with orbital angular momentum quantum numbers $l_1$ and $l_2$ | $0 = S$ $1 = P$ $2 = D$ $3 = F$ etc. |
| $M_L$ | Resultant orbital magnetic quantum number; describes orientation of angular momentum vector | $M_L = \Sigma \, (m_l)_i$ $2L + 1$ possible values | |
| $S$ | Resultant spin quantum number; determines magnitude of spin angular momentum | $S = s_1 + s_2, s_1 + s_2 - 1, \ldots, \lvert s_1 - s_2 \rvert$ for two electrons with spin quantum numbers $s_1$ and $s_2$ | For $2S + 1$: $0 = $ singlet $1 = $ doublet $2 = $ triplet etc. |
| $M_S$ | Resultant spin magnetic quantum number; describes orientation of spin angular momentum vector | $M_S = \Sigma(m_s)_i$ $2S + 1$ possible values | |
| $J$ | Total angular momentum quantum number; determines magnitude of total angular momentum of entire atom | $J = L + S, L + S - 1, \ldots, \lvert L - S \rvert$ | |
| $M_J$ | Resultant total magnetic quantum number; describes orientation of total angular momentum vector | $M_J = J, J - 1, \ldots, -J$ $2J + 1$ possible values | |

$^2S_{1/2}$. Here for the single $3s$ valence electron, $L = l = 0$, $S = s = \frac{1}{2}$, $J = L + S = \frac{1}{2}$.

For atoms with several valence electrons, a given configuration can give rise to many states. With carbon ($1s^2\ 2s^2\ 2p^2$), for example, there are two open-shell $p$ electrons to consider. One $2p$ electron has $n = 2$, $l = 1$, three possible values of $m_l$, and two possible values of $m_s$, a total of six possible sets of quantum numbers. For the second $2p$ electron (same $n$ and $l$) there are five possible sets because of the exclusion principle. For both electrons, there are $(6 \times 5)/2 = 15$ possible assignments since the electrons are indistinguishable. Since $l_1 = l_2 = 1$, $L$ can be 2, 1, or 0, and there are $D$, $P$, and $S$ terms.

Table 7-3 gives the possible arrangements and the resulting terms for the carbon ground electronic configuration. Finding the terms is a matter of picking out those multiplets that can give rise to all the states in the table. We can see that there must be a $D$ term because there are states with $M_L = \pm 2$. For this term, the spins must be paired by the exclusion principle ($S = 0$); thus the $D$ term must be a singlet term. The $D$ term is composed of five states with $M_L = +2, 1, 0, -1, -2$ and $M_S = 0$. Similarly, there must be a triplet term because of the presence of the values $M_S = \pm 1$. The latter values appear only in association with $M_L = \pm 1$. Hence this is a triplet $P$ term ($L = 1$, $S = 1$) composed of nine states with $M_L = +1, 0, -1$ and $M_S = +1, 0, -1$. The remaining term has $L = 0$, $S = 0$ and is thus a $^1S$ term. Note that all $L$ values up to the maximum are represented. Because the electrons are indistinguishable, it is impossible in some cases to assign a specific microstate to a particular term. Hence states that differ only in the spins of the electrons are associated with more than one term as can be seen in Table 7-3. The $J$ values associated with

**TABLE 7-3**
Origin of terms for the carbon atom
$$1s^2\ 2s^2\ 2p^2:\ l_1 = l_2 = 1$$

| $m_{l_1}$ | $m_{l_2}$ | $M_L$ | $m_{s1}$ | $m_{s2}$ | $M_S$ | Term |
|---|---|---|---|---|---|---|
| 1 | 1 | 2 | $\frac{1}{2}$ | $-\frac{1}{2}$ | 0 | $^1D$ |
| 1 | 0 | 1 | $\frac{1}{2}$ | $\frac{1}{2}$ | 1 | $^3P$ |
|   |   |   | $\frac{1}{2}$ | $-\frac{1}{2}$ | 0 | $^1D,\ ^3P$ |
|   |   |   | $-\frac{1}{2}$ | $\frac{1}{2}$ | 0 | |
|   |   |   | $-\frac{1}{2}$ | $-\frac{1}{2}$ | $-1$ | $^3P$ |
| 1 | $-1$ | 0 | $\frac{1}{2}$ | $\frac{1}{2}$ | 1 | $^3P$ |
|   |   |   | $-\frac{1}{2}$ | $-\frac{1}{2}$ | $-1$ | $^3P$ |
|   |   |   | $\frac{1}{2}$ | $-\frac{1}{2}$ | 0 | |
|   |   |   | $-\frac{1}{2}$ | $\frac{1}{2}$ | 0 | $^1D,\ ^3P,\ ^1S$ |
| 0 | 0 | 0 | $\frac{1}{2}$ | $-\frac{1}{2}$ | 0 | |
| 0 | $-1$ | $-1$ | $\frac{1}{2}$ | $\frac{1}{2}$ | 1 | $^3P$ |
|   |   |   | $\frac{1}{2}$ | $-\frac{1}{2}$ | 0 | $^1D,\ ^3P$ |
|   |   |   | $-\frac{1}{2}$ | $\frac{1}{2}$ | 0 | |
|   |   |   | $-\frac{1}{2}$ | $-\frac{1}{2}$ | $-1$ | $^3P$ |
| $-1$ | $-1$ | $-2$ | $\frac{1}{2}$ | $-\frac{1}{2}$ | 0 | $^1D$ |

the terms for carbon are $J = 2$ for the $D$ term, $J = 2$, 1, 0 for the $^3P$ term, and $J = 0$ for the $^1S$ term. The complete set of term symbols for the ground electronic configuration of carbon is thus $^1D_2$, $^3P_2$, $^3P_1$, $^3P_0$, and $^1S_0$. Term symbols for almost all practical configurations have been tabulated in E. U. Condon and G. H. Short-ley, *The Theory of Atomic Spectra* (Cambridge University Press, Cambridge, 1963).

States with unlike term symbols differ in energy because of different degrees of electronic repulsion and correlation. The lowest-energy state for an atom with several possibilities can be determined with the aid of **Hund's rules:**

1. The highest multiplicity term within a configuration is of lowest energy.
2. For terms of the same multiplicity, the highest $L$ value has the lowest energy ($D < P < S$).
3. For subshells that are less than half-filled, the minimum $J$-value state is of lower energy than higher $J$-value states. For subshells that are more than half-filled, the state of maximum $J$ value is the lowest energy.

For carbon, the ground electronic configuration has the following energy order: $^3P_0 < {}^3P_1 < {}^3P_2 < {}^1D_2 < {}^1S_0$. The $^3P$ states are quite close together in energy since they differ only in the $J$ values.

Excited-state term symbols are determined in an analogous manner. Excitation of Na, for example, can result in promotion of the valence electron to the $3p$

orbital. In this case $n = 3$, $S = \frac{1}{2}$, $2S + 1 = 2$, $L = 1$ ($P$ term), $J = \frac{3}{2}, \frac{1}{2}$. There are two closely spaced levels in the excited term of sodium with term symbols $^2P_{1/2}$ and $^2P_{3/2}$. The state with $J = \frac{1}{2}$ is of slightly lower energy than the state with $J = \frac{3}{2}$ (Hund's rule 3). The well-known yellow sodium D lines result from transitions from this pair of excited levels to the ground level as shown below.

$$^2P_{1/2} \rightarrow {}^2S_{1/2} \qquad 5896\ \text{Å (D}_1\ \text{line)}$$

$$^2P_{3/2} \rightarrow {}^2S_{3/2} \qquad 5890\ \text{Å (D}_2\ \text{line)}$$

This type of term splitting (same $L$ but different $J$) is called **fine structure.**

Additional term symbols and transitions in the sodium atom are illustrated in the energy-level diagram of Figure 7-11a. The energy-level diagram for the calcium atom is shown in Figure 7-11b.

### Selection Rules and Atomic Spectra

Only a fraction of the total possible number of transitions between states are observed in practice. It is possible to derive from quantum mechanics **selection rules** which tell us the transitions that are allowed (see Appendix F). *Allowed transitions* are those that occur with high probability and give reasonably intense lines. *Forbidden transitions*, on the other hand, are those that occur with low probability and are thus weak in intensity. We shall consider here only electric dipole tran-

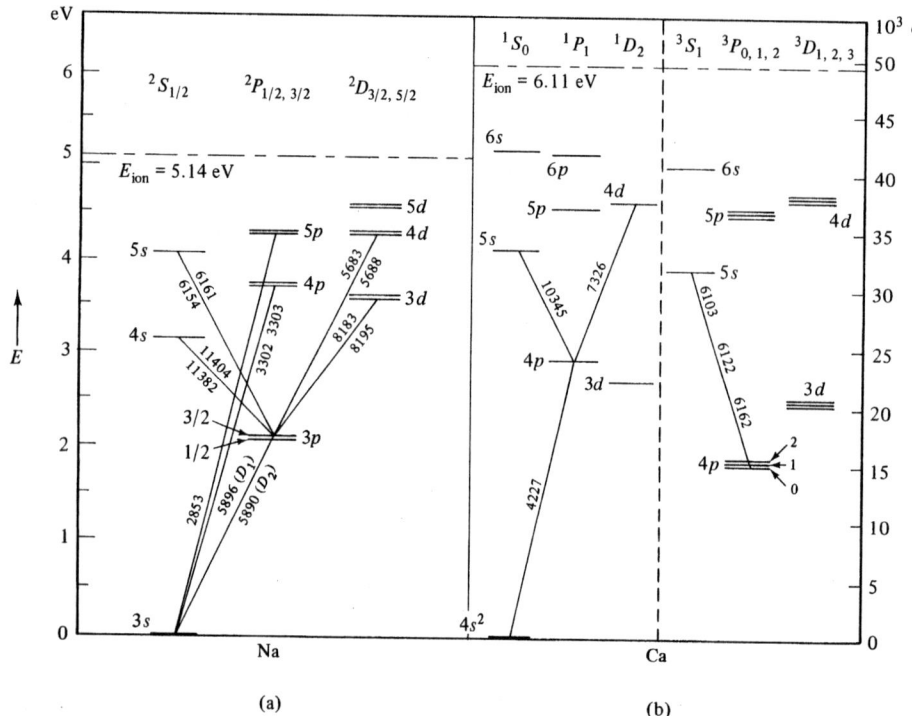

(a)   (b)

**FIGURE 7-11** Energy-level diagram for sodium (a) and calcium (b). The height of each line corresponds to the energy in eV (left scale) relative to the ground level. The ionization limit is the top dashed line (5.14 eV for Na and 6.11 eV for Ca). The right scale expresses the energy in wavenumbers relative to the ground level. Allowed transitions and their wavelengths (Å) are shown by the lines connecting levels. The term symbols are shown at the top of each diagram.

sitions. Transitions that are dipole forbidden may occur because they are electric quadrupole or magnetic dipole allowed. Furthermore, the approximation is made that the transitions involve only single electrons; that is, there are no changes in the orbitals of all the other electrons or mixing of states. This is not strictly correct, but the selection rule model works fairly well for describing spectra and for calculating the intensities of atomic lines.

For single-photon transitions the **general selection rules** that must be strictly obeyed for an allowed transition are:

1. The parities of the upper and lower level must be different.
2. $\Delta l = \pm 1$.
3. $\Delta J = 0$ or $\pm 1$, but $J = 0$ to $J = 0$ is forbidden.

In addition, there are **special selection rules** that depend on which coupling scheme (*LS* or *jj*) prevails. For *LS* coupling and single-photon transitions, these special selection rules are:

1. $\Delta S = 0$.
2. $\Delta L = 0$, or $\pm 1$, but $L = 0$ to $L = 0$ is forbidden.

For two-photon processes (see Chapter 11), only transitions between levels that have the *same* parity are

allowed. For *LS* coupling, the change in *J* is restricted to $|\Delta J| \leq 2$; the change in *L* is limited to $|\Delta L| \leq 2$. For alkali metal atoms the two-photon selection rules allow *only* $\Delta L = 0$ and $|\Delta L| = 2$.

The selection rules noted above can also account for the existence of long-lived atomic states with energies above the ground state. Such states are often called **metastable states**. These cannot lose their energy by emitting a photon because the transition required is forbidden. For example, the $^3P_0$ state of mercury cannot decay to the $^1S_0$ ground state because the transition is forbidden by general selection rule 3 ($J = 0$ to $J = 0$ is forbidden). Metastable atoms are often quite useful carriers of energy in sparks, plasmas, and flames.

Some of the allowed single-photon transitions in Na are shown in Figure 7-11a. Note that the allowed transitions are determined by the selection rule $\Delta l = \pm 1$. The emission spectrum of sodium (and other alkali metals) consists of three major series of spectral lines. Transitions between *P* states and the 3*S* ground state form the *principal series*. Transitions from excited *S* states to one of the 3*P* states form the *sharp series*, while those from *D* states to one of the 3*P* states form the *diffuse series*. For hydrogen and alkali metals the spectral series converge at the ionization limit. For more complex atoms different series may converge to different limits. The alkaline earth elements give spectra quite different from the alkali metals. From the energy-level diagram for Ca in Figure 7-11b, it can be seen that the resonance line at 422.7 nm is a single line and not a doublet as was the case with Na. Note also that the triplet terms do not give resonance lines at all, as the transition to the ground state is spin forbidden. The $4p$ triplet terms are, in fact, metastable.

If an electron other than the outermost is excited in an atom or if two electrons are excited simultaneously, the energy of the resulting state may be high enough that it can pass into a ground-state ion and a free electron with excess kinetic energy. Such states are said to be **autoionizing states**. They result from spectral lines with upper terms above the normal ionization limit. These lines are often observed to be broadened due to the very short lifetime of the upper state (see the next section). Spectral lines from such states may also have anomalous intensities. The existence of a level above the normal ionization limit implies that the recombination of a free electron of the appropriate energy with the ion has an enhanced probability. High current densities in electrical discharges are, in fact, often found to enhance lines which show autoionization because of recombination followed by emission.

### Additional Splitting Effects

An additional splitting of terms can occur because of magnetic coupling of the spin and orbital motion of the electrons in atoms with the nuclear spin. Such splitting results in **hyperfine structure**. For Na the $^2S_{1/2}$ ground level is split into two close levels, which leads to additional splitting of each D line. For Na, the hyperfine splitting is on the order of 0.02 Å and not readily detected. Other atoms, such as Cu and In, exhibit larger hyperfine splittings that are important in analytical applications.

When several isotopes of an element are present, a splitting of atomic lines can result from the **isotope shift**. The resonance lines of $^6$Li and $^7$Li are separated by about 0.16 Å, enough to determine their concentration ratios by atomic absorption.

In external fields atoms undergo additional term splitting. The splitting of spectral lines in a magnetic field is known as the **Zeeman effect**. The quantum number $M_J$ (see Table 7-2) describes the orientations (projections) of the $J$ vector; it assumes values from $+J$ to $-J$ that differ by unity. In the absence of a magnetic field, states with the same values of $L$, $S$, and $J$ but with different $M_J$ values are all of equivalent energy and are said to be **degenerate states**. For carbon, for example, the five states with the term symbol $^1D_2$ are degenerate. Similarly, for calcium (see Figure 7-11b), the three states with the term symbol $^1P_1$ are degenerate. In a magnetic field, however, there are slight energy differences corresponding to the different orientations of the $J$ vector

with respect to the field. Figure 7-12a illustrates the energy splittings for several different values of $J$. It can be shown theoretically that the selection rule for $M_J$ is $\Delta M_J = 0, \pm 1$. Singlet terms give rise to the **normal Zeeman effect**, in which the splitting between components is the same in different states of the atom, as illustrated in Figure 7-12b for Ca. For the normal Zeeman effect, the magnitude of the splitting is directly proportional to the magnetic field strength.

For terms with $S > 0$, the Zeeman splitting between components is no longer the same in different states, and this **anomalous Zeeman effect** can give rise to a more complicated splitting pattern, as illustrated in Figure 7-12c for the Na $D_1$ lines. The magnitude of the splitting in this case is a complex function of the field strength. The Zeeman effect is used for background correction in atomic absorption spectrophotometry (see Section 10-3).

In the presence of an electric field spectral lines can also be split into closely spaced components by the **Stark effect**. For hydrogen the Stark effect removes the degeneracy between states of the same $n$, but different $l$, producing an effect in which the splitting is proportional to the field strength (*first-order Stark effect*). In more complicated atoms there is no degeneracy between states of different $l$, and the splitting is proportional to the square of the electric field (*quadratic Stark effect*). Line broadening due to the Stark effect is important in arcs, sparks, and plasmas, as discussed in the next section. The Stark effect is widely used in analyzing microwave spectra of molecules.

### Statistical Weights and Partition Functions

We have seen in the preceding section that atomic states differing only in their $M_J$ values are degenerate. The **statistical weight** of a level is the number of degenerate quantum states making up that level. It is important in calculations of population distributions and spectral line radiances because we must account for the degeneracy of a particular level by weighting it according to the numbers of quantum states it contains. An electronic state with total angular momentum quantum number $J$ has a statistical weight $g$ given by

$$g = 2J + 1 \qquad (7\text{-}12)$$

The $^2P_{1/2}$ excited state of Na has a statistical weight of 2 and the $^2P_{3/2}$ state has a statistical weight of 4. In some cases fine structure splitting is ignored and the statistical weight refers to the multiplet or term as a whole. In this case $g = (2L + 1)(2S + 1)$. For the Na $^2P$ term,

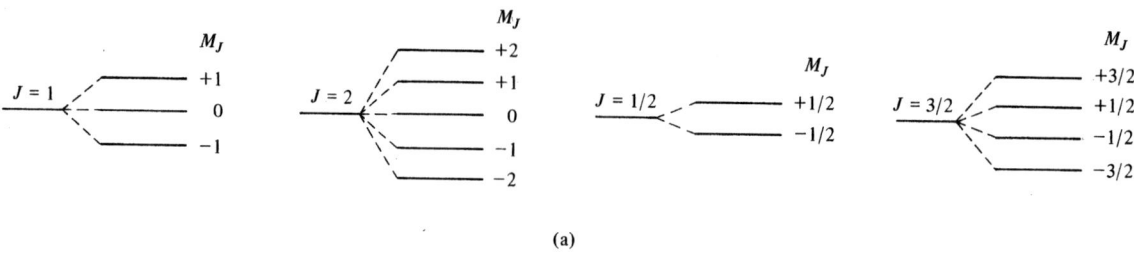

(a)

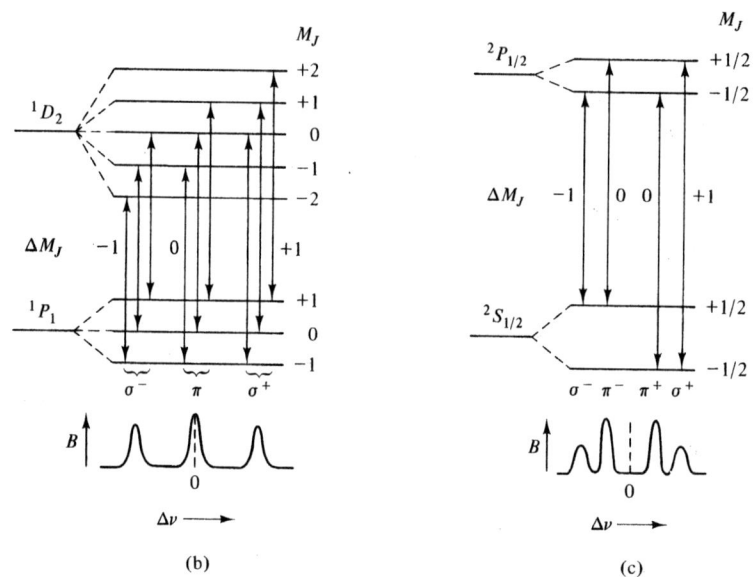

(b)                                        (c)

**FIGURE 7-12** Zeeman splitting of spectral lines. In (a) the energy levels corresponding to different $M_J$ values are shown for 4 values of $J$ in the absence (to the left of the dashed lines) and in the presence of a magnetic field (to the right). The normal Zeeman splitting pattern for the Ca 7326-Å line is shown in (b). The upper and lower levels are split by equal amounts in the normal Zeeman effect giving rise to three lines corresponding to the three allowed values of $\Delta M_J$. Components with $\Delta M_J = 0$ are called $\pi$ components, while those with $\Delta M_J = \pm 1$ are called $\sigma$ components. The $\sigma^-$ transition occurs at lower energy than the $\sigma^+$ transition. In (c) the anomalous Zeeman effect is illustrated for the Na $D_1$ line. Here the upper and lower levels are split by different amounts so that the components with the same $\Delta M_J$ are no longer of equivalent energy. This gives rise to four lines. The same effect is observed for the Na $D_2$ line ($^2P_{3/2} \rightarrow {}^2S_{1/2}$).

ignoring the fine structure splitting gives rise to a statistical weight of 6 for the doublet as a whole.

The **internal partition function** for an atom is defined as

$$Z(T) = \sum_{i=0}^{\infty} g_i e^{-E_i/kT} \qquad (7\text{-}13)$$

where $g_i$ is the statistical weight and $E_i$ is the energy of the $i$th level. For the ground state $E_0 = 0$. The partition function or "sum-over-states" is thus a weighted sum

of the number of states available ($g_i$ weighted by the Boltzmann factor $e^{-E_i/kT}$). It is the function needed to obtain the statistical mechanical properties of a system, as we have seen earlier in this chapter with dissociation and ionization constants. As shown in Chapter 2 (equation 2-14), the partition function is also needed in the normalization of the Boltzmann distribution to allow calculation of the fractional population of an energy level. As equation 7-13 shows, partition functions depend on temperature.

One problem in calculating the partition functions

for atoms is that the number of energy levels is infinite since the principal quantum number $n$ is unlimited. Since the Boltzmann factors in equation 7-13 are limited to values below the ionization limit, the partition function would also be infinite. In practice this dilemma is avoided by assuming an upper cutoff of the summation; the sum is bounded because with a very large orbital radius the electron begins to interact with other atoms or the walls of the container. Fortunately, if the atom has only a single ground level and a lowest excited level at least 2 eV above the ground level, the contributions of higher levels to $Z(T)$ are negligible if $T < 5000$ K. In these cases we make the assumption that $Z(T) \approx g_0$, where $g_0$ is the statistical weight of the ground state. In cases where the ground state is a multiplet with fine-structure splitting, however, the assumption above is not valid and other terms must be included. Partition functions have been calculated for several atoms at temperatures existing in common media [see, e.g., L. deGalan, R. Smith, and J. D. Winefordner, *Spectrochim. Acta, 23B,* 521 (1968)].

## 7-4 SPECTRAL LINE PROFILES

When atomic emission, absorption, or fluorescence spectra are recorded, narrow spectral lines are obtained. With ordinary spectrometers, the widths of the lines obtained are determined not by the atomic system, but by the properties of the spectrometer employed (slit function and spectral bandpass). With very high resolution monochromators or with Fabry–Perot interferometers, the actual widths and profiles of atomic spectral lines can be recorded. The finite widths obtained are the result of a variety of line-broadening phenomena. These processes give rise to a **spectral distribution** or **spectral profile** of photons which we shall call $S_\nu$ or $S_\lambda$. The quantity $S_\nu\,d\nu$ or $S_\lambda\,d\lambda$ can be interpreted as the fraction of photons with frequencies in the interval $\nu$ to $\nu + d\nu$ or with wavelengths in the interval $\lambda$ to $\lambda + d\lambda$. The spectral distribution function is normalized by

$$\int_{\text{line}} S_\nu\,d\nu = 1 \quad \text{or} \quad \int_{\text{line}} S_\lambda\,d\lambda = 1$$

We shall deal most often with the spectral distribution in terms of frequency, because line-broadening expressions are simpler and easier to interpret than in wavelength units. The distribution $S_\nu$ has the units of time or $\text{Hz}^{-1}$; it can be converted to $S_\lambda$ in (length)$^{-1}$ units (e.g., $\text{nm}^{-1}$) by $S_\lambda = S_\nu(c/\lambda_m^2)$, where $\lambda_m$ is the peak wavelength.

In thermal equilibrium, the forward rate of a microscopic process must equal the reverse rate of that same process, which is known as the **principle of detailed balancing**. This principle allows us to state that the photons emitted and absorbed from a continuous radiation field in equilibrium have the same spectral distribution, namely $S_\nu$. Thus we shall consider here the factors that contribute to the distribution and keep in mind that the results apply equally well to absorption and emission as long as equilibrium conditions prevail.

### Lifetime Broadening

Consider a two-level atomic system undergoing the various radiational and nonradiational processes shown in Figure 7-13. Because of emission and absorption from the radiation field and collisional processes, states $j$ and $i$ have finite lifetimes, and this gives rise to uncertainties in the energies of both states according to the Heisenberg relationship $\Delta E \Delta t \approx h/2\pi$. Because photons emitted or absorbed by the atomic system have frequencies determined by $\nu = |E_j - E_i|/h$, uncertainties in the energies of the states give rise to a frequency distribution of photons.

The upper-state lifetime $\tau_j$ is determined by radiative processes (spontaneous and stimulated emission) and collisional deactivation. In fact, the excited-state lifetime is the reciprocal of the sum of the rate constants for deactivation as given below.

$$\tau_j = (A_{ji} + B_{ji}U_\nu + k_j)^{-1}$$

In most spectroscopic experiments, except for some with high-power lasers, we can neglect stimulated emission, $(A_{ji} + k_j) \gg B_{ji}U_\nu$, and this equation becomes $\tau_j = (A_{ji} + k_j)^{-1}$. The finite upper-state lifetime leads to an

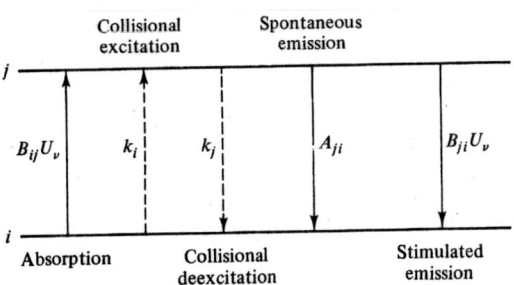

**FIGURE 7-13** Excitation/deexcitation processes in a two-level atom. Radiational processes are absorption, spontaneous emission, and stimulated emission. Nonradiational processes are collisional excitation and deexcitation. The rate constants for these processes are given in the diagram. The radiation field has energy density $U_\nu$ at frequency $\nu_{ij}$.

uncertainty in its energy $\Delta E_j$ given by

$$\Delta E_j = \frac{h}{2\pi\tau_j} = \frac{h(A_{ji} + B_{ji}U_v + k_j)}{2\pi}$$

where $\tau_j$ is substituted for $\Delta t$ in the Heisenberg relationship.

The lower-state lifetime $\tau_i$ is determined by the rate of absorption and collisional excitation as shown below.

$$\tau_i = (B_{ij}U_v + k_i)^{-1}$$

The finite lower-state lifetime leads to an uncertainty in its energy $\Delta E_i$ given by

$$\Delta E_i = \frac{h}{2\pi\tau_i} = \frac{h(B_{ij}U_v + k_i)}{2\pi}$$

The total uncertainty in frequency (the half-intensity line width or just half-width) due to lifetime effects $\Delta v_l$ is the sum of that due to the upper state and that due to the lower state:

$$
\begin{aligned}
\Delta v_l &= \frac{\Delta E_i + \Delta E_j}{h} \\
&= \frac{A_{ji} + B_{ij}U_v + B_{ji}U_v + k_i + k_j}{2\pi}
\end{aligned}
\tag{7-14}
$$

When more than two levels are involved, additional collisional and radiative excitation and deactivation processes must be considered. If the lower level is not the ground level, coupling to yet lower levels must also be considered. In this case the expressions above include a summation of $A$ coefficients. The two-level system is most closely approximated by the lowest-energy resonance line of an atom (transition from the ground state to the lowest excited state). Even here we must often consider the presence of a third level, as in the case of the Na $^2P_{3/2}$, $^2P_{1/2}$ levels. For resonance lines, the lifetime of the lower level can be considered very long compared to the excited-state lifetime. Hence in equation 7-14, $\Delta E_j \gg \Delta E_i$. Also, the field-induced rates (absorption and stimulated emission) are often negligible unless laser fields are used. Equation 7-14 then simplifies to

$$\Delta v_l = \frac{A_{ji} + k_j}{2\pi} \tag{7-15}$$

*Natural Broadening.*   The first term in equation 7-15 ($A/2\pi$) is often called the **natural line width** $\Delta v_N$.

It results from the natural or **radiative lifetime** $\tau_r$ given by

$$\tau_r = A_{ji}^{-1} \tag{7-16}$$

The broadening due to the radiative lifetime is thus called **natural broadening** and $\Delta v_N$ is

$$\Delta v_N = \frac{1}{2\pi\tau_r} = \frac{A_{ji}}{2\pi} \tag{7-17}$$

For many atoms the Einstein transition probability for the first resonance line is on the order of $10^8$ s$^{-1}$. The natural lifetime is thus on the order of $\tau_r \approx 10^{-8}$ s and $\Delta v_N \approx 1.6 \times 10^7$ Hz. In wavelength units the half-intensity width $\Delta\lambda_N = \Delta v_N \lambda_m^2/c$, where $\lambda_m$ is the wavelength of maximum intensity. This leads for Na ($\lambda_m = 5890$ Å), for example, to $\Delta\lambda_N \approx 2 \times 10^{-4}$ Å or 0.2 mÅ. Under normal conditions in media used in analyses, natural broadening is negligible compared to other effects.

Spontaneous emission of photons leads to an exponential time decay of the excited-state population. To determine the frequency distribution $S_v$ of the radiation emitted, it is necessary to convert from a time-domain description to a frequency-domain description through Fourier transformation. The Fourier transform of an exponentially damped sine wave is a Lorentzian function. Thus we are not surprised that the spectral profile that results from natural broadening is a Lorentzian or dispersion function. Mathematically, the normalized spectral profile due to natural broadening $S_{vN}$ is

$$S_{vN} = \frac{2/(\pi \Delta v_N)}{1 + [2(v_m - v)/\Delta v_N]^2} \tag{7-18}$$

where $v_m$ is the frequency at the line center. The Lorentzian profile is symmetric with respect to the line center.

*Collisional Broadening.*   The second term in equation 7-15 ($k_j/2\pi$) results from collisional deactivation of the excited state. Collisions which leave the atom in a different energy level as shown in Figure 7-13 are called **diabatic collisions**. In addition to such state-changing collisions, there can also be collisions which leave the atom in the same energy level. Such collisions are termed **adiabatic collisions**. The amount of broadening caused by collisions increases with the concentration of collision partners (perturbers). Consequently, collision broadening is sometimes called **pressure broadening**. Two types of collision partners can be distinguished. When collisions are between the atom and partners that are a different species (atoms, molecules, or ions), the broadening is called **foreign gas broadening** or **Lorentz broadening**. When collisions are between like atoms, the broadening is called **Holtzmark broadening** or **resonance broadening**.

Diabatic collisions reduce the lifetime of the initial state and give rise to an exponential decay of the excited-state population with time. The diabatic collision lifetime $\tau_d$ is given by

$$\tau_d = k_j^{-1} \tag{7-19}$$

and the diabatic collision half-width $\Delta\nu_d$ is

$$\Delta\nu_d = \frac{k_j}{2\pi} \tag{7-20}$$

Note that the collisional deactivation rate constant $k_j$ (s$^{-1}$) may be replaced by $k_j' n_x$, where $k_j'$ is the second-order rate constant (cm$^3$ s$^{-1}$) and $n_x$ is the number density of collision partners $x$ (cm$^{-3}$). The spectral line profile for a line broadened by diabatic collisions is Lorentzian and of the form of equation 7-18 with $\Delta\nu_N$ replaced by $\Delta\nu_d$. The values of $\Delta\nu_d$ are highly media dependent. For Na in flames containing Ar as a diluent, $\Delta\lambda_d$ is as small as 0.3 mÅ. For flames diluted with N$_2$, a much more effective collision partner, $\Delta\lambda_d$ can be as high as 10 mÅ (0.01 Å). In flames diabatic collision broadening is often, but not always, small compared to adiabatic collision broadening.

Adiabatic collisions are more difficult to treat theoretically, although extensive classical, semiclassical, and quantum mechanical treatments have been made. In the classical sense an adiabatic collision results in a change in phase of the atomic oscillator. Such phase-changing collisions can be characterized by a correlation time $\tau_c$, which can be interpreted as the average time between two phase-changing collisions. The more frequent the collisions, the shorter the correlation time. The half-width of the spectral distribution that results from adiabatic collisions $\Delta\nu_a$ is then

$$\Delta\nu_a = \frac{1}{\pi\tau_c} \tag{7-21}$$

and the spectral profile is Lorentzian as in equation 7-18 with $\Delta\nu_a$ replacing $\Delta\nu_N$. The reciprocal of the correlation time is the average number of effective collisions per unit time. For conditions of thermal equilibrium $\tau_c^{-1}$ is

$$\tau_c^{-1} = \bar{\nu}_r \sigma_a n_x \tag{7-22}$$

where $\bar{\nu}_r$ is the average relative velocity of the collision partners (cm s$^{-1}$), $\sigma_a$ is the *optical cross section* for *adiabatic collision broadening* (cm$^2$), and $n_x$ is the density of perturbers. Here we define $\sigma_a$ as $\pi(r_1 + r_2)^2$, where $r_1$ and $r_2$ are the radii of the collision partners. From the kinetic theory of gases, the average relative velocity can be expressed as $\bar{\nu}_r = (8kT/\pi\mu)^{1/2}$, where $\mu$ is the reduced mass of the collision partners in grams. If we use this value in equation 7-22 and substitute $\tau_c$ into equation 7-21, the adiabatic collision half-width is expressed as

$$\Delta\nu_a = \left(\frac{8kT}{\pi^3\mu}\right)^{1/2} \sigma_a n_x \tag{7-23}$$

For Na in a 2500-K flame diluted with N$_2$, the value of $\Delta\lambda_a$ is $\approx 35$ mÅ. This value greatly exceeds the natural half-width and is greater than the diabatic collision half-width. In addition to broadening the lines, adiabatic collisions can also cause a small shift in the line center (usually toward the red) and an asymmetry in the far line wings.

*Total Lorentzian Profile.* If natural and collisional broadening are assumed to be mutually independent, the resulting half-intensity width of the Lorentzian profile $\Delta\nu_L$ is $\Delta\nu_L = \Delta\nu_N + \Delta\nu_C$, where $\Delta\nu_C$ is the half-intensity width from both adiabatic and diabatic collisions. The latter is not the simple sum of $\Delta\nu_d$ and $\Delta\nu_a$, because the two types of collisions are not mutually independent. The total spectral profile $S_{\nu L}$ is then a Lorentzian of the form

$$S_{\nu L} = \frac{2/(\pi \, \Delta\nu_L)}{1 + [2(\nu_m - \nu)/\Delta\nu_L]^2} \tag{7-24}$$

Here we have neglected any shifts in the line center or asymmetry caused by adiabatic collisions. In flames and plasmas, it is usually found that $\Delta\nu_C \gg \Delta\nu_N$, so $\Delta\nu_L \approx \Delta\nu_C$.

## Doppler Broadening

The Doppler effect is an additional source of broadening in most atomic spectroscopic experiments. It results because of a statistical distribution of velocities of the emitting or absorbing atoms along the observation path. Because atoms are in motion with respect to the observation line, the Doppler effect causes a statistical distribution in the frequencies observed that is directly related to the velocity distribution. In thermal equilibrium the distribution of velocities is given by Maxwell's law, which expresses the fraction, $f(v_x) \, dv_x$, of particles moving along the $x$ axis that have velocities in the interval $v_x$ to $v_x + dv_x$. Maxwell's law predicts a Gaussian distribution of velocities given by

$$f(v_x)\,dv_x = \left(\frac{m}{2\pi kT}\right)^{3/2} e^{-mv_x^2/kT}\,dv_x$$

where $m$ is the mass of the particle, $T$ is its temperature, and $k$ is Boltzmann's constant. If the velocity of an atom does not change while radiating, the profile of a Doppler-broadened line $S_{vD}$ is also Gaussian with a half-width $\Delta v_D$:

$$S_{vD} = \frac{2\sqrt{\ln 2}}{\Delta v_D \sqrt{\pi}} e^{-4(\ln 2)(v - v_m)^2/(\Delta v_D)^2} \qquad (7\text{-}25)$$

The maximum value of the Doppler-broadened spectral profile $(S_{vD})_m$ occurs when $v = v_m$ and is given by

$$(S_{vD})_m = \frac{2\sqrt{\ln 2}}{\Delta v_D \sqrt{\pi}} = \frac{0.939}{\Delta v_D} \qquad (7\text{-}26)$$

The half-width $\Delta v_D$ is given by

$$\Delta v_D = 2\left[\frac{2(\ln 2)kT}{m}\right]^{1/2} \frac{v_m}{c} \qquad (7\text{-}27)$$

where $c$ is the speed of light. If we express $m$ by the atomic or molecular weight of the particle $M = mN$, where $N$ is Avogadro's number, and evaluate the constants in equation 7-27, the result is

$$\frac{\Delta v_D}{v_m} = \frac{\Delta \lambda_D}{\lambda_m} = 7.16 \times 10^{-7} \frac{\sqrt{T}}{\sqrt{M}} \qquad (7\text{-}28)$$

For atoms in flames $\Delta \lambda_D$ is typically $10^{-1}$ to $10^{-2}$ Å. For example, for sodium atoms with $M = 23$ g mol$^{-1}$, $T = 2500$ K, and $\lambda = 5890$ Å, a Doppler half-intensity width of $\Delta \lambda_D = 4.5 \times 10^{-2}$ Å (45 mÅ) is found from equation 7-28.

We have assumed in equations 7-25 to 7-28 that there is no change in the velocity of the species while it is radiating. This is equivalent to assuming that the mean free path between collisions is long compared to the central wavelength of the line. While this is ordinarily the case in flames, in other media the assumption may not be valid. Under these conditions one may observe a reduction in the Doppler width (collisional line narrowing) and a non-Gaussian profile.

We have also assumed in the treatment above that the spectral lines are homogeneously broadened. **Homogeneous broadening** occurs when all atoms within a specified time interval have an equal probability of absorbing or emitting any frequency within the absorption line. **Inhomogeneous broadening** occurs when different atoms are responsible for different parts of the line profile. Natural broadening is always homogeneous, but collision broadening can be inhomogeneous if the time interval for absorption of a photon is short compared to the duration of the disturbance of the atomic energy levels. The distinction between these two types of broadening is important when intense, monochromatic laser beams interact with atomic systems. In this case only those atoms having a velocity component in the beam direction that satisfies the Doppler relationship will contribute to the absorption and induced emission. Such inhomogeneous broadening can be used to generate Doppler-free spectra (see Section 17-4).

### Other Causes of Line Broadening

In addition to the sources of line broadening described above, others may be operable under special conditions. **Stark broadening** results from perturbations of the atomic system by ions, electrons, or molecules with a permanent dipole moment. It is generally negligible in flames, but can be a significant broadening source in sparks and plasmas with a high amount of ionization. Stark broadening of hydrogen emission lines is frequently used to measure electron densities in sparks and plasmas. Broadening proportional to the square of the electric field strength (quadratic Stark broadening) can be significant in sparks and plasmas. **Radiation** or **power broadening** can occur when strong radiation fields (i.e., high-powered lasers) are applied to an atomic system. High rates of induced absorption and emission lead to a reduction in the radiative lifetime which is a function of the energy density of the field (see equation 7-14). Such broadening of the Na D lines has been observed in flames irradiated by pulsed dye laser radiation. It has been predicted that under very strong, narrow bandwidth fields, a splitting rather than a broadening of the lines should occur (ac Stark effect). This effect is well known in microwave spectroscopy, but is difficult to observe in flames.

An apparent broadening occurs when a high-intensity laser is scanned across an atomic absorption line. This **saturation broadening** is not the result of change in the profile $S_v$ itself, but the result of a decrease in the population difference between the two levels under strong fields when the laser is tuned to the line center. Because the fluorescence or absorption increases less near the line center than in the wings with increasing laser power, an apparent broadening is seen under conditions of saturation.

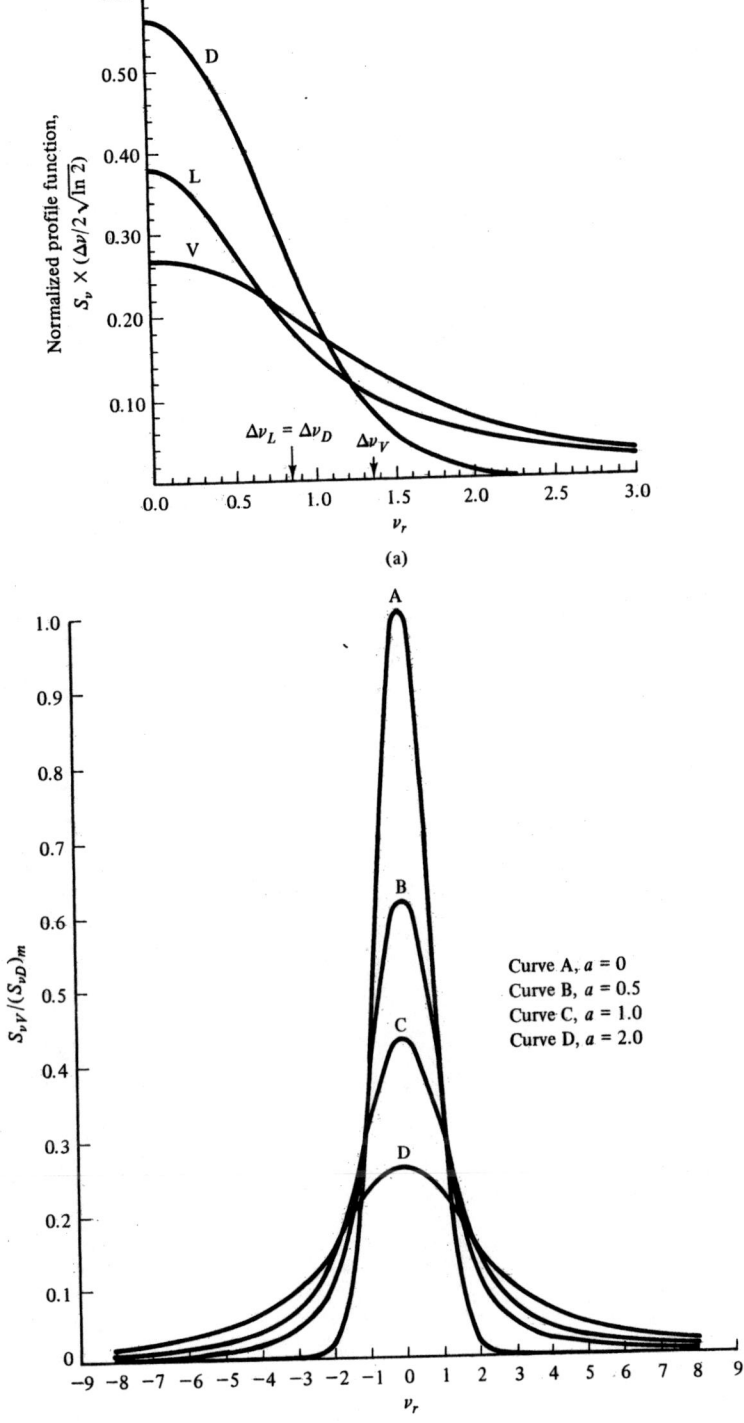

**FIGURE 7-14** (a) Normalized spectral profiles vs. relative frequency $\nu_r$ for pure Doppler (D), pure Lorentzian (L), and Voigt (V) distributions for $\Delta\nu_L = \Delta\nu_D$. The profiles are symmetric with respect to the line center. (b) Voigt profiles for various values of the $a$-parameter. Here $S_{\nu V}$ is normalized so that for pure Doppler broadening ($a = 0$), $S_{\nu V} = 1$ at $\nu_m$. Note that the Lorentz profile becomes dominant in the line wings. In fact, for $\Delta\nu_L = \Delta\nu_D$, at only two half-widths from the line center, the Lorentz profile function is 1000 times that of the Doppler function. At three half-widths, it is $10^9$ times the Doppler function.

## Overall Line Profiles

The overall profiles of most atomic spectral lines are neither purely Gaussian nor purely Lorentzian, but rather a combination known as a **Voigt profile**, $S_{\nu V}$. For simplicity, we shall assume that collisional and Doppler broadening are independent processes. Also we shall neglect any line shifts and asymmetry. Under these conditions the combined line profile is a *convolution* of the normalized Lorentzian and Gaussian profiles given by

$$S_{\nu V} = (S_{\nu D})_m \delta(a, \nu_r) \qquad (7\text{-}29)$$

The quantity $\delta(a, v_r)$ is the Voigt integral given by

$$\delta(a, v_r) = \frac{a}{\pi} \int_{-\infty}^{+\infty} \frac{e^{-y^2}}{a^2 + (v_r - y)^2} dy \qquad (7\text{-}30)$$

where $y$ is an integration variable. The **a-parameter** or **damping constant** $a$ is related to the ratio of the Lorentz half-width to the Doppler half-width by

$$a = \sqrt{\ln 2} \; \frac{\Delta v_L}{\Delta v_D} = 0.84 \frac{\Delta v_L}{\Delta v_D}$$

The parameter $v_r$ is the relative frequency compared to $\Delta v_D$ and is given by

$$v_r = \frac{2\sqrt{\ln 2}\,(v - v_m)}{\Delta v_D} = \frac{1.68(v - v_m)}{\Delta v_D}$$

In the limit as $a \to 0$, equation 7-29 simplifies to a pure Gaussian profile (equation 7-25), and as $a \to \infty$, to a pure Lorentzian profile (equation 7-24).

The Voigt distribution function (equation 7-29) is fully determined by the $a$-parameter, $v_m$, and $\Delta v_D$. We shall see in the next section that the $a$-parameter plays an important role in describing the intensity of a spectral line as a function of the atom density.

Numerical values of the Voigt integral, the ratio of the normalized Voigt profile to the normalized Doppler profile, can be found as a function of the $a$-parameter and $v_r$ in many tabulations. Within 10%, the following approximate expressions hold for the Voigt integral at the line center ($v_r = 0$):

$$\delta(a, 0) \approx (1 + 1.2a)^{-1} \qquad \text{for } 0 < a < 2$$

and

$$\delta(a, 0) \approx \frac{0.56}{a} \qquad \text{for } a > 2$$

Note that the Voigt integral and the $a$-parameter are identical if profiles are expressed in wavelength or in frequency units.

One cannot find the half-intensity width $\Delta v_V$ of the Voigt profile by simple addition of $\Delta v_L$ and $\Delta v_D$. Within about 1% the following empirical expression holds for all $a$-parameter values.

$$\Delta v_V \approx \frac{\Delta v_L}{2} + \left[ \left( \frac{\Delta v_L}{2} \right)^2 + (\Delta v_D)^2 \right]^{1/2}$$

Figure 7-14a shows an example of normalized Lorentzian, Gaussian, and Voigt profiles for $\Delta v_L = \Delta v_D$ ($a = 0.84$). It can be seen that the Lorentzian profile becomes a dominant part of the Voigt function in the line wings. For values of $a < 1$, as shown, the Voigt function resembles a Gaussian profile near the line center. Figure 7-14b shows Voigt profiles for various values of the $a$-parameter.

The convolution process implicit in the Voigt profile is just one way to account for the simultaneous action of various broadening processes. It assumes that the natural, diabatic, and adiabatic profiles are Lorentzian functions and that Doppler broadening is a Gaussian function. The major assumption is that the Gaussian and Lorentzian components are mutually independent. The mutual independence of the various broadening processes has been questioned by several authors. Quantum mechanical theories of line broadening have also been developed. These do not require the foregoing assumption, and are formally at least more satisfactory. However, they do require an interaction potential to account for the broadening effects before numerical calculations become feasible. The form of the interaction potential is not known and thus its inclusion also involves several assumptions. Experimental results of line broadening studies in flames have shown that the line core and the near wings are adequately described by the Voigt profile. However, the decrease in intensity in the far line wings has been found to deviate from the $v_r^{-2}$ dependence predicted by the Voigt profile. Similar results have been obtained in gas cells at pressures between $10^{-2}$ and 1 atm.

## 7-5 SPECTRAL LINE INTENSITIES

The intensities of atomic emission, absorption, and fluorescence lines are functions of many variables. We shall consider here general expressions for line intensities as a function of atomic population densities. In later chapters that deal with specific atomic spectrochemical methods, these expressions are expanded to include specific experimental and instrumental parameters.

### Thermal Emission

We shall find here the radiance of a spectral line emitted by a collection of atoms in thermal equilibrium. In Chapter 2 (equation 2-15), the radiant power of emission was shown to be a function of the population density of excited atoms, the number of photons emitted per second by each excited atom (the Einstein A coefficient), the energy of each photon $hv$, and the observed volume element $V$ of the emitting system. If we consider emission from atoms with upper level $j$ and lower level $i$, equation 2-15 becomes, after substitution of the Boltzmann distribution (equation 2-14),

$$\Phi_E = \frac{A_{ji}h\nu_{ji}g_j V n_M e^{-E_j/kT}}{Z(T)} \tag{7-31}$$

where $n_M$ is the total number of free atoms per cm$^3$. This simple equation holds at low population densities.

At high population densities, a significant fraction of the photons emitted by excited atoms is absorbed by atoms in the lower level. This phenomenon of **self-absorption** is particularly a problem when the lower level is the ground level (resonance transition) because the majority of atoms are present as ground-state atoms even in very hot environments. Self-absorption could be included in equation 7-31 by multiplying $\Phi_E$ by a factor that describes the fraction of emitted photons that escape the emission source without being absorbed by lower-level atoms. However, we will take here an alternative approach that includes the self-absorption phenomenon in a natural way.

The spectral radiance $B_{\lambda E}$ of any thermal emitter at temperature $T$ is related to that of a blackbody radiator $B_\lambda^b$ at the same wavelength and temperature by its emissivity $\varepsilon(\lambda)$ (see Section 4-1). Thus

$$B_{\lambda E} = B_\lambda^b \varepsilon(\lambda)$$

For a radiator in thermal equilibrium, the absorptance $\alpha(\lambda)$ equals the emissivity at the same wavelength $[\alpha(\lambda) = \varepsilon(\lambda)]$. Hence

$$B_{\lambda E} = B_\lambda^b \alpha(\lambda)$$

The spectral absorptance $\alpha(\lambda)$, often called simply the *absorption factor*, is related to the spectral transmittance of the medium by $\alpha(\lambda) = 1 - T(\lambda)$.

The total line radiance $B_E$ emitted by the thermal radiator is found by integrating $B_{\lambda E}$ over the entire line:

$$B_E = \int_{\text{line}} B_{\lambda E}\, d\lambda = \int_{\text{line}} B_\lambda^b \alpha(\lambda)\, d\lambda \tag{7-32}$$

Since atomic lines are very narrow, the spectral radiance of a blackbody is essentially constant over the line profile; thus $B_\lambda^b$ can be taken outside the integral and replaced by $B_{\lambda m}^b$, the blackbody spectral radiance at the line center $\lambda_m$:

$$B_E = B_{\lambda m}^b \int_{\text{line}} \alpha(\lambda)\, d\lambda = B_{\lambda m}^b A_t \tag{7-33}$$

where $A_t$ is called the **integral** or **total absorption** and is given by $A_t = \int_{\text{line}} \alpha(\lambda)\, d\lambda$; it has the units of length

(m, cm, nm). Thus the line radiance can be calculated if $A_t$ can be evaluated, since $B_{\lambda m}^b$ is known from Planck's law.

The absorption factor $\alpha(\lambda)$ is related to the atomic absorption coefficient $k(\lambda)$, in cm$^{-1}$ or m$^{-1}$, by Beer's law:

$$\alpha(\lambda) = 1 - e^{-k(\lambda)l} \tag{7-34}$$

where $l$ is the pathlength (cm or m). The atomic absorption coefficient can be written in terms of the population density $n_i$ (atoms cm$^{-3}$) of the lower level and Einstein coefficients or oscillator strengths as discussed in Appendix F. The most useful relationships for this discussion are equations F-26 and F-28, which are repeated here:

$$k(\lambda) = \frac{A_{ji}\lambda_m^4 g_j S_\lambda n_i}{8\pi c g_i} \tag{7-35}$$

$$k(\lambda) = \frac{e^2 \lambda_m^2 S_\lambda n_i f_{ij}}{4\varepsilon_0 m_e c^2} = 8.82 \times 10^{-13} \lambda_m^2 S_\lambda n_i f_{ij} \tag{7-36}$$

where $A_{ji}$ is the Einstein coefficient for spontaneous emission, $S_\lambda$ is the spectral profile function (we will omit the subscript $V$ for Voigt profile for simplicity), $h$ is Planck's constant, the $g$ values are the statistical weights, $\varepsilon_0$ is the permittivity of free space, $m_e$ is the mass of the electron, $c$ is the speed of light, and $f_{ij}$ is the absorption oscillator strength. The constant in equation 7-36 has the units of cm with $S_\lambda$ in cm$^{-1}$ and $\lambda$ in cm.

The spectral profile can also be expressed in terms of $(S_{\lambda D})_m$, the peak value of the Doppler profile, and $\delta(a, v_r)$, the Voigt integral, or

$$S_\lambda = (S_{\lambda D})_m \delta(a, v_r)$$

where $(S_{\lambda D})_m$ is given by $2\sqrt{\ln 2}/(\Delta\lambda_D \sqrt{\pi})$ (equation 7-26). Using this equation in equation 7-35 yields

$$k(\lambda) = \frac{(S_{\lambda D})_m \delta(a, v_r) A_{ji} \lambda_m^4 g_j n_i}{8\pi c g_i}$$

This latter equation is often written as

$$k(\lambda) = k_D^0 \delta(a, v_r)$$

where $k_D^0$ is the peak absorption coefficient for pure Doppler broadening given by

$$k_D^0 = \frac{2\sqrt{\ln 2}\, A_{ji} \lambda_m^4 g_j n_i}{8\pi c g_i\, \Delta\lambda_D \sqrt{\pi}} = \frac{A_{ji} \lambda_m^3 g_j n_i m^{1/2}}{8\pi^{3/2} g_i (2kT)^{1/2}}$$

An approximation can be made to evaluate the line radiance if the product of $k(\lambda)$ and $l$ in equation 7-34 is much smaller than unity, which usually occurs at low values of $n_i$. In this case the system is said to be **optically thin** and we can approximate $1 - e^{-k(\lambda)l}$ as $k(\lambda)l$. The line radiance is then

$$(B_E)_{k(\lambda)l \to 0} = B_{\lambda m}^b \int_{\text{line}} k(\lambda)l \, d\lambda \qquad (7\text{-}37)$$

The integral in equation 7-37 can be evaluated exactly since the only wavelength-dependent term is $S_\lambda$ and $\int_{\text{line}} S_\lambda \, d\lambda = 1$ because of normalization. Thus, under optically thin conditions, the integral absorption is given by

$$(A_t)_{k(\lambda)l \to 0} = \int_{\text{line}} k(\lambda)l \, d\lambda = \frac{A_{ji}\lambda_m^4 g_j n_i l}{8\pi g_i c} \qquad (7\text{-}38)$$

Again evaluating the constants in equation 7-38 and expressing $A_{ji}$ in terms of $f_{ij}$, we find $(A_t)_{k(\lambda)l \to 0} = 8.82 \times 10^{-13}\lambda_m^2 n_i f_{ij} l$ in cm.

If we assume that Wien's law (equation 4-7) can be used to approximate $B_{\lambda m}^b$ (stimulated emission assumed negligible), the line radiance is

$$(B_E)_{k(\lambda)l \to 0} = \frac{A_{ji}hcg_j n_i l e^{-hc/\lambda_m kT}}{4\pi\lambda_m g_i} \qquad (7\text{-}39)$$

Note that for the optically thin case, the line radiance is independent of the line shape and width and directly proportional to $n_i l$. If level $i$ is an excited level, $n_i$ is given by $n_i = n_M g_i e^{-E_i/kT}/Z(T)$. When substituted into equation 7-39, the line radiance becomes

$$(B_E)_{k(\lambda)l \to 0} = \frac{A_{ji}hcg_j n_M l e^{-E_j/kT}}{4\pi\lambda_m Z(T)}$$

or $\hspace{9cm} (7\text{-}40)$

$$(B_E)_{k(\lambda)l \to 0} = \frac{e^2 hg_i f_{ij} n_M l e^{-E_j/kT}}{2\varepsilon_0 m_e \lambda_m^3 Z(T)}$$

where $Z(T)$ is the partition function and $E_j$ is the excitation energy ($E_j = E_i + hc/\lambda_m$). Equation 7-40 also holds if level $i$ is the ground level and the transition is a resonance transition. Furthermore, for this latter case, if $T < 5000$ K and $E_j > 2$ eV, $n_M \approx n_0$ [also, $Z(T)$ in the second form of equation 7-40 becomes $g_0$].

Note that the first form of equation 7-40 is identical to the simplified equation 7-31 considered earlier. Here the line radiance in equation 7-40 can be converted to total radiant power $\Phi_E$ by multiplying by the cross-sectional area of the cell (area $\times l = V$) and the total solid angle over which emission occurs ($4\pi$ sr).

In the case of high optical thickness [$k(\lambda)l \gg 1$], an asymptotic expression can be derived for the integral absorption $A_t$. The result is

$$(A_t)_{k(\lambda)l \to \infty} = \left( \frac{e^2}{4\varepsilon_0 m_e c^2} \lambda_m^2 \, \Delta\lambda_L n_i f_{ij} l \right)^{1/2} \qquad (7\text{-}41)$$

Thus for high atomic populations [$k(\lambda)l \gg 1$], $B_E = B_{\lambda m}^b A_t$ shows a square-root dependence on population density $n_i$. Note that the value of $A_t$ and thus $B_E$ at a given density depends on the Lorentzian component of the Voigt profile $\Delta\lambda_L$. This can be understood in terms of self-absorption of the emitted radiation. At low concentrations the entire line grows with concentration as shown in Figure 7-15a. When the atomic population is high enough, a significant fraction of the emitted radiation is absorbed by atoms in the lower level. Under strong self-absorption, the center of the line profile reaches the limit given by $B_{\lambda m}^b$ [i.e., $\alpha(\lambda) = 1$]. Further growth in $A_t$ with increasing concentration then occurs mainly by expansion of the wings of the profile, which are dominated by the Lorentzian component. Thus the asymptote reached does not depend on the Doppler component. As shown in Figure 7-15b, however, the intermediate regions of the curve of growth depend on the Doppler component through the $a$-parameter. As discussed in later chapters, these *curves of growth* are closely related to analytical curves.

In media where the temperature is not homogeneous, strong resonance lines can have a dip in the center because of **self-reversal**. In arcs, sparks, plasmas, and many flames, the hot central region may be surrounded by a cooler outer zone that also contains metal vapor. Under strong self-absorption, emission lines from the hot inner core are broadened (see Figure 7-15a) compared to the absorption profile of atoms in the cooler outer zone. Hence the line center can be more strongly absorbed in this outer region, giving rise to a dip surrounded by two maxima. It is often stated that the emission profile from the inner core is broadened relative to that in the outer zone due to greater Doppler broadening. However, computer modeling has shown that the broadening effects of self-absorption are mainly responsible. Self-reversal is also found in many gas-discharge lamps used as line sources for absorption and fluorescence.

In some cases a self-reversal dip is found even under uniform temperature conditions. This is attributed to the phenomenon of **radiation diffusion**, where a photon generated in the source is absorbed and re-emitted many times in succession. Calculations have

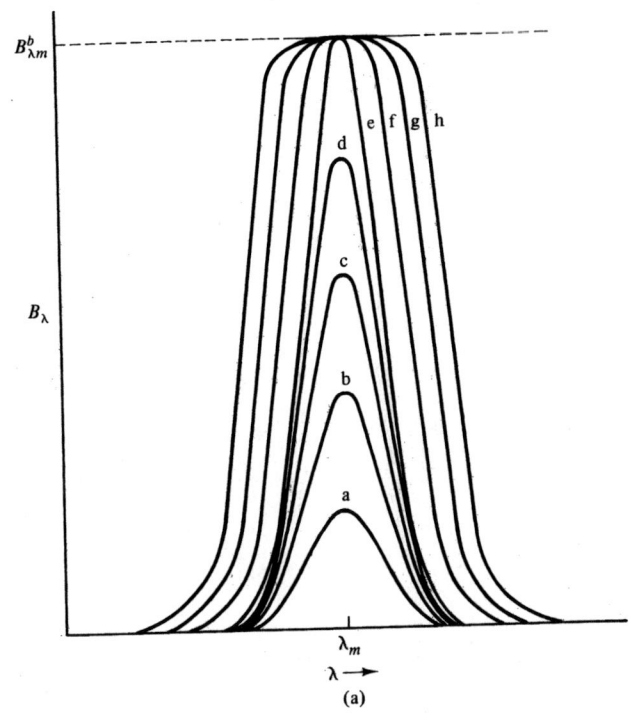

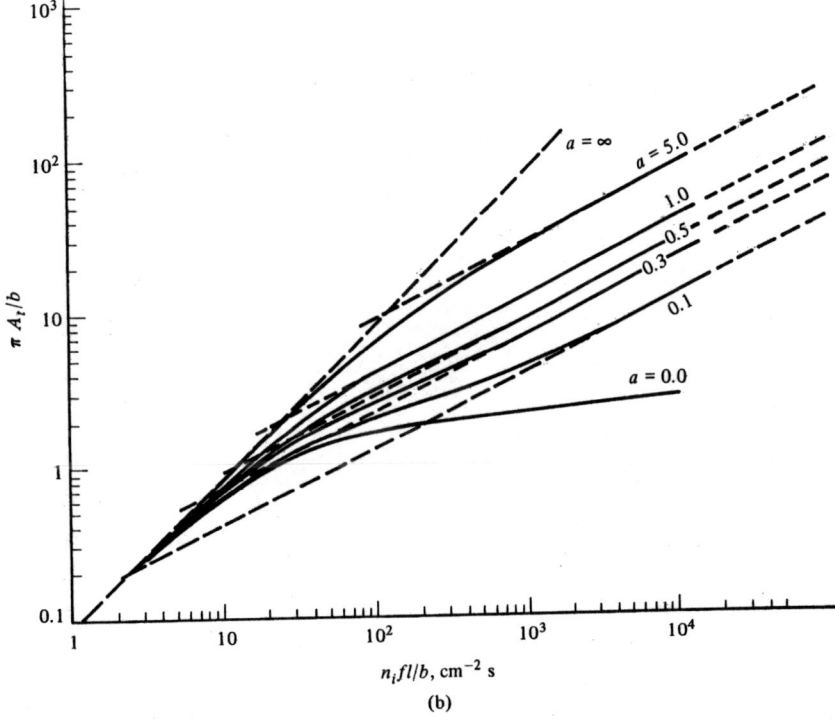

**FIGURE 7-15** Self-absorption and the curves of growth. In (a) spectral line profiles are shown as a function of increasing atomic concentration (a → h). Note that the center of the line reaches the limit given by $B^b_{\lambda m}$ at high atom densities. In (b) the growth curves are shown for different values of the $a$-parameter. A plot of $A_t$ (or some multiple of $A_t$) vs. $n_i$ (or some multiple of $n_i$) on a log-log scale is a curve of growth. For low and high atom densities all curves reach the same limiting slopes. In the intermediate regions the curve shapes depend on values of the $a$-parameter. In (b) the parameter $b = \pi \, \Delta\nu_D \sqrt{\ln 2}$.

shown that resonance lines can be reversed under conditions where deactivation by photon emission is more probable than deactivation by collisions (high quantum efficiencies). When most of the photons are converted to heat before reaching the source boundaries, we speak of **radiation trapping** or **radiation imprisonment**.

## Absorption

In the case of atomic absorption, the measurement result is expressed as the absorption factor $\alpha = 1 - T$ or the absorbance $A = -\log T = -\log (1 - \alpha)$. The transmittance $T$ is given by $\Phi/\Phi_0$, where $\Phi$ is the transmitted radiant power and $\Phi_0$ is the incident radiant power. The absorption factor $\alpha$ is the radiant power absorbed ($\Phi_0 - \Phi$) divided by the incident radiant power $\Phi_0$, or

$$\alpha = \frac{\Phi_0 - \Phi}{\Phi_0}$$

The transmitted spectral radiant power at wavelength $\lambda$ is given by Beer's law as $\Phi_\lambda = (\Phi_\lambda)_0 e^{-k(\lambda)l}$ and the total radiant power transmitted is $\Phi_0 = \int (\Phi_\lambda)_0 \, d\lambda$. Hence we can express $\alpha$ as

$$\alpha = \frac{\int (\Phi_\lambda)_0 (1 - e^{-k(\lambda)l}) \, d\lambda}{\int (\Phi_\lambda)_0 \, d\lambda} \qquad (7\text{-}42)$$

where the integral extends over the entire wavelength interval for which the incident spectral radiant power $(\Phi_\lambda)_0$ exists or is measured. In the ensuing discussion it is assumed that the absorption coefficient $k(\lambda)$ is independent of source radiant power $(\Phi_\lambda)_0$. In the case of irradiation by saturating laser beams, this assumption is invalid.

There are two different cases to be considered in evaluating $\alpha$ and $A$: the case of a narrow line source and that of a continuum source. For a narrow spectral line source (e.g., hollow cathode lamp), the spectral profile of the source is much narrower than that of the absorption coefficient, as shown in Figure 7-16a. Here $k(\lambda)$ is essentially constant over the source profile and equal to its maximum value $k_m$. Thus the absorption factor for a line source $\alpha_L$ is given by

$$\alpha_L = \frac{\int (\Phi_\lambda)_0 (1 - e^{-k_m l}) \, d\lambda}{\int (\Phi_\lambda)_0 \, d\lambda} = 1 - e^{-k_m l} \qquad (7\text{-}43)$$

The maximum absorption coefficient can readily be evaluated. It is worthwhile to define an effective width $\Delta\lambda_{\text{eff}}$ as the width of an equivalent rectangular profile with the same peak value and same area as $S_\lambda$. Thus

$$\Delta\lambda_{\text{eff}} = \frac{\int_{\text{line}} S_\lambda \, d\lambda}{S_{\lambda m}} = \frac{1}{S_{\lambda m}} \qquad (7\text{-}44)$$

where $S_{\lambda m}$ is the maximum value of $S_\lambda$. The maximum value of $k(\lambda)$ is then found from equation 7-35 or 7-36 by substituting $S_{\lambda m}$ for the profile function $S_\lambda$, and $1/\Delta\lambda_{\text{eff}}$ for $S_{\lambda m}$.

$$k_m = \frac{A_{ji}\lambda_m^4 g_j n_i}{8\pi c g_i \, \Delta\lambda_{\text{eff}}} = \frac{e^2 \lambda_m^2 n_i f_{ij}}{4\varepsilon_0 m_e c^2 \, \Delta\lambda_{\text{eff}}} \qquad (7\text{-}45)$$

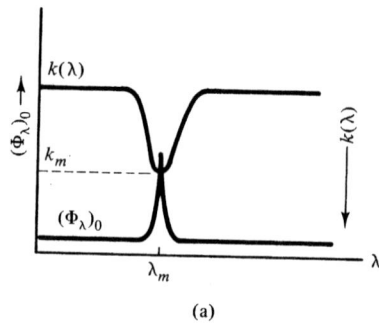

(a)

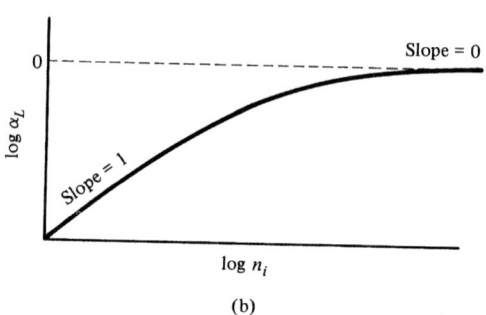

(b)

**FIGURE 7-16** Spectral profiles of source and atomic absorption coefficient (a) for a narrow line source and log-log plot of absorption factor $\alpha_L$ vs. atom density $n_i$ (b). The absorption coefficient increases top to bottom in (a), while the incident spectral radiant power increase bottom to top. Note that $k(\lambda)$ is essentially constant over the entire source profile and equal to its maximum value $k_m$.

Note that $k_m$ can also be expressed as $k_m = k_D^0 \delta(a, 0)$. By substituting the second form in equation 7-45 for $k_m$ into equation 7-43, we find

$$\alpha_L = 1 - \exp\left(\frac{-e^2 \lambda_m^2 n_i f_{ij} l}{4\varepsilon_0 m_e c^2 \, \Delta\lambda_{\text{eff}}}\right) \qquad (7\text{-}46)$$

For optically thin conditions ($n_i l \to 0$), series expansion of the exponential term in equation 7-46 gives, after evaluation of the constants,

$$(\alpha_L)_{n_i l \to 0} = \frac{8.82 \times 10^{-13} \lambda_m^2 n_i f_{ij} l}{\Delta\lambda_{\text{eff}}} \qquad (7\text{-}47)$$

Hence the absorption factor for a line source is related linearly to atom density $n_i$ for small values of $n_i$. At high atom densities $\alpha_L$ approaches its limiting value of unity as shown in log-log form in Figure 7-16b.

If absorbance is measured the peak absorbance, $A_L$, is given by $A_L = -\log(1 - \alpha_L) = 0.434 k_m l$ and from equation 7-45

$$A_L = 0.434 \frac{e^2 \lambda_m^2 n_i f_{ij} l}{(4\varepsilon_0 m_e c^2 \, \Delta\lambda_{\text{eff}})} \qquad (7\text{-}48)$$

After evaluating the constants equation 7-48 becomes

$$A_L = \frac{3.83 \times 10^{-13}\, n_i f_{ij} l}{\Delta \lambda_{\text{eff}}} \qquad (7\text{-}49)$$

Note here that $A_L$ is predicted to be strictly a linear function of $n_i$. In practice nonlinearity occurs because of the variation of $k(\lambda)$ over the source width (polychromatic radiation), because of stray radiation, and because not all light rays travel the same path $l$ through the absorbing gas. Unresolved hyperfine structure of the atomic line can also cause deviations from linearity, because of different $k(\lambda)$ values for the hyperfine components.

In the case of a continuum source (e.g., xenon arc lamp), the incident spectral profile is much broader than the absorption profile, as shown in Figure 7-17a. In fact, the incident spectral profile detected is usually determined by the spectral bandpass $s$ of the monochromator that follows the atomic vapor cell, since the monochromator bandpass is much larger than the width of the absorption profile.

To find the absorption factor $\alpha_C$ for a continuum source, we evaluate equation 7-42, where now the re-

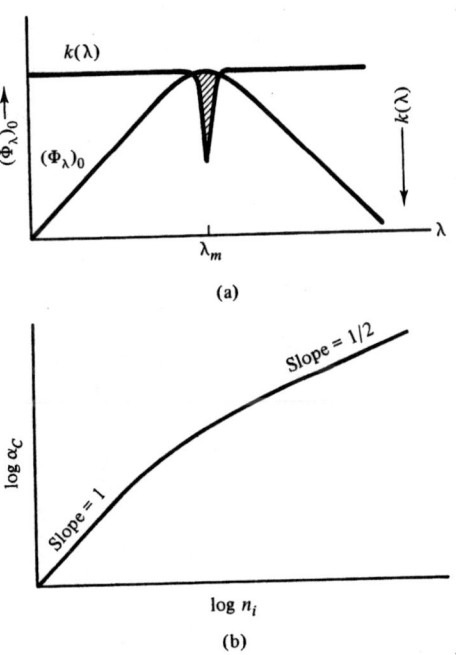

**FIGURE 7-17** Spectral profiles (a) and dependency of absorption factor $\alpha_C$ on atomic population (b) for a continuum source. The source spectral profile in (a) is determined by the monochromator slit function $t(\lambda)$ and the half-width is the monochromator bandpass $s$. Note in (a) that $(\Phi_\lambda)_0$ is essentially constant over the absorption profile. The hatched area in (a) represents the wavelengths where absorption can occur.

gion over which $(\Phi_\lambda)_0$ exists is determined by the spectral bandpass $s$ of the monochromator. We will assume that $(\Phi_\lambda)_0$ is constant over $s$ and equal to $(\Phi_{\lambda m})_0$, the spectral radiant power at the absorption line center. Since the bandpass $s$ is much larger than the absorption line width, we can express $\alpha_C$ as

$$\alpha_C = \frac{(\Phi_{\lambda m})_0 \int (1 - e^{-k(\lambda)l})\, d\lambda}{(\Phi_{\lambda m})_0 \int d\lambda} \qquad (7\text{-}50)$$

where all integrals are taken over $\lambda_0 \pm s$. Since $k(\lambda)$ is finite only over the absorption line width, the integral in the numerator is just $A_t$, the integral absorption. The integral in the denominator is just $s$, the spectral bandpass. Hence

$$\alpha_C = \frac{A_t}{s} \qquad (7\text{-}51)$$

Note that $A_t$ has the limiting values given by equation 7-38 for optically thin conditions and by equation 7-41 for optically thick conditions. A plot of log $\alpha_C$ vs. log $n_i$, for a given bandpass $s$, is shown in Figure 7-17b and has the same shape as the growth curves of Figure 7-15. For optically thin conditions, evaluating the constants in $A_t$ in equation 7-51 gives

$$(\alpha_C)_{n_i l \to 0} = \frac{8.82 \times 10^{-13} \lambda_m^2 n_i f_{ij} l}{s} \qquad (7\text{-}52)$$

This expression differs from that for a line source only in the replacement of $\Delta\lambda_{\text{eff}}$ with the monochromator spectral bandpass $s$. Because $s$ is usually much larger than $\Delta\lambda_{\text{eff}}$, $\alpha_L$ is usually larger than $\alpha_C$. This lower calibration sensitivity for continuum source AA occurs both because the integral absorption is obtained rather than the peak absorption, and because much of the incident radiant power $\Phi_0$ is not capable of being absorbed. The absorbance for a continuum source $A_C$ is only linear with $n_i$ at low values of $n_i$, unless $s$ is much smaller than the absorption profile width.

**Atomic Fluorescence**

Many of the concepts and expressions described earlier in this section apply as well to atomic fluorescence (AF) because fluorescence involves absorption of incident radiation as in AA, emission of fluorescence radiation, and possible self-absorption of the emitted radiation as in AE. Although there are many types of AF transitions, we will restrict the discussion here primarily to resonance fluorescence, which is the most widely used type. Other AF excitation/deexcitation schemes are considered in Chapter 11.

Let us consider again the two-level atomic system depicted in Figure 7-13. For AF the absorption rate $B_{ij}U_\nu n_i$ generally exceeds the collisional excitation rate $k_i n_i$ or modulation is used to discriminate against the continuous thermal emission. Hence we are not surprised that the strength of an AF line is directly proportional to the absorbed radiant power $\Phi_A$, where

$$\Phi_A = \Phi_0 - \Phi = \int (\Phi_\lambda)_0 \left(1 - e^{-k(\lambda)l}\right) d\lambda \qquad (7\text{-}53)$$

Upon excitation there are two competing paths (for the simple two-level system) for deexcitation: emission and collisional deexcitation. We shall also assume here that a conventional line or continuum source (and not a laser) is employed for excitation. Thus stimulated emission is assumed negligible and only spontaneous emission is considered. (Laser-excited AF is discussed in Chapter 11.) With these assumptions, we can calculate the efficiency of the fluorescence process and relate it to atomic parameters. The **fluorescence quantum efficiency** or **fluorescence quantum yield** $\phi$ is defined as the fraction of absorbed photons that are emitted as fluorescence photons of the appropriate wavelength. This fraction is equivalent to the probability that an atom excited by absorption will undergo the appropriate transition and emit a fluorescence photon. The **fluorescence power yield** $Y$ is the fraction of the absorbed radiant power emitted as fluorescence radiant power at the appropriate wavelength. Since radiant power is equal to flux (photons s$^{-1}$) multiplied by the photon energy, $Y$ is related to $\phi$ by $Y = \phi(h\nu'/h\nu) = \phi(\lambda/\lambda')$, where $\nu'$ and $\lambda'$ are the fluorescence frequency and wavelength, and $\nu$ and $\lambda$ are those of absorption. For a resonance transition $\lambda' = \lambda$ and hence $Y = \phi$.

The fluorescence quantum efficiency is just the rate constant for spontaneous emission divided by the sum of the rate constants for all the deexcitation pathways. For the atomic system of Figure 7-13 the efficiency $\phi$ is given by

$$\phi = \frac{A_{ji}}{A_{ji} + k_j} \qquad (7\text{-}54)$$

This equation assumes negligible stimulated emission and only a single collisional deexcitation pathway. For systems involving more than two levels, equation 7-54 can be modified by including in the sum in the denominator the rate constants for the additional radiative and nonradiative pathways. Since the radiative lifetime $\tau_r$ of the upper level as given by equation 7-16 is $\tau_r = A_{ji}^{-1}$ and its observed lifetime $\tau$ is $\tau = (A_{ji} + k_j)^{-1}$, the fluorescence quantum efficiency can

also be written

$$\phi = \frac{\tau}{\tau_r} \qquad (7\text{-}55)$$

The total fluorescence radiant power produced, $\Phi_F'$, in the volume element $V$ by absorption over pathlength $l$ is given by

$$\Phi_F' = \Phi_A Y \qquad (7\text{-}56)$$

Only a portion of this total fluorescence is collected and impingent on the detector. The fraction that strikes the detector depends on the shape of the sample container (e.g., the flame or plasma shape), the geometry of illumination, and the solid angle and fraction of the volume element illuminated and viewed as defined by the emission optics and emission wavelength selector. For simplicity we will assume that the shape of the atomizer is a parallelepiped that is fully illuminated by the excitation radiation. We will also assume that all the fluorescence from the viewing face is collected and measured.

Let us consider first the case of continuum source excitation where $(\Phi_\lambda)_0$ in equation 7-53 can be considered to be constant over the absorption line and equal to $(\Phi_{\lambda m})_0$ the spectral radiant power at the line center. Hence we can express $\Phi_A$ in the general fluorescence equation 7-56 as

$$\Phi_A = (\Phi_{\lambda m})_0 \int \left(1 - e^{-k(\lambda)l}\right) d\lambda = (\Phi_{\lambda m})_0 A_t \qquad (7\text{-}57)$$

Substituting this result into equation 7-56 gives

$$\Phi_{FC}' = (\Phi_{\lambda m})_0 Y A_t \qquad (7\text{-}58)$$

For optically thin conditions, equation 7-38 gives $A_t$. If we convert from $A_{ji}$ to $f_{ij}$ in equation 7-38 and substitute the resulting $A_t$ expression into equation 7-58, the result is

$$(\Phi_{FC}')_{n_i l \to 0} = \frac{e^2 \lambda_m^2 l (\Phi_{\lambda m})_0 Y n_i f_{ij}}{4\varepsilon_0 m_e c^2} \qquad (7\text{-}59)$$

Evaluation of the constants gives

$$(\Phi_{FC}')_{n_i l \to 0} = 8.82 \times 10^{-13} \lambda_m^2 (\Phi_{\lambda m})_0 Y n_i f_{ij} l \qquad (7\text{-}60)$$

where the constant has the units cm. Note that under optically thin conditions, the fluorescent radiant power is directly proportional to $n_i$ and to the source spectral radiant power. Equation 7-59 is often expressed in terms of the source spectral radiance by including the solid angle of collection of the source radiation and the area of the source viewed.

At large atomic concentrations (i.e., high $n_i$), $A_t$ in equation 7-58 varies as the square root of $n_i$ (see equation 7-41). Thus under these conditions, $\Phi'_{FC}$ is proportional to $\sqrt{n_i}$.

If the source is a line source that emits at $\lambda_m$, from equation 7-43 we have

$$\Phi_A = \Phi_0\alpha_L = \Phi_0(1 - e^{-k_m l})$$

where $\Phi_0$ is the incident source radiant power for the source line providing excitation. If this result is substituted into equation 7-56, the radiant power of fluorescence with line source excitation is obtained:

$$\Phi'_{FL} = \Phi_0\alpha_L Y \qquad (7\text{-}61)$$

Under optically thin conditions, $\alpha_L$ is given by $k_m l$, where $k_m$ is defined in equation 7-45. Thus

$$(\Phi'_{FL})_{nil\to 0} = \frac{e^2\lambda_m^2\Phi_0 Y l n_i f_{ij}}{4\varepsilon_0 m_e c^2 \,\Delta\lambda_{\text{eff}}} \qquad (7\text{-}62)$$

After evaluating the constants the latter expression becomes

$$(\Phi'_{FL})_{nil\to 0} = \frac{8.82 \times 10^{-13}\lambda_m^2\Phi_0 Y l n_i f_{ij}}{\Delta\lambda_{\text{eff}}} \qquad (7\text{-}63)$$

where again the constant has the units cm. The observed radiant power of fluorescence is predicted to be directly proportional to $n_i$ and to the integrated spectral radiant power of the source. By comparing equation 7-62 to equation 7-59 for the continuum source, we see that if $\Phi_0/\Delta\lambda_{\text{eff}} \gg (\Phi_{\lambda m})_0$, line source AF should produce more fluorescence radiant power than continuum source AF.

When $n_i$ is large, $\alpha_L$ in equation 7-61 approaches its limiting value of unity. Thus $\Phi'_{FC}$ becomes independent of $n_i$ because all the excitation radiation is absorbed.

Even under idealized conditions, only a portion of the fluorescent radiation in the volume element excited can be observed because of self-absorption (i.e., the fluorescence photons are absorbed by other analyte atoms before they pass out of the atomizer). Thus the observed total fluorescence $\Phi_F$ is given by

$$\Phi_F = \Phi'_F f_s \qquad (7\text{-}64)$$

where $f_s$ is the fraction of fluorescent photons that leave the sample container without being reabsorbed. The prime symbol distinguishes the fluorescence radiant power produced in volume element $V$ from that observed.

We will consider only the qualitative effects of self-absorption on resonance fluorescence growth curves. Exact expressions have been derived, but they are complex and highly dependent on cell geometry. As a first approximation, the self-absorption factor $f_s$ exhibits the proportionality

$$f_s \propto \frac{\int (1 - e^{-k(\lambda)L})\,d\lambda}{\int k(\lambda)L\,d\lambda} = \frac{A'_t}{(A'_t)_{niL\to 0}} \qquad (7\text{-}65)$$

where $A'_t$ is the integral absorption of the fluorescence beam over emission distance $L$ and $(A'_t)_{niL\to 0}$ is the integral absorption in the optically thin limit. Here $L$ is the width of the illuminating beam and of the atomizer. We can see that at low population densities, since $e^{-k(\lambda)L} \approx 1 - k(\lambda)L$, $f_s \to 1$, while at high population densities $f_s \propto 1/\sqrt{n_i}$ since in this case $A'_t \propto \sqrt{n_i}$ (see equation 7-41) and $(A'_t)_{niL\to 0} \propto n_i$.

At low population densities $\Phi_F = \Phi'_F$ and $\Phi_{FL}$ and $\Phi_{FC}$ are proportional to $n_i$. At high population densities with continuum source illumination, $\Phi_{FC}$ becomes independent of $n_i$ because $\Phi'_{FC}$ or $A_t$ is proportional to $\sqrt{n_i}$, while $f_s$ is proportional to $1/\sqrt{n_i}$. A log-log plot of $\Phi_{FC}$ vs. $n_i$ should then exhibit a limiting slope of unity at low concentrations and a slope of zero at high concentrations, as shown in Figure 7-18.

With a line source and high population densities, $\Phi_{FL}$ is predicted to be proportional to $1/\sqrt{n_i}$ because $\Phi'_{FL}$ or $\alpha_L$ is independent of $n_i$ while $f_s$ is proportional to $1/\sqrt{n_i}$. Thus log-log plots of $\Phi_{FL}$ vs. $n_i$ should show limiting slopes of 1 at low concentrations and $-\frac{1}{2}$ at high concentrations, as can also be seen in Figure 7-18.

We can summarize the dependencies of emission, absorption, and fluorescence line strengths on the lower-state population density $n_i$ as in Table 7-4. Note that under optically thin conditions all techniques give linear relationships.

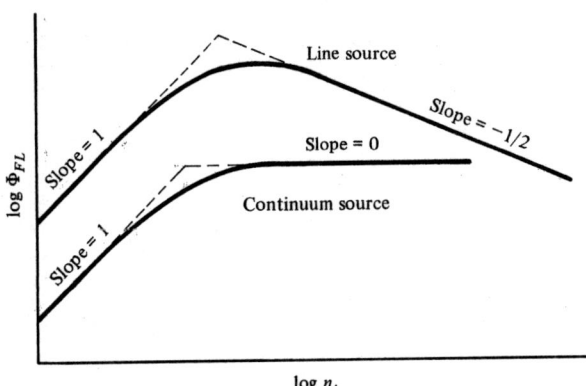

**FIGURE 7-18** Growth curves for atomic fluorescence for continuum source and line source excitation.

**TABLE 7-4**

Dependencies of line strengths on atomic population densities[a]

| | AE | AAL | AAC | AFL | AFC |
|---|---|---|---|---|---|
| Optically thin limit | $B_E \propto n_i$ | $\alpha_L \propto n_i$ | $\alpha_C \propto n_i$ | $\Phi_{FL} \propto n_i$ | $\Phi_{FC} \propto n_i$ |
| | | $A_L \propto n_i$ | $A_C \propto n_i$ | | |
| High density limit | $B_E \propto \sqrt{n_i}$ | $\alpha_L \approx 1$ | $\alpha_C \propto \sqrt{n_i}$ | $\Phi_{FL} \propto 1/\sqrt{n_i}$ | $\Phi_{FC} \neq f(n_i)$ |
| | | $A_L \propto n_i$ | | | |

[a]AE, Atomic emission; AAL, line source atomic absorption; AAC, continuum source atomic absorption; AFL, line source atomic fluorescence; AFC, continuum source atomic fluorescence.

## PROBLEMS

**7-1.** Find the ground-state term symbols for the following atoms and show how you arrived at them.

Ne, B, Li, Ar, Ca, Al

**7-2.** The total pressure of sodium (atoms plus ions) in a 2800-K flame is $1.0 \times 10^{-8}$ atm. The ionization energy of Na is 5.14 eV, and the equilibrium constant for ionization is $K_i = 7.4 \times 10^{-8}$ atm. Find the fraction of Na present as nonionized atoms for the following conditions.
   (a) All electrons in the flame arise from Na ionization.
   (b) The partial pressure of electrons is buffered at $10^{-7}$ atm by adding potassium.

**7-3.** Two solutions used in flame emission spectrometry with an air–$C_2H_2$ flame contain identical amounts of calcium. Solution 1 is a standard, and solution 2 is the sample. For each of the measurement conditions below, compare the **atomic** ground-state number densities with solutions 1 and 2 in a qualitative manner $[(n_0)_1 = (n_0)_2, (n_0)_1 > (n_0)_2,$ relative populations indeterminant, etc.]. Justify your answers.
   (a) The flow rate of air, the nebulizing gas, is higher when solution 2 is sprayed than when solution 1 is sprayed.
   (b) The viscosity of solution 2 is greater than that of solution 1.
   (c) Solution 2 contains 1000 µg mL$^{-1}$ KCl in addition to calcium.
   (d) Both solutions contain a large excess of phosphate. However, solution 2 was made 0.01 $M$ in EDTA prior to introducing it into the flame.

**7-4.** Give a short (one or two sentence) *scientific* reason why the statements below are true.
   (a) Spectral lines in flames often have half-widths (full widths at half maximum intensity) on the order of 0.005 to 0.01 nm, whereas natural broadening is predicted to give a line width on the order of $2 \times 10^{-5}$ nm.
   (b) A monochromator with a spectral bandpass < 0.1 nm is usually found in atomic emission systems, while in atomic absorption systems the monochromator often has a spectral bandpass in the range 0.2 to 1.0 nm.

**7-5.** For the Ca 422.7-nm resonance line in a flame at 3000 K, calculate the Doppler half-width in nanometers.

**7-6.** For Na atoms in a 2500 K flame diluted with $N_2$ at atmospheric pressure, the value of the adiabatic collision half-width for the 589.0 nm D line is approximately 35 mÅ. Find the optical cross section for adiabatic collision broadening $\sigma_a$ in Å$^2$.

**7-7.** Oscillator strengths for absorption transitions are often tabulated as $gf$ values, where $g$ is the statistical weight of the lower level and $f$ is the emission oscillator strength. For the Na $^2S_{1/2} \rightarrow {}^2P_{3/2}$ transition at 5890 Å, $gf = 1.3$. Find the value of the Einstein transition probability (see Appendix F for the relationship between $A$ and $f$), and the radiative lifetime of the upper level.

**7-8.** Consider flame atomic spectrometric measurements of Cu at 324.7 nm in a typical stoichiometric $N_2O/C_2H_2$ flame. (See Table 8-3 for temperature and stoichiometry information.) We will be concerned with transitions between the first excited state and the ground state at $\lambda = 324.7$ nm. Assume that the solution transport rate is 5.0 cm$^3$ min$^{-1}$, $\varepsilon_a = 0.030$, $\varepsilon_n = 0.050$, $Q = 10$ L min$^{-1}$, $Z(T) = g_0$, $A_{10} = 0.95 \times 10^8$ s$^{-1}$, $g_0 = 2$, and $g_1 = 4$. The test solution is a 2.0 µg mL$^{-1}$ aqueous solution of Cu. Calculate the following values.
   (a) The values of $n_0$ and $n_1$.
   (b) The total radiant power emitted in 1.0 cm$^3$, $\Phi_E$.
   (c) The radiance due to Cu emission from a 1.0 cm$^3$ cube of atomic vapor, $B_E$.

**7-9.** For the same system and conditions as in problem 7-8, calculate the following values.
   (a) The natural, Doppler, collisional, total Lorentzian, and total (Voigt) half-widths. (Calculate in Hz and in nm, and assume that adiabatic collisions are dominant and that the average atomic mass of the species colliding with Cu is 23 g, the number density of perturbers is $2.3 \times 10^{18}$ cm$^{-3}$, and the collision cross section is 300 Å$^2$.)

**(b)** The *a* parameter.

**(c)** The value of the Voigt integral at the maximum.

**(d)** The maximum value of the Doppler-broadened profile in $Hz^{-1}$ and $nm^{-1}$.

**(e)** The maximum value of the Voigt profile in $Hz^{-1}$ and $nm^{-1}$.

**(f)** The effective half-width in nm.

**(g)** The peak absorption coefficient for pure Doppler broadening and for the Voigt profile.

**(h)** The fraction absorbed $\alpha_L$ and the absorbance with a line source, $A_L$, with a burner of 5 cm pathlength.

**(i)** The total fluorescence radiant power emitted from a 1.0-$cm^3$ cube of atomic vapor if the incident radiant power is 1.0 mW and the power yield is 0.2.

**7-10.** The accompanying figure is a partial term diagram for Hg. Use the information in the diagram to answer the following questions.

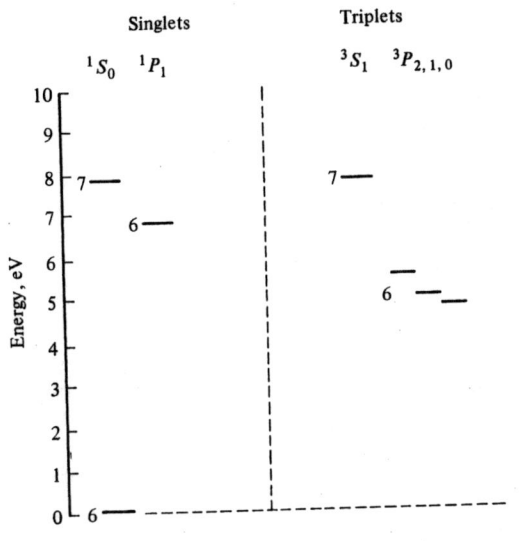

**(a)** List the allowed transitions and justify according to the selection rules.

**(b)** If an external continuum source is used to observe the absorption of Hg atoms in a room-temperature vapor, list the transitions that will be seen if the selection rules are strictly obeyed.

**7-11.** If the $^3P_{2,1,0}$ states of Hg are thermally populated in a flame at 2700 K, calculate the ratio of the populations of the $^3P_2$ and $^3P_0$ levels. The splitting is 6398 $cm^{-1}$.

**7-12.** Compare continuous and noncontinuous atomizers.

**7-13.** Discuss why the "optimum" temperature might be different for AA, AE, or AF measurements of a given element.

**7-14.** Why do the shapes of AA and AF calibration curves differ between line and continuum source excitation?

**7-15.** For CaOH, the dissociation energy $E_d = 4.3$ eV. Find the dissociation constant $K_d$ at 1000, 1500, and 2000 K. You can assume that the ratio $Z_{Ca}Z_{OH}/Z_{CaOH} = 1$.

**7-16.** For LiOH, the dissociation energy $E_d = 103$ kcal $mol^{-1}$. If the partition function ratio is unity, find $K_d$ at 2000 K.

# REFERENCES

The following are reference books and reviews dealing with atomic structure, atomic spectra, and atomic energy levels.

1. H. G. Kuhn, *Atomic Spectra*, Academic Press, New York, 1962. A very readable book on atomic spectra.

2. E. U. Condon and G. H. Shortley, *The Theory of Atomic Spectra*, Cambridge University Press, London, 1967. One of the classics on the quantum mechanical treatment of atomic spectra. Particularly good on term symbols and coupling modes.

3. G. Herzberg, *Atomic Spectra and Atomic Structure*, Dover, New York, 1944. A classic.

4. J. J. Devlin, "Origins of Atomic Spectra," in *Analytical Emission Spectroscopy*, vol. 1, pt. II, E. L. Grove, ed., Marcel Dekker, New York, 1971, pp. 73–130.

5. A. C. G. Mitchell and M. W. Zemansky, *Resonance Radiation and Excited Atoms*, Cambridge University Press (New York and London), 1961. The classic work on atomic spectroscopy.

The following are references having a great deal of data on energy levels, transition probabilities, and wavelengths of spectral lines.

6. S. Bashkin and J. O. Stoner, Jr., *Atomic Energy Levels and Grotrian Diagrams*, vols. 1–4, North-Holland, Amsterdam, 1975–1982.

7. C. E. Moore, *Atomic Energy Levels*, vols. I–III, NBS Circular 467 (vol. I), NSDRS-NBS-35 (vols. II, III), National Bureau of Standards, Washington, D.C., 1949 (vol. I), 1970 (vols. II, III).

8. C. H. Corliss and W. R. Bozman, *Experimental Transition Probabilities for Spectral Lines of Seventy Elements*, NBS Monograph 53, National Bureau of Standards, Washington, D.C., 1962.

9. W. L. Wiese, M. W. Smith, and B. M. Glennon, *Atomic Transition Probabilities, Hydrogen through Neon*, NSDRS-NBS-4, vol. I, National Bureau of Standards, Washington, D.C., 1969.

10. W. L. Wiese, M. W. Smith, and B. M. Miles, *Atomic Transition Probabilities, Sodium through Calcium,* NSDRS-NBS-22, vol. II, National Bureau of Standards, Washington, D.C., 1969.

11. J. Reader, C. H. Corliss, W. L. Wiese, and G. A. Martin, *Wavelengths and Transition Probabilities for Atoms and Atomic Ions,* NSDRS-NBS-68, National Bureau of Standards, Washington, D.C., 1980.

12. W. F. Meggers, C. H. Corliss, and B. F. Scribner, *Tables of Spectral Line Intensities,* pt. I, *Arranged by Element;* pt. II, *Arranged by Wavelength,* U.S. Government Printing Office, Washington, D.C., 1961.

13. G. R. Harrison, *Wavelength Tables,* MIT Press, Cambridge, Mass., 1969.

14. A. N. Zaidel, V. K. Prokof'ev, S. M. Raiskii, V. A. Slavnyi, and E. Ya. Shreider, *Tables of Spectral Lines,* Plenum Press, New York, 1970.

The references below deal with the principles of analytical atomic spectroscopy.

15. C. Th. J. Alkemade, Tj. Hollander, W. Snelleman, and P. J. Th. Zeegers, *Metal Vapors in Flames,* Pergamon Press, Elmsford, N.Y., 1982. Everything you wanted to know about line broadening, intensities of spectral lines, and anything fundamental in flames.

16. C. Th. J. Alkemade and R. Herrmann, *Fundamentals of Analytical Flame Spectroscopy,* Halsted Press, New York, 1979. An excellent discussion of flame spectrometry from a fundamental, yet qualitative viewpoint. Particularly good discussion of interferences in atomic spectrometry.

17. P. W. J. M. Boumans, *Theory of Spectrochemical Excitation,* Adam Hilger, Bristol, England, 1966. Another classic reference on atomic spectrometric theory.

18. G. F. Kirkbright and M. Sargent, *Atomic Absorption and Fluorescence Spectroscopy,* Academic Press, London, 1974. Includes a chapter on spectroscopic theory and chapters on the theory of AA and AF measurements. Also includes a chapter on introduction of liquids into flames.

19. C. Th. J. Alkemade and P. J. Th. Zeegers, "Excitation and De-excitation Processes in Flames," in *Spectrochemical Methods of Analysis,* J. D. Winefordner, ed., Wiley-Interscience, New York, 1971, pp. 3–125. An excellent treatment of the processes occurring in flames from a fundamental point of view.

20. T. J. Vickers and J. D. Winefordner, "Flame Spectrometry," in *Analytical Emission Spectroscopy,* vol. 1, pt. II, E. L. Grove, ed., Marcel Dekker, New York, 1972.

The following deal with more specialized topics of concern in this chapter.

21. P. J. Th. Zeegers, R. Smith, and J. D. Winefordner, "Shapes of Analytical Curves in Flame Spectroscopy," *Anal. Chem.,* 40, 13 (1968).

22. J. D. Winefordner, V. Svoboda, and L. J. Cline, "Critical Comparison of Flame Spectrometric Methods," *CRC Crit. Rev. Anal. Chem.,* 1, 233 (1970).

23. J. D. Winefordner, "Spectroscopic Methods," in *Trace Analysis: Spectroscopic Methods for Elements,* chap. IV, J. D. Winefordner, ed., Wiley-Interscience, New York, 1976.

24. A. Syty, "Developments in Methods of Sample Injection and Atomization," *CRC Crit. Rev. Anal. Chem.,* 4, 155 (1974).

25. E. J. Meehan, "Optical Methods: Emission and Absorption of Radiant Energy," in *Treatise on Analytical Chemistry,* 2nd ed., pt. 1, vol. 7, chap. 1, I. M. Kolthoff and P. J. Elving, eds., Wiley, New York, 1981.

26. R. F. Browner and A. W. Boorn, "Sample Introduction: The Achilles Heel of Atomic Spectroscopy," *Anal. Chem.,* 56, 787A (1984). This article and the following one are excellent discussions of the problems and techniques of sample introduction in atomic spectrometry.

27. R. F. Browner and A. W. Boorn, "Sample Introduction Techniques for Atomic Spectroscopy," *Anal. Chem.,* 56, 875A (1984).

28. R. K. Skogerboe and S. J. Freeland, "Experimental Characterization of Aerosol Production, Transport, Vaporization, and Atomization Systems: Part I. Factors Controlling Aspiration Rates; Part 2. Factors Controlling Aerosol Size Distributions Produced," *Appl. Spectrosc.,* 39, 916 (pt. I), 920 (pt. II) (1985).

29. R. K. Skogerboe and S. J. Freeland, "Effect of Solution Composition on the Physical Characteristics of Aerosols Produced by Nebulization," *Appl. Spectrosc.,* 39, 925 (1985).

30. L. deGalan, "New Directions in Optical Atomic Spectrometry," *Anal. Chem.,* 58, 697A (1986).

31. M. L. Parsons, B. W. Smith, and G. E. Bentley, *Handbook of Flame Spectroscopy,* Plenum Press, New York, 1975.

32. R. W. B. Pearse and A. G. Gaydon, *The Identification of Molecular Spectra,* 3rd ed., Chapman & Hall, London, 1963.

33. "Atomic Absorption, Atomic Fluorescence, and Flame Emission Spectrometry." This review appears every two years in the Fundamental Reviews issue of *Analytical Chemistry.*

# CHAPTER 8

# Flame and Plasma Atomic Emission Spectrometry

Emission spectrometry has great potential as a qualitative and quantitative tool since all elements can be made to emit characteristic spectra under the appropriate circumstances. Unfortunately, no single excitation source exists that can excite all the elements in an optimum manner. This chapter considers emission spectroscopy with flame and plasma excitation. These two techniques are considered together because instrumental and sample introduction considerations are similar.

Flame emission is the first of the atomic spectrochemical techniques that we will consider. It is not only the simplest atomic spectrochemical method, but also the oldest; the first observations of the colors produced by metallic salts in flames date back to the eighteenth century. The pioneering work of Kirchhoff and Bunsen in the mid-1800s led to a general recognition of the power of flame emission for the qualitative identification of many elements. In fact, these workers discovered the elements cesium, thallium, indium, and gallium by spectrochemical methods. Quantitative flame emission was not fully established until the 1920s, when the Swedish plant physiologist Lundegardh developed quantitative methods for studying plant metabolism. Commercial interest soon followed, and the first commercially available flame emission instrument was introduced in the mid-1930s by Siemens and Zeiss in Europe. Flame emission rapidly grew to become an important analytical technique, particularly after the introduction of Gilbert's total consumption burner by Beckman Instruments in 1951.

In the 1940s interest developed in materials that were difficult to analyze by flame emission methods. The war years provided great impetus for the development of more energetic excitation sources than common flames. Hence arc and spark discharges became important sources particularly for the analysis of solids and refractory materials. We discuss arc and spark techniques in detail in Chapter 9.

The atomic absorption (AA) technique, discussed in Chapter 10, was introduced as a new flame spectrometric method in 1955, and commercial instruments for AA became widely available in the early 1960s. The growth of the AA method, and particularly the growth in commercial AA instruments, soon relegated flame emission to a lesser role. At present flame emission is little used except for easily excited elements such as alkali metals.

Interest in new excitation sources increased dramatically in the 1960s with the development of stable high-frequency and dc plasma devices. Like the combustion flame, plasma devices are relatively easy and convenient to operate and capable of high precision.

Flame emission remains useful in determining alkali metals and other easy-to-excite elements. However, common flames are not very energetic and this limits the detectability of elements with high excitation energies. Because plasma sources are much more energetic than common flames, they can provide higher atomization efficiencies, high degrees of excitation, and lower detection limits for many elements. Following the introduction of commercial inductively coupled and dc plasmas in the mid-1970s, these sources became rapidly accepted as analytical tools. Plasma sources now dominate the sales of emission spectrometers; most of the research on emission sources deals with plasma devices.

This chapter begins with a consideration of the ideal source for emission spectrochemical analysis. The characteristics of the ideal source provide a basis for determining the strengths and weaknesses of the real sources that are used. Flames are next considered as excitation sources. Their structures, combustion processes, temperatures, and spectral background emission are important not only in flame emission, but also in flame AA and flame atomic fluorescence (AF). Plasma atomic emission sources, including the inductively coupled plasma, the dc plasma, and the microwave plasma, are then described and their characteristics compared to those of the ideal source. Complete atomic emission spectrometers useful in single- and multielement determinations are introduced and described. To gain some feeling for how the readout signal in emission spectrometry depends on instrumental parameters and analyte characteristics, mathematical expressions are developed for the signals obtained, and signal-to-noise considerations in emission spectrometry are discussed. The $S/N$ expressions are useful in predicting conditions for achieving high precision or low detection limits. The factors that influence linearity, accuracy, precision, and detection limits are developed next. This chapter concludes by describing how the practical methodology, discussed in Chapter 6, is adapted to meet the specific requirements of plasma and flame emission methods. The determination of Na and K in blood serum is used to illustrate several practical aspects of flame emission spectrometry. Similarly, the application of plasma emission methods to water testing and marine sediment analysis illustrates some of the considerations with these sources.

## 8-1  THE IDEAL ATOMIC EMISSION SPECTROMETRIC SYSTEM

The analytical chemist has long sought a single instrumental technique that would be capable of providing qualitative and quantitative information about the ma-

jor, minor, and trace constituents of a sample. Because all elements emit characteristic spectral lines whose radiances are proportional to concentration, at least over a limited range, emission spectroscopy is potentially a very powerful elemental analysis technique. In this section we investigate the properties and characteristics of the ideal atomic emission spectrometric system in order to place the actual excitation sources and spectrometers discussed later into proper perspective.

### Information Desired from Emission Spectra

What type of information would the ideal emission system provide? Since we are dealing with techniques capable of atomizing samples, we will obtain information concerning mainly the atomic composition of the sample; we will leave it to molecular spectrometric techniques to give us information concerning the molecular makeup of the sample. Information that would allow identification of the elements in the sample is certainly desirable. Thus the ideal emission system should be capable of rapid, qualitative elemental analysis so that samples with unknown constituents can be identified and rapid comparisons made of different samples. With the prevailing interest in trace constituents, the technique should cover a wide concentration range (pg mL$^{-1}$ to mg ml$^{-1}$, or nine orders of magnitude, would be desirable).

For quantitative determinations of a single element, we now demand that trace methods be capable of accuracies and relative standard deviations of a few percent. Beyond this, however, an ideal system would be capable of true, simultaneous multielement determinations with similar precision and accuracy. We can imagine for elemental profiling purposes (moon rocks or pollution studies, for example) that it would be desirable to determine as many as 50 elements simultaneously. Although not all situations are this demanding, the ability to determine one element at a time or many elements simultaneously would give the technique great flexibility. These lofty ideals are not satisfied at present by any existing emission system. However, modern emission sources and spectrometers are quite versatile tools that can give a wealth of information.

### Characteristics of the Ideal Emission Source

In emission spectrometry, the "emission" source actually serves a dual purpose. First, it must provide the necessary energy to atomize the analyte with high efficiency (ideally, 100% atomization efficiency). As we have seen in Chapter 7, this is no easy task. Second, the source must supply the energy needed to excite the element or elements of interest. Thus the ideal emission

**TABLE 8-1**
Characteristics of the ideal atomic emission source

1. Complete atomization of all elements
2. Controllable excitation energy
3. Sufficient excitation energy to excite all elements
4. Inert chemical environment
5. No background
6. Accept solutions, gases, or solids
7. Tolerant to various solution conditions and solvents
8. Simultaneous multielement analysis
9. Reproducible atomization and excitation conditions
10. Accurate and precise analytical results
11. Inexpensive to purchase and maintain
12. Ease of operation

source should be an ideal atomizer as well as an ideal excitation source. Table 8-1 lists a few of the characteristics of such an ideal emission source. In addition to the primary features discussed above, we want the source to have an excitation energy that can be controlled by the user. This would allow optimization of source conditions for the determination of easy-to-excite elements as well as those with high excitation energies. Atomization and excitation should occur in an inert chemical environment to minimize compound formation and background emission. With sufficient energy and an inert environment, the chemical, physical, and spectral interferences that plague many techniques should be minimal. We would also like the source to be able to accept samples as solids, liquids, and gases with minimal (ideally no) sample pretreatment. It should be tolerant of samples that contain organic solvents, are viscous, or contain particulates. The source should maintain reproducible atomization and excitation conditions in order to give precise and accurate single element and multielement results. Finally, the source should be inexpensive to purchase and maintain, and quite simple to operate. Needless to say, the real sources that will be considered in this chapter do not meet all the criteria listed in Table 8-1. However, as we shall see, some sources approach these ideals and, indeed, this is a major reason for their popularity.

**Atomizer Temperature**

Atomizer temperature not only influences the population of ground-state atoms through the atomization efficiency, but also the fraction of excited atoms through the Boltzmann factor $e^{-E_i/kT}$. Table 8-2 gives the Boltzmann factor for several different atomizer temperatures and excitation energies. If the upper and lower levels are of equal statistical weight, the factors shown are equal to the fraction of atoms excited. It should be noted that in highly ionized plasmas, even though the Boltzmann factors are relatively high, the absolute neutral atom populations can be low because of significant ionization.

## 8-2 FLAME ATOMIC EMISSION SOURCES

The flames used in spectrochemical analysis are hot, chemical flames in which a highly exoergic, autocatalytic chemical reaction takes place in the gas phase between a fuel (hydrogen, acetylene, propane) and an oxidant (oxygen, air, nitrous oxide). After the flame is ignited, it propagates rapidly and spontaneously through the combustible mixture. Flames are made stationary by means of a burner which supplies the reacting gases and supports the combustion reaction. The fuel and oxidant are continuously supplied to the burner and ignited on top of the burner head. The flames are usually **premixed flames,** in which the fuel and oxidant are mixed prior to the region where combustion takes place, but they can be **unpremixed flames,** in which the fuel gas and oxidant are first mixed in the combustion region itself. It is extremely important to control gas flow rates carefully. Normally, the fuel and oxidant pressures from compressed gas cylinders are regulated by a two-stage regulator. Instruments that use flames also have pressure gauges or flow meters preceeding the nebulizer/ burner for fine-adjustment purposes.

In flame emission, the sample to be analyzed is usually nebulized into the high-temperature environment

**TABLE 8-2**
Boltzmann factors for several excitation energies and atomizer temperatures

| Resonance line wavelength, $\lambda_m$ (nm) | $E_i$, J (eV) | $e^{-E_i/kT}$ | | | | |
|---|---|---|---|---|---|---|
| | | 2000 K | 3000 K | 4000 K | 5000 K | 6000 K |
| 200 | $9.9 \times 10^{-19}$ (6.2) | $2.6 \times 10^{-16}$ | $4.1 \times 10^{-11}$ | $1.6 \times 10^{-8}$ | $5.9 \times 10^{-7}$ | $6.4 \times 10^{-6}$ |
| 300 | $6.6 \times 10^{-19}$ (4.1) | $4.1 \times 10^{-11}$ | $1.2 \times 10^{-7}$ | $6.4 \times 10^{-6}$ | $7.0 \times 10^{-5}$ | $3.5 \times 10^{-4}$ |
| 400 | $5.0 \times 10^{-19}$ (3.1) | $1.4 \times 10^{-8}$ | $5.7 \times 10^{-6}$ | $1.2 \times 10^{-4}$ | $7.3 \times 10^{-4}$ | $2.4 \times 10^{-3}$ |
| 500 | $4.0 \times 10^{-19}$ (2.5) | $5.1 \times 10^{-7}$ | $6.4 \times 10^{-5}$ | $7.1 \times 10^{-4}$ | $3.0 \times 10^{-3}$ | $8.0 \times 10^{-3}$ |
| 600 | $3.3 \times 10^{-19}$ (2.1) | $6.4 \times 10^{-6}$ | $3.5 \times 10^{-4}$ | $2.5 \times 10^{-3}$ | $8.3 \times 10^{-3}$ | $1.9 \times 10^{-2}$ |
| 700 | $2.8 \times 10^{-19}$ (1.8) | $3.9 \times 10^{-5}$ | $1.2 \times 10^{-3}$ | $6.3 \times 10^{-3}$ | $1.7 \times 10^{-2}$ | $3.5 \times 10^{-2}$ |

of the flame, where desolvation, volatilization, and atomization take place (recall Figure 7-2). In the vapor phase, molecules, free atoms, and ions are produced. Emission from excited atoms (occasionally molecules or ions) is then measured and related to the identity of the sample species and/or their concentrations.

## Properties of Flames

The premixed flames most often used in spectrochemical analysis are generally, but not always, **laminar flames** in which the gases exhibit laminar flow characteristics. Most often the flames burn in the open air and are thus **atmospheric pressure flames.** When the flame gases burn they expand substantially and do work ($p \, dV$), which consumes energy that can no longer heat the gases. Thermal expansion also increases the dissociation of molecules such as $H_2O$, $CO_2$, and $O_2$, formed by combustion reactions, and the energy used for dissociation is unavailable for heating the gases. Both of these effects limit the maximum achievable temperatures in atmospheric pressure chemical flames to around 5000 K. More typical are flames with temperatures in the range 1500 to 3000 K, as described below.

*Flame Structure.* The three zones that are observable in a premixed, laminar flame are illustrated in Figure 8-1. In the **primary combustion zone**, sometimes called the **inner cone**, the fuel/oxidant mixture is ignited, and combustion reactions proceed rapidly to completion or as far as the available oxidant allows. The gases proceed through this primary combustion zone (0.01 to 0.1 mm in thickness in atmospheric pressure, premixed flames) in about 10 μs. The reactions that occur in this zone are not in thermodynamic equilibrium. Because of this, no meaningful single temperature can be assigned to the primary combustion zone. Although some atomic lines produced by chemiluminescent processes are observed in this region, the inner cone is rarely used in analytical methods.

The **interzonal region**, sometimes called the **interconal zone**, lies immediately above the primary combustion zone. Here, local thermodynamic equilibrium exists, and the region is fairly homogeneous in composition and temperature. The flame gases reach their maximum temperature in the interzonal region, and the rise velocity is approximately constant (distance above burner head is proportional to time in this region). Gases traverse the interzonal region in a few milliseconds. For the foregoing reasons, most analytical observations are made here.

The interzonal region is externally bounded by the **secondary combustion zone,** which is much more diffuse and less well defined than the primary combustion zone.

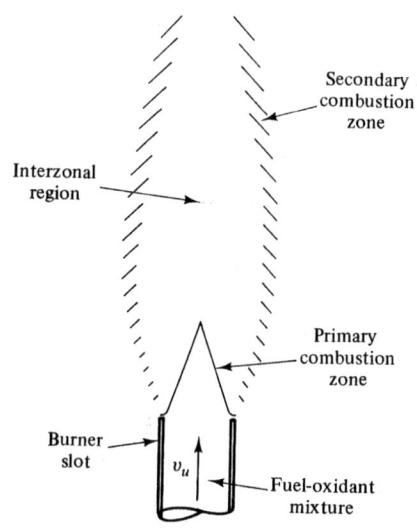

**FIGURE 8-1** Structure of premixed, laminar flame. The primary combustion zone, the interzonal region and the secondary combustion zone are indicated. The unburned gases enter the region of combustion with velocity $v_u$. The magnitude of the burning velocity is the component of $v_u$ perpendicular to the flame front (primary combustion zone). The burning velocity is a fundamental parameter of the gas mixture and is important in determining the flame shape and stability.

Since the secondary combustion zone is in contact with the atmosphere, oxygen and nitrogen from the air can penetrate and cause additional reactions here. When the flame has an insufficient supply of oxidant, the secondary combustion zone becomes more distinct, and the temperature can rise due to the heat generated by secondary combustion reactions. The secondary combustion zone can stabilize the flame, but it also produces background radiation that can interfere with the measurement of analyte emission.

*Flame Types and Temperatures.* The most widely used flames in flame emission are those fueled with hydrogen or acetylene. Propane is also used as a fuel in some simple Na or K analyzers, and cyanogen and carbon monoxide have occasionally been employed. Air, oxygen, and nitrous oxide are the most widely used oxidants. Table 8-3 lists the combustion reactions, the theoretical temperatures, and the maximum burning velocities for several common gas mixtures. The temperatures are calculated based on ideal adiabatic conditions and full thermodynamic equilibrium. Experimental temperatures depend highly on the conditions used (gas flow rates, fuel-to-oxidant mole ratios, burner design, position in flame, presence or absence of liquid droplets, etc.). They are often sub-

**TABLE 8-3**
Properties of common flames

| Fuel–oxidant | Combustion reaction[a] | Theoretical stoichiometric temperature (K) | Maximum burning velocity (cm s$^{-1}$) |
|---|---|---|---|
| 1. $C_3H_8$–air | $C_3H_8 + 5O_2 + 20N_2 \rightarrow$ $3CO_2 + 4H_2O + 20N_2$ | 2267 | 39–43 |
| 2. $H_2$–air | $2H_2 + O_2 + 4N_2 \rightarrow$ $2H_2O + 4N_2$ | 2380 | 300–440 |
| 3. $C_2H_2$–air | $C_2H_2 + O_2 + 4N_2 \rightarrow$ $2CO + H_2 + 4N_2$ | 2540 | 158–266 |
| 4. $H_2$–$O_2$ | $2H_2 + O_2 \rightarrow 2H_2O$ | 3080 | 900–1400 |
| 5. $C_3H_8$–$O_2$ | $C_3H_8 + 5O_2 \rightarrow$ $3CO_2 + 4H_2O$ | 3094 | 370–390 |
| 6. $C_2H_2$–$N_2O$ | $C_2H_2 + 5N_2O \rightarrow$ $2CO_2 + H_2O + 5N_2$ | 3150 | 285 |
| 7. $C_2H_2$–$O_2$ | $C_2H_2 + O_2 \rightarrow$ $2CO_2 + H_2$ | 3342 | 1100–2480 |

SOURCE: C. Th. J. Alkemade and R. Herrmann, *Fundamentals of Flame Spectroscopy*, Halsted Press, New York, 1979.
[a]$N_2$ is included in the air mixture reactions so that the stoichiometry can be discerned from the air composition.

stantially lower than theoretical values. For example, a temperature $\approx 2950$ K has been obtained for a fuel-rich $N_2O$–$C_2H_2$ flame, which is lower than the theoretical stoichiometric value of 3150 K. The oxycyanogen flame (not listed in Table 8-3) has occasionally been employed for analytical purposes. Its temperature has been experimentally determined to be in excess of 4500 K, one of the hottest flames known. With aspiration of aqueous solutions, the temperature decreases due to the energy expended for solvent evaporation.

Note from Table 8-3 that the use of oxygen in place of air increases the theoretical flame temperature by 700 to 800 K. This is because the inert $N_2$ from the air absorbs heat. The theoretical maximum temperatures of the flames listed in Table 8-3 are often slightly higher (20 to 100 K) than the stoichiometric temperatures and, for acetylene flames, are often achieved under fuel-rich conditions because the combustion products (CO or $CO_2$) are partly dissociated, which withdraws heat from the flame gases. For hydrogen flames, the maximum temperature occurs very near the stoichiometric fuel-to-oxidant ratio.

The burning velocities listed in Table 8-3 are critical flame parameters. Flames achieve stability only within a certain region of gas flow rates. If gas rise velocities do not exceed the burning velocity, the flame propagates inside the burner, resulting in a flashback condition. As the gas flow is increased, the rise velocity soon exceeds the burning velocity at all points, and the flame rises until it reaches a point above the burner where the rise and burning velocities are just equal.

This is the region of a stable flame. At yet higher flow rates the flame continues to rise and blows off the burner.

Flames that burn with air as the oxidant can be premixed because of the low burning velocities (see Table 8-3). The $N_2O$–$C_2H_2$ flame can likewise be premixed without problems. However, mixtures of oxygen and acetylene require a great deal of care to burn safely after premixing. Oxygen and hydrogen, if diluted with argon or helium, can be premixed.

*Flame Background.* Background emission can be the limiting factor in determining the precision of analytical measurements or the limits of detection. The blank measurement in flame emission spectrometry can compensate for the background emission only if it does not change between sample and blank determinations. This implies that the background emission does not fluctuate with time or with slight variations in conditions, such as fuel and oxidant flow rates, and that the background emission is not influenced by concomitants in the sample that are not present in the blank. If analyte emission occurs in a spectral region of high background emission, spectral overlap can occur, and any fluctuations in background emission will degrade precision and the limits of detection. For these reasons it is important to examine flame background emission in some detail. We shall consider here only background emission produced by the flame itself; concomitants in the sample can, of course, also give rise to background emission.

Most often analyte emission is measured in the region above the primary combustion zone. The background emission observed is from both the secondary combustion zone and the interzonal region itself, unless a separated flame is used in which case only the latter region need be considered. Background emission spectra for three common analytical flames are shown in Figure 8-2. There are two types of flame background that are commonly observed: spectral continua and band spectra. Band spectra are emitted by molecular species. Hydrogen flames show intense emission from OH radicals with bandheads at 281.1, 306.4 (strongest), and 342.8 nm. These radicals are formed from reactions such as

$$O + H_2 \rightarrow H + OH$$

$$H + O_2 \rightarrow O + OH$$

Oxygen–hydrogen flames also show the Schumann–Runge $O_2$ bands in the region 250 to 400 nm. Hydrocarbon flames not only produce the bands seen in hydrogen flames, but also bands from such hydrocarbon radicals as CH, which has bandheads at 431.5, 390.0, and 314.3 nm. The primary combustion zone also contains emission from $C_2$, the Swan bands at 474, 516, and 563 nm. Oxygen–acetylene flames also produce some CO bands in the region 205 to 245 nm. In fuel-rich nitrous oxide–acetylene flames, emission from CN, $C_2$, CH, and NH can be seen in the region 300 to 700 nm.

The continuum can be caused by blackbody emission from hot solids (e.g., incandescent carbon particles) or from unquantized recombination reactions. In $O_2$–$H_2$ flames the reaction

$$H + OH \rightarrow H_2O + h\nu$$

produces continuum emission in the blue to the near-UV region. In acetylene flames the reactions

$$CO + O \rightarrow CO_2 + h\nu$$

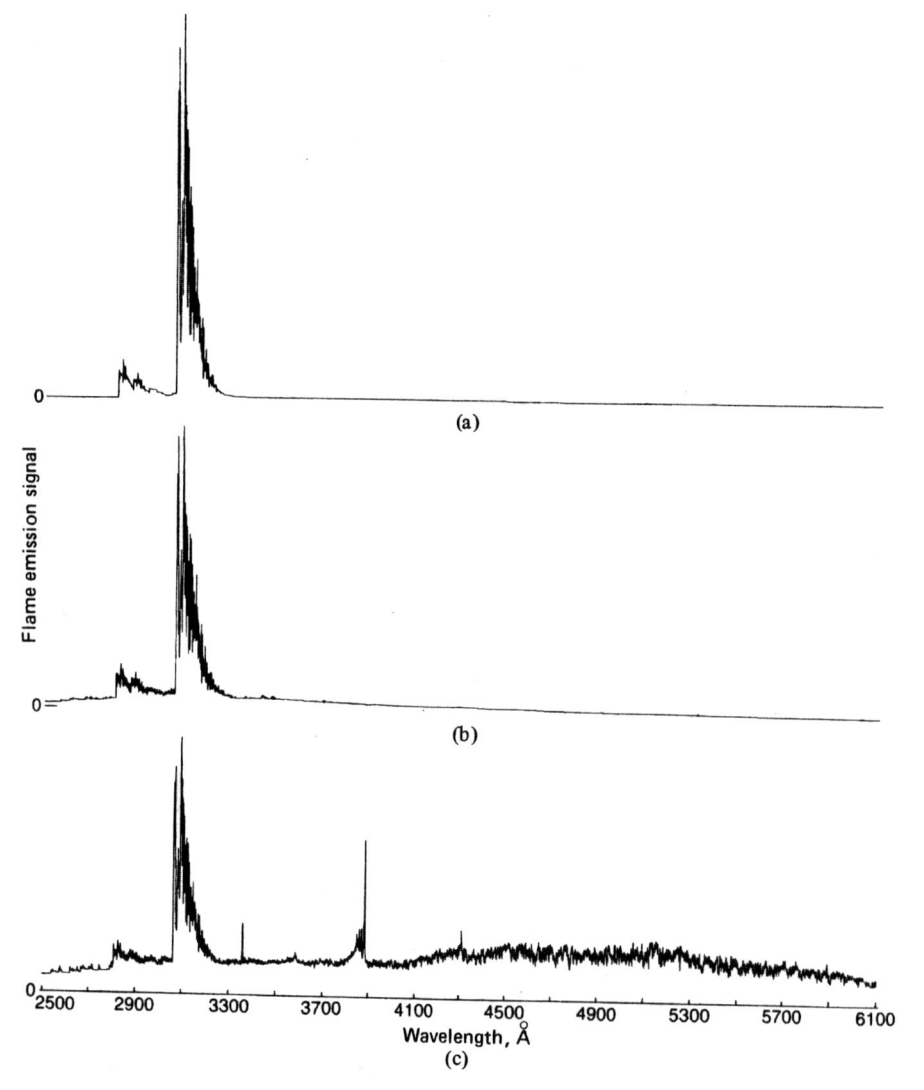

**FIGURE 8-2** Background emission spectra for common analytical flames: (a) oxygen–hydrogen; (b) oxygen–acetylene; (c) nitrous oxide–acetylene. The vertical sensitivities for the three spectra are not the same. (From D. G. Peters, J. M. Hayes, and G. M. Hieftje, *Chemical Separations and Measurements: Theory and Practice of Analytical Chemistry.* Copyright © by W. B. Saunders, Co., New York, 1974. Reprinted with permission of CBS College Publishing.)

and

$$NO + O \rightarrow NO_2 + h\nu$$

also produce continuum emission. When flames contain a high concentration of an easily ionized metal, the radiative recombination of ions and electrons produces an increase in background emission according to

$$M^+ + e^- \rightarrow M + h\nu$$

From Figure 8-2 it can readily be seen that the background emission is lower and simpler in flames fueled with hydrogen. The exact intensity of the background emission is a strong function of the fuel-to-oxidant ratio, gas purity, the type of burner, the type of solvent being sprayed, and the observation region in the flame. When the spectrometer resolution is low, the molecular bands shown in Figure 8-2 may appear diffuse and more like continua than bands. In addition, concomitants in the sample that form hard to dissociate molecular species can give rise to additional emission bands, and atomized concomitants can produce atomic lines.

## Sample Introduction into Flames

Solution samples are usually introduced into flames by means of a nebulizer/burner system. We will consider here several nebulizer/burner systems. Nebulizers and the principles of atom formation with this type of sample introduction were described in Section 7-1.

*Premixed Burners.* Most modern commercial flame spectrometers use a premixed burner system based on a pneumatic concentric nebulizer and spray chamber, as shown in Figure 8-3. In research applications, other types of nebulizers (e.g., crossed-flow and ultrasonic) have been employed with a nebulizing chamber, as discussed in Section 7-1. Because the gas flows are laminar in a premixed burner, a smooth steady flame is produced with low flame flicker.

In Figure 8-4, several types of burner heads that are used with spray chamber nebulizers are shown. In addition to the types shown, capillary burners form the burner head with stainless steel capillary tubes that can be positioned in different ways to change the shape and pathlength of the flame produced. The Alkemade burner is very useful in fundamental studies. It consists of an inner flame into which analyte solution is sprayed, an outer or mantle flame of identical gas composition to the inner flame, but without analyte, and a sheath of inert gas surrounding both flames. The mantle flame helps to maintain a homogeneous radial temperature,

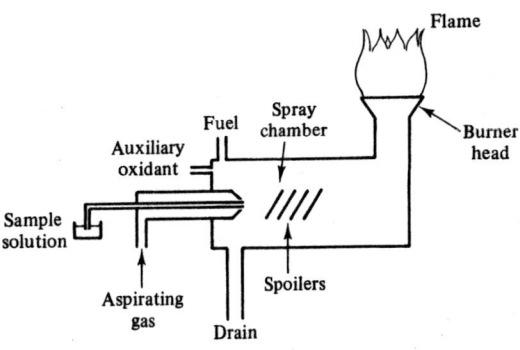

**FIGURE 8-3** Premixed, spray chamber nebulizer/burner. The sample is aspirated and nebulized into the spray chamber by a concentric nebulizer with oxidant as the nebulizing gas. Often aerosol modifiers (spoilers) are used to break up larger droplets. All but the finest spray particles condense and are drained away (see Chapter 7). The sample spray is mixed in the spray chamber with the fuel and additional oxidant gas to control the fuel-to-oxidant ratio. Only the small (5–20 μm)-diameter particles enter the flame. Typical solution flow rates are 2 to 5 mL min$^{-1}$.

while the sheath reduces air entrainment, and renders the flame stable against room drafts. Inert-gas sheathing can also reduce background emission and minimize flame flicker due to external air movement. The Alkemade burner is usually employed with a relatively low temperature flame ($O_2$–Ar–$H_2$ or air–$C_2H_2$); it does not have the high atomization efficiency typical of most analytical flame systems.

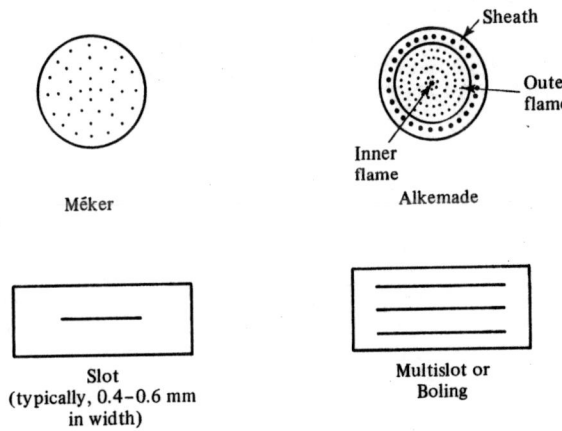

**FIGURE 8-4** Types of premixed burner heads. A top view of the burner head is shown. The Méker burner has many small holes compared to the common Bunsen burner which has only one large hole. The Alkemade burner is a cylindrical Méker burner. It produces a stable flame with a homogeneous radial temperature distribution. Slot burners have longer pathlengths than circular burners. They are often used for emission and almost always used in AAS.

*Total Consumption Burners.* Many older flame emission spectrometers use the **total consumption burner**, often called a **direct injection burner**, shown in Figure 8-5. With this burner, all the sample mist is injected into the flame ($\varepsilon_n = 1$), including the large droplets that would be removed with spray chamber-premixed burner designs. The residence time for the large droplets is often not long enough to ensure complete desolvation and vaporization. Hence desolvation efficiencies are often significantly less than unity. The larger amount of water entering the flame (1 to 2 mL min$^{-1}$) compared to chamber burners cools the flame upon evaporation. The total consumption burner is also subject to salt encrustation at the burner tip which can distort the flame.

The unpremixed flames produced by total consumption burners are **turbulent flames** or **diffusion flames**. Since fuel and oxidant are mixed above the burner tip, highly explosive gas mixtures with high burning velocities, such as $O_2–H_2$ and $O_2–C_2H_2$, can safely be burned, and combustible samples (i.e., gasoline) can safely be aspirated.

Unpremixed, turbulent flames have much less distinct structure than do premixed flames. The primary combustion zone can still be distinguished, but it is usually much more diffuse because of turbulent mixing. These flames are acoustically noisier than premixed flames because of the irregularity of the flame front and its erratic, changing velocity and direction. There is an interzonal region which attains some degree of thermodynamic equilibrium, but the turbulent flame is not at all homogeneous in its properties; the secondary combustion zone is often not easily recognized. Turbulent flames are usually characterized by higher rise velocities

and lower temperatures than theoretically achievable. Along with the shortness of the flame, these characteristics are responsible for the short analyte residence times, which can increase nonspectral interferences.

## Atomization and Excitation Characteristics

The choice of what flame to use in a given application depends on many factors. For flame emission, it is desirable to choose a flame consistent with high atomization efficiency, freedom from interferences, and sufficient excitation energy for the analyte. The hotter flames generally produce higher background emission, more ionization, and richer sample spectra with subsequent spectral overlap problems. Although hydrocarbon flames produce more background emission than hydrogen flames, they also have lower burning velocities and can provide longer residence times for analyte species. Hydrocarbon flames also produce, under fuel-rich conditions, reducing carbon-containing radicals (C, $C_2$, CN, CH), which can minimize the formation of refractory oxides with metals. The $N_2O–C_2H_2$ flame is particularly useful in this respect. Table 8-4 shows experimentally measured free atom fractions ($\beta_a$ values) for several elements in air–$C_2H_2$ and $N_2O–C_2H_2$ flames. Many of the elements that form stable oxides (Al, Ba, Mn, Sn) and hydroxides (Ca, Ba, Mg) are more highly dissociated (higher $\beta_a$ values) when nitrous oxide is employed.

The thermal energy ($kT$) in typical flames, 2000 to 4000 K, ranges from $2.8 \times 10^{-20}$ J, which is 0.17 eV, to $5.5 \times 10^{-20}$ J, which is 0.35 eV. As can be seen in Table 8-2, the Boltzmann factors and hence the fraction of atoms excited is quite low for UV transitions

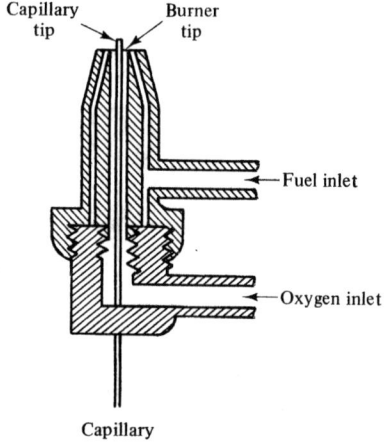

**FIGURE 8-5** Total consumption nebulizer/burner. The pneumatic concentric nebulizer aspirates and nebulizes the solution at the nebulizer tip. The fuel flows through the outer concentric ring around the burner tip.

**TABLE 8-4**

Atomization efficiencies ($\beta_a$ values) in acetylene flames

| Element | $C_2H_2$–air flame | $C_2H_2$–$N_2O$ flame |
|---------|--------------------|------------------------|
| Al | $<10^{-5}$ | 0.42–0.59 |
| Ba | 0.0011–0.009 | 0.15[a]–0.30[a] |
| Ca | 0.066–0.14 | 0.69[a]–1.4[a] |
| Cu | 0.87–1.00[b] | 1.00[b] |
| Fe | 0.38–0.66 | — |
| Li | 0.21–0.26[a] | 0.44[a] |
| Mg | 0.59–0.84 | 1.5–2.3 |
| Mn | 0.45–0.93 | 0.76–0.80 |
| Na | 0.50–1.00[a] | 0.33[a]–0.65[a] |
| Sn | $<10^{-4}$–0.078 | 0.71–0.76 |
| Zn | 0.45–1.10 | 0.91–1.00 |

SOURCE: J. D. Winefordner, S. G. Schulman, and T. C. O'Haver, *Luminescence Spectrometry in Analytical Chemistry*, Wiley-Interscience, New York, 1972.

[a]Ionization suppressor added.

[b]Some methods were relative to Cu for which $\beta_a = 1.00$ was assigned.

with flames. Even for visible transitions in the range 500 to 700 nm, the Boltzmann factors are 0.1% or less for normal flame temperatures.

For AA, the considerations in choosing a flame are similar to those for flame emission, except that excitation energy is not a concern. For AF, quenching by flame constituents must also be considered. Many workers have used relatively cool flames for fluorescence measurements. Flames such as $H_2-O_2-Ar$ produce high fluorescence yields for alkali metals (see Chapter 11).

### Comparison to the Ideal Source

Let us now compare the characteristics of flames to those of our ideal emission source (Table 8-1). As can be seen from Table 8-2, common chemical flames do not possess sufficient energy to excite all elements. In fact, only the "easy-to-excite" elements are excited to an appreciable extent in common flames. The energy available in flames is not readily controlled by the user. Fuel and oxidant flow rates can be varied, but only over a limited range. Flames are certainly not inert, high-temperature media; instead, they are highly reactive systems. Reactions of analyte atoms with flame gas combustion products or reactive intermediates can reduce the free-atom fraction and lead to low atomization efficiencies. Flames also produce significant amounts of background radiation that must be compensated for in order to obtain accurate and precise determinations. Flames are normally limited to solution samples, although they have been used for gases and, under unusual circumstances, solids. Because of the limited thermal energies available, flames are less than ideal sources for multielement determinations. Nevertheless, flames are convenient, reproducible, easy to use, and inexpensive. For this reason they are still highly useful atomizers for AA and AF. The plasma excitation sources described in the next section have supplanted flames for atomic emission determinations of most elements.

## 8-3 PLASMA ATOMIC EMISSION SOURCES

A plasma is a hot, partially ionized gas. In recent years, electrically generated plasmas have become widely available for atomic emission spectrometry. These sources produce flamelike plasmas with significantly higher gas temperatures and less reactive chemical environments compared to flames. Samples can be introduced into the high-temperature regions of these plasmas much as they are introduced into flames, as convenient solution aerosols. Thus the formation of free atoms occurs by the processes discussed in Chapter 7, and many of the same nebulization devices are used to produce aerosols.

The plasmas are energized with high-frequency electromagnetic fields (RF or microwave energy) or with direct current. When combined with a high-quality spectrometer, plasma sources provide many of the characteristics of the ideal source. The RF, inductively coupled plasma is the most prevalent of the plasma devices and is discussed first. Then various microwave plasmas and the dc plasma are described. Whenever appropriate, the characteristics of these sources are compared to those of combustion flames and to our ideal atomic emission source.

### Inductively Coupled Plasmas

The **inductively coupled plasma** (ICP) was first studied as an analytical emission source in the early 1960s by Fassel and coworkers in the United States (Iowa State University) and by Greenfield and coworkers in England (Albright and Wilson Ltd.). Unfortunately, this new source was introduced at a time when flame atomic absorption spectrometry was undergoing explosive growth and gaining widespread acceptance as an analytical tool. At the same time the established atomic emission techniques (arc and spark excitation) were experiencing sharp declines in general usage. Thus it took what now seems an incredibly long time for the ICP technique to gain acceptance in the analytical community. In fact, the first commercial ICP spectrometer was not introduced until 1975, more than 10 years after the first description of ICP sources. Since that time, however, there has been phenomenal growth in the number of ICP spectrometers in scientific laboratories and in analytical service laboratories.

*ICP Operation.* The main body of the inductively coupled plasma torch consists of a quartz tube (15 to 30 mm in diameter) surrounded by an induction coil that is connected to a high-frequency generator as shown in Figure 8-6a. The generator is operated at frequencies of 4 to 50 MHz (27 MHz is common) and at output powers of 1 to 5 kW. An inert gas, usually argon (Ar), flows through the tube. This flow of gas acts as the support gas for the plasma and as the coolant for the quartz tube. To form the plasma, the spark from a Tesla coil is used to produce "seed" electrons and ions in the region of the induction coil. Once the Ar conducts, the plasma forms spontaneously if the flow patterns are proper inside the tube. Figure 8-6b shows a more detailed diagram of the high-frequency currents and magnetic fields inside the quartz tube. The induced current, composed of ions, and electrons flowing in a closed circular path, heats the support gas to a temperature on the order of 9000 to 10,000 K and sustains the ionization necessary for achieving a stable plasma.

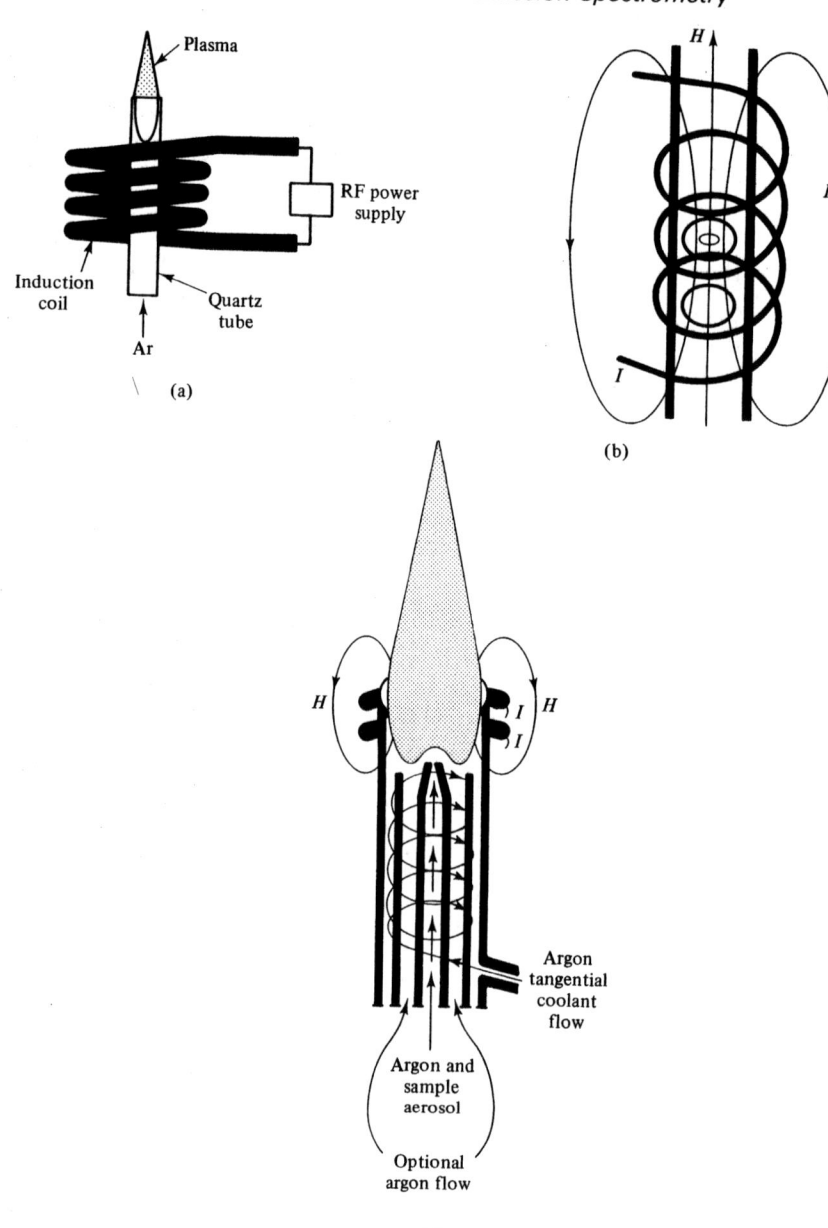

(a)

(b)

(c)

**FIGURE 8-6** Operation of ICP. In the schematic diagram (a), an induction coil surrounds a quartz tube through which an inert gas flows. The coil is connected to an RF generator. When the Ar is "seeded" with electrons it becomes conductive and the plasma forms. In (b) the magnetic fields ($H$) and eddy currents ($I$) inside the quartz tube are shown. The high frequency alternating currents in the induction coil generate magnetic fields with their lines of force located parallel to the tube. The seed ions and electrons are accelerated in a circular flow or eddy current. When the current in the induction coil reverses direction, the magnetic field and eddy current also reverses. The accelerated ions and electrons collide with support gas atoms and cause additional ionization along with intense ohmic heating. In (c) the configuration of a complete ICP torch is shown. Sample is introduced with Ar carrier gas through the central tube. A high flow rate of Ar is introduced tangentially for thermal isolation by vortex stabilization. The third flow of Ar support gas is optional in most designs.

Two different argon flows are used. A relatively low flow velocity, about 1 L min$^{-1}$, is used to transport the sample to the plasma. A much higher flow velocity, typically 10 L min$^{-1}$, is introduced tangentially as shown in Figure 8-6c. The latter flow of Ar thermally isolates the plasma from the outer quartz confinement tube and prevents overheating. The plasma itself is located near the exit end of the concentric tubes.

*Sample Introduction.*  One of the early problems with the ICP was introduction of the sample. If the plasma is operated at frequencies of about 5 MHz, a plasma shape much like a teardrop is obtained, as shown in Figure 8-7a. The sample tends to avoid the high-temperature region, where the resistance due to

gas expansion is the highest. Such a shape thus results in inefficient heating of the sample. The annular plasma, obtained at higher frequencies (Figure 8-7b), gives rise to efficient desolvation, volatilization, and excitation of the sample. This annular-shaped plasma looks much like a doughnut when viewed from below. The shape results from the skin-depth effect of induction heating. As the frequency of the RF generator is increased, the eddy current paths move closer to the outer or skin region of the plasma. The greatest heating occurs in the region of high eddy currents. The central region of the plasma (the doughnut hole) is thus somewhat cooler, which reduces the resistance to sample introduction.

Because of efficient desolvation and volatilization in the ICP, the most widely used mode of sample in-

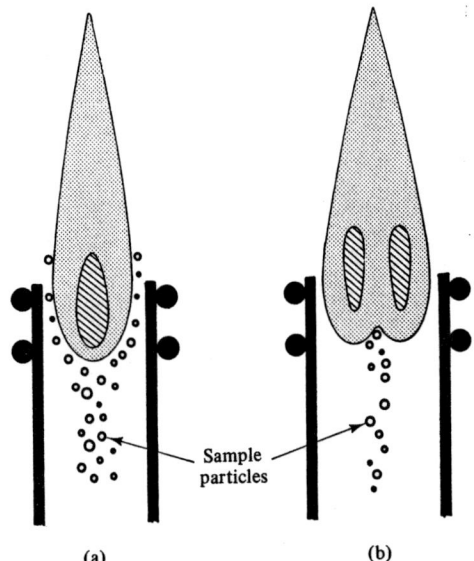

(a)        (b)

**FIGURE 8-7** Plasma shapes at 5 MHz (a) and 27 MHz (b). With the teardrop shape in (a), the sample particles do not penetrate the high-temperature region shown by the crosshatching. When operated at higher frequencies ($\approx$ 27 MHz), the plasma assumes an annular or toroidal shape (b). Sample particles travel through a narrow axial channel surrounded by the high-temperature core.

troduction is direct introduction of the aerosol from a spray-chamber nebulizer (see Chapter 7). The crossed-flow, pneumatic nebulizer is most common, although ultrasonic nebulizers are also employed. Samples have also been introduced into ICP sources as vapors from electrothermal atomization devices and hydride generators (see Section 10-2), or from chromatographic columns. This versatility of sample introduction is one of the major reasons for the success of the ICP in practical analyses.

*ICP Characteristics.* Some of the most important characteristics of the ICP are listed in Table 8-5. Gas temperatures in the axial channel can be as high as 7000 to 8000 K, while sample residence times in the plasma are usually 2 to 3 ms. This combination leads to nearly complete solute vaporization and a high atomization efficiency. Matrix and interelement effects are quite low, although not always insignificant. The inert chemical environment means that free atoms should have relatively long lifetimes in the plasma. The high electron densities found in argon ICPs (see below) means that the total number of electrons is not significantly altered by the addition of easily ionized elements. Hence ionization interference effects are small. It has been found in some cases that the addition of easily ionized elements influences the spatial profile of emission from analyte neutral atoms and ions. The absence of molec-

ular species should lead to high fluorescence yields since Ar has a small quenching cross section. In the normal observation region, the population density of free analyte atoms in the axial channel can be quite high. Since the population density is much lower in the hot Ar sheath than in the central channel, self-absorption effects are small compared to many sources.

The region of highest gas temperature, the so-called plasma core, is located inside the induction coil; it extends a few millimeters above the coil. This core region appears to the eye as a bright white, nontransparent zone. Spectrally, the core is characterized by an intense continuum and the atomic lines of Ar. The continuum is typical of ion–electron recombination reactions and *bremsstrahlung* (continuum radiation that results from the slowing or stopping of charged particles). Because of the continuum, the core region is of limited analytical utility. A short distance above the core region, the plasma becomes slightly transparent, and the continuum emission is reduced by several orders of magnitude over that in the core. This second zone extends from 1 to 3 cm above the induction coil. It is in this zone that the highest $S/N$'s are observed for most analytes. Background in this second zone consists primarily of Ar lines, OH band emission, and some other molecular bands, as can be seen in the spectra shown in Figure 8-8. Because there is a large temperature gradient in this second region, emission is found from elements with a wide range of excitation energies. Above this second zone, the "tail flame" region can be observed when easily excited analytes are introduced into the plasma. Temperatures in this third zone are similar to those in ordinary combustion flames. The ICP thus consists of a very high temperature core for efficient atomization of the sample, a secondary zone where many analytes are observed, and a low-temperature "tail flame" that can be used for easily excited elements.

The excitation mechanisms active in the ICP are not completely understood at this time. There is now general agreement that the plasma in argon is not in local thermodynamic equilibrium (see Chapter 7 for the LTE model), at least in the region used for analytical measurements. Thus different values are found for the

**TABLE 8-5**
Characteristics of the ICP

1. High temperatures
2. Long residence times
3. High electron number densities (few ionization interferences)
4. Free atoms formed in nearly chemically inert environment
5. Molecular species absent or present at very low levels
6. Optically thin
7. No electrodes
8. No explosive gases

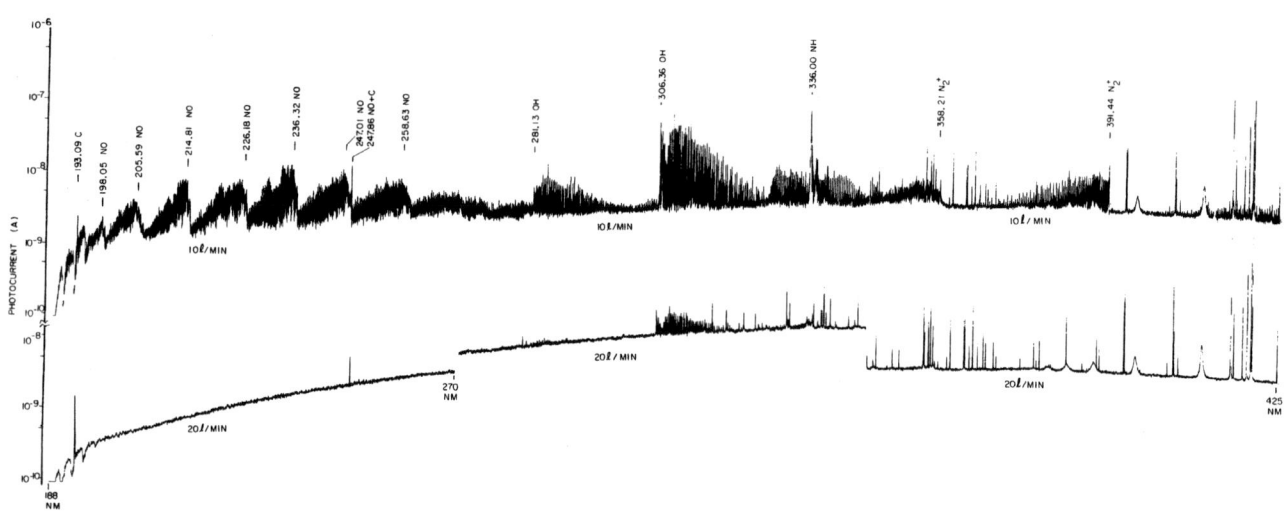

**FIGURE 8-8** Background emission spectra from the argon ICP with deionized water being aspirated into the plasma. The upper recording was taken under conditions favoring molecular band emission, while the lower recording was taken under conditions minimizing molecular band emission. The flow rates listed are those for the outer, tangential argon flow. All other conditions were the same for the two recordings. Note the presence of Ar lines, some OH band emission, and some other molecular bands, particularly in the upper trace. In the lower recording most of the molecular bands have disappeared and there are mainly Ar lines. (Reprinted with permission from R. K. Winge, V. A. Fassel, V. J. Peterson, and M. A. Floyd, *Inductively Coupled Plasma-Atomic Emission Spectroscopy: An Atlas of Spectral Information*, Elsevier, Amsterdam, 1985.)

several measures of plasma temperature (gas kinetic temperature, excitation temperature, ionization temperature, etc.). Electron number densities are quite high in a pure Ar ICP; measurements based on Stark broadening indicate values in the range $10^{14}$ to $10^{16}$ cm$^{-3}$. These high values explain the lack of ionization interferences. The electron, and hence argon ion, number densities are, however, higher than predicted by the spectroscopically measured excitation temperatures. The values found for excitation temperatures increase as the upper level of the transition used increases in energy. In other words, the overpopulation of an energy level increases with energy. It has also been found that ion line-to-atom line intensity ratios for a given element are significantly higher than expected from the LTE model.

Several excitation models have been presented for the argon ICP. In one model a high population of argon metastable levels is used to explain the observed ion/atom intensity ratios. Another model explains the overpopulation of high energy levels, particularly argon metastable levels, by a radiation-trapping model (see Section 7-5). Yet a third model uses ambipolar diffusion to explain observations made in the analytical region based on conditions present in the region inside the load coil. To date the most comprehensive model is a collisional-radiative model that includes radiation trapping and transport processes.

If we compare the ICP to the ideal emission source characterized in Table 8-1, we can see that it meets many of the criteria listed. While the ICP energy is not readily controlled, it is capable of exciting a wide range of elements within the observation zone. This significantly enhances the simultaneous, multielement capabilities of the source. The ICP does have sufficient energy to excite all common elements of interest. Atomization and excitation occur in an inert argon atmosphere, which lessens the probabilities for chemical interferences. Solutions and gases can be injected directly into the plasma, but solids are analyzed only with some difficulty. Direct insertion probes made from carbon or graphite have been used to introduce solids. A spark discharge is used in one commercial system to vaporize solids for introduction into the ICP (see Chapter 9). The ICP is a reproducible source and capable of producing accurate and precise results (see later). ICP sources are less than ideal, however, in several categories. First, there are spectral overlap interferences, even though these are minimal in many cases. Second, ICP emission systems are not inexpensive either to purchase or to maintain. Complete systems begin in the $50,000 range and can be significantly more expensive. Operating costs are not inconsiderable, particularly with the large flow rates of argon required; and maintenance costs can be substantial. Finally, the ICP is not a simple source to operate. Considerable training is required to become an efficient and knowledgeable user of ICP emission systems.

In addition to their uses as spectrochemical sources, ICPs have recently been used as ionization sources for mass spectrometry (MS). Solution aerosols can be introduced into an ICP, and a few of the positive ions extracted through an orifice into a differentially pumped vacuum system containing a quadrupole mass spectrometer. Trace element and isotopic composition data on solution samples can be obtained with the ICP/MS technique. Solids can be introduced into the ICP/MS system by various volatilization techniques.

## Microwave Plasmas

Although they have not been studied as extensively as ICPs, microwave plasmas have been used for many emission spectrometric applications. These sources are generally operated at lower powers than ICP sources and, of course, at microwave frequencies.

*Plasma Types.* Two different types of microwave plasmas have been employed: the capacitively coupled, coaxial waveguide plasma and the electrodeless, microwave cavity plasma, or **microwave-induced plasma** (MIP). Both of these designs are shown in Figure 8-9. Like the ICP, a microwave plasma is initiated by providing "seed" electrons from the spark of a Tesla coil. The electrons oscillate in the microwave field and gain sufficient kinetic energy to ionize the support gas by collisions. A microwave frequency of 2450 MHz is usually used because of the commercial availability of microwave generators of this frequency. To obtain efficient power transfer to the plasma, it is necessary to tune the coupling element to match the load impedance. With cavity designs, the plasma itself contributes significantly to the system impedance. Large amounts of sample or solvent vapor can result in considerable changes in plasma impedance and thus coupling efficiency. If such changes due to the sample are too large, the plasma may be extinguished. The capacitively coupled plasma shows better tolerance for the introduction of foreign materials.

Capacitively coupled microwave plasmas are usually operated with argon or nitrogen as support gases. Typical gas flow rates are in the range 4 to 8 L min$^{-1}$.

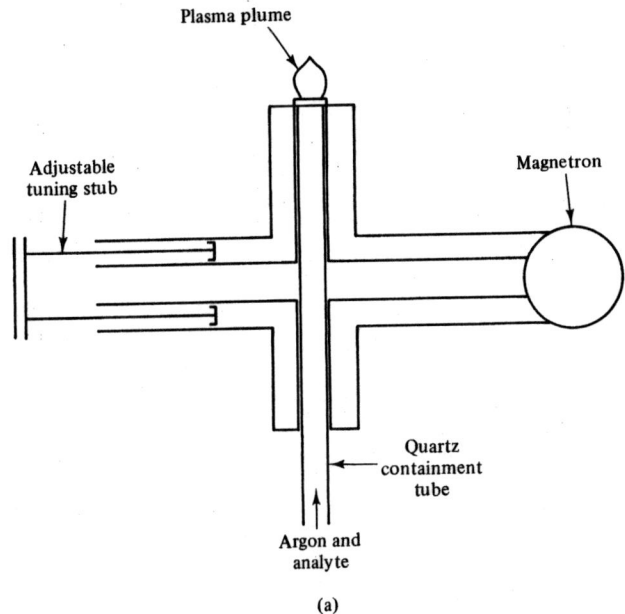

(a)

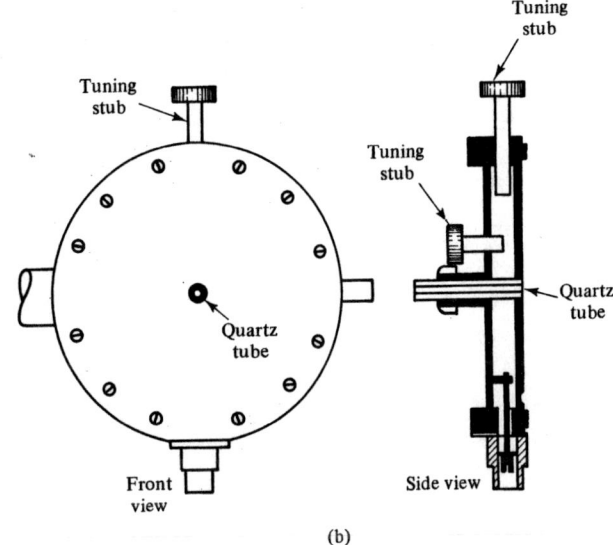

(b)

**FIGURE 8-9** Coaxial waveguide microwave plasma (a) and cavity-tuned plasma (b). In (a) the microwave power from a magnetron (microwave power tube) is conducted through a coaxial waveguide to the tip of a coaxial electrode. In (b) the microwave energy is coupled to the gas stream in the containment tube by an external cavity. Microwave plasmas can also be sustained by surface wave propagation (surfatron).

Microwave powers of 200 to 1000 W at 2450 MHz are employed. Microwave-induced plasmas have been generated with argon, helium, or nitrogen. These typically employ 25 to 250 W of microwave power. Of the two plasma types, the microwave-induced plasma has been used more extensively, primarily because of its low power requirements and atmospheric-pressure operation.

*Sample Introduction.* Microwave plasmas are normally operated at substantially lower applied powers than ICP devices. Although advantageous in terms of expense and electrical noise, the low power levels do not produce sufficiently energetic plasmas for efficient desolvation and volatilization of solution samples. In addition, plasma stability can be affected seriously when solutions are injected directly. For this reason, microwave plasmas, particularly MIPs, have been used most successfully with vapor-phase sample introduction. For example, MIPs have been extensively employed as detectors for gas chromatography. Also, generation of a gaseous compound of the analyte, such as a hydride or chloride, has been used to introduce samples into the plasma (see Section 10-1). Direct spraying of solution samples into a He MIP is a particular problem since pneumatic nebulizers will not operate with He at the flow rates needed for stable plasma operation. Pneumatic nebulization has been employed with Ar plasmas after desolvation of the droplets by techniques such as that shown in Figure 8-10. Ultrasonic nebulization and desolvation have also been employed successfully for introducing aqueous samples.

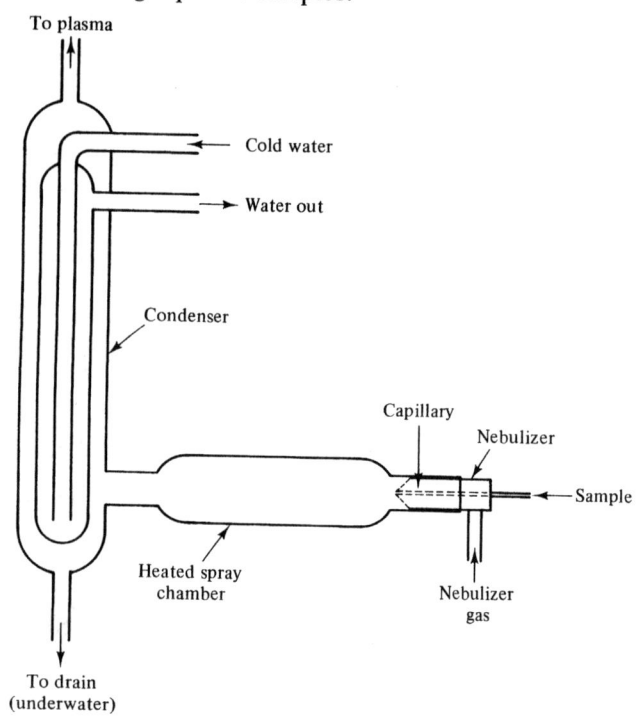

**FIGURE 8-10** Nebulizer with desolvation system. Solutions are sprayed with a pneumatic nebulizer into a heated chamber where solvent is vaporized. A condenser cooled to about 10°C causes the solvent to condense. Solvent flows out the drain and desolvated sample is carried away to the source by the nebulizer gas. Nearly complete desolvation can be achieved with overall efficiencies of 35%.

An attractive sample introduction approach is to use an electrothermal atomizer to desolvate, vaporize, and atomize the analyte prior to injection and excitation in the plasma. Filament or strip atomizers made from Pt, W, or Ta have been used. Also, a microarc discharge has been employed to atomize the analyte into the support gas stream of an MIP. Clearly, sample introduction difficulties have been primarily responsible for the lower popularity of MIPs than ICPs, and have hindered the development of commercial instrumentation based on MIP sources.

*Microwave Plasma Characteristics.* Thermodynamic equilibrium is usually not attained with microwave plasmas, and lower gas temperatures are generally reported than those with ICPs. Gas kinetic temperatures on the order of 2000 to 3000 K have been determined. As a result, atomization efficiencies are only slightly better than those of common flames. Excitation temperatures, however, are usually higher than gas temperatures (typically, 4000 to 5000 K in a MIP), and emission from high-excitation-energy elements has been reported. When used as gas chromatographic detectors, for example, MIPs can readily excite such elements as C, H, N, P, B, O, and the halogens. The background emission spectra produced by Ar and He MIPs are generally simple and of relatively low intensity, although continuum emission is observed in the UV. Atomization and excitation in a MIP are conducted in an inert gas environment, but interferences can be more troublesome than with ICPs because of the lower temperatures. Sample introduction, however, remains the largest problem that must be overcome if MIP devices are to become widely used in routine analyses.

## DC Plasmas

**Direct-current plasmas** (DCPs), also called **plasma jets,** are also extensively used in emission spectroscopy. Commercial instrumentation based on DCPs is available and this has undoubtedly stimulated the use of these devices. In a DCP, a flow of partially ionized gas is forced out of a small orifice at high velocity. The plasma is produced by a dc discharge, similar to a dc arc (see Section 9-1), struck between two or more electrodes. To obtain high current densities, the discharge is made to constrict by cooling the outer regions of the plasma with a high-velocity gas stream. The latter makes the outer regions of the arc less conductive and decreases the current channel area.

Early dc plasmas were made from ring-shaped carbon electrodes and an upper cathode. They had the disadvantage that observations were made in a region where the background continuum due to electron–ion recombination was relatively intense. In the early 1970s the background problem was partially overcome by a new electrode geometry which positioned the anode and cathode at an angle of 30° to one another. The plasma then took the shape of an upside-down V. Aerosol was sprayed from between the two electrodes, and emission was observed above the apex of the arc. This observation zone avoided the strongly emitting plasma core and resulted in higher signal-to-background ratios.

The third major development in DCPs occurred with the introduction of a commercial three-electrode system by Spectrametrics, Inc. This three-electrode DCP is shown in Figure 8-11. Here two independent plasma jets have a single common cathode, and the overall plasma burns in the form of an upside down Y. A flow of argon gas is directed over each electrode at a velocity high enough to cool the ceramic sleeves and prevent their melting. The sample is also nebulized with Ar gas, and the total Ar consumption is about 8 L min$^{-1}$.

Experiments have shown that although excitation temperatures can reach 6000 K in this DCP, sample volatilization is not complete because residence times in the plasma are relatively short. This can be troublesome with samples that contain difficult-to-volatilize materials. Electron number densities in the three-electrode DCP are quite similar to those in the ICP. The major advantages of the three-electrode DCP are its good stability, its low power requirements (<1 kW), its ability to handle organic as well as aqueous solutions,

and its acceptance of solutions with relatively high solids content. The small region where optimum line-to-background ratios occur is the principal disadvantage of this DCP. With a conventional spectrometer, a much larger region is ordinarily viewed (5 to 10 mm slit height) unless optics are used to magnify the source image. For this reason echelle spectrometers are most often used with this DCP. Different elements are found to have their optimum line-to-background ratios in different spatial regions, which makes the selection of compromise conditions for multielement analysis, particularly near the limits of detection, somewhat difficult. It has also been found that plasma properties change when high concentrations of easily ionized materials (e.g., alkali metals) are introduced. Like the ICP, a DCP spectrometer is neither inexpensive to purchase or maintain. For an operator to use DCP systems effectively requires considerable training. Despite these problems, the three-electrode DCP is a versatile and useful source especially when a broad range of analytical matrices is encountered.

## 8-4 FLAME AND PLASMA EMISSION SPECTROMETERS

In addition to the excitation source and sample introduction system, an emission spectrometer must contain a wavelength selection device with entrance optics, one or more transducers, an electronic processing system, and a readout system, as shown in the block diagram of Figure 8-12. Although the trend in recent years has been toward multichannel spectrometers, single-channel systems are still widely used, and, for flame emission, simple filter photometers can sometimes be sufficient. We consider first the optical components of the emission spectrometer system; later in this section the electronic components and readout system are described along with computer-controlled emission systems.

### Wavelength Selection

The entire optical train in an emission spectrometer should (1) collect as much radiation as possible from the excitation source while selecting a well-defined volume element and solid angle for observation; (2) isolate the desired emission line(s) from unwanted lines and continuous background; (3) focus radiation of the selected wavelengths on the photosensitive transducer(s); and (4) provide modulation capabilities if desired for *S/N* enhancement. It is also highly desirable for the spectrometer system to provide some means for measuring and correcting for the background emission at or

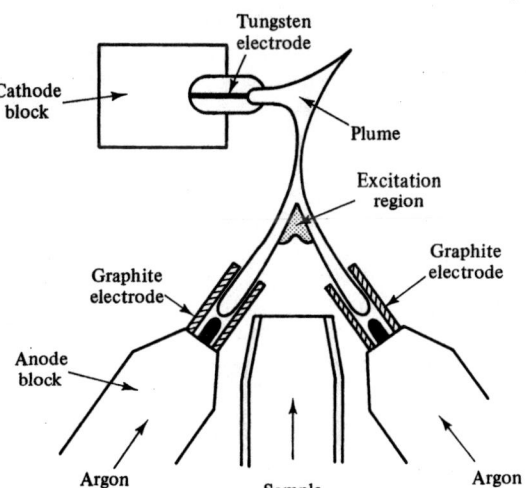

**FIGURE 8-11** Three-electrode dc plasma. Two separate dc plasmas operate with a common tungsten cathode. Sample is introduced as an aerosol from between the two graphite anodes. Desolvation, vaporization, and excitation occur in the region beneath the strongly emitting plasma core. Observation of emission in this region avoids much of the plasma background.

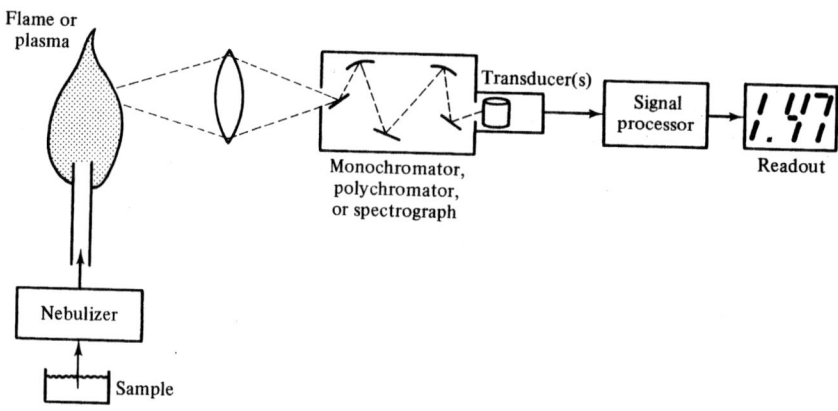

**FIGURE 8-12** Flame or plasma emission spectrometer. The sample to be analyzed is nebulized (converted to a liquid aerosol or mist) into the high-temperature environment of the flame or plasma atomizer, where the droplets are desolvated and solid particles are vaporized. In the vapor phase, molecules, free atoms, and ions are produced. The emission from excited atoms or ions (occasionally, excited molecules) is directed to the wavelength selection device, which isolates radiation characteristic of the analyte. The transducer converts the isolated radiation into an electrical signal which is processed and displayed on an appropriate readout device.

near the analyte line. We will look first at the entrance optics, which play an important role in the first goal above.

*Entrance Optics.* Typically, a lens or mirror is used to focus an image of the source (flame or plasma) onto the wavelength selection device. With dispersive devices, the F/n of the optical element and its position are chosen to match the monochromator (or polychromator) solid angle ($\Omega$) of collection. For 1:1 imaging, the F/n is one-half the F/n of the monochromator, and the excitation source and entrance slit are placed $2f$ ($f$ is the focal length of the lens or mirror) from the entrance optic. With lenses, chromatic aberrations can cause the throughput to vary with wavelength. The entrance slit dimensions determine the portion of the source that is observed. With flames the viewing window is readily adjusted by mounting the burner on a translational stage and moving the flame vertically with respect to the optical element. With plasmas it is usually easier to adjust the optics or the spectrometer than to move the plasma. In some systems a concave mirror is placed behind the excitation source with its center of curvature located within the source to increase the solid angle of collection.

If ac or lock-in detection is to be employed, a chopper must be placed in the optical train unless wavelength or sample modulation is employed. In most cases, a mechanical chopper is positioned between the source and the spectrometer entrance slit (see Section 4-5).

*Single-Channel Systems.* Single-channel instruments are very common in both flame and plasma emission spectrometry. For routine flame emission determinations (e.g., Na or K in blood serum), interference filter photometers are often employed. Because of their low resolution, filter photometers are used only where spectral interferences are unlikely. The major advantages of filter instruments are their low cost, small size, and high light throughput. Several commercial flame photometers (see Section 8-7) designed specifically for clinical applications are based on interference filters.

The grating monochromators discussed in Chapter 3 are most often used in single-channel instruments. The most versatile monochromators are capable of moderate to high resolution (< 0.1 nm bandpass) and have variable slit widths to allow optimization of the S/N (see the next section). Automatic scanning is highly desirable to allow the background emission to be obtained on either or both sides of the analytical line. Alternatively, wavelength modulation (see Section 4-5) can be employed for automatic background subtraction. With single-channel systems, the response time of the measurement electronics can limit the maximum scan rates achievable. If more than a few elements are to be determined, or the analysis wavelengths are widely separated, linear scanning can lead to intolerable analysis times and excessive sample consumption.

*Sequential Multielement Instruments.* Several different types of sequential spectrometers are employed for emission spectrometry. Slew-scan monochromators can provide a significant savings in analysis time and sample consumption over conventional linear scanning monochromators. Slew-scan monochromators can be programmed to scan slowly over or to momentarily stop at wavelength regions of interest, but to move rapidly (slew) through uninteresting regions. Both conventional grating monochromators and echelle grating monochromators can be made to slew-scan. Of course, for developmental work in choosing appropriate analytical lines or in investigating spectral characteristics, it is desirable to have both linear scanning and slew-scanning capabilities. For example, one commercial echelle spectrometer used with a dc plasma allows linear scanning over long- or short-wavelength regions as well as slew scanning for multielement determinations. The spectrometer can slew from 190 to 900 nm in less than 3 s. Typical slewing times between lines are 1.5 s or less. Slew-scan systems also allow simple background correction methods to be used. The spectrometer can be programmed to measure the background intensity on either or both sides of the analytical line.

***Multichannel Spectrometers.*** For true simultaneous multielement determinations, instruments capable of monitoring at multiple wavelengths in parallel are required. In the past spectrographs with photographic detection were commonly used. Photographic emulsions, however, have limited linearity, are difficult to calibrate, must be developed, and must be subjected to densitometry to obtain intensity information. For these reasons, photographic emulsions are used now primarily for qualitative spectral observations. Modern multichannel emission systems use either polychromators with photomultiplier detectors or spectrographs with array detectors.

A modern direct-reading spectrometer used with an ICP source is shown in Figure 8-13. A commercial instrument of this design has provision for more than 60 exit slits and photomultiplier tubes; smaller 48- and 36-channel versions are available. All channels can be

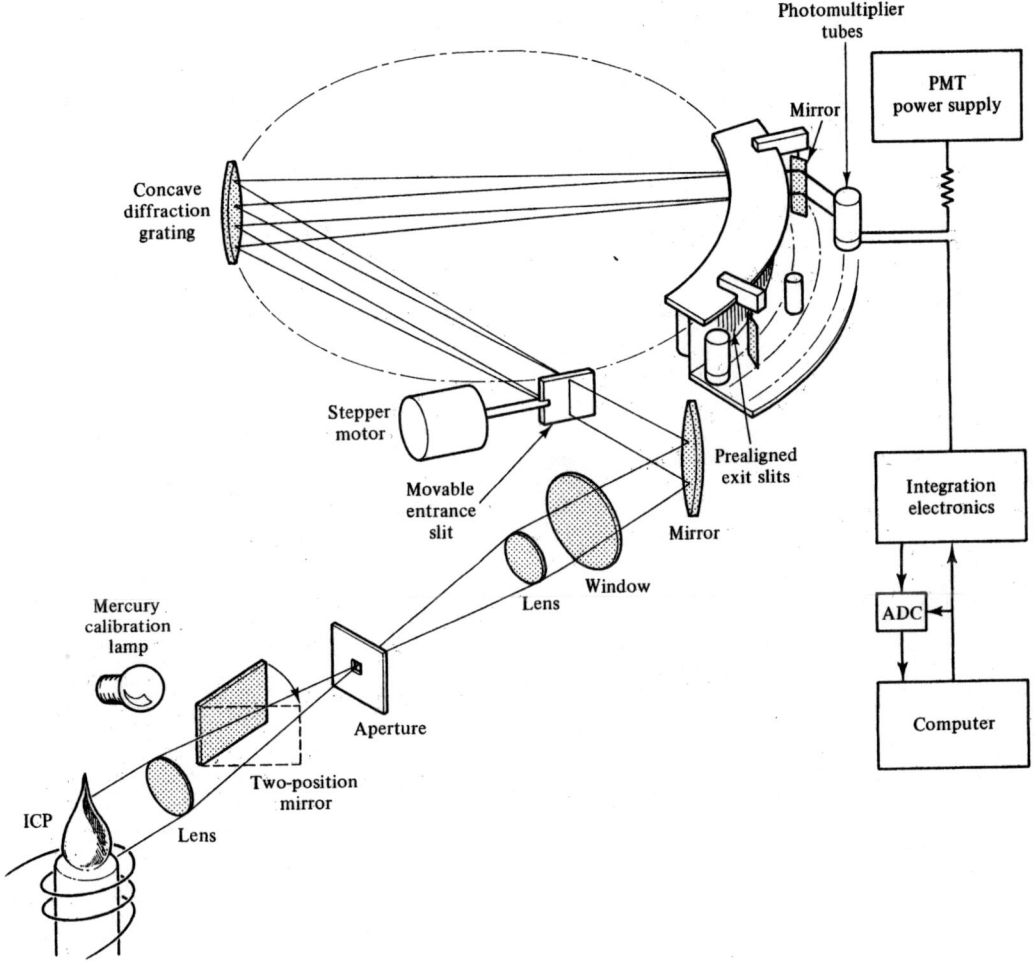

**FIGURE 8-13** Direct-reading ICP emission spectrometer. The polychromator is of the Paschen–Runge design. It uses a concave diffraction grating and produces a spectrum in focus around a Rowland circle. Separate exit slits isolate each spectral line, and a separate photomultiplier tube converts the optical information into electrical signals.

dedicated to analytes or some can be used for the measurement of background. In some cases multiple lines of the same element can be used for analysis. Alignment of the individual exit slits in a direct reader is extremely tedious and time consuming. Hence analysis lines are added or changed only very rarely. Unfortunately, alignment also depends on temperature and even the refractive index of the air. Some spectrometers use one channel to monitor an intense spectral line from a hollow cathode or mercury lamp. A servo system continuously adjusts the position of the grating or exit-slit assembly to keep this reference line centered. In other systems quartz refractor plates in front of each slit allow

small adjustments to be made in the wavelengths passing through the slit. Because of the difficulties in aligning the exit slits, direct-reading spectrometers are not very flexible systems. Many commercial instruments, however, provide an extra observation port to which a scanning monochromator can be added for additional versatility (the $n + 1$ channel).

Background correction is also difficult with direct-reading spectrometers. Quartz refractor plates which scan a small wavelength interval around the analytical line are often used in many background correction schemes (see Figure 4-34). One refractor plate can be used behind the entrance slit to scan all channels or

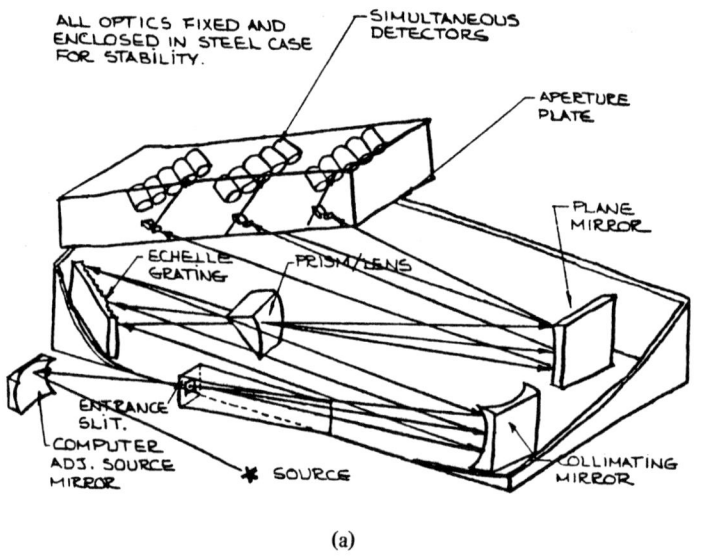

(a)

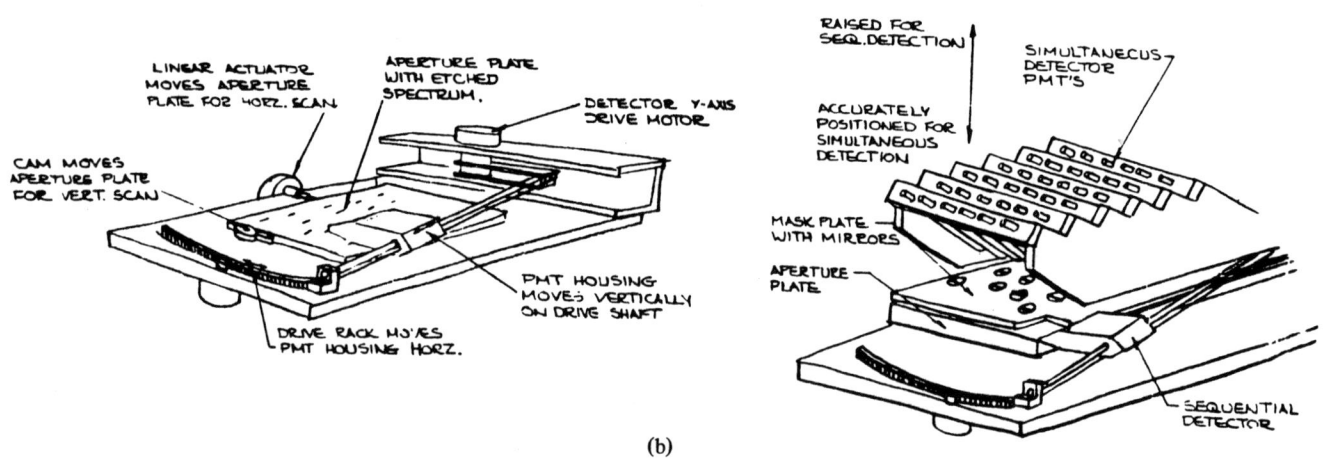

(b)

**FIGURE 8-14** Direct-reading echelle spectrometer. In the simultaneous analysis mode (a), 45 photomultiplier tubes are positioned above the exit slit plate. In the sequential mode (b), a single photomultiplier detector is moved to locations behind a preset aperture plate. (Courtesy of Leeman Labs.)

refractor plates in front of each exit slit can be used to scan each channel individually. The entrance slit can also be oscillated which effectively scans all the slit images dispersed in the focal plane. However, these can add complexity to a system already difficult to use. The high cost and low optical speed of these spectrometers are additional considerations. Signal-to-noise ratio calculations have shown that for a fixed total measurement time the multichannel techniques provide the highest $S/N$'s. Slew-scan techniques approach multichannel techniques only when few lines must be measured.

Echelle spectrometers are also commonly used as direct readers. Recall that the echelle grating with a prism order sorter disperses the spectrum in two dimensions. Hence multiple exit slits and photomultipliers can be placed in the two-dimensional focal plane as illustrated in Figure 8-14. Excellent resolution can be obtained in a compact spectrometer by using an echelle grating. One commercial instrument based on the design of Figure 8-14 allows operation in the sequential mode (Figure 8-14b) as well as the true simultaneous mode (Figure 8-14a). In either mode the exit slit plate can be moved off the analysis line to enable background correction or to allow the selected line to be scanned. One advantage of the fixed optics approach shown in Figure 8-14 is that the grating is never moved and the optical components are mounted in a thermally and vibrationally isolated housing for maximum stability. The throughput of echelle polychromators or monochromators can be less than for conventional grating systems because smaller slit heights are used and only one order is monitored.

Another approach to multichannel systems is to use a spectrograph with an array detector. At present there is no perfectly suitable solid-state replacement for the photographic plate in emission applications. Devices such as diode arrays are too small (1 in. long or less) to achieve both the desired resolution and the desired wavelength coverage for multielement determinations. They are also not as sensitive as photomultiplier tubes unless operated with an intensifier. One recent spectrometer combines the direct reader approach with a diode array detector. A grating coarsely disperses the radiation, and several wavelength regions are allowed to pass through a mask to an echelle grating. The latter further disperses the radiation from each of the selected wavelength windows onto the array. As a result, several wavelength regions (a few nm each) appear side by side on the array. A different mask is required for a different selection of elements, but since the masks are coarse openings, they are not expensive.

Some of the newer solid-state devices, such as the charge-coupled device and the charge-injection device, appear to have considerable potential in high-resolution spectroscopic applications, particularly with crossed-dispersion, echelle grating spectrographs, but at present they are quite expensive and difficult to use. With two-dimensional arrays, it is possible to intercept the entire two-dimensional spectrum generated by the echelle spectrograph.

Multiplex spectrometers (see Chapter 4), such as Hadamard and Fourier transform spectrometers, have also been proposed for multielement emission spectrometry. These techniques have been widely used in the IR spectral region, where detector noise is the limiting source. They have not been very successful in the UV-visible region, where quantum noise or flicker noise is often the limiting noise source.

## Transducers and Electronic Components

Photomultiplier tubes are used as transducers in the majority of emission spectrometers, although some simple filter photometers for flame emission have used vacuum phototubes or photodiodes. Since the radiant flux emitted by atoms (ions) can vary over several orders of magnitude, it is important that the transducer, the signal processing electronics, and the readout system have a wide dynamic range. Most emission sources are also somewhat noisy. Because photomultipliers are rapid response detectors, some signal averaging is usually needed for $S/N$ improvement. Instruments that are dc-based have used current-to-voltage converters with dc amplifiers (if needed) to obtain voltages in the range 0.1 to 10 V. Typically, these instruments use low-pass filtering with time constants of a few seconds. Another approach that is often used with dc-based plasma emission systems is to use a linear integrator with an integration time of 1 to 30 s. For an $n$-channel direct reader, $n$ integrators are therefore required for averaging. In one commercial 60-channel system, 60 operational amplifier integrators are used. To increase the dynamic range, the integration capacitor is usually charged to full scale and discharged several times during the measurement. The processing system then counts the number of full-scale integration cycles and adds this to the final integrator output. The integrator sensitivity (value of integration capacitor) is different for weak lines than for strong lines. Hence the electronics for each channel are somewhat customized. The output of each integrator is sent to a multiplexed analog-to-digital conversion system for conversion to the digital domain. A computer system stores the acquired data for each channel. In another approach, the PMT outputs are converted to voltages by operational amplifier circuits. The voltage output of each channel is then acquired with an integrating analog-to-digital converter, such as a voltage-to-frequency-to-digital converter. In either case computers are required to store and analyze the large amounts of data obtained.

With modulation (source, sample, or wavelength), current-to-voltage conversion followed by either ac amplification and demodulation or by lock-in amplification is employed. Except for specific clinical analyzers, it is difficult to find new commercial instruments that are dedicated to flame emission determinations. Many flame AA instruments (see Chapter 10) have provision for flame AE measurements. Often a mechanical chopper is activated to modulate the flame emission signal at the modulation frequency of the AA hollow cathode lamp (not employed for emission) so that the same ac electronics can be used. AA instruments are not always well suited for AE measurements because automatic scanning is usually not provided and their resolution may not be sufficient.

### Computer Control

Nearly all recently manufactured emission spectrometer systems include a microprocessor or a microcomputer system. The degree of computer involvement in the experiment varies from model to model. In some cases the computer merely acquires the emission data, computes calibration data from standards, calculates the concentrations of unknowns from stored calibration equations, and outputs the data to a printer or video display. Some programs allow emission data to be corrected for matrix effects by including various empirical correction factors. Others allow background correction to be implemented under program control. Many instruments also provide an automatic sampler and in some cases the computer can control the sampler.

With some instruments the computer takes a more active role in the experiment. For example, many sequential or slew-scanned monochromators are placed under computer control so that all scanning functions (e.g., scan range, scan rate, analysis wavelengths, and integration time for each element) can be programmed. This permits routines to be developed for various determinations and stored in a library for later reuse. Ideally, one would also like to control the excitation source via the computer. Parameters such as gas flow rates, nebulizer flow rates, observation windows, and many others are amenable to control by the computer. Placing as many parameters as possible under computer control allows self-optimization routines to be used to develop optimum analysis conditions as discussed in Section 8-5.

## 8-5 SIGNAL AND NOISE CONSIDERATIONS

Expressions for the readout signal, the noise, and the signal-to-noise ratio ($S/N$) can give insights into the factors that influence these quantities. Such expressions can be used as guidelines in optimizing measurement precision. By combining several equations considered previously, an expression is developed here for the readout voltage in flame and plasma emission spectrometry. The theoretical equation for the $S/N$ that was developed in Chapter 5 is then modified to a form appropriate for emission measurements, and limiting noise sources are identified. Optimization of $S/N$'s and the influence of experimental variables on detection limits are then discussed.

### Readout Signals

Let us consider the flame or plasma emission spectrometer of Figure 8-12 and expand the equation for the readout voltage $E_E$ given in equation 4-34 and repeated here:

$$E_E = mG \int B_{\lambda E} Y(\lambda) R(\lambda) \, d\lambda \qquad (8\text{-}1)$$

where $G$ is the gain of the signal modifier (V A$^{-1}$), $m$ is the transducer gain, $B_{\lambda E}$ is the spectral radiance due to analyte emission, $Y(\lambda)$ is the monochromator (or polychromator) throughput factor, and $R(\lambda)$ is the radiant cathode responsivity of the transducer. We consider here the normal emission case where a narrow spectral line is monitored ($\Delta \lambda_{\text{eff}} \ll s$). The transition monitored is assumed to be a resonance transition from the first excited level (level 1) to the ground level (level 0) and self-absorption is presumed negligible. The spectrometer throughput factor, given by equation 3-70, and $R(\lambda)$ are constant over the narrow emission line width. Thus

$$E_E = mGR(\lambda)WH\Omega T_{\text{op}} \int B_{\lambda E} \, d\lambda \qquad (8\text{-}2)$$

One-to-one imaging is assumed and $T_{\text{op}}$ includes the entrance optics transmission factor. The integral in equation 8-2 has been shown in equation 7-32 to be the total line radiance, $B_E$. Thus we can write

$$E_E = mGR(\lambda)WH\Omega T_{\text{op}} B_E \qquad (8\text{-}3)$$

If we assume that self-absorption is negligible (optically thin limit), we can use equation 7-39 for the total line radiance, which gives

$$E_E = mGR(\lambda)WH\frac{\Omega}{4\pi}T_{op}E_1A_{10}\frac{g_1}{g_0}n_0e^{-E_1/kT} \qquad (8\text{-}4)$$

Here we have made use of the relationship $E_1 = hc/\lambda_m$ for a resonance transition in order to express the exponential in equation 8-4 in terms of the upper-state energy. For sources that are not in thermal equilibrium, $T$ refers specifically to the excitation temperature. The ground-state number density $n_0$ can be expressed in terms of the original solution concentration $c$ by equation 7-9, where we assume that $Z(T) \approx g_0$. If we substitute equation 7-9 into equation 8-4, the result is

$$E_E = 6 \times 10^{17}c\frac{F\varepsilon_a l}{Qe_f}WH\frac{\Omega}{4\pi}T_{op}mGR(\lambda)E_1A_{10}\frac{g_1}{g_0}e^{-E_1/kT} \qquad (8\text{-}5)$$

Equation 8-5 expresses the readout signal $E_E$ in terms of atomizer factors $(F\varepsilon_a l/Qe_f)$, instrumental factors $[WH(\Omega/4\pi)T_{op}mGR(\lambda)]$, analyte spectroscopic factors and source excitation temperature $[E_1A_{10}(g_1/g_0)e^{-E_1/kT}]$. Note that source modulation by a mechanical chopper between the source and the spectrometer would reduce $E_E$ by 50%. Equation 8-5 predicts that $E_E$ will be directly proportional to analyte concentration for a particular analyte and transition (e.g., specific values of $E_1$, $A_{10}$, $g_1$, and $g_0$) under certain conditions. If the instrumental factors are constant, $E_E \propto B_E$. If self-absorption is negligible and the source temperature is constant, $B_E \propto n_0$, and if the atomizer factors are constant, $n_0 \propto c$. Thus $E_E \propto c$ under these conditions. Source temperature is a critical experimental variable because it influences the population of ground-state atoms through $\varepsilon_a$ and $e_f$ as well as the fraction of atoms excited through the Boltzmann factor $e^{-E_1/kT}$ (see Table 8-2). For ionic emission, the values of $\varepsilon_a$, $E_1$, $A_{10}$, $g_1$, and $g_0$ correspond to those of the ion.

Let us roughly compare flames and plasmas with the aid of equation 8-5. The gas flow rates ($Q$ values) are roughly the same for the two media, but the solution flow rate $F$ is usually lower in plasmas because of the low flow rate nebulizers used. However, with plasmas, the values of the expansion factor $e_f$ and the Boltzmann factor $e^{-E_1/kT}$ are significantly higher because of the higher gas kinetic and excitation temperatures. The larger Boltzmann factors along with the larger atomization efficiencies are usually more than enough to yield higher readout signals in plasma emission.

The background emission signal is given from equation 4-36 as

$$E_{bE} = mG\int B_{\lambda bE}Y(\lambda)R(\lambda)\,d\lambda \qquad (8\text{-}6)$$

If we substitute equation 3-70 for $Y(\lambda)$, approximate the background emission near the spectral line as a continuum, assume that $B_{\lambda bE}$ and $R(\lambda)$ are constant over $\lambda_m \pm s$, and note that $\int t(\lambda)\,d\lambda = s_g = R_dW$, the result is

$$E_{bE} = mGR(\lambda)W^2H\Omega T_{op}R_dB_{\lambda bE} \qquad (8\text{-}7)$$

## Signal-to-Noise Expressions

The analytical signal $E_E$ in flame or plasma emission is extracted from the total signal $E_{tE}$ by subtraction of the blank signal $E_{bk}$ (equation 2-22), where the blank includes contributions from background emission and from dark signals. The rms noise $\sigma_t$ in $E_{tE}$ was given in equation 5-26 for voltage output or in equation 5-27 in terms of cathodic currents. The resulting $S/N$ expressions for emission spectrometry are summarized in Table 8-6, where $E$ is used as a subscript to indicate emission $(E_S = E_E, E_{bE} = E_B)$.

If the $S/N$ is plotted vs. analyte concentration (or vs. $E_E$ or $i_E$), curves similar in shape to those shown in Figure 5-6 are obtained. At low analyte concentrations,

---

**TABLE 8-6**

Signal-to-noise ratio expressions for flame and plasma emission[a]

| | |
|---|---|
| Voltage output: | $\dfrac{S}{N} = \dfrac{E_E}{[mGK(E_E + E_{bE} + E_d) + (\xi_E E_E)^2 + (\chi_E E_{bE})^2 + \{(\sigma_d)_{ex}\}^2 + (\sigma_{ar})^2]^{1/2}}$   (8-8) |
| Cathodic currents: | $\dfrac{S}{N} = \dfrac{i_E}{[K(i_E + i_{bE} + i_d) + (\xi_E i_E)^2 + (\chi_E i_{bE})^2 + \{(\sigma_d)_{ex}/mG\}^2 + (\sigma_{ar}/mG)^2]^{1/2}}$   (8-9) |

[a] $E_E(i_E)$, analyte emission voltage (cathodic current) [V(A)]; $E_{bE}(i_{bE})$, background emission voltage (cathodic current) [V(A)]; $E_d(i_d)$, dark voltage (cathodic current) [V(A)]; $m$, transducer gain (dimensionless); $G$, signal processor gain (V A$^{-1}$); $K$, bandwidth constant (C s$^{-1}$); $\xi_E$, analyte emission flicker factor (dimensionless); $\chi_E$, background emission flicker factor (dimensionless); $(\sigma_d)_{ex}$, rms dark current excess noise (V); $\sigma_{ar}$, rms amplifier-readout noise (V).

the noise is limited by such factors as background emission flicker or shot noise and dark current noise which are independent of $E_E$. Thus at low analyte concentrations, the $S/N$ is directly proportional to concentration ($S/N \propto E_E$ or $i_E$). Usually, background emission levels are high enough in emission spectrometry that dark current noise and amplifier-readout noise are negligible ($E_{bE} \gg E_d$). At intermediate analyte concentrations, the first term in the denominator of the equations in Table 8-6 (shot noise term) can become dominant with $E_E \gg (E_{bE} + E_d)$. When analyte emission shot noise dominates, the $S/N$ is proportional to the square root of concentration [$S/N \propto (E_E)^{1/2}$ or $(i_E)^{1/2}$]. At high concentrations, the analyte emission flicker noise term dominates and the $S/N$ becomes independent of concentration [$S/N \neq f(E_E$ or $i_E)$].

### Signal-to-Noise Optimization

The equations of Table 8-6 can give us insight into how experimental variables influence the $S/N$ and how the $S/N$ can be maximized. If measurement precision is limited by random noise, maximizing the $S/N$ minimizes the relative standard deviation (RSD) of the measurements. Let us first consider low analyte concentrations where the contribution of analyte emission shot and flicker noise is negligible. For the rest of this discussion we will assume that excess dark current noise and amplifier-readout noise can also be neglected, and that the gains ($mG$) are adjusted so that measurements are not readout resolution limited. In this case, equation 8-8 reduces for low concentrations to

$$\frac{S}{N} \approx \frac{E_E}{[mGK(E_{bE} + E_d) + (\chi_E E_{bE})^2]^{1/2}} \quad (8\text{-}10)$$

To improve the $S/N$ at low concentrations, the signal-to-background ratio $E_E/E_{bE}$ can be increased, the bandwidth constant $K$ can be reduced, and the background emission flicker factor $\chi_E$ can be reduced. Equation 8-5 gives the dependence of $E_E$ on experimental variables. Unfortunately, many of the factors that increase the analyte emission signal $E_E$ also increase the background emission signal $E_{bE}$.

We could substitute the detailed expressions for $E_E$ and $E_{bE}$ (equations 8-5 and 8-7) into equation 8-10, to provide a $S/N$ expression that indicates the dependence of the $S/N$ on all variables. Use of such an expression for optimization is difficult since the dependence of the magnitude of some of the variables on other variables is not known or the variables are interrelated.

For example, changing the fuel-to-oxidant ratio affects not only $Q$, but also $T$, which in turn affects $\varepsilon_a$, $e_f$, $B_{bE}$, and $\chi_E$. In practice, the $S/N$ is usually measured experimentally as a function of the variable of interest with theoretical expressions as a guide. Some variables show optimum values while others do not. The most important instrumental variables are monochromator slit width, $W$, noise bandwidth $\Delta f$ (included in $K$), and certain atomizer factors (gas flow rates, nebulizer variables, height of observation, etc.).

Increasing the monochromator slit width increases $E_E$ and $E_{bE}$. Since $E_E \propto W$ while $E_{bE} \propto W^2$, from equation 8-10 $S/N \propto W/[k_1(W^2 + k_2) + k_3W^4]^{1/2}$, where the $k$'s are constants and the background emission is assumed to be a continuum. At very narrow slit widths, dark current shot noise predominates and the $S/N$ is proportional to $W$ because the noise is independent of $W$. As the slits become wider, background emission shot noise becomes appreciable and the $S/N$ becomes independent of $W$. Eventually, background emission flicker noise dominates, and the $S/N$ is proportional to $W^{-1}$. A typical plot of $S/N$ vs. $W$ shows a maximum at a particular optimum value of slit width $W_{op}$. If the background emission includes lines from concomitants, the $S/N$ can improve dramatically if the slit width is decreased enough to eliminate an intense background line.

At low analyte concentrations and small slit widths, when background emission flicker is low, equation 8-10 predicts that decreasing the noise equivalent bandpass $\Delta f$ and thus $K$ increases the $S/N$. Since the background emission flicker noise often shows a $1/f$ behavior at low frequencies, decreasing $\Delta f$ will eventually cause background emission flicker noise to predominate and the $S/N$ becomes independent of $\Delta f$ or may even decrease. With flame emission, an optimum $\Delta f$ on the order of 0.1 to 1 Hz is normally found experimentally.

With a mechanical chopper and source modulation, the background emission signal and noise are carried along with the emission signal so that synchronous ac detection provides no advantage if background emission flicker noise is the limiting noise source. Improvements in the $S/N$ at low concentrations and the detection limit have been achieved with wavelength and sample modulation since in this case the background emission signal and low-frequency background emission noise are not carried.

Gas flow rates, nebulizer parameters, slit width, slit height, and observation height are adjusted for maximum signal or $S/N$, although the effects of these variables are very difficult to predict theoretically. With flames, the type of fuel and oxidant and their flow rates can influence both $E_E$ and $E_{bE}$ by changing flame temperature, solution flow rate, atomization efficiency, and other factors. Optimum values are often found experimentally. With plasmas, gas flow rates, the solution flow rate, and the applied RF power influence both $E_E$ and $E_{bE}$. Similarly, in flames and plasmas the optimum observation height must be found experimentally since $E_E$, $E_{bE}$, and $\chi$ vary with position.

Background emission flicker noise can result from variations in atomizer conditions or in the sample introduction system. The flicker factor depends on noise equivalent bandpass and many other variables. For a given situation, $\chi_E$ is determined experimentally. For flames and plasmas, typical values are 0.3 to 3% for $\Delta f$ in the range 0.1 to 1 Hz.

At high analyte concentrations, analyte emission flicker often becomes the dominant noise source. Under these conditions, equation 8-8 reduces to

$$\frac{S}{N} \approx \frac{1}{\xi_E} \qquad (8\text{-}11)$$

The only way to improve the $S/N$ is to reduce the magnitude of the analyte emission flicker factor $\xi_E$. The fluctuations in analyte emission are caused mainly by variations in the free-atom population in the viewed region of the flame or plasma and variations in excitation conditions. Sometimes the analyte emission noise is called *nebulizer noise* because it is believed to be due to fluctuations in the sample introduction rate. As a first approximation $\xi_E$ is assumed independent of analyte concentration so that the $S/N$ becomes independent of $c$ at high concentrations. At concentrations where self-absorption is significant, fluctuations in self-absorption may cause the relative flicker noise and hence $\xi_E$ to increase. Again $\xi_E$ is determined experimentally and is typically 0.3 to 3% in flames or plasmas for $\Delta f$ values in the range 0.1 to 1 Hz.

## 8-6 PERFORMANCE CHARACTERISTICS

There are several ways to evaluate the performance of an emission spectrometer or a particular analytical method. Among the factors that are important in characterizing performance are the linearity of calibration curves, the accuracy of determinations, the precision of repetitive determinations, and the experimentally achievable limits of detection. In this section these characteristics are discussed, and the factors leading to nonideal behavior are described. Because solution samples are most often employed with flame and plasma sources,

we will assume that determinations are carried out with solutions. Experimental methods for evaluating and improving performance are introduced.

### Linearity

Equation 8-4 gives the general relationship between readout signal $E_E$ and ground-state number density $n_0$ in all emission techniques. In addition, equation 8-5 gives the relationship between $E_E$ and solution concentration $c$ for techniques that introduce samples by continuous nebulization. The proportionality between readout signal and concentration is invariably established by means of a calibration curve. Here $E_E$ values are obtained for several standards of known concentrations, and $E_E$ is plotted vs. $c$ or a log-log plot is made. For many reasons it is desirable that such calibration curves be linear or at least follow predicted relationships. We expect to obtain linear calibration curves if the emission intensity is proportional to $n_0$, if $n_0$ is proportional to $c$, and if instrumental and source factors are constant from one standard to the next. Verification of linearity is then experimental evidence that the instrumental system is performing as expected.

A major cause of nonlinearity when resonance lines are used is self-absorption at high concentrations. Self-absorption causes the slope of a log-log plot of $E_E$ vs. $c$ to deviate from its low concentration value of 1 and approach its high concentration limiting value of ½ (see Figure 7-15). In flame emission, the onset of nonlinearity due to self-absorption typically occurs in the concentration range 10 to 100 $\mu g$ $mL^{-1}$. The onset of self-absorption in ICPs often occurs at higher concentrations than in combustion flames. Although high concentrations can give lower calibration curve slopes, and thus lower sensitivity, useful analytical results can still be obtained in the region where self-absorption occurs. Thus nonlinearity at high concentrations should be expected and not taken as an instrumental failure unless the theoretical relationship is not obeyed.

At low concentrations, ionization of the analyte can cause nonlinearity (see Section 7-2). With flames it is common to use an ionization buffer with easily ionized elements to overcome this effect. With ICP and DCP sources the high electron densities tend to buffer the plasma against changes in the extent of ionization with concentration. Also, with ICP sources ionic emission lines are often used for easily ionized elements and these are often less susceptible than neutral atom lines to additional ionization. Changes in solution flow rate, atomization efficiency, and atomizer temperature with

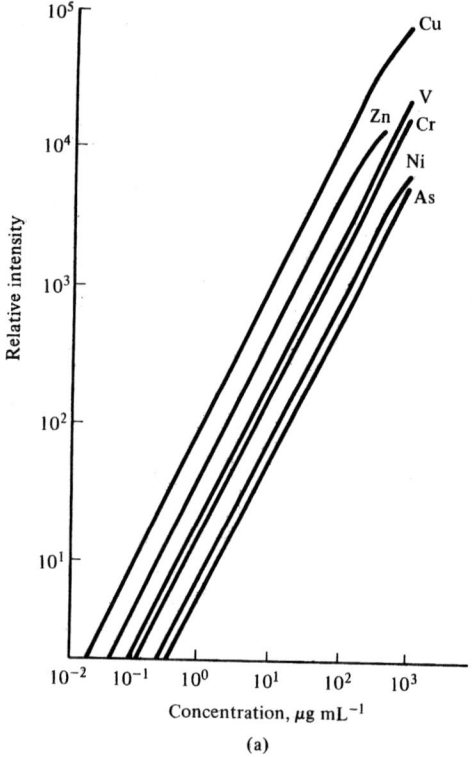

(a)

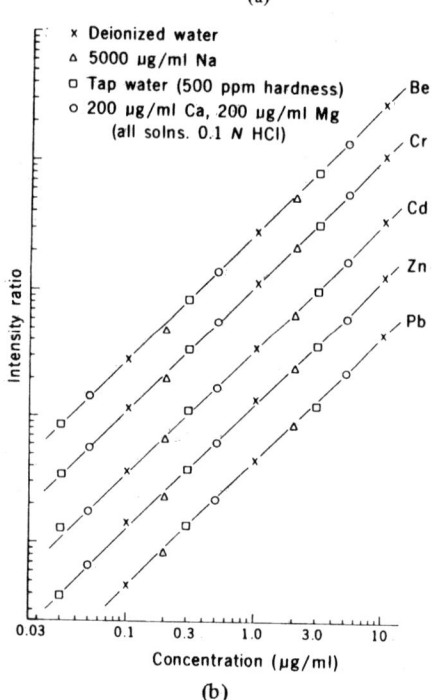

x Deionized water
△ 5000 µg/ml Na
□ Tap water (500 ppm hardness)
○ 200 µg/ml Ca, 200 µg/ml Mg
   (all solns. 0.1 N HCl)

(b)

**FIGURE 8-15**. Calibration curves obtained by ICP atomic emission spectrometry. In (a), working curves for 6 elements are shown. For most elements such curves are linear over a concentration range that covers four to six orders of magnitude. In (b), the lack of interelement effects is shown. For the calibration curves, some solutions were prepared in deionized water, while in other cases tap water and high-salt-concentration matrices were used. A yttrium line at 242.2 nm was used as an internal standard. [(b) With permission from V. A. Fassel, *Science, 202,* 183 (1978). Copyright 1978 by the AAAS.]

analyte concentration are also common causes of calibration curve nonlinearity.

For simultaneous determinations involving multiple elements, compromise source conditions must be established. Such conditions may not be optimum for any one element, but best for the group of elements as a whole. Because linear ranges may be quite different from one element to another, it is important to establish these through calibration curves obtained with multielement standard mixtures. Spectral interferences among the multiple analytes can also be a cause of nonlinearity and can be detected with multielement standards. With modern instruments, it is unusual for there to be nonlinearities in the transducer or amplifier-readout system.

Flame emission techniques often show linearity over two or three decades in concentration. ICP and DCP sources can show remarkable ranges of linearity, often four to five decades in concentration. The large linear range of the ICP is one of its major attractions. Often, major, minor, and trace constituents can be determined without the dilution or sample preconcentration steps needed in flame techniques, for example. Some typical calibration curves showing the excellent linearity of the ICP and its lack of interelement effects are shown in Figure 8-15.

### Precision

Precision is almost always evaluated experimentally by measuring replicate samples as discussed in Chapter 6. The most complete but most time-consuming procedure is to carry the samples through the entire analytical scheme, including for each the preparation of standards, the blank, and the calibration curve. Calibration curve uncertainty can also be estimated statistically as described in Chapter 6.

Relative standard deviations in flame emission are typically on the order of a few percent, although special differential scale expansion methods have been used to reduce the RSD to the level 0.1 to 0.3%. The reproducibility of experimental variables such as solution flow rate $F$, atomization efficiency $\varepsilon_a$, pathlength $l$, gas flow rate $Q$, flame gas expansion factor $e_f$, and flame temperature $T$ often determine the precision. Flame temperature is a particularly critical factor. If we naively assume that excitation of the analyte is the only step influenced by temperature, we can get an idea of how fluctuations in $T$ affect precision. If we assume that the measured signal $E_E$ depends on temperature as $E_E = k'e^{-E_i/kT}$ and that $k'$ is independent of temperature (not really true), then

$$\sigma_E = \frac{E_1}{kT}\frac{\sigma_T}{T}$$

Since $E_1/kT$ is on the order of 10 in flames, the RSD in temperature $\sigma_T/T$ should be less than 0.1% for the RSD in $E_E$ to be less than 1%. Hence in a 2500-K flame, $\sigma_T$ would need to be $\pm 2.5$ K for the RSD in $E_E$ to be 1%. In reality, the influence of flame temperature is much more complex since it also affects desolvation, solute volatilization, compound dissociation, ionization, and flame background. Because of the many variables affected, flame temperature fluctuations can give rise to larger or smaller values of the RSD than predicted by excitation considerations alone.

Signal-to-noise ratio considerations with plasma emission spectrometers are very similar to those for flame emission systems, although the expressions can be considerably more complex when multielement determinations are included. It is not uncommon with ICP and DCP sources to achieve relative standard deviations on standards in the range 1 to 5% and on real samples in the range 1 to 10%. With many of the sources considered here, the stability of the source itself can limit precision through its effect on temperature and indirectly on atomization and excitation efficiencies.

Of course, the ultimate limit of precision of the measurement step is imposed by random noise, as described in Section 8-5. The goal of an experimentalist should always be to improve the precision of all the noninstrumental steps until electrical noise dominates. Then the *S/N* optimization methods mentioned previously can be used to advantage and reflected in the precision of the entire procedure.

### Accuracy

The accuracy of determinations must necessarily be evaluated by inference since the analyte concentration is unknown in an actual sample. Usually, periodic determinations of the concentrations of standard reference materials or samples of known composition with matrices similar to the analytical sample are used to estimate accuracy and uncover errors in procedure. In certain areas, such as clinical chemistry, preanalyzed blood serum samples with normal and abnormal analyte concentrations are available for evaluation. With other sample types, standards certified by the National Bureau of Standards are used. If the errors are outside the expected or desired range, a systematic search can be made for their source and for ways to minimize them.

For new analytical methods, samples should also be analyzed by independent methods when they are available.

An optimized emission spectrometer and analytical procedure should be capable of precision (random error)-limited accuracy. Since RSDs are typically in the range 1 to 5% for many procedures, accuracies in this range are often the goal. Precision-limited accuracy can be achieved, of course, only if systematic errors are minimized or eliminated.

With external standards, the accuracy depends directly on standard preparations and how well the standards approach the ideal standard (see Chapter 6). Incorrectly prepared standards lead to calibration errors which are usually seen as a deviation from expected calibration curve behavior or as an error when samples of known concentration are analyzed. If there is doubt as to the reliability of the standards, they should be analyzed by independent methods.

Less easily detected and corrected are the blank and analyte interferences discussed in Section 7-2. Blank or additive spectral interferences are generally detected by using blank solutions that contain the suspected interferent(s) at varying concentration levels and by changing the monochromator spectral bandpass. The high excitation energies of the plasma sources can give rise to very rich emission spectra and make spectral overlap problems more severe than with flames. Stray radiation can result from the scattering of intense spectral lines by grating imperfections or by other components in the spectrometer, or it can result from overlapping orders. In any of these cases a spectral interference can result. For these reasons, moderate-to high-resolution spectrometers are preferred in plasma emission methods. However, the 0.75-m spectrometers used with many ICP sources may not possess sufficient resolution for trace-element determinations in complex samples. The echelle grating spectrometers possess higher resolution in a moderate-sized instrument. With many plasma sources, background emission from the plasma continuum and from molecular bands is also dependent on the composition of the sample. Because of the difficulty of applying background correction methods with many multichannel spectrometers, such changes in the background can lead to inaccurate results. These effects often seem more severe in multielement determinations because so many elements are being determined simultaneously under compromise conditions. Elaborate correction procedures have been established to compensate for additive interference effects with multichannel spectrometers. These typically involve multichannel measurements on a series of single-element standards. All responses are then used to construct an interference matrix that accounts for the additive effects. Computerized systems are, of course, required to acquire and process the large amount of data. For the ICP, extensive tabulations of interelement effects and interelement corrections are available from manufacturers of ICP instruments. Most require a fairly good knowledge of the major sample constituents.

Nonspecific analyte or multiplicative interferences can be distinguished from specific analyte interferences by determining a different analyte in the same sample. A nonspecific effect of HCl on solution viscosity, surface tension, density, or vapor pressure, for example, alters the flow rate of solution into the atomizer and influences different analytes in the same manner. A specific interference of HCl on dissociation or ionization should not affect two different analytes in exactly the same way. Excitation interferences can be discovered by using several analyte lines of different excitation energies. Since temperature changes influence the fraction excited through the Boltzmann factor ($e^{-E_i/kt}$), the percentage change in signal should be different for lines of differing $E_i$ values. An excitation interference can also be distinguished from a nonspecific interference by observing the same line in absorption and emission. An ionization interference by a concomitant can be recognized by simultaneous emission measurements of an atomic and an ion line. Because of the high temperatures, many of the analyte specific interferences found so commonly in flames are negligible or reduced to tolerable proportions in plasma sources. For example, the Ca neutral atom line was found to undergo only a 2 to 3% enhancement when the Na concentration was changed from 0 to 0.7 wt % in an ICP source. Similarly, the depressant effect of $PO_4^{3-}$ on Ca was found to be negligibly small in the ICP. Enhancements of atomic and ionic emission have been found, however, in the DCP when high concentrations of alkali metals were present. These have been attributed to changes in the plasma properties and not to ionization suppression.

Some interferences can be distinguished by plotting the measured analyte signal vs. interferent concentration and noting the shape of the plot. In other cases, measuring the analyte emission as a function of the height of observation in the flame or plasma can give insight into the type of interference. Once the type of interference has been discovered, procedures can be designed to compensate or correct for the resulting errors as discussed in Chapter 6.

**TABLE 8-7**

Some flame emission detection limits

| Element | Wavelength (nm) | Flame[a] | DL ($\mu$g mL$^{-1}$) |
|---|---|---|---|
| Ag | 328.1 | N/A | 0.02 |
| Al | 396.2 | N/A | 0.005 |
| Au | 267.6 | N/A | 0.5 |
| Ba | 553.6 | O/A | 0.002 |
| Bi | 223.1 | O/A | 2 |
| Ca | 422.7 | N/A | 0.0001 |
| Cd | 326.1 | N/A | 2 |
| Co | 345.4 | N/A | 0.05 |
| Cr | 425.4 | N/A | 0.005 |
| Cs | 852.1 | O/A | 0.008 |
| Cu | 327.4 | N/A | 0.01 |
| Fe | 372.0 | N/A | 0.05 |
| Ga | 417.2 | N/A | 0.01 |
| In | 451.1 | N/A | 0.2 |
| K | 766.5 | O/A | 0.003 |
| Li | 670.8 | N/A | 0.000003 |
| Mg | 285.2 | N/A | 0.005 |
| Mo | 390.3 | N/A | 0.1 |
| Na | 589.0 | O/A | 0.0001 |
| Ni | 341.5 | N/A | 0.6 |
| Pb | 405.8 | N/A | 0.2 |
| Rb | 780.0 | O/A | 0.002 |
| Sn | 284.0 | N/A | 0.3 |
| Sr | 460.7 | N/A | 0.0002 |
| Ti | 399.9 | N/A | 0.2 |
| Tl | 377.6 | N/A | 0.02 |
| V | 437.9 | N/A | 0.2 |
| Zn | 213.8 | O/A | 50 |

[a] N/A, Nitrous oxide–acetylene; O/A, oxygen–acetylene.

## Detection Limits

The detection limit (DL) for an element on a given instrument is quite useful as a figure of merit since it defines the lower limit of the concentration range that can be used. Detection limits are evaluated as discussed in Chapter 6 by measuring the calibration curve slope $m$ and the blank standard deviation ($DL = ks_{bk}/m$). It is frequently found that DLs calculated from standards prepared in a matrix that duplicates that of the analytical sample as closely as feasible are significantly higher than those obtained on standards prepared in a pure solvent. Since the latter are usually reported by instrument manufacturers, they should be used with some caution. In any case, DLs obtained experimentally or reported by others can often determine whether or not a particular determination is even feasible. Small differences in DLs between workers or instruments are usually insignificant, while differences of one or more orders of magnitude are usually significant.

The best detection limits reported to date for flame emission methods are listed in Table 8-7. These were obtained with several different instruments, nebulizer-burner types, flame types, and measurement electronics. The DLs reported do give an excellent indication of the elements for which flame emission is highly sensitive. As would be expected, the alkali metals and alkaline earths have quite low DLs by flame emission spectrometry, whereas elements with UV resonance lines (e.g., Zn) and small Boltzmann factors have poor DLs.

Detection limits are reported in Table 8-8 for ICP and DCP sources. These are the best detection limits obtained to date with pneumatic nebulization and research-grade instrumentation. In general it can be seen that the ICP achieves superior powers of detection compared to the DCP, but often by less than one order of magnitude. For many elements a comparison with Table 8-7 reveals that ICP detection limits are significantly lower than those reported by flame emission. The alkali metals are the major exceptions. For some elements with the ICP source, detection limits have been reported in matrices other than distilled water. These have revealed that powers of detection do not significantly deteriorate in some "real-life" matrices. Detection limits can vary significantly, however, with the spectrometer quality, throughput, and the type of nebulizer. In many cases, particularly for the ICP, DLs obtained in practice with commercial instruments may be significantly worse (e.g., by a factor of 10 or more) than the values shown in Table 8-8. With ultrasonic nebulization and research-grade instrumentation, the best DLs obtained can be a factor of 10 lower than those given in Table 8-8.

If random noise in the blank limits measurement

**TABLE 8-8**
ICP and DCP detection limits

| Element | ICP (ng mL$^{-1}$) | DCP (ng mL$^{-1}$) |
|---|---|---|
| Ag | 0.2 | 2 |
| Al | 0.2 | 2 |
| As | 2 | 45 |
| Au | 0.9 | 3 |
| B | 0.1 | 5 |
| Ba | 0.01 | 2 |
| Be | 0.003 | 0.5 |
| Bi | 10 | 75 |
| Ca | 0.0001 | 0.2 |
| Cd | 0.07 | 0.5 |
| Co | 0.1 | 1 |
| Cr | 0.08 | 1 |
| Cu | 0.04 | 2 |
| Er | 1 | 5 |
| Eu | 0.06 | NA |
| Fe | 0.09 | 3 |
| Ga | 0.6 | 38 |
| Ge | 0.5 | NA |
| In | 0.4 | 38 |
| La | 0.1 | 2 |
| Mg | 0.003 | 0.2 |
| Mn | 0.01 | 0.5 |
| Mo | 0.2 | 0.5 |
| Na | 0.1 | 0.05 |
| Ni | 0.2 | 2 |
| P | 15 | 75 |
| Pb | 1 | 23 |
| Pt | 0.9 | 26 |
| Sb | 10 | 3 |
| Sc | 0.4 | NA |
| Si | 2 | 15 |
| Sr | 0.002 | 2 |
| Te | 15 | 75 |
| Th | 3 | NA |
| Ti | 0.03 | 1 |
| V | 0.06 | 8 |
| W | 0.8 | 30 |
| Zn | 0.1 | 2 |
| Zr | 0.06 | 8 |

precision near the DL, S/N theory can be used to predict DLs and their variation with experimental variables. Thus the discussion in Section 8-5 on optimizing the S/N at low analyte concentrations applies to optimizing the DL.

## 8-7 METHODOLOGY AND APPLICATIONS

The analytical sample presented to the flame or plasma source is most often in the form of a solution. Hence the bulk sample, if it is not initially in solution form, must first be dissolved and/or decomposed. Next, sample cleanup methods are used, if necessary, for separation of the analyte, removal of primary interferences, or preconcentration of the analyte. Finally, the analytical sample itself is prepared along with the appropriate blank and standard solutions. The exact treatment of the samples and standards depends on the specific method to be utilized. The most common sample dissolution and separation methods employed in atomic spectrochemical methods are reviewed in Appendix E.

A major advantage of the emission spectrometric methods is their ability to accept samples without a great deal of sample pretreatment. Because of the low detection limits of the plasma emission methods, preconcentration methods are not often needed. Although some of the procedures described in Appendix E may be needed for sample dissolution, in general, extensive separation and sample treatment steps are not required with these sources. Flame emission methods more often require sample preconcentration and the addition of chemicals such as ionization buffers and complexing or releasing agents.

In addition to the sample treatment discussed above, emission spectrochemical methods usually require a choice of instrument and conditions as well as the specific methodology to be employed (external standard, internal standard, standard addition method, etc.). The particular solvent to be used may also be an important factor because of sample insolubility in aqueous media. Samples in organic matrices, for example, may dictate that a DCP or flame technique be used, since ICPs are less tolerant of the solvent. This section begins by considering the identification and selection of the emission lines that will be employed in quantitative methods. Then we consider briefly some of the most common procedures that are used in emission spectroscopy and the trade-offs involved in choosing a specific method. Finally, a few specific applications are presented to illustrate how the techniques are used in practical situations.

## Identification and Selection of Analytical Lines

In order to use emission spectroscopy the analysis line(s) must be identified and selected. For qualitative analysis, wavelength identification is the primary task, while with quantitative methods selection of the appropriate line is of paramount importance.

*Wavelength Identification.* With flame emission the scarcity of lines makes wavelength identification quite straightforward. Since plasma sources produce much richer spectra, identification can be more difficult. Scanning spectrometers can simplify the task of wavelength identification. However, errors in the grating drive system and the wavelength readout should be assessed frequently by recording spectra from hollow cathode lamps or other standard spectral sources (e.g., low-pressure Hg arc lamps). The presence of spectral multiplets often greatly assists assignment of lines to elements because of their fingerprint nature and the stability of their intensity ratios under various conditions. Once a particular wavelength has been assigned, it is useful to determine whether the assignment is correct by using standards of different concentrations. With multichannel spectrometers, the positions of the various exit slits are often preset at the factory and not readily changed by the user. The exact wavelength regions passed by each slit can often be varied by refractor plate adjustments. These are usually made while radiation from a stable, narrow line source, such as a hollow cathode lamp, is incident on the spectrometer. The refractor plate position is adjusted to give a maximum photomultiplier tube output signal.

*Selection of Analysis Lines.* With flame emission it is often an easy task to pick the exact emission line to be used for the analysis. For most elements, the resonance lines are the most sensitive in flames, and they are thus used in the majority of flame methods. With plasma sources, the choice is more difficult because several emission lines from neutral atoms or ions of the same element may appear useful in quantitative determinations. These often vary widely in relative intensities and in the abundance of nearby, potentially interfering lines from other elements. With some direct readers, of course, the selection may have been substantially narrowed by the initial setup of the exit slits and detectors. Reference books, atlases of spectral lines, journal articles, and application notes from instrument manufacturers can be extremely helpful in choosing a good plasma analysis line. These often give linear ranges and potential interferences for a given emission line with a particular source/spectrometer combination. Often the expected concentration range will dictate whether to use a neutral atom resonance line, an ion line, or a line arising from transitions between excited atomic states. Resonance lines are, of course, most useful for trace constituents, but they are also susceptible to self-absorption at high concentrations. Lines of lower relative intensities are often used for minor and major constituents. If an internal standard is being used (see later), the availability of a suitable internal standard line also influences the choice of analysis line. In many cases two or three lines are monitored during the initial method development stages, and the most suitable of these is chosen for final use. With some computerized direct readers, it is possible to use more than one line in the final analysis procedure and obtain results that are the average of several lines. Also, more than one line can be used to extend the concentration range if samples vary widely in analyte concentration.

If multiple elements are to be determined with a scanning or slew-scan monochromator, it is usually desirable that the lines chosen be near each other in wavelength. This minimizes the time required to scan or slew between lines and allows the same detector to be used for all the lines. The specific source and spectrometer often plays a major role in choosing an appropriate line.

### Analytical Procedures

Most of the analytical methodology discussed in Chapter 6 has been applied to emission spectrochemical analyses. With flame emission, external standard methods are most common. In the usual analytical procedure, the flame emission spectrometer is first optimized for maximum signal or $S/N$ as described in Section 8-5. Then the analytical signal $E_E$ (total signal minus blank) is obtained for a series of standards of known analyte concentration $c_s$, preferably with added constituents to simulate the matrix of the analytical sample. The blank should be a reagent blank that contains all reagents used in the procedure at identical concentrations as in the sample. A calibration curve is obtained by plotting $E_E$ vs. $c_s$ or log $E_E$ vs. log $c_s$. The net signal is then measured for the analytical sample, and the analyte concentration obtained from the calibration curve or its mathematical equivalent.

With plasma emission, external standard methods are also used, but they require careful control over source conditions if high accuracy and precision are to be achieved. Because line intensities are highly dependent on source parameters, small changes between standards and samples can cause large errors.

*Internal Standard Methods.* The internal standard method (see Section 6-5) is widely used in emission spectrometry, particularly in flame emission determinations of Na and K, as discussed later in this section. In internal standard methods, a known concentration of the reference element is added to all samples, standards, and blanks, and the ratio of the intensity of the analyte to that of the reference element is obtained. If we consider the equation relating readout signal to ground-state number density (equation 8-4), we can see that a number of factors cancel or are in constant proportion when the ratio of two line intensities is obtained. Equation 8-12 expresses the measured voltage signal ratio as a function of the analyte number density, $(n_0)_a$, the reference element number density, $(n_0)_r$, and the difference in upper state energies $(E_r - E_a)$ for both transitions being resonance transitions.

$$\frac{(E_E)_a}{(E_E)_r} = \frac{k(n_0)_a e^{-(E_r - E_a)/kT}}{(n_0)_r} \qquad (8\text{-}12)$$

In order to make effective use of the internal standard method, the two line intensities should be measured simultaneously. With a direct-reading spectrometer or a spectrograph with multichannel detection, simultaneous acquisition of intensities is readily achieved. The internal standard method is, however, difficult to implement with a slew-scan or linear scan monochromator since measurements are necessarily taken at different times.

There are several requirements for a good internal standard element. First, the internal standard element should not be present in the samples or standards so that a known concentration can be added. This requirement is often difficult to meet with complex samples. In some cases a constituent already present in the samples at a fixed level can be used as an internal standard. Second, as can be seen from equation 8-12, the excitation energies of the two transitions should be similar to compensate for changes in source conditions (temperature, electron density, etc.). Third, factors that relate gas-phase concentrations to solution or solid concentrations should be affected similarly by changes in conditions. For solution samples and standards, for example, this requirement means that the two elements should have similar ionization potentials, similar chemical compound formation tendencies, and similar volatilities so that changes in atomization efficiencies are compensated. Changes in solution introduction rates should be well compensated by the internal standard method.

It is rare to find an internal standard element that meets all the requirements given. Instead compromises must be made, particularly with multielement determinations, where a single internal standard often serves for several analytes. The internal standard method can, however, be quite successful in reducing variations due to changes in source temperature, source position, and sample introduction rate.

*Standard Additions.* The method of standard additions is also frequently used in emission spectrometry for solution samples (see Section 6-5). Usually, the method of multiple additions is preferred because of its built-in check of linearity. Standard additions have also been applied to multielement determinations. Here a multielement standard is used to make the additions. Concentrated standards are added in small volumes to avoid changing the matrix of the analytical sample.

### Applications

Emission spectrochemical analysis has been used for a wide variety of sample types and almost every stable element. Here we consider one practical application of flame emission and two applications that use plasma excitation.

### Determination of Na and K in Blood Serum.

Although the use of flame emission has been decreasing in recent years, it is still widely used in clinical methods. Even here, however, many laboratories have switched to the use of ion selective electrodes, particularly for determining Ca in serum. Flame emission results give the total calcium present, which is not as clinically useful as the free, ionized calcium. For Na and K, flame emission is still widely used.

The normal ranges in serum are 135 to 142 meq $L^{-1}$ (3105 to 3266 $\mu g\ mL^{-1}$) for Na and 3.5 to 5.0 meq $L^{-1}$ (137 to 195 $\mu g\ mL^{-1}$) for K. Typically, samples are diluted 1:1000 with a Li internal standard solution (usually, $\approx$ 15 meq $L^{-1}$) before introduction to the flame spectrometer. Simple dilution produces a solution that can be readily nebulized without additional treatment (deproteinization or ashing).

To take full advantage of the internal standard method, a three-channel photometer, such as that shown in Figure 8-16, is used for simultaneous ratio recording. This ratio recording system outputs directly the ratio of the Na signal to the Li internal standard signal and the ratio of the K signal to the Li signal. Still other commercial systems allow Na and K to be measured relative to a Li internal standard and Li to be measured relative to a K internal standard. Because of the low excitation energy needed for alkali metals and to avoid ionization, air–propane flames are commonly used.

Standard solutions containing both Na and K are prepared. Usually, these range from 100 to 160 meq $L^{-1}$ of $Na^+$ and 2 to 8 meq $L^{-1}$ of $K^+$. Many clinical analyzers provide automatic dilutions so that standards and serum samples are identically diluted with the internal standard. A blank containing only the internal standard is used for the zero adjust. Often only a one- or two-point calibration is used. Control serum samples with known amounts of Na and K are frequently analyzed for accuracy assessment. Typical reproducibilities are $\approx$ 1 meq $L^{-1}$ for Na and 0.1 meq $L^{-1}$ for K.

### Water Testing with an ICP Spectrometer.

Several water resource laboratories use ICP emission spectrometry for testing the quality of various ground- and surface-water samples. In one method [J. R. Garbarino and H. E. Taylor, *Spectrochim. Acta, 38B*, 323 (1983)], 17 elements (Ba, Be, Ca, Cd, Co, Cu, Fe, Li, Mg, Mn, Mo, Na, Pb, Si, Sr, V, and Zn) were determined simultaneously with an automated spectrometer. A computer provided acquisition, processing, and storage of data generated by the spectrometer system. In addition, the computer controlled an automatic sampler and an automatic standardization system for introducing the external standards. A rotating refractor plate (spectrum shifter) was used for dynamic background corrections. Interelement corrections were applied to analytes that suffered spectral interferences. One particular standard was analyzed frequently for quality control of the analytical results. Approximately 100 samples could be analyzed daily for the elements listed above; the element profiles were automatically entered into a data-base management system on a laboratory computer.

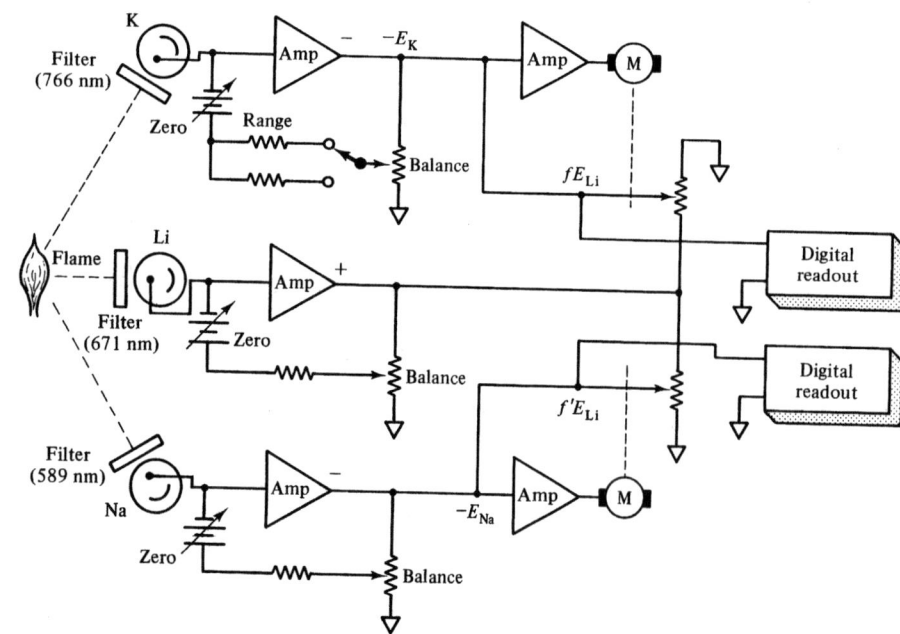

**FIGURE 8-16** Three-channel filter photometer for Na and K determinations. Lithium (central channel) is used as an internal standard. A zero adjust made with only the internal standard being aspirated allows automatic subtraction of flame background and dark current. When the upper servo system is at balance, $fE_{Li} - E_K = 0$. The readout displays the fraction $f = E_K/E_{Li}$. At balance, the lower readout displays the fraction $f' = E_{Na}/E_{Li}$.

*Marine Sediment Analysis with a DCP Spectrometer.* Marine sediments taken near offshore drilling installations often show heavy metal accumulations. These can be monitored with emission spectrometry. In one study [W. C. Grogan, *Spectrochim. Acta, 38B,* 357 (1983)], a DCP system was used because of its ability to handle solutions with a high content of dissolved solids. Sediments were digested with $HNO_3$ and diluted. A high concentration set was analyzed and the trace elements, Cd, Cr, Cu, Ni, Pb, V, and Zn determined. Further dilution was necessary in order to determine the major elements Ba, Ca, Fe, and Sr. Standards were matrix matched as closely as possible. The high concentrations of Ca were found to result in stray radiation errors for all elements except Cd. The errors were reduced to tolerable levels by adding Ca to all standards. In a related study, shellfish extracts were analyzed by the same method. Stray radiation errors were found to arise from both Ca and Mg. Correction equations were used to subtract the contribution of stray radiation. Excellent agreement between DCP results and certified values were found when standard materials of similar matrices were analyzed.

## PROBLEMS

**8-1.** Give a brief scientific reason why the following statements are true.
   **(a)** Ionization interferences are usually not as severe in the inductively coupled plasma (ICP) as in flames.
   **(b)** Echelle-grating monochromators or polychromators are increasingly being found in commercial ICP atomic emission systems.

**8-2.** The flame emission determination of Na in a biologically derived fluid is being considered. Burners and gases for all common flames listed in Table 8-3 are available. Discuss the most important factors that will influence the choice of which flame to use. Include the advantages and disadvantages of low and high temperatures for this determination.

**8-3.** Compare and contrast free-atom formation from a solution sample for a $N_2O-C_2H_2$ flame and an ICP. List all of the steps involved and draw conclusions as to which excitation source is most efficient for each step. Justify.

**8-4.** Compare and contrast the DCP and the ICP for the following sample types.
   **(a)** An aqueous solution sample.
   **(b)** A sample in an organic solvent.
   **(c)** A sample with a high particulate content.

**8-5.** Consider the list of ICP characteristics given in Table 8-5. Do these apply to the MIP and DCP as well? For each item in the table state whether it is applicable to these sources and why or why not.

**8-6.** Contrast $H_2$ and $C_2H_2$ flames in terms of background and atomization characteristics.

**8-7.** Calculate the $S/N$ and $S/B$ in an emission measurement limited by background emission shot and flicker noise with a slit width of 100 μm if $E_E = 0.010$ V, $E_{bE} = 0.050$ V, $mGK = 4.0 \times 10^{-6}$ V, and $\chi_E = 0.010$. Calculate the $S/N$ and $S/B$ at slit widths of 200 and 400 μm assuming that the background emission is a continuum. Justify the observed dependence on slit width.

**8-8.** Consider the flame emission measurement of a 50-μg $mL^{-1}$ Cu solution at 324.7 nm in a 2000-K flame. The atomizer and analyte spectroscopic factors are $e_f =$ 10, $F = 5.0$ mL $min^{-1}$, $\varepsilon_a = 0.03$, $Q = 10$ L $min^{-1}$, $A_{10} = 0.95 \times 10^8$ $s^{-1}$, $g_0 = 2$, and $g_1 = 4$. The pathlength $l = 1$ cm. A monochromator is used with a slit width $W$ of 50 μm, a slit height $H$ of 5 mm, a solid angle $\Omega$ of 0.02 sr, and an optics transmission factor $T_{op} = 0.6$. A PMT is used with a gain $m$ of $5 \times 10^5$ and a cathodic responsivity $R(\lambda)$ of 15 mA $W^{-1}$. The current-to-voltage converter has a transfer function of $10^7$ V $A^{-1}$. Calculate the analyte emission signal voltage $E_E$ under these conditions.

**8-9.** What types of data suggest that local thermodynamic equilibrium does not exist in an ICP?

**8-10.** Compare and contrast the thermal energy ($kT$) in a flame of 3500 K to that in the axial channel of an ICP at 7500 K.

**8-11.** Discuss the atomizer and excitation parameters that would be influenced by a sudden increase in fuel gas flow rate in a flame emission determination. How would these parameters change, and what would be the effect on the emission signal for Na in an air–$H_2$ flame? If the oxidant is also the nebulizing gas, what parameters would be affected by a sudden increase in oxidant flow rate? Will the Na emission signal increase or decrease or is the result unpredictable?

**8-12.** In an atomic emission spectrometry determination, the concentration of the analyte in the sample solution is in the region of the calibration curve where self-absorption is noted with standards. Discuss the various options that are available to obtain accurate analytical results.

**8-13.** What would be the effect on the $S/N$ of working in the self-absorption region? If dilution by a factor of at least 20 is needed to put the sample emission signal into the linear region, is it better from an $S/N$ standpoint to work in the nonlinear region or to dilute the sample? Assume that signal shot noise is limiting for both cases.

**8-14.** Discuss how the electron number density might be measured in a particular spatial region of an ICP.

# REFERENCES

The following references deal with flames and flame emission spectroscopy.

1. C. Th. J. Alkemade, Tj. Hollander, W. Snelleman, and P. J. Th. Zeegers, *Metal Vapors in Flames,* Pergamon Press, Elmsford, N.Y., 1982.

2. C. Th. J. Alkemade, and R. Herrmann, *Fundamentals of Analytical Flame Spectroscopy,* Halsted Press, New York, 1979.

3. C. Th. J. Alkemade and P. J. Th. Zeegers, "Excitation and De-excitation Processes in Flames," in *Spectrochemical Methods of Analysis,* J. D. Winefordner, ed., Wiley-Interscience, New York, 1971, pp. 3–125. An excellent treatment of the processes occurring in flames from a fundamental point of view.

4. R. Mavrodineanu, *Analytical Flame Spectroscopy—Selected Topics,* Springer-Verlag, New York, 1970.

5. A. Syty, "Flame Emission Spectroscopy," *in Treatise on Analytical Chemistry,* 2nd ed., pt. 1, vol. 7, chap. 7, I. M. Kolthoff and P. J. Elving, eds., Wiley-Interscience, New York, 1981.

6. A. G. Gaydon, *The Spectroscopy of Flames,* 2nd ed., Halsted Press, New York, 1974. An excellent introduction to flame phenomena from a spectroscopic viewpoint.

7. A. G. Gaydon and H. G. Wolfhard, *Flames, Their Structure, Radiation and Temperature,* 3rd ed., Chapman & Hall, London, 1971. A widely used book on the fundamental properties of flames.

8. J. A. Dean and T. C. Rains, eds., *Flame Emission and Atomic Absorption Spectrometry,* Vols. I–III, Marcel Dekker, New York, 1969, 1971, 1975. A multiple volume work written by experts in the field.

9. R. Mavrodineanu and H. Boiteux, *Flame Spectroscopy,* Wiley, New York, 1965.

The following references deal with plasmas and plasma spectroscopy.

10. V. A. Fassel and R. N. Kniseley, "Inductively Coupled Plasma Optical Emission Spectroscopy," *Anal. Chem., 46,* 1110A, 1115A (1974). Among the early ICP papers.

11. V. A. Fassel, "Simultaneous or Sequential Determination of the Elements at All Concentration Levels—the Renaissance of an Old Approach," *Anal. Chem., 51,* 1290A (1979). A review stressing the unique aspects of the ICP in multielement analysis.

12. R. K. Skogerboe and G. N. Coleman, "Microwave Plasma Emission Spectrometry," *Anal. Chem., 48,* 611A (1976).

13. "Plasma Spectrochemistry III," *Spectrochim. Acta, 41B,* 1–196 (1986). This entire issue of *Spectrochimica Acta* is devoted to proceedings of a conference on plasma spectrochemistry. Articles from earlier conferences may be found in "Plasma Spectrochemistry II," *Spectrochim. Acta, 40B,* 1–412 (1984) and "Plasma Spectrochemistry I," *Spectrochim. Acta, 38B,* 1–445 (1983).

14. H. R. Griem, *Plasma Spectroscopy,* McGraw-Hill, New York, 1964. A leading book on plasma diagnostics and physics from a spectroscopic perspective.

15. R. K. Winge, V. A. Fassel, V. J. Peterson, and M. A. Floyd, *Inductively Coupled Plasma-Atomic Emission Spectroscopy: An Atlas of Spectral Information,* Elsevier, New York, 1985.

16. M. L. Parsons, A. Forster, and D. Anderson, *An Atlas of Spectral Interferences in ICP Spectroscopy,* Plenum Press, New York, 1980.

17. R. M. Barnes, "Sample Preparation and Presentation in Inductively Coupled Plasma Spectrometry," *Spectroscopy, 1*(5), 24 (1986).

18. R. S. Houk, "Mass Spectrometry of Inductively Coupled Plasmas," *Anal. Chem., 58,* 97A (1986).

19. P. W. J. M. Boumans, ed., *Inductively Coupled Plasma Emission Spectroscopy,* pts. I and II, Wiley-Interscience, New York, 1987. Covers the fundamentals, instrumentation, methodology, and applications of ICP emission spectroscopy.

20. A. Montaser and D. W. Golightly, eds., *Inductively Coupled Plasmas in Analytical Atomic Spectrometry,* VCH Publishers, Inc., New York, 1987. Sixteen chapters written by experts on ICP devices and their applications.

The following are general references on emission spectroscopy and spectrochemical excitation (see also Chapter 7 references).

21. P. W. J. M. Boumans, *Theory of Spectrochemical Excitation,* Adam Hilger, Bristol, England, 1966.

22. C. Veillon, "Optical Atomic Spectroscopic Methods," in *Trace Analysis: Spectroscopic Methods for Elements,* chap. VI, J. D. Winefordner, ed., Wiley-Interscience, New York, 1976.

23. R. M. Barnes, ed., *Emission Spectroscopy,* Dowden, Hutchinson & Ross, Stroudsburg, Pa., 1976. This book is a collection of the "classic" papers in the field of emission spectrochemical analysis.

24. "Emission Spectrometry." A review of progress in the field that appears once every two years in the Fundamental Reviews issue of *Analytical Chemistry.*

# CHAPTER 9

# Arc and Spark Emission Spectrometry

The story of emission spectroscopy would certainly be incomplete without a consideration of arc and spark excitation. These sources were primarily responsible for the "coming of age" of emission spectroscopy. Before arcs and sparks became available, only flames were widely used analytically to excite atomic emission. With arcs and sparks it became possible to excite nearly all the stable elements in the periodic table. Many of the instrumentation developments that were made with arc and spark sources were carried over directly for use with the plasma sources that became popular in the 1970s.

Arc and spark discharges have been used as excitation sources for qualitative and quantitative emission spectrometry since the 1920s. Many new developments in arc and spark excitation occurred during the war years of the 1940s, particularly during the Manhattan Project. Commercial instrumentation became available during this period, and use of the photomultiplier tube instead of the photographic plate increased dramatically. Photoelectric detection greatly increased the throughput of quantitative determinations and led to the development of many new quantitative procedures.

With the dc arc some 70 to 80 elements have been excited. The major use of the arc is in qualitative and semiquantitative analysis, because the precision of quantitative determinations is not as high as desirable. The high-voltage spark is even more energetic than the arc; even the rare gases and the halogens can be excited in a spark discharge. The spark is capable of higher precision than the dc arc, and is applied primarily in quantitative determinations. Although plasmas have recently become the most widely used excitation sources for emission spectrometry, arcs and sparks are still used, particularly for the analysis of solid samples. In fact, as we shall see, one commercial instrument uses a spark discharge to vaporize solids prior to ICP excitation.

We saw in Chapter 8 that the ICP and DCP approach our ideal atomic emission source. However, there are many problems to be solved with plasmas. For example, the plasma sources are less than ideal when it comes to dealing with solid samples or samples in organic matrices. Thus much research is currently being devoted to the development of new excitation sources for atomic emission spectrometry.

This chapter begins by describing the arc and spark discharges that have proven useful in analytical emission spectrometry. These sources are compared to the ideal atomic emission source developed in Chapter 8. Several "miscellaneous" excitation sources are then considered. These include laser devices and reduced pressure discharges. A brief discussion of instrumentation for arc and spark spectroscopy follows with some consideration of the photographic instruments that are still found in many laboratories. The final section considers methodology and applications and emphasizes the analysis of solid materials by arc and spark methods.

## 9-1 ARC EXCITATION SOURCES

In this section we consider the characteristics, advantages, and limitations of various types of arc discharges, such as *dc arcs, ac arcs, controlled-atmosphere arcs,* and *gas-stabilized arcs.* Recent developments in arc excitation sources are outlined.

As applied to spectrochemical analysis, an arc is an electrical discharge between two or more conducting electrodes. One of the electrodes normally contains the sample as a powder, a solid mixture, or a solution residue. Emission intensities are normally integrated, either photographically or electronically, throughout the complete time of arcing, called the **burn.** Arcs can be free-burning, in air or an inert-gas atmosphere, or they can be gas stabilized. Free-burning arcs have been most often employed for spectrochemical analysis. Three distinct types of arcs are in use: the dc arc, the ac arc, and the intermittent or unidirectional arc.

### Free-Burning DC Arcs

The dc arc is the most commonly employed type of arc in spectrochemical analysis. It is conventionally characterized by its good detectability and poor precision. Although ionization can be substantial in an arc discharge, the arc produces a preponderance of neutral atom lines. In fact, neutral atom lines are often called **arc lines** or designated as type (I) lines in spectroscopic nomenclature. Thus an Ar(I) line means a line of neutral argon.

The dc arc consists of a continuous discharge of 1 to 30 A between a pair of metal or graphite electrodes. A simplified electrical circuit diagram is shown in Figure 9-1. Although a resistor could be used to control the arc current, an inductor is more common because of the considerable energy loss (through heating) in the resistor. The arc displays negative resistance properties because an increase in arc current results in a decrease in the voltage drop across the arc gap and a decrease in arc resistance, as can be seen in Figure 9-2. Current control of the arc is necessary because of the negative resistance characteristics; the current would increase without limit as the arc became more conductive. Precise current control helps to achieve a smooth burn and produces more reproducible emission intensities. To

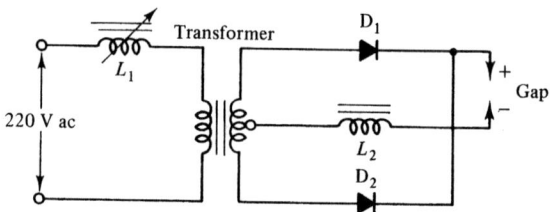

**FIGURE 9-1** Power supply for dc arc. Variable inductor $L_1$ is used to control the arc current, while inductor $L_2$ filters the pulsating direct current produced by diodes $D_1$ and $D_2$. The arc current is typically 1 to 30 A, while the gap spacing is 1 to 20 mm.

achieve better current regulation, the supply voltage should be much larger than the voltage drop across the arc. Because of this, arcs are usually powered from 220 V ac lines. Note from Figure 9-2 that the voltage does not change linearly from one electrode to the other, but drops substantially near each electrode. The **cathode fall** and **anode fall** regions occur because of an accumulation of electrons near the anode and positive ions near the cathode (localized space charges).

The most common electrode material is graphite, although occasionally metallic samples are themselves machined into an appropriate shape and used as electrodes. Graphite is inexpensive, available in high purity, and resistant to attack by most reagents. It is also a refractory material and this enables high-boiling materials to be vaporized. Most often the samples to be analyzed are solids; powders, chips, and filings are common. Samples are generally introduced into the arc by evaporation from a cup-shaped lower electrode similar in shape to one of those shown in Figure 9-3. Solid

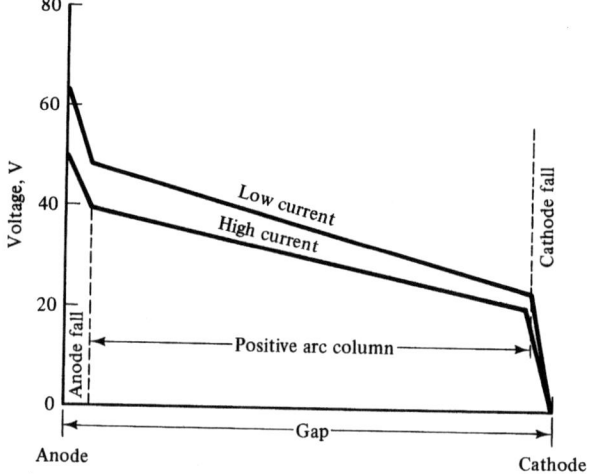

**FIGURE 9-2** Voltage profile from anode to cathode in a dc arc. Higher arc current results in lower resistance and lower voltage drop. Near each electrode a space charge exists, which is responsible for the anode and cathode fall regions.

**FIGURE 9-3** Graphite electrode shapes for dc arc. Most electrodes are made from 0.242- or 0.120-in.-diameter rods. The cup shapes are used for solid samples, while the conical shape is used as the counter electrode. This conical shape helps to stabilize the arc positionally.

samples are packed into the electrode. Alternatively, solution samples can be deposited on a graphite electrode and evaporated to dryness; the resulting residue is then analyzed in the arc. The upper, counter electrode often has a rounded, cone-shaped tip, as shown in Figure 9-3. To ignite the arc, the two electrodes are either brought into momentary contact, or a low-current spark igniter is used to provide initial ionization. The arc is sustained by thermal ionization of the gap material and by the supply of electrons and ions from the electrodes.

In the United States the arc is usually operated with the sample electrode as the anode and the counter electrode as the cathode. Cathodic sampling is most often employed in Europe. With anodic sampling, there is an upward action of the field on ionized materials. Only a fairly low concentration of ionized material exists in the arc column and little vapor escapes by sideward diffusion. With cathodic excitation, ionized vapors are subjected to downward forces in the arc column. The result is a low concentration in the column, and an accumulation of metal particles at the cathode, the so-called "cathode layer." Cathodic excitation is sometimes used to obtain low absolute detection limits due to enhanced emission in the cathode layer. However, intense background emission is also found in the cathode layer region, and signal-to-background ratios may be no better than for anodic sampling. Typical transit times for atoms in free-burning arcs are on the order of a few milliseconds.

Although the dc arc is not in exact thermodynamic equilibrium, localized regions can be considered to be in equilibrium. Thus the intensities of spectral lines are determined by the local excitation temperatures and can be calculated from the Boltzmann and Saha equations (see Chapter 7). Arc temperatures are generally in the range 3000 to 8000 K and depend almost linearly on the ionization potential of the material in the gap region. At a constant current, the energy dissipated, and thus the arc temperature, is proportional to the resistance of the arc plasma. When easily ionized materials

are present, the electron densities in the gap are high, the resistance between the electrodes is low and thus the temperature is low. Similarly, high ionization potential materials lead to high temperatures. This dependence of arc temperature on the nature of the sample is quite undesirable and often leads to severe matrix effects. The arc temperature also varies significantly in the axial direction. Higher temperatures are found in the anode and cathode fall regions than in the arc column itself. In the radial direction, the temperature is a maximum in the current channel and rapidly decreases with increasing distance. The low-temperature outer regions of the arc can contain high densities of ground-state atoms, which often leads to severe problems with self-absorption and self-reversal. This occurs because the radiation emitted in the high-temperature channel must pass through the outer fringes of the arc before it reaches the spectrometer entrance slit.

Another feature of a dc arc discharge is **selective volatilization**, which occurs because the electrodes are only slowly heated by the arc. Thus the most volatile materials vaporize first, followed by higher-boiling materials, as shown in Figure 9-4. Often samples are completely burned in a dc arc analysis. This typically requires a few minutes of burn time for normal samples. Because of drastic temperature changes that occur during the burn, line intensities can vary significantly with the sample matrix. Also, spectral interferences can readily occur because of selective volatilization. For example, in a complex sample, spectral interferences are more

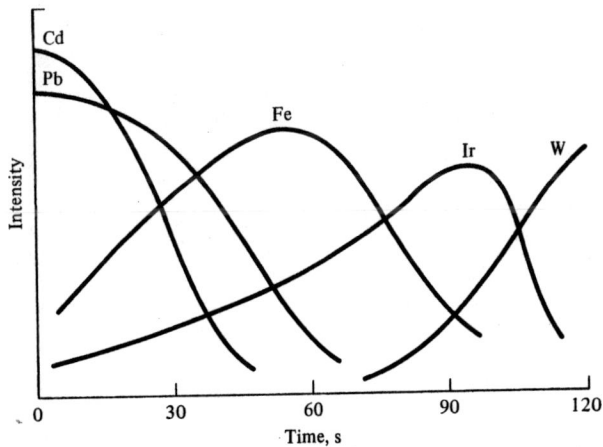

**FIGURE 9-4** Selective volatilization in a dc arc discharge. Volatile elements (Cd, Pb, Zn, As, Hg, etc.) vaporize completely during the early stages of the burn, while more refractory elements (W, Ir, Mo, Pt, etc.) are not significantly vaporized until later in the burn. Volatilization orders are difficult to predict because of nonuniform heating and changes in chemical composition during the burn.

likely to occur late in the burn when refractory materials with rich spectra are volatilized.

Selective volatilization can also be used to advantage to enhance line-to-background ratios and minimize interferences if the exposure period (integration time) is selected correctly. A low-boiling element such as gallium is sometimes intentionally added to the sample in the *carrier distillation* method. Although the exact mechanism is not understood, when the low-boiling element volatilizes it is thought to carry with it other easily volatilized elements, which can separate them from refractory materials. Selective volatilization is often reduced by adding a **spectrochemical buffer** to the sample. These are usually volatile, low-ionization-potential materials such as alkali and alkaline earth carbonates or halides. In many cases, graphite powder is added to the buffer to increase conductivity. An excess of buffer tends to keep the arc temperature constant and nearly independent of sample composition. Dilution with the buffer, of course, degrades detection limits.

The dc arc emits band and continuum radiation, which contribute to background. Band emission occurs from molecules and radicals that are stable in the cooler fringes of the arc. For example, in air, emission from the cyanogen radical (CN), formed from burning carbon electrodes in the presence of nitrogen, is quite intense. The CN band system can render the entire spectral region between 360.0 and 420.0 nm completely unusable for analytical work. Continuum emission can occur from the hot electrodes (blackbody radiation), from incandescent particles in the arc, and from radiative recombination and bremsstrahlung. One final problem with the free-burning dc arc is **arc wander**, which occurs because the arc column contacts the electrodes only at very small spots. Most of the volatilization of sample and electrode material occurs at these cathode and anode spots, which tend to move erratically over the electrode surface, causing different portions of the electrode to be sampled. The arc is also positionally unstable because of thermal currents that are generated during the burn. Arc wander is a major contributor to the poor reproducibility usually associated with the dc arc.

## Other Types of Arcs

### Controlled-Atmosphere and Gas-Stabilized Arcs.

To eliminate the CN band emission, arcs have been operated in atmospheres of Ar or He. Such environments also reduce continuum emission, provide higher temperatures, and lower the electrode consumption rate. The higher arc temperatures can enhance both neutral atom and ion emission. The increased line intensities and the lower background can

provide substantial improvements in detection limits. When operated in pure Ar atmospheres, the arc consumes electrodes and samples at extremely slow rates, which can result in long analysis times. For this reason a mixture of 80% Ar and 20% $O_2$ is often used. The added $O_2$ increases the rate of sample consumption without increasing CN emission.

The Stallwood jet shown in Figure 9-5 is commonly used for controlled-atmosphere arc excitation. The particular gas mixture used establishes the excitation conditions. The gas flow helps to reduce arc wander, which improves analysis precision. The gas stream also removes the absorbing vapor in the outer fringes of the arc, which reduces self-absorption. The cooling of the sample electrode helps to reduce selective volatilization.

### AC and Intermittent Arcs.

Two types of ac arcs have been employed. The high-voltage ac arc is operated at 2000 to 4000 V, while low-voltage ac arcs employ 100 to 400 V. In contrast to the continuous dc arc, the ac arc extinguishes at the end of each half-cycle. High-voltage arcs reignite automatically when the applied voltage exceeds the breakdown voltage of the gas. Low-voltage arcs are reignited on each half-cycle by a low-current spark discharge. An intermittent arc is similar to an ac arc except that the electrode polarity does not

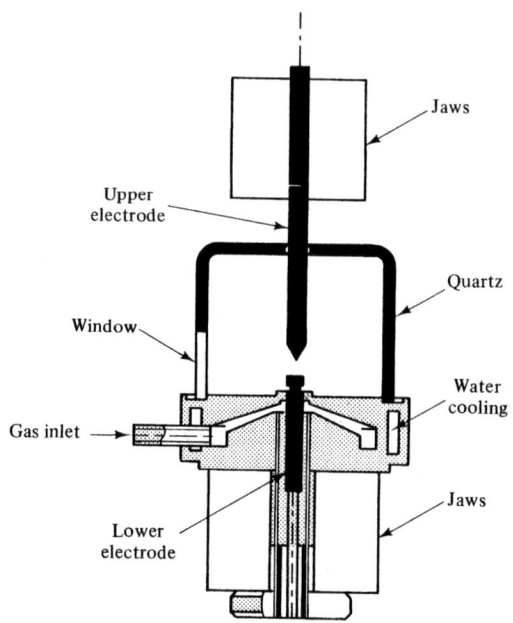

**FIGURE 9-5** Stallwood jet cross section. The sample electrode is placed in a water-cooled holder, and a mixture of Ar and $O_2$ flows upward around the electrode as it burns. The entire electrode assembly is encased in a quartz envelope which excludes ambient air and creates a slight positive pressure of Ar and $O_2$.

change on each half-cycle. In all cases the extinguishing and reignition of the arc causes a different portion of the electrode to be sampled on each cycle; this provides greater precision than the erratic, slow-moving dc arc. The ac arc is, however, less sensitive than the dc arc because of the periodic changes in electrode polarity. The arc behaves like a cathode layer arc on one half-cycle and like an arc with anodic sampling on the other. Sample volatilization is more efficient when the sample electrode is anodic. On the cathodic half-cycle, signal-to-background ratios are reduced because there is less sample vaporization. With the same current magnitude and arc dimensions, there is less electrode heating in an ac arc than in a dc arc. This can be a disadvantage for difficult-to-volatilize materials, but an advantage with metallic electrodes that might melt under continuous arcing. The unidirectional, intermittent arc maintains the sample electrode as the anode on both half-cycles. It thus has the sensitivity of the dc arc, while maintaining the high precision of the ac arc.

### Arcs as Ideal Emission Sources

Much of the recent work with arc excitation has involved fundamental studies of the processes leading to atomic emission. For example, studies have been made of the effects of various inert and reactive gases in arc formation, reproducibility and excitation. Various new geometries such as U-shaped arcs and V-shaped arcs have been investigated. Several workers have also studied the influence of a variety of external parameters, such as magnetic fields and rotating electric fields on arc behavior.

Since the arc suffers from many nonidealities in its vaporization, atomization, and excitation roles, some research has been aimed at separating the various functions. For example, arcs have been used to vaporize samples prior to their introduction into plasma excitation sources. The microarc discharge shown in Figure 9-6 has been used for introducing small volumes of solution samples into MIPs and ICPs. The microarc/plasma combination separates in time the processes of desolvation and volatilization from those of atomization and excitation so that these can be separately optimized. Introduction of samples with the microarc does not seem to degrade the precision and sensitivity of the plasma source. Matrix interferences are not significantly different from those normally experienced with the plasma.

As can be inferred from the discussion above, arcs are far from meeting the criteria established for our ideal emission source. Arcs do not provide the controllable excitation energy desired. In fact, the energy dissipated in the arc depends highly on the sample constituents and their ionization potentials. The arc discharge does not occur in an inert chemical environment, and arcs are subject to severe matrix effects. Free-burning arcs often suffer from severe self-absorption and even self-reversal of analytically useful lines. Background emission can be severe unless controlled-atmosphere excitation is used. Although an arc is capable of exciting most elements, the reproducibility of the atomization and excitation conditions is not high. For these reasons the use of arc discharges in emission spectroscopy has decreased dramatically in recent years, although we may find more uses being made of arc discharges in systems that separate sampling and excitation.

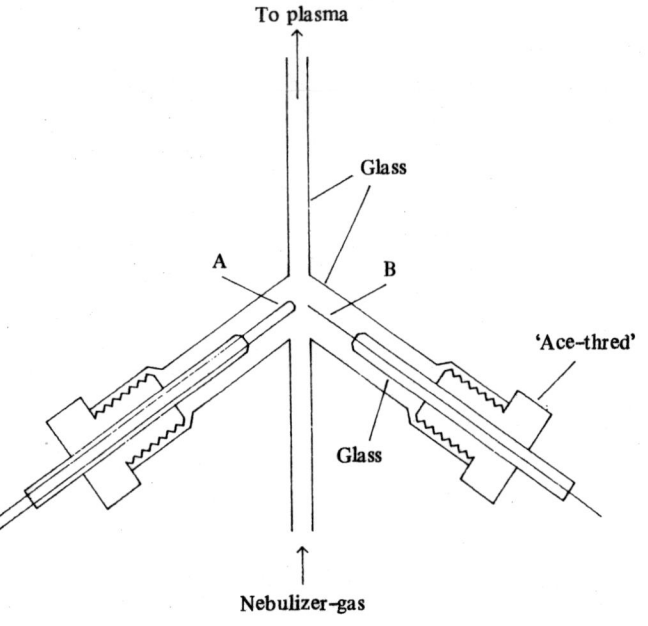

**FIGURE 9-6** Diagram of a micro-arc vaporizer for plasma atomic emission spectrometry. The arc is formed between a tungsten cathode loop (A) and a stainless steel anode (B). A small volume of sample (1 to 10 μL) is deposited on the cathode loop with a syringe. A high-voltage, low-current pulsating dc discharge (1500 V, 20 mA) desolvates the sample and vaporizes it into the nebulizing gas stream of the ICP or MIP. [From J. P. Keilsohn, R. D. Deutsch, and G. M. Hieftje, *Appl. Spectrosc.*, 37, 101 (1983). Reprinted by permission of *Applied Spectroscopy*.]

## 9-2   HIGH-VOLTAGE SPARKS AND OTHER EMISSION SOURCES

High-voltage spark discharges have long been used as excitation sources in emission spectrometry, especially in the ferrous metal industry. These sources are characterized by the high precision with which they can vaporize and excite many solid samples. The high-voltage spark discharge is an intermittent discharge rather than a continuous discharge like the dc arc. A single spark normally lasts a few microseconds. Electrode material is sampled many times during sparking in order to improve reproducibility. Because of the intermittent nature of the spark and the resulting time-dependent atomic emission, the spark discharge is an extremely complicated excitation source. The combination of the complex excitation processes and the inhomogeneities associated with many solid samples often leads to "matrix effects" with spark discharges. We have only recently begun to understand how sampling and excitation occur with high-voltage spark discharges. Fundamental mechanistic studies of spark excitation have led to several recent developments, including stabilized sparks and controlled-waveform discharges. In addition to sparks, many other sources have been proposed and studied for use in emission spectrometry. We consider here two sources that use lasers to vaporize samples, several low-pressure discharges, and two types of sources under development.

### The High-Voltage Spark Discharge

The sampling process in a high-voltage spark discharge is more favorable than that in a dc arc discharge. An arc tends to burn for some time at one or a few spots, which leads to selective and erratic volatilization. With a spark, electrode material is sampled by many successive discharges, striking different spots on the surface. This random, multiple sampling helps to improve the precision of spark emission measurements because of averaging over multiple spots. However, the spark is also not an ideal sampling device. The surface of the material sampled tends to change with sparking time, which leads to changes in spark spectra with time. This so-called "sparking-off" effect is often overcome by postponing measurements until a reproducible point in the sampling process has been reached. Such a prespark or preburn time is typically 1 or 2 minutes.

Although the spark discharge is an extremely energetic source with high peak currents and high power densities, limits of detection are usually higher than with dc arc discharges because states of high internal energy are produced, and the extent of ionization is higher. Also, very little sample is ordinarily consumed during sparking. Doubly ionized elements can produce quite intense spectra in a high-voltage spark discharge. In fact, ionization is so common in spark excitation that ion lines are often simply called **spark lines**. In spectroscopic terminology ion lines are designated as type (II), or type (III) lines, where the Roman numeral indicates the state of ionization plus one. Thus a singly ionized element produces a type (II) line, a doubly ionized element a type (III) line, and so on. The larger number of ionized and excited states produced in sparks results in spectra that are more complex and less intense than those produced by dc arcs.

*Spark Sources.*   Simplified diagrams of two high-voltage spark source designs are shown in Figure 9-7. To regulate the breakdown of the analytical spark gap, a control gap (Figure 9-7a) or an electronic switch (thyratron tube in Figure 9-7b) is used. A fixed control gap allows the spark to fire only after a particular capacitor voltage is reached (voltage-thresholded spark), while a rotary gap can be used for time-thresholded operation. The thyraton-triggered source of Figure 9-7b can be operated in either a voltage- or a time-thresholded mode. Once the analytical gap breaks down, a high current-density discharge channel is formed between the two electrodes. The duration of a single spark is generally a few microseconds. Conditions are arranged so that several hundred breakdowns occur per second, and the radiation emitted by several thousand individual sparks is integrated to improve precision. In spark excitation, the sample electrode is arranged to be initially the cathode because vaporization of cathodic electrodes is more rapid than that of anodic electrodes.

The sources shown in Figure 9-7 both operate by overvolting the analytical gap. Walters and coworkers have described an entirely different approach to spark source design in which gap breakdown occurs at the open circuit end of a transmission line connected to a radio-frequency power supply. Once gap breakdown occurs, a separate current source is used to inject current into the conducting gap. An arbitrary current waveform can be injected into this quarter-wave source because the gap ignition circuitry is completely separated from the current supply circuitry.

Samples for spark excitation are most often machined into the proper shape to serve as electrodes. In some cases they can be solutions or solution residues. Conducting samples are often ground flat and used as one electrode with the other being a pointed graphite electrode (point-to-plane configuration). Powders can be mixed with graphite and pressed into pellets, which then serve as the planar electrode. Solutions have been introduced directly in the form of an aerosol from a pneumatic nebulizer both with and without prior de-

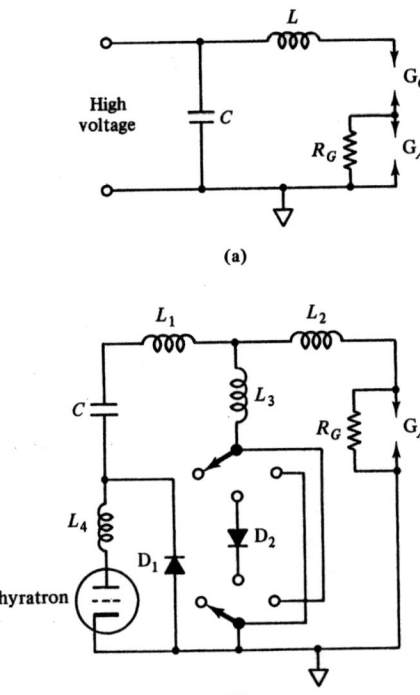

(a)

(b)

**FIGURE 9-7** Simplified circuit diagrams of common high-voltage spark source designs. In (a) a high-voltage power supply charges capacitor $C$ to a voltage in the range 5 to 20 kV. The control gap $G_c$ is used to regulate the breakdown of the analytical gap $G_A$. The control gap is arranged to breakdown at a higher voltage than the analytical gap by electrode spacing and gap environment. When the control gap becomes conductive, the capacitor discharges through the analytical spark gap. The analytical gap is formed from metallic or graphite electrodes spaced a few millimeters apart. The discharge can be made unidirectional by putting a diode in parallel with the analytical gap. In another modification, the fixed control gap is replaced by a rotary gap, which can synchronize the firing of the spark to the power line frequency. In (b) an adjustable waveform thyratron triggered source is shown. The anode of the thyratron, connected to $L_4$, holds off the full capacitor voltage until a trigger pulse, applied to the thyratron grid, causes the thyratron to conduct. When the thyratron is nonconductive, diode $D_1$ carries the current. Diode $D_2$ is used to control the direction of the current through the analytical gap. Several different discharge current waveforms can be produced with this source. Since an independent trigger source is used, either voltage or time thresholding can be chosen.

solvation. More common are rotating disk electrodes, which contact the solution sample and transfer some to the analytical gap, or porous cups in which the sample reaches the surface of the electrode via capillary action.

Evaporation of a solution onto the surface of a copper, silver, or treated graphite electrode is also common. With either solutions or solids, a very small amount of sample is vaporized by each individual spark ($<$ 1 $\mu$g). Because of this, matrix and interelement effects are substantially lower than those encountered with dc arcs. Selective volatilization is also not as bothersome as with the dc arc because of the short violent sampling and the low duty cycle (short on/off time ratio) of the spark. The latter keeps the electrodes relatively cool, which further reduces selective volatilization.

In recent years much work has been done to produce positionally stable spark discharges. While the random sampling position of a normal spark is sometimes advantageous to average inhomogeneities, surface information is lost in this process. Positionally stable discharges would also allow spatial resolution of the emission from the spark. Hence regions of high background emission or intense continuum emission could be masked so that radiation from them does not reach the spectrometer. At present the design of positionally stable discharges has centered around directing inert gas flows from one electrode to the other and on control of electrode shape. Rotation of the sample electrode also aids stability.

In the past, the spark source was usually operated in analytical applications so that a damped oscillatory discharge was obtained as shown in Figure 9-8. Values of circuit resistance, inductance, and capacitance were changed to vary the discharge waveform. Damping of the resonant circuit occurs because of losses in the analytical gap, in the switch or control gap, and in the circuit resistance. Because of the temporal variations

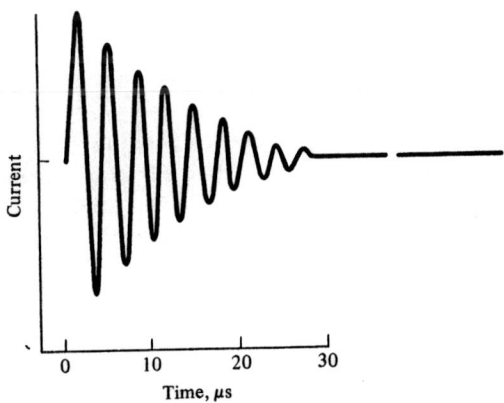

**FIGURE 9-8** Current waveforms for a single high-voltage spark discharge. The oscillation frequency is $[(1/LC) - (R^2/4L^2)]^{1/2}$. If $R$ is in ohms, $L$ in henrys, and $C$ in farads, the frequency is in hertz. Typically, the frequency is about 1 MHz. Peak currents range from 500 to 2000 A, but the average current is only a few amperes or less.

in the discharge current, excitation conditions in such a spark were highly time dependent. Today, however, most spark sources are unidirectional. Such sources give higher precision because of lower drift. Some of the newer spark source designs allow quite elegant control over the discharge waveform. The quarter-wave source mentioned earlier allows arbitrary waveforms to be used. Further investigations of tailored waveforms for high-voltage spark discharges are warranted.

### Processes in Spark Formation and Decay.

The processes that occur during the buildup and decay of a high-voltage spark are extremely complex. It is only in recent years that we have begun to understand some of these processes and to take advantage of this knowledge. We will present here a qualitative description of the stages of spark formation and decay. For more extensive treatments, the reader should consult the references for this chapter.

We will assume here that a triggered, non-quarter-wave spark source is used. After the source has been triggered, the capacitor and gap voltage begin to increase. Typically this voltage buildup stage lasts from 100 to 500 ns. Initial spark formation begins with the formation of a plasma in the gap. This plasma consists of a thin streamer of charge carriers which cross the gap and define the current channel between the electrodes. Emission from ionized atmospheric species and continuum background are observed in the early stages of a spark. During these stages the channel is anchored to the cathode, and a cathode spot forms. Current is then conducted through the spark channel. This stage usually lasts about 100 $\mu$s.

The mechanism of energy transfer from the current channel to the electrode surface is not entirely clear, but positive-ion bombardment (sputtering), positive-ion neutralization, neutral-atom condensation, and the return of previously ejected electrons to the electrode have been proposed. Bulk heating also occurs, which leads to some surface melting. Charge transfer (neutralization) appears to be the dominant mechanism in some sparks. Collisional excitation and ionization of electrode material may occur in the space-charge region near the electrode surface or material may be vaporized directly in excited or ionic states. During the sampling stage, intense continuum emission is observed near the cathode as the current increases. Electrode vapor then begins to propagate into the channel, where it interacts with plasma material.

At the current peak, the temperature in the discharge channel may reach 40,000 K, and the intensities of ion lines are quite high with respect to neutral atom lines. (Because of the temporal and spatial inhomogeneity it is probably improper to attempt to associate any single temperature to a spark plasma.) Ionic lines tend to show intensity-time variations that follow roughly the current waveform. Neutral atom lines tend to reach their maximum intensities well after the current peaks. When the current decreases, ion–electron recombination can occur along with various decay processes to produce lower-energy states. Energy transfer from metastable species has also been proposed as a means of producing excited analyte atoms. Neutral atom emission is most pronounced in the radial wings of the discharge as the current is decreasing.

After the current ceases, emission disappears and a postdischarge toroidal structure is formed. This torus has been observed by Schlieren photography. The exact mechanism of torus formation and decay are not well understood, but the torus has been shown to contain ground-state analyte atoms as well as ions. Reexcitation of the torus material has been suggested as a way of separating spark sampling from analyte excitation. The postdischarge vapor cloud can exist several hundred microseconds after the current ceases. Typically, there are 100 $\mu$s to 10 ms between discharges. The processes above are then repeated for the total number of sparks used. It has been found that the breakdown voltage for later discharges in a train of sparks is lower than that for single discharges or low-repetition-rate discharges. It is reasonable to presume that ions and metastable species from previous discharges are present at the beginning stages of later discharges and responsible for this reduced breakdown voltage.

### Time-Resolved Spectroscopy.

As discussed above, ionic and atomic analyte emission can vary dramatically with time during a spark discharge. In some cases, the background emission and the analyte emission show significantly different intensity-time behavior. In such cases, time-resolved measurements of the analyte emission signal, rather than integration throughout a complete train of sparks, have been shown to improve the signal-to-background ratio substantially.

Several different approaches to time-resolved measurements have been proposed. Usually, it is desirable to obtain microsecond or better resolution. Time resolution can be accomplished optically by rotating a mirror so that the spark image is swept across the spectrometer slit at the desired time. Alternatively, with electronically triggered spark sources, gated-integration techniques (e.g., boxcar integration) can be used for single wavelength measurements over a particular time window. It is also possible to gate an image intensifier tube so that multiwavelength measurements can be made.

Much of the fundamental information regarding behavior of electrode material with time has been gathered using time-resolved spectroscopy. Several workers have also demonstrated the enhancement in signal-to-noise ratios that can be achieved with such techniques. Unfortunately, only a very small portion of the elegant work done by various researchers in the past 15 years has found its way into commercial developments.

*Other Uses of Sparks.* Although the most common use of spark discharges is in the analysis of solid samples, recent studies have shown that linear calibration curves over several decades in concentration can be obtained with solution samples introduced after desolvation. A positionally stabilized, electronic-triggered spark source was used. Analyte emission was optimized relative to the background emission by gated-integration techniques.

Spark discharges have been used as detectors for gas and liquid chromatography. The spark discharge is easy to adapt for GC detection since the sample is already in the gas phase. Since it totally atomizes the GC effluent, the spark is free from any structural effects. The analyte emission observed can be directly correlated with concentration. Detection of nonmetals, such as C, B, N, O, and S, was demonstrated. In fact, it was possible to calculate the empirical formulas of several organic materials as they eluted from the GC column. Spark discharges have not been employed as successfully for detection of LC eluates.

Like arcs, sparks have also been used for introducing samples into other excitation sources. Because solid samples are not easily introduced into plasmas, the spark discharge is used in one commercial instrument to vaporize solids into the support gas stream of an ICP. The spark carries out the sampling and vaporization of the solid material and the ICP finishes off the atomization and provides the excitation energy. Such a combination utilizes the high precision of spark sampling and the high efficiency of the ICP for atomization and excitation.

*Conclusions.* The high-voltage spark is less than ideal in many of its characteristics. Its high energy produces quite complex spectra with significant background. Solution samples are not as conveniently handled as they are with ICP and DCP sources. Spark sources are neither inexpensive to purchase nor inexpensive to maintain. Simplicity of operation is not a feature of traditional spark sources.

Considerable progress has been made in recent years in understanding the mechanisms involved in energy transfer and excitation of analyte species. Electronically triggered spark sources have been developed in which the current waveform can be varied from pulsed to unidirectional to fully oscillatory. This can provide a degree of control over spark conditions not previously

feasible. The application of modern computer technology to spark instrumentation should also lead to many improvements. Although developments in plasma sources have preoccupied manufacturers of emission spectroscopy instrumentation, these sources have not solved the problems associated with analyzing solid materials. It is here that the spark discharge still has an important role to play.

## Miscellaneous Excitation Sources

Because no emission source truly meets all the criteria established for the ideal source, much current research is devoted to the development of new sources and the study of their characteristics. We will consider here only a few of the most useful or promising sources.

*Laser Microprobe.* A high-powered, pulsed laser beam (ruby, Nd:glass, Nd:YAG, flashlamp-pumped dye) when focused to a small-diameter spot (5 to 50 μm) is capable of vaporizing solid materials, even if the sample is not electrically conductive. When the laser radiation strikes the surface, it produces a high-temperature vapor plume of atoms, ions, and molecules. The most successful applications of this principle have used the laser for vaporization and a spark discharge for excitation, as shown in Figure 9-9. The laser and spark are combined with a microscope objective for viewing the target, selecting the desired sampling area, and focusing

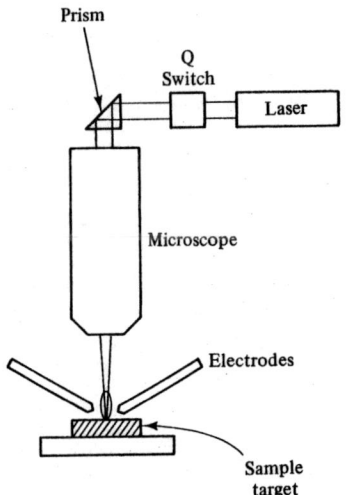

**FIGURE 9-9** Diagram of laser microprobe. Radiation from a *Q*-switched laser is focused onto a target with a conventional microscope. The laser energy causes a small volume of the sample to vaporize. A low-voltage spark discharge ignites spontaneously when ions from the sample pass between the graphite electrodes a few millimeters from the surface.

the laser beam. Indeed, a commercial **laser microprobe** which used photographic detection was developed for analysis by this technique. With cross excitation by a spark, the spectrum obtained is determined primarily by the parameters of the spark source and not by the characteristics of the laser. Such an instrument has achieved quite spectacular results in determining trace elements in selected portions of single blood cells, for example. Unfortunately, laser microprobes are no longer being manufactured in the United States. With modern array detector and computer technology, such an approach would seem quite attractive.

### Laser-Induced Breakdown.

A newer, related technique, known as **laser-induced breakdown spectroscopy** (LIBS), uses a laser to cause dielectric breakdown of a gas located at the laser focal point. The fundamental frequency (1.06 $\mu$m) from a Nd:YAG laser can produce power densities of several MW cm$^{-3}$, enough power to break down most gases, including ambient air. The small spark that results produces atoms and ions in excited states that radiate for a few microseconds following the nanosecond laser pulse. Time-resolved techniques have been used to discriminate against the intense continuum that is produced during and immediately following the laser pulse. The technique has been used to analyze dust particles in the ambient air. In addition, sample cells have been designed so that gaseous samples or aerosols can be introduced into an inert-gas (e.g., Ar) atmosphere prior to laser spark formation. Solution samples have been introduced as either dry aerosols, formed by desolvating the mist from a pneumatic nebulizer, or isolated droplets without desolvation. Although the LIBS technique is quite new, it shows considerable promise for spectrochemical analysis.

### Reduced Pressure Discharges.

Gas discharges operated at reduced pressures have been well studied by physicists. Some are beginning to be used in spectrochemical analysis. The **glow discharge** is generated between two planar electrodes in a cylindrical glass tube, which is filled with gas to a pressure of a few torr. A dc voltage source and a current-controlling series resistance are connected to the electrodes. Typically, a normal glow discharge occurs with currents of $10^{-4}$ to $10^{-2}$ A and voltage drops of a few hundred volts across the electrodes. The presence of metallic electrodes in the tube produces the spectrum of the metals present. In the late 1960s Grimm took advantage of this and introduced a new type of glow discharge in which conductive solid samples could be readily inserted for analysis, as illustrated in Figure 9-10. In the *Grimm discharge*, Ar is used as the filler gas at pressures of 10 to 12 torr. This discharge has achieved detection limits in the $\mu$g g$^{-1}$ range for many elements. Grimm discharge sources are commercially available.

The **hollow cathode discharge** has also been em-

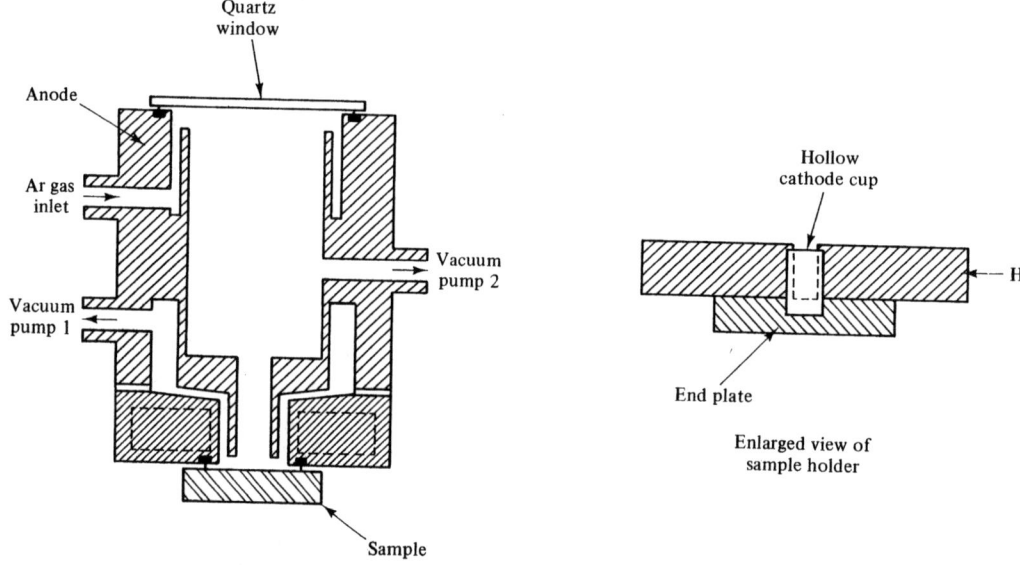

**FIGURE 9-10** Grimm discharge for emission spectrochemical analysis. Solid samples are placed in the hollow cathode cup. The solid sample is clamped to the cathode block and serves as the active cathode surface. Ar is used as the filter gas at pressures of 10 to 12 torr, and currents are 0.1 to 0.2 A. Positive Ar ions formed by collisions with energetic electrons sputter the sample from the surface. Sample vapor then enters the discharge region where excitation and emission occur.

ployed in atomic emission spectrometry in addition to its widespread use as a spectral source in atomic absorption and fluorescence. As discussed in Section 4-2, the low gas temperatures in a hollow cathode tube result in very narrow spectral lines, which are, of course, advantageous in emission techniques. Demountable hollow cathodes are required for emission spectrometry so that the cathode material can be changed. Unfortunately, the hollow cathode tube must be brought to atmospheric pressure each time the cathode is changed. The time-consuming evacuation and gas-filling operations make this emission source less than ideal for rapid, high-throughput determinations. Hollow cathode discharges are advantageous, however, for solid samples and for the analysis of refractory materials. Like the glow discharge, the primary mechanism for sample introduction is sputtering. Since the cathode remains cool, sputtering can be accomplished without significant selective volatilization. Detection limits for many metals are better than those obtained with atmospheric pressure discharges. Extremely good detection limits have also been achieved for nonmetallic elements, such as sulfur, phosphorus, and the halogens.

Low-pressure discharges have also been studied as atomization sources for solid samples with atomic absorption or fluorescence detection. One company markets a low-pressure discharge attachment for atomic absorption spectrophotometers. The rate of sample vaporization and atomic number density above the solid sample surface are enhanced significantly by directing gas jets at the surface during the discharge.

*Exploding Thin Films.*   Pulsed electrical discharges, such as sparks, are not convenient to use for nonconducting samples. Exploding conductors, however, have been shown by Sacks and coworkers to be quite useful for analyzing such samples. While exploding wires and foils have also been used, thin films have been shown to be more reproducible and convenient.

The film is formed by vacuum depositing a metal vapor on a flat, nonconducting substrate such as polyethylene. The film, which may be as thin as 25 nm, is exploded by discharging several hundred joules of energy through the high film resistance under a reduced-pressure inert-gas atmosphere. Apparently, the atmosphere near the film ionizes to initiate breakdown. Analytical applications are usually carried out by totally vaporizing the film.

The exploding thin-film technique has been shown to be capable of vaporizing materials with extremely high boiling points. For example, ZrC particulates have been completely vaporized even though the material boils at over 5000 K. Further mechanistic and analytical

studies of this promising method of excitation are under way.

*Theta Pinch Discharge.*   A theta pinch is created by pulsing a high magnetic field around a diffuse plasma. The field compresses and heats the plasma. The plasma can be created by any of several convenient methods such as a microwave discharge or a spark discharge. Such pinched discharges are used in nuclear fusion experiments. They can achieve temperatures an order of magnitude higher than those in sparks or exploding thin films. Although research on the analytical potential of such discharges has only begun, they are promising discharges for sampling nonconducting solid materials.

## 9-3  INSTRUMENTATION AND PERFORMANCE CHARACTERISTICS

The optical and electronic instrumentation used with arc and spark emission is nearly identical to that discussed in Chapter 8 in conjunction with flame and plasma emission. Thus direct-reading spectrometers are used for simultaneous, multielement determinations, while slew-scanned monochromators can be used for sequential determinations. Integration techniques are invariably used to increase signal-to-noise ratios and decrease the effects of source instability. Many of the older emission instruments, however, use photographic detection, and a brief discussion of such systems is included here. We also consider the performance characteristics of arc and spark emission systems.

### Photographic Detection for Arc and Spark Emission

The most common of the older instruments for arc and/or spark emission used spectrographs with photographic emulsions to detect the radiant energy emitted. Even with photoelectric detection, the photographic emulsion remains an excellent qualitative detector to view an entire region of the spectrum at once. The use of photographic emulsions for qualitative identification is considered in Section 9-4. Here we focus on the use of photographic emulsions for quantitative measurements.

When a photographic emulsion is exposed to radiant energy and later subjected to development (see Section 4-4), we obtain a photographic *response*, which is normally measured as the density of silver (weight of silver produced per unit area). The response is related to the radiant **exposure** (irradiance integrated over the exposure time $t$). After development of the plate in an appropriate darkroom, the density $D$ (similar to absorbance) is measured with a microphotometer called a **densitometer**. To calculate $D$, the densitometer measures the transmittance of a narrow beam of light through the exposed region of interest on the plate. Photoelectric detectors (phototubes or photomultiplier tubes) are commonly used. The 100% $T$ signal ($E_r = E_{rt} - E_d$) is obtained by moving the light beam to a clear area on the plate. The signal representing the transmitted flux ($E_s = E_{st} - E_d$) is obtained by passing the beam through the exposed region of interest. The density is then related to the measured signals by the equation

$$D = -\log T = \log \frac{E_r}{E_s} = \log \frac{E_{rt} - E_d}{E_{st} - E_d} \quad (9\text{-}1)$$

The relationship between density and relative exposure is established by means of a calibration curve known as a Hurter and Driffield or **H and D curve**, as shown in Figure 9-11. The slope of the linear region, known as the **gamma** of the emulsion, is a measure of the contrast. Unfortunately, gamma is also wavelength dependent, so that the calibration should be carried out at several wavelengths over the region of interest.

A widely used method to obtain a known series of relative exposures is the rotating-step sector method. Here a source with many lines, such as an Fe arc, is

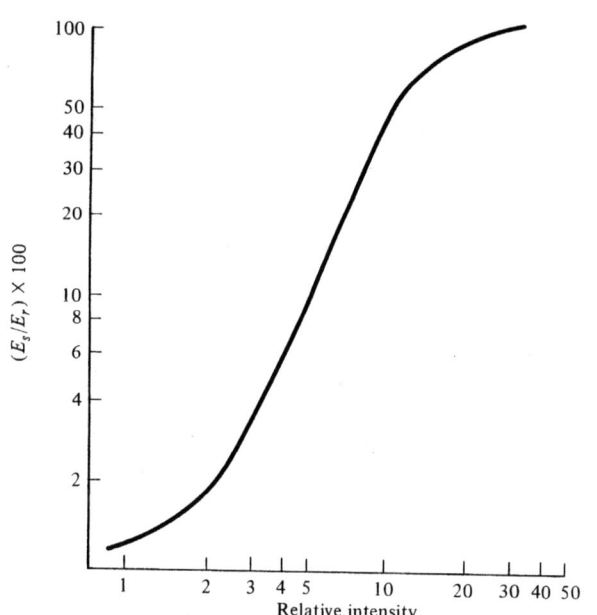

**FIGURE 9-11** H and D characteristic curve for photographic emulsion. At moderate exposures a linear relationship is obtained. The linear region of the curve is called the latitude, while the slope of the linear region is the gamma of the emulsion. At low exposures the emulsion shows inertia, while at high exposures photographic saturation becomes apparent.

used so that the calibration can be made over a large wavelength range. The sector, containing four to seven steps, is rotated rapidly in front of the entrance slit of the spectrograph. It is cut so that a series of exposures reaches the plate stepwise upon rotation. Many other techniques have been proposed to vary the exposure in a known manner.

The procedure for quantitative photographic work can be seen to be quite involved. First, the photographic emulsion is calibrated. Next, the spectra of interest are obtained and the plate is removed to a darkroom for development. After this, the plate is taken to a densitometer for measurement of the densities of the lines of interest. The densities obtained are then converted to relative intensities via the emulsion calibration curve. Of course in analytical spectroscopy, the relative intensities must be related to concentrations through an additional calibration curve, the working curve.

## Performance Characteristics

The most commonly employed sample types with arc and spark sources are solid samples, which makes it difficult to compare results obtained with these sources directly to those obtained with plasmas and flames on solution samples. These sources must expend part of their energy in sampling and vaporizing the solid materials to be excited.

Equation 8-4 gives the general relationship between readout signal $E_E$ and ground-state number density $n_0$ in all emission techniques. For solid samples introduced into arcs or sparks, we must modify equation 8-4 by introducing the relationship between ground-state number density and solid-phase concentration. Unfortunately, the latter relationship is extremely complex because it depends on the homogeneity of the sample, the area sampled by the arc or spark, the efficiency of vaporization, which may be time and matrix dependent because of selective volatilization, and many other factors. Linear ranges obtained thus depend highly on the type of sample and matrix. It is not uncommon to obtain working curves with arc discharges that bend back over at high concentrations because of self-absorption, self-reversal, photographic saturation, or other causes. For solid samples, just obtaining appropriate standard materials for calibration curves and testing can be a problem in itself. For powders that are to be mixed with spectrochemical buffers, pure standards can be added in different amounts. For metallic samples and alloys several standards are available in certified form from the NBS.

Signal-to-noise considerations with arc and spark emission spectrometers are very similar to those discussed in Section 8-5 for flame and plasma emission systems, although the expressions can be considerably more complex when solid sampling is included. In practice, precision is almost always evaluated experimentally by measuring replicate samples. With arcs and sparks, the stability of the source itself can limit precision through its effect on temperature and indirectly on atomization and excitation efficiencies. Similarly, precision can be limited by sample homogeneity. High-voltage spark excitation can under ideal circumstances give precision similar to that of an ICP (1 to 5% RSDs with standards and 1 to 10% with real samples). By contrast with free-burning dc arcs and solid samples that show selective volatilization, RSDs of better than 10% are rare.

Assessment of accuracy is usually a more difficult experimental task, since accuracy depends highly on the quality of the standards used and on the absence of interferences. For new analytical methods, samples should be analyzed by independent methods when they are available. Again, standard reference materials available from NBS should be used to assess accuracy if needed. In some cases only semiquantitative results are necessary. An entire semiquantitative methodology has been developed for use with dc arc excitation and photographic detection, as described in the next section.

Interferences can be a major problem in trace analysis procedures, particularly with solid samples. Solid samples are often extremely complex matrices with widely varying compositions. This can greatly complicate the preparation of standards and lead to severe matrix effects. Sample inhomogeneity is also a major problem with solid materials. Solids can, of course, be dissolved by the methods described in Appendix E, and the resulting solution analyzed by ICP or DCP methods. However, dissolution steps may introduce contaminants and be excessively time consuming when results are needed rapidly. Solids are most often determined by arc or spark excitation methods, which are well suited for direct analysis.

Detection limits in arc and spark emission also depend on a number of variables, such as sample matrix, source conditions, optical factors, and detector factors (photographic plate vs. photoelectric detection). Thus detection limits are not often meaningful because of the difficulty in reproducing them on other instruments or in other laboratories. Instead, Table 9-1 gives detection limits as ranges for several elements with a dc arc. Because solid samples are usually employed these ranges are given in wt %.

With spark discharge excitation, detection limits are usually somewhat poorer than those shown in Table 9-1, as discussed previously. However, some 40 ele-

**TABLE 9-1**
Approximate detection limits for dc arc

| Approx. DL (wt %) | Elements |
|---|---|
| $<10^{-4}$ | Li, Na, Cu, Ag |
| $10^{-4}$–$10^{-3}$ | K, Rb, Cs, Be, Mg, Ca, Y, La, Ti, Zr, V, Cr, Mo, Mn, Fe, Ru, Co, Rh, Ni, Au, Zn, Cd, B, Al, Ga, In, Tl, Ge, Sn, Pb, Nd, Eu, Th, Dy, Pd, Ho, Er, Tm, Yb, Lu |
| $10^{-3}$–$10^{-2}$ | Hf, Nb, Ta, W, Re, Os, Ir, Pt, Hg, Si, P, As, Sb, Bi, F, Th, U |
| $10^{-2}$–$10^{-1}$ | Se, Te, Ce, Sm, Gd |

ments can be detected in the range $10^{-4}$ to $10^{-2}$ wt %, with the majority being in the upper half of that range.

## 9-4 METHODOLOGY AND APPLICATIONS

Dc arc emission spectroscopy is useful in qualitative and semiquantitative analysis as well as in quantitative determinations. Hence this section begins by discussing this methodology. The general quantitative schemes that have proven useful for arcs and/or sparks are then described. The applications presented include the use of dc arc emission for samples of geological origin and the use of the high-voltage spark for the analysis of ferrous metals.

### Qualitative and Semiquantitative Methods

The dc arc is particularly useful in qualitative analysis because of its excellent detectability and its ability to excite so many elements. To establish the presence or absence of an element, two or three sensitive lines are usually employed. More than one line is usually necessary for positive identification because of the possibility of spectral interferences, stray radiation, and background emission at the wavelength chosen. The most sensitive lines of the element(s) of interest are normally found in standard wavelength tables, such as the MIT tables. Most tables give relative intensities of lines in the dc arc and high-voltage spark and the wavelengths of these lines; some tables are arranged by element, while others are arranged by wavelength.

When spectra are recorded photographically, a reference spectrum is usually recorded on each photographic plate to facilitate in wavelength identification. Typically, the spectrum from a low-current iron arc is used for this purpose. The Fe spectrum contains a multitude of intense, narrow lines throughout the UV-visible region. Each photographic instrument should have available a standard plate that contains the Fe spectrum and wavelength indications for the most intense lines. Identification of unknown samples is then made by comparison of the sample and reference spectra to the standard spectrum.

Because of the time involved in using photographic detection for quantitative measurements, several semiquantitative methods have been developed to provide more rapid results. In the simplest case, the method of **quantitative estimates**, the arc spectrum of the sample is recorded photographically, and the plate developed. The wavelengths to be used are then identified and a visual estimate of film blackness on a scale of 1 to 10 is made at each of the analytical wavelengths. Published sensitivity factors are then used with the blackness estimate to give a rough ($\pm$ 50 to 100%) estimate of the weight percent of the element(s).

In the semiquantitative method of Harvey, a weighed amount of sample is mixed with graphite and loaded into the cavity of an electrode. The sample is completely burned in the dc arc under prescribed conditions, and the photographic plate is developed. Line-to-background ratios are then measured on a densitometer and used along with published sensitivity factors to calculate the percent analyte in the sample. Since the method uses the background emission as an internal standard, careful control of conditions is necessary. When used correctly, it is capable of accuracies on the order of $\pm$ 30 to 50%. Neither emulsion calibration curves nor working curves are needed.

## Quantitative Methods

The considerations for true quantitative determinations are quite similar to those discussed in Chapter 8 for plasma and flame emission except that sample dissolution is usually not required. First the analysis lines must be selected and tested for interferences during the methods development stage. Then the specific methodology to be employed must be chosen.

With traditional arc and spark sources, it is often quite difficult to maintain reproducible source conditions. Emission intensities are also highly sample dependent, particularly with solid samples. This often means that matrix matching of the standards to the sample is required if external standards are to be used. Unfortunately, this means a prior knowledge of the sample composition. Hence quantitative arc and spark determinations most often involve the use of internal standards.

With the new generation spark sources (quarter-wave, positionally stabilized, etc.) discussed previously, it is possible to achieve highly reproducible source conditions. By combining such sources with time- and spatial-resolution methods, it should be possible to achieve much higher precision and accuracy than obtainable previously. Indeed, one commercial system (Shimadzu, Inc.) now uses time-resolved techniques for emission-intensity measurements.

In addition to the criteria given in Section 8-7 for choosing an appropriate internal standard, certain other factors must be considered with arc and spark excitation. For solid samples, the analyte and the internal standard should be similar in chemical properties so that their rates of vaporization are similar. If detection is photographic, the line pair should be close in wavelength and of similar intensity to minimize emulsion calibration errors. If a good internal standard is used, the internal standard method can be quite successful in reducing variations due to changes in source temperature and position (e.g., arc wander).

The method of standard additions is not often used in arc and spark spectroscopy, although it has been applied to solid samples. The major disadvantage, when applied to solids, is the time required to weigh and mix the standards and samples.

## Applications

Arc and spark spectrochemical analyses have been applied to many different analytes and sample types. We will consider here one common application of the dc arc and one widely used high-voltage spark method.

*Analysis of Silicate Minerals and Rocks.* The dc arc has found major application in the analysis of minerals, soils, rocks, and meteorites. We will discuss here a general scheme for these materials, the scheme of Ahrens, which provides a good compromise between analysis speed and accuracy. The method takes advantage of selective volatilization to reduce matrix effects.

Nearly all silicate rocks and minerals contain alkali metals at concentration levels exceeding 0.5%. The alkali elements distill rapidly in the dc arc, and because of their low ionization potentials, they exert a strong influence on arc characteristics (e.g., temperature and electron density). The method of Ahrens divides the elements to be determined into three groups: the alkali elements themselves, a group of volatile elements, and an involatile group. The alkali elements and the volatile group (Pb, Ga, Ag, Cu, Tl, Zn, Sn, Ge, In, Bi, Sb, and As) are determined during the alkali distillation phase. The sodium present in the rock or mineral is used as the internal standard for the other alkali elements. Added lithium is used as the internal standard for sodium. For the volatile group, indium added as the oxide is used as the internal standard because its volatility is in the middle of the group and indium is normally present at undetectable levels in such samples.

The emission signals from the involatile group (V, Ni, Co, Zr, Cr, Y, Sc, La, Nd, Sr, Ba, Mo, B, Be, Ti, Ca, Fe, Mg, Si, and Mn) reach their maximum intensities in the arc well after the alkali metals have distilled. For determinations of these elements, samples are mixed with graphite, and the arc is burned until complete consumption has occurred. Photographic measurements are begun after the alkali metals have distilled as indicated by a change in the color and visual appearance of the arc. Relative standard deviations on the order of 2 to 5% have been achieved with careful attention to details.

*Determination of Metals in Steels.* The high-voltage spark has been extensively used in the ferrous metals industry. Spark spectrometers are often found on the floors of large foundrys, where they can be used to give rapid results. Direct-reading spectrometers are often used for multielement determinations. It is important in the steel industry to determine the composition of a melt rapidly enough that it can be adjusted before the melt is poured. Hence samples of the melt are taken, cooled, and sparked in the point-to-plane method. Major constituents can be used as internal standards or an added element can be employed. Routinely, more than 20 elements are monitored. Although some companies have switched to ICP excitation, spark excitation is likely to remain important in the steel industry because of the need for minimal sample preparation, the speed with which results can be obtained, and the high precision of the spark.

## PROBLEMS

**9-1.** A grating spectrograph has an angular dispersion of $1.33 \times 10^{-3}$ rad nm$^{-1}$ in the first order and a focal length of 1.5 m.
  **(a)** What is the reciprocal linear dispersion in the first order?
  **(b)** Find the focal plane separation in mm between the 3099.9- and 3100.7- Å components of the iron triplet.
  **(c)** Repeat parts (a) and (b) for a spectrograph with an angular dispersion of $1.25 \times 10^{-3}$ rad nm$^{-1}$ and a focal length of 2.5 m.

**9-2.** What must the resolution of a spectrograph be to separate the iron triplet components at 3099.9 and 3100.3 Å?

**9-3.** It is desired to analyze a very inhomogeneous solid disk sample. Which excitation source (dc arc or high-voltage spark) would be better suited for this analysis? Explain and justify.

**9-4.** A dc arc method for the determination of Cu in a powder sample is being developed. The 3247-Å line has been chosen. Photographic detection is all that is available. An internal standard must be chosen. List the characteristics that a good internal standard for this determination would have.

**9-5.** Choose an appropriate emission spectrochemical analysis technique for the following determinations. Consider all the flame and plasma methods of Chapter 8 as well as the arc and spark methods of this chapter. Explain and justify your choices. Consider sample preparation time, accuracy, and precision in your answer.
  **(a)** Determination of Mo in a steel sample.
  **(b)** Determination of Cu, V, Fe, W, and Si in a deposit from a boiler.
  **(c)** Determination of Ca in a hard-water sample.
  **(d)** Determination of 12 elements in a river water sample.
  **(e)** Determination of Cr, Fe, Ni, and Co in a crude oil sample.

**9-6.** Describe the advantages and disadvantages caused by selective volatilization in a dc arc discharge.

**9-7.** What advantages are provided by controlling the atmosphere around dc arcs?

**9-8.** Compare high-voltage spark sources to dc arc sources in terms of spectra produced and performance characteristics.

**9-9.** What are the advantages of a unidirectional high-voltage spark compared to a damped oscillatory discharge waveform?

**9-10.** Discuss how one might measure the spectroscopic temperature present in an observed region of a dc arc discharge. What additional considerations are necessary with a high-voltage spark discharge?

## REFERENCES

The following are general references on the theory and principles of emission spectroscopy (see also Chapters 7 and 8 references).

1. P. W. J. M. Boumans, "Excitation of Spectra," in *Analytical Emission Spectroscopy*, vol. 1., pt. II, E. L. Grove, ed., Marcel Dekker, New York, 1972. An excellent dis-

cussion of arc and spark sources from the mechanistic point of view.

2. P. W. J. M. Boumans, *Theory of Spectrochemical Excitation*, Adam Hilger, Bristol, England, 1966.

3. R. M. Barnes, ed., *Emission Spectroscopy*, Dowden, Hutchinson & Ross, Stroudsburg, Pa., 1976. This book is a collection of the "classic" papers in the field of emission spectrochemical analysis.

4. R. D. Sacks, "Emission Spectroscopy," in *Treatise on Analytical Chemistry*, 2nd ed., pt. 1, vol. 7, chap. 6, I. M. Kolthoff and P. J. Elving, eds., Wiley-Interscience, New York, 1981. An excellent treatment of the principles and instrumentation for arc, spark, and plasma emission methods.

5. M. Slavin, *Emission Spectrochemical Analysis*, Wiley-Interscience, New York, 1971.

The following deal specifically with arc and/or spark excitation for emission spectrochemical analysis.

6. A. Scheeline, "High Voltage Discharges: Diagnostics and Opportunities," *Prog. Analy. At. Spectrosc.*, 7, 21 (1984).

7. J. P. Walters, "The Formation and Growth of a Stabilized Spark Discharge," *Appl. Spectrosc.*, 26, 323 (1972). An excellent summary.

8. J. P. Walters and H. V. Malmstadt, "Emission Characteristics and Sensitivity in a High-Voltage Spark Discharge," *Anal. Chem.*, 37, 1477 (1965). A classic.

9. R. D. Sacks and J. P. Walters, "Short-Time, Spatially-Resolved Radiation Processes in a High Voltage Spark Discharge," *Anal. Chem.*, 42, 61 (1970).

10. J. P. Walters, "Source Parameters and Excitation in a Spark Discharge," *Appl. Spectrosc.*, 26, 17 (1972).

11. C. E. Harvey, *Spectrochemical Procedures*, Applied Research Laboratories, 1950. An older book on dc arc procedures, including semiquantitative methods.

12. L. H. Ahrens and S. R. Taylor, *Spectrochemical Analysis*, Addison-Wesley, Reading, Mass., 1961. A classic book on the dc arc for emission spectrochemical analysis.

The following deal with more specialized topics discussed in this chapter.

13. W. F. Meggers, C. H. Corliss, and B. F. Scribner, *Tables of Spectral Line Intensities*, pt. I, *Arranged by Element*; pt. II, *Arranged by Wavelength*, U.S. Government Printing Office, Washington, D.C., 1961.

14. J. Reader, C. H. Corliss, W. L. Wiese, and G. A. Martin, *Wavelengths and Transition Probabilities for Atoms and Atomic Ions*, NSDRS-NBS-68, National Bureau of Standards, Washington, D.C., 1980.

15. D. A. Cremers, L. J. Radziemski, and T. R. Loree, "Spectrochemical Analysis of Liquids Using the Laser Spark," *Appl. Spectrosc.*, 38, 721 (1984).

16. Yu. P. Razier, *Laser-Induced Discharge Phenomena*, Consultants Bureau, New York, 1977.

17. R. W. B. Pearse and A. G. Gaydon, *The Identification of Molecular Spectra*, 3rd ed., Chapman & Hall, London, 1963.

# CHAPTER 10

# Atomic Absorption Spectrophotometry

Atomic absorption spectrophotometry (AAS) has become the most widely used single-element technique for the determination of metals. It is based on the absorption of radiation by neutral, ground-state atoms produced by an atomizer.

The phenomenon of atomic absorption was noted by Wollaston and Fraunhofer and explained by Kirchhoff and Bunsen in the nineteenth century. They observed dark lines in the solar spectrum. In 1912, Malinowski measured the absorbance of mercury vapor as a function of mercury concentration. Paschen in 1914 developed the hollow cathode lamp (HCL), the present primary source for AAS, but did not use this source with flames. Despite these early studies, the use of AAS was restricted to astrophysics applications until the 1950s, except for the determination of mercury in laboratory air. In 1955, Walsh in Australia and Alkemade and Milatz in Holland proposed and demonstrated analytical flame AAS, in which radiation from a line source, such as an HCL, is absorbed by atomic vapor in a flame. The technique of flame AAS became accepted in the 1960s after the introduction of the first commercial flame AA spectrophotometer in 1959.

New developments soon contributed to the enhanced growth of AAS. The nitrous oxide–acetylene flame, introduced by Willis in 1965, extended the range of elements determined by flame AAS. In the 1960s, L'vov and later Woodriff, Massman, and West pioneered the development of electrothermal atomizers for AAS. By the 1970s, commercial electrothermal atomizers became available and accepted as alternative atomization sources to flames in AAS. Electrothermal atomizers provide superior detection limits for many elements. Since the early 1970s, AA instrumentation has been further refined by the development of more intense line sources, by the introduction of instrumentation for background absorption correction, and by the incorporation of microcomputers.

The typical instrumental configuration for an AA spectrophotometer is shown in Figure 10-1. The source is usually an HCL (Figure 4-7), although an electrodeless discharge lamp (EDL) is often used for some elements (As, Se, and Te) where the radiant power output of HCLs is low. The lamps are usually modulated at a 50% duty cycle by operation from a pulsed power supply. (Mechanical choppers were used in early commercial instruments.) Normally, 1:1 imaging is used where lens L1 focuses the radiation in the center of the atomizer and L2 refocuses the radiation on the monochromator entrance slit. Thus the F/n of the lenses is usually one-half that of the monochromator, which has an F/n in the range F/6 to F/10. Mirrors are also used

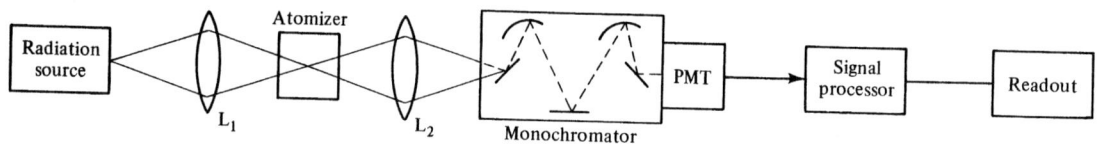

**FIGURE 10-1** Single-beam atomic absorption spectrophotometer. Radiation from a line source is focused on the atomic vapor produced by the atomizer and then directed to the monochromator, where the atomic line of interest is isolated. The attenuation of the source radiation by the analyte atomic vapor is detected by the PMT. The signal is appropriately processed to yield an absorbance readout signal.

for focusing in some commercial instruments. The atomizer is typically a flame or an electrothermal device.

A monochromator with a spectral bandpass of 0.1 to 2 nm is required to isolate the resonance line of the element to be determined from lamp impurity and filler gas lines, and from the atomizer background emission. The signal processor converts the photoanodic current from the PMT into a voltage. Amplification and demodulation circuitry extract the amplitude information from the carrier waveform. Hardware or software logarithmic conversion is used to provide direct absorbance readout on a digital meter, video display, video terminal, or printer. Normally, the noise equivalent bandpass of the electronics is adjustable by the user, who selects a time constant or integration time. A continuum source with a high-resolution echelle monochromator can also be used in place of a line source and a conventional monochromator. Continuum source AA and several other options and refinements are discussed in Section 10-3.

This chapter begins by considering the flame and electrothermal atomizers that are commonly used in AAS. Specialized sample introduction and atomization schemes are also reviewed. Next, signal and $S/N$ expressions are presented and used to illustrate how critical experimental variables affect performance characteristics. A discussion of various instrumental options and configurations follows. The chapter concludes with a description of performance characteristics, methodology, and several common applications.

## 10-1 ATOMIZERS

For AAS, the ideal atomizer would provide complete atomization ($\beta_a = 1$) of the element of interest irrespective of the sample matrix. For the lowest detection limits, the atomic vapor should not be highly diluted by the atomizer gas so that a large ground-state neutral atom population is produced. Excitation of the analyte and other species should be minimal so that analyte and background emission noise is small. Although not ideal, flame and electrothermal atomizers have gained ac-

ceptance as reliable atomizers for AAS in many situations.

### Flame Atomizers

Combustion flames are still the most popular atomization sources for AAS. In most commercial AA spectrometers, a premixed, chamber-type nebulizer burner system is employed. The nebulizer is usually based on the concentric, pneumatic design (Figure 7-4a). Nebulizer parts are fabricated with robust metals (e.g., Pt-Ir alloy, Ta, Pt) to enhance the chemical resistance to acidic solutions and other corrosive mixtures. Variable-flow-rate nebulizers allow solution flow rates to be reduced for solutions that would normally yield high absorbances. Invariably, a slot burner head (Figure 8-4) is used to provide a long pathlength (e.g., 5 to 10 cm) for absorption. Special burner heads with wider slots or three slots are available to minimize clogging with solutions of high salt content.

The most popular flames are air–$C_2H_2$ and $N_2O$–$C_2H_2$. The nitrous oxide–acetylene flame provides higher atomization efficiencies (see Table 8-4 for a comparison of $\beta_a$ values) and thus better detection limits for refractory elements, such as Si, Al, Sc, Ti, V, Zr, and the rare earths. In addition, the hotter and more reducing $N_2O$–$C_2H_2$ flame minimizes interference effects (e.g., solute vaporization interferences) for many elements such as Mg and Ca. An air–$H_2$ flame or nitrogen–$H_2$ entrained air flame is sometimes used for easy-to-atomize elements such as As and Se with absorption lines in the far-UV region.

The burner control unit is designed for convenient and safe burner operation. Flow controllers with flow meters allow adjustment of the fuel/oxidant ratio, which is critical for some elements. With Cr, for example, a fuel-rich air–$C_2H_2$ flame yields the best atomization efficiency and detection limit.

Due to potential flashback problems, a $N_2O$–$C_2H_2$ flame cannot be ignited directly. First an air–$C_2H_2$ flame is lighted and then the $N_2O$ flow is turned on and increased as the air flow is simultaneously decreased to zero. The procedure is reversed before extinguishing the flame. Automatic gas flow modules implement this procedure without operator assistance. The slot length and width for a $N_2O-C_2H_2$ burner head (typically, a 5-cm slot) is different than for an air–$C_2H_2$ burner head (typically, a 10-cm slot) to prevent flashback with the higher-burning-velocity flame (see Section 8-2). Burner interlocks are often provided to prevent use of the incorrect burner head. Many instruments provide push-button flame ignition through activation of a small starter flame. An optical sensor is often used to monitor the flame. If the flame extinguishes accidentally, the gas flows are automatically shut down.

The burner is mounted on translational stages to allow the flame to be positioned so that the focused line source radiation passes through the middle of the flame at the desired burner height to maximize the absorbance. Usually, the burner can be rotated to reduce the pathlength for cases in which the absorbance is too large.

## Electrothermal Atomizers

With electrothermal atomizers, a discrete sample is deposited, and the atomizer is electrically heated to produce a transient cloud of atomic vapor. The absorption of this atomic vapor is probed to determine the amount of analyte in the sample. In 1961, L'vov demonstrated that electrothermal atomizers in an AA spectrometer could yield absolute detection limits in the picogram range and concentration detection limits over 100 times better than achieved with flame AA. This potential for extremely low detection limits provided the impetus to understand, develop, and refine electrothermal atomizers for routine analytical use.

The reason for the superior detectability provided by electrothermal atomizers is readily seen. For a typical flame atomizer with a chamber nebulization system, the solution flow rate is about 5 mL min$^{-1}$ and the nebulization efficiency ($\varepsilon_n$) is 5%. Thus the actual rate of sample delivery to the flame is about 4 μL s$^{-1}$. For a typical flame with a 10-L min$^{-1}$ total gas flow rate and a gas expansion factor of 10, the analyte atomic vapor produced from the sample is diluted in 100 L min$^{-1}$ or 1.6 L s$^{-1}$ of hot flame gases. Thus every 4 μL of sample is diluted in 1.6 L of flame gases.

By contrast, the bulk of a 5-μL sample is atomized in about 1 s into a furnace volume of about 2 mL in an electrothermal atomizer. Thus the dilution factor is about 1000 times less or the atomic number density of the analyte produced from a given solution is about 1000 times greater in a furnace atomizer compared to a flame.

This increased number density comes with a penalty; the concentration of other gaseous components from the sample matrix is correspondingly larger. Thus interference effects can be more severe in a furnace compared to a flame. As we shall see, much of the development of electrothermal devices, methodology, and ancillary equipment has been aimed at eliminating or minimizing these interferences.

Commercial electrothermal atomizers are essentially small, electrically heated tubular furnaces, as shown in Figure 10-2a. The furnace is usually a graphite tube that is 1 to 3 cm in length and 3 to 8 mm in diameter. The furnace assembly is usually mounted on translational stages in place of the nebulizer-burner, and its position is adjusted so that the radiation from the line source is directed through the tube. A hole in the top of the tube allows typically 5 to 50 μL of sample, blank, or standard solution to be injected manually with a syringe or with an automatic injector. Each end of the furnace tube is connected to a high-current, programmable power supply through water-cooled contacts. Because the electrical resistance of the furnace tube is small (i.e., milliohms), the power supply must be capable of delivering several hundred amperes at 10 to 12 V.

Typically, the furnace is heated in three stages, as illustrated in Figure 10-2b, in which the temperature of the furnace is progressively increased by passing larger currents through the atomizer tube. The first step is the *drying* or *desolvation step*, in which a sufficient current causes the furnace temperature to be increased and maintained at about 110°C. During this stage, the solvent is evaporated, leaving a solid residue in the furnace. The temperature is chosen to evaporate the solvent as rapidly as possible without spattering.

The second step is the *ash step*, in which the power supply current is increased so that the furnace temperature is raised, typically to 350–1200°C. During this stage, organic matter in the sample is ashed or converted to $H_2O$ and $CO_2$, and volatile inorganic components are vaporized. The temperature should ideally be set high enough to remove all volatile components without loss of the analyte.

The final step is the *atomization step*. The current through the atomizer tube is increased so that the tube temperature is 2000 to 3000°C. During this period the sample is vaporized and atomized to produce atomic vapor that is probed by the source. The atomic vapor is produced rapidly and diffuses out of the observation zone to produce a transient, peak-shaped response.

The current (related to the tube temperature) and the duration of each step are adjustable so that conditions can be optimized for a given element, type of sample, and sample size. Typically, the whole sequence takes 45 to 90 s. The dry and ash steps are typically 20 to 45 s in duration, while the atomize step is 3 to 10 s in length. Sometimes a fourth "clean" step involving a

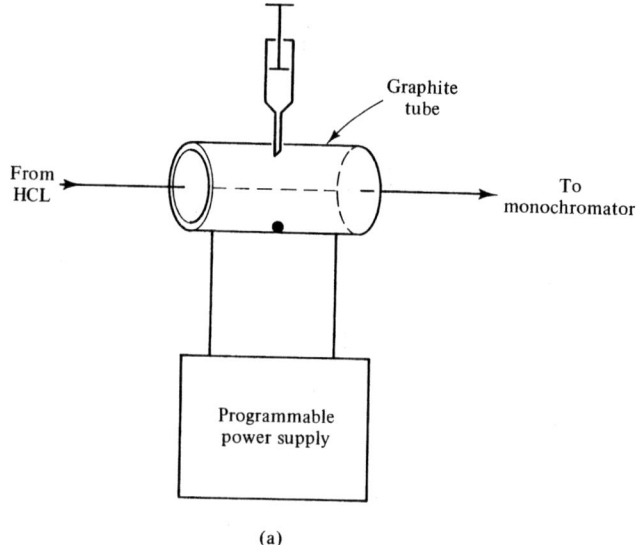

(a)

**FIGURE 10-2** Design and use of electrothermal atomizers. In (a) an electrothermal atomizer system is shown to be a small furnace tube heated by passing current through it from a programmable power supply. The sample is deposited into the tube with a syringe or injector and then the tube is heated in stages as shown in (b). The dry and ash steps remove water and organic or volatile inorganic matter, respectively. The atomization step produces a pulse of atomic vapor that is probed by the radiation beam from the HCL as shown in (c). Many instruments sense and hold the peak absorbance value for readout.

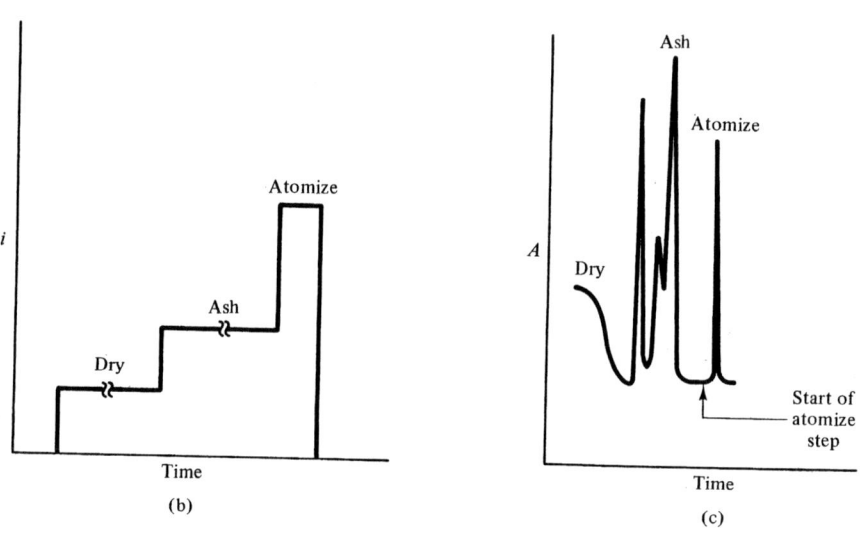

(b)

(c)

higher current and tube temperature is employed after the atomize step to remove any remaining sample residue in the tube. Alternatively, the atomize step can be repeated before the next sample is run.

During operation, the furnace tube is continually bathed in an inert gas such as argon or nitrogen. The inert gas protects the atomizer surface from oxidation. The gas also transports the analyte atoms from the surface to the center of the atomizer and removes all gaseous substances produced during the operational sequence. The vapor atmosphere inside the furnace tube can be more inert than in flames due to the absence of flame gases and combustion products; oxide formation can be reduced and atomization efficiencies can be higher.

As indicated above, the furnace is usually made of some form of carbon in a tube shape. The original L'vov design and some commercial designs are shown in Figure 10-3. Metals with a high melting point, such as tantalum, tungsten, or platinum, have also been em-

ployed as construction materials. Many other shapes and configurations for electrothermal atomizers have been used. These include carbon cups, carbon filaments and braids, tantalum strips, and tungsten loops. Today, carbon furnace tubes are coated with pyrolytic carbon, which prevents the sample from soaking into the graphite and reduces sample loss due to diffusion through porous graphite walls. A small amount of methane can be added to the sheath gas to reduce the rate of deterioration of the coating.

The heating characteristics of the furnace are critical for obtaining reliable analytical results. Faster heating rates during the atomization step cause the atomic population to build up rapidly before the atomic vapor diffuses out of the observation zone. This creates a higher instantaneous atomic concentration and larger absorbances. In current commercial furnaces, the heating rate is in excess of $1000°C\ s^{-1}$.

The control of the temperature during the various

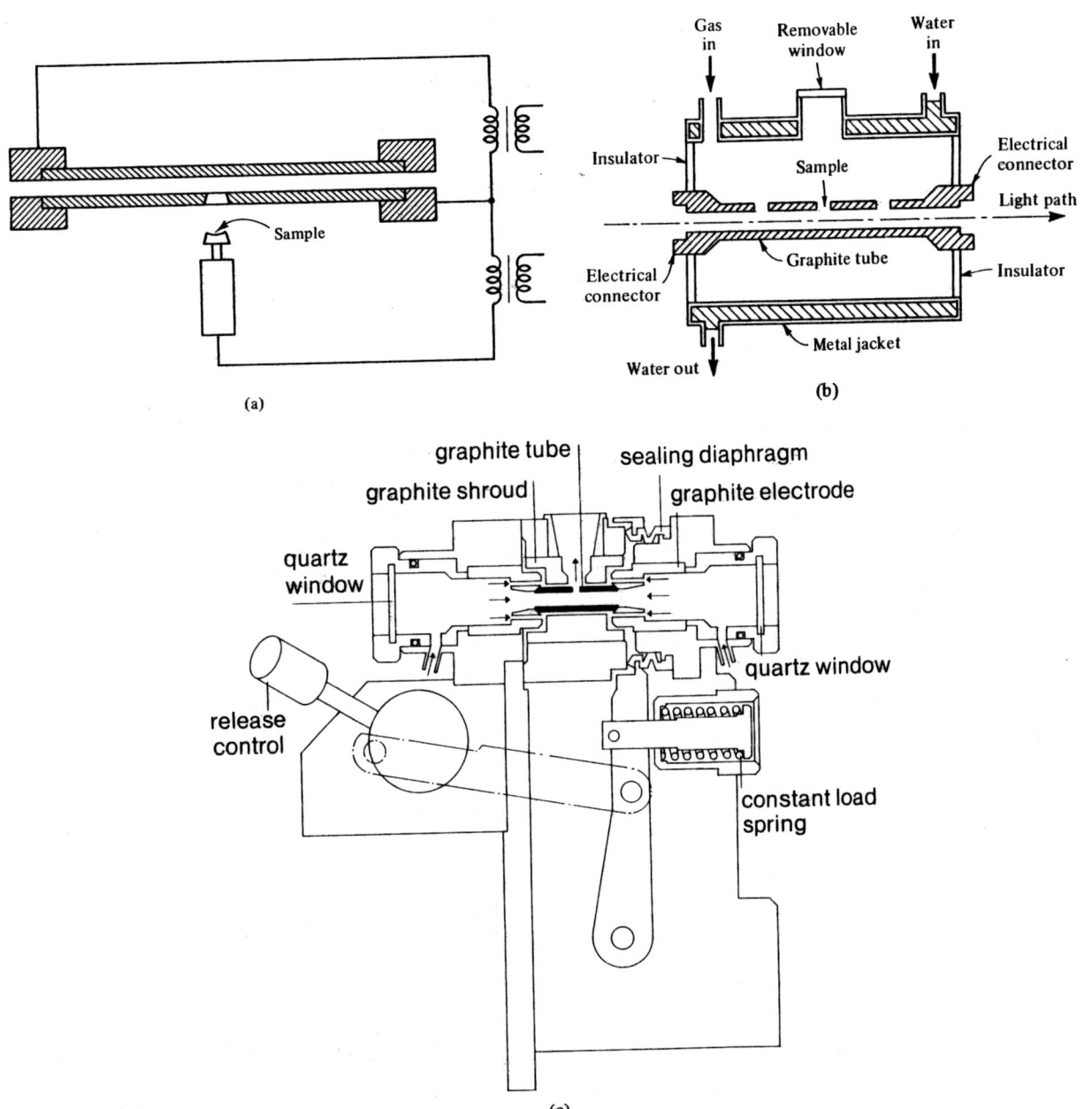

**FIGURE 10-3** Designs of electrothermal furnace atomizers. The original L'vov design is shown in (a). In (b) the heated graphite atomizer (HGA) is shown whereas in (c), the design of the graphite tube atomizer (GTA) is illustrated. [(b) Courtesy of Perkin-Elmer Corporation; (c) courtesy of Varian Associates, Inc., Instrument Group.]

cycle steps is critical for reproducibility. In early furnace designs, constant power was provided in each portion of the cycle. Constant power does not ensure a constant temperature. Also, for low final atomization temperatures, less power is applied with a resulting slower, nonlinear rise to the final temperature. Many modern furnace designs use an optical temperature sensor that

views the outside of the tube. This allows the user to choose the atomization temperature. With feedback control, the preselected temperature is rapidly achieved and held for the duration of the atomization cycle.

Although the power to the furnace is stepped essentially instantaneously to its selected value during the atomization step, it takes a finite time for the furnace

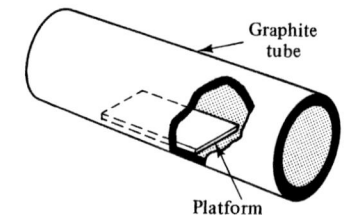

**FIGURE 10-4** Modified L'vov platform.

temperature to reach its equilibrium value. When the wall of the furnace on which the sample is deposited reaches a critical temperature, the analyte vaporizes off the surface. The vaporization temperature depends on the analyte, the analyte concentration, and the sample matrix. The gas inside the tube is at a lower temperature than the furnace walls, so that atomized analyte atoms may suffer compound formation after vaporization.

To alleviate these problems, which ultimately cause interference effects, a modified L'vov platform is now often used. As shown in Figure 10-4, the L'vov platform is a thin graphite plate placed at the bottom of the furnace onto which the sample is deposited. The graphite plate is heated primarily by radiation from the walls, so that the temperature of the plate (and hence the sample temperature) lags the tube wall and gas-phase temperature. Conditions can be adjusted so that the analyte is atomized after the wall and gas phase have reached near steady-state temperatures. The platform enhances the efficiency of dissociation of analyte molecules and reduces vapor-phase-related interferences. Similar improvements have been achieved by different designs of the graphite tube (e.g., varying the thickness along the optical axis).

## Other Sample Introduction and Atomization Techniques

The majority of AAS determinations are carried out by continuous aspiration of solution into a flame with a pneumatic nebulizer or by discrete sampling with an electrothermal atomizer. However, many other techniques have been developed for sample introduction into flames or for atomization in special circumstances (e.g., sample-limited conditions, improvement of detection limits). These specialized techniques are reviewed below in the context of AA detection; however, they can also be applied in atomic emission and fluorescence spectrometry.

*Other Flame Techniques.* With adjustable nebulizers it is possible to turn down the aspiration rate to zero and use a pump to provide the driving force for solution flow. Although this is not common, a high-quality pump can provide a more constant solution flow rate than that achieved with aspiration and may be advantageous for solutions with variable viscosity.

The flow rates used in HPLC are compatible with aspiration rates of flame AA nebulizers. Thus it is relatively simple to use a flame AA as a detector for HPLC. This makes it possible to separate different species of a given metal on a column and specifically detect the metal-containing species as they elute. An HPLC can also be used as an on-line preconcentration tool. Here a large volume of sample solution is passed through a column containing an ion-exchange resin under conditions where the metal ion is retained. After the preconcentration step, the metal is stripped off the column in a small volume of a suitable reagent and sent to the flame through the nebulizer.

Sample modulation (see Section 4-5) allows automatic referencing of sample information to that of a blank. In flame AAS, sample modulation has been implemented by rapidly alternating (up to 70 Hz) between the flow of sample and blank solution or aerosol into the burner chamber. For example, an automated three-way valve can be used to switch sample and blank solution alternately into a single nebulizer.

Typically, 2 to 5 mL of sample solution is required with standard nebulizers to achieve a steady-state absorbance signal and acquire two or three measurements. In some cases the amount of sample solution may be limited. Several techniques have been developed for the introduction of discrete micro samples into flames and a few of these are illustrated in Figure 10-5. These methods produce a peak-shaped signal and the peak absorbance or peak area is taken as the analytical signal. In the pulse nebulization method, a special accessory (Figure 10-5a) allows a discrete volume (10 to 100 μL) of sample to be injected into the aspiration tube.

Flow injection techniques are gaining popularity for sample introduction with AAS. Here a sample solution is loaded into the loop of an HPLC-type sample loop injection valve (Figure 6-12) with a syringe or inexpensive pump or by aspiration (Figure 10-5b). The sample loop volume can vary from microliters to milliliters. While the sample is being loaded, the carrier stream solvent is pumped through the bypass valve of the sample injection valve into the flame nebulizer. After the sample is loaded, the valve is switched to the inject position, the sample loop is placed on-line, and the sample is carried as a plug into the nebulizer to produce a peak-shaped response. This method can be used for small sample volumes and for routine analysis to increase sample throughput.

Some techniques for handling small sample volumes totally bypass the use of the nebulizer. Here a small volume (e.g., microliters) of sample such as blood

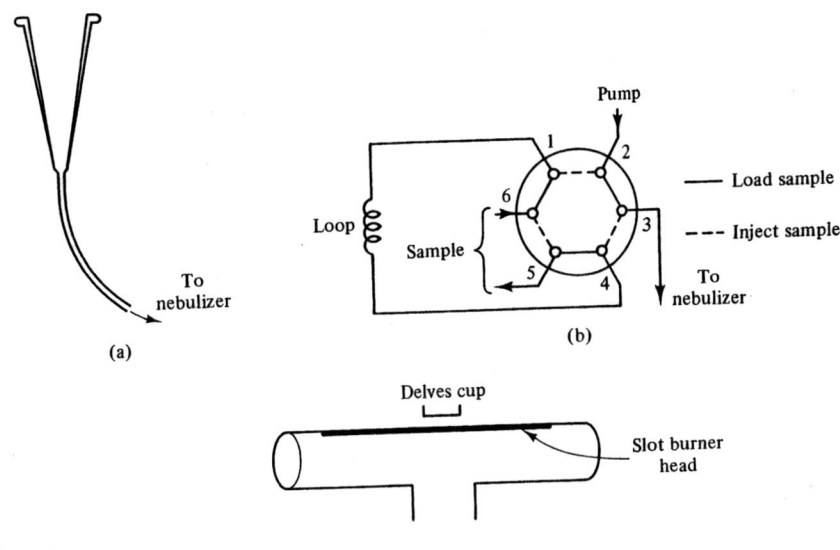

(a)

(b)

(c)

**FIGURE 10-5** Methods of introducing discrete samples. In (a) a small conical cup is attached to the nebulizer aspiration capillary tube. In (b) an HPLC type injection valve is used to inject a volume of sample determined by the loop volume. In (c) the sample is placed in a cup that is heated by the flame.

is pipetted into a sampling device (Figure 10-5c). The sampling device can be a platinum wire loop, a tantalum boat or cup, or a stainless steel rod with a cavity. Usually, the sample is dried at about 100°C to drive off the solvent, which leaves a solid residue in the sampling device. A mechanical positioning apparatus then places the sampling device into the flame, where the sample is atomized. The source radiation beam probes the transient cloud of atomic vapor that is produced above the sampling device. Because the sampling device is at a lower temperature than the flame, this technique is used primarily for volatile elements such as Pb, Cd, As, Bi, and Zn. Note that this technique is similar to electrothermal atomization except that the energy to heat the sample is provided by a flame instead of by electrical energy. The use of the sampling cup technique has decreased since the advent and acceptance of electrothermal atomizers.

Samples can also be introduced into the flame in the gaseous state. Of course, the element must be converted to a chemical form that is volatile at or near room temperature. Elements such as As and Se are easily converted in solution to hydrides that can then be swept out of solution into the atomizer. Reaction of some metal ions with appropriate complexing agents produces volatile metal complexes. For example, Cu, Fe, Be, and Cr form volatile trifluroacetylacetonates. The volatile species can be introduced directly into the flame through the nebulizer tube or a port into the spray chamber. Alternatively, the vapor phase sample can be directed to a quartz tube heated by a flame or even into the furnace tube of an electrothermal atomizer. Some researchers have used a GC to separate and volatilize metal containing species with a flame AA apparatus as a metal specific detector.

*Hydride Techniques.* The flame AA detection limit for some elements, such as As and Se, is only on the order of 1 $\mu$g mL$^{-1}$ and not adequate for determining environmental levels of these species. Elements such as As, Se, Sb, Bi, Ge, Sn, Te, and Pb form hydrides in acidic solutions with NaBH$_4$. For example, the reaction of borohydride with As(III) to form arsine is described by

$$3BH_4^- + 3H^+ + 4H_3AsO_3 \rightarrow 3H_3BO_3 + 4AsH_3 + 3H_2O$$

The hydrides formed can be swept out of solution into an atomizer as shown in Figure 10-6; a peak-shaped response is obtained. When basic borohydride is added to an acidic solution, excess hydrogen is also produced according to

$$BH_4^- + 3H_2O + H^+ \rightarrow H_3BO_3 + 4H_2$$

There are two primary advantages to the hydride generation technique. First, the analyte is removed from the sample matrix, which reduces the potential for interferences. Second, detection limits are improved to the ng/mL level or below because all the analyte in a 1 to 50-mL sample is introduced into the atomizer in a few seconds (i.e., the atomic number density during the peak is much greater than with nebulization techniques).

Various atomizers have been used with hydride generation. These include cool combustion flames (e.g., H$_2$–air or Ar–H$_2$–entrained air) and heated quartz tubes.

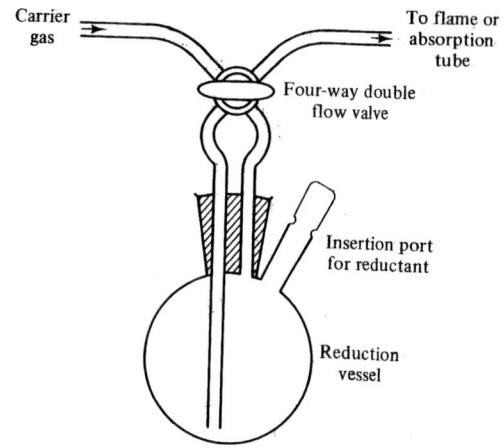

**FIGURE 10-6** Apparatus for generating hydrides or elemental mercury. The apparatus shown allows the sample and reductant to be mixed. The volatile analyte species (hydride or $Hg^0$) is swept out of solution by a carrier gas and carried to the atomizer or absorption cell. Most manufacturers of AA spectrophotometers provide hydride and cold vapor accessories for their instruments.

The quartz tubes can be heated directly with a flame or wrapped with nichrome wire and heated electrically. For best detection limits, quartz tube atomizers are preferred since problems due to flame background emission noise and absorption are eliminated, and the residence time of the atomic vapor in the optical path is longer. Initially, it was believed that the atomization of hydrides in quartz tubes was a simple thermal process. However, quartz tubes are only heated to about 900°C, well below the temperature needed to atomize hydrides in electrothermal atomizers. Actually, the decomposition of hydrides to the elemental form is believed to involve H radicals that are produced when $H_2$ (formed in the borohydride reduction reaction) adsorbs on the surface of the heated quartz tube and is converted to hydrogen radicals. The hydrogen radicals react with the hydride and extract hydrogen atoms ($MH_x + xH\cdot \rightarrow M + xH_2$).

The efficiency of atomization can vary as the surface of the quartz tube deteriorates. Thus some researchers have used a "flame-in-a-tube" atomizer. Here the hydride is mixed with $H_2$ and $O_2$ (or air) before the mixture reaches the observation quartz tube atomizer. A flame is sustained inside the quartz tube to produce a continuous supply of H radicals for atomization of the hydrides.

The hydride has also been trapped in a balloon or a liquid-nitrogen trap and later released to the atomizer. In the latter case, the hydride is expelled by heating the trap. The trapping technique is advantageous if the hydride formation reaction is relatively slow.

*Cold Vapor Mercury Techniques.* Mercury is unique among the metallic elements because of its appreciable vapor pressure at ambient temperatures. Mercuric ions in solutions can be reduced by $SnCl_2$ ($Sn^{2+} + Hg^{2+} \rightarrow Sn^{4+} + Hg^0$) or $NaBH_4$ to elemental mercury. The mercury is swept out of solution with a carrier gas (e.g., $N_2$) into a long-path glass absorption tube cell where the atomic absorption of the mercury atoms at 253.7 nm is measured. An apparatus similar to that shown in Figure 10-6 for hydride generation can be used.

As with hydride methods, the cold vapor method removes the analyte from the sample matrix, concentrates all the mercury in the analytical sample into a small plug of carrier gas, and provides a long residence time. Detection limits depend on the apparatus, but vary from sub-ng $mL^{-1}$ to pg $mL^{-1}$ concentrations. Water mist entering the observation cell can lead to false absorbance signals due to scattering. Thus it is common to remove the water mist with a trap or to vaporize the water mist by heating the mercury-containing carrier gas and the observation tube to slightly above 100°C.

## 10-2 SIGNAL AND NOISE EXPRESSIONS

In this section the general expressions for the analyte absorption signal developed in Chapters 2 to 4 are combined with expressions for absorption factors from Chapter 7. The complete readout expressions provide insight into how experimental variables and characteristics of a particular element affect the absorption signals observed. General $S/N$ expressions from Chapter 5 are refined to apply specifically to AA measurements.

### Readout Expressions

The most important equations from previous chapters on absorption measurements are summarized in Table 10-1. In some cases the equations are simplified to the specific form applicable to conventional flame AA measurements. The measured transmittance $T'$ (equation 10-4) is obtained from the three readout signals defined in equations 10-1 to 10-3. Note that the readout signal expressions are simplified from the general expressions (equations 2-28 to 2-31) for two reasons. First, analyte and background luminescence (fluorescence) are usually negligible in flame AA measurements. Second, with source modulation, dark current, background emission, and analyte emission signals are not carried by the modulation waveform, as illustrated in Figure 10-7 (i.e., their mean values with synchronous detection are zero at the readout device). Thus the

**TABLE 10-1**
Signal and readout equations for flame AAS

| | | |
|---|---|---|
| 0% $T$ signal[a] | $E_{0t} = E_{bE} + E_d = 0$ | (10-1) |
| Total reference signal[b] | $E_{rt} = E_r + E_{0t} = E_r$ | (10-2) |
| Total sample sample[c] | $E_{st} = E_s + E_E + E_{0t} = E_s$ | (10-3) |
| Measured transmittance | $T' = \dfrac{E_{st} - E_{0t}}{E_{rt} - E_{0t}} = \dfrac{E_s}{E_r} = T$ | (10-4) |

Radiant power incident on PMT with flame off[d]

General:
$$\Phi_0 = \int_0^\infty (\Phi_\lambda)_0 \, d\lambda = \int_0^\infty B_\lambda Y(\lambda) \, d\lambda \qquad (10\text{-}5)$$

Line:
$$\Phi_0 = WH\Omega T_{op}B \qquad (10\text{-}6)$$

Continuum:
$$\Phi_0 = WH\Omega T_{op}sB_\lambda \qquad (10\text{-}7)$$

Reference signal[e]

General:
$$E_r = mG \int_0^\infty B_\lambda Y(\lambda) T_r R(\lambda) \, d\lambda \qquad (10\text{-}8)$$

Line:
$$E_r = mGR(\lambda)WH\Omega T_{op}BT_r \qquad (10\text{-}9)$$

Continuum:
$$E_r = mGR(\lambda)WH\Omega T_{op}sB_\lambda T_r \qquad (10\text{-}10)$$

Sample signal[f]

General:
$$E_s = mG \int_0^\infty B_\lambda Y(\lambda) T_s R(\lambda) \, d\lambda \qquad (10\text{-}11)$$

Line:
$$E_s = mGR(\lambda)WH\Omega T_{op}BT_s \qquad (10\text{-}12)$$

Continuum:
$$E_s = mGR(\lambda)WH\Omega T_{op}sB_\lambda T_s \qquad (10\text{-}13)$$

| | | |
|---|---|---|
| Sample transmission factor | $T_s = T_r(1 - \alpha)$ | (10-14) |
| Spectral absorptance | $\alpha(\lambda) = 1 - e^{-k(\lambda)l}$ | (10-15) |

[a] Measured with source radiation blocked and flame on; background fluorescence assumed negligible here and for total sample and reference measurements; with source modulation and synchronous detection the mean value of $E_{0t}$ is 0.

[b] Measured with the source on and blank aspirating into the flame; with source modulation and synchronous detection the mean value of $E_{rt}$ is $E_r$.

[c] Measured with the source on and an analyte solution aspirating into the flame; analyte fluorescence assumed negligible; with source modulation and synchronous detection, the mean value of $E_{st}$ is $E_s$.

[d] $Y(\lambda)$ and $T_{op}$ account for the optical efficiency of components in the monochromator and of the external optical components between the source and monochromator. In equation 10-6 and subsequent equations for a line source, $T_{op}$ is assumed to be constant over the source line width. In equation 10-7 and subsequent equations for a continuum source, $T_{op}$ and $B_\lambda$ are assumed constant over the spectral bandpass of the monochromator.

[e] In equations 10-9 and 10-10, $T_r$ and $R(\lambda)$ are assumed to be constant over the source line width or spectral bandpass, respectively.

[f] In equation 10-12, $T_s$ and $R(\lambda)$ are assumed to be constant over the line source width.

measured transmittance is the true transmittance or the ratio of the lamp signal passed by the sample to that passed by the blank $T' = T = E_s/E_r$, and the true absorbance A is obtained from the measured transmittance ($A = -\log T$). It is important to realize that $E_d$, $E_{bE}$, and $E_E$ can be measured before synchronous detection, and noise in these signals is still present at the modulation frequency.

With modern AA spectrometers, the 0% $T$ signal is not directly measured (i.e., the signal modifier electronics are internally adjusted so that $E_{0t} = 0$ V or $T = 0$ when the lamp radiation is blocked). All commercial AA spectrometers provide direct absorbance readout as one readout option and in some cases as the only readout option. Here the sample and reference signals are converted to sample and reference readout signals displayed in absorbance units (see equations 2-32 to 2-34). Thus the reference absorbance signal with the blank aspirating ($A_r$) and the sample absorbance with the sample aspirating ($A_s$) are used to calculate the analyte absorbance ($A = A_s - A_r$).

The radiant power passed by the monochromator

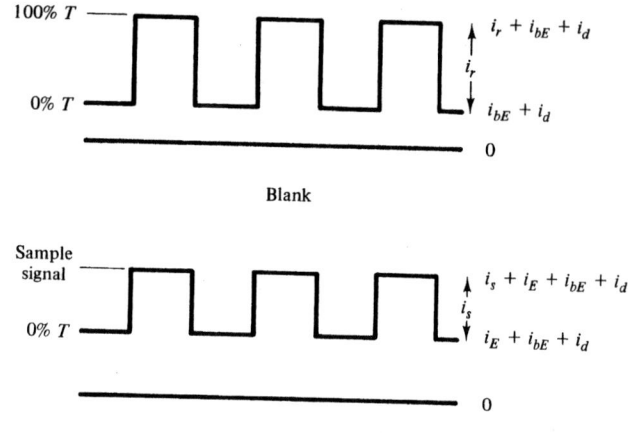

**FIGURE 10-7** Modulation signals observed in AAS referenced to photocathodic current. With the blank aspirating, the amplitude of the source-modulated signal at the PMT photocathode or photoanode is proportional to the source radiation passed even if the background emission signal varies between samples and standards. When an analyte solution is aspirated into the flame, background and analyte emission are not carried and are observed as dc signals. However, the amplitude of the modulated signal is proportional to the amount of lamp radiation transmitted by the sample regardless of the amount of analyte emission.

exit slit ($\Phi_0$) and impingent on the PMT when the flame is off (typically, $10^{-10}$ to $10^{-12}$ W) is given by equation 10-5. It is a function of the lamp spectral radiance and the collection and transmission characteristics of the monochromator and of the external optical components. Equations 10-6 and 10-7 are the specific expressions for line and continuum sources.

Part of the incident radiant power is absorbed when blank or analyte solutions are aspirated into the flame. The general equations and specific equations that apply with line and continuum sources for the reference and sample signals are given by equations 10-8 to 10-13.

The attenuation of the lamp signal with sample or blank aspirating is a function of the sample and reference transmission factors ($T_s$ and $T_r$). The reference transmission factor is less than 1 due to absorption by flame species or blank species (primarily molecular species), to scattering by particles, and to reflection losses caused by refractive index differences between air and the flame. The sample transmission factor is controlled by the same factors plus the analyte absorptance ($\alpha$) as shown in equation 10-14.

With a proper blank, $T_r$ is the same for the sample and blank measurements so that $T = E_s/E_r = T_s/T_r =$

($1 - \alpha$) and due solely to analyte absorption. It is also assumed that all instrumental conditions, such as the source radiance, the solution flow rate, and gas flow rates, are held constant between the reference and blank measurements.

With equation 10-15 we can calculate the analyte spectral absorptance from the analyte absorption coefficient [$k(\lambda)$] and the pathlength ($l$). We can now use the results developed in Sections 7-1 and 7-5 to write specific equations for $\alpha$ and $A$ with line or continuum source AAS.

For a source line width much smaller than the absorption line width, $k(\lambda)$ is considered constant over the line profile and equal to its maximum value ($k_m$). Thus the analyte absorptance with a line source ($\alpha_L$) is equal to $\alpha(\lambda_m)$ as given by equation 7-43. In this case

$$
\begin{aligned}
A_L &= -\log T = -\log (1 - \alpha_L) \\
&= -\log (e^{-k_m l}) = 0.434 k_m l
\end{aligned}
\tag{10-16}
$$

Expressing $k_m$ in terms of fundamental constants, the number density of the state from which absorption occurs ($n_i$), the source wavelength ($\lambda_m$), and effective half-width ($\Delta\lambda_{\text{eff}}$) (see equation 7-45), we obtain equation 7-48, which is repeated here:

$$
A_L = \frac{0.434(e^2\lambda_m^2 n_i f_{ij} l)}{4\varepsilon_0 m_e c^2 \, \Delta\lambda_{\text{eff}}}
\tag{10-17}
$$

Equation 10-17 may be found in the literature in various forms. With the appropriate conversion equations (see Appendix F), $A_{ji}$ and $B_{ij}$ can be substituted for $f_{ij}$. Note that $(\Delta\lambda_{\text{eff}})^{-1}$ is just the maximum value of the absorption line profile ($S_\lambda$) or $S_{\lambda m}$. From equation 7-29, $S_{\lambda m} = (S_{\lambda D})_m \delta(a, 0)$, where $(S_{\lambda D})_m$ is the maximum value of a purely Doppler broadened profile (equation 7-26) and $\delta(a, 0)$ is the value of the Voigt integral at the wavelength of maximum absorption.

For most AA determinations, the transition from the ground state to the first excited state is employed (i.e., $i = 0$, $j = 1$). Under these conditions, $n_i = n_0$ and equation 7-9 can be used for $n_0$ in equation 10-17, with the result that

$$
A_L = \frac{(2.60 \times 10^{17})F\varepsilon_a c_M e^2 \lambda_m^2 f_{01} l}{4Q e_f \varepsilon_0 m_e c^2 \, \Delta\lambda_{\text{eff}}}
\tag{10-18}
$$

where $c_M$ is the molar concentration of the analyte and it is assumed that $Z(T) = g_0$. If we substitute for fundamental constants, the result is

$$
A_L = \frac{(2.30 \times 10^5)F\varepsilon_a c_M \lambda_m^2 f_{01} l}{Q e_f \, \Delta\lambda_{\text{eff}}}
\tag{10-19}
$$

where $\lambda_m$ and $\Delta\lambda_{\text{eff}}$ are expressed in cm. Equation 10-19 indicates how the measured absorbance depends on nebulizer-flame parameters ($F$, $\varepsilon_a$, $l$, $Q$, $e_f$), the analyte ($\lambda_m$, $f_{01}$), and a combination of flame and analyte parameters ($\Delta\lambda_{\text{eff}}$). It predicts a linear relationship between the analyte concentration and measured absorbance. As discussed in more detail in Section 10-4, calibration curves become nonlinear at high absorbances partially because $k(\lambda)$ does vary slightly over the finite source profile.

With a continuum source, the absorption coefficient and hence analyte spectral absorptance varies over the wavelength range passed by the monochromator ($\lambda_0 \pm s$). If the width of the absorption line profile is less than this wavelength range, the absorbance with a continuum source ($A_C$) is given by

$$A_C = -\log T = -\log (1 - \alpha_C) \qquad (10\text{-}20)$$
$$= -\log \left(1 - \frac{A_t}{s}\right)$$

where $\alpha_C$ is the absorptance with a continuum source given by equation 7-51 and $A_t$ is the integral absorption. Because $A_t$ involves an integral of $1 - e^{-k(\lambda)l}$ over the absorption profile with $k(\lambda)$ dependent on $\lambda$, a simple expression cannot be derived. In the limit of small $A$, $\alpha_C$ is proportional to $n_0$ (see Section 7-5) and equations 10-17 to 10-19 apply if $s$ is substituted for $\Delta\lambda_{\text{eff}}$. For small absorbances, $A_C$ is less than $A_L$ by the factor $\Delta\lambda_{\text{eff}}/s$. For higher absorbances, the relationship between $A_C$

and $n_0$ and hence analyte concentration is nonlinear because $\alpha_C \propto n_0^{1/2} \propto c_M^{1/2}$. As the spectral bandpass is decreased, the linearity of the calibration curve improves.

The readout expressions presented can be adapted for other atomizers and sample introduction techniques. With electrothermal atomizers, the absorption signal is time dependent because $n_0$, and thus $k(\lambda)$, varies with time as the finite amount of sample is atomized.

## Signal-to-Noise Expressions

The expressions for the relative standard deviation (RSD) in absorbance ($\sigma_A/A$) developed in Section 5-5 apply to AAS. General and specific equations for AAS are summarized in Table 10-2. Equations 10-21 to 10-23 relate the standard deviation (SD) in absorbance ($\sigma_A$) and $\sigma_A/A$ to the SD in the total sample signal ($\sigma_{st}$).

Equation 10-24 shows that $\sigma_{st}$ is due to shot and flicker noise in the transmitted source signal and 0% $T$ noise as discussed previously (equation 5-46), and in addition, to analyte emission noise ($\sigma_E$). The lamp signal shot and signal flicker noise are calculated from equations 10-25 and 10-26, respectively. The noise with the source turned off and the blank aspirating, the 0% $T$ noise ($\sigma_{0t}$), is seen from equation 10-27 to be due to background emission noise ($\sigma_{bE}$), dark current noise ($\sigma_d$), and amplifier-readout noise ($\sigma_{ar}$). (See equation 5-55 for an expanded form of this equation.)

The RSD in $E_s$ due to signal flicker noise is spec-

## TABLE 10-2
Signal-to-noise expressions for AAS[a]

| | |
|---|---|
| $\sigma_A = \dfrac{0.43\sigma_T}{T}$ | (10-21) |
| $\sigma_T = \dfrac{\sigma_{st}}{E_r}$ | (10-22) |
| $\dfrac{\sigma_A}{A} = \dfrac{-\sigma_{st}}{TE_r \ln T}$ | (10-23) |
| $\sigma_{st} = [(\sigma_s)_s^2 + (\sigma_s)_f^2 + \sigma_{0t}^2 + \sigma_E^2]^{1/2}$ | (10-24) |
| $(\sigma_s)_s = mGKTE_r$ | (10-25) |
| $(\sigma_s)_f = \xi_s TE_r$ | (10-26) |
| $\sigma_{0t} = (\sigma_{bE}^2 + \sigma_d^2 + \sigma_{ar}^2)^{1/2}$ | (10-27) |
| $\xi_s = [\xi_1^2 + \xi_2^2 + (\xi_A^*)^2]^{1/2}$ | (10-28) |
| $\xi_A^* = 2.3\xi_A A$ | (10-29) |
| $\dfrac{\sigma_A}{A} = \left\{(E_r \ln T)^{-2}\left[KmGE_rT^{-1} + E_r^2(\xi_1^2 + \xi_2^2) + \left(\dfrac{\sigma_{0t}}{T}\right)^2 + \left(\dfrac{\sigma_E}{T}\right)^2\right] + \xi_A^2\right\}^{1/2}$ | (10-30) |
| $\dfrac{\sigma_A}{A} = \left\{(-i_r \ln T)^{-2}\left[Ki_rT^{-1} + i_r^2(\xi_1^2 + \xi_2^2) + \left(\dfrac{\sigma_E}{mGT}\right)^2 + \left(\dfrac{\sigma_{0t}}{mGT}\right)^2\right] + \xi_A^2\right\}^{1/2}$ | (10-31) |

[a] $\xi_s$, Sample flicker factor (dimensionless); $\xi_1$, source flicker factor (dimensionless); $\xi_2$, atomizer transmission factor (dimensionless); $\xi_A$, analyte absorption flicker factor (dimensionless); $\sigma_E$, analyte emission noise (V).

ified by the sample flicker factor $\xi_s$. For AAS, the signal flicker noise arises from three sources, as specified by three flicker factors (equation 10-28). Here $\xi_1$ is the source flicker factor and $\xi_2$ is the flame or atomizer transmission flicker factor. The latter flicker factor accounts for the fluctuating transmission properties of the flame or other atomizer. The third flicker factor ($\xi_A$ or $\xi_A^*$) is unique to AAS and is denoted the *analyte absorption flicker factor*. It can be written in two forms that are related according to equation 10-29. Analyte absorption flicker noise is due to fluctuations in the measured analyte neutral ground state number density. It is believed to be due to fluctuations in sample introduction or in atomization characteristics that cause $n_0$ to vary. Thus $\xi_A$ is the RSD in $n_0$ for a given time constant or integration time. If $A \propto n_0 \propto c$, $\xi_A$ is expected to be independent of $A$ while $\sigma_A$ due to analyte absorption flicker noise is proportional to $A$. Note that $\xi_A^*$ specifies the RSD in $E_s$ due to analyte absorption flicker noise.

Equations 10-30 and 10-31 are the final expressions for $\sigma_A/A$ that account for all noise sources. Although these equations are written in terms of the signals observed with transmittance readout, they apply equally to direct absorbance readout since readout resolution and noise added by logarithmic conversion are usually negligible.

Figure 10-8 shows plots of $\sigma_A/A$ vs. $A$ for five limiting cases in which one of the five noise terms in equation 10-30 is dominant. Curves a to c show how precision varies with $A$ for the three standard limiting cases discussed in Section 5-5 in which 0% $T$ noise, signal shot noise, or signal flicker noise (source or atomizer transmission noise) are limiting. Analyte emission noise can be dominant at higher analyte concentrations or absorbances when the analyte emission signal is significant. The relative contribution of analyte emission noise ($\sigma_E$) to $\sigma_A/A$ increases faster than 0% $T$ noise (curve d) because $\sigma_E$ increases with analyte concentration, whereas $\sigma_{0t}$ is independent of $A$. Curve e shows

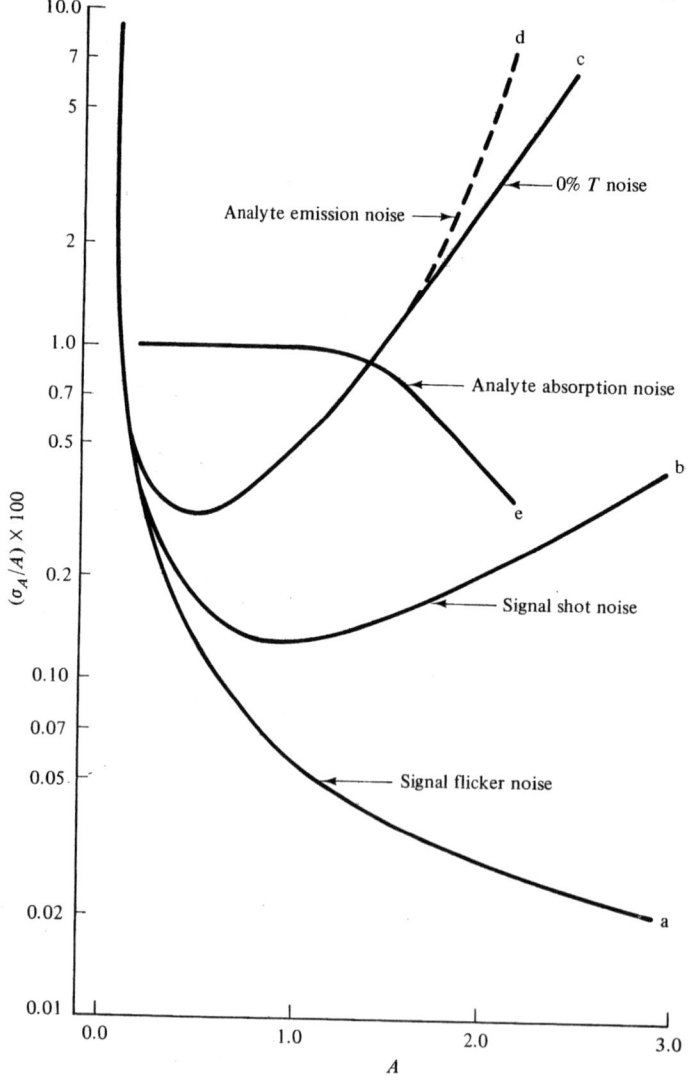

**FIGURE 10-8** Precision plots for limiting cases in flame AAS. In curve a, source and flame transmission noise are limiting; in curve b, signal shot noise is limiting; in curve c, 0% $T$ noise is limiting; in curve d, analyte emission noise is limiting; in curve e, analyte absorption flicker noise is limiting. For curves a, b, and c it is assumed that $\sigma_{st}/E_r = 10^{-3}$ at $A = 0$.

that analyte absorption flicker noise is usually dominant at moderate absorbances and the precision is often constant. At higher absorbances, the contribution of analyte absorption flicker noise to the RSD decreases because of negative deviation in the calibration curve, as illustrated in Figure 10-9. Although $\sigma_A/A$ decreases, the RSD for determining the analyte concentration does not improve.

Although the limiting noise sources in a given analysis depend on the element, source, type of flame, optical throughput, and other conditions, some general statements can be made. For most elements, $\sigma_A/A$ is determined by different types of noise in different absorbance regions. From the detection limit to about $A = 0.1$ to 0.2, the precision constantly improves with increasing $A$ and signal shot noise, source flicker noise, or flame transmission noise are limiting. Often the magnitudes of these three types of noise are within a factor of 2 to 3 of each other and typically $\xi_1$ and $\xi_2$ equal 2 to $10 \times 10^{-4}$. However, for elements with resonance lines at wavelengths shorter than 230 nm (e.g., As, Zn, Cd, Pb, Se, Te, Sb, Sn, Bi, and Ir), where flame absorption is significant, flame transmission noise is usually limiting and $\xi_2 = 2 - 3 \times 10^{-3}$. The absolute rms noise due to signal shot or flicker noise decreases with increasing $A$ since these noises are proportional to $T^{1/2}$ and $T$, respectively.

The absolute noise in $E_s$ due to analyte absorption flicker noise is proportional to $A$ and often becomes dominant at $A = 0.1$ to 0.2 and remains so up to $A = 1.0$ to 1.5. Typically, $\xi_A = 0.5$ to 1.0% for a 1-s integration time. Over this region, $\sigma_A/A$ is relatively constant if the calibration curve is relatively linear or constantly decreases if the calibration curve bends off. The analyte absorption flicker factor at higher absorbances can be estimated from

$$\xi_A = \xi_A' \frac{mc}{A} \qquad (10\text{-}32)$$

where $m$ is slope of calibration curve at $A$ and $\xi_A'$ is the value of $\xi_A$ at $A = 0.2$. At an absorbance of 0.2, $\xi_A$ can be easily evaluated because it is often the limiting noise source and calibration curves are usually linear.

Often for absorbances greater than 1 to 1.5, 0% $T$ noise and analyte emission noise become dominant and precision decreases. This region is not very useful for analyses because nonlinearity in the calibration curve is often severe at absorbances greater than 1.0. The relative 0% $T$ noise ($\sigma_{0t}/E_r$) is typically 2 to $200 \times 10^{-5}$. With modern AA spectrophotometers dark current noise and amplifier-readout noise are usually negligible compared to background emission noise. Digital readout with a resolution of 0.0001 A.U. prevents readout resolution from being the factor limiting precision. Thus 0% $T$ noise, if significant, is usually limited by background emission noise, which is larger and more important at lower absorbances if the background emission signal is large at the analysis wavelength (e.g., near the OH radical bands or in the visible region) or where the lamp signal ($i_r$) is small. Analyte emission noise is most likely to be the limiting noise source for elements with $\lambda > 350$ nm such as alkali metals, rare earths, some alkaline earths, and group 3B metals because the analyte emission signal is larger. Both background and analyte emission noise can be due to shot or flicker noise. Because the analyte and background emission signals are not carried by the modulation waveform (see Section 5-6), the contribution from $1/f$ flicker noise in these signals over the measurement bandpass centered at the modulation frequency is reduced.

Equation 10-31 indicates that the source radiance [e.g., lamp current or type of lamp (EDL vs. HCL)] and spectral bandpass should be increased such that $i_r$ is large enough that signal shot noise, $\sigma_{0t}$ and $\sigma_E$ are

$(\sigma_A/A)_{A-0.2} = (0.02/0.2) = 0.1$

$(\sigma_A/A)_{A-0.32} = (0.02/0.32) = 0.062$

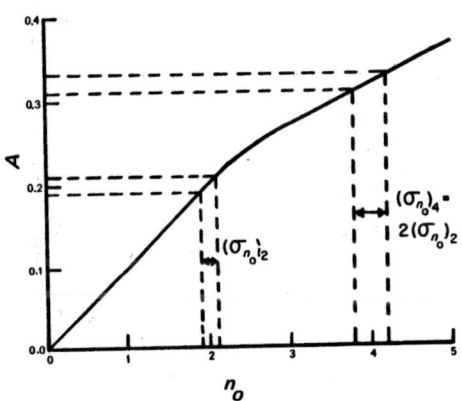

**FIGURE 10-9** Effect of calibration curve nonlinearity on analyte absorption flicker noise. At an absorbance of 0.2, a 10% relative fluctuation in the neutral ground-state number density ($n_0$) causes a 10% RSD in $A$ and $\xi_A = 0.1$. When the analyte concentration is doubled, $A$ increases only to 0.32 because of nonlinearity. The same RSD in $n_0$ causes a smaller RSD in $A$ because the calibration slope is one-half of the initial value. Thus $\xi_A = 0.062$ at this absorbance. [From N. W. Bower and J. D. Ingle, Jr., *Anal. Chem.*, *49*, 575 (1977), copyright 1977, with permission of the American Chemical Society.]

negligible at the highest absorbances to be measured. Note that $i_r$ is proportional to $\Phi_0$, as described by equations 10-5 to 10-7. In addition, $K$ (i.e., $\Delta f$) should be reduced so that signal shot noise is negligible. The source flicker factor may depend on lamp current. Generally, $\sigma_{0t}$, $\sigma_E$, and $\xi_2$ are less for cooler flames and leaner flames. Typical precision plots are shown in Figure 10-10. They show the limiting regions and the dominance of analyte absorption flicker noise at moderate absorbances.

The $\sigma_A/A$ expressions can easily be modified for other atomizers or sample introduction techniques. With electrothermal atomizers, background emission noise is not due to the flame but to blackbody emission from the furnace walls and usually important only in the visible region. The analyte absorption flicker factor may be higher since it accounts for reproducibility in sample introduction and atomization conditions. The time constant with electrothermal atomization must be smaller (e.g., 0.1 s) than with flame atomization to prevent distortion of the transient absorption signals. This increases the absolute rms noise due to most noise sources compared to flame measurements. The smaller size of the furnace or extra apertures used to restrict blackbody emission can reduce $\Phi_0$ and $i_r$ and increase the relative signal shot noise.

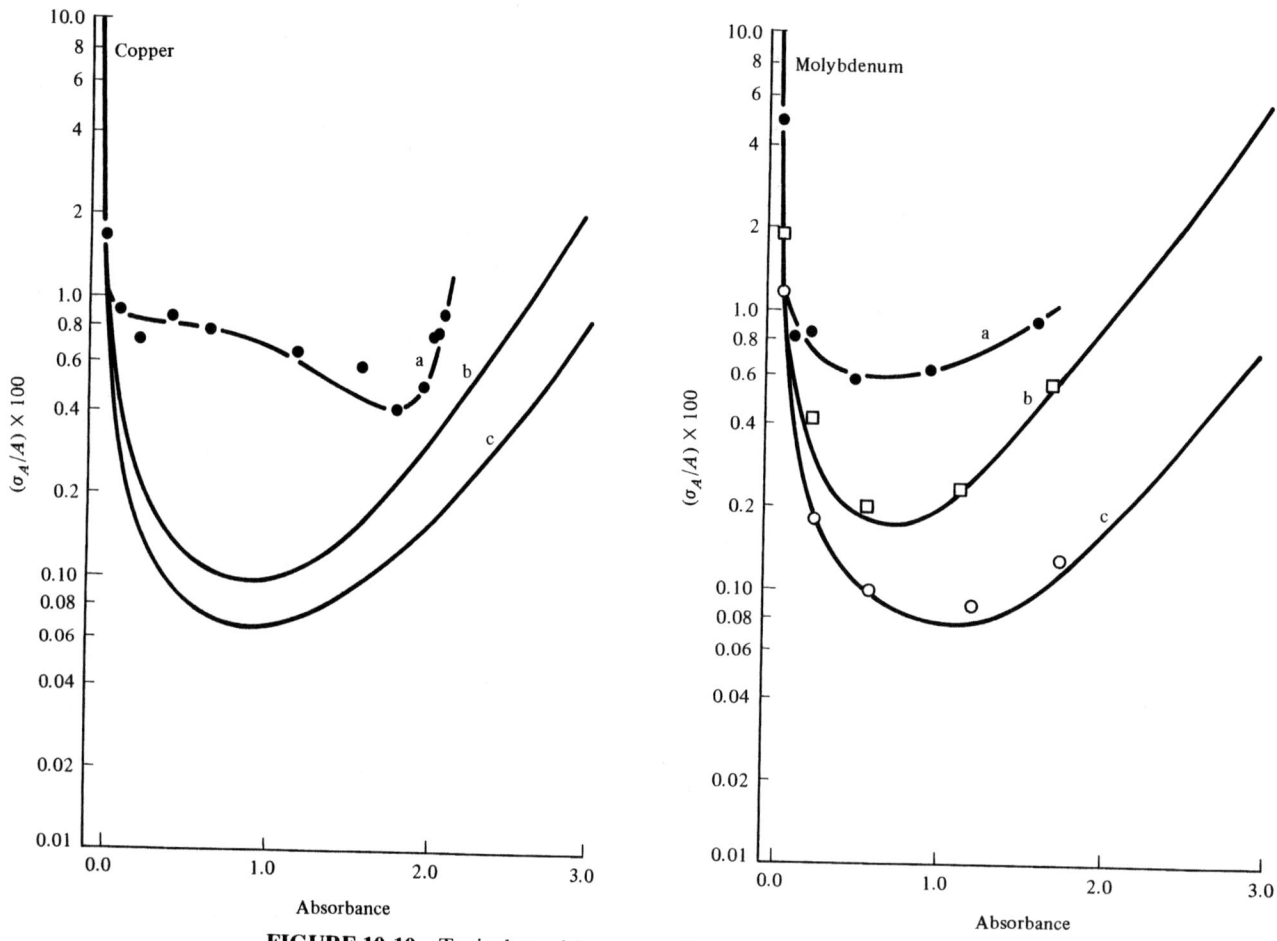

**FIGURE 10-10**  Typical precision curves for two elements in AAS. Curve a in both figures represents the experimental values of $\sigma_A/A$ measured with analyte solutions aspirating. In the curves labeled b and c in both figures, the absorbance is varied by placing filters in the light beam. In curve b the blank is aspirating while in curve c, the flame is off. Clearly, analyte absorption noise is dominant at moderate absorbances because the RSD is worse at the same value of $i_s$ (i.e., $A$) when analyte is present and responsible for the absorption. Comparison of curves b and c illustrates that flame background emission noise is greater than dark current and amplifier-readout noise. The rapid decrease in precision for Cu at high concentrations is due to analyte emission noise. (From N. W. Bower, Ph.D. thesis, Oregon State University, 1978, with permission of author.)

## 10-3 INSTRUMENTATION

The general design characteristics of a single-beam AA spectrophotometer were described earlier in this chapter. This section is concerned with various instrumental features and configurations that enhance the capabilities of AA spectrophotometers.

### Common Features of Commercial AA Spectrophotometers

Most commercial AA spectrophotometers provide several features for convenience. Normal operation requires that the PMT bias voltage be adjusted to vary the PMT gain ($m$) during the process of setting the readout absorbance to near zero with blank aspirating. With an *auto zero option*, the instrument automatically performs this adjustment when the appropriate front panel switch is activated. Thus when sample is aspirated, the readout absorbance is nearly the true sample absorbance. For best accuracy, the blank absorbance should be recorded and subtracted from the measured sample absorbance.

A scale expansion control allows the measured absorbance to be multiplied by a known factor before display on the readout device. This is useful if measurements are limited by readout resolution rather than noise. Scale expansion also allows readout directly in concentration units. Here the scale expansion factor is adjusted so that the number displayed on the readout device is equal to the concentration of the standard. Thus when a sample is aspirated, the concentration is directly displayed.

To use direct concentration readout, the calibration curve must be linear or nonlinearity must be compensated for. In older commercial instruments, correction circuitry was provided to linearize the calibration curve. Today linearization is usually based on software that implements curve-fitting techniques. Curve-correction techniques must be used with caution because the linearization algorithm used may not provide correct fitting of all analytical curves. Additional standards should be analyzed to check the validity of the linearization scheme.

When electrothermal atomizers are used, a *peak mode option* is convenient. Here the maximum absorbance value during the atomization step is electronically held and displayed. Calculation and reporting of the peak area is also common. Peak area measurements can be more accurate than peak height measurements if the analyte atomization profile is changed by the presence of concomitants. Ideally, the peak area for a given mass of analyte is independent of the width of the atomization peak if the total amount of analyte atomized is constant. Particularly for concentrations near the detection limit, precision in the peak area mode can be worse than for the peak height mode because the *S/N* for measuring points in the wings of the atomization profile is poor.

As mentioned in Chapter 8, many AA spectrometers provide an emission mode to allow flame AE determinations. In this mode, the AA source is turned off and a mechanical chopper is activated to modulate the emission signal at the same frequency that the AA lamp is modulated so that the AA electronics can process the analyte emission signal.

The incorporation of a microprocessor into an AA spectrophotometer allows more convenient use of the features discussed above and provides additional versatility and automation capabilities. Computerized systems are used for entry of experimental conditions, data acquisition, calculations, and presentation of results. With many systems, the operator can enter through a keyboard experimental conditions such as the values of the wavelength, slit width, lamp current, integration time, and gas flow rates, and in the case of electrothermal atomizers, dry, ash, and atomization parameters. These values may even be stored in memory so that the proper conditions are adjusted rapidly and reliably without operator attention once the element is chosen.

Data presentation can be as simple as a report of the stored absorbance values for all solutions before or after blank correction. With electrothermal atomization, peak heights and areas can be reported. If repetitive measurements can be programmed, measurement precision can be evaluated from the reported means and standard deviations. A sequence of blank, standard, and sample measurements can often be programmed. After this sequence, absorbance data from the standards are fitted to a calibration curve, even if nonlinear, from which the analyte concentration in the samples is calculated and reported.

Automatic samplers are available as accessories to enhance automation and allow unattended measurements (see Section 6-6). They allow the aspiration tube to be dipped sequentially into a series of standard and sample solutions loaded into cups on a sample turntable. Automatic samplers and dispensers are also available for electrothermal atomizers. They can increase the precision of the critical step of placement of the sample into the furnace tube. Provision must be made for cleaning the dispenser tube between samples, standards, and blanks. One manufacturer uses a nebulizer to spray a controlled amount of sample into the furnace tube. Standard addition or addition of matrix modifiers can also be automated.

Often, printers are used to provide a hard copy

of the experimental conditions, absorbance data, and statistical values. Video terminals or monitors provide faster display of this information as well as of calibration curves. Video displays are especially useful with electrothermal atomization since the time profile of the absorbance peak, and in some cases the temperature profile, can be viewed. Changes in peak shape or position can be indicative of interference problems.

The use of flow injection techniques (see Sections 6-6 and 10-1) with flame AA detection is expected to increase as a means of fast automated sampling. Flow injection techniques can be used for on-line sample dilution, for standard additions, or for introduction of releasing agents and/or ionization buffers.

### Double-Beam Systems

Most manufacturers of AA spectrophotometers provide double-beam (DB) models based on a design similar to that shown in Figure 10-11. The primary advantage of a DB system is that it continually compensates for source drift and flicker noise (lowers $\xi_1$). With a single-beam (SB) system, source drift between the sample and blank measurements can cause error. With a double-beam system, the warm-up time for source lamps is also reduced since measurements can be made before the lamp stabilizes. However, many SB spectrophotometers provide two to six separate HCL sockets and power supplies, at added expense, so that additional HCLs are already warmed-up when they are needed.

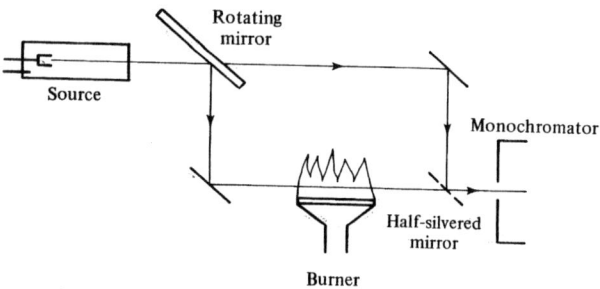

**FIGURE 10-11**  Double-beam optical system. A rotating mirror is used to direct the source beam in an alternating fashion through the atomizer or around the atomizer. The split beams are recombined and focused on the monochromator entrance slit. The electronics are configured to process separately the two signals that are 180° out of phase. Usually, the final signal displayed on the readout device represents the $-\log$ of the ratio of the sample beam signal to the reference beam signal. In some systems, a beam splitter is located at the position of the rotating mirror, and a rotating mirror is located in the position of beam combiner in the diagram above.

With a double-beam system, the reference beam does not pass through a "reference atomizer" to correct for blank absorption. Also, there is no compensation for drift and noise in the flame transmission characteristics. The added optical components reduce the measured photon signal levels (reduction in $i_r$), which reduces precision if signal shot noise or 0% $T$ noise is dominant. Finally, a DB system does not compensate for changes in the profile of the lamp emission line, which can influence the measured absorbance.

### Background Correction

In addition to the analyte, both the flame itself and concomitant species introduced into the flame (or an electrothermal atomizer) can absorb the source radiation. Flame absorption below 220 nm is serious, as shown in Figure 10-12. However, a water blank can compensate for this flame absorption.

Although absorption of the HCL analyte line by other atomized elements is rare in AAS, scattering and background absorption by molecular species originating from the sample can occur, particularly with electrothermal atomizers. This can cause measured absorbances, and hence predicted analyte concentrations, to be too high unless a proper blank can be prepared [i.e., the reference transmission factor ($T_r$) is different for the samples and standards]. Absorption by molecular species such as NaCl occurs over a broad wavelength range, as shown in Figure 10-13. Several molecular absorption background correction schemes have been developed and incorporated into AA spectrophotometers. Most of these schemes involve measurement of the total absorbance ($A_t$), which accounts for analyte and background absorption, and the background correction absorbance ($A_{bc}$), which ideally accounts for background absorption at the analyte wavelength. The corrected analyte absorbance ($A_c$) is then calculated from

$$A_c = A_t - A_{bc} \qquad (10\text{-}33)$$

In general one would expect detection limits with background correction to be worse than those obtained without background correction and background absorption. First, there is added imprecision due to the measurement of $A_{bc}$. Second, if background absorption is large, the effective lamp signal ($i_r$) for both measurements is reduced. This increases the effective blank noise if signal shot noise or 0% $T$ noise becomes the limiting noise source. Also the atomizer transmission noise increases due to fluctuations in the background absorbance.

Various correction schemes obtain $A_t$ and $A_{bc}$ in different ways. The specific correction method can af-

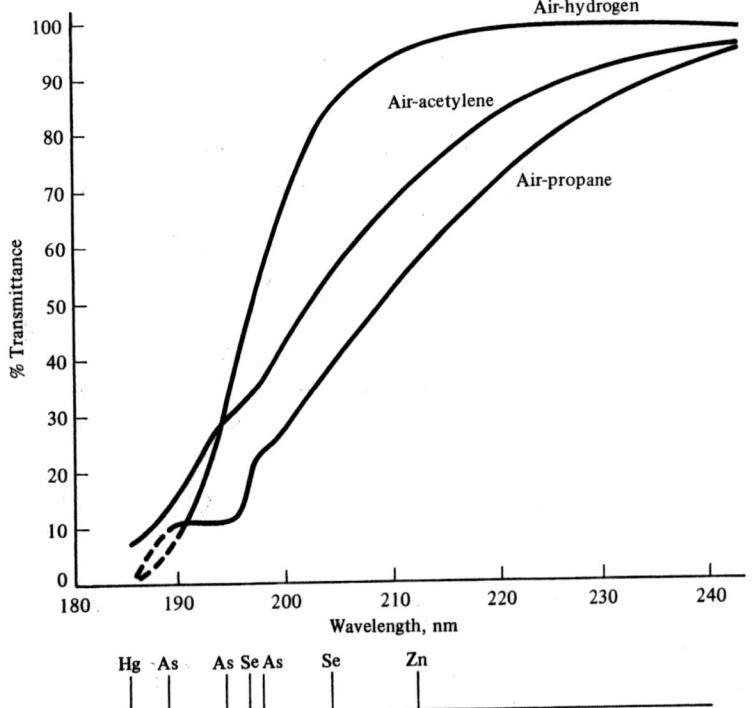

**FIGURE 10-12** Absorption by several flames in the short-wavelength region. The analysis wavelengths for several elements in this wavelength region are indicated. (Adapted from J. Ramirez-Munoz, *Atomic-Absorption Spectroscopy and Analysis by Atomic-Absorption Flame Photometry,* Elsevier, Amsterdam, 1968, with permission of publisher.)

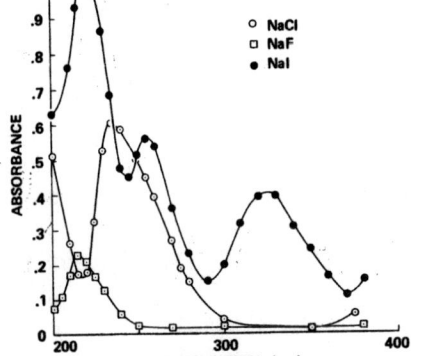

**FIGURE 10-13** Molecular absorption spectra of NaCl, NaF, and NaI in an electrothermal atomizer [5 μL of 0.1% (w/v) solutions]. (From B. R. Culver and T. Surles, *Anal. Chem.*, **47**, 920 (1975), copyright 1975, with permission of the American Chemical Society.)

fect the accuracy of the correction and the shape and slope of the corrected absorbance calibration curve. The total measured absorbance is the sum of the true analyte absorbance ($A_a$) and the background absorbance ($A_b$), or

$$A_t = A_a + A_b \qquad (10\text{-}34)$$

The measurement of the background correction absorbance yields an absorbance with contributions from the background ($A_b'$) and the analyte ($A_a'$) or

$$A_{bc} = A_b' + A_a' \qquad (10\text{-}35)$$

If equations 10-34 and 10-35 are substituted into equation 10-33, the result is

$$A_c = (A_a + A_b) - (A_b' + A_a') \qquad (10\text{-}36)$$

Ideally, $A_b = A_b'$, so that the background absorbance is totally compensated by the background correction absorbance measurement. The background absorbance correction is in error if this condition is not true. The contribution of analyte absorption to $A_{bc}$ is some fraction of $A_a$ or $A_a' = fA_a$. Ideally, $f = A_a' = 0$, so that $A_c = A_a$ if $A_b = A_b'$ and the corrected absorbance calibration curve is identical to the normal calibration curve in the absence of background absorption. If $f$ is significantly different from zero, but independent of analyte concentration, the calibration curve for corrected absorbance has the same shape as that of the normal calibration curve but is lowered by a fixed fraction $(1 - f)$, as illustrated in Figure 10-14a. The lower calibration slope degrades the detection limit. If $f$ increases at higher analyte concentrations, the shape of the calibration curve is changed and roll-over can occur, as illustrated in Figure 10-14b. Thus two very different analyte concentrations can yield the same absorbance (i.e., the analytical curve is double valued).

We will now consider the background correction

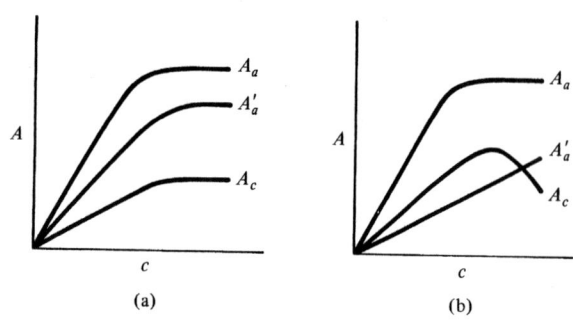

(a)                    (b)

**FIGURE 10-14** Effect of background correction on calibration curves. Here background absorption is assumed negligible. Calibration curves are shown for the normal analyte absorbance ($A_a$), the analyte absorbance measured in the correction absorbance measurement ($A_a'$), and the corrected absorbance ($A_c$). In (a), $A_a'$ is a constant fraction of $A_a$ and the curve for $A_c$ is lowered by a fixed fraction relative to the curve for $A_a$. In (b), the curve for $A_a$ reaches a plateau at high analyte concentrations but $A_a'$ is proportional to the analyte concentration over the range shown. Note that the corrected absorbance calibration curve rolls over at high analyte concentrations where $A_a$ is constant but $A_a'$ is increasing. If background absorption is significant and constant for all analyte concentrations, all points in the curves for $A_a$, $A_a'$, and $A_c$ would be raised by $A_b$, $A_b'$, and ($A_b - A_b'$), respectively.

schemes (two-line, continuum source, Zeeman, and Smith–Hieftje) developed for line source AAS. Background absorption correction with continuum source AAS is discussed later in the chapter.

*Two-Line Background Correction Method.*
Although not used much today, the two-line method

for correction of background absorption was the first method proposed. First $A_t$ was measured in the normal way at the analyte wavelength with a blank measurement. Second $A_{bc}$ and an appropriate blank absorbance were measured with a second line (filler gas or nonresonance) from the same HCL or a different HCL. In this method, the second line should be near the analyte wavelength but outside the spectral bandpass used for the measurement at the analyte wavelength. Normally, $A_a'$ is zero, but the correction is accurate only if the background absorbance is the same at both wavelengths ($A_b = A_b'$). Thus a uniform background absorption is assumed between the two measured wavelengths. It is often difficult to find an intense line near the analyte wavelength. Also, the method is tedious and subject to error. In the time interval between the two measurements during which experiment conditions (e.g., wavelength) must be readjusted, experimental conditions can drift.

*Continuum Source Background Correction.*
The second background correction technique developed and now available as an option with most AA spectrophotometers is the continuum lamp technique illustrated in Figure 10-15. With a beam splitter, radiation from a deuterium or hydrogen arc lamp (or $H_2$- or $D_2$-filled HCL) and from the analyte HCL are directed through the atomizer. The lamps are pulsed at different frequencies or out of phase, so that the signal processing electronics can distinguish and process a separate absorption signal measured with each lamp. With the HCL, $A_t$ is measured, whereas the signal measured with the $D_2$ lamp is $A_{bc}$. Typically, the instrument automatically calculates and reports the corrected absorbance by taking the difference (equation 10-33). It is assumed that

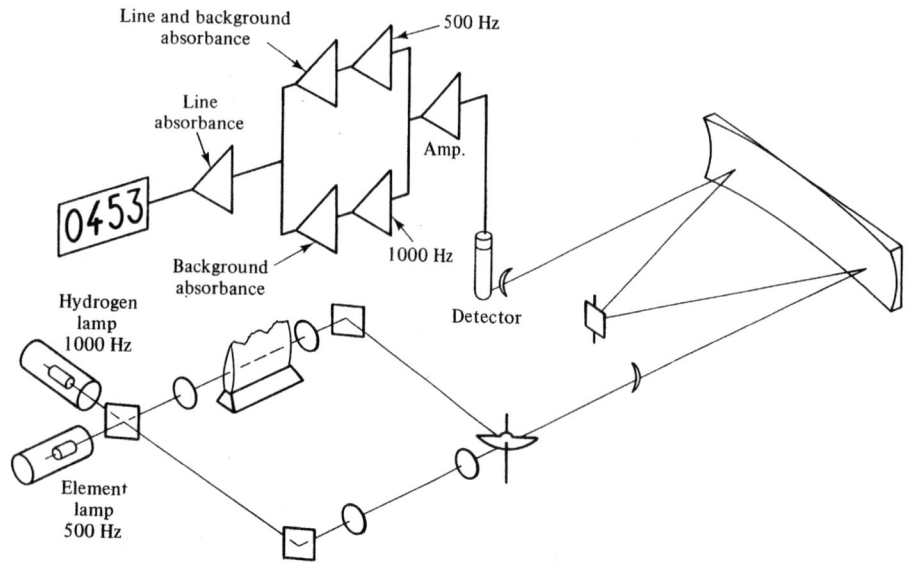

**FIGURE 10-15** Optical diagram for continuum source background correction. Background is corrected by two sources operated double beam and pulsed at different frequencies. (Courtesy of Thermo Jarrell Ash, Inc.)

the background absorbance is constant over the spectral bandpass and thus equal to the background absorbance at the HCL wavelength ($A_b = A_b'$). It is also assumed that analyte absorption of the continuum radiation is insignificant ($A_a' = 0$) because the analyte absorption profile is very narrow compared to the spectral bandpass over which continuum source radiation is passed (typically, 0.1 to 1 nm).

Despite the success of the $D_2$ lamp correction system, there are some limitations. First, the correction can degrade detection limits because the radiance of $D_2$ or $H_2$ lamps is weak at longer wavelengths (see Figure 4-5). The signal processing electronics usually require that the reference signals for both sources be the same. Because $\Phi_0$ for the $D_2$ lamp is often less than for the analyte HCL, the lamp current for the analyte HCL is often reduced below its normal value. This degrades precision if signal shot noise or 0% $T$ noise becomes significant. A tungsten lamp can be used in place of the $D_2$ lamp at higher wavelengths to reduce the problem. We could envisage a system in which the HCL and continuum lamp signals were separately processed in way that did not require reduction of the HCL intensity. The precision of the background correction measurement could still be poorer.

The correction can be in error if the background absorption over the spectral bandpass is nonuniform or structured. For example, it is possible for atomic absorption by lines from other elements within the spectral bandpass to increase the background absorbance measured even though this type of absorption does not affect the analyte absorbance measured with the HCL. Thus the method can overcorrect or undercorrect for background absorption.

The alignment of two sources so that their beams pass exactly through the same volume element of atomizer is tedious. If the lamps are misaligned, the cor-

rection can be erroneous since the background absorption probed by each source is different.

The $D_2$ background correction method can work with background absorbances greater than 1.0 and has little effect on the calibration curve. Detection limits are degraded by at least a factor of 2; the degradation depends on the wavelength and amount of background absorbance. With a single-beam instrument, the continuum source correction can be implemented manually by sequentially measuring the absorbances with the HCL lamp and $D_2$ lamp in position.

*Zeeman Background Correction.* Zeeman atomic absorption (ZAA) correction is available with some commercial instruments. Here a magnetic field splits normally degenerate spectral lines into components with different polarization characteristics. Elemental and background absorption are distinguished based on the magnetic and polarization characteristics of the lines.

There are several ways to implement ZAA correction. Either the source or the atomizer can be placed in a strong magnetic field. The magnetic field can be parallel or perpendicular to the light path and generated with a permanent magnetic or an electromagnet (ac or dc).

To illustrate the principle of the ZAA technique, we will first consider *analyte-shifted ZAA correction*, where the atomizer is placed in a transverse dc magnetic field as shown in Figure 10-16. We will assume that the normal Zeeman splitting applies (see Figure 7-12b). The analyte absorption lines are split into $\pi$ and $\sigma$ components that are polarized parallel and perpendicular to the magnetic field, respectively. The wavelength of the $\pi$ component remains at the zero field wavelength while the $\sigma$ components are shifted away from this central wavelength, as shown in Figure 10-17a. With one

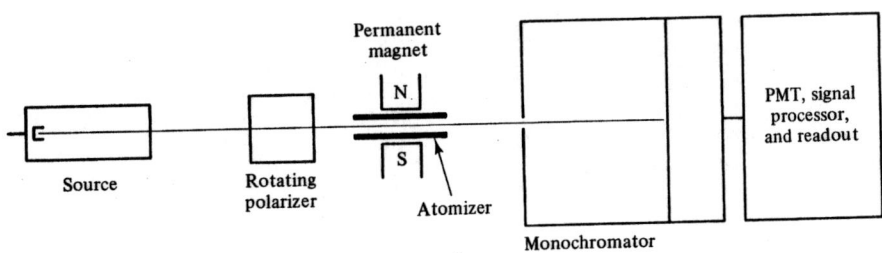

**FIGURE 10-16** Zeeman atomic absorption background absorption correction. A dc magnetic field perpendicular to the optical axis splits the analyte absorption lines into $\pi$ and $\sigma$ polarization components. A rotating polarizer allows the $\pi$ and $\sigma$ components of the source light beam to pass through the atomizer alternately. The signal processor measures the absorbance of each polarization component and reports the difference in absorbance of the $\pi$ and $\sigma$ components.

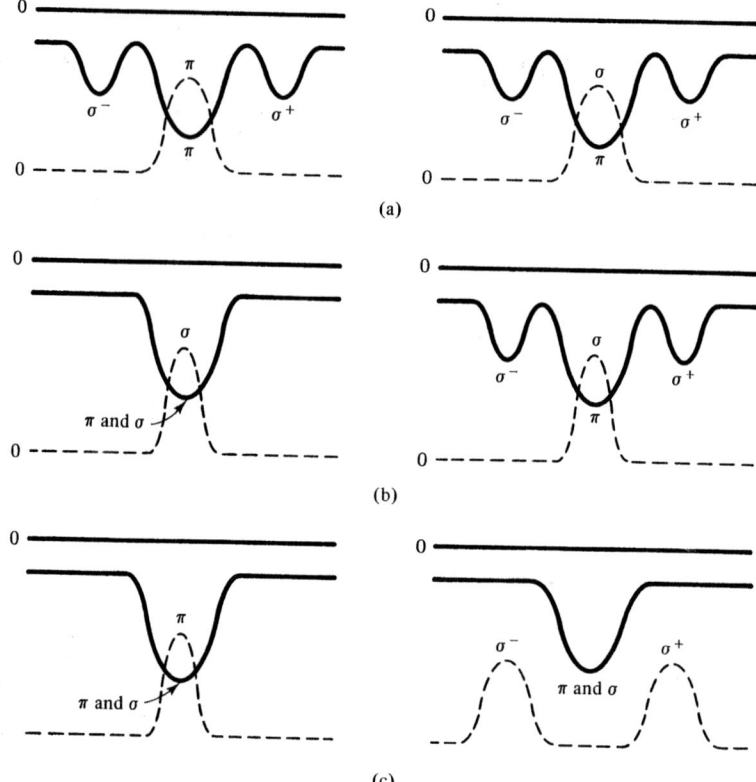

**FIGURE 10-17** Source profiles (dashed lines) and absorption profiles (solid lines) with different Zeeman configurations. In (a) a dc magnetic field splits the absorption σ components continually and the absorption of the source σ and π components is alternately measured. In (b) the atomizer is placed in an ac magnetic field. The absorbance is measured at zero field and when the field is at its maximum value. In (c), the source σ and π components are split by a dc magnetic field and are alternately passed through the atomizer.

orientation of the rotating polarizer, the lamp emission π component is transmitted by the polarizer and is absorbed by the unshifted π component of the analyte. Thus $A_t$ or both analyte and background absorption at the normal analyte wavelength are measured. In the second cycle, the source σ component is transmitted and $A_{bc}$ is measured. The lamp σ component is not absorbed by the unshifted analyte π component because its polarization is different. The shifted analyte absorption σ components ($σ^+$ and $σ^-$) are at different wavelengths and ideally absorb no source radiation, as illustrated in Figure 10-17a. Thus under ideal circumstances only background absorption at the analyte wavelength is measured by σ source radiation ($A_b = A_b'$ and $A_a' = 0$). Subtraction of the two absorbances measured in each cycle yields the corrected absorbance.

If the atomizer is placed in an ac transverse magnetic field, a static linear polarizer is oriented between the atomizer and monochromator to transmit only the source σ components for all measurements. As shown in Figure 10-17b, $A_t$ is measured when the field is off because the σ absorption components are unshifted. When the field is on, $A_{bc}$ is measured with ideally no analyte absorption by the shifted absorption σ components.

With either ac or dc techniques, it is less convenient to place a flame between the magnetic poles than it is to place an electrothermal atomizer. Flame at-

omizers require more space and thus a larger and more powerful magnet to obtain the same field strength across the atomization region. Longitudinal fields can be used with electrothermal atomizers. In this case the absorption π component is polarized along the optical axis and does not absorb the π component of the source radiation which is in another plane. Hence a linear polarizer is not needed to isolate the source σ component.

With *source-shifted* ZAA *correction*, the source is placed in a dc magnetic field and a polarization modulator (an electro-optic- or magneto-optic-based variable-phase retardation plate, see Section 3-3) is used to rotate the plane of polarization of the two components of the source beam back and forth by 90°. The beam is then passed through a static linear polarizer. This method of alternating the polarization component transmitted is often superior to the use of a rotating polarizer (e.g., two linear polarizers mounted on a rotating wheel with their polarization axes oriented 90° with respect to each other). With a rotating polarizer, it can be difficult to align both polarization components along the same optical axis. Often, additional optical components are required with a rotating polarizer to compensate for differences in the intensity of the two polarization components.

As illustrated in Figure 10-17c for source-shifted ZAA, $A_t$ is measured when the π component has the proper plane of rotation to be transmitted by the static linear polarizer. The unshifted source $\pi$ component is absorbed by the $\pi$ component of the analyte and by molecular species. When the plane of rotation is altered 90°, the wavelength-shifted source $\sigma$ components are transmitted and ideally absorbed only by the background species.

Source Zeeman correction can be used with any kind of atomizer. However, special magnetically stabilized lamps are required. Conventional HCLs cannot be used. Often, EDLs or glow discharge lamps are employed.

Many elements produce the anomalous Zeeman effect (see Figure 7-12c), and this complicates ZAA correction because there are several $\sigma$ and $\pi$ components. Both polarization components are split from the central wavelength with the $\sigma$ components undergoing wider splitting. With hyperfine structure and isotope effects, the $\sigma$ and $\pi$ profiles can be broadened considerably. The shift of the $\pi$ component away from the field free central wavelength reduces the analyte absorption coefficient (the calibration sensitivity) and is dependent on the field strength. This shift is not a problem with analyte-shifted ZAA correction with an ac field since $A_t$ is measured when the field is zero.

There are several advantages of Zeeman techniques over the continuum-source-based correction method. With a single source, normal source intensities can be used at all wavelengths without the alignment or differential drift problems experienced with two sources. In general, background correction is more accurate ($A_b = A_b'$) if the background absorption is structured over the spectral bandpass because the background absorbance is measured exactly at or very near the wavelength that the analyte absorbance is measured.

The Zeeman technique involves effectively a double-beam in-time measurement. The difference in two absorption measurements is continually calculated at the frequency that the polarization components are modulated. Thus the contributions from $1/f$ source flicker noise and atomizer transmission flicker noise are reduced, which can improve detection limits if these noise sources are limiting.

There are some disadvantages of ZAA correction systems compared to conventional AA measurements without correction. First, the implementation is rather complex and expensive. With analyte-shifted Zeeman systems, the background absorption bands can also experience Zeeman shifts and reduce the accuracy of the correction ($A_b' \neq A_b$). For optimized correction, the magnetic field strength should be adjusted for each element, and this is not possible with some commercial systems. The S/N and detection limit can be worse if signal shot noise or 0% T noise become the limiting

noise source because the source intensity is reduced due to transmission losses in polarizer components and the use of only a fraction of the field-free source intensity for each measurement.

The calibration slope for the corrected absorbance may be reduced a factor of 2 to 3 over uncorrected measurements (i.e., worse detection limits) and calibration curves can exhibit greater nonlinearity and even roll-over. With typical field strengths of about 1 tesla (kilogauss), the shifts in polarization components are 1 to 10 mÅ. The wings of the $\sigma$ component profiles are not totally shifted away from the $\pi$-component profiles, and some analyte absorption can be seen in the measurement of $A_{bc}$ (i.e., $A_a' \neq 0$). This effect is worse with anomalous splitting, significant hyperfine structure, and isotope effects. Absorption in the line wings causes $A_c$ to be less than $A_a$ and roll-over at absorbances of 0.5 to 1.0 for most elements. The calibration sensitivity is also reduced for cases where the $\pi$ component is shifted significantly from the zero-field wavelength. Detection limits can be better or worse than those obtained without background correction, depending on the element, the limiting noise sources, and the configuration for correction used.

*Pulsed Hollow Cathode Lamp Background Correction.* The newest commercially available background correction system is the Smith–Hieftje (S-H) system. As shown in Figure 10-18a, a HCL is pulsed at a normal current (6 to 20 mA) for 9.7 ms and a high current (100 to 500 mA) for 0.3 ms during each period of the modulation cycle. During the low-current part of the waveform, $A_t$ due to analyte and background absorption is measured. When the short-duration high-current pulse is passed through the HCL, $A_{bc}$ is sampled. The HCL must be capable of stable operation when pulsed to high currents.

The emission line from the HCL is considerably broadened and somewhat self-reversed at the higher current, as shown in Figure 10-18b. Because the average absorption coefficient over the line profile is much less, the measured analyte absorbance ($A_a'$) for a given analyte concentration is much smaller, as seen in Figure 10-18c. If the background absorbance is constant over the broadened emission profile, $A_b = A_b'$. The instrument automatically calculates the absorbance difference ($A_c = A_t - A_{bc}$).

As with ZAA correction systems, one HCL is employed so that the technique works at all wavelengths, alignment is simple, and there is compensation for source flicker noise and atomizer transmission flicker noise. Additional optical components, such as beam splitter or polarizers, which reduce the measured source intensity and can degrade the S/N are not required.

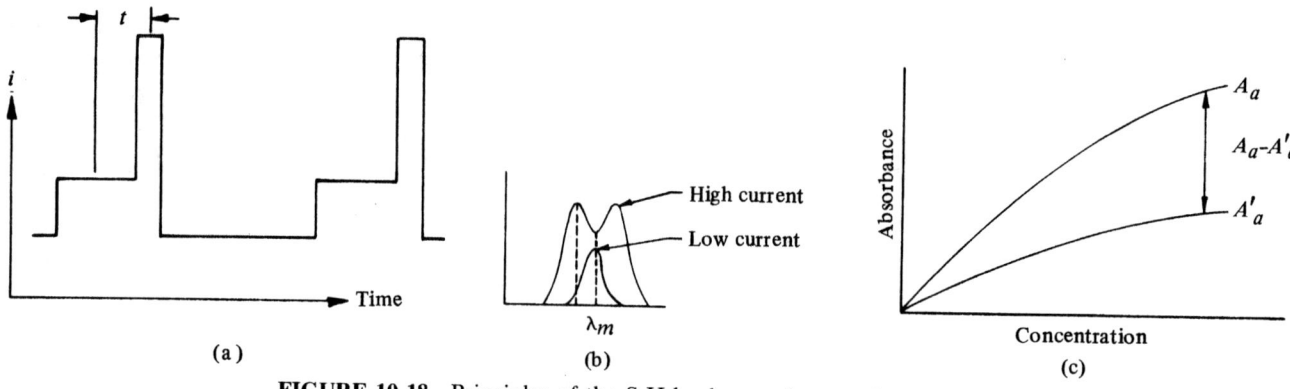

**FIGURE 10-18** Principles of the S-H background correction system. In (a) the current waveform used to drive the HCL is shown. The time between low-current and high-current segments ($t$) is only 4.5 ms, indicating the temporal proximity of background correction. In (b) the idealized emission profiles of the HCL at low and high lamp current are shown. In (c) the analyte absorbance calibration curves are shown for the low and high current measurements. [Adapted from S. B. Smith, Jr., and G. M. Hieftje, *Appl. Spectrosc.*, *37*, 419 (1983), with permission of the Society for Applied Spectroscopy.]

Overall the technique is relatively simple, inexpensive, and can be used with any atomizer.

The background correction accuracy is usually very good ($A_b' = A_b$), even for background absorbances greater than 2, unless the background absorbance over the broadened emission profile (about 0.01 Å) is structured. The primary disadvantage is a reduced calibration sensitivity (10 to 80%, depending on the element) due to analyte absorption during the high-current period ($A_a' \neq 0$). Roll-over in calibration curves at very high absorbances can occur but is less severe than with ZAA systems. Detection limits are typically worse by a factor of 2, and in some cases, five to six times worse than those obtained without correction.

### Multielement Spectrophotometers

Of the three basic atomic spectrometric techniques, AA is the least well suited for simultaneous multielement measurements. With multiple single-element line sources, all beams must be combined into one beam to pass through the atomizer along the same optical axis. This can be accomplished with beam splitters or with fiber optics. However, with very many line sources the alignment is tedious and the reduced intensity (50% loss per beam splitter) can ultimately reduce precision (increase the relative importance of signal shot noise and 0% *T* noise). Also, a separate power supply is needed for each single-element HCL. These problems can be somewhat alleviated by the use of multielement HCLs or by the use of a continuum source as discussed later in this section. A grating can also be used in reverse to combine the beams from several HCLs (i.e., wavelengths) into one beam.

A second problem is the limited linear dynamic range of AAS (approximately two to three orders of magnitude for a given analytical line). With one sample dilution, it may be difficult to adjust the concentration of all elements to be determined to be both above the detection limit and in the linear range.

The optimum conditions (e.g., type of flame, fuel/oxidant ratio, burner position) for single-element determinations do vary. Thus a third problem is the need to choose compromise conditions in which the detection limit and calibration curve linearity for some elements is degraded. This may be more critical with electrothermal atomizers, where optimum ashing and atomization temperatures vary considerably.

*Line Source Multielement AA.* Several designs for simultaneous multielement AA spectrometers with line sources for 2 to 10 elements have been proposed which differ in the way the spectral signals at different wavelengths are resolved and detected. In one general approach, a direct-reader arrangement with a separate slit and PMT for each element is employed. Alternatively, the radiation transmitted through several exit slits in the focal plane can be directed to one PMT. The absorption signals from each element are distinguished with a *time multiplex approach*. Here the HCL for each element is pulsed on at a separate time or the slits are sequentially blocked and unblocked with a rotating wheel with apertures. The data acquisition electronics are synchronized to the lamp pulsing or slit openings.

Some researchers have used a multichannel detector (e.g., diode array) in the focal plane. There is some compromise between wavelength coverage and resolution due to the limited size of these detectors. The limited dynamic range of these detectors also causes problems. Ideally, the reference signal for each detector element corresponding to a specific analyte line should be adjusted to be near the saturation value of the array elements for the integration time used. In reality this is difficult because the source radiance of different HCLs varies greatly. For weaker analysis lines, the relative amplifier-readout noise ($\sigma_{ar}/E_r$) is greater because $E_r$ is less and precision degrades. Different integration times can be used to alleviate this problem and different analysis lines can be used to extend the dynamic range.

The only commercial simultaneous multielement instruments available are dual-channel spectrophotometers. The radiation from two HCLs is passed through the flame and then directed to a beam splitter. The transmitted and reflected components are directed to two different monochromators. The radiant power isolated by each monochromator is detected and processed separately. Such instruments increase sample throughout by a factor of 2 and allow internal standard methods to be used. Dual-line or continuum source background absorption correction schemes can be implemented. In addition, two analysis lines can be used for one element to yield an overall greater dynamic range.

Computerized commercial AA spectrophotometers are available that allow automated sequential multielement analysis. The operator initially chooses new or previously stored values for experimental conditions such as wavelength, slit width, lamp position, lamp current, integration time, and gas flow rates for each element. Once initiated, the instrument readies itself for determination of the first element and performs the requested absorption measurements. This sequence of setting up and measurement is automatically reported in a sequential manner for each element (up to six elements) without operator assistance.

Although these instruments do not provide the high throughput and reduced sample consumption of a true simultaneous multielement instrument, they speed up the analysis and allow optimization of conditions for each element. Normally, the burner position or type of burner head cannot be optimized automatically in commercial spectrophotometers.

Several researchers have developed sequential multielement AA spectrometers based on rapid-scan or slew-scan monochromators or on an image dissector tube (IDT) (see Section 4-4). In the latter case, the IDT is placed in the focal plane of a spectrograph to select different wavelengths sequentially. As with the use of diode arrays, there is a trade-off between wavelength coverage and resolution because of the size of the active detector element or array.

*Continuum Source Multielement AA.*    The most successful design of a simultaneous multielement AA spectrophotometer is based on a continuum source with an echelle polychromator. A high-intensity Xe arc lamp is used as the single source, which alleviates the expense, complexity, and source alignment problem experienced with using multiple line sources. An echelle monochromator or polychromator is required to obtain a spectral bandpass on the order of the atomic absorption line width. For simultaneous analysis, a bank of PMTs are positioned behind appropriate exit slits in the two-dimensional focal plane of an echelle polychromator to intercept the transmitted source radiation corresponding to the analysis wavelength for each element. Sequential systems are based on a scanning echelle monochromator.

There are some disadvantages of continuum source AA compared to conventional line source AAS. Because the spectral bandpass is still greater than HCL line profile widths ($s > \Delta\lambda_{eff}$, see Section 10-2), the initial calibration slopes are less, and nonlinearity at higher absorbances is more pronounced. Characteristic concentrations (i.e., the concentration yielding an absorbance of 0.0044) are typically two to four times poorer.

The measured lamp radiant power ($\Phi_0$) can be less because of a smaller lamp radiance over the absorption line width compared to a line source or because of lower optical throughput. The latter reason is believed to be dominant. Typically, smaller slit widths and particularly slit heights (100 $\mu$m) are used with echelle monochromators and polychromators. Also, the source radiation for a given wavelength is spread over many orders and only one order is intercepted by the exit slit.

The lower measured radiant power and hence lower reference signal ($i_r$) increase the blank noise as signal shot noise or 0% $T$ noise become more significant. The combined effect of reduced calibration sensitivity and increased blank noise degrades detectability. Above 250 nm, detection limits are 2 to 5 times poorer and can be 10 to 20 times poorer for analysis wavelengths shorter than 250 nm.

Most continuum source echelle AA spectrophotometers are modified for wavelength modulation by placing a refractor plate behind the entrance slit (see Figure 4-34). This provides background absorption correction and discrimination against 1/$f$ source and atomizer transmission noise. The wavelength modulation depth is typically about 0.1 Å, and background correction is accurate if the background absorption is constant and unstructured over this wavelength interval.

With continuum source AA, it is also possible to make measurements in the wings of the absorption profile, where the absorption coefficient is less. This can extend the dynamic range of the calibration curve to higher concentrations.

## 10-4 PERFORMANCE CHARACTERISTICS

We now consider linearity, precision, accuracy, and detection limits in some detail. Emphasis is placed on how instrumental factors affect these performance characteristics.

### Linearity

Most AA calibration curves exhibit nonlinearity at absorbances greater than 0.5 to 1.0 and flatten out at absorbances of 1.5 to 2.0. There are several primary reasons for this nonlinearity. In flame and electrothermal atomizers, absorption line widths are typically 5 to 50 mÅ due to collisional and Doppler broadening, and the wavelength of peak absorption is often shifted to the red by up to 0.03 Å. The emission line widths from HCLs and EDLs are smaller because temperatures and pressures of these lamps are lower than those of the heated atomizer. At normal HCL currents (6 to 20 mA peak), source line widths are primarily Doppler broadened and typically 1 to 10 mÅ. Therefore, although emission line widths are much smaller, there is still some variation of the absorption coefficient over the source line width, and the maxima of the source and absorption profiles may not coincide. These effects are illustrated in Figure 10-19a. This situation is complicated by hyperfine structure and isotope effects, as shown in Figure 10-19b. Because of fine structure, other closely spaced analyte lines that are within the monochromator spectral bandpass can also cause nonlinearity because they are absorbed to different degrees.

Source line broadening considerations are often important in the selection of the HCL current. The HCL current can be adjusted in a range determined by the minimum current at which the HCL will operate and the maximum current specified by the manufacturer. Above this maximum current, the lamp may operate erratically, and the lifetime is considerably shortened. The HCL emission profiles for all elements are broadened at very high currents (> 100 mA) due to higher temperatures and number densities. This phenomenon is the basis of the S-H background correction system discussed previously. For volatile elements such as Cd, Ag, Ni, Pb, and Zn, significant broadening and even

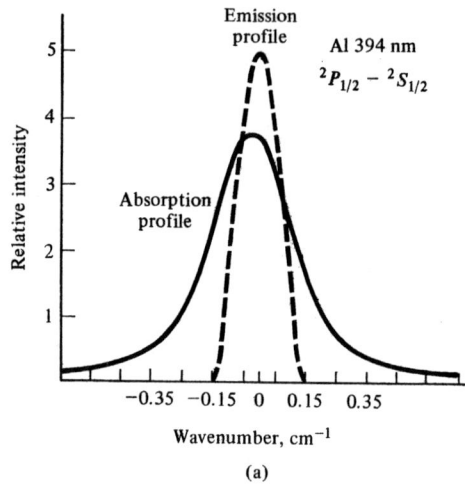

(a)

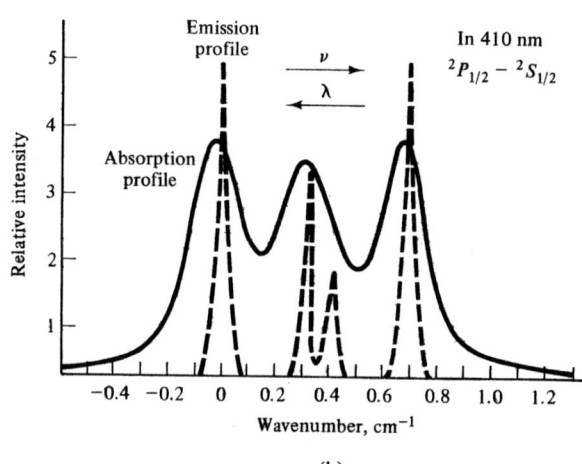

(b)

**FIGURE 10-19** HCL emission and flame absorption profiles. In (a) the width of the source profile is seen to be less than that of the absorption profile of Al in a $N_2O-C_2H_2$ flame. However the absorption coefficient does vary slightly over the source profile. The peak of the absorption profile is slightly shifted to the red due to collisional broadening. In (b) the emission and absorption profiles of In are seen to be actually composed of several hyperfine structure components. In this case, the measured absorbance depends upon the relative absorption of each source component of variable intensity by each corresponding absorption component with a different peak absorption coefficient. In the figures, $0.1 \text{ cm}^{-1} \approx 16 \text{ mÅ}$. [Adapted with permission from H. C. Wagenar and L. de Galan, *Spectrochim. Acta, 28B,* 157 (1973), Pergamon Press, Ltd., Oxford.]

self-reversal occurs at moderate currents (5 to 20 mA). These line profile effects increase the variation of $k(\lambda)$ over the absorption profile, reduce the initial calibration slope, and enhance nonlinearity as illustrated in Figure 10-20 for Cd.

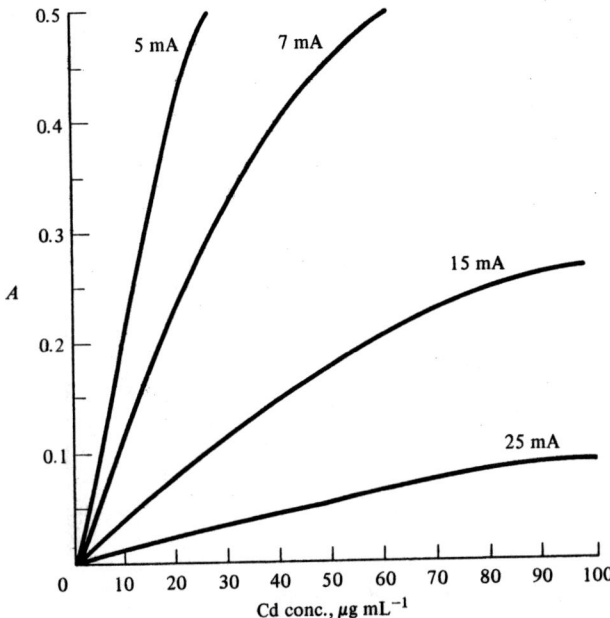

**FIGURE 10-20** Calibration curves for cadmium for different HCL currents. The increase in nonlinearity with increasing HCL current is more severe than for most elements. [Adapted with permission from B. J. Russell, J. P. Shelton, and A. Walsh, *Spectrochim. Acta*, *8*, 317 (1957), Pergamon Press, Ltd., Oxford.]

Stray radiation is a second major cause of nonlinearity. Traditionally, stray radiation is radiation outside of $\lambda_0 \pm s$ that is passed by the monochromator. In AAS, source radiation from lines of wavelengths far from the analysis line may be partially passed. However, in AAS, stray radiation usually refers to radiation from other source lines, due to impurities or the filler gas, that is within the spectral bandpass of the monochromator. The contribution of these unabsorbed lines (or in some cases less efficiently absorbed analyte lines) to the total radiant power passed by the monochromator can become significant at high analyte concentrations, where the analyte resonance line is highly attenuated.

As illustrated in Figure 10-21, reduction of the spectral bandpass can eliminate nonabsorbed lines, increase the initial calibration sensitivity, and reduce nonlinearity. Of course, unabsorbed lines at the analyte wavelength or closer than the resolution of the monochromator (0.2 to 0.005 nm) cannot be eliminated. Also, for many HCLs, the line emission is superimposed on top of a weak continuum whose relative contribution (typically, 0.1 to 2%) can be minimized, but not eliminated, by reducing the spectral bandpass. The effect of nonabsorbed lines and the slit width on the amount of stray radiation can depend on lamp current, which can alter the relative intensities of the source analyte and unabsorbed lines.

The shape of the calibration curve can sometimes be described by (see derivation of equation 13-53)

$$A' = -\log (T + r) \qquad (10\text{-}37)$$

where $T$ is the ideal transmittance when $r = 0$ and $r$ is the fraction of the reference readout signal due to unabsorbed radiation passed by the monochromator. The fraction $r$ can be measured at high analyte concentrations, where $T \ll r$ and is typically on the order of

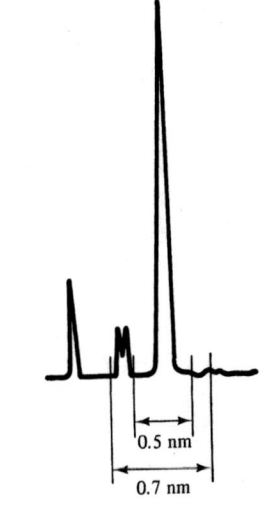

(a)

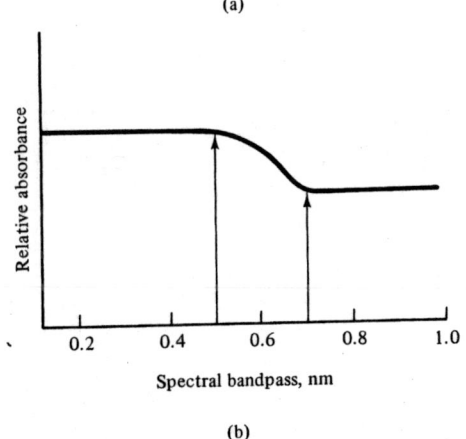

(b)

**FIGURE 10-21** Effect of slit width on absorbance in AAS. In (a) the spectrum of a HCL lamp in the region near the intense analyte resonance line is shown. Several other less intense nonabsorbed lines are nearby. In (b) the absorbance for a given analyte solution is seen to decrease if the spectral bandpass is larger than 0.5 nm because the nonabsorbed lines are passed by the monochromator. With elements such as Ce, Co, Cd, Ir, La, Nb, Os, Pd, Sb, Tb, Ni, U, W, and Rb, the proximity of nonabsorbed lines to the HCL resonance line makes adjustment of the spectral bandpass more critical.

1%. Equation 10-37 is often used as the fitting equation for implementing curve correction. Alternatively, curve fitting based on a polynomial equation may be more accurate.

One assumption used in the derivation of Beer's law is that the concentration of the absorbers is uniform. Thus a third major cause of nonlinearity in AA calibration curves is variation of the analyte atom density and even the absorption profile both laterally and vertically over the cross section of the source beam viewed by the monochromator. In the extreme, some of the sampled radiation could bypass the atomization zone (e.g., the flame) entirely. Also, the pathlengths of different rays on and off the optical axis can vary. Both effects cause negative deviations and are minimized by using higher F/n optics and focusing the beam to a small spot in the center of the flame or electrothermal atomizer. It would appear that collimation of the incident beam would be advantageous. This is uncommon because of the increased optical complexity.

As with other atomic spectrometric techniques (see Section 7-1), nonlinearity occurs if the sample flow rate ($F$) or the overall atomization efficiency ($\varepsilon_a$) varies with analyte concentration. With proper precalibration, instrumental nonlinearity is usually negligible. With modulation, analyte emission does not directly cause nonlinearity. However, high analyte emission or blackbody radiation from electrothermal atomizers can saturate the PMT or electronics and yield erroneous results. Analyte fluorescence could cause nonlinearity at high analyte concentrations, but the effect is usually negligible because of quenching in the flames used for AA and the relatively high F/n collection optics.

### Precision

Measurement precision is evaluated experimentally by making repetitive measurements on samples and standards. For a given solution, precision is normally limited by noise as discussed in Section 10-2. At absorbances larger than 0.1 or 0.2, the RSD is typically 0.3 to 1.0% with flame atomization and 1 to 5% with electrothermal atomization. At high absorbances in flames, precision may decrease if 0% $T$ noise or analyte emission noise become significant. As always, the precision of the entire analytical scheme (i.e., including calibration and measurement of all solutions) is somewhat lower than the measurement step precision.

For analyte concentrations greater than about 50 times the detection limit, analyte flicker noise is often limiting in both AE and AA and $\xi_A \approx \xi_E$. Thus the precision in both techniques is controlled by the same factors, the fluctuations in $n_0$ due to variations in atomization efficiency, temperature, and solution gas flow rates. Atomic absorption spectrophotometry is less susceptible to fluctuations in excitation conditions (i.e., the Boltzmann factor is less critical). This fact is often overemphasized since temperature fluctuations also affect the atomization efficiency (see Section 8-5).

### Accuracy

The evaluation of the accuracy of AA measurements is similar in many respects to that discussed for AE (Section 8-6). Under proper conditions, accuracy can be limited by random error and noise to the range 0.5 to 5%. Highest accuracy is obtained by bracketing sample measurements with measurements of standards of approximately the same analyte concentration.

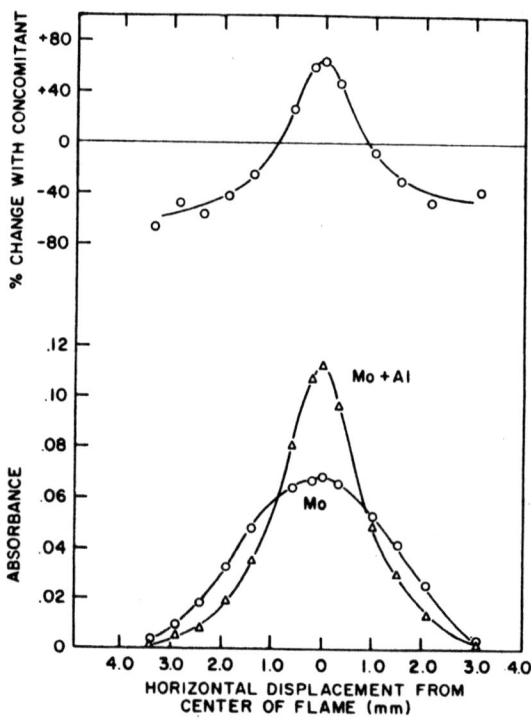

**FIGURE 10-22** Lateral diffusion effect of Al on Mo in an argon-shielded nitrous oxide-acetylene flame. The absorption was measured with source beam perpendicular to the burner slot. The presence of Al enhances the absorption of Mo in the center of the flame, but depresses it in the edges. The presence of aluminum is thought to delay the volatilization of molybdenum-containing species until they are higher in the flame. Thus there is less time for free analyte atoms to diffuse laterally out of the center of the flame in the presence of Al than in its absence. At a constant observation height, the Mo absorption thus appears enhanced in the center and depressed in the edges. [From A. C. West, V. A. Fassel, R. N. Kniseley, *Anal. Chem.*, *45*, 1586 (1973), copyright 1973, with permission of the American Chemical Society.]

Systematic errors can arise due to analyte and spectral interferences, as discussed in Section 7-2. Analyte interferences, except for excitation interferences, affect flame AA and flame AE equally. When flame AA is compared to AE in hotter, more inert, and electron-rich plasmas, AA is subject to worse volatilization, dissociation, and ionization interferences. Lateral diffusion interferences are unique to flame AA with $N_2O-C_2H_2$ flames and are illustrated in Figure 10-22.

Generally, analyte interference effects are worse and less well characterized with electrothermal atomizers than with flames. The chemical form of the analyte after solvent evaporation (e.g., PbO or $PbSO_4$) as well as the salts in the matrix can greatly influence the atomization efficiency. The profile and the time of appearance of the analyte absorption peak are also affected, as illustrated in Figure 10-23. Vaporization interferences can be worse than with flames because the salt particles formed are much larger than those formed from a nebulized droplet. Dissociation interferences can be worse because of the higher number density of all species. However, sample transport and ionization interferences are less important with furnaces.

In AE, interference due to lines of other elements near or at the analyte emission wavelength are the major cause of spectral interference. By contrast, for AAS atomic spectral interferences due to overlap of atomic absorption profiles is possible (e.g., between the 403.307-nm Mn line and the 403.298-nm Ga line), but rare. However, spectral interference due to scattering and molecular absorption can be serious. Such interferences are expected for analytes with lines in the UV, with solutions of high salt content, and are more likely with electrothermal atomization. Background absorption correction is necessary in these cases.

## Characteristic Concentrations and Detection Limits

Typical characteristic concentrations and detection limits (DL) for flame and electrothermal AAS are listed in Table 10-3. Clearly, most metals can be determined at concentrations lower than 1 μg mL$^{-1}$. Electrothermal atomizers are not hot enough to determine a few high-melting-point metals, such as W, Ta, and Nb.

The initial calibration slope ($m$) is related to the characteristic concentration ($m_A$) by $m = 0.0044/m_A$ (see Section 6-3). Thus $m_A$ can be used to compare the calibration slopes of different elements or of two atomization techniques for the same element. It can be seen in Table 10-3 that characteristic concentrations are usually a factor of 10 to 500 better with electrothermal atomization except for a few refractory elements (Al, Mo).

If the DL is defined as the concentration yielding an absorbance equal to twice the blank standard deviation ($s_{bk}$) in absorbance units (A.U.), DL = $2s_{bk}/m$. In terms of $m_A$, DL = $s_{bk}m_A/(2.2 \times 10^{-3})$. Thus DL = $m_A$ if $s_{bk}$ = 2.2 × 10$^{-3}$ A.U. and DL = 0.1 $m_A$ if $s_{bk}$ = 2.2 × 10$^{-4}$ A.U. For flame AA, most DLs reported in Table 10-3 are about $\frac{1}{10}$ of $m_A$, implying that a typical blank noise is about 2 × 10$^{-4}$ A.U. For elements with resonance lines at wavelengths shorter than 250 nm, such as Zn and Cd, the blank noise is

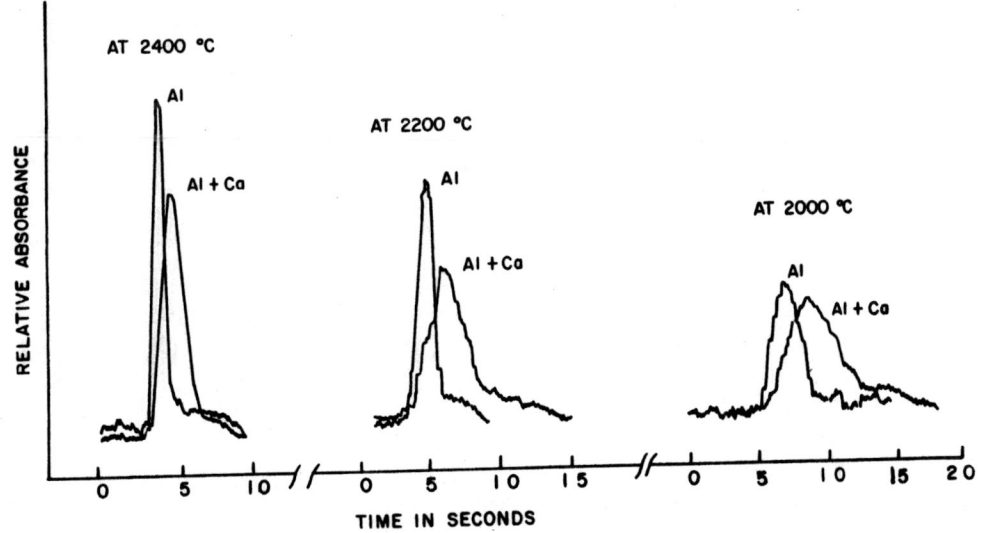

**FIGURE 10-23** Effect of interferent and temperature on atomization peak shapes. The presence of Ca is seen to delay and broaden the Al atomization peak. [From J. Y. Hwang and G. P. Thomas, *Am. Lab.*, 6(8), 55 (1974), with permission of publisher.]

**TABLE 10-3**

Typical detection limits and characteristic concentrations for flame and electrothermal AAS

| Element | Wavelength (nm) | Flame | | Electrothermal[a] | | | |
|---|---|---|---|---|---|---|---|
| | | $m_A$ (ng mL$^{-1}$) | DL (ng mL$^{-1}$) | $m_A$ (pg) | $m_A$ (ng mL$^{-1}$) | DL (pg) | DL (ng mL$^{-1}$) |
| Ag | 328.1 | 30 | 3 | 2 | 0.1 | 0.2 | 0.01 |
| Al | 309.3 | 340 | 30 | 10 | 0.5 | 2 | 0.1 |
| As | 193.7 | 500 | 200 | 20 | 1 | 10 | 0.5 |
| Au | 242.8 | 80 | 20 | 10 | 0.5 | 10 | 0.5 |
| B | 249.8 | 7000 | 2000 | 1000 | 50 | — | — |
| Ba | 553.6 | 160 | 20 | 20 | 1 | 5 | 0.25 |
| Be | 234.9 | 20 | 2 | 1 | 0.05 | 0.6 | 0.03 |
| Bi | 223.1 | 160 | 30 | 12 | 0.6 | 2 | 0.1 |
| Ca | 422.7 | 20 | 1 | 1 | 0.05 | 0.5 | 0.25 |
| Cd | 228.8 | 9 | 1 | 0.5 | 0.025 | 0.2 | 0.01 |
| Co | 240.7 | 40 | 4 | 10 | 0.5 | 6 | 0.3 |
| Cr | 357.9 | 40 | 4 | 6 | 0.3 | 0.6 | 0.03 |
| Cs | 852.1 | 100 | 20 | 6 | 0.3 | — | — |
| Cu | 324.8 | 25 | 2 | 3 | 0.15 | 1 | 0.05 |
| Eu | 459.4 | 300 | 40 | 16 | 0.8 | 2 | 0.1 |
| Fe | 248.3 | 50 | 6 | 10 | 0.5 | 5 | 0.25 |
| Ga | 287.4 | 1100 | 50 | 40 | 2 | 20 | 1 |
| Ge | 265.2 | 1500 | 200 | 100 | 5 | 50 | 2.5 |
| Hg | 253.6 | 4000 | 500 | 200 | 10 | 100 | 5 |
| K | 766.5 | 20 | 2 | 5 | 0.25 | 1 | 0.05 |
| Li | 670.8 | 20 | 1 | 10 | 0.5 | 5 | 0.25 |
| Mg | 285.2 | 3 | 0.2 | 0.4 | 0.02 | 0.04 | 0.002 |
| Mn | 278.5 | 20 | 2 | 1 | 0.05 | 0.2 | 0.01 |
| Mo | 313.3 | 20 | 5 | 20 | 1 | 5 | 0.5 |
| Na | 589.0 | 5 | 0.2 | 1 | 0.05 | 0.4 | 0.02 |
| Ni | 232.0 | 50 | 3 | 20 | 1 | 10 | 0.5 |
| Pb | 217.0 | 60 | 8 | 5 | 0.25 | 2 | 0.1 |
| Pd | 244.8 | 100 | 20 | 20 | 1 | 20 | 1 |
| Pt | 266.0 | 1000 | 50 | 100 | 5 | 50 | 2.5 |
| Rb | 780.0 | 40 | 5 | 15 | 0.8 | 10 | 0.5 |
| Sb | 217.6 | 260 | 20 | 30 | 1.5 | 10 | 0.5 |
| Se | 196.0 | 260 | 200 | 40 | 2 | 10 | 0.5 |
| Si | 251.6 | 1500 | 20 | 80 | 4 | 20 | 1 |
| Sn | 235.5 | 720 | 15 | 40 | 2 | 100 | 5 |
| Sr | 460.7 | 50 | 4 | 20 | 1 | 10 | 0.5 |
| Ti | 276.8 | 350 | 20 | 100 | 5 | 20 | 1 |
| V | 318.4 | 500 | 25 | 100 | 5 | 20 | 1 |
| Zn | 213.9 | 10 | 1 | 0.2 | 0.02 | 0.1 | 0.005 |

[a]Sample size for electrothermal AA is 20 μL.

generally greater due to higher flame transmission noise. Blank noise is also greater for elements with weak HCL emission lines, such as As, Se, and Te, because the relative signal shot noise is greater. For those elements, EDLs are usually used in place of HCLs to overcome this effect. EDLs have 10 to 50 times the radiance of HCLs for some analysis lines.

The flame DLs reported are obtained after extensive optimization of instrumental parameters and with pure sample solutions, new lamps, and 3 to 10-s integration times. With older lamps or smaller integration times, the blank noise and hence DLs can be two or three times worse.

Reported flame DLs by different manufacturers are quite similar, probably because of comparable nebulizer-burner systems and sources. Flame AA DLs are less affected by the sample matrix than are flame or plasma AE DLs. In flame AA, the blank noise is limited by source and flame transmission flicker noise or lamp shot noise, which are affected only slightly by the sample matrix. Also, depression of the calibration slope by matrix components is rarely greater than a factor of 2 for flame AAS.

Detection limits achieved with electrothermal atomizers are more variable between manufacturers because they depend on the atomizer design and atomization conditions. Often, absolute or mass DLs and $m_A$ values are reported (e.g., in pg). Thus $m_A$ and the DL depend on the sample volume used and do not necessarily improve linearly with sample volume. For most elements, the DLs with electrothermal atomization are a factor of 10 to 500 better than achieved with flames, due to the larger calibration curve slopes (higher $n_0$). The blank noise is usually slightly larger with electrothermal atomization because smaller time constants are employed to avoid distortion of the peak-shaped signals. Detection limits in complex samples can be degraded significantly, due to reduced atomization efficiency (lower $m$) or higher atomizer transmission noise in the case of significant background absorption.

In general, electrothermal AA DLs are best for more volatile elements and often below 1 pg. For less volatile elements such as Al, Ba, alkaline earth metals, and rare earth metals, $\beta_a$ is lower in a furnace and DLs are in the range 1 to 100 pg. Characteristic concentrations and DLs are also higher for elements that form nonvolatile carbides (Ba, B, Ca, Mo, W, V, and Zr) with the carbon of the atomizer or carbon in the organics in the sample matrix. The furnace tube can be coated with one of these carbide-forming metals to reduce the loss in calibration sensitivity for other carbide-forming elements.

Some workers have demonstrated that furnaces are hot enough to excite emission from some of the more easily excited elements (Al, Cr, Cu, Mn) with DLs near 1 pg, as with AAS. However, DLs for other elements are significantly worse than with electrothermal AA.

In general, flame AA DLs (Table 10-3) are better than flame AE DLs (Table 8-6) except for easy-to-excite elements (alkali metals). The Boltzmann factor in flames is very small for elements with only UV resonance lines. For many elements, flame AA DLs (Table 10-3) and ICP emission DLs (Table 8-8) are within an order of magnitude but significantly worse than electrothermal AA DLs. The ICP does provide the best DLs for many refractory elements and the multielement capabilities and dynamic range of plasma emission techniques must be considered. For As, Bi, Se, and Te, hydride techniques provide the best DLs, while the cold vapor method yields the best DL for Hg.

## 10-5 METHODOLOGY AND APPLICATIONS

The analytical sample for AAS is usually in the form of a solution. Thus for solid samples (e.g., geological, biological), dissolution procedures must first be applied, as discussed in Appendix E. If detection limits are inadequate, preconcentration or a different sample

introduction or atomization technique may be required. Separation procedures are sometimes necessary if interferences cause serious systematic errors that cannot be eliminated by simpler means.

### Flame Atomization

With flame AA most analyses are first attempted with external standards. Selection of optimum experimental conditions for flame AA is usually straightforward. All manufacturers provide a "cookbook" with suggested experimental conditions for most elements. The type of flame, the type of source, the fuel-to-oxidant ratio, the analysis wavelength, the lamp current, the spectral bandpass, and the observation height are specified. Fine adjustments may be necessary. Usually, the burner position and the fuel-to-oxidant ratio are optimized by maximizing the absorbance while aspirating a solution yielding an absorbance of about 0.5. This normally optimizes precision at moderate analyte concentrations. For measurements near or at the DL, the ratio of the calibration sensitivity to the blank noise should be optimized. Generally, the smallest HCL current and slit width that ensure that measurements are not signal shot noise or background emission noise limited are best (i.e., $\sigma_A$ is independent of $\Phi_0$). Higher lamp currents or slit widths can cause lower calibration sensitivity and nonlinearity for some elements, as discussed in Section 10-4.

Flame AA cookbooks generally list a number of possible lines that can be used for analysis. An example is shown in Table 10-4 for Cu. Normally, the absorption line corresponding to a transition from the ground state to the first excited state is used because it has the largest transition probability (oscillator strength) and hence the smallest characteristic concentration or the largest calibration curve slope. Also, the corresponding transition line from the source usually has the highest radiance. Other less sensitive lines may yield better linearity at high concentrations, eliminate the need for excessive

**TABLE 10-4**
Analysis lines for copper

| $\lambda$ (nm) | $f^a$ | $m_A{}^b$ ($\mu g\ mL^{-1}$) | DL ($\mu g\ mL^{-1}$) |
|---|---|---|---|
| 324.8 | 0.74 | 0.04 | 0.002 |
| 327.4 | 0.38 | 0.1 | |
| 217.9 | 0.011 | 0.6 | |
| 222.6 | 0.004 | 2.0 | |
| 249.2 | — | 10 | |
| 244.2 | — | 40 | |

$^a f$ is the oscillator strength.
$^b$ In an air–$C_2H_2$ flame.

dilution, increase the *S/N* due to less background emission noise or flame transmission noise, or reduce susceptibility to background absorption interferences.

Potential interferences for flame AA are identified in AA cookbooks. With flame AA, spectral interferences are usually minimal. Background absorption correction is sometimes required for elements with resonance lines below 300 nm or for solutions of high salt content.

Several options are available if analyte interferences that depress or enhance the analyte absorbance are observed. First, the type of flame and flame conditions can be adjusted to minimize analyte interferences. For some elements ionization buffers and releasing agents are required; these can also improve linearity of calibration curves. Test solutions can be made to contain 1 mg mL$^{-1}$ of Cs or another easy to ionize element to prevent ionization interferences for other elements, such as alkali metals, alkaline earth metals, rare earth metals, Al, Au, and Si. Lanthanum at the 1-mg mL$^{-1}$ level is effective for minimizing solute vaporization interferences for some elements, such as alkaline earth metals, and is also a good ionization buffer.

Usually, the foregoing steps are adequate with flames. For special cases, standard addition procedures can be employed. They should be used with care because many AA calibration curves exhibit subtle nonlinearity at moderate absorbances. Alternatively, matrix-matched standards or separation of interferences may be required. For some elements the flame atomization efficiency varies slightly when sample solutions contain more than 1% acid. Therefore, standards should be prepared in the same concentration of acid. Internal standard methods are rarely used and quite inconvenient except with dual-channel instruments. They can improve precision slightly if slow variations in atomization conditions are limiting and can compensate for analyte interferences if an internal standard can be found that suffers the same interference effects as the analyte.

### Electrothermal Atomization

Electrothermal AA complements flame AA and is most suitable when flame AA DLs are inadequate or sample volumes are limited. Electrothermal atomizers are more expensive than nebulizer-burners, although operating costs due to flame gases are less. Compared to flame AA measurements, electrothermal AA has worse precision for measurements well above the detection limit, suffers from greater interference effects, requires more operator skill, and has a lower sample throughput. Additional time-consuming dilution steps may actually be required for some samples because of the excellent detectability.

Electrothermal AA is useful when low DLs or special sample considerations are critical. Particularly for analyte concentrations near or below 10 μg L$^{-1}$, the precision of flame AA is poor or the analyte concentration is below the DL. With electrothermal AA these low analyte concentrations can often be measured directly without time-consuming preconcentration steps.

The ability to analyze small sample sizes down to 1 μL is useful for sample-limited situations, such as in some clinical assays. Although special techniques are available for introduction of submilliliter samples into flames (see Section 10-2), the flame AA DL in terms of concentration is generally worse than obtained with continuous aspiration of 1 mL or more of sample.

Time-consuming sample preparation steps can be bypassed by placing viscous solutions (oil, blood) or solid samples (plastics, paper) directly into the furnace atomizer. However, solid samples do present special problems. It must be possible for the atomizer to melt the sample to prevent blockage of the light beam and to vaporize all the sample to prevent memory effects. For the highest accuracy, solid standards similar to the sample in composition should be used. The atomization characteristics for a given mass of analyte can be quite different if it is deposited in solution form rather than in a solid matrix. Although it is easy to weigh small solid samples accurately, it is more difficult to transfer them faithfully to the atomizer. Also, sample homogeneity is of concern for small sample sizes.

For elements such as Hg, Se, Sb, Bi, and Sn, electrothermal atomizers provide better AA DLs than do flames. However, the cold vapor method is best for low Hg concentrations because the DL is lower and losses of volatile mercury in the ashing step are avoided. Similarly, hydride techniques are usually preferred for low concentrations of As, Se, Sb, Bi, and Sn because of better detection limits. These alternative AA techniques do require sample volumes of 1 mL or more.

As for flame AA, cookbook settings are available as an initial guide for setting the time interval and temperature for the drying, ashing, and atomization steps. Selection and optimization of the analysis wavelength, type of lamp, lamp current, and spectral bandpass are very similar to flame AA. Fine tuning of the drying, ashing, and atomization conditions is quite critical to achieve maximum atomization efficiency and minimal interference effects because the optimum settings are very matrix dependent.

Many potential interferences are identified in electrothermal AA cookbooks. The major spectral interference is absorption by molecular species, which are covolatilized with the analyte. This type of interference can be detected by making measurements with and without background correction. Although background

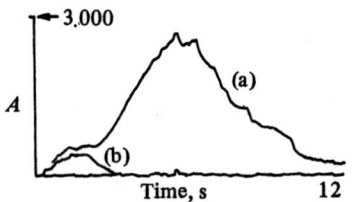

←— 3.000

A

(a)

(b)

**Time, s**        12

**FIGURE 10-24** Determination of 5 μg L$^{-1}$ of
Cd in 2% (w/v) NaCl with electrothermal AA.
In (a) the broad peak is due primarily to background absorption by NaCl. In (b) the absorbance is background corrected (S/H system) and
the atomization peak due to Cd is clearly seen.
[Adapted from S. B. Smith, Jr., and G. M. Hieftje,
*Appl. Spectrosc., 37,* 419 (1983), with permission
of Society for Applied Spectroscopy.]

correction can result in loss of detectability, it may be
essential for obtaining reasonable accuracy, as shown
in Figure 10-24. Analyte interferences are due to matrix
components which alter the volatilization characteristics
of the analyte, or vapor-phase reactions which involve
the analyte. Different peak shapes for standards and
samples are indicative of analyte interferences.

The first step in minimizing interference effects is
adjustment of atomization conditions. The ashing temperature and time should be adjusted to volatilize selectively as many matrix species as possible without loss
of the analyte. The atomization temperature should be
high enough to atomize the analyte completely and
maintain atomization. Excessive atomization temperatures can vaporize less volatile matrix components which
are potential spectral interferences and reduce the lifetime of the atomizer tube. The L'vov platform is also
useful for minimizing vapor-phase interferences.

Newer designs for electrothermal atomizers may
reduce some of the limitations of present commercial
designs. Heating of the furnace with a capacitive discharge increases heating rates by an order of magnitude
or greater. This reduces the time for atomization, increases calibration sensitivity, and minimizes spectral
and nonspectral interference effects. With probe furnaces, the sample is placed into a furnace that has already reached the atomization temperature. Again interference effects are reduced. One manufacturer uses
a tungsten filament instead of the traditional carbon
tube. Interference effects are claimed to be reduced
because the tungsten atomizer can be heated more rapidly (up to 6000°C s$^{-1}$) and is less porous than graphite.

Sample preparation techniques can also aid in
minimizing interferences. Generally oxyacids (sulfuric,
phosphoric) should be used in such procedures instead
of HCl. Chlorides of several elements, including Cu,
Cd, Zn, and Ge, are more volatile than other chemical
forms. Thus when high concentrations of chloride are

present, it is more difficult to eliminate interfering components because lower ash temperatures must be employed to prevent volatilization of the analyte.

Addition of matrix modifiers to the sample is advantageous in many cases. Matrix modifiers can be used
to increase the volatility of selected matrix components
such as chloride so that they can be removed at a lower
ash temperature. For example, ammonium acetate
(NH$_4$OAc) is often added to samples that contain a
large amount of NaCl. The NH$_4$Cl and NaOAc formed
have boiling points below 900°C compared to about
1400°C for NaCl. Thus they can be removed at a lower
ash temperature to minimize background absorption
due to undissociated NaCl present in the atomize step.
Some matrix modifiers, such as phosphates, are added
to stabilize volatile elements so that they remain in the
atomizer until after the matrix has been removed. Nickel
is used with Se, Te, and As to increase the permissible
ash temperature by about 700°C.

Standard addition procedures are often used with
electrothermal atomizers for cases in which the calibration slope is altered by the sample matrix. If all else
fails, it may be necessary to remove interferences with
separation techniques.

Determination of very low concentrations with
electrothermal AA presents special difficulties. Memory effects can be serious; if so, care must be taken to
remove all remnants from the previous sample. With
small sample sizes, contamination problems are enhanced especially for ubiquitous elements, such as Na,
K, Ca, Mg, and Zn. For example, one grain of pollen
that is 10 μm in diameter contains about 10 pg of Mg,
which is about 100 times the DL.

Both memory effects and analyte interferences effects depend on the condition of the atomization tube.
The pyrolytic coating deteriorates with use. The rate of
deterioration depends on the type of samples analyzed
and the atomization temperatures used. The furnace
tube should be replaced typically after 50 to 100 firings,
or when the shape of atomization peaks significantly
change.

## Applications

Atomic absorption spectrophotometry has been used
to determine metals and some nonmetals in almost every
conceivable type of sample. Many standard procedures
for water analysis (river, ground, and lake water; seawater; industrial effluents; beverages) are based on AAS.
Except for water samples with a high salt content, analysis is usually straightforward. Suspended matter is removed by filtration before dissolved metals are determined. Sorbed metals are released prior to filtration
and determination by acidification to 1%(v/v) HNO$_3$ or

by acid digestion. Abundant elements (Na, K, Ca, Mg, Fe) and many trace transition metals can usually be determined directly with flame AA. However, to determine many metals which are present at concentrations lower than 10 μg L$^{-1}$, particularly in seawater, electrothermal AA and/or preconcentration steps (ion exchange, solvent extraction) are necessary.

The use of AAS in the metallurgical and mining industries is common for analysis of metals, alloys, geochemical samples, and electroplating solutions. Flame AAS is adequate for determination of most elements at concentrations of 0.001% (w/w) or above in solid samples after dissolution by acid digestion or fusion. For lower concentrations, electrothermal AAS may be required.

Analysis of petroleum products presents special sample preparation difficulties. Oils and gasoline are often diluted by a factor of 2 to 10 with suitable solvents (heptane, xylene, methyl isobutyl ketone) so they can be aspirated into a flame. Usually, organometallic standards soluble in the solvent are used because the atomization efficiency, and hence the calibration slope, are different between aqueous solutions of inorganic metal standards and organic solutions of organic metal standards. Background absorption correction is often needed. These considerations make the standard addition technique a useful tool and also apply to situations where metals are extracted from aqueous solutions into an organic solvent before aspiration into a flame.

For biological and clinical samples, it is critical to remove the organic matrix by digestion or some other method before analysis. For example, serum protein clogs burners and causes high background absorbance and residues with electrothermal atomizers. Deproteinization with trichloroacetic acid is common. Dilution is simpler if detectability is adequate. With most biological samples, a flame ionization buffer is needed because of the high content of Na and K.

## PROBLEMS

**10-1.** Multielement hollow cathode lamps are sometimes used in atomic absorption (AA) spectrophotometry. For example, a brass hollow cathode produces lines of Zn and Cu in addition to filler gas lines. Zinc is often determined in AA using its 213.9-nm resonance line. In a brass hollow cathode, the Cu lines at 216.5 and 217.9 nm are fairly near the Zn line.
   **(a)** If the only monochromator available has a spectral bandpass of 5 nm, what effect will the Cu lines from the hollow cathode have on AA calibration curves if (i) there is no Cu in the standards, and (ii) the standards contain a constant amount of Cu?
   **(b)** If another monochromator is available with a reciprocal linear dispersion of 2 nm mm$^{-1}$ and slits variable from 50 μm to 2 mm, what slit width should be used to avoid interference from the Cu lines?

**10-2.** Is the following statement true or/and false? *Criticize* or *defend* the statement, but begin with a definite opinion. Use equations, diagrams, words or any other information in the answer.

> Flame atomic absorption (AA) spectrometry would be a much better absolute technique than flame emission spectrometry. That is, you can carry out determinations without standards. The fraction absorbed with a line source $\alpha_L$ is a simple function of concentration in AA since at low concentrations ($\alpha_L = k_m l = Kn_0 = K'c$), where $k_m$ is the atomic absorption coefficient, $l$ is the pathlength, $K$ and $K'$ are constants, $n_0$ is the ground-state number density, and $c$ is the solution concentration. The constants $K$ and $K'$ are known from theory. In flame emission, however, all the instrumental factors (mono-

chromator factors, transducer factors, amplifier-readout factors) come into play and render absolute measurements almost impossible.

**10-3.** The determination of Pb in a brass sample was carried out with AA spectrophotometry and the method of standard additions. The original sample was diluted to 50 mL after dissolution. This solution was introduced into the AA spectrometer and a transmittance of 0.380 was obtained. To the original solution 20 μL of a 10.0 μg mL$^{-1}$ Pb standard was then added. The transmittance of this solution was 0.263. Find the concentration of Pb in the original sample. What assumption has been made in order to use a single standard addition? Is there any built-in check on this assumption in the procedure outlined above?

**10-4.** Compare and contrast the continuum source, Zeeman, and Smith–Hieftje methods of background correction in AA spectrophotometry. Consider in your discussion the range of applicability (i.e., number of elements), the accuracy of the correction, the ease of use, the expense, the possibility of errors being introduced, and any other factors you deem important.

**10-5.** Discuss the general applicability of AA spectrophotometry for multielement determinations. Compare AA to ICP emission spectrometry for the simultaneous determination of multiple elements.

**10-6.** In the early days of AA spectrophotometry various claims as to instrumental errors and interferences were made. It was often stated, for example, that AAS is less susceptible to changes in flame temperature than AES because the large ground-state population measured in AAS changes only slightly with temperature, while the excited-state population changes drastically with temperature. For example, consider 1000 atoms in the ground state and 1 atom in the excited state. If the temperature changes such that 2 atoms are now in the excited state and 999 in the ground state, the AES signal should change by 100%, while the AAS signal will change by only 0.1%. Discuss why this argument is oversimplified and what temperature-dependent factors it neglects. In actual practice, are AAS and AES much different in their behavior with temperature?

**10-7.** One reason that combination flame emission/atomic absorption instruments have not been great successes is that the monochromator requirements in the two techniques differ substantially.

(a) Discuss the function of the monochromator in the two techniques. What role does the monochromator play in reducing interferences in the two methods?

(b) Your laboratory is considering the purchase of a monochromator. Under consideration are an expensive, high-resolution, variable-slit-width system (smallest spectral bandpass = 0.5 Å) and a less expensive, moderate-resolution, fixed-slit-width system (spectral bandpass = 10 Å). Which system would you choose if your laboratory was only going to do atomic absorption measurements? Why?

**10-8.** What are the advantages and disadvantages of double-beam systems relative to single-beam systems in AAS?

**10-9.** Discuss the major factors involved in the decision to use flame or electrothermal atomization in AAS.

**10-10.** Why do hydride techniques for As and Se and the cold vapor technique for Hg yield better AA detection limits than conventional flame AAS with solution nebulization?

**10-11.** Why do different flames yield different AA calibration slopes for a given element?

**10-12.** Discuss the limiting noise sources in flame AA at low, moderate, and high absorbances. How is your answer affected by the selection of the slit width and HCL current in a given determination?

**10-13.** Why is background absorption correction more important with electrothermal atomization than with flame atomization?

**10-14.** Why can the monochromator slit width affect the linearity of AA calibration curves? Consider both line and continuum sources in your answer.

**10-15.** Consider the flame AA determination of Cu at 324.7 nm. The solution flow rate is 6.5 mL min$^{-1}$, $\varepsilon_a = 0.02$, $e_f = 12$, $Q = 9.3$ L min$^{-1}$, $Z(T) = g_0 = 2$, $g_1 = 4$, $\Delta\lambda_{eff} = 4 \times 10^{-3}$ nm, $l = 10$ cm, and $A_{10} = 0.95 \times 10^8$ s$^{-1}$. The test solution is a 5.0-$\mu$g mL$^{-1}$ aqueous Cu solution. The flame temperature is 2800 K. Find the absorption factor with a line source, $\alpha_L$, and the absorbance $A_L$.

**10-16.** Why is the ash step very critical with electrothermal atomizers?

**10-17.** Contrast analyte-shifted and source-shifted Zeeman AA correction techniques.

**10-18.** The flame AA detection limit for Cu is found to be independent of the spectral bandpass over the bandpass range 0.2 to 1.0 nm. What noise sources can be limiting precision?

# REFERENCES

The following books, chapters, articles, and reviews are general references devoted entirely or substantially to atomic absorption spectrophotometry.

1. W. J. Price, *Spectrochemical Analysis by Atomic Absorption*, Heyden, London, 1979.

2. L. Ebdon, *An Introduction to Atomic Absorption Spectroscopy: A Self-Teaching Approach*, Heyden, London, 1982.

3. "Atomic Absorption Spectroscopy, Past, Present, and Future," *Spectrochim. Acta, 36B*, (5 and 7) (1981). Two issues devoted to AA.

4. J. W. Robinson, "Atomic Absorption Spectroscopy," in *Treatise on Analytical Chemistry*, 2nd ed., vol. 1, pt. I, chap. 8, P. J. Elving and I. M. Kolthoff, eds., Wiley-Interscience, New York, 1981.

5. J. C. Van Loon, *Analytical Atomic Absorption Spectroscopy: Selected Methods*, Academic Press, New York, 1980.

6. W. Slavin, "Atomic Absorption Spectroscopy—The Present, Past and Future," *Anal. Chem., 54*, 685A (1982).

7. M. Pinta, *Atomic Absorption Spectrometry, Vol. 2. Application to Chemical Analysis*, 2nd ed., Masson, Paris, 1980.

8. C. Th. J. Alkemade and R. Herrmann, *Fundamentals of Analytical Flame Spectroscopy*, Wiley, New York, 1979.

9. B. V. L'vov, *Atomic Absorption Spectrochemical Analysis*, American Elsevier, New York, 1970.

10. J. W. Robinson, *Atomic Absorption Spectroscopy*, 2nd ed., Marcel Dekker, New York, 1975.

11. B. Welz, *Atomic Absorption Spectroscopy*, Verlag Chemie, Deerfield Beach, Fla., 1976.

12. A. Asha, *CRC Handbook of Atomic Absorption Analysis*, CRC Press, Boca Raton, Fla., 1984.

13. A. Walsh, "Atomic Absorption Spectroscopy—Stagnant or Pregnant," *Anal. Chem., 49*, 1269A (1977).

14. J. E. Cantle, *Atomic Absorption Spectrometry*, Elsevier, Amsterdam, 1982.

15. M. Salvin, *Atomic Absorption Spectroscopy*, 2nd ed., Wiley, New York, 1979.

16. K. C. Thompson and R. J. Reynolds, *Atomic Absorption, Fluorescence, and Flame Emission Spectroscopy: A Practical Approach*, 2nd ed., Wiley, New York, 1978.

17. G. F. Kirkbright and M. Sargent, *Atomic Absorption and Fluorescence Spectroscopy*, Academic Press, London, 1974.

18. T. J. Vickers, "Atomic Fluorescence and Atomic Absorption Spectroscopy," in *Physical Methods in Modern Chemical Analysis*, T. Kuwana (ed.), vol. 1, Academic Press, New York, 1978, p. 192.

The following references are concerned specifically with electrothermal AA.

19. S. R. Koirtyohann and M. L. Kaiser, "Furnace Atomic Absorption—A Method Approaching Maturity," *Anal. Chem.*, **54**, 1115A (1982).

20. W. Slavin, *Graphite Furnace Source Book*, Perkin-Elmer Corp., Ridgefield, Conn., 1984.

21. R. E. Sturgeon, "Factors Affecting Atomization and Measurement in Graphite Furnace Atomic Absorption Spectrometry," *Anal. Chem.*, **49**, 1225A (1977).

22. C. W. Fuller, *Electrothermal Atomization for Atomic Absorption Spectrometry*, Royal Society of Chemistry, London, 1977.

Background absorption correction is discussed in the following articles.

23. L. DeGalan and M. T. C. de Loos-Vollebregt, "Roll-Over of Analytical Curves in Atomic Absorption Spectrometry Arising from Background Correction with Pulsed Hollow-Cathode Lamps," *Spectrochim. Acta.*, **34B**, 1011 (1984).

24. J. J. Cotera and H. L. Kahn, "Background Correction in AAS," *Am. Lab.*, 193, November (1982).

25. S. D. Brown, "Zeeman Effect-Based Background Correction in Atomic Absorption Spectrometry," *Anal. Chem.*, **49**, 1269A (1977).

26. R. Stephens, "Zeeman Modulated Atomic Absorption Spectroscopy," *CRC Crit. Rev. Anal. Chem.*, **9**, 167 (1980).

Articles on more specific topics in AA are given below.

27. M. L. Parsons and P. M. McElfresh, "Comparison of Theoretical and Experimental Limits of Detection in Atomic Absorption Spectrometry Using Air-Acetylene and Nitrous Oxide–Acetylene Flames," *Appl. Spectrosc.*, **26**, 972 (1972).

28. M. L. Parsons, B. W. Smith, and P. M. McElfresh, "On the Selection of Analysis Lines in Atomic Absorption Spectrometry," *Appl. Spectrosc.*, **27**, 971 (1973).

29. N. W. Bower and J. D. Ingle, Jr., "Precision of Flame Atomic Absorption Measurements of Arsenic, Cadmium, Calcium, Cobalt, Iron, Magnesium, Molybdenum, Nickel, and Zinc," *Anal. Chem.*, **49**, 574 (1977).

30. T. C. O'Haver, "Continuum-Source Atomic-Absorption Spectrometry: Past, Present, and Future Prospects," *Analyst*, *109*, 211 (1984).

31. T. Nakahara, "Applications of Hydride Generation Techniques in Atomic Absorption, Atomic Fluorescence, and Plasma Atomic Emission Spectroscopy," *Prog. Anal. At. Spectrosc.*, *6*, 163 (1983).

# CHAPTER 11

# Atomic Fluorescence Spectrometry

Atomic fluorescence (AF) involves the emission of photons from an atomic vapor that has been excited by photon absorption. The phenomenon of atomic fluorescence was studied in the late nineteenth and early twentieth centuries by physicists who observed fluorescence from several elements (e.g., Na, Hg, Cd, Tl) in heated cells and flames. Starting in 1956, Alkemade used AF to study physical and chemical processes in flames and in 1962 suggested the use of AF in chemical analysis. In 1964, Winefordner and Vickers proposed and demonstrated AF flame spectrometry as a new analytical method. After 1964, atomic fluorescence spectrometry (AFS) was extensively studied and refined, particularly by Winefordner's group in the United States and West's group in England.

Much of the research on AFS as an analytical technique has centered around the development of suitable intense sources. We have already seen (Section 7-5) that at low analyte concentrations, a very small fraction of the excitation source radiant power is absorbed. A fraction of this absorbed radiant power is converted to fluorescence as determined by the fluorescence power yield ($Y$). Finally, a small fraction of the fluorescence emitted over $4\pi$ sr is actually collected and detected. Thus, in many experiments, the fluorescence radiant power detected may be $10^{-6}$ to $10^{-9}$ or an even smaller fraction of the excitation radiant power. This small analyte fluorescence signal must be distinguished from various background signals and noise.

In Section 11-1 we consider the types of fluorescence that can be observed. Next, instrumental design and various instrumental configurations and components are discussed. The ideal characteristics of instrumental components are compared to the characteristics of components actually used. This is followed by a review of signal and noise expressions that provides insight into how instrumental variables and atomic properties affect the signals observed and the linearity and precision of AFS. The phenomenon of saturated fluorescence is also presented. The chapter concludes by summarizing performance characteristics and applications of AF measurements.

## 11-1 TYPES OF FLUORESCENCE

There are five basic types of fluorescence: resonance fluorescence, direct-line fluorescence, stepwise-line fluorescence, sensitized fluorescence, and multiphoton fluorescence. These are summarized in Figure 11-1. Some specific examples are given in Table 11-1.

In **resonance fluorescence** (Figure 11-1a and b),

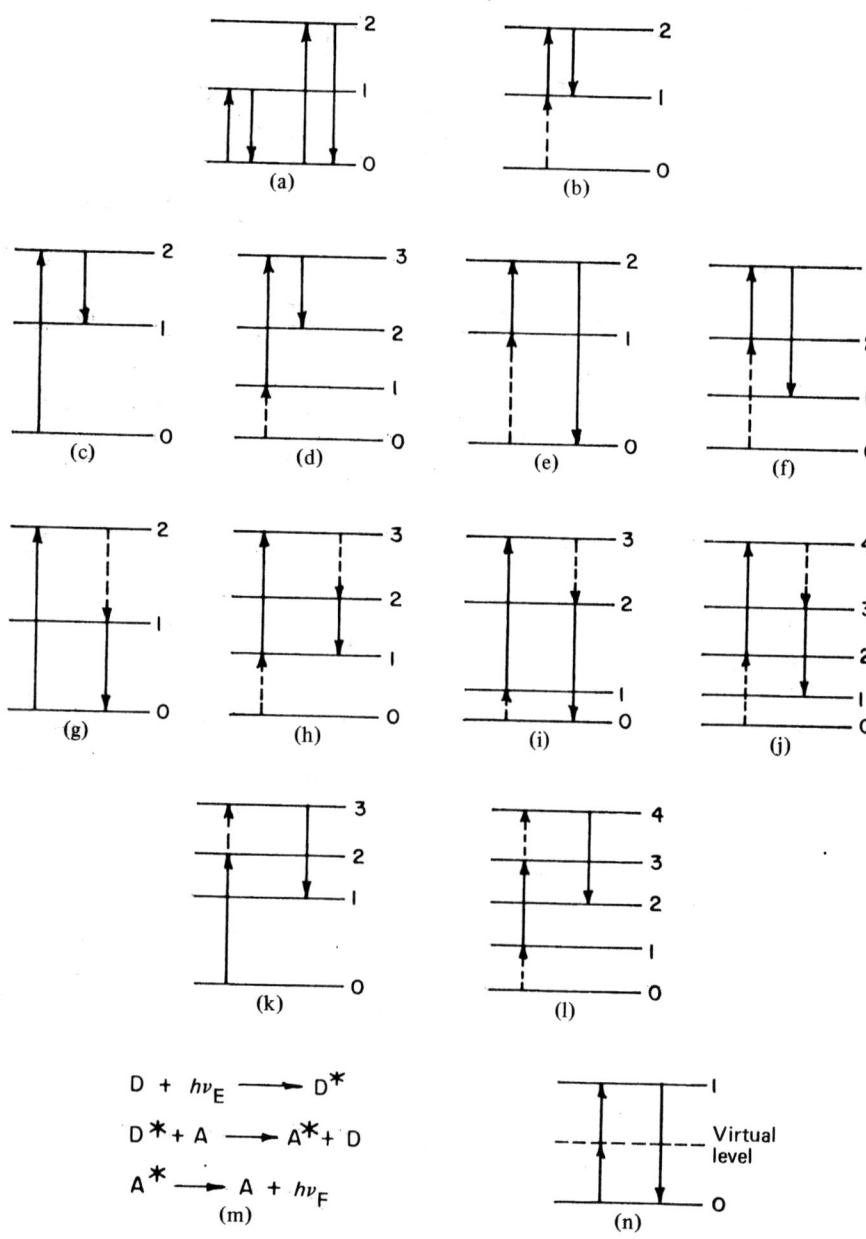

**FIGURE 11-1** Types of atomic fluorescence transitions (the spacings between atomic levels is not indicative of any specific atom): (a) resonance fluorescence (either process); (b) excited-state resonance fluorescence; (c) Stokes direct-line fluorescence; (d) excited-state Stokes direct-line fluorescence; (e) anti-Stokes direct-line fluorescence; (f) excited-state anti-Stokes direct-line fluorescence; (g) Stokes stepwise-line fluorescence; (h) excited-state Stokes stepwise-line fluorescence; (i) anti-Stokes stepwise-line fluorescence; (j) excited-state anti-Stokes stepwise-line fluorescence; (k) thermally assisted Stokes or anti-Stokes stepwise-line fluorescence (depending on whether the absorbed radiation has shorter or longer wavelengths than the fluorescent radiation); (l) excited-state thermally assisted Stokes or anti-Stokes stepwise-line fluorescence (depending on whether the absorbed radiation has shorter or longer wavelengths than the fluorescence radiation); (m) sensitized fluorescence (D = donor; D* = excited donor; A = acceptor; A* = excited acceptor; $h\nu_E$ = exciting radiation; $h\nu_F$ = fluorescence radiation); (n) two-photon excitation fluorescence (multiphoton processes involving more than two photons are even less probable than the two-photon process). [From N. Omenetto and J. D. Winefordner, *Appl. Spectrosc.*, **26**, 555 (1972). Reproduced by permission of the Society for Applied Spectroscopy.]

**TABLE 11-1**

Examples of fluorescence types

| Type | Element | $\lambda_{ex}$ (nm) | $\lambda_{em}$ (nm) |
|---|---|---|---|
| Resonance | Zn | 213.9 | 213.9 |
| Excited-state resonance | Ga | 417.2 | 417.2 |
| Stokes direct-line | Pb | 283.3 | 504.8 |
| Excited-state Stokes direct-line | Sn | 270.7 | 333.1 |
| Anti-Stokes direct-line | In | 451.1 | 410.2 |
| Stokes stepwise-line | Na | 330.3 | 589.0 |
| Excited-state Stokes stepwise-line | Pb | 283.3 | 368.4 |
| Thermally assisted anti-Stokes stepwise-line | Cr | 359.4 | 357.9 |

the same upper and lower levels are involved in the excitation-deexcitation process so that the absorption and emission wavelengths are the same. **Direct-line fluorescence** (Figure 11-1c to f) results when the same upper level is involved in excitation or emission, whereas in **stepwise-line fluorescence** (Figure 11-1g to l), different upper levels are involved in the excitation-deactivation process. Both direct-line and stepwise-line fluorescence are grouped in the category of **nonresonance fluorescence**. **Sensitized fluorescence** (Figure 11-1m) occurs when a donor species excited by photon absorption transfers energy to an acceptor atom which radiatively deactivates. Finally, **multiphoton fluorescence** (Figure 11-1n) results when two or more photons promote an atom to an excited state which then emits a photon.

The intermediate levels may be virtual or real, and the excitation photon energies may be the same or different. The selection rules for multiphoton excitation are different than for single-photon absorption (see Section 7-3).

Different types of resonance, direct-line, and step-wise-line fluorescence can be distinguished. If the upper and lower states involved are both excited states, the fluorescence process is said to be **excited-state fluorescence** (Figure 11-1b, d, f, h, j, and l). The lower state is usually a low-lying metastable level that is thermally populated. The adjectives **Stokes** and **anti-Stokes** denote that the excitation wavelength is less than or greater than the emission wavelength, respectively. If the excitation process involves radiative excitation followed by further thermal excitation, the process is **thermally assisted** (Figure 11-1k and l).

Most analytical work has involved normal resonance fluorescence because the transition probabilities and the source radiances with conventional line sources are often the greatest. Sensitized and multiphoton fluorescence have found little analytical use because of the low fluorescence radiances produced. With nonresonance fluorescence, wavelength selection can be used to resolve scattering from fluorescence because the excitation and emission wavelengths are different.

## 11-2 INSTRUMENTATION

Figure 11-2 shows a block diagram of a typical single-element AF spectrometer or photometer. The atomizer can be a flame, plasma, electrothermal device, or a special-purpose atomizer (e.g., heated quartz cell). The excitation source can be a conventional line or continuum source, a laser, or an ICP. Normally, a pulsed source (e.g., pulsed HCL or laser) or a continuous source

with a mechanical chopper provides modulated excitation radiation.

Excitation optics focus the source radiation onto the desired volume element in the atomizer. With conventional sources, the collection efficiency of the excitation optics is critical; the solid angle of collection should be large. This consideration does not apply to lasers because of the directionality of the light beam. A mirror can be placed on the opposite side of the atomizer from the source to reflect transmitted excitation radiation back onto the atomizer.

The emission collection-detection system may be dispersive (monochromator-based) or nondispersive (filters or no filter). In either case, wavelength selection is usually required to isolate the analyte fluorescence line from background signals (emission and fluorescence from the atomizer and concomitants). For nonresonance fluorescence, the wavelength selector allows the analyte fluorescence signal to be distinguished from the source scattering signal; for resonance fluorescence, such a distinction based on wavelength is not possible. A portion of the total fluorescence excited within the atomizer is collected and imaged on the monochromator entrance slit or on an aperture preceding the detector in nondispersive systems. The goal is to maximize the solid angle of collection and the volume element viewed. With dispersive systems, the F/n of the collection optics should be matched to the F/n of the monochromator. To minimize the collection of scattered source radiation, the fluorescence is usually viewed at an angle of 90° with respect to the excitation axis. The angle between the excitation and emission axes can be any angle except 180°, as this angle would result in direct viewing of the excitation source radiation. Additional baffles restrict the collection and detection of source radiation that is reflected or scattered from external components.

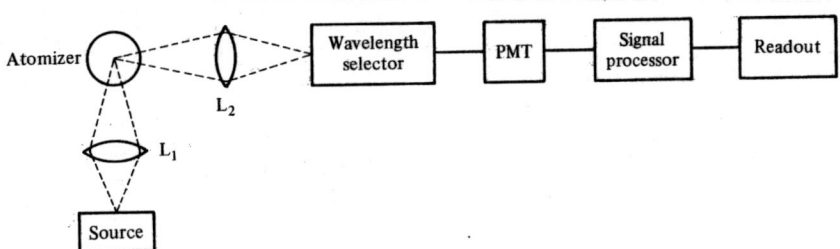

**FIGURE 11-2** Single-beam atomic fluorescence spectrometer. Source radiation is focused onto the atomizer. A fraction of the fluorescence photons produced are collected and imaged onto the wavelength selector and ultimately a PMT. The resulting photoanodic current is electronically processed to yield a fluorescence readout signal. In dispersive systems, a monochromator is used to isolate the analyte fluorescence from background emission and fluorescence from other species. With nondispersive systems a filter is used or a solar blind PMT is sometimes used without a filter.

Because of the low light levels to be detected, a PMT is normally the detector although multichannel solid-state detectors and image dissector tubes (IDTs) have been used with spectrographs. In some cases a *solar blind* PMT (response from 180 to 320 nm) has been used without a filter.

The signal processor typically includes either a synchronous ac amplifier (lock-in) if the source duty cycle is 50%, or a gated integrator (boxcar integrator) for pulsed, low-duty cycle sources. Conventional read-out devices include digital meters, analog recorders, and printers.

The relationship between the measured fluorescence signal and the instrumental and analyte parameters is discussed in detail in the next section along with $S/N$ considerations. To provide a framework to discuss the different instrumental components and designs in this section, we first note that at low analyte concentrations (see Section 7-5) the fluorescence signal detected is proportional to the source radiance (ignoring saturation effects for now), the analyte number density, the fluorescence quantum efficiency, and the emission collection efficiency. Thus, to maximize the detected fluorescence signal, these quantities should be optimized.

Maximizing the fluorescence signal does not tell the whole story because detectability is ultimately a $S/N$ problem. The small fluorescence signal must be distinguished from background signals and noise due to emission or fluorescence from other species or scattering of the excitation radiation.

## Excitation Sources

The ideal source for AFS would be stable and provide an extremely high radiance at the excitation wavelengths for the elements of interest. The most popular conventional line sources have been hollow cathode lamps (HCLs), electrodeless discharge lamps (EDLs), and metal vapor discharge lamps. Normally, a separate source is required for each element, although some multielement lamps are available. If the line source profile is narrower than the analyte absorption profile in the atomizer, it is desired to maximize the integrated line radiance.

Metal vapor lamps were used early in the development of AFS, but are not used much today because they are available for only the most volatile elements (Cd, Zn, Hg, Ga, In, Tl, Na, and K) and they are difficult to pulse electronically. When these lamps are operated at high currents to increase the total line radiance, the source profile becomes significantly broadened and self-reversed. The amount of source radiation emitted and absorbed by the analyte, and hence the fluorescence signal, can actually decrease. The Cd, Zn, Hg, and Tl discharge lamps are the most useful, and flame AF detection limits (DLs) below 1 ng mL$^{-1}$ have been obtained.

Microwave-powered EDLs have been the most popular AF excitation sources because they provide relatively clean spectra and higher radiances than most other conventional sources. Their use for AF has decreased in the last few years. Although EDLs can be prepared for most elements, they are available from commercial sources for only a few elements. Their preparation and optimization is still tedious and difficult.

Hollow cathode lamps have long been used and still are used as excitation sources even though their radiances are usually lower than those provided by other sources. They are available for most elements. For AFS, HCLs are now used exclusively in a pulsed mode in which the peak current is typically 0.1 to 1 A, and the duty cycle is 1 to 10%. Pulsed operation increases the peak radiance 10 to 100 times relative to the steady-state radiance with dc current operation. At high currents, the line profile is significantly broadened so that the radiant power absorbed does not increase in direct proportion to the integrated line radiance. Boosted output (high intensity) HCLs can provide up to 10 times the radiances of normal shielded HCLs.

A high radiance continuum source would be ideal for AFS because only one source would be needed for all elements for single-element or multielement measurements. No presently available continuum source provides sufficiently high radiance over the entire wavelength region needed for AF. The most suitable continuum excitation source has been a 150 to 500-W Xe high-pressure arc lamp run in a continuous mode. Usually, lamps with an integral elliptical or parabolic reflector (e.g., EIMAC, see Section 4-2) have been employed because they collect and focus or collimate more than $2\pi$ sr of the radiation emitted from the lamp. The spectral radiance is lower than that of a conventional line source. However, the radiant power is absorbed over the entire absorption profile rather than over just the narrower emission profile of the line source. The total radiance absorbed is generally less than that obtained with an EDL. There is a substantial decrease in radiance for wavelengths shorter than 250 nm.

Lasers are excellent sources for AFS in many respects because they can provide irradiances many orders of magnitude higher than available with conventional sources. Both continuous wave and pulsed dye lasers have been used to provide tunable radiation over the visible region. Pulsed dye lasers have found the most use because they can be tuned to wavelengths as low as $\approx$220 nm (N$_2$ laser pump) or $\approx$180 nm (Nd:YAG laser pump) with frequency doubling. The spectral bandwidth can be adjusted from 1 to less than $10^{-4}$ nm (much narrower than absorption line widths), and the beam can be focused to a very small spot or expanded to any cross section desired. There is a considerable loss in source irradiance for small line widths. The very high peak irradiances ($10^4$ to $10^8$ W cm$^{-2}$) of pulsed dye lasers can cause saturation, which provides added advantages, as discussed in Section 11-3. The high irradiance allows the use of weaker nonresonance transitions, and thus wavelength discrimination against source scattering. Unfortunately, the use of laser excitation in routine applications is hindered by the cost and operational complexity of pulsed, tunable lasers.

More recently an ICP has been used as a versatile excitation source in conjunction with another atomizer (e.g., flame or plasma). The excitation wavelength is chosen and matched to the analyte by aspirating a high concentration of the analyte (e.g., 20 mg mL$^{-1}$) into the source ICP. The ICP excitation source is characterized by high radiance, long-term stability, narrow line widths, and freedom from self-reversal. As with lasers, the primary disadvantages are the cost and complexity of the source and its operation.

## Atomizers and Sample Introduction

As for other atomic spectrometric techniques, the ideal atomizer would have a large and stable overall nebulization-atomization efficiency (see equation 7-9) so that $n_0$ (or $n_1$ for excited-state fluorescence), and hence the fraction of absorbed radiation, is significant even for small analyte concentrations. Thus minimal dilution of the atomic vapor in hot atomizer gases is desirable. The atomization efficiency ($\varepsilon_a$) should ideally be unity and independent of the sample matrix to prevent analyte interferences. To reduce noise, the atomizer should cause minimal thermal excitation of emission from the analyte and other species.

In particular for AF, the atomizer should provide an environment that maximizes the fluorescence power yield (minimal nonradiative deactivation of the excited state). For excited-state or thermally assisted fluorescence, the atomizer should provide enough energy to populate the appropriate state significantly. Measurements can be limited by noise in the signal due to scattering of excitation radiation from unvaporized particles in the atomizer. Thus the atomizer temperature and residence times should be high enough to provide complete vaporization. Rayleigh scattering from atoms or molecules cannot be eliminated and determines the fundamental background noise limit for resonance fluorescence measurements.

Most AF measurements have been made with flame atomizers by continuous sample introduction with pneumatic nebulizers (occasionally, ultrasonic nebulizers). Although early measurements were made with total consumption burners, today only chamber-type burners are employed because of more complete vaporization and atomization of the smaller droplets introduced. The burner head is usually of the Méker or capillary design (Section 8-2) with a circular or square cross section. Provision is often made for sheathing the flame with an inert gas such as Ar, which prevents entrainment of the surrounding air. This provides a separated flame in which the secondary reaction zone is distinct from the interconal zone (see Section 8-1) and reduces background emission and quenching in some cases.

Selection of the type of flame involves compromises among atomization efficiency, background emission, and quenching characteristics. The general order for quenching efficiency is Ar < H$_2$ < H$_2$O < N$_2$ < CO < O$_2$ < CO$_2$. In early work H$_2$-based flames were used because of their reduced quenching efficiencies and low background emission. Detection limits can be improved by an order of magnitude by using a mixture of O$_2$ and Ar in place of air to reduce quenching from N$_2$. Unfortunately, these cooler flames provide inadequate atomization efficiency for refractory elements, and unvaporized solute particles can be a problem.

The trend has been to go to air–C$_2$H$_2$ and N$_2$O–C$_2$H$_2$ flames, as used in flame AAS. These increase the atomization efficiency for refractory elements and minimize analyte interferences and scattering. The loss in signal due to lower quantum efficiencies (i.e., quenching by CO, CO$_2$) and the increase in background emission signals and noise compared to cooler H$_2$ flames can be somewhat offset with higher radiance sources such as lasers.

More recently, the ICP has been used as an atomizer for AFS. Plasmas generally provide better vaporization and atomization efficiency and more freedom from analyte interferences than flames. Quantum efficiencies are usually higher than in hydrocarbon flames because of the inert Ar atmosphere (lower quenching). The background emission is highly dependent on the viewing region, but generally is greater than in flames. Usually, the viewing region is higher above the induction coils than used in AES to minimize background emission; excitation temperatures in the range 3000 to 3500 K exist in the region viewed. For some elements fluorescence from ions is observed. The higher temperature compared to flames enhances the usefulness of excited-state and thermally assisted fluorescence.

To date, electrothermal atomizers have been used sparingly in AFS. The advantages (e.g., small sample size, low absolute DLs) and disadvantages (e.g., significant analyte interferences, background absorption by molecular species) are similar to those discussed in contrasting flame and electrothermal AAS (see Section 10-5). The inert atmosphere reduces quenching problems, scattering can be worse, and blackbody emission in the visible region is a problem.

Alternate sample introduction and atomization techniques such as those used with AAS (Section 10-2) can be employed with AFS in particular cases. For example, elements such as As, Se, and Te have been determined by AFS with hydride introduction into cool flames or heated cells. Mercury has been determined by cold vapor AF. Mercuric ion is reduced to elemental Hg, which is carried to a quartz cell or vented into the atmosphere with excitation by the intense 254-nm line from a low-pressure Hg arc lamp.

### Wavelength Selection and Signal Processing

The requirements for wavelength selection in AFS are source dependent. With conventional line source or laser excitation, the "self-monochromating" nature of AF can be used to advantage. With conventional line sources, lines due to impurities or the filler gas do not cause interference or nonlinearity unless they excite significant fluorescence from concomitants or produce significant scattering within the wavelength region selected for viewing. Therefore, small-F/n and low-dispersion monochromators or nondispersive systems can be employed. With the latter, the collection efficiency (product of the volume element viewed, the solid angle viewed, and the optical transmission efficiency) can be 10 to 100 times greater than with dispersive systems.

With continuum source excitation, a medium-resolution monochromator is usually required to distinguish against the scattering of source radiation and the fluorescence excited from other species over the entire UV-visible region. In contrast to line source excitation, scattering is observed for all wavelengths in the spectral bandpass even though a small fraction of this wavelength range is useful for excitation.

Source modulation is generally employed to distinguish against background and analyte emission, which are not carried by the modulation waveform. Low-duty-cycle modulation provides S/N advantages, as discussed in Section 11-3. Background fluorescence and scattering at the monitored emission wavelength are carried by the modulation waveform and are also proportional to the source radiance and emission collection efficiency. Often, with continuum source excitation, wavelength modulation with a refractor plate (see Figure 4-34) is used alone or in conjunction with source modulation to discriminate against continuum background fluorescence and scattering.

### Multielement Instrumentation

The multielement capabilities of AFS have always been a major attraction. A single continuum source can provide many excitation wavelengths, several conventional line sources can be arranged around the atomizer, or a dye laser can be sequentially tuned to different wavelengths.

Sequential multielement AF systems have been based on dispersive designs, in which case a conventional or slew-scan monochromator is employed, or on nondispersive designs. In the early 1970s, Technicon briefly marketed a nondispersive, sequential, multielement flame AF spectrometer based on rotating filters, as shown in Figure 11-3. Sequential measurements can also be carried out with a spectrograph and image dissector tube.

Many simultaneous multielement AF spectrometer designs have been tested. The majority are based on multiple line sources. In the **time-division multiplex** approach, the hollow cathode lamps (HCLs), one for each element, are pulsed at the same frequency with a low duty cycle such that the pulses are interweaved (i.e., a given lamp is pulsed on only when the other lamps are in their off-period). The fluorescence from all ex-

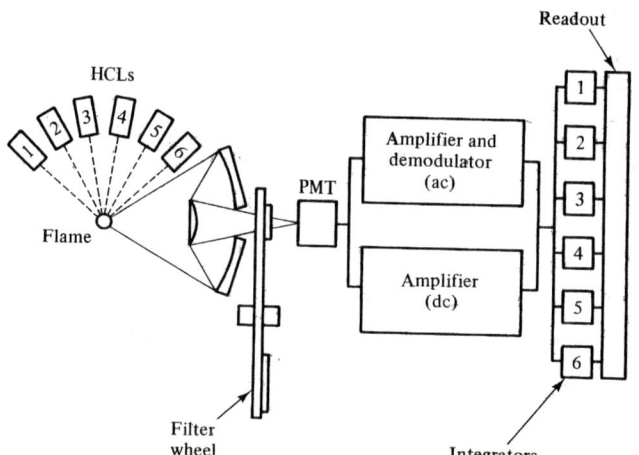

**FIGURE 11-3** Schematic diagram of Technicon AFS-6 multichannel AF spectrometer. Radiation from six different pulsed HCLs is focused on the flame. An inverse Cassegrain mirror system is used to collect fluorescence radiation with a solid angle of 0.82 sr. The collected fluorescence radiation passes through a rotating filter wheel with six filters and is incident on a single PMT. Each lamp is pulsed at 500 Hz only when the proper filter is in place. The demodulated signals for each element are switched at the proper time to six integrators. The dc signal can also be selected to measure the atomic emission signal for an element.

cited elements is imaged on one PMT. The fluorescence photoanodic current signals corresponding to each element are distinguished by demultiplexing the information encoded in different time slots. Thus the signal processor is synchronized to the lamp pulsing circuitry; it directs the pulse height information for each element to a separate information channel. For *S/N* enhancement, the signals obtained from many lamp pulses are averaged. Although the time-division multiplex method is not a true simultaneous method, *n* elements can be determined in the same time as one element. The collection optics can be nondispersive with a broadband UV filter and conventional PMT or with a solar blind PMT. With dispersive time multiplex spectrometers, a multiple-exit-slit monochromator is employed. The radiation exiting all slits is collected and imaged on one PMT.

Another approach is to modulate the line source for each element at a different frequency, the **frequency multiplex** approach. The fluorescence signals for each element are viewed by one PMT and distinguished by frequency demultiplexing with lock-in amplifiers tuned to each frequency or by Fourier transformation of a digitized record of the signal information.

The only commercial AFS system available at present is a multielement nondispersive system with ICP atomization. As shown in Figure 11-4, a separate HCL source-interference filter-PMT module is used for each element. Up to 12 of these modules are arranged around the plasma. The HCLs are pulsed on at separate times with a low duty cycle, as with the time-division multiplex method. However, demultiplexing is not required, as there is a separate detector and synchronized signal processor for each element. An extended plasma source with a long sleeve above the induction coil allows viewing higher in the plasma than is used for plasma AE measurements.

Simultaneous multielement AF spectrometers with line or continuum source excitation can also be based on direct readers or spectrographs with silicon-intensified target vidicons (SITs). The low throughput of direct readers is not favorable for AF measurements, whereas the high electronic readout noise of SITs degrades DLs.

## 11-3  SIGNAL AND NOISE EXPRESSIONS

In this section, readout and *S/N* expressions are developed for AF measurements. These equations provide a quantitative description of the effect of instrumental and analyte parameters on the magnitude of the analyte fluorescence signal and the precision with which it can be measured.

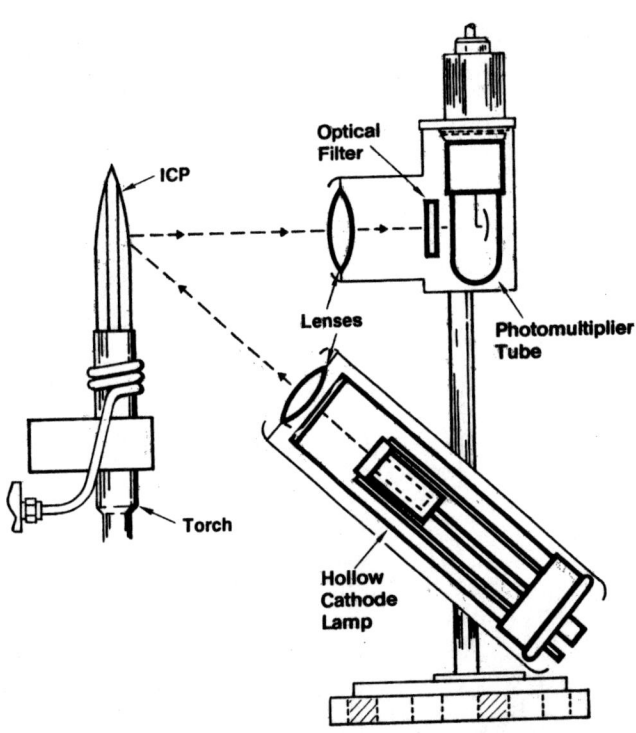

**FIGURE 11-4**  Diagram of source-detector module for a filter-based, multielement ICP-AF spectrometer. A number of these modules circle the plasma torch. Radiation from a given HCL is focused on the plasma. A fraction of the AF produced is collected and passed through an interference filter to the PMT for that module. The lamp in each module is pulsed on at a separate time so that the AF from only one element is excited at a time. The detection electronics are synchronously gated to the pulsing. (Courtesy of Baird Corp.)

## Signal Expressions

The most important signal expressions for AFS are summarized in Table 11-2. The measured analyte fluorescence signal ($E_F'$) is obtained from the difference in the total sample fluorescence signal, $E_{tF}$, and the blank signal, $E_{bk}$ (equation 11-1). The blank signal (equation 11-2) is due to dark current ($E_d$), background emission ($E_{bE}$), background fluorescence ($E_{bF}$), and scattering ($E_{sc}$). The total analyte fluorescence signal (equation 11-3) includes contributions from all blank components, analyte fluorescence ($E_F$), and analyte emission ($E_E$). With source modulation the mean value of $E_E$ is zero. If, additionally, the blank measurement compensates for scattering and background fluorescence, the measured analyte fluorescence signal equals the true analyte fluorescence signal ($E_F' = E_F$).

*Variable Dependence.*  In Section 7-5, general expressions were developed for the total fluorescence

**TABLE 11-2**

Signal expressions for AFS[a,b]

$$E'_F = E_{tF} - E_{bk} \tag{11-1}$$

$$E_{bk} = E_d + E_{bE} + E_{bF} + E_{sc} \tag{11-2}$$

$$E_{tF} = E_F + E_E + E_{bk} \tag{11-3}$$

$$E_F = mGR(\lambda')(\Phi_F)_0 \tag{11-4}$$

$$(\Phi_F)_0 = B_F Y(\lambda') \tag{11-5}$$

$$Y(\lambda') = hl^*\Omega'T'_{op} \tag{11-6}$$

$$B_F = \frac{\Phi_A^* Y f_s}{4\pi h l^*} \tag{11-7}$$

$$\Phi_A^* = hL \int (E_\lambda)_0 f_{pr} \ \alpha^*(\lambda) \ d\lambda \tag{11-8}$$

$$\alpha^*(\lambda) = 1 - e^{-k(\lambda)l^*} \tag{11-9}$$

$$f_{pr} = e^{-k(\lambda)l_{pr}} \tag{11-10}$$

$$E_F = mGR(\lambda')\frac{\Omega'}{4\pi}T'_{op}Yf_s hL \int_{\text{line}} (E_\lambda)_0 f_{pr} \alpha^*(\lambda) \ d\lambda \tag{11-11}$$

At low analyte number densities:

$$E_{FL} = mGR(\lambda')Y\frac{\Omega'}{4\pi} T'_{op}hLl^*E_0 k_m \tag{11-12}$$

$$E_{FC} = mGR(\lambda')Y\frac{\Omega'}{4\pi} T'_{op}hL(E_{\lambda m})_0 A_t^* \tag{11-13}$$

$$E_{FL} = \frac{mGR(\lambda')\Omega'T'_{op}E_0 l^* Lhe^2 \lambda_m n_i f_{ij}Y}{16\pi\varepsilon_0 m_e^2 c^2 \ \Delta\lambda_{\text{eff}}} \tag{11-14}$$

$$E_{FL} = \frac{(3.4\times10^3)mGR(\lambda')\Omega'T'_{op}E_0 hLl^*\lambda_m^2 f_{01}YF\varepsilon_a c}{\Delta\lambda_{\text{eff}}Qe_f} \tag{11-15}$$

[a]Equations 11-1 to 11-3 are specific forms of equations 2-23 to 2-25. With source modulation, the mean values of $E_d$, $E_{bE}$, and $E_E$ are zero. With sample modulation, the mean values of $E_d$, $E_{bE}$, $E_{sc}$, and $E_{bF}$ are zero with a good blank. With line source wavelength modulation, the mean values of $E_d$, $E_{bE}$, and $E_{bF}$ are zero if the background is a continuum. The same is true for continuum source wavelength modulation, but in addition the mean of $E_{sc}$ averages to zero. It is assumed that the bandpass of the wavelength selector is much greater than the width of the fluorescence profile. Here a prime is used to distinguish the emission wavelength or quantities related to this wavelength from the excitation wavelength or quantities related to the excitation wavelength. Both $Y(\lambda')$ and $T'_{op}$ are assumed to account for the optical efficiency of all optical components between the atomizer and the detector.

[b]$(\Phi_F)_0$, Analyte fluorescence radiant power incident on detector (W); $B_F$, radiance of analyte fluorescence from the viewed volume element ($V^*$) (W cm$^{-2}$ sr$^{-1}$); $Y(\lambda')$, throughput factor for emission optics and wavelength selector (cm$^2$ sr); $\Phi_A^*$, radiant power absorbed by analyte in $V^*$ (W); $f_s$, self-absorption factor (dimensionless); $f_{pr}$, prefilter-effect factor (dimensionless); $l_{pr}$, prefilter pathlength (cm); $L_{pt}$, postfilter pathlength (cm).

radiant power produced in volume element V in the atomizer. Here we will take into account the effect of the geometry of illumination and the collection optics on the portion of total fluorescence actually observed by the detector and on the shape of the calibration plot.

Equation 11-4 shows that $E_F$ is determined by the PMT and signal modifier gains, the PMT responsivity, and the analyte fluorescence radiant power incident on the PMT [$(\Phi_F)_0$]. Similar expressions apply for the relationship between different components of the background readout signals and their respective radiant powers.

To derive the equation for the fluorescence radiant power detected, consider the idealized model of

Figure 11-5 in which the fluorescence is collected at 90° to the excitation axis. The volume element $V$ in which atoms are radiationally excited is determined by the rectangular cross section of a collimated beam ($L \times h$) and the pathlength of the atomizer ($l$). Normally, the emission optics restrict the pathlength and the volume element of excitation and observation: these are denoted $l^*$ and $V^*$, respectively ($l^* < l$, $V^* < V$). We will assume for simplicity that 1:1 imaging is used to focus the fluorescence from the atomizer onto a limiting emission aperture, between the flame and the detector, of width $l^*$. Thus with an emission monochromator, $l^*$ is equal to the entrance slit width ($W$). In nondispersive systems, $l^*$ is determined by the width of the aperture in front of the detector. The height of the emission aperture is assumed to be equal to or greater than $h$.

We can consider the volume element $V^*$ as a source of fluorescence radiation characterized by line radiance $B_F$. The radiant power observed by the PMT is the product of $B_F$ and the emission optics throughput factor, $Y(\lambda')$ (equation 11-5). From equation 11-6, $Y(\lambda')$ is determined by the area of the viewed face of $V^*$ ($hl^*$), the collection solid angle of the emission optics ($\Omega'$), and the optical efficiency of the emission optics ($T'_{op}$).

The radiance observed from $V^*$ is calculated from equation 11-7. Here the product of the absorbed radiant power in $V^*$ ($\Phi_A^*$) and the fraction of this radiation converted to fluorescence, or the fluorescence power yield ($Y$), give the total fluorescence emitted in $V^*$. This radiant power is converted into radiance by dividing by $4\pi hl^*$. The self-absorption factor $f_s$ accounts for

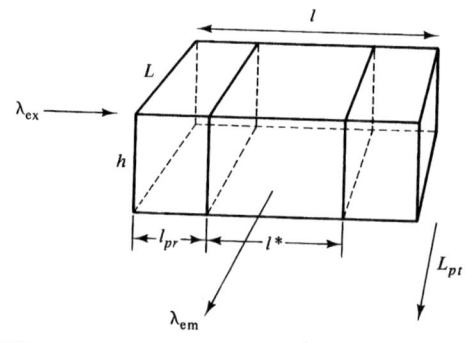

**FIGURE 11-5** Geometry for fluorescence excitation and collection. The excitation beam of cross-sectional area $L \times h$ passes through the atomizer of length $l$. Thus the volume element of excitation $V$ equals $Llh$. The emission optics define the width $l^*$ over which fluorescence is excited and viewed. Hence fluorescence is collected from volume element $V^* = Lhl^*$ through the front face of $V^*$ with an area of $hl^*$. The excitation photons must pass through distance $l_{pr}$ before reaching $V^*$. Fluorescence photons must travel a distance $L_{pt}$ from the viewed face of $V^*$ before leaving the atomizer.

the fraction of the emitted photons that are reabsorbed before passing out of the viewed face of the atomizer. It is understood that $Y$ varies with the particular $i$ and $j$ levels involved in the transition.

The radiant power absorbed in $V^*$ is given by equation 11-8. It is similar in form to equation 7-58 except that the source intensity is expressed in terms of its spectral irradiance $(E_\lambda)_0$, where $hL(E_\lambda)_0 = (\Phi_\lambda)_0$ and the prefilter-effect factor, $f_{pr}$, is added. The spectral absorptance in $V^*$ [$\alpha^*(\lambda)$] is given by equation 11-9. The prefilter factor accounts for the fraction of the excitation radiation that is absorbed before reaching the viewed volume element and is calculated from equation 11-10. Thus the product of $(E_\lambda)_0$ and $f_{pr}$ is the source spectral irradiance actually seen by atoms in $V^*$. We will assume that there is no absorption of the excitation or the fluorescence radiation by other species and that reflection or scattering losses at the atomizer-atmosphere interface are negligible. Additional factors could be included to account for these effects or flame absorption, which can be significant below 250 nm.

Equation 11-11 is the final general form for the analyte fluorescence readout voltage obtained by combining equations 11-4 to 11-10. Equations 11-12 and 11-13 are the limiting forms of equation 11-11 for low number densities with line and continuum source excitation, respectively. Note that $f_{pr} = f_s = 1$ under these conditions. For line source excitation, the fraction of the source radiation absorbed is $k_m l^*$ if it is assumed that the absorption coefficient is constant over the source profile and equal to the maximum value $k_m$. Note that $E_F$ is proportional to the volume element illuminated and viewed ($V^* = hLl^*$). For continuum source excitation, the irradiance is assumed to be constant over the absorption profile and equal to the value at the wavelength of maximum absorption, $(E_{\lambda m})_0$. In this case the absorbed radiation in $V^*$ is equal to the product of $(E_{\lambda m})_0$ and $A_t^*$ (the integral absorption in $V^*$).

If we substitute equation 7-45 for $k_m$ in equation 11-12, the result is equation 11-14, from which we see that $E_{FL}$ is related to the number density of the state from which excitation occurs ($n_i$), the absorption oscillator strength ($f_{ij}$), and the effective width of the absorption profile ($\Delta\lambda_{eff}$). If we assume normal resonance fluorescence between the ground state and the first excited state ($n_i = n_0$) and $Z(T) = g_0$, use equation 7-9 for $n_0$ in equation 11-14, and evaluate the constants, the result is equation 11-15. Expressions similar to equations 11-14 and 11-15 can be written for continuum source excitation, if $E_0/\Delta\lambda_{eff}$ is replaced by $(E_{\lambda m})_0$. Thus at low analyte concentrations, the measured analyte fluorescence signal is proportional to the analyte concentration ($c$), instrumental factors [$mGR(\lambda')\Omega' T'_{op}E_0 hLl^*$], atomizer factors ($F\varepsilon_a/Qe_f$), and analyte spectroscopic factors ($\lambda_m^2 f_{01}Y/\Delta\lambda_{eff}$).

For a given element and fluorescence transition (specific excitation and emission wavelength), the fluorescence signal varies with the instrumental configuration. This is due primarily to the type of atomizer (e.g., $F$, $\varepsilon_a$, $Y$), the type of source and excitation optics [e.g., $E_0$ or $(E_\lambda)_0$, $h$, $L$, and the solid angle of collection for conventional sources], and the type of emission collection system (e.g., $l^*$ and $\Omega'$). Both $\Omega'$ and $l^*$ are considerably larger with nondispersive systems than with dispersive systems. For a given element, spectrometer, and atomizer, the fluorescence signal varies with the specific excitation and emission transitions utilized. From equation 11-14, these differences are seen to be due primarily to the magnitudes of $E_0$, $n_i$, $f_{ij}$, and $Y$. With excited-state fluorescence, the number density of the lower excited state is usually lower than the ground-state number density unless this state is very close to the ground state (e.g., Ga, Tl, In, Pb, Sn). In the case of thermally assisted, stepwise fluorescence, only a fraction of the number density of the upper state of excitation is raised thermally to the state from which fluorescence occurs. When fluorescence occurs from higher excited states, there are numerous radiative and nonradiative pathways for deexcitation. Thus fluorescence can occur at two or more wavelengths, and the fluorescence power yield for a particular transition may be less than for resonance fluorescence from the first excited state. For elements with complex atomic spectra, the fluorescence is distributed among wavelengths yielding low $Y$ values for a particular transition. In some cases it is possible to monitor several emission wavelengths simultaneously.

*Linearity.* In equation 11-11, three terms depend on the analyte number density or

$$E_F \propto f_s \int_{\text{line}} f_{pr}\alpha^*(\lambda) \, d\lambda \tag{11-16}$$

Nonlinearity occurs when primary and secondary absorption effects become significant. *Primary absorption effects* are due to significant absorption of the excitation radiation in the region before the viewed volume element (i.e., the prefilter effect) and within the viewed volume element $V^*$. At high analyte concentrations, $f_{pr}$ is less than unity. With continuum source excitation, the integral of the spectral absorptance over the absorption profile is not proportional to $n_0$ at high concentrations. Similarly, with line source excitation, the absorptance over the source profile is no longer proportional to $n_0$ at high analyte concentrations. Use of high lamp currents with HCLs or metal discharge lamps can broaden the source profile and enhance nonlinearity. *Secondary absorption effects* are due to significant absorption of the fluorescence within $V^*$ or in the post-filter region. Because AF lines are very narrow, secondary absorption is almost always due to the analyte. Hence, it is called self-absorption. Secondary absorption causes $f_s$ to be less than unity at high analyte number densities.

We now consider the effect of each of the foregoing three terms on $E_F$ and on the calibration plots for resonance fluorescence. First, ignoring the effect of $f_s$ and $f_{pr}$, we have already noted in Section 7-5 that for large $n_0$, the fluorescence signal exponentially approaches a limiting value with line source excitation when all of the available radiation is absorbed. With continuum source excitation, the integral absorption and hence the fluorescence signal are proportional to $n_0^{1/2}$ at high analyte concentrations because absorption still occurs in the wings of the absorption profile even though effectively all the source radiation at the maximum of the absorption profile has been absorbed. In either case, it is important to remember that $E_F$ is proportional to the amount of radiation absorbed ($\Phi_A^*$). When $n_0$ is large, $\Phi_A^*$ and hence $E_F$ are no longer proportional to $n_0$.

As also discussed in Section 7-5 (equation 7-65), $f_s$ for photons emitted and absorbed within the excited and viewed region is proportional to $n_0^{-1/2}$ for large $n_0$. This, combined with the dependence of the fraction of absorbed radiation on $n_0$, causes the calibration curve to roll over at high $n_0$ with line source excitation and to reach a plateau with continuum source excitation (see Figure 7-18). Absorption by atoms in the nonilluminated postfilter region causes further attenuation of the observed fluorescence signal and affects the magnitude of $f_s$. The manner in which $f_s$ and the calibration curve shape depend on the analyte concentration is complex and highly influenced by the geometry of illumination and viewing.

The prefilter-effect factor decreases exponentially to zero as $n_0$ is increased with line source excitation [equation 11-10 with $k(\lambda) = k_m$]. Thus in the limit of infinite analyte concentration, all the excitation radiation is absorbed before reaching the viewed volume element $V^*$, and the observed fluorescence signal approaches zero. Combined with the previous two effects, the prefilter effect causes additional nonlinearity at moderate analyte concentrations and increased roll-over at high concentrations. With continuum source excitation, the prefilter effect causes similar behavior, although the effect is more complex and less severe because $k(\lambda)$ and $f_{pr}$ vary over the absorption profile. An AF calibration plot is shown in Figure 11-6, where it is compared to AA and AE calibration plots.

In a real system, the exact shape of the calibration

curve can be predicted only if all geometric factors as well as analyte parameters (e.g., the $a$-parameter) are known. Prefilter effects are minimized by decreasing $l_{pr}$, by increasing the fraction of the excited volume element viewed ($l^* \to l$), or by using front-surface illumination (exciting and viewing from the same surface of the atomizer). Self-absorption is negligible for nonresonance fluorescence.

## Saturated Fluorescence

Up to this point, we have assumed that the excitation source does not significantly alter the distribution of analyte states. With very high irradiances from focused pulsed dye laser beams, it is possible to deplete significantly the population of the state from which excitation occurs.

To understand this effect, we will consider the simple case of resonance fluorescence between the ground state of population $n_0$ and first excited state of population $n_1$, where the number density of higher levels (e.g., $n_2$) is not significant ($n_M = n_0 + n_1$). We shall also assume the following: (1) quasi-continuum laser excitation so that the irradiance is constant over the analyte absorption profile; (2) the high source irradiance does not affect the analyte absorption profile, the temperature, or the velocity distribution in the atomizer; (3) the excitation beam stimulates only absorption (i.e., the laser does not enhance ionization); (4) the beam intensity is uniform across its profile; and (5) the analyte number density is low enough that the self-

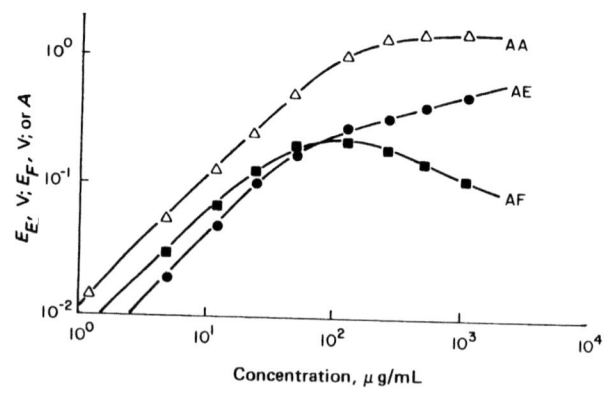

**FIGURE 11-6** Calibration plots for Cu with AA, AE, and AF in an air–$H_2$ flame with HCL excitation. The nonlinearity and rollover in the AF calibration plot are due to primary absorption effects, including the prefilter effect, and secondary absorption effects (self-absorption in this case). [From N. W. Bower and J. D. Ingle, Jr., *Appl. Spectrosc., 35,* 317 (1981), with permission of the Society for Applied Spectroscopy.]

absorption effect, the prefilter effect, and any attenuation of the source radiation over $l^*$ are negligible.

The equations that describe saturated fluorescence for a two-level system are summarized in Table 11-3. Equation 11-17 shows that the total fluorescence radiant power emitted from $V^*$ ($\Phi_F$) is determined by the number of atoms in the excited state ($n_1 V^*$), the Einstein coefficient for spontaneous emission ($A_{10}$), and the energy per photon ($h\nu$). Under steady-state conditions the rate of absorption and the rate of deactivation by collisions, spontaneous emission, and stimulated emission are balanced, as shown in equation 11-18. Here we assume that thermal excitation is negligible and note that $(E_\lambda)_0/c$ is the spectral energy density $(U_\lambda)_0$. Equation 11-18 can be rearranged to solve for the ratio of the number densities of the two states (equation 11-19). For high source irradiances, the second term on the right side of equation 11-19 vanishes, which indicates that the maximum value of $(g_0 n_1/g_1 n_0)$ is unity, the rates of stimulated absorption and stimulated emission are balanced, and there is no net absorption. We can define the saturation spectral irradiance, $(E_\lambda^s)_0$, to be the value of $(E_\lambda)_0$ for which $n_1$ (or the measured fluorescence signal) is 50% of its maximum value, as shown in equation 11-21. For $g_0 = g_1$, this saturation spectral irradiance corresponds to one-fourth of the total atoms in the excited state. For complete saturation, one-half of the atoms are in the excited state. Combining equations 11-17 and 11-19 to 11-21 yields equation 11-22, which predicts how $\Phi_F$ depends

on various constants, the total analyte number density, and the source irradiance.

For low source radiances (conventional sources), equation 11-22 simplifies to equation 11-23, which is similar in form to the equations developed in Table 11-2. It predicts that $\Phi_F$ is proportional to the source radiance, the fluorescence power yield, and the absorption coefficient. As the source irradiance is increased, $\Phi_F$ is no longer proportional to the irradiance and, in the limit (equation 11-24), $\Phi_F$ is independent of $Y$, $(E_\lambda)_0$, and $B_{01}$.

Experimentally, saturated AF has been demonstrated with pulsed dye lasers, which provide irradiances from $10^4$ to $10^7$ W cm$^{-2}$ nm$^{-1}$. An example for thallium is shown in Figure 11-7. Measurements under saturation can provide several advantages. The analyte fluorescence signal is independent of source fluctuations and quenching species. At higher number densities, linearity is improved because primary absorption effects and self-absorption are negligible. Self-absorption still occurs over the nonilluminated distance between the viewed volume element and the edge of the atomizer.

The exact dependence of $\Phi_F$ on $(E_\lambda)_0$ can be more complicated than presented here, particularly for small laser bandwidths, because some of the assumptions made are not always valid (e.g., saturation only occurs in the

**TABLE 11-3**

Equations for saturated fluorescence

$$\Phi_F = h\nu A_{10} n_1 V^* \tag{11-17}$$

$$\frac{B_{01}(E_\lambda)_0 n_0}{c} = \left[ k_{10} + A_{10} + \frac{B_{10}(E_\lambda)_0}{c} \right] n_1 \tag{11-18}$$

$$\frac{g_0 n_1}{g_1 n_0} = \left[ 1 + \frac{A_{10} c g_1}{Y B_{01}(E_\lambda)_0 g_0} \right]^{-1} \tag{11-19}$$

$$n_1 = n_M \left( 1 + \frac{n_0}{n_1} \right) \tag{11-20}$$

$$(E_\lambda^s)_0 = \frac{A_{10} c}{Y B_{01}} \frac{g_1}{(g_0 + g_1)} \tag{11-21}$$

$$\Phi_F = h\nu A_{10} V^* n_M \frac{g_1}{(g_0 + g_1)} \left[ 1 + \frac{(E_\lambda^s)_0}{(E_\lambda)_0} \right]^{-1} \tag{11-22}$$

If $(E_\lambda)_0 \ll (E_\lambda^s)_0$,

$$\Phi_F = \frac{h\nu}{c} V^* n_M (E_\lambda)_0 Y B_{01} \tag{11-23}$$

If $(E_\lambda)_0 \gg (E_\lambda^s)_0$,

$$\Phi_F = h\nu A_{10} V^* n_M \frac{g_1}{(g_0 + g_1)} \tag{11-24}$$

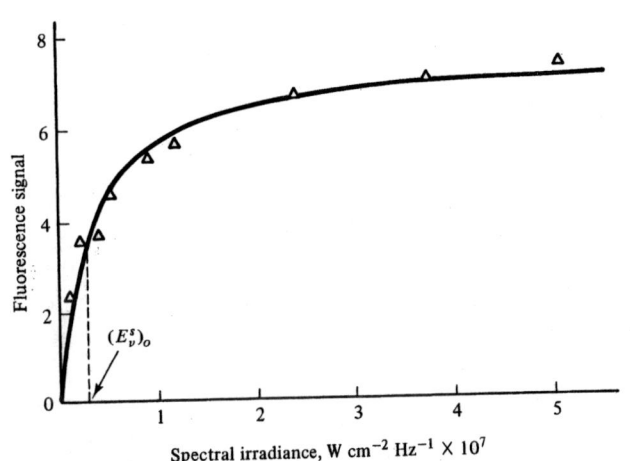

**FIGURE 11-7** Saturation curve for thallium fluorescence in an air–H$_2$ flame. The dependence of the nonresonance direct-line fluorescence signal at 535.0 nm on spectral irradiance with laser excitation at 377.5 nm is illustrated. At high irradiances, the signal reaches a plateau due to saturation effects. The irradiance yielding one-half the maximum signal, the saturation spectral irradiance, is $2.8 \times 10^{-8}$ W cm$^{-2}$ Hz$^{-1}$ ($2.9 \times 10^4$ W cm$^{-2}$ nm$^{-1}$). [Adapted with permission from D. R. De Olivares and G. M. Hieftje, *Spectrochim. Acta, 36B*, 1059 (1981), Pergamon Press, Ltd., Oxford.]

center of the beam profile where the irradiance is greatest). However, the qualitative behavior described is still applicable.

## Signal-to-Noise Expressions

The general noise and *S/N* expressions for luminescence measurements presented in Section 5-4 (equations 5-25 and 5-31), converted to the specific forms applicable to AFS, are presented in Table 11-4. The noise in the total fluorescence signal (equation 11-25) is due to shot and flicker noise in the analyte fluorescence signal, analyte emission noise, and blank noise. The blank noise (equation 11-26) is due to amplifier-readout noise, dark current noise, and noise in the three types of background signals (background emission, background fluorescence, and scattering).

At low analyte concentrations, blank noise is dominant and equation 11-28 applies. The *S/N* is proportional to $i_F$ (see Figure 5-6). Note that there are shot and flicker components in all three background signals. The flicker noise in the background fluorescence and scattering signals is assumed to be due to source flicker noise.

At higher analyte concentrations, the analyte fluorescence or emission signals are much greater than the background signals, so that shot or flicker noise in these signals becomes limiting. The *S/N* is proportional to $i_F^{1/2}$ if analyte fluorescence shot noise is dominant or independent of $i_F$ if flicker noise in the analyte fluorescence signal is dominant (see Figure 5-6). The flicker noise in the analyte fluorescence signal is due to source flicker noise and *analyte fluorescence flicker noise*. The latter noise is characterized by the analyte fluorescence flicker factor ($\xi_F$), which is related to $\xi_E$ and $\xi_A$. In other

words, analyte fluorescence flicker noise is due to fluctations in the number density of analyte species in the viewed volume element of the atomizer.

The strategy for optimizing the *S/N* depends on the limiting noise sources. To improve the DL or the precision for low analyte concentrations, $i_F/\sigma_{bk}$ should be optimized. Dark current noise or amplifier-readout noise is likely to be limiting only with dispersive systems, low radiance sources, and cool, low background flames. Under these conditions, the *S/N* is proportional to $i_F$. This signal can be increased with controllable instrumental factors such as $\Omega'$, $(E_\lambda)_0$, or $l^*$. These are identified in equation 11-5. With ICP excitation or atomization, RF noise can increase the electronic noise. This noise can be minimized by appropriate filtering techniques.

If background signal noise is limiting, the situation becomes more complex. Here instrumental variables, such as the viewing position in the atomizer, gas flow rates, $(E_\lambda)_0$, $\Omega'$, and $l^*$, should be adjusted to optimize the *S/B*. If the analyte fluorescence and the background signals are both increased by changing the magnitude of a variable, the *S/N* reaches a limiting best value at the point that flicker noise in the background signals becomes dominant.

Background emission noise is more likely to be limiting with high background atomizers and nondispersive systems (i.e., large $l^*$ and $\Omega'$). With dispersive systems ($l^* \propto W$), $i_{bE}$ is proportional to $W^2$ if the background is a continuum. Thus for large $W$, the *S/N* becomes independent of $W$ if background emission shot noise is limiting and decreases with $W$ once background emission flicker noise becomes dominant. Use of a low-duty-cycle pulsed light source and synchronous gated detection decreases the relative magnitude of back-

**TABLE 11-4**

Signal-to-noise expressions for AFS[a]

$$\sigma_{tF} = [(\sigma_F)_s^2 + (\sigma_F)_f^2 + \sigma_E^2 + \sigma_{bk}^2]^{1/2} \tag{11-25}$$

$$\sigma_{bk} = (\sigma_{bE}^2 + \sigma_d^2 + \sigma_{ar}^2 + \sigma_{bF}^2 + \sigma_{sc}^2)^{1/2} \tag{11-26}$$

$$\frac{S}{N} = \frac{E_F}{\sigma_{tF}} \tag{11-27}$$

At low analyte concentrations

$$\frac{S}{N} = \frac{i_F}{\left\{ K(i_d + i_{sc} + i_{bE} + i_{bF}) + \xi_1^2(i_{bF} + i_{sc})^2 + (\chi i_{bE})^2 + \left[\frac{(\sigma_d)_{ex}}{mG}\right]^2 + \left(\frac{\sigma_{ar}}{mG}\right)^2 \right\}^{1/2}} \tag{11-28}$$

At high analyte concentrations

$$\frac{S}{N} = \frac{i_F}{[K(i_F + i_E) + (\xi_1 i_F)^2 + (\xi_F i_F)^2 + (\xi_E i_E)^2]^{1/2}} \tag{11-29}$$

[a] $\sigma_{tF}$, Noise in total fluorescence signal (V); $(\sigma_F)_s$, analyte fluorescence shot noise (V); $(\sigma_F)_f$, analyte fluorescence flicker noise (V); $\sigma_E$, analyte emission noise (V); $\sigma_{bk}$, blank noise (V); $\sigma_{sc}$, scattering noise (V); $\sigma_{bF}$, background fluorescence noise (V); $\xi_1$, source flicker factor (dimensionless); $\xi_F$, analyte fluorescence flicker factor (dimensionless); $\xi_E$, analyte emission flicker factor (dimensionless); $\chi$, background emission flicker factor (dimensionless).

ground emission noise because the average background emission signal relative to $i_F$ is reduced (see Section 5-6). Source, wavelength, or sample modulation discriminate against $1/f$ background emission flicker noise.

With high lamp radiances and large values of $l^*$ and $\Omega'$, the excitation-source-dependent background signals become limiting and background fluorescence or scattering noise becomes dominant. Once the point is reached that flicker noise in these signals is limiting, further increases in the source radiance or emission collection efficiency provide no improvement or even decrease the $S/N$. Background fluorescence can be due to molecular species (e.g., CH, OH, $PH_3$, CaOH) or atoms, particularly with continuum source excitation. Scattering of excitation radiation by small particles is proportional to $(\lambda_{ex})^{-4}$ and thus more likely in the UV region. With continuum source excitation, the scattering signal is proportional to $W^2$ and $1/f$ noise in the background fluorescence and scattering signals can be discriminated against with sample or wavelength modulation.

Although resonance fluorescence is most commonly used, some of the other fluorescence types can provide $S/N$ advantages. In particular, scattering and noise due to scatter can be eliminated if nonresonance fluorescence is used. Here it is possible to pick an emission wavelength where background emission noise is low. In many cases, the $S/N$ and the DL of excited-state resonance fluorescence or various types of nonresonance fluorescence can be poorer than that for normal resonance fluorescence if the signal decreases more than the noise. The reasons for lower signals in these cases were discussed earlier in this section.

At higher analyte concentrations, blank noise is negligible and the noise carried by analyte signals is dominant. Analyte fluorescence shot noise can be limiting at moderate concentrations in which case $i_F$ should be increased [larger $hL(E_\lambda)_0$, $\Omega'$, or $l^*$) or the electronic bandwidth ($K$) should be reduced to improve precision. At higher analyte concentrations, flicker noise in the analyte fluorescence signal is usually limiting. With high blank noise, the signal-shot-noise-limiting region is not often observed (i.e., when the analytical signal-carried noise becomes greater than the blank noise, signal shot noise is negligible relative to signal flicker noise).

The total flicker noise in the analyte fluorescence signal is due to source flicker noise and analyte fluorescence flicker noise. With conventional sources, this latter noise source is usually dominant, and the precision is atomizer limited. Typically, $\xi_F$ is 0.5 to 1% with a 1- to 10-s time constant or integration time, whereas $\xi_1$ is about $10^{-3}$. With pulsed lasers, the pulse-to-pulse intensity variations (often 10%) significantly increase $\xi_1$. In this case, a double-beam system (see Section

6-5) can be used to compensate for source intensity variations as long as there is no saturation.

In the region of substantial calibration plot nonlinearity at high analyte concentrations, the $S/N$ often decreases, as shown in Figure 11-8. This is due to increased analyte fluorescence flicker noise (increase in $\xi_F$) because of fluctuations in pre- and postfilter absorption or decreased $i_F$ in the roll-over region with line source excitation. In the visible region, analyte emission noise can also become significant at high concentrations. Its effect is reduced by increasing the source radiance or by using low-duty-cycle, pulsed sources with synchronous, gated detection. Source modulation provides discrimination against $1/f$ analyte emission flicker noise.

Under conditions near or in saturation, scattering or background fluorescence noise is usually limiting at low concentrations. At higher concentrations, source flicker noise in the analyte fluorescence signal is reduced, $\xi_F$ due to fluctuations in the pre- or postfilter absorption is reduced, and analyte emission is negligible. Very high source radiance can worsen the $S/N$ because the scattering signal and noise increase in proportion to $(E_\lambda)_0$, whereas $i_F$ becomes independent of $(E_\lambda)_0$.

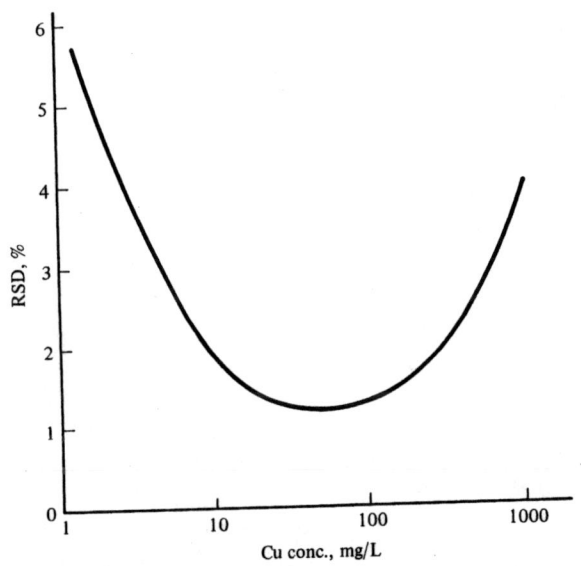

**FIGURE 11-8** Precision curve for AF. At low concentrations the $S/N$ is limited by blank noise. As the analyte concentration and AF signal increase, analyte fluorescence shot noise or flicker noise become dominant. At very high analyte concentrations, precision decreases due to analyte emission noise or noise due to fluctuations in pre- and postfilter absorption. (Adapted from S. Ghaffari, Ph.D. thesis, Oregon State University, 1984, with permission of author.)

## 11-4 PERFORMANCE CHARACTERISTICS AND APPLICATIONS

### Linearity

The linear range for atomic fluorescence measurements can vary from three to eight orders of magnitude. Larger ranges are obtained with elements for which the DL is very low. As discussed in Section 11-3, nonlinearity occurs when primary absorption and secondary absorption effects are significant. Generally, this occurs when the analyte concentration reaches the range 10 to 100 mg L$^{-1}$. Nonlinearity is readily observed when the absorbance across the atomizer pathlength ($l$) is about 0.1 and is severe when the absorbance approaches unity. The largest linear ranges are often obtained with laser excitation because the DLs are the lowest. Also, under saturation conditions, the upper range of linearity is extended as previously described. As for other atomic spectrometric techniques, nonlinearity can arise due to changes in solution flow rate or atomization efficiency with analyte concentration (i.e., ionization and dissociation equilibria).

Measurements can be made in the nonlinear region if it is well characterized with standards. Use of the plateau or roll-over region, often observed with line source excitation, is not suggested.

### Precision and Accuracy

As for AE and AA, the precision of AF measurements at analyte concentrations significantly above the DL is usually limited to 0.5 to 2% by noise from the atomization system (analyte flicker noise). The overall precision depends on the reproducibility of all steps in the analytical procedure.

Accuracies of 0.5 to 5% can be obtained in the absence of systematic errors. Analyte interferences and the techniques for minimizing analyte interferences (e.g., use of releasing agents or ionization buffers) in AF are similar to those for AE and AA and are primarily atomizer dependent. Concomitants that alter the fluorescence quantum yield are an added source of analyte interference. Here standard addition techniques can be useful. Background absorption interferences could be significant but have not been addressed in the literature.

With source modulation, blank spectral interferences due to concomitant emission are eliminated. Thus only background fluorescence and scattering usually cause the measured analytical signals to be too high without a good blank measurement. Uncompensated atomic fluorescence due to concomitants is rare with dispersive systems, particularly with line source excitation. Both continuum source excitation and nondispersive systems (filter or no filter with a solar blind PMT wavelength selection) enhance the probability of exciting and viewing AF from other species.

The importance of scattering interferences with resonance fluorescence is highly dependent on the sample matrix, the atomizer (e.g., efficiency of volatilization) and the wavelength. It can be compensated with continuum source excitation by wavelength modulation. The scattering component can be estimated with line source excitation by measuring the signal at a wavelength close to the excitation wavelength.

### Detection Limits

Detection limits in AFS are extremely source and atomizer dependent. Some of the best flame AF DLs obtained with line source, continuum source, laser, and plasma ICP excitation are given in Table 11-5. Also listed are DLs obtained with ICP plasma atomization and pulsed HCL excitation.

Spectacular DLs well below 1 ng mL$^{-1}$ have been obtained for Ag, Cd, and Zn with line source excitation and flame atomization in cool flames with nebulized samples. Similarly, excellent DLs are obtained for As, Se, Sb, and Te with hydride sample introduction or for Hg with cold vapor techniques. In general, DLs with continuum source excitation are worse than those obtained with line source excitation except for Cr, Mg, Pt, and Sr. Surprisingly to date, high-radiance, laser excitation has not provided significant improvements in detectability for many elements for which conventional source excitation work well. At present, most pulsed dye lasers do not provide wavelengths near 200 nm to excite elements such as Zn, Cd, As, Se, and Te. However, laser excitation has made it possible to determine many refractory elements with DLs below 1 μg mL$^{-1}$ and often down to 1 ng mL$^{-1}$ in the N$_2$O–C$_2$H$_2$ flame or the ICP. The high radiance allows less efficient nonresonance transitions to be used to eliminate scattering problems. Initial studies have shown that the ICP is a good source of excitation radiation for some elements.

The ICP has proved to be an excellent atomization source for AFS. The Baird system (ICP atomization, pneumatic nebulization with aerosol desolvation, and pulsed boosted-output HCL excitation) yields DLs comparable to or better than flame AAS and ICP-AES with pneumatic nebulization for many nonrefractory elements. For some refractory elements, DLs are up to a factor of 10 worse than those obtained with ICP-AES. Excellent DLs have also been achieved with ICP atomization with another ICP as the excitation source. Finally, DLs in the range 0.4 to 7 ng mL$^{-1}$ have been obtained with ICP atomization and tunable pulsed dye laser excitation for several elements (Al, B, Ba, Ca, Mo, Pb, Si, Sn, Ti, Tl, V, Y, Zr).

**TABLE 11-5**
Detection limits for AFS (ng mL$^{-1}$)

| Element | Line source[a] | Continuum source[a] | Laser[a] | ICP[b] | HCL[c] |
|---|---|---|---|---|---|
| Ag | 0.1 | 0.7 | 4 | 3 | 0.1 |
| Al | 70 | 200 | 0.6 | — | 5 |
| As | 70[0.1] | — | — | — | 15 |
| Au | 5 | 150 | — | — | 0.3 |
| Ba | — | — | 2 | — | 50 |
| Be | 8 | 15 | — | — | 0.2 |
| Bi | 2 | 10 | 3 | — | 2 |
| Ca | 0.3 | 40 | 0.08 | 4 | 0.4 |
| Cd | 0.001 | 1 | 8 | 0.7 | 0.1 |
| Ce | — | — | 500 | — | — |
| Co | 2 | 15 | 200 | 11 | 2 |
| Cr | 0.3 | 1.5 | 1 | 2 | 0.6 |
| Cu | 0.3 | 1.5 | 1 | 2 | 0.2 |
| Eu | — | — | 20 | — | — |
| Fe | 0.6 | 10 | 30 | 6 | 0.3 |
| Ga | 7 | 140 | 0.9 | — | — |
| Gd | — | — | 800 | — | — |
| Ge | 140 | 1800 | — | — | 400 |
| Hg | 0.2[0.002] | — | — | — | 5 |
| In | 10 | 25 | 0.2 | — | 20 |
| Li | — | — | 0.5 | — | 0.4 |
| Mg | 0.1 | 0.1 | 0.2 | 0.09 | 0.3 |
| Mn | 0.5 | 2 | 0.4 | 2 | 1 |
| Mo | 200 | 100 | 12 | — | 8 |
| Na | 100,000 | 8 | <0.1 | — | 0.3 |
| Nb | — | — | 1,500 | — | — |
| Ni | 1 | 25 | 2 | 50 | 0.4 |
| Os | — | — | 150,000 | — | — |
| Pb | 10 | 10 | 13 | — | 5 |
| Pr | — | — | 1,000 | — | — |
| Pt | 150 | 700 | — | — | 0.3 |
| Rh | 3,000 | — | 100 | — | — |
| Ru | 5,000 | — | 500 | — | 250 |
| Sb | [0.1] | — | 50 | — | — |
| Sc | — | — | 10 | — | 9 |
| Se | 36[0.06] | — | — | — | 40 |
| Si | 600 | — | — | — | 200 |
| Sn | 30 | 20 | — | — | 2 |
| Sr | 30 | 0.9 | 0.3 | — | 4 |
| Te | 5[0.08] | 790 | — | — | 25 |
| Ti | — | 200 | 2 | — | 30 |
| Tl | 0.6 | 6 | 4 | — | 25 |
| V | 70 | 30 | 30 | — | 25 |
| Zn | 0.02 | 0.5 | — | 0.5 | 0.1 |

[a]Conventional line excitation source, continuum excitation source, or laser excitation source with flame atomization; DLs in brackets with hydride or cold vapor generation.
[b]ICP as excitation source with flame atomization.
[c]Pulsed HCL excitation source with ICP atomization (Baird system).

## Applications

*Analytical.* In research laboratories, AFS has been used to analyze a wide variety of sample types. It should be useful for any of the types of samples for which AAS or AES has been employed. Until recently, the routine use of AFS has been hampered by the lack of commercial instrumentation.

*Localized Diagnostics.* The characteristics of AFS make it an excellent tool for obtaining fundamental information about conditions in atomizers. In contrast to AE and AA, one can directly probe a specific volume element in an atomizer defined by the intersection of the excitation beam and the optical path of detection. This allows three-dimensional profiles of characteristics to be obtained. Important information obtained with AF includes temperatures, absolute number densities of electrons, ions, atoms, and flame species, quantum yields, and nonradiative rate constants. In many cases, laser-saturated fluorescence has been particularly useful for obtaining diagnostic information. For example, equation 11-24 shows that atomic number densities can be obtained by measuring the absolute fluorescence radiant power under saturated conditions if accurate values of $V^*$ and $A_{10}$ are known. Note that the measurement is independent of the source radiance and quantum yield.

*Resonance Detector.* A resonance detector (resonance monochromator) is based on atomic fluorescence and used to isolate atomic lines in AAS. As shown in Figure 11-9, all the radiation from the source passes through the atomizer into the resonance detector. The resonance detector is some type of container which produces and maintains an atomic population of the element to be determined. The fluorescence excited is collected and directed through a filter to a PMT. The fluorescence signal detected is directly proportional to the intensity of the excitation beam.

The atomic cloud for resonance detection has been produced in several ways; in early designs, a heating element was placed in a block of a volatile metal in an evacuated chamber. Cathodic sputtering as used in HCLs is applicable to most elements. More recently, flames or plasmas have been used as convenient resonance detectors. Here the resonance detector is "tuned" to a specific element by aspiration of that element into the atomizer.

Resonance detectors provide several potential advantages: response only to the change in intensity of the incident analyte resonance line, discrimination against nonabsorbed lines from the excitation source which do not stimulate fluorescence, a narrow bandpass limited by the width of the absorption line (typically 0.003 Å), no mechanical selection of the wavelength, and no drift

in the wavelength selection. It is also possible to modulate the detection signal by modulating the atomic vapor. Two or more resonance detectors can be put in tandem for multielement measurements.

Resonance detection has not been employed commercially for several reasons. Although the efficiency of transmission into the resonance detector is high, only a small portion of the incident radiant power is converted to fluorescence and collected for viewing. Therefore, the measured signal can be much smaller than observed by wavelength selection with a monochromator. This increases the relative signal shot noise and can degrade detection limits. Noise added by fluctuations in the atomic populations in the detector chamber can also degrade precision. Resonance detection is most efficient for elements with simple spectra.

A variation of resonance devices not based on fluorescence detection involves the use of atomic vapor cells as selective modulators. For example, the emission radiation from an ICP can be directed through an atomic vapor cell, containing the element of interest, before it is imaged on the monochromator or polychromator entrance slit. By pulsing the atomic vapor on and off, only the analyte emission radiation is modulated and detected by a synchronous detection system. Emission by other species at wavelengths within the spectral bandpass but not overlapping the modulator analyte absorption profile is not affected and discriminated against. A similar approach can be used with atomic absorption spectrophotometry.

## Overview

The future of AFS is difficult to predict. The acceptance of a new technique for routine use requires that it offer significant advantages over established techniques and that it be commercialized and promoted by an instrument manufacturer. Certainly, the acceptance of AFS

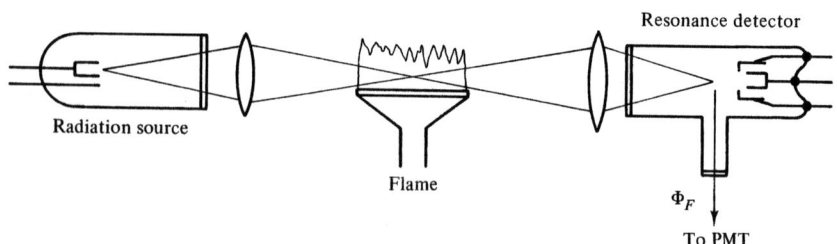

**FIGURE 11-9** Diagram of a resonance detector used for detection of atomic absorption signals. Here the resonance detector is a chamber in which a cloud of atomic vapor is produced by sputtering. The intensity of the source radiation transmitted through the atomizer is directly proportional to the AF signal excited within the resonance detector and viewed by the PMT.

has been hampered by the lack of support by instrument companies until 1982.

The most often cited advantages of AFS include low DLs, large linear dynamic ranges, multielement capabilities, simplicity, and freedom from spectral interferences. Overall DLs are not generally better and often worse than those reported for flame AA, electrothermal AA, or plasma AE except for a few elements. It is recognized in terms of minimizing analyte and scattering interferences, $C_2H_2$ flames or plasmas are generally the best atomization sources for AFS as they are in AAS and AES. Detection limits with these atomizers are often much higher than the best reported in cooler flames.

In terms of dynamic range and multielement capability, atomic fluorescence spectrometry is superior to AAS, but comparable to plasma AES. Nondispersive AF systems based on conventional excitation sources can be relatively simple and inexpensive. However, AF systems based on monochromators or laser excitation can be at least as expensive and complex as many AAS or AES systems. Spectral interference problems are generally less severe for AFS than for AES, but comparable to those experienced with flame AAS.

## PROBLEMS

**11-1.** The following components are available for flame atomic spectroscopy experiments.

> Flame (nebulizer/burner system), F
> Lenses (various focal lengths), L
> Hollow cathode lamp (high intensity), HCL
> Chopper (1-kHz mechanical chopper), C
> Monochromator with photomultiplier tube attachment, M/PMT
> Signal processing (ac) and readout system, SP

Draw block diagrams of the setups you would put together for **(a)** flame atomic emission, **(b)** flame atomic absorption, and **(c)** flame atomic fluorescence. Indicate carefully in your diagrams the positions of the various components and the geometry of the source/flame/monochromator arrangement.

**11-2.** Critically compare flame atomic absorption, atomic fluorescence, and atomic emission as to detection limits for elements of various excitation energies, range of linearity of calibration curves, susceptibility to spectral, chemical, and physical interferences, and any other features.

**11-3.** Consider the flame spectrometric techniques of atomic emission (AE), atomic absorption (AA), and atomic fluorescence (AF). The possible sources of noise in these techniques are flicker noise in the radiation source, flame transmission flicker noise, flicker noise in the flame background emission, analyte flicker noise, shot noise in the PMT photoanodic current, Johnson noise in the load (feedback) resistor, amplifier noise, dark current noise, and readout resolution uncertainty. For each of the three techniques, one source of noise tends to predominate near the limit of detection ($S/N = 2$). Give the most likely predominant noise source(s) for AE, for AA, and for AF near the detection limit (note that these are different for the three techniques) and justify why the other sources can be neglected.

**11-4.** Compare the multielement capabilities of flame AFS with that of AAS and AES. If it is desired to construct a single spectrometer capable of simultaneous or near-simultaneous determinations of most of the elements using two of the three flame techniques, which two would you choose? Explain and justify. If you were not limited to flames but could also pick plasma atomizers, which two techniques would you choose, and why?

**11-5.** Discuss the potential interferences in AFS that are not present in AAS and AES. Why is AFS considered more susceptible to environmental effects (matrix and flame conditions) than the other techniques?

**11-6.** Indicate the differences between the following types of atomic fluorescence: direct-line and stepwise-line, Stokes and anti-Stokes, normal and excited-state.

**11-7.** For a particular atom, only two levels need to be considered for the resonance transition. The Einstein coefficient ($A_{10}$) for that atom is known to be $1.3 \times 10^8 \, s^{-1}$, and the quantum efficiency in a certain flame is 0.09. What is the observed lifetime of the upper level?

**11-8.** Why is an excitation monochromator not normally employed in AF measurements?

**11-9.** Filters are often used as wavelength selectors for AF measurements. However, monochromators are always used for AA measurements and usually used for AE measurements. Why?

**11-10.** Contrast the reasons for nonlinearity of calibration curves at high analyte concentrations for AA, AE, and AF.

**11-11.** Discuss which of the three flame spectrometric techniques (AE, AA, or AF) would be most suitable for determination absolute concentrations in flames. Give your reasoning.

**11-12.** Discuss the advantages and disadvantages of saturated AF measurements.

**11-13.** For the Na $^2P_{1/2} \rightarrow {}^2S_{1/2}$ transition at 589.6 nm in a $H_2-O_2$ flame diluted with Ar, the fluorescence power yield $Y = 0.63$ under a certain set of conditions. If the Einstein coefficient $A_{10} = 1.0 \times 10^8 \, s^{-1}$, find the saturation spectral irradiance. If the viewed volume is 5.0 mm$^3$ and the total atomic population is $5.0 \times 10^{10}$ atoms cm$^{-3}$, calculate the maximum fluorescence radiant power possible under complete saturation.

**11-14.** At low analyte concentrations, the $S/N$ of an AF measurement is independent of the incident radiant power. What noise sources could be limiting the $S/N$? How would your answer differ if the above were true at much higher analyte concentrations?

**11-15.** Discuss two methods (or more if you can think of them) to measure the fluorescence quantum yield $\phi$ of a resonance transition in a particular flame. Compare the two methods as to potential accuracy and ease of application.

# REFERENCES

1. J. D. Winefordner, S. G. Schulman, and T. C. O'Haver, *Luminescence Spectrometry in Analytical Chemistry,* Wiley-Interscience, New York, 1972.

2. G. F. Kirkbright and M. Sargent, *Atomic Absorption and Fluorescence Spectroscopy,* Academic Press, London, 1974.

3. V. Sychra, V. Svoboda, and I. Rubeska, *Atomic Fluorescence Spectroscopy,* Van Nostrand Reinhold, Wokingham, Berkshire, England, 1975.

4. J. D. Winefordner, "Atomic Fluorescence Spectrometry: Past, Present, and Future," in *Recent Advances in Analytical Spectroscopy,* Pergamon Press, Oxford, 1982, pp. 151–164.

5. A. H. Ullman, "Multielement Atomic Fluorescence Spectrometry," *Prog. Anal. At. Spectrosc., 3,* 187 (1980).

6. N. Omenetto and J. D. Winefordner, "Atomic Fluorescence Spectrometry: Basic Principles and Applications," *Prog. Anal. At. Spectrosc., 2,* 1 (1979).

7. J. C. Van Loon, "Atomic Fluorescence Spectroscopy—Present Status and Future Prospects," *Anal. Chem., 53,* 333A (1981).

8. J. D. Winefordner, "Principles, Methodologies, and Applications of Atomic Fluorescence Spectroscopy," *J. Chem. Educ., 55,* 72–78 (1978).

9. T. S. West, "Atomic-Fluorescence and Atomic-Absorption Spectrometry for Chemical Analysis," *Analyst, 99,* 886 (1974).

10. R. F. Browner, "Atomic-Fluorescence Spectrometry as an Analytical Technique," *Analyst, 99,* 617 (1974).

11. C. Th. J. Alkemade, Tj. Hollander, W. Snelleman, and P. J. Th. Zeegers, *Metal Vapors in Flames,* Pergamon Press, Oxford, 1982.

12. T. J. Vickers, "Atomic Fluorescence and Atomic Absorption Spectroscopy," in *Physical Methods in Modern Chemical Analysis,* T. Kuwana, ed., vol. 1, Academic Press, New York, 1978, p. 192.

13. C. Th. J. Alkemade and R. Herrmann, *Fundamentals of Analytical Flame Spectroscopy,* Wiley, New York, 1979.

14. R. Smith, "Flame Fluorescence Spectrometry," chap. 4 in *Spectrochemical Methods of Analysis,* J. D. Winefordner, ed., Wiley-Interscience, New York, 1971.

15. N. W. Bower and J. D. Ingle, Jr., "Experimental and Theoretical Comparison of the Precision of Atomic Absorption, Fluorescence, and Emission Measurements," *Appl. Spectrosc., 35,* 317 (1981).

# CHAPTER 12

# Introduction to Molecular Spectroscopy

Molecular spectroscopy differs from atomic spectroscopy in many aspects. With molecules, the bonding of atoms gives rise to rotational and vibrational energy levels. Thus rotational and vibrational transitions occur in addition to electronic transitions. These added transitions expand the wavelength range of interest from the UV through the microwave region. The greater number of excited states increases the complexity of spectra. Mixtures of molecules are more difficult to determine by electronic spectral characteristics because of the breadth of molecular electronic bands compared to atomic lines. The large number of bands or lines in vibrational or rotational spectra also make determination of species in mixtures more difficult.

In many respects the variety possible in molecular spectroscopy is greater because of the numerous different kinds of molecules compared to the approximately 100 different types of atoms. Also, molecular spectroscopy can be conducted in the gas, liquid, or solid phase, with spectral characteristics changing significantly with phase. The utilization of chemical reactions to enhance selectivity plays a much larger role in molecular spectroscopy than in atomic spectroscopy.

We can calculate natural, collisional, and Doppler line widths for molecules in the gas phase as described in Section 7-3 for atomic electronic transitions. How-

ever, this information is generally useful only for describing line widths in pure rotational spectra in the gas phase; collisional broadening is usually dominant. Most molecular vibrational and electronic spectroscopy in analytical chemistry is applied to solutions where molecular interactions cause rotational detail to be lost. The width of vibrational bands is determined by the large range of rotational transitions possible between two vibrational states. Similarly, the breadth of electronic absorption bands is determined by the multitude of vibrational transitions possible between two electronic states. In solution the vibrational detail is usually either considerably blurred or not even observed.

In this chapter the origins of rotational and vibrational spectra are first reviewed in terms of the energy levels and transitions possible. Next, factors that determine the characteristics of electronic absorption spectra of diatomic molecules are presented. This includes a discussion of molecular orbital theory, molecular quantum numbers, and term symbols, selection rules, and band shapes. This discussion of electronic absorption spectra is then extended to polyatomic molecules with an emphasis on the symbolism used to denote electronic states and on the types of spectra observed for different classes of molecules. Section 12-6 reviews the origins and characteristics of luminescence spectra. The various deactivation processes for electronically excited molecules, including quenching, fluorescence and phosphorescence, luminescence quantum efficiencies and lifetimes, and polarization of luminescence are now discussed. The material in this chapter lays the foundation for later discussions of specific molecular spectrochemical techniques such as UV-visible molecular absorption spectrophotometry (Chapter 13), infrared spectrophotometry (Chapter 14), molecular luminescence spectrometry (Chapter 15), and molecular scattering methods (Chapter 16).

## 12-1 MOLECULAR SPECTRA

There are three basic types of optical spectra that we can observe for molecules: (1) electronic or vibronic spectra, which involve transitions between a specific vibrational and rotational level of one electronic state and a vibrational and rotational level of another electronic state; (2) vibrational or vibrational-rotational spectra, which involve transitions from the rotational levels of one vibrational level to the rotational levels of another vibrational level in the same electronic state; and (3) rotational spectra, where the transitions are between rotational levels of the same vibrational level of the same electronic state. The vibrational and rotational levels of two electronic states A and B of a molecule are illustrated in Figure 1-3. Purely rotational spectra are normally observed in the microwave region of the spectrum, vibration-rotation spectra in the infrared region, and electronic spectra in the ultraviolet, visible, and near-IR wavelength regions.

As discussed in Appendix F, quantum mechanics through the Schrödinger equation can help us predict wavefunctions, energy levels, selection rules, and the strengths of transitions (e.g., transition probabilities). Exact analytical solutions to the Schrödinger equation are not possible except with a few simple model systems (e.g., harmonic oscillator, rigid rotor). However, these simple model systems provide a means to understand complex real molecules. We will be concerned here only with the results of these quantum mechanical derivations.

As a first approximation (the *Born–Oppenheimer approximation*) the wavefunction for a molecule is assumed to be the product of three independent wavefunctions describing the electronic, vibrational, and rotational energy levels. In this case the energy levels associated with each type of transition can be initially treated separately, and the total energy $E_t$ of a molecule in a given state is approximately equal to the sum of its electronic energy ($E_e$), its vibrational energy ($E_v$), and its rotational energy ($E_r$ or $E_J$).

## 12-2 ROTATIONAL SPECTRA

To describe rotational spectra of linear molecules, the *rigid rotor model* is used as a first approximation. Here a dumbbell model is a system consisting of two mass points $m_1$ and $m_2$ connected by a massless rod of length $r$. If the Schrödinger equation is solved for a rotating linear system, the total angular momentum is found to be quantized in units of $h/2\pi$, and the energies of the rotational levels are given by equation 12-2. (Refer to Table 12-1 for all equations in this discussion.) Absorption lines are observed in the microwave region of the electromagnetic spectrum. Rotational transitions occur only for molecules with a permanent dipole moment (i.e., not homonuclear diatomics). For linear molecules, the selection rule is $\Delta J = \pm 1$, where $J$ is the rotational angular momentum quantum number. The allowed rotational transitions produce a series of equally spaced lines of wavenumber equal to $\bar{\nu} = 2\bar{B}(J + 1)$, where $\bar{B}$ is the rotational constant. Note that the spacing between energy levels increases with increasing $J$.

The degeneracy of rotational energy levels is $(2J + 1)$. For small $J$, $E_J \ll kT$. Thus at room temperature, the Boltzmann distribution (equation 2-14) predicts that typically 30 to 50 rotational levels are appreciably populated. The relative intensity of the rotational lines depends on the population and degeneracy of each rotational level.

In the simple rigid rotor model, it is assumed that molecular dimensions are independent of molecular vibrations and undisturbed by molecular rotation. If the centrifugal stretching of bond lengths is considered, the moment of inertia $I$ is dependent on $J$. For linear mol-

**TABLE 12-1**

Equations for rotational spectroscopy[a]

| Rigid rotor model | |
|---|---|
| Angular momentum: $P = I\omega = \dfrac{J(J + 1)h}{2\pi}$ | (12-1) |
| Energy levels: $E_J = \bar{B}J(J + 1)$ in cm$^{-1}$<br>$\quad\quad\quad\quad\quad E_J = BJ(J + 1)$ in J | (12-2) |
| Rotational constant: $\bar{B} = \dfrac{h}{8\pi^2 Ic}$ in cm$^{-1}$ | |
| $B = \dfrac{h^2}{8\pi^2 I}$ in J | |
| Moment of inertia: $I = \Sigma m_i r_i^2$ | |

| Nonrigid rotor model | |
|---|---|
| Energy levels: $E_J = \bar{B}J(J + 1) - \bar{D}J^2(J + 1)^2$ | (12-3) |

[a] $m_i$, Mass of $i$th atom (g); $r_i$, distance of $i$th atom from center of gravity (cm); $\omega$, angular velocity of rotation (rad s$^{-1}$); $J$, angular momentum quantum number, 0, 1, 2, . . .; $\bar{D}$, centrifugal distortion constant (cm$^{-1}$); typically, $\bar{D} < 10^{-4}\bar{B}$ and $\bar{B} = 0.1 - 10$ cm$^{-1}$.

ecules, this *nonrigid rotor model* results in the subtraction of a small term as shown in equation 12-3. Here the spacing between successive lines is seen to decrease with increasing $J$.

Application of an electric field splits and shifts the degenerate rotational levels (the Stark effect). In this case the angular momentum quantized along the direction of the applied field must be considered and additional selection rules are applicable.

Hyperfine structure in rotational levels is caused by the nuclear spin interacting with the angular momentum. Here the selection rules must take into account both $J$ and nuclear spin quantum numbers.

For tetratomic molecules that are symmetric tops, exact rotational energy levels can be calculated. Energy-level expressions are more complex and must consider the angular momentum contributions along different axes. Selection rules must be broadened to include $\Delta J = 0$ in addition to $\Delta J = \pm 1$. Also, additional selection rules exist for the angular momentum along a unique axis. Molecules of analytical interest are usually more complex. However, the equations derived for simple molecules provide a basis to describe the rotational transitions observed.

In the gas phase at pressures above 1 torr, collisional broadening determines the line shape and half-width of rotational lines. The line shape is Lorentzian with typical half-widths of $10^{-3}$, $10^{-4}$, and $10^{-5}$ cm$^{-1}$ at 0.1, 0.01, and 0.001 atm, respectively. For gases at higher pressures, the rotational energies are blurred due

to more frequent collisions. In condensed phases, rotational structure is not observed. In liquids, rotational energies become effectively nonquantized due to molecular collisions more frequent ($10^{12}$ to $10^{13}$ s$^{-1}$) than the period of rotation ($10^{-10}$ s). In solids rotations are totally restricted. Hence microwave rotational spectroscopy is little used for analytical purposes.

## 12-3  VIBRATIONAL SPECTRA

### Pure Vibrational Transitions

Molecular vibrations of diatomic molecules may be described, as a first approximation, by the **harmonic oscillator model**. A simple harmonic oscillator is a mechanical system consisting of a point mass $m$ connected to a massless spring. The mass is under action of a restoring force proportional to the displacement of the particle from its equilibrium position and the force constant $k$ of the spring. The potential energy $V$ is proportional to the square of the displacement. The frequency of vibration $\nu$ is given by $\nu = (k/m)^{1/2}/(2\pi)$.

Likewise for a diatomic molecule, $V$ is a function of the displacement $q$ of the internuclear distance $r$ of the atoms from their equilibrium position $r_e$ (or of a particle of reduced mass $\mu$ from the center of gravity). This potential energy function is parabolic in shape as shown in equation 12-4. (Refer to Table 12-2 for all equations in this discussion.) Use of this potential en-

**TABLE 12-2**

Equations for vibrational spectroscopy[a]

| | Harmonic oscillator model | |
|---|---|---|
| Potential energy: | $V = \frac{1}{2}kq^2$ | (12-4) |
| Energy levels: | $E_v = \bar{\nu}(v + \frac{1}{2})$   in cm$^{-1}$   for $v = 0, 1, 2, \ldots$ | (12-5) |
| | $E_v = h\nu(v + \frac{1}{2})$   in J (ergs) | |
| Wavenumber or frequency: | $\bar{\nu} = \dfrac{\nu}{c} = \dfrac{(k/\mu)^{1/2}}{2\pi c}$; typically, $\bar{\nu} = 200$ to 3500 cm$^{-1}$ | |
| | $\nu = \dfrac{(k/\mu)^{1/2}}{2\pi}$ | (12-6) |
| Reduced mass: | $\mu = \dfrac{m_1 m_2}{m_1 + m_2}$ | (12-7) |

| | Anharmonic oscillator model | |
|---|---|---|
| Morse potential energy: | $V = \bar{D}_e|1 - e^{-\beta q}|^2$ | (12-8) |
| Energy levels: | $E_v = \bar{\nu}(v + \frac{1}{2}) - \bar{\nu}\chi_e(v + \frac{1}{2})^2$   for $v = 0, 1, 2, \ldots$ | (12-9) |

| | Rotational-vibration transitions | |
|---|---|---|
| $E_{vJ} = \bar{\nu}(v + \frac{1}{2}) + \bar{B}_v J(J + 1)$   for $v = 0, 1, 2, \ldots$ and $J = 0, 1, 2, \ldots$ | | (12-10) |

[a] $k$, Force constant (N m$^{-1}$)(typically, 200 to 2000); $q = r - r_e$, change in internuclear distance from its equilibrium value (cm); $r$, internuclear distance (cm); $r_e$, equilibrium internuclear distance (cm); $v$, vibrational quantum number, 0, 1, 2, . . .; $\beta$, molecular constant (cm$^{-1}$); $\chi_e$, anharmonicity constant (dimensionless): typically, 0.002 to 0.02; $\bar{D}_e$, dissociation energy (cm$^{-1}$); $\bar{B}_v$, rotational constant dependent on $v$ (cm$^{-1}$).

ergy function in the Schrödinger equation results in the quantized energy levels described by equation 12-5. Vibrational absorption transitions are observed in the mid- or far-infrared portion of the spectrum.

The actual potential energy of vibrations fits the parabolic function fairly well only near the equilibrium internuclear distance. The Morse potential function shown in equation 12-8 more closely resembles the potential energy of vibrations in a molecule for all internuclear distances. The shape of this function and the observed vibrational levels for a diatomic are shown in Figure 12-1. If the Morse potential function is used in the Schrödinger equation, the solution is more complex. Equation 12-9 is an approximate solution for the energy levels that result from this **anharmonic oscillator model**. The second term in equation 12-9 leads to the observed convergence of energy levels at large values of $v$ near the dissociation limit of the molecule. The harmonic oscillator model predicts that all energy levels are equally spaced by $\bar{v}$ and does not allow for bond rupture.

The selection rule for transitions between vibrational levels is $\Delta v = 1$. In addition, a change in dipole moment must occur so that homonuclear diatomic molecules exhibit no vibrational transitions in IR absorption spectroscopy. Weaker transitions called *overtones* are sometimes observed. These correspond to $\Delta v = 2$ or 3, and their frequencies are less than two or three times the fundamental frequency ($\Delta v = 1$) because of anharmonicity.

Typical energy spacings for vibrational levels are

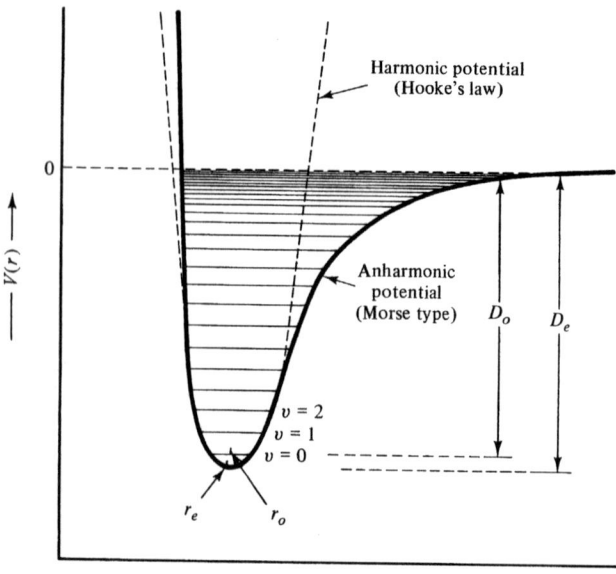

**FIGURE 12-1**   The harmonic (Hooke's law) and anharmonic (Morse-type) potentials for a diatomic molecule. Here $D_e$ is the dissociation energy relative to the minimum of the potential curve and $D_0$ is the measured dissociation energy relative to the zero-point vibrational energy.

on the order of $10^{-20}$ J. From the Boltzmann distribution, it can be shown that at room temperature typically 1% or less of the molecules are in excited states in the absence of external radiation. Thus most absorption transitions observed at room temperature are from the $v = 0$ level to the $v = 1$ level.

### Rotation-Vibrational Transitions

When vibrational spectra of gaseous diatomic molecules are observed under high-resolution conditions, each band is found to contain a large number of closely spaced components. Because of this, molecular spectra are often referred to as *band spectra*. The structure observed can be explained in terms of excitation of rotational motion during a vibrational transition. The form of such a vibration-rotation spectrum can be predicted from the energy levels of a vibrating-rotating molecule. In the case of simple diatomic molecules with no anharmonicity or centrifugal corrections, equation 12-10 gives the energies of the possible transitions in wavenumbers. Additional terms involving $(v + \frac{1}{2})^2$ and $[J(J + 1)]^2$ can be added to equation 12-10 to fit the anharmonic oscillator and nonrigid rotor models. These correction terms are less important than the rotational-vibrational coupling term. Redefining $\bar{B}$ as $\bar{B}_v$ signifies the dependence of the rotational constant on $v$. This occurs because the internuclear distance $r$ varies with the vibrational level.

A vibrational absorption transition from $v$ to $v + 1$ gives rise to three sets of lines called branches, which are shown in Figure 12-2. The lower-frequency $P$ branch corresponds to transitions with $\Delta v = 1$ and $\Delta J = -1$. The higher-frequency $R$ branch corresponds to transitions with $\Delta v = 1$ and $\Delta J = +1$, while the $Q$ branch arises from a transition with $\Delta v = 1$ and $\Delta J = 0$. The $Q$ branch is not observed for diatomics except those with an odd number of electrons (e.g., NO). The relative intensities of the components of the $R$ and $P$ bands in absorption spectra are governed by the population and the degeneracies of the various rotational levels in the ground vibrational state.

Analysis of vibration-rotation spectra can give force constants for bonds and also rotational constants. In condensed phases, the rotational structure is blurred by frequent molecular collisions and IR vibrational bands are broad with no rotational structure. Under these conditions, typical half-widths are 5 to 20 cm$^{-1}$. These bands can be further broadened by intermolecular interactions such as H bonding.

Polyatomic molecules give rise to much more complex vibrational motions (e.g., stretching, bending) than diatomics. For linear polyatomics with $N$ atoms, there are $3N - 5$ normal modes of vibrations, while nonlinear polyatomics have $3N - 6$ normal modes. Each mode

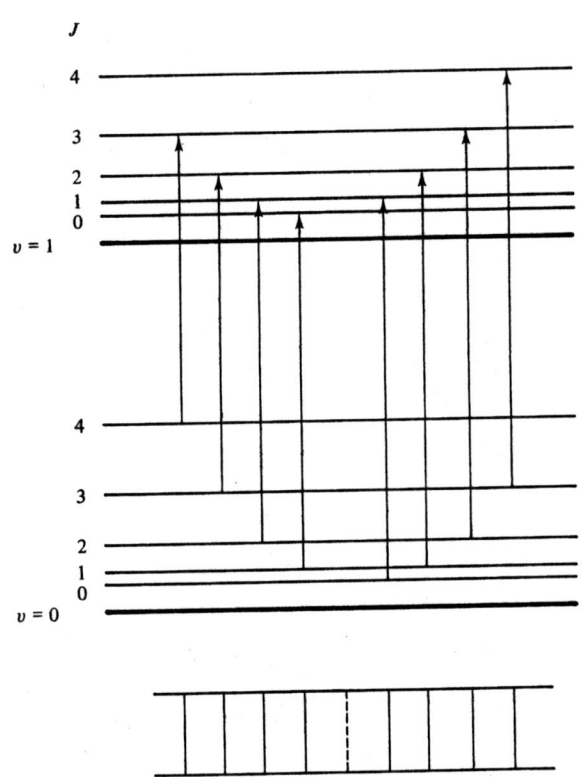

**FIGURE 12-2** Vibrational-rotational energy level diagram for a diatomic molecule. For the lines in the $R$ branch, $\Delta J = 1$, while for the $P$ branch, $\Delta J = -1$. The $Q$ branch is not usually observed but corresponds to $\Delta J = 0$. In condensed phases, the individual rotational transitions are not observed and the IR band for the $v = 0$ to $v = 1$ transition is the envelope of the probable rotational transitions.

can be considered as an independent harmonic oscillator (neglecting anharmonicities). Often the predicted $3N - 6$ (or $3N - 5$) bands are not observed in absorption vibrational spectra because some transitions are forbidden or two or more modes may be degenerate and thus have the same vibrational frequency. Additional vibrational bands can occur due to overtones.

Vibrational spectra are often characteristic of various functional groups in a molecule and frequently used for qualitative analysis, as discussed in Chapter 14. As for diatomics, rotational structure is not observed in the condensed phases.

## 12-4 ELECTRONIC ABSORPTION SPECTRA OF DIATOMIC MOLECULES

Our discussion of electronic spectra begins with a review of molecular quantum numbers and electronic states for diatomic molecules. Here we assume a knowledge of atomic quantum numbers (e.g., $n$, $l$, $m_l$, $m_s$, and $s$), which are reviewed in Section 7-3.

### Electronic States

*Molecular Orbitals and Quantum Numbers.* As a first approximation we will assume that molecular orbitals (MOs) are composed of linear combinations of the atomic orbitals (AOs) of the separate atoms making up the molecule. In a diatomic molecule, we redefine the atomic electronic angular momentum to be along the molecular axis or bond where now the $l$ components are not degenerate but represent different energies. For electrons of given $n$ and $l$ quantum numbers, but different $m_l$ values, the quantum number for the axial component of the orbital angular momentum $\lambda$ is employed. The quantum number $\lambda$ can take on values of $|m_l| = \{l, l - 1, \ldots, 0\}$. The MOs with $\lambda = 0, 1,$ and $2$ are called $\sigma$, $\pi$, and $\delta$ MOs, respectively. The $\sigma$ MOs are not degenerate, while the $\pi$ and $\delta$ MOs are degenerate, because $m_l$ can be positive or negative for $l \geq 1$.

The MOs formed by combinations of AOs in diatomics are shown in Figure 12-3. Note that A and B are used as subscripts on the AO symbols to indicate the nucleus associated with the particular orbital. The symbols used to represent the MOs depend on the situation and are discussed below.

First, we will consider homonuclear diatomics, in which case the AOs that combine are equivalent and have the same energy. Here the A and B subscripts can be dropped. The MOs are symbolized here as $nl\lambda$, but in other texts $\lambda nl$ may be used. The appropriate symbol for $\lambda$ in the MO designation is subscripted with $g$ or $u$. The subscripts stand for the German words *gerade* ($g$) and *ungerade* ($u$) and indicate that the MO is either symmetric (even) or asymmetric (odd), respectively, relative to inversion through the molecular center.

Each two AOs that combine form two MOs designated bonding and antibonding; the latter are denoted with the superscript *. For example, two $1s$ AOs with $m_l = 0$ form the MOs that originate from the linear combinations $1s + 1s$ and $1s - 1s$. Since $\lambda = 0$ for both AOs, these are $\sigma$ MOs designated $1s\sigma_g$ and $1s\sigma_u$. For $2p$ AOs, $\lambda = 0$ or $1$, so that $2p\sigma$ and $2p\pi$ MOs are formed.

For heteronuclear diatomics, three different types of MO symbols are shown in Figure 12-3. The first type of symbolism is similar to that used for homonuclear diatomics (i.e., $nl\lambda$). Here the $s$ or $p$ is subscripted with A or B to indicate that the MO has primarily the character of one of the two nonequivalent AOs. The latter types of symbols use letters ($z$, $y$, $x$, . . .) or numbers ($1$, $2$, $3$, . . .) to indicate the energetic order of the $\sigma$ and $\pi$ MOs with no direct reference to the AOs. The letter notation developed by Mulliken starts with the

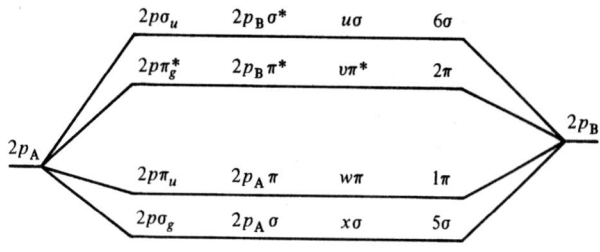

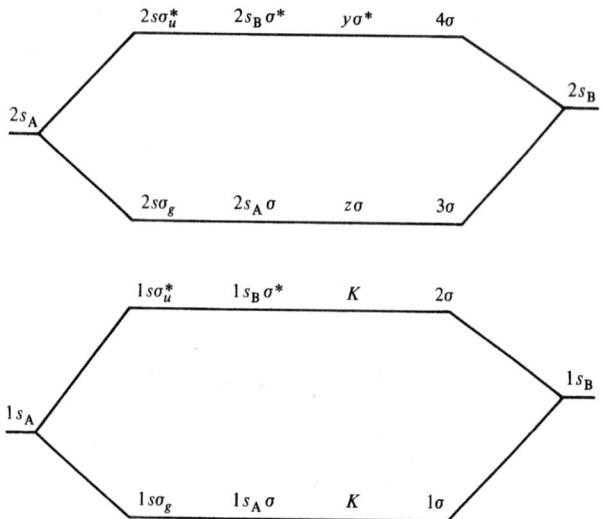

**FIGURE 12-3** Formation of molecular orbitals from atomic orbitals. Often the $2p\pi$ MO is lower in energy than the $2p\sigma$ orbital.

$2s\sigma$ orbital, where $K$ is used to represent the filled $1s\sigma$ MOs. Mulliken's notation is normally used where the atomic number of the atoms in the molecule differ by only one or two units (e.g., NO, CO). The number notation is used for atoms with unlike nuclei such as hydrides (e.g., LiH), where the electronic energy is determined largely by the heavier atom. For such cases, the orbitals formed with inner-shell electrons are primarily AOs or nonbonding MOs and only one or two orbitals are true MOs responsible for bonding. In naming states of diatomics, it is also common to call the ground state the X state and to label the excited states A, B, C, and so on.

The ordering of the MOs according to energy is often different than shown in Figure 12-3. It depends on the atoms involved and the internuclear distance, and usually must be determined experimentally. It is especially different for diatomics with shorter bonds such as $H_2$ and hydrides or heteronuclear diatomics containing atoms from different periods.

In a similar way to atoms, the ground-state electron configuration of a molecule is determined by successively filling the MOs from lower to higher energy

with electrons. The total number of electrons involved is the sum of the number of electrons from each atom. The bonding MOs involve attraction and the overlap of the two AOs, while the antibonding orbitals result in repulsion and a nodal plane corresponding to zero electron density between the atoms. A filled bonding MO increases stability of the molecular state, whereas a filled antibonding MO contributes instability. The **bond order** is one-half the difference between the number of bonding and antibonding electrons. If the bond order is $\frac{1}{2}$ or greater (or if the number of bonding electrons is greater than the number of antibonding electrons), the molecular state is stable. If the bond order is 0 or less, instability is predicted.

*Term Symbols.* For a given electronic configuration of a molecule, there are often several electronic states possible as described by term symbols. The designation requires that a few molecular quantum numbers be defined. In a molecule, the angular momentum of each AO can be combined to form various MO angular momentum states characterized by the **total axial electronic angular quantum number** $\Lambda$. Here $\Lambda = |\Sigma \lambda_i|$ and we symbolize $\Lambda = 0, 1, 2,$ and 3 by $\Sigma$, $\Pi$, $\Delta$, and $\Phi$, respectively. For two $\sigma$ orbitals ($\lambda = 0$), $\Lambda = 0$ and only $\Sigma$ states are possible. For two $\pi$ orbitals ($\lambda = 1$), $\Lambda = 0$ or 2 because the individual $\lambda$ may have the same or opposite signs (1 or $-1$). Thus both $\Sigma$ and $\Delta$ states are possible.

In atoms, the spins of all electrons couple to give a **resultant** or **total spin quantum number** $S$. In molecules, we can couple the spin of all electrons in both atoms comprising the molecule to yield the total spin quantum number $S$, where $S = \Sigma s_i$. In a manner analogous to the atomic term symbol, the term symbol designation of a molecular electronic state is given as $(^{2S+1}\Lambda)$, where $2S + 1$ is the multiplicity.

The quantum number for the axial component of the spin angular momentum is designated $\Sigma$ ($\Sigma = +S$, $S - 1$, to $-S$) and the multiplicity is $2\Sigma + 1$. The quantum number for the total axial angular momentum (due to the sum of the spin and orbital components) is $\Omega = |\Lambda + \Sigma|$, where $\Lambda$ and $\Sigma$ may have the same or opposite directions along the internuclear axis. Often, $\Omega$ is used as a subscript for the term symbol. Thus for the case in which $\Lambda = 1$ and $S = 1$, $\Sigma = 1, 0,$ or $-1$, $\Omega = 2, 0,$ or 1, and the possible states are $^3\Pi_2$, $^3\Pi_1$, and $^3\Pi_0$.

For $\Sigma$ states of homonuclear molecules, it is common to subscript the term symbol with $g$ or $u$ instead of $\Omega$ to indicate whether there is an even or odd number of electrons in the $u$ molecular orbitals. The $\Sigma$ states of homonuclear or heteronuclear molecules are further distinguished by being symmetric ($\Sigma^+$ state) with respect to a reflection in the plane of symmetry axis or antisymmetric ($\Sigma^-$ state) with respect to this reflection. ($\sigma$ orbitals give only $+$ states, while $\pi$ and $\delta$ orbitals can lead to either $+$ or $-$ states.)

The ground-state configuration of homonuclear diatomic molecules can be found in a manner similar to that for the configurations of many-electron atoms. With atoms, electrons are put into hydrogen-like AOs, paying attention to the Pauli principle and Hund's rules. For homonuclear diatomic molecules, the electrons are put into $H_2^+$-like MO's formed hypothetically from AOs. Again the Pauli principle must be followed, and Hund's rules tell us the lowest-energy state is that of the highest multiplicity.

For $H_2$, for example, both electrons go into the $1s\sigma_g$ orbital with opposite spins. The electron configuration for the ground state of $H_2$ is thus $(1s\sigma_g)^2$. Since $\Lambda = 0$ and $S = 0$, the ground-state term symbol is $^1\Sigma_g^+$. The bond order is 1. For $Li_2$ (6 electrons), the electron configuration is $(1s\sigma_g)^2(1s\sigma_u^*)^2(2s\sigma_g)^2$. Again $\Lambda = 0$, $S = 0$, and the ground state is $^1\Sigma_g^+$ and the bond order is 1. For $B_2$ (10 electrons) the ground-state configuration is $(1s\sigma_g)^2(1s\sigma_u^*)^2(2s\sigma_g)^2(2s\sigma_u^*)^2(2p\pi_u)^2$ or $KKLL (2p\pi_u)^2$. In the latter case, $KK$ and $LL$ indicate filled $1s$ and $2s$ bonding and antibonding MOs. For this configuration $\Lambda = 1 + 1 = 2$ or $\Lambda = 1 - 1 = 0$. For the $\Sigma$ states ($\Lambda = 0$), the electrons can be opposite spins or the same spins, giving rise to the two states $^1\Sigma_g^+$ and $^3\Sigma_g^-$. The $\Delta$ state ($\Lambda = 2$) is forced to be a singlet state by the Pauli principle. It is, in fact, a $^1\Delta_g$ state. The ground state, $^3\Sigma_g^-$, is the state of maximum multiplicity.

Electronic term symbols for excited-state electron configurations of homonuclear diatomics are determined in a similar manner. For example, consider excited $H_2$ in which one electron has been raised from the ground state to a higher bonding MO. For the $(1s\sigma_g)^1(2s\sigma_g)^1$ electronic configuration, $\Lambda = 0$ and $S = 0$ or 1, which yields term symbols $^1\Sigma_g^+$ and $^3\Sigma_g^+$ for the singlet and triplet states, respectively. For the $(1s\sigma_g)^1(2p\sigma_g)^1$ electronic configuration, the possible values of

$\Lambda$, $S$, and $\Omega$ are the same, but the overall state is an odd since an even $1s$ state and odd $2p$ state are combined. Thus the term symbols are $^1\Sigma_u^+$ and $^3\Sigma_u^+$. If the excited electron is promoted to the $2p\pi$ orbital, the electronic configuration is $(1s\sigma_g)^1(2p\pi_u)^1$. Now $\Lambda = 1$, and $S = 0$ or 1. The possible term symbols are $^1\Pi_g$ and $^3\Pi_u$.

The electronic states for heteronuclear diatomics are obtained in a similar fashion, although the symbolism and energy ordering of the MOs may be different. For diatomics with similar nuclei, the Mulliken symbolism is often used. For example, CO (14 electrons) has a ground-state electronic configuration of $KK(z\sigma)^2(y\sigma)^2(w\pi)^4(x\pi)^2$ with a $^1\Sigma^+$ ground electronic state. The $z\sigma$, $x\sigma$, and $w\pi$ MOs are bonding MOs and the $y\sigma$ MO is a nonbonding MO yielding a bond order of 3 (a triple bond).

If the atomic numbers of the atoms differ by more than two units, the electronic energy is determined largely by the heavier atom. The orbitals formed from inner-shell electrons are primarily atomic in character. Thus not all valence-shell electrons contribute significantly to the MOs. Usually, the numbered MO symbolism is employed.

For HF (10 electrons) the ground-state electron configuration is $(1\sigma)^2(2\sigma)^2(3\sigma)^2(1\pi)^4$, and the term symbol is $^1\Sigma^+$. Here the $1\sigma$ orbital contains atomic inner-shell electrons, and the $2\sigma$ and $1\pi$ orbitals are nonbonding AOs. Only the $3\sigma$ orbital is an effective MO resulting in the HF single bond.

The term symbols resulting from different combinations of equivalent electrons (i.e., same $n$ and $l$ quantum numbers) and nonequivalent electrons are summarized in Table 12-3.

### Electronic Transitions

In electronic absorption transitions, electrons are promoted to empty or partially filled MOs. These transitions between ground and excited electronic states are symbolized by specifying the term symbols of the elec-

**TABLE 12-3**
Electronic term symbols resulting from equivalent and nonequivalent electrons

| Equivalent electrons | Electronic terms | Nonequivalent electrons | Electronic terms |
|---|---|---|---|
| $\sigma^2$ | $^1\Sigma^+$ | $\sigma\sigma$ | $^1\Sigma^+, {}^3\Sigma^+$ |
| $\pi^2$ | $^1\Sigma^+, {}^3\Sigma^+, {}^1\Delta$ | $\sigma\pi$ | $^1\Pi, {}^3\Pi$ |
| $\delta^2$ | $^1\Sigma^+, {}^3\Sigma^+, {}^1\Gamma$ | $\sigma\delta$ | $^1\Delta, {}^3\Delta$ |
| $\pi^4$ | $^1\Sigma^+$ | $\pi\pi$ | $^1\Sigma^+, {}^3\Sigma^+, {}^1\Sigma^-, {}^1\Delta, {}^3\Delta$ |
| | | $\pi\delta$ | $^1\Pi, {}^3\Pi, {}^1\Phi, {}^3\Phi$ |

tronic states involved in the transitions. The probability and hence intensity of absorption (or emission) for a specific transition between two electronic states is related to the transition moment ($R$) as described in Appendix F. The Born–Oppenheimer approximation allows us to assume that the electronic wavefunctions are independent of nuclear motions (i.e., vibrations and rotations) because the nuclei are fixed during an electronic transition. Thus the total time-independent wavefunction ($\psi_t$) for a given electronic state can be written as

$$\psi_t = \psi_e\psi_s\psi_v\psi_r \qquad (12\text{-}11)$$

where the subscripts signify electronic, spin, vibrational, and rotational wavefunctions, respectively. With this approximation for the wavefunctions of the upper and lower electronic states involved in the transition and the assumption that the total first-order dipole moment ($\mu$) can be separated into nuclear ($\mu_n$) and electronic ($\mu_e$) dipole moments, the transition moment for a given electronic band from equation A6-2 becomes

$$R = R_e O_s O_v O_r \qquad (12\text{-}12)$$

where   $R_e = \int \psi_e'\mu_e\psi_e'' d\tau_e$ = electronic transition moment

$O_s = \int \psi_s'\psi_s'' \, d\tau_s$ = electronic spin overlap integral

$O_v = \int \psi_v'\psi_v'' \, d\tau_v$ = vibrational overlap integral

$O_r = \int \psi_r'\psi_r'' \, d\tau_r$ = rotational overlap integral

Here a prime and double prime signify wavefunctions of the upper and lower electronic states, respectively, and the superscript *, indicating the complex conjugate, for the upper-state wavefunction is not used for simplicity.

*Selection Rules.*   Selection rules for electronic transitions in diatomics are based on $R_e$ or $O_s$ being zero (forbidden transition) or finite (allowed transition) as a first approximation. This can be determined from the symmetry properties of the electronic states involved in a transition. If spin-orbit coupling is weak (light nuclei) the following selection rules apply:

$\Delta\Lambda = 0, \pm 1$

$\Delta S = 0$

$u \leftrightarrow g$   Homonuclear diatomics   (Laporte rule)

$\Sigma^+ \leftrightarrow \Sigma^+, \; \Sigma^- \leftrightarrow \Sigma^-$   (Parity rule)

Allowed transitions between various electronic angular momentum states of homonuclear and heteronuclear diatomics are summarized in Table 12-4. The electronic transition moment determines the intensity of the band as a whole. The electronic spin overlap integral vanishes unless $\psi_s' = \psi_s''$ or $\Delta S = 0$. Thus transitions are permitted only between states of the same multiplicity. In the case of strong spin-orbital interactions that are encountered for diatomics with heavier nuclei, the rules are replaced by $\Delta\Omega = 0, \pm 1$, except that $\Delta\Omega \neq 0$ if $\Omega'' = \Omega' = 0$.

*Band Intensities.*   The breadth and shape of an electronic absorption band is determined by the vibrational overlap integral. That is, $R$ is finite only where $O_v$ is finite. The square of the vibrational overlap integral is called the **Franck–Condon factor**. The Franck–Condon principle states that an electronic transition is rapid ($10^{-15}$ s) with respect to the nuclear motions so that a vibronic transition occurs to a vibrational state in the upper electronic state such that the positions and momenta of the nuclei are essentially the same as in the initial vibrational state of the lower electronic state. Thus the magnitude of the vibrational overlap is determined by the vertical coincidence of the maxima and minima of the vibrational wavefunctions of the two electronic states involved in the transition. If the probability of finding the nuclei at the same internuclear separation is the same for both states, the probability of a vibronic transition is large.

This principle is illustrated in Figure 12-4. The vibrational levels and wavefunctions in the potential wells for the ground and first excited states are shown for a typical diatomic in which the internuclear distance for the excited state ($r_e'$) is somewhat larger than the internuclear distance for the ground electronic state ($r_e''$). At room temperature, most molecules in the ground electronic state are in the ground vibrational level ($v'' = 0$). The most intense transitions occur from the $v'' = 0$ state to higher $v'$ states in the upper electronic level because the overlap of the vibrational wave-

**TABLE 12-4**

Allowed electronic transitions between various states

| Heteronuclear diatomics | Homonuclear diatomics |
|---|---|
| $\Sigma^+ \leftrightarrow \Sigma^+$ | $\Sigma_g^+ \leftrightarrow \Sigma_u^+$ |
| $\Sigma^- \leftrightarrow \Sigma^-$ | $\Sigma_g^- \leftrightarrow \Sigma_u^-$ |
| $\Pi \leftrightarrow \Sigma^+$ | $\Pi_g \leftrightarrow \Sigma_u^+, \Pi_u \leftrightarrow \Sigma_g^+$ |
| $\Pi \leftrightarrow \Sigma^-$ | $\Pi_g \leftrightarrow \Sigma_u^-, \Pi_u \leftrightarrow \Sigma_g^-$ |
| $\Pi \leftrightarrow \Pi$ | $\Pi_g \leftrightarrow \Pi_u$ |
| $\Pi \leftrightarrow \Delta$ | $\Pi_g \leftrightarrow \Delta_u, \Pi_u \leftrightarrow \Delta_g$ |
| $\Delta \leftrightarrow \Delta$ | $\Delta_g \leftrightarrow \Delta_u$ |

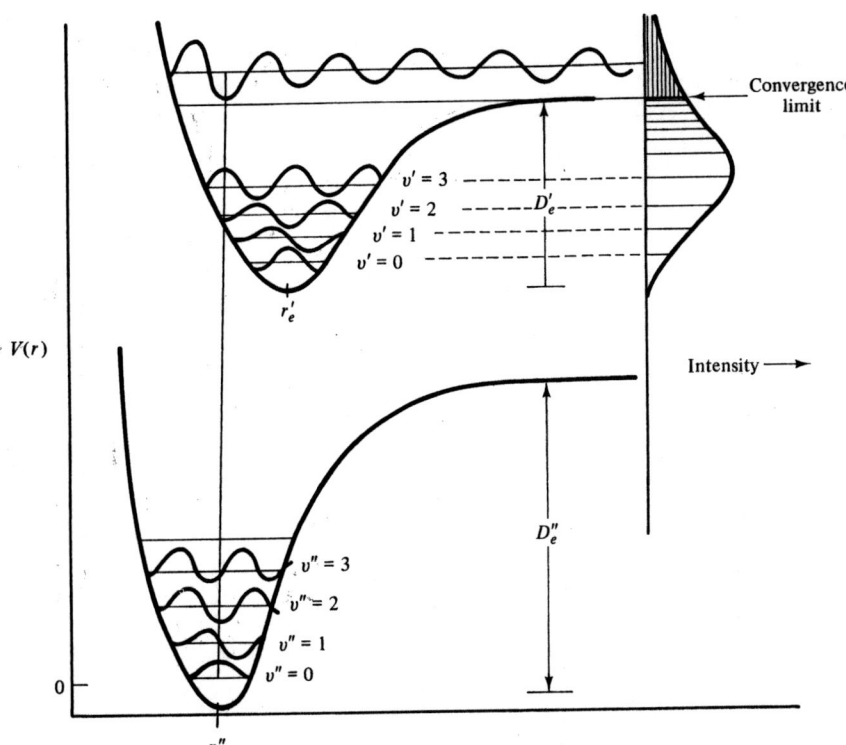

**FIGURE 12-4** Vibronic transitions between two electronic states. The intensity of a given vibronic transition depends on the overlap of the vibrational wavefunctions of the states $v''$ and $v'$ involved in the transition.

functions is greater. Figure 12-4 is drawn so that the $(v'' = 0) \rightarrow (v' = 3)$ transition is most intense. Transitions to vibrational states above or below the $v' = 3$ state are less intense because there is less probability that the nuclei will have the same internuclear distance as in the ground state. The intensity of a given vibronic transition (specific values of $v'$ and $v''$) is proportional to the square of the vibrational overlap integral for the specific $v''$ and $v'$ states involved. The vibrational overlap integral integrated over all space (all possible vibrational states or values of $v'$ and $v''$) is effectively the summation of the probabilities for all specific vibronic transitions.

Note that $\Delta v = v' - v''$ can assume any integer value $(0, \pm 1, \pm 2, \ldots)$, so that there is no specific selection rule such as the $\Delta v = \pm 1$ rule in normal IR vibrational spectroscopy. Excitation to the unbounded region above the discrete vibrational states in the upper electronic state (the convergence limit) leads to nondiscrete or continuous absorption and causes dissociation of the bond. For cases in which $r''_e = r'_e$, the $(v' = 0) \rightarrow (v' = 0)$ or (0–0) transition is strongest. If $r''_e \ll r'_e$, dissociation is more likely since overlap of the wavefunctions occurs only with the uppermost vibrational levels of the excited electronic state.

Rotational levels are associated with each vibrational state. For a given vibronic transition, the intensities of the possible rotational transitions are determined by the appropriate rotational overlap integral. For diatomics the general selection rule $\Delta J = \pm 1$ applies without the restriction of a permanent dipole moment. Since many rotational levels are populated, a vibronic band consists of many rotational transitions and P and R branches can be observed. If $\Lambda \neq 0$ ($\Pi$ and $\Delta$ states), transitions with $\Delta J = 0$ are also allowed and a Q band may be observed within a vibronic band.

The term **band** is used to describe a specific vibronic transition. A **band system** is an ensemble of bands associated with the same electronic transition. A **hot band** applies to transitions from an excited vibrational state in the lower electronic state to a lower vibrational state in the upper electronic state ($v' < v''$ and $v'' > 0$). A series of bands having the same value of ($v' - v''$) is called a **sequence**, while a **progression** applies to a set of bands with either the same $v'$ or $v''$.

*Band Shapes.* The total energy difference ($\Delta E_t$) for a transition between specific vibrational and rotational levels in the upper and lower electronic states ($v'', J''$) $\rightarrow$ ($v', J'$) is given by

$$\Delta E_t = \Delta E_e + \Delta E_v + \Delta E_r \qquad (12\text{-}13)$$

and thus is due to the changes in electronic, vibrational, and rotational energy. If we use the energy levels derived for a harmonic oscillator and rigid rotor (equations 12-2 and 12-5, respectively) for the vibrational and rotational energies in the two electronic states, as a first approximation, $\Delta E_t$ in wavenumbers is given by

$$\Delta E_t = \Delta \bar{v}_e + \bar{v}'(v' + \tfrac{1}{2}) - \bar{v}''(v'' + \tfrac{1}{2})$$
$$+ B'_v J'(J' + 1) - B''_v[J''(J'' + 1)] \quad (12\text{-}14)$$

where $\Delta \bar{v}_e$ is the difference in electronic energy in wavenumbers between the two states relative to the minima of the potential-energy curves. More complex expressions can be written for diatomics if anharmonicity and centrifugal distortion are considered.

Qualitatively, equation 12-14 makes it clear that the wavelength of absorption is determined primarily by the energy difference in the electronic states ($\Delta \bar{v}_e \gg \bar{v} > \bar{B}_v$). An electronic absorption band or band system consists of many bands due to the different vibronic transitions that are probable. This vibronic structure is observed for simple molecules in the gas phase at low temperatures and pressures. At higher pressures or temperatures in the gas phase or in condensed phases, the electronic absorption band is often a broad envelope of the possible vibronic transitions as shown in Figure 12-5. Thus the breadth of the electronic band, or the range of $\Delta E_t$ for which absorption is significant, is controlled by the range of vibronic transitions (values of $v'$ and $v''$) for which the Franck–Condon factor is significant. At room temperature, often only the $v'' = 0$ vibrational state is significantly populated. Thus the wavelength of maximum absorption is determined by the vibronic transition from $v'' = 0$ level to the vibrational level $v'$ in the upper electronic state yielding the maximum overlap of wavefunctions.

Each vibronic band is composed of numerous rotational lines due to specific rotational transitions. Thus the rotational component of equation 12-14 determines the shape and width of a given vibronic band. As in IR vibrational spectroscopy, the breadth of the vibrational band is controlled by the range of rotational states populated in the $v''$ vibrational level. At moderate temperatures or pressures, the rotational structure is often blurred and only the envelope of the rotational transitions is observed, as shown in Figure 12-5. As mentioned above, at higher temperatures or pressures in the gas phase or in condensed phases, usually no rotational structure and little or no vibrational structure are observed.

## 12-5 ELECTRONIC ABSORPTION SPECTRA OF POLYATOMIC MOLECULES

The description of electronic configurations, electronic states, and spectra for diatomic molecules forms a framework to discuss polyatomic molecules. However, the added complexity due to three or more bonded atoms and the lack of simple models makes it more difficult to treat electronic absorption transitions theoretically. A qualitative treatment will suffice for our purposes and is presented here.

Except for linear triatomics, much of the symbolism used for diatomics is no longer valid. The loss of simple axial symmetry means that quantum numbers for orbital angular momentum along the internuclear axis are in general not useful.

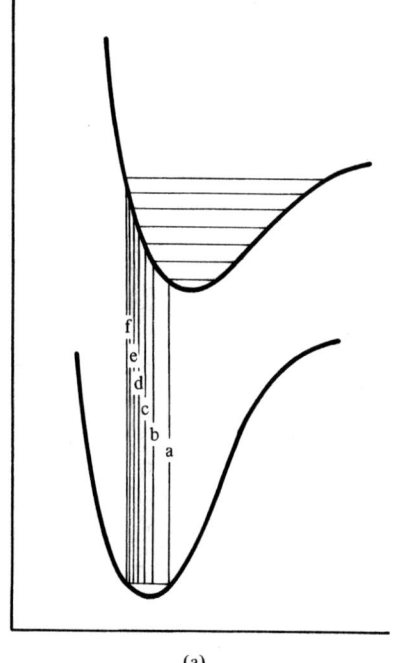

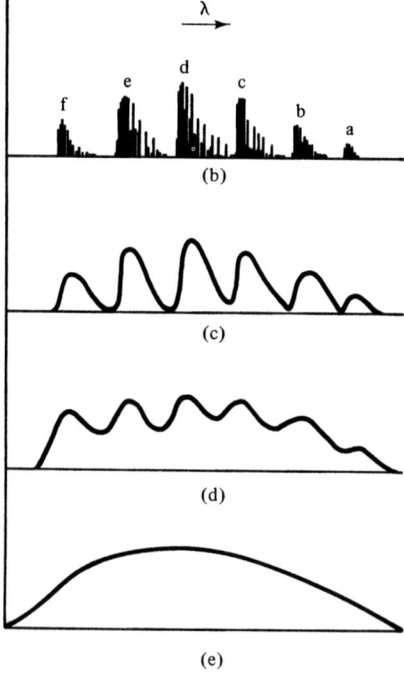

**FIGURE 12-5** Effect of collisions and intermolecular forces on an idealized absorption band: (a) potential energy curves involved in the transition; (b) absorption spectrum of the dilute vapor; (c) absorption spectrum of the vapor at moderate pressures (rotational structure smeared out); (d) absorption spectrum in liquid with weak intermolecular forces (vibrational structure present, but less pronounced than in vapor and somewhat shifted in frequency); (e) absorption spectrum in liquid with strong intermolecular forces (vibrational structure smeared out).

Because one atom can be bonded to two or more atoms in polyatomics, MOs are often formed from combinations of three or more AOs. The familiar concept of hybridization (e.g., $sp$, $sp^2$, $sp^3$) is often employed. Although not considered in our previous discussion, it is often necessary even for some diatomics to consider mixing of AOs before formation of MOs.

### Electronic States and Transitions

For many polyatomic molecules, it is still possible to consider that transitions involve $\sigma$ and $\pi$ bonding and antibonding MOs and nonbonding ($n$) MOs as shown in Figure 12-6. Typical characteristics and Mulliken's symbolism for these transitions are given in Table 12-5. In Mulliken's symbolism, $N$ denotes the ground state and $V$, $Q$, and $R$ denote excited states. As shown in Table 12-5, $N \rightarrow V$ transitions occur from bonding to antibonding orbitals and are strongly allowed. The $N \rightarrow Q$ transitions involve promotion of an electron in a nonbonding orbital to an antibonding orbital and are usually weaker than $N \rightarrow V$ transitions. Of the $N \rightarrow Q$ transitions, the $n \rightarrow \sigma^*$ transitions are generally more intense than $n \rightarrow \pi^*$ transitions. Both $N \rightarrow Q$ and $N \rightarrow V$ transitions are called **sub-Rydberg transitions**.

Transitions from bonding or nonbonding orbitals to very high energy states near the ionization limit are called **Rydberg transitions** (denoted $N \rightarrow R$). These transitions are observed in the vacuum UV region with conventional single-photon spectroscopy.

Other symbols are used to represent electronic states in polyatomics. These include **Platt symbols**, for which $A$ denotes the ground state, $B$ excited states involved in highly allowed transitions, and $L$ and $C$ excited states involved in partially forbidden transitions. Particularly for organic molecules, it is also common to represent electronic states as singlet states ($S_0$, $S_1$, $S_2$, . . .) and triplet states ($T_1$, $T_2$, . . .), where the subscript indicates the energy ordering. Usually, the ground electronic state is the lowest-lying singlet state ($S_0$) because electrons are all paired in the ground-state electronic configuration.

Finally, electronic states can be specified with group theory notation. Here it is necessary to identify the point group to which the molecule belongs. The symbols for the symmetry species or irreducible representations of the point group are then used to denote the electronic states. For example, water belongs to the $C_{2v}$ point group and the electronic states are denoted $A_1$, $A_2$, $B_1$, and $B_2$. The MOs are specified with the corresponding lowercase symbols. Thus the electron configuration for $H_2O$ is $(1a_1)^2(2a_1)^2(1b_2)^2(3a_1)^2(1b_1)^2$.

The group theory notation is even used for diatomics which belong to the $C_{\infty v}$ point group. Here the $\Sigma^+$, $\Sigma^-$, $\Pi$, $\Delta$, and $\Phi$ term symbols normally used correspond directly to the $A_1$, $A_2$, $E_1$, $E_2$, and $E_3$ symmetry species, respectively.

A further complication arises in polyatomics because the ground-state and excited-state molecular configurations may be different and described by different point groups. For example, $CS_2$ is linear ($D_{\infty h}$) in the ground electronic state but is bent ($C_{2v}$) in the first two excited states.

Once the electronic states are identified, the allowed transitions can be determined by group theory techniques. Because $\Lambda$ is no longer a good quantum

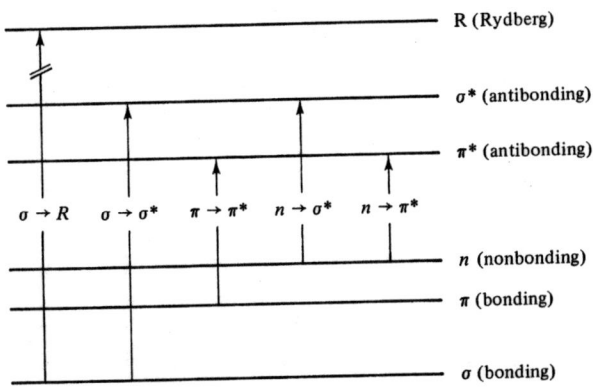

**FIGURE 12-6** Electronic transitions between $\sigma$, $\pi$, and $n$ orbitals.

*TABLE 12-5*
Characteristics of electronic transitions between $\sigma$, $n$, and $\pi$ orbitals

| Transition | Mulliken's designation | $\lambda$ (nm) | $\varepsilon$ (L mol$^{-1}$ cm$^{-1}$) | Examples |
|---|---|---|---|---|
| $\sigma \rightarrow \sigma^*$ | $N \rightarrow V$ | <200 | — | Saturated hydrocarbons |
| $\pi \rightarrow \pi^*$ | $N \rightarrow V$ | 200–500 | $\approx 10^4$ | Alkenes, alkynes, aromatics |
| $n \rightarrow \sigma^*$ | $N \rightarrow Q$ | 160–260 | $10^2 - 10^3$ | $H_2O$, $CH_3OH$, $CH_3Cl$, $CH_3NH_2$ |
| $n \rightarrow \pi^*$ | $N \rightarrow Q$ | 250–600 | $10^1 - 10^2$ | Carbonyls, nitro, nitrate, carboxyl |

number, selection rules based on $\Omega$ are no longer valid. The selection rule based on spin ($\Delta S = 0$) is still good as a first approximation. Many violations to this rule occur because of appreciable spin-orbit coupling. In general, there must be a change in symmetry for the two electronic states involved in a transition.

### Electronic Spectra

Electronic absorption spectra are correlated to the molecular structure of polyatomics in several ways. Often, spectral features are correlated to transitions that are localized in a given bond or group in the molecule.

Groups that are responsible for absorption are called **chromophores** and are often isolated double bonds. The absorption characteristics of common chromophores are listed in Table 12-6.

The MO approach is still used in some cases but it is usually localized to particular bonds in the molecule. For example, for $H_2CO$ the ground-state electron configuration can be written as

$$(1s_C)^2(1s_O)^2(2s_O)^2(\sigma_{CH})^2(\sigma_{CH'})^2(\sigma_{CO})^2(\pi_{CO})^2(n_O)^2$$

Electrons are associated with AOs of C and O, bonding and antibonding MOs of the CH and CO groups, and

**TABLE 12-6**
Chromophores and auxophores

| Group | Example | $\bar{\nu}$ ($10^3$ cm$^{-1}$) | $\lambda$ (nm) | $\varepsilon$ (L mol$^{-1}$ cm$^{-1}$) |
|---|---|---|---|---|
| C=C | $H_2C$=$CH_2$ | 55 | 182 | 250 |
| | | 57.3 | 174 | 16,000 |
| | | 58.6 | 170 | 16,500 |
| | | 62 | 162 | 10,000 |
| C≡C | H—C≡C—$CH_2$—$CH_3$ | 58 | 172 | 2,500 |
| C=O | $H_2CO$ | 34 | 295 | 10 |
| | | 54 | 185 | Strong |
| C=S | $CH_3$—C(S)—$CH_3$ | 22 | 460 | Weak |
| —$NO_2$ | $CH_3$—$NO_2$ | 36 | 277 | 10 |
| | | 47.5 | 210 | 10,000 |
| —N=N— | $CH_3$—N=N—$CH_3$ | 28.8 | 347 | 15 |
| | | >38.5 | <260 | Strong |
| (benzene) | | 39 | 255 | 200 |
| | | 50 | 200 | 6,300 |
| | | 55.5 | 180 | 100,000 |
| —Cl | $CH_3Cl$ | 58 | 172 | — |
| —Br | $CH_3Br$ | 49 | 204 | 1,800 |
| —I | $CH_3I$ | 38.8 | 258 | — |
| —OH | $CH_3OH$ | 49.7 | 201 | 1,200 |
| | | 55 | 183 | 200 |
| | | 67 | 150 | 1,900 |
| —SH | $C_2H_5SH$ | 43 | 232 | 160 |
| —$NH_2$ | $CH_3NH_2$ | 46.5 | 215 | 580 |
| | | 52.5 | 190 | 3,200 |
| —S— | $CH_3$—S—$CH_3$ | 44 | 228 | 620 |
| | | 46.5 | 215 | 700 |
| | | 49.3 | 203 | 2,300 |
| C=C—C=C | $H_2C$=CH—CH=$CH_2$ | 48 | 209 | 25,000 |
| (naphthalene) | | 32 | 211 | 250 |
| | | 37 | 270 | 5,000 |
| | | 45 | 221 | 100,000 |
| (anthracene) | | 28 | 360 | 6,000 |
| | | 40 | 250 | 150,000 |
| O=⬡=O | | 23 | 440 | 20 |
| | | 34 | 300 | 1,000 |
| | | 40 | 250 | 15,000 |

nonbonding orbitals of O. The three $\sigma$ bonds to the carbon atom are formed from $sp^2$ trigonal hybrid orbitals and the $\pi$ bond is formed from the remaining $p$ orbital of C and an O $p$ orbital. Only the valence electrons in the $\pi_{CO}$ and $n_O$ orbitals are involved in transitions to empty $\pi_{CO}^*$ and $\sigma_{CO}^*$ antibonding orbitals.

The absorption characteristics of a molecule can often be predicted from the chromophores present in a molecule. Similarly, the absorption spectrum of an unknown can be used to identify the presence of certain functional groups in a molecule. Chromophores separated by a —CH$_2$— behave almost independently. If there are two chromophores in a given molecule which absorb at the same wavelength, the total molar absorptivity is approximately the sum of the individual molar absorptivities. The absence of absorption bands in the region 200 to 800 nm is a good indication of the absence of chromophores and thus a saturated organic molecule or an inorganic molecule with no double bonds.

Shifts in the wavelength of absorption of chromophores to longer or shorter wavelengths are denoted **bathochromic** and **hypsochromic** shifts, respectively. An increase or decrease in molar absorptivity is denoted a **hyperchromic** or **hypochromic** effect, respectively. Conjugation (i.e., C=C—C=C, C=C—C=O, Ph—C=C) usually causes a hyperchromic effect and a bathochromic shift.

**Auxophores** are groups which do not absorb themselves, but when conjugated to chromophores cause a bathochromic shift and a hyperchromic effect by inductive or resonance effects. Examples are —OH, —Br, and —NH$_2$. Usually, they have at least one pair of $n$ electrons which interact to lower the energy of the $\pi^*$ orbital associated with the chromophore.

In solution, solvent–solute interactions often affect the absorption wavelengths of chromophores. Peaks from $n \rightarrow \pi^*$ transitions often suffer a hypsochromic shift as the polarity of the solvent increases. This behavior is attributed to the increased solvation of the nonbonding pair, which lowers the energy of the $n$ orbital. In contrast, a bathochromic shift with increased solvent polarity is often observed for $\pi \rightarrow \pi^*$ transitions.

Absorption by inorganic molecules is often due to transitions involving unfilled $d$ and $f$ orbitals on the metals or to charge-transfer transitions. In transition metal complexes, the bonding between $d$ orbitals of the metal and orbitals from the ligand breaks the degeneracy of the five $d$ orbitals. Thus transitions are between lower-energy filled $d$ orbitals and higher-energy unfilled $d$ orbitals. The type of splitting of $d$ orbitals depends on the molecular geometry (e.g., octahedral, tetrahedral) and the magnitude of the splitting depends on the particular ligand and is quantitatively treated with crystal field theory or ligand field theory. Generally, transition metal complex absorption transitions are in the visible region of the spectrum. An example is the blue $Cu(NH_3)_4^{2+}$ complex.

Lanthanide and actinide ions have narrow absorption bands due to transitions to empty $f$ orbitals. The spectra are little affected by the solvent or anions present because the $f$ orbitals are well shielded.

Charge-transfer absorption occurs with many inorganic ions where the components of the complex have donor-acceptor properties. These transitions have very high molar absorptivities in the visible region and occur when an electron in an orbital associated with a donor which has low electron affinity is transferred to an orbital associated with an acceptor of high electron affinity. An example is the Fe(III) thiocyanate complex where, in the excited state, Fe(II) and SCN$\cdot$ are the predominant species. Common anions which undergo charge-transfer transitions are $SO_4^{2-}$, $NO_3^-$, $I_3^-$, $CrO_4^{2-}$, and $MnO_4^-$. For the latter two anions, the absorption transition involves a transfer of an O electron to an empty metal orbital.

## Electronic Band Shapes and Intensities

Molecular absorption electronic spectra can be composed of one band or several bands, depending on the types and number of chromophores in a molecule. Separate absorption bands can often be observed due to transitions from the ground state to different electronic excited states (e.g., $S_0 \rightarrow S_1$, $S_0 \rightarrow S_2$). As for diatomic molecules (Section 12-4), the shapes and intensities of bands are determined by the transition moment and the Franck–Condon factors.

Experimentally, the absorption transition strength at a particular wavelength is evaluated from the molar absorptivity ($\varepsilon$). The molar absorptivity or the *integrated molar absorptivity* over the whole band ($\bar{\varepsilon}$) can be used to calculate oscillator strengths, transition probabilities, and Einstein coefficients as detailed in Appendix F. An oscillator strength ($f$) value of 1 corresponds to a very strong absorption with a molar absorptivity at the wavelength of maximum absorption ($\varepsilon_m$) of about $10^5$ L mol$^{-1}$ cm$^{-1}$. Typical values of $f$ are $10^{-8}$ to $10^{-6}$ for the rare earth ions, $10^{-4}$ for aquo complexes of transition metal ions, 0.03 for $MnO_4^-$, and nearly unity for organic dyestuffs.

The particle in a box or free electron model has also been applied to electronic transitions. Consider the electron, which undergoes a transition to be a particle in a box in which the positive charge is symmetrically distributed about the midpoint. It can be shown that the transition probability $(R^2)$ in J cm$^3$ is given by

$$R^2 = \frac{e^2 a^2}{\pi^5 \varepsilon_0} = (9.46 \times 10^{-24})a^2 \quad (12\text{-}15)$$

where $e$ is the charge of an electron $(1.6 \times 10^{-19}$ C) and $a$ equals the width of box or size of molecule in meters. Substitution of equation 12-15 into equation F-39 yields

$$\bar{\varepsilon} = \frac{8Ne^2 a^2 v_m}{6909 \pi^2 \varepsilon_0 hc} \quad (12\text{-}16)$$

where $N$ is Avogadro's number. This equation indicates that $\bar{\varepsilon}$ is related to the size of the molecule or more correctly to the part of the molecule over which the electron involved in the transition resides. This explains the increased molar absorptivity in conjugated systems. The model also predicts the bathochromic shift observed with conjugation.

## 12-6   LUMINESCENCE SPECTRA

Up to this point we have been concerned about how molecules are promoted to excited electronic states by absorption of electromagnetic radiation. Now we will investigate the ways in which an excited molecule can dissipate its excess energy and return to the ground electronic state. The two general modes of deactivation involve nonradiative and radiative processes. In nonradiative deactivation, the excess electronic energy is converted to translational, rotational, or vibrational energy with no emission of radiation. In contrast, the radiative dissipation process involves emission of a photon. Most analytical luminescence determinations are carried out in solutions or frozen solids, and the following discussion is directed to these applications.

### Processes of Deactivation

Part of the energy-level diagram for a hypothetical aromatic molecule is shown in a **Jablonski diagram** in Figure 12-7. The various activation and deactivation processes are indicated. The absorption or excitation process (a) is very rapid and on the order of the period of oscillation of the electric field of a visible photon $(10^{-15}$ s). Absorption transitions from the ground vibrational level of the ground singlet electronic state $(S_0)$ to different vibrational levels in the first and second excited electronic states $(S_1$ and $S_2)$ are shown.

Molecules in excited vibrational states rapidly dissipate their excess vibrational energy and relax to the ground vibrational level in a given electronic state. The energy goes into thermal or vibrational motion of the solvent molecules in condensed phases. This nonradiational process is denoted **vibrational relaxation**. Normally, vibrational relaxation proceeds in a stepwise fashion $(\Delta v = 1)$ in which one vibrational quantum is lost per collision. This typically takes $10^{-11}$ to $10^{-10}$ s. Since a typical vibrational period is $10^{-13}$ s, many vibrations occur before the excess vibrational energy is lost.

The crossover between two states of the same mul-

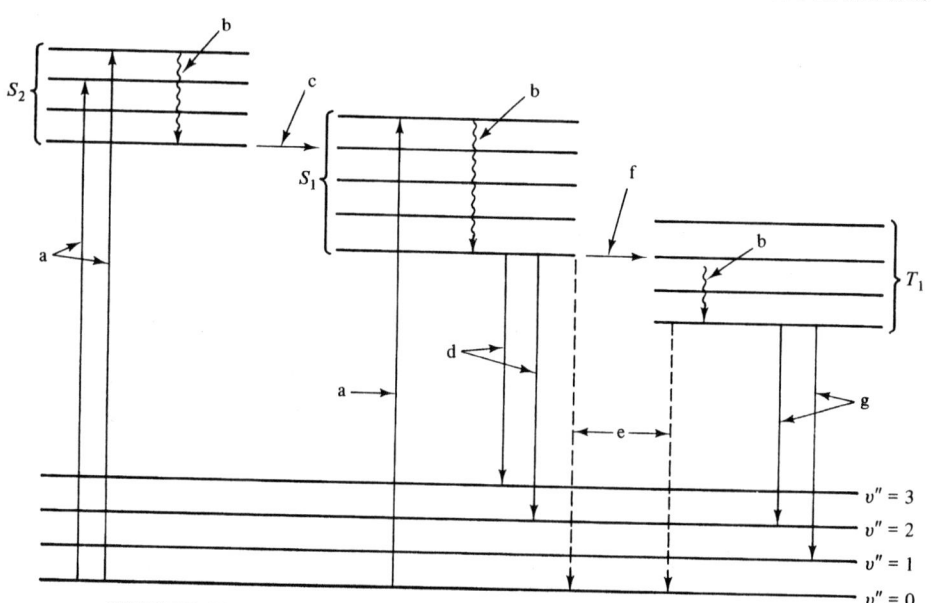

**FIGURE 12-7**   Deactivation processes for an excited molecule. a, absorption; b, vibrational relaxation; c, internal conversion; d, fluorescence; e, external conversion; f, intersystem crossing; g, phosphorescence.

tiplicity is a nonradiative electron state transition called **internal conversion**. This is likely to occur when the potential energy curves for two electronic states cross such that the lower vibrational levels of the higher electronic state are approximately the same energy as higher vibrational levels of the lower electronic singlet state. Internal conversion ultimately results in the conversion of excess electronic energy to excess vibrational energy.

Internal conversion can occur between excited states (e.g., $S_2 \rightarrow S_1$) or between the first excited electronic state and the ground electronic state ($S_1 \rightarrow S_0$). Generally, internal conversion between excited electronic states is rapid ($10^{-12}$ s). Internal conversion from the $S_1$ to the $S_0$ state depends on the molecule but is often less efficient if there is a wide energy separation between $S_1$ and $S_0$, so that there is no overlap of the potential energy wells. After internal conversion, the excess vibrational energy is rapidly dissipated through vibrational relaxation to the ground vibrational level of the lower electronic state.

**Fluorescence** is a radiational transition between electronic states of the same multiplicity. For most molecules, the electrons are paired in the ground state so that fluorescence involves a singlet-singlet transition. Because internal conversion to $S_1$ and vibrational relaxation are more rapid processes than fluorescence, fluorescence usually occurs from the ground vibrational state of $S_1$ to various vibrational levels in $S_0$ ($S_1 \rightarrow S_0 + h\nu$). For this reason, only one fluorescence band is normally observed even if absorption to different excited singlet states occurs. Typically, fluorescence requires $10^{-10}$ to $10^{-6}$ s to occur. Fluorescence usually appears at longer wavelengths than absorption because absorption transitions are to higher excited electronic states or to higher vibrational levels in the $S_1$ manifold. This is illustrated in Figure 12-7. Fluorescence can occur from higher electronic states in rare instances. Azulene and its derivatives exhibit $S_2 \rightarrow S_0$ fluorescence.

The term **external conversion** refers to nonradiative processes in which excited states transfer their excess energy to other species, such as solvent or solute molecules. One primary mechanism of external conversion is **dynamic quenching**. It involves the nonradiational transfer of energy from excited species to other molecules during collisions. Therefore, the rate of dynamic quenching is reduced by cooling the sample.

Absorption transitions to triplet states are forbidden by symmetry ($\Delta S = 0$), although weak absorption is possible in some molecules. The triplet state can also be populated from excited singlet states by a process denoted **intersystem crossing**, which is a crossover between electronic states similar to internal conversion except that the states have different multiplicities (usually $S_1 \rightarrow T_1$).

After intersystem crossing, a molecule in the $T_1$ state deactivates by vibrational relaxation to the ground vibrational level of $T_1$. Normally, the triplet state deactivates by external conversion or intersystem crossing to the ground state ($T_1 \rightarrow S_0$). The triplet state can also deactivate by emission of a photon. This radiational deactivation process between electronic states of different multiplicity is called **phosphorescence** (typically, $T_1 \rightarrow S_0 + h\nu$). Usually, phosphorescence takes $10^{-4}$ to $10^4$ s to occur because the process is spin forbidden. Thus phosphorescence is usually observed only if external conversion is reduced by cooling or other techniques. The wavelengths of phosphorescence for a given molecule are generally longer than those for fluorescence because the energy of $T_1$ is less than $S_1$, due to the electrons being unpaired and in different molecular orbitals.

Fluorescence caused by direct excitation to the $S_1$ state or internal conversion to the $S_1$ state is called more precisely **prompt fluorescence**. **Delayed fluorescence** has a longer lifetime than prompt fluorescence because $S_1$ is populated by indirect mechanisms. For example, in *E-type delayed fluorescence*, $S_1$ is populated by thermally assisted intersystem crossing ($T_1 \rightarrow S_1$) back from a triplet state originally derived from an $S_1$ state. *P-type delayed fluorescence* involves a bimolecular reaction between triplet states followed by triplet-triplet annihilation ($T_1 + T_1 \rightarrow S_0 + S_1$). The term **photoluminescence** is used to describe any emission of photons after photon excitation. For molecules, photoluminescence includes prompt and delayed fluorescence as well as phosphorescence.

If the energy of the excitation photon is greater than the convergence limit of the excited electronic state, a bond is ruptured after absorption. This process is called **dissociation**. If a bond is ruptured after internal conversion the process is called **predissociation**. Predissociation or dissociation is more likely in molecules that absorb at wavelengths shorter than 200 nm (200 nm corresponds to 140 kcal mol$^{-1}$).

Excited singlet or triplet states can also be depopulated by photochemical reactions. This includes reactions of excited states with solvent or solute molecules. Excited states are distinctly different from the ground state in polarity and reactivity.

### Quantum Efficiencies and Power Yields

The efficiencies and hence intensities of fluorescence and phosphorescence depend on the relative competition between radiative and nonradiative routes of deactivation. Different ways of expressing luminescence efficiencies are summarized in Table 12-7. From equation 12-17, the **luminescence quantum efficiency** or **lumi-**

**TABLE 12-7**

Expressions for luminescence efficiencies[a]

| | |
|---|---|
| Luminescence quantum efficiency: $\phi_L = \dfrac{\Phi_{L,p}}{\Phi_{A,p}}$ | (12-17) |
| Fluorescence quantum efficiency: $\phi_F = \dfrac{\Phi_{F,p}}{\Phi_{A,p}}$ | (12-18) |
| Phosphorescence quantum efficiency: $\phi_P = \dfrac{\Phi_{P,p}}{\Phi_{A,p}}$ | (12-19) |
| Luminescence power yield: $Y_L = \dfrac{\Phi_L}{\Phi_A}$ | (12-20) |
| Spectral luminescence quantum efficiency: $\phi_{\nu L} = \dfrac{\Phi_{\nu L,p}}{\Phi_{A,p}} = \phi_L S'_\nu$ | (12-21) |
| Spectral luminescence power yield: $Y_{\nu L} = \dfrac{\Phi_{\nu L}}{\Phi_A}$ | (12-22) |

[a]The subscripts to the radiant power ($\Phi$) have the following meanings: $L$, due to luminescence; $F$, due to fluorescence; $P$, due to phosphorescence; $A$, absorbed; $p$, in photons per second; $\nu L$, luminescence over the frequency interval $\nu$ to $\nu + d\nu$. $S'_\nu$ is the emission band profile in hertz.

nescence quantum yield ($\phi_L$) is the ratio of the luminescence radiant power to the absorbed radiant power where the radiant powers are expressed in photons per second. Thus $\phi_L$ indicates the fraction of the absorbed photons which are converted to luminescence photons ($0 \leq \phi_L \leq 1$). Equations 12-18 and 12-19 are specific expressions for the **fluorescence quantum efficiency** and the **phosphorescence quantum efficiency**, respectively. Here and elsewhere we will use the subscripts $F$ and $P$ to distinguish fluorescence from phosphorescence. The **luminescence power yield** (equation 12-20) is defined in an analogous fashion to the luminescence quantum efficiency (equation 12-17) except that the radiant power of absorbed and emitted radiation is expressed in terms of watts instead of photons per second.

The **spectral luminescence quantum efficiency** ($\phi_{\nu L}$) is the fraction of the absorbed photons that results in luminescence over a frequency interval $\nu$ to $\nu + d\nu$, as defined in equation 12-21. This equation also shows that the spectral luminescence quantum efficiency can be expressed in terms of the luminescence quantum efficiency and the emission band profile ($S'_\nu$) which is normalized so that $\int S'_\nu \, d\nu' = 1$. In this discussion a prime is used to distinguish an emission frequency or function from excitation quantities. Note that

$$\phi_L = \int \phi_{\nu L} \, d\nu' = \phi_L \int S'_\nu \, d\nu'$$

If $S'_\nu$ is a Gaussian in shape,

$$\phi_L = 1.0645(\phi_{\nu L})_m \, \Delta\nu' \qquad (12\text{-}23)$$

where the subscript $m$ denotes at the wavelength of maximum luminescence and $\Delta\nu'$ is the half-width of the luminescence band (Hz).

The corresponding **luminescence spectral power yield** ($Y_{\nu L}$) is the ratio of radiant power of luminescence over frequency interval $\nu$ to $\nu + d\nu$ to $\Phi_A$, as shown in equation 12-22. If excitation occurs over a small frequency interval centered at $\nu$ and emission is observed at $\nu'$,

$$Y_{\nu L} = \frac{\Phi_{\nu L}}{\Phi_A} = \frac{\Phi_{\nu L,p} h\nu'}{\Phi_{A,p} h\nu}$$

With this relationship and equation 12-21, we see that

$$Y_{\nu L} = \phi_{\nu L} \frac{\nu'}{\nu} = S'_\nu \phi_L \frac{\nu'}{\nu} \qquad (12\text{-}24)$$

The total luminescence power yield for a given excitation frequency is the integral of equation 12-24 over the emission band, or

$$Y_L = \phi_L \int S'_\nu \frac{\nu'}{\nu} \, d\nu' \qquad (12\text{-}25)$$

As a rough approximation,

$$Y_L \approx \phi_L \frac{\nu'_m}{\nu_m} \qquad (12\text{-}26)$$

where the subscript $m$ denotes the frequency of maximum absorption or emission.

Often $\phi_L$ is totally independent of the excitation wavelength. However, in some cases, $\phi_L$ can vary as much as 20% as the excitation wavelength is changed. Abnormal behavior occurs where the simple model presented is not adequate. For example, fluorescence can arise from two different excited singlet states.

Note that $Y_L < \phi_L$ and that $Y_L$ always depends on the excitation frequency (see equation 12-25). Even if $\phi_L = 1$, some of the excess electronic energy is converted to heat before luminescence occurs by internal conversion and vibrational relaxation. As the excitation frequency is increased, $Y_L$ decreases because a greater fraction of the absorbed radiant power is converted to heat.

*Fluorescence Quantum Efficiency.* The fluorescence quantum efficiency is related to the rate of absorption and the rate of deactivation of the first excited singlet state. If it is assumed that all processes are first order with respect to number densities of $S_0$ and $S_1$ ($n_{S_0}$ and $n_{S_1}$ in molecules per cm³, respectively), the rate of absorption is $k_A n_{S_0}$ and the rate of deactivation is $(k_F + k_{nr})n_{S_1}$, where $k_A$, $k_F$, and $k_{nr}$ are the first-order rate constants in s⁻¹ for absorption, fluorescence, and nonradiative deactivation, respectively. Thus the rate of change of the number density of the $S_1$ state is given by

$$\frac{dn_{S_1}}{dt} = k_A n_{S_0} - (k_F + k_{nr})n_{S_1} \qquad (12\text{-}27)$$

In this discussion it is assumed that stimulated emission, dissociation, predissociation, and photochemical reactions are negligible.

If the **fluorophore** of interest (i.e., the analyte molecule with fluorescence properties) is contained in a sample volume $V$ that is fully illuminated with radiation of constant intensity, a steady-state concentration of $S_1$ is rapidly achieved ($dn_{S_1}/dt = 0$). Thus, from equation 12-27,

$$n_{S_1} = \frac{n_{S_0}k_A}{k_F + k_{nr}} \qquad (12\text{-}28)$$

The rates of photon absorption and fluorescence emission are given by

$$\Phi_{A,p} = k_A n_{S_0} V \qquad (12\text{-}29)$$

$$\Phi_{F,p} = k_F n_{S_1} V \qquad (12\text{-}30)$$

Use of the foregoing three relationships in equation 12-17 yields

$$\phi_F = \frac{k_F}{k_F + k_{nr}} \qquad (12\text{-}31)$$

If $k_{nr} \gg k_F$, $S_1$ is deactivated by nonradiative processes before the molecule has a chance to fluorescence, $\phi_F$

is small, and detection of fluorescence is difficult. The quantum efficiency of a molecule is appreciable only if $k_F$ is comparable to or greater than $k_{nr}$ since this increases the probability of emission before nonradiative deactivation.

Typically, $k_F$ is $10^6$ to $10^9$ s⁻¹. To understand the factors that affect the magnitude of $k_{nr}$, we can break it up into its components:

$$k_{nr} = k_{ec} + k_{ic} + k_{isc} \qquad (12\text{-}32)$$

where $k_{ec}$, $k_{ic}$, and $k_{isc}$ are the first-order rate constants for external conversion, internal conversion ($S_1 \rightarrow S_0$), and intersystem crossing ($S_1 \rightarrow T_1$), respectively. The magnitude of these rate constants is highly dependent on the structure and environment of the molecule. The rate constant for internal conversion to the ground state varies from typically $10^5$ to $10^7$ s⁻¹. For organic molecules, $k_{isc}$ is typically $10^6$ to $10^9$ s⁻¹. In inorganic systems with high-atomic-number metal ions relative to C, H, and N, spin-orbit coupling constants are greater and $k_{isc}$ is typically $10^9$ to $10^{12}$ s⁻¹.

The rate of fluorescence photon emission from equations 12-28, 12-30, and 12-31 can be expressed as

$$\Phi_{F,p} = n_{S_0}k_A\phi_F V \qquad (12\text{-}33)$$

Thus the fluorescence intensity is proportional to the ground-state population of the fluorophore, the rate of absorption, the fluorescence quantum efficiency, and the volume element of the sample illuminated. Note that $k_A$ is proportional to $U_v B_{0j}$ (equation 4-8). Hence the fluorescence intensity is proportional to the intensity of the excitation beam [with ordinary (nonlaser sources)] and the Einstein coefficient for absorption or the transition probability for absorption.

The rate constant for fluorescence emission, $k_F$, is, in fact, $A_{10}$, the Einstein coefficient for spontaneous emission. Thus $k_F$ and $\phi_F$ are more likely to be large if the transition probability between $S_1$ and $S_0$ is large. Because absorption to $S_1$ and fluorescence involve the same electronic states, both $k_F$ and $k_A$ ($S_0 \rightarrow S_1$) are larger in molecules for which the molar absorptivity ($\epsilon$) for absorption to $S_1$ is large. From equation 12-33, significant fluorescence is more likely to be observed from molecules for which $\epsilon$ ($S_0 \rightarrow S_1$) is large. As an approximation, $k_F = 3 \times 10^{12} \epsilon_m \Delta\lambda/\lambda_m^4$, where $\lambda_m$ is the wavelength of maximum absorption ($S_0 \rightarrow S_1$) in nm, $\Delta\lambda$ is the half-width of the band in nm, and $\epsilon_m$ is the molar absorptivity at the maximum.

*Phosphorescence Quantum Efficiency.* The phosphorescence quantum efficiency (equation 12-19) depends on the rate that the triplet state ($T_1$) is populated by intersystem crossing and the rate of deactivation $T_1$. Note that $(\phi_p + \phi_F) \leq 1$.

Under conditions of constant illumination, a steady-state population of $T_1$ is produced $(dn_{T_1}/dt = 0)$. Thus the rate of production of $T_1$ equals the rate of deactivation of $T_1$ or

$$k_{isc}n_{S_1} = (k_P + k'_{nr})n_{T_1} \qquad (12\text{-}34)$$

where $k_P$ and $k'_{nr}$ are the first-order rate constants for phosphorescence and nonradiative decay of $T_1$, respectively. It is assumed that $T_1 \rightarrow S_1$ intersystem crossing is negligible. From equation 12-34 the steady-state triplet concentration is given by

$$n_{T_1} = \frac{k_{isc}n_{S_1}}{k_P + k'_{nr}} \qquad (12\text{-}35)$$

The rate of phosphorescence photon emission from volume element $V$ is

$$\Phi_{P,p} = k_P n_{T_1} V = \frac{k_P k_{isc} n_{S_1} V}{k_P + k'_{nr}} \qquad (12\text{-}36)$$

Use of equations 12-28, 12-29, and 12-36 in equation 12-19 yields

$$\phi_P = \left(\frac{k_{isc}}{k_F + k_{nr}}\right)\left(\frac{k_P}{k_P + k'_{nr}}\right) \qquad (12\text{-}37)$$

Thus $\phi_P$ is the product of two factors, the fraction of the absorbed photons that produce triplet states and the fraction of the triplet molecules that undergo phosphorescence. The former factor is often denoted the *quantum efficiency of triplet formation* $(\phi_{isc})$.

Phosphorescence is favored for molecules and environmental conditions in which intersystem crossing is favorable $(k_{isc} > k_F + k_{ec} + k_{ic})$. Thus, if $k_{isc} > k_F$, phosphorescence is favored over fluorescence.

Even if intersystem crossing is efficient, phosphorescence is not usually observed because nonradiative decay of $T_1$ occurs before phosphorescence occurs (i.e., $k'_{nr} > k_P$). Nonradiative decay of $T_1$ is due to external conversion and intersystem crossing from $T_1$ to $S_0$, or

$$k'_{nr} = k'_{ec} + k'_{isc} \qquad (12\text{-}38)$$

where $k'_{ec}$ and $k'_{isc}$ are the first-order rate constants for these processes. Typically, $k'_{ec} \gg k'_{isc}$, so that phosphorescence is usually observed only if $k'_{ec}$ is substantially reduced by cooling the sample to a rigid glass in liquid $N_2$. Substantial reduction of $k'_{ec}$ is required because typically $k_P < 10^4$ s$^{-1}$, due to the fact that the

$T_1 \rightarrow S_0$ transition is spin forbidden. Thus temperature is usually a more critical variable for phosphorescence than fluorescence. Cooling can also enhance $\phi_P$ because the quantum efficiency of triplet formation is increased by reducing dynamic quenching of $S_1$. Dynamic quenching can also be reduced by immobilizing the molecule on a solid support or by protecting the molecule in solution (inside cavities or micelles).

## Luminescence Lifetimes

If the excitation source is turned off instantaneously, the concentrations of $S_1$ and $T_1$, and hence the fluorescence and phosphorescence signals, decay. Because the excited states are often deactivated by first-order processes, the decay in either type of luminescence signal can be described by an exponential:

$$\Phi_L(t) = \Phi_L^0 e^{-t/\tau_L} \qquad (12\text{-}39)$$

where $\Phi_L^0$ is the luminescence radiant power at the time the excitation source is shut off and $\tau_L$ is the luminescence lifetime. The luminescence lifetime is defined as the time for the luminescence signal to decay to $1/e$ of its initial value. Sometimes the luminescence lifetime is expressed as a half-life, $(\tau_L)_{1/2}$, which is the time for decay to one-half of the initial intensity. It is easily related to $\tau_L$ [i.e., $(\tau_L)_{1/2} = 0.69\tau_L$].

The lifetimes for fluorescence and phosphorescence are related to the rate constants for deactivation with the following equations:

$$\tau_F = (k_F + k_{nr})^{-1} \qquad (12\text{-}40)$$

$$\tau_P = (k_P + k'_{nr})^{-1} \qquad (12\text{-}41)$$

In the absence of nonradiative decay, the radiative, natural, or inherent lifetimes (denoted with a subscript $r$) are given by

$$(\tau_F)_r = \frac{1}{k_F} \qquad (12\text{-}42)$$

$$(\tau_P)_r = \frac{1}{k_P} \qquad (12\text{-}43)$$

Usually, $k_F$ is $10^5$ to $10^8$ s$^{-1}$, so that $(\tau_F)_r$ is $10^{-5}$ to $10^{-8}$ s. Because $k_P < 10^5$ s$^{-1}$, $(\tau_P)_r > 10^{-5}$ s. For organic systems $k_P$'s are rarely larger than $10^2$ s$^{-1}$, whereas for inorganic complexes, $k_P$ is most commonly $10^4$ to $10^2$ s$^{-1}$, but can be as low as $10^{-1}$ s$^{-1}$.

The equations above indicate that nonradiative decay decreases the steady-state fluorescence and phosphorescence intensities and lifetimes by the same factors [i.e., $\tau_F$ and $\phi_F$ are proportional to $(k_F + k_{nr})^{-1}$ and $\tau_p$ and $\phi_P$ are proportional to $(k_P + k'_{nr})^{-1}$]. If the ratio of equation 12-40 to equation 12-42 is compared to equation 12-31, we note that $\phi_F = \tau_F/(\tau_F)_r$.

## Quenching and Excited-State Reactions

The luminescence signals observed can be reduced by the presence of concomitants through several mechanisms. Reabsorption of emitted radiation by other species (or even other analyte molecules) is sometimes called **trivial quenching**. We will refer to this process as secondary absorption to distinguish it from mechanisms that cause nonradiative deactivation before photon emission as discussed below.

*Dynamic Quenching.* Quenching normally refers to nonradiative energy transfer from excited species to other molecules. **Dynamic quenching** or collisional quenching requires contact between the excited lumophore and the quenching species, the quencher (Q). The rate of quenching is diffusion controlled and depends on the temperature and viscosity of the solution. The quencher concentration must be high enough that the probability of collision between the analyte and quencher is significant during the lifetime of the excited species.

If external conversion is controlled by collisions with a single quencher, a second-order rate process, the rate constant for external conversion, is given by

$$k_{ec} = k_q[Q] \qquad (12\text{-}44)$$

where $k_q$ is the second-order rate constant in L mol$^{-1}$ s$^{-1}$ for quenching and self quenching is assumed negligible. From equations 12-31 and 12-32, the fluorescence quantum efficiency in the absence of quenching $(\phi_F^0)$ is given by

$$\phi_F^0 = \frac{k_F}{k_F + k_{ic} + k_{isc}} \qquad (12\text{-}45)$$

whereas with quenching

$$\phi_F = \frac{k_F}{k_F + k_{ic} + k_{isc} + k_q[Q]} \qquad (12\text{-}46)$$

By ratioing equations 12-45 and 12-46 we obtain the Stern–Volmer equation

$$\frac{\phi_F^0}{\phi_F} = 1 + K_q[Q] \qquad (12\text{-}47)$$

where $K_q$ is the Stern–Volmer quenching constant defined as $K_q = k_q/(k_F + k_{ic} + k_{isc})$ in L mol$^{-1}$ and the quencher concentration is expressed in terms of molarity. Rearrangement yields

$$\phi_F^{-1} = (\phi_F^0)^{-1} + \frac{K_q[Q]}{\phi_F^0} \qquad (12\text{-}48)$$

Since the observed fluorescence signal is proportional to the fluorescence quantum efficiency, a plot of the reciprocal of the observed fluorescence signal vs. the quencher concentration should yield a straight line. From the intercept and slope, the Stern–Volmer constant can be determined. Deviations from the ideal behavior are sometimes observed when the extent of quenching is large.

Note that $K_q^{-1}$ is equal to the quencher concentration where the fluorescence quantum efficiency or fluorescence intensity is reduced by a factor of 2 relative to the value with a quencher concentration of zero. Also, the fluorescence quantum efficiency and lifetime are reduced the same amount by dynamic quenching. In fact, $K_q$ is given by

$$K_q = k_q \tau_F^0 \qquad (12\text{-}49)$$

where $\tau_F^0$ is the fluorescence lifetime in the absence of the quencher. Similar Stern–Volmer expressions apply to phosphorescence measurements where $\phi_P$ replaces $\phi_F$.

In aqueous solutions at room temperature, the bimolecular collision rate is about $10^{10}$ L mol$^{-1}$ s$^{-1}$. Hence, if all collisional encounters result in quenching, we can estimate that the maximum value of $k_q$ is about $10^{10}$ L mol$^{-1}$ s$^{-1}$. For a fluorescence lifetime of 1 ns, equation 12-49 shows that the Stern–Volmer constant is approximately 10 L mol$^{-1}$. Substituting this result in equation 12-47 shows that dynamic quenching of fluorescence is usually negligible when the quencher concentration is below 1 mM. With a lifetime of 1 ms, such as is typical for phosphorescence, $K_q$ can be as high as $10^7$ L mol$^{-1}$. Clearly, phosphorescence is totally quenched by an efficient quencher at 1 mM concentrations.

Dissolved oxygen is a very efficient quencher for long-lived triplet states, and its equilibrium concentration is typically about 1 mM in many solvents. For this reason it is usually necessary to deoxygenate solutions or solidify the sample to eliminate diffusion for successful phosphorescence measurements. Oxygen is also a good quencher for some fluorophores, such as aromatic hydrocarbons, particularly if the fluorescence lifetime is greater than 10 ns.

*Static Quenching.* Another form of quenching is **static quenching** in which the quencher and the fluorophore in the ground state form a stable complex (i.e., the *dark complex*). Fluorescence is only observed from the unbound fluorophore. The decrease in fluorescence intensity is described by equation 12-47, where $K_q$ is the formation constant (i.e., $K_q = [F \cdot Q]/[F][Q]$ and F is the fluorophore). The lifetime is not affected in this case; measurement of the lifetime provides a means of distinguishing between dynamic and static quenching. More complex equations can be written to describe situations in which both types of quenching are significant.

*Long-Range Quenching.* Energy transfer can also occur between molecules without collisions. This type of nonradiational deactivation is called **long-range quenching** or **Förster quenching**. It can be considered to be due to dipole-dipole coupling between a donor (the excited lumophore) and an acceptor (the quencher). The rate of energy transfer ($k_T$) to a specific acceptor is given by

$$k_T = \tau_D^{-1} \left( \frac{R_0}{R} \right)^6 \tag{12-50}$$

where $\tau_D$ is the luminescence lifetime of the donor, $R$ is the average distance between the donor and acceptor molecules, and $R_0$ is the Förster distance. Note that for $R = R_0$, the efficiency of energy transfer is 50% or the luminescence efficiency is 50% of what it would be in the absence of the acceptor species. The Förster distance for efficient quenchers is often in the range 20 to 50 Å and can be as large as 100 Å. Efficient long-range energy transfer is favored in situations where the emission spectrum of the donor and the absorption spectrum of the acceptor overlap and the molar absorptivity of the donor is relatively high in the overlap region. The rate of energy transfer increases with acceptor concentration as the average distance between molecules decreases. The dependence of the degree of quenching on the quencher concentration is complex and does not follow the Stern–Volmer model.

In both dynamic and long-range quenching, the quencher is promoted to an excited state. Often, the quencher deactivates by nonradiational processes. However, in some cases, the quencher can luminesce. This luminescence is termed **sensitized luminescence**. The overall quantum efficiency of sensitized luminescence can be greater than the luminescence quantum efficiency of the original lumophore if the efficiency of energy transfer is high and the luminescence efficiency of the quencher is higher than that of the species originally excited by photon absorption.

*Excited-State Reactions.* Molecules in excited states can react with other molecules to form complexes as follows:

$$S_1 + Q \to S_1 \cdot Q$$

The excited complex is called an **exciplex** if Q is different from the analyte molecule and an **excimer** (an excited-state dimer) if Q is a ground-state analyte molecule. Often the complexes are charge-transfer in nature. These excited complexes can dissipate their excess energy by releasing heat to the solvent, by dissociating into solvated ions, or by emission of a photon. Luminescence from singlet or triplet excimers is generally shifted red with respect to the luminescence of the original excited state. Excimer formation is likely only at relatively high analyte concentrations (i.e., greater than 1 mM).

### Band Shapes

A fluorescence emission band is broad because many vibronic transitions are possible from the ground vibrational level of $S_1$ ($v' = 0$) to many vibrational levels in $S_0$, as illustrated in Figure 12-8. As for absorption, the breadth of the emission band is determined by the range of vibrational levels ($v''$) over which the vibrational overlap integral is significant. Often, the fluorescence spectrum is a mirror image of the absorption spectrum, as also shown in Figure 12-8. This occurs if the vibrational levels in $S_0$ and $S_1$ are similar. Thus if the Franck–Condon factor for the $v'' = 0$ to $v' = 2$ absorption transition is largest, the corresponding emission transition from $v' = 0$ to $v'' = 2$ is the strongest. The wavelength of maximum emission is greater because the energy difference between $v' = 0$ and $v'' = 2$ is less than between $v'' = 0$ and $v' = 2$.

There are many exceptions to the mirror image rule such as when the molecular geometry of $S_0$ and $S_1$ are different or if the fluorescence is from an excimer. Also, more than one fluorescence band is observed in molecules such as biphenyl, where fluorescence originates from different parts of the molecule.

Usually, one phosphorescence band is observed and the breadth is determined by the range of probable transitions from the ground vibrational level of $T_1$ to different vibrational levels in $S_0$. The mirror image rule does not apply because of the very different nature of singlet and triplet states.

### Structural Effects

The luminescence efficiency of a molecule depends on its structure and on the environment in which the luminescence is measured. Although it is difficult to pre-

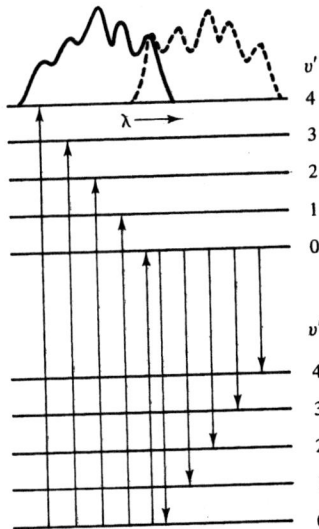

**FIGURE 12-8** Relationship between absorption and fluorescence. The absorption and fluorescence emission spectra are represented by the solid and dashed lines, respectively. The fluorescence transitions occur generally at longer wavelengths because the energy differences are less. In solution the vibronic detail is often not present and only a broad band is observed.

dict if a molecule exhibits luminescence, some general rules can be stated. It should be noted that there are always exceptions to these rules.

The nature of the lowest-lying excited singlet ($S_1$) is critical in determining the luminescence behavior of a molecule because fluorescence and intersystem crossing usually occur from this state. In organic molecules, the transitions between $S_0$ and $S_1$ can involve $\pi-\pi^*$ transitions ($\pi,\pi^*$ excited states) or $n-\pi^*$ transitions ($n,\pi^*$ excited states). Similarly, the triplet state ($T_1$) can be a $\pi,\pi^*$ or an $n,\pi^*$ excited state.

The most efficient fluorescence usually involves $\pi-\pi^*$ transitions because the transition probability is high (i.e., large $\varepsilon$, $k_A$, and $k_F$ and small $\tau_F$). For $n,\pi^*$ states, the fluorescence transition probability is lower because of the low degree of overlap between $n$ and $\pi^*$ orbitals. This makes fluorescence less favorable. However, the rate of intersystem crossing is usually enhanced because the energy difference between the singlet and triplet is smaller and the degree of spin-orbit coupling is greater. This can result in high phosphorescence yields. The rate of intersystem crossing is generally 1000 times faster between states of different electronic origin ($S_1(n,\pi^*) \rightarrow T_1(\pi,\pi^*)$ or $S_1(\pi,\pi^*) \rightarrow T_1(n,\pi^*)$). Phosphorescence from an $n, \pi^*$ triplet tends to be short lived and more efficient than transitions from a $\pi,\pi^*$ triplet.

With the discussion above in mind, we can make the following general observations:

1. Luminescence is generally not observed from saturated hydrocarbons as there are no $\pi$ or $n$ electrons (Weak fluorescence is sometimes observed in the vacuum UV due to $\sigma-\sigma^*$ transitions.)

2. Luminescence is rarely observed from nonaromatic hydrocarbons that have some double bonds ($\pi$ electrons). Weak fluorescence in the UV is observed for some aliphatic carbonyl compounds involving $n-\pi^*$ transitions. Some highly conjugated, but nonaromatic hydrocarbons (e.g., β-carotene and vitamin A) do exhibit substantial fluorescence due to $\pi-\pi^*$ transitions. Diacetyl exhibits significant phosphorescence even at room temperature.

3. Many aromatic hydrocarbons are intensely fluorescent because they possess a low-lying $\pi-\pi^*$ singlet state. The energy required for excitation is often low enough to prevent bond disruption. Phosphorescence is less likely without atoms providing $n$ electrons or substitutent groups that promote intersystem crossing.

4. Phosphorescence is often favorable in aromatic molecules containing carbonyl groups (e.g., benzophenones) or heteroatoms such as nitrogen (e.g., pyrimidine, pyrazine). These groups introduce nonbonding electrons and the possibility of $n-\pi^*$ transitions. The resulting increased rate of intersystem crossing generally reduces fluorescence intensities. Strong fluorescence is observed for some heterocyclic molecules (tryptophan and others that include the indole ring moiety in which nitrogen is not part of the aromatic ring system) because the $\pi,\pi^*$ state is lower in energy than the $n,\pi^*$ state.

5. Substituents attached to aromatic rings can dramatically influence quantum efficiencies and emission wavelengths as summarized in Table 12-8. The groups often influence the nature of the lowest-lying excited state ($n,\pi^*$ or $\pi,\pi^*$). In general, all groups except those having little effect on $\phi_F$, cause a red (bathochromic) shift of the fluorescence emission. The table shows that electron-donating groups (ortho-para directing groups) such as —OH in general increase $\phi_F$ relative to the parent compound. By contrast, electron-withdrawing groups (metadirecting groups) such as

**TABLE 12-8**

Effect of substituents on luminescence of aromatic compounds[a]

| Substituent | $\phi_F$ | $\phi_P$ |
|---|---|---|
| Alkyl | Slight | Increase |
| Hydroxyl, methoxyl | Increase | Increase |
| Carboxyl, keto | Large decrease | Large increase |
| Nitro, nitroso | Decrease | Increase |
| Primary, secondary, or tertiary amine | Increase | Increase |
| Sulfhydryl | Decrease | — |
| Sulfonic acid | Slight | — |
| Halogen | Decrease | Increase |
| Cyanide | Increase | — |

[a]Effect on $\phi_F$ and $\phi_P$ relative to the parent compound.

—$NO_2$ decrease $\phi_F$ by introducing a low-lying $n,\pi^*$ state.

6. The effect of halide substituents is specifically known as the **internal heavy atom effect**. Heavy atoms perturb the electron spins and enhance state mixing. This increases the rates of $S_1 \rightarrow T_1$ intersystem crossing, $T_1 \rightarrow S_0$ phosphorescence, and $T_1 \rightarrow S_0$ intersystem crossing. Because the third effect is often minor, the net effect is a decrease in $\phi_F$, an increase in $\phi_P$, and a decrease in $\tau_P$. This dramatic effect is illustrated in Table 12-9, where the ratio $\phi_P/\phi_F$ varies from 0.60 to >1000 as the halide changes from fluoride to iodide.

7. There is often an increase in $\phi_F$, a decrease in $\phi_P$, and a bathochromic shift in the emission bands as the size of the ring system and the extent of conjugation increases (see Table 12-10). For a given number of aromatic rings, the linear ring molecules usually fluoresce at longer wavelengths than the corresponding nonlinear molecules.

8. Luminescence is favored in molecules with rigid planar structures. These characteristics increase the interaction and conjugation of the π-electron system with a resultant decrease in the interaction with the solvent, internal conversion from $S_1 \rightarrow S_0$, and external conversion. For example, fluorescein is very fluorescent, while phenolphthalein is nonfluorescent. The only difference is the oxygen

fluorescein            phenolphthalein

bridge, which forces planarity. When aryl groups are separated by an alkene group, the more planar trans isomer is usually more fluorescent than the nonplanar cis isomer. The nonplanarity forced by steric hindrance is also manifested by the lower fluorescence quantum efficiency of hexamethylbenzene relative to less substituted benzenes.

9. For molecules consisting of two or more aromatic ring systems separated by alkyl groups, the fluorescence characteristics are sometimes those of the independent aromatic groups. However, in some cases, emission from one ring system results from excitation and energy transfer from another ring system in the molecule.

**TABLE 12-9**

Internal heavy-atom effect illustrated for 1-substituted naphthalenes[a]

| Compound | $\phi_F$ | $\lambda_F$ (nm) | $\phi_P$ | $\lambda_P$ (nm) | $\tau_P$ (s) | $k_{isc}$ (s) |
|---|---|---|---|---|---|---|
| Naphthalene | 0.55 | 325 | 0.051 | 469.5 | 2.6 | $1 \times 10^5$ |
| 1-Fluoronaphthalene | 0.84 | 316 | 0.056 | 473 | 1.5 | $2 \times 10^5$ |
| 1-Chloronaphthalene | 0.058 | 319 | 0.30 | 483 | 0.29 | $1.5 \times 10^7$ |
| 1-Bromonaphthalene | 0.0016 | 320 | 0.27 | 483 | 0.018 | $5 \times 10^8$ |
| 1-Iodonaphthalene | <0.0005 | — | 0.38 | 480 | 0.002 | $<3 \times 10^9$ |

[a]Measurements in ethanol–ether at 77 K.

**TABLE 12-10**
Fluorescence of linear aromatics in EPA at 77K[a]

| Compound | | $\Phi_F$ | $\lambda_{ex}$ (nm) | $\lambda_{em}$ (nm) | $\phi_p$ | $\tau_p$ (s) |
|---|---|---|---|---|---|---|
| Benzene | | 0.11 | 205 | 278 | 0.26 | 7 |
| Naphthalene | | 0.29 | 286 | 321 | 0.1 | 2.6 |
| Anthracene | | 0.46 | 365 | 400 | <0.01 | 0.04 |
| Naphthacene | | 0.60 | 390 | 480 | — | — |

[a]EPA is a mixture of ethanol, isopropanol, and ether.

10. Fluorescence from metals usually occurs only in organometallic complexes, particularly rigid metal chelates. Often, the ligand is nonfluorescent. The complex exhibits fluorescence if the lowest-lying singlet state of the ligand is changed from an $n,\pi^*$ to a $\pi,\pi^*$ state. Sometimes fluorescence involves charge-transfer transitions with metal $d$ electrons and ligand orbitals. In some rare earths, luminescence is due to $f$–$f$ transitions of metal electrons.

## Environmental Effects

Environmental factors such a temperature, solvent, pH, and the presence of other species can have a profound affect on the luminescence characteristics of a given molecule. These factors can affect the rate constants of luminescence and nonradiational deactivation or the nature of the lowest-lying excited state.

*Temperature.* In general, increasing the temperature decreases luminescence efficiencies because the rate of dynamic quenching is increased. The effect is much more dramatic for phosphorescence measurements. Delayed fluorescence (E-type) intensities can actually increase at higher temperatures because the rate of thermally assisted $T_1 \rightarrow S_1$ intersystem crossing is enhanced.

*Solvent.* The viscosity, polarity, and hydrogen-bonding characteristics of the solvent can significantly affect luminescence characteristics. In some cases, luminescence efficiencies increase with solvent viscosity due to the reduced rate of bimolecular collisions and the rate of dynamic quenching.

The solvent polarity and hydrogen-bonding char-

acteristics are critical because they affect the nature of the excited state. There is often a rapid ($10^{-11}$ to $10^{-12}$ s) reorientation of solvent molecules around the excited species that occurs before photon emission. Thus the energies of the excited state during emission and the ground state immediately following emission can be different than they were at the time of absorption. For $\pi-\pi^*$ transitions, the excited state is often more polar and basic than the ground state. Increasing the solvent polarity or protic nature decreases the energy of the excited state more so than that of the ground state, with a resultant red shift in the wavelengths of luminescence. By contrast, the excited state is less polar than the ground state for $n-\pi^*$ transitions and a blue shift occurs with increasing solvent polarity or hydrogen-bond-forming capability. In polar or protic solvents, the lowest-lying singlet state can switch from being $n,\pi^*$ to $\pi,\pi^*$ if these states are relatively close in energy. This explains why some heterocyclic or carbonyl-containing compounds fluoresce weakly or not at all in nonpolar, aprotic solvents, but appreciably in polar, protic solvents. The overall luminescence quantum efficiency is often less in hydrogen-bonding solvents, due to an increased rate of $S_1 \rightarrow S_0$ internal conversion.

*pH Effects.* The pH of solutions in protic solvents can be critical for aromatic molecules with acidic or basic functional groups (e.g., phenols, amines). In some cases, only the protonated or unprotonated form of the acid or base may be fluorescent. For example, many phenols are fluorescent only in the nonionized form. The fluorescence of amine containing compounds can decrease in acidic solution as $-NH_3^+$ forms and withdraws electrons from the ring system.

The p$K_a$ of the excited state can be a factor of 4 to 9 lower than that of the ground state. For example, the p$K_a$ of the ground state of 2-naphthol is 9.5, while the $pK_a$ of the excited state is 3.1. Different fluorescence excitation and emission spectra are observed for 2-naphthol and the 2-naphtholate anion. Fluorescence from the unprotonated form is observed at pHs much less than 9.5 because after excitation of 2-naphthol, rapid deprotonization to the anion form occurs before emission.

*External Heavy Atom Effect.* We have already observed how other solutes can affect luminescence efficiencies through different quenching mechanisms. Solvents or solutes containing heavy atoms such as haloalkanes (e.g., propyl bromide), alkali halide salts (e.g., NaBr), or some heavy metal ion salts (e.g., $TlNO_3$) can increase the rate of intersystem crossing and phosphorescence yields. For these cases, the effect is termed the **external heavy atom effect**. This effect is to be distinguished from the internal heavy atom effect, which involves heavy atoms in the molecule of interest.

### Polarization of Luminescence

When a fluorophore is illuminated with polarized light, the molecules whose absorption transition dipoles are parallel to the electric vector of the excitation beam are selectively excited. The resultant fluorescence emission is partially polarized. This phenomenon is called **photoselection**. The degree of polarization is determined by the random distribution of the orientation of molecules and the angle between the absorption and emission dipoles. The emission is depolarized by a number of phenomena. Rotational diffusion is one major cause. Radiationless transfer of energy among fluorophores can also cause depolarization. This depolarization mechanism is normally important only for analyte concentratins above 0.01 M.

In a fluorescence polarization experiment, one measures either the degree of polarization ($p$) or the anisotropy ($r$), defined as

$$p = \frac{\Phi_\parallel - \Phi_\perp}{\Phi_\parallel + \Phi_\perp} \quad (12\text{-}51)$$

$$r = \frac{\Phi_\parallel - \Phi_\perp}{\Phi_\parallel + 2\Phi_\perp} \quad (12\text{-}52)$$

where $\Phi_\parallel$ and $\Phi_\perp$ are the measured emission radiant power of the parallel (vertical or $\pi$) and perpendicular (horizontal or $\sigma$) polarization components, respectively. For completely polarized light, $\Phi_\perp = 0$ and $p = r = 1.0$. For unpolarized light, $\Phi_\perp = \Phi_\parallel$ and $p = r = 0$. For intermediate values, $p$ and $r$ are not equal

but are related by

$$p = \frac{3r}{2 + r} \quad \text{or} \quad r = \frac{2p}{3 - p} \quad (12\text{-}53)$$

We will use anisotropy in the following discussion because many theoretical expressions are simpler when expressed in terms of this parameter.

The measured anisotropy ($r$), assuming that rotational diffusion is the only significant process resulting in loss of anisotropy, is given by one form of the Perrin equation,

$$r = \frac{r_0}{1 + \tau_F/\phi_r} \quad (12\text{-}54)$$

where $r_0$ is the intrinsic anisotropy or the anisotropy in the absence of rotational diffusion (a rigid fluorophore), $\tau_F$ is the fluorescence lifetime, and $\phi_r$ is the rotational correlation or relaxation time. For $\tau_F \ll \phi_r$, $r \approx r_0$ and there is little depolarization because the fluorophore finishes emitting before the molecule has a chance to rotate (i.e., for the emission dipole to reorientate). Similarly, if $\tau_F \gg \phi_r$, there is almost complete depolarization ($r \approx 0$) because the orientation of the molecule changes significantly before it can emit. Typically, $r_0$ is no greater than 0.4 and can be as low as $-0.2$ (or $p_0$ varies from 0.5 to $-0.33$).

The rotational relaxation time for a spherical molecule is given by

$$\phi_r = \frac{\eta \overline{V}}{RT} \quad (12\text{-}55)$$

where $\eta$ is the viscosity of solution in poise (P) or $g\ cm^{-1}\ s^{-1}$, $\overline{V}$ is the partial molar volume of the molecule ($cm^3\ mol^{-1}$), $R$ is the gas constant ($8.3 \times 10^7\ erg\ mol^{-1}\ K^{-1}$), and $T$ is the absolute temperature. Combining equations 12-54 and 12-55 and taking the reciprocal yield

$$\frac{1}{r} = \frac{1}{r_0} + \frac{\tau_F RT}{\eta \overline{V} r_0} \quad (12\text{-}56)$$

In a Perrin plot, $1/r$ is plotted vs. $T/\eta$. The intercept at infinite viscosity yields the intrinsic anisotropy. The molar volume is given by

$$\overline{V} = \frac{M}{\rho} \quad (12\text{-}57)$$

where $\rho$ is the density ($g\ cm^{-3}$) and $M$ is the molecular weight of the molecule (g mol$^{-1}$). Thus

$$\phi_r = -\frac{\eta M}{\rho RT} \qquad (12\text{-}58)$$

For a typical protein, $\rho = 1.37$ g cm$^{-3}$, and at 25°C, $RT = 2.43 \times 10^{10}$ erg mol$^{-1}$. Thus

$$\phi_r = (3.0 \times 10^{-11})\eta M \qquad (12\text{-}59)$$

For water, $\eta = 0.01$, so that the rotational correlation time in nanoseconds is given by

$$\phi_r = (3 \times 10^{-4})M$$

For a relatively small molecule ($M = 333$ g mol$^{-1}$) with $\tau_F = 10$ ns, $\phi_r = 0.1$ ns, $\tau_F/\phi_r = 100$, and $r = 0.01 r_0$. Thus the fluorescence is almost completely depolarized because a small molecule rotates many times during the fluorescence lifetime. With a large molecule ($M = 33,300$ g mol$^{-1}$) and the same lifetime, $\phi_r = 10$ ns, $\tau_F/\phi_r = 1$ and $r = r_0/2$. Thus the anisotropy can be as high as 0.2 if $r_0 = 0.4$.

The time decay of the anisotropy can also be measured. In a pulse excitation experiment, the total emission intensity often decreases exponentially, as indicated by equation 12-39. Independent of the intensity decay, the anisotropy decays with time according to

$$r(t) = r_0 e^{-t/\phi_r} \qquad (12\text{-}60)$$

Thus, as the molecule rotates after the excitation pulse with parallel polarization, the polarization decreases as

the $\Phi_\perp$ component increases relative to the $\Phi_\parallel$ component. Both $\Phi_\perp$ and $\Phi_\parallel$ decrease with time also with the same exponential form as equation 12-39, whereas the difference $D(t) = \Phi_\parallel(t) - \Phi_\perp(t)$ decreases as

$$D(t) = (\Phi_t)_0 r_0 e^{-t/\tau_F - t/\phi_r} \qquad (12\text{-}61)$$

where $(\Phi_t)$ is the total initial emission radiant power ($\Phi_t = \Phi_\parallel + 2\Phi_\perp$) and the subscript 0 indicates the value at time zero. The decay of the total fluorescence, polarization components, and $D(t)$ is illustrated in Figure 12-9.

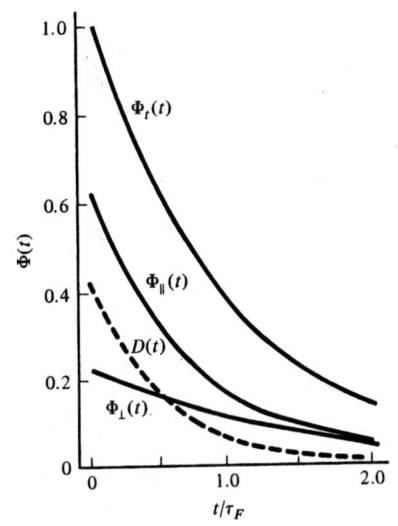

**FIGURE 12-9** Time-resolved decays of fluorescence anisotropy. It is assumed that $r_0 = 0.4$ and $\tau_F = \phi_r$.

## PROBLEMS

**12-1.** For a molecule the rate constants for fluorescence and nonradiative decay are $1.00 \times 10^9$ and $9 \times 10^9$ s$^{-1}$, respectively. Calculate the fluorescence quantum efficiency and lifetime. Also calculate the radiative lifetime.

**12-2.** From the results of problem 12-1, calculate the fluorescence power yield if the wavelengths of maximum excitation and emission are 300 and 400 nm, respectively.

**12-3.** Calculate the frequencies of the two lowest rotational transitions of ICl in MHz and cm$^{-1}$. Assume that the equilibrium internuclear distance is 2.32 Å and that centrifugal distortion is negligible.

**12-4.** Calculate the oscillator strength, the integrated molar absorptivity, and the transition probability for the $n \to \pi^*$ transition for acetaldehyde from the following

data: $\lambda_m = 209$ nm, $\varepsilon_m = 15$ L mol$^{-1}$ cm$^{-1}$, and $\Delta\bar{\nu} = 600$ cm$^{-1}$ (see Appendix F).

**12-5.** Determine the electronic configuration and bond order for the ground state of $N_2$. What are the possible term symbols?

**12-6.** Why do the half-widths of electronic absorption bands differ?

**12-7.** The fluorescence lifetime without quenching is 6.0 ns while with a quencher concentration of $1.00 \times 10^{-2}$ M, the fluorescence lifetime is 2.0 ns. Assume the rate constant for intersystem crossing is $1.00 \times 10^8$ s$^{-1}$ and that internal conversion from $S_1$ to $S_0$ is negligible. Calculate the Stern–Volmer quenching constant ($K_q$), the rate constant for fluorescence, and the quenching rate constant ($k_q$).

**12-8.** For a molecule, $k_F = 1.0 \times 10^8$ s$^{-1}$, $k_{isc} =$

$1.0 \times 10^7 \, s^{-1}$, $k_{ic} = 1.0 \times 10^4 \, s^{-1}$ and $k_q$ for a specific quencher equals $1.0 \times 10^{10} \, L \, mol^{-1} \, s^{-1}$. Calculate the fluorescence quantum efficiency with no quencher and with a quencher concentration of $1.0 \times 10^{-2} \, M$.

**12-9.** The quantum yield $\phi_L$ of a luminescence process (fluorescence or phosphorescence) is expressed as the rate constant of the luminescence process divided by the sum of the rate constants for all deactivation processes (i.e., it is the relative rate of the luminescence process).

(a) A molecule is excited into the first excited singlet state $S_1$ and fluorescence is observed ($S_1 \rightarrow S_0 + h\nu$). The deactivation rate constants for the various processes are: $k_F = 2.0 \times 10^8 \, s^{-1}$ (fluorescence), $k_{isc} = 2.0 \times 10^8 \, s^{-1}$ (intersystem crossing $S_1 \rightarrow T_1$), and $k_{ec} = 5 \times 10^7 \, s^{-1}$ (external conversion $S_1 \rightarrow S_0$). Internal conversion to $S_0$ is negligible. Find the fluorescence quantum yield $\phi_F$ and lifetime of the $S_1$ state.

(b) Phosphorescence from the triplet state is also observed ($T_1 \rightarrow S_0 + h\nu$). If $k_P = 0.70 \, s^{-1}$ (phosphorescence) and $k'_{ec}$ and $0.20 \, s^{-1}$ (external conversion $T_1 \rightarrow S_0$), find the quantum yield or phosphorescence $\phi_P$ and the lifetime of the triplet state.

**12-10.** For a molecule, the phosphorescence lifetime without quenching is $1.0 \, s$, while with a quencher population of $10^{-2} \, M$ the phosphorescence lifetime is $0.10 \, s$. Calculate the rate constant for phosphorescence as-

suming that $k_{isc} \gg (k_F + k_{ec} + k_{ic})$. Calculate the rate constant for external conversion of the triplet state ($k'_{ec}$), the quenching rate constant for the triplet state ($k'_q$), and the Stern–Volmer quenching constant. Assume that intersystem crossing from $T_1$ to $S_0$ is negligible.

**12-11.** In molecular fluorescence spectrometry a quantum efficiency of unity is almost never achieved. Describe three processes which are responsible for this.

**12-12.** Calculate the degree of polarization and the anisotropy for a molecule in which the fluorescence signals for the parallel and perpendicular components are $0.70 \, V$ and $0.50 \, V$, respectively.

**12-13.** For fluorescence excited with parallel polarized radiation, why does the parallel component of emission decay more rapidly than the perpendicular component?

**12-14.** Calculate the bond force constant for a C—H vibrational stretch in a molecule that occurs at $3000 \, cm^{-1}$.

**12-15.** A Perrin plot of the reciprocal of the anisotropy vs. the ratio of the absolute temperature to viscosity is made for a solution of a fluorescent protein of fluorescence lifetime 10 ns by varying the solvent. The intercept is 3.33 and the slope is $1.84 \times 10^{-5}$ $K^{-1} \, g^{-1} \, cm^{-1} \, s^{-1}$. Determine the intrinsic anisotropy, the specific molar volume of the protein, and its rotational correlation time at 25°C if the solution viscosity is 0.012 P.

## REFERENCES

The following are general references to molecular spectroscopy.

1. W. A. Guillory, *Introduction to Molecular Structure and Spectroscopy*, Allyn and Bacon, Boston, 1977.
2. C. N. Banwell, *Fundamentals of Molecular Spectroscopy*, 2nd ed., McGraw-Hill, Maidenhead, Berkshire, England, 1972.
3. G. M. Barrow, *Introduction to Molecular Spectroscopy*, McGraw-Hill, New York, 1962.
4. G. Herzberg, *Diatomic Molecules*, vol. I, Van Nostrand Reinhold, New York, 1950.
5. G. Herzberg, *The Spectra and Structures of Simple Free Radicals*, Cornell University Press, Ithaca, N.Y., 1971.
6. G. Herzberg, *Molecular Spectra and Molecular Structure: Infrared and Raman Spectra of Polyatomic Molecules*, vol. 2, D. Van Nostrand, Princeton, N.J., 1945.
7. E. B. Wilson, Jr., J. C. Decius, and P. C. Cross, *Molecular Vibrations*, McGraw-Hill, New York, 1955.
8. G. Herzberg, *Molecular Spectra and Molecular Structure: Electronic Spectra of Polyatomic Molecules*, vol. 3, D. Van Nostrand, New York, 1967.

9. J. I. Steinfield, *Molecules and Radiation: An Introduction to Molecular Spectroscopy*, 2nd ed., MIT Press, Cambridge, Mass., 1985.
10. R. Chang, *Basic Principles of Spectroscopy*, R. E. Krieger, Melbourne, 1971.
11. R. P. Bouman, *Absorption Spectroscopy*, Wiley, New York, 1962.

The following references deal with molecular luminescence spectroscopy.

12. E. L. Wehry, "Effects of Molecular Structure and Molecular Environment on Fluorescence," Chap. 3 in *Practical Fluorescence: Theory, Methods, and Techniques*, G. G. Guilbault, ed., Marcel Dekker, New York, 1973.
13. S. G. Schulman, "Luminescence Spectroscopy: An Overview," chap. 1 in *Molecular Luminescence Spectroscopy: Methods and Applications*, pt. I, S. G. Schulman, ed., Wiley, New York, 1985.
14. J. D. Winefordner, S. G. Schulman, and T. C. O'Haver, *Luminescence Spectrometry in Analytical Chemistry*, Wiley-Interscience, New York, 1972, pp. 32–81.
15. R. S. Becker, *Theory and Interpretation of Fluorescence*

*and Phosphorescence*, Wiley-Interscience, New York, 1969.

16. W. R. Seitz, "Luminescence Spectrometry (Fluorimetry and Phosphorimetry)," in *Treatise on Analytical Chemistry*, 2nd ed., I. M. Kolthoff and P. J. Elving, eds., pt. I, vol. 7, Wiley-Interscience, New York, 1981, pp. 161–204.

17. C. A. Parker, *Photoluminescence of Solutions*, Elsevier, Amsterdam, 1968.

18. S. Udenfriend, *Fluorescence Assay of Biology and Medicine*, vol. II, chap. 1, Academic Press, New York, 1969.

19. J. R. Lakowicz, *Principles of Fluorescence Spectroscopy*, Plenum Press, New York, 1983.

# CHAPTER 13

# Ultraviolet and Visible Molecular Absorption Spectrophotometry

Ultraviolet-visible molecular absorption spectrophotometry (often called light absorption spectrophotometry or just UV-visible spectrophotometry) has a long and continuing history of use in analytical chemistry. This technique is based on measuring the absorption of near-UV or visible radiation by molecules. Radiation in this wavelength region causes electronic transitions at wavelengths characteristic of the molecular structure of the molecule, as detailed in Chapter 12.

The term *spectrophotometry*, rather than *spectrometry*, is used for absorption measurements because they involve measuring, either simultaneously or sequentially, the ratio of the radiant power transmitted through a sample solution to that transmitted by a blank solution. This ratio, the transmittance $(T)$, is used to calculate the absorbance $(A = -\log T)$. Under suitable conditions, the absorbance is directly proportional to the analyte concentration through Beer's law $(A = abc)$.

Quantitative molecular absorption measurements were first demonstrated in the nineteenth century. The intensities of the colors of standard and sample solutions were matched by adjusting the thickness of the solution through which the light passed. These are denoted **colorimetric** techniques because the human eye is the detector. For molecular absorption measurements in the visible region, the modifier colorimetric is still often used instead of spectrophotometric. The introduction of the Beckman DU spectrophotometer with UV capability and double-beam recording spectrophotometers by Cary in the 1940s stimulated the development of new spectrophotometric methods.

In this chapter we first consider the general design characteristics of UV-visible spectrophotometers including some commercial examples. Next signal and $S/N$ expressions are introduced to provide an understanding of the instrumental and chemical factors that affect the measurement accuracy and precision. This is followed by a discussion of the causes of nonlinear calibration curves and the methodology and performance characteristics. The chapter concludes with a survey of the broad range of applications in which UV-visible spectrophotometry is used.

## 13-1 INSTRUMENTATION

Spectrophotometers or photometers can be based on single detectors or multichannel detectors, configured for single-beam (SB) or double-beam (DB) measurements, and designed for fixed-wavelength measurements or for acquiring complete absorption spectra.

Some of these basic design features are illustrated in Figure 13-1.

The single-detector approach is the most popular, in which case a particular wavelength band of source radiation is transmitted by the wavelength selector to the sample cell(s) and finally to the detector, as shown in Figure 13-1a and b. With SB instruments (Figure 13-1a), one beam passes through one cell and is incident on the detector. In DB instruments (Figure 13-1b), the beam from the wavelength selector is split, with a rotating sectored mirror, into two beams, which alternately pass through a sample cell and a reference cell. The sample and reference beams are then recombined to be incident on a single detector. Filter-based instruments (photometers) are always intended for fixed-wavelength measurements, while instruments based on monochromators (spectrophotometers) may also be configured for scanning spectra. The double-beam in-time approach illustrated in Figure 13-1b is generally preferred over the double-beam in-space approach (see Figure 4-41a) in scanning instruments because of the difficulty in matching two detectors. However, a sep-

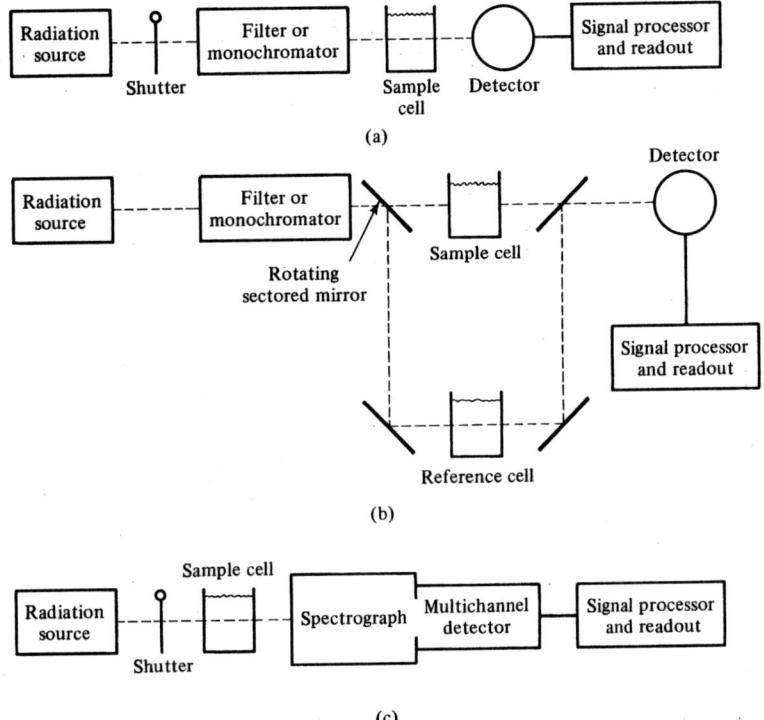

(a)

(b)

(c)

**FIGURE 13-1** Block diagrams of different types of UV-visible molecular absorption spectrophotometers. In (a) a single-detector, single-beam spectrophotometer is shown. A filter of monochromator selects a particular wavelength range of source radiation to be incident on the sample cell in the sample compartment. The transmitted radiant power is detected with a vacuum phototube, a photomultiplier tube, or a photodiode and converted to a readout signal with a signal processor. In (b) a single-detector, double-beam in-time spectrophotometer is shown. The radiation beam from the wavelength selector is split and alternately passes through the sample and reference cells and then recombined to be incident on the detector. The amplitude information proportional to the radiant power passed by the sample and reference cells is extracted and processed to yield a readout signal proportional to the transmittance or absorbance. In (c) a single-beam multichannel spectrophotometer is illustrated. The source radiation is directly impingent on the sample cell. The transmitted radiant power is directed to the entrance slit of a spectrograph where the radiation is dispersed onto a multichannel detector in the focal plane. After a suitable integration time, the signals accumulated by the individual detector elements are rapidly interrogated, digitized, stored, and processed to yield an absorption spectrum.

arate detector for the sample and reference beam is utilized for some fixed-wavelength instruments such as HPLC absorption detectors.

A different and newer approach is illustrated in Figure 13-1c, which shows the design of a single-beam, multichannel spectrophotometer. Note that the continuum source radiation first passes through the sample cell and then is dispersed by a spectrograph. A multichannel detector, a linear diode array, senses the transmitted radiation to allow the simultaneous detection of many spectral elements (i.e., small wavelength bands) rather than detection of one spectral element at a time as occurs with single-detector instruments. Thus absorption spectra can be acquired rapidly and then displayed, or the absorbances at particular wavelengths can be reported. Multichannel spectrophotometers can also be configured as double-beam in-time spectrophotometers.

In any spectrophotometer or photometer, the detector signal in the absence of source radiation must be measured and subtracted or compensated to define complete absorption or 0% *T*. Therefore, a shutter or chopper placed between the source and detector is used to block the source radiation momentarily and establish the 0% *T* condition.

### Instrumental Components

We now consider the characteristics of the various components of UV-visible spectrophotometers in some de-

tail. The components used to extend measurements into the near-infrared (NIR) region are also discussed.

*Sources.* The radiation source is usually a tungsten or tungsten-halogen lamp to provide adequate continuum radiation over the entire visible and NIR region. For UV capability, a $D_2$ or $H_2$ arc is required. Provision is made for manually or automatically switching the proper lamp into position. The arc lamp is used for absorption measurements below about 350 nm, where its spectral radiance is greater than that of the tungsten lamp. A $D_2$ lamp is generally preferred over an $H_2$ lamp because its radiance is about three times greater. For simplicity, a $D_2$ or $H_2$ lamp can serve as a single source for the entire UV-visible region, although the radiance in the visible region is less than with a tungsten lamp (see Figure 4-5c) and sometimes less stable. In some instruments the source radiation is modulated with a mechanical device (a chopper or a rotating sectored mirror). A well-regulated and stable lamp power supply is critical for providing a constant radiance; in some instruments, the lamp radiance is stabilized by optical feedback (see Section 6-5).

*Wavelength Selection.* A portion of the source radiation is collected and directed to the wavelength selector or the sample cell(s) with lenses or mirrors. With single-detector instruments, the wavelength selector is normally placed before rather than after the sample cell to minimize heating or photodecomposition

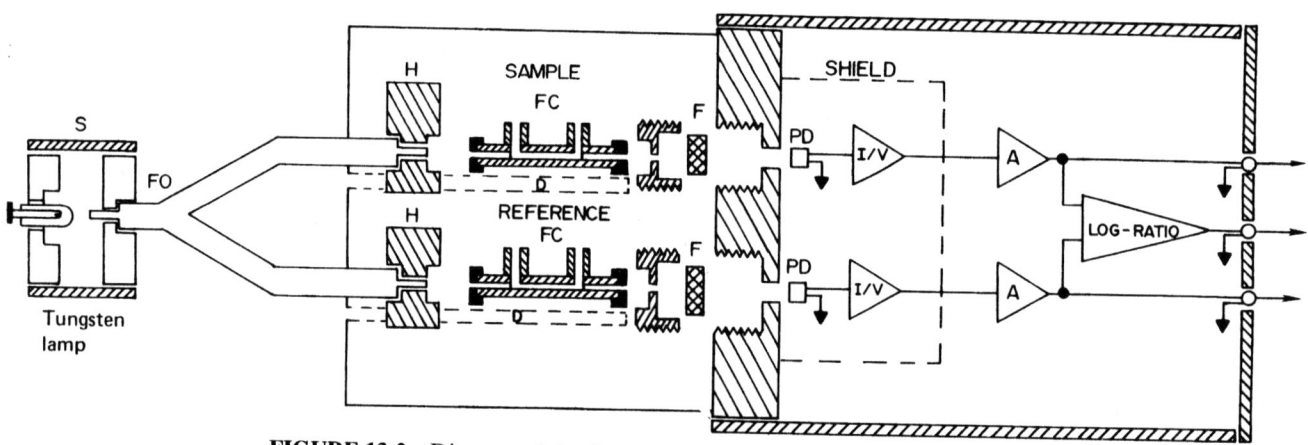

**FIGURE 13-2** Diagram of the design used in a commercial double-beam in-space photometer (Alpkem Corp). The radiation from a miniature tungsten lamp is split into two beams with a randomized, bifurcated fiber optic bundle (FO) and directed to the sample and reference flow cells (FC). Two identical filters (F) isolate the desired wavelength range in each beam. The radiation transmitted by each filter is converted to a current and then a voltage by separate photodiodes (PD) and current-to-voltage converters (*I/V*). Finally, the two signals are processed by log-ratio amplifier to yield a readout voltage proportional to the absorbance. [Adapted from C. J. Patton and S. R. Crouch, *Anal. Chim. Acta, 179,* 189 (1986), with permission of Elsevier Science Publishers.]

of the sample by radiation at wavelengths not monitored. If these effects are not significant, the wavelength selector can be placed after the sample cell. With multichannel spectrophotometers, the multichannel detector must be located at the focal plane of the spectrograph; the sample cell is placed before the wavelength selector and illuminated with white radiation.

Simple *photometers* such as that shown in Figure

13-2 employ interference filters for wavelength selection. Provision is made for easy changing of the filters. Inexpensive photometers are useful for routine analysis in which a large spectral range can be tolerated. Often a set of filters with half-widths of 10 to 20 nm is provided with central wavelengths corresponding to the analysis wavelengths of species commonly determined.

In *single-detector spectrophotometers*, an image of

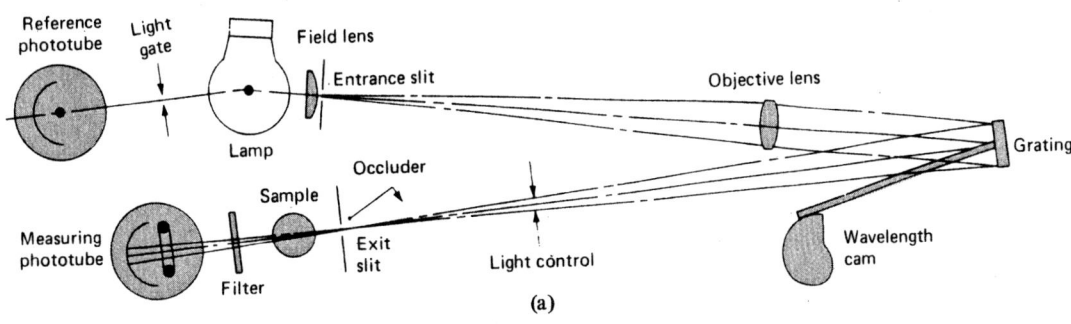

(a)

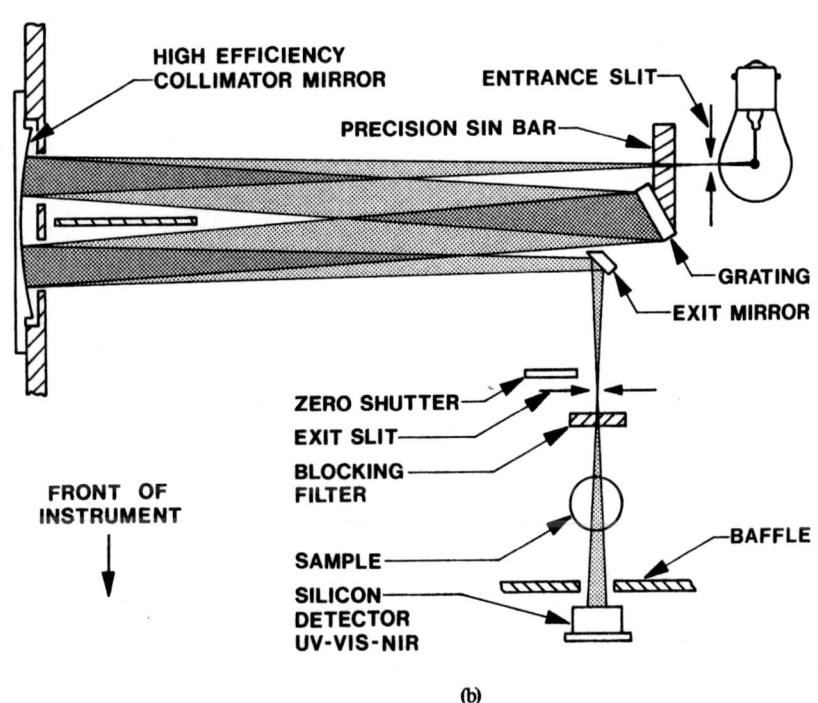

(b)

**FIGURE 13-3** Two examples of simple single-beam spectrophotometers. (a) The Spectronic 20. A vee-shaped slit (light control) is used to adjust the radiant power incident on the sample cell. The occluder blocks the light beam when the sample cell is removed. (b) The Turner 320. The monochromator is based on the Ebert design, and the lamp current is varied to adjust the radiant power incident on the sample cell. [(a) Courtesy of Milton Roy Co., Analytical Division, Rochester, NY, (b) courtesy of Sequoia-Turner Corp., Mt. View, Calif.]

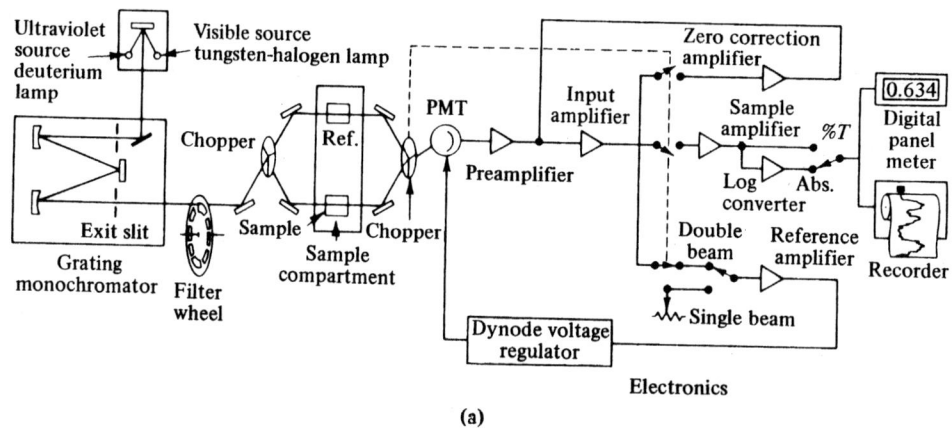

**(a)**

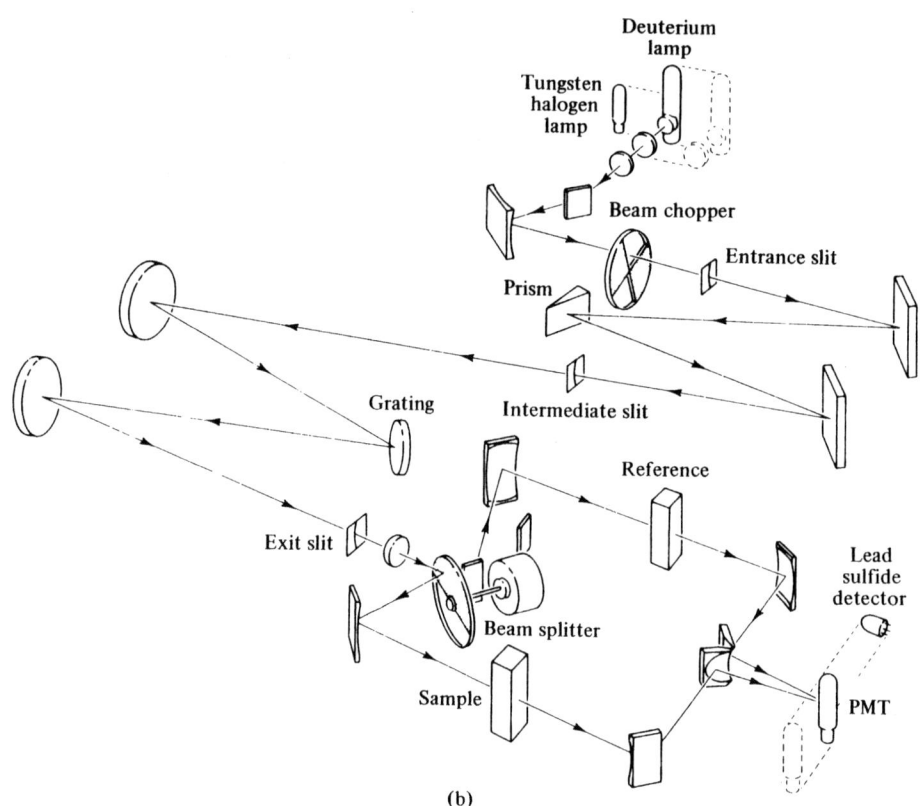

**(b)**

FIGURE 13-4 Schematic diagrams of double-beam in-time scanning spectrophotometers. In (a), the optics and signal processing electronics of a spectrophotometer based on dual-source and single-pass monochromator design are emphasized (Varian 634). Note that separate amplifiers synchronized to the chopper are used to process the 0% *T*, sample, and reference signals. The reference signal amplifier is in a feedback loop with the photomultiplier tube detector to maintain the reference signal as a constant value during scanning. A filter wheel before the monochromator entrance slit reduces stray radiation. The diagrams in (b) and (c) illustrate the optical paths in spectrophotometers based on a double monochromator (Cary 17) and a dual-pass monochromator (Varian 2200), respectively. In (b), the source radiation is dispersed by a prism and a grating monochromator in series. In contrast, the spectrophotometer in (c) utilizes the same grating (different portions) to disperse the radiation twice. (Courtesy of Varian Associates, Inc., Instrument Group.)

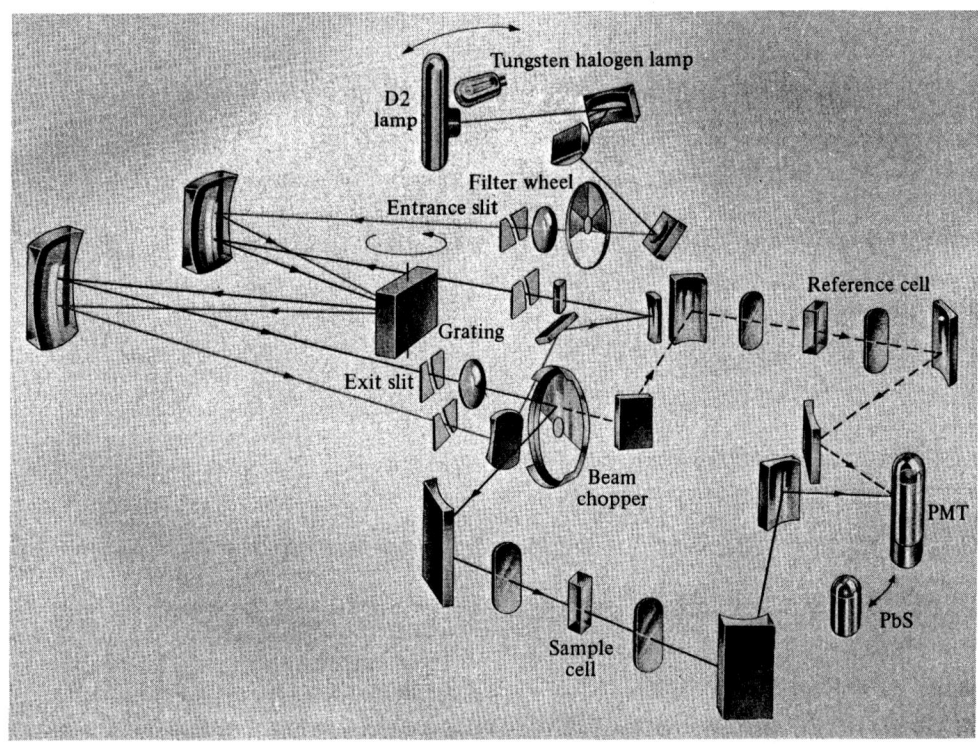

Tungsten halogen lamp

D2 lamp

Filter wheel

Entrance slit

Reference cell

Grating

Exit slit

Beam chopper

PMT

PbS

Sample cell

(c)

**FIGURE 13-4** (*continued*)

the source is focused on the entrance slit of the mono-chromator with an F/n that matches that of the disper-sive device. A low-dispersion monochromator with a fixed slit width (typically, a 2- to 20-nm spectral band-pass) and manual wavelength adjustment is used for such simple single-beam spectrophotometers as those shown in Figure 13-3. Higher-quality SB and DB spec-trophotometers are based on medium-dispersion mono-chromators. Depending on the instrument, the spectral bandpass can be varied from 10 nm down to as low as 0.05 nm. For instruments intended for only fixed-wave-length measurements, slew scanning is often provided to allow rapid adjustment of the analysis wavelength.

*Scanning spectrophotometers* usually provide a wide range of scanning rates (from 1 up to as large as 2000 nm min$^{-1}$). Some typical commercial double-beam scanning spectrophotometers are shown in Figure 13-4. Rapid-scan spectrophotometers based on an os-cillating optical component in the monochromator (see Section 4-6) are available to acquire spectra in 1 s or less.

Many fixed-wavelength and scanning spectropho-tometers allow manual or automatic placement of filters in the optical train to reduce stray radiation at wave-lengths significantly different from the monochromator

wavelength setting (see Figure 13-4a). Sometimes a double monochromator (Figure 13-4b) or a double-pass monochromator (Figure 13-4c) is used to reduce stray radiation.

With multichannel spectrophotometers (Figure 13-5), the wavelength resolution is determined by the size of the pixels in the diode array. Therefore, the entrance slit of the spectrograph is fixed at a slit width approximately equal to the width of an individual diode element; resolution of the order of 1 to 2 nm is common.

*Sample Compartment and Sample Cells.* The sample compartment is a light-tight compartment with a lid that has provision for securely holding one or more sample cells. The radiant power transmitted by the wavelength selector (or from the source in multichannel spectrometers) is directed to the sample cell(s) in the sample compartment. In some spectrophotometers, ad-ditional optics are used to collimate the light beam, to focus the light in the center of the sample cell, or in the case of double-beam instruments, to split and direct the light beam to the sample and reference cells. The change in the width of the beam across the sample cell should be minimal, so that all rays travel approximately the same pathlength. Often, additional lenses or mirrors

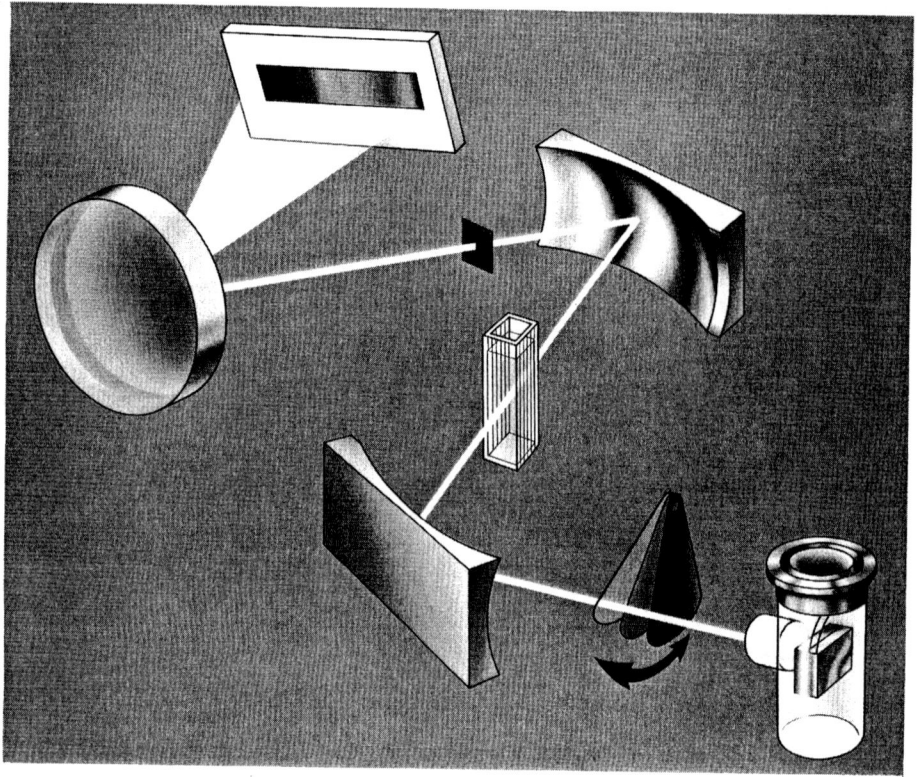

(a)

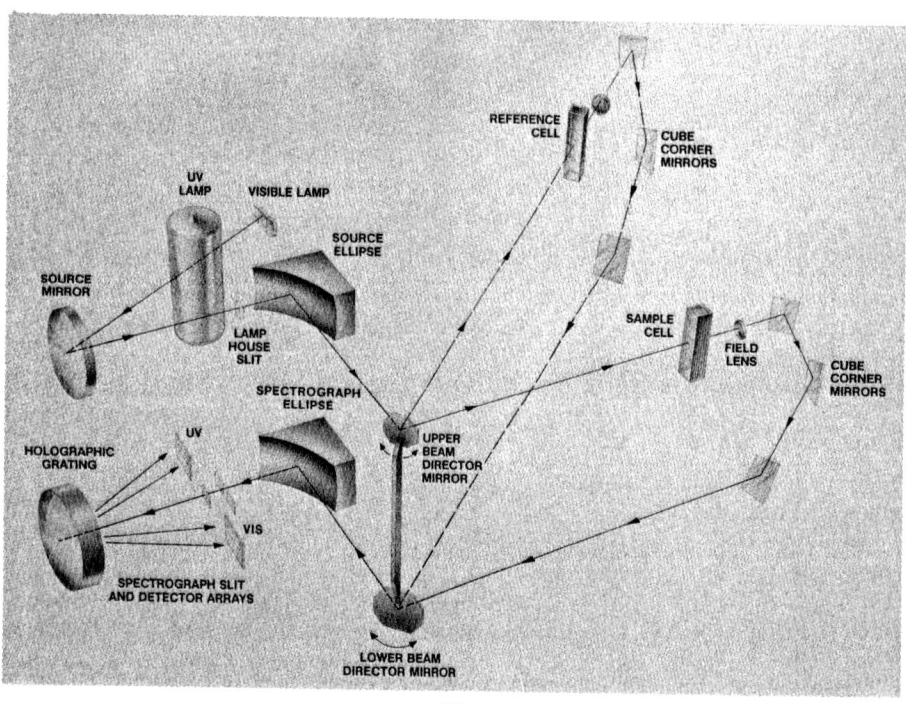

(b)

**FIGURE 13-5** Multichannel spectrophotometers. In (a) the HP 8451A single-beam diode array spectrophotometer is shown. One mirror is used to direct the source radiation ($D_2$ lamp) through the same cell, and a second mirror focuses the transmitted radiation on the spectrograph entrance slit. A concave holographic grating disperses the radiation on a 315-element diode array. The wavelength coverage is 190 to 820 nm with a resolution of 2 nm. In (b) a HP8450A double-beam diode array spectrophotometer is illustrated. The upper beam director mirror directs the radiation from two sources (the radiation from the tungsten lamp passes through the $D_2$ lamp) sequentially to up to five sample cells, of which one normally contains the reference solution. Additional mirrors and the lower beam director mirror divert and image the transmitted radiation onto the spectrograph entrance slit. The radiation in the regions 200 to 400 nm and 400 to 800 nm is separately diffracted and directed by two holographic gratings on one substrate to two 311-element diode arrays. The resolution is 1 nm in the UV region and 2 nm in the visible region. (Courtesy of Hewlett-Packard Co.)

direct the light transmitted through the sample cell(s) to the detector (or to the spectrograph entrance slit in multichannel spectrophotometers).

It is critical that the sample cell holder allow easy insertion and removal of sample cells and that the sample cell be reproducibly located. A multiple sample cell holder based on rotary or linear movement allows several samples to be loaded into the sample compartment and rapid manual or automatic positioning of the desired sample cell into the light beam. Provision is often made for controlling the sample cell temperature by circulating constant temperature water through the cell holder. Alternatively, thermoelectric devices are used for temperature control. A magnetic stirrer below the sample cell allows mixing of solutions added to the sample cell.

A wide variety of sample cell types (i.e., construction materials, pathlengths, and shapes) are available. Common construction materials include normal glass, quartz, fused silica, or various plastics. (See Appendix B for transmission characteristics.) The more expensive quartz or fused silica sample cells are required for measurements at wavelengths shorter than 300 nm, where normal glass and most plastics absorb significantly.

Some common cell shapes and designs are shown in Figure 13-6. The standard spectrophotometer cell shown in Figure 13-6a has a 1-cm internal pathlength and a 4.5-mL internal volume. Semimicro and micro cells (Figure 13-6b) with a 1-cm pathlength but a smaller internal volume are available for sample-limited situations. Sample cells can be purchased with pathlengths from 1 mm to 5 cm. Short pathlengths are useful for highly absorbing solutions. Sample cells with path-

lengths up to 10 cm, often of cylindrical shape (Figure 13-6c), are provided for absorption measurements of weakly absorbing solutions or gases. In simple spectrometers or photometers, an inexpensive test tube-shaped sample cell (Figure 13-6d) is often used.

With standard sample cells, test solutions are added to the cell manually, or in some cases, automatically with electrically controlled dispensers. Different sample cells can be used for different solutions. The sample cell must be emptied or evacuated and then rinsed before the next sample is added.

Sample throughput can be increased by using various types of flow cells (Figure 13-6e) in which the filling and cleaning operations are easily automated. Test and rinse solutions are pumped to the cell or drawn through the cell and then stopped for the absorption measurement. The pumping time is adjusted to ensure that the previous solution has been completely removed and replaced by the next sample solution. This approach is used with automatic samplers to provide a highly automated sequential batch analyzer (see Section 6-6).

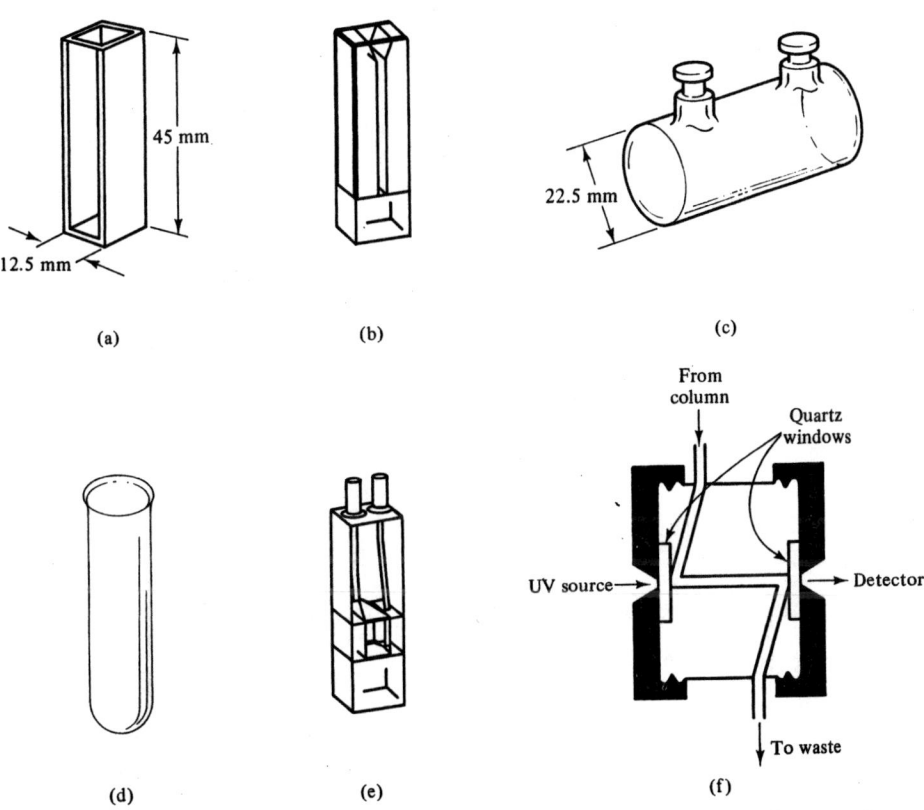

**FIGURE 13-6** Sample cell types. In (a) a standard spectrophotometric cell with a 1-cm pathlength is shown. In (b) a micro-sample cell allows smaller solution volumes to be used. Cylindrical cells as illustrated in (c) provide long pathlengths. Simple spectrophotometers often employ test tube–shaped cells as shown in (d). For automated sample delivery, flow cells as shown in (e) are used. In (f) a zee-shaped flow cell suitable for use in HPLC detection is illustrated.

Flow cells can also be used to make absorption measurements on flowing streams. With analyzers based on segmented or unsegmented flow (see Sections 6-6 and 13-5), sample plugs are mixed with reagents and pumped to the flow cell. The flow rate and data acquisition time are controlled so that the absorbance due to a given sample is measured at the proper time.

Specially designed flow cells are available for use in HPLC absorption detectors (Figure 13-6f). Here the sample cell volume (typically 0.5 to 2 μL) and flow path design are critical to prevent significant broadening of the eluting bands.

*Detectors.* Vacuum phototubes and photodiodes are used as detectors in many simple photometers and spectrophotometers with a large spectral bandpass. With higher-resolution single-detector spectrophotometers, the light throughput is less; a PMT is normally used as the optical transducer to provide a higher $S/N$. To enhance the $S/N$ in the region 600 to 1000 nm, a PMT with an extended red response is required. A few commercial spectrophotometers allow measurements further into the NIR region. Provision is made for temporarily replacing the PMT detector with a PbS detector to extend the useful wavelength range to about 2.5 μm (see Figure 13-4b). Often, a different grating is used for the NIR region.

Multichannel spectrophotometers based on diode arrays are normally configured to cover the range 190 to 850 nm. The detector elements provide good responsivity to wavelengths as long as 1100 to 1200 nm.

*Signal Processor and Readout.* The detector output signal contains information about the radiant power transmitted by the sample or reference solutions. This signal must be processed by hardware and/or software to extract the desired information and to convert it to a form suitable for display by the readout device.

In all spectrophotometers, some analog signal processing is required. With single-detector spectrophotometers or photometers, an operational amplifier (op amp), in the current-to-voltage ($I–V$) configuration, transforms the photoanodic or photodiode current into a voltage of suitable amplitude. With noncomputerized instruments, further signal processing is usually accomplished with analog circuitry, although digital displays are common. With computerized, single-detector instruments, the amplified voltage signals are digitized by an analog-to-digital converter (ADC). The digital data are stored in memory and further processed by software. Multichannel spectrophotometers are always computer-based, due to the large amount of data generated in a short time. Here the spectral information from typically 300 to 1000 photodiodes must be read out, amplified, digitized, and stored in several milliseconds (see Figure 4-26).

With noncomputerized single-beam spectrophotometers intended for fixed-wavelength measurements, the voltage from the $I–V$ converter, sometimes after further voltage amplification, is directly displayed on a readout device for transmittance readout. For direct absorbance readout, it is necessary to convert the photosignal into a voltage proportional to the absorbance with logarithmic circuitry (see Section 2-5) before display. The readout device in either case is typically an analog or digital voltmeter. If the source is modulated, synchronous detection electronics are required to extract the amplitude information before display or logarithmic conversion.

**TABLE 13-1**

Readout signals in absorption measurements

| Readout signal | Solution | Shutter | Equation | |
|---|---|---|---|---|
| 0% $T$ signal[a] | Blank | Closed | $E_{0t} = E_d$ | (13-1) |
| Total reference signal[b] | Blank | Open | $E_{rt} = E_r + E_{0t}$ | (13-2) |
| Total sample signal[b] | Analyte | Open | $E_{st} = E_s + E_{0t}$ | (13-3) |
| Reference logarithmic signal[c] | Blank | Open | $E_{lr} = -k' \log \dfrac{E_r}{k''}$ | (13-5) |
| Sample logarithmic signal[c] | Analyte | Open | $E_{ls} = -k' \log \dfrac{E_s}{k''}$ | (13-6) |

[a]The dark current signal ($E_d$) is due to dark current and offsets.

[b]The lamp signals are due to the source radiation transmitted by the reference solution ($E_r$) and the sample solution ($E_s$).

[c]$k'$ is the logarithmic conversion factor [V (A.U.)$^{-1}$] and $k''$ is the reference voltage for which $E_{lr} = 0$ V.

Signal processing with noncomputerized double-beam spectrophotometers is more complex. The sample and reference signal amplitudes are first extracted from the modulated signal waveform. Then the ratio or logarithm of the ratio of the sample and reference signals is computed. For scanning instruments, the spectral data ($A$ or $T$ vs. $\lambda$) are plotted in real time on an analog recorder.

With microcomputer-based spectrophotometers, most of the computation steps discussed above (calculation of $T$ and $A$ and extraction of amplitude information from modulated waveforms) are performed on digitized and stored reference and sample signal data by software. The readout device can be a digital meter, plotter, a printer, a video monitor, or a combination of these devices.

In all spectrophotometers, the noise equivalent bandpass ($\Delta f$) is controlled by some limiting time constant or integration time which is set to about 1 s on simple instruments or is user controllable over the range 0.01 to 10 s with more sophisticated instruments. With multichannel spectrophotometers, $\Delta f$ is usually determined by the integration time set by the manufacturer, which ensures near saturation of the diodes experiencing the highest intensity.

## Readout Considerations

The reference signal ($E_r$) and sample signal ($E_s$) must be extracted from the readout signals to calculate the transmittance or absorbance. The transmittance can be calculated from the three readout signals defined in Table 13-1 and equation 2-27, which is repeated here:

$$T' = \frac{E_{st} - E_{0t}}{E_{rt} - E_{0t}} \qquad (13\text{-}4)$$

At room temperature blackbody emission from the sample cell or solutions in the UV-visible region is negligible and analyte and background luminescence signals are usually insignificant compared to the lamp signals ($E_r$ and $E_s$). Thus subtraction of $E_{0t}$, or $E_d$ in this case, from the total reference and sample signals yields $E_r$ and $E_s$, respectively, and allows calculation of the true transmittance ($T' = E_s/E_r = T$) and absorbance ($A = -\log T$).

For instruments that use analog logarithmic circuitry to provide direct absorbance readout, the absorbance is calculated from the sample and reference logarithmic readout signals defined in Table 13-1 and the equation

$$A = A_s - A_r = \frac{E_{ls} - E_{lr}}{k'} \qquad (13\text{-}7)$$

where $k'$ is the conversion factor in volts per absorbance unit (A.U., a dimensionless quantity). Here the dark signal must be negligible, subtracted, or otherwise compensated before logarithmic conversion.

In all spectrophotometers and photometers, there is provision for compensating for the dark signal and, in most, for adjusting the magnitude of the reference signal. The reference signal is wavelength dependent because the lamp radiance, wavelength selector transmission efficiency, and detector responsivity vary with wavelength. Accuracy can be limited by resolution if the reference signal is significantly less than the full-scale capability of the readout device in the transmittance mode or of the ADC in computerized instruments. Also, the accuracy of analog computational circuits (ratio, logarithmic) is often best over a certain range of input voltages. Therefore, the reference signal at each wavelength is usually manually or automatically adjusted to a value that uses the full resolution capabilities of the readout device and provides good computational accuracy. The adjustment of the reference signal and the compensation for the dark signal are implemented by different means in different instruments, as we shall see.

*Fixed-Wavelength Spectrophotometers.* With simple spectrophotometers or photometers of single-beam design, the readout scale is calibrated in transmittance or absorbance units. Thus the readout signals are manually adjusted to define and utilize these scales once the wavelength is selected.

For transmittance readout, the signals $E_{rt}$ and $E_{0t}$ are adjusted to correspond to the full-scale reading ($E_{fs}$) and zero reading of the readout device, respectively, as shown in Figure 13-7. A 0% $T$ control knob is first used to adjust the readout signal to zero on the scale when

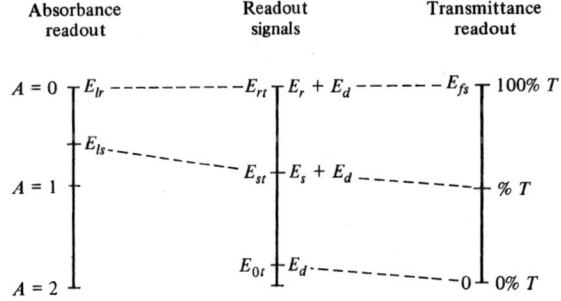

**FIGURE 13-7** Adjustment of the readout scales. With transmittance readout, the total reference signal is adjusted to the full-scale setting of the readout device ($E_{fs}$) and the 0% $T$ signal is adjusted to the zero setting of the readout device. With direct absorbance readout, the logarithm of the reference signal is adjusted to the zero setting of the readout device.

a shutter is activated to block all source light from reaching the detector. The 0% $T$ control allows a variable suppression voltage or current to be input at an appropriate point in the signal processor circuitry to null out the dark signal before display. Next, with the shutter open and the blank in the sample cell, a 100% $T$ control knob (often called the zero absorbance or gain control) is used to make the readout device display 100% $T$.

Different approaches are employed to make the reference signal adjustment. The radiant power incident on the sample cell can be varied by adjusting the lamp radiance (i.e., the lamp current) or the fraction of the lamp radiance that reaches the cell (i.e., with a variable aperture such as the monochromator slit width or a movable vee-shaped slit). Alternatively, the PMT gain ($m$) or signal processor gain ($G$) is adjusted so that $E_{rt} = E_{fs}$. Finally, a potentiometer can be used to select the fraction of the signal processor output to be sent to the readout device. In some cases, the 0% $T$ and 100% $T$ controls are interactive and must be alternately adjusted until the readout scale is properly defined.

After these adjustments, the analyte solution is placed in the sample cell, and the resulting reading is directly in transmittance units. Note that equation 13-4 is effectively used to calculate $T$, although $E_{0t}$ and $E_{rt}$ are set to 0 and 100 units, respectively, rather than measured. From the transmittance measured, the absorbance is manually calculated ($A = -\log T$). Alternatively, instruments with analog readout also include a second scale labeled in nonlinear absorbance units so that the absorbance can be estimated directly.

For instruments with analog logarithmic circuits, the procedure is similar except that no external 0% $T$ adjustment is made because $A$ would be off-scale with the shutter closed (i.e., $A = \infty$ for 0% $T$). In this case, the readout with blank solution in the sample cell is adjusted for 0 A.U. with a zero absorbance control knob that performs the same function as the 100% $T$ control for transmittance readout.

In more sophisticated spectrophotometers, the readout adjustment procedure is simplified. With an *auto zero* option, the adjustment of the blank signal to 100% $T$ or 0 A.U. is automatically performed when the appropriate front-panel switch is activated. In some instruments, a front-panel 0% $T$ control is not needed because the 0% $T$ signal is quite stable, internally compensated, and independent of the 100% $T$ adjustment. This is often the case when the source is modulated, as illustrated in Figure 13-8a. An internal adjustment may be required periodically to ensure that the transmittance or the input signal to analog logarithmic circuits is zero when the light source is blocked.

For double-beam instruments, manual adjustment of the limits of the readout scale is unnecessary because

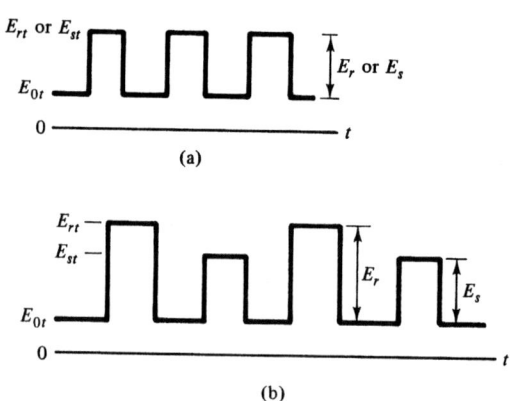

**FIGURE 13-8** Modulation waveforms. In (a) the signal waveform from a single-beam instrument with source modulation is shown at a point before demodulation. The amplitude is proportional to the reference or sample signal depending on the solution in the sample cell. The dark signal does not contribute to the demodulated output signal. In (b) the signal waveform from a double-beam in-time spectrophotometer is illustrated. The signal processor samples and stores the $E_{rt}$, $E_{st}$, and $E_{0t}$ signals at the proper time and computes $E_s$ ($E_s = E_{st} - E_{0t}$), $E_r$ ($E_r = E_{rt} - E_{0t}$), and finally $T = E_s/E_r$ or $A = -\log(E_s/E_r)$.

the reference and sample signals are processed simultaneously to obtain continuous computation and display of $T$ or $A$. With the double-beam in-time design, the sample and reference signals are processed to compensate for the dark signal (see Figure 13-8b); a front-panel 0% $T$ adjustment is not required. Usually, feedback circuitry is used to compare and adjust the reference signal continuously to a preset value.

The optical transmission characteristics of the optical components, including the sample cells, in the sample and reference beams always differ slightly. Therefore, it is first necessary to fill the sample and reference cells with blank solution and adjust the readout for 100% $T$ or 0 A.U. with a zero control knob or an automatic zero switch. This control allows $E_s$ to be multiplied by a suitable factor so that it is equal to $E_r$. After this adjustment, the sample is placed in the sample cell, and $T$ or $A$ is directly displayed.

With computerized instruments, the scale setup procedure is not required, although the reference signal is usually adjusted first to a nominal value to utilize the full range of the ADC. The reference and sample signals (and the 0% $T$ signal if needed) are stored and later used to calculate $T$ or $A$ with software.

*Scanning and Multichannel Spectrophotometers.* It is tedious to obtain spectra with noncomputerized single-beam spectrophotometers because $T$ or

$A$ is not directly calculated and $E_r$ varies during scanning. To generate a point-by-point spectrum, it would be necessary to adjust the readout to 100% $T$ or 0 A.U. with blank solution, replace the blank solution with sample solution, read $T$ or $A$, change the wavelength, and repeat the reference adjustment and sample reading procedure for each wavelength. Alternatively, a separate scan of $E_r$ vs. $\lambda$ with the blank solution and of $E_s$ vs. $\lambda$ with the analyte solution could be obtained and followed by a point-by-point calculation of $T$ or $A$ from the values of $E_r$ and $E_s$ at each wavelength. Here the accuracy can be limited by readout resolution at wavelengths where $E_r$ is small due to a lower lamp radiance or detector responsivity.

Traditionally, absorption spectra have been obtained with scanning double-beam in-time spectrometers (often called *ratio-recording spectrophotometers*). During scanning, $T$ or $A$ is continually displayed, and the reference voltage is maintained at a constant value for best readout resolution and computational accuracy. In the **auto slit** mode, the slit width is adjusted during the scan to maintain the incident radiant power at a constant value by feedback circuitry or a slit program. This mode of operation results in a constant $S/N$ at all wavelengths but a variable wavelength resolution. In the **auto gain** mode, the PMT gain is continually decreased or increased as the wavelength is changed to maintain a constant photoanodic current. Although the resolution remains constant throughout the spectrum, the $S/N$ is degraded where the lamp radiance or detector responsivity is lower.

Ideally, a spectrum with blank solution in both the sample and reference cells would yield zero absorbance at all wavelengths once the absorbance is set to zero at one wavelength. In reality, there is some variation in the baseline absorbance due to differences in the wavelength dependence of the reflectance and transmission characteristics of the different optical components and the sample cells in the sample and reference beams. Thus some type of baseline compensation scheme is often provided. In older instruments, a multipot system was used to adjust the baseline to zero in a number of separate wavelength regions.

Although not common, some manufacturers have used a sample modulation approach for double-beam measurements (see Section 4-6). Here the sample and reference cells are secured in an oscillating carriage which alternately places the sample and reference cells into the optical path. The detector output is similar to that of double-beam in-time spectrophotometers (see Figure 13-8) although the modulation frequency is typically 2 to 4 Hz. The sample modulation design provides double-beam information without having to split the original optical beam into two separate beams. Thus

the optical path for the reference and sample beams is identical except for the differences in transmission characteristics of the cells.

Computerization of scanning spectrophotometers enhances their performance and capabilities. The essential feature is that the reference intensity spectrum ($E_r$ vs. $\lambda$) and the sample intensity spectrum ($E_s$ vs. $\lambda$) are digitized and stored. This allows spectra to be obtained conveniently with single-beam instruments of simpler optical design. Because the reference and sample intensity spectra are obtained at separate times, it is critical that the light-source radiance remain constant. Computer calculation of $T$ or $A$ essentially eliminates computational error and nonlinearity. Very high scan rates (e.g., 1000 nm/min) can be used with such spectrophotometers because the data can be digitized rapidly and displayed later. With conventional scanning instruments, the response time of the computational circuitry or the analog recorder limits the maximum scan rate. For baseline compensation a spectrum of blank versus blank in double-beam instruments can be stored and later subtracted from the sample absorption spectrum to account for differences in the wavelength dependence of the optical efficiencies of the optical components in the sample and reference paths.

As with noncomputerized scanning instruments, there is usually some provision for adjusting $E_r$ to a relatively constant value during the scan. Either auto slit or auto gain modes can be used. In one commercial single-beam scanning instrument, filters with different transmission characteristics are automatically positioned into the optical path during the scan to maintain the incident radiant power at a relatively constant value.

Both SB and DB multichannel spectrophotometers provide the information and calculation options of scanning computerized single-detector spectrophotometers. However, the spectral data can be obtained much more rapidly because all wavelengths are measured simultaneously. Mechanical simplicity and wavelength accuracy are added advantages, particularly in SB designs, where an automatic shutter is the only moving part.

### Other Features

*Display Options.* With both fixed-wavelength and scanning instruments, provision is often made for displaying a portion of the transmittance or absorbance range on the readout device (e.g., 100 to 90% $T$ or 0.4 to 0.5 A.U.) to enhance the readout resolution. With noncomputerized instruments, the display range must be selected before the scan is made. In contrast, computerized instruments allow the user to display at leisure the acquired and stored data in different formats. Thus any particular wavelength region of the total spectrum acquired can be displayed with any desired readout scale. Scale expansion is useful up to the point that the display is limited by wavelength resolution, the resolution of the ADC, or the noise in the digitized signals. Computer-based instruments also allow successive spectra to be stored and later superimposed for comparison or added, subtracted, and normalized before display. A plot of log $A$ vs. λ can also be generated easily with appropriate software. This type of data presentation weights the relative change in absorbance with wavelength at low and high absorbances equally. If Beer's law is followed, log $A$ spectra at different concentrations are parallel and shifted from each other by log $bc$.

A *concentration readout* mode is available with many instruments as a convenience feature. This is really a direct absorbance readout mode in which the voltage or number corresponding to the absorbance is multiplied by a suitable factor before display. A standard is used to adjust the readout to correspond to the analyte concentration, after which the analyte concentration of samples is directly reported. The multiplication factor is adjusted with a potentiometer in noncomputerized instruments or software-selected in computerized instruments.

*Dual-Wavelength Spectrophotometers.* The difference in the absorbances measured at two wavelengths [$\Delta A = A(\lambda_1) - A(\lambda_2)$] is a useful quantity in some applications, as discussed in Section 13-5. Specialized dual-wavelength (DW) spectrophotometers are available to calculate and display $\Delta A$. They are configured to pass the radiation at the two wavelengths alternately through the sample cell (i.e., double-channel in-time). The signals are processed much like a double-beam in-time spectrophotometer, except that $\Delta A$ is calculated. Wavelength selection is based on a conventional monochromator with a vibrating refractor plate located near the exit slit (Figure 4-34) or a dual-grating monochromator. A *dual-grating monochromator* contains two stacked gratings that are independently controlled. The radiation selected by each grating passes through a different portion of the same exit slit or exits at a slightly different angle. A mechanical chopper is configured to intercept the beams such that at a given time, one beam is blocked while the other is directed to the sample cell. To generate a DW spectrum, either one wavelength is scanned with the other wavelength fixed or both wavelengths are scanned simultaneously with a constant wavelength difference. With microcomputer-based scanning instruments or multichannel spectrophotometers, $\Delta A$ or a DW spectrum can be calculated from the stored spectral absorbance data.

*Derivative Spectra.* A first or higher derivative of the absorbance spectrum (e.g., $dA/d\lambda$, $d^2A/d\lambda^2$) is often useful for accentuating spectral details such as absorption band maxima and shoulders on bands, as discussed in Section 13-5. With noncomputerized instruments, the first derivative is obtained with electronic derivative circuits which output a response proportional to $dA/dt$ during a scan. The time derivative is proportional to $dA/d\lambda$ through the dependence of λ on the time $t$ as determined by the wavelength scan rate. A derivative spectrum can also be obtained with a DW spectrophotometer by scanning both wavelengths simultaneously with a constant wavelength difference [$\Delta A/\Delta\lambda \approx dA/d\lambda$ if $\Delta\lambda$ is small] or by wavelength modulation during a scan (see Figure 4-34d). Finally, computerized instruments allow derivative spectra to be calculated from the stored absorption spectrum with appropriate software.

Taking the derivative also accentuates the noise as $dA/d\lambda$ due to noise is often greater than that due to the change in absorptivity with wavelength. Therefore, smoothing routines are often built into computerized instruments.

*Probe-Type Photometers.* Normally, the sample is brought to the sample cell in the instrument. Fiber optics are making possible a new generation of instruments for in situ measurements. Fiber optics can be used to direct light between the sample and the instrument. Figure 13-9 shows an example of a fiber optic probe photometer that allows absorption measurements directly in sample containers.

*Microcomputer Enhancements.* We have already noted that microcomputer-based instruments have profoundly enhanced the power and ease of use of spectrophotometers in terms of data acquisition, storage, and display. In addition, control and calculation options are greatly increased. Often, a keyboard allows user selection of wavelength, scanning parameters (e.g., scan rate and range), slit width, type of lamp, time constant or integration time, or number of repetitive measurements or scans. Programming and storage of repetitive scans or absorbance values at one or more wavelengths provides $S/N$ enhancement and allows means, standard deviations, and other statistical data to be calculated.

It is often possible to measure and store the absorbance data from a series of standards along with the standard concentrations. The calibration data are then fitted to generate a calibration function which is subsequently used to calculate and report analyte concentrations from the stored sample absorbances. The ab-

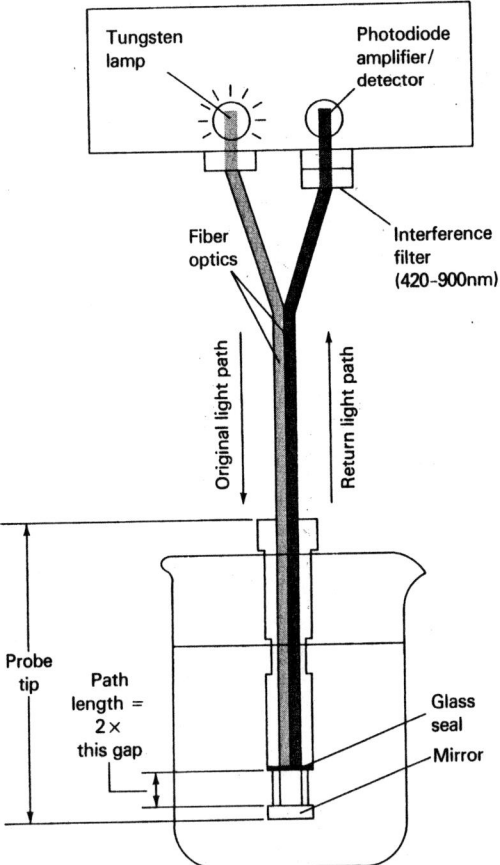

**FIGURE 13-9** Probe-type photometer. Light from a tungsten lamp is directed into the sample solution through a fiber optic. The light passes through the solution and is reflected by a mirror back to a second fiber optic. This returning light passes through an interference filter to a detector where the signal is processed as in a conventional photometer. (Courtesy of Brinkmann Instruments, Inc., Westbury, N.Y.)

sorbance data at a number of preselected wavelengths may be printed or used in fitting routines for multicomponent analysis.

Many software packages are also available for specialized applications such as kinetic studies or chromatographic detection. More detail is presented in Section 13-5.

## 13-2  SIGNAL AND NOISE EXPRESSIONS

In this section we use and adapt equations developed in Chapters 2 to 5 to express quantitatively the relationship of absorption readout signals and $S/N$ to instrumental and analyte parameters.

### Readout Expressions

Important signal expressions for molecular absorption measurements are summarized in Table 13-2. Although most expressions are written for conventional spectrophotometers with PMT detection, they can easily be adapted to other situations.

The source spectral radiant power $[(\Phi_\lambda)_0]$ and total radiant power $(\Phi_0)$ incident on the sample cell(s) depend on the source radiance $(B_\lambda)$ and the collection and transmission characteristics of the wavelength selector and other optical components $[Y(\lambda)]$, as shown in equations 13-8 and 13-9. Equations 13-10 and 13-11 apply when a monochromator is the wavelength selector; in the latter equation, $B_\lambda$ is assumed to be constant over $\lambda_0 \pm s$.

Absorption, reflection, and scattering by the sample cell and its contents attenuate the incident radiation, as illustrated in Figure 13-10. The attenuation by a sample cell filled with blank solution at a given wavelength is characterized by the reference transmission factor $(T_r)$. From equation 13-12, $T_r$ is seen to be due to the reflectance losses at the two cell walls $[(1 - \rho_c)^2]$, absorption and scattering within the cell walls $(1 - \alpha_c)^2$, and absorption by the solvent and nonanalyte species $(1 - \alpha_b)$.

The reflection loss at one cell wall is determined by the reflectance at the air–cell wall $(\rho_1)$ and the cell wall–solution interfaces $(\rho_2)$ (equation 13-13). For a typical glass or silica (quartz) cell filled with an aqueous solution, the Fresnel equation (equation 3-8) indicates that $\rho_1$ and $\rho_2$ are typically 4 and 0.3%, respectively. Thus $T_r$, considering only reflection losses, is about 91.5% (we will ignore multiple reflections for the time being). The exact value depends on the wavelength-dependent refractive indices of the cell wall material and solution.

The absorption and scattering by the cell walls depend on the type, thickness, and quality (e.g., imperfections) of the cell wall material (see Appendix B). Normally, $\alpha_c$ is insignificant in the visible region and becomes significant in the UV or NIR region at a wavelength dependent on the cell material.

Absorption and scattering by the blank solution are wavelength dependent and determined by the cell pathlength, the solvent absorptivity, the concentration and absorptivity of all nonanalyte species, and the size distribution and number of scatters. For wavelengths shorter than 190 nm, a factor accounting for absorption by $O_2$ should be added to the equations. Purging the entire optical path with $N_2$ can eliminate this absorption effect.

Equation 13-14 shows that the transmission factor with analyte solution in the sample cell $(T_s)$ is deter-

## TABLE 13-2
### Signal expressions for molecular absorption spectrophotometry[a,b]

| | |
|---|---|
| Incident radiant power | General:<br>$$(\Phi_\lambda)_0 = B_\lambda Y(\lambda) \tag{13-8}$$<br>$$\Phi_0 = \int_0^\infty (\Phi_\lambda)_0 \, d\lambda = \int_0^\infty B_\lambda Y(\lambda) \, d\lambda \tag{13-9}$$<br>With monochromator:<br>$$(\Phi_\lambda)_0 = WH\Omega B_\lambda T_{op} t(\lambda) \tag{13-10}$$<br>$$\Phi_0 = WH\Omega T_{op} s B_\lambda = W^2 H\Omega B_\lambda T_{op} R_d \tag{13-11}$$ |
| Reference transmission factor | $$T_r = \frac{(\Phi_\lambda)_r}{(\Phi_\lambda)_0} = (1 - \rho_c)^2 (1 - \alpha_c)^2 (1 - \alpha_b) \tag{13-12}$$<br>$$1 - \rho_c = (1 - \rho_1)(1 - \rho_2) \tag{13-13}$$ |
| Sample transmission factor | $$T_s = \frac{(\Phi_\lambda)_s}{(\Phi_\lambda)_0} = T_r[1 - \alpha(\lambda)] \tag{13-14}$$<br>$$1 - \alpha(\lambda) = 10^{-\varepsilon(\lambda)bc} = 10^{-A(\lambda)} = T(\lambda) \tag{13-15}$$ |
| Reference readout signal | General:<br>$$E_r = mG \int_0^\infty (\Phi_\lambda)_r R(\lambda) \, d\lambda = mG \int_0^\infty B_\lambda Y(\lambda) T_r R(\lambda) \, d\lambda = mG i_r \tag{13-16}$$<br>With monochromator:<br>$$E_r = mGWH\Omega \int_{\lambda_0 - s}^{\lambda_0 + s} B_\lambda T_{op} t(\lambda) R(\lambda) T_r \, d\lambda \tag{13-17}$$<br>$$E_r = mGR(\lambda) WH\Omega T_{op} s B_\lambda T_r \tag{13-18}$$ |
| Sample readout signal | General:<br>$$E_s = mG \int_0^\infty (\Phi_\lambda)_s R(\lambda) \, d\lambda = mG \int_0^\infty B_\lambda Y(\lambda) T_s R(\lambda) \, d\lambda = mG i_s \tag{13-19}$$<br>With monochromator:<br>$$E_s = mGWH\Omega \int_{\lambda_0 - s}^{\lambda_0 + s} B_\lambda T_{op} t(\lambda) R(\lambda) T_s \, d\lambda \tag{13-20}$$<br>$$E_s = mGR(\lambda) WH\Omega T_{op} s B_\lambda T_s \tag{13-21}$$ |
| Transmittance expressions | General:<br>$$T = \frac{E_s}{E_r} = \frac{\int (\Phi_\lambda)_s R(\lambda) \, d\lambda}{\int (\Phi_\lambda)_r R(\lambda) \, d\lambda} = \frac{\int B_\lambda Y(\lambda) T_s R(\lambda) \, d\lambda}{\int B_\lambda Y(\lambda) T_r R(\lambda) \, d\lambda} \tag{13-22}$$<br>With monochromator:<br>$$T = \frac{\int_{\lambda_0 - s}^{\lambda_0 + s} B_\lambda T_{op} t(\lambda) R(\lambda) T_s \, d\lambda}{\int_{\lambda_0 - s}^{\lambda_0 + s} B_\lambda T_{op} t(\lambda) R(\lambda) T_r \, d\lambda} \tag{13-23}$$<br>$$T = \frac{\int_{\lambda_0 - s}^{\lambda_0 + s} t(\lambda) 10^{-\varepsilon(\lambda)bc} \, d\lambda}{\int_{\lambda_0 - s}^{\lambda_0 + s} t(\lambda) \, d\lambda} = \frac{\int_{\lambda_0 - s}^{\lambda_0 + s} t(\lambda) 10^{-\varepsilon(\lambda)bc} \, d\lambda}{s} \tag{13-24}$$<br>$$T = T(\lambda_0) = 10^{-\varepsilon(\lambda_0)bc} = 1 - \alpha(\lambda_0) \tag{13-25}$$ |

[a] $Y(\lambda)$ and $T_{op}$ account for the optical efficiency of all components used for wavelength selection and of the external optical components between the source and the detector. Therefore, the incident (spectral) radiant power in these equations actually represents the radiant power incident on the detector when the sample cell is removed. When the wavelength selector is before the sample cell, the radiant power incident on the sample cell can be greater because of losses due to optical components between the sample cell and detector. In equations 13-11, 13-18, and 13-21, all wavelength-dependent quantities are assumed to be constant over $\lambda_0 \pm s$.

[b] $(\Phi_\lambda)_0$, Incident spectral radiant power (W nm$^{-1}$); $\Phi_0$, incident radiant power over the wavelength range passed by the wavelength selector (W); $Y(\lambda)$, wavelength selector throughput factor [$WH\Omega T_{op} t(\lambda)$ for a monochromator (cm$^2$ sr)] $B_\lambda$, source spectral radiance (W cm$^{-2}$ sr$^{-1}$ nm$^{-1}$); $T_r$, reference transmission factor with blank or reference solution in the sample cell (dimensionless); $(\Phi_\lambda)_r$, spectral radiant power passed by reference cell (W nm$^{-1}$); $\rho_c$, fraction of radiant power reflected in passing through one cell wall (dimensionless); $\alpha_b$, fraction of radiant power absorbed by nonanalyte species or scattered in the reference solution (solvent, reagents, and impurities) (dimensionless); $\alpha_c$, fraction of radiant power lost due to absorption or scattering by a cell wall (dimensionless); $\rho_1$, spectral reflectance for air–glass interface (dimensionless); $\rho_2$, spectral reflectance for glass–solution interface (dimensionless); $T_s$, sample transmission factor with analyte solution in the sample cell (dimensionless); $(\Phi_\lambda)_s$, spectral radiant power passed by analyte solution (W nm$^{-1}$); $\alpha(\lambda)$, spectral absorptance due to the analyte (dimensionless); $R(\lambda)$, photomultiplier photocathodic responsivity (A W$^{-1}$); $i_r$, reference photocathodic current, (A); $G$, gain of signal processor (V A$^{-1}$); $i_s$, sample photocathodic current (A).

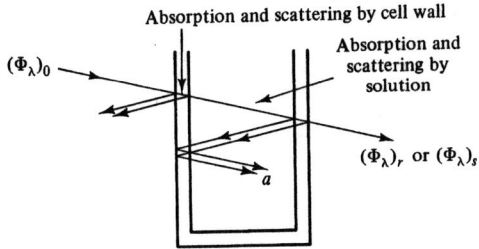

Absorption and scattering by cell wall

Absorption and scattering by solution

$(\Phi_\lambda)_0$

$(\Phi_\lambda)_r$ or $(\Phi_\lambda)_s$

$a$

**FIGURE 13-10** Radiation losses in a sample cell. The incident radiation $(\Phi_\lambda)_0$ is attenuated at both cell walls by reflection at the air–wall and wall–solution interfaces and by scattering and absorption within the cell wall. Inside the cell walls, losses occur due to scattering and absorption by the solvent and blank species, and for standard or sample solutions, also because of absorption by the analyte. The radiant power transmitted by the reference solution $(\Phi_\lambda)_r$ and the sample solution $(\Phi_\lambda)_s$ is determined from equations 13-12 and 13-14, respectively. The incident ray is not drawn perpendicular to the cell wall, as it normally is, to illustrate the reflections that occur. The effect of the multiple reflected rays (a) is discussed in Section 13-3.

mined by the same factors that affect $T_r$ plus naturally the analyte spectral absorptance $[\alpha(\lambda)]$. The analyte spectral absorptance at a given wavelength is related to the analyte molar concentration and molar absorptivity $[\varepsilon(\lambda)]$ at that wavelength with equation 13-15. Here $T(\lambda)$ and $A(\lambda)$ are the spectral transmittance and absorbance due solely to the analyte ($T_r = 1$), respectively. Often in the literature, the transmittance of the solution, ignoring cell wall effects, is called the **internal transmittance** and is equal to $T(\lambda)(1 - \alpha_b)$.

The reference readout signal ($E_r$) is determined by the spectral radiant power passed by the reference solution $[(\Phi_\lambda)_r]$, the PMT responsivity, the PMT gain, and the signal processor gain, as shown in equation 13-16. Equations 13-17 and 13-18 are the more specific forms that apply when a monochromator is used. In equation 13-18, all wavelength-dependent quantities are assumed to be constant over $\lambda_0 + s$. Equations 13-19 to 13-21 are the corresponding equations that relate the sample readout signal to the radiant power passed by an analyte solution $[(\Phi_\lambda)_s]$. In these equations, $T_r$ is replaced by $T_s$.

Equations 13-16 to 13-18 make it clear why the reference signal is highly instrument and wavelength dependent. For a given analysis wavelength, the reference signal is instrument dependent because of differences in the lamp radiance, the optical efficiency of the wavelength selector ($WH\Omega T_{op}s$), the detector responsivity $[R(\lambda)]$, and the gains ($mG$) among instruments. For a given instrument with a constant bandpass

(e.g., $W$) and gains ($mG$), $E_r$ varies typically about two orders of magnitude over the range 200 to 800 nm because of the dependence of $T_{op}$ and $T_r$, and especially $B_\lambda$ and $R(\lambda)$, on wavelength.

Equation 13-22 defines the measured transmittance as the ratio of equation 3-16 to equation 3-19, where $m$ and $G$ are assumed to remain constant for the sample and reference measurements. For a monochromator-based instrument, equation 13-23 applies where $W$, $H$, and $\Omega$ are assumed to remain constant for both measurements. Finally, in equation 13-24, all wavelength-dependent quantities $[B_\lambda, T_{op}, R(\lambda), T_r]$ except $\varepsilon(\lambda)$ and $t(\lambda)$ are assumed to be constant over $\lambda_0 \pm s$ and identical for the sample and reference measurements.

If $\varepsilon(\lambda)$ is constant over the wavelength range passed by the wavelength selector (e.g., $\lambda_0 \pm s$ for a monochromator-based system), equations 13-22 to 13-24 simplify to equation 13-25 and $A$ is proportional to $c$. Strictly, this is valid only if the incident radiation is monochromatic (i.e., infinitesimally small spectral bandpass). Because the bandpass is finite in real instruments, the measured $T$ or $A$ only approximates the spectral transmittance or absorbance at the wavelength setting ($\lambda_0$) of the monchromator. For further discussions we will drop the spectral designation ($\lambda$) and write Beer's law, in the normal form ($A = -\log T = \varepsilon bc$). In doing so we realize that $A$, $T$, and $\varepsilon$ are effective values determined by a weighted average of the corresponding spectral quantities $[T(\lambda), A(\lambda),$ and $\varepsilon(\lambda)]$ over $\lambda_0 \pm s$. Moreover, the proportionality between $A$ and $c$ is only an approximation, as we discuss in more detail in Section 13-3.

The equations in Table 13-2 apply to multichannel spectrophotometers with only minor modifications. For a diode array instrument, the sample and reference signal expressions are applicable to a given detector element in the array, where the wavelength range observed is determined by the width of a diode-array pixel and the dispersion of the spectrograph. As with vacuum phototubes and photodiode detectors, the detector gain ($m$) is unity. Because wavelength selection occurs after the continuum radiation passes through the sample cell, $(\Phi_\lambda)_0$ represents the effective incident spectral radiant power (i.e., the spectral radiant power incident on a detector element with the sample cell removed).

## Signal-to-Noise Expressions

In this section we discuss the factors that affect the precision with which the absorbance of a given solution can be measured. In other words, if we make repetitive absorption measurements on the same solution, what factors affect the measured standard deviation? We will see that noise, readout resolution, and cell positioning reproducibility must all be considered.

*Theory.* The expressions for the relative standard deviation (RSD) in absorbance ($\sigma_A/A$) developed in Section 5-5 apply to molecular absorption spectrophotometry. General and specific equations are summarized in Table 13-3. Equations 13-26 to 13-28 relate the standard deviation (SD) in absorbance ($\sigma_A$) and $\sigma_A/A$ to the SD in the total sample signal ($\sigma_{st}$). In this discussion, we will neglect the small contribution of the precision of the blank or 0% $T$ measurements to the magnitude of $\sigma_A/A$.

Equation 13-29 shows that $\sigma_{st}$ is due to shot and flicker noise in the transmitted source signal and 0% $T$ noise, as discussed previously (equation 5-46). The lamp signal shot noise and signal flicker noise are calculated from equations 13-30 and 13-31, respectively. The noise with the source radiation blocked or 0% $T$ noise ($\sigma_{0t}$) is seen from equation 13-32 to be due to dark current noise ($\sigma_d$) and an amplifier-readout noise ($\sigma_{ar}$).

The RSD in $E_s$ due to signal flicker noise is specified by the sample flicker factor $\xi_s$. Signal flicker noise arises from two sources as specified by two flicker factors (equation 13-33). Here $\xi_1$ is the *source flicker factor* and $\xi_2$ is the *cell transmission flicker factor*.

Cell transmission noise is not noise in the normal sense, as it is not observed as fluctuations during the measurement of a given test solution in a conventional sample cell. The transmission flicker factor is a measure of random variation in the reference signal due to the sample cell being removed and replaced in the optical path between measurements. Thus it accounts for variations in the reference transmission factor ($T_r$) due to sample cell placement. A random variation is observed because the incident light beam is imaged onto a slightly different portion of the cell walls every time the cell is moved; the reflection, transmission, and scattering characteristics of the sample cell vary spatially. The position of the sample cell can also affect the portion of the active surface of the detector that is illuminated and hence the effective detector responsivity for each

**TABLE 13-3**

Signal-to-noise expressions for molecular absorption spectrophotometry[a]

$$\sigma_A = \frac{0.43\,\sigma_T}{T} \tag{13-26}$$

$$\sigma_T = \frac{\sigma_{st}}{E_r} \tag{13-27}$$

$$\frac{\sigma_A}{A} = \frac{-\sigma_{st}}{TE_r \ln T} \tag{13-28}$$

$$\sigma_{st} = [(\sigma_s)_s^2 + (\sigma_s)_f^2 + \sigma_{0t}^2]^{1/2} \tag{13-29}$$

$$(\sigma_s)_s = (mGKTE_r)^{1/2} \tag{13-30}$$

$$(\sigma_s)_f = \xi_s TE_r \tag{13-31}$$

$$\sigma_{0t} = (\sigma_d^2 + \sigma_{ar}^2)^{1/2} = [mGKE_d + (\sigma_d)_{ex}^2 + \sigma_{ar}^2]^{1/2} \tag{13-32}$$

$$\xi_s = (\xi_1^2 + \xi_2^2)^{1/2} \tag{13-33}$$

$$\frac{\sigma_A}{A} = (E_r \ln T)^{-1}\left[KmGE_rT^{-1} + E_r^2(\xi_1^2 + \xi_2^2) + \left(\frac{\sigma_{0t}}{T}\right)^2\right]^{1/2} \tag{13-34}$$

$$\frac{\sigma_A}{A} = (-i_r \ln T)^{-1}\left[Ki_rT^{-1} + i_r^2(\xi_1^2 + \xi_2^2) + \left(\frac{\sigma_{0t}}{mGT}\right)^2\right]^{1/2} \tag{13-35}$$

$$\frac{\sigma_A}{A} = (-i_r \ln T)^{-1}\left[Ki_rT^{-1} + i_r^2(\xi_1^2 + \xi_2^2) + \left(\frac{\sigma_{0t}i_r}{TE_{fs}}\right)^2\right]^{1/2} \tag{13-36}$$

$$\sigma_{bk} = \frac{0.43}{i_r}\left[Ki_r + (\xi_s i_r)^2 + \left(\frac{\sigma_{0t}i_r}{E_{fs}}\right)^2\right]^{1/2} \tag{13-37}$$

[a] $\sigma_T$, Standard deviation (SD) in the transmittance (dimensionless); $\sigma_A$, SD in the absorbance (dimensionless); $\sigma_{st}$, SD in total sample signal (V); $(\sigma_s)_s$, SD due to signal shot noise (V); $(\sigma_s)_f$, SD due to signal flicker noise (V); $\sigma_{0t}$, SD due to 0% $T$ noise (V); $K$, the bandwidth constant (A); $\xi_s$, sample flicker factor (dimensionless); $E_{fs}$, full-scale or 100% voltage for readout device or ADC (V); $\xi_1$, source flicker factor or relative standard deviation (RSD) of the source spectral radiance over the measurement bandwidth due to flicker noise (dimensionless); $\xi_2$, cell transmission flicker factor or RSD in the transmission of the sample cell filled with reference solution due to positioning differences (dimensionless) (note that $\xi_2$ is independent of $\Delta f$ and $K$); $\sigma_{bk}$, SD in blank the absorbance (A.U.).

measurement. With flow cells, transmission flicker noise can be caused by pump pulsations, bubbles, temperature-dependent refractive index variations, or incomplete mixing of the components of the mobile phase.

Equations 13-34 to 13-36 are the final expressions for $\sigma_A/A$ that account for all noise sources. In all these equations, the three terms in the brackets represent the contributions from signal shot noise, signal flicker noise, and 0% $T$ noise, respectively. Equation 13-36 takes into account that the gains ($mG$) are usually adjusted so that $E_r$ equals the full-scale setting ($E_{fs}$) of the readout device or of the ADC in computerized instruments.

If direct absorbance readout is provided by analog logarithmic circuitry, the same equations apply in most cases. Here $E_{fs}$ represents the input voltage to the computational circuitry, which corresponds to 0 A.U. If measurements are limited by readout resolution, a term accounting for the uncertainty in absorbance units due to readout resolution is added (see equations 5-59 and 5-60).

The blank noise ($\sigma_{bk}$) in absorbance units can be estimated with equation 13-37 (repeat of equation 6-13). The detection limit (DL) can be estimated from this value and DL $= k\sigma_{bk}/m$, where $k$ is the confidence factor and $m$ is the calibration slope ($\varepsilon b$).

### Effect of Variables.

The limiting noise sources in a given absorption measurement depend on the instrument, the instrumental parameters, the sample cell, and the blank and analyte absorbances. Equation 13-36 makes it clear that the measurement precision is critically dependent on the reference photocathodic current ($i_r$), the source and cell transmission flicker factors ($\xi_1$ and $\xi_2$), the relative 0% $T$ noise ($\sigma_{0t}/E_{fs}$), and the transmittance.

The reference photocathodic current can vary typically from $10^{-12}$ to $10^{-6}$ A, depending on the instrument and the instrumental conditions ($\lambda_0$, $W$) chosen for a given analysis. For a given instrument, $i_r$ can easily vary two to four orders of magnitude with changes in the slit width or analysis wavelength. Among instruments set to a given analysis wavelength, $i_r$ varies greatly due to differences in the incident radiant power as determined by the specific characteristics of the light source (e.g., $B_\lambda$), the wavelength selector and imaging optics (e.g., $WH\Omega T_{op}s$), and the detector [e.g., $R(\lambda)$]. The transmission characteristics of the specific sample cell and blank solution also influence the magnitude of the reference signal.

The source flicker factor is typically $10^{-3}$ to $10^{-5}$ and dependent on the type and age of the lamp and the quality of the lamp power supply. It is often nearer the lower limit for single-beam instruments using optical feedback or for double-beam instruments because $E_s$

and $E_r$ are constantly ratioed. Formally, source flicker noise is due to nonfundamental fluctuations in the source radiance. The effective source flicker noise can be greater, due to mechanical vibrations or varying magnetic fields due to choppers and motors such as those used in double-beam instruments. Vibrations or nonreproducible imaging of the sample or reference beams onto the detector active surface results in additional noise if the detector responsivity in spatially dependent.

The cell transmission flicker factor varies from $10^{-2}$ to $10^{-4}$. It depends on the quality of the sample cell, the condition of the sample cell (e.g., scratches, cleanliness of the surfaces), and the ability to position the sample cell reproducibly. Generally, $\xi_2$ is larger with test tube-type cells, where the rotation of the cell is not easily controlled. Dual-wavelength measurement techniques compensate for both cell transmission and source flicker noise.

With low-resolution readout devices, 0% $T$ noise is often limited by readout resolution ($\sigma_{0t} = \sigma_r$). The relative readout resolution is dependent on the characteristics of the readout device or ADC and can vary from 0.2% $T$ or 0.002 A.U. with analog readout (meters or recorders) to 0.001% $T$ or $10^{-5}$ A.U. with high-resolution digital readout or ADCs. In modern spectrophotometers with sufficient readout resolution, the limiting 0% $T$ noise is usually dark current noise ($\sigma_d > \sigma_a$). Dark current noise is more likely to be the limiting noise source under conditions where $i_r$ is smaller. Source modulation allows discrimination against $1/f$ dark current noise or amplifier-readout noise.

The shapes of the plots of $\sigma_A/A$ vs. $A$ for the three limiting cases (one type of noise term dominant) were discussed previously (Figure 5-7). In all cases, $\sigma_A/A$ decreases with increasing $A$ up to an absorbance of about 0.43, after which $\sigma_A/A$ continually increases if 0% $T$ noise is dominant. If signal shot noise is limiting, the increase in $\sigma_A/A$ is delayed until the absorbance is about 0.87 and is more gradual. Precision continually improves with increasing $A$ if signal flicker noise is dominant. The latter behavior is also observed if readout resolution is limiting in instruments that provide direct absorbance readout with analog logarithmic circuitry.

For real instruments, one or more noise sources may be dominant in different absorbance regions and the optimum absorbance with respect to precision usually lies between about 0.43 and 1 to 3. The contribution of signal shot noise and 0% $T$ noise to $\sigma_A/A$ vary as $T^{-1/2}$ and $T^{-1}$, respectively. Therefore, if signal flicker noise is dominant at low absorbances, one or both of these two types of noise will limit the precision at higher absorbances and ultimately cause the precision to degrade. Noise independent of the lamp signal (0% $T$ noise) always eventually becomes limiting at some higher absorbance where the lamp signal and hence noise sources related to the lamp signal become small. These limiting situations may not be observed because the maximum absorbance obtainable is limited by the display range of the readout device or by excessive nonlinearity due to stray radiation or other factors (see Section 13-3). Note also that is always a transition region between the limiting regions over which two types of noise are sig-

nificant (e.g., signal flicker and shot noise or signal shot noise and 0% $T$ noise).

Simple single-beam spectrophotometers and photometers often have a large spectral bandpass, and thus provide a large light throughput (large $\Phi_0$ and $i_r$); they often use lower-resolution readout devices (0.1% $T$ or 0.001 A.U. or worse). For these cases, noise may not be obvious such that measurements are readout reso-

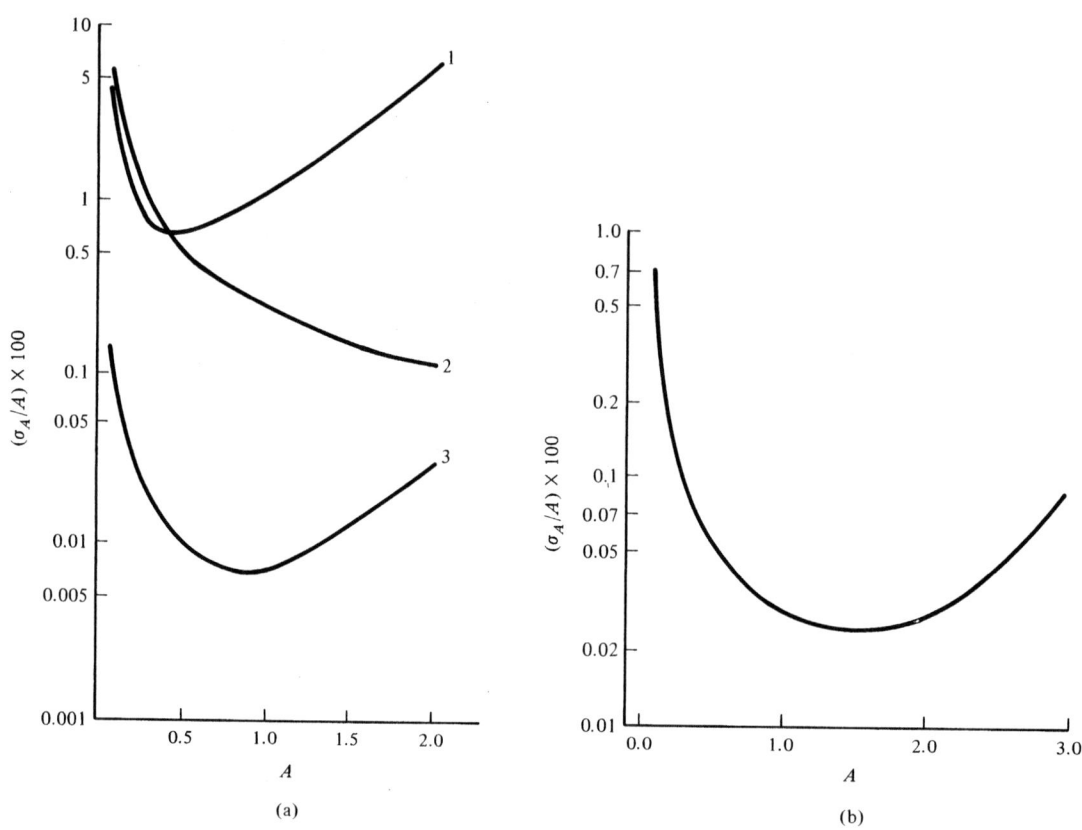

(a)

(b)

**FIGURE 13-11** Precision plots for two spectrophotometers. In (a) the dependence of the precision ($\sigma_A/A$) on absorbance for a simple spectrophotometer (Turner 330) is shown. In curve 1, the precision is limited by the readout resolution of the spectrophotometer's analog meter (0.2% $T$). To obtain curves 2 and 3, a digital meter with a resolution of 0.001% $T$ was employed so that readout resolution is not limiting. In curve 2, the sample cell was moved between measurements and precision is totally limited by cell transmission flicker noise. For curve 3, the cell was kept stationary. In this case source flicker noise limits precision up to about $A$ = 1 after which 0% $T$ becomes dominant. Signal shot noise is insignificant at all absorbances. The precision plot for a high quality double beam spectrophotometer (Cary 118C) is shown in (b). Here source flicker noise is limiting at low absorbances. As the absorbance increases, source flicker noise and signal shot noise become comparable in magnitude and eventually signal shot noise becomes dominant at moderate absorbances. With further increases in absorbance, the signal shot noise decreases further in magnitude and becomes comparable and eventually insignificant compared to the dark current noise at high absorbances. [(a) Adapted from J. D. Ingle, Jr., *Anal. Chim. Acta, 88,* 131 (1977), with permission of Elsevier Science Publishers.]

lution limited (a 0% $T$ noise) at all absorbances. The precision characteristics of a simple spectrophotometer are illustrated in Figure 13-11a.

With higher-quality instruments, the light throughput is often less (i.e., smaller $i_r$), and the readout resolution is $10^{-4}$ $T$ or A.U. or better. With such spectrophotometers, noise fluctuations or cell positioning imprecision can normally be observed. Often source or transmission flicker noise is limiting at the detection limit and lower absorbances, while signal shot noise or 0% $T$ noise become dominant at moderate or high absorbances. The absorbance at which the precision starts to decrease is largely a function of the reference photocathodic current. At wavelengths where the source radiance and detector responsivity are high (e.g., 400 to 600 nm), $i_r$ is large and signal flicker noise dominates up to high absorbances. The reference photocathodic current becomes small at short wavelengths near 200 nm because the source radiance is low and often at long wavelengths (greater than 650 nm) because the detector responsivity of many PMTs is low. Thus, for these wavelength regions, signal shot noise or dark current noise is more likely to dominate, particularly at high absorbances. This behavior also occurs when $i_r$ is small due to narrow slit widths or a large blank absorbance ($T_r \ll 1$). The precision characteristics of a double-beam spectrophotometer are illustrated in Figure 13-11b.

For NIR measurements with PbS detectors, the detector noise is considerably higher than for PMTs, PTs, or PDs, such that 0% $T$ noise limits the precision at all absorbances. With multichannel detectors 0% $T$ noise is often limiting at all absorbances and is due to electronic readout noise. The relative 0% $T$ noise ($\sigma_{0r}/E_{fs}$) is typically 0.01 to 0.05% for diode signals near saturation.

For instruments in which the slit width can be controlled, precision is improved by increasing the slit width to increase $i_r$ until measurements become limited by signal flicker noise or readout resolution or the wavelength resolution is insufficient. Source flicker noise, signal shot noise, and dark current noise can be decreased by increasing the time constant or integration time (i.e., reducing $K$ or $\Delta f$). The $1/f$ nature of source flicker noise precludes reducing its magnitude significantly for noise bandwidths smaller than 0.1 Hz.

*Scale Expansion.* With low-resolution readout devices, the readout resolution often limits the precision [$\sigma_A/A = (\sigma_r/E_{fs})/(-T \ln T)$], which causes a large uncertainty in the measured absorbance for low absorbances and also for high absorbances with transmittance readout. To utilize the inherent capabilities of an instrument fully, the relative readout resolution ($\sigma_r/E_{fs}$)

should be improved until noise is observed. This can be accomplished by substituting a high-resolution readout device (smaller $\sigma_r$) or by using scale expansion techniques.

Before the advent of high-resolution digital readout devices and ADCs, scale expansion was a popular technique for increasing precision. It is still useful for some fixed-wavelength instruments and for displaying spectra. Scale expansion allows a portion of the transmittance scale to be displayed on a readout device. This is often accomplished by calibrated multiplication and suppression of the signal processor output signal before it is sent to the readout device. For example, consider the case where a 1-V full-scale meter with a resolution of 1 mV is used to display the range 0 to 100% $T$. The relative readout resolution is $10^{-3}/1 = 10^{-3}$, or 0.1% $T$. If the readout signal is amplified by a factor of exactly 10 and then 9 V is subtracted before display, the readout scale now represents 100 to 90% $T$, $E_{fs}$ is effectively increased to 10 V, and the readout resolution is 0.01% $T$. If 8 V is subtracted, the readout scale represents 90 to 80% $T$. Note that $\sigma_A/A$ is proportional to $\sigma_r/E_{fs}$ only if readout resolution is limiting. In the example above, the precision at a given absorbance is improved by a factor of 10 only if readout resolution is limiting before and after expansion.

For instruments that use logarithmic circuitry to provide direct absorbance readout, similar electronic manipulation can be used to improve readout resolution by displaying a smaller portion of the absorbance scale (e.g., 1.0 to 1.5 A.U. instead of 0.0 to 3.0 A.U.). In some instruments, the signal processor gain is automatically increased by a factor of 10 when the transmittance falls below 10% $T$ to enhance the readout resolution for small sample signals. With computer-based or multichannel spectrophotometers, one can choose to display a portion of the transmittance or absorbance scale to accentuate detail; however, the uncertainty due to readout is actually limited by the readout resolution of the ADC.

In the past it was common to use analyte solutions of known concentration and hence transmittance to set one or both ends of a transmittance readout scale. For example, with a standard solution of 90% $T$ in the sample cell, the 100% $T$ control can be used to adjust the readout to 100% $T$. Next, the 0% $T$ control can be used to adjust the readout to 0% $T$ with a standard solution normally yielding 80% $T$. Now sample solutions with transmittances in the range 90 to 80% $T$ can be measured with 10 times the readout resolution under normal conditions. Thus this *chemical* scale expansion technique (often called differential or high-precision spectrophotometry) can also provide the same advantages as electronic scale expansion. However, this tedious technique is rarely used in modern instruments with high-resolution digital readout, electronic scale expansion, and/or direct absorbance readout.

## 13-3 APPARENT DEVIATIONS FROM BEER'S LAW

From Beer's law, a calibration plot of $A$ vs. $c$ from measurements on a series of standards should be linear with an intercept of zero. However, calibration curves are sometimes found to be nonlinear or have a nonzero intercept. These effects are rarely due to Beer's law being invalid, but rather are a consequence of measurement conditions in which the assumptions used to derive Beer's law (see Table 3-1) are not valid. The terms **positive deviation** and **negative deviation** from linearity are used to describe nonlinear calibration curves that bend toward or away from the concentration axis, respectively. Hence in the case of a negative deviation, the calibration slope decreases with increasing analyte concentration. We now discuss the specific instrumental and chemical effects that cause such apparent deviations from Beer's law.

### Nonzero Intercept

A nonzero intercept is usually due to an improper blank measurement or adjustment or to nonequivalent measurement conditions for the blank and standard solutions. The blank measurement should account for all the attenuation of the incident beam by the sample cell filled with standard solution except for that due to the analyte. This implies that the reference transmission factor $T_r$ (equation 13-12) should be identical for the standard and blank measurements. Differences can occur because of an improper blank [i.e., the composition of the blank solution is different from the matrix of the standard solutions such that the absorption by nonanalyte species ($\alpha_b$) or the refractive index which affects the reflection loss ($\rho_r$) is different].

If different and nonequivalent sample cells are used for blank and standard measurements, differences in cell wall absorption and scattering ($\alpha_c$) or reflection properties ($\rho_c$) can cause a nonzero intercept. With double-beam instruments, differences in cell properties are adjusted by zeroing the absorbance with blank in both cells. In single-beam instruments, either one sample cell should be used for all solutions or a measurement should be made with the blank solution in each cell. Correction factors can be developed to account for sample cells with different transmission characteristics. Drifts in the lamp radiance, PMT or signal processor gain, or detector responsivity can also cause apparent nonzero intercepts. For example, if the lamp radiance is constantly

decreasing in a single-beam instrument and the blank measurement is always made before the standard measurement, the transmittance calculated will always be too low; this yields a positive intercept corresponding to the absorbance change due to drift between the blank and sample measurement.

Random error can also result in a nonzero intercept. When calibration data are fit by least-squares methods, the intercept is normally not zero. Errors in preparation of standards can cause similar effects. Statistical methods can be used to evaluate the uncertainty in the intercept and determine if it differs significantly from zero.

### Nonlinearity due to Chemical Equilibria

If several species absorb at a given wavelength, the additive form of Beer's law can be used:

$$A = \sum_{i=1}^{n} A_i = b \sum_{i=1}^{n} \varepsilon_i c_i \tag{13-38}$$

where the subscript $i$ identifies that quantity for the $i$th absorbing species. For the moment we will assume that the blank compensates for the absorbance due to concomitants, but consider that the analyte can exist in several chemical forms in solution, which may be in equilibrium.

If only one of the chemical forms absorbs at the monitored wavelength, we can rewrite Beer's law as

$$A = \varepsilon b f c_a \tag{13-39}$$

where $c_a$ is the total or analytical concentration (the total concentration of analyte placed into solution) and $f$ is the fraction of $c_a$ that exists in the absorbing form. Thus we see that the "$c$" in Beer's law is not necessarily the analyte concentration but more precisely the concentration of the absorbing form of the analyte. Clearly, a linear calibration curve is observed only if $f$ is independent of the analytical concentration ($c_a$). A negative or positive deviation in the calibration curve is observed if $f$ decreases or increases with increasing $c_a$, respectively.

The situation becomes more complex if two or more chemical forms of the analyte absorb at the analysis wavelength. Consider an equilibrium between a monomer (M) and a dimer (D) in solution ($2M \rightleftharpoons D$) with an equilibrium constant $K$ ($K = c_D/c_M^2$). The total absorbance due to both species from equation 13-38 is given by

$$A = b(\varepsilon_D c_D + \varepsilon_M c_M) \tag{13-40}$$

The analytical concentration $(c_a)$ is found from

$$c_a = c_M + 2c_D = c_M + 2Kc_M^2 \qquad (13\text{-}41)$$

If equation 13-41 is solved for $c_M$ and $c_D$ and the results are substituted into equation 13-40, we obtain

$$A = \frac{b}{2}\left[\varepsilon_D c_a + (2\varepsilon_M - \varepsilon_D)\frac{(8Kc_a + 1)^{1/2} - 1}{4K}\right]$$

$$(13\text{-}42)$$

If $\varepsilon_D = 2\varepsilon_M$, the second term vanishes and a linear calibration curve is obtained. In other cases, the contribution of the second term in equation 13-42 to the total absorbance decreases with increasing analytical concentration and causes a positive or negative deviation depending on the sign of the quantity $(2\varepsilon_M - \varepsilon_D)$. The fraction of the total analytical concentration in the dimer form $(c_D/c_a)$ increases with increasing $c_a$. Thus if $\varepsilon_D > 2\varepsilon_M$, a positive deviation is observed because the decrease in the absorbance due to the loss of two monomer molecules is more than compensated by the gain in absorbance due to the formation of one dimer molecule. By contrast, a negative deviation occurs if $\varepsilon_D < 2\varepsilon_M$ because the increase in dimer absorbance does not compensate for the loss in monomer absorbance. Negative and positive deviations due to dimer formation are observed with methylene blue when the absorbance is monitored at 664 and 600 nm, respectively.

Many types of equilibria involving the analyte are possible, and the shape of the calibration plot depends on the particular case. For example, if a metal is determined by the absorbance of a metal complex, the total absorbance can be due to a number of complexes involving different numbers of ligands associated with the metal or polynuclear complexes. Each complex can have a different spectrum and thus different molar absorptivities at a given wavelength. The relative fraction of each metal complex can vary with the total metal concentration. Sometimes it is possible to choose the initial ligand concentration so that only one metal complex forms in a significant amount. Alternatively, with a sufficient excess of ligand, the fraction of each complex formed can be independent of the initial metal concentration.

Many equilibria involving analyte species are pH dependent. A classic example is Cr(VI) in solution $(Cr_2O_7^{2-} + H_2O \rightleftharpoons 2H^+ + 2CrO_4^{2-})$. The relative fraction of yellow chromate to orange dichromate varies with the Cr(VI) concentration and the pH; positive or negative deviations can be observed depending on the monitored wavelength. For such pH-dependent equi-

libria, the pH can be adjusted or buffered to force the equilibrium predominantly in one direction. For example, at high pH ($\geq 12$), Cr(VI) exists primarily as yellow chromate, and a linear calibration curve is observed at an analysis wavelength of 370 nm. At very low pHs, nonlinearity is observed at most wavelengths because of equilibria between several absorbing species (e.g., $HCr_2O_7^-$, $H_2CrO_4$, chromium sulfate complexes).

Equilibria involving metal complexes or the acid and base forms of a species are also pH dependent. The latter case is illustrated in Figure 13-12. Absorption spectra of the indicator phenol red are shown as a function of pH. The plot of absorbance vs. total indicator concentration can be made linear by adjusting the pH of all solutions to a value that results in the indicator being totally unprotonated or protonated or that maintains a constant ratio for the concentration of the two forms for any total indicator concentration.

Note in Figure 13-12 that the absorbance is independent of pH at about 495 nm. This wavelength is called the **isobestic** or **isoabsorptive** point. An isobestic point is observed when two absorbing species in equilibrium have the same molar absorptivity at one wavelength. In fact, the observation of an isobestic point is a criterion to prove the existence of two interconvertible absorbing forms of a species with overlapping spectra. The isobestic point is an analytically useful wavelength because a linear calibration curve is obtained at this wavelength without controlling such solution conditions as pH.

## Nonlinearity due to Other Chemical Effects

Nonlinearity can also occur if the analyte molar absorptivity is dependent on the analyte concentration. Such effects are usually minor and occur at relatively high concentrations ($> 10^{-2}$ M). Differences in solute–solvent interactions, solute–solute interactions, or hydrogen bonding at high concentrations can change the chemical or electrostatic environment and hence the absorptivity of the analyte.

Changes in the solution refractive index $(\eta)$ with analyte concentration can also cause nonlinearity by altering the position, size, or solid angle of the image transmitted to the detector, the reflectance loss at the cell wall–solution interface $(\rho_2)$, of the analyte molar absorptivity. In the last case, $\varepsilon$ in Beer's law is replaced by $\varepsilon\eta/(\eta^2 + 2)^2$.

## Nonlinearity due to Polychromatic Radiation

We have repeatedly emphasized that Beer's law is valid only for monochromatic radiation. To illustrate the ef-

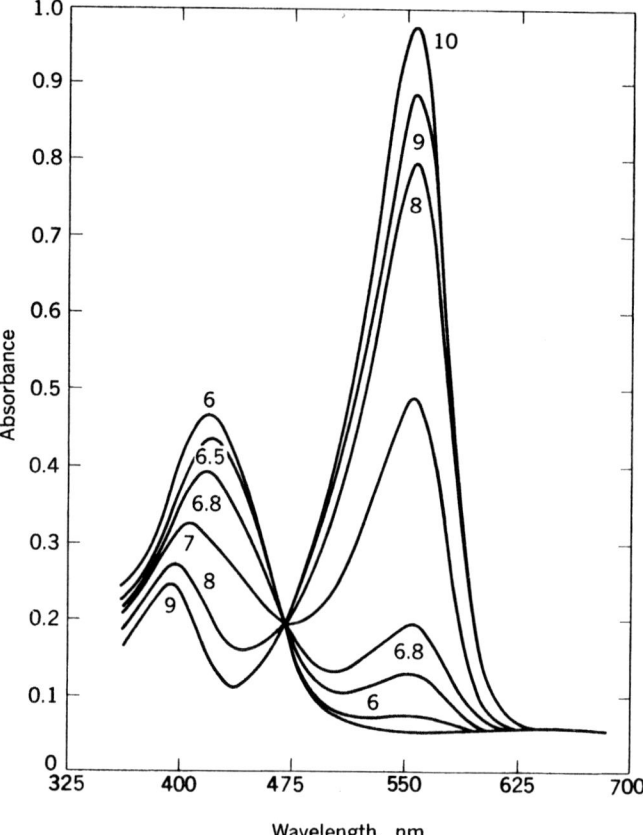

**FIGURE 13-12** Absorption spectra of phenol red at various pH values. The absorbance at a given wavelength is given by $A = b(\varepsilon_u c_u + \varepsilon_p c_p)$ where the subscripts $p$ and $u$ indicate that quantity for the protonated and unprotonated forms of phenol red. At most wavelengths, $A$ is seen to depend on pH because $c_u$ and $c_p$ vary with pH. At the isobestic point ($\lambda \approx$ 495 nm), $\varepsilon_u = \varepsilon_p$ and $A$ is proportional to the total indicator concentration, but $A$ is independent of the pH and the fraction in each form. (From G. W. Ewing, *Instrumental Methods of Analysis*, McGraw-Hill Book Co., New York, 1985, with permission of publisher.)

fect of nonmonochromatic, heterochromatic, or polychromatic radiation in real measurements, we first consider a simple example in which the sample cell is illuminated with two monochromatic lines of wavelengths $\lambda_1$ and $\lambda_2$. In this case, the integrals in equation 13-22 reduce to summations of the quantities at two wavelengths. If we assume that Beer's law is followed for each wavelength (equation 13-25), equation 13-22 simplifies to

$$T = \frac{E_1 10^{-\varepsilon_1 bc} + E_2 10^{-\varepsilon_2 bc}}{E_1 + E_2} \qquad (13\text{-}43)$$

where $E_1$ and $E_2$ are the reference readout signals observed at wavelengths $\lambda_1$ and $\lambda_2$, respectively, when measured separately. In further discussions, we will assume that $\lambda_1$ is the desired wavelength and that the radiation at $\lambda_2$ represents the undesired contribution. If we let $r = E_2/E_1$, equation 13-43 becomes

$$T = \frac{10^{-\varepsilon_1 bc}}{1 + r} \times [1 + r \times 10^{(\varepsilon_1 - \varepsilon_2)bc}] \qquad (13\text{-}44)$$

Note that $r$ is the weighting factor for the undesired contribution and accounts for differences in the source radiance, wavelength selector optical efficiency, refer-

ence transmission factor, or detector responsivity at the two wavelengths. From equation 13-44, the absorbance is given by

$$A = \varepsilon_1 bc + \log(1 + r) - \log[1 + r \times 10^{(\varepsilon_1 - \varepsilon_2)bc}]$$

$$(13\text{-}45)$$

If this equation is differentiated with respect to $bc$, we find that the effective absorptivity ($\varepsilon$) is given by

$$\varepsilon = \varepsilon_1 - \frac{r(\varepsilon_1 - \varepsilon_2)10^{(\varepsilon_1 - \varepsilon_2)bc}}{1 + r \times 10^{(\varepsilon_1 - \varepsilon_2)bc}} \qquad (13\text{-}46)$$

Clearly, $A$ is not proportional to $c$ and $\varepsilon$ is not independent of $c$ except when $r$ equals 0 or $\infty$ (monochromatic radiation) or $\varepsilon_1 = \varepsilon_2$.

Two limiting cases are of interest. When $bc \rightarrow 0$, equation 13-46 simplifies

$$\varepsilon = \frac{\varepsilon_1 + r\varepsilon_2}{1 + r} \qquad (13\text{-}47)$$

and linearity is observed with an effective molar absorptivity that is a weighted average of $\varepsilon_1$ and $\varepsilon_2$. For large absorbance ($bc \rightarrow \infty$) with $\varepsilon_1 \gg \varepsilon_2$, we find that

$\varepsilon = \varepsilon_2$. Hence the least absorbed wavelength controls the calibration slope because effectively all the radiation at $\lambda_1$ is absorbed, and further increases can only attenuate the radiation at $\lambda_2$.

These results pinpoint the two major effects of polychromatic radiation. First, the calibration slope at low analyte concentrations is lower than it would be at the wavelength of highest molar absorptivity (i.e., $\varepsilon_1$ is greater than $\varepsilon$ given by equation 13-47). Second, negative deviations occur at larger absorbances (i.e., $\varepsilon$ decreases with increasing $c$ from the value given by equation 13-47 to $\varepsilon_2$). The magnitude of these two effects depends on the difference between $\varepsilon_1$ and $\varepsilon_2$ and the value of $r$ as illustrated in Figure 13-13.

The same type of treatment can be extended to an infinite number of wavelengths as would be the case with most instruments that use a continuum source. Here the wavelength range incident on the sample is determined by the transmission characteristics of the wavelength selector. To evaluate the integrals in equation 13-22, it is necessary to describe functionally the wavelength dependence of all quantities in the integrals (e.g., lamp radiance, molar absorptivity). Even if this difficult task is accomplished, it is not usually possible to obtain an analytical function for the dependence of $A$ on $c$ or other variables. However, qualitatively, the effects of polychromatic radiation on the shape of the calibration curve are the same as described for the simple two-wavelength example.

Numerical integrations have been carried out with equations similar to equation 13-24 where only the wavelength dependencies of the molar absorptivity and the monochromator slit function are considered. Often,

the slit function is modeled as a triangular function (equation 3-60), and the absorption band shape [dependence of $\varepsilon(\lambda)$ on $\lambda$] is described as a Gaussian function (equation A-3). The results of such studies are now summarized.

First consider the case where the monochromator wavelength ($\lambda_0$) is set to the wavelength of maximum absorption ($\lambda_m$) of an absorption band of true half-width $\Delta\lambda$. With an infinitesimally small spectral bandpass or monochromatic radiation, we can measure the absorbance at the maximum ($A_m$), calculate the molar absorptivity at the maximum ($\varepsilon_m$), determine the true half-width, and obtain a linear calibration curve. We are concerned with how a finite spectral bandpass ($s$) affects the observed absorbance ($A$), the calculated effective molar absorptivity ($\varepsilon$), and the linearity. It is convenient to express the results in terms of how $A/A_m$ or $\varepsilon/\varepsilon_m$ vary with the bandwidth ratio ($s/\Delta\lambda$). As seen in Figure 13-14a, the observed absorbance with a finite bandpass is less than $A_m$. However, the calibration curve is reasonably linear up to 1.5 A.U. if $s/\Delta\lambda \leqslant 0.43$. With $s/\Delta\lambda = 0.43$, the effective molar absorptivity is about 92% of $\varepsilon_m$. For a spectral bandpass such that $s/\Delta\lambda < 0.1$, the observed absorbance is within 0.4% of $A_m$ as shown in Figure 13-14b. Figure 13-14 also illustrates that the observed absorbance is much less than $A_m$ and serious negative deviations occur when the spectral bandpass approaches the true half-width or is greater.

From these results we can develop some practical rules of thumb. The monochromator spectral bandpass or the filter half-width should be less than 1/10 of the true half-width of the absorption band when the wavelength is set to the peak maximum. This ensures good

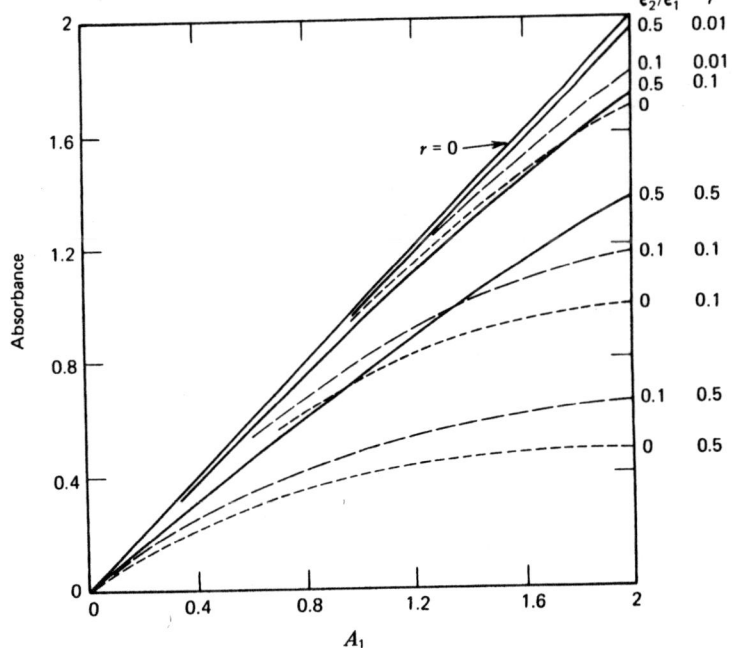

**FIGURE 13-13** Observed absorbance vs. the absorbance at $\lambda_1$ ($A_1$). The plots show how the measured absorbance depends on the ratio of the molar absorptivities ($\varepsilon_2/\varepsilon_1$) and of the reference signals ($r = E_2/E_1$) at the two monitored wavelengths. The reduction in the slope and the degree of negative deviation are enhanced when $r$ or the difference between $\varepsilon_1$ and $\varepsilon_2$ is greater. (Adapted from E. J. Meehan, "Fundamentals of Spectrophotometry," in *Treatise on Analytical Chemistry*, 2nd ed., part I, vol. 7, I. M. Kolthoff and P. J. Elving, eds., Wiley, New York, 1981, p. 73, by permission of John Wiley & Sons, Inc., New York, copyright © 1981.)

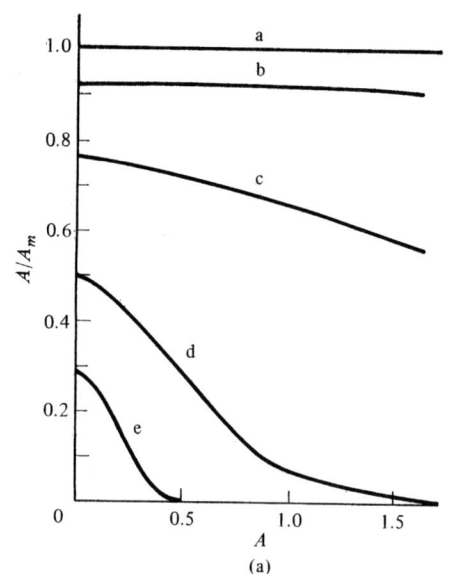

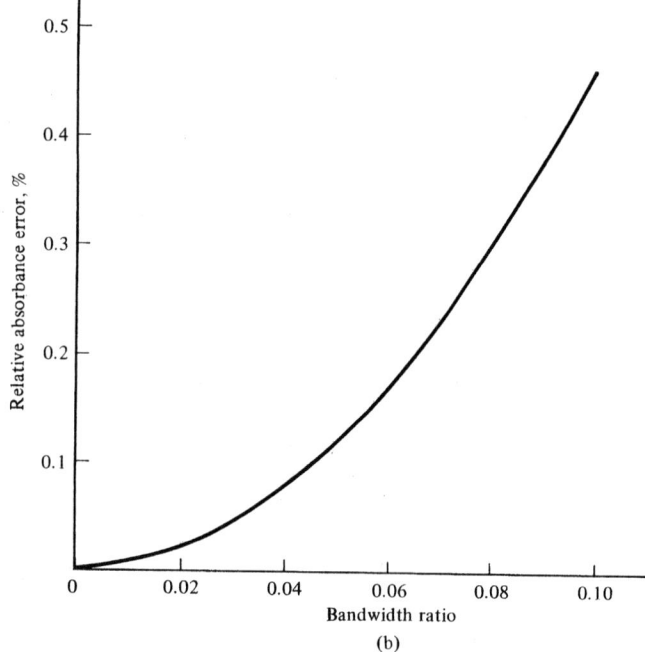

**FIGURE 13-14** Effect of polychromatic radiation on the measured absorbance. In (a) the ratio of the observed absorbance to the absorbance at the maximum of the absorption band is shown as a function of the observed absorbance and the bandwidth ratio ($s/\Delta\lambda$). The values of ($s/\Delta\lambda$) are 0, 0.43, 0.85, 1.70, and 2.55 in curves a, b, c, d, and e, respectively. In (b), the relative absorbance error (($A_m - A)/A_m$) is shown as a function of the bandwidth ratio. [(a) Adapted from S. Broderson, *J. Opt. Soc. Am.*, **44**, 22 (1954), with permission of the publisher.]

linearity and a calibration slope and effective molar absorptivity which are within better than 0.5% of the maximum values. For 0.1% accuracy ($\varepsilon/\varepsilon_m = 99.9\%$), the spectral bandpass should be reduced so that $s/\Delta\lambda \le 0.05$.

When measurements are made on the side of an absorption band ($\lambda_0 \ne \lambda_m$), the requirements are more strict because $\varepsilon(\lambda)$ varies more radically over $\lambda_0 \pm s$. As a rule of thumb, the spectral bandpass should be 1 nm or less to prevent significant nonlinearity and decreased calibration sensitivity.

Half-widths of many absorption bands are in the range 50 to 100. Thus a spectral bandpass of 5 nm or less is usually adequate. However, there are many exceptions to this rule. For example, the absorption bands of rare earth ions are quite narrow. Significant nonlinearity due to polychromatic radiation is observable with a spectral bandpass greater than 1 nm, as illustrated in Figure 13-15. In general, absorption bands in the UV region are narrower than those in the visible region. In the gas phase, absorption bands exhibit vibronic structure as discussed in Section 12-4, such that a spectral bandpass less than 1 nm is necessary to record faithfully the absorption spectrum and to make quantitative measurements.

## Nonlinearity due to Stray Radiation

Stray radiation or stray light occurs when the wavelength selector passes wavelengths outside its nominal range. For a monochromator, the amount of stray radiation is normally expressed as the stray radiation fraction ($f$) or percent ($\% f$), defined as

$$f = \frac{\Phi_{SR}}{\Phi_0} \qquad (13\text{-}48)$$

where $\Phi_{SR}$ is the stray radiant power and $\Phi_0$ is the radiant power output within $\lambda_0 \pm s$ (see Section 3-5). The fraction $f$ depends on the wavelength and the slit width. In the literature, $f$ is sometimes defined as $\Phi_{SR}/(\Phi_0 + \Phi_{SR})$, which is usually close to the value of $f$ defined by equation 13-48 because $\Phi_{SR}$ is usually much less than $\Phi_0$.

For a spectrophotometer, the reference signal due to stray radiation ($E_{SR}$) is given by

$$E_{SR} = mG \int_0^\infty (\Phi_\lambda)_{SR} T_r R(\lambda) \, d\lambda \qquad (13\text{-}49)$$

where $(\Phi_\lambda)_{SR}$ is the spectral radiant power due to stray radiation. The stray radiation readout fraction ($r$) or the relative contribution of stray radiation to the ideal

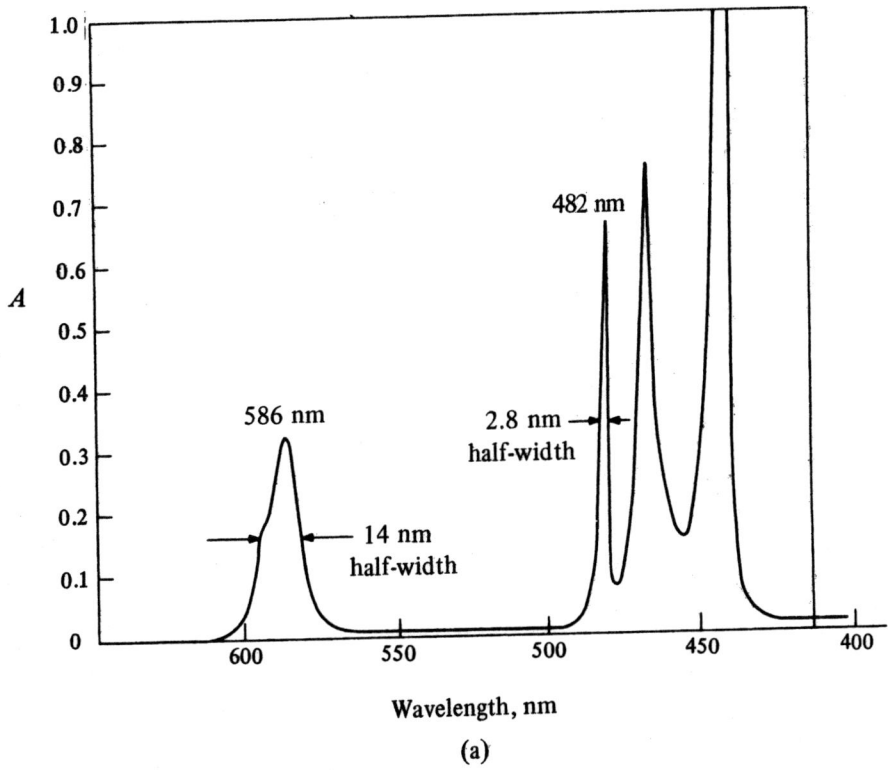

(a)

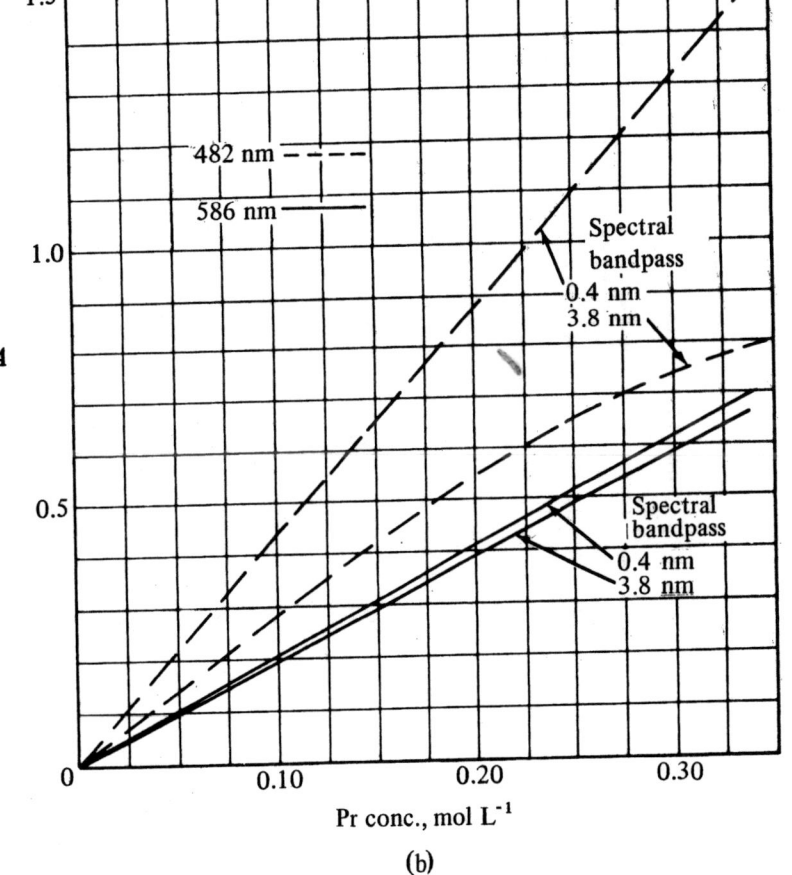

**FIGURE 13-15** Effect of the spectral bandpass on the linearity of the calibration curve for praseodymium. In (a), the absorption spectrum of 0.25 M $Pr^{3+}$ 1M HCl is shown to possess several narrow absorption bands. The calibration curves in (b) are for measurements at 482 and 586 nm with spectral bandpasses of 0.4 and 3.8 nm. For the narrow band at 483 nm, there is a significant decrease in the initial slope and negative deviation at higher concentrations with a 3.8 nm spectral bandpass (i.e., $s/\Delta\lambda = 1.36 \geq 0.1$). (Adapted from the manual for the model EU-701 UV-visible single-beam spectrophotometer, 1969, courtesy of McPherson, a division of Schoeffel Instrument Corp.)

(b)

reference signal ($E_r$ from equations 13-16 to 13-18) can be defined as

$$r = \frac{E_{SR}}{E_r} \qquad (13\text{-}50)$$

In a spectrophotometer, room light leaks into the sample compartment or detector housing can also contribute to the total stray radiation signal.

The presence of stray radiation usually causes the measured transmittance to be larger than it should be because the majority of the stray radiation is at wavelengths less strongly absorbed by the analyte than radiation within the bandpass. As the analyte concentration increases, this effect progressively worsens and causes a negative deviation in the calibration curve because the stray radiant power transmitted becomes a larger fraction of the total transmitted radiation power.

For simplicity we will first assume that the stray radiation is not absorbed by the analyte. Thus the readout signal due to the stray radiation component is identical for the reference and analyte solutions. In this case, the measured transmittance ($T'$) is given by

$$T' = \frac{E_s + E_{SR}}{E_r + E_{SR}} \qquad (13\text{-}51)$$

If we divide the numerator and denominator by $E_r$ and use the definition for $r$ (equation 13-50), we obtain

$$T' = \frac{T + r}{1 + r} \approx T + r \qquad (13\text{-}52)$$

where $T$ is the ideal transmittance when $r = 0$. The second form of equation 13-52 is normally a good approximation since $r \ll 1$. Note that equations 13-51 and 13-52 are just limiting forms of equations 13-43 and 13-44, respectively, where $E_s = E_1$, $E_{SR} = E_2$, and $\varepsilon_2 = 0$. The measured absorbance ($A'$) is given by

$$A' = -\log T' = -\log(T + r) + \log(1 + r)$$
$$\approx -\log(T + r) \qquad (13\text{-}53)$$

Figure 13-16 shows calibration plots for different amounts of stray radiation. Negative deviations become significant when $T$ approaches $r$ and $A'$ reaches an asymptotic limit of $-\log[r/(1 + r)] \sim -\log r$ when $T \ll r$ in equation 13-53. From equation 13-53 it can be shown that the measured absorbance is low by 1% when $r \sim (T^{0.99} - T)$. Thus to measure absorbances up to 3 with 1% accuracy, $r$ should be less than 0.0072%.

The foregoing treatment is an oversimplification because the analyte also absorbs radiation at wavelengths outside the range selected by the monochromator or filter. The absorption of some of the stray radiation by the analyte reduces the contribution of stray radiation to the sample readout signal and efficiently causes $r$ to decrease with increasing analyte concentration. Thus the error and nonlinearity due to stray radiation can be less than predicted by equations 13-51 to 13-53.

If the analysis wavelength is not set to the wavelength of maximum absorption, it is possible that the stray radiation is more strongly absorbed than radiation

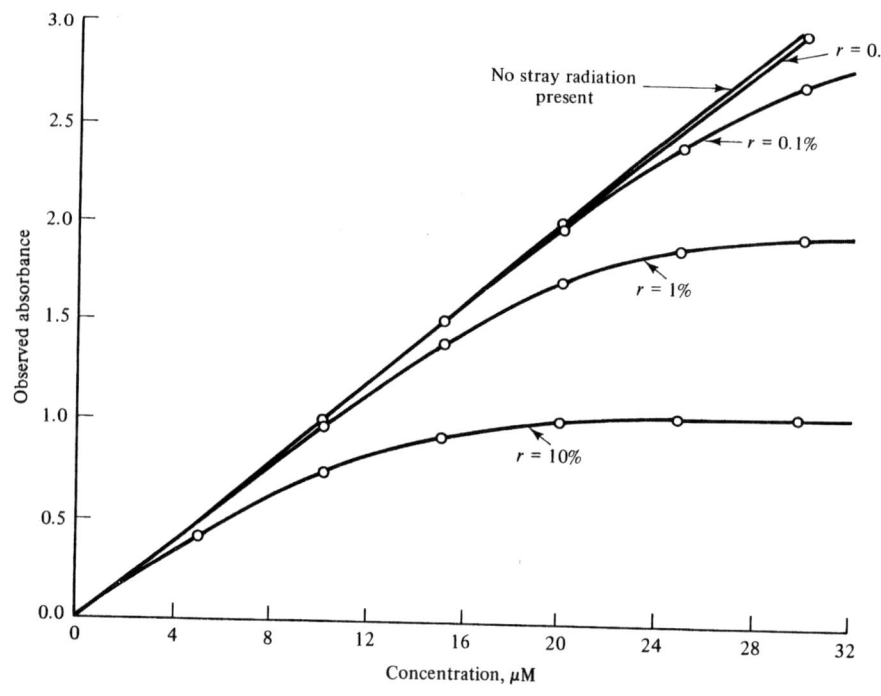

**FIGURE 13-16** Effect of stray radiation. Calibration plots derived from equation 13-53 are shown for different values of the stray radiation readout fraction ($r$) where $\varepsilon b = 10^5$ L mol$^{-1}$. As $r$ increases in value, the degree of negative deviation increases and the maximum limiting value of $A'$ decreases.

within the bandpass. In this case, at higher absorbances the absorbance measured can be too high and a positive deviation occurs.

Standard procedures are available to evaluate the stray radiation readout fraction. Often they are based on measuring the absorbance of a concentrated solution or a cutoff filter at a wavelength where the true absorbance is large enough that $T \ll r$. The observed absorbance is used to calculate $r$ ($A' = -\log r$). For example, $NaNO_2$ (50 g $L^{-1}$), NaI (10 g $L^{-1}$), or NaBr (10 g $L^{-1}$) are used to evaluate the stray radiation at 340, 220, and 200 nm, respectively. Such procedures assume that stray radiation is the only cause of nonlinearity and usually underestimate the effective stray radiation in a particular analysis because the test solution or filter absorbs more of the stray radiation than the analyte of interest.

Note that $r$ evaluated with the procedures above is different than the stray radiation fraction $f$ because $r$ depends on the differences in the reference transmission factor and the detector responsivity for radiation within the bandpass $[(\Phi_\lambda)_0]$ and outside the bandpass $[(\Phi_\lambda)_{SR}]$. Consider a simple case where the source emits at only two wavelengths ($\lambda_1$ and $\lambda_2$) with equal radiance. If $\lambda_1$ is selected for analysis and $f = 1\%$, then 1% of the radiation at $\lambda_2$ is incident on the sample cell as the stray radiation component. If the detector responsivity is 10 times greater at $\lambda_2$, $r = 0.1$.

Stray radiation is more significant at analysis wavelengths where the product of the source radiance and detector responsivity $[B_\lambda R(\lambda)]$ is low compared to the average for the rest of the wavelength range. Normally, $r$ is largest from 200 to 220 nm with a $D_2$ lamp and 340 to 400 nm with a tungsten lamp because the stray radiation arises from wavelength regions where the lamp radiance is greater. The stray radiation readout fraction is often greater in the NIR region than the visible region because $R(\lambda)$ is smaller for longer wavelengths. Stray radiation can be reduced by using absorption band filters in conjunction with the monochromator, double monochromators, or holographic gratings. The first two options cannot be used with diode array–based spectrophotometers, for which stray radiation effects can be particularly troublesome because of the large aperture in the focal plane.

### Other Instrumental Causes of Nonlinearity

There are numerous minor causes of nonlinearity. Most are significant only for applications requiring very high accuracy.

*Variability in Pathlength.* Beer's law is based on the assumption that all rays in the incident beam traverse the same thickness of the absorbing medium. This assumption is false if the cell walls are not parallel or flat or if the incident beam is not collimated. To illustrate this effect, consider the simple case where half of the rays in the incident beam experience pathlength $b_1$ and the other half experience $b_2$. The measured transmittance is given by

$$T = \frac{10^{-\epsilon b_1 c} + 10^{-\epsilon b_2 c}}{2} \quad (13\text{-}54)$$

Equation 13-54 is very similar in form to equation 13-43, which describes the case of two wavelengths being absorbed with different molar absorptivities. Thus the measured $T$ or $A$ values are between those that would be measured separately with each pathlength. If the difference in the pathlengths is great, a negative deviation is observed at higher absorbances where the rays traveling the shorter pathlength dominate the transmitted radiant power.

Pathlength effects are usually most important for test tube–shaped cells, as illustrated in Figure 13-17a. Even if the width of the incident beam is one-half the cell diameter, the effective pathlength and measured absorbances are only about 2.5% low relative to that expected from the sample cell diameter and nonlinearity is insignificant even up to absorbances near 2. With rectangular and cylindrical (flat window) sample cells, the pathlength for different rays varies primarily because the incident beam is never perfectly collimated,

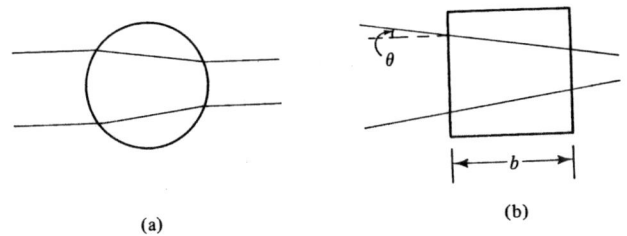

(a)                          (b)

**FIGURE 13-17** Examples of variability in pathlength. Top views are shown of the incident radiation entering and traversing a test tube–shaped sample cell in (a) and a standard spectrophotometer cell in (b). For the test tube–shaped sample cell, the pathlength varies across the beam cross section even if the incident beam is collimated. In addition, rays impingent on different portions of the sample cell are refracted differently and experience different pathlengths. If the rays in the incident beam are not parallel as shown in (b), the effective pathlength is greater than $b$ and varies with the angle of the incident ray. The maximum angle of divergence ($\theta$) of an incident ray with respect to the normal to the cell wall is shown.

as shown in Figure 13-17b. However, even with a moderate divergence angle of 13.5° for the incident beam, the effective pathlength and measured absorbance are only 0.5% too high and nonlinearity is insignificant.

### Multiple Reflections.

In our previous discussion of reflection losses at the cell walls, we ignored the effect of multiply reflected rays. About 4% of the radiation reaching the rear cell wall–air interface is reflected back into the cell and passes through the solution. Most of this reflected radiation passes out of the cell; however, about 4% of the radiant power is reflected off the front cell wall–air interface and then passes back through the solution and out the rear cell wall to the detector. This twice-reflected ray (see Figure 13-10) experiences a pathlength of $3b$ and corresponds to about $(0.04)^2 = 0.16\%$ of the directly transmitted beam. The fraction of the radiation experiencing a pathlength of $3b$ can be greater if some of the once-reflected beam that passes out of the front cell wall is reflected back into the cell from such surfaces as lenses or the monochromator exit slit. The contribution from rays undergoing even more reflections is usually insignificant.

The net result of multiple reflections is that the measured absorbance is too large because part of the monitored radiation is absorbed more strongly due to the enhanced pathlength. Thus multiple reflections are just a special case of pathlength variability. The absolute error ($\Delta A$) is approximately given by

$$\Delta A = 0.434F(1 - T^2) \qquad (13\text{-}55)$$

where $F$ is the fraction of the total transmitted radiant power due to the multiple reflected beam. If $F$ is assumed to be 0.16%, multiple reflections cause positive errors of about 0.3% at low absorbances. The absolute error approaches a constant value of about 0.0007 A.U. at higher absorbances because the multiply reflected beam is effectively completely absorbed.

### Circular Dichroism.

When molecules are optically active, the molar absorptivity depends on the polarization of the incident radiation (see Sections 3-2 and 13-5). As with polychromatic radiation, negative deviations can result when different components of the incident beam are absorbed to different degrees. Thus the absorbance is given by an expression similar to equation 13-45 except that now $\varepsilon_1$ and $\varepsilon_2$ represent the molar absorptivities at a given wavelength for the $d$ and $l$ circularly polarized components and $r$ is the ratio of the incident radiant power for the two polarization components. The output radiation from a monochromator is usually partially polarized by the grating.

Linearity can be achieved by placing a polarizer in the optical beam to produce radiation polarized in only one direction. In solution, the relative difference in $\varepsilon$ for the two polarization components rarely exceeds 10%; a negative deviation of only 0.6% or less is observed even at an absorbance of 2. Nonlinearity due to circular dichroism can be more significant in solids or liquid crystals.

### Fluorescence.

At higher analyte concentrations, a significant portion of the incident radiation is absorbed. A fraction of this absorbed radiation is converted to fluorescence and observed by the detector. Thus the observed sample radiant power is higher than expected, which causes the measured transmittance to be too high or the absorbance to be too low. The measured transmittance is given by an equation similar to that for stray radiation (equation 13-52) except that $E_{SR}$ is replaced by $E_F$, the readout signal due to analyte fluorescence, and $E_{SR}$ in the denominator equals zero. As the analyte concentration increases, $E_s$ decreases while $E_F$ increases, which causes negative deviation in the calibration curve when $E_F$ becomes significant relative to $E_s$.

Generally, analyte fluorescence is negligible even compared to the dark signal and does not cause nonlinearity for three primary reasons. First, the fluorescence quantum efficiency of most species is very small. Second, the source radiance is much lower than that used for fluorescence measurements. Third, the solid angle of collection between the sample cell and detector is usually smaller than that used in fluorescence spectrometers.

Because analyte fluorescence occurs at longer wavelengths than the excitation wavelength, it can be eliminated by placing an appropriate filter between the sample cell and detector. In multichannel spectrophotometers, the fluorescence signal is isolated from the absorption signal because the wavelength selector is between the sample cell and the detector. However, because the sample is illuminated by all source wavelengths, fluorescence can occur at the wavelength of the monitored absorption band due to excitation at wavelengths of other absorption bands at lower wavelengths.

Background fluorescence from concomitants is rarely significant but can cause effects similar to those just described for analyte fluorescence. In this case, the contribution from background fluorescence is constant and independent of the analyte concentration unless the analyte absorbs the background fluorescence. If the species responsible for background fluorescence are in the blank, equation 13-51 applies if $E_{SR}$ is replaced by $E_{bF}$, the background fluorescence readout signal.

*Miscellaneous Causes.* Nonideal performance of the detector (e.g., hysteresis), the signal processor, or the readout device can affect linearity. Thus the characteristics of these components must be such that the transmittance readout signals are directly proportional to the radiant power incident on the detector. With direct absorbance readout, the accuracy and linearity of logarithmic conversion can be limiting. Incorrect setting of 0% $T$ on the readout scale or improper subtraction of the dark current signal from the total readout signals can also cause nonlinearity.

## 13-4 METHODOLOGY AND PERFORMANCE CHARACTERISTICS

In this section we explore the factors that influence the use and performance characteristics of quantitative molecular absorption spectrophotometry. Usually the analyte concentration in the analytical sample is evaluated from the sample absorbance and a calibration curve. The calibration curve is most often generated from absorption measurements on external standards.

### General Considerations in Quantitative Analysis

*Sample Treatment.* Sample treatment varies considerably with the application. Because most determinations are carried out on solution samples, solids must be dissolved. Sometimes the analyte can be determined directly from the absorbance of its original chemical form. More commonly, derivatization is necessary because the absorption bands of the analyte in its original chemical form possess inadequate absorptivity or are in an unsuitable wavelength range. Derivatization involves converting the analyte to a strongly absorbing form by a selective analytical reaction with appropriate reagents. The absorbance of a reaction product is related to the original analyte concentration by external calibration. Usually, an excess of the reagent(s) is added, and the analytical reaction is allowed to proceed to equilibrium before measurements are made. Alternatively, the rate of the chemical reaction can be related to the analyte concentration, or the analyte can be titrated with a standard reagent that reacts with it with known stoichiometry. The reaction between the analytical sample or standards and reagents can be carried out directly in the sample cell or in another suitable container. In the latter case, an aliquot of the reaction mixture is transferred to the sample cell after the reaction is complete. In such specialized applications as titrations and kinetics studies and determinations, absorption measurements are made prior to complete conversion of the analyte to its absorbing form as discussed in Section 13-5. Often separation steps (e.g., filtering, solvent extraction) are necessary to eliminate interferents, and in some applications, spectrophotometry is used as the detection technique for chromatography (see Section 13-5).

*Choice of Solution Conditions.* In analytical procedures, one must first determine the absorbing form of the analyte to be used. The nature of the absorbing species greatly influences the range of concentrations to be measured, the calibration sensitivity, the detection limit, the precision, the accuracy, and the linearity. For trace analysis, the absorptivity of the absorbing species must be large enough to yield a good detectability and precision. The chemical nature of the absorbing species affects how it interacts with potential interferents or takes part in equilibria that can cause nonlinearity. Similarly, the spectral characteristics of the absorbing species determine the potential for spectral overlap with other species.

The solvent and such solution conditions as pH, ionic strength, and temperature should be optimized and controlled to minimize interferences and nonlinearity due to equilibria and to maximize the time stability of the absorbing form (i.e., to prevent decomposition due to oxidation, photodecomposition, or other mechanisms). When an analytical reaction is used to form an absorbing derivative, the reagent concentration(s) must also be optimized to achieve the goals noted above. Additionally, the time and temperature required to form the absorbing product, the order of addition of reagents, and the reaction of concomitants with the reagent must be considered. Masking reagents are sometimes added to block the reaction of concomitants with the analytical reagent. Ideally, the conversion of the analyte to the measured absorbing form should be quantitative. Otherwise, the efficiency of conversion should be constant irrespective of the concentration of the analyte or the concomitants. When solvent extraction is employed, the effect of the type of solvent on the extraction efficiency for the analyte and concomitants, the molar absorptivity, and the spectral positions of the absorption bands should be evaluated.

*Choice of Wavelength and Spectral Bandpass.* For standard quantitative procedures, the absorption spectrum of the absorbing form of the analyte is available in the literature, and the wavelength ($\lambda_0$), a suitable spectral bandpass, and solution conditions (e.g., reagent concentrations) are specified. In developing a quantitative method for determining the analyte concentration, the choice of the absorbing species and ultimately the particular absorption band and wavelength to use for analysis is critical as discussed above. To make these decisions, UV-visible absorption spectra of all relevant species (the analyte and/or its absorbing form, the reagent(s), and potential interferents) are obtained with a scanning or multichannel spectrophotometer. For accurate recording of the true absorption band shapes and widths, the spectral bandpass is adjusted to be much less than the half-width of the narrowest absorption band in the spectrum. From the spectrum of the absorbing form of the analyte, the wavelength of maximum absorption ($\lambda_m$) and half-width ($\Delta\lambda$) of all absorption bands should be characterized.

When several absorption bands of suitable absorptivity are present, the absorption band selected should favor wavelength regions that correspond to relatively high source radiance, high detector responsivity, and a high reference transmission factor (i.e., minimal absorption by the sample cell and blank solution) to maximize the reference photocathodic current ($i_r$) and hence the precision. If possible, the wavelength region selected for analysis should not overlap absorption bands of concomitants in the sample or of any reagents used. Although it is possible to compensate for reagent absorbance with a proper reagent blank, it is best to choose a band and analysis wavelength where reagent absorption is minimal because the exact amount of excess reagent after reaction with the analyte is usually unknown.

Often, $\lambda_0$ is chosen to be equal to $\lambda_m$ for the absorption band utilized. This provides the largest calibration curve slope and minimizes nonlinearity due to polychromatic radiation. There are two primary reasons to choose an analysis wavelength on the side of a band. First, $\lambda_m$ may be outside the wavelength range of the spectrophotometer or photometer used for analysis or at the extremes of the wavelength range where the $S/N$ may be poor. Second, there may be significant reagent or interferent absorption at the band maximum which cannot be compensated with a reagent blank. A large blank absorption can also reduce $i_r$ and hence the $S/N$.

If possible, the slit width should be adjusted so that $s < \Delta\lambda/10$ when $\lambda_0 = \lambda_m$ to minimize nonlinearity due to polychromatic radiation. If measurements must be made on the size of absorption bands, it is best to determine experimentally the spectral bandpass that provides the desired linearity. Too small a slit width reduces $i_r$ to the point that the precision is decreased.

*Choice of Instrument.* If several spectrophotometers are available, the type of instrument to use depends on the relative importance of cost, simplicity, precision, accuracy, and detectability. For established routine analytical procedures involving determination of the same species in a large number of samples, a simple inexpensive photometer is often adequate. Spectrophotometers provide more wavelength versatility which may be needed when many different types of species must be determined or when new procedures are developed. Less expensive spectrophotometers with a fixed spectral bandpass are suitable for many routine procedures. More expensive and higher-quality spectrophotometers often provide a wider range of usable wavelengths and better detection limits. Linearity can also be better because of lower stray radiation figures and the ability to adjust the spectral bandpass to a small enough value for narrow absorption bands. Computerized spectrophotometers which provide relatively fast scanning capabilities, and particularly multichannel spectrophotometers, increase sample throughput for cases where many spectra or multiple wavelength information is required. Instruments with automated sample treatment and sample introduction capabilities (see Sections 6-6 and 13-5) provide high sample throughput and allow unattended operation for routine determinations.

Single-beam instruments are often less complex and expensive than double-beam instruments and provide higher radiation throughput (i.e., large $\Phi_0$ and $i_r$) and hence higher $S/N$ because fewer optical components are in the light beam. The NBS uses SB spectrophotometers for very high accuracy absorption measurements and certification of absolute absorbance standards.

The choice between DB and SB designs is not as clear as in the past. Formerly, a double-beam spectrophotometer was necessary for obtaining spectra and often better compensated for source flicker noise and drift because the sample and reference signals were continually ratioed. Also, sample throughput could be higher because it was not necessary to alternate manual sample and reference measurements. Some modern computerized SB spectrophotometers (single-detector and multichannel) allow spectra to be acquired. High-stability lamp power supplies and/or optical feedback stabilization minimize lamp flicker noise and drift and allow several sample measurements to be made after one initial reference measurement.

Most manufacturers of spectrophotometers provide the values for a standard list of specifications shown in Table 13-4 that can be used in choosing a spectrophotometer. The values of these reported parameters are obtained by applying standardized testing procedures. The **wavelength range** is particularly useful for deciding if UV or NIR capability is required. The **wavelength accuracy** is evaluated by comparing the wavelength setting of the monochromator to the known wavelength at the maxima of emission lines from a low-pressure Hg source or from the $D_2$ lamp in the spectrophotometer (e.g., 656.1 nm) or absorption bands of test solutions or filters (e.g., didymium glass filters). The **wavelength repeatability** is a measure of the precision of the above measurement. The spectral bandpass value(s) or range indicate the usefulness of the instrument for narrow absorption bands. The stray radiation percentage is evaluated at specific wavelengths and spectral bandpasses with strongly absorbing solutions or filters as described in Section 13-3. **Photometric accuracy** is determined by comparing the measured absorbance to the certified absorbance of accepted standards (e.g., NBS filters), whereas **photometric reproducibility** is the precision obtained for repetitive measurements on these standards. **Photometric linearity** indicates the ability of the spectrophotometer to provide linearity for a set of standards known to follow Beer's law. The **baseline noise level** usually indicates the rms or peak-to-peak noise in absorbance units at 0 A.U. (blank in the sample cell) at a specified wavelength and spectral bandpass. The **baseline stability** specifies the drift in the 0 A.U. signal in A.U. per hour after a specified warm-up time. The adjustable range of the noise equivalent bandpass is indicated by the time-constant or response-time (time for response to 98% of the final value) range available. The **photometric range** indicates the transmittance and absorbance range displayed on the readout device or over which the instrument can be used. For scanning or multichannel spectrophotometers, the **baseline flatness** indicates the average deviation from 0 A.U. for a blank vs. blank spectrum and may be specified with or without baseline compensation.

## Performance Characteristics

We now consider several important performance characteristics, including precision, detection limit, linearity, and accuracy.

*Precision.* The precision of determining the analyte concentration in a sample can be no better than the sample measurement precision (the precision with which the absorbance of a given sample solution is measured). The sample measurement precision is determined by readout resolution, noise, sample cell placement reproducibility, and the sample absorbance. It can be estimated from $S/N$ considerations as discussed in Section 13-3 or experimentally determined from repetitive measurements. The instrumental specification of photometric reproducibility provides some of this information at selected absorbance values and wavelengths. With a linear calibration curve, the relative standard deviation (RSD) in determining a concentration ($\sigma_c/c$) is equal to the RSD in measuring the sample absorbance ($\sigma_A/A$) if this is the factor limiting the random error.

If possible, the analyte concentrations in the standards and sample should be adjusted so that their absorbances are in the range where measurement precision is best. The measurement precision of simple photometers and spectrophotometers, with low-resolution readout devices (e.g., analog meters), is often limited to 1 to 2% by readout resolution for absorbances in the range 0.2 to 0.9. For higher-quality instruments with adequate readout resolution, noise or cell positioning imprecision is often dominant; the RSD can be 0.1% or better for absorbances greater than about 0.1

**TABLE 13-4**
Spectrophotometer specifications

| Specification | Value[a] | |
|---|---|---|
| 1. Wavelength range (nm) | 185–3200 | 340–1000 |
| 2. Wavelength accuracy (nm) | 0.2 (0.8 in NIR) | 2 |
| 3. Wavelength repeatability (nm) | 0.05 (0.2 in NIR) | 1 |
| 4. Spectral bandpass (bandwidth) or resolution (nm) | 0.05–5 (0.2–20 in NIR) | 10 |
| 5. Stray radiation (light) (%) | <0.000012 (<0.002 in NIR) | 0.3 |
| 6. Photometric accuracy (A.U.) | ±0.003 at 1 | ±0.003 at 0.4 |
| 7. Photometric repeatability | ±0.001 A.U. | — |
| 8. Photometric linearity | — | 1% $T$ |
| 9. Baseline noise level (A.U.) | <0.0002 | 0.001 |
| 10. Baseline or absorbance zero stability (drift) (A.U. h$^{-1}$) | <0.0005 | 0.003 |
| 11. Response time or time constant (s) | 0.2–10 | — |
| 12. Photometric range (A.U.) | −5 to 5 | 0 to 2 |
| 13. Baseline flatness (A.U.) | 0.001 (0.02 in NIR) | — |

[a]The values in the first column are for a high-quality, double-beam spectrophotometer (Perkin-Elmer Lambda 9). The values in the second column refer to a less expensive, simple, single-beam spectrophotometer with a fixed spectral bandpass (Bausch and Lomb Spectronic 21MV).

A.U. This precision is maintained up to 1 to 3 A.U., depending on the limiting noise sources. For a given analysis wavelength, the slit width is the primary variable that the user can vary to change the shape of the precision curve.

Except for measurements at low absorbances near the detection limit, the overall precision of a spectrophotometric determination is often limited by the reproducibility of the sample acquisition, sample treatment, sample placement, or calibration steps. The overall RSD can range from 0.3 to 5%, depending on the situation.

*Detection Limit.* The detection limit (DL) is determined by the blank standard deviation ($s_{bk}$) and the slope of the calibration curve (DL $= ks_{bk}/\varepsilon b$). The value of $s_{bk}$ depends on the instrument and instrumental conditions, such as the analysis wavelength and the slit width. It typically varies from $10^{-3}$ to $10^{-5}$ A.U. Thus for a solution of a very strongly absorbing species ($\varepsilon = 10^5$ L mol$^{-1}$ cm$^{-1}$) in a 1-cm-pathlength cell, the DL is typically in the range $2 \times 10^{-10}$ to $2 \times 10^{-8}$ M.

The DL can be improved by preconcentration techniques, by longer-pathlength cells, or by reducing the blank noise. If signal shot noise, dark current noise, or amplifier noise determines the SD, the slit width should be increased to reduce the blank noise. In general, instrumental conditions should be adjusted to minimize the ratio of $s_{bk}$ to $\varepsilon$. Note that if the wavelength or slit width is adjusted to a value such that $\varepsilon$ is not equal to the value at the maximum of the strongest absorbing band ($\varepsilon_m$), nonlinearity can occur at higher absorbances. If sample cell placement limits precision, the sample cell should be left in the sample cell holder while solutions are added and removed. Readout resolution should be improved if it limits the DL.

The rms noise observed with the blank in the sample cell (i.e., the baseline noise specification) is often equated to the blank SD and used to calculate the DL. This method often yields a falsely low value. If sample cell positioning reproducibility is important, a better estimate of $s_{bk}$ is obtained by making repetitive measurements on a blank by removing and replacing the sample cell between each measurement. If the blank absorbs significantly, a more realistic estimate of $s_{bk}$ is obtained from measurements on separate reagent blank solutions carried through the complete preparation procedure. The practical detection limit in complex real samples may be even higher than calculated above, due to the variability of uncompensated blank absorbance from sample to sample. In some cases, the practical value of $s_{bk}$ may be $10^{-2}$ A.U. or greater. When a spectrophotometer is used as an HPLC detector, the DL estimated from the baseline noise is often a more realistic estimate of the practical DL because the analyte is ideally isolated from other absorbing species before detection.

*Linearity.* The factors affecting linearity were presented in Section 13-3. The actual range of linearity is best confirmed by experimental measurements. As described previously, the natural half-width of the absorption band can be used to estimate the monochromator spectral bandpass or filter half-width needed to minimize deviations due to polychromatic radiation. Stray radiation specifications give rough guidance to estimate the expected nonlinearity due to stray radiation; this often underestimates the effect of stray radiation. Nonlinearity due to chemical equilibria or another chemical effect can be distinguished by measuring the same solutions in cells of different pathlengths. Instrumental effects are generally worse for a given solution in longer-pathlength cells where the absorbance is greater, whereas nonlinearity due to chemical effects is independent of pathlength and dependent on concentration. If measurements must be made in the nonlinear portion of the calibration curve, enough standards should be measured to ensure that the calibration function is accurately known.

In general, sample solutions should be diluted so that the absorbance is 2 or lower. Extending the absorbance measurements to 3 A.U. increases the linear dynamic range by only 50%. The additional factor-of-10 reduction in the transmittance greatly increases the probability of nonlinearity due to instrumental effects and often results in poorer precision due to the dominance of signal shot noise or 0% $T$ noise. In some applications, high dilution factors cannot be used because a small analyte absorbance is measured superimposed on top of a large background absorbance or scattering signal. In this case, instrumental characteristics such as stray radiation must be appropriate to obtain linearity.

*Accuracy.* To obtain accurate results with external standardization, a standard and sample of the same analyte concentration should yield the same measured absorbance. This is not usually the case because of random errors (as discussed above) and systematic errors. Such causes of systematic errors as blank and analyte interferences are discussed below.

Blank interferences often limit the accuracy of spectrophotometric measurements and are usually spectral interferences, which cause the contribution of the reference transmission factor ($T_r$) to the sample transmission factor ($T_s$) (see equation 13-14) to be different for the sample and standards. Thus spectral interference is caused by concomitants or particles present in the analytical sample, but absent from the standards, that absorb or scatter at the analysis wavelength (i.e., that affect $\alpha_b$), that fluoresce significantly, that change the reflection loss as the cell wall-solution interface (i.e., that affect $\rho_c$), or that alter the focusing of the beam image on the detector. When an analytical reaction is used to convert the analyte to a strongly absorbing form, blank spectral interference also arises from species in the sample, which do not necessarily absorb themselves, but which react with the analytical reagent(s) to form a species absorbing at the analysis wavelength.

Most analyte interferences are nonspectral in nature and caused by species in the sample, but not the standards, which affect the concentration of the absorbing form of the analyte or the molar absorptivity of the absorbing form. Thus nonspectral interference occurs when the sample matrix causes any of the factors which affect the linearity or slope of the calibration curve (see Section 13-3) to be different between the standards and the sample. Concomitants can affect the equilibria between different forms of the analyte, the fraction of the analyte in the absorbing form, the electrostatic environment of the analyte, and thus the molar absorptivity, and the relative contribution of polychromatic or stray radiation to the measurement. Also, the conversion efficiency of the analytical reaction can be altered by species that interact with the analyte and prevent its reaction with the reagents or that consume a significant fraction of the reagents.

Interference effects are reduced by a careful choice of the analysis wavelength, spectral bandpass, and/or analytical reaction to achieve selectivity for the analyte. Careful control and choice of such solution conditions as reagent concentrations, the type of solvent, pH, and ionic strength can enhance selectivity and prevent shifts in equilibria due to species only in the sample. Separation of the analyte from some interferents by solvent extraction or another technique is sometimes essential. Filtering is often recommended to reduce potential scattering interferences. An internal blank is occasionally useful when analytical reactions are employed. Here the absorption due to species in the original sample which do not react with the analytical reagent(s) can be compensated with a blank that is the sample reaction mixture minus a critical reagent needed for the analytical reaction to proceed. In some cases, an internal blank can be made by adding another reagent which blocks the formation of the absorbing product. Standard addition techniques can be used when analyte interferences affect the slope of the calibration plot. Additional selectivity can be gained by using titration, reaction-rate, derivative, or dual-wavelength techniques, as discussed in Section 13-5.

Factors other than interference effects that can cause systematic errors include calibration errors (e.g., improper standards), sampling errors, contamination and loss, decomposition, and different measurement conditions (e.g., pH, temperature, sample cells with different pathlengths or transmission characteristics) for the samples and standards.

*Absolute Accuracy.* Up to this point, we have been concerned primarily with the accuracy of determinations with external calibration. A determination can be accurate even if the measured absorbance is not the true absorbance as long as the calibration accounts for nonideal behavior. In several applications it is necessary to know the true absorbance. The instrumental specification of photometric accuracy is a measure of the ability of the instrument to measure absorbance values accurately.

From Beer's law ($A = \varepsilon bc$), either $\varepsilon$, $b$, or $c$ can be determined if the absolute absorbance is measured accurately and two of the other three quantities are known. For example, the molar absorptivity can be determined from the absorbance if the pathlength and analyte concentration are accurately known. An accurate value of the molar absorptivity is needed for calibration of $b$ or $c$ or for fundamental studies relating measured absorption parameters to fundamental quantities (e.g., transition probability) or molecular structure. Similarly, the pathlength of a sample cell can be determined from the absorbance if a solution with a known analyte concentration and molar absorptivity can be prepared. Sample cells with pathlengths accurate to $10^{-4}$ cm are available from the NBS. To determine $b$ or $\varepsilon$ accurately, extreme care must be exercised in preparing solutions of known analyte concentration.

In analytical applications, the sample analyte concentration can be directly calculated from the measured sample absorbance and known values of $b$ and $\varepsilon$ without resorting to external standardization. This is common where the purity of compounds must be assayed or with clinical enzymatic assays for which pure standards for enzymes or proteins are not readily available.

When absolute measurements are made, experimental conditions must be adjusted to ensure linearity. For very high accuracy work, the more subtle causes of nonlinearity and systematic errors such as multiple reflections and variable pathlengths must be considered. An angle of incidence 3% from the normal causes a 0.1% positive error. The preparation of the blank solution and measurement of the blank must be conducted in a way that the reference transmission factor ($T_r$) is the same for the sample and blank solutions. Thus $\alpha_b$, $\rho_c$, and $\alpha_c$ in equation 13-12 must be the same for the reference and sample measurements. Additionally, the wavelength setting of the monochromator must be accurately calibrated. Errors in wavelength calibration of 1 to 2 nm can cause absolute errors in absorbance of 0.1 to 1%, depending on the spectral bandpass and the half-width of the absorption band.

## 13-5 APPLICATIONS

In this section we stress the use of spectrophotometry as a quantitative tool in fundamental and analytical applications. The section begins with a brief discussion of qualitative analysis with spectrophotometry.

### Qualitative Analysis

The spectral position of an absorption band is indicative of the presence or absence of certain structural features or functional groups in a molecule as discussed in Section 12-5 (see Table 12-6 in particular). Also, compilations of UV-visible absorption spectra of many compounds are available to compare to the absorption spectrum of a pure unknown. Usually, a match between the reference and unknown spectra is not sufficient proof of the identity of the compound because the positions and intensities of the few absorption bands are not greatly affected by minor differences in structure, particularly for large molecules. Thus spectrophotometry is not considered a major tool for qualitative analysis; such techniques as NMR, IR, and mass spectrometry are more often employed for positive identification.

### Fundamental Applications

Spectrophotometry is a major tool for studying chemical equilibria and kinetics. Often it is possible to choose appropriate wavelengths to monitor the absorbance(s) of one or more reactants, products, or intermediates in the presence of other species. The concentrations are then determined by applying Beer's law and known molar absorptivities.

For equilibrium studies, known concentrations of reactants are mixed, and the absorption spectrum of the reaction mixture is obtained after equilibrium is reached. The final concentrations of reactants and products are determined from the measured absorbances at selected wavelengths and stoichiometric relationships; these are then used to calculate equilibrium constants. An example is presented later.

Spectrophotometry is also used to determine the stoichiometry of reactions and, in particular, of metal complexes. The dependence of the solution absorbance on the ratio of the metal ion (M) and ligand (L) analytical concentrations is used to determine the molar ratio $l/m$ for the reaction $m\text{M} \text{ to } l\text{L} \rightleftharpoons \text{M}_m\text{L}_l$. Some of

the most common techniques are discussed below. For simplicity, we will assume that only one complex is formed and that only the complex absorbs at the monitored wavelength. More sophisticated data manipulation techniques are available for situations in which the foregoing assumptions are invalid.

In the **molar ratio method**, the absorbances of a series of solutions with different ligand-to-metal ion concentration ratios are measured; the metal ion analytical concentration ($c_M$) is kept constant while the ligand analytical concentration ($c_L$) is varied. As shown in Figure 13-18a, the value of $c_L/c_M$ at the break point in the plot of $A$ vs. $c_L/c_M$ equals $l/m$.

For the **continuous variations** or **Job's method**, the absorbance of a series of solutions is measured in which the mole fraction $X$ of the ligand is varied from 1 to 0 as shown in Figure 13-18b; the total number of moles of L and M is held constant. The mole fraction yielding

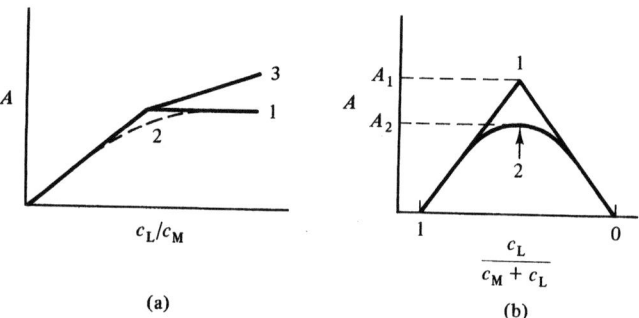

(a)     (b)

**FIGURE 13-18** Techniques for determining the stoichiometry of a complex. (a) and (b) show plots of the absorbance data obtained with the molar ratio and continuous variations methods, respectively. In both figures, curves 1 and 2 represent the cases where the dissociation of the complex is negligible and significant, respectively, and only the metal complex is assumed to absorb at the monitored wavelength. In curve 1 of (a), the absorbance reaches a plateau when the molar ratio of ligand to metal equals the actual ratio in the complex. The solution absorbance does not increase with further increases in the ligand concentration because the moles of metal ion limit the moles of metal complex formed. In curve 1 of (b), the maximum absorbance is observed when the mole fraction of ligand equals the mole fraction of ligand in the metal complex. If dissociation of the metal complex is significant, the behavior shown in curve 2 of both figures is observed and extrapolation of the linear portions can be used to locate the position of the break point or the maximum more accurately. If the ligand additionally absorbs with a different molar absorptivity [curve 3 in (a)], the slopes of the two segments are different.

the maximum absorbance $X_m$ is related to the stoichiometric ratio in the complex by $l/m = X_m/(1 - X_m)$. In both the molar ratio and continuous variation methods, negligible dissociation of the complex is often assumed; extrapolation techniques can be used if the dissociation is not excessive as shown in the figure.

The **Mollard method** involves the measurement of the absorbance of only two solutions. One measurement is made with metal ion analytical concentration $c_M$ and a large excess of ligand to yield absorbance $A_M$. The second absorption measurement is made on a solution with an excess of metal ion compared to the ligand analytical concentration $c_L$ to yield absorbance $A_L$. The amount of excess reactant in both cases is adjusted so that dissociation of the complex is negligible and the concentration of the metal complex is determined by the concentration of limiting species. Under these conditions, $l/m = (c_L A_M)/(c_M A_L)$.

A variation of the Mollard method is the **slope ratio method**, in which a series of solutions with excess L and different $c_M$ and with excess M and different $c_L$ are prepared. The ratio of the slopes of plots of $A_M$ vs. $c_M$ and $A_L$ vs. $c_L$ is equal to $l/m$. Curvature of the plots indicates that dissociation of the complex is significant.

Numerous schemes have been proposed to determine equilibrium constants from the data acquired with the foregoing methods. For example, consider the continuous variations method (Figure 13-18b) applied to a simple 1:1 metal complex ($l = m = 1$). If $A_1$ and $A_2$ are the extrapolated and measured absorbances at the maximum, respectively, the equilibrium concentrations (denoted by brackets) are given by

$$[ML] = c_M \frac{A_2}{A_1}$$

$$[M] = [L] = c_M \left(1 - \frac{A_2}{A_1}\right)$$

where $c_M$ is the initial (analytical) metal ion concentration. Thus the formation constant, $K = [ML]/[M][L]$, is calculated from

$$K = \frac{A_2/A_1}{c_M(1 - A_2/A_1)^2}$$

In kinetics studies, spectrophotometry is used to monitor the disappearance of reactants or the formation of products or intermediates. If Beer's law is valid for the absorbing species, the rate can be calculated from the rate of change in absorbance [$dc/dt = (dA/dt)/\varepsilon b$]. Thus absorbance monitoring provides a tool to determine rate laws and rate constants and to elucidate the mechanisms of reactions. Diode array spectrophotom-

eters have become powerful tools for kinetics studies because complete absorption spectra can be recorded throughout the course of the reaction.

Some other fundamental applications include the determination of molecular parameters and the molecular weight. The shape, width, and wavelength of the absorption bands of a pure compound can be used to elucidate the energy-level structure of the molecule, as discussed in Section 12-4. The integrated absorbance can be used to calculate fundamental quantities such as the transition probability, as reviewed in Appendix F. If a compound is derivatized with a reagent possessing a chromophore of known molar absorptivity that is not affected by the coupling, the molecular weight (MW) can be calculated from the measured absorbance and the formula

$$MW = \frac{\varepsilon cb}{A} \tag{13-56}$$

where $c$ is concentration of the species in g $L^{-1}$.

## Conventional Quantitative Determinations

As discussed in Section 13-4, the analyte is usually determined by measuring the absorbance of an absorbing product after equilibrium has been reached for the reaction of the analyte with selective reagents. Many analytical reactions have been proposed for a host of species, and often numerous derivatization reactions are available for a given analyte. Only a small percentage of all developed procedures are accepted and used routinely.

In certain applications, the absorbance of the analyte is measured directly provided that it is the predominant absorbing species at the analysis wavelength or if it is separated from potential absorbing species before detection. For example, nitrate in natural water is determined by measuring the absorbance at 220 nm. This technique is normally used only as a screening procedure because dissolved organic matter can also absorb at the same wavelength. Often a second absorbance measurement is made at 270 nm, where nitrate does not absorb and is used to correct for interferences from organic species.

## Determination of Inorganic Species

Molecular absorption spectrophotometry is widely used to determine metals, cationic species, anionic species, and complex ions. Although atomic spectrometric techniques are most commonly used to determine metals because of the selectivity and detectability provided, molecular absorption spectrophotometry is still used in selected applications. Molecular spectrophotometric instrumentation is usually less expensive and can be made more portable for field studies or highly automated for high sample throughput. Spectrophotometric techniques can also be used to determine specific oxidation states of metal ions. The following two examples illustrate molecular spectrophotometric determinations of metals based on the selective formation of a metal complex between a metal cation and an organic ligand and a redox reaction between the metal cation and an organic reagent.

The determination of iron is often based on the formation of the tris complex of 1,10-phenanthroline and ferrous iron. The absorbance of the orange-red complex formed is measured at 510 nm. Hydroxylamine is used to reduce ferric iron to determine total iron or to minimize interference by oxidizing species. Because Cu(I) also forms a strong complex with phenanthroline, reduction must be carried with a suitable reducing agent and pH such that Cu(II) is not also reduced. Spectral interferences from excessive amounts of organic concomitants are eliminated by prior sample digestion with acid. Interference from high concentrations of other transition metal ions can be a problem. To prevent interference from other metals that form chelates with the reagent that do not absorb at the analysis wavelength, a sufficient excess of reagent should be used. In some cases it is necessary to eliminate interferences by oxidation of all the iron to the ferrous state, acidification with HCl, extraction of $FeCl_3$ into isopropyl alcohol, back extraction of the iron into water, and reduction to Fe(II) before the analytical reaction is run.

The spectrophotometric determination of chromium is usually based on the oxidation of the reagent 1,5-diphenylcarbohydrazide (diphenylcarbazide) by Cr(VI) in acidic solution. The reaction is

$$2CrO_4^{2-} + 3H_4L + 8H^+ \rightarrow$$

$$Cr^{III}(HL)_2 + Cr^{3+} + H_2L + 8H_2O \qquad (13\text{-}57)$$

where $H_4L$ is the diphenylcarbazide and $H_2L$ is the reduced form of the reagent, diphenylcarbazone. The red-purple product is a chelate of Cr(III) and diphenylcarbazone and its absorbance is measured at 550 nm. The direct reaction of $H_2L$ with $Cr^{3+}$ in aqueous solution is kinetically slow. To determine total chromium, the sample is first oxidized with hypobromite or permanganate in alkaline solution; then the excess oxidant is destroyed with sodium azide. Interference from organics can be eliminated by prior acid digestion. The reaction is quite selective, although high concentrations of Hg, V, Mo, Fe, and Cu can interfere; these metals also form complexes with diphenylcarbazone. The latter four metals can be removed by extraction of their cupferrates into chloroform before reagent addition.

Molecular absorption spectrophotometry is widely applied in the determination of many inorganic anions; some common examples are summarized in Table 13-5. Although ion chromatographic techniques are available for many of the same determinations, spectrophotometric procedures provide higher sample throughput and can easily be automated.

Many of the analytical reactions used for inorganic anions are quite complex and often involve a series of reactions. For example, the molybdenum blue method for phosphate involves the reaction of orthophosphate with molybdate in acid solution to form yellow 12-molybdophosphoric acid (12-MPA); as

$$12Na_2MoO_4 + H_2PO_4^- + 24H^+ \rightleftharpoons$$

$$[H_2PMo_{12}O_{40}]^- + 24Na^+ + 12H_2O \qquad (13\text{-}58)$$

The 12-MPA is then reduced by ascorbic acid to produce molybdenum blue, which absorbs and is measured at 880 nm. An acid hydrolysis pretreatment step is required to determine polyphosphates, whereas a persulfate digestion is necessary to release orthophosphate from organophosphates. High concentrations of silicate or arsenate can interfere because they also form heteropolymolybdates.

**TABLE 13-5**

Spectrophotometric determination of inorganic anions

| Species | Reagents |
|---------|----------|
| $CN^-$ | $Cl_2$, pyridine, pyrazolone |
| $NH_3$ | Phenol, NaOCl |
| $PO_4^{3-}$ | $Na_2MoO_4$, ascorbic acid |
| $Cl^-$ | $Hg(SCN)_2$, $Fe^{3+}$ |
| $NO_3^-$ | Cd, sulfanilic acid, chromotropic acid |
| $NO_2^-$ | Sulfanilamide, *N*-1-naphthlethylenediamine dihydrochloride |

*Determination of Organic Species.* The analytical reactions used for the determination of many organic species can be organized around the specific functional group in the molecule that reacts with the reagent(s). Table 13-6 lists some common reagents used for different functional groups and typical molar absorptivities for the absorbing products that are measured. Many types of reactions are used including substitution, addition, elimination, and rearrangement reactions. The specificity and efficiency of the analytical reaction and the molar absorptivity depend on the type and placement of other functional groups in the molecule. Because many of these reactions are selective for only a type of functional group, they are used when the analyte is the primary species in the sample with the reactive moiety.

In the clinical area, organics are often determined as the substrate in very selective enzymatic reactions. Many assays involve the oxidation of the substrate by the coenzyme nicotinamide adenine dinucleotide ($NAD^+$). The concentration of the substrate is determined from the absorbance measured at 340 nm due to NADH, the reduced form of $NAD^+$. For example, the determination of L-lactate is based on the following reaction:

$$\text{L-lactate} + NAD^+ \underset{\text{LDH}}{\rightleftharpoons} \text{pyruvate} + NADH + H^+$$

(13-59)

where LDH is the enzyme L-lactate dehydrogenase. The amount of the substrate L-lactate is determined by measuring the absorbance due to the NADH produced at 340 nm. Because the equilibrium lies far to the left, the efficiency of the reaction can be improved by adding hydrazine to trap the pyruvate formed through hydrazone formation. Although the reaction is quite selective, such acids as glycolic and L-lactic acid can also be oxidized and cause interference at high concentrations.

Some clinical assays are based on the enzyme-catalyzed reduction of the substrate with NADH, in which case the decrease in absorbance due to the oxidation of NADH to $NAD^+$ is monitored. It is also possible to couple the NADH formed in one reaction to another reaction to form a product which has a higher molar absorptivity than NADH or to monitor other reaction products. If the $NAD^+$/NADH couple is not directly involved in the substrate reaction, one or more enzymatic reactions can be coupled to the original substrate reaction such that NADH is eventually produced. For example, the determination of glucose is often based on the following reaction sequence:

$$\text{glucose} + ATP \underset{}{\overset{\text{HK}}{\rightleftharpoons}} \text{glucose 6-phosphate} + ADP$$

(13-60)

$$\text{glucose 6-phosphate} + NADP^+ \overset{\text{G-6PDH}}{\rightleftharpoons}$$

$$NADPH + \text{6-phosphoglucono-}\delta\text{-lactone} + H^+$$

(13-61)

where the enzyme hexokinase (HK) first catalyzes the phosphorylation of glucose with the coenzyme ATP (adenosine triphosphate), which is converted to ADP (adenosine diphosphate). The second reaction is the indicator reaction in which the glucose 6-phosphate produced in the first reaction is oxidized by the coenzyme $NADP^+$ (nicotinamide adenine dinucleotide phosphate) and catalyzed by the enzyme glucose-6-phos-

**TABLE 13-6**
Spectrophotometric determination of organic species based on the type of function group

| Functional group | Reagent | Molar absorptivity ($L\ mol^{-1}\ cm^{-1}$) | Absorption maxima (nm) |
|---|---|---|---|
| Acid | Pinacyanol | $2–5 \times 10^4$ | 620 |
| Carbonyl | Dinitrophenylhydrazine | $1 \times 10^4$ | 480 |
| | Chromotropic acid | $2 \times 10^4$ | 570 |
| | 3-Methyl-2-benzothiazolone hydrazone | $7 \times 10^4$ | 635 670 |
| Ester | Hydroxamic acid | $0.8–1.2 \times 10^3$ | 530 |
| Olefin (cyclic) | Phenyl azide | $5 \times 10^4$ | 515 |
| Peroxide | Ferrous thiocyanate | $5 \times 10^3$ | 525 |
| Phenol | 4-Aminoantipyrine | $1 \times 10^4$ | 460 |
| Pyrrole | Dimethylamino-benzaldehyde | $5–9 \times 10^4$ | 520–580 |
| Sulfonate | Methylene blue | $9 \times 10^4$ | 650 |

phate dehydrogenase (G-6PDH) to produce the reduced form of the coenzyme (NADPH). This species is similar to NADH and can be monitored at 340 nm.

## Multicomponent Determinations

The additivity of absorbances from several species is the basis for determining several analytes in one sample. For two species $x$ and $y$, the absorbance of the sample is measured at two wavelengths and

$$A_1 = b(\varepsilon_{x1}c_x + \varepsilon_{y1}c_y) \tag{13-62}$$

$$A_2 = b(\varepsilon_{x2}c_x + \varepsilon_{y2}c_y) \tag{13-63}$$

where the subscripts 1 and 2 indicate that quantity measured at $\lambda_1$ and $\lambda_2$, respectively. These two equations can be solved for $c_x$ and $c_y$ in terms of measured or known quantities:

$$c_x = \frac{\varepsilon_{y2}A_1 - \varepsilon_{y1}A_2}{b(\varepsilon_{y2}\varepsilon_{x1} - \varepsilon_{y1}\varepsilon_{x2})} = \alpha_1 A_1 - \alpha_2 A_2 \tag{13-64}$$

$$c_y = \frac{\varepsilon_{x1}A_2 - \varepsilon_{y1}A_2}{b(\varepsilon_{y2}\varepsilon_{x1} - \varepsilon_{y1}\varepsilon_{x2})} = \beta_1 A_1 - \beta_2 A_2 \tag{13-65}$$

where $\varepsilon_{x1} > \varepsilon_{x2}$ and $\varepsilon_{y2} > \varepsilon_{y1}$. The second form is more convenient for calculations because the coefficients can be calculated from the pathlength and molar absorptivities. Conditions must be arranged so that Beer's law applies to both species and that the species behave independently from each other. Highest accuracy is obtained by choosing wavelengths for which the molar absorptivity difference at each wavelength is greatest.

The treatment above can be extended to three or more analytes in a mixture. Thus for $n$ analytes it is necessary to measure the absorbance of the mixture at $n$ wavelengths. From $n$ simultaneous equations and knowledge of the molar absorptivity of each analyte, the concentration of all species can be determined. In general, the accuracy decreases as the number of components and thus the number of measurements increases.

The advent of rapid scanning computerized spectrophotometers and diode array spectrophotometers has greatly enhanced the feasibility and use of multicomponent absorption measurements. With these instruments, both data acquisition and post analysis calculations can be accomplished automatically and rapidly. With the sophisticated software now available, the spectra from all components to be determined can be stored in computer memory. After acquisition of the sample spectrum, matrix techniques are used to find the weighting factors to generate a synthetic spectrum from the stored component spectra that matches the sample spectrum. From the weighting factors, the concentrations of all components are determined. This approach can yield higher accuracy as the absorbance values from more than $n$ wavelengths are used to determine $n$ species.

## Spectrophotometric Titrations

A **titration** is the process by which the quantity of an analyte (An) is determined by adding a standard solution of a titrant (T) with which the analyte reacts in a known and stoichiometric manner. A detector (manual or automatic) is required to signal when the amount of titrant added is chemically equivalent to the amount of analyte present. This point of chemical equivalence is called the **equivalence point,** while that actually measured is denoted the **endpoint.** If a good endpoint detector and reaction are used, the accuracy of a titration depends predominantly on the accuracy with which the titrant concentration is known. The latter is usually determined by titration against a primary standard substance.

In **spectrophotometric titrations**, the spectrophotometer serves as the detector, monitoring the sample transmittance or absorbance at a suitable wavelength during the addition of increments of the titrant. The spectrophotometer sample cell is the titration vessel. For a titration reaction of the form

$$\text{An} + \text{T} \rightleftharpoons \text{P} \tag{13-66}$$

where P is the product, the absorption due one or more of the species involved is monitored. Instrumental titration methods can also be based on fluorescence or chemiluminescence detection as discussed in Sections 15-3 and 15-5.

The plot of the absorbance versus titrant volume is called a **spectrophotometric titration curve** and many shapes are possible, depending on the absorbing species, as shown in Figure 13-19. Normally, the absorbance values plotted are corrected for dilution by the titrant by multiplying the measured absorbance values by $(V_T + V_{An})/V_{An}$, where $V_{An}$ and $V_T$ are the volumes of the sample containing the analyte and the titrant, respectively. Ideally, the endpoint is indicated by a sharp change in absorbance at the intersection of the two straight-line regions of differing slopes on either side of the equivalence point region. If the titration or indicator reaction is not quantitative near the equivalence point, a rounded intersection results. The linear segments before and after the endpoint can be extrapolated to locate the endpoint if conditions are adjusted so that Beer's law applies.

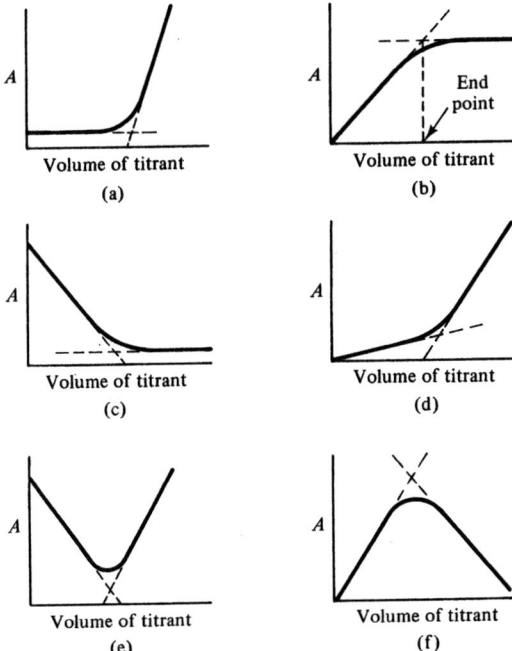

**FIGURE 13-19** Shapes of spectrophotometric titration curves. The curve in (a) is typical of a titration where only the titrant (T) absorbs [titration of As(III) with bromine] or where an indicator reacts with excess T. The curve in (b) is characteristic of a titration reaction in which the only the product absorbs (e.g., titration of Cu(II) with EDTA). When an absorbing analyte is converted to a nonabsorbing product with a nonabsorbing titrant, a curve such as that shown in (c) is observed. In curves (d) to (f), two species involved in the titration reaction absorb. In figure (e) an absorbing analyte is converted to a nonabsorbing product by an absorbing titrant. The curve in (d) could result if both the titrant and product absorb but the analyte does not absorb or if two different metal complexes with different molar absorptivities are formed at the monitored wavelength during the titration of a nonabsorbing metal with a nonabsorbing ligand. The latter situation could also produce a titration curve of the shape shown in (f). In all the cases above, the rounding near the endpoint is due to the titration reaction not going to completion. The dashed extrapolated portions indicate the behavior observed if the titration reaction goes to completion.

There are several advantages of spectrophotometric titrations compared to titrations with visual endpoint detection. The determination of the endpoint can be more precise because the spectrophotometer can better detect smaller changes in color shade or absorbance than the eye. Titrations can be carried out in turbid or colored solutions, which make visual endpoint detection difficult. Titration reactions with unfavorable

equilibrium constants (i.e., where the color change at the endpoint in not sharp) can be employed because measurements taken away from the endpoint can be used to extrapolate to the true endpoint. These advantages allow titrations to be extended to lower analyte concentrations ($\leqslant 10^{-4}$ M).

Because spectrophotometric titrations are based on a relative measurement, in contrast to conventional spectrophotometric measurements, it is not necessary to measure the absolute absorbance or to calibrate the absorbance scale with standards. The presence of other species which absorb or scatter does not cause interference. Concomitants that react with the titrant do interfere. Analytical reactions can be used in cases where neither the analyte nor its reaction product absorb if a suitable absorbing indicator is available to react with the excess titrant. A sufficient excess of indicator must be added to provide a reasonable linear region beyond the equivalence point.

Commercial spectrophotometric titration systems are available which include a spectrophotometer or photometer, a titration vessel, an automated buret, and often specialized data handling and plotting capabilities. A diagram of a commercial titration accessory is shown in Figure 13-20. Typically, the titration vessel has a volume of 5 to 100 mL and is stirred. In computerized systems, the absorbance vs. titrant volume data are stored during the titration, corrected for dilution, and later plotted. Software allows the endpoint to be located by extrapolation or by differentiation (i.e., the endpoint is indicated by where $dA/dV_T$ is at a maximum).

Spectrophotometric titrators are commonly used for the routine determination of species based on classical acid-base, redox, and metal complex titration reactions. Spectrophotometric titration curves provide much of the same information as the molar ratio method if the absorbance is corrected for dilution effects (compare Figures 13-18a and 13-19a). Thus spectrophotometric titrators are commonly used to acquire data to determine stoichiometric coefficients and equilibrium constants.

### Reaction-Rate Methods of Determination

In **kinetics-based methods** or **reaction-rate methods**, the rate of the analytical reaction is measured and related to the analyte concentration. Normally, the initial rate in the first few percent of the reaction is monitored. Numerous instrumental techniques can be used to provide a reaction monitor signal that tracks the formation of a product or the disappearance of a reactant. When a reaction is monitored spectrophotometrically, the rate of change of absorbance is measured. Thus the calibration curve is a plot of the rate of change of absorb-

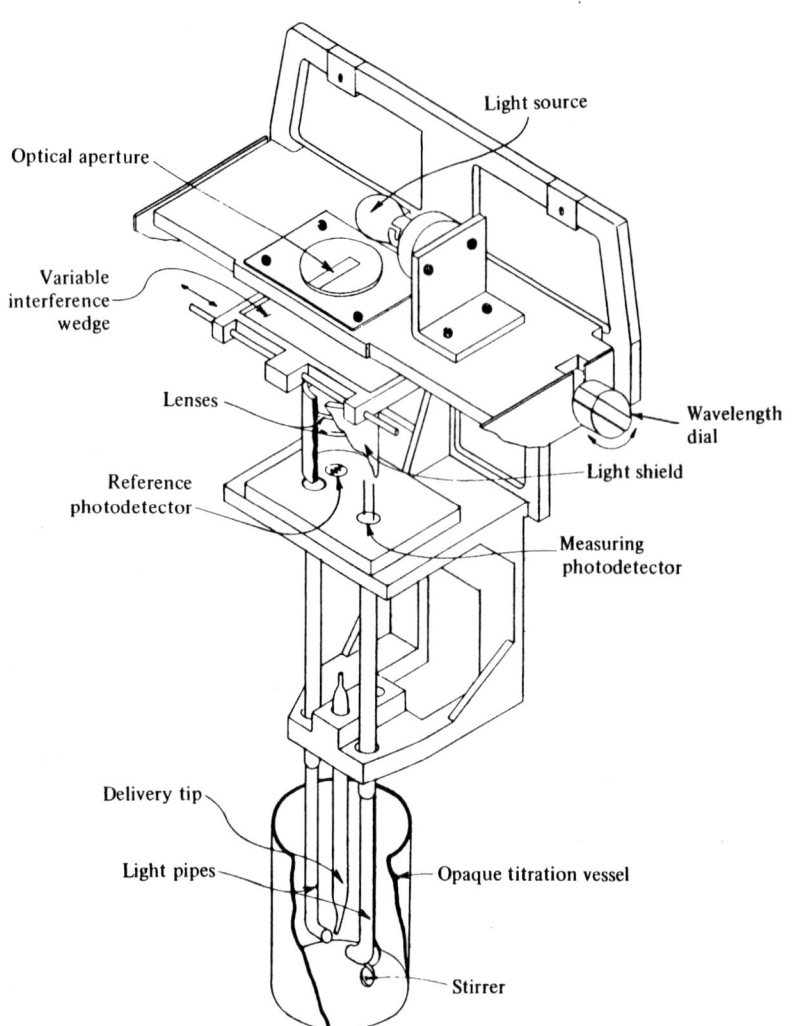

**FIGURE 13-20** Fisher photometric accessory for automatic photometric titrations. Note that fiber optics are used to direct the light beam through the titration vessel. (Courtesy of Fisher Scientific Co.)

ance, or a related quantity, versus the analyte concentration of standards.

Reaction-rate methods offer several unique advantages compared to conventional equilibrium-based measurements in which the analytical reaction is allowed to proceed to equilibrium before measurements are made. There can be a considerable time savings for reactions that require several minutes or more to reach equilibrium. Kinetics methods can be more selective (greater freedom from interferences) for two primary reasons. First, a kinetics measurement is a relative measurement and only changes in the reaction monitor signal are measured. Thus nonreacting species or instrumental factors that contribute to the absolute magnitude of the reaction monitor signal (e.g., the absolute absorbance), and that may vary from sample to sample, but not during the reaction, do not interfere as they would in an equilibrium-based method. Second, conditions can often be arranged so that the kinetic contribution of only one species is significant over the measurement period. For example, in the case of a very fast

initial interfering reaction, the analytical rate measurement can be made after the interfering reaction has reached equilibrium. Because of this greater specificity, kinetics measurements can be faster than equilibrium measurements if time-consuming separations steps can be avoided. If the reaction is quite specific and there is no rate when the analyte is absent, further time savings can be realized by omitting frequent blank measurements. Nonstoichiometry, unstable reaction products or reagents, or side reactions make some analytical reactions unsuitable or difficult to use for equilibrium-based determinations. However, the initial rates of such reactions can often be related to the analyte concentration. Finally, catalysts can be uniquely determined since they affect the rate but not the equilibrium concentrations.

The limitations of kinetics-based methods should also be understood. The reaction rate must be in a suitable range and a half-life of a few milliseconds to tens of minutes is typical. The lower limit is determined by the time necessary to mix the reagents, whereas the upper limit is restricted by slow drifts in the instrumental monitoring system and the need to have a reasonable analysis time. Reaction conditions such as temperature, pH, and reagent concentrations must be carefully controlled as they often have a greater effect on the rate than on the final equilibrium concentrations. Detection limits and the $S/N$ can be lower compared to equilibrium-based methods because only a portion of the reaction is used in the measurement (i.e., the change in absorbance measured is much less than the final absorbance at equilibrium). Also, the noise equivalent bandpass must be large enough to prevent distortion of the changing reaction monitor signal. However, the increased selectivity against variable and uncompensated absorbance by other species can result in practical detection limits comparable to or even better than those achieved with equilibrium-based methods. Concomitants which alter the rate of reaction cause analyte interference. Species which react with the reagents at rates comparable to the analyte to form products that absorb at the monitored wavelength result in a blank interference unless a suitable blank rate can be determined.

Reaction and measurement conditions are often arranged so that the rate law is simple and the measured rate, or another appropriate quantity, is directly proportional to the analyte concentration. Most commonly, the reagent concentrations are in a sufficient excess that the reaction is pseudo-first order with respect to the analyte in the initial stages of the reaction.

To make rate measurements reproducible and convenient, a spectrophotometer should have provision for mixing the sample and reagents in the sample cell, temperature control, and suitable hardware and/or software for timing the acquisition of absorption data and computing the rate or a related quantity. Mixing in the sample cell with magnetic stirring bars is suitable for measuring rates of reactions with half-lifes of a few minutes or longer. Specialized accessories or instruments based on *stopped-flow* mixing extend the range down to half-lifes on the order of milliseconds. Pressure-jump, temperature-jump, or flash photolysis techniques can be used for even faster reactions. Some instruments allow the reactions in several sample cells to be monitored simultaneously. The sample cells are sequentially positioned in the light beam in a repetitive manner at a sufficient rate to allow absorbance vs. time data to be acquired in a faithful manner for all cells. Highly automated instruments are available for high-sample-throughput routine measurements.

Rate information is extracted from the time-varying reaction monitor signal in several ways. With modern computer-based instruments, curve-fitting techniques are applied to the stored data. With suitable software, the region over which the rate is constant (pseudo-zero order) can be determined and the initial rate calculated and reported. Several other rate computational approaches are utilized and can be implemented through software or analog or digital circuitry. In the **variable-time method**, the amount of time required for the reaction-monitor signal to change by a predetermined amount ($\Delta S$) is measured. The change in the reaction monitor signal in a predetermined time interval ($\Delta t$) is measured in the **fixed-time method**. In the **derivative method**, the instantaneous slope of the reaction monitor signal is measured. The fixed-time and variable-time methods are integral methods in which the measured quantity ($\Delta S/\Delta t$) approximates the true instantaneous rate ($dS/dt$). If $\Delta t$ is sufficiently small and the measurement is made very early in the reaction (pseudo-zero order conditions), the initial rate is measured. Even if there is significant curvature in the reaction monitor signal over the measurement time $\Delta t$, $\Delta S/\Delta t$ is still linearly related to the analyte concentration under suitable conditions (e.g., first order for the analyte for fixed-time methods).

Several techniques have been developed to compensate for imprecision in rate measurements caused by variations in the rate constant due to run-to-run changes in the magnitude of variables that affect the rate constant. For first-order reactions, precision can be improved by making rate measurements near the half-life, by computing the ratio of the rate measured at two different times during the reaction, or by estimating the equilibrium position from reaction monitor data taken over several half-lifes. It is also possible to correct for changes in variable values by measuring the magnitude of the variable for a given run and applying a suitable correction factor.

Kinetics-based measurements with absorption monitoring are usually the method of choice for assaying enzymes in clinical samples. For example, the enzymes L-lactate dehydrogenase and hexokinase are determined by measuring the rate of increase in absorbance due to the formation of NADH or NADPH in reactions described by equation 13-59 and equations 13-60 and 13-61, respectively. In this case, the substrate concentrations are adjusted to be large enough that pseudo-zero-order kinetics prevail. These same reactions are the basis for the determination of the substrates L-lactate and glucose. For substrate determinations, conditions are arranged such that the substrate concentration is low enough that the reaction is first order in the substrate.

Inorganic and organic species are determined by measuring the rate of formation of absorbing products in noncatalytic reactions which are often used for equilibrium-based measurements. For example, phosphate can be determined by measuring the initial absorbance change due to the formation of molybdenum blue (the reduction of 12-MPA formed in reaction described in equation 13-58). The reaction is pseudo-first order in phosphate if the reagent concentrations are sufficiently large. Inorganic catalysts may be determined with similar procedures.

## Other Quantitative Techniques and Uses

*Derivative Techniques.* Some spectrophotometers allow the first derivative or higher-order derivatives of the absorption spectrum to be calculated and displayed (e.g., $dA/d\lambda$ vs. $\lambda$) by the techniques discussed in Section 13-1. As shown in Figure 13-21, derivative spectra can be used to emphasize and distinguish overlapping bands, to locate the maximum of a broad band, or to extract quantitative information when the analyte absorption band is overlapped by a concomitant absorption band that is not compensated with the blank measurement.

Derivative spectrophotometry is particularly advantageous when a narrow analyte absorption band is superimposed on top of a relatively flat or gradually changing absorbance due to turbidity or overlap with the side of a broad interferent absorption band. The derivative spectrum emphasizes the contribution from the analyte for which the change in absorbance with wavelength is much greater than that for the background (i.e., the contribution from the background is a flat baseline). The difference between the maximum and minimum values of the derivative is proportional to the analyte concentration even in the presence of an uncompensated and variable background absorbance.

*Dual-Wavelength Techniques.* Some spectrophotometers can be configured to output the difference in absorbance ($\Delta A$) between two wavelengths, as discussed in Section 13-1. This dual-wavelength (DW) option is particularly useful when it is difficult to prepare a blank that compensates for absorption by other species or scattering. The beams corresponding to $\lambda_1$ and $\lambda_2$ are normally selected to be the sample and reference beams, respectively. Often, $\lambda_1$ is adjusted to a wavelength corresponding to analyte absorption while $\lambda_2$ is set to a wavelength where the analyte does not absorb (e.g., the base of the analyte absorption band) or an isobestic point. A spectrum is obtained by keeping $\lambda_2$ constant and scanning only $\lambda_1$ over the analyte absorption band. Thus the sample is used as a blank and $\Delta A$ is used as the analytical signal proportional to the analyte concentration.

Dual-wavelength measurements find widespread use in biochemical applications where absorption measurements must often be made in cloudy or turbid suspensions. It is difficult to observe a small analyte absorption band in a conventional absorption spectrum when it is superimposed on top of a large and relatively constant background absorbance baseline. In kinetics studies or measurements, the reference wavelength can be set to a value for which the absorbance does not change during the course of the reaction. Dual-wavelength measurements can even compensate for concomitant absorption when the analyte and concomitant bands are of about the same breadth and seriously overlap, as shown in Figure 13-22.

In addition to discrimination against spectral interferences, DW techniques afford additional advantages. There is some compensation for sample cell errors. The light beams for both wavelengths follow identical light paths through one sample cell and strike the same portion of the photocathode of the PMT. Thus systematic errors due to differences in the transmission characteristics of different cells, the positioning of the cell, or the refractive index of different samples are compensated to the extent that they are independent of wavelength. Random error due to sample cell placement can also be reduced.

Depending on the situation, the precision of DW measurements can be poorer than conventional single-wavelength measurements. Also, stray radiation and polychromatic radiation can cause more serious nonlinearity and even roll-over in the calibration curve.

*Difference Spectra Techniques.* In difference spectroscopy, the difference between spectra is displayed or the absorbance values at a specific wavelength are subtracted for two samples of slightly different composition or physical state. Because common features in the two spectra cancel, the difference spectrum accentuates the subtle differences in absorption due to variations in solvent, pH, temperature, or analyte or reagent concentration. For example, the difference spectrum of some barbiturates in 0.45 M NaOH and a pH 10 buffer over the range 230 to 280 nm is highly characteristic. When only one component is present, the difference in absorbance at 260 nm is useful for quantitative determinations.

Difference spectra can be obtained by placing the two test solutions in the reference and sample cells of a double-beam spectrophotometer or by taking the difference between two spectra that are successively stored. Because small absorbance differences are often measured, the spectrophotometer must be capable of high photometric precision and accuracy, sample cells must be closely matched, and sample cell placement must be reproducible.

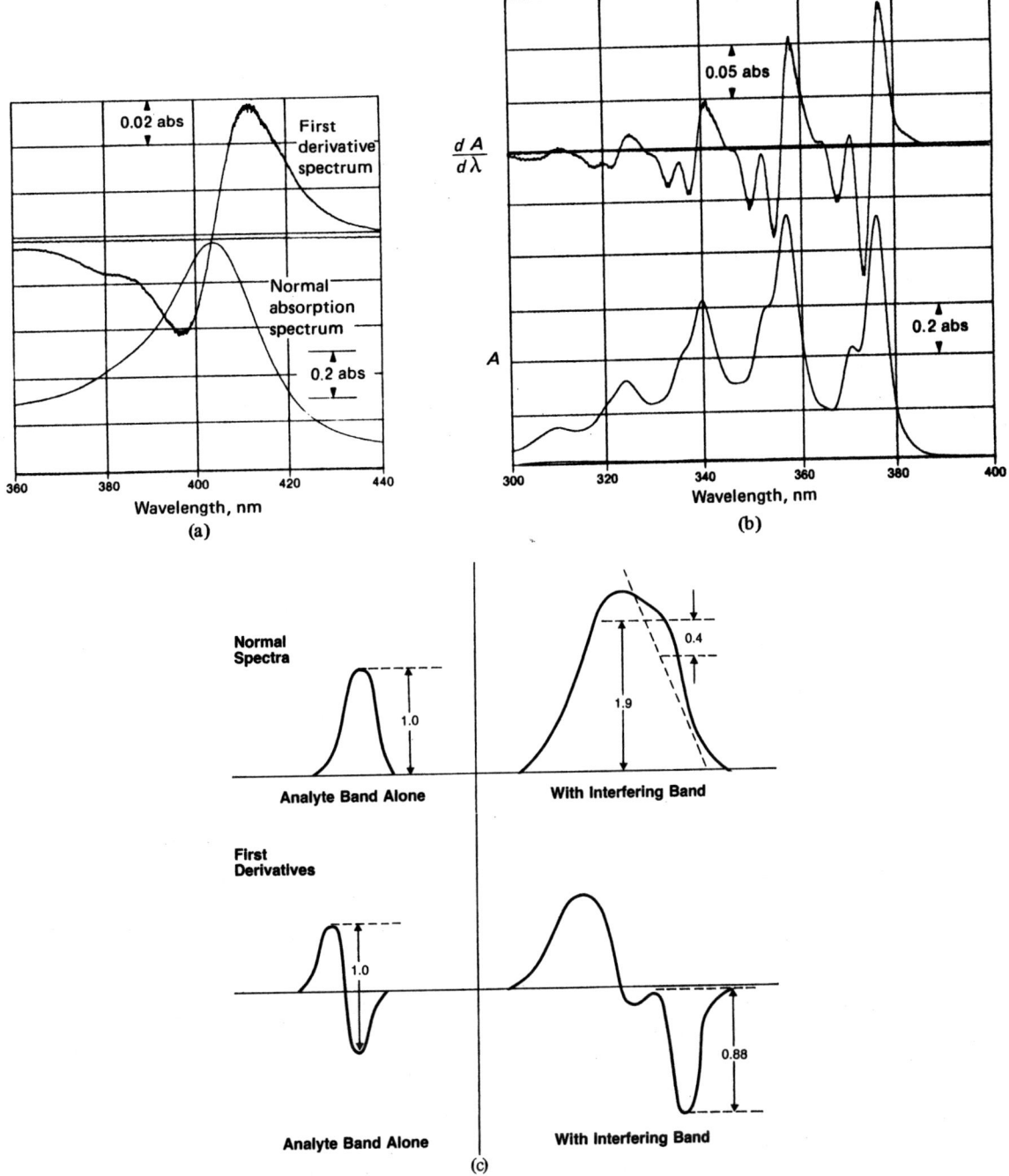

**FIGURE 13-21** Applications of derivative spectra. In (a), the wavelength of maximum absorbance of the Soret band of bovine methemoglobin is easier to detect with the derivative spectrum. The absorbance and derivative spectra of anthracene in (b) demonstrate that the overlapping bands are more emphasized in the derivative mode. (c) illustrates that overlapping absorption bands from different species can be more readily observed. In the normal absorption spectrum, it is very difficult to estimate the contribution from the analyte band which is almost totally obscured by the interfering band. In the derivative spectrum of the mixture, the analyte concentration can be estimated within 12% by using the difference between the adjacent maximum and minimum as the analytical signal. [(a) and (b) from *Varian Instru. Appl.*, 7 (2), 11 (1973), courtesy of Varian Associates, Inc., Instrument Group; (c) adapted from T. C. O'Haver, *Anal. Chem.*, *51*, 91A (1979), copyright 1979, with permission of the American Chemical Society.]

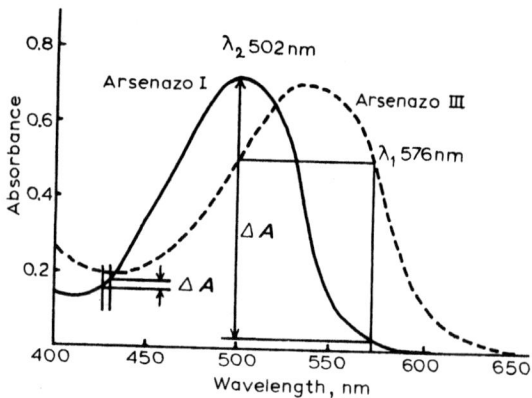

**FIGURE 13-22** Compensation for spectral interference with dual-wavelength measurements. Arsenazo I can be determined in the presence of arsenazo III by proper selection of the two wavelengths at which the absorbance difference is calculated. For the wavelengths shown, the absorbances of arsenazo III are equivalent such that its contribution to $\Delta A$ is zero for any concentration. Thus $\Delta A$ is directly proportional to the concentration of arsenazo I. [From S. Shibata, K. Goto, and Y. Ishiguro, *Anal. Chim. Acta, 67,* 305 (1972), with permission of Elsevier Science Publishers.]

## Chromatographic Applications

Presently UV-visible spectrophotometry is the most popular detection technique for high-performance liquid chromatography (HPLC). This popularity is due in part to the fact that UV-visible absorption is a reasonably universal phenomenon for the majority of compounds and provides good-to-excellent detection limits (e.g., 1 ng) in most cases.

Many spectrophotometers and photometers can be converted to HPLC detectors by installing a suitable flow cell in the sample cell holder. The performance characteristics are often poorer than those of instruments designed specifically as HPLC absorption detectors. The characteristics of the incident beam in conventional spectrophotometers are not usually compatible with the smaller cross section of most HPLC flow cells (typically, 1 mm²). The cell dimensions restrict the area and solid angle of the source viewed. The reduction in effective radiant power can degrade the *S/N*.

The optics of HPLC absorption detectors are designed to maximize the radiant power throughput through the flow cell. Often special heat sinks or water jackets around the sample cell stabilize the temperature and therefore minimize drift and noise due to temperature-dependent refractive index changes. Continuum or line sources are used. Most HPLC absorption detectors are intended for fixed-wavelength measurements and are based on a double-beam in-space design to minimize

light-source drift and flicker noise. The baseline rms noise can be as low as $10^{-5}$ A.U. Often the reference cell is filled only with air. In other cases, a fraction of the mobile phase is diverted at a point before the injector to the reference cell. This arrangement provides some compensation for absorption by the mobile phase. This compensation is critical if the mobile phase absorbs significantly, particularly if the baseline absorbance changes during a chromatographic run, as can occur with gradient programming.

The most established version of photometric detectors is based on a low-pressure mercury arc lamp (see Figure 13-23). Many detectors are configured for measurements only at the 254-nm line of mercury, which is easily isolated from other Hg lines by an interference filter. This wavelength is fortuitous for several reasons: Its radiance is high, it is long enough that many organic solvents do not absorb significantly, and it is in a region where many aromatic and heterocyclic compounds absorb strongly even though the wavelength may not correspond to the wavelength where maximum absorption occurs. With a line source, nonlinearity due to polychromatic radiation is essentially eliminated if only one line is passed by the filter.

Different filters are used to isolate other mercury lines (e.g., 313, 365, 405, 436, 546, 578 nm), for cases in which the absorption at 254 nm is inadequate for the compounds of interest. Also, a phosphor attachment between the source and sample cells can be used to convert shorter wavelengths to longer wavelengths. Conversion of 254-nm radiation to 280-nm radiation is common. The emission band of phosphors is typically 10 nm in width, and an interference filter with a smaller half-width is often used to minimize nonlinearity due

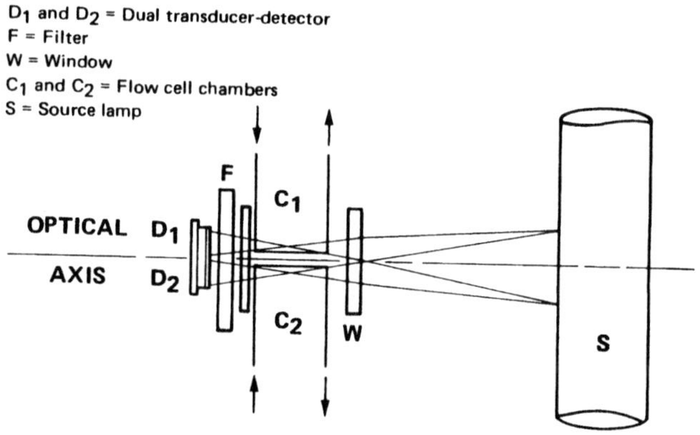

$D_1$ and $D_2$ = Dual transducer-detector
F = Filter
W = Window
$C_1$ and $C_2$ = Flow cell chambers
S = Source lamp

**FIGURE 13-23** Schematic of a fixed-wavelength HPLC absorption detector. Radiation from a Hg arc lamp is directed simultaneously through a sample and reference flow cell, a filter to isolate the desired Hg line (usually 254 nm), and finally to a sample and reference detector. (Courtesy of LDC/Milton Roy.)

to polychromatic radiation. The radiance provided by other mercury lines, especially phosphors, is lower than that at 254 nm; the baseline noise is often greater. Other line sources such as Cd lamps (229 and 326 nm) and Zn lamps (208 and 214 nm) are used in particular where the compounds of interest only absorb at wavelengths shorter than 254 nm.

In dual-channel absorption detectors, radiation from two wavelengths emitted by a line source passes through the flow cell and is detected simultaneously by separate filter-detector modules. Monitoring two wavelengths simultaneously (a common wavelength pair is 254 and 280 nm) is useful for cases in which monitoring at only one wavelength does not provide adequate absorption for all the compounds of interest. The ratio of the absorbance at the two wavelengths can be calculated and displayed as the chromatographic signal. This response ratio is useful for confirming peak identity because it is independent of the analyte concentration, as shown in Figure 13-24.

More expensive and complex absorption detectors are based on monochromators and conventional continuum sources (e.g., tungsten and $D_2$ lamps). The spectral bandpass is typically in the range 2 to 10 nm; it cannot be as low as for some conventional spectropho-

tometers because the lower throughput would reduce the radiant power to a level that would increase the baseline noise. The wavelength versatility provides several advantages. The wavelength of optimum response for the compound(s) can be found and then used in determinations. The wavelength is chosen both to maximize the absorbance and to minimize the response from other species whose chromatographic peaks overlap the analyte peak. Many compounds only absorb strongly near 200 nm. On the other hand, spectral interference is often less at longer wavelengths. If chromatographic peaks are sufficiently separated in time, it is possible to program the wavelength to the value that is optimal for each peak during the chromatographic run. In some cases, the flow can be stopped and an absorption spectrum obtained.

Many commercial diode array absorption detectors are now available. Spectra can be obtained in as little time as 0.01 s during the chromatographic run. The obvious advantage of such detectors is that absorption data at all wavelengths are taken and stored. The optimum wavelength(s) for plotting can be evaluated after the chromatographic run. With appropriate software, the user can display multidimensional absorption data, as shown in Figure 13-25. A more conventional chromatogram is obtained by plotting the absorbance at one wavelength vs. time. Alternatively, the absorbance plotted can be for different wavelengths in different portions of the chromatogram to optimize the calibration slope and selectivity for each compound.

If the compounds of interest do not absorb sufficiently, a post-column reaction system may be needed. The eluted compounds are mixed with a reagent that converts the species to a strongly absorbing product before the eluant is sent to the detector. Ninhydrin is a common post-column reagent for amino acids, and the absorbance is monitored at 440 and/or 570 nm. Metal

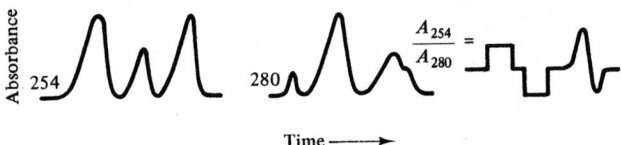

**FIGURE 13-24** The response ratio technique. Chromatograms are shown as a function of the absorbance at 254 and 280 nm and as the ratio of absorbances at these two wavelengths. A square peak indicates a constant ratio. Lack of a square peak indicates a response from two or more compounds.

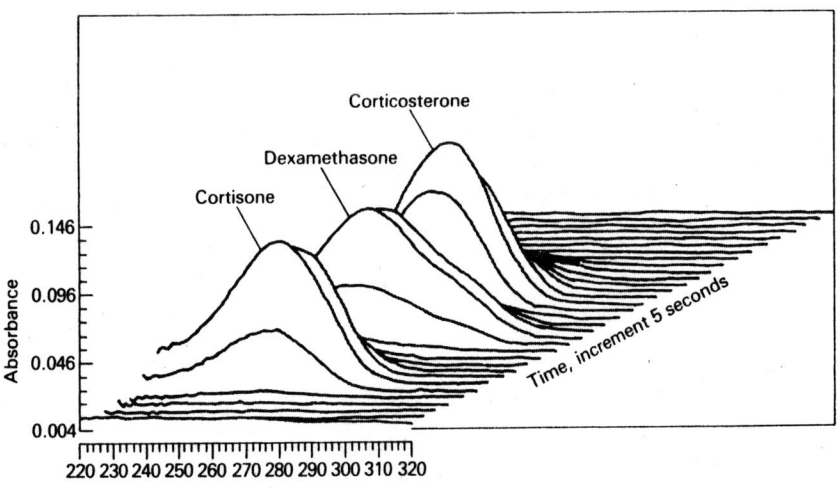

**FIGURE 13-25** Absorption spectra taken with a diode array absorption detector during a chromatographic run of three steroids. The spectra are taken 5 s apart. (Courtesy of Hewlett-Packard Company.)

ions can be separated on an ion exchange column and detected after post-column reaction with the general complexing agent 4-(2-pyridylazo)-resorcinol (PAR).

Special *gel scanning* attachments are available for some spectrophotometers. The sample components are first separated on a gel, and then the gel is sprayed with an appropriate staining reagent. Next, the gel is mounted in a special transport assembly in the sample compartment that orientates the gel perpendicular to the light beam. The gel is moved at a constant rate past the incident beam along the axis of development to produce a plot of absorbance vs. distance. The retention distances are then used to identify the separated components.

### Automated Measurements

As discussed in Section 6-6, automated batch or continuous analyzers are most commonly based on spectrophotometric detection. They are widely used in clinical and water analysis laboratories where a few species must be determined in a large number of similar samples. Most routine spectrophotometric determinations require that the analyte be converted to an absorbing product with a suitable selective analytical reaction. Hence the automated control of sample introduction, addition of reagents, mixing, incubation time, and sample cleanup can greatly increase sample throughput and precision and reduce manual labor and blunders. Automated systems can be configured for equilibrium-based measurements or kinetics-based measurements, as is common with enzymatic-based reactions in clinical assays.

In segmented flow analyzers, the time between mixing of the sample and reagents and the absorption measurement of the product formed is relatively long such that the reaction often reaches equilibrium. In unsegmented or flow injection analysis (FIA) systems, conditions are often such that the reaction does not go to completion before detection occurs. If the reaction is first order or pseudo-first order, the extent of reaction for a given reaction time is independent of the analyte concentration. Thus the absorbance is proportional to initial analyte concentration; the measurement is effectively a one-point kinetics-based measurement. It is also possible to configure FIA systems to carry out pseudo titrations.

### Reflectance Measurements

The absorption properties of opaque materials cannot be measured directly in conventional UV-visible spectrophotometers. In such cases, reflectance spectrometry is a useful technique. The reflectance of a material is correlated to the absorption properties because the reflectance is lower at wavelengths where the absorption is higher. Reflectance techniques are commonly used with materials such as paints, textiles, plastics, and inks. They can provide semiquantitative information about the color characteristics of samples. Reflectance is a complex function not only of the absorption characteristics of the sample but the nature of the surface (e.g., particle size, shape, and orientation).

Special attachments are available for many spectrophotometers for either specular or diffuse reflectance measurements. They are similar in design to the reflectance attachments for the IR region (see Figures 14-23 and 14-24).

### Optical Rotatory Dispersion and Circular Dichroism Measurements

The instrumentation for optical rotary dispersion (ORD) and circular dichroism (CD) measurements is quite similar to that used for molecular absorption spectrophotometry. The principles for ORD and CD measurements are reviewed in Section 3-2.

Spectropolarimetry involves measuring the angle of rotation of the plane of linearly polarized radiation by an optically active species. The dependence of the optical rotation on wavelength is termed the *optical rotatory dispersion*. As shown in Figure 13-26, radiation over a small wavelength band is converted to plane-polarized radiation with a linear polarizer. This radiation passes through a sample cell containing a solution of the optically active compound of interest and then through a second linear polarizer called the analyzer. The polarizer or analyzer axis is orientated with a solution not exhibiting optical activity to achieve a reference detector signal that defines an optical rotation angle ($\alpha$) of zero degrees. The reference signal can be defined as the null or minimum signal achieved with crossed polarizers or some other signal relative to the null point. After the sample is added to the sample cell, the analyzer or polarizer is rotated to compensate for the rotation caused by the sample and restore the detector signal to the original reference value. The necessary rotation angle is taken as the optical rotation of the sample.

In recording spectropolarimeters, the position of the analyzer or polarizer is automatically adjusted with a mechanical servo feedback network during wavelength scanning to maintain a null signal. Often the polarizer is mechanically rotated back and forth through a small angle to produce an ac detector signal. The analyzer rotation is adjusted to achieve a detector signal that is symmetrical around the minimum signal. The oscillation of the plane of polarization of the incident

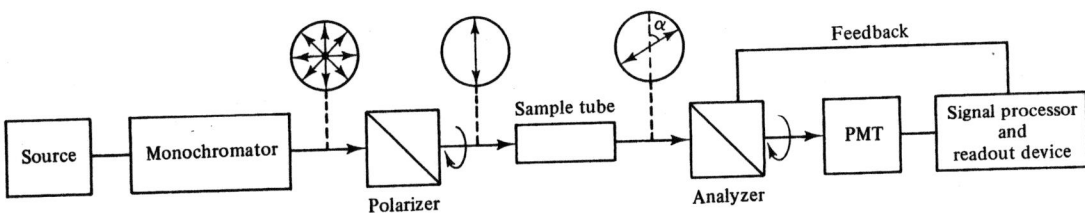

**FIGURE 13-26** Block diagram of a spectropolarimeter. Source radiation over a given wavelength range is selected by a monochromator and then linearly polarized by a polarizer. This radiation passes through the sample tube, a second polarizer called the analyzer, and is incident on a PMT. The analyzer is often initially oriented to produce the minimum detector signal with a blank solution. With an optically active sample in the sample tube, the optical rotation is determined from the angle it is necessary to rotate the analyzer axis to reestablish the null detector signal.

radiation can also be achieved with an electro-optic polarizer such as a Pockels cell (see Figure 3-15) placed between the polarizer and analyzer and driven with an ac voltage. In some spectropolarimeters, another electro-optic polarizer is employed to adjust the plane of polarization for the null signal. The dc voltage applied to the device is related to the optical rotation of the sample.

Measurement of the *circular dichroism* (CD) is quite similar to conventional absorption measurements. The primary difference is that the radiation incident on the sample cell must be circularly polarized; a circular polarizer is placed between the monochromator and the sample cell and is sometimes available as an attachment for spectrophotometers. A circular polarizer is constructed from the combination of a linear polarizer and a 90° retarder such as quarter-wave plate or a Fresnel rhomb (see Figure 3-13). It is necessary to measure the absorbance with both *d*- and *l*-circularly polarized radiation and calculate the difference. This difference is in turn used to calculate the *molar ellipicity* (see equation 3-28), which is the normal measure of CD.

In single-beam in-time instruments, the axis of the linear polarizer or the retarder is mechanically rotated between two orientations that are 90° different to illuminate the sample cell in an alternate fashion with *d* and *l* circularly polarized radiation. The instrument automatically calculates the difference in absorbance between the two components. It is often necessary to compensate for the difference in the incident intensities of the two polarization components. The rotation of the plane of the linear polarized beam can also be electrically modulated with an electro-optic device such as a Pockels cell or a magneto-optic device such as a Faraday cell. Double-beam in-space designs are often based on splitting the incident beam into both polarization components, which are each separately directed through two equivalent sample cells.

Both CD and ORD measurements are employed primary in fundamental studies to provide structural information about optically active compounds. Most applications are in the biochemical area (e.g., amino acids, proteins).

## PROBLEMS

**13-1.** Describe the major differences between a nonlinearity due to stray radiation and that due to polychromatic radiation. What are the major causes of stray radiation in a monochromator? What steps are taken in many monochromators to reduce the stray radiation figure?

**13-2.** The slit width of the monochromator in a UV-visible spectrophotometer is extremely important since it determines throughput, and thus signal-to-noise ratio (*S/N*), as well as spectral bandpass, and thus resolution.

(a) The nonlinearity due to polychromatic radiation is usually negligible if the spectral bandpass $s$ is much smaller than the full width at half maximum (FWHM) of an absorption band. As a rule of thumb $s < 0.1$ FWHM for negligible deviations. The absorption band of $Pr^{3+}$ at 482 nm has a FWHM of 3.0 nm. What slit width should be used to avoid a nonlinearity with a monochromator that has a $R_d$ value of 4.0 nm mm$^{-1}$?

(b) A different absorption band of $Pr^{3+}$ has a FWHM value of 13.0 nm. For the same monochromator as in part (a), what a slit width should be used to avoid a nonlinearity due to polychromatic radiation?

**(c)** What is the radiant power "cost" of doubling spectrophotometric resolution? Can the loss in radiant power be overcome by increasing slit height?

**13-3.** A UV spectrophotometer is operated under signal-shot-noise-limited conditions. The signal-to-noise ratio (*S/N*) under these conditions is proportional to the square root of the reference beam photocurrent $i_r$ ($i_r$ is directly proportional to the reference beam radiant power). A given measurement has a *S/N* of 15 when the slit width of the monochromator is 35 μm. What slit width would be required to give a *S/N* of 50? of 100? By what factor will the resolution be degraded if the slits are opened to give a *S/N* of 100?

**13-4.** Draw the expected spectrophotometric titration curves for the following cases and indicate the equivalence points.

**(a)** The analyte absorbs, but neither the titrant nor the product absorb.

**(b)** The product absorbs, but neither the analyte nor the titrant absorb.

**(c)** Both the analyte and the titrant absorb, but $\varepsilon_{analyte} > \varepsilon_{titrant}$.

**(d)** Both the analyte and the titrant absorb, but $\varepsilon_{titrant} > \varepsilon_{analyte}$.

**(e)** Both the product and the titrant absorb, but $\varepsilon_{product} > \varepsilon_{titrant}$.

**(f)** Both the product and the titrant absorb, but $\varepsilon_{titrant} > \varepsilon_{product}$.

**13-5.** You are in charge of selecting a new UV-visible *scanning* spectrophotometer for your industrial research laboratory. The laboratory mostly does spectral scanning to ascertain if certain impurities are present in one of their products. If they find the impurity, they use the spectrophotometer at a fixed wavelength to accurately determine the concentration of the impurity. At a recent instrument exhibit, you note that there are two different types that seem to fit your needs. A new microprocessor-controlled, single-beam spectrophotometer sells for $15,000. It scans the spectrum of the reference cuvette and stores that in memory. It then scans the spectrum of the sample cuvette and automatically obtains the ratio (i.e., transmittance) and calculates the absorbance. The second system to fit your needs is a true double-beam system (also microprocessor-controlled) that sells for $25,000. Compare the advantages and limitations of these two types of spectrophotometers for your application. Recommend to your supervisor one of the two systems. How would your recommendation change if you also consider a diode array spectrophotometer with 2-nm resolution that sells for $15,000?

**13-6.** A new reaction-rate method has been proposed that measures the initial rate $v_0$ and the rate at some later time during the reaction $v_t$. It is proposed that these two rate measurements, for a first- or pseudo-first-order reaction will allow the analyte initial concentration $[A]_0$ to be obtained independent of the rate constant for the reaction. Consider the reaction to be

$$A + R \xrightarrow{k} P$$

where A is the analyte, R is the reagent, $k$ is the rate constant, and P is the product. Show *mathematically* that the combination of $v_0$ and $v_t$ measurements allow $[A]_0$ to be obtained independent of rate constant $k$.

**13-7.** The following are several of the most important specifications that manufacturers give for UV-visible absorption spectrophotometers.

1. Stray radiation (fraction of light outside bandpass, but reaching the exit slit)
2. Photometric accuracy (accuracy of absorption measurements)
3. Photometric precision (precision of absorption measurements)
4. Baseline stability (drift of 100% *T* with time)
5. Baseline flatness (deviation of 100% *T* with wavelength)
6. Wavelength reproducibility (precision of setting wavelength)
7. Wavelength scanning accuracy
8. Spectral bandpass

For the following applications, discuss which of the specifications given above are most important for achieving optimal results (high accuracy and precision) and which are of little or no importance.

**(a)** A reaction-rate method for glucose based on the reaction of glucose with glucose oxidase and subsequent conversion of the product $H_2O_2$ to a colored absorbing species.

**(b)** A study to determine the formation constant of a metal-ligand complex in which the ligand absorbs at one wavelength and the complex at another. The equilibrium is monitored by spectral scanning for various analytical concentrations of metal and ligand.

**(c)** A determination in which the spectrophotometer is to be used as a high-performance liquid chromatography detector at a fixed wavelength.

**13-8.** The following are errors or problems encountered in spectrophotometric applications. They can all be remedied by a change in instrumental or chemical conditions. State why each error or problem occurs and explain in a few brief sentences how the problem or error can be corrected.

**(a)** A UV-visible spectrophotometric determination was found to have a precision of 2%, which was identical to the absorbance reading error on the strip-chart recorded readout.

**(b)** In a biological kinetics study, one could be required to work at high absorbances ($A = 3$ to $4$) to monitor a reactant in a reaction mixture. Dilution of the sample cannot be used because it changes the reaction mechanism. It is noted that absorbance is nearly independent of concentration at these absorbances.

**13-9.** Give the major reasons that the following methods or techniques are becoming increasingly useful and popular in instrumental analysis. Give the major advantages over alternatives.

(a) Enzyme-catalyzed reactions to determine substrates.

(b) UV-visible spectrophotometers with diode array detectors.

(c) Double-beam spectrophotometers with an option for slit programming or photomultiplier voltage programming.

(d) Double monochromators for UV-visible spectrophotometers.

(e) Flow injection analysis with colorimetric detection.

**13-10.** The simultaneous determination of titanium and vanadium, each as their peroxide complex, can be done in steel. Preliminary measurements indicated that 1.00 mg of Ti in 50 mL with $H_2O_2$ gave an absorbance of 0.269 at 400 nm and 0.134 at 460 nm in a 1.00-cm cell. Under similar conditions, 1.00 mg of V in 50 mL gave an absorbance of 0.057 at 400 nm and 0.091 at 460 nm. For a 1.000-g steel sample treated with $H_2O_2$ and ultimately diluted to 50 mL, the absorbance at 400 nm was 0.172 and that at 460 nm was 0.116 in the same 1.00-cm cell. Calculate the percent titanium and vanadium from these absorbance values.

**13-11.** Calculate the measured absorbance for $2.00 \times 10^{-5}$ M solution of an analyte molecule with a molar absorptivity of $1.00 \times 10^5$ L mol$^{-1}$ cm$^{-1}$ in a 1.00-cm-pathlength cell. The stray radiation readout fraction ($r$) is 2%.

**13-12.** Consider a solution of a weak acid HA which has a dissociation constant $K_a$. The undissociated acid HA has a molar absorptivity of $\varepsilon_{HA}$ and the anion $A^-$ has a molar absorptivity of $\varepsilon_{A^-}$. The analytical concentration of HA is designed as $c_a$. Derive an expression for the absorbance of a solution of this acid in terms of $c_a$, $K_a$, $[H^+]$, $\varepsilon_{HA}$, and $\varepsilon_{A^-}$. Indicate how the Beer's law plot depends on the relative magnitude of $\varepsilon_{HA}$ compared to $\varepsilon_{A^-}$, the pH, and $c_a$. Assume that the hydrogen ion concentration is independently adjusted by another buffer.

**13-13.** For a given analyte solution, indicate what a plot of $\sigma_A/A$ versus the slit width $W$ of the monochromator would look like. Justify the shape of the curve. Assume that the slit width is varied from 0 to some large value.

**13-14.** Calculate the effective pathlength of a cell of width $b$ in which the angle of incidence is 1° from the normal.

**13-15.** For a molecular absorption spectrophotometer, $i_r = 1.00 \times 10^{-10}$ A, $\xi_s = 1.00 \times 10^{-3}$, $\sigma_{0t} = (\sigma_d)_{ex} = 1 \times 10^{-14}$ A. Calculate the detection limit for species X where $\varepsilon b = 1.00 \times 10^4$ L mol$^{-1}$. Assume in your

calculation that $K = 4.0 \times 10^{-19}$ A and $mG = 1.0 \times 10^{12}$ V A$^{-1}$.

**13-16.** A solution contains 35.3 µg mL$^{-1}$ Cr. The Cr is present only as dichromate ($Cr_2O_7^{2-}$) in 1.0 M $H_2SO_4$. The transmittance of this solution was 0.762 at 440 nm in a 1-cm-pathlength cell. What is the molar absorptivity of dichromate at 440 nm?

**13-17.** If the relative standard deviation in concentration is given by

$$\frac{\sigma_c}{c} = \frac{-\sigma_T}{T \ln T}$$

where $\sigma_T$ is the standard deviation in transmittance, show that the optimum precision occurs at 36.8% $T$. Assume that $\sigma_T$ is constant and independent of concentration.

**13-18.** If absorption measurements are made under signal-shot-noise-limited conditions, the relative standard deviation in concentration is given by

$$\frac{\sigma_c}{c} = \frac{-kT^{-1/2}}{\ln T}$$

where $k$ is a constant for a given reference current and noise equivalent bandpass. Show that the optimum precision occurs at 13.5% $T$ in this case.

**13-19.** In a flow injection analysis experiment, a sample of $1.5 \times 10^{-4}$ M $Cu^{2+}$ is injected into a flowing stream containing a buffered reagent. As the sample is carried downstream, it mixes and reacts with the reagent in a reaction coil. The product is then detected spectrophotometrically. Identify as many variables of the system as you can. Indicate those which it would be difficult to vary experimentally and any you would not wish to vary.

**13-20.** Several replicate experiments are carried out using the FIA system described in problem 13-19. The average height of the resultant peaks is $h$ and the standard deviation is $\sigma$.

(a) What would be the effect of optimizing the response function $R = h/\sigma$?

(b) Would it be better to optimize $R = h/(\sigma + 1)$?

**13-21.** From the following data, indicate whether the nonlinearity in the calibration curve is due to instrumental causes or chemical equilibria involving the analyte.

(a) The absorbance at which nonlinearity occurs changes from 0.5 to 1.0 when the cell pathlength is doubled.

(b) The absorbance at which nonlinearity occurs is independent of cell pathlength.

**13-22.** Radiation at two wavelengths is incident on a sample. The reference signal for both wavelengths is identical, but the molar absorptivities are $1.0 \times 10^4$ and $1.0 \times 10^5$ L mol$^{-1}$ cm$^{-1}$. Calculate the absorbance at analyte concentrations of 1, 10, 20, and 30 µM in a 1-cm-pathlength cell.

# REFERENCES

The following are general references to spectrophotometry.

1. E. J. Meehan, "Fundamentals of Spectrophotometry," chap. 2 in *Treatise on Analytical Chemistry*, P. J. Elving and I. M. Kolthoff, eds., 2nd ed., pt. I, vol. 7, Wiley-Interscience, New York, 1981.

2. G. F. Lothian, *Absorption Spectrophotometry*, 2nd ed., Hilger and Watts, London, 1958.

3. C. Burgess and A. Knowles, *Standards in Absorption Spectrometry*, Chapman & Hall, London, 1981.

4. R. P. Bauman, *Absorption Spectroscopy*, Wiley, New York, 1962.

5. K. L. Cheng, "Absorptimetry," in *Spectrochemical Methods of Analysis*, J. D. Winefordner, ed., Wiley-Interscience, New York, 1971.

6. H. H. Jaffe and M. Orchin, *Theory and Applications of Ultraviolet Spectroscopy*, Wiley, New York, 1962.

7. P. A. St. John, "Optical Molecular Spectroscopic Methods," chap. VIIA in *Trace Analysis: Spectroscopic Methods for Elements*, J. D. Winefordner, ed., Wiley-Interscience, New York, 1976.

8. K. L. Cheng, "Spectrophotometry and Fluorometry," in *Trace Analysis: Physical Methods*, G. H. Morrison, ed., Wiley-Interscience, New York, 1965, p. 161.

9. "Ultraviolet and Light Absorption Spectrometry," *Anal. Chem.* This review article appears every two years in the Fundamental Reviews issue of *Analytical Chemistry*. Recent advances in chemistry and instrumentation as well as specific applications are covered.

Spectrophotometric methods for specific species are discussed in the following references.

10. E. B. Sandell and H. Onishi, *Photometric Determination of Traces of Metals*, pt. I, 4th ed., Wiley-Interscience, New York, 1978.

11. E. B. Sandell, *Colormetric Determination of Traces of Metals*, 3rd ed., Wiley-Interscience, New York, 1959.

12. D. F. Boltz and J. A. Howell, eds., *Colorimetric Determination of Nonmetals*, 2nd ed., Wiley-Interscience, New York, 1978.

13. F. D. Snell, *Photometric and Fluorometric Methods of Analysis: Metals*, Wiley-Interscience, New York, 1978.

14. F. D. Snell, *Photometric and Fluorometric Methods of Analysis: Non-metals*, Wiley-Interscience, New York, 1981.

15. L. C. Thomas and G. J. Chamberlin, *Colorimetric Chemical Analytical Methods*, 9th ed., Wiley, New York, 1980.

16. M. Pesez and J. Bartos, *Colorimetric and Fluorimetric Analysis of Organic Compounds and Drugs*, Marcel Dekker, New York, 1974.

17. F. T. Weiss, *Determination of Organic Compounds: Methods and Procedures*, Wiley-Interscience, New York, 1970.

18. S. Siggia, ed., *Instrumental Methods of Organic Functional Group Analysis*, Wiley-Interscience, New York, 1972.

19. K. A. Connors, *Reaction Mechanisms in Organic Analytical Chemistry*, Wiley-Interscience, New York, 1973.

20. S. Siggia and J. Gordon Hanna, *Quantitative Organic Analysis via Functional Groups*, 4th ed., Wiley-Interscience, New York, 1979.

21. H. Freiser, "Reactive Groups as Reagents: Introduction and Inorganic Applications," in *Treatise on Analytical Chemistry*, P. J. Elving and I. M. Kolthoff, eds., 2nd ed., pt. I, vol. 3, Wiley-Interscience, New York, 1983, p. 395.

22. J. G. Hanna, "Reactive Groups as Reagents: Organic Applications," in *Treatise on Analytical Chemistry*, P. J. Elving and I. M. Kolthoff, eds., 2nd ed., pt. I, vol. 3, Wiley-Interscience, New York, 1983, p. 507.

23. P. J. Elving and I. M. Kolthoff, eds., *Treatise on Analytical Chemistry*, pt. II, 1st ed., Wiley-Interscience, New York. Seventeen volumes in this part of the series contain detailed discussions of spectrophotometric and other determination methods for specific elements and compounds.

Performance characteristics are discussed in the following references.

24. L. D. Rothman, S. R. Crouch, and J. D. Ingle, Jr., "Theoretical and Experimental Investigation of Factors Affecting Precision in Molecular Absorption Spectrophotometry," *Anal. Chem.*, *47*, 1226 (1975).

25. J. D. Ingle, Jr., "Precision Characteristics of Simple Spectrophotometers," *Anal. Chim. Acta*, *88*, 131 (1977).

26. *Tentative Definitions of Terms and Symbols Relating to Molecular Spectroscopy*, ASTM publication E131-68, American Society for Testing and Materials, Philadelphia, 1968.

27. L. S. Goldring, R. C. Hawes, G. H. Hare, A. O. Beckman, and M. E. Stickney, "Anomalies in Extinction Coefficient Measurements," *Anal. Chem.*, *25*, 869 (1953).

28. W. Kaye, "Stray Light Ratio Measurements," *Anal. Chem.*, *53*, 2201 (1981).

29. K. L. Ratzlaff and D. F. S. Natusch, "Theoretical Assessment of Accuracy in Dual Wavelength Spectrophotometric Measurements," *Anal. Chem.*, *51*, 1209 (1979).

Reaction rate and enzymatic measurements are discussed in the following references.

30. H. V. Malmstadt, C. J. Delaney, and E. A. Cordos, "Reaction-Rate Methods of Analysis," *CRC Crit. Rev. Anal. Chem.*, *2*, 559 (1972).

31. H. A. Mottola, "Catalytic and Differential Rate Methods," *CRC Crit. Rev. Anal. Chem.*, *4*, 229 (1975).

32. H. L. Pardue, "Applications of Kinetics to Automated Quantitative Analysis," in *Advances in Analytical Instrumentation*, C. N. Reilley and F. W. McLafferty, eds., vol. 7, p. 141, Wiley-Interscience, New York, 1968.

33. H. B. Mark, Jr., "Reaction Rate Methods in Analysis," *Talanta*, *14*, 717 (1972).

34. H. V. Malmstadt, E. A. Cordos, and C. J. Delaney, "Automated Reaction Rate Measurements," *Anal. Chem.*, *44*, 26A (October 1972).

35. H. V. Malmstadt, C. J. Delaney, and E. A. Cordos, "Instrumentation for Rate Determinations," *Anal. Chem.*, *44*, 79A (October 1972).

36. H. B. Mark, Jr., and G. A. Rechnitz, *Kinetics in Analytical Chemistry*, Wiley-Interscience, New York, 1968.

37. G. G. Guilbault, *Handbook of Enzymatic Methods of Analysis*, Marcel Dekker, New York, 1976.

38. H. L. Pardue, "A Comprehensive Classification of Kinetic Methods of Analysis Used in Clinical Chemistry," *Clin. Chem.*, *23*, 2189 (1977).

39. S. R. Crouch, "Applications of Computer Circuitry and Techniques to Instrumentation for Chemical Analysis," in *Computers in Chemistry and Instrumentation*, vol. 3, J. S. Mattson, H. C. MacDonald, and H. B. Mark, eds., Marcel Dekker, New York, 1973, p. 107.

40. P. W. Carr and L. D. Bowers, *Immobilized Enzymes in Analytical and Clinical Chemistry*, Wiley, New York, 1980.

Other specialized applications and instrumentation are reviewed in the following references.

41. J. B. Headridge, *Photometric Titrations*, Pergamon Press, Oxford, 1961.

42. R. L. Stevenson, "UV-VIS Absorption Detectors for HPLC," Chap 2 in *Liquid Chromatography Detectors*, T. M. Vickery, ed., Marcel Dekker, New York, 1983.

43. D. G. Jones, "Photodiode Array Detectors in UV-VIS Spectroscopy," *Anal. Chem.*, *57*, pt. I, 1057A; pt. II, 1207A (1985).

44. R. W. Frei, H. Jansen, and V. A. Th. Brinkman, "Postcolumn Reaction Detectors for HPLC," *Anal. Chem.*, *57*, 1524A (1985).

45. W. C. Johnson, Jr., "Circular Dichroism and its Empirical Application to Biopolymers," in *Methods of Biochemical Analysis*, D. Glick, ed., vol. 31, Wiley-Interscience, New York, 1985, p. 61.

46. J. Růžička and E. H. H. Hansen, *Flow Injection Analysis*, Wiley, New York, 1981.

47. R. W. Frei and J. F. Lawrence, eds., *Chemical Derivatization in Analytical Chemistry*, Plenum Press, New York, Vol. 1, 1981; Vol. 2, 1982. These volumes have chapters covering reaction detectors in liquid chromatography, pre- and postcolumn derivatization techniques, and flow injection analysis.

# CHAPTER 14

# Infrared Spectrometry

Infrared (IR) spectrometry deals with the interaction of infrared radiation with matter. The IR spectrum of a compound can provide important information about its chemical nature and molecular structure. Most commonly, the spectrum is obtained by measuring the absorption of IR radiation, although infrared emission and reflection are also used. Infrared spectrometry finds its widest applicability in the analysis of organic materials, but it is also useful for polyatomic inorganic molecules and for organometallic compounds.

The infrared region of the electromagnetic spectrum is generally considered to lie in the wavelength range from 770 nm to 1000 μm; the corresponding wave number range is from 12,900 to 10 cm⁻¹ (see Table 1-1). Radiant energies in the IR region are on the order of the energies of vibrational transitions. Hence IR spectroscopy is one branch of vibrational spectroscopy. The IR region is often further subdivided into three subregions. The **near-infrared** region (nearest to the visible) extends from 770 nm to 2.5 μm (12,900 to 4000 cm⁻¹), the **mid-infrared** region from 2.5 to 50 μm (4000 to 200 cm⁻¹), and the **far-infrared** region from 50 to 1000 μm (200 to 10 cm⁻¹).

Nearly all molecules absorb infrared radiation, the exceptions being homonuclear diatomics such as $O_2$, $N_2$, and $H_2$. The IR spectrum of a polyatomic molecular can be quite complex because of the many possible vibrational transitions and the existence of overtones and sum and difference bands. However, IR absorption bands are often quite sharp and characteristic of certain groups within the molecule. The IR spectrum for a given molecule is also unique to that molecule and thus highly useful in compound identification. Detailed analysis of IR absorption bands can be of great assistance in determining the molecular structure of a compound, particularly when used in conjunction with another structure-sensitive technique, such as nuclear magnetic resonance spectrometry or mass spectrometry.

Infrared spectrometry is also useful for quantitative analysis, although it is considerably more difficult to achieve accurate and precise results with IR spectrometry than with UV-visible methods. Beer's law provides the basis for quantitative IR methods as it does in UV-visible spectrophotometry. However, nonlinearities and the presence of overlapping absorption bands can lead to severe problems in quantitative applications of IR absorption.

This chapter is concerned primarily with infrared absorption spectrometry, because of its widespread utility compared to emission and reflection spectrometry. We begin our study by considering the basis of IR absorption, the selection rules and the types of transitions observed. We also consider the instrumentation used for both conventional IR absorption and Fourier transform IR (FTIR) spectrometry, in particular the various optical materials that must be used in the infrared region. Because many different sample preparation techniques are used in obtaining IR spectra, these methods are discussed here in some detail prior to illustrating some of the qualitative and quantitative applications of IR absorption spectrometry. While most of the discussion here deals with absorption in the mid-IR region, a section is included on near-IR and far-IR absorption. This chapter concludes with a brief consideration of infrared emission and reflection methods and some of their applications.

## 14-1 BASIS OF INFRARED ABSORPTION

The IR absorption spectrum can be obtained with gas-phase or with condensed-phase molecules. For gas-phase molecules vibration-rotation spectra are observed (see Section 12-3), while in condensed phases, the rotational structure is lost. For most routine analytical applications of infrared spectrometry, spectra are obtained with condensed-phase samples. Hence the discussion here centers around the vibrational transitions observed with molecules present as pure liquids, as solutions, or in the solid state.

### Requirements for Infrared Absorption

Selection rules for infrared absorption can be determined by evaluating the transition moment $R$ as given in equation F-2. For vibrational motion the electric dipole moment $\mu$ can be expressed as

$$\mu = \mu_0 + (r - r_e)\left(\frac{\partial\mu}{\partial r}\right)_0 + \frac{1}{2}(r - r_e)^2\left(\frac{\partial^2\mu}{\partial r^2}\right)_0 + \cdots$$

$$(14\text{-}1)$$

where $\mu_0$ is the permanent dipole moment, $r$ is the internuclear distance, and $r_e$ is the equilibrium bond distance. If we neglect all but the first two terms in equation 14-1 and substitute for $\mu$ into equation F-2, the result is

$$R = \int \psi_i^* \left[\mu_0 + (r - r_e)\left(\frac{\partial\mu}{\partial r}\right)_0\right]\psi_j\, d\tau \qquad (14\text{-}2)$$

which reduces to

$$R = \int \psi_i^* \left[(r - r_e)\left(\frac{\partial\mu}{\partial r}\right)_0\right]\psi_j\, d\tau \qquad (14\text{-}3)$$

since $\mu_0$ is a constant and $\int \psi_i^*\psi_j\, d\tau = 0$ because of orthogonality.

From equation 14-3 it is clear that there must be a *change* in dipole moment during the vibration in order for a molecule to absorb infrared radiation. The selection rules predict that the *fundamental* absorption will occur with $\Delta v = \pm 1$, with much weaker overtone absorption corresponding to $\Delta v = \pm 2$, and so on. The intensity of a vibrational band in the infrared region depends on the square of the transition moment (see Appendix F) and thus to the square of the dipole moment derivative from equation 14-3. A molecule with a small permanent dipole moment may have a large dipole moment derivative, and vice versa. Carbon monoxide, for example, has a very small dipole moment of only 0.11 D, but is a good infrared absorber because of a large dipole moment derivative. For homonuclear diatomics, however, $\mu = 0$ for all internuclear separations, and no IR absorption occurs. Vibrational modes which do not involve a change in dipole moment are said to be **infrared-inactive**.

It should be noted that the molar absorptivities of even the most strongly absorbing groups in the infrared region are one to three orders of magnitude lower than molar absorptivities of the most intense electronic transitions in the ultraviolet and visible. Thus highly polar bonds such as C—Cl and C—F, which have large dipole moment derivatives, may have molar absorptivities of only 100 to 1000 L mol$^{-1}$ cm$^{-1}$, whereas highly absorbing transitions in the UV and visible region may have absorptivities on the order of 10,000 to 100,000 L mol$^{-1}$ cm$^{-1}$.

### Number of Vibrational Modes

As discussed in Section 12-3, there are $3N - 6$ fundamental vibrational modes for a nonlinear molecule, while a linear molecule has $3N - 5$ vibrational modes. Thus water, being a nonlinear molecule, should have three fundamental modes of vibration. These modes are illustrated in Figure 14-1a. A permanent dipole moment in a molecule arises because of the separation of charges. In the $H_2O$ molecule, the H atoms can be considered to have a partial + charge and the O atom a partial − charge. Each of the vibrational modes illustrated in Figure 14-1a gives rise to a change in dipole moment and thus is **infrared-active**. Note that other motions in the $H_2O$ molecule are actually rotations since the oxygen atom is free to rotate. The $H_2O$ molecule

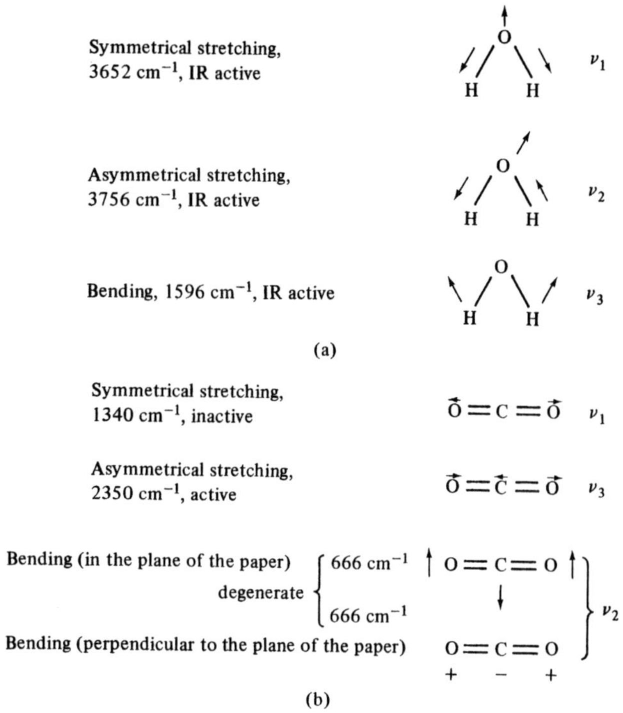

Symmetrical stretching, 3652 cm$^{-1}$, IR active   $\nu_1$

Asymmetrical stretching, 3756 cm$^{-1}$, IR active   $\nu_2$

Bending, 1596 cm$^{-1}$, IR active   $\nu_3$

(a)

Symmetrical stretching, 1340 cm$^{-1}$, inactive   $\vec{O} = C = \vec{O}$   $\nu_1$

Asymmetrical stretching, 2350 cm$^{-1}$, active   $\vec{O} = \vec{C} = \vec{O}$   $\nu_3$

Bending (in the plane of the paper) degenerate
$\left\{ \begin{array}{l} 666 \text{ cm}^{-1} \quad \uparrow O = C = O \uparrow \\ 666 \text{ cm}^{-1} \quad \downarrow \end{array} \right\}$

Bending (perpendicular to the plane of the paper)   $O = C = O$
$+ \quad - \quad +$   $\nu_2$

(b)

**FIGURE 14-1** Illustration of vibrational modes in $H_2O$ (a) and $CO_2$ (b). For a nonlinear molecule ($H_2O$), there are $3N - 6$ vibrational modes, while for a linear molecule ($CO_2$), there are $3N - 5$ modes. Note that all the $H_2O$ modes are IR active, while with $CO_2$, the symmetric stretch is IR inactive. The $\nu_n$ symbolism for labeling the frequencies of the vibrations is discussed in the text. This should not be confused with the $v_n$ symbolism that is used to label the vibrational energy levels.

shows three fundamental absorption bands in the IR region.

Carbon dioxide, on the other hand, is a linear molecule and thus has four fundamental vibrational modes, as illustrated in Figure 14-1b. The symmetric stretch in $CO_2$ does not give rise to a dipole moment change and thus is infrared-inactive. The other three modes are infrared-active, but the two bending modes are degenerate and absorb at the same frequency. Thus $CO_2$ shows two fundamental absorption bands in the IR region, one due to the asymmetric stretch and the other to the degenerate bending modes.

Also introduced in Figure 14-1 is the $\nu_n$ symbolism that is used to label the frequencies of the various fundamental vibrations. By convention, the highest frequency totally symmetric vibration is labeled $\nu_1$, the next-highest frequency symmetric vibration $\nu_2$, and so on. After assigning all the symmetric vibrations, the highest-frequency assymmetric vibration is counted next. Then the rest of the asymmetric vibrations are numbered in the order of decreasing frequency. The bending

vibration of a linear molecule, which is labeled $\nu_2$, is an exception to this rule.

In addition to the vibrations illustrated in Figure 14-1, more complex molecules can undergo several additional motions, as illustrated in Figure 14-2a for the methylene group. Ring compounds can undergo a stretching motion, called a *breathing* vibration, that involves the entire ring as shown in Figure 14-2b.

Although the number of expected vibrations can be calculated from the number of degrees of freedom of the molecule, polyatomic molecules often given fewer, or in some cases more, absorption bands than predicted. The observed number of absorption bands may be less than predicted if some vibrations are infrared-inactive, if vibrations are degenerate, if absorption is very weak, if bands are unresolved instrumentally, or if some vibrations occur outside the range of the spectrophotometer. The number of observed absorption bands may be greater than predicted if there are overtones as discussed previously, combination bands, or difference bands. Combination bands result when the absorption of a photon excites two vibrational modes simultaneously. If the absorption frequencies of the two independent vibrations are $\nu_1$ and $\nu_2$ a combination band is sometimes found at frequency $\nu_1 + \nu_2$. Similarly, a

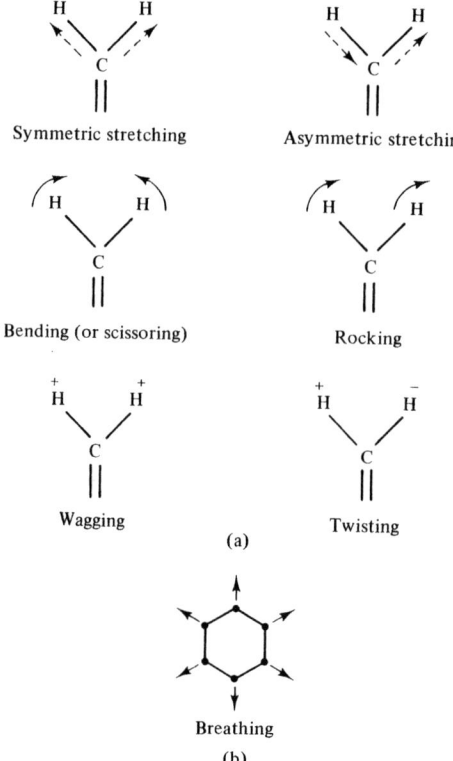

Symmetric stretching

Asymmetric stretching

Bending (or scissoring)

Rocking

Wagging

Twisting

(a)

Breathing

(b)

**FIGURE 14-2** Vibrational modes for a methylene group (a) and breathing vibration for a ring compound (b).

difference band is the result of a molecule in one excited vibrational level absorbing energy and being promoted to another excited vibrational level. All of these non-fundamental transitions are less intense than absorption bands due to fundamental vibrations as discussed below.

## Group Frequencies

With certain functional or structural groups, it has been found that their vibrational frequencies are nearly independent of the rest of the molecule. For example, the stretching vibration of the carbonyl group in various aldehydes and ketones is almost always observed in the range 1650 to 1740 cm$^{-1}$. Such frequencies are characteristic of the functional or structural group involved and are thus known as **group frequencies**. The presence of various group vibrations in the infrared spectrum is of great assistance in identifying the absorbing molecule.

For many groups involving only two atoms, the approximate frequency of the fundamental vibration can be calculated from a simple harmonic oscillator model as shown in Section 12-3 (see Table 12-2). Thus we can write the vibrational frequency in wave numbers as

$$\bar{\nu} = \frac{(k/\mu)^{1/2}}{2\pi c} \qquad (14-4)$$

where $k$ is the force constant of the bond and $\mu$ is the reduced mass of the two atoms. If $k$ is expressed in N cm$^{-1}$ and the masses in atomic mass units, evaluation of the constants in equation 14-4 gives

$$\bar{\nu} = 1302 \left(\frac{k}{\mu}\right)^{1/2} \qquad (14-5)$$

Force constants have been calculated for many bonds based on measurement of the fundamental vibrational frequency and equation 14-5. In some cases force constants can be obtained or calculated independently and used to predict a particular vibrational frequency. For a carbonyl group, for example, the force constant is approximately 12 N cm$^{-1}$, which predicts that the stretching vibration should occur at approximately 1722 cm$^{-1}$, in good agreement with the observed frequency range. Arguments based on changes in mass are often used to predict changes in vibrational frequency upon substitution of one atom for another. For example, in a molecule or group XY, if atom Z is substituted for atom Y, the ratio of the vibrational frequencies is related to the reduced masses by

$$\frac{\bar{\nu}_{XY}}{\bar{\nu}_{XZ}} = \sqrt{\frac{\mu_{XZ}}{\mu_{XY}}}$$

as long as factors affecting the force constant $k$ are equal. This equation does work well for isotopic substitutions. In other cases, however, changing one atom for another often affects bond strengths, polarities, and lengths as well as mass, which invalidates the assumptions leading to the equation above.

Calculations using equation 14-4 show that for most groups of interest, characteristic frequencies of stretching vibrations should lie in the region 4000 to 1000 cm$^{-1}$. In practice, the region from 4000 to 1300 cm$^{-1}$ is often called the **group frequency region**. Figure 14-3 gives the approximate location of various group vibrations in the IR spectrum. It can be seen that functional

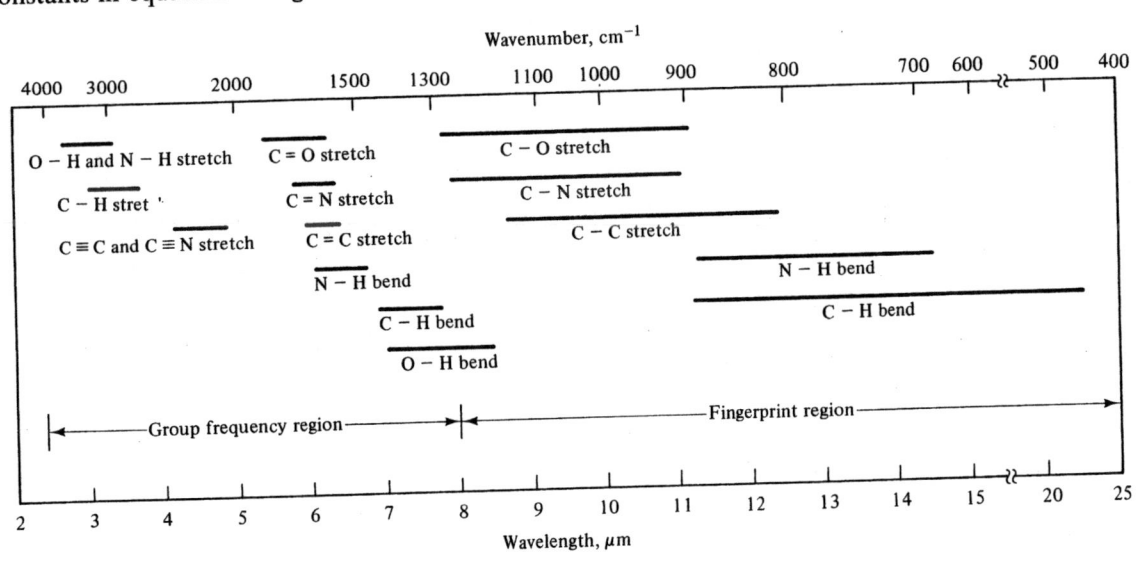

**FIGURE 14-3** Frequencies of various group vibrations in the group frequency region and in the fingerprint region.

group absorption is only slightly affected by the composition of the rest of the molecule in the group frequency region.

In the region from $\approx 1300$ to $400 \text{ cm}^{-1}$, vibrational frequencies are affected by the entire molecule, as the broader ranges for group absorptions in Figure 14-3 illustrate. Since absorption in this **fingerprint region** is characteristic of the molecule as a whole, this region finds widespread use for identification purposes by comparison with library spectra, as discussed in Section 14-4.

### Vibrational Coupling

The energy of a particular vibrational mode, and thus its frequency, may be influenced by the presence of other vibrations in the molecule through vibrational coupling. Strong coupling can occur between two stretching vibrations, for example, if the two vibrations have an atom in common. Interaction also occurs between two bending vibrations with a common bond between the groups. Coupling is strongest when the energies of the isolated vibrations are approximately equal. If the groups are separated by two or more bonds, little or no coupling occurs.

As a result of vibrational coupling, the position of an absorption band cannot be specified exactly. For example, the coupling of the C—O stretching vibration with nearby C—C and C—H vibrations leads to a difference of $20 \text{ cm}^{-1}$ in the C—O stretch in methanol and ethanol. Although coupling leads to uncertainty in group vibrational frequencies, it also makes the infrared spectrum unique for a given molecule and is one of the major reasons that IR spectrometry is so valuable in qualitative analysis.

In order for vibrational coupling to occur several conditions must be fulfilled. First, both vibrations must have the same symmetry and approximately the same frequency. Second, there must be an appreciable interaction between the groups responsible. This in turn implies that the two groups are near to each other in the molecule and that a vibrational energy transfer mechanism exists.

When an overtone or combination frequency interacts with a fundamental vibration, the interaction is called **Fermi resonance**. Fermi first proposed such a mechanism to explain the splitting of the $1340\text{-cm}^{-1}$ fundamental of $CO_2$. In this case Fermi resonance gives rise to two bands in the Raman spectrum where only one is predicted. For $CO_2$, an overtone of the degenerate bending vibrations ($2\nu_2$) is at $2 \times 666 \text{ cm}^{-1} = 1332 \text{ cm}^{-1}$. This frequency is very close to that of the symmetric stretching vibration ($\nu_1$) at $1340 \text{ cm}^{-1}$. The resonance interaction raises the frequency of one of the vibrations and lowers that of the other. In quantum mechanical terms the resonance interaction mixes the energies of the two initial states. The separation and relative intensities of the bands depend on nearness of the two unperturbed frequencies. The interaction of the CH stretch in aldehydes at about $2800 \text{ cm}^{-1}$ with the first overtone of the $1400 \text{ cm}^{-1}$ in-plane CH bending vibration is another example of Fermi resonance. This interaction gives a doublet in the region 2700 to 2900 $\text{cm}^{-1}$.

Fermi resonance is not always easy to detect. It should be suspected whenever a normally single band appears as a doublet. In complex molecules, it can sometimes be detected by deuteration or by using various solvents. In compounds containing methyl groups, partial deuteration has been used to circumvent Fermi resonance between the CH deformation overtone and the CH stretching vibration. The frequencies are shifted enough by the isotope effect that Fermi resonance no longer occurs.

## 14-2 INSTRUMENTATION

Three distinct types of instruments employed for IR absorption spectrometry are considered in this section. Dispersive instruments with a monochromator are used in the mid-IR region for spectral scanning and quantitative analysis. Fourier transform IR systems are widely applied in the far-IR region and becoming quite popular for mid-IR spectrometry. Nondispersive instruments that use filters for wavelength selection or an infrared-absorbing gas in the detection system are often used for gas analysis at specific wavelengths.

### Dispersive IR Spectrophotometers

Modern dispersive infrared spectrophotometers are invariably double-beam instruments, but many allow single-beam operation via a front-panel switch. Single-beam instruments are not very practical in the IR region because of the absorption of IR radiation by atmospheric $H_2O$ and $CO_2$, as shown in Figure 14-4. Double-beam operation compensates for atmospheric absorption, for the wavelength dependence of the source spectral radiance, the optical efficiency of the mirrors and grating, and the detector responsivity, and for source and detector instability, which can be serious in the IR region. Also, all solvents absorb in the infrared region, which makes single-beam operation somewhat impractical even with storage of the reference spectrum.

*Dispersive Spectrophotometer Designs.* A majority of the dispersive IR spectrophotometers as of this writing use the optical null principle illustrated in Figure 14-5. Ratio recording systems, similar to those

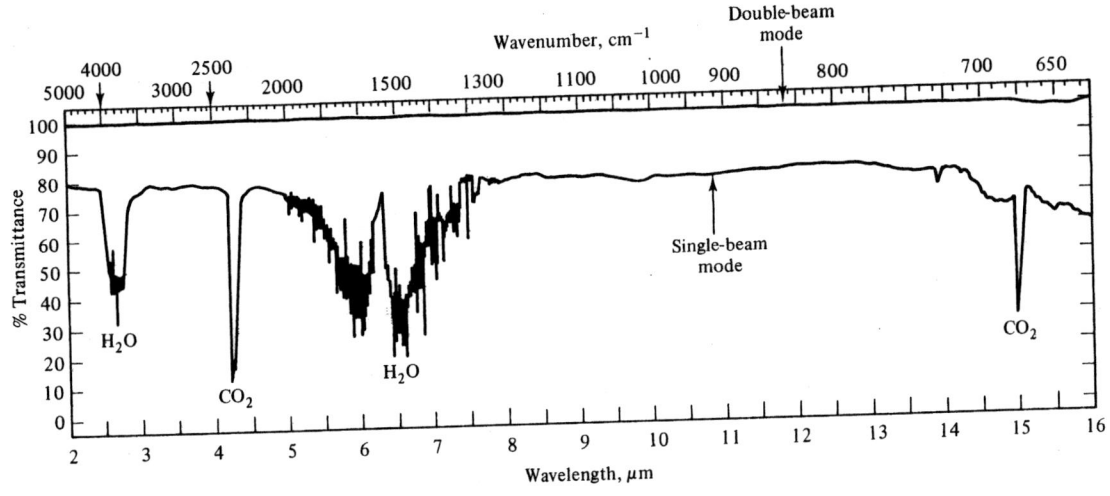

**FIGURE 14-4** Single- and double-beam spectra of atmospheric water vapor and $CO_2$. In the lower, single-beam trace, the absorption of atmospheric gases is apparent. The top, double-beam trace shows that the reference beam compensates nearly perfectly for this absorption and allows a stable 100% $T$ baseline to be obtained.

used in the UV and visible regions, are used with IR spectrophotometers, but until recently were found only in the more expensive research instruments. Several differences can be noted when the schematic of Figure 14-5 is compared to the corresponding diagram for a UV-visible spectrophotometer (see Figure 13-4). First, with infrared spectrophotometers, the sample cell is usually placed in front of the monochromator to minimize the effects of infrared emission and stray radiation from the cell compartment. In the UV-visible region, the sample cell is often placed after the monochromator to minimize photodecomposition of the sample. The sample and reference beams are usually chopped at relatively low frequencies (5 to 30 Hz) with IR spectrophotometers because of the slow response of most IR detectors. Modulation also allows discrimination against IR emission by materials near the detector and detector noise with a $1/f$ nature.

In inexpensive infrared spectrophotometers, the recorder pen and the attenuator are mechanically coupled as shown in Figure 14-5a. In more versatile systems, the pen is driven by a separate servomechanism. A potentiometer with a fixed voltage across it is coupled to the attenuator so that the attenuator position controls the wiper of the potentiometer. The voltage on the wiper is proportional to the transmittance. This voltage is the input to a normal servo recorder. By using several different fixed voltages across the potentiometer, any portion of the transmittance scale may be expanded to full scale on the recorder.

Ratio recording systems are less frequently used in IR spectrophotometers because of the slow response times of the thermal detectors used and the high gains required. With the improvements in electronic ampli-

fiers and their reduction in price, these systems have gained in popularity in recent years. Optical null systems read out directly in transmittance. Ratio recording spectrometers can read out in transmittance or absorbance. Because of the present popularity of FTIR systems, not many dispersive spectrometers have been introduced recently.

The radiant power emitted by infrared sources varies dramatically over the wavelength region covered by most spectrophotometers. Slit programming is used in modern IR instruments to compensate for these variations and thus maintain a nearly constant radiant power level at the detector. Either a cam is used to vary the slit width as a function of wavelength in a predetermined manner or the reference beam detector signal is monitored and compared to a preset desired level. In the latter case, any difference is amplified and used to drive a slit control motor, which opens or closes the slits to minimize the difference signal. The cam method is inexpensive, but not very flexible since the program cannot be varied as components age or a new source is installed. The slit servo method allows the instrument to adapt to changing conditions.

Because of slit programming, the resolution of IR spectrophotometers degrades in wavelength regions where the source intensity is low or where strong absorption occurs in the reference beam. Regions of high solvent absorption, for example, are often accompanied by wide slits and low resolution. Care must often be taken in interpreting IR spectra in such wavelength regions.

*Components of Dispersive Infrared Spectrophotometers.* Sources and detectors for IR spec-

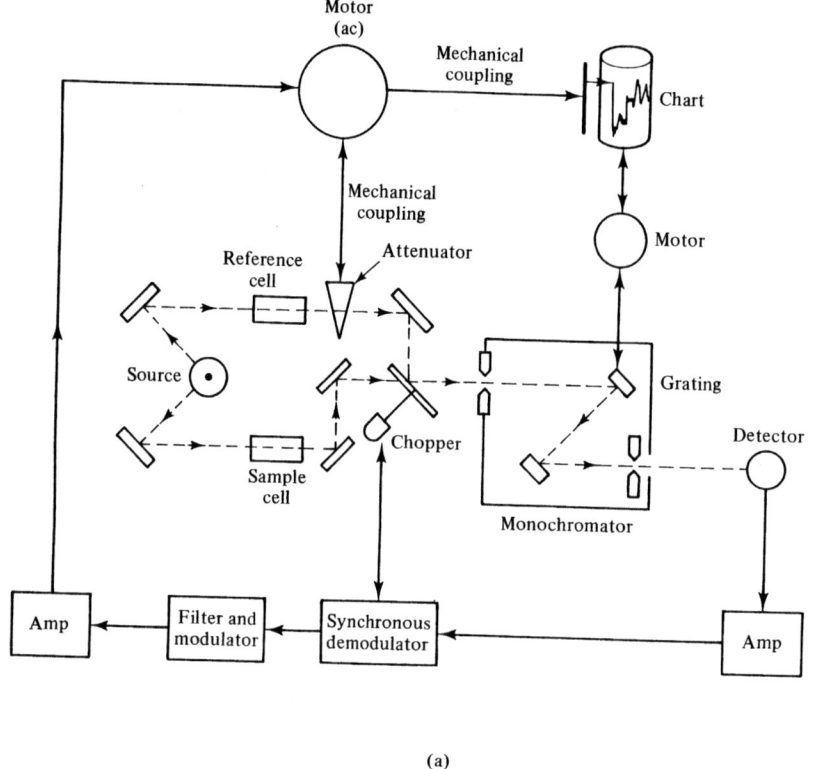

(a)

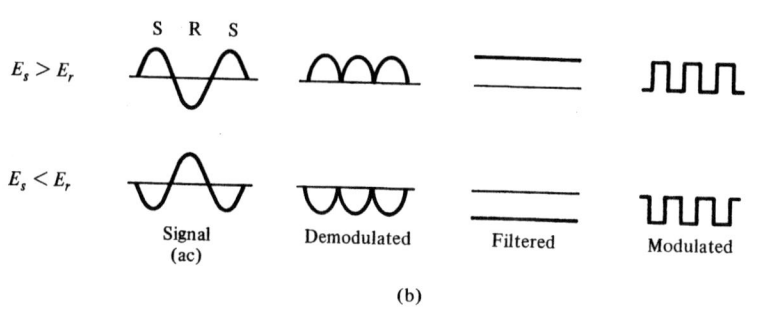

(b)

**FIGURE 14-5** Optical null double-beam spectrophotometer for infrared absorption measurements. In (a), the radiation from the source is split into two beams. One beam passes through the reference cell and the other through the sample cell. The beams are recombined and chopped prior to passing alternately through the monochromator and striking the detector. Any difference in the intensity of the beams causes an ac signal from the detector, which is amplified and synchronously demodulated (b). The sign of the dc output of the synchronous demodulator (lock-in amplifier) depends on the phase of the ac demodulator input, which in turn depends on which beam is more intense. The dc output can be further amplified to drive a dc motor which is connected to a beam attenuator and the pen on the recorder. Alternatively, the dc signal can be remodulated by an electrical chopper and applied to one winding of an ac motor. The other winding is attached to the same signal that drives the electrical chopper. In either case, if the reference beam intensity is higher than the sample beam intensity, as in the lower ac signal in (b), the motor turns in the appropriate direction to move the attenuator further into the reference beam and reduce its intensity. Similarly, if the sample beam intensity is higher, as in the upper ac signal in (b), the attenuator is pulled back so as to increase the reference beam intensity.

trometry were discussed in Sections 4-2 and 4-4, respectively. Nernst glowers and globars (see Table 4-2) are typically used as sources for general-purpose mid-IR instruments and can also be used in the far-IR region to wavelengths of $\approx 50$ μm. Some inexpensive mid-IR instruments use nichrome or rhodium wire sources electrically heated to $\approx 1100$ K. These are less intense than the globar or Nernst glower, but they have longer lifetimes. Tungsten filament lamps can be used in the near-IR region. In the very far-IR region, high-pressure mercury arc lamps can be used to wavelengths of $\approx 300$ μm, although they provide relatively low radiance.

Thermocouples are the most widely used detectors in dispersive mid-infrared instruments, although bolometers and Golay cells are sometimes used. For the far-IR region, the Golay (pneumatic) detector is

more efficient. In the near-IR region, photon energies are high enough that photon detectors can be used. Photoconductive cells, such as CdS, PbS, and PbSe, are often employed in this region. A multichannel IR spectrophotometer has been constructed that uses an IR spectrograph and a linear pyroelectric detector array. At this time such a system is still quite expensive.

Prisms were commonly used in older infrared instruments for wavelength isolation. Glass and quartz cannot be used because of excessive absorption at wavelengths longer than 3 μm. Hence prisms were made of ionic crystals such as NaCl, LiF or $CaF_2$. Unfortunately, with NaCl great care has to be taken to protect the prism from moisture since it is highly hygroscopic. Modern IR spectrophotometers employ reflection gratings that are similar to those used in the UV-visible region, although the groove spacing is greater (e.g., 120 grooves mm$^{-1}$). Reciprocal linear dispersion values in the tens of nanometers per millimeter range are common. Replica gratings are now available made from various plastics. To reduce the effect of overlapping orders and stray radiation, filters or a preceding prism are employed. Because a blazed grating is only efficient over a limited wavelength range, two or more gratings are often used with several filters to scan a wide region.

Lenses are rarely used in the IR region. Instead, mirrors are used to focus and collimate the radiation. Mirrors are generally made from Pyrex or another material with a low coefficient of thermal expansion. Front surfaces are usually coated with a vacuum-deposited thin metal film of aluminum, silver, or gold. Aluminum is the most widely used material, although silver and gold have higher reflectivities. Aluminized mirrors retain their reflectivities for long periods of time, while silver surfaces tend to tarnish when exposed to substances normally found in laboratory atmospheres.

Window materials are a particular problem in IR spectrometry. Windows are used for sample cells and to permit various compartments of the spectrophotometer to be isolated from the environment. Window materials must be transparent to IR radiation over the wavelength range of the instrument, and inert to the various chemicals they will contact. They must be capable of being shaped, ground, and polished to the desired optical quality.

Crystalline alkali halides, such as NaCl, KBr, CsBr, and CsI, are most widely employed as window materials (see Appendix B) because of their high transmittances in the IR region. The long-wavelength limits for many IR materials are listed in Table B-3. In the very far-IR region, quartz becomes transparent again at wavelengths greater than 50 μm. The Irtran series of window materials (see Appendix B) is prepared by sintering various types of powders under high pressure and temperature. These have nearly identical optical properties as single crystals of the same substance. Since water is present in nearly every organic solvent and as a vapor in the laboratory environment, the solubility of the window material in water is of particular interest. Water is also a desirable solvent for many IR studies. The water solubilities of common window materials is also listed in Table B-3.

*Features of Dispersive Spectrophotometers.* Dispersive IR spectrophotometers are available with a wide range of performance features and prices. Inexpensive instruments that cover the mid-IR region may have resolution figures in the range 1 to 3 cm$^{-1}$. More expensive systems often have options available for extending the range to the far-infrared region and may have resolution figures in the range 0.1 to 0.5 cm$^{-1}$. The optics of a commercially available, inexpensive infrared spectrophotometer are shown in Figure 14-6.

Newer IR spectrophotometers contain microcom-

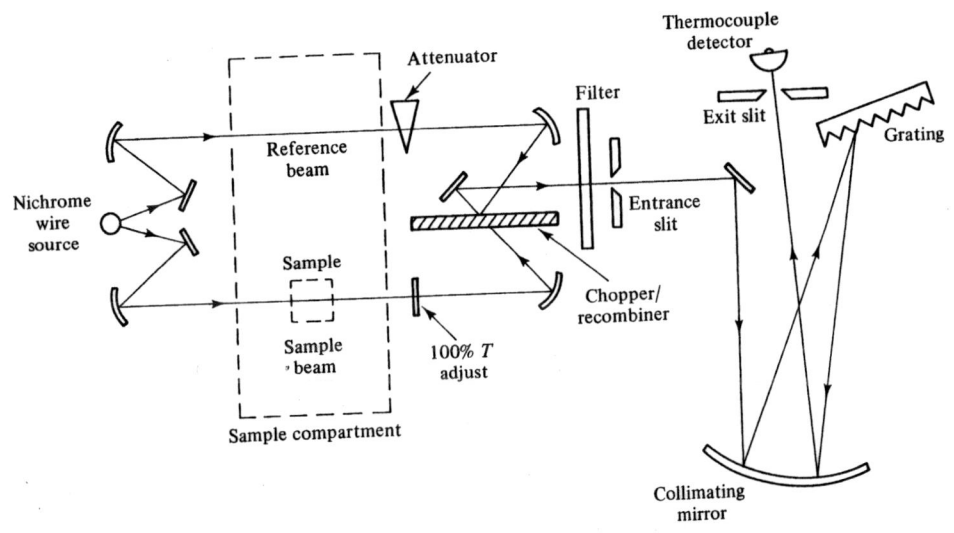

**FIGURE 14-6** Optical diagram of Beckman Acculab infrared spectrophotometer. This system is of the optical null type. Radiation from the nichrome wire source is split into two beams and sent through the sample and the reference cells. The reference beam attenuator is controlled so that the reference beam intensity matches that of the sample beam. The two beams are recombined and alternately pass through the monochromator to a thermocouple detector. The recorder pen is mechanically linked to the reference beam attenuator. (Courtesy of Beckman Instruments, Inc.)

puters to aid in measurement automation. In inexpensive instruments the microcomputer may store the slit program, provide wavelength calibration modes, control monochromator scanning, and calculate absorbance values. In addition to these features, more expensive instruments provide for storage of spectra, automatic background subtraction, and signal averaging. Software packages to aid in quantitative analysis and qualitative identification of compounds are available.

## Fourier Transform IR Spectrometers

Fourier transform infrared spectrometers are commercially available from several instrument manufacturers. Some of these are of limited wavelength coverage (e.g., far-IR or mid-IR range), while others are capable of obtaining spectra over an extended wavelength range (e.g., UV to far-IR region). The majority of the commercial instruments are based on the Michelson inter-

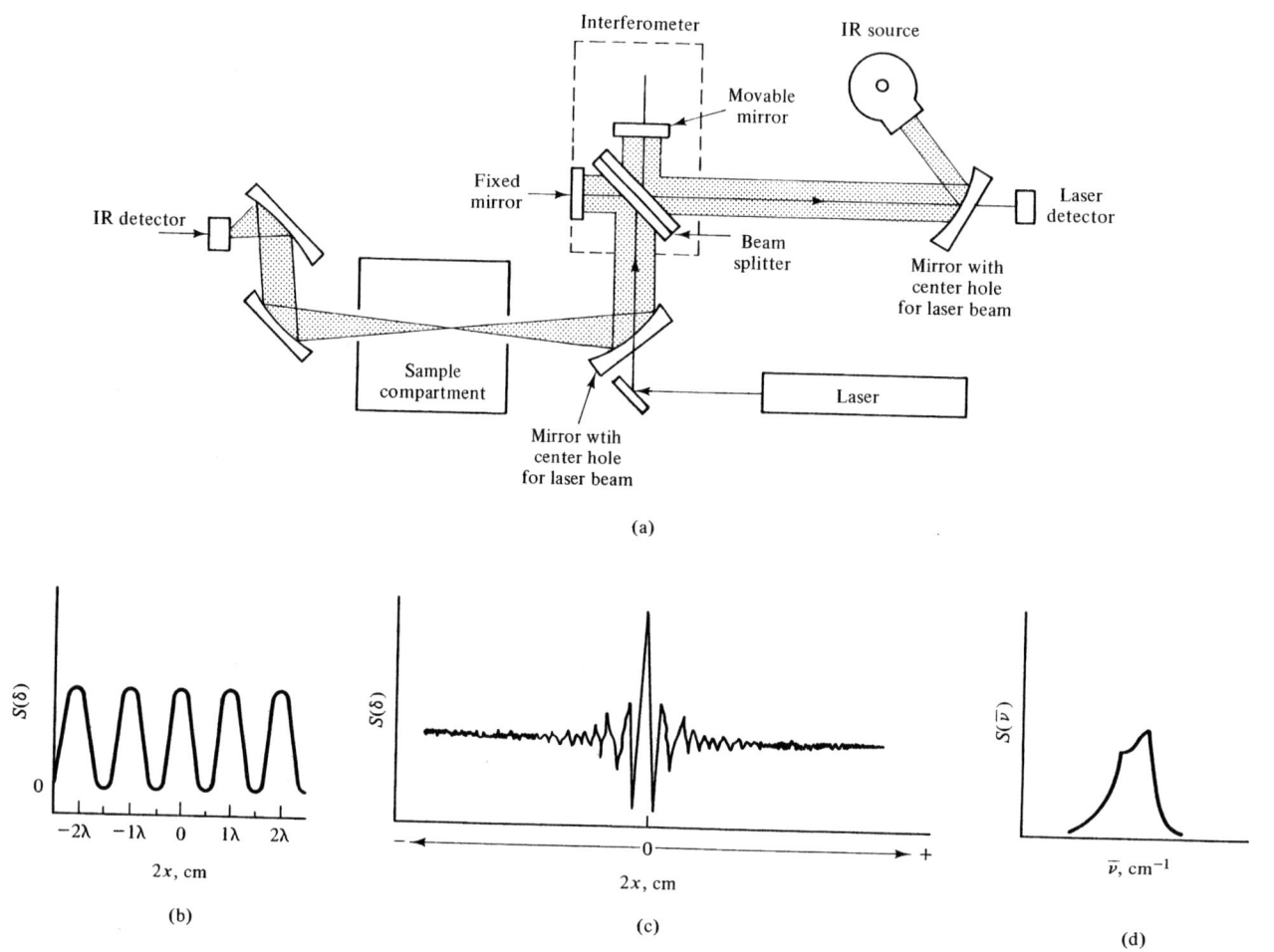

**FIGURE 14-7** Single-beam FTIR spectrometer of Michelson interferometer type. In one arm of the interferometer (a), the IR source radiation travels through the beam splitter to the fixed mirror back to the beam splitter through the sample and to the detector. In the other arm, the IR source radiation travels to the beam splitter to the movable mirror, back through the beam splitter to the sample and to the detector. The difference in pathlengths of the two beams is the retardation $2x$. A He-Ne laser is used as a monochromatic reference source. The laser beam is sent through the interferometer in the opposite direction to that of the IR beam. The interference pattern observed at the laser detector is shown in (b). It is a cosine wave that varies from maximum amplitude to minimum amplitude every $\frac{1}{4}\lambda$ movement of the movable mirror, which corresponds to a pathlength difference of $\lambda/2$. The interferogram in (c) corresponds to that of a continuum source. The spectrum (d) of the continuum source is obtained after Fourier transformation.

ferometer discussed in Section 3-6. The reader may wish to review this section before proceeding. We consider here some of the design and performance features of FTIR spectrometers.

*Interferometer Systems.* Instruments for far-IR absorption spectrometry are most often single-beam designs, while mid-IR spectrometers are most often double-beam systems. Figure 14-7 shows a typical arrangement for a single-beam FTIR system. Note that the sample is placed after the Michelson interferometer. Because of the modulation characteristics of interferometry, infrared emission and stray radiation from the sample compartment appear as a dc component of the interferogram and are inherently discriminated against. Since a computer system is a necessary part of the FTIR spectrometer, single-beam systems acquire, calculate, and store the reference spectrum, usually versus air as a blank. Later the sample spectrum is obtained and true sample absorbances are then calculated.

A double-beam spectrometer for the mid-IR re-gion is illustrated in Figure 14-8. The mirrors directing the interferometer beam through the sample and reference cells are oscillated rapidly compared to the movement of the interferometer mirror so that sample and reference information can be obtained at each mirror position. The double-beam design thus provides compensation for source and detector drifts.

Moving the interferometer mirror without any tilt is a particular problem in FTIR systems. In the far-IR region, where wavelengths are long, mirror drive mechanisms can be relatively crude. Often the mirror is positioned by means of a motor-driven piston which moves through a cylinder with normal grease as a lubricant. For mid- and near-IR operation, however, a more precise and sophisticated drive mechanism is required. Usually, the mirror mount is floated on an air cushion held within stainless steel sleeves and driven by a voice coil similar to that in a loudspeaker. Very high resolution systems require tilt-compensation devices and some even stop the mirror at each sampling point to adjust its planarity prior to measurement.

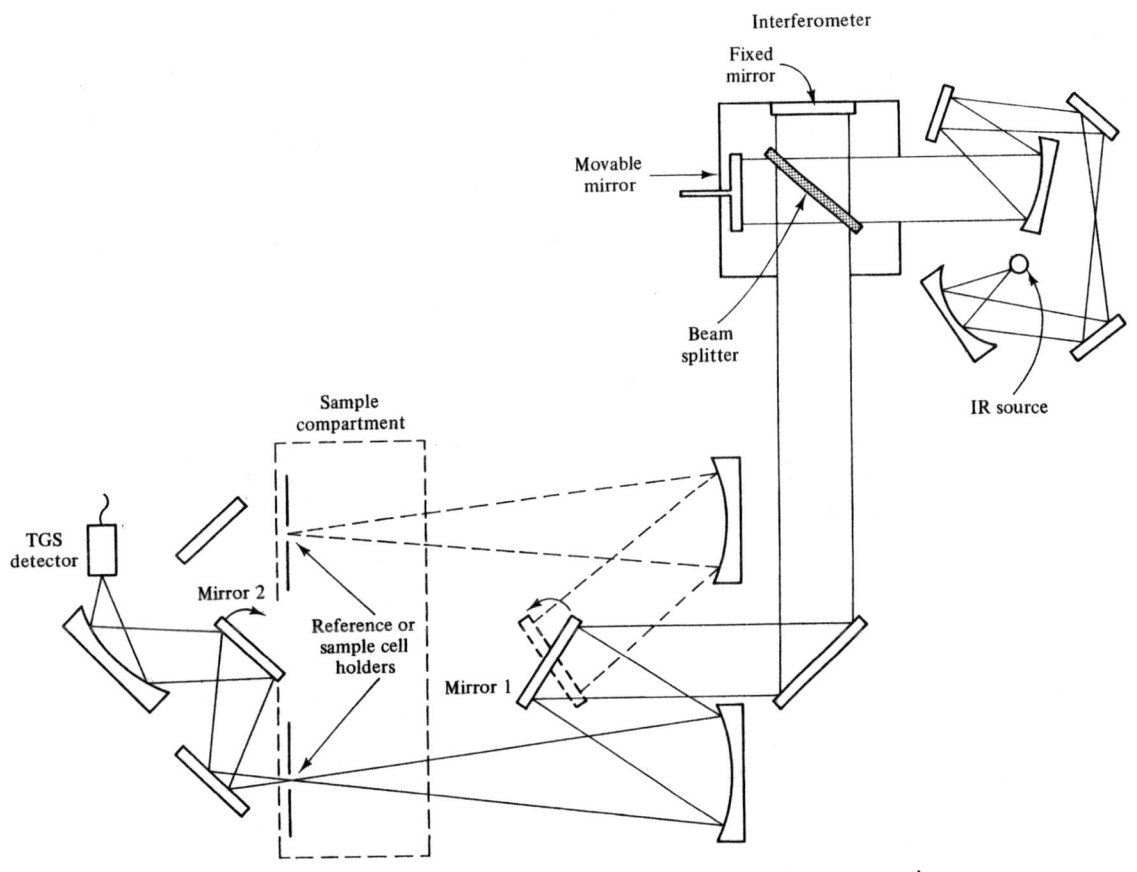

**FIGURE 14-8** Double-beam FTIR spectrometer. The beam emerging from the interferometer strikes mirror 1, which, in one position, directs it through the reference cell, and in the other position, directs it through the sample cell. Mirror 2, which is synchronized to mirror 1, alternately directs the reference beam and the sample beam to the detector.

Modern FTIR systems also provide for automatic location of the zero path difference position and for sampling of the interferogram at precisely spaced intervals. This is accomplished by the three-interferometer system shown in Figure 14-9. The triple interferometer design yields very high precision in determining IR wavelengths. Data collection at exactly the same mirror position on every scan allows coherent signal averaging for *S/N* improvement.

*Components of FTIR Spectrometers.* Sources for FTIR are identical to those used in dispersive IR spectrophotometers. Thermal detectors are, however, too slow in response for the modulation frequencies normally obtained. For far-IR spectroscopy, the Golay pneumatic detector and the triglycine sulfate (TGS) pyroelectric detector are most common. Only the py-

roelectric detector has sufficient response time for rapid-scanning applications.

Pyroelectric detectors are also widely used in mid-infrared instruments. Although thermocouples have $D^*$ values an order of magnitude larger than pyroelectric detectors (see Table 4-5) at low modulation frequencies, the TGS detector has better responsivity at high modulation frequencies (e.g., 1 kHz). Mercury-doped germanium bolometers and silicon bolometers have been used for very high sensitivity mid-infrared measurements.

In the far-infrared region, beam splitters are most commonly made from thin films of Mylar (polyethylene terephthalate). Unfortunately, Mylar strongly absorbs in the mid-IR region and is not mechanically strong enough to make into films of the necessary thickness ($<1$ μm). Thus mid-infrared beam splitters are normally thin films of germanium or silicon deposited on

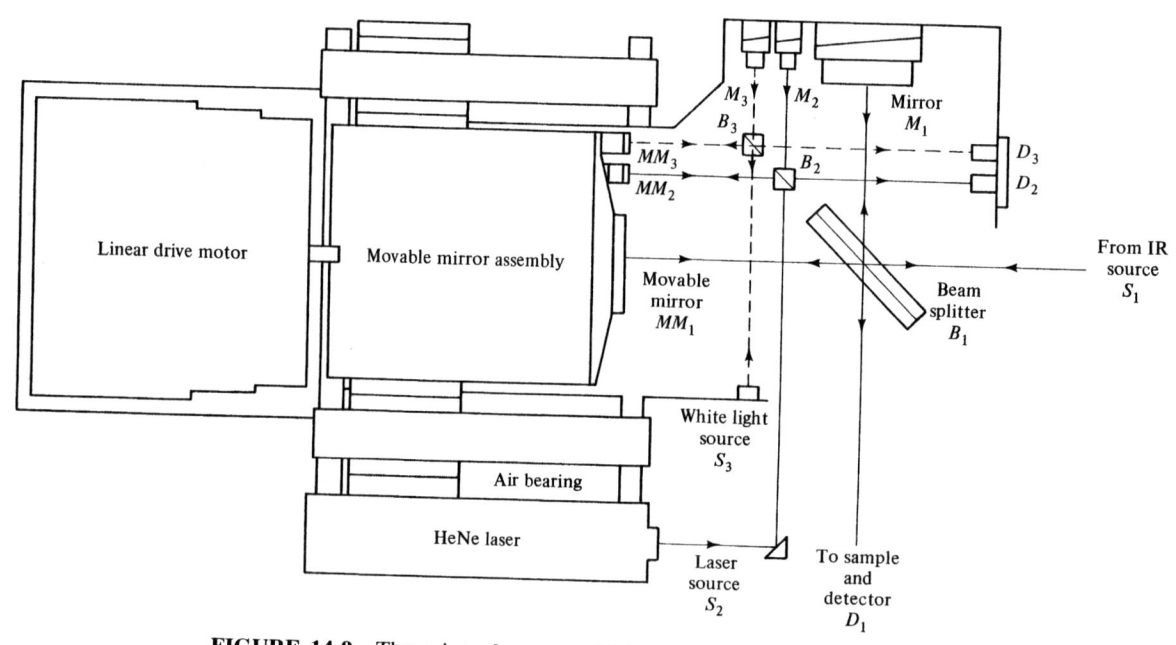

**FIGURE 14-9** Three-interferometer FTIR spectrometer. The main interferometer consists of the components labeled with subscript 1. Interferometer 2 (subscript 2) is a laser-fringe reference system, while interferometer 3 is a white-light interferometer. The movable mirrors (*MM*) of all three interferometers are driven by the same mirror drive assembly. Interferometer 2 has a He-Ne laser source. The interferogram of this monochromatic source is a single-frequency cosine wave. Sampling of the interferogram is done on each zero crossing of the reference interferogram. The white-light interferometer uses a polychromatic source which gives an interferogram with a very intense output at zero path difference, since this is the only point where the cosine waves from all the different frequencies are in phase. The maximum in the white-light interferogram is used to signal the start of data acquisition for each mirror scan. The zero path difference position of the white-light interferometer is normally displaced slightly with respect to the main interferometer so that sampling begins shortly before the maximum in the main interferometer.

flat alkali halide substrates. A layer of $Fe_2O_3$ on quartz or $CaF_2$ has been used in the near-IR region.

Computer system requirements for FTIR spectrometers vary dramatically with the wavelength coverage and the resolution desired. Low-resolution ($5\ cm^{-1}$) far-IR spectrometers normally acquire less than 1000 points per interferogram, whereas moderate resolution ($0.5\ cm^{-1}$) mid-IR systems may acquire as many as 32,000 points. The recent dramatic decrease in the price of computer memory and disk storage has reduced the expense of the computer system for FTIR spectrometers.

*Features of Commercial FTIR Systems.* There has been a clear trend in recent years toward the development of low-cost FTIR instrumentation. The least expensive FTIR systems are intended for use in one spectral region, such as the mid-IR or far-IR. These are often priced to compete with the lower-priced dispersive systems (e.g., $12,000 to 15,000). Typically, these benchtop systems provide resolution values of a few wavenumbers. More versatile, and more expensive, instruments are available to cover a wider wavelength range and provide higher resolution. One commercial instrument system, for example, can produce spectra from the UV to the far-IR region with a few changes in sources, detectors, and beam splitters. The resolution provided varies from $0.5\ cm^{-1}$ with moderate expense systems to less than $0.1\ cm^{-1}$ with research-quality instruments.

In the far-IR region, Fourier transform spectrometers are clearly superior to conventional, dispersive systems in terms of $S/N$, resolution, and scanning time. In the mid-IR region, FTIR systems allow the spectrum to be obtained with higher $S/N$ for comparable analysis time compared to dispersive systems. Conversely, the spectrum can be obtained in a much shorter time with nearly equivalent $S/N$. For example, a moderate resolution (few wavenumbers) spectrum can be obtained with comparable $S/N$ to a conventional IR spectrum in a few seconds. Fourier transform IR clearly shows superior characteristics for applications where transient species or time-varying samples are involved (kinetic studies or GC/IR), for measurements where energy is limited (far-IR region or highly absorbing samples), for measurements at very high resolution, and for measurements where excellent wavenumber reproducibility and accuracy are required. Wavenumber reproducibility is particularly important for accurate background subtraction.

In recent years, there has been an emphasis on increasing the versatility of FTIR systems through providing new accessories for spectrometers. Most FTIR manufacturers now provide microsampling accessories

(see the next section) and various reflectance attachments (see Section 14-7). Some even have accessories for such new techniques as FTIR photoacoustic spectrometry (see Section 17-1).

## Nondispersive IR Instruments

There are many routine applications of infrared spectrometry where it is desired to obtain quantitative information about one constituent or a few species in a mixture instead of obtaining a complete spectral scan. Such applications include monitoring a single constituent in a process stream, and atmospheric gas analyses where a few important gases are routinely determined. For these types of determinations, the dispersive spectrophotometers and FTIR systems just described are overly complex, too expensive, and often not rugged enough. Single-beam, nondispersive spectrometers can be used in such applications for accurate quantitative determinations. Among the nondispersive spectrometers used for these purposes are filter photometers, filter-wedge spectrophotometers, and infrared gas analyzers. We consider here each of these nondispersive spectrometer types.

*Filter Photometers.* Several simple IR photometers are available that use filters for wavelength selection. Figure 14-10 shows an example of a commercial IR gas analyzer designed around a filter photometer.

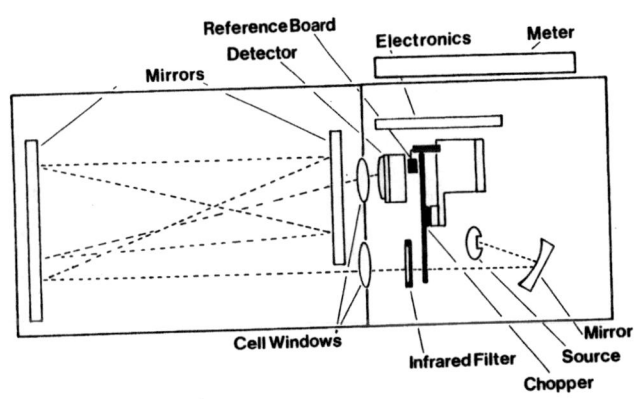

**FIGURE 14-10** Portable, infrared filter photometer for gas analysis. A nichrome wire source is used with a pyroelectric detector. Interchangeable interference filters are available for the determination of specific gases. Gaseous samples are introduced into the cell by means of a battery-powered pump. A folded-pathlength design allows the cell pathlength to be set from 0.5 to 13.5 m at the factory. (Courtesy of the Foxboro Company.)

Interference filters are available at wavelengths that correspond to transitions in industrially important gases.

Filter photometers with long-pathlength cells, such as the instrument shown in Figure 14-10, can detect a few parts per million of the selected compound. Among the gases that can be detected with readily available filters are phosgene, hydrogen cyanide, acrylonitrile, carbon monoxide, and chlorinated hydrocarbons. The instrument is battery operated and weighs less than 20 lb. Since the filters are expensive and somewhat inconvenient to change, this instrument is best suited for determination of a single constituent or for infrequently monitoring a few constituents.

*Dielectric Filter Spectrometers.* Circular wedge dielectric filters are available (see Appendix C) in which the transmitted frequency varies continuously and uniformly on a disk-shaped substrate. These dielectric filters can be used like a monochromator to select wavelengths. The spectrum can be scanned by slowly rotating the disk. A spectrometer based on such a filter is shown in Figure 14-11. With microprocessor control over wavelength selection and data collection, the spectrometer of Figure 14-11 can be programmed to determine absorbances at multiple wavelengths. Calibration data (absorptivities, slopes of working curves, etc.) can be stored in computer memory and used to determine the concentrations of the various components automatically. Note that the sample area allows solid, liquid, or gaseous samples to be used. Such spectrometers are considerably more expensive than fixed-wavelength filter instruments, but much more versatile.

*Infrared Gas Analyzers.* There are several photometers available that employ no wavelength selection device at all, as illustrated in Figure 14-12. These nondispersive gas analyzers are very simple, rugged, and inexpensive. Selectivity is achieved by filling the detector compartment with the analyte gas so that the detector absorbs and responds only to radiation in the wavelength region absorbed by the analyte gas.

Another type of gas analyzer uses a double-beam arrangement. Radiation in the reference beam passes through both the sample cell and a reference cell filled with the analyte gas. Radiation in the sample beam passes through the sample cell only. The two beams strike bolometers that are arms of a Wheatstone bridge. The null balance point of the bridge is related to the analyte concentration. Interferents can be troublesome in the latter approach if they have absorption spectra that overlap that of the analyte. Interference effects can be minimized by putting the interfering substance at high concentration in a cell that precedes the sample cell. This removes radiation at the wavelengths that the interferent can absorb.

*Special-Purpose Spectrometers.* Not surprisingly, tunable IR lasers have been used to make special-purpose IR spectrometers. The lasers that have been used at present are diode lasers and parametric oscillators (see Section 4-3). Diode lasers have very narrow bandwidths (e.g., 1 to $10^{-6}$ cm$^{-1}$) and relatively high irradiances, but their tuning ranges are quite narrow (20 to 50 cm$^{-1}$). The parametric oscillator is tunable over a wider range, but produces a broader spectral

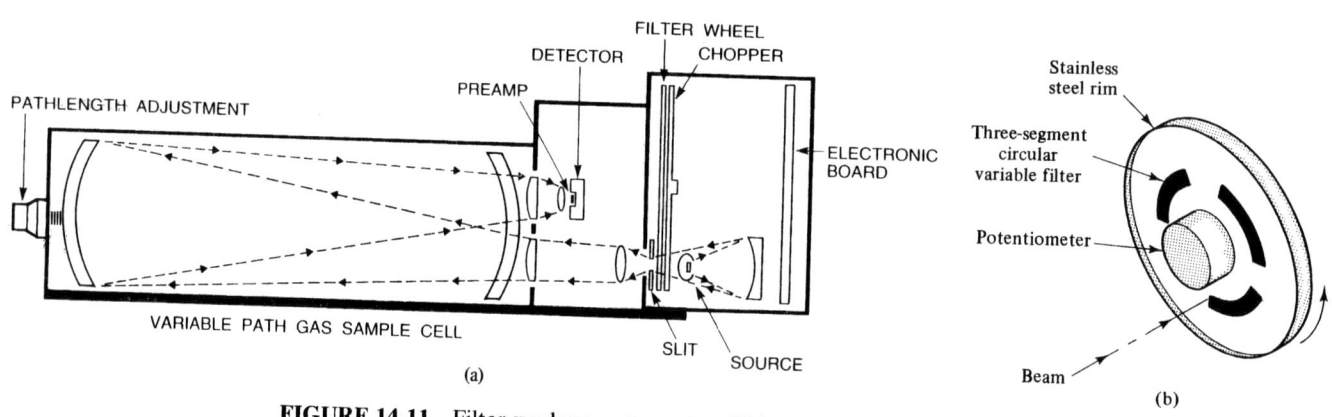

(a)

(b)

**FIGURE 14-11** Filter-wedge spectrometer. This spectrometer (a) covers the region 2.5 to 14.5 μm. The wavelength selection device is a circular variable filter wheel (b). A microprocessor controls the motor that turns the filter wheel and thus determines the wavelength. The system has scanning capabilities. The cell pathlength is adjustable in the range from 0.75 to 20.25 m. An integral air pump allows ready introduction of samples. [(a), Courtesy of the Foxboro Company]

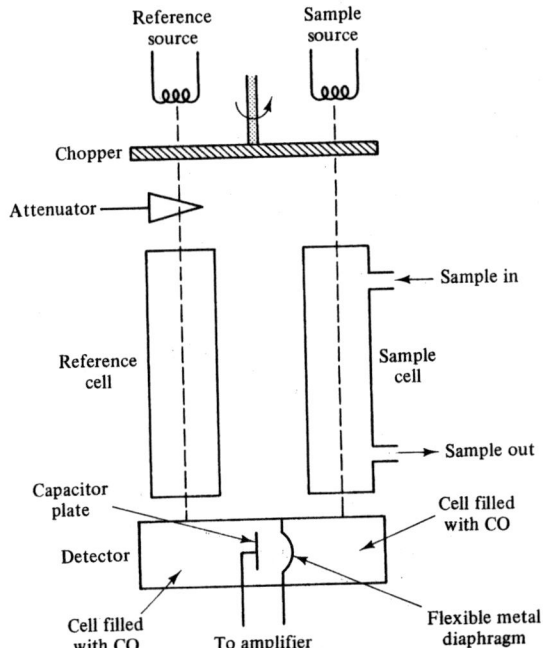

**FIGURE 14-12** Nondispersive gas analyzer. The reference cell is filled with a nonabsorbing gas (CO in this example), while the sample flows through the sample cell. Selectivity is achieved by filling both compartments of the detector cell with the gas being analyzed. A thin metal diaphragm separates the two sections and serves as one plate of a capacitor. If the analyte gas is absent in the sample cell, both compartments are equally heated by the two sources of infrared radiation. If the sample cell contains the analyte gas, the beam on the right is attenuated at the specific wavelengths where the detector gas absorbs. Thus the right-hand detector compartment becomes cooler than the left, and the diaphragm moves to the right. Since the capacitance of two plates, separated by an insulator, depends on the distance between them, the capacitance changes whenever there is analyte gas in the sample cell. This capacitance change can be measured electronically or used to drive a beam attenuator into the reference beam so that its intensity matches the sample beam.

output. At present such laser-based spectrometers are limited to fixed-wavelength applications.

Special-purpose spectrometers for detecting single constituents such as an important atmospheric gas have been constructed using IR LEDs as "pseudo-mono-chromatic" sources. Spin-flip Raman lasers have also been used. These are pumped by a $CO_2$ laser and tuned by varying the magnetic field strength. This laser can be tuned over more than 800 cm$^{-1}$. Spectral bandwidths < 0.05 cm$^{-1}$ have been achieved. A spin-flip Raman

laser spectrometer has been mounted in a balloon and used to study NO concentrations in the stratosphere.

## 14-3  SAMPLE PREPARATION TECHNIQUES

The preparation of samples for infrared spectrometry is often the most challenging task in obtaining an IR spectrum. Since almost all substances absorb IR radiation at some wavelengths, cell window materials, cell pathlengths, and solvents must be carefully chosen for the wavelength region and sample of interest. However, infrared spectroscopists have developed a large number of sample handling techniques that allow spectra to be obtained on a wide variety of sample types, including gases, pure liquids, solutions, films, or solids.

### Gas Samples

A gas sample cell consists of a cylinder of glass or sometimes a metal. The cell is closed at both ends with an appropriate window material and equipped with valves or stopcocks for introduction of the sample. For routine spectra, cells with ≈ 10 cm pathlength are commercially available. Long-pathlength cells can be used to study dilute or weakly absorbing samples. Multipass cells are, however, more compact and efficient and are often used instead of long-pathlength cells. With multipass cells, mirrors are used so that the beam makes several passes through the sample before exiting the cell. Effective pathlengths of greater than 10 m are readily achieved.

When it is desired to resolve the rotational structure of the sample, the spectrum must be taken at reduced pressure. Hence the cell must be capable of being evacuated. For quantitative determinations with light molecules, the cell is sometimes pressurized in order to broaden the rotational structure and allow simpler measurement.

### Liquid Samples

Liquid samples are often used in IR spectrometry. Most often these are solution samples, but occasionally they are pure liquids or even gases dissolved in liquid solvents.

*Pure Liquids.*  Infrared spectra are often taken on pure liquid samples, particularly when sample amounts are limited or suitable solvents are not available. With pure liquids, the sample is most often in the form of a thin film, which has a sufficiently short pathlength to give reasonable absorbances.

The most common technique to obtain qualitative spectra with nonvolatile liquids is to place a drop of the

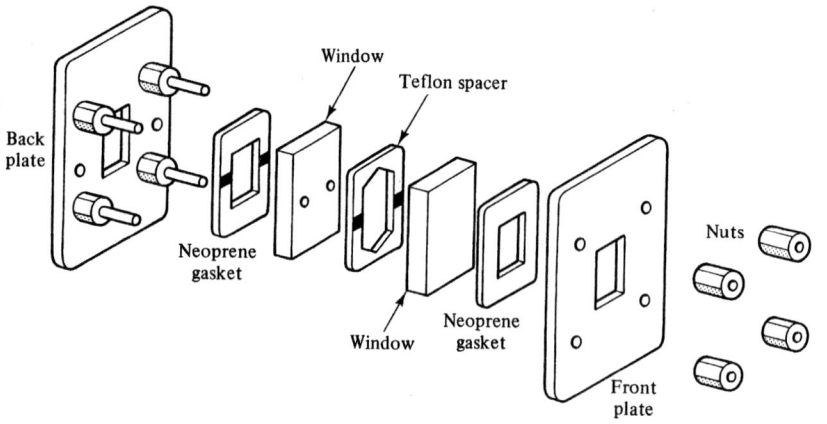

**FIGURE 14-13** Demountable liquid cell for infrared spectrometry. Teflon spacers are used to vary the cell pathlength in the range 0.015 to 1 mm.

pure liquid between two NaCl plates which are then clamped together in a demountable cell such as that shown in Figure 14-13. Spacers can be used to increase the pathlength over that formed by capillary action. Obviously, the thickness cannot be very precisely controlled by this technique, and hence it is used only for qualitative studies.

*Solution Samples.* The first problem when using solution samples for IR spectrometry is to find a suitable solvent. As shown in Figure 14-14, all common solvents absorb in the IR region, but there are wavelength regions where each solvent is somewhat transparent. Of course, solvent transparency is not the only concern. The sample must also be soluble to the desired extent in the solvent and the solvent must be chemically inert toward the sample. For qualitative analyses solution concentrations $\approx 10\%$ are usually satisfactory.

The most commonly used solvents are carbon tetrachloride, carbon disulfide, and chloroform. Water is infrequently employed, not only because it is a strong infrared absorber in certain regions (see Figure 14-4), but also because it dissolves the most common window

materials. Organic solvents must also be dry if water-soluble window materials are employed. When it is necessary to use aqueous solutions, as it often is with inorganic and biological samples, the insoluble window materials listed in Table B-3 must be used. Fourier transform IR techniques are often used with aqueous solutions, because excellent compensation for solvent absorption can be obtained.

To obtain spectra for the qualitative analysis of nonvolatile solutions, a demountable sample cell, such as that shown in Figure 14-13, is often used. For volatile samples or when the cell thickness must be controlled as in quantitative methods, sealed sample cells are necessary. Sealed cells are either of fixed pathlength design or variable in pathlength. Sealed cells are available with pathlengths of 0.1 to 1 mm. For quantitative analysis this allows the determination of many species in the concentration range 0.05 to 10%.

The thicknesses of narrow path IR cells are often determined by an interference fringe method. The transmittance spectrum of an empty cell is recorded against air in the reference beam. The reflected radiation from the two walls of the cell interferes with the transmitted beam to produce an interference pattern similar to that shown in Figure 14-15; note that the cell is essentially a Fabry–Perot cavity. If $n$ interference maxima are observed in a wavenumber region $\Delta\bar{\nu}$, the cell thickness $b$ can be calculated from

$$b = \frac{n}{2\Delta\bar{\nu}} \qquad (14\text{-}6)$$

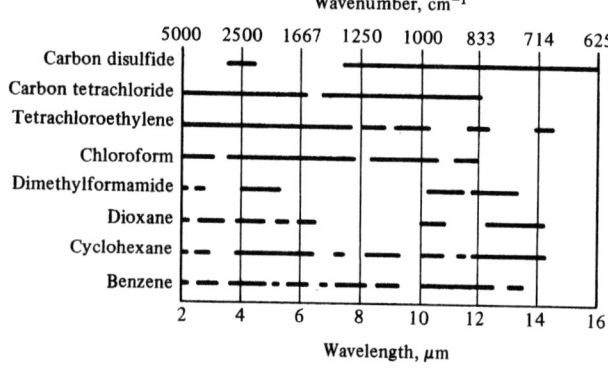

**FIGURE 14-14** Regions of transparency for common infrared solvents. The horizontal lines indicate regions where the solvent transmits at least 25% of the incident radiation in a 1-mm cell.

as shown in Figure 14-15. Measurements should be made with flat, freshly polished windows to obtain clean fringes. Cell thicknesses should be measured fairly often for quantitative work since erosion of the window materials can cause thicknesses to change. Interference fringes are not normally seen in cells that are filled with liquid because the refractive index difference between the

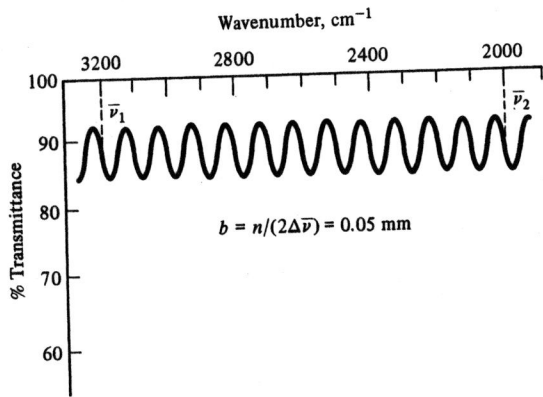

**FIGURE 14-15** Interference fringe method for determination of cell thickness. Note that 12 interference maxima are found in the region 3200 to 2000 cm$^{-1}$. The cell pathlength is thus 6/1200 cm or 0.05 mm.

window material and the liquid is small, which reduces the amount of radiation reflected at the interface.

## Solid Samples

In cases where suitable solvents do not exist or where it is too time consuming to search for an appropriate solvent, solid samples can be used in the form of mineral-oil mulls or as KBr pellets. In both cases the sample must be finely ground so that the particle size is smaller than the wavelength of the IR radiation. If this is not the case a significant fraction of the incident radiation can be lost to scattering processes.

*Mulls.* The mineral oil that is most often used for mulling is Nujol, a highly refined mixture of saturated hydrocarbons. Nujol shows strong absorption in the region near 3000 cm$^{-1}$ due to its C—H stretching vibration and absorption in the 1400-cm$^{-1}$ region due to its C—H bending vibrations; elsewhere, it is relatively transparent. If Nujol absorption is severe in a region of interest, chlorinated (hexachlorobutadiene) or fluorinated (Fluorolube) oils can be used.

Mulls are normally prepared by grinding a few milligrams of the powered sample with a mortar and pestle or with pulverizing equipment. A few drops of the mineral oil are then added. Grinding is continued in the presence of the oil until a smooth paste is obtained. Each particle should be separately coated with a thin layer of liquid. A small amount of the resulting paste is then spread between two NaCl plates, and the spectrum of the thin layer obtained. With too small an amount of liquid, the particles are not evenly coated and separated, and scattering results. With too much oil, the absorption bands of the liquid will be too strong.

Satisfactory mulls are often obtained with a good deal of practice.

*Salt Pellets.* In the KBr pellet method, a milligram or less of the finely ground ($< 2 \mu m$ particle size) sample is mixed with $\approx 100$ mg of dry KBr powder in a mortar, ball mill, or miniature ball mill (e.g., a Wig-L-Bug). The mixture is then compressed in a die to form a transparent pellet, and the pellet is mounted in a suitable holder. Pellets that are made in simple dies tend to become cloudy because of air entrapment. With more sophisticated pellet-making facilities, the die assembly can be evacuated to remove entrapped or occluded air in the mixture. Properly made pellets are quite clear and the KBr is transparent in the infrared region out to $\approx 25 \mu m$.

There are numerous problems associated with the KBr pellet technique. Many substances tend to react with KBr under pressure or even while mixing. Thus, with unknown samples it is usually wise to obtain a spectrum of the material in a mull as well for comparison purposes. In addition, KBr is quite hygroscopic and the spectra obtained are difficult to reproduce. The latter problem results from the difficulty of obtaining uniform particle sizes during grinding and from variations in pressure from one pellet to another.

While mulls and pellets are satisfactory for qualitative analysis, neither technique is well suited for quantitative analysis. Since the thickness of mulls is not readily determined, an internal standard is almost always used for quantitative determinations. Internal standards are also used in quantitative analysis with pellets, although these can be used directly if the thickness is known and the sample and the KBr are carefully weighed.

The infrared spectra of some solid samples can also be obtained by infrared internal reflection spectroscopy, as discussed in Section 14-7.

## Microsampling Devices

With normal infrared techniques it is easy to obtain spectra for qualitative analyses on a few milligrams of material. Quantitative analyses can be accurately carried out with 20 mg or more of sample. In cases where smaller samples must be analyzed, special micromethods are employed.

For liquids, special microcells are available in which there is a minimum of dead space in the cell; nearly all the liquid is contained between the windows. Sample volumes on the order of 1 to 10 $\mu L$ can provide reasonable spectra.

For very small solid samples (e.g., $< 1$ mg) a beam-condensing system is needed to concentrate the

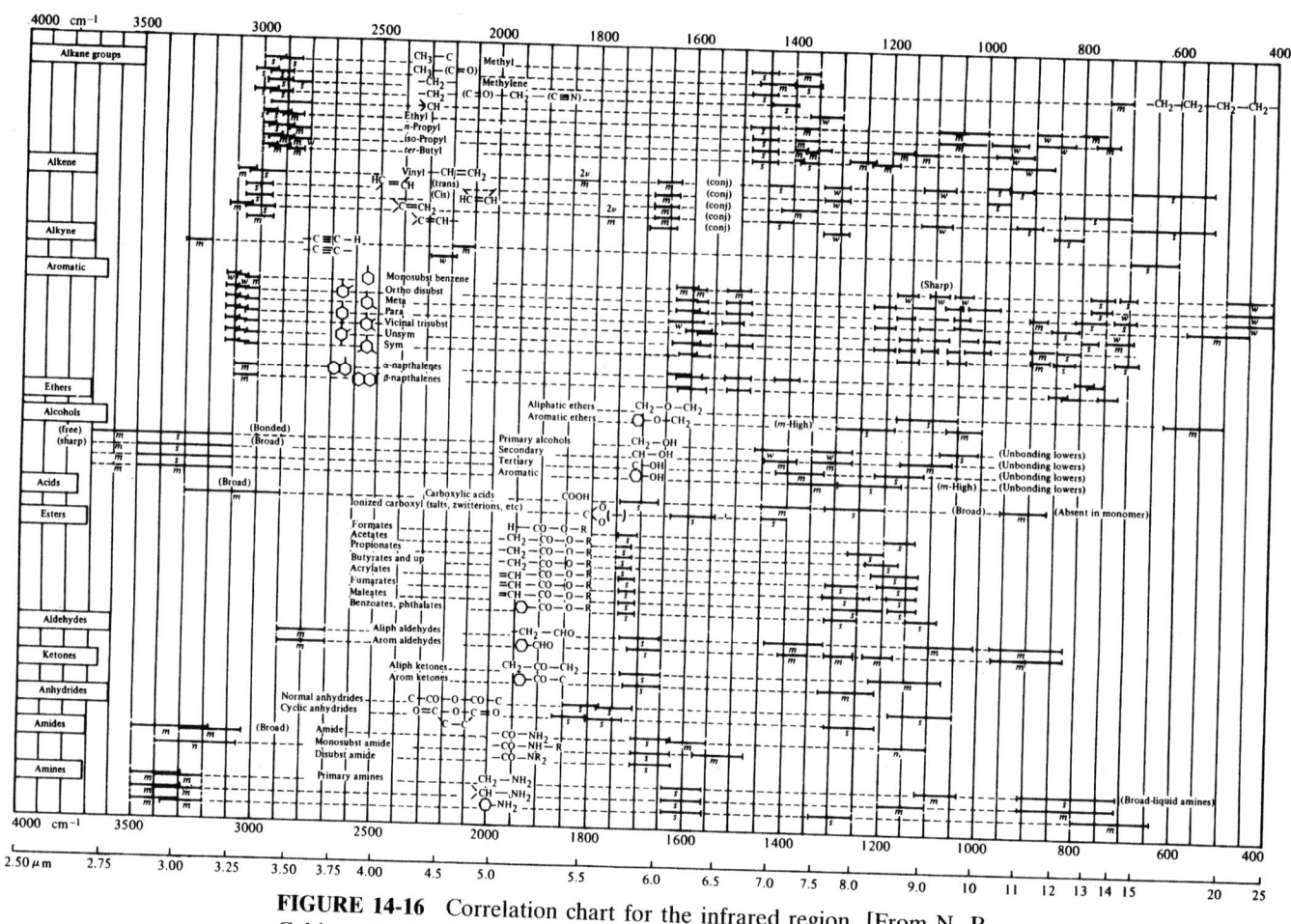

**FIGURE 14-16** Correlation chart for the infrared region. [From N. B. Colthup, *J. Opt. Soc. Am., 40*, 397, (1950). With permission.]

IR beam through a small aperture. Commercial beam-condensers are available. The simplest arrangement uses KBr or AgCl lenses mounted on a bed that can be inserted into the sampling area of a normal spectrometer. An off-axis elliptical mirror microscope is also available commercially. Mirror systems avoid the chromatic aberrations that plague lens systems. With such attachments, solid microsamples can be handled by preparing minute pressed disks. Pellets as small as 0.5 mm in diameter have been used to obtain spectra. Nonvolatile liquids can be dissolved in a volatile solvent and the solution used to wet finely ground KBr powder. The volatile solvent is then removed and a microdisk is pressed. Volatile liquids are transferred into suitable microcells that fit in the beam-condensing system.

**Infrared microscopy** is a technique that is undergoing rapid growth after remaining stagnant for about 25 years. Nearly every FTIR manufacturer makes a microscope attachment for their spectrometer. In addition, stand-alone systems are available based on dispersive spectrometers. Most of these allow visible lighting of the sample so that precise alignment can be done

through the microscope. Most infrared microscopes allow collection of data in either the transmittance mode or the reflectance mode (see Section 14-7). Spectra can be obtained on samples, such as fibers, with diameters as small as 10 μm.

## 14-4 QUALITATIVE ANALYSIS AND STRUCTURE DETERMINATION

The major uses of IR spectrometry are for qualitative identification of compounds and for the determination of molecular structure. Both of these applications rely on quite similar methodology. In structure determinations, it is important that the compound being studied be a *pure* compound so that the correct interpretation of the IR spectrum can be made. Information from IR spectrometry is normally combined with that from several other techniques (e.g., mass spectrometry, NMR spectrometry) to deduce the structure. For qualitative identifications, samples need not be pure compounds, but can be relatively simple mixtures. Information con-

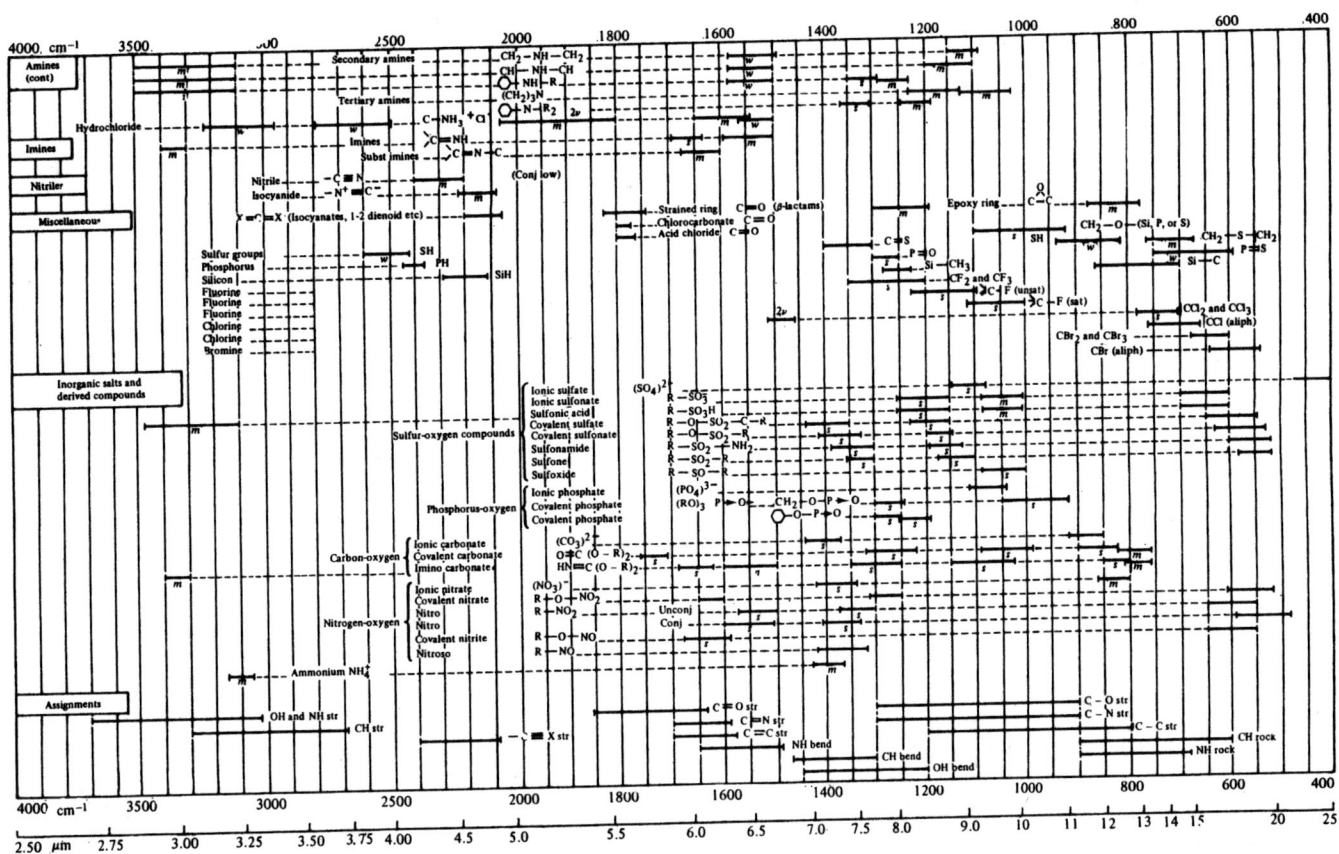

**FIGURE 14-16** *(continued)*

cerning functional groups in the unknown molecule can be obtained from the IR spectrum and used in conjunction with spectra of known compounds to aid in identifying the unknown compound. Since the procedures used in identifying compounds and in obtaining structural information are essentially identical, both applications are considered here.

### Correlation Charts

Qualitative applications of IR spectrometry begin with identifying the functional groups in molecules responsible for infrared absorption. The group frequency region (see Figure 14-3) is the most useful for this purpose. Because of the widespread use of IR spectrometry for qualitative purposes, a good deal of empirical information is available concerning group frequencies. The most comprehensive of these are **correlation charts** such as that shown in Figure 14-16. This type of chart not only contains all the group frequency information, but also is organized according to the type of compound in which the group is found.

Usually, after the spectrum is obtained, the strongest absorption bands are selected, and the wave-

numbers or wavelengths of the absorption maxima are identified. A correlation chart or a similar aid is then used to attempt to identify the groups responsible for the bands. After the strongly absorbing bands have been assigned, identification of the moderate and weak absorption bands is then attempted. A complete assignment of all the bands in the spectrum, even for pure compounds, is usually impossible because of the presence of combination and overtone bands.

For qualitative analysis, final identification is most often done by comparison of the spectrum obtained to reference spectra available in library collections (see below). Any additional information, such as boiling point, melting point, and other spectroscopic data, is frequently of great assistance in an unambiguous identification. For structural determinations, the IR spectrum is often only one piece of evidence among several. Infrared spectral evidence is frequently strengthened by calculations of the number of expected infrared active modes based on assumed structures. Such calculations use the symmetry of the molecule and the techniques of group theory or normal coordinate analysis. Any ambiguities that arise in interpretations can often be resolved with the aid of mass spectrometry, NMR spectrometry, Raman spectrometry, isotopic substitution methods, or a variety of other techniques.

As implied above, it is seldom possible to establish the identification or structure of a compound from correlation charts alone. Instead, such charts provide indications that a certain group is present or absent in a molecule and thus guidance as to what to select for more detailed study. Ambiguities can arise because of overlapping group frequencies, combination and overtone bands, the sample preparation technique, and instrumental factors.

## Spectral Collections and Search Systems

Extensive collections of infrared spectra are available from several sources for use in spectral matching and searching procedures. These collections are now so large ($> 100,000$ spectra) that special retrieval techniques must be used. In the simplest of these, edge-punched cards are manually sorted to obtain the spectrum of interest. Infrared data are presented in these collections in several different formats. For manual searching, replica spectra are presented in notebook format. Spectra are also available on microfilm and microfiche.

In recent years, computer storage of reference IR spectra has become popular, due to the speed with which computer searches can be accomplished. In the Sadtler system, the library compounds are coded first according to the position of the strongest absorption band. Next, the region from 4000 to 2100 cm$^{-1}$ is divided into ten 200-cm$^{-1}$ subregions, and all bands with % $T$ values $< 60$ are coded by position. Finally, the region from 2100 to 400 cm$^{-1}$ is divided into 100-cm$^{-1}$ subregions, and all intense bands are similarly coded.

With a computerized IR instrument, the spectrum of the unknown compound can be automatically coded in a manner that conforms to the Sadtler system. Alternatively, with older instruments, the spectrum can be manually coded. The search algorithm first identifies those library compounds that have the same strongest absorption band as the unknown. Further narrowing of the possibilities can then be done with the additional coded bands. Alternatively, the spectra of candidate compounds can be viewed on a terminal, or the Sadtler numbers obtained from the search can be used to obtain the full spectra from the library. Several manufacturers of Fourier transform IR spectrometers now include the Sadtler software package with their data systems.

Some of the newer microcomputer-based, conventional IR spectrophotometers now include a more limited set of reference spectra. The computer collects absorption band location and relative intensity information to compare automatically with the spectra stored in the data base. With the microcomputer systems available with most general-purpose IR spectrometers, several thousand reference spectra can be stored for matching purposes. Such computer-based search systems are bound to improve greatly in searching speed, efficiency, and power in the future.

## Applications

There are many applications of infrared spectrometry for identifying organic compounds and structural features of molecules. Many of these are adequately covered in several of the references for this chapter or in books on organic structural elucidation.

With the advent of FTIR spectrometry, it is becoming more common to use IR spectrometry to monitor the effluents of chromatographic columns. For this application, IR methods are highly selective and can provide identification of the eluting species. The example given in Figure 14-17 shows the use of FTIR spectrometry to monitor the spectrum of a GC effluent. The spectrum was obtained after injecting 0.2 µL of a sample containing the analyte at the 1-µg ml$^{-1}$ level.

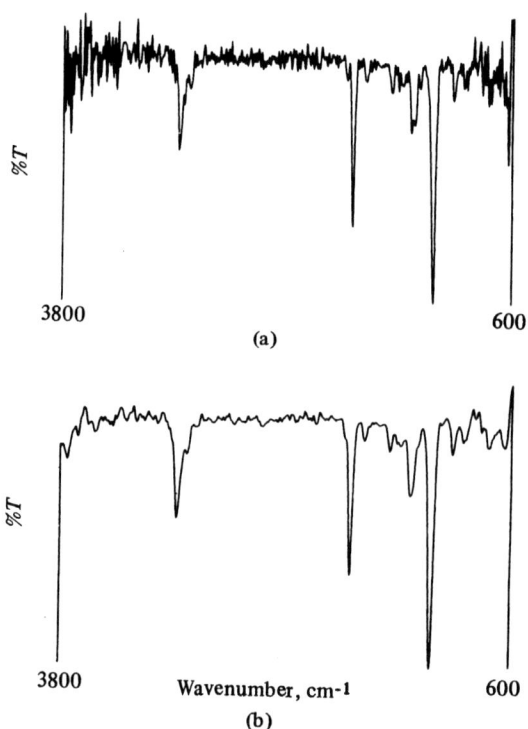

**FIGURE 14-17** FTIR spectrum of isobutyl methacrylate separated by gas chromatography; the infrared spectrum was obtained during the elution of the analyte. In (a) the spectrum is shown without smoothing, while in (b) smoothing was used to improve the *S/N*. (With permission from P. R. Griffiths, *Transform Techniques in Chemistry*, Plenum Press, New York, 1978.)

The cell for most GC-IR experiments is a long (25 to 100 cm), narrow (few mm), gold-coated, flow-through gas cell, known as a light pipe. The infrared beam undergoes several reflections as it traverses the cell. The cell volume must be small in order to prevent mixing of the separated components in the light pipe. The long pathlength is necessary to obtain adequate absorbances for the low concentrations measured. Infrared methods have also been used to monitor liquid chromatographic effluents.

## 14-5 QUANTITATIVE INFRARED ABSORPTION SPECTROPHOTOMETRY

Although IR spectrophotometry is used most often for qualitative analysis and structural determinations, quantitative IR measurements are, nonetheless, quite important. We shall first consider the characteristics of IR absorption and IR instrumentation that combine to make routine quantitative uses more difficult than routine uses of UV-visible spectrophotometry. These include the need for frequent calibrations, nonlinearities, and the difficulty in measuring the analyte absorbance. Next, the precision of IR absorption measurements is considered briefly. This section concludes with several examples of quantitative infrared absorption measurements.

Quantitative applications of IR spectrometry are based on Beer's law. Because of the extremely thin cells used and their deterioration with age, frequent pathlength calibrations are necessary when using liquid cells for quantitative IR methods. Such calibrations are often done by measuring interference fringes (see Figure 14-15). To avoid such pathlength calibrations, external standards can be used in the same cell used to measured analyte absorbance; the pathlength is then taken into account in the slope of the resulting calibration curve. Again, because liquid cells deteriorate and the pathlength varies with time, more frequent calibrations must be made than with UV-visible spectrophotometry. Usually, gas cells are of sufficient pathlength that a direct pathlength measurement is possible.

### Nonlinearities

Nonlinearities arise from similar causes to those considered in Section 13-3 for UV-visible spectrophotometry. Nonlinearity due to polychromatic radiation is much more common in IR spectrophotometry than in the UV-visible region. This is particularly true when the sample is in the gas phase at reduced pressure, and rotational fine structure is present. Such absorption lines can be as narrow as a few tenths of a wavenumber. Fortu-

nately, at higher pressures such lines are substantially broadened and can be made to merge into a broad envelope. With some quantitative IR techniques for gases, higher pressures are intentionally used to broaden the lines in order to obtain adherence to Beer's law.

In the liquid phase, absorption bands are seldom narrower than 3 or 4 $cm^{-1}$ and often wider than 10 $cm^{-1}$. As a rule of thumb recall from Section 13-3 that the spectrophotometer spectral bandpass $s$ should be less than one-tenth the FWHM of the absorption band for 0.5% accuracy. With modern grating IR spectrophotometers and many FTIR spectrometers, spectral bandpasses are often small enough to measure absorbances accurately. However, older prism instruments often possessed insufficient resolution and produced nonlinearities due to polychromatic radiation. Even with modern spectrophotometers, however, double-beam systems with automatic slit servo mechanisms can give rise to such deviations in spectral regions where the source intensity is low or the detector is unresponsive. Thus in quantitative infrared spectrometry, even with modern instruments, molar absorptivity data obtained on one instrument may not transfer accurately to another, and calibration curves made with external standards are almost always necessary.

Stray radiation, extraneous radiation, and scattered radiation can also be serious problems in IR spectrophotometry, particularly with dispersive spectrometers. Stray radiation inside the monochromator poses an identical problem in IR spectrophotometers as with UV-visible instruments. However, because of atmospheric absorption in the IR region, the effects of stray radiation may appear even though the sample absorbance itself is small. For this reason it is difficult to obtain exact cancellation for atmospheric absorption with some samples even though compensation is achieved without the sample. Dispersive spectrometers usually use filter-grating arrangements to minimize stray radiation. With FTIR systems, the stray radiation arises from aliasing which can be effectively avoided by electronic filtering.

Extraneous radiation is also more often a problem in the infrared region than in the UV-visible region, especially with dispersive spectrometers. Here, any IR radiation not emanating from the primary absorption source is considered extraneous radiation. Although the modulation schemes of modern IR spectrophotometers are designed to reduce the effects of IR emission and other sources of extraneous radiation, total elimination of these effects is not possible. Hence measurement of high absorbances ($A > 2$) is not recommended with dispersive systems. With FTIR spectrometers, extraneous radiation merely increases the dc level of the interferogram and is effectively discriminated against in the data processing. With FTIR and solvent absorption subtraction methods, accurate absorbance measurements have been obtained on samples with a solvent absorbance greater than 2.0.

Scattered radiation in the sample cell area is a major reason, along with pathlength variations, why mull techniques are not useful in quantitative analysis. Although the salt pellet technique can be used in quantitative analysis, scattered radiation can also be a major contributor to nonlinearity with this method.

## Measurement of Absorption

Even the measurement of the absorbance of a single component can be somewhat difficult in the IR region. Because the solvent is quite likely to absorb radiation at the analytical wavelength and overlapping bands from concomitants in the sample are common, background absorption can be much more severe than in the UV-visible region. Also, IR absorption measurements are often made by scanning through the wavelength region of interest. For these reasons, the strategy used in the UV-visible region for obtaining absorbances, the so-called "cell-in/cell-out" method, is not very practical for IR measurements. In this method, the 100% *T* or 0 *A* setting is made with a reference solution, and the absorbance of the sample is obtained versus this reference setting. In the infrared region, a more effective method to measure absorbances is the **baseline method** described below.

*Baseline Method.* In the baseline method a line is drawn to represent the baseline of the absorption band. Usually, the method illustrated in Figure 14-18 is used to estimate the baseline. The choice of the points of minimum absorption is often unclear, particularly with overlapping bands. The major guiding principle is to try to approximate the baseline to be the absorbance at the analytical wavelength if the analyte band were absent.

The baseline method assumes that the background varies linearly with wavelength in the region of the absorption band. This assumption is most likely to hold when there are few overlapping bands present, and the chosen baseline is nearly horizontal. If a band from a concomitant directly overlaps the analyte band, the baseline method does not work. When such severe background absorption is present, it is often wise to choose another analytical wavelength where this assumption may be valid. Similarly, it is preferable to work with low absorbance baselines than with those drawn at higher absorbances. Often, the analyst must resort to trial-and-error methods on synthetic mixtures in order to choose an appropriate baseline.

There are several advantages of the baseline method

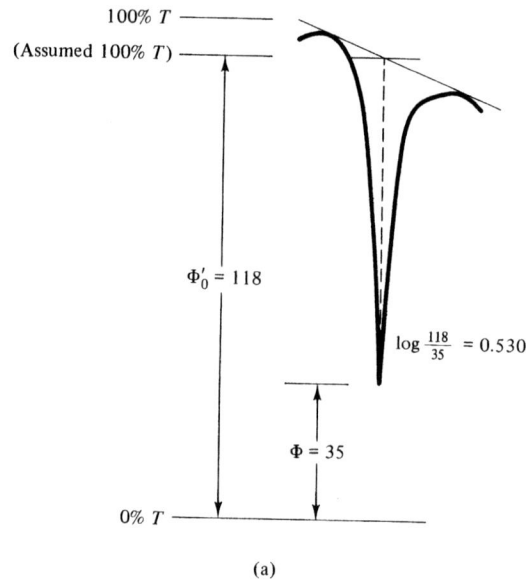

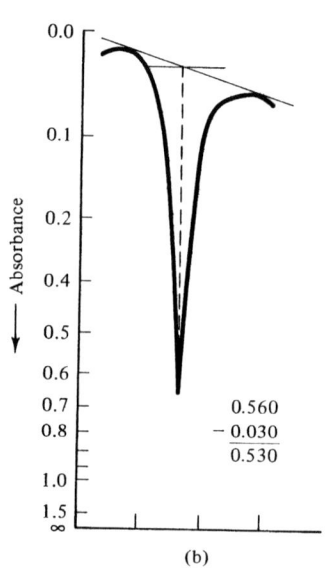

**FIGURE 14-18** Baseline method for measuring absorbances in quantitative IR spectrometry. The 100% *T* (a) or 0 *A* (b) reading is established by drawing a straight line between points of minimum absorption surrounding the absorption band. In some automated systems readings are taken at several wavelengths on either side of the absorption band, and a linear regression is used to obtain the absorbance or transmittance at the analytical wavelength.

over other methods employed in IR spectrometry. First, the baseline is established with the same cell and the same solution that is used for the analyte absorption. This compensates for cell absorption and cell reflection losses as well as background absorption by the sample. Second, since measurements are made on a spectrum obtained by scanning, there are no critical wavelength settings or adjustments to be made. Finally, the method is quite rapid and relatively easy to automate with computer-controlled spectrometers.

*Measurement of Integrated Absorbance.* Quantitative applications of IR spectrometry are normally based on measurements made at the absorption maximum as just described. However, there are applications that use the integrated absorbance defined below:

$$\int_{band} A \, dv = bc \int_{band} a \, dv \qquad (14\text{-}7)$$

The integrated absorbance and the integrated absorption coefficient are related to the total probability for absorption as discussed in Appendix F. For example, consider the measurement of the total aliphatic C—H content of a sample from the $2900\text{-cm}^{-1}$ C—H band. The shape and position of this band vary from compound to compound, but the integrated absorbance is proportional to the number of C—H groups in the molecule. The integrated absorbance is also much less sensitive to changes in the monochromator spectral bandpass than the peak absorbance.

With earlier instruments it was necessary to carry out the integration graphically or by planimetry. With modern computer-controlled systems, the calculation of integrated absorbance is often included in the software or firmware that accompanies the instrument.

## Precision of IR Absorption Measurements

The precision of IR absorption measurements can be treated in a similar manner to that of UV-visible absorption (see Sections 5-5 and 13-2) measurements. For dispersive IR spectrometers, signal expressions are the same as those given in Table 13-2, while signal-to-noise ratio expressions are identical to those in Table 13-3. Because of the lower radiance of IR sources and the higher noise levels of IR thermal detectors compared to UV-visible sources and detectors, signal-carried shot or flicker noise is not observed. Thus the limiting source of noise is usually detector noise (Johnson noise), which is independent of the illumination level. Hence the RSD in absorbance, $\sigma_A/A$, often conforms to equation 5-53, which is repeated here:

$$\frac{\sigma_A}{A} = \frac{\sigma_{0t}}{E_r} \times \frac{-1}{T \ln T} \qquad (14\text{-}8)$$

where $\sigma_{0t}$ is the 0% $T$ noise and $E_r$ is the reference readout signal. For thermal detectors $\sigma_{0t}$ is normally

equal to the detector noise. This equation assumes that there is no uncertainty in measuring the 0 or 100% $T$ signals and a constant uncertainty in measuring the transmittance, $\sigma_T = \sigma_{0t}/E_r$, that is independent of the transmittance. Under these conditions the optimum % $T$ for minimizing the RSD in absorbance is 36.8 ($A = 0.434$), and the RSD in absorbance follows curve $a$ of Figure 5-7. Since the minimum RSD in absorbance is a rather broad function of $T$, transmittances between 20 and 60% provide nearly optimum results.

In actuality, because of background absorption and overlapping bands there may be a considerable uncertainty in the 100% $T$ value used to calculate the sample absorbance as shown in the baseline method described previously. Thus in practice infrared measurements often show RSD values in the range 5 to 10%, particularly with complex sample matrices.

Equation 14-8 also gives us some guidance in improving the $S/N$ for a given IR absorption measurement. Since $\sigma_{0t}$ is usually constant for a given IR detector, the $S/N$ in measuring a given $T$ ($A/\sigma_A$) can be improved by reducing the relative 0% $T$ noise by increasing $E_r$. This is normally accomplished by increasing the slit width. Equation 13-18 shows that $E_r$ is proportional to $W^2$ since $s$ is proportional to $W$. Normally, $E_r$ is set equal to the full-scale setting (100% $T$) of the instrument. When $W$ is increased, the gain of the electronics $G$ is decreased to keep $E_r$ constant. In any case the $S/N$ usually improves with $W^2$ with dispersive IR spectrophotometers. It should be noted that although the $S/N$ improves with increasing slit width, the resolution, of course, degrades as the slits are opened. The throughput can be much larger with filter-based, nondispersive spectrometers, which allows lower values of electronic gain to be used. Baseline noise levels as low as $10^{-4}$ A.U. are possible.

Fourier transform IR spectrometers show the multiplex advantage where the $S/N$ of spectra taken with the same resolution and same measurement time is $n^{1/2}$ times greater than that obtained on a grating instrument, where $n$ is the number of resolution elements (see Section 5-6). Michelson interferometers also show increased throughput over grating spectrometers (Jacquinot's advantage). Let us now briefly estimate the magnitude of the throughput advantage. The throughput factor at a particular wavelength $\Upsilon$ (cm² sr) is the product of the solid angle of the spectrometer ($\Omega$), the area of the source viewed ($A_s$) and the transmission factor ($T_{op}$) of the optical system (see Section 3-6) as shown by

$$\Upsilon = \Omega A_s T_{op} \qquad (14\text{-}9)$$

For a grating spectrometer, the throughput factor

$Y_G$ can be written

$$Y_G = \frac{WHA_G T_{op}}{f^2} = \frac{sHA_G T_{op}}{f^2 R_d} \quad (14\text{-}10)$$

where $A_G$ is the projected grating area, $s$ is the spectral bandpass, $f$ is the focal length of the monochromator, and $R_d$ is the reciprocal linear dispersion. The experimental resolving power of the grating monochromator $R_G = \bar{\nu}/s$, where $s$ is in wavenumbers. Using this relationship and expressing $R_d$ in terms of the angular dispersion of the grating $D_a$ [$R_d = 1/(fD_a)$], equation 14-10 becomes

$$Y_G = \frac{\bar{\nu} HA_G D_a T_{op}}{f R_G} \quad (14\text{-}11)$$

For a Michelson interferometer, if the detector area is not limiting, the solid angle is $2\pi/R_I$, where $R_I$ is the resolving power of the interferometer. The throughput factor of the interferometer $Y_I$ is then

$$Y_I = \frac{2\pi A_I T'_{op}}{R_I} \quad (14\text{-}12)$$

where $A_I$ is the area of the limiting the mirror in the interferometer and $T'_{op}$ is the optics transmission factor. By taking the ratio of equation 14-12 to equation 14-11, the value of Jacquinot's advantage can be expressed as

$$\frac{Y_I}{Y_G} = \frac{2\pi A_I T'_{op} f R_G}{R_I \bar{\nu} HA_G D_a T_{op}} \quad (14\text{-}13)$$

For a rough estimation, let us assume that the spectrometers have nearly equal limiting areas and optics transmission factors ($A_I T'_{op} \approx A_G T_{op}$), and the same resolving powers ($R_G = R_I$). Furthermore, if the diffraction angle $\beta$ is near $45°$, $D_a\bar{\nu} \approx \tan\beta \approx 1$. With these assumptions, equation 14-13 becomes

$$\frac{Y_I}{Y_G} = \frac{2\pi f}{H}$$

The ratio $H/f$ seldom exceeds 1/30, even for a high-throughput monochromator. Assuming that $f/H = 30$, the throughput advantage of the interferometer is approximately $60\pi$ or about 200. In practice, Jacquinot's advantage is wavenumber dependent and varies from 10 to over 200 in the mid-IR region.

The combination of the multiplex advantage and the throughput advantage should give interferometers enormous improvements in $S/N$ over dispersive spec-

trometers in the mid-IR region. In practice, the thermocouple detectors used with dispersive systems can be one to two orders of magnitude more responsive than the TGS detectors used for FTIR. This can offset significantly the multiplex and throughput advantages.

## Applications

Two examples are given here to illustrate the procedures used in quantitative IR absorption. In the first, the analysis of a three-component mixture is described, while the second example discusses the use of IR absorption for the measurement of reaction rates.

*Quantitative Mixture Analysis.* Quantitative IR absorption is often used in the analysis of mixtures of organic compounds. The determination of *o*-xylene, *m*-xylene, and *p*-xylene is a typical example. The spectrum of the mixture in the region 700 to 800 cm$^{-1}$ is shown in Figure 14-19. The frequencies chosen for the determination were 795 cm$^{-1}$ (*A* due mostly to *p*-xylene), 768 cm$^{-1}$ (*m*-xylene), and 741 cm$^{-1}$ (*o*-xylene). Spectra were obtained for standards of each component in cyclohexane, and the product of absorptivity and pathlength was calculated at each frequency. The following

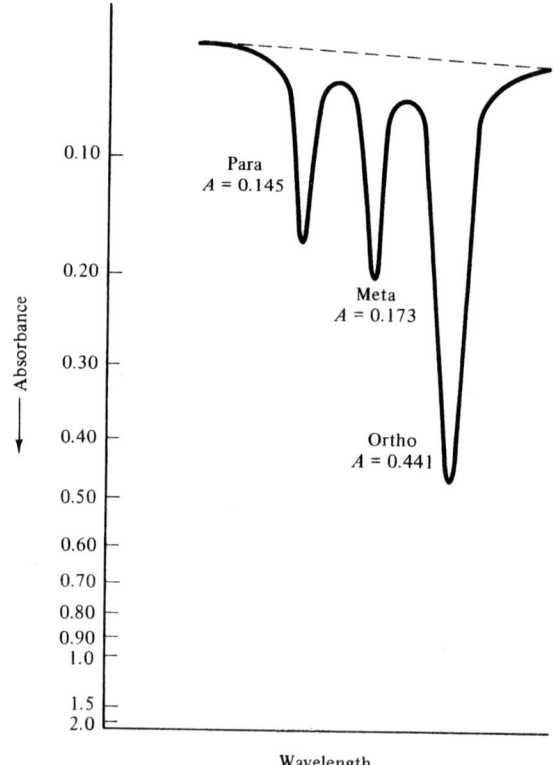

**FIGURE 14-19** IR spectrum of a mixture of xylenes in cyclohexane.

simultaneous equations result from the additivity of absorbances:

$$A_{795} = 1.506c_p + 0.048c_m + 0.000c_o$$

$$A_{768} = 0.025c_p + 1.440c_m + 0.000c_o$$

$$A_{741} = 0.032c_p + 0.033c_m + 2.405c_o$$

where the numbers are $ab$ values in L g$^{-1}$ at each frequency and the $c$ values are concentrations in g L$^{-1}$.

As shown in Figure 14-19, the baseline method is used to obtain the absorbances at each frequency. With these measured absorbances, the simultaneous equations can be solved for the three concentrations to yield $c_p = 0.095$ g L$^{-1}$, $c_m = 0.118$ g L$^{-1}$, and $c_o = 0.180$ g L$^{-1}$. The accuracy of the simultaneous determination is better than 10%.

*Reaction-Rate Studies.* If a reacting molecule or product has an absorption band that is separated cleanly from other bands, IR absorption can be extremely valuable in kinetics studies. Unfortunately, because of the special thin cells usually used in conventional IR methods, reactions are not conveniently carried out in the cell as they can be in UV-visible studies.

In one study, the reaction between phenyl isocyanate and various alcohols was investigated. The reaction was carried out in a vessel suspended in a constant-temperature bath. At the desired times, a sample was removed from the mixture with a syringe and transferred to an IR cell of known thickness. The absorbance of the isocyanate band at 2260 to 2270 cm$^{-1}$ was measured by scanning a narrow region surrounding the band. Several such samples were withdrawn at various times during the 5- to 7-h period over which the reaction was studied. A calibration curve had previously been prepared from solutions of known phenyl isocyanate concentration. The absorbance values were then converted to concentration and used to test various rate laws; the reaction was found to be first order in phenyl isocyanate and first order in the alcohol.

Obviously, the reactions that can be studied by procedures similar to that described must be very slow so that changes do not occur during the solution transfer and scanning step. Thus conventional IR spectrometry is not very convenient for rapid reactions. Also, it is difficult to thermostat the cell compartments in most conventional IR instruments. Often, the geometry used in conventional IR systems (sample cell in front of the monochromator) leads to considerable heating of the sample unless precautions are taken. Fourier transform IR spectrometry has been used successfully to study fairly rapid reactions, as discussed previously.

## 14-6 NEAR-INFRARED AND FAR-INFRARED ABSORPTION

The techniques and applications of near-infrared (NIR) and far-infrared (FIR) spectrometry are quite different from those discussed above for conventional, mid-IR spectrometry. We briefly discuss here a few of the features that make these spectral regions quite useful to the chemist.

### Near-Infrared Spectrometry

Near-infrared spectrometry shows some similarities to UV-visible spectrophotometry and some to mid-IR spectrometry. Indeed, as was discussed in Chapter 13, the spectrophotometers used in this region (0.77 to 2.5 μm) are often combined UV-visible-NIR spectrophotometers. Although a few electronic transitions can occur in the wavelength range 0.77 to 1.0 μm, the majority of the absorption bands observed are due to overtones (or combinations) of fundamental bands that occur in the region 3 to 6 μm, usually hydrogen-stretching vibrations. Hence the region bears a strong relationship spectrally to the mid-IR.

*Instrumentation and Techniques.* The instrumentation used in the NIR region has been described in Chapter 13. Tungsten lamps are used as sources, and the PbS photoconductive cell is the usual detector. Several commercial instruments intended primarily for the UV-visible region allow near-infrared operation.

Sample preparation and handling techniques are not nearly as troublesome as those used in the mid-IR region and are often identical to the techniques used in the UV-visible region. Similarly, sample cells are usually the same quartz or fused silica cells used in the region 200 to 770 nm. Cell pathlengths vary between 0.1 and 10 cm. Because there are few instrumental or sample handling difficulties, NIR molar absorptivities are much more reliable than molar absorptivities obtained in the mid-IR region. Consequently, data obtained in one laboratory can often be used reliably by another.

Several solvents are suitable for use in the NIR region, although solvents containing O–H, N–H, and C–H groups are usually avoided. Figure 14-20 shows the regions that are useful for several common solvents. Note that CCl$_4$ is transparent throughout the NIR region, and CS$_2$ absorbs only weakly in a narrow region near 2.22 μm.

*Spectral Information.* Because of the nature of the absorption bands observed in the NIR region, the spectral information obtained is not nearly as charac-

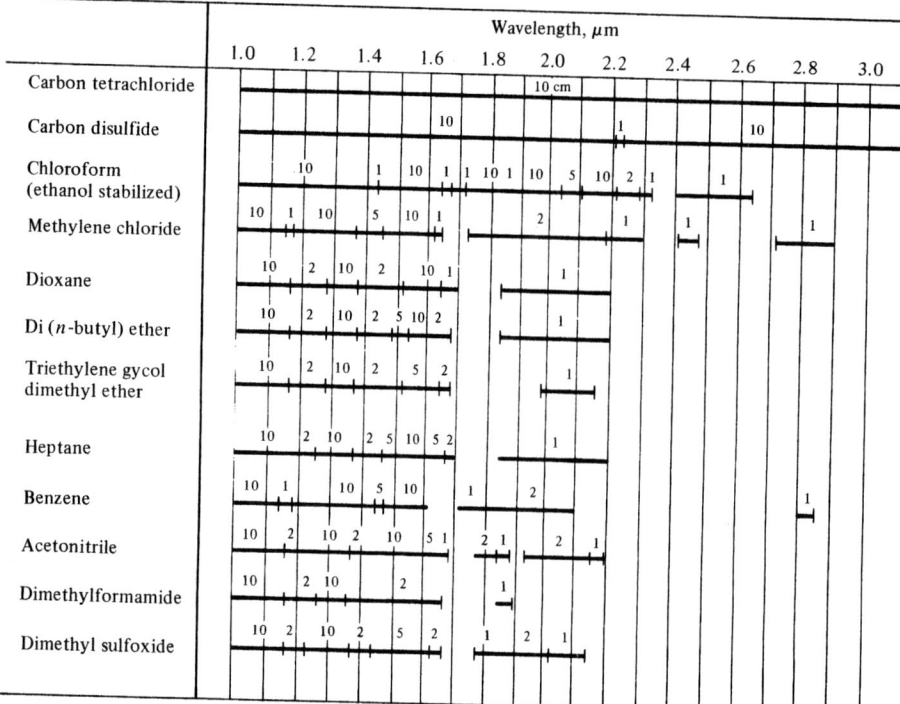

**FIGURE 14-20** Characteristics of several common solvents in the near-infrared region. The solid lines indicate useful regions of low solvent absorption; the numbers indicate the maximum desirable pathlengths in cm. (Reprinted with permission from R. F. Goddu, "Near Infrared Spectrophotometry," in *Advances in Analytical Chemistry and Instrumentation*, C. N. Reilley, ed., vol. I, Wiley, New York, 1960. Copyright © John Wiley & Sons, Inc., 1960.)

teristic of a particular molecule as information obtained in the fingerprint region of the mid-IR. Hence NIR spectrometry is not as generally useful for qualitative analysis as is mid-IR spectrometry. The greatest utility of NIR spectra is in the detection and subsequent quantitative determination of functional groups that contain unique hydrogen atoms.

The functional group correlations normally made in the NIR region are summarized in Figure 14-21. Note that molar absorptivities are also shown for many of the functional groups. Because of the reliability of molar absorptivity data in this region, it is often used, in conjunction with band location, for qualitative identification of the absorbing group as well as for quantitative analysis.

*Applications.* In contrast to mid-IR spectrophotometry, NIR spectrophotometry is most widely used for *quantitative* organic funcitonal-group analysis. However, the NIR region has also been used for qualitative analyses and studies of hydrogen-bonding, solute–solvent interactions, organometallic compounds, and inorganic compounds.

As an example, consider the qualitative identification of primary, secondary, and tertiary amines. As can be seen from Figure 14-21, these amines have quite different NIR spectra. Primary amines have two strong absorption bands between 2.85 and 3.05 μm, due to fundamental N—H stretching vibrations, a corresponding overtone doublet between 1.45 and 1.55 μm, a com-

bination band near 2.0 μm, and a single absorption band near 1.0 μm. Secondary amines have only singlet bands and no bands near 2.0 μm. Tertiary amines, of course, have none of the N—H bands mentioned above, although they show NIR absorption due to C—H and other hydrogen-containing groups. Near-infrared spectrometry thus provides a very clear way to distinguish the three amine types.

Quantitative applications of NIR spectrometry include the determination of alcohols, phenols, hydroperoxides, organic acids, and water through measurement of the fundamental O—H stretching vibration in the region 2.7 to 3.0 μm and the determination of esters, ketones, and acids through measurement of the first overtone of the carbonyl group at 2.8 to 3.0 μm. Detection limits for compounds with "reasonable" molar absorptivities (e.g., $> 1.0$ L mol$^{-1}$ cm$^{-1}$) are usually less than 0.1%.

**Far-Infrared Spectrometry**

The far-infrared region (50 to 1000 μm) has long been considered a valuable region for obtaining chemical information. In the past, however, far-IR spectrometry was frought with experimental difficulties. There are very few sources that can be used in this region, and these are notoriously weak in intensity. Grating instruments have relatively low throughput in the far-IR region, particularly with the order-sorting filters that are necessary to prevent radiation diffracted from higher grating orders from reaching the detector. It is not surprising then that FTIR spectrometry was first applied in the far-IR region. The high throughput of interferometers and the rather low mechanical tolerances required in this region have made it relatively easy to obtain FTIR spectra with an adequate signal-to-noise ratio.

*Instrumentation and Techniques.* Almost all FIR studies are now carried out with FTIR spectrometers (see Section 14-2). Sources in the far-IR region are typically mercury arc lamps, and detectors are usually the Golay pneumatic detector or the TGS pyroelectric detector.

Solution sampling in the far-IR region is hampered by the lack of suitable solvents. Since absorptivities are low in this region, it is desirable to use long-pathlength cells, which only compounds the problem of solvent transparency. Hydrocarbons, such as cyclohexane and benzene, are among the most useful, while $CCl_4$, $CS_2$, and $CHCl_3$ are also used. Sample cells are often made with polyethylene windows, but quartz has been used in the region 50 to 330 $\mu$m.

Solid samples are generally preferred to solution samples in the far-IR region. Scattering is not nearly as serious in this region as it is in the mid-IR since the scattered radiation intensity depends inversely on the fourth power of the wavelength. Spectra have been obtained on samples dispersed in mulls and in polyethylene and wax matrices. Pressed polyethylene plates are commonly used with solids.

*Spectral Information and Applications.* The far-IR region can provide unique information. For example, the fundamental vibrations of many organometallic and inorganic molecules fall in this region due to the heavy atoms and weak bonds in these molecules. In addition, lattice vibrations of crystalline materials occur in this region, and electron valence/conduction band transitions in semiconductors often correspond to far-IR wavelengths.

Far-infrared spectrometry has been applied to many different chemical problems. For example, low-frequency metal-ligand vibrational modes have been studied by FIR techniques. Because such low-frequency vibrations were difficult to study in the past, the spectra of many complexes were incorrectly assigned. Such studies have now become almost routine with FTIR instruments and have allowed inorganic chemists to assign such low-frequency modes correctly. As an illustration of the utility of FIR spectrometry in studying inorganic complexes, Figure 14-22 shows the FIR spectrum of a copper complex obtained in a Nujol mull matrix. The absorption bands are sharp and the scat-

tering background is much less than seen in the mid-IR region with mulls.

Far-infrared spectrometry has been widely used in structural studies of organosulfur, organophosphorus, and organometallic compounds. In addition, the far-IR region has been used in qualitative analysis as a supplement to mid-IR studies. Many closely related compounds, such as isomers and polymers of different chain lengths, have quite similar mid-IR spectra, but markedly different FIR spectra.

It is clear that far-infrared spectrometry can provide unique and valuable spectrochemical information. The widespread use of FTIR instruments is leading to an explosion of interest in this spectral region and a reevaluation of its relative importance.

## 14-7 INFRARED REFLECTION AND EMISSION

Infrared reflection techniques, particularly the attenuated total reflectance (ATR) method, are widely applicable. The ATR method is frequently used to obtain spectra of difficult samples, such as polymer films, rubber, food, and various resins. Although the ATR phenomenon was first observed by Newton, it was not widely applied to IR spectroscopy until the 1960s. Specular and diffuse reflection methods are also used in the infrared region. Infrared emission spectrometry is not used nearly as often or as routinely as IR absorption spectrometry. However, IR emission can provide some unique information and is quite useful for certain studies and samples. In this section we first consider IR reflection methods and ATR in particular and then discuss the principles of IR emission methods.

### Infrared Reflectance Methods

Most routine IR spectroscopy is carried out in the absorption mode, on the beam transmitted through the sample. However, information about the sample can also be obtained by studying the radiation reflected from it.

We will consider here the three general types of reflectance phenomena that are used: specular, diffuse, and internal reflection. This last method provides more chemical information, while specular and diffuse reflectance can provide optical information.

*Specular and Diffuse Reflectance.* As discussed in Section 3-1, **specular reflectance** occurs from smooth surfaces, while **diffuse reflectance** occurs from surfaces with irregularities. With many solid surfaces it is desirable to obtain reflectance information as a func-

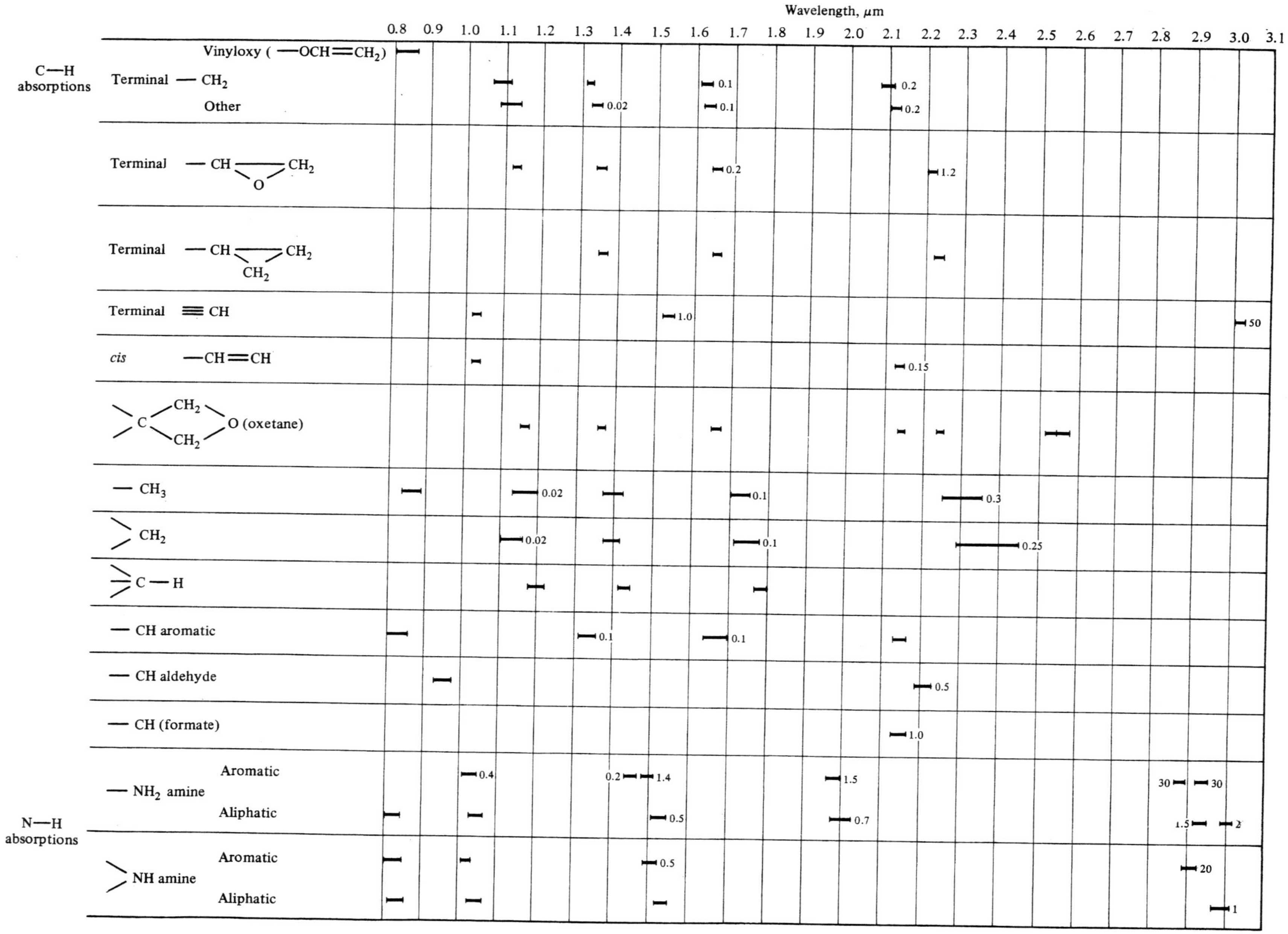

Wavelength, μm

Axis values: 0.8 0.9 1.0 1.1 1.2 1.3 1.4 1.5 1.6 1.7 1.8 1.9 2.0 2.1 2.2 2.3 2.4 2.5 2.6 2.7 2.8 2.9 3.0 3.1

**C—H absorptions**

Vinyloxy ( —OCH=CH₂ )

Terminal —CH₂ — 0.2 (2.1), 0.1 (1.65), 0.02 (1.38)

Other — 0.2 (2.1), 0.1 (1.65)

Terminal —CH (with CH₂ O, oxetane type) — 1.2 (2.2), 0.2 (1.65)

Terminal —CH CH₂ CH₂

Terminal ≡CH — 50 (3.0), 1.0 (1.55)

cis —CH=CH — 0.15 (2.1)

$\begin{array}{c} C \\ CH_2 \\ CH_2 \end{array}$ O (oxetane) — 0.2 (2.22)

—CH₃ — 0.3 (2.35), 0.1 (1.7), 0.02 (1.2)

CH₂ — 0.25 (2.4), 0.1 (1.75), 0.02 (1.2)

C—H

—CH aromatic — 0.1 (2.1), 0.1 (1.65), 0.1 (1.3)

—CH aldehyde — 0.5 (2.2)

—CH (formate) — 1.0 (2.1)

**N—H absorptions**

—NH₂ amine
  Aromatic — 30 (2.85), 1.5 (1.97), 1.4 (1.5), 0.2 (1.5), 0.4 (1.0)
  Aliphatic — 30 (2.9), 2 (3.0), 1.5 (2.85), 0.7 (1.97), 0.5 (1.5)

NH amine
  Aromatic — 20 (2.88), 0.5 (1.5)
  Aliphatic — 1 (2.95)

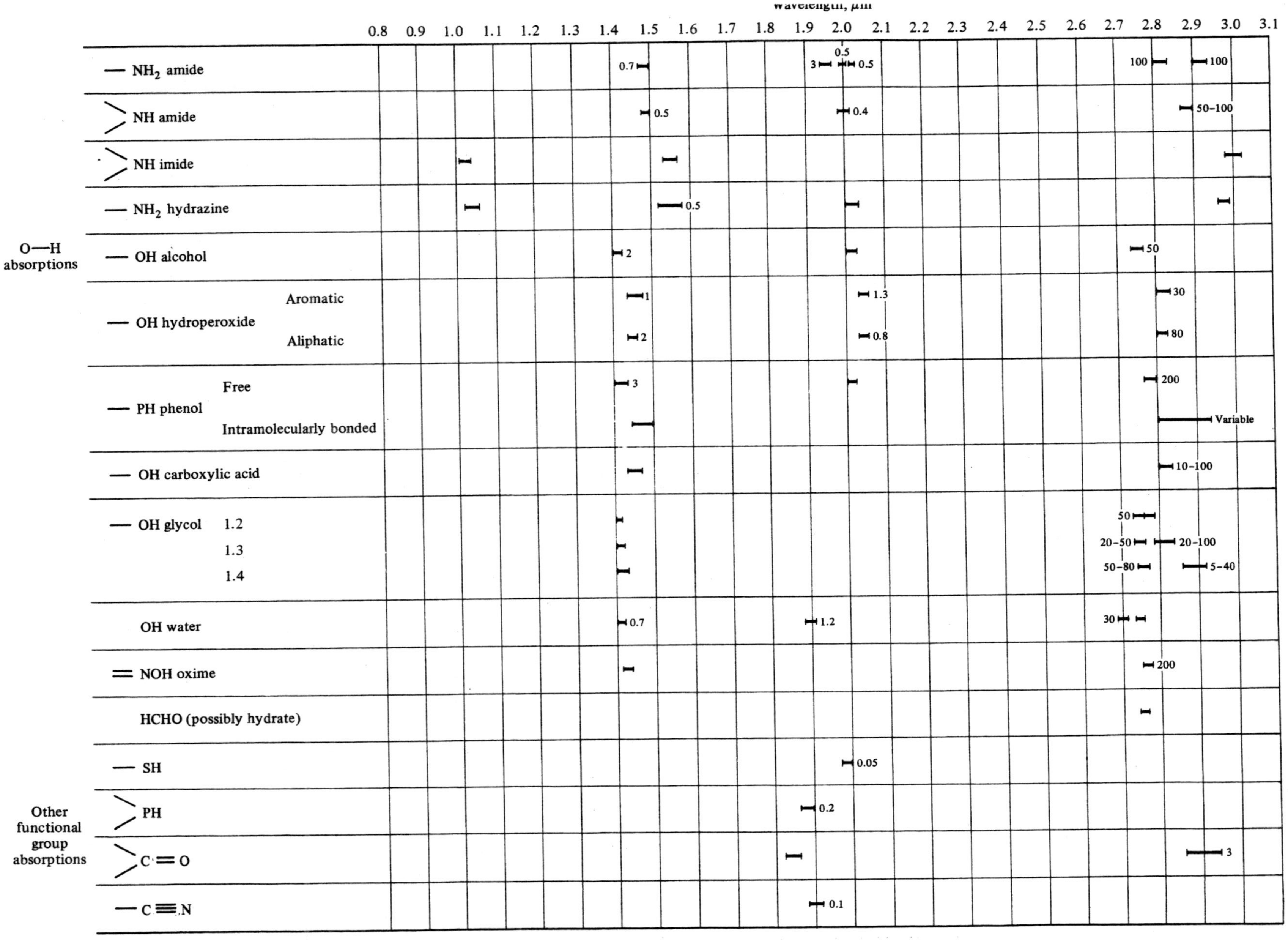

**FIGURE 14-21** Near-infrared functional group correlations. The numbers shown near the band locations are approximate molar absorptivities. (Reprinted with permission from R. F. Goddu, "Near Infrared Spectrophotometry," in *Advances in Analytical Chemistry and Instrumentation*, C. N. Reilley, ed., vol. I, Wiley, New York, 1960. Copyright © John Wiley & Sons, Inc., 1960.)

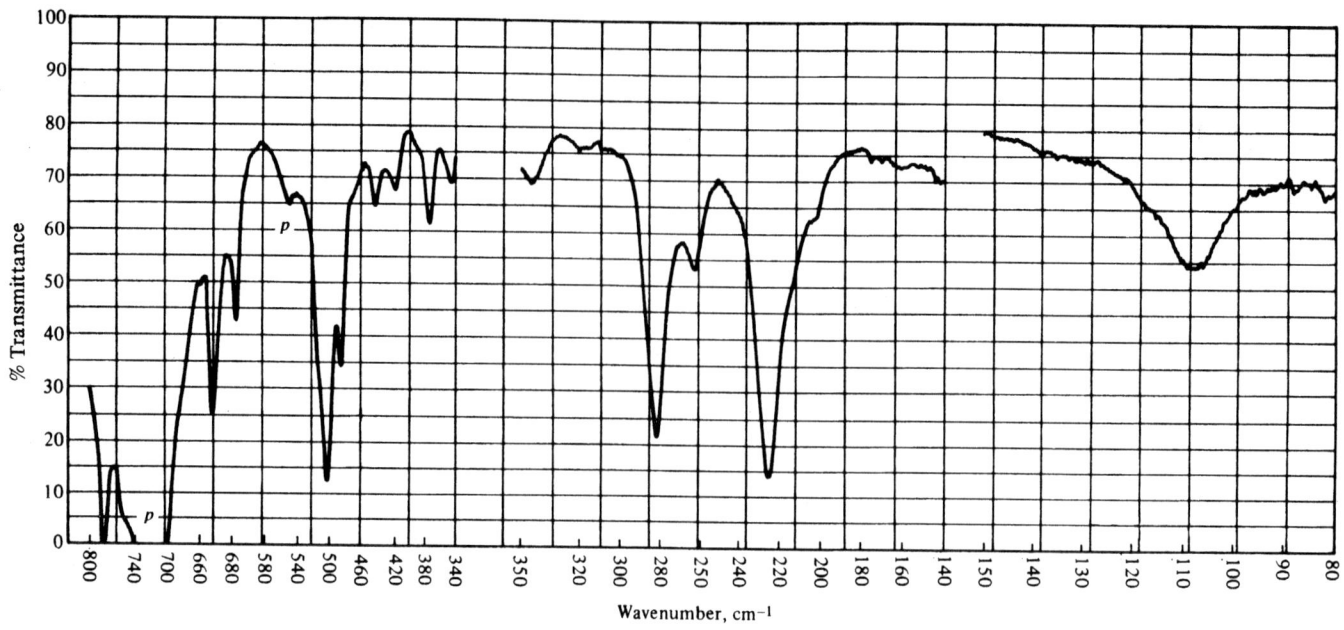

**FIGURE 14-22**  Far-infrared spectrum of the copper complex of 1-hydroxy-2-carbethoxycylcooctene. A Nujol mull was pressed between polyethylene plates for this spectrum. The bands marked p are due to absorption by the polyethylene. [Courtesy of Beckman Instruments (adapted).]

tion of wavelength, incident angle, and incident beam polarization.

A simple reflectance attachment of the type available commercially for single beam IR spectrophotometers is illustrated in Figure 14-23. The spectrum is recorded with the sample in place and then with a reference mirror substituted for the sample. The ratio of the two spectra is the reflectance of the sample relative to the reference. Reflectance measurements can also be made with double-beam spectrophotometers by placing an identical attachment with the reference mirror in the reference beam. This alleviates the need to do point-by-point ratio calculations to obtain the reflectance spectrum. Microreflectance attachments have also been constructed with suitable optics for condensing the beam.

Specular reflectance measurements as a function

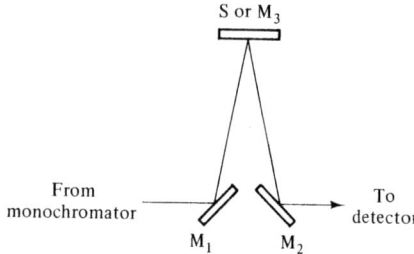

**FIGURE 14-23**  Specular reflectance measurements with a single-beam spectrophotometer. Mirror $M_1$ deflects the beam to the reflecting sample S or to a reference mirror, $M_3$. Mirror $M_2$ returns the beam to its original path.

of incident angle can be made with an optical arrangement that rotates the source and source mirror relative to the sample axis. Insertion of polarizers into the beam allows reflectance to be obtained as a function of the incident beam polarization.

Reflectance measurements are also used to determine the thickness of thin films deposited on solids or grown on solid surfaces. Here rays reflected directly from the surface interfere with those reflected from the interface between the film and the substrate. The number of interference fringes in a certain wavenumber interval is used to obtain the film thickness in a manner similar to that used to obtain sample cell pathlengths (see Figure 14-15).

Diffuse reflectance methods are used to study irregular surfaces. With such surfaces the reflected radiance is nearly independent of the viewing angle. Such measurements are often made with an **integrating sphere**, similar to that shown in Figure 14-24. Usually, with this technique the radiant power reflected from the sample is obtained as a function of the wavelength of the IR beam. A reference reflector is then placed in the sphere and its spectrum obtained. The reflectance spectrum of the sample relative to the reference is then calculated by ratioing. In the past few years, diffuse reflectance FTIR spectrometry has been shown to be quite useful for the analysis of solids.

*Internal Reflection Methods.*  The phenomenon of total internal reflection was discussed in Section

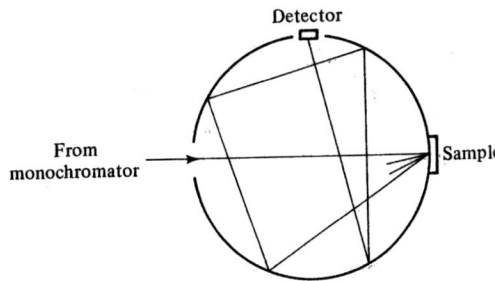

**FIGURE 14-24** Diffuse reflectance measurement with an integrating sphere. The inside of the sphere is coated with a nearly perfect diffuse reflector (e.g., $BaSO_4$ or MgO for the UV-visible-near IR region and flowers of sulfur for working into the mid-IR region). For measurements farther into the IR region, integrating spheres are not coated. Radiation reflected from the sample in all directions is rereflected by the sphere and eventually detected, lost out the entrance window, reflected again by the sample or absorbed. Because most of the radiation experiences multiple reflections in random directions, it becomes totally integrated (uniformly distributed within the spherical cavity). Once the radiation has been integrated, the effects of angular response and inhomogeneities are averaged, and the detector senses a radiant power proportional to the total radiation reflected from the sample.

3-1. In ATR, although the beam is totally reflected at the interface, radiation does penetrate a small distance into the medium of lower refractive index. This penetrating radiation is called the **evanescent wave**. If the less dense medium is capable of absorbing the IR radiation, the reflected beam is attenuated at characteristic wavelengths corresponding to absorption bands. This phenomenon is known as **attenuated total reflection**.

In ATR spectroscopy the sample is placed tightly against the surface of a prism or an internal reflection crystal. Figure 14-25 shows an experimental arrangement for obtaining ATR spectra. With the appropriate incident angle, the IR beam undergoes multiple internal reflections before it passes out of the ATR crystal. Attenuation due to absorption can take place at each reflection. The prism or multireflection crystal must be nonabsorbing and chemically inert.

The refractive indices of the crystal and the sample are critical factors in determining the spectrum obtained. The effective penetration depth $d_p$ depends on the wavelength of the beam, the refractive indices of the two media and the beam angle; it can be calculated from

$$d_p = \frac{\lambda_c}{\{2\pi\,[\sin^2\theta - (\eta_s/\eta_c)^2]\}^{1/2}} \qquad (14\text{-}14)$$

where $\lambda_c$ is the wavelength in the crystal ($\lambda/\eta_c$), $\theta$ is the angle of incidence, and $\eta_s$ and $\eta_c$ are the refractive indices of the sample and crystal, respectively. Note that the closer the two refractive indices are to each other, the deeper the evanescent wave penetrates. This, of course, leads to greater attenuation by the sample and thus higher contrast in the spectrum. However, near strong absorption bands the refractive index can change rapidly; this may give rise to regions where the refractive index of the sample exceeds that of the crystal and the radiation is no longer internally reflected. This

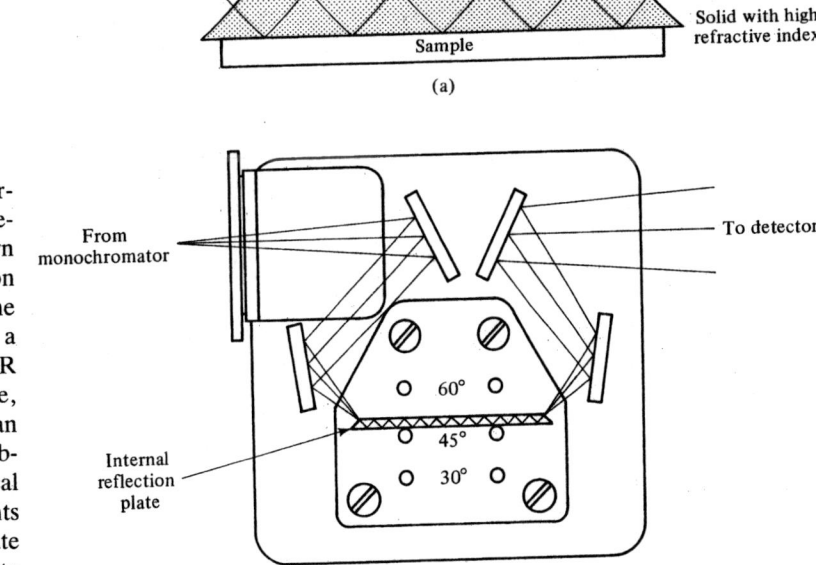

**FIGURE 14-25** Experimental arrangement for attenuated total reflection. In (a) a solid sample is shown mounted on an internal reflection crystal of high refractive index. The crystal can be a plate as shown or a prism. The materials used as ATR crystals include KRS-5, AgCl, Ge, Si, and the Irtrans. Solid samples can be pressed against the crystal to obtain optical contact. In (b) a typical attachment for ATR measurements is shown. The internal reflection plate can be positioned in the holder to provide angles of 30°, 45°, or 60°.

distorts the spectrum unless the sample thickness is less than the penetration depth of the evanescent wave. The wavelength dependence of $d_p$ has an effect on the spectrum in that longer-wavelength absorption bands are more intense than in the corresponding absorption spectrum.

Note that the penetration depth can be changed by changing the crystal material, the angle of incidence, or both. It is thus possible to obtain a depth profile of a surface using ATR spectroscopy. In practice a multireflection crystal with a 45° incidence angle can accommodate most routine samples.

One of the major features of ATR spectroscopy is that absorption spectra can be obtained on a wide variety of sample types with minimal preparation. In addition to polished solid slabs, samples of fabrics, threads, yard, or fibers can be pressed against the ATR crystal. Powders, pastes, or suspensions can also be accommodated. Viscous liquids can be spread on the ATR crystal, and cells are available for obtaining ATR spectra of less viscous liquids. There are even ATR flow cells available. Water solutions can be used if the crystal is water insoluble.

Because the attenuation in ATR takes place in a very thin layer, surface layers and gases adsorbed on surfaces can readily be studied. If the sample thickness exceeds the beam penetration depth, variations in thickness do not affect the intensity of the bands obtained. However, the ATR band intensity is proportional to concentration in accordance with Beer's law, so that useful quantitative measurements can be made.

The spectra obtained with ATR methods are quite similar to IR absorption spectra except as noted above. Thus standard IR spectra and correlation charts can be used for qualitative analysis. There are, however, some significant differences in the peak shapes and relative intensities obtained in ATR and those obtained by IR absorption methods. In ATR the peak shape is not only a function of absorption coefficient, but also a function of refractive index. This can introduce asymmetries in the band profile. Also, the orientation of the sample on the ATR crystal can influence the band shapes and relative intensities.

The ATR method has been applied to many sample types. It is convenient, and ATR attachments are relatively inexpensive. They are available for nearly all commercial IR spectrophotometers.

## Infrared Emission Spectroscopy

Infrared emission spectroscopy was quite important in the early part of the twentieth century in measurements of the spectral distribution of blackbody radiation. The experimental verification of Planck's law was, in fact,

based to a large extent on IR emission measurements. Following this period, however, IR emission spectrometry was nearly dormant until after World War II, when scientists and engineers again became interested in thermal radiation from rocket exhausts, flames, hot gases, and other sources. In recent years, the widespread availability of FTIR instruments has led to a resurgence of interest in IR emission, and new applications are now appearing regularly. Today IR emission is a useful technique for gaseous samples, for some solids and for spectral distribution studies of new materials and reference IR sources. The luminescence of materials has also been measured in the IR region (IR fluorescence). However, luminescence measurements are quite difficult because the long lifetimes enhance the probability of radiationless decay processes.

*Studies of IR Sources.* Infrared emission spectrometry is still used in studying the emittance of IR radiation sources and the emissivity of various materials. In these studies radiation from the emitting sample is collected and imaged on the entrance slit of an IR spectrometer. Usually, the radiometric quantity desired is the spectral emittance or spectral emissivity. In order to obtain the true emittance spectrum, a calibrated spectrometer must be used or the spectrum obtained must be compared to that of a blackbody or calibrated source at the same temperature and under the same instrumental conditions. Care must be taken to correct for absorption by atmospheric $H_2O$ and $CO_2$.

Double-beam methods have been used to measure spectral emissivities of materials directly. In a true double-beam system, the source under study is placed in one beam of an optical null spectrometer and a reference source or blackbody is placed in the other. The two beams are initially balanced with the same source in both beams. In the semi-double-beam method, spectra of the reference and the sample are recorded in turn relative to the usual source of an optical null spectrometer and then compared. In both cases computation is simplified over single-beam methods because absorption due to atmospheric gases is canceled.

*Solid, Liquid, and Gaseous Samples.* Upon thermal excitation, molecules that absorb in the IR region can also emit IR radiation. Because transitions can be observed between excited vibrational levels, IR emission spectrometry can provide information that is different from that provided by IR absorption spectrometry. The major drawbacks to analytical IR emission spectrometry are the low radiances emitted by many samples at reasonable temperatures and the difficulty in extracting the desired analyte emission from the background IR radiation emitted by concomitants in the sample and the sample surroundings.

The use of FTIR instruments is particularly attractive for the study of weak IR emitters or of samples at low temperatures because of the throughput and multiplex advantages. In fact, FTIR systems have proven very useful in IR emission studies of planetary and stellar radiation sources, which are very weak emitters. With such sources, multichannel averaging techniques are usually employed for further *S/N* improvement. With modulation techniques FTIR systems can compensate for IR emission by the sample surroundings (cells, holders, etc.) With computers it is relatively straightforward to correct emission spectra for atmospheric absorption or emission from concomitants.

Conducting solid samples can be heated for IR emission studies by an electrical current. An inert atmosphere is used to prevent oxidation. Alternatively, solids can be dissolved in suitable solvents which are evaporated onto a salt plate. The plate is then heated electrically near the spectrometer entrance slit, and the IR emission is monitored. In one interesting application, pesticides such as malathion, DDT, and dieldrin were identified by IR emission on salt plates in quantities in the range 1 to 10 μg.

The IR thermal emission spectra of liquids can also be obtained. Usually, temperatures below a few hundred degrees are used to prevent extensive evaporation. The liquid sample can be placed in a heated sample cell of the type usually used for IR absorption. Because of low radiances, the FTIR approach is usually used for IR emission studies of liquids. Fourier transform IR emission spectrometry has also been applied as a detector for constituents separated by liquid chromatography.

With gaseous samples the IR emission spectra can sometimes be obtained without heating, particularly with FTIR methods. Ammonia gas, for example, has been detected in clouds. Similarly, air pollutants have been detected remotely in plumes from industrial plants. In one case an interferometer was mounted on a reflecting telescope, and $CO_2$ and $SO_2$ were detected several hundred feet from an industrial smokestack.

Rapid scanning FTIR emission measurements have also been used to study transient species in shock tubes, electrical discharges, rocket exhausts and flames. GC effluents have also been detected. Although such species can be studied by IR absorption spectrometry, IR emission is simpler because it requires no external IR radiation source, and difficult optical alignment is avoided. Because of its simplicity and its potential for obtaining valuable information, IR emission, particularly with FTIR methods, is certain to be more widely used in the future.

## PROBLEMS

**14-1.** Find the wavelengths (μm) and the wavenumbers (cm$^{-1}$) for each of the following infrared frequencies:
   **(a)** $7.031 \times 10^{13}$ Hz.
   **(b)** $1.034 \times 10^{14}$ Hz.
   **(c)** $2.897 \times 10^{13}$ Hz.

**14-2.** The mid-infrared region of the spectrum is usually considered to be 2.5 to 50 μm. What frequency range in Hz does this region encompass?

**14-3.** What frequencies (Hz) are found in the far-infrared region (50 to 1000 μm)?

**14-4.** Use the simple harmonic oscillator model for the vibration between two atoms and calculate the vibrational frequencies (cm$^{-1}$) and wavelengths (μm) expected for the following bonds.
   **(a)** The C—C bond in ethane ($k = 4.5$ N cm$^{-1}$).
   **(b)** The C—C bond in benzene ($k = 7.6$ N cm$^{-1}$).
   **(c)** The C—N bond in $CH_3CN$ ($k = 17.5$ N cm$^{-1}$).
   **(d)** The C—H bond in ethane ($k = 5.1$ N cm$^{-1}$).

   Compare the calculations with the group frequency ranges given in Figure 14-3.

**14-5.** Interference fringes were obtained for three empty cells to determine their pathlengths. Find the pathlengths of the cells if the number of fringes between 1200 and 1000 cm$^{-1}$ was:
   **(a)** 13.
   **(b)** 6.
   **(c)** 22.

**14-6.** The following are typical reasons for instrumental errors in IR spectrophotometry. Some can be corrected by proper instrument adjustment, some are compensated in double-beam systems, and others can be eliminated by appropriate calibrations. For each potential source of error suggest first how it could be detected and second how it could be corrected.
   **(a)** Amplifier gain is set too high.
   **(b)** Beam attenuator (optical null system) is nonlinear.
   **(c)** Stray radiation levels are high in some wavelength regions.
   **(d)** The solvent absorbs strongly in a certain region of the spectrum.
   **(e)** The spectrum is scanned too rapidly.

**14-7.** An infrared spectrum was acquired for a high-boiling liquid with an empirical formula of $C_9H_{10}O$. Band intensities were described as weak (w), medium (m), and strong (s). Major absorption bands were found at 3000 (w), 1690 (s), 1610 (w), 1575 (w), 1360 (s), 1300 (m), 1250 (s), 1077 (w), 1048 (w), 980 (m), 760 (s), and 725 (m) cm$^{-1}$. Draw as many conclusions as feasible as to the identity of the compound.

**14-8.** What distances must the mirror be driven in an FTIR spectrometer in order to achieve 0.05-, 0.5-, and 4.0-cm$^{-1}$ resolution?

**14-9.** Criticize or defend the following quotation. Use as many specifics as you can to justify whether or not you agree with the statement.

Over most of the mid-infrared spectral range, Fourier transform instruments appear to have signal-to-noise ratios that are better than those of a good-quality dispersive instrument by more than an order of magnitude. However, the Fourier transform method offers little or no advantage (other than shortened analysis time) over a good-quality grating spectrophotometer for routine qualitative applications in the mid-infrared region. Furthermore, it suffers by comparison in terms of high initial cost and substantial maintenance problems. The latter arise because the quality of a Fourier transform spectrum degrades much more rapidly with instrument maladjustment than does a spectrum produced by a grating instrument.

**14-10.** In scanning the infrared absorption spectrum of a solution to identify a new product in a reaction mixture, the slits open very wide in the spectral region in which the product should absorb. The source is still quite intense in this region and the detector is still quite responsive. Give possible reasons why this occurs and what errors it might cause. State how the problem might be corrected.

**14-11.** Why are nondispersive IR instruments often used for the determination of gases instead of dispersive IR spectrophotometers?

**14-12.** Contrast the use of mulls and salt pellets for the infrared analysis of solids.

**14-13.** The mid-IR region is the most common region used in IR spectrophotometry. Discuss the unique information provided and special instrumental requirements for measurements in the near-IR and far-IR regions.

**14-14.** What is the purpose of an integrating sphere in diffuse reflectance measurements?

**14-15.** Contrast conventional IR absorption measurements with attenuated total reflection measurements.

**14-16.** In an ATR measurement, the sample has a refractive index of 1.37 at 3.9 μm. The ATR crystal was AgCl, which has a refractive index of 2.00 at this wavelength. If the angle of incidence is 45°, what is the effective penetration depth of the evanescent wave? How would this change if the angle of incidence was 60°?

**14-17.** The X—H fundamental vibration in a molecule containing the X—H functional group occurs at 3010 cm$^{-1}$. If deuterium were substituted for hydrogen, at what wavenumber would you expect the X—D vibration?

**14-18.** Single-beam instruments are quite useful for UV-visible absorption measurements, but are not often used in the IR region. Explain.

**14-19.** The C≡N stretching vibration in HCN occurs at 2006 cm$^{-1}$. What is the force constant of the C≡N bond?

**14-20.** In the mid-IR region, the *S/N* of absorption measurements is rarely limited by signal shot noise as it often is in the UV-visible region. Discuss the reasons for this.

**14-21.** In IR spectrophotometers, the source is usually modulated at 30 Hz or 15 Hz. Discuss the advantages and the disadvantages of using these modulation frequencies.

# REFERENCES

The following are general references dealing with infrared spectroscopy.

1. W. J. Potts, *Chemical Infrared Spectroscopy*, vol. 1, Wiley, New York, 1963. A classic reference filled with practical information.

2. A. L. Smith, "Infrared Spectroscopy," in *Treatise on Analytical Chemistry*, 2nd ed., pt. 1, vol. 7, chap. 5, I. M. Kolthoff and P. J. Elving, eds., Wiley-Interscience, New York, 1981. A fairly recent treatment of IR methods.

3. A. L. Smith, *Applied Infrared Spectroscopy*, Wiley, New York, 1979. One of the newer books with comprehensive IR coverage.

4. J. E. Stewart, *Infrared Spectroscopy: Experimental Methods and Techniques*, Marcel Dekker, New York,

1970. An excellent book on IR instrumentation, including electronics and optics.

5. D. N. Kendall, *Applied Infrared Spectroscopy*, Reinhold, London, 1966.

6. G. M. Barrow, *Introduction to Molecular Spectroscopy*, McGraw-Hill, New York, 1962.

7. N. B. Colthup, L. H. Daly, and S. E. Wiberley, *Introduction to Infrared and Raman Spectroscopy*, 2nd ed., Academic Press, New York, 1975.

8. R. T. Conley, *Infrared Spectroscopy*, 2nd ed., Allyn and Bacon, Boston, 1972.

9. G. Herzberg, *Molecular Spectra and Molecular Structure II: Infrared and Raman Spectra of Polyatomic Molecules*, Van Nostrand, New York, 1945.

10. A. L. Smith, "Infrared Spectroscopy," in *Handbook of*

*Spectroscopy*, vol. II, J. W. Robinson, ed., CRC Press, Boca Raton, Fla., 1974.

The following references deal specifically with Fourier transform infrared methods.

11. P. R. Griffiths, ed., *Transform Techniques in Chemistry*, Plenum Press, New York, 1978. Covers FTIR as well as FT mass spectrometry, FT NMR, and other transform methods.

12. P. R. Griffiths and J. A. deHaseth, *Fourier Transform Infrared Spectroscopy*, Wiley, New York, 1986. A comprehensive treatment of FTIR theory and practice.

13. P. R. Griffiths, *Chemical Infrared Fourier Transform Spectroscopy*, Wiley, New York, 1975.

14. R. J. Bell, *Introductory Fourier Transform Spectroscopy*, Academic Press, New York, 1972.

15. J. Ferraro and L. Basile, eds., *Fourier Transform Infrared Spectroscopy*, Academic Press, New York, 1979.

16. A. E. Martin, "Infrared Interferometric Spectrometers," in *Vibrational Spectra and Structure*, vol. 8, J. R. Durig, ed., Elsevier, Amsterdam, 1980.

17. J. R. Durig, *Analytical Applications of FT-IR to Molecular and Biological Systems*, D. Reidel, Dordrecht, The Netherlands, 1980. This book presents the proceedings of a NATA study institute in 1979.

18. G. A. Vanasse, ed., *Spectrometric Techniques*, vols. I–IV, Academic Press, New York, 1977–1985. There are chapters in this set that deal with interferometers, FTIR, diode lasers for IR, and many other topics of interest.

The following references deal with specific topics of concern in this chapter.

19. J. F. Butler, K. W. Nill, A. W. Mantz, and R. S. Eng, "Applications of Tunable-Diode Laser IR Spectroscopy to Chemical Analysis," in *New Applications of Lasers to Chemistry*, G. M. Hieftje, ed., American Chemical Society, Washington, D.C., 1978.

20. N. J. Harrick, *Internal Reflection Spectroscopy*, Wiley-Interscience, New York, 1967.

21. G. Kortum, *Reflectance Spectroscopy*, Springer, New York, 1969.

22. D. L. Wetzel, "Near-Infrared Reflectance Analysis," *Anal. Chem.*, 55, 1165A (1983).

23. B. R. Buchanan and D. E. Honigs, "Advances in Near-Infrared Spectroscopy," *Spectroscopy*, 1(7) (1986).

# CHAPTER 15

# Molecular Luminescence Spectrometry

Molecular luminescence spectrometry is concerned with the measurement of photon emission from molecules. Analytical luminescence spectrometry can be based on *photoluminescence* or *chemiluminescence*. *Fluorescence* and *phosphorescence* spectrometry are types of photoluminescence spectrometry in which excitation occurs by absorption of photons. Fluorescence involves emission between states of the same multiplicity, usually singlet to singlet; phosphorescence involves radiational transitions between states of different multiplicity, usually triplet to singlet. In chemiluminescence spectrometry, emission occurs from excited states produced by a chemical reaction.

The phenomenon of fluorescence has been known for many centuries. In 1852, Stokes introduced the term "fluorescence" and later identified what is now known as the Stokes shift, in which the emitted radiation is of a longer wavelength than that of the exciting radiation. He even used two colored filters to select the excitation and emission wavelengths. In the 1860s, Stokes proposed and demonstrated fluorometric analysis. By the 1930s, phototubes and photomultipliers replaced photographic plates as radiation detectors and allowed more convenient and lower-light-level measurements. The first commercial spectrofluorometer, based on a design by Bowman, Caulfield, and Udenfriend, was introduced in 1955 by the American Instrument Company. Since that time, instrumentation has been continually improved, and the number of applications has grown enormously. The excellent detection limits, simplicity, and selectivity of fluorescence has spurred the development and adoption of fluorometry as a major technique in chemistry, biochemistry, biology and medicine.

Phosphorescence was first recognized in 1568 by Cellini. The concept that phosphorescence involves absorption and remission of light was established in the seventeenth century, about two centuries before the realization that it applied to fluorescent materials as well. In the nineteenth century, Edmond Becquerel developed the first **phosphoroscope,** a mechanical device allowing the sequential excitation and viewing of phosphorescence. This allowed the duration and exponential decay of phosphorescence to be observed. In 1894, James Dewar observed phosphorescence from solutions cooled by liquid air. Lewis and Kasha in 1944 proposed that phosphorescence could be used to identify organic compounds. The first paper on quantitative phosphorimetry by Kiers, Britt, and Wentworth did not appear until 1957. Important advances in phosphorescence instrumentation were reported by Winefordner's group in the 1960s. More recently, there has been considerable activity in the area of room-temperature phosphorescence.

The use of chemiluminescence as an analytical technique has emerged relatively recently compared to photoluminescence techniques. Chemiluminescence from living organisms such as fireflies (bioluminescence) has been observed since antiquity. Chemiluminescence due to oxidation by oxygen or peroxides of phophorous, halogens, and organic species was observed by many workers late in the eighteenth century and the nineteenth century. The first uses of chemiluminescence in analytical chemistry date back to the 1930s, in which chemiluminescent species were used as endpoint indicators in titration reactions. Quantitative techniques based on relating the chemiluminescence intensity to the analyte concentration have only appeared in the last two decades.

This chapter begins with a discussion of molecular luminescence instrumentation. Next signal and *S/N* expressions are developed and used to discuss optimization of measurement conditions. The next three sections separately address the special considerations, methodology, performance characteristics, and applications of fluorescence, phosphorescence, and chemiluminescence measurements. Section 15-6 deals with more specialized applications of luminescence measurements, including lifetime and polarization measurements.

## 15-1 INSTRUMENTATION

The design of luminescence spectrometers and photometers is profoundly affected by the very low radiant power normally detected (often picowatts or less). By contrast, the transmitted radiant power in molecular absorption measurements is usually in the range 1 to 1000 nW. The small analyte luminescence signal must be distinguished from background signals, the dark signal, and noise in these signals.

In photoluminescence measurements, the luminescence signal is proportional to the radiant power absorbed by the analyte (see equations 2-17 to 2-21). Usually, analyte concentrations are in the range 1 μM to 1 pM for which the absorbed radiant power and the luminescence signal are linearly related to concentration. For these low analyte concentrations, only a small fraction of the incident radiation is absorbed (typically $10^{-6}$ to 1%). The luminescence quantum efficiency determines the fraction of the absorbed radiant power that is converted to luminescence. Although the quantum efficiency approaches unity in favorable cases, it is often considerably less. Because luminescence is isotropic, only a small fraction of the luminescence emitted in all directions is collected and imaged on the detector. Usu-

ally, only a fraction of the wavelength range of emission is monitored.

For many of the reasons noted above, chemiluminescence signals are also quite small. However, for this type of luminescence, the efficiency of the chemiluminescence reaction, rather than the source intensity, determines the number of excited molecules produced.

In this discussion of luminescence instrumentation, we emphasize molecular fluorescence spectrometry, the most widely used luminescence technique. However, many of the design principles apply equally to phosphorescence and chemiluminescence spectrometry. The special instrumental requirements for the latter techniques are discussed in Sections 15-4 and 15-5.

The basic component schematic for photoluminescence measurements is shown in Figure 15-1. The source and excitation wavelength selector and the auxiliary optics define the characteristics of the excitation or primary beam incident on the sample. The emission optics and wavelength selector determine the characteristics of the emission or secondary beam that is incident on the detector. Note that the emission is normally collected at 90° with respect to the excitation axis. Other configurations are discussed later in this section. For chemiluminescence measurements, the excitation source and wavelength selector are not required.

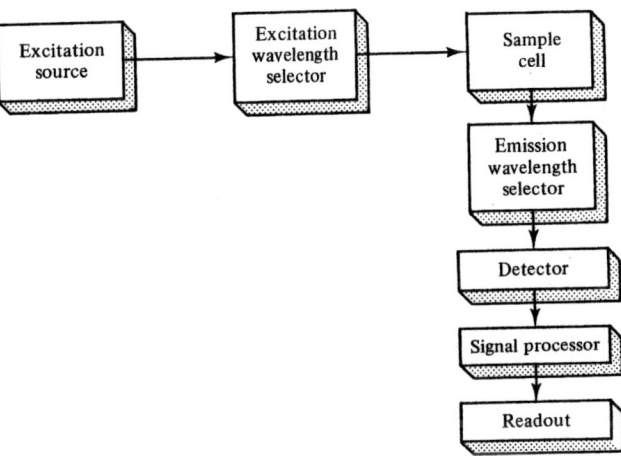

**FIGURE 15-1** Block diagram of a molecular photoluminescence spectrometer. Radiation from the excitation source is imaged onto the excitation wavelength selector, which isolates a particular wavelength band to be incident on the sample cell. A portion of the luminescence emitted is normally collected at right angles to the excitation axis and imaged onto the emission wavelength selector. The portion of the luminescence radiant power over the emission wavelength range selected is sensed by the detector. The detector signal is processed to yield a readout signal proportional to the luminescence.

If only filters are used for wavelength selection, the instrument is designated a **fluorometer.** In a **spectrofluorometer,** monochromators are employed. Typical optical configurations for both types of instruments are shown in Figure 15-2. Hybrid instruments using one filter and one monochromator are available; they are usually still referred to as spectrofluorometers. The most common hybrid configuration is based on an excitation monochromator and an emission filter. Note that the spellings *fluorimeter* and *fluorimetry* are commonly used in Europe. For phosphorescence measurements, the spelling **phosphorimetry** is always used. In this book we adopt the common U.S. spellings of *fluorometry* and *phosphorimetry.* **Fluorometry** refers to fluorometric measurements conducted with either a fluorometer or a spectrofluorometer.

### Excitation Sources

Under most conditions, the photoluminescence signal is directly proportional to the incident radiant power. Therefore, the ideal excitation source should provide a

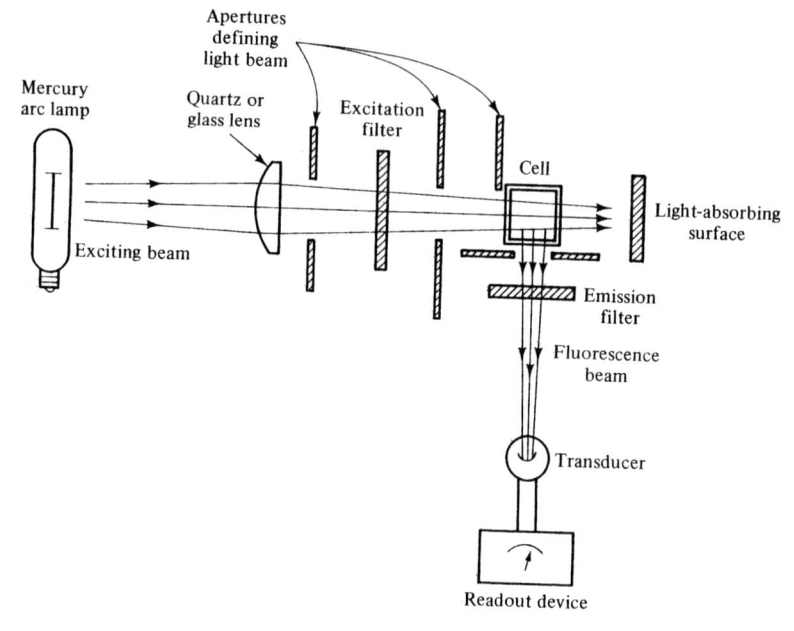

(a)

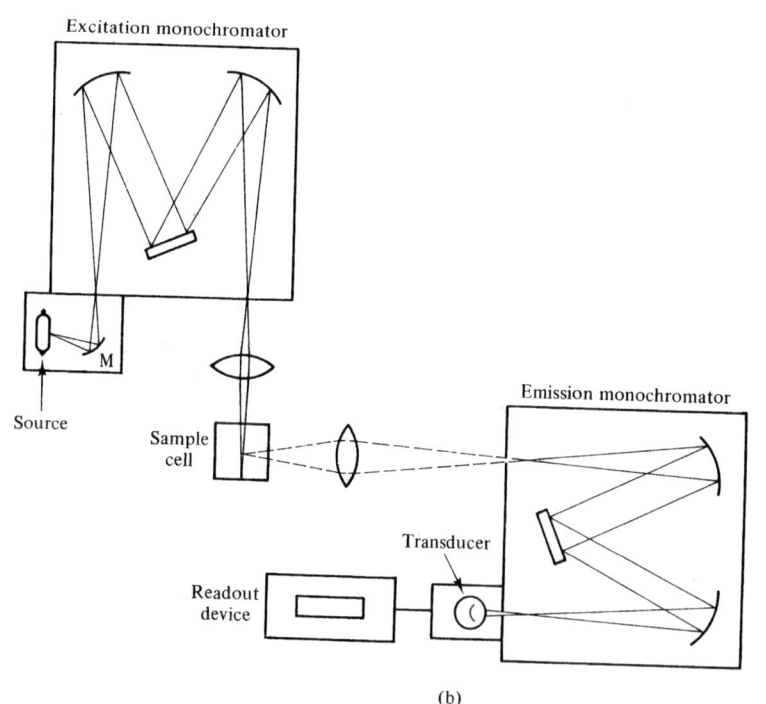

(b)

**FIGURE 15-2** Optical diagrams of a typical fluorometer and a typical single-beam spectrofluorometer. For the fluorometer shown in (a), apertures are used to define the width of the excitation beam entering the sample cell and the width of the emission beam viewed by the detector. With the spectrofluorometer shown in (b), a mirror is used to focus an image of the source onto the entrance slit of the excitation monochromator. One or more lenses, between the exit slit of the excitation monochromator and the cell, are used to image a narrow beam of excitation radiation into the cell. Another lens or combination of lenses collects and images the sample fluorescence onto the entrance slit of the emission monochromator.

stable, high radiance or irradiance at the excitation wavelengths of interest. For conventional sources, appropriate lenses or mirrors are used to maximize the solid angle and the area of source radiation collected and to focus an image of the source onto an aperture or into the sample cell. Elliptical reflectors provide the most efficient radiation collection (over $2\pi$ sr), although the lamp image magnification is large. The reflector and lamp can be separate units or one integral unit (see Figure 4-4e).

For simple fluorometers, a low-pressure mercury arc lamp is the most common source. The intense lines at 254, 312, and 365 nm are suitable excitation wavelengths for many molecules. These lamps can also be coated with phosphors to shift the wavelength to longer values (e.g., 254 to 280 nm is common) with a significant decrease in radiance.

In commercial spectrofluorometers, the most common source is a 75- to 450-W high-pressure Xe arc lamp (see Figure 4-4). High-pressure Hg and Xe-Hg lamps are also employed. The radiance at the Hg emission wavelengths is much greater than that produced by Xe at these wavelengths (see Figure 4-6). Arc lamps require a large power supply capable of providing 15 to 30 V at 5 to 20 A. Often the supply is configured to adjust the voltage across the arc to maintain a constant arc current and a more stable radiance. In arc lamps, there is a tendency, particularly during warm-up, for the arc to change position between the electrodes; this is denoted **arc wander.** Arc wandering can cause sudden variations in the observed luminescence signal, especially if the image of the arc is focused on a small slit aperture. Sometimes arc wander can be reduced by placing the arc lamp in a varying magnetic field (e.g., from a magnetic stirrer). To prevent danger from toxic ozone generated by high-pressure lamps, the lamp housing should be sealed or have provision for de-ozonating or venting the contents out of the laboratory. Pulsed Xe arc lamps are gaining popularity as luminescence excitation sources. They provide a larger relative radiance below 300 nm compared to arc lamps operated in the dc mode and can be used for time-resolved phosphorescence measurements. Deuterium lamps can be employed for excitation at wavelengths lower than 300 nm, where the radiance of dc powered Xe and Hg arc lamps falls off dramatically.

It might be expected that the much greater irradiance of lasers compared to conventional sources would be ideal for molecular photoluminescence measurements. Additionally, no excitation wavelength selector is required. However, lasers are presently not used in commercial systems except those designed for lifetime measurements. Detection limits with laser excitation are often no better than those obtained with high-intensity conventional sources if background lumines-

cence or scattering flicker noise is dominant at the detection limit. Compared to conventional sources, lasers are more expensive, more complex, more difficult to maintain (i.e., more downtime), and often less stable. Additionally, photodecomposition can be a problem with high excitation irradiance. Pulsed lasers and Xe lamps required more complex and expensive signal processors (e.g., boxcar integrators).

Laser-based spectrofluorometers have been constructed, however, in many research laboratories. The most popular lasers are the pulsed nitrogen laser (337.1-nm line), the argon laser (488.8- and 514.6-nm lines), and tunable dye lasers. The wavelength provided by the $N_2$ laser is suitable for excitation of many species. The wavelengths provided by the Ar ion laser are somewhat limiting, as the fraction of molecules that are excited to fluorescence at these wavelengths is small. Dye lasers are the most versatile laser source because the excitation wavelength can be optimized for the molecule of interest. Pulsed dye lasers with pulse widths less than 1 ns are well suited for lifetime measurements.

## Wavelength-Selection Devices

The central wavelength and bandpass of the excitation and emission wavelength selectors (e.g., filters or monochromators) are chosen to maximize the fluorescence signal and to minimize the background signal. To prevent the direct viewing of elastically scattered source radiation, the wavelength ranges passed by the excitation and emission wavelength selectors should not overlap. Increasing the excitation bandpass with continuum sources has the benefit of increasing the incident radiant power. Often, the excitation bandpass is larger than one-tenth the half-width of the absorption band. At first thought, one might expect this condition to cause nonlinearity due to polychromatic radiation as it does in absorption spectrometry. However, as we will see in the next section, the amount of radiation power absorbed by the analyte for sufficiently low analyte concentrations is proportional to the analyte concentration even if the molar absorptivity varies over the excitation bandpass. Small excitation bandpasses can provide more selectivity by restricting the excitation of fluorescence from concomitants.

Because the emission band is usually broad (20 to 100 nm), increasing the emission bandpass results in a larger fluorescence signal. This is tempered by the need to distinguish the analyte fluorescence from background fluorescence and Raman scattering. At low analyte concentrations or in the absence of self-absorption, the profile of a fluorescence band is independent of concentration. Therefore, a large emission bandpass does not cause nonlinearity even though the fluorescence profile varies over the bandpass.

With the foregoing considerations in mind, we should not be surprised that filters are often used as high-throughput wavelength selectors. The throughput with a filter can be 10 to 100 times greater than with a monochromator because of the larger bandpass and faster collection optics (e.g., F/1 to F/2) that can be used. Absorption band and cutoff filters transmit a broader wavelength range and are adequate for isolating one or two lines from a low-pressure Hg source. These types of filters can be used to transmit most of the wavelength range over which analyte fluorescence occurs. The narrower bandpass of interference filters provides more selectivity compared to absorption filters. The throughput is less because of this and the transmittance is lower at the maximum.

When filters are used, they should be selected for low-fluorescence characteristics. Fluorescence generated in the excitation filter can be scattered and detected if a portion of the wavelength range of filter emission overlaps the emission wavelength range viewed. Similarly, scattered excitation radiation that reaches the emission filter may be blocked, but it can still generate fluorescence seen by the detector. Emission interference filters should be blocked to eliminate second-order effects. Where extra selectivity is required, two or more filters are combined to define a narrower bandpass.

Normally, medium-resolution monochromators (reciprocal linear dispersion of 2 to 10 nm mm$^{-1}$) are used in spectrofluorometers. A wavelength resolution of 1 nm is more than adequate for most applications. For quantitative analysis relatively large slit widths (0.5 to 2 mm) and spectral bandpasses (4 to 20 nm) are employed to increase the detected fluorescence radiant power. Large slit widths increase the area of the excitation source or the volume element of the sample viewed. Larger values of spectral bandpass increase the range of excitation wavelengths incident on the sample or the range of emission wavelengths viewed. Low-F/n monochromators (typically F/3 to F/4.5) are used to increase the solid angles of collection of excitation and emission radiation. For maximum optical efficiency, the blaze wavelength for the excitation monochromator is typically in the region 300 to 400 nm, while that for the emission monochromator is 450 to 550 nm.

Low stray radiation is critical in fluorescence measurements. A small fraction of the source radiation at longer wavelengths can be passed by the excitation monochromator as a stray radiation component, scattered by the sample, transmitted by the emission monochromator or filter, and detected. More important, elastically scattered excitation radiation can be passed as a stray radiation component of the emission monochromator. Stray radiation effects are reduced by using holographic gratings, double monochromators, or an additional short-wavelength cutoff filter in the emission beam. In some cases, an IR absorbing filter (e.g., long-wavelength cutoff filter or a cell containing water) is placed between the source and excitation monochromator or filter to block some of the intense IR radiation emitted by high-pressure arc lamps. This reduces sample heating and potential thermal damage to filters, slits, and so on.

## Sample Compartment and Sample Cells

The sample compartment contains a suitable cell holder that allows simple and reproducible positioning of the sample cell. It is very critical that the access lid and the rest of the sample compartment be perfectly lighttight. As with molecular absorption instruments, there may be provision for holding and positioning several sample cells, for controlling the cell temperature, and for stirring the contents of the cell.

Some of the optical components for imaging excitation radiation or collecting the emission radiation may also be in the sample compartment. Polarizers are sometimes added, as discussed in Section 15-6. A light trap or blackened surface is often used to trap or absorb the transmitted excitation radiation. A fraction of this radiation can be scattered or reflected by surfaces inside the sample compartment and contribute to the signal seen by the detector. A manual or automatic shutter, located either in the sample compartment or at some other point in the excitation beam, is useful for measurements of species that undergo photodecomposition. The shutter is opened only during the period of measurement, so that sample irradiation time is minimized.

Several different types of sample cells are used. Test-tube-shaped cells are adequate for simple fluorometers. More often, a standard square sample cell (1 × 1 × 4.5 cm) is employed (see Figure 13-6d); they are polished on all sides, unlike some absorption cells, which need only be polished on two opposite sides. Although normal glass is adequate for some measurements, high-quality fused synthetic silica is the best construction material. It provides high transmission for lower-UV-excitation wavelengths and exhibits the least fluorescence.

Micro sample cells (see Figure 13-6b) are available for sample-limited situations. Low-volume fluorescence flow cells are used for fluorescence detection in flowing streams (e.g., continuous flow analyzers and HPLC detectors). The types of fluorescence cells described above are also used for solution room-temperature phosphorescence measurements and solution chemiluminescence measurements. Specialized sample cells and sampling handling are required for low-temperature luminescence measurements.

## Cell Geometry

The three basic excitation-emission geometries for sample cell excitation and viewing are illustrated in Figure 15-3. The right-angle geometry is most common because at this angle, the elastic scattering signal is the least intense (see Chapter 16). Also, the contribution of cell wall photoluminescence is reduced because the cell walls directly illuminated by the excitation beam are not directly viewed. Additional apertures or baffles adjacent to the excitation and emission cell walls are often used to restrict the section of the cell walls that are illuminated or observed. With the right-angle configuration, the luminescence from a small volume element in the middle of the cell is normally viewed. This presents a problem in highly absorbing solutions because a significant portion of the excitation radiation is absorbed before it reaches the central portion of the cell (prefilter effect) or a significant portion of the luminescence can be absorbed before exiting from the sample cell (postfilter effect). These effects are described quantitatively in the next section.

The frontal geometry is useful for solutions of high absorbance or opaque solids because the excited and viewed volume element is moved next to a common cell wall; the attenuation of the luminescence signal by absorbers is reduced. The in-line geometry is rarely employed because the emission wavelength selector must isolate the intense transmitted excitation radiation from

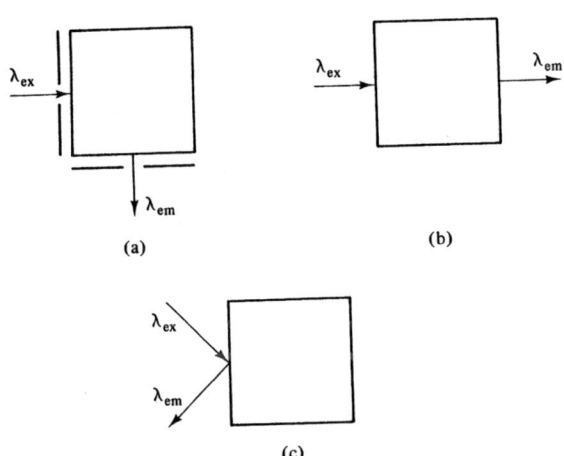

**FIGURE 15-3** Three common cell geometries. In the right-angle configuration (a), luminescence is viewed at 90° with respect to the excitation axis. Mirrors can be placed at the cell walls opposite the excitation and viewing walls to increase the detected luminescence signal by about 50%. In (b) the in-line geometry is illustrated. The luminescence is viewed along the excitation axis. The same cell wall is illuminated and viewed in the frontal geometry (c).

the weak luminescence signal. This configuration does allow absorption spectra to be acquired if the excitation and emission wavelengths are identical. For both the frontal and in-line geometries, an illuminated cell wall is directly viewed.

## Detectors

The low luminescence radiant power normally observed precludes the use of most detectors. Therefore, a photomultiplier tube (PMT) is normally used for fixed-wavelength measurements or slow-scanning experiments. It should be selected for low dark current and dark current noise. Complete emission spectra can be acquired rapidly with a multichannel detector interfaced to an emission spectrograph. This is particularly useful for kinetic studies and chemiluminescence measurements where the concentration of the luminescent species (and thus the luminescence signal) is changing with time. Both intensified diode arrays and silicon-intensified target vidicons (SITs) have been used. Some commercial instruments have provision for installing these devices and spectrofluorometers with dedicated intensified diode arrays are becoming available. To date, multichannel spectrofluorometers have not achieved the popularity of multichannel spectrophotometers.

## Data Processing, Manipulation, and Readout

Usually, the photoanodic current from the PMT is first converted to voltage with an operational amplifier in the current-to-voltage configuration. Often after further voltage amplification, the readout signal is displayed on an analog or digital voltmeter; a servo recorder is required to display spectra. In newer microcomputer-based instruments, the voltage signal is digitized and stored in memory for later manipulation and display on video terminals, printers, or digital plotters.

A zero control is usually provided for adjusting the readout signal to near zero with blank in the sample cell. The signal processing and readout system must be designed to accommodate the large dynamic range of fluorescence signals. Often the electronic or PMT gain can be adjusted by exactly factors of 10 over three to five orders of magnitude.

High-quality analog electronics with low drift and noise are critical. For signal-to-noise ratio ($S/N$) enhancement, the time constant or integration time is adjustable over the range 0.1 to 10 s. If the background signal levels are low, photon counting signal processing can provide an $S/N$ advantage (see Section 5-6); some commercial instruments have a photon counting option.

Also in this case, cooling the PMT to reduce the dark current signal and noise can further enhance the *S/N*.

In some spectrofluorometers, the excitation source radiation is modulated with a mechanical chopper and processed with synchronous detection electronics. This processing discriminates against the dark signal and $1/f$ dark current and amplifier noise. Other specialized signal processing techniques are discussed in later sections.

As for spectrophotometers, incorporation of microcomputers has greatly enhanced the capabilities and convenience of molecular luminescence measurements. Keyboard control of scanning parameters (scan rate and wavelength range), slit widths, and time constants or integration times is often provided. Repetitive measurements or scans can be programmed for *S/N* enhancement. Common data manipulation and display options applied to the stored luminescence data include blank signal subtraction, calculation and display of difference spectra or derivative spectra, fitting of calibration data and calculation of the analyte concentration in samples, calculation of statistical information, and smoothing of spectra. Because the data are stored, high scan rates can be used without distortion by the response time of analog recorders. Later any desired portion (wavelength interval) of the spectrum can be displayed with the desired scale expansion. Software packages are available for specialized applications, including kinetic studies and measurements, HPLC detection, and analysis of mixtures.

## Data Presentation

*Fixed-Wavelength Measurements.* For most quantitative work, the required data are the blank-corrected luminescence signals for the samples and standards. The excitation wavelength ($\lambda_{ex}$) and emission wavelength ($\lambda_{em}$) are fixed.

*Excitation and Emission Spectra.* The photoluminescence properties of a molecule are characterized by its fluorescence and phosphorescence quantum efficiencies and the excitation and emission spectra. The **excitation spectrum** is a plot of the luminescence signal vs. excitation wavelength with a constant emission wavelength and bandpass. It is used to determine the best excitation wavelength for analysis and is related to the absorption spectrum of the analyte. A plot of the luminescence signal vs. the emission wavelength with a constant excitation wavelength and bandpass is denoted an **emission spectrum**. This spectrum indicates the best emission wavelengths for analysis. A given molecule has one excitation spectrum and two emission spectra,

as illustrated in Figure 15-4. For chemiluminescence, only the emission spectra are relevant.

*Total Luminescence Spectra.* A more complete picture of the spectral characteristics of a luminescence molecule can be obtained by plotting a **total luminescence spectrum,** as shown in Figure 15-5. It is seen to be a three-dimensional plot or a contour map of the luminescence signal as a function of both the excitation and emission wavelengths. The data acquired are often termed the **excitation-emission matrix.**

The acquisition and display of such a plot can be accomplished with a computerized single-channel spectrofluorometer. This procedure requires acquiring and storing a large number of emission scans with the excitation wavelength incremented to a new value for each scan. The data acquisition time can be reduced by using an emission spectrograph and a multichannel detector to obtain each emission spectrum more rapidly. The **video fluorometer** design shown in Figure 15-6 provides a rapid and novel means of acquiring the data for total luminescence spectra. Once the data for a three-dimensional spectrum are acquired, a conventional excitation or emission spectrum can be plotted by selecting the portion of the data to display.

*Synchronous Spectra.* Some spectrofluorometers allow both the excitation and emission mono-

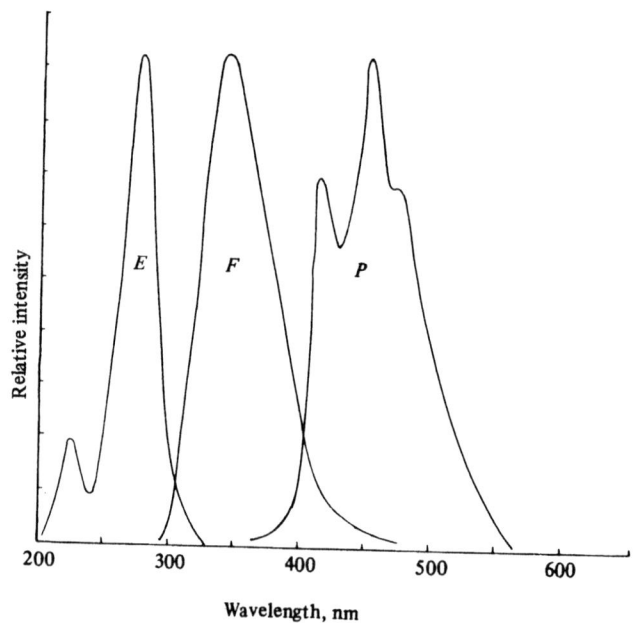

**FIGURE 15-4** The excitation (*E*), fluorescence (*F*), and phosphorescence (*P*) spectra of the amino acid tryptophan. (Reprinted from G. G. Guilbault, *Practical Fluorescence*, Marcel Dekker, New York, 1973, p. 164; courtesy of Marcel Dekker, Inc.)

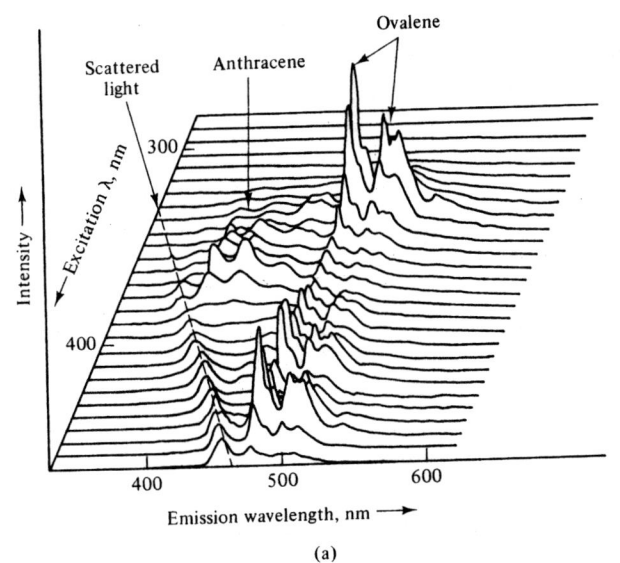

(a)

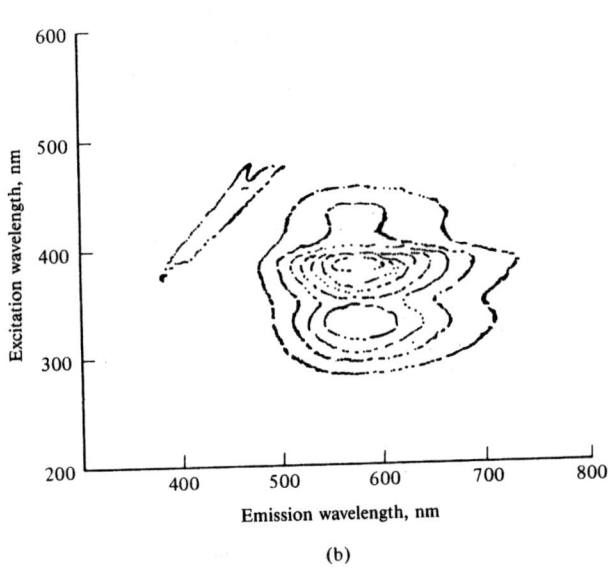

(b)

**FIGURE 15-5** Total luminescence spectra. In (a) the total fluorescence spectrum of a mixture of anthracene and ovalene is represented as a three-dimensional plot. Each horizontal slice represents a separate emission scan with a fixed excitation wavelength. A diagonal slice parallel to the excitation axis represents an excitation spectrum. The signals along the dashed line are due to elastically scattered excitation radiation. The total fluorescence spectrum for 8-hydroxybenzo[a]pyrene is depicted in (b) as a contour plot, sometimes called a fluorogram. Each line represents a particular magnitude of the luminescence signal. The oblong region is due to scattered radiation. [(a) From Y. Talmi et al., *Anal. Chem., 50,* 936A (1978); (b) adapted from J. H. Rho and J. L. Stewart, *Anal. Chem., 50,* 620 (1978); copyright 1978, with permission of the American Chemical Society.]

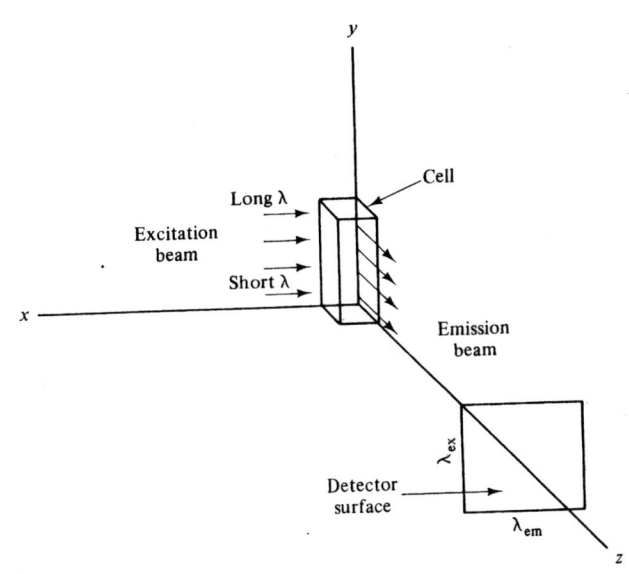

**FIGURE 15-6** Principles of operation of a video fluorometer. Compared to a conventional spectrofluorometer, the sample cell is illuminated simultaneously with all appropriate excitation wavelengths. This is accomplished by rotating the excitation monochromator 90° so that the entrance slit is horizontal. The exit slit is removed leaving a slot in the focal plane. Thus images of different wavelengths are focused at different vertical positions in the sample in the *xy* plane. An image of the cuvette is focused on the entrance slit of a spectrograph with a silicon intensified target vidicon (SIT) detector. Emission radiation is dispersed in the *xz* plane. Each horizontal slice in the two-dimensional image focused on the detector represents an emission spectrum with an excitation wavelength corresponding to a vertical position in the cell.

chromator to be simultaneously scanned at identical rates with a fixed-wavelength difference ($\Delta\lambda = \lambda_{em} - \lambda_{ex}$). The resulting spectrum is called a **synchronous spectrum.** A fluorescence signal is observed only at wavelengths where both excitation and emission occur for a wavelength separation $\Delta\lambda$, as illustrated in Figure 15-7. The magnitude of $\Delta\lambda$ is critical. In general, the simplicity of the spectrum increases as $\Delta\lambda$ is decreased. A synchronous spectrum can be generated from the data for a total luminescence spectrum by software selection of $\Delta\lambda$. By scanning the emission monochromator at a slightly faster rate than the excitation monochromator, a synchronous spectrum with a constant energy difference ($\Delta\nu$) can be obtained. Ideally, only one peak is obtained when $\Delta\nu$ is set to the Stokes shift of the fluorophore of interest.

***Derivative and Wavelength Modulation Spectra.*** In **wavelength modulation spectrometry,** the

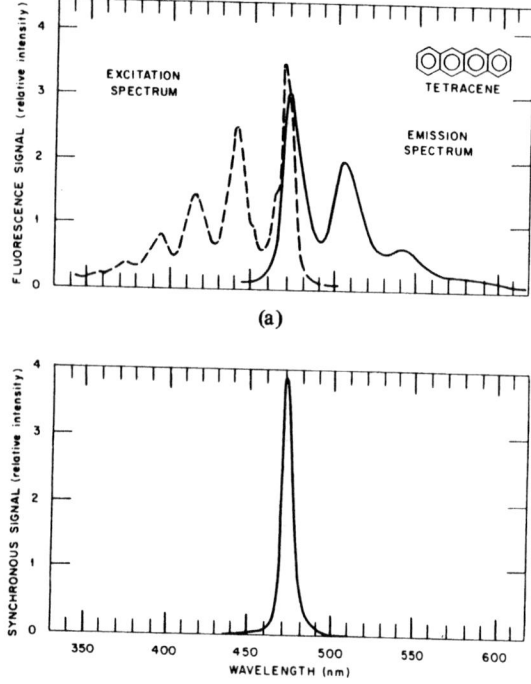

**(a)**

**(b)**

**FIGURE 15-7** Synchronous spectra. The excitation spectrum (dashed-line) and emission spectrum of tetracene is shown in (a). In (b) the synchronous spectrum is illustrated. Here the excitation and emission monochromators are scanned simultaneously with a fixed-wavelength difference of 3 nm. [From T. Vo-Dinh, *Anal. Chem.*, *50*, 396 (1978), copyright 1978, with permission of the American Chemical Society.]

wavelength to which the monochromator is tuned is rapidly scanned back and forth over a wavelength interval called the **wavelength modulation interval**, $\Delta\lambda$. This is accomplished by oscillating an optical component (e.g., the grating or an interference filter) or a refractor plate near the monochromator exit slit (Figure 4-34). The average or central wavelength can be fixed or scanned to produce a spectrum. This technique produces results similar to those obtained with the dual-wavelength technique employed in spectrophotometry. In photoluminescence spectrometry, either the excitation or the emission wavelength can be modulated.

Wavelength modulation produces a detector signal with a component whose magnitude is related to the change in signal over $\Delta\lambda$. Normally, a lock-in amplifier extracts the amplitude information. As shown in Figure 15-8, wavelength modulation over a relatively large interval can be used to discriminate against spectral overlap interferences.

As in spectrophotometry, *derivative spectra* are used to accentuate subtle spectral details, particularly in mixtures. Modulation of the emission wavelength

with a small $\Delta\lambda$ provides derivative information; the amplitude of the first and second harmonic signals is proportional to the first and second derivatives, respectively (see Figure 4-34). The first derivative can also be obtained by analog differentiation of the emission signal during scanning. Any order derivative can be obtained by applying computer software-based numerical techniques to stored spectral data. The $S/N$ of derivative spectra can be worse than that of normal spectra because noise is more emphasized than the rate of change of the signal.

## Compensation and Correction Techniques

Source stability, the wavelength dependence of the efficiency of optical components, and the absorption characteristics of the test solution affect the magnitude of luminescence signals, calibration curve linearity, or the magnitude and shape of luminescence spectra. Many instruments incorporate additional hardware and software to compensate or correct for these effects.

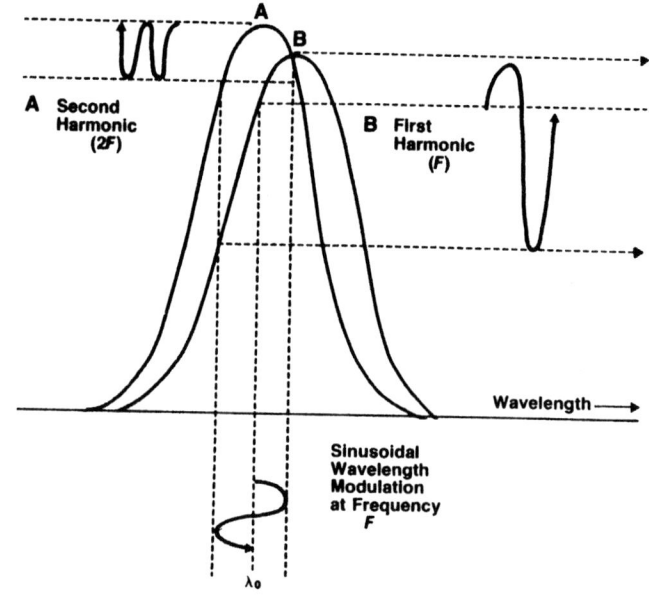

**FIGURE 15-8** Wavelength modulation for discrimination against spectral interferences. The central wavelength ($\lambda_0$) of modulation is adjusted so that $\lambda_0$ corresponds to the maximum of the band for species A. The modulation wavelength interval ($\Delta\lambda$) is situated on one side of the band for species B. Under these conditions, band A generates a second harmonic signal, while band B generates a predominantly first harmonic signal. This approach can be applied to the excitation or emission wavelength. [Adapted from T. C. O'Haver, *Anal. Chem.*, *51*, 91A (1979), copyright 1979 with permission of the American Chemical Society.]

*Source Intensity Compensation.* The drift and flicker noise of arc lamps can be significant. To correct for these nonfundamental temporal changes in lamp radiance, the single-beam instrumentation described earlier in this section is upgraded to a double-beam design. Several possible arrangements are illustrated in Figure 15-9. They are all based on obtaining a reference signal that tracks the source intensity. For the designs shown in Figure 15-9a and b, the reference signal is obtained by directly monitoring a portion of the excitation radiation with a separate detector or the sample fluorescence detector. Another approach (Figure 15-9c) involves diverting a portion of the excitation beam to a sample cell filled with a standard solution of the analyte of interest. The resulting fluorescence is detected and used to generate a reference signal. A reference signal can also be obtained by monitoring the transmitted radiant power. This technique is appropriate if sample absorption is negligible.

Most commonly the sample fluorescence signal is continually ratioed to the reference signal to compensate for source fluctuations and drift. Analog ratio circuitry or division by a computer can be employed. Alternatively, the reference signal is compared to a preset voltage. The difference signal is used in a feedback network to correct for source radiance variations. Either

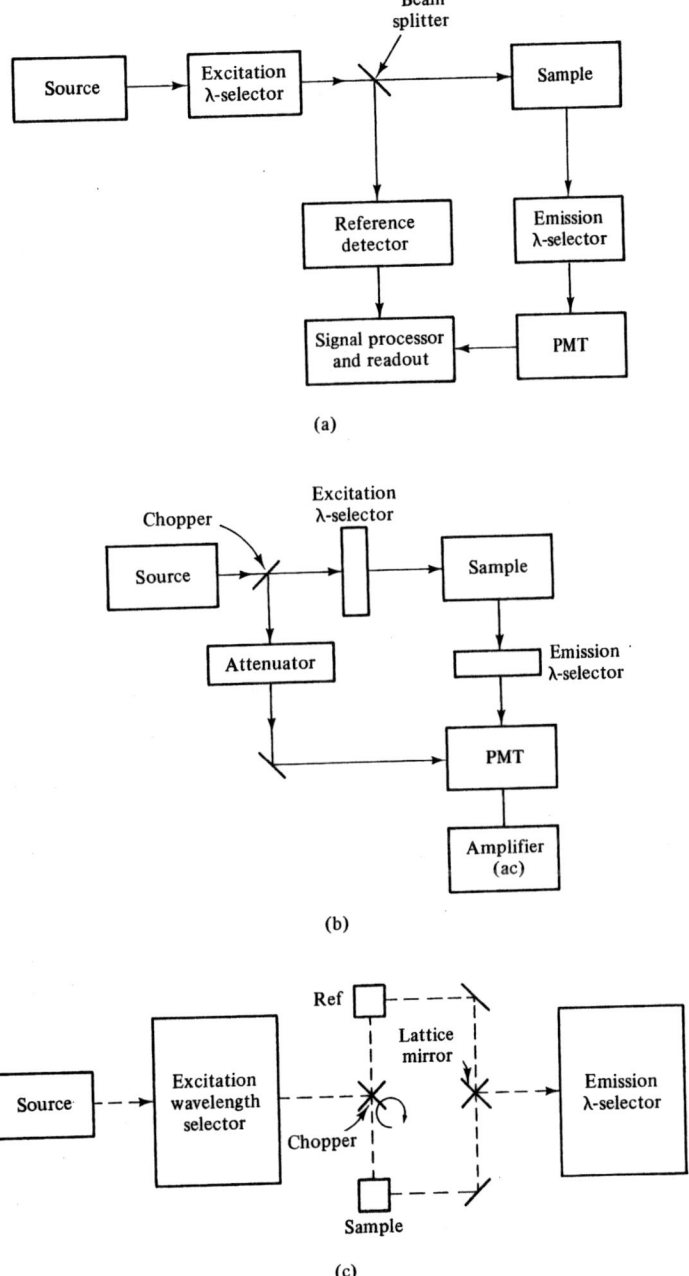

**FIGURE 15-9** Double-beam designs for source compensation or correction. In (a) a double-beam in-space design is shown. A small portion of the radiation passed by the excitation monochromator is reflected by a beam splitter to a reference detector (e.g., PMT, PD, or PT). In some instruments, the beam splitter is placed before the excitation wavelength selector, or the source is directly viewed from a different angle by the reference detector. The latter approach has the disadvantage that the total source radiance may vary differently than the source radiance over a selected wavelength range. In (b) a double-beam in-time approach is illustrated. The source radiation is directed alternately to the sample and to the detector. Thus the detector views the sample fluorescence and the excitation source at different times. The reference signal must be substantially attenuated so that it is approximately equal to the sample fluorescence signal. In (c), the excitation beam is alternately impingent on a sample and reference cell. The fluorescence from both cells is directed through one wavelength selector to one PMT.

the electrical power to the source can be adjusted to maintain constant source radiance (optical feedback) or the sample PMT bias voltage is adjusted to maintain the product of the sample PMT gain and source radiance ($mB_\lambda$) at a constant value (dynode feedback). The null balance method is used in some simple fluorometers. A light attenuator in the reference beam is adjusted until the sample fluorescence and reference signals are identical. The dial controlling the attenuator is calibrated in reference signal units which are proportional to the sample fluorescence signal at null.

Deriving a reference signal from a standard (Figure 15-9c) compensates not only for source radiance fluctuations, but also for changes in the sample fluorescence due to temperature fluctuations that are matched in the two cells. In addition, this configuration allows the difference in the fluorescence signals between the sample and reference solutions to be computed on-line. The difference emission spectrum cancels common features in the spectra of two solutions. It can be used for automated subtraction of blank signal components (e.g., scattering signals or fluorescence of the cell, solvent, or solvent impurities) or for emphasizing the difference in fluorescence characteristics of two similar samples.

*Corrected Excitation Spectra.* Excitation spectra taken with single-beam instruments are uncorrected for the wavelength dependence of the source radiance and the efficiency of the excitation wavelength selector and optical components. Hence the uncorrected excitation spectrum for a given analyte can vary significantly among instruments. These instrumental distortions can be corrected on-line during scanning with a double-beam design such as that used for source intensity compensation (Figure 15-9a). However, for the correction to be accurate, the detector must have a wavelength-independent responsivity. In this case, a plot of the ratio of the luminescence signal to the reference signal vs. excitation wavelength yields a spectrum with a shape that would be obtained if the spectral radiant power incident on the sample was the same at all wavelengths. Thus a properly corrected excitation spectrum indicates how efficiently the analyte is excited at different wavelengths. With PMTs, PTs, or PDs, the reference signal can be quite different at two excitation wavelengths even if the ratio of the radiant power impingent on the sample and the reference detector is identical at both wavelengths.

In commercial instruments, the reference detector is either a thermal detector or a quantum counter. With a thermal detector (see Section 4-3), the correction signal is proportional to total energy in watts at a given wavelength. The slow response speed and poor detectivity of thermal detectors compared to photoelectric

detectors can be limiting in this application. The relative responsivity can also vary several percent over the UV-visible range. The large wavelength range of the responsivity can also be a source of error if the reference radiation beam contains stray radiation in the IR region.

More commonly a quantum counter is employed, as illustrated in Figure 15-10. Physically, a **quantum counter** is a sample cell filled with a concentrated solution of an highly efficient fluorophore. A portion of the excitation beam excites the fluorescence, which is monitored by a PMT. All excitation wavelengths are converted to the same spectral distribution such that the dependence of the PMT responsivity on wavelength is not of concern. The absorption spectrum and concentration of the fluorophore are such that all the excitation radiation is absorbed. If the emission spectrum and quantum efficiency are independent of wavelength, the reference fluorescence signal is directly proportional to source radiant power in photons per second impingent on the sample.

Rhodamine B is, by far, the most widely used compound in quantum counters. At concentrations of 3 to 8 g $L^{-1}$ in glycerol or ethylene glycol, the fluores-

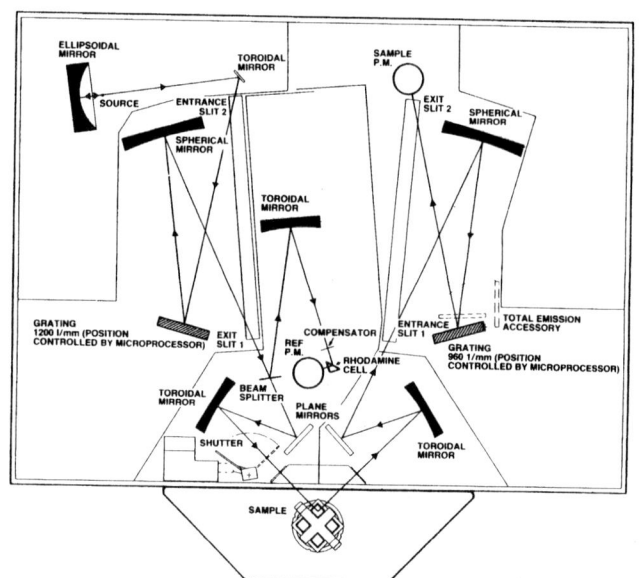

**FIGURE 15-10** Optical diagram of a Perkin-Elmer Model LS-5 spectrofluorometer. Part of the excitation beam from exit slit 1 is reflected by the beam splitter into the quantum counter, the triangular cross section cell containing rhodamine B. The fluorescence excited is detected by the reference PMT. This reference signal is processed and divided into the sample fluorescence signal observed by the sample PMT to provide corrected excitation spectra. Note that a shutter is provided to block the excitation beam between measurements. (Courtesy of Perkin-Elmer Corp.)

cence spectral quantum efficiency over 610 to 620 nm is constant to ±2% for excitation wavelengths in the range 350 to 600 nm and to ±5% down to 250 nm.

The difference between a corrected and uncorrected excitation spectrum is shown in Figure 15-11. A corrected excitation spectrum and the absorption spectrum are often identical under proper conditions. Ideally, the corrected excitation spectrum reflects the wavelength dependence of excitation efficiency which is proportional to the molar absorptivity. The excitation and absorption spectra will differ if significant self-absorption occurs or if the quantum efficiency is dependent on the excitation wavelength. For the highest-accuracy corrections, factors must be included to account for the wavelength dependence of the transmittance and reflectance of the beam splitter and the transmittance of the cell walls of the quantum counter and the sample cell.

*Corrected Emission Spectra.* Uncorrected emission spectra of a given compound differ among instruments because of the wavelength dependence of the efficiency of the emission wavelength selector and

optics and of the responsivity of the PMT. The shape of a properly corrected emission spectrum reflects only how the luminescence efficiency varies with the emission wavelength, as shown in Figure 15-11. Correction is more difficult than for excitation spectra because a real-time correction signal cannot be obtained. Normally, the transfer function of the emission monochromator–PMT combination is determined with a calibrated light source of known spectral radiance (see Section 4-1) or with a compound or series of compounds with known corrected emission spectra. Alternatively, the output spectral radiant power of the excitation source and monochromator system can be calibrated with a quantum counter. After calibration, a reflector or scatterer, with a wavelength-independent reflectivity, is placed in the sample position. The excitation and emission monochromator are set to the same wavelength and scanned simultaneously at the same rate to determine the calibration function. The uncorrected emission spectral data for each resolution element are multiplied before display by the predetermined correction factor for that spectral element. The correction is easily accomplished by software in computerized instruments where the correction factors for each wavelength are stored in memory. In older instruments, the correction factors were programmed into a multitap potentiometer ganged to the emission monochromator wavelength drive.

*Absorption-Corrected Measurements and Spectra.* At high analyte concentrations, significant absorption of the primary (excitation) beam or the secondary (emission) beam by the analyte causes calibration curve nonlinearity and distorts the shape of the excitation and emission spectra. Significant primary or secondary absorption by concomitants attenuates the analyte fluorescence and distorts the spectra. Correction factors for these absorption effects can be obtained by simultaneously measuring the absorbance of the test solution. Correction factors can also be obtained by the **cell-shift method.** In this case the luminescence signals from different volume elements in the sample cell, which experience the primary and secondary absorption effects to different degrees, are measured. The origin of the correction factors is discussed in Section 15-2.

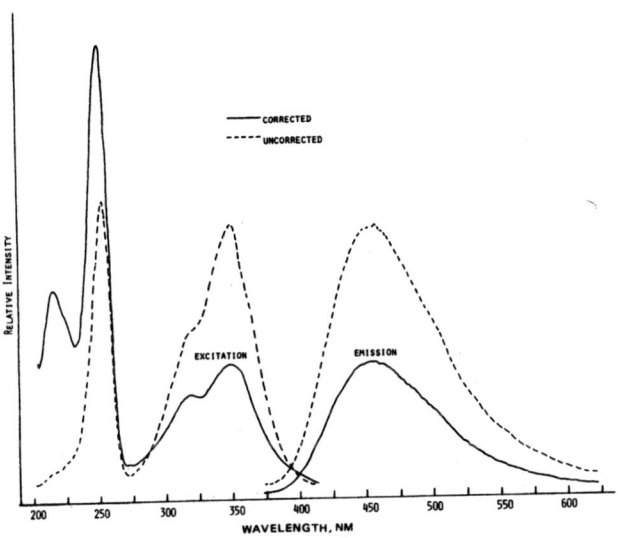

**FIGURE 15-11** Corrected and uncorrected spectra for quinine sulfate in 0.2 M $H_2SO_4$. The relative height of the bands in the corrected excitation spectrum below 300 nm are greater than those above 300 nm because the molar absorptivity for these high-energy bands is greater. The relative intensity of the lower wavelength bands is less in the uncorrected spectrum because the source radiance is less. (From *Natl. Bur. Stand. (U.S.) Tech. Note 584*, p. 55, December 1971.)

## 15-2 SIGNAL AND NOISE EXPRESSIONS

Readout signal expressions are first developed in this section to indicate how the luminescence signal and the shape of the calibration curve depend on instrumental and analyte parameters. Next, the precision and detection limit of luminescence measurements are discussed with the aid of *S/N* expressions.

**TABLE 15-1**

Signal expressions for photoluminescence spectrometry[a]

$$E_L = E_{tL} - E_{bk} \tag{15-1}$$

$$E_{bk} = E_d + E_{bL} + E_{sc} \tag{15-2}$$

$$E_{tL} = E_L + E_{bk} \tag{15-3}$$

$$E_L = mG \int (\Phi_{\lambda L})_0 R(\lambda') \, d\lambda' \tag{15-4}$$

$$\Phi_A^* = \int \Phi_{\lambda A}^* \, d\lambda \tag{15-5}$$

$$\Phi_{\lambda A}^* = (\Phi_\lambda)_0 T_c f_{pr} \alpha_t^*(\lambda) f_a \tag{15-6}$$

$$(\Phi_\lambda)_0 = B_\lambda Y(\lambda) \tag{15-7}$$

$$Y(\lambda) = A_s \Omega_s T_{ex}(\lambda) \tag{15-8}$$

$$T_c = (1 - \alpha_c)(1 - \rho_c) \tag{15-9}$$

$$f_{pr} = 10^{-A_t^{pr}(\lambda)} \tag{15-10}$$

$$\alpha_t^*(\lambda) = 1 - 10^{-A_t^*(\lambda)} \tag{15-11}$$

$$f_a = \frac{A_a^*(\lambda)}{A_t^*(\lambda)} \tag{15-12}$$

$$\Phi_A^* = A_s \Omega_s \int B_\lambda T_{ex}(\lambda) T_c f_{pr} \alpha_t^*(\lambda) f_a \, d\lambda \tag{15-13}$$

$$\Phi_A^* = \Omega_{ex} WHs T_{op} B_\lambda f_{pr} T_c \alpha_t^*(\lambda) f_a \tag{15-14}$$

$$\Phi_A^* = \Omega_s A_s T_i T_f B f_{pr} T_c \alpha_t^*(\lambda) f_a \tag{15-15}$$

$$B_{\lambda L} = \frac{\Phi_A^* Y_{\lambda L} f_s T_c'}{4\pi h b^*} \tag{15-16}$$

$$T_c' = (1 - \rho_c')(1 - \alpha_c') \tag{15-17}$$

$$f_s = 10^{-A_t^{pr}(\lambda')} \frac{1 - 10^{-A_t^w(\lambda')}}{2.3 A_t^w(\lambda')} \tag{15-18}$$

## Signal Expressions

*Readout Voltage Signals.* The most important signal expressions for photoluminescence spectrometry are summarized in Table 15-1. The analyte luminescence signal ($E_L$) is obtained from the difference in the total sample luminescence signal ($E_{tL}$) and the blank signal ($E_{bk}$) (equation 15-1). The blank signal (equation 15-2) is due to dark current ($E_d$), background luminescence ($E_{bL}$), and scattering ($E_{sc}$). The total luminescence signal (equation 15-3) includes contributions from all blank components and analyte luminescence ($E_L$).

Molecular analyte or background thermal emission in the UV-visible region is considered insignificant at or below room temperature. Chemiluminescence signals are also negligible except in special circumstances. Both the analyte and background photoluminescence signals can be due to fluorescence or phosphorescence. Fluorescence and phosphorescence can usually be distinguished from each other by wavelength and time discrimination. Under the conditions normally used for fluorescence measurements, the triplet state is rapidly deactivated by nonradiative processes, and the analyte or background phosphorescence signal is negligible. Thus $E_L = E_F$ and $E_{bL} = E_{bF}$, where the subscript $F$ denotes

**TABLE 15-1** Cont.

$$(\Phi_{\lambda L})_0 = B_{\lambda L} Y(\lambda') \tag{15-19}$$

$$Y(\lambda') = b^* h \Omega' T_{\text{em}}(\lambda') \tag{15-20}$$

$$(\Phi_{\lambda L})_0 = \Phi_A^* Y_{\lambda L} f_s T_c' T_{\text{em}}(\lambda') \frac{\Omega'}{4\pi} \tag{15-21}$$

$$E_L = mG\Phi_A^* \frac{\Omega'}{4\pi} \int Y_{\lambda L} f_s T_c' T_{\text{em}}(\lambda') R(\lambda') \, d\lambda' \tag{15-22}$$

$$E_L = mGA_s \Omega_s \int B_\lambda T_{\text{ex}}(\lambda) f_{pr} T_c \alpha_t^*(\lambda) f_a \, d\lambda \frac{\Omega'}{4\pi} \int Y_{\lambda L} f_s T_c' T_{\text{em}}(\lambda') R(\lambda') \, d\lambda' \tag{15-23}$$

$$E_L = mGR(\lambda') W H \Omega_{\text{ex}} s T_{\text{op}} B_\lambda f_{pr} T_c \alpha_t^*(\lambda) f_a Y_{\lambda L} f_s T_c' s' T_{\text{op}}' \frac{\Omega'}{4\pi} \tag{15-24}$$

$$E_L = mGR(\lambda') W H \Omega_{\text{ex}} s T_{\text{op}} B_\lambda T_c (2.303 A_a^*) Y_{\lambda L} T_c' s' T_{\text{op}}' \frac{\Omega'}{4\pi} \tag{15-25}$$

$$E_L = mGR(\lambda') W H \Omega_{\text{ex}} s T_{\text{op}} B_\lambda T_c (2.303 \varepsilon b^* c) Y_L T_c' T_{\text{op}}' \frac{s'}{1.06 \Delta\lambda'} \frac{\Omega'}{4\pi} \tag{15-26}$$

[a]$(\Phi_{\lambda L})_0$, Luminescence spectral radiant power incident on the PMT (W); $B_{\lambda L}$, luminescence spectral radiance observed from $V^*$ (W cm$^{-2}$ sr$^{-1}$ nm$^{-1}$); $Y(\lambda)$ and $Y(\lambda')$, excitation and emission throughput factor, respectively; $T_{\text{ex}}(\lambda)$ and $T_{\text{em}}(\lambda)$, excitation and emission transmission factor, respectively (dimensionless); $\Omega_s$ and $\Omega'$, solid angle of collection of source excitation radiation and of luminescence from the sample cell, respectively (sr); $T_c$ and $T_c'$, excitation and emission cell wall transmission, respectively (dimensionless); $\Phi_A^*$ and $\Phi_{\lambda A}^*$, radiant power and spectral radiant power absorbed by the analyte in $V^*$, respectively (W$_c$ or W nm$^{-1}$); $Y_{\lambda L}$, luminescence spectral power yield (nm$^{-1}$); $f_s$, secondary absorption factor (dimensionless); $\rho_c$ and $\rho_c'$, reflectance of cell wall at excitation and emission wavelengths, respectively (dimensionless); $\alpha_c$ and $\alpha_c'$, absorptance of cell wall at excitation and emission wavelengths, respectively (dimensionless); $A_t^{pr}(\lambda)$ and $A_t^*(\lambda)$, total solution absorbance at an excitation wavelength in the prefilter region and in $V^*$, respectively (dimensionless); $A_t^w(\lambda')$ and $A_t^{pt}(\lambda')$, total solution absorbance at an emission wavelength on cell wall and in the postfilter region, respectively (dimensionless); $(\Phi_\lambda)_0$, source spectral radiant power incident on cell wall (W nm$^{-1}$); $f_{pr}$, prefilter factor (dimensionless); $f_a$, fraction of excitation radiation absorbed by the analyte (dimensionless); $B_\lambda$, source spectral radiance (W cm$^{-2}$ sr$^{-1}$ nm$^{-1}$); $A_s$, source area viewed (cm$^2$); $\Omega_{\text{ex}}$, solid angle of collection of excitation monochromator (sr); $W$ and $W'$, slit width of excitation and emission monochromator, respectively (cm); $H$ and $H'$, slit height of excitation and emission monochromator, respectively (cm); $s$ and $s'$, spectral bandpass of excitation and emission monochromator, respectively, (nm); $A_a^*(\lambda)$, analyte absorbance over $b^*$ (dimensionless); $T_{\text{op}}$ and $T_{\text{op}}'$, transmission efficiency of optical components in the excitation beam (between the source and the cell) and the emission beam (between the cell and the detector), respectively (dimensionless); $\Delta\lambda'$, half-width of the emission band (nm); $\varepsilon$, analyte molar absorptivity (L mol$^{-1}$ cm$^{-1}$); $b_{pr}$ and $b_{pt}$, prefilter and postfilter pathlength, respectively (cm); $b^*$, pathlength of illumination and observation (cm); $w$, width of the excitation beam in the center of the cell (cm); $V^*$, volume element of illumination and observation (cm$^3$).

fluorescence. With source modulation, the mean value of $E_d$ is zero. Longer-lived phosphorescence signals are distinguished from fluorescence and scattering signals by temporally separating the excitation and observation steps. Hence with time discrimination, $E_{sc} = 0$, $E_L = E_P$, and $E_{bL} = E_{bP}$, where the subscript $P$ denotes phosphorescence. In chemiluminescence (CL) measurements, there is no external excitation source, no fluorescence, phosphorescence, or scattering; thus $E_L = E_{CL}$, and $E_{bL} = E_{bCL}$.

Equation 15-4 shows that $E_L$ is determined by the PMT and signal modifier gains, the PMT responsivity, and the analyte luminescence spectral radiant power incident on the PMT [$(\Phi_{\lambda L})_0$]. Similar expressions apply for the relationship between different components of the background readout signals and their respective spectral radiant powers. Here and elsewhere a prime is used to distinguish the emission wavelengths or quantities related to the emission wavelength from excitation wavelengths or related quantities.

*Initial Assumptions.* Now we derive equations relating the luminescence radiant power seen by the detector [$(\Phi_{\lambda L})_0$] to instrumental and solution variables. First, we calculate the radiant power absorbed by the analyte. Then we find the fraction of the absorbed radiant power that is converted to luminescence and collected by the emission wavelength selector. We consider the idealized model of Figure 15-12, in which the luminescence is collected at 90° with respect to the excitation axis. The volume element $V$ in which molecules are radiationally excited is determined by the rectan-

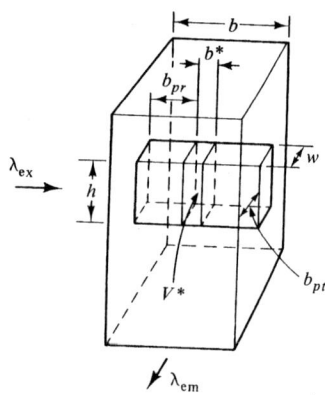

**FIGURE 15-12** Geometry for photoluminescence excitation and collection. The collimated excitation beam of cross-sectional area $w \times h$ passes through the sample cell of pathlength $b$. The volume element of excitation $V$ thus equals $wbh$. The emission optics define the width $b^*$ over which luminescence is excited and viewed. Hence luminescence is collected from volume element $V^* = wb^*h$ through the front face of $V^*$ with an area of $hb^*$. The excitation photons must pass through distance $b_{pr}$ before reaching $V^*$. The emitted photons must travel a distance $b_{pt}$ from the viewed face of $V^*$ before leaving the sample cell.

gular cross section of a collimated beam ($w \times h$) and the cell pathlength ($b$). Normally, some aperture in the emission beam restricts the pathlength and the volume element in which both excitation and observation occurs; these are denoted $b^*$ and $V^*$, respectively ($b^* < b$, $V^* < V$). The height of the emission aperture is initially assumed to be equal to or greater than $h$.

*Excitation Variables.* The analyte luminescence signal is directly proportional to the radiant power absorbed by the analyte in $V^*$ ($\Phi_A^*$). Thus the first step is to calculate the spectral radiant power absorbed by the analyte in the excited and viewed volume element ($\Phi_{\lambda A}^*$) and to integrate over the range of excitation wavelengths incident on the sample (equation 15-5). Equation 15-6 shows that $\Phi_{\lambda A}^*$ is proportional to the incident source spectral radiant power [$(\Phi_\lambda)_0$], the excitation cell wall transmission factor ($T_c$), the prefilter-effect factor ($f_{pr}$), the total spectral absorptance ($\alpha_t^*$), and the fraction of the total spectral radiant power absorbed by the analyte ($f_a$).

For a conventional source, the excitation spectral radiant power is a function (equation 15-7) of the source spectral radiance ($B_\lambda$) and the throughput factor for the excitation wavelength selector and auxiliary optics [$Y(\lambda)$]. The excitation throughput factor is given by equation 15-8, where $A_s$ is the source area viewed and imaged

into the cell as determined by some limiting aperture, $\Omega_s$ is the solid angle of the source viewed, and $T_{ex}(\lambda)$ is the transmission factor of the emission optics. The last factor accounts for the efficiency with which a given excitation wavelength is transmitted between the source and sample cell, and thus the range of source wavelengths passed from a continuum source.

The product of $(\Phi_\lambda)_0$, $T_c$, and $f_{pr}$ is the source spectral radiant power actually seen by the molecules in the viewed volume element. The excitation cell wall transmission factor is determined by the cell wall absorptance ($\alpha_c$) and the cell wall reflectance ($\rho_c$) as shown in equation 15-9. Typically, $T_c$ is about 96% and is controlled by the reflectance at the air–cell wall interface. The prefilter factor accounts for the fraction of the excitation radiation that is absorbed in the prefilter region. Thus $f_{pr}$ is the transmittance over $b_{pr}$ ($T_{pr}$) and is determined by equation 15-10, where $A_t^{pr}(\lambda)$ is the total absorbance in this region.

The fraction of the radiant power incident on $V^*$ that is absorbed at a given wavelength in $V^*$ is the total spectral absorptance, $\alpha_t^*(\lambda)$. From equation 15-11 it is determined by the total spectral absorbance in $V^*$ [$A_t^*(\lambda)$]. Note that the right side of equation 15-11 is just $1 - T_t^*$, where $T_t^*$ is the transmittance across $b^*$. Only the fraction of the total absorbed radiation power absorbed by the analyte results in analyte luminescence. Equation 15-12 shows that this fraction is determined by the ratio of the analyte spectral absorbance in $V^*$ [$A_a^*(\lambda)$] to the total spectral absorbance in $V^*$ [$A_t^*(\lambda)$].

By combining equations 15-5 to 15-8, we obtain equation 15-13 for the total radiant power absorbed by the analyte in $V^*$. We now examine what equation 15-13 tells us about instrumental variables and different instrumental configurations. The excitation radiant power incident on the cell is determined by the type of source and excitation wavelength selector. The most critical excitation instrumental variables are $A_s$, $\Omega_s$, $B_\lambda$, and $T_{ex}(\lambda)$.

In a spectrofluorometer, the excitation transmission factor is determined by $T_{ex}(\lambda) = T_{op}t(\lambda)$ and the source is usually a high-pressure arc lamp. Thus the range of wavelengths over which equation 15-13 is integrated is determined by the excitation spectral bandpass ($\lambda_0 \pm s$). The excitation throughput factor (equation 15-4) can be specifically written (see equation 3-70) as $Y(\lambda) = WH\Omega_{ex}T_{op}t(\lambda)$. We assume that the source entrance optics are properly configured to match the F/n (or solid angle $\Omega_{ex}$) of the excitation monochromator (see Figure 15-13) and that the image of the source focused on the entrance slit is larger than the slit dimensions. In that case the product $A_s\Omega_s$ is equal to $WH\Omega_{ex}$. If the source imaging is 1:1, $A_s = WH$ and

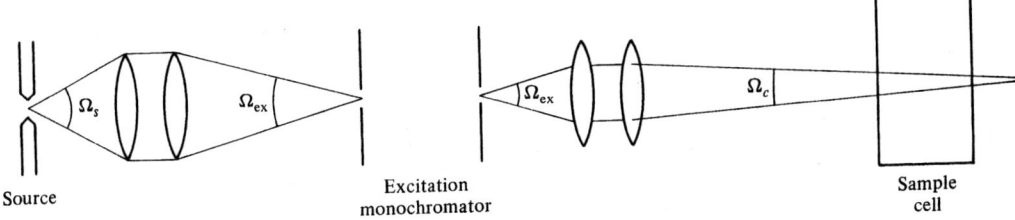

Source  Excitation  Sample
monochromator  cell

**FIGURE 15-13** Excitation considerations. A lens or mirror or combination of optical components collects the source radiation with solid angle $\Omega_s$ and focuses an image of the source onto the entrance slit of the excitation monochromator. The imaging solid angle matches the solid angle of collection of the monochromator ($\Omega_{ex}$). Overfilling the monochromator would increase the stray radiation and decrease the optical throughput. The two solid angles are related by $\Omega_s = m^2\Omega_{ex}$ where $m$ is the magnification. The area of the source image focused at the entrance slit is $m^2$ times the original source area. Often the dimensions of the source arc image are larger than the entrance slit dimensions, and the source area seen by the entrance slit ($A_s$) is given by $WH/m^2$. Thus for a given monochromator and entrance slit dimensions, $\Omega_s A_s = \Omega_{ex}WH$. Hence collecting the source radiation at a larger solid angle than the acceptance solid angle of the monochromator generally provides no greater radiation throughput because the area of the source viewed is reduced correspondingly. If the source dimensions are smaller than the entrance slit dimensions, the throughput can be increased by choosing $\Omega_s > \Omega_{ex}$ up to the point that the entrance slit is filled with the magnified source image. When an elliptical reflector is employed to collect the source radiation (the source at one focus and the entrance slit at the other focus), the collection solid angle can be greater than $2\pi$. However the image is highly magnified such that a significant increase in throughput is not realized. Similar considerations apply to imaging the radiation leaving the exit slit onto the cell. As shown, the solid angle of incidence onto the cell ($\Omega_c$) is often made less than $\Omega_{ex}$ to reduce beam divergence and create a beam that is more nearly collimated. In this case, the size of the image in the cell is magnified and greater than $WH$.

$\Omega_s = \Omega_{ex}$. The dimensions of the source arc or its image at the entrance slit or some other aperture (an aperture mask at the cell wall) may limit the maximum magnitude of $A_s$.

As is also shown in Figure 15-13, the cross-sectional area of the image in the viewed volume element ($wh$) depends on the imaging between the excitation monochromator exit slit and the sample cell. To achieve a more collimated beam within the cell, the F/n of incidence is often increased, with the result that the image size in the cell is greater than $WH$.

Equation 15-14 is a limiting form of equation 15-13 which applies when an excitation monochromator is employed and all wavelength-dependent quantities are constant over the excitation spectral bandpass. Because the excitation bandpass $s$ is proportional to $W$, the absorbed radiant power is proportional to $W^2$.

For a fluorometer, the source radiation is collected and imaged into the sample cell with optical components. The excitation transmission factor is given by $T_{ex}(\lambda) = T_iT_f(\lambda)$, where $T_i$ is the transmission factor for all imaging optics between the source and the sample cell and $T_f(\lambda)$ is the spectral transmittance of the excitation filter. Typically, a low-pressure Hg arc lamp is the source. The combination of some limiting excitation aperture between the source and sample cell and the type of imaging (i.e., magnification) determines the actual source area ($A_s$) that is viewed and imaged into the sample cell and the cross-sectional area of the beam within the cell. The source radiation is usually collected with fast F/1 or F/2 optics so that the solid angle of collection ($\Omega_s$) is relatively large. Normally, the excitation filter isolates a single source line. In this case the integration in equation 15-13 is not necessary, $B_\lambda$ is replaced by the line radiance $B$, and all other quantities are evaluated at the source line wavelength, as shown in equation 15-15. If two or more source lines are transmitted, $\Phi_A^*$ is determined by evaluating the right side of equation 15-15 for each excitation wavelength and then summing the results. With a phosphor-coated lamp, equation 15-13 applies, and the limits of integration are determined by the combination of the wavelength range of the source emission profile and the spectral transmittance of the excitation filter.

With laser excitation, all wavelength-dependent quantities can usually be considered to be constant over the line width, and the total absorbed power is determined by an equation similar to equation 15-15. In this case we replace $\Omega_s A_s B T_i T_f(\lambda)$ by $\Phi_0$, the total radiant power measured at the cell wall.

*Emission Variables.* We can consider the volume element $V^*$ as a source of luminescence characterized by analyte spectral radiance $B_{\lambda L}$. This spectral radiance, referenced to just outside the viewed cell wall, is calculated from equation 15-16. The product of $\Phi_A^*$ and the luminescence spectral power yield ($Y_{\lambda L}$) equals the total spectral luminescence radiant power emitted in $V^*$. This radiant power is converted into a radiance by dividing by $4\pi h b^*$. The secondary absorption factor $f_s$ accounts for the absorption of the photons between their point of emission in $V^*$ and the viewed cell face. Finally, $T_c'$ is the emission cell wall transmission factor (typically, 96%) as controlled (equation 15-17) by the emission cell wall absorptance ($\alpha_c'$) and the cell wall reflectance ($\rho_c'$).

The secondary absorption factor $f_s$ is the product of two factors, as shown in equation 15-18. The first exponential factor equals the transmittance in the postabsorption region ($T_t^{p'}$) between the viewed face of $V^*$ and the emission cell wall. It can be calculated if the total spectral absorbance [$A_t^{p'}(\lambda')$] over the postfilter pathlength ($b_{pt}$) is known. The second factor in equation 15-18 accounts for the absorption of emitted photons within $V^*$ and depends on the total solution absorbance [$A_t^w(\lambda')$] at the emission wavelengths across the width of the exciting beam ($w$). This factor is more complex because different photons travel different pathlengths within $V^*$ depending on where they are emitted.

The spectral radiant power observed by the PMT is the product of $B_{\lambda L}$ and the emission optics throughput factor, $Y(\lambda')$ (equation 15-19). From equation 15-20, $Y(\lambda')$ is determined by the area of the viewed face of $V^*$ ($h b^*$), the collection solid angle of the emission optics ($\Omega'$), and the emission optics transmission factor [$T_{em}(\lambda')$]. This last factor accounts for the efficiency with which a given wavelength is transmitted between the sample cell and the detector; it thus determines the range of emission wavelengths passed to the detector (assumed to be a PMT).

Substituting equations 15-16 and 15-20 into equation 15-19 yields equation 15-21 for the luminescence spectral radiant power incident on the PMT. The integral of equation 15-21 over the wavelength interval passed by the emission wavelength selector is the total

luminescence radiant power incident on the detector. When this result is substituted into equation 15-4, we obtain equation 15-22. We will now use this equation to discuss emission parameters.

The most critical emission instrumental parameters are $b^*$, $T_{em}(\lambda')$, and $\Omega'$. The pathlength $b^*$ is critical because it affects the magnitude of $\Phi_A^*$ or the size of the volume element from which luminescence is collected. The latter two factors influence what fraction of the luminescence emitted over a broad wavelength range and in all directions is actually collected and observed.

With an emission monochromator, the emission transmission factor in equation 15-20 is given by $T_{em}(\lambda') = T_{op} t(\lambda')$. The emission monochromator spectral bandpass ($s'$) determines the wavelength range over which $t(\lambda')$ is finite, equation 15-22 is integrated, and emission is viewed. As for the excitation monochromator (see Figure 15-13), the F/n or solid angle ($\Omega_{em}$) of the emission monochromator should be matched. For a given magnification ($m'$) of the optics between the cell and emission monochromator, $b^* = W'/m'$, $\Omega' = \Omega_{em}(m')^2$, and $h' = H'/m'$, where $W'$ and $H'$ are the emission monochromator slit width and slit height and $h'$ is the height of the volume element viewed in the sample cell. Hence the overall throughput is determined by the F/n of the emission monochromator. Collecting the fluorescence with a solid angle larger than $\Omega_{em}$ results in a smaller volume element (i.e., smaller $b^*$) being viewed for a given slit width. Often $h' \geq h$ and the height of the excitation beam rather than $H'$ controls the size of $V^*$. However, if the magnification is large enough that $h' < h$, $H'$ controls the height of the volume element viewed, $h$ is replaced by $h'$ in equations 15-16 and 15-20, and $\Phi_A^*$ is correspondingly reduced by the factor $h'/h$.

For a fluorometer, $T_{em}(\lambda') = T_i' T_f(\lambda')$, where $T_i'$ is the transmission factor for all imaging components used with the filter and $T_f(\lambda')$ is the spectral transmittance of the filter. The latter factor determines the wavelength range over which equation 15-22 is integrated. The magnitude of $b^*$ is determined by some aperture in the emission beam. The product $b^*\Omega'$ is usually much larger than that with an emission monochromator. It is possible to collect fluorescence over a pathlength of about 0.5 cm with an F/n of 1 to 2 and to intercept the complete image with the PMT photocathode.

In a spectrofluorometer, the viewed volume element is normally only a small fraction of the total volume element in which excitation occurs, as illustrated in Figure 15-14a. As shown in Figure 15-14b, the viewed pathlength $b^*$ and $V^*$ can be substantially increased by using a horizontal slit arrangement.

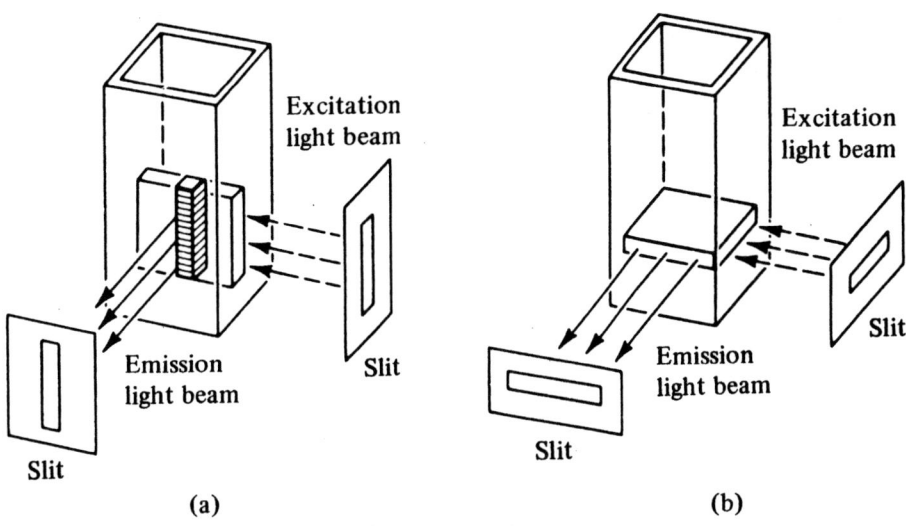

**FIGURE 15-14** Slit arrangements in spectrofluorometers. The normal vertical slit configuration is shown in (a). The excited and viewed volume element is in the center of the cell. In (b) a horizontal slit arrangement is shown. The monochromators are rotated 90° so that the slit heights, rather than the slit widths, determine $b^*$ and $w$. This configuration increases $V^*$ and hence the luminescence signal by a factor of 5 to 30. (Courtesy of Perkin-Elmer Corp.)

*Final Signal Expressions.* By combining equations 15-13 and 15-22, we obtain equation 15-23 for the photoluminescence signal. Equations 15-24 to 15-26 are limiting cases of this general equation for a spectrofluorometer in which it is assumed that all wavelength-dependent quantities are constant over the excitation and emission bandpasses. It is further assumed in equation 15-25 that the absorbances at the excitation and emission wavelengths are small such that $f_{pr} = f_s = 1$ and $\alpha_t^* f_a = 2.303 A_a^*$. Finally, in equation 15-26 we substitute $\varepsilon b^* c$ for $A_a^*$, where $c$ and $\varepsilon$ are the analyte concentration and molar absorptivity. It is further assumed that measurements are made at the maximum of an emission band that is Gaussian in shape. In this case we can substitute $[Y_L/(1.06\Delta\lambda')]$ for $Y_{\lambda L}$, where $Y_L$ is the luminescence power yield and $\Delta\lambda'$ is the half-width of the emission band. Note that the photoluminescence signal is directly proportional to the analyte concentration. Because $s'$ and $b^*$ are proportional to $W'$, $E_L$ is proportional to $(W')^2$. With 1:1 imaging in the emission beam, $b^* = W'$ and $\Omega' = \Omega_{em}$.

## Refinements of Signal Expressions

The equations in Table 15-1 tell us how instrumental variables affect the luminescence signal. They should be considered semiquantitative in terms of predicting the absolute value of the luminescence signal. The equations could be refined to account for some assumptions that have been made.

We first note that the region of excitation and viewing ($V^*$) is not a perfect parallelepiped. The excitation beam is not perfectly collimated such that $b_{pr}$ and $b^*$ vary for different excitation rays. The irradiance of the excitation beam often varies across its cross sec-

tion. The solid angle of luminescence collected and the emission pathlengths, $w$ and $b_{pt}$, vary for photons collected from different regions in $V^*$.

We have also ignored some other factors, such as stray radiation and elastic scattering by particles in the sample solution. Reflection of excitation and emission radiation from the cell walls opposite the excitation and emission cell walls, respectively, will slightly increase the observed luminescence signal. For some molecules at high concentration, **reemission** must be considered. Here analyte molecules in the emission beam absorb luminescence photons emitted by other analyte molecules; a fraction of this absorbed radiation is reemitted as luminescence.

In the literature, the luminescence radiance is often calculated by dividing the total radiant power of luminescence by $4\pi h b^* \eta^2$ rather than $4\pi h b^*$ (see equation 15-16), where $\eta$ is the refractive index of the test solution. The refractive index factor accounts for the fact that emission rays, not parallel to the normal to the emission cell wall, are bent away from the normal when passing from a higher refractive medium (the solution) to a lower refractive medium (air). Because the emission radiation is expanded into a larger solid angle after passing through the cell wall, the actual solid angle of collection from $V^*$ is less than $\Omega'$ by $\eta^{-2}$. However, the refractive index effect also changes the magnification of the image focused on the entrance slit of the monochromator. Thus the viewed volume element becomes larger in proportion to $\eta^2$; this counteracts the reduction in solid angle. It has been shown that under some experimental conditions, the efficiency of collection of fluorescence from a given fluorophore in solvents of different refractive index does not vary significantly.

## Solution Variables

The dependence of the luminescence signal on analyte concentration is complex. If we concern ourselves only with those factors that depend on the absorption by the analyte or concomitants, equation 15-24 indicates that

$$E_L \propto f_{pr}\alpha_t^*(\lambda)f_a f_s \qquad (15\text{-}27)$$

If we expand the terms with the aid of equations 15-10 to 15-12 and equation 15-18, we obtain

$$E_L \propto (10^{-A_t^{pr}})(1 - 10^{-A_t^*})\left(\frac{A_a^*}{A_t^*}\right)\frac{1 - 10^{-A_t^w}}{2.303A_t^w}(10^{-A_t^{pt}}) \qquad (15\text{-}28)$$

For simplicity we have dropped the $(\lambda)$ and $(\lambda')$ designations.

All the exponential terms can be expanded according to $10^{-x} = 1 - 2.303x + (2.303x^2)/2! + \cdots$. For small $x$, $10^{-x} \approx 1$ and $(1 - 10^{-x}) \approx 2.303x$ to 1% or better if $x$ is less than 0.0044 and 0.0091, respectively. Hence for small values of absorbance at the excitation and emission wavelengths, $f_{pr} = f_s \approx 1$ and $\alpha_t^* f_a \approx 2.3A_a^*(\lambda)$, in which case equation 15-28 simplifies to

$$E_L \propto A_a^* \propto \varepsilon(\lambda)b^*c \qquad (15\text{-}29)$$

Note that under these conditions, $E_L \propto c$ even if $\varepsilon(\lambda)$ varies over the range of excitation wavelengths incident on the sample [i.e., $E_L \propto b^*c \int \varepsilon(\lambda)\,d\lambda$].

As a rule of thumb, the absorbance of the sample solution in a 1-cm-pathlength cell should be less than 0.005 at both the excitation and emission wavelengths to achieve linearity within 1% or to realize less than 1% attenuation by concomitant absorption. This assumes that luminescence measurements are made in a 1-cm cell with $V^*$ in the center of the cell. For higher solution absorbances, either nonlinearity occurs or concomitant absorption is significant.

Significant absorption of the excitation beam radiation is denoted a **primary absorption effect**, whereas significant absorption of the emission beam is designated a **secondary absorption effect**. Together these two effects are often called the **inner filter effect**. The primary absorption effect is due both to significant absorption of excitation radiation in the prefilter region, the **prefilter effect**, and in $V^*$. Similarly, the secondary absorption effect is due to the significant absorption of emission radiation in $V^*$ and in the postfilter region, the **postfilter effect**. Secondary absorption due to the analyte is specifically denoted **self-absorption**.

*Linearity.* To understand primary and secondary absorption effects, we will first consider the effect of analyte absorption on the linearity of calibration curves by assuming that absorption by concomitants is negligible. In this case equation 15-28 reduces to

$$E_L \propto 10^{-A_a^{pr}}(1 - 10^{-A_a^*})\frac{1 - 10^{-A_a^w}}{2.303A_a^w}(10^{-A_a^{pt}}) \qquad (15\text{-}30)$$

where the subscript $a$ denotes that the absorbance in each region is due only to the analyte.

Let us first consider the second term, accounting for the fraction of the radiation absorbed within $V^*$. At low analyte concentrations (i.e., $A_a^* < 0.01$), $\alpha_a^* = 2.303A_a^*$ and $E_L \propto c$. As $c$ increases, $\alpha_a^*$ exponentially approaches unity when all the radiation is absorbed. Note that the relationship between luminescence signal and analyte concentration is inherently nonlinear. The absorbance, but not the absorption (the fraction absorbed), is proportional to concentration.

The prefilter absorption effect enhances the nonlinearity. At very high concentrations, the first term in equation 15-30 ($f_{pr}$) decreases more rapidly than $\alpha_a^*$ increases, and the calibration curve rolls over. In the limit of very high analyte concentration, the luminescence signal approaches zero because all the excitation radiation is absorbed before it reaches $V^*$. In spectrofluorometers, the prefilter effect usually has the dominant effect on linearity because $b_{pr} > b^*$.

Figure 15-15 illustrates the nonlinearity caused by primary absorption effects due to the analyte. Figure 15-16 shows how the prefilter factor and the fraction of radiation absorbed are related to the sample absorbance across the full pathlength of the cell.

The secondary absorption effect, self-absorption in this case, causes further nonlinearity. Its effect is usually minor compared to primary absorption effects. Self-absorption occurs for some molecules at emission wavelengths on the lower-wavelength side of the emission band if the excitation and emission spectra overlap. If overlap occurs, the molar absorptivity is lower than that at the maximum of the excitation band.

*Spectral Interference due to Concomitants.* Significant primary or secondary absorption by concomitants causes a spectral interference. The luminescence signal for a given analyte concentration is lower compared to the signal in the absence of absorbing concomitants. For example, if $A_t^{pr} = 0.1$, $A_t^* = 0.025$, $A_a^* = 0.001$, and secondary absorption is negligible, the luminescence signal is 77% of what it would be in the absence of concomitant absorption.

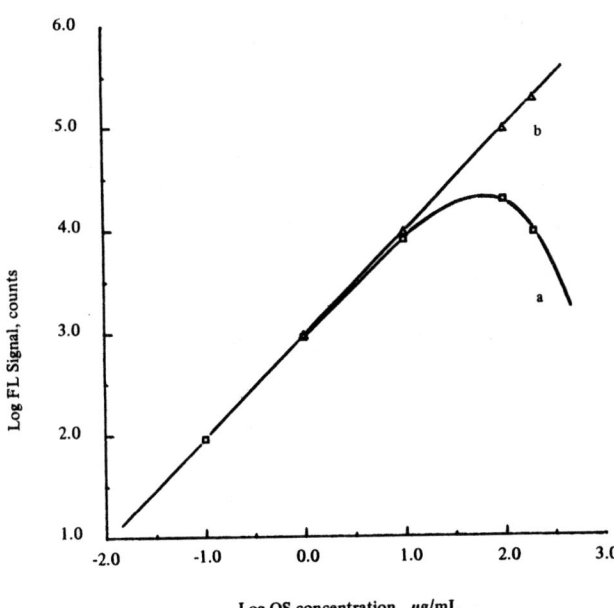

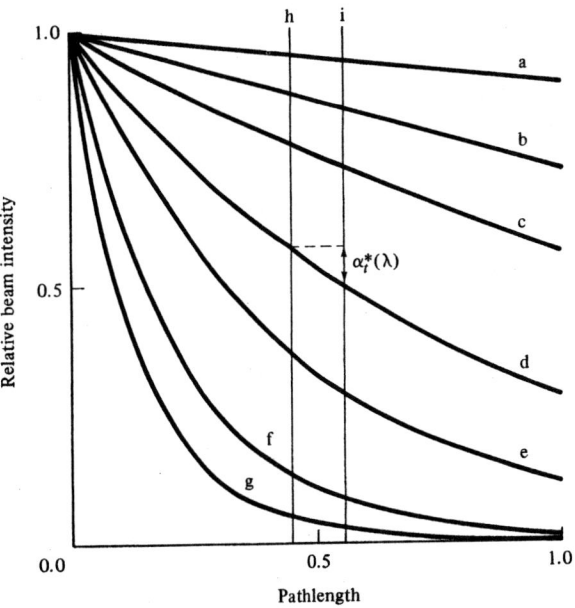

**FIGURE 15-15** Fluorescence calibration curve for quinine sulfate (QS). Curve a is the actual calibration plot. The calibration is linear to within 1% up to 1 μg mL$^{-1}$ because $f_{pr} = f_s = 1$ and $\alpha_a^* = 2.303 A_a^*$. Above 1 μg mL$^{-1}$, nonlinearity due to primary absorption effects is observed because $f_{pr} < 1$ and $\alpha_a^* = (1 - 10_a^{-A})$. Above about 100 μg mL$^{-1}$, the prefilter effect causes roll-over. In curve b, the signals have been corrected for the attenuation due to primary absorption and linearity is restored. Conditions: $\lambda_{ex} = 365$ nm; $\lambda_{em} = 450$ nm. (From M. C. Yappert, Ph.D. thesis, Oregon State University, 1985, with permission of author.)

**FIGURE 15-16** Relative excitation beam intensity as a function of distance traversed through the cell for different sample transmittances: (a) 80% $T$; (b) 65% $T$; (c) 50% $T$; (d) 25% $T$; (e) 10% $T$; (f) 1% $T$; (g) 0.1% $T$. The section between the two vertical lines labeled h and i represents $b^*$, the pathlength over which luminescence is viewed. The difference in transmittance between h and i is the fraction of the excitation radiation absorbed within $V^*$ as shown for curve d. The pathlength from the front of the cell to h is the prefilter pathlength. Thus the transmittance at point h is the prefilter factor $(f_{pr})$ or the fraction of the incident radiation that reaches the viewed volume element.

## Corrected Measurements

The equations developed help us understand the correction schemes described in Section 15-1.

*Source Intensity Correction.* The luminescence signal is proportional to $B_\lambda$. In source intensity correction, the instrument is configured to keep $B_\lambda$ constant or to divide $E_L$ by a reference signal proportional to $B_\lambda$.

*Corrected Excitation Spectra.* During an excitation scan, all emission variables are held constant. From Equation 15-26, $B_\lambda$, $T_{op}$, $T_c$, and ε vary with excitation wavelength. With an instrument configured for corrected excitation spectra, $E_L$ is ratioed to a reference signal obtained from a reference detector (see Figure 15-9a). The reference signal is proportional to $B_{\lambda,p}T_{op}$ with a quantum counter and to $B_\lambda T_{op}$ with a thermal detector. With a quantum counter, the cor-

rected excitation spectrum is a plot of relative quanta per second per nanometer vs. $\lambda_{ex}$. If the quantum efficiency is independent of the excitation wavelength, primary absorption effects are negligible, and the optical characteristics of the beam splitter are independent of the excitation wavelength or are compensated, the shape of the spectrum reflects the dependence of the molar absorptivity on wavelength; it has a shape identical to the absorption spectrum. If a thermal detector is used as the reference detector, the reference signal is proportional to energy rather than quanta. The excitation spectra can be normalized to the shape obtained with a quantum counter by multiplying the reference signal by $\lambda_{ex}$.

*Corrected Emission Spectra.* Excitation variables are held constant during an emission scan. From equation 15-25, $R(\lambda')$, $Y_{\lambda L}$, $T_c$, and $T'_{op}$ are dependent on the emission wavelength. For corrected emission spectra, the instrument is configured to multiply $E_L$ by a factor that is proportional to $[R(\lambda')T'_{op}]^{-1}$ or $[K(\lambda')T'_{op}]^{-1}$. The corrected emission spectrum thus indicates the true shape of the emission spectrum (i.e., how $Y_{\lambda L}$ or $\phi_{\lambda L}$ depends upon $\lambda_{em}$). Usually, $T'_c$ is assumed to be independent of $\lambda_{em}$.

*Absorption Corrected Measurements.* For absorption corrected measurements, the observed luminescence signal $E'_L$ is divided by a primary absorption correction factor ($f_p$) and a secondary absorption correction factor ($f_s$):

$$E^c_L = \frac{E'_L}{f_p f_s} \qquad (15\text{-}31)$$

The corrected luminescence signal ($E^c_L$) represents the value that would be obtained if primary or secondary absorption were insignificant. This equation can be used to correct for nonlinearity due to significant analyte absorption or spectral interference due to significant concomitant absorption. We have already derived the secondary absorption correction factor (equation 15-18). The primary absorption correction factor is given by

$$f_p = \frac{f_{pr}\alpha^*_t f_a}{(f_{pr}\alpha^*_t f_a)_{A(\lambda)\to 0}} \qquad (15\text{-}32)$$
$$= \frac{10^{-A^{pr}_t}(1 - 10^{-A'_t})}{2.303 A^*_t}$$

Usually, $f_s$ and $f_p$ are evaluated by measuring the sample absorbances across the total pathlength of the cell ($b$) at both the excitation ($A_t$) and emission wavelengths ($A'_t$). The absorbances in the prefilter, $V^*$, and postfilter regions are calculated from the known pathlengths as shown below:

$$A^{pr}_t = A_t \frac{b_{pr}}{b}$$

$$A^*_t = \frac{A_t b^*}{b}$$

$$A^w_t = A'_t \frac{w}{b}$$

$$A^{pt}_t = A'_t \frac{b_{pt}}{b}$$

In the cell-shift method, $f_p$ is calculated from the values of $E_L$ observed from a given $V^*$ at two positions along the excitation axis (i.e., different $b_{pr}$). Similarly, $f_s$ is determined from measurements of a given $V^*$ at

two different positions along the emission axis (i.e., different $b_{pt}$).

## Saturation Effects

With laser excitation, it is possible to approach saturation conditions for fluorescence in which a significant fraction of the analyte molecules is in the excited state. It can be shown that the fluorescence signal is related to the incident irradiance $(E_p)_0$ in photons cm$^{-2}$ s$^{-1}$ by

$$E_F \propto ck_F \left[ 1 + \frac{k_F}{\sigma(\lambda)(E_p)_0\phi_F} \right]^{-1} \qquad (15\text{-}33)$$

where $k_F$ is the rate constant for fluorescence deactivation in s$^{-1}$ and $\sigma(\lambda)$ is the absorption cross section (cm$^{-2}$); $\sigma(\lambda)$ is related to the molar absorptivity (see Table F-1 and equation F-17) by

$$\sigma(\lambda) = \frac{2.303\varepsilon(\lambda)}{N} = 3.82 \times 10^{-21}\varepsilon(\lambda) \qquad (15\text{-}34)$$

where $N$ is Avogadro's number.

Equation 15-33 predicts that a plot of $E_F$ vs. incident irradiance will reach a plateau. At low irradiances such as achieved with conventional sources, the second term in the brackets in equation 15-33 dominates and $E_L \propto (E_p)_0\phi_F\varepsilon(\lambda)$, as we have discussed. At very high irradiances, the second term vanishes. Under these conditions the fluorescence signal is independent of the source irradiance and the fluorescence quantum efficiency because the rates of stimulated absorption and emission are balanced. Nonradiational deactivation becomes negligible. The absorbance decreases with increasing irradiance because the ground-state singlet population is reduced.

Working under saturated conditions provides several advantages: $E_F$ is independent of the source intensity fluctuations, $E_F$ is independent of the environment of the molecule (e.g., the effect of quenchers), and nonlinearity due to primary absorption effects and due to secondary absorption effects within $V^*$ are reduced or eliminated.

Routine fluorescence measurements are not made under saturated conditions. The disadvantages of the expense and complexity of laser excitation and the likelihood of photodecomposition outweigh the potential advantages in most applications. Also fluorescence may only be saturated in the most intense, central portion of the laser beam. Additionally, nonlinearity can result if fluorescence photons produced within $V^*$ are reabsorbed by molecules outside of $V^*$ and cause additional fluorescence.

## Signal-to-Noise Expressions

The general noise and *S/N* expressions for luminescence measurements presented in Section 5-4 (equations 5-25 and 5-31), converted to the specific forms applicable to molecular photoluminescence spectrometry, are presented in Table 15-2. The noise in the total luminescence signal (equation 15-35) is due to shot and flicker noise in the analyte luminescence signal and blank noise. The blank noise (equation 15-36) is due to amplifier-readout noise, dark current noise, and noise in the background luminescence and scattering signals.

At low analyte concentrations, blank noise is dominant and equation 15-38 applies. The *S/N* is proportional to $i_L$ (see Figure 5-6). Note that there are shot and flicker components in the background luminescence and scattering signals; the flicker noise is assumed to be due to source flicker noise.

At higher analyte concentrations, the analyte luminescence is much greater than the background signals, so that analyte luminescence shot or flicker noise becomes limiting (equation 15-39). The *S/N* is proportional to $i_L^{1/2}$ if analyte luminescence shot noise is dominant or independent of $i_L$ if flicker noise in the analyte luminescence signal is dominant (see Figure 5-6). It is assumed that imprecision due to cell positioning or sample introduction is negligible.

The strategy for optimization of the *S/N* depends on the limiting noise sources. To improve the DL or the precision for low analyte concentrations, $i_L/\sigma_{bk}$ should be maximized. Dark current noise or amplifier-readout noise is likely to be limiting only with small bandpasses, low radiance sources, or samples with low background luminescence. Under these conditions, the *S/N* is proportional to $i_L$. Equations 15-22 to 15-26 identify the instrumental factors that can be changed. For a given instrument, the user has control of the excitation and emission wavelengths and bandpasses. For a fluorometer, the bandpasses are determined by the transmission characteristics of the filters chosen. In monochromator-based systems, the excitation and emission slit widths control both the bandpasses and the size of $V^*$. The excitation wavelength should be chosen to maximize the magnitude of $B_\lambda \varepsilon$, whereas the emission wavelength is chosen to maximize the product $Y_{\lambda L}R(\lambda')$. The *S/N* varies from instrument to instrument because of differences in the magnitudes of $A_s$, $\Omega_s$, $B_\lambda$, $T_{ex}(\lambda)$, $T_{em}(\lambda')$, $\Omega'$, and $b^*$. High-quality analog signal processing or photon counting signal processing (often with cooling of the PMT) is used to minimize $\sigma_{ar}$ and $\sigma_d$.

Noise in the background signal is likely to be limiting when the incident radiant power, $b^*$, or the efficiency and wavelength range of luminescence collection is large. For a given instrument and experimental conditions, the background signal and noise are larger for samples containing a higher concentration of luminescing concomitants or scatterers. If noise in the background signal is limiting, instrumental conditions are adjusted to maximize the signal-to-background ratio (*S/B*). Most instrumental factors affect the analytical and background signals identically (i.e., the background signals are proportional to the incident radiant power, the size of the viewed volume element, the efficiency with which luminescence is collected, etc.). To obtain the best *S/B* and *S/N*, it may be necessary to adjust the excitation and emission wavelengths to values other than those producing the largest analytical signal. If the analyte luminescence and the background signals are both increased by changing the magnitude of a variable such as the excitation or emission slit widths, the *S/N* reaches

***TABLE 15-2***
Signal-to-noise expressions for molecular photoluminescence spectrometry

$$\sigma_{iL} = [(\sigma_L)_s^2 + (\sigma_L)_f^2 + \sigma_{bk}^2]^{1/2} \tag{15-35}$$

$$\sigma_{bk} = (\sigma_d^2 + \sigma_{ar}^2 + \sigma_{bL}^2 + \sigma_{sc}^2)^{1/2} \tag{15-36}$$

$$\frac{S}{N} = \frac{E_L}{\sigma_{iL}} \tag{15-37}$$

At low analyte concentrations:

$$\frac{S}{N} = \frac{i_L}{\{K(i_d + i_{sc} + i_{bL}) + \xi_i^2(i_{bL} + i_{sc})^2 + [(\sigma_d)_{ex}/mG]^2 + (\sigma_{ar}/mG)^2\}^{1/2}} \tag{15-38}$$

At high analyte concentrations:

$$\frac{S}{N} = \frac{i_L}{[Ki_L + (\xi_i i_L)^2]^{1/2}} \tag{15-39}$$

a limiting best value at the point that flicker noise in the background signals becomes dominant.

As the analyte concentration is increased, blank noise becomes negligible and noise carried by the analyte signal dominates. Analyte luminescence shot noise can be limiting at moderate concentrations in which case $i_L$ should be increased as for the blank noise case or the electronic bandwidth ($K$) should be reduced to improve precision. At yet higher analyte concentrations, flicker noise in the analyte luminescence signal is usually limiting. With high blank noise, the signal-shot-noise-limiting region is often not observed (i.e., when the analytical signal-carried noise becomes greater than the blank noise, the signal shot noise is negligible relative to signal flicker noise). The source flicker factor ($\xi_1$) with conventional sources is typically 0.003 to 0.03 with a 1- to 10-s time constant or integration time. Double-beam systems compensate for source intensity fluctuations and reduce the source flicker factor.

At even higher analyte concentrations, where the prefilter effect causes the analyte luminescence signal to decrease with increasing concentration, analyte luminescence shot noise or even blank noise can again become dominant. Because the background signals are also attenuated, dark current and amplifier-readout noise are relatively more significant.

A common spectrofluorometer specification is the *S/N* for the Raman band of common solvents under specified conditions (e.g., for water, $\lambda_{ex}$ = 350 nm, $\lambda_{em}$ = 397 nm, $s = s'$ = 5 to 10 nm, $\tau$ = 1 s). This specification is most useful for comparing different spectrofluorometers. The Raman signal is proportional to the same instrumental parameters as the luminescence. Thus the *S/N* of the Raman band is higher for instruments with larger source radiances and excitation and emission throughput factors. For good spectrometers the *S/N* of the Raman band is in the range 30 to 100. The *S/N* of the Raman band has little relevance to fluorometric measurements in which background luminescence noise limits the *S/N*.

### Detection Limits

For efficient luminescent molecules ($\phi_L \geq 0.5$), the detection limit (DL) with a good photoluminescence instrument in a low-background test solution can be in the range 1 to 10 pM. Detection limits in this range are 100 to 1000 times lower than those achieved for the same molecules with absorption measurements. The DL for quinine sulfate ($\phi_F$ = 0.55) is often quoted as a specification for commercial spectrofluorometers and fluorometers. It is typically below 1 ng mL$^{-1}$ and often near 1 pg mL$^{-1}$.

We can estimate the DLs achievable in photolu-

minescence by calculating typical values of the calibration curve slope ($m$) and the blank noise ($s_{bk}$) under favorable conditions. From equation 15-26, the slope referenced to the PMT photocathode is $E_L/mGc$ or

$$m \approx (0.5)\Phi_0 \varepsilon b^* Y_L \frac{s'}{\Delta\lambda'} \frac{\Omega'}{4\pi} R(\lambda') \quad (15\text{-}40)$$

In this equation $\Phi_0$ is the incident radiant power ($\Phi_0 = WH\Omega_{ex}sT_{op}B_\lambda$), and we estimate that ($T_c T'_c T'_{op}/1.06$) = 0.22. With a typical spectrofluorometer and excitation wavelength and bandpass, $\Phi_0$ is in the range 0.01 to 10 mW. For calculation purposes, we will assume the following typical values for $\Phi_0$ and cell and emission parameters:

$$\Phi_0 = 1 \text{ mW}$$

$$b = 1.0 \text{ cm}$$

$$b^* = 0.1 \text{ cm}$$

$$\frac{s'}{\Delta\lambda} = 0.1$$

$$\frac{\Omega'}{4\pi} = 0.004 \text{ (F/4 system)}$$

$$R(\lambda') = 0.050 \text{ AW}^{-1}$$

Note that $(b^*/b)(s'/\Delta\lambda')(\Omega'/4\pi) = 4 \times 10^{-5}$ or only about 0.004% of the luminescence in the cell is collected by the emission monochromator. Substituting these values into equation 15-40 yields

$$m \approx 1 \times 10^{-9}\varepsilon Y_L \quad \text{A M}^{-1} \quad (15\text{-}41)$$

The blank noise is given by the denominator of equation 15-38. If background signals are low ($i_B < i_d$), dark current shot noise is limiting and

$$\sigma_{bk} = (Ki_d)^{1/2} \quad (15\text{-}42)$$

if excess dark current noise and amplifier-readout noise are negligible. A typical value of $i_d$ is $10^{-15}$ A. With a bandwidth constant $K$ of $1 \times 10^{-19}$ A ($\tau$ = 1 s, $\alpha$ = 0.27, $\Delta f$ = 0.25 Hz), $\sigma_{bk}$ is $1 \times 10^{-17}$ A. Substituting this result and equation 15-41 into the DL equation (DL = $ks_{bk}/m$) and using a confidence factor ($k$) of 2 yields

$$\text{DL} = \frac{2 \times 10^{-8}}{\varepsilon Y_L} \quad (15\text{-}43)$$

For $\varepsilon = 1 \times 10^5$ L mol$^{-1}$ cm$^{-1}$ and $Y_L$ = 0.5, the calculated DL is 0.4 pM.

The DL calculated above corresponds to an analyte absorbance of $4 \times 10^{-8}$ in a 1-cm-pathlength cell. In conventional absorption measurements, the noise limited DL is reached when the analyte absorbance is about $10^{-4}$. At this point, signal flicker noise limits our ability to distinguish two large signals (the sample and reference signal) which differ by only about 0.01%. For photoluminescence measurements of appropriate molecules, the difference between the total signal and the blank signal can still be distinguished even when the net analyte absorption is very small.

Theoretically, the DL calculated above could be lowered substantially. By cooling the PMT and using photon counting, the dark current noise, and thus the blank noise, can readily be reduced by an order of magnitude. The slope of the calibration curve could be increased two orders of magnitude with an emission filter (i.e., collection of a larger fraction of the total luminescence emitted) and another two orders of magnitude by increasing the incident radiant power (i.e., laser excitation). Thus it would seem that femtomolar or even attomolar DLs are possible.

The best experimental DLs for efficient fluorophores such as fluorescein are about 0.1 pM. Such detection limits have been obtained with laser excitation with the analyte dissolved in prepurified solvents. Experimental DLs are much larger than the theoretical DLs in the femtomolar or attomolar range predicted by the discussion above because scattering or background luminescence noise becomes limiting when experimental variables are adjusted to maximize the calibration slope. For a given fluorophore, the DL becomes independent of the values of instrumental variables (see equation 15-40) that affect $m$ when background flicker noise is limiting (i.e., $\sigma_{bk} \propto i_B \propto \Phi_0, s', \Omega'$, etc.).

To understand the magnitude of the background signal, it is useful to define a *blank equivalent concentration* (BEC). It is the concentration of the analyte that yields a luminescence signal equal to the dark signal-corrected background signal (BEC = $i_B/m$). For purified solvents, and efficient fluorophores, the BEC is rarely lower than about 10 pM. Under these conditions, the experimental DL is usually no better than about 0.1 pM. At this concentration, the analyte luminescence signal is at most 1% of the blank signal ($S/B = 0.01$) and the RSD in the blank signal is at best about 1%. In complex samples, the BEC is much higher due to luminescing concomitants. For example, the BEC for human serum diluted 1:100 can be 1000 times greater than that for water.

The discussion above indicates that the DL for a given analyte must be experimentally determined for a given instrument. The value obtained for the analyte in a pure solvent must be used with caution. First, the

blank noise in samples can be much greater. Second, the calibration slope can be less due to quenching. Third, the calibration curve can be nonlinear at subnanomolar concentrations due to contamination problems, as illustrated in Figure 15-17. Finally, the practical DL can be limited by the variability of the uncompensated blank component from sample to sample rather than noise.

## 15-3 MOLECULAR FLUORESCENCE SPECTROMETRY

Molecular fluorescence spectrometry has become established as a routine technique in many specialized applications. For some species, molecular fluorescence spectrometry can yield lower DLs and greater selectivity than molecular absorption spectrometry and is the preferred technique when these characteristics are paramount. However, it should be noted that the instrumentation and theory are more complex.

Although fluorescence can be observed from almost any molecule with a sufficiently intense excitation beam, only a small fraction of molecules exhibit fluorescence characteristics (e.g., high $\phi_F$) that are favorable for analytical purposes. This makes fluorescence

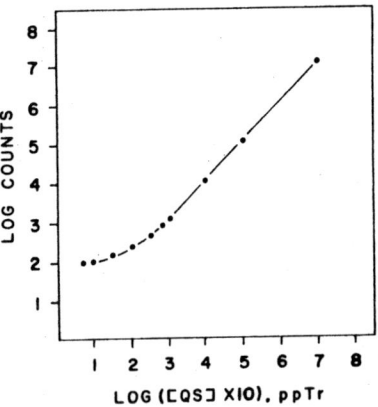

**FIGURE 15-17** Fluorescence calibration curve for quinine sulfate (QS). Here the fluorescence signal is given in counts and the concentration in part per trillion (pptr) or pg mL$^{-1}$. The large dynamic range of fluorescence is illustrated. However, below about 100 pg mL$^{-1}$, nonlinearity is observed (the fluorescence signal is greater than expected). This behavior is believed to be due to contamination problems (e.g., desorption of fluorescent contaminants from glassware or the cell walls). Under these conditions, the calibration slope at higher concentrations cannot be used to evaluate the DL. [From J. D. Ingle, Jr., and R. L. Wilson, *Anal. Chem., 48*, 1641 (1976), copyright 1976, with permission of the American Chemical Society.]

less universal than absorption techniques, but more selective. Even if many species in a sample fluoresce significantly, selectivity is often enhanced by an appropriate choice of excitation and emission wavelengths. Ultimately, the selectivity is limited by the presence of background signals. Electrical noise in the dark and background signals usually controls the detection limits achievable.

### Methodology

Most fluorometric determinations are carried out in solution with external standardization. Therefore, solid samples must be dissolved. In specialized applications to be discussed later in this section, gas or solid phase fluorescence measurements are useful.

In many applications, minimal sample treatment is required because the native fluorescence properties of the analyte are used. For cases in which the analyte is not fluorescent or the fluorescence quantum efficiency is inadequate, derivatization reactions are used to convert the analyte into a product with good fluorescence characteristics. The fluorescence signal of the product is related to the analyte concentration. Most commonly, the analytical reaction is allowed to reach equilibrium before measurements are made. As with molecular absorption spectrometry, titrations or kinetics-based measurements can be carried out by monitoring a fluorescent product or reactant involved in a selective reaction.

Such solution conditions as ionic strength, pH, viscosity, and temperature must be carefully controlled, as they can affect the molar absorptivity, the quantum yield, and the fraction of the analyte in the fluorescent chemical form. When a derivatization reaction is employed, reagent concentrations are adjusted to maximize the fluorescence signal and the conversion efficiency of the analytical reaction. The background fluorescence from reagents, or contaminants in the reagents, must be considered during optimization.

In general, the rate of external conversion increases with increasing temperature because the collisional rate increases; this reduces $\phi_F$. Temperature coefficients are typically 1 to 2% per °C, although they can be much higher. Thus a temperature-controlled sample cell holder is recommended to ensure that standards and samples are at the same temperature.

Photodecomposition or photochemical reactions involving the excited state cause the fluorescence signal to decrease during exposure to excitation radiation. The degree of photodecomposition is proportional to the incident radiant power and the exposure time. If photodecomposition is significant, the incident radiant power can be reduced (e.g., smaller excitation slit width). The

potential decrease in the $S/N$ must be acceptable. Alternatively, a shutter can be used so that the sample is exposed to excitation radiation only during the measurement time.

For quantitative work, a spectrofluorometer or fluorometer may be employed. Fluorometers are less expensive and can provide better detection limits if the source radiant power is reasonably high. With line source excitation, the available excitation wavelengths may not suitably match the absorption spectrum of the analyte. For developmental work, spectrofluorometers are invaluable. Excitation and emission spectra of a standard indicate the wavelengths and bandpasses that maximize the signal from the fluorophore or the types and characteristics of filters to use in a fluorometer. Spectra of the sample can be compared to those of a standard to delineate the nature of the background fluorescence and scattering signals and to assess primary and secondary absorption by concomitants. Wavelengths and values of bandpass can be chosen to optimize the DL, $S/B$, or $S/N$ for a particular type of sample.

### Background Signals

Because background signals and noise are critical, it is important to understand their origin. Common sources of background signals are listed in Table 15-3. The origin and characteristics of the elastic scattering and Raman scattering signals are discussed in detail in Chapter 16.

*Elastic Scattering and Reflection Signals.* The wavelengths of the scattering signals can be predicted as shown in Figure 15-18. The elastic scattering signal, as well as the reflection signal, occurs over the wavelength range passed by the excitation wavelength selector. Ideally, these can be totally eliminated if the excitation and emission bandpasses do not overlap. In reality some of the scattered or reflected radiation is passed by the emission wavelength selector as a stray radiation component. Also, excitation radiation outside

***TABLE 15-3***
Sources of background signals

---

1. Elastic scattering of excitation radiation by the solution, sample cell, or optical components
2. Reflection of excitation radiation by the sample cell walls, optical components, or other surfaces within the sample compartment
3. Raman scattering by the solvent
4. Stray radiation due to room light leaks
5. Fluorescence from the solvent and concomitants (dissolved or adsorbed on particles or sample cell walls)
6. Luminescence from the sample cell walls or optical components (lenses, filters)

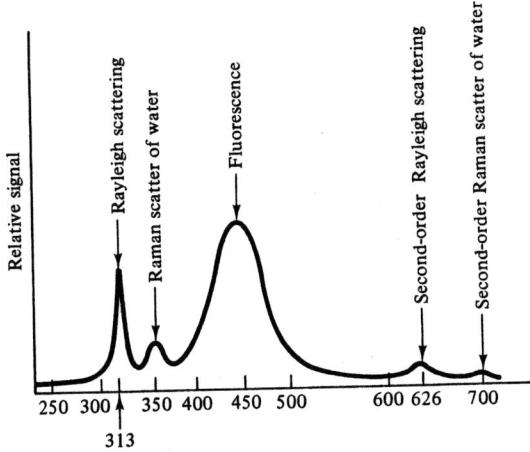

**FIGURE 15-18** Hypothetical fluorescence spectrum of a species in water excited by the 313-nm mercury line. The hypothetical molecule is assumed to have a fluorescence maximum at 450 nm (e.g., quinine sulfate in acidic solution). The elastic scattering signal appears in the first order at 313 nm and in the second order at 626 nm. The Raman band for water has a maximum at 350 nm and the second-order Raman band is observed at 700 nm with 313-nm excitation.

the nominal bandpass of the excitation wavelength selector (stray excitation radiation) and within the emission wavelength selector bandpass can be scattered or reflected and observed. Scattering from molecules and particles within $V^*$ is inevitable. Background signals resulting from reflection, elastic scattering from outside $V^*$, and light leaks are minimized by proper design of the sample compartment (e.g., baffles and light traps) and the geometry of illumination and observation.

Scratches, imperfections, or dust particles on cell walls or optical components can enhance the scattering signal. Scattering from particles in the test solution can be reduced by filtering. Scattering by molecules, however, cannot be eliminated because of its fundamental nature. The intensity of Rayleigh scattering is proportional to the sixth power of the radius of a molecule and $\lambda_{ex}^{-4}$. Thus Rayleigh scattering intensities are lower with excitation in the visible region than in the ultraviolet region.

*Raman Scattering.* Although the Raman effect is weak (i.e., a typical absorption cross section is $10^{-16}$ cm$^2$, whereas a typical Raman cross section is $10^{-29}$ cm$^2$), the Raman band for the most abundant species, the solvent molecule, is readily detected in good spectrofluorometers. The Stokes Raman band appears at a frequency shifted lower by one vibrational quantum from the excitation frequency. The maximum wave-

length of the Raman band ($\lambda_R$) and the Raman wavelength shift ($\Delta\lambda_R$) in nanometers are given by

$$\lambda_R = \frac{\lambda_{ex}}{1 - \lambda_{ex}\,\Delta\bar{\nu}} \qquad (15\text{-}44)$$

$$\Delta\lambda_R = \lambda_R - \lambda_{ex} = \frac{(\lambda_{ex})^2\,\Delta\bar{\nu}}{1 - \lambda_{ex}\,\Delta\bar{\nu}} \qquad (15\text{-}45)$$

where $\Delta\bar{\nu}$ is the Raman shift in nm$^{-1}$. The Raman shifts for common solvents are 0.339 $\mu$m$^{-1}$ (water), 0.292 and 0.143 $\mu$m$^{-1}$ (ethanol), and 0.287 $\mu$m$^{-1}$ (cyclohexane). From equation 15-44, the Raman band for water excited at 313 nm appears at 350 nm (see Figure 15-18). In many cases, the *Stokes shift* for the fluorophore ($\lambda_{em} - \lambda_{ex}$) is much greater than the Raman shift. Thus, wavelength selection usually provides discrimination against the Raman band of the solvent. The width of the Raman band is determined by the excitation bandpass.

*Background Luminescence.* Background luminescence is the most troublesome component of the background signal because it can occur over the same wavelength range as the analyte fluorescence. The magnitude of the background fluorescence signal from solution components with $V^*$ can be quite variable, as it depends on the nature and the concentrations of the concomitants. Background fluorescence from the solvent or reagents is often due to contamination. Thus the solvents and reagents used should be of high purity or further purified by the user. The sample cell and glassware should be scrupulously cleaned to prevent desorption of substances from surfaces. Soaps should be avoided, as they often contain fluorescent components. Cleaning with concentrated acids or strong oxidants (e.g., a $KMnO_4$ solution followed by an HCl rinse) is usually recommended. In general, plastic materials should be avoided (e.g., Tygon is a notorious source of fluorescence contaminants).

The background luminescence (fluorescence or phosphorescence) from cell walls or from optical components depends on the characteristics of the optical material used as illustrated in Figure 15-19; high-grade synthetic silica is the preferred optical material. Background luminescence from glass or silica surfaces directly illuminated by excitation radiation can be eliminated by proper design. By contrast, background luminescence from viewed optical surfaces that is stimulated by scattered or reflected excitation radiation is much more difficult to eliminate. It is also possible for background luminescence from nonviewed optical surfaces to be scattered or reflected into the emission beam.

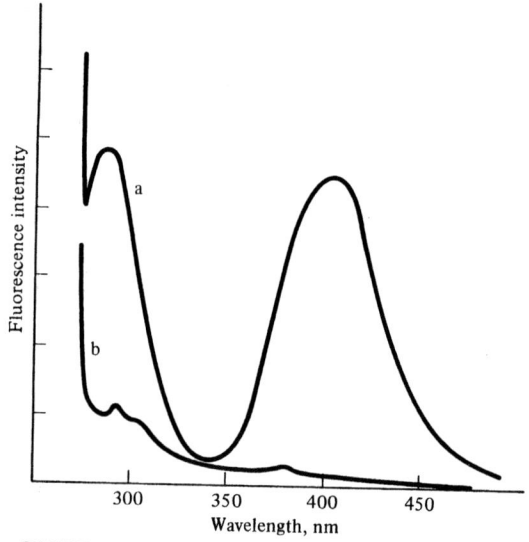

**FIGURE 15-19** Background luminescence of cuvettes. Spectral-grade cyclohexane excited by light at 250 nm; curve a, in fused-quartz cuvette; curve b, in synthetic silica cuvette. (Adapted from C. A. Parker, *Proceedings of SAC Conference, Nottingham,* P. W. Schallis, ed., W. Heffer and Sons, Ltd., 1965, p. 208, with permission of Royal Society for Chemistry.)

## Performance Characteristics

*Precision and Detection Limits.* Measurement precision at concentrations well above the detection limit (DL) is typically in the range 0.5 to 2% and is limited by source flicker noise and drift. The reproducibility of sample acquisition, sample treatment, and calibration must additionally be considered.

As discussed in section 15-2, DLs for efficient fluorophores can be in the range 0.1 pM to 1 nM, depending on the instrument (i.e., excitation radiant power, volume element viewed, efficiency of emission collection) and the magnitude of the background signal and noise. Preconcentration and separation of species causing background fluorescence noise are often employed to improve the DL. Estimation of the DL from the noise in the reagent blank signal can yield a falsely low value due to the noise or variability of uncompensated background signals from concomitants in the sample.

With a spectrofluorometer, the DL can be improved by replacing the emission monochromator with an emission filter if background signal flicker noise is not the dominant noise source. Unfortunately, this option is not usually available with commercial spectrofluorometers. However, it is possible to adjust the emission monochromator wavelength setting to zero (zero order) and insert an emission filter in the emission beam to discriminate against scattering signals. Although the

grating efficiency at zero order is significantly less than in the first order, a significant throughput advantage is often realized due to the larger wavelength range of emission viewed. One commercial spectrofluorometer allows rapid insertion of a mirror in front of the grating to increase the throughput with an emission filter.

*Accuracy.* In addition to the random errors discussed above, systematic errors due to blank and analyte interferences can limit the accuracy. With external standardization, an equivalent analyte concentration in the standard and sample should yield the same analyte fluorescence signal. Concomitants or particles in the sample can attenuate excitation or emission radiation, cause additional fluorescence or scattering, or quench the analyte fluorescence.

Blank interferences are usually spectral interferences due to background fluorescence from concomitants in the sample, but not in the reagent blank, which have native fluorescence or are also converted to a fluorescent product by reacting with the analytical reagent. The potential interferent or its derivative will interfere if its excitation and emission spectra are similar to those of the analyte or its derivative. Scatterers absent in the reagent blank can also cause the total measured fluorescence signal to be too high.

Analyte interferences can be both spectral or nonspectral in nature. Species in the sample, but not in the standards, can alter the absolute value of the fluorescence signal observed from the analyte. When derivatization is necessary, concomitants in the sample can change the conversion efficiency of the analytical reaction.

Nonspectral analyte interference is due to chemical equilibria involving the analyte in the ground state or quenching. In the former case, the fraction of the analyte that exists in the fluorescing form can vary with concomitant concentration. As discussed in Section 12-6, this is sometimes called "static quenching." Quenching usually refers to dynamic or long-range quenching in which a species in the sample increases the rate of nonradiational deactivation and thus lowers the luminescence quantum efficiency. For some analytes, $O_2$ is a good quencher. In such cases, the solutions should be deoxygenated or all standard and sample solutions should have the same $O_2$ content. Bulk properties such as temperature and viscosity should be identical for standards and samples, as they affect $\phi_F$ through their influence on the rate of collisional quenching. Although not common, the presence of dynamic quenchers can be confirmed by measuring the fluorescence lifetime of the analyte in the sample; a decrease in lifetime relative to the standard is indicative of the presence of quenchers. Concomitants can also affect the electrostatic environment of the molecule and change $\varepsilon$ or $Y_{\lambda F}$.

Spectral analyte interference is caused by concomitants that absorb significant amounts of excitation or emission radiation. This can be compensated as discussed in Section 15-2 (equation 15-31). Presently, commercial instruments do not provide on-line correction for absorption. However, the sample absorbance at the excitation and emission wavelength can be measured in a spectrophotometer to ascertain if there is significant primary or secondary absorption (i.e., $A > 0.01$ over $b_{pr} + b^*$ or $A' > 0.01$ over $w + b_{pt}$).

Fluorescence measurements are less dependent than absorption measurements on uncompensated concomitant absorption. Consider a case where the analyte and concomitant absorbances are each 0.001. In absorption spectrometry, the absolute error would be 100% if the blank did not compensate for the concomitant absorption. The concomitant absorption would cause less than 0.1% error in fluorescence measurements. Even with relatively high background absorption, the absolute error caused by uncompensated background absorbance is less with fluorescence measurements than with absorption measurements.

Blank fluorescence interferences are minimized by careful selection of the excitation and emission bandpass and wavelength. In some cases, separation from fluorescent concomitants is mandatory. More specialized techniques are sometimes used to enhance the $S/B$ or DL or to permit the determination of several fluorescent species in one sample. These include derivative, synchronous, and wavelength modulation techniques (see Section 15-1), methods that sharpen the structure of the excitation and emission spectra (see later in this section), and lifetime discrimination (see Section 15-6).

Analyte interference due to absorption, scattering, or quenching can often be eliminated by simple dilution if the DL is adequate. Standard additions can also be useful in these cases.

*Two-photon excited fluorescence* involves the simultaneous absorption of two photons to reach an excited state whose energy is the sum of the energies of the two photons. Because the transition probability is much less than that for one-photon absorption, laser sources are required (i.e., a two-photon cross section is typically $10^{-50}$ cm$^4$ s molecule$^{-1}$ photon$^{-1}$). Selectivity can be enhanced because different selection rules apply for two-photon processes. Because the fluorescence occurs at wavelengths lower than the excitation wavelength, Rayleigh and Raman scattering signals can be virtually eliminated. Also, primary absorption effects are usually minimal at the longer excitation wavelengths. To date, the DLs achieved with two-photon excited fluorescence are considerably poorer than those obtained with conventional spectrofluorometry.

*Linearity.* The linear range of molecular fluorescence can be four to six orders of magnitude for efficient fluorophores with low detection limits. As described in Section 15-2, nonlinearity begins to occur at analyte concentrations where primary absorption becomes significant (i.e., the absorbance over the prefilter and viewed pathlength, $b_{pr} + b^*$, is greater than 0.01). Secondary or self-absorption is usually insignificant at emission wavelengths near the maximum of the emission band, but it can distort the emission spectrum on the short-wavelength side. Self-quenching (i.e., collisional deactivation of an excited analyte molecule by a ground-state analyte molecule) or dimer formation can be significant at relatively high concentration (e.g., $> 1$ mM). Contamination can cause nonlinearity at very low concentrations (see Figure 15-17).

Other potential causes of nonlinearity are similar to those in molecular absorption measurements (see Section 13-3). Chemical equilibria involving the analyte can cause significant nonlinearity. Usually, one chemical form is fluorescent. However, to achieve linearity the fraction of the analyte in the monitored chemical form must be independent of the total analyte or analytical concentration.

Changes in solution bulk properties with analyte concentration or instrumental factors (e.g., polychromatic excitation radiation, stray radiation, variability in pathlengths of different excitation rays, multiple reflections, or dichroism) can cause nonlinearity in absorption measurements. These sources of nonlinearity are not usually significant in fluorescence measurements because they only become important at high analyte concentrations, where primary absorption effects usually dominate.

Measurements can be made in the nonlinear region if the shape of the calibration curve is established with a sufficient number of standards. Precision can be poor in the region yielding maximum signal before rollover because the fluorescence signal changes only slightly with concentration (i.e., the calibration slope approaches zero). It is also possible to linearize the calibration curve with correction factors (see equation 15-31 and Figure 15-15).

With proper design, the characteristics of the PMT, signal processor, and readout device should not limit linearity. If photon counting signal processing is employed, nonlinearity due to pulse overlap can occur at high incident light levels.

Generally, it is preferred to make measurements in the linear region of the calibration curve. This can be accomplished by simple dilution or by changing the excitation wavelength to one where the absorbance is less. Nonlinearity due to prefilter effects is minimized by using microcells or by moving the cell to reduce the prefilter pathlength (i.e., reducing $b_{pr}$).

## Qualitative Analysis

Fluorescence spectrometry is not considered a major structural or qualitative tool for solution measurements. Often molecules with subtle structural differences have similar spectral characteristics. Also, excitation and emission bands are relatively broad in solution at room temperature. Libraries of fluorescence excitation and emission spectra of many molecules are available. These reference spectra can be compared to sample spectra to suggest the possible presence or to confirm the absence of certain species. Only corrected spectra should be compared as the shapes of uncorrected spectra are highly instrument dependent. Care must be taken that spectra are acquired at a low enough analyte concentrations that inner filter effects do not distort the spectral characteristics.

Fluorescence spectrometry has proved a valuable forensic tool for oil spill identification. For example, the source of an oil spill can be confirmed by comparing the emission spectrum of the environmental sample to the emission spectrum of the suspected source of oil. This fingerprinting is made possible by the high degree of vibrational structure of polycyclic hydrocarbons in oil.

## Fundamental Applications

Fluorometry is used to study chemical equilibria and kinetics in much the same way as spectrophotometry (see Section 13-5). Thus common applications include determination of equilibrium constants and reaction stoichiometry (e.g., mole ratio method, method of continuous variations), rates of reactions ($dc/dt \propto dE_F/dt$), rate constants, and reaction mechanisms. The enhanced selectivity makes it easier to monitor one species involved in the reaction. Reactions can be studied at lower reactant concentrations because of the improved detection limits.

The shape and spectral position of bands in the excitation and emission spectra of a molecule yield information about its molecular structure and energy level structure. As we have already noted, the corrected excitation spectrum is often equivalent to a conventional absorption spectrum. However, the excitation spectrum can be obtained at much lower concentration if $\phi_F$ is favorable; this is advantageous if a compound has limited solubility.

Fluorescent probes or tags can be covalently linked or physically bound to specific sites in molecules such as proteins. The magnitude of the observed fluorescence signal provides information about energy transfer processes, the distances between reactive sites, or the polarity of the protein in the vicinity of the binding site. The nonfluorescent tag aminonaphthalene sulfonic acid becomes highly fluorescent when bound to proteins. It can be used to titrate the number of available binding sites.

## Quantum Efficiency Determinations

The fluorescence quantum efficiency (yield) or power yield is also of fundamental importance. The magnitude and variation in magnitude of these quantities with solution conditions provide information about mechanisms and rate constants of deactivation and quenching. The determination of the absolute quantum efficiency or power yield is a challenging experimental problem. Here we review briefly some of the techniques utilized. Depending on the specific technique, either the quantum yield or power yield is determined. The other quantity can be calculated from the determined quantity, as discussed in Section 12-6.

To determine the fluorescence quantum efficiency or power yield directly, it is necessary to measure the total radiant power of fluorescence and the total radiant power absorbed by a given amount of fluorophore. The ratio of these quantities is $\phi_F$ or $Y_F$ (see Table 12-7). The quantum yield can be evaluated with chemical actinometry. A **chemical actinometer** is a device for measuring the absolute radiant power absorbed by a solution by determining the photolysis products of a photochemical reaction of known yield. For example, a sufficiently concentrated solution of potassium ferrioxalate absorbs all radiation below 490 nm. The absorbed radiation reduces ferrioxalate to ferrous oxalate with near-unity conversion efficiency. The number of moles of ferrous ion produced during a given exposure time is determined by titration or by another technique.

For quantum efficiency determinations, the sample cell is totally surrounded (except for small windows to transmit the excitation radiation) by a chemical actinometer solution. This yields a measurement of the total fluorescence radiant power emitted after correction for the window losses. The same actinometer is then used to determine the incident radiant power.

Calorimetric methods provide a simple and elegant method of indirectly determining the luminescence power yield. The efficiency with which the luminescent species ($x$) converts absorbed radiation into heat is measured. This is compared to the efficiency with which a nonluminescent, photochemically inert reference species ($r$) converts absorbed radiation into heat:

$$Y_F = 1 - \frac{H_x}{H_r} \frac{\Phi_{0,r}}{\Phi_{0,x}} \frac{\alpha_r}{\alpha_x} \qquad (15\text{-}46)$$

where $H$ is a signal proportional to the heat produced corrected for the solvent blank, $\Phi_0$ is the incident radiant power, and $\alpha$ is the fraction of the incident radiation absorbed. Traditionally, $H$ is determined by conventional calorimetry (e.g., a Dewar flask or a thermostatted compartment with a temperature transducer to measure $\Delta T$). The response $H$ has also been measured with photoacoustic and thermal lensing techniques (see Chapter 17).

Other methods for the determination of $\phi_F$ utilize a more standard spectrofluorometer setup. The fluorescence signal measured with a spectrofluorometer is only a relative signal. Moreover, only a fraction of the total fluorescence emitted is collected and detected. By completely characterizing all geometric and instrumental factors (see Table 15-1), calibrating the wavelength selection-detection system, and making additional absolute measurements of the incident and transmitted radiant power, we could eventually calculate the total radiant power absorbed and fluoresced. Such absolute calibration is exceedingly difficult. Therefore, relative methods based on some type of reference species are usually employed.

The Weber–Teale approach involves ratioing the fluorescence flux seen by the detector to the photon flux measured with a scatterer (e.g., glycogen) in the sample cell. Under suitable conditions (e.g., the solutions have identical absorbance), the scattering solution is considered an ideal emitter (i.e., a yield of unity). Complications arise due to polarization effects and differences in the spatial distribution of scattering. In this and other methods, integrating spheres and quantum counters are often used to minimize these effects.

The most popular indirect method for determining quantum efficiency is the Parker–Rees method. Here $\phi_F$ is evaluated by comparing the corrected emission spectra of the analyte ($x$) and a reference fluorescent species ($r$) of known $\phi_F$:

$$\phi_{F,x} = \phi_{F,r} \frac{A_r}{A_x} \frac{\Phi_{r,p}}{\Phi_{x,p}} \frac{\int E_x \, d\lambda'}{\int E_r \, d\lambda'} \qquad (15\text{-}47)$$

where $A$ is the absorbance at the excitation wavelength, $\Phi_p$ is the relative incident radiant power (photons s$^{-1}$), and $\int E \, d\lambda'$ is the integral of the corrected emission band (relative quanta per second). Additional factors may be included to account for other differences in

excitation or emission variables if different excitation or emission wavelengths or different solvents (e.g., refractive index effects) are used.

## Quantitative Analysis

*Inorganic Species.* Although many fluorometric methods for metal ions have been proposed, few of these methods are used routinely. Atomic spectrometric techniques are more universal and in general provide better detection limits and selectivity.

Only a few unchelated metal ions, primarily lanthanides [Ce(III), Tb(III), Eu(III)] and actinides [$UO_2^+$, Th(I)], exhibit appreciable fluorescence in solution. The fluorescence, or actually phosphorescence in some cases, is due to $d$–$f$ and $f$–$f$ electronic transitions, is rather long-lived ($\tau_L \approx 1$ ms), and often consists of several narrow emission bands. Detection limits for some elements are as low as 1 ng mL$^{-1}$.

Inorganic reagents can increase the luminescence of metal ions. The quantum efficiency for uranyl ion, for example, is greatly enhanced in concentrated phosphoric acid. In concentrated HCl and HBr, several nonrare earth elements [e.g., Tl(I), Sn(II), Pb(II), As(III)] luminesce. The luminescence is often significantly increased at very low temperatures (e.g., 77 K or lower). Rare earths can also be excited to luminescence by energy transfer from excited states of aromatic carbonyl compounds produced by photon excitation.

The majority of inorganic fluorometric determinations are based on the formation of fluorescent chelates with nonfluorescent ligands. The chelating reagents shown in Table 15-4, or their derivatives, are commonly used. Selectivity is often limited by the specificity of the chelation reaction. The best DLs are often obtained with low atomic number, diamagnetic metal ions ($Be^{2+}$, $Mg^{2+}$, $Ca^{2+}$, $Al^{3+}$), which are less likely to promote intersystem crossing. The fluorometric DL for $Al^{3+}$ with several reagents is in the range 0.1 to 1 ng mL$^{-1}$, better than that achieved with most atomic spectrometric techniques.

The fluorometric determination of selenium in biological materials with 3,3'-diaminobenzidine, of boron with 4'-chloro-2-hydroxy-4-methyloxybenzophenone, and of beryllium with morin (2',3,4',5,7-pentahydroxy-flavone) are still popular methods in specialized applications.

Some proposed fluorometric determinations are based on the inorganic ion acting as a reducing or oxidizing reagent or a quencher. These methods have generally proved to be less selective than more direct fluorometric techniques.

As in spectrophotometry, reaction-rate methods can yield added selectivity by discriminating against

**TABLE 15-4**
Some common chelating agents and fluorescent chelates

| Name | Ligand structure | Chelate structure |
|---|---|---|
| Benzoin | (keto form) | |
| 2,2'-Dihydroxyazobenzene | | |
| 2-Hydroxy-3-naphthoic acid | | |
| 8-Hydroxyquinoline | | |
| Salicylidene-2-aminophenol | | |

steady-state signals (uncompensated background fluorescence) or species that react at rates significantly different from the analyte. For example, $Al^{3+}$ can be determined down to 0.1 ng mL$^{-1}$ by fluorometric monitoring of the rate of formation of its chelate with 2,4,2'-trihydroxyazobenzene-5'-sulfonic acid. Many kinetics-based determinations have been proposed in which the analyte metal ion acts as a catalyst.

Metal ions, particularly rare earth ions, can also be fluorometrically determined if they are incorporated into a host solid matrix by precipitation, crystallization, or solid-state reactions. The fluorometric determination of uranium fused in a NaF matrix has been widely used for decades. In a solid matrix, the absorption and emission bands become much narrower. With a tunable dye laser, it is possible to achieve high selectivity by exciting a specific absorption line of a given rare earth in a particular environment or crystallographic site. Detection limits for many lanthanide ions in $CaF_2$ are in the range 1 to 100 pg mL$^{-1}$.

Nonfluorescent ions can be determined by **probe ion** techniques. Here the analyte is incorporated into a solid matrix where it forms associates with the fluorescent rare earth ions. The associates experience a changed crystal field splitting and the spectral bands are shifted. It is then possible to obtain single-site spectra of only rare earth associates with the same detection limits as achieved for rare earth ions directly. Sample preparation can be tedious, and cryogenic facilities are usually used to cool the sample for narrowing the bands.

*Organic Species.* Fluorometric methods of analysis are used routinely for the determination of many types of organic molecules. They are employed extensively in the clinical and biochemical area; however, environmental applications are also quite important.

Clinical analysts are often required to determine very low concentrations of analytes (often $1 \mu g\ L^{-1}$ or less) and to work with small samples (e.g., serum from a newborn). The detection limits and selectivity provided by fluorescence are thus well suited for clinical assays. Before measurement, some form of sample cleanup is usually required. Red blood cells are usually removed from blood by coagulation and centrifugation before enzymatic assays. For other species, deproteinization of the blood serum or urine is also required to reduce background absorption and fluorescence from proteins and to release analytes that are bound to proteins. Solvent extraction or simple column cleanup procedures are used to separate the analyte from the majority of the sample constituents. Tissue samples are usually homogenized in a blender, and then the homogenate is filtered or extracted with a suitable solvent.

Several biochemical compounds exhibit significant native fluorescence at room temperature as shown in Table 15-5. Thus their determination requires only a separation from potential interferents. The residues of highly fluorescent tryptophan account for about 90% of the fluorescence from proteins.

Many selective derivative reactions have been developed for nonfluorescent organics or organic molecules with inadequate natural fluorescence properties. Some common reagents are listed in Table 15-6.

The detection limit of enzymatic assays is often improved by using fluorometric instead of spectrophotometric monitoring. Substrates can be determined by equilibrium-based or kinetics-based procedures; enzyme activity must be assayed by reaction-rate methods. Often, the same chemistry used in spectrophotometric monitoring is employed in cases where the coenzymes NADH or NADPH are involved in the reaction (see equations 13-59 to 13-61). The fluorescence quantum efficiencies of the reduced coenzymes are about 2%. To improve the DL, the NADH formed can be coupled to another reaction which produces a product that is highly fluorescent. Alternatively, other reactions can

**TABLE 15-5**

Types of organic compounds with native fluorescence

| Class | Examples |
|---|---|
| Aromatic amino acids | Tyrosine, tryptophan, phenylalanine |
| Vitamins | A, $B_2$ (riboflavin), $B_6$ (pyridoxine), $B_{12}$, E, folic acid |
| Flavins | Flavin mononucleotide (FMN), flavin dinucleotide (FAD) |
| Reduced pyridine nucleotides | NADH, NADPH |
| Cathecholamines | Dopamine, norepineprine, epinephrine (adrenalin) |
| Porphyrins | Chlorophylls |
| Steroids | Estrogen |
| Drugs | Quinine, salicylates, tetracyclines, barbiturates |

be coupled to a product of the initial enzymatic reaction to produce a fluorescent product.

Most fluorometric enzymatic assays are carried out in solution. However, special front-surface techniques have been developed to monitor reactions on solid surfaces. The reagents can be adsorbed or immobilized on the solid surface. A drop of sample is placed on the active surface.

Fluorescence is used in environmental applications for detecting certain aromatic hydrocarbons. The most extensive applications are for determining polycyclic aromatic hydrocarbons (PAH) such as chrysene, perylene, pyrene, fluorene, and 1,2-benzofluorene. These compounds are released into the environment by fossil-fuel burning, and many are carcinogenic.

*Automated Measurements.* Automated batch or continuous flow analyzers are available for fluorometric determinations. They are similar to the systems described in Section 6-6 except that the spectrophotometer or photometer is replaced with a spectrofluorometer or fluorometer with a flow cell. Their use

**TABLE 15-6**

Common fluorometric reagents for organics

| Analyte | Reagent |
|---|---|
| Estrogens | $H_2SO_4$ |
| Corticosteroids | $H_2SO_4$ |
| Amino acids | *o*-Phthaldehyde (condensation) |
| Thiamine | $Fe(CN)_6^{3-}$ or $Hg^{2+}$ (oxidation to thiochrome) |
| Ascorbic acid | Phenyldiamine (condensation) |

is prevalent in clinical applications where high sample throughput is critical. Background due to desorption of fluorescent species from plastic tubing must be considered. The centrifugal analyzer (see Figure 6-10) can also be configured for fluorescence measurements.

## Multicomponent Analysis

Often a situation is encountered in which it is necessary to determine one or more species in samples containing several fluorophores with overlapping excitation and emission spectra. Several techniques which do not rely on separations have been proposed. The derivative and synchronous spectra techniques were discussed in Section 15-1; these are often useful for mixtures. In addition, there are methods based on simultaneous equations and low-temperature measurements that have proven useful for mixtures.

*Simultaneous Equations.* By making measurements at *n* sets of excitation and emission wavelengths, one can determine *n* components by techniques similar to those used in multicomponent spectrophotometric determinations. The fluorescence signal for a given set of wavelengths is the summation of the products of the concentration of each species and the calibration slope for that species. The latter factors are determined by measuring the fluorescence signal for pure standards of each species. The concentration of each species is then determined by the method of simultaneous equations, often using matrix techniques. The excitation-emission matrix, acquired to obtain total luminescence spectra, is a convenient data set for this technique.

*Low-Temperature Techniques.* Normally, electronic bandwidths are determined by **inhomogeneous broadening**, spectral broadening due to the dif-

ferent microenvironments experienced by different analyte molecules. In a rigid inert matrix, the environmental heterogeneity is reduced and it is possible to approach bandwidths limited by **homogeneous broadening**, the inherent contribution to bandwidths determined by the Heisenberg uncertainty principle and other factors. With this in mind, it is not surprising that researchers have incorporated samples into a solid matrix to sharpen the spectra for multicomponent determinations. The sharpening of the vibronic structural detail in both the excitation and emission spectra makes it possible to excite (with lasers) and observe fluorescence more selectively by tuning into specific vibronic bands of the analyte. Under certain conditions, it is possible to excite molecules only in a particular type of microenvironment, a procedure denoted **site-selection spectroscopy, fluorescence line narrowing**, or **energy selective spectroscopy**.

Samples can be dissolved in some organic solvents and cooled to liquid-nitrogen temperatures (77 K) to form a clear transparent glass. This technique is used primarily in phosphorescence measurements (see Section 15-4). At very low temperatures achievable with the cryogenic facilities (4 K), spectral detail is greatly enhanced as illustrated in Figure 15-20. Bandwidths of 0.01 to 0.1 nm are achievable.

The sharpening of the fluorescence spectra is known as the **Shpol'skii effect.** Most of the work in this area has involved PAHs dissolved and frozen in *n*-alkanes, especially *n*-heptane, to form a polycrystalline solid. The nature of the spectra and the sharpness of the bands depend critically on the solvent. Optimum resolution is achieved when there is a close match between the length of the alkane carbon skeleton and the longest dimension of the aromatic solute molecule. Cooling provides the added benefit of enhancing the quantum efficiency due to reduced collisional deactivation.

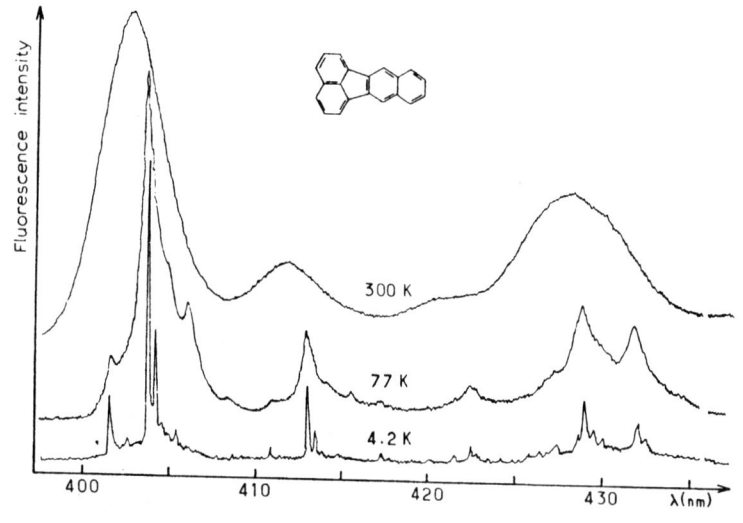

**FIGURE 15-20** Fluorescence spectra of benzo[k]fluoranthene in *n*-heptane. As the temperature is lowered, the emission bands become narrower. [From J. M. Colin, G. Vion, M. Lamotte, and J. Joussot-Dubien, *J. Chromatogr., 204,* 135 (1981), with permission of Elsevier Science Publishers.]

Other techniques can be used to prepare samples in inert matrices. In **matrix isolation techniques,** the sample is vacuum co-deposited with $N_2$ or a Shpol'skii solvent on a cold finger at 10 to 15 K. Samples can also be dissolved and solidified in plastic matrices.

## Spectrofluorometric Titrations

Fluorescence indicators are used in titrations where colorimetric indicators lack sensitivity or are difficult to observe, as in highly turbid or colored solutions. Titrations with visual endpoint detection are carried out in a dark room or a viewing box with illumination by a UV lamp. In instrumental fluorometric titrations, the titration vessel is located in the compartment of a fluorometer or spectrofluorometer. Acid-base indicators are based on changes in $\phi_F$ for the acid and base forms of the indicator. In precipitation titrations, the difference in $\phi_F$ for the indicator in solution as opposed to adsorbed on the precipitate is exploited. Redox titration indicators exhibit different fluorescence characteristics in different oxidation states. Indicators for complex formation titrations, often called *metallofluorescent indicators*, usually form fluorescent complexes with the untitrated metal ion.

## Fluorescence Chromatographic Detection

Fluorescence has long been used as a detection method in thin-layer chromatography (TLC). For nonfluorescent compounds, fluorescent derivatives are prepared prior to separation or the plates are sprayed with derivatization reagents after separation. For qualitative analysis, the plate is observed under UV light from a Hg arc lamp. For quantitative measurements, attachments for commercial spectrofluorometers are available to translate the TLC plate past the excitation probe beam to obtain a plot of fluorescence signal vs. distance. Front-surface viewing is employed.

Fluorometric detectors are now widely used in HPLC. A fluorometer or spectrofluorometer can be converted to an HPLC detector by replacing the standard sample cell with a micro-flow cell. Flow cells are often constructed from narrow-bore fused silica tubing. To prevent extra broadening, the internal volume of the cell should be less than 20 μL. There can be a considerable loss in the absolute fluorescence photon flux observed by the detector because the dimensions of the viewed volume element are reduced [i.e., the cross-sectional area of illumination ($wh$) and the viewed pathlength ($b^*$) are reduced relative to a 1-cm-pathlength cell]. Because the viewed volume element is closer to the cell walls, scattering, reflection, and cell wall luminescence signals are more difficult to discriminate

against. The loss in absolute calibration sensitivity and increased background signals and noise often degrade detection limits.

Because of the difficulties noted above, many companies market dedicated HPLC fluorescence detectors specifically designed to illuminate and collect fluorescence radiation from microvolume cells. Fluorometers often provide adequate selectivity and higher-radiation throughput than monochromator-based detectors. Specially designed emission selection systems as shown in Figure 15-21 are used to enhance the emission signal detected. Although spectrofluorometers may yield poorer detection limits than fluorometers, the excitation and emission wavelengths can be changed at appropriate times during the chromatographic run to optimize conditions for specific compounds.

Researchers have designed specialized "windowless" flow cells to reduce background signals from silica cell walls. In one design, eluate in drop form is illuminated as it flows out of the HPLC column. In another design, the eluate is confined in a sheath of solvent.

Laser excitation is well suited for fluorometric

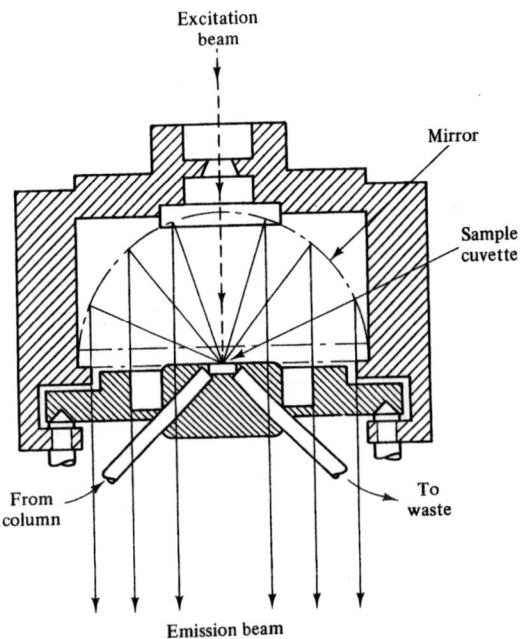

**FIGURE 15-21** Diagram of the "$2\pi$ steradian cuvette" for fluorescence HPLC detection. Excitation radiation from an excitation monochromator is impingent on the sample cell. The mirror collects front-surface fluorescence emitted over $2\pi$ sr and directs it to an emission filter and PMT. A $D_2$ lamp is used for excitation because it has a higher radiance in the region 200 to 250 nm, where many organic species have high molar absorptivities. (Courtesy of ABI Analytical, Kratos Division.)

HPLC detection. One can take advantage of the ability to focus an intense spot into a small cell volume. Absolute detection limits of about 1 pg have been obtained.

HPLC with fluorometric detection has been shown to be particularly useful for samples containing a mixture of native fluorophores such as PAHs. For nonfluorescent molecules, either precolumn or postcolumn derivatization has been used. The determination of amino acids is an important primary example. Precolumn derivatization is often based on labeling reactions. The labeling reagent is also fluorescent, but is separated from the labeled analytes. Dansyl chloride (5-dimethyl-aminonapthalene-1-sulfoenyl chloride) reacts with primary and secondary amino groups (and also phenolic hydroxyl groups) to form highly fluorescent derivatives. The reagent o-phthalaldehyde reacts with amino acids and amino sugars in the presence of a reducing agent to form a fluorescent condensation product. This reaction is fast enough to use for postcolumn derivatization.

### Vapor-Phase Fluorescence

Fluorescence measurements in the gas phase are not as common because many fluorescent compounds are not volatile. However, fluorescence detection in gas chromatography has been demonstrated. Several important air pollutants, including $NO_2$ and $SO_2$, can be selectively detected by their fluorescence.

Samples can be cooled to translational temperatures less than 1 K by supersonic expansion of a gas through an orifice into a vacuum (see Figure 17-7b). Under these conditions, fluorophores in mixtures can be determined selectively, as the spectral bands are very narrow; resolution is often an order of magnitude better than achieved with low-temperature measurements in the solid phase.

Recent studies have demonstrated the analytical potential of sensitized fluorescence in the gas phase. Energy transfer occurs from metastable species such as $Ar^*$ and $N_2^*$ produced in a discharge.

### Immunoassays

**Immunoassay** is based on competitive binding between the analyte in a sample and a labeled analyte and a binding molecule. Usually, the analyte is an antigen and the binding molecule is an antibody. Antigen-antibody binding is extremely specific. The concentration ratio of the sample analyte to tagged analyte bound to a limiting amount of antibody increases with increasing sample analyte concentration. Thus the displacement of the tagged antigen by the sample analyte is determined by measuring a physical or chemical property associated with the labeled antigen. In **heterogeneous assays,** the free and bound antigen are physically separated by solid-phase adsorption, precipitation, or chromatography before measurement. On the other hand, there is no prior partitioning in **homogeneous assays.**

Radioimmunoassay (RIA) heterogeneous techniques were the first developed. Detection limits of $10^{-12}$ to $10^{-15}$ M have been demonstrated. More recently, there has been a tendency to move away from radiotracers because of safety and ecological considerations. The expense and shelf life of isotopic labels are also of concern. Of the other types, fluorescent labels have received the most attention. Many types of fluorescence immunoassay have been developed.

The most widely used fluorescent probes have been derivatives of fluorescein–isothiocyanate. The isocyanate group is bound through reaction with amino groups on the antigen. For heterogeneous immunoassays, the bound antigen is separated from the unbound antigen before the fluorescence is measured. A calibration curve is prepared in which the fluorescence signal from a mixture of labeled antigen, antibody, and unlabeled antigen standard is plotted vs. the standard antigen concentration. The fluorescence signal decreases in a nonlinear fashion with increasing unlabeled antigen because a larger fraction of the labeled antigen is displaced by the unlabeled antigen. Homogeneous assays are based on differences in the fluorescence characteristics of the bound and unbound fluorescent tag, such as quenching or enhancement, polarization, lifetime, and resonance energy transfer. Some homogeneous immunoassays involve labeling the antigen with a substrate which is converted to a fluorescence product by enzymatic hydrolysis. Conditions are arranged so that only the unbound antigen can undergo the enzymatic reaction.

### Remote Sensing

In situ, laser-excited fluorescence of atmospheric pollutants has been demonstrated. Remote sensing at long distances is limited by the small solid angle of collection. Laser sources are often used in such studies (see Section 16-5).

Remote sensing in solutions is a new and developing area. Single fibers, two fibers, or bifurcated fiber bundles have been employed to direct excitation radiation to the sample and return the fluorescence generated to a wavelength selector-detection system. The sampling step is eliminated and in situ determinations can be made in locations normally difficult to access (e.g., underwater, groundwater wells, inside blood vessels). This type of remote sensing is also advantageous for measuring samples in hazardous locations (e.g., chemically corrosive or potentially explosive environments); the expensive and delicate components are located in a safe environment.

Species with native fluorescence can be monitored directly (e.g., chlorophyll in lakes). For nonfluorescent species it is necessary to modify the distal end of the fiber to respond to the analyte. Such devices are called **optrodes.** A selective reagent is immobilized or confined by a membrane at the tip of the fiber. The analyte selectively reacts with the reagent to form a fluorescent derivative. Optrodes have also been constructed to measure pH or temperature by incorporating at the optrode tip a fluorescent substance whose fluorescence signal depends on pH or temperature in a known manner.

## 15-4 MOLECULAR PHOSPHORESCENCE SPECTROMETRY

The instrumentation, methodology, and applications of molecular phosphorescence are similar to molecular fluorescence in many respects. Differences arise primarily because the phosphorescence emission is from a long-lived triplet state, whereas fluorescence occurs from the short-lived singlet state. Two basic conditions must be met for the quantum efficiency of phosphorescence ($\phi_P$) to be large enough that phosphorescence measurements are feasible (see Section 12-6). First, intersystem crossing must be a favorable process in the analyte molecule or the environment of the molecule must be changed to make it favorable. Second, the triplet state must be protected to prevent significant quenching before phosphorescence can occur. Other considerations can also be critical. Usually, discrimination techniques in addition to wavelength selection are needed to prevent interference from scattering and fluorescence signals. In solutions, $O_2$ quenching is much more serious than for fluorescence measurements. We shall now see how the considerations noted above profoundly affect sample preparation and handling and some instrumental components.

### Low-Temperature Phosphorescence

Traditionally, phosphorescence measurements have been made on samples dissolved in organic solvents which form clear rigid glasses at liquid $N_2$ temperatures (77 K). The rigid matrix minimizes collisional quenching. The sample cell is usually a small (1 to 3 mm i.d.) synthetic silica tube. After it is filled with sample, it is immersed into a silica Dewar flask which is filled with liquid coolant as shown in Figure 15-22. Liquid $N_2$ is almost always used as the coolant. The dimensions of the tube involve a compromise between too small a

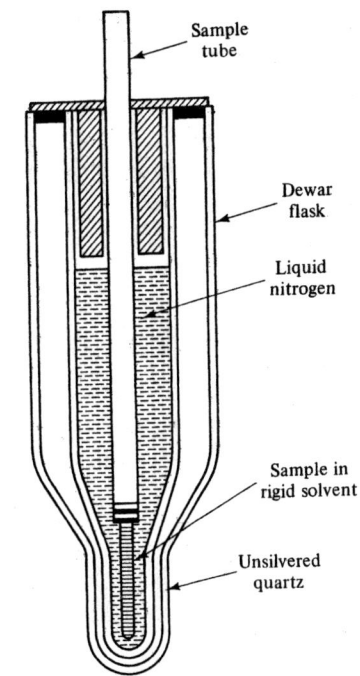

**FIGURE 15-22** Dewar vessel and cell for low-temperature phosphorescence measurements. The optical path goes through an unsilvered part of the Dewar.

diameter, which limits $V^*$ and the $S/N$, and too large a diameter, which leads to poor solidification and cracking of the glass. Solidification of solutions through conductive cooling is also employed. A sample capillary tube is filled with sample solution and placed in contact with metal that is cooled with liquid nitrogen.

Often, the sample tube is connected to a spinning apparatus to average out optical inhomogeneities and to minimize the effect of variations in sampling position. The rotation can improve precision and detection limits by an order of magnitude. It permits reasonable measurements under conditions where a cracked glass or snow is formed.

Selection of the proper solvent is critical. The analyte should be readily soluble in the solvent at low temperatures. The solvent should form a clear rigid glass at 77 K and have low phosphorescence background. The most frequently used solvent for nonpolar compounds is known as EPA, a mixture of ethanol, isopentane, and diethyl ether in the proportions by volume of 2:5:5. Other solvents (ethanol) or mixtures of solvents (e.g., ethanol–methanol, isopropyl alcohol–isopentane) are sometimes employed. Solvents should be purified to remove aromatic and heterocyclic compounds that phosphoresce. A solvent mixture denoted IEPA is 10 parts of EPA to 1 part of methyl iodide. The methyl iodide increases $\phi_P$ for many compounds due to the external heavy atom effect. Although not extensively used, glassy plastics such as polymethyl methacrylate have also been employed as a solid support matrix for phosphorescence measurements.

The requirement of cryogenic conditions has somewhat limited the use of low-temperature phosphorescence (LTP) measurements. Also, the filling and cleaning of long narrow cells can be tedious; about 5 minutes per sample is required.

### Room-Temperature Phosphorescence

Because of the time required for LTP measurements, the development of room-temperature phosphorescence (RTP) techniques has received considerable attention. Phosphorescence at room temperature is feasible if the sample is deposited on a solid support or dissolved in a micellar or cyclodextrin solution.

*Solid Substrate–Based Measurements.* The first analytical RTP work involved front surface phosphorescence measurements of samples deposited on solid substrates. Common substrates include cellulosic supports (filter paper disks), inorganic substrates (silica gel, alumina, chalk), and organic substrates (sodium acetate, sucrose). The binding of the analyte to the solid substrate provides the rigidity required to reduce collisional quenching of the triplet state. Matrix modifiers are often mixed with the sample or codeposited with the sample to enhance the phosphorescence. Bases or acid salts promote ionization; the binding to the substrate is often stronger, with a resultant increase in $\phi_P$, when the analyte is in its ionized form. Inorganic salts or organics containing heavy atoms (e.g., NaI, $AgNO_3$, lead acetate) are used as matrix modifiers to promote intersystem crossing. Normally, a drying step is required to remove $H_2O$, which competes with the analyte for bonding sites on the support material and enhances transport of oxygen into the sample matrix. In general, moisture reduces $\phi_P$. Background phosphorescence from the substrate is a problem for most substrates. Various treatments to reduce the background effects have been unsuccessful. Overall, the choice of the optimum substrate, solvent, and matrix modifiers is still somewhat empirical.

The sample preparation procedure used for RTP analysis based on a filter paper substrate is illustrated schematically in Figure 15-23. Somewhat similar procedures are used for silica gel plates. For powdered substrates, the solvent, sample, and powder are mixed, dried, ground, and placed in a cavity in a plate that can be mounted in the sample compartment. In the packed flow cell technique, a mixture of crushed quartz and paper lint scraped from filter paper is packed into a narrow-bore quartz tube. The sample and reagents are injected with a syringe.

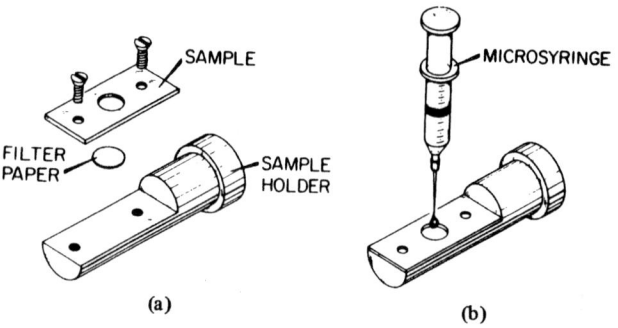

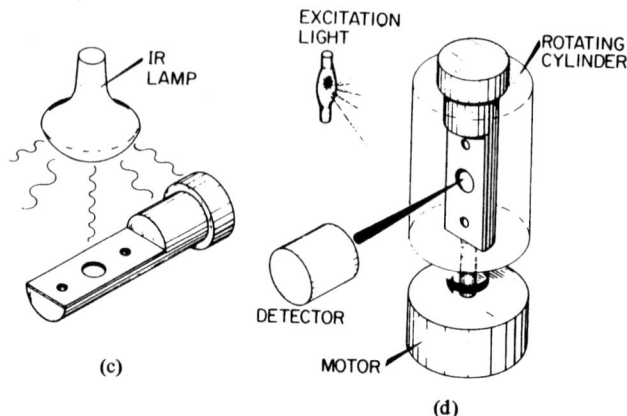

**FIGURE 15-23** Schematic illustration of the procedure for RTP analysis using a filter paper substrate. In (a) a filter paper disk is placed in the sample holder. Next, the filter paper is spotted with solution delivered by a syringe as shown in (b). After the sample is dried (c), the sample holder is mounted in the sample compartment of the phosphorimeter as illustrated in (d) and the RTP measurement is made. (Adapted from T. Vo-Dinh, *Room Temperature Phosphorimetry for Chemical Analysis*, p. 99, by permission of John Wiley and Sons, Inc., New York, copyright © 1984.)

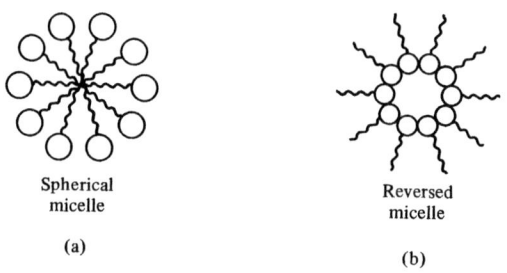

**FIGURE 15-24** Pictorial representation of the aggregation of micelles. In an aqueous solvent, (a) the aggregate has a polar boundary and a nonpolar core due to repulsion of the polar heads (circles in diagram) and association of the hydrophobic tails. The arrangement is reversed in nonpolar solvents as shown in (b).

*Micelle- and Cyclodextrin-Based Measurements.* More recently, solution-based RTP techniques have been developed. Micellar solutions have been found to be particularly attractive for RTP measurements. Molecules having a nonpolar hydrophobic group attached to a polar hydrophilic group are known as surfactants or detergents. The nonpolar group is usually a hydrocarbon chain ("tail"), and the polar group ("head") can be either an ionic or a neutral compound. Above a certain concentration known as the *critical micelle concentration* (CMC), these surfactants dynamically associate in solutions to form colloidal aggregates called **micelles**. The type of aggregation of the surfactants depends on the chemical structure of the surfactants and the nature of the solvent, as shown in Figure 15-24.

Phosphorescence from many organic molecules is observed at room temperature in solutions containing detergents (sodium lauryl sulfate is common) at concentrations above the CMC. In a typical sample preparation procedure, a small amount of sample is mixed with a micellar solution. An ultrasonic bath often aids mixing. The sample is usually spiked with salts containing heavy atom ions [Tl(I), Pb(II)] to enhance $\phi_P$ (external heavy atom effect). Next, the sample solution is transferred to a container where it is deoxygenated

**TABLE 15-7**
Summary of some analytical features of different RTP methods

| RTP method | Advantages | Disadvantages |
|---|---|---|
| Filter paper | Simple<br>Rapid (5–10 min)<br>Inexpensive substrate material<br>Versatile (easily treated with chemical agents)<br>Gives best results for the widest variety of compounds<br>Can be combined with paper chromatography<br>Can be automated easily | Sensitive to humidity<br>Has a strong background emission |
| Silica gel chromatoplates | Simple<br>Relatively rapid (15–30 min)<br>Selective for certain compounds<br>Can be combined with thin layer chromatography<br>Less sensitive to humidity than paper | Less applicable to many compounds than filter paper<br>More susceptible to oxygen quenching<br>Requires substrate materials that cost more than filter paper |
| Powder substrates<br>Sodium acetate<br>Polymer mixtures<br>Inorganic materials | Selective for a few compounds<br>Almost insensitive to humidity<br>Has minimal background interference[a] | Requires more elaborate sample preparation than filter paper<br>Long assay time (25–65 min)<br>Limited applicability to many compounds<br>Lower sensitivity (than with filter paper) |
| Packed flow-through | Can analyze liquid samples directly<br>Can be combined with liquid chromatography<br>Can be automated | Elaborate sample preparation procedures<br>Very sensitive to humidity |
| Micellar solutions | Can analyze aqueous samples<br>Increases solubility of nonpolar aromatics<br>Can be automated<br>Can be directly interfaced with high-performance liquid chromatography | Elaborate sample preparation procedures<br>Requires deoxygenation process (sensitive to oxygen)<br>Relatively long assay time (1 h) |

SOURCE: T. Vo-Dinh, *Room Temperature Phosphorimetry for Chemical Analysis*, p. 119; by permission of John Wiley & Sons, Inc., New York, copyright © 1984.
[a]Applies to the method using sodium acetate.

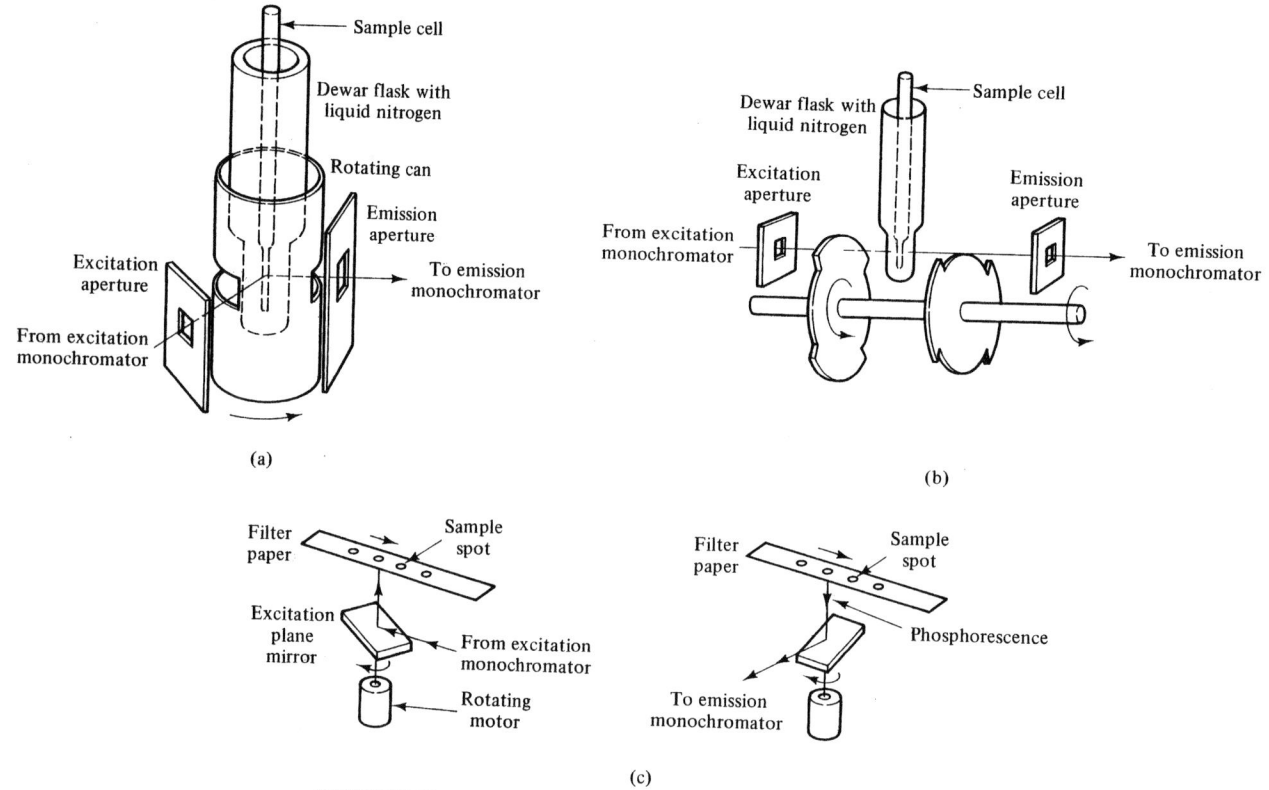

**FIGURE 15-25** Phosphoroscope designs. In (a) the rotating can or cylinder design is shown. The hollow cylinder has one or more apertures equally spaced on the circumference. During part of the rotation cycle, the excitation radiation passes through an aperture and strikes the sample cell while the emission radiation is blocked from reaching the emission monochromator. As the can rotates, it reaches a position where the excitation radiation is blocked, but the emission radiation can pass through an aperture and be detected. It is suitable for the right angle excitation emission geometry. The Becquerel or rotating disk design (b) provides the same type of out-of-phase excitation and viewing with an in-line excitation-emission geometry. The cam-shaped disks can be replaced by circular disks with apertures. In both (a) and (b) the shaft is driven by a variable-speed motor and the sample is in a Dewar flask for LTP measurements. The Dewar flask can be replaced by a normal cell for micelle-based RTP measurements or by a suitable holder for solid substrate-based RTP measurements (see Figure 15-23b). A rotating mirror phosphoroscope is shown in (c). The filter paper strip can be moved to allow viewing of different sample spots.

by purging with nitrogen and applying a vacuum to the solution. Finally, the sample solution is transferred to a sample cell for excitation and observation of the phosphorescence.

The incorporation of the analyte molecule into the core of a micelle imparts certain advantages:

1. The structural conformation of the micelles protects the phosphorescing triplet state of the analytes from external quenchers.

2. The orientational constraint decreases vibrational deactivation.

3. The altered microenvironment can provide favorable polarity and acid/base equilibria for enhanced phosphorescence quantum efficiency.

4. The micellar solutions can improve the detection limit for hydrophobic species in aqueous solution by increasing their solubility.

5. The proximity of interacting species (phosphors and heavy atoms) is increased, and can result in a more efficient spin-orbit coupling.

Although deoxygenation is needed, it can provide an advantage. The oxygenated solution containing the analyte produces little or no phosphorescence because of the efficient quenching by $O_2$. Hence this solution can serve as a blank that accounts for scattering signals and luminescence signals that are not dependent on $O_2$.

Cyclodextrins are doughnut-shaped polymerized sugar molecules. They form inclusion complexes with molecules in solution. Phosphorescence and fluorescence are enhanced for molecules included in the cavity. In contrast to micellar solutions, some phosphorescence is observed without deaeration. However, deoxygenation does increase the phosphorescence signal. Heavy atom-containing reagents (e.g., 1,2-dibromoethane) are added so that they are incorporated into the cavity with the analyte to promote formation of the triplet state. Selectivity can be greater than in micellar solutions, due to the stringent size limits imposed by the inclusion process.

*Sensitized RTP Measurements.* In sensitized RTP techniques, a weakly or nonphosphorescent analyte (donor) is excited and transfers its triplet-state energy to the triplet state of an acceptor with a good phoshorescence quantum efficiency. The phosphorescence signal from the acceptor is measured. Good acceptors are bromonaphthalenes and diacetyl which exhibit significant phosphorescence in solutions even at room temperature. To obtain specificity, chromatographic separation normally precedes mixing of the donor and acceptor and detection. The donor can also be determined from the quenching of its phosphorescence by the acceptor.

The advantages and limitations of different RTP methods are summarized in Table 15-7.

## Instrumentation

*Phosphoroscopes.* The primary difference between many fluorescence spectrometers and phosphorescence spectrometers is the incorporation of a phosphoroscope. A **phosphoroscope** is a mechanical device that allows out-of-phase excitation and observation of phosphorescence from the sample. Several types are shown in Figure 15-25. The rotating can type (Figure 15-25a) is available as an attachment for some commercial spectrofluorometers. The phosphoroscope eliminates short-lived scattering and fluorescence signals. Scattering and reflection signals can be much greater from frozen samples in small tubes or solid substrate surfaces.

The operation of a phosphoroscope is characterized by three time periods, as shown in Figure 15-26. The fraction ($\alpha$) of the total phosphorescence observed (i.e., that observed without a phosphoroscope) is given by

$$\alpha = \frac{\tau_P \exp(-t_d/\tau_P)[1 - \exp(t_e/\tau_P)]^2}{t_c[1 - \exp(-t_c/\tau_P)]} \quad (15\text{-}48)$$

To maximize the phosphorescence flux from a given phosphor, it is important to adjust the cycle time so that the decay of the phosphorescence is small during the delay and observation time. If $\tau_P \gg t_c$, $\alpha$ is independent of $\tau$ and the cycle time and

$$\alpha = \left(\frac{t_e}{t_c}\right)^2 \quad (15\text{-}49)$$

The maximum rotation rate of rotating can and disk phosphoroscopes is about 6000 rpm, which yields a minimum $t_c$ of about 10 ms. Thus phosphoroscopes are generally useful for $\tau_P > 1$ ms.

*Pulse Source, Gated Detection.* Time resolution can also be achieved with pulse excitation and electronically gated detection. The sample is repetitively excited with radiation from a pulsed or chopped light source to populate the triplet state. The resulting phosphorescence from a given excitation pulse is observed after a suitable delay time measured from the end of the excitation pulse. Often, a boxcar integrator is used to allow control of both the delay time and the period

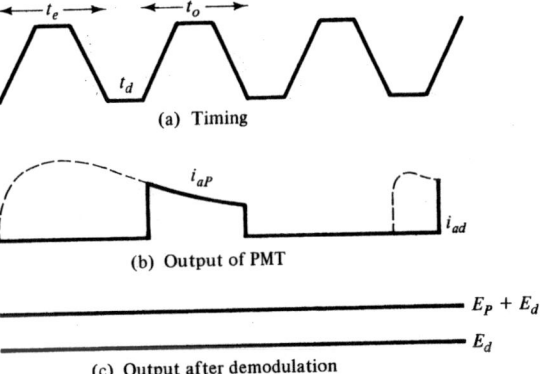

(a) Timing

(b) Output of PMT

(c) Output after demodulation

**FIGURE 15-26** Timing diagram for a phosphoroscope. In (a) the phosphoroscope is characterized by $t_e$, the exposure time, $t_d$, a delay time, and $t_o$, the observation time. The total cycle time $t_c$ is equal to $t_e + t_d + t_o$. The decay of the phosphorescence signal ($E_p$) during the observation period can be monitored on an oscilloscope (b). For convenience, the signal processing electronics can be configured to demodulate the signal and output a readout signal proportional to the average signal during the observation time as shown in (c).

of observation and to average the phosphorescence signals from many pulses. In some cases, the PMT bias voltage is turned off during the excitation pulse to prevent overloading the PMT with the scattered radiation signal.

In early studies, a continuous Xe arc lamp and a mechanical chopper were employed. More recently, the preferred source is a pulsed Xe arc flash lamp which yields pulse widths of 1 to 10 μs. This allows phosphors with $\tau_P$ values below 0.1 ms to be measured and offers more convenient control of the delay and observation times. The *S/N* is better than with chopped systems because the peak irradiance is much higher than the average irradiance for continuum lamps. Pulsed lasers have also been used.

### Methodology

The methodology used in phosphorimetry is similar in many respects to fluorometry except for sample preparation and presentation. Most analyses are based on species with native phosphorescence; derivative reactions are rarely used. Reagents containing heavy atoms are often employed to increase $\phi_P$. These heavy-atom matrix modifiers must be used with care because the lifetime is also reduced (i.e., α in equation 15-48 is reduced for a given $t_c$).

Wavelength and bandpass optimization, derivative techniques, and synchronous scanning are employed to improve the *S/N* or the selectivity and to carry out multicomponent determinations as in fluorometry. Time discrimination is an added dimension of selectivity and *S/N* enhancement. The delay and observation times can be adjusted to discriminate against other phosphors with half-lifes much shorter or greater than that of the analyte.

### Performance Characteristics

The measurement precision of phosphorescence for concentrations well above the DL can be 1 to 2%. However, sampling irreproducibility often limits sample-to-sample precision to 5 to 10%. In low-temperature phosphorescence measurements, the *S/N* and DL can be less due to the smaller $V^*$, the loss of radiation in passing through many interfaces, and convection flicker noise (from radiation passing through liquid $N_2$ with bubbles and refractive index changes).

The factors limiting accuracy are much the same as for fluorescence measurements. The primary blank interference is uncompensated background phosphorescence from concomitants. Analyte interference can be due to concomitants causing quenching or absorbing of a significant fraction of the primary and secondary

radiation. The later effect is smaller because of the smaller cell pathlengths in LTP and especially in RTP with front-surface solid-substrate measurements.

For good phosphors, the best DLs are generally in the range 1 to 10 nM with LTP and micelle-based RTP. These DLs are significantly worse than for the best fluorophores. However, DLs for many analytes are better with phosphorescence than fluorescence if $\phi_P \gg \phi_F$ for the analyte. Detection limits with solid substrate RTP are typically 1 to 10 ng, although values near 1 pg have been achieved with some compounds. For a 2.5-μL sample size, a DL of 1 ng corresponds to 2 μM for a compound with a molecular weight of 200.

The linear range is typically two to four orders of magnitude depending on the DL. Nonlinearity occurs at higher concentrations when the fraction of radiation absorbed is no longer proportional to the analyte concentration or prefilter effects occur in solution RTP.

### Applications

The fact that phosphorescence is rarer than fluorescence restricts its applicability, but enhances selectivity. It has found use in situations where excellent (e.g., nanomolar) but not spectacular (e.g., picomolar), DLs are required and selectivity is paramount.

Phosphorimetry has been used primarily for the determination of organics in environmental applications (e.g., PAHs) and in the drug and pharmaceutical field (e.g., theophylline, sulfonamides). The range of analytes is demonstrated by the list of data in Table 15-8. Solution RTP can be used for HPLC detection.

Phosphorimetric determination of inorganic species is rare (other than for rare earths, as discussed in Section 15-3). However, some metal ions can be determined by forming phosphorescent complexes with nonphosphorescent ligands.

At this time, phosphorimetry is not used as routinely as fluorometry. The development of RTP methods could change this situation. Much research is being conducted to improve RTP methods and to make them adaptable to flow systems and HPLC detection.

### 15-5 CHEMILUMINESCENCE

Chemiluminescence (CL) techniques are based on the fact that in a few reactions, a significant fraction of intermediates or products are produced in excited electronic states. In other words, part of the chemical energy released in the reaction is used to produce excited species. The emission of photons from the excited molecules is measured. Often, it is possible to arrange reaction conditions so that the luminescence signal is re-

**TABLE 15-8**
Phosphorescence data for some organic compounds[a]

| Compound | Solvent | $\lambda_{ex}$ (nm) | $\lambda_{em}$ (nm) | $\tau_P$ (s) | DL | Compound | Solvent | $\lambda_{ex}$ (nm) | $\lambda_{em}$ (nm) | $\tau_P$ (s) | DL |
|---|---|---|---|---|---|---|---|---|---|---|---|
| Adenine | WM | 278 | 406 | 2.9 | 0.02 | Pyridine | EtOH | 310 | 440 | 1.4 | 0.0001 |
|  | RTP | 290 | 470 | — | 4.1 | Pyridoxine HCl | EtOH | 291 | 425 | — | 0.008 |
| 6-Amino-6-methyl- | WM | 321 | 456 | 0.66 | 0.0002 | Salicyclic acid | EtOH | 315 | 430 | 6.2 | 0.05 |
| mercaptopurine |  |  |  |  |  |  | RTP | 320 | 470 | — | 0.7 |
| Anthracene | EtOH | 300 | 462 | — | 0.05 | Sulfamerazine | EtOH | 280 | 405 | 0.7 | 0.0001 |
|  | EPA | 240 | 380 | 2.1 | 0.10 | Sulfamethazine | EtOH | 280 | 410 | 0.8 | 0.0001 |
| Aspirin | EtOH | 310 | 430 | 5.3 | 0.007 | Sulfanilamide | EtOH | 297 | 411 | 2.9 | 0.012 |
| Benzoic acid | EPA | 240 | 400 | 2.4 | 0.005 |  | RTP | 267 | 426 | — | 3 |
| Caffeine | EtOH | 285 | 440 | 2.0 | 0.2 | Sulfapyridine | EtOH | 310 | 440 | 1.4 | 0.0001 |
| Cocaine HCl | EtOH | 240 | 400 | 2.7 | 0.01 | Tryptophan | EtOH | 295 | 440 | 1.5 | 0.002 |
|  | RTP | 285 | 460 | — | 0.04 |  | RTP | 280 | 448 | — | 4 |
| Codeine | EtOH | 270 | 505 | 0.3 | 0.01 | Vanillin | EtOH | 332 | 519 | — | 0.1 |
| DDT | EtOH | 270 | 420 | 0.2 | 0.007 |  |  |  |  |  |  |
| Diacetylsulfanamide | EtOH | 280 | 405 | 1.3 | 0.001 |  |  |  |  |  |  |

[a] For LTP measurements the solvent is indicated (WM = water/methanol in 9:1 ratio) and the DL is in $\mu g \ mL^{-1}$. For RTP measurements on different support materials the DL is in ng.

lated to the analyte concentration. Note that unlike photoluminescence measurements, no external radiation source is required.

## Principles

Three common reaction sequences used to produce CL are the following:

$$A + B \rightarrow I + I^*$$
$$\downarrow$$
$$I + h\nu \rightarrow P \qquad (15\text{-}50a)$$

$$A + B \rightarrow P + P^*$$
$$\downarrow$$
$$P \rightarrow h\nu \qquad (15\text{-}50b)$$

$$I^*(\text{or } P^*) + F \rightarrow I(\text{or } P) + F^*$$
$$\downarrow$$
$$F + h\nu \qquad (15\text{-}50c)$$

where A and B are reactants, I is an intermediate, P is the product, and F is a fluorescence acceptor. We will assume in this discussion that A is the species converted to the luminescing species (i.e., A is the CL precursor), while B represents other reagents or species that influence the reaction. Reaction sequences 15-50a and 15-50b illustrate that the luminescing species can be an intermediate or a product in the reaction. Reaction 15-50c represents **sensitized CL** in which the excited species produced in one of the first two reaction sequences transfers its energy to an efficient fluorophore acceptor which than fluoresces.

For CL to occur, three general conditions must be met. First there must be sufficient energy to produce an excited state. Thus the reaction must be sufficiently exothermic such that

$$-\Delta G \geq \frac{hc}{\lambda_{ex}} \qquad (15\text{-}51)$$

where $\Delta G$ is the free-energy change (kcal $mol^{-1}$) of the reaction and $\lambda_{ex}$ is the long-wavelength limit in nanometers for excitation of the luminescing species. For CL emission in the visible region, $-\Delta G$ must be 40 to 70 kcal $mol^{-1}$. Second, there must be a favorable reaction pathway to produce the excited state. In most reactions, the ratio of I* to I (or P* to P) produced by mechanisms 15-50a or 15-50b is exceedingly small and no CL is observed. Also, other intermediates or products may be produced by different pathways, with only one pathway leading to the excited species and CL. Third, photon emission must be a favorable deactivation process for direct CL. For sensitized CL, both the efficiency of energy transfer from the excited species to the donor and the fluorescence efficiency of the donor must be good.

The chemiluminescence quantum efficiency ($\phi_{CL}$) is defined as

$$\phi_{CL} = \frac{\text{number of photons emitted}}{\text{number of molecules reacted}} \qquad (15\text{-}52)$$

It can be represented as a product of two efficiencies

or

$$\phi_{CL} = \phi_{ex}\phi_L \qquad (15\text{-}53)$$

where $\phi_{ex}$ is the efficiency of production of the excited species [the fraction of the reacting molecules (A) that produces an excited molecule] and $\phi_L$ is luminescence efficiency (usually, the fluorescence quantum efficiency) of the luminescing species. Matching of the fluorescence emission spectrum of the product and the CL spectrum of the reaction is good evidence that the product is the luminescing species. For sensitized CL, $\phi_{ex}$ also accounts for the efficiency of energy transfer.

The CL radiant power in photons per second at time $t$ $[\Phi_{CL,p}(t)]$ is given by

$$\Phi_{CL,p}(t) = \phi_{CL}\,\frac{-dn_A(t)V}{dt} \qquad (15\text{-}54)$$

where $-dn_A(t)/dt$ is the rate of the reaction (i.e., the rate of disappearance of the CL precursor in molecules $cm^{-3}\,s^{-1}$) and $V$ is the volume of the CL mixture viewed. Thus after the reactants are mixed, the CL signal eventually decays with time as the reagents are consumed, as illustrated in Figure 15-27. The shape of the intensity vs. time curve may also be dependent on $\phi_{CL}$ if it changes with time. For example, products formed in the reaction may inhibit the reaction pathway leading to excited species (change $\phi_{ex}$) or may quench the luminescence (change $\phi_L$). Although most CL reactions are carried out in one phase (gas or solution), heterogeneous CL reactions can be employed (e.g., a gas in contact with a solution or solid, or a species in solution in contact with immoblized reagents). For such cases, the CL signal depends on both the reaction kinetics and the rate of mass transfer (e.g., diffusion). **Electrochemiluminescence** is a particular type of chemiluminescence that results from electrochemical reactions involving species generated at electrodes and is discussed in more detail later in this section.

The background above makes it clear that the analytical performance of CL techniques is governed primarily by the chemistry of the reactions utilized and the luminescence characteristics of the emitting species. Careful control of the reaction conditions is mandatory. Excellent detection limits are possible if $\phi_{CL}$ is reasonably large. Compared to photoluminescence measurements, there are no scattering or background photoluminescence signals since no external source is required. The CL signal must be distinguished from only dark current noise in some cases. In other cases, the CL signal due to the analyte must be distinguished from a background CL signal observed with no analyte. This background signal is often determined by contaminants in the reagents and solvents. Selectivity and linearity are very dependent on the reaction and reaction conditions chosen. As for photoluminescence measurements, absorption of emission radiation by the analyte, products, or concomitants can cause nonlinearity or spectral interference.

In some CL reactions (usually in the gas phase), the analyte is the CL precursor. For many solution CL applications, the analyte is not the CL precursor. Rather, the analyte is a necessary reagent for the CL reaction to occur or a species that affects the rate or efficiency of the CL reaction (e.g., $\phi_{ex}$). For the latter case, the analyte can be an activator (e.g., catalyst), an inhibitor, or an acceptor. The increase or decrease in the CL signal caused by the presence of the analyte is measured. In some cases where the analyte cannot be directly involved in a CL reaction, it is possible to use coupling reactions. Here the analyte reacts with suitable reagents to produce a product which is a reactant in a CL reaction.

### Instrumentation

One attractive feature of CL techniques is the simplicity of the instrumentation. All that is required is a sample cell, provision for mixing the sample and reagents, a detector, and a signal processor-readout system. Except to obtain spectra, no wavelength selection device is usually required because the CL arises from one species; other species affect the intensity of CL, but not usually the spectral distribution of the light emitted. Filters are sometimes used for gas-phase CL where concomitants can result in CL with a different spectral distribution.

Most commercial CL photometers for solution measurements are batch or discrete-sampling instruments, as shown in Figure 15-28. The sample and reagents are mixed rapidly and the CL intensity vs. time

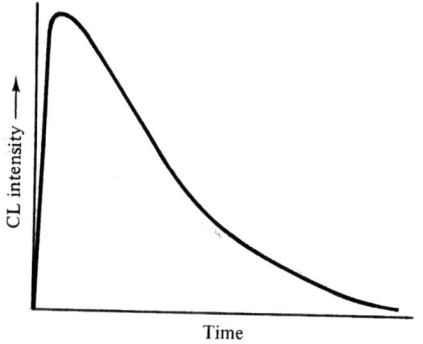

**FIGURE 15-27** Hypothetical curve showing CL intensity as a function of time after the reagents are mixed to initiate the reaction. The initial increase in intensity is due to the finite time for mixing of reagents or an induction period in the reaction. The decay of the signal is caused by the consumption of the reagents and changes in the CL quantum efficiency with time.

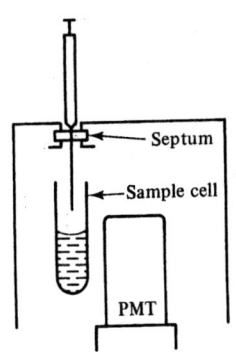

**FIGURE 15-28** Diagram of simple CL photometer. The sample and some reagents are dispensed into the sample cell. The final reagent is injected with a syringe through a septum to initiate the CL reaction. The resultant CL is monitored with a PMT, the photoanodic current is converted to a voltage, and the voltage is displayed on a readout device or digitized and stored for later data manipulation and display. Often an automatic syringe is employed to achieve a more reproducible injection speed and volume and synchronization of data collection to the initiation of the reaction. Although the force of injection provides mixing, a magnetic stirring bar and stirrer are sometimes used to achieve better mixing. Often, the temperature of the sample cell is controlled.

time period, or the integral (area) of the entire peak. Stopped-flow systems are best for very fast CL reactions. Here the sample and reagents are rapidly forced into a mixing chamber and then to an observation cell. The centrifugal analyzer (see Figure 6-10) can also be used for batch CL measurements. Liquid scintillation counters have been used for CL measurements, although they do not normally have provision for automatic injection.

In continuous flow CL methods, the sample and reagents are continually pumped and combined at tees and sent to a flow cell or mixed and observed in an integral reactor-flow cell as shown in Figure 15-29. A steady-state signal is achieved when the cell is totally filled with the reaction mixture. The signal represents the integrated output over the residence time of the reaction mixture in the cell.

Compared to the batch method, the continuous flow method yields a signal that is easier to measure and depends less on the mixing characteristics for fast reactions. With a steady-state signal, it is also possible to scan a monochromator and obtain a CL spectrum. However, reagent consumption is greater, reaction kinetics information is not obtained, and only a portion of the total CL is observed for slow reactions. The flow system components can also be used for flow injection analysis or configured as a postcolumn CL detection system for HPLC.

For fast CL reactions, it is difficult to obtain spectra with batch systems because the CL intensity changes during the scanning period. A point-by-point CL spec-

profile is measured (see Figure 15-27). Depending on the situation, the analytical signal can be taken as the peak signal, the signal after some fixed delay time from the point of mixing, the integral of the signal over some

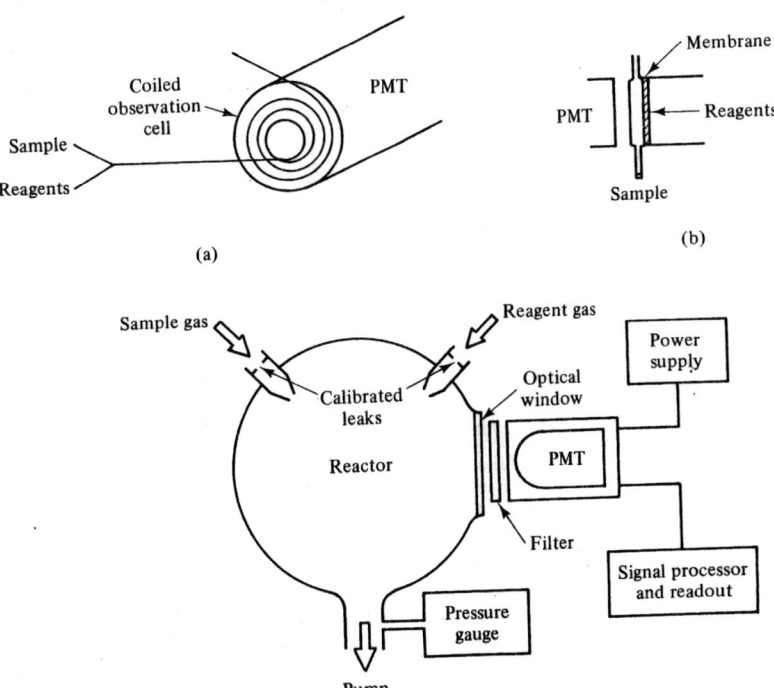

(a)

(b)

(c)

**FIGURE 15-29** Continuous flow CL photometers. In (a) the sample and reagents are combined at a tee and sent to a flow cell for observation of the CL. Mixing aids are often employed. The design in (b) is also used for solutions. In this case, the reagent is forced through a membrane into the flowing stream of the sample. A reaction chamber for CL determination of gases is illustrated in (c).

trum can be collected by incrementing the wavelength between each run. Multichannel detectors, such as intensified diode arrays, have proven useful in this application. Monochromators or spectrographs are not used for quantitative measurements because of the loss in radiation throughput.

Because CL signals are often very weak, the efficient collection and coupling of the CL radiation to the detector are critical. The sample cell is placed as close to the detector as possible. Additional mirrors, lenses, and other devices are used to increase the solid angle of collection. Often, dark current noise limits the detection limit, in which case cooling the PMT and photon counting signal processing enhance the DL and S/N.

## Solution Chemiluminescence

*Chemiluminescence Systems.* Some of the most used CL reagents are shown in Figure 15-30. In most analytical techniques, the CL reagent is oxidized in basic (pH 9-11) solutions. Common oxidants include $H_2O_2$, hypochlorite ion, or ferricyanide. The oxidation of luminol involves the formation of 3-aminophthalate,

as shown below. This product has been identified as the luminescing species.

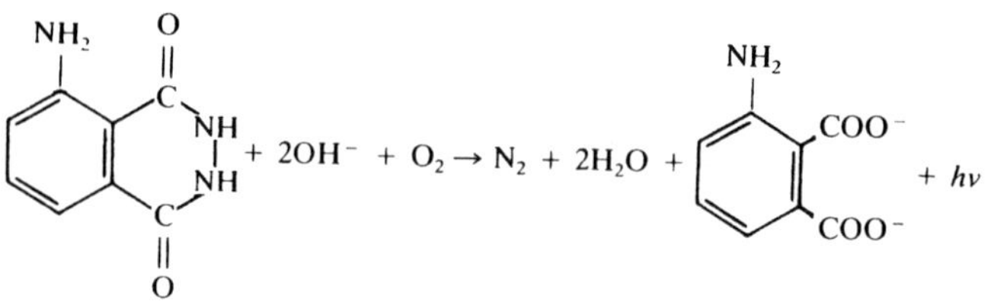

For the lucigenin reaction,

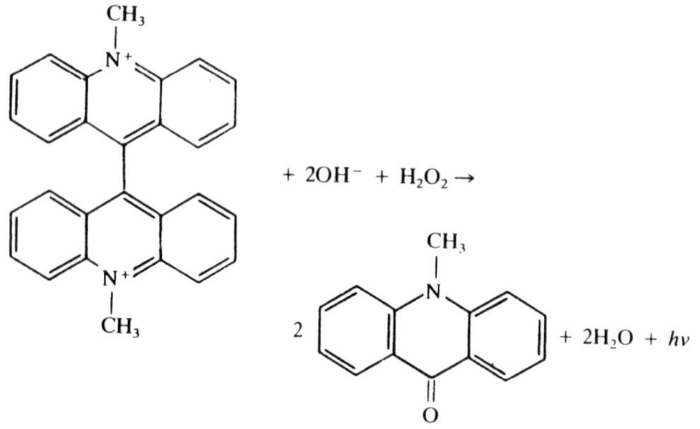

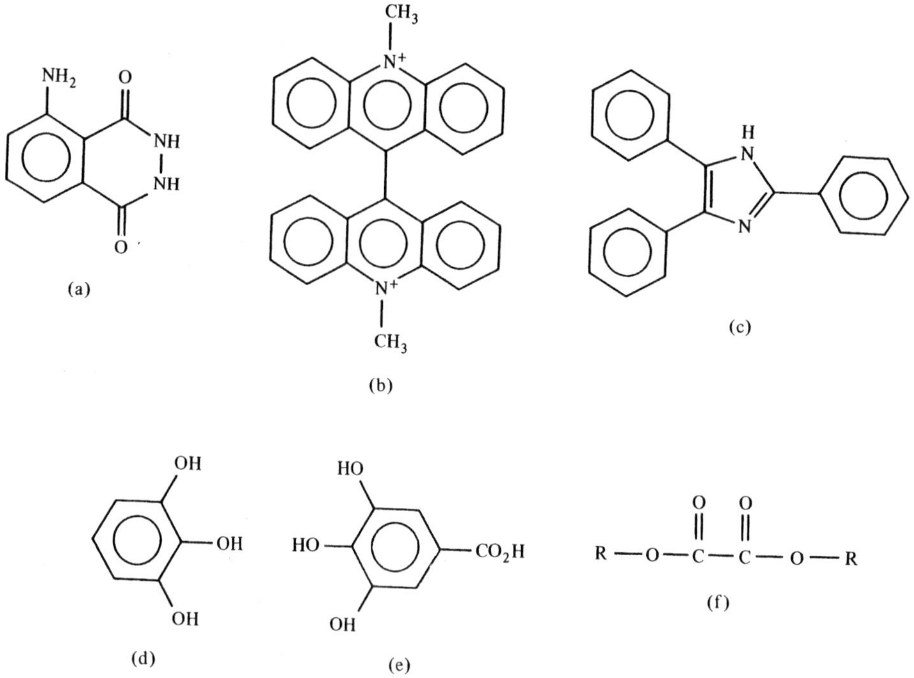

**FIGURE 15-30** Common chemiluminescence reagents: (a) luminol (5-amino-2,3-dihydrophthalazine-1,4-dione); (b) lucigenin (bis-N-methylacridinium nitrate; (c) lophine (2,4, 5-triphenylimidazole); (d) pyrogallol (1,2,3-benzenetriol); (e) gallic acid (3,4,5-trihydroxybenzoic acid); (f) peroxyoxalate derivatives.

there is still controversy as to whether the luminescing species is one of the products (*N*-methylacridone) or an intermediate. Chemiluminescence measurements involving peroxyoxalates employ sensitized CL.

$$ArO-\underset{\underset{O}{\|}}{C}-\underset{\underset{O}{\|}}{C}-OAr + H_2O_2 \longrightarrow \underset{\underset{O}{\diagdown C}-\underset{O}{C \diagup}}{\overset{O-O}{\overset{|\quad|}{}}} + 2ArOH$$

$$\underset{\underset{O}{\diagdown C}-\underset{O}{C \diagup}}{\overset{O-O}{\overset{|\quad|}{}}} + F \longrightarrow \left[\underset{\underset{O}{\diagdown C}-\underset{O}{C \diagup}}{\overset{O-O}{\overset{|\,\overline{\cdot}\,|\,F\!+}{}}}\right]$$

$$\left[\underset{\underset{O}{\diagdown C}-\underset{O}{C \diagup}}{\overset{O-O}{\overset{|\,\overline{\cdot}\,|\,F\!+}{}}}\right] \longrightarrow F^* + 2CO_2$$

$$F^* \longrightarrow F + h\nu$$

The proposed excited cyclic $C_2O_4$ intermediate transfers its energy to an efficient fluorophore (F). Most peroxyoxalate CL is carried out in organic solvents.

### Chemiluminescence Indicators.

The first analytical use of solution CL involved CL indicators. As for fluorescence indicators, they are useful for titrations of colored or turbid solutions in which colorimetric endpoints are difficult to observe. Their use is based on the initiation of CL due to the change in solution composition that occurs at the equivalence point. For example, in the titration of an acid with a base with added $H_2O_2$, CL from the indicators luminol, lucigenin, or lophine occurs only after the acid is completely titrated and the solution pH is high enough for the CL reactions to proceed. Redox titrations can be carried out with luminol as the indicator and hypochlorite or hypobromite as the titrant. Chemiluminescence appears at the equivalence point when excess oxidant is present. Another CL indicator is siloxene, a polymeric silicon oxide ($Si_6H_6O_3$ units). It is used for redox titrations with an oxidant as the titrant in acidic solutions.

### Inorganic Determinations.

Most of the CL reagents identified in Figure 15-30 have been used for determination of various inorganic species, primarily transition metal ions. In most cases, the metal ions enhance the rate of the reaction and the observed CL signal with $H_2O_2$ as the oxidant. Detection limits below

1 ng mL$^{-1}$ and as low as 10 pg mL$^{-1}$ have been obtained for several metals, including Fe(II), Cu(II), Co(II), Cr(III), and Cr(VI). Detection limits and specificity depend on the CL reagent. Although reaction conditions can be somewhat optimized for a given metal ion, additional steps are often necessary to enhance the selectivity. The determination of Cr(III) with luminol can be made quite specific by determining the CL signal immediately after adding EDTA to the sample. The EDTA masks most metal ions, but the complexation reaction with Cr(III) takes hours. In other cases, rapid separation schemes have been developed to isolate the analyte from the major interfering species.

In the absence of $H_2O_2$, the luminol system is much more selective. Thus oxidants such as hypochlorite or ferricyanide can be determined selectively in alkaline luminol solutions. With $O_2$ as the oxidant, Fe(II) can be determined through its enhancement of the luminol CL reaction. Under proper conditions, Cr(VI) can be determined selectively in alkaline lophine solutions with $H_2O_2$ as the oxidant. The linear range extends from 0.1 ng mL$^{-1}$ to 100 mg mL$^{-1}$.

The determination of hypochlorite can also be based on the following simple CL reaction:

$$H_2O_2 + OCl^- \rightarrow O_2 + H_2O + Cl^-$$

Excited singlet oxygen is responsible for the red CL.

### Organic Determinations.

Several types of organic species are determined by their effect on some of the common CL reactions discussed above. Metal-containing compounds such as hemes and vitamin B12 enhance the rate of the luminol CL reaction. Formaldehyde is determined by its enhancement of the CL reaction between gallic acid and $H_2O_2$. The lucigenin reaction involves some reduction steps. This makes possible the CL determination of reducing species such as glucose, ascorbic acid, and uric acid in the absence of $H_2O_2$. Some organic analytes can be determined by sensitized CL with the peroxyoxalate reaction. The overall CL efficiency with acceptors such as rubene or some anthracenes can be greater than 20%. Postcolumn detection schemes have been developed for the determination of separated dansyl or fluorescamine derivatives of amino acids and steroids, porphyrins, and polycyclic hydrocarbons. Attomole DLs have been achieved in some cases.

Clinical assays of substrates and enzymes can be performed by coupling to CL reactions. These assays often involve an initial oxidase enzymes–catalyzed reaction in which $H_2O_2$ is a product. The amount of $H_2O_2$ produced is monitored with the luminol CL or the peroxyoxalate CL reaction. Often, it is necessary to run the enzymatic and indicator reactions sequentially because the optimum pH for the two reactions differs.

In most of the applications above, the CL precursor is a reagent. More recently it has been shown that some analytes, in particular organics with polyphenolic groups (e.g., humic acid, tannin), can be determined from the CL produced by their oxidation with permanganate.

*Other Applications.* The use of chemiluminescence tags for heterogeneous immunoassays is rapidly increasing. Weak CL has been noted for the autooxidation of hydrocarbons, fuels, food oils, blood, and polymers. This CL is often enhanced by oxygen or heating. In some cases, the intensity of CL can be related to the condition or age of the product.

### Bioluminescence

Bioluminescence (BL) is a subset of CL in which biological or enzymatic reactions are involved. The CL efficiency of some BL reactions is greater than 50%.

The most famous BL reaction is the firefly reaction:

$$LH_2 + E + ATP \xrightarrow{Mg^{2+}} E:LH_2:AMP + PP$$

$$E:LH_2:AMP + O_2 \longrightarrow E + L{=}O + CO_2$$

$$+ AMP + light (\lambda_m = 562\,nm)$$

where $LH_2$ is firefly luciferin, $L{=}O$ is oxyluciferin, E is firefly luciferase, PP is pyrophosphate, ATP is adenosine triphosphate, and AMP is adenosine monophosphate. The BL originates from excited oxyluciferin produced in the second step of the reaction.

Although the reaction can be used to determine $Mg^{2+}$ or $O_2$, it is primary used to determine ATP. Several commercial CL photometers are marketed with the application of biomass determinations in mind. The DL for ATP can be as low as $10^{-14}$ g for a 10-$\mu$L sample. There is often a good correlation between the dry weight of living cells and the amount of ATP. Clinical assays are based on coupling reactions which form ATP to the firefly reaction.

Bacterial BL is based on the following reaction:

$$FMNH_2 + O_2 + R{-}CHO \xrightarrow{luciferase} FMN$$

$$+ RCOOH + H_2 + h\nu$$

where $FMNH_2$ is the reduced form of flavin mononucleotide (FMN) and RCHO is a long-chain aldehyde. Clinical assays are based on coupling this indicator re-

action to other reactions involving the formation of $FMNH_2$. Most applications involve coupling to reactions that form NADH or NADPH; these can be coupled to the bacterial BL reaction with the following reaction:

$$NADH + FMN \xrightarrow{NADH:FMN\ oxidoreductase} NAD + FMNH_2$$

### Gas-Phase Chemiluminescence

Several commercial CL analyzers are available for the determination of nitrous oxide and ozone. They are based on the following CL reactions:

$$NO + O_3 \longrightarrow NO_2 + O_2$$

$$2O_3 + C_2H_4 \longrightarrow 2HCHO + 2O_2$$

For the reaction involving NO, the emitting species is excited $NO_2$. Total nitrous oxides can be determined by passing the sample airstream over a hot carbon bed to reduce $NO_2$ to NO ($NO_2 + C \rightarrow CO + NO$). For ozone, ethylene is the CL reagent.

Ozone can also be determined from its CL reaction with rhodamine B adsorbed on silica gel or with a solution of gallic acid and rhodamine B. In the latter case, sensitized CL is employed.

We also note that some emission in flames involves CL excitation. Sulfur is determined by the CL of its excited dimer ($S + S \rightarrow S_2^*$) and phosphorus from emission of chemiexcited HPO ($H + PO \rightarrow HPO^*$).

### Electrochemiluminescence

Emission of light from solutions undergoing electrolysis is termed electrochemiluminescence or electrogenerated luminescence. For example, luminescence is observed from some aromatic hydrocarbons (e.g., anthracenes, rubene), in DMF or acetonitrile, when subjected to an alternating potential at the working electrode. During the cathodic and anodic cycles, anion radicals ($R^-$) and cation radicals ($R^+$) are produced at the electrode surface. Luminescence originates from excited species produced by an annihilation reaction:

$$R^- + R^+ \longrightarrow R^* + R$$

The luminescence spectrum corresponds to the fluorescence spectrum of the parent compound. Excited species can also be produced by oxidation of hydrocarbon anion radicals generated by oxidants generated at the electrode. Formation of excimers that luminesce is also possible.

To date, analytical applications of electrochemiluminescence are rare. A linear relationship between the luminescence signal and analyte concentration has been demonstrated for some molecules.

## 15-6 LIFETIME AND POLARIZATION MEASUREMENTS

In this final section we discuss the instrumentation and applications of luminescence lifetime and polarization measurements.

### Lifetime Measurements

The luminescence lifetime is an important characteristic of a molecule and its environment. Lifetime measurements provide information that is useful in fundamental studies of energy transfer and quenching. In analytical applications, the luminescence lifetime is an additional parameter that can be used to enhance selectivity or permit multicomponent determinations.

Several experimental approaches are used to obtain lifetime data. The primary goal of any of these approaches is to obtain data representing the time dependence of the decay of the luminescence. For a single component sample in which the excited state decays by first-order processes, the luminescence time-decay profile is described by

$$E_L(t) = E_L^0 e^{-t/\tau_L} \tag{15-55}$$

where $E_L^0$ is the initial luminescence signal and $\tau_L$ is the luminescence lifetime. Thus, from $E_L(t)$ vs. time data, the lifetime is the time for the luminescence signal to decay to $1/e$ of its initial value. It can be determined by fitting the decay data to equation 15-55.

Before the widespread availability of laboratory computers, lifetimes were commonly determined by graphical procedures. For example, by taking the natural logarithm of both sides of equation 15-55, we obtain

$$\ln E_L(t) = \ln E_L^0 - \frac{t}{\tau_L} \tag{15-56}$$

Hence the negative reciprocal of the slope of a semilogarithmic plot of $\ln E_L(t)$ vs. $t$ yields the lifetime. This **log slope method** assumes that the luminescence signal decays to zero for long times (i.e., the contribution of the dark signal, background luminescence signals, and electronic offsets is insignificant compared to the analyte luminescence signal).

More sophisticated data-reduction techniques are available to extract the lifetime information and to compensate for the residual baseline signal. For example, in the **Guggenheim method**, the natural logarithm of the difference in the luminescence signal between pairs of data points along the decay curve, separated by a constant time interval, is plotted vs. time. The lifetime is obtained from the negative reciprocal of the slope.

For multicomponent mixtures, the decay curve can be fit to a weighted sum of exponential terms to obtain the lifetime of each component. This approach is also used for fitting a nonexponential decay from one species even though the lifetime calculated may have little physical meaning.

We will now discuss some of the primary approaches used to obtain lifetime data. They are based on time resolution, phase resolution, and correlation techniques.

*Pulse Excitation Method.* In the pulse excitation method, the analyte is excited with a short-duration pulse of radiation, and the decay of the resultant luminescence signal after the excitation pulse is monitored. Thus the method involves time resolution. Either a pulsed source is used or the radiation from a continuous source is gated with an external modulator. Data from many excitation pulses are summed for signal-averaging purposes. If the excitation pulse width ($t_p$) is much shorter than the luminescence lifetime ($\tau_L$), the true decay of the luminescence is observed as shown in Figure 15-31a. The log slope method is illustrated in Figure 15-31b.

Phosphorescence lifetime measurements are reasonably straightforward because phosphorescence lifetimes are relatively long. The pulsed Xe arc lamps used in the pulse source-gated detection technique usually provide a sufficiently short pulse. With mechanical shutter arrangements (phosphoroscopes), the excitation and observation steps are completely separated, and the

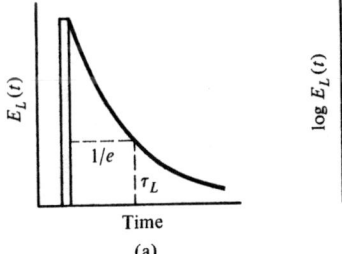

**FIGURE 15-31** Pulse lifetime measurements. In (a) the decay of luminescence after an ideal short, rectangular pulse of radiation is illustrated. The luminescence lifetime is the time for the signal to decay to $1/e$ of its initial value. In the log slope method (b), the lifetime is determined from the slope.

first part of the decay curve is not observed (see Figure 15-26).

Fluorescence lifetime measurements are more difficult because the lifetimes are usually five to seven orders of magnitude smaller than phosphorescence lifetimes. To observe the true decay of the fluorescence signal, the excitation pulse width should be on the order of nanoseconds or even smaller. Moreover, the response time of the detector, signal processor, and readout device must be sufficiently small to prevent distortion of the true decay curve.

Most available pulsed light sources such as nitrogen or deuterium flash lamps or pulsed lasers yield pulse widths of several nanoseconds' duration. Only mode-locked lasers provide picosecond-width pulses. With nanosecond excitation pulse widths, the decay curve represents a composite or convolution of the source and fluorescence decay as illustrated in Figure 15-32, and the following equation applies:

$$E_F'(t) = \int_0^t E_r(t')E_F(t - t') \, dt' \qquad (15\text{-}57)$$

where $E_r(t')$ is a reference signal related to the time profile of the excitation pulse, $E_F'(t)$ is the measured decay curve, and $t'$ is a dummy integration variable $(t > t')$. The true fluorescence decay curve is extracted from the experimental decay curve by deconvolution methods. This requires that the convolution integral be solved for $E_F(t)$. Thus it is necessary to know the excitation pulse profile, which is experimentally obtained by measuring the signal from a scattering solution. It is

critical that the shape of the excitation profile be reproducible. With deconvolution procedures it is possible to measure a 1-ns lifetime with a 2-ns excitation pulse width.

Detection systems with subnanosecond response times are difficult to construct. Special fast-response PMTs are employed. The type of dynode biasing and wiring are critical for a fast response. The decay curve could be directly displayed on a oscilloscope or digitized (e.g., transient recorder) for later display. Because these devices do not provide time resolution of 1 ns, it is more common to use a pulse sampling method (also called the stroboscopic method) in which the fluorescence signal over a small time window is sampled and stored after a controlled delay period relative to the excitation pulse. The signal from different points of the decay curve are measured by varying the delay time. Thus the time-resolved decay is reconstructed from the data. The sampling can be accomplished with a boxcar integrator or a sampling oscilloscope. Alternatively, the PMT bias voltage can be pulsed to a high value for a short period (typically, 0.2 to 1 ns) to increase the PMT gain at some point along the fluorescence decay curve. The signal resulting from the gain pulse is digitized and stored.

*Time-Correlated Single-Photon Method.* The time-correlated single-photon (TCSP) method is the most widely used time-resolution method for lifetime measurements. As shown in Figure 15-33, the detection system measures the time between the excitation pulse and the arrival of the first fluorescence photon. The time of photon emission is random but weighted according to the decay kinetics. By measuring the emission time for a large number of excitation pulses, the fluorescence decay curve is reconstructed. To obtain the true decay curve, it is necessary to arrange conditions (e.g., source irradiance) such that the probability of observing more than a single luminescence photon per excitation pulse is very small. In practice an average count rate of 0.01 to 0.05 photon per excitation pulse is used. This technique provides a resolution of about 0.2 ns. Typically, $10^3$ to $10^6$ counts are accumulated in the peak channel. With a low repetition rate and pulsed light source, several hours are required to accumulate the data.

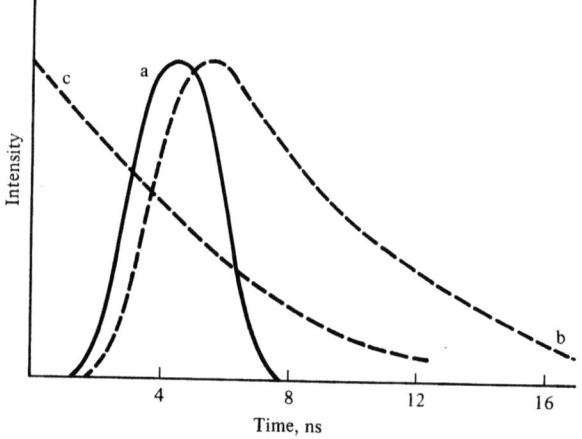

**FIGURE 15-32** Source and decay profiles. The source time profile (curve a) and the measured fluorescence decay curve (curve b) are shown. Because $t_p \approx \tau_F$, the initial part of the decay curve is distorted. The true fluorescence decay curve (curve c) is obtained by deconvolving the contribution of the excitation pulse.

*Phase-Resolved Methods.* Phase-resolved, phase modulation, or phase-sensitive lifetime measurements are based on the use of a continuous, sinusoidally modulated excitation source and phase-sensitive detection, as shown in Figure 15-34. A reference signal $[E_r(t)]$ that tracks the time-dependent excitation radiance is obtained by directing part of the excitation beam to a scattering solution and monitoring the scattering

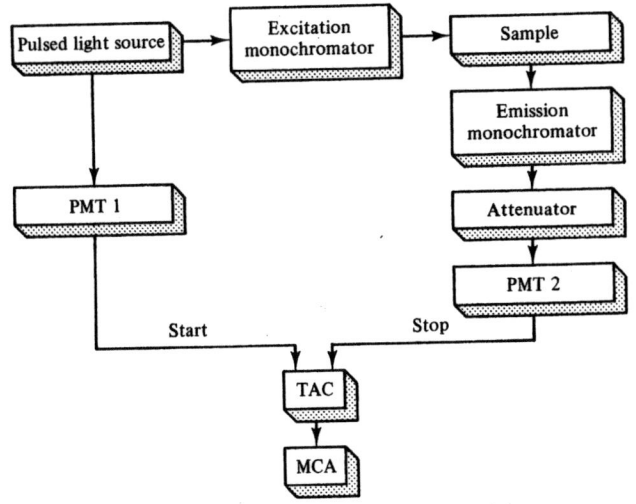

**FIGURE 15-33** Block diagram of apparatus for time-correlated single-photon lifetime measurements. When the light source is fired, a photon striking PMT 1 starts the time-to-amplitude converter (TAC). The TAC is stopped when a fluorescence photon strikes PMT 2. If no photon is detected in a preset time interval, the TAC is reset to zero. The TAC outputs a voltage pulse with an amplitude proportional to the elapsed time between starting and stopping. The output pulse is sorted by a multichannel analyzer (MCA) according to its amplitude. After many excitation pulses, each channel of the MCA indicates the number of pulses within a given small amplitude interval and thus the number of photons that were emitted in a given elapsed time.

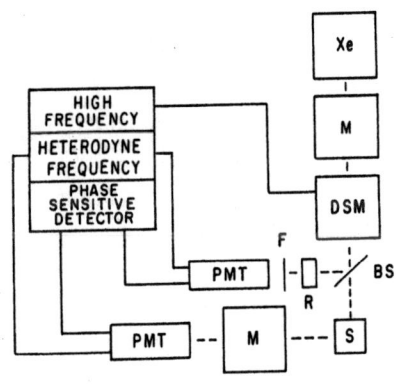

**FIGURE 15-34** Block diagram of a phase fluorometer with phase-sensitive detection. Xe, xenon lamp; M, monochromator; DSM, Debye–Sears modulator; BS, beam splitter; F, optical filter; S, sample; PMT, photomultiplier tube; R, reference solution. [From J. R. Lakowicz and H. Cherek, *J. Biochem. Biophys. Methods*, **5**, 19 (1981), with permission of Elsevier Science Publishers B. V. (Biomedical Division).]

signal. In this case

$$E_r(t) = (E_r)_{dc}(1 + d_m \sin \omega t) \qquad (15\text{-}58)$$

where $(E_r)_{dc}$ is the magnitude of the dc component, $d_m$ is the modulation depth of the ac component (i.e., the ratio of the ac to dc amplitude of the excitation radiance), and $\omega$ is the angular modulation frequency ($\omega = 2\pi f$, where $f$ is the linear modulation frequency).

If the modulation period is on the order or smaller than the fluorescence lifetime, the resulting fluorescence signal is demodulated and phase shifted as shown in Figure 15-35. The time-dependent fluorescence signal [$E_F(t)$] is described by

$$E_F(t) = (E_F)_{dc}[1 + d_m m \sin (\omega t - \phi)] \qquad (15\text{-}59)$$

where $(E_F)_{dc}$ is the dc component of the fluorescence signal, $\phi$ is the phase shift, and $m$ is the demodulation factor. The demodulation factor is given by

$$m = \cos \phi \qquad (15\text{-}60)$$

For an exponential decay, the fluorescence lifetime $\tau_F$ can be calculated from the phase shift or demodulation factor as follows:

$$\tau_F = \frac{\tan \phi}{\omega} \qquad (15\text{-}61)$$

$$\tau_F = \frac{(m^{-2} - 1)^{1/2}}{\omega} \qquad (15\text{-}62)$$

Both $\phi$ and $m$ are measured relative to the signal from a scattering solution or a reference fluorophore of known lifetime. In the former case, $m$ is the ratio of the modulation depth of the ac waveform of the fluorophore to that for a scattering solution.

The excitation beam is usually modulated with an electro-optic modulator (e.g., Kerr or Pockels cell) or a Debye–Sears acousto-optical modulator (see Section 3-3). Modulation frequencies vary typically from 10 to 200 MHz. The modulated fluorescence signal is extracted by synchronous detection with a lock-in amplifier. Heterodyning techniques are used to shift high-frequency signals to low-frequency signals ($\approx 10$ Hz), which are easier to process. For example, the PMT bias voltage and hence the PMT gain is modulated at frequency $\omega + \Delta\omega$ to produce a heterodyne signal at $\Delta\omega$ which is processed. Resolution as good as 1 to 100 ps is possible. Lifetime measurements require only a few minutes at most rather than hours as with the time-correlated single-photon method.

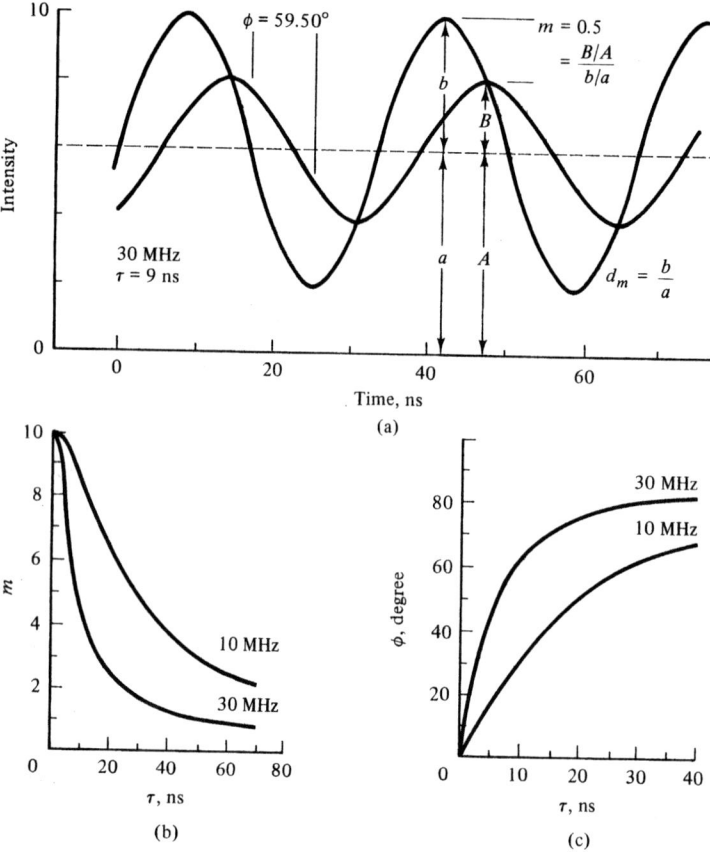

**FIGURE 15-35** Phase modulation technique. In (a) the phase shift and demodulation of the fluorescence signal relative to a scattering signal are illustrated. A modulation frequency of 30 MHz and a lifetime of 9 ns are assumed. The dependence of the demodulation factor ($m$) and the phase shift ($\phi$) on the lifetime for two modulation frequencies is shown in (b) and (c). [Adapted from J. R. Lakowicz, *Fluorescence Spectroscopy*, Plenum Press, New York, 1983, with permission of publisher.]

*Other Lifetime Measurement Methods.* In the correlation fluorometer approach, the signal from a pulsed mode-locked laser, detected by a reference detector, and the fluorescence signal are multiplied with a microwave mixer. The resultant signal is time averaged with a low-pass filter. A plot of the correlation function vs. delay time yields the fluorescence decay curve. The delay time for the reference signal is varied optically by changing the distance of the reference detector from the laser. Correlation techniques based on optical mixing or excited-state population mixing have also been employed.

Time-resolution methods, such as the pulse source method, involve measuring the impulse response function of the fluorescence species in the time domain. In the phase modulation method, the frequency response function of the fluorophore is monitored in the frequency domain. Because the source is modulated at one frequency, the frequency response at one frequency is measured at a given time.

The complete frequency response function can be obtained by exciting with a source that is modulated at all frequencies and monitoring the frequency response of the fluorophore with a spectrum analyzer. The amplitude of the fluorophore signal is attenuated at high frequencies because the excited-state population cannot follow the excitation source. The Fourier transform of the power spectrum obtained is the impulse response function or the fluorescence decay curve. The roll-off of the power spectrum at higher frequencies is Lorentzian if the fluorescence decay is exponential. Thus the lifetime can be obtained from the reciprocal of the Lorentzian half-width without resorting to Fourier transformation. Note that the power spectrum of a scattering solution will be flat if the source is modulated equally at all frequencies. Thus white noise in the frequency domain corresponds to a infinitely narrow pulse in the time domain (i.e., the Fourier transform of a pulse is flat).

Experimentally, this approach has been implemented by using a CW laser as the source. The laser mode noise results in a laser output that is amplitude modulated at discrete frequencies separated by an amount equal to the laser's mode spacing. For a typical 1-m laser, the modes oscillate at frequencies from 150 MHz up to as high as 4.5 GHz with a mode spacing of 150 MHz. The power spectrum of a scattering signal and the fluorescence signal elicited is measured with a microwave analyzer. The ratio of these two spectra is taken to produce a normalized power spectrum that accounts for the different amplitudes of the laser modes. Lifetimes are extracted from this power spectrum as discussed above.

*Analytical Applications.* In phosphorescence measurements, time discrimination of the phosphorescence signal from scattering and fluorescence signals is standard practice as discussed in Section 15-4. Instruments configured for pulse source time-resolved or phase-resolved fluorescence lifetime measurements have been used in analytical applications to increase selectivity. In time-resolved measurements, scattering signals are eliminated by delaying observation until the source intensity has decayed to zero. Background fluorescence from short-lived species can be reduced by choosing a delay time such that the fluorescence from short-lived species has decayed to zero. This works best if $\tau_F$ for the analyte is greater than 10 ns. Multicomponent determinations are implemented by measuring the fluorescence signal at different delay times.

With phase-resolved instruments it is possible to tune into one species by adjusting the phase angle of delay of the reference signal to be equal to $\phi$ in equation 15-59. The fluorescence signal from one component can be nulled out as shown in Figure 15-36. Multicomponent analysis is accomplished by adjusting the reference phase angle for each species or by measuring the fluorescence signal at different phase angles and solving for the component concentrations by the method of simultaneous equations.

Lifetime measurements can also be used to determine the free and bound fractions of a species labeled with a fluorescence tag if the binding of the species to a macromolecule significantly changes the lifetime of the tag. This approach has proved useful for phase-resolved fluoroimmunoassays.

## Polarization Techniques

In Section 12-6 we discussed how fluorescence emission is partially polarized when excited with polarized light. Rotational diffusion results in depolarization of the emission radiation. For small molecules, the depolarization is essentially complete. However, for very large molecules with large rotational correlation times, the degree of polarization or anisotropy of the emission radiation can be measured. We now consider the instrumentation and applications of polarization measurements.

*Instrumentation.* A spectrofluorometer is configured for polarization measurements by placing polarizers in the excitation and emission beams. The polarizer in the excitation beam is normally adjusted for parallel or vertical polarization. Often, emission polarizers are located to view the fluorescence emission from both cell walls that are parallel to the excitation beam.

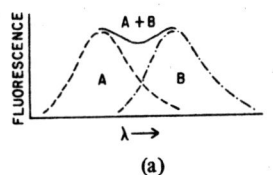

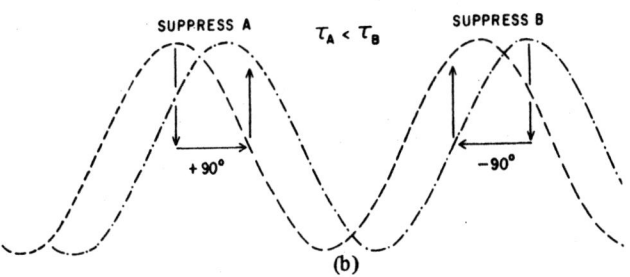

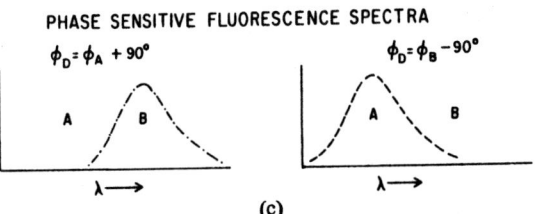

**FIGURE 15-36** Nulling with phase modulation spectrometry. The fluorescence emission spectra of components A and B are seen to overlap in (a). In (b) the modulated signals for A and B are shown. By adjusting the reference detector phase angle to $\pm 90°$ relative to $\phi$ for one component, its signal is nulled out and only the emission signal for the other component is observed (c). [From J. R. Lakowicz and H. Cherek, *J. Biochem. Biophys. Methods*, 5, 19 (1981), with permission of Elsevier Science Publishers B. V. (Biomedical Division).]

A separate wavelength selector, detector, and signal processor are used for each emission beam. The polarization axes of the emission polarizers are oriented 90° with respect to each other to allow simultaneous measurement of the parallel and perpendicular components.

The anisotropy ($r$) is calculated from the ratio of the polarization signals ($R = \Phi_\parallel / \Phi_\perp$) and the equation

$$r = \frac{R - 1}{R + 2} \tag{15-63}$$

It is usually necessary to correct $R$ for the different transmission efficiencies of the emission monochromator(s) for vertically and horizontally polarized radiation.

The decay of anisotropy or dynamic anisotropy has been measured with the pulse excitation or phase modulation methods. In the latter case, the phase difference between the polarization components is measured and ultimately related to *r*.

*Applications.* Fluorescence polarization is used in many fundamental studies of the hydrodynamic properties, size, conformation, and shape of macromolecules. It has also been used to quantify protein denaturation, protein–ligand association reactions, and rotational rates of proteins. The time-dependent decay of anisotropy provides additional information about the diffusive motions of a fluorophore.

The primary analytical use of polarization measurements is in fluorescence homogeneous immunoassays. The anisotropy of a tagged antigen increases substantially when it is bound to a high-molecular-weight antibody because of the increased rotational correlation time. Thus the anisotropy of a mixture of antibody, tagged antigen, and analyte antigen decreases with increasing analyte concentration because more tagged antigen is displaced.

Polarizers have been used to reduce the elastic scattering signal. Only the parallel polarization component of the excitation beam is scattered. Thus a polarizer in the excitation or emission beam that is oriented to pass only the perpendicular polarized component can significantly reduce the viewed scattering signal.

Fluorescence-detected circular dichroism has also gained popularity. Here the sample is excited with circularly polarized light. The corrected excitation spectrum yields the same information as a conventional absorption-detected circular dichroism measurement. However, measurements can be made at lower concentrations if $\phi_F$ is significant.

# PROBLEMS

**15-1.** Calculate the peak wavelength of the Raman band of water with excitation at 365 nm.

**15-2.** Discuss the major reasons why molecular phosphorescence spectrometry has not been widely applied as an analytical tool. Will the development of room-temperature phosphorescence methods be likely to change this situation drastically? Why or why not?

**15-3.** A method for obtaining molecular fluorescence results that are independent of quenching has been proposed [G. M. Hieftje and G. R. Haugen, *Anal. Chim. Acta, 123,* 255 (1981)]. The fluorescence signal $E_F$ is normally related to the concentration of analyte by $E_F = k\phi_F c$, where $k$ is a constant and $\phi_F$ is the quantum yield of fluorescence. Because the quantum yield of fluorescence is the ratio of the observed lifetime of fluorescence $\tau_F$ to the radiative lifetime $(\tau_F)_r$, the proposed method measures both the fluorescence signal $E_F$ and the fluorescence lifetime $\tau_F$. Show mathematically that the ratio $E_F/\tau_F$ is independent of quantum yield.

**15-4.** Describe the major experimental differences between the following types of molecular fluorescence spectra. Tell which monochromator(s) are scanned (emission or excitation), which are kept at a constant wavelength, if any, and the relationship of the spectrum obtained to the molecular absorption spectrum, if any.
   **(a)** The excitation spectrum.
   **(b)** The emission spectrum.
   **(c)** The synchronous fluorescence spectrum.
   **(d)** The excitation/emission matrix spectrum.

**15-5.** Discuss why a molecular fluorescence calibration plot becomes nonlinear at high concentrations.

**15-6.** Describe the instrumental differences between molecular fluorescence and molecular phosphorescence spectrometry.

**15-7.** Contrast chemiluminescence and photoluminescence techniques in terms of the factors that control the selectivity.

**15-8.** Saturated fluorescence can be advantageous since the fluorescence signal becomes independent of quenchers and prefilter effects. Consider an analyte molecule with a quantum efficiency of 0.10, a molar absorptivity of $1.0 \times 10^4$ L mol$^{-1}$ cm$^{-1}$, and a fluorescence rate constant of $1.0 \times 10^8$ s$^{-1}$. The analyte concentration in the sample cell is 1.0 pM.
   **(a)** Calculate the source irradiance incident on the sample required to reach 50% of saturation.
   **(b)** How does this irradiance compare to that provided by a conventional arc lamp–excitation monochromator configuration ($\approx$ 10 mW cm$^{-2}$ over 10 nm centered at 300 nm)?
   **(c)** Assume that 25% ($0.50 \times 50\%$) of the analyte molecules are excited at a given time. Calculate the total number of fluorescence photons emitted per second in a viewed volume element of $1.0 \times 10^{-3}$ cm$^3$.

**15-9.** In Section 15-2, equations were presented to estimate the detection limit in photoluminescence measurements that are limited by dark current shot noise. Assume that the slope of the calibration curve for the analyte referenced to the photocathode is given by equation 15-41 and that the molar absorptivity is $1.0 \times 10^5$ L mol$^{-1}$ cm$^{-1}$ and the power yield is 0.50. The blank equivalent concentration (BEC) is 1.0 nM. Calculate the following assuming that the noise bandwidth constant ($K$) is $1.0 \times 10^{-19}$ A and the source flicker factor is 0.0050.

(a) The detection limit when limited by background luminescence shot noise. How does this detection limit compare to that calculated with a typical value of dark current shot noise?

(b) What is the BEC for which background luminescence shot and flicker noise are equal?

**15-10.** In a pulse excitation experiment, the fluorescence signal decays to one-tenth of its initial value in 10 ns. Calculate the fluorescence lifetime. Calculate the $S/B$ for measurements at 2 and 10 ns after the excitation pulse. Assume that the $S/B$ is 1.0 at zero time and that the background luminesence has a lifetime of 1.0 ns.

**15-11.** Contrast molecular absorption and fluorescence measurements with respect to selectivity and detection limits.

**15-12.** In a fluorescence polarization immunoassay, why does the anisotropy decrease at the analyte concentration increases?

**15-13.** Fluorescence measurements are made in a 1-cm-path-length cell under conditions that $b_{pr} = b_{pt} = 0.40$ cm and $b^* = 0.10$ cm. The absorbance of the analyte is

$0.10$ $\mu M^{-1}$ $cm^{-1}$ and the fluorescence signal for $0.01$ $\mu M$ of the analyte is 1.0 mV.

(a) Calculate the primary absorption correction factor and the observed fluorescence signal at analyte concentrations of 0.1, 1, and 10 $\mu M$.

(b) Calculate the error in the measurement of an analyte concentration of 0.10 $\mu M$ caused by secondary absorption by concomitants if the absorbance at the emission wavelength over 1 cm is 0.10 and 1.0.

**15-14.** Discuss the use of derivative and wavelength modulation techniques for photoluminescence measurements.

**15-15.** Calculate the fraction of the total phosphorescence signal observed with a phosphoroscope for analytes with phosphorescence lifetimes of 10 and 100 ms with a delay time of 2 ms and exposure and observation times of 3 ms.

**15-16.** Phase modulation measurements are made with a source modulation frequency of 30 MHz. The analyte fluorescence signal is phase shifted 60° relative to a scattering solution. Calculate the fluorescence lifetime and the demodulation factor ($m$).

## REFERENCES

The following references deal with molecular luminescence spectroscopy.

1. G. G. Guilbault, *Practical Fluorescence: Theory, Methods, and Techniques*, Marcel Dekker, New York, 1973.

2. S. G. Schulman, ed., *Molecular Luminescence Spectroscopy: Methods and Applications, Part I*, Wiley, New York, 1985.

3. J. D. Winefordner, S. G. Schulman, and T. C. O'Haver, *Luminescence Spectrometry in Analytical Chemistry*, Wiley-Interscience, New York, 1972.

4. R. S. Becker, *Theory and Interpretation of Fluorescence and Phosphorescence*, Wiley-Interscience, New York, 1969.

5. W. R. Seitz, "Luminescence Spectrometry (Fluorimetry and Phosphorimetry)," in *Treatise on Analytical Chemistry*, 2nd ed., I. M. Kolthoff and P. J. Elving, eds., pt. I., vol. 7, Wiley-Interscience, New York, 1981, pp. 161–204.

6. C. A. Parker, *Photoluminescence of Solutions*, Elsevier, Amsterdam, 1968.

7. S. Udenfriend, *Fluorescence Assay of Biology and Medicine*, Academic Press, New York, vol. I, 1962; vol. II, 1969.

8. J. R. Lakowicz, *Principles of Fluorescence Spectroscopy*, Plenum Press, New York, 1983.

9. J. N. Miller, *Standards in Fluorescence Spectrometry*, Chapman & Hall, London, 1981.

10. E. L. Wehry, ed., *Modern Fluorescence Spectroscopy*, Plenum Press, New York, vol. 1, 1976; vol. 2, 1976; vol. 3, 1981; vol. 4, 1981. This series includes many excellent review chapters on recent advances in techniques and applications of molecular luminescence in such areas as modulation, synchronous, and derivative techniques, laser excitation, chemiluminescence, clinical and biochemical applications, reaction-rate methods, fluoroimmunoassays, array detectors, probe ion techniques, and low-temperature measurements.

11. E. J. Bowen, ed., *Luminescence in Chemistry*, Van Nostrand, London, 1968.

12. D. M. Hercules, ed., *Fluorescence and Phosphorescence Analysis*, Wiley-Interscience, New York, 1965.

13. A. J. Pesce, C. J. Rosen, and T. L. Pasby, *Fluorescence Spectroscopy: An Introduction for Biology and Medicine*, Marcel Dekker, New York, 1971.

14. S. G. Schulman, *Fluorescence and Phosphorescence Spectroscopy: Physiochemical Principles and Practice*, Pergamon Press, Oxford, 1977.

15. C. E. White and R. J. Argauer, *Fluorescence Analysis: A Practical Approach*, Mercel Dekker, New York, 1970.

16. P. A. St. John, "Fluorometric Methods for Traces of Elements," chap. VII-B in *Trace Analysis*, J. D. Winefordner, ed., Wiley-Interscience, New York, 1976.

17. B. L. Van Duuren and T. L. Chan, "Fluorescence Spectrometry," chap. 7 in *Spectrochemical Methods of Anal-*

*ysis*, J. D. Winefordner, ed., Wiley-Interscience, New York, 1971.

18. T. C. O'Haver, "The Development of Luminescence Spectrometry as an Analytical Tool," *J. Chem. Educ.*, 55, 423 (1978).

19. F. Van Geel, E. Voigtman, and J. D. Winefordner, "General Intensity Expression in Molecular Luminescence Spectrometry," *Appl. Spectrosc.*, 38, 228 (1986).

20. I. M. Warner, G. Patonay, and M. P. Thomas, "Multidimensional Luminescence Measurements," *Anal. Chem.*, 57, 463A (1985).

21. K. D. Mielenz, ed., "Measurement of Photoluminescence," vol. 3 in *Optical Radiation Measurements*, F. Grum and C. J. Bartleson, eds., Academic Press, New York, 1982. This reference includes excellent chapters on correction of excitation and emission spectra and quantum yield measurements.

22. "Molecular Fluorescence, Phosphorescence, and Chemiluminescence Spectrometry," *Anal. Chem.* This review article appears once every two years in the Fundamental Reviews issue of *Analytical Chemistry*. Applications as well as recent advances in instrumentation and chemistry are covered.

The following references deal with phase-resolved and lifetime measurements.

23. L. McGown and F. Bright, "Phase-Resolved Fluorescence in Chemical Analysis," *CRC Crit. Rev. Anal. Chem.*, 18, 245 (1987).

24. J. N. Demas, *Excited State Lifetime Measurements*, Academic Press, New York, 1983.

25. G. M. Hieftje and G. R. Haugen, "Correlation-Based Approaches to Time-Resolved Fluorimetry," *Anal. Chem.*, 53, 755A (1981).

26. T. D. Harris and F. E. Lytle, "Analytical Applications of Laser Absorption and Emission Spectroscopy," chap. 7 in *Ultrasensitive Laser Spectroscopy*, D. S. Kliger, ed., Academic Press, New York, 1983.

27. D. V. O'Connor and D. Phillips, *Time-Correlated Single Photon Counting*, Academic Press, London, 1979.

The following references deal specifically with phosphorimetry.

28. T. Vo-Dinh, *Room Temperature Phosphorimetry for Chemical Analysis*, Wiley-Interscience, New York, 1984.

29. M. Zander, *Phosphorimetry*, Academic Press, New York, 1968.

30. R. J. Hurtubise, "Phosphorimetry," *Anal. Chem.*, 55, 669A (1983).

The following references deal with chemiluminescence.

31. W. R. Seitz, "Chemiluminescence and Bioluminescence Analysis: Fundamentals and Biochemical Applications," *CRC Crit. Rev. Anal. Chem.*, 13, 1 (1981).

32. V. Isacsson and G. Wettermark, "Chemiluminescence in Analytical Chemistry," *Anal. Chim. Acta, 68*, 339 (1974).

33. D. B. Paul, "Recent Analytical Development Using Chemiluminescence in Solution," *Talanta, 25*, 337 (1978).

34. E. Schram and P. E. Stanley, eds., *Proceeding of the International Symposium on Analytical Applications of Bioluminescence and Chemiluminescence*, State Printing and Publishing, Westlake, Calif., 1979.

35. M. Cormier, D. M. Hercules, and J. Lee, eds., *Bioluminescence and Chemiluminescence*, Plenum Press, New York, 1971.

36. W. R. Seitz and M. P. Neary, "Chemiluminescence and Bioluminescence," *Anal. Chem., 46*, 188A (1974).

37. M. A. Deluca and W. D. McElroy, eds., *Bioluminescence and Chemiluminescence* Academic Press, New York, 1981.

38. M. A. Deluca, ed., "Bioluminescence and Chemiluminescence," in *Methods of Enzymology*, vol. 57, Academic Press, New York, 1978.

39. A. K. Campbell, M. B. Hallett, and I. Weeks, "Chemiluminescence as an Analytical Tool in Cell Biology and Medicine," in *Methods of Biochemical Analysis*, D. Glick, ed., vol. 31, Wiley-Interscience, New York, 1985, p. 317.

40. L. J. Kricka and G. H. G. Thorpe, "Chemiluminescent and Bioluminescent Methods in Analytical Chemistry," *Analyst, 108*, 1274 (1983).

41. L. J. Kricka, P. E. Stanley, G. H. G. Thorpe, and J. P. Whitehead, eds., *Analytical Applications of Bioluminescence and Chemiluminescence*, Academic Press, London, 1984.

The following references deal with fluorescence HPLC detection.

42. D. C. Shelly and I. M. Warner, "Fluorescence Detectors in High-Performance Liquid Chromatography," chap. 3 in *Liquid Chromatography Detectors*, T. M. Vickery, ed., Marcel Dekker, New York, 1983.

43. A. T. Rhys Williams, *Fluorescence Detection in Liquid Chromatography*, Perkin-Elmer, Buckinghamshire, England, 1980.

The following references deal the absorption-corrected luminescence measurements.

44. J. F. Holland, R. E. Teets, P. M. Kelly, and A. Timnick, "Correction of Right-Angle Fluorescence Measurements for the Absorption of Excitation Radiation," *Anal. Chem., 49*, 706 (1977).

45. D. R. Christman, S. R. Crouch, and A. Timnick, "Automated Instrument for Absorption-Corrected Molecular Fluorescence Measurements by the Cell Shift Method," *Anal. Chem., 53*, 276 (1981).

46. K. Adamsons, J. E. Sell, J. F. Holland, and A. Timnick, "Cell Rotation for Absorption Corrected Measurements in Molecular Fluorescence," *Am. Lab., 16*(11), 16 (1984). This article includes a review of the different techniques used for absorption correction.

Other specific applications of molecular luminescence are discussed in the following references.

47. F. D. Snell, *Photometric and Fluorometric Methods of Analysis: Metals*, Wiley-Interscience, New York, 1981.

48. F. D. Snell, *Photometric and Fluorometric Methods of Analysis: Non-metals*, Wiley-Interscience, New York, 1981.

49. P. A. St. John, W. J. McCarthy, and J. D. Winefordner, "Applications of Signal-to-Noise Theory in Molecular Spectrometry," *Anal. Chem.*, *38*, 1828 (1966).

50. F. E. Lytle, "Solution Luminescence of Metal Complexes," *Appl. Spectrosc.*, *24*, 314 (1970).

51. E. M. Chait and R. C. Ebersole, "Clinical Analysis— A Perspective on Chromatographic and Immunoassay Technology," *Anal. Chem.*, *53*, 682A (1981).

52. E. Soini and I. Hemmila, "Fluorimmunoassay: Present Status and Key Problems," *Clin. Chem.*, *25*, 353 (1979).

53. I. M. Warner and L. B. McGown, "Recent Advances in Multicomponent Fluorescence," *CRC Crit. Rev. Anal. Chem.*, *13*, 155 (1982).

54. C. G. deLima, "The Shopol'skii Effect as an Analytical Tool," *CRC Crit. Rev. Anal. Chem.*, *16*, 177 (1986).

55. W. R. Seitz, "Fluorescence Derivatization," *CRC Crit. Rev. Anal. Chem.*, *8*, 367 (1980).

# CHAPTER 16

# Molecular Scattering Methods

In previous chapters we have considered the absorption, emission, or luminescence of electromagnetic radiation by atoms and molecules. When radiation interacts with matter, however, several additional processes can occur (recall Figure 1-1). The radiation can, for example, be refracted, reflected, or scattered. While the first two processes do not find many direct analytical applications, the scattering of electromagnetic radiation forms the basis for several analytical techniques, including Raman spectroscopy.

In this chapter the principles of radiation scattering are first considered in order to put the Raman effect and various other scattering phenomena into an overall perspective. Many scattering processes involve no change in the energy of the scattered beam compared to the incident beam. On the surface, these processes appear merely to randomize the direction of an incident beam of photons. In actuality, however, the details of the various scattering processes are quite complex; only a qualitative discussion of the theory of radiation scattering is given here.

Following this general introduction to radiation scattering, Raman spectroscopy is considered. The Raman effect is a scattering phenomenon that *does* involve a net energy change between the scattered beam and

the incident beam. The Raman effect can be treated in most aspects with classical theories. We discuss the theory here qualitatively in order to develop selection rules for Raman transitions. The instrumentation used to observe Raman bands is next described. Also considered are resonance Raman methods, nonlinear Raman methods, such as coherent anti-Stokes Raman, and various applications of Raman spectroscopy.

The next section considers the analytically useful molecular scattering methods known as turbidimetry and nephelometry. The chapter concludes with a description of various laser light scattering methods and includes a discussion of remote sensing with laser sources.

## 16-1 PRINCIPLES OF RADIATION SCATTERING

The general principles of radiation scattering were introduced in Section 2-3. We concern ourselves here with two general classes of scattering phenomena: elastic scattering, in which the scattered radiation is of the same frequency as the incident radiation, and inelastic scattering, in which the scattered radiation is of a different frequency.

**TABLE 16-1**
Scattering classes

| Type of scattering | Refractive index requirement[a] | Size requirement[b] |
|---|---|---|
| Rayleigh | $|(\eta_r - 1)| \ll 1$ | $d_s < 0.05 \lambda$ |
| Debye | $|(\eta_r - 1)| \approx 0.1$ | $0.05 \lambda < d_s < \lambda$ |
| Mie | $(\eta_r - 1) \gg 0$ | $d_s > \lambda$ |

[a] $\eta_r$, Relative refractive index, $\eta_s/\eta_m$; $\eta_s$, refractive index of scatterer; $\eta_m$, refractive index of medium.
[b] $d_s$, Major dimension of scatterer.

## Elastic Scattering

In Section 2-3 **elastic scattering** was classified into three types: **Rayleigh scattering, Debye scattering** and **Mie scattering**. It should be noted that this classification is not universally followed. Debye scattering is often referred to as Rayleigh–Debye scattering and is sometimes omitted as a class. The type of scattering observed depends on the refractive index of the particle relative to the surrounding medium and the dimensions of the particle relative to the wavelength of the incident radiation. Approximate criteria for the types of scattering are listed in Table 16-1.

Radiation scattering, like reflection and refraction, has its origins in the induced secondary emission of particles in the path of an incident beam of radiation. Scattering occurs only if the particles have dimensions on the order of or smaller than the incident beam wavelength and they are randomly distributed in a medium of refractive index different from their own. In other cases only reflection and refraction are observed.

From a classical viewpoint, let us first consider the sample to be a collection of spherical molecules with polarizable electron clouds illuminated with electromagnetic radiation of frequency $\nu$. The incident radiation has oscillating electric and magnetic fields that can interact strongly with the electron clouds of the analyte molecules. If these molecules are polarizable, charges are induced that oscillate at the frequency of the field as shown schematically for a single molecule in Figure 16-1.

The oscillating dipole induced in the analyte molecule produces a field of its own which also oscillates at frequency $\nu$. This field acts as a source of radiation itself so that radiation of the same frequency as the incident beam is sent in all directions from the particle.

From a quantum mechanical viewpoint, elastic scattering can be considered to be the removal of energy from the incident beam and the subsequent reemission of that energy. We can view scattering in terms of transitions between energy levels. A molecule in the ground state absorbs an incident photon and reaches an intermediate level called a **virtual level** because it does not correspond to a real energy level of the atom or molecule. Despite this, there is a finite transition probability connecting the ground state to a virtual state. The closer the virtual state is to a real state, the higher the probability of absorption. The molecule returns to the ground state emitting a photon of the same energy that was absorbed. Elastic scattering and resonance fluorescence are similar processes, except that the level reached upon absorption is virtual in the scattering case and real in the resonance fluorescence case. Scattering occurs essentially instantaneously, while the upper level in fluorescence has a measurable lifetime.

Let us now examine in more detail the types of elastic scattering.

*Rayleigh Scattering.* Rayleigh scattering is characteristic of scattering from such particles as atoms and molecules where the dimensions of the scatterer are much smaller than the incident beam wavelength

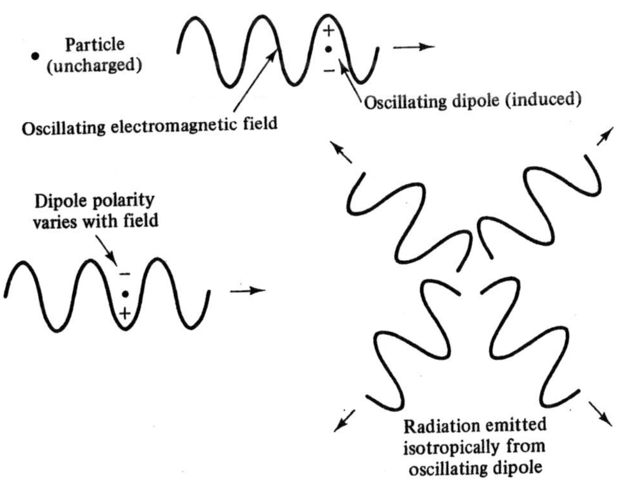

**FIGURE 16-1** Schematic diagram illustrating elastic scattering. The oscillating dipole induced in the particle behaves as a secondary source to produce scattered radiation of the same frequency as that incident on the particle.

Particle (uncharged)

Oscillating electromagnetic field

Oscillating dipole (induced)

Dipole polarity varies with field

Radiation emitted isotropically from oscillating dipole

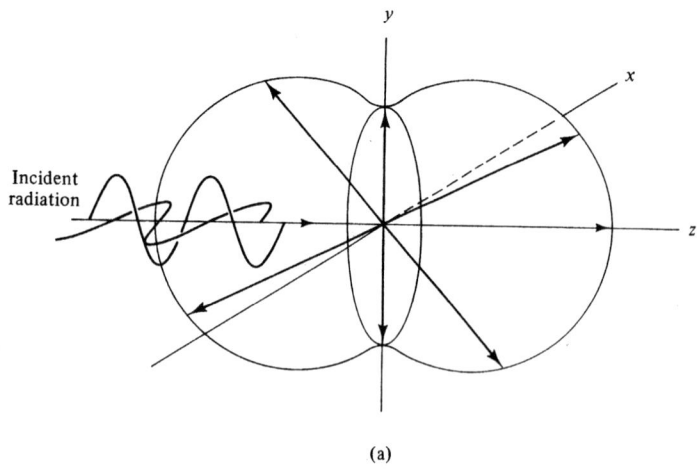

(a)

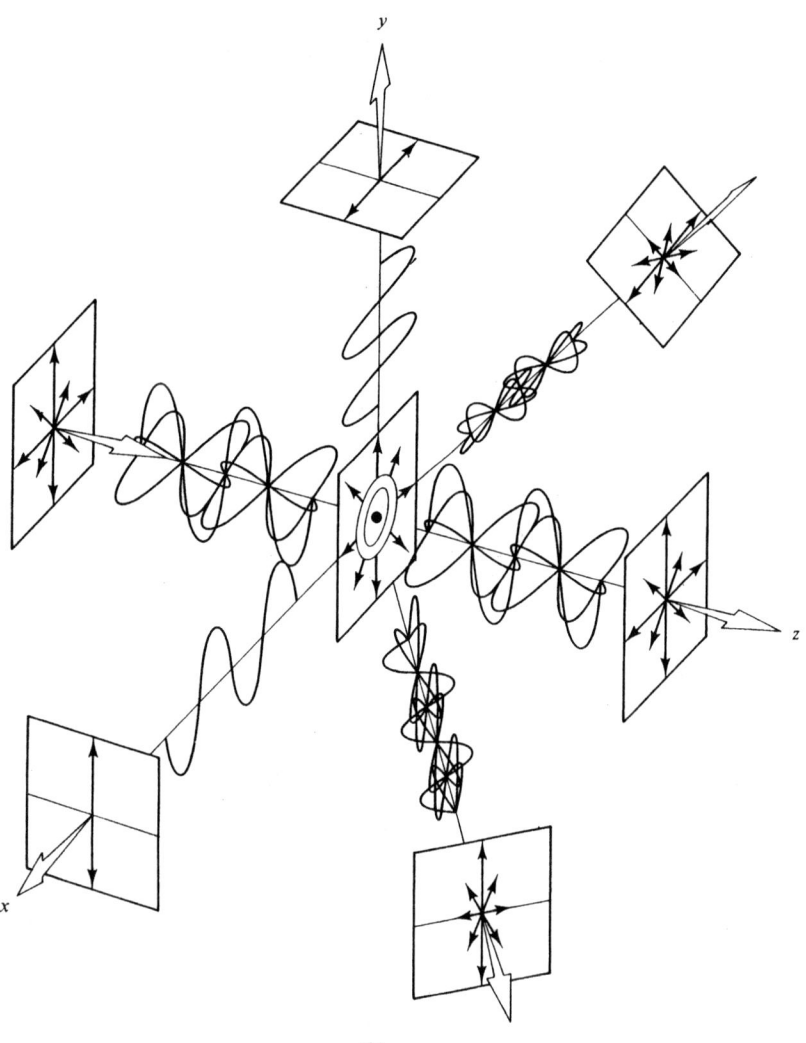

(b)

**FIGURE 16-2** Scattering of unpolarized monochromatic light by a small spherical particle. The particle is shown at the origin of the axes. In (a) the scattered radiation intensity at a given angle is indicated by the length of the vectors. The envelope represents the cross sections of scattering in the *yz* and the *xy* planes. In (b) the state of polarization of the scattered rays is indicated by the vectors. Radiation scattered in either direction along the *z* axis is unpolarized; off that axis it is partially polarized. When the direction of observation is at 90° to the direction of propagation, the radiation is completely linearly polarized.

(typical molecular dimensions are on the order of nanometers). In the case of Rayleigh scattering, we can consider the scatterer to be a point source of secondary emission.

Lord Rayleigh was the first to investigate the dependence of the scattered radiation intensity on the incident wavelength. For a single particle, a symmetrical intensity distribution of the type shown in Figure 16-2 is obtained with unpolarized incident radiation. The irradiance of the scattered radiation at angle $\theta$, $(E_{sc})_\theta$, is given by

$$(E_{sc})_\theta = \frac{8\pi^4(\alpha')^2(1 + \cos^2\theta)E_0}{\lambda^4 d^2} \quad (16\text{-}1)$$

where $\alpha'$ is the polarizability of the particle in $m^3$, $\lambda$ is the wavelength of the incident radiation, $\theta$ is the angle between the incident and scattered ray, $E_0$ is the incident beam irradiance, and $d$ is the distance from the center of scattering to the detector. The polarizability is a measure of how efficiently a given incident frequency induces a dipole in the particle. Because the polarizability varies roughly with the volume of the scattering particle, equation 16-1 predicts that the scattered radiation intensity increases with increasing particle size. Thus, in a sample that contains particles of various sizes, the larger particles tend to contribute most heavily to the scattering.

Equation 16-1 also predicts that the scattered radiation intensity varies inversely with the fourth power of the wavelength (directly with $\nu^4$). This dependence on wavelength is responsible for such natural phenomena as the blue color of the sky and the red color of the sun at sunset. The sky appears blue because short wavelengths of sunlight are efficiently scattered by small dust particles and water vapor in the atmosphere. The red color of the sun as seen through haze, smoke, or fog results from the efficient transmission of the longer wavelengths of sunlight relative to the shorter wavelengths which are effectively scattered. Note also from Figure 16-2 that with unpolarized incident radiation, the scattered radiation at 90° is linearly polarized.

*Large-Particle Scattering.* The comprehensive theory of large-particle (Debye and Mie) scattering was formulated by Mie. In Debye scattering we can no longer consider the particles to be point sources. Instead, centers of scattering are found in various areas of the particle. When such particles scatter radiation, the scattering centers are far enough apart that some interference is likely between the rays emitted from one area of the particle and those from another. This leads to an intensity distribution that is quite different from the fairly uniform distribution characteristic of Rayleigh scatter-

ing. Figure 16-3a illustrates the interference that can occur in large-particle scattering. As a result the intensity distribution pattern shifts to one showing predominantly forward scattering as depicted by the center scattering envelope in Figure 16-3b. Note that the envelope is similar to Rayleigh scattering in the forward direction, but quite different in the reverse direction.

Mie scattering occurs from large particles of relatively high refractive index. Here, not only are there fixed phase relationships between waves scattered from different parts of the same particle, but there are also distortions of the electric field of the incident and scattered radiation. The result is the complex scattering envelope shown in the inner part of Figure 16-3b. Note again the similarity to Rayleigh scattering in the forward direction and the complicated angular dependence in the reverse direction.

*Scattering from Liquids and Solutions.* Pure liquids and solutions of small molecules are generally poor light scatterers. Because they contain a fairly regular array of molecules, lateral destructive interference occurs; this virtually ensures that most of the incident radiation is refracted or reflected at the surface. Very weak scattering can occur, however, because of inhomogeneities (localized density and concentration variations) produced by thermal forces. These lead to some volume elements with more particles than the average and some with less. Each volume element that differs from the average can act briefly as a local scattering center of polarizability $\Delta\alpha'$, where $\Delta\alpha' = \alpha'_{local} - \alpha'_{av}$. For such a case we can replace $(\alpha')^2$ in equation 16-1 with $(\Delta\alpha')^2$. Since the latter is very small, scattering is expected to be quite weak.

If the particles in solution are large molecules, such as polymers or sols, the scattered radiation can be primarily the result of concentration fluctuations which increase with increasing concentration. With molecules that have dimensions between approximately 1.5 and 40 nm, the scattering is still of the Rayleigh type.

## Inelastic Scattering

Two types of **inelastic scattering** can be distinguished. In Raman scattering, relatively large frequency shifts occur that are independent of scattering angle. **Raman scattering** is caused by rotational and vibrational transitions in molecules. In addition, scattering can occur with a relatively small frequency shift that varies with the scattering angle. This type of scattering, caused by thermal fluctuations in the medium, is known as **Brillouin scattering**. Because Brillouin scattering requires a high-resolution spectrometer (e.g., Fabry–Perot in-

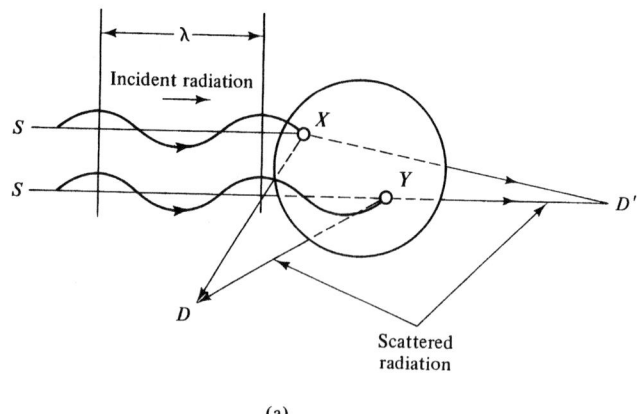

(a)

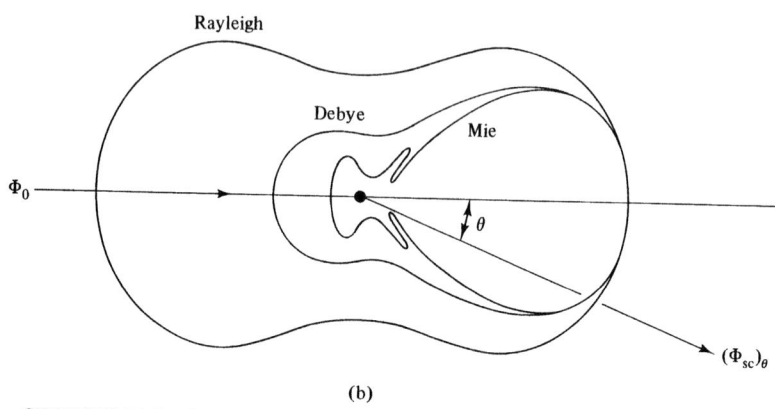

(b)

**FIGURE 16-3**    Scattering from large particles. In (a) the interference that occurs in the scattering from a large particle is illustrated. Note that the rays emitted in a backward direction are highly susceptible to destructive interference because of the large path differences possible. Thus scattered radiation originating at centers $X$ and $Y$ and observed at point $D$ is likely to show destructive interference effects because there is a large path difference between the route $SXD$ and the route $SYD$. In the forward direction, there is a much smaller path difference to detector position $D'$. Hence rays traveling the route $SXD'$ and those arriving via $SYD'$ are more likely to constructively interfere. In (b) the distributions of scattered radiation intensity in the plane of propagation are shown for the three scattering types. In Debye scattering, the highest intensity is observed in the forward scattering direction because of destructive interference in the backward direction. Mie scattering is much more complex and diffraction effects appear in the angular distribution diagram. Note that the scattering angle $\theta$ is taken as the angle between the incident and scattered rays. Thus $\theta$ is 0° for forward scattering and 180° for scattering in the backward direction.

terferometer) to observe and has few analytical applications, it is discussed only briefly in Section 16-4.

In discussing Rayleigh scattering, it was assumed that the polarizability $\alpha'$ of the scattering molecule remains constant. A changing polarizability gives rise to a scattered radiation intensity that changes correspondingly. A molecule can undergo a change in polarizability during one of its normal modes of vibration. A normal mode that involves such a change in polarizability is said to be a **Raman-active mode**. The resulting changes in the amplitude of the scattered radiation are illustrated in Figure 16-4. For a monochromatic incident beam, the amplitude-modulated waveform is shown in Section 16-2 to have three frequency components. One component corresponds to the Rayleigh scattered radiation and occurs at the frequency of the monochromatic source; the remaining two components correspond to Raman scattered radiation and occur shifted

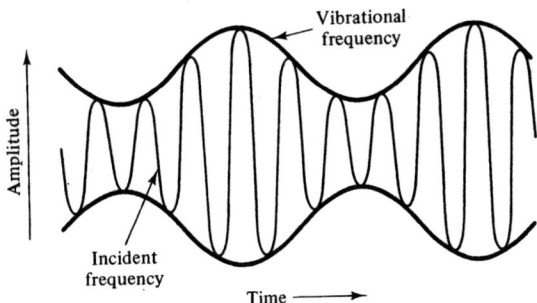

**FIGURE 16-4** Amplitude–time waveform of scattered radiation from molecule undergoing a periodic change in polarizability during a vibration.

to higher and lower frequencies than the incident frequency by an amount equal to the frequency of the vibration.

From a more quantum mechanical viewpoint we can consider the Raman effect to involve virtual states, as shown in the simplified energy-level diagram of Figure 16-5a. Molecules here are considered to be either in the ground vibrational state ($v = 0$) or in the lowest excited vibrational state ($v = 1$). Absorption of a photon of energy $h\nu_{ex}$ can give rise to Rayleigh scattering and to Stokes and anti-Stokes Raman scattering as shown in the spectrum of Figure 16-5b. The frequency shifts between the incident radiation and the Raman scattered radiation thus correspond to vibrational energy levels. Hence we expect that Raman spectrometry will yield

information similar to IR spectrophotometry. Because of differences in the selection rules, however, the shifts observed do not always correspond to observable IR bands. Note also that the shifts can be observed with incident radiation that is in the visible region of the spectrum, which can be a distinct advantage. In Section 16-2 we consider the Raman effect in more detail and make a more comprehensive comparison between Raman and IR spectroscopy.

## 16-2 RAMAN SPECTROSCOPY

The Raman effect was discovered by Sir C. V. Raman, an Indian physicist, in 1928. A great deal of interest was generated by the discovery as evidenced by many early applications and the awarding of the Nobel Prize to Raman in 1930. Since then, Raman spectroscopy has become a powerful tool for characterizing the structures of molecules. However, until the laser became the dominant radiation source for Raman measurements, Raman spectroscopy was not often used for chemical analysis. With the low-pressure mercury arc, used prior to the advent of the laser, the intensities of Raman bands were very low, which made it necessary to use high concentrations. Also, with the mercury arc source, much larger solution volumes were needed than for IR spectrophotometry. Today, lasers are used almost exclusively as sources, and Raman methods have become feasible alternatives to IR methods for structure elu-

**FIGURE 16-5** Energy-level diagram illustrating Raman scattering (a) and resulting Raman spectrum (b). In (a), molecules in the ground vibrational state ($v = 0$) can absorb a photon of energy $h\nu_{ex}$ and reemit a photon of energy $h(\nu_{ex} - \nu_v)$. Molecules in a vibrationally excited state can scatter inelastically and return to the ground state, producing a Raman effect with energy $h(\nu_{ex} + \nu_v)$. The lower-frequency transition is called Stokes scattering, while the higher-frequency transition is called anti-Stokes scattering. We will often label the Stokes frequency $\nu_s$ and the anti-Stokes frequency $\nu_a$. If the system is in thermal equilibrium, the equilibrium populations of the ground and excited states follow a Boltzmann distribution. Because the ground-state population is greater than that of the excited state, the Stokes lines are more intense than the anti-Stokes lines (b).

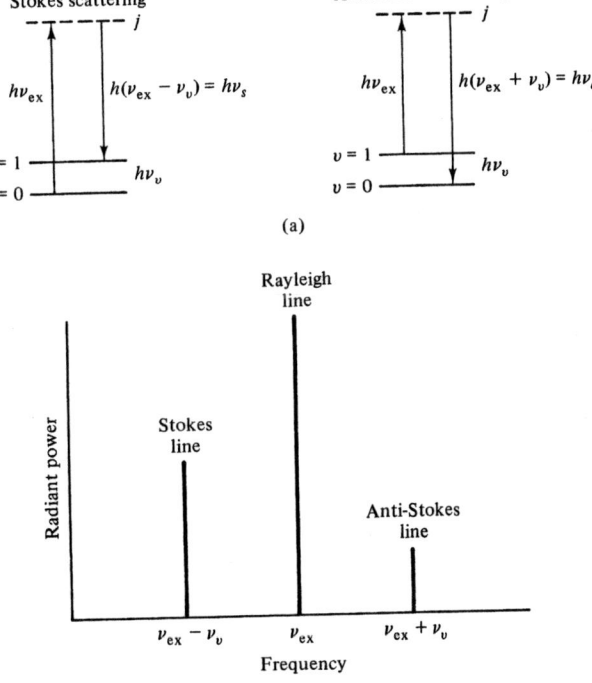

cidation, multicomponent qualitative analysis, and quantitative determinations of minor and trace constituents.

Although Raman scattering can be used to obtain information about rotational, electronic, and vibrational transitions, we will be concerned primarily with its use in vibrational spectroscopy.

## Theory

We treat here briefly the classical theory of the Raman effect. Let us consider a beam of exciting frequency $\nu_{ex}$ to be incident on a sample. The electric field **E** associated with the beam can be written

$$\mathbf{E} = \mathbf{E}_m \cos (2\pi\nu_{ex}t) \qquad (16\text{-}2)$$

where $\mathbf{E}_m$ is the amplitude of the wave. When this oscillating field interacts with the polarizable electron clouds of the sample molecules, it induces a dipole moment $\mu_{in}$ given by

$$\mu_{in} = \alpha\mathbf{E} = \alpha\mathbf{E}_m \cos (2\pi\nu_{ex}t) \qquad (16\text{-}3)$$

where $\mu_{in}$ has the units of C m, and $\alpha$ is the polarizability of the sample molecules in $J^{-1} C^2 m^2$. To obtain the polarizability $\alpha'$ in cubic meters, $\alpha$ must be divided by $4\pi\varepsilon_0$ (i.e., $\alpha = 4\pi\varepsilon_0\alpha'$).

The Raman effect results from the interaction of the polarizability with the normal modes of vibration of the molecules. The polarizability varies with internuclear separation around its equilibrium value $\alpha_0$ according to

$$\alpha = \alpha_0 + (r - r_e)\left(\frac{\partial\alpha}{\partial r}\right)_e + \cdots \qquad (16\text{-}4)$$

where $\alpha_0$ is the polarizability of the molecule at the equilibrium bond distance $r_e$, and $r$ is the internuclear separation. The subscript $e$ indicates evaluation at the equilibrium position. The change in internuclear distance varies with the frequency of the vibration $\nu_v$ according to

$$r - r_e = r_m \cos (2\pi\nu_v t) \qquad (16\text{-}5)$$

where $r_m$ is the maximum internuclear separation relative to the equilibrium position. If we substitute equation 16-5 into equation 16-4, we obtain

$$\alpha = \alpha_0 + \left(\frac{\partial\alpha}{\partial r}\right)_e r_m \cos (2\pi\nu_v t) \qquad (16\text{-}6)$$

If we substitute equation 16-6 into equation 16-3, we

obtain an expression for the induced dipole moment $\mu_{in}$:

$$\mu_{in} = \alpha_0\mathbf{E}_m \cos (2\pi\nu_{ex}t)$$
$$+ \mathbf{E}_m r_m \left(\frac{\partial\alpha}{\partial r}\right)_e \cos (2\pi\nu_v t) \cos (2\pi\nu_{ex}t) \qquad (16\text{-}7)$$

If we now use the identity for the product of two cosines, $\cos x \cdot \cos y = [\cos (x + y) + \cos (x - y)]/2$, the latter equation becomes

$$\mu_{in} = \alpha_0\mathbf{E}_m \cos (2\pi\nu_{ex}t) + \frac{\mathbf{E}_m}{2}r_m \left(\frac{\partial\alpha}{\partial r}\right)_e \cos 2\pi(\nu_{ex} + \nu_v)t$$
$$+ \frac{\mathbf{E}_m}{2}r_m \left(\frac{\partial\alpha}{\partial r}\right)_e \cos 2\pi(\nu_{ex} - \nu_v)t \qquad (16\text{-}8)$$

Here the first term represents the Rayleigh scattering (frequency $\nu_{ex}$), while the second and third terms represent the anti-Stokes and Stokes Raman scattering, respectively. Although this classical treatment predicts that there will be Raman scattered radiation at higher and lower frequencies than the exciting radiation, it fails to account for the difference in intensity between the Stokes and anti-Stokes lines. Quantum mechanical treatments, which take into account the relative populations of the energy levels, can accurately predict the intensity ratios of the Stokes and anti-Stokes Raman lines.

*Requirements for Raman Scattering.* The selection rules for Raman scattering are determined by evaluating the transition moment, as described for IR spectroscopy in Section 14-1. This leads to an equation analogous to equation 14-2 with the polarizability substituted for the dipole moment:

$$R = \int \psi_i^* \left[\alpha_0 + (r - r_e)\left(\frac{\partial\alpha}{\partial r}\right)_e\right] \psi_j \, d\tau \qquad (16\text{-}9)$$

which reduces to

$$R = \int \psi_i^* \left[(r - r_e)\left(\frac{\partial\alpha}{\partial r}\right)_e\right] \psi_j \, d\tau \qquad (16\text{-}10)$$

From equation 16-10 it is clear that there must be a *change in polarizability* during the vibration in order for Raman scattering to occur. The selection rules further predict that Raman lines, corresponding to fundamental modes, occur with $\Delta v = \pm 1$. Just as in IR spectroscopy, overtone transitions, which are much weaker, appear at $\Delta v = \pm 2$.

We can now explain the difference between Raman and IR selection rules. In IR spectrometry, in order for the integral in equation 14-3 to be nonzero, that is, for the transition to be allowed, there must be a change in *dipole moment* during the vibration. For Raman spectroscopy, on the other hand, there must be a change in *polarizability* during the vibration. An easy way to determine whether a transition is allowed or not in IR and Raman spectroscopy is to use group theory and the symmetry properties of the states and the operator (dipole moment or polarizability) involved. For integrals such as those in equations 14-3 and 16-10 to be nonzero, they must be totally symmetric. That is, the product of the ground state, the operator and the excited state symmetry must be totally symmetric (i.e., totally invariant under all symmetry operations). This leads to the conclusion that allowed IR transitions generally involve unsymmetrical vibrations. By contrast, Raman active vibrations include those that are themselves totally symmetric.

The differences between Raman and IR activity arise from the different symmetry properties of the dipole moment and the polarizability operators. For molecules with a center of symmetry, these differences lead to the conclusion that there are no IR active transitions in common with Raman active transitions, the **mutual exclusion principle**. For example, the symmetric stretching mode of $CO_2$ is IR inactive because there is no dipole moment change during the vibration (recall Figure 14-1). On the other hand, the polarizability varies during the vibration, which leads to Raman activity. For the asymmetric stretch of $CO_2$, the dipole moment changes during the vibration. However, as the polarizability of one of the C—O bonds increases as it lengthens, that of the other decreases, and overall, there is no change. Thus the asymmetric stretching vibration of $CO_2$ is Raman inactive.

For noncentrosymmetric molecules, there are many cases in which the mutual exclusion principle still holds. However, many other vibrational modes may be both Raman and IR active. One interesting case is that of $C_1$ symmetry, in which there is no symmetry. Here all vibrations are both Raman and IR active. In cases where vibrations are Raman and IR active, the intensities observed for the same vibration may be quite different.

*Raman Intensities.* It is difficult to write a simple expression relating the Raman intensity to molecular parameters. However, the following equation gives some important factors that influence the radiant power of Raman scattering $\Phi_R$:

$$\Phi_R \propto \sigma(\nu_{ex})\nu_{ex}^4 E_0 n_i e^{-E_i/kT} \qquad (16\text{-}11)$$

where $\sigma(\nu_{ex})$ is the Raman cross section in $cm^2$, $n_i$ is the number density in state $i$, and the exponential term is the Boltzmann factor for state $i$. Typically, $\sigma(\nu_{ex})$ is

on the order of $10^{-29}$ $cm^2$ for a good Raman scatterer. Fundamentally, the strength of a Raman line depends on the square of the polarizability derivative ($\sigma(\nu_{ex})$ is proportional to $[(d\alpha/dr)_e]^2$). In actual practice, the intensities measured depend on *averages* of the polarizability derivatives for isotropic systems (e.g., liquids and gases). For a sample such as a single crystal, the radiant power of Raman-scattered radiation depends experimentally on the specific crystal orientation.

Raman intensities also are directly proportional to the source irradiance $E_0$, as shown in equation 16-11. For this reason and for several others discussed later, laser sources have replaced conventional sources in modern spectrometers. The radiant power of Raman-scattered radiation is usually directly proportional to the concentration of the active species. Because of this, quantitative analysis with Raman spectrometry has become quite popular.

The dependence of the detected Raman scattering radiant power on instrumental parameters such as source irradiance, the emission collection efficiency, and concentration is quite similar to that observed in molecular luminescence spectrometry (see Table 15-1). The major difference is that the Raman cross section replaces the $\varepsilon Y_L$ term in the luminescence expression. Thus many of the techniques used in luminescence spectrometry for maximizing the analytical signal and for optimizing the signal-to-noise ratio (see Chapter 15) are also applicable with Raman spectrometry.

A typical Raman spectrum is shown in Figure 16-6. It is common in Raman spectrometry to calibrate the

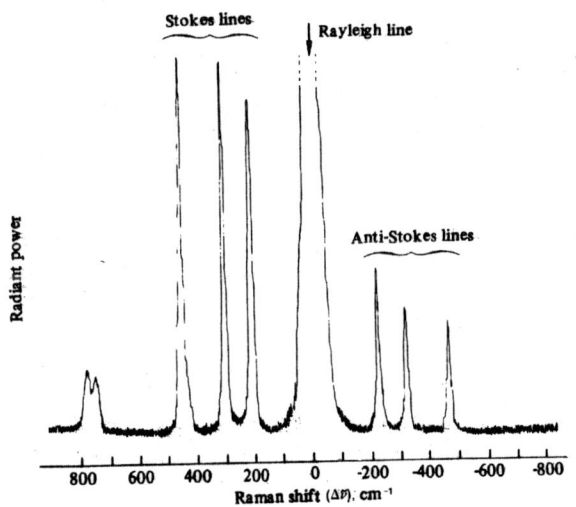

**FIGURE 16-6** Raman spectrum of pure carbon tetrachloride. This spectrum was obtained with an He-Ne laser and 3 μL of sample. The Raman shift ($\Delta\bar{\nu}$) is the difference in wavenumbers between the Rayleigh line and the Raman line. [Redrawn with permission from B. J. Bulkin, *J. Chem. Educ., 46*, A781 (1969).]

horizontal axis in wavenumbers (cm⁻¹). Hence spectra are usually obtained as scattering magnitude vs. the Raman shift in cm⁻¹. Note that the Raman bands are considerably lower in intensity than the Rayleigh line. Also note that the Stokes lines are much more intense than the anti-Stokes lines.

*Depolarization Ratios.* The wavenumbers of the Raman shifts observed in molecules provide qualitative and structural information similar to that observed in IR spectrophotometry. The radiant power of Raman scattering provides quantitative information as discussed above. Raman measurements can, however, yield an additional factor, called the **depolarization ratio**, that is useful in structure elucidation. If the incident beam is polarized, as it is with a CW laser source, the Raman-scattered radiation can be polarized to various degrees that depend on the nature of the active vibration.

The depolarization ratio $\rho$ is defined as

$$\rho = \frac{(\Phi_R)_\perp}{(\Phi_R)_\parallel} \qquad (16\text{-}12)$$

where $(\Phi_R)_\perp$ is the Raman radiant power polarized perpendicular to the polarization of the original beam and $(\Phi_R)_\parallel$ is that polarized parallel to the original beam. Experimentally, it is obtained by inserting an analyzer prism (see Section 3-5) between the sample and the monochromator. The prism is first oriented to pass radiation polarized parallel to the original beam and then to pass that polarized perpendicular. The ratio of the intensities obtained with the two orientations is then $\rho$.

The depolarization ratio can give information about the symmetry of the vibration involved. For example, if the molecule is approximately spherical and the vibration is totally symmetric, the incident beam polarization is maintained; the depolarization ratio would be very small in this case. On the other hand, if the vibration distorts the symmetry, or if the molecule is not symmetric to begin with, a significant depolarization can occur. From scattering theory, it is predicted that for nonsymmetric vibrations $\rho = 0.75$, while for symmetric vibrations $\rho < 0.75$. With $CCl_4$, for example, the 459-cm⁻¹ Raman line has a value of $\rho = 0.005$. This line arises from the totally symmetric breathing vibration; the four chlorine atoms move simultaneously away and toward the central carbon atom. The lines at 218 and 314 cm⁻¹, however, arise from nonsymmetric vibrations; for these, $\rho \approx 0.75$.

## Instrumentation

Raman spectroscopy has been greatly influenced by instrumentation developments. Commercial laser-excited Raman spectrometers with high-quality monochroma-

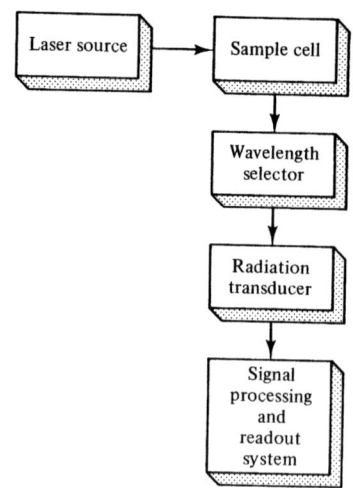

**FIGURE 16-7** Block diagram of a Raman spectrometer. The laser source radiation is directed into a sample cell. The Raman scattering is normally observed at right angles to avoid directly viewing the source radiation. A wavelength selector isolates the spectral region of interest. The radiation transducer converts the radiant power or photon flux into a dc electrical signal or a count rate.

tors became available in the 1960s. Since that time significant improvements have been made in lasers, in gratings for monochromators, and in detection systems. The new spectrometers can observe far weaker signals than previous generations, and some can obtain spectra of transient species.

A normal Raman spectrometer consists of the laser source, a sample cell, a wavelength selector, a radiation transducer, and an appropriate signal processor and readout device, as illustrated in Figure 16-7. Because these components are critical to the nature and quality of the spectra obtained, they are discussed in detail below.

*Sources.* The early laser-excited Raman spectrometers used the He–Ne laser. However, virtually all normal Raman instruments today use the powerful (e.g., > 1 W) argon and krypton ion lasers with outputs in the blue or green region of the spectrum. The most common laser wavelengths used for exciting Raman spectra are listed in Table 16-2. Since the scattered

**TABLE 16-2**

Laser excitation wavelengths

| Laser type | Wavelengths (nm) |
|---|---|
| He-Ne | 632.8 |
| Ar⁺ | 488.0, 514.5 |
| Kr⁺ | 530.9, 647.1 |

intensity increases as $(\nu_{ex})^4$, $Ar^+$ and $Kr^+$ lasers can provide a significant improvement in Raman intensity over the He–Ne laser. In addition, the most commonly used detectors have higher responsivities in the blue or green than in the red.

The excitation wavelength for Raman spectrometry must be chosen carefully. Photodecomposition of samples can be a severe problem with short-wavelength excitation. In addition, fluorescence can also occur with some samples. Colored samples can absorb either the incident radiation or the Raman-scattered radiation. Hence a compromise is often made in selecting the excitation wavelength. The best wavelength would produce high Raman intensities, but minimal photodecomposition, fluorescence, and, except for resonance Raman spectrometry (discussed later), minimal absorption.

Tunable lasers are also used for exciting Raman spectra. Often these are CW dye lasers pumped by powerful $Ar^+$ or $Kr^+$ lasers. These sources allow optimization of the excitation wavelength. In addition, they are highly useful for resonance Raman spectroscopy (see later). A potentially important source is the synchronously pumped dye laser. This is produced by a mode-locked $Ar^+$ laser pumping a dye laser with a

capacity dumping accessory (see Section 4-3). By frequency doubling the output, tunable UV radiation can be produced. The short (e.g., picosecond) pulses also allow time-resolution techniques to be employed to reject fluorescence.

*Sample Cells and Cell Configurations.* Raman spectra can be readily obtained with gas, liquid, and solid samples. Since the exciting and scattered radiation is in the visible (occasionally ultraviolet) region, cell materials, windows, and optical materials can be made of glass or quartz. With laser sources, many different sample cells and configurations can be used.

The most common cells for liquids are capillary tubes (typically, 1.5 mm o.d.) in which pure liquids or solutions are contained. Two widely used sample cell configurations for micro samples are shown in Figure 16-8. Cell configurations have also been described for macro samples. Flow-through cells, while not commonly used for Raman spectra, have been described for examining liquid samples without realignment.

Another popular cell design, shown in Figure 16-9, is the rotating cell, originally developed for ob-

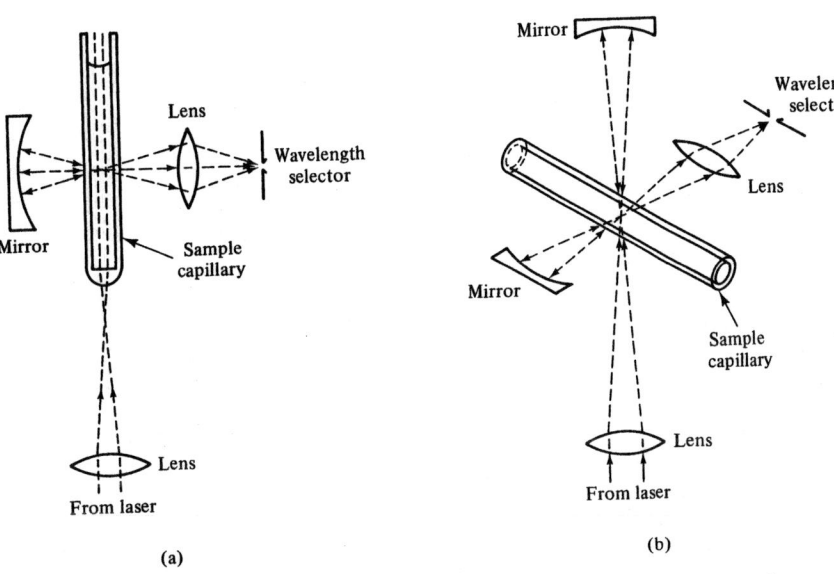

(a)  (b)

**FIGURE 16-8** Cell configurations for Raman spectra of micro samples. In (a) the laser beam enters through the bottom of the capillary cell. The cell bottom serves as a lens to collimate the laser radiation. The scattered radiation is collected at 90° to the excitation beam and focused with a lens onto the wavelength selection device. Note that this arrangement provides a long pathlength for excitation and viewing along the monochromator slit width. Sample volumes can be as small as 0.04 μL. In (b) the laser beam enters the cell from the side and is reflected by a mirror to traverse the cell again. Transverse excitation simplifies alignment and allows the used of open end capillary tubes. The scattered radiation is again collected at right angles to the laser beam. With this geometry, sample volumes can be as small as 0.008 μL.

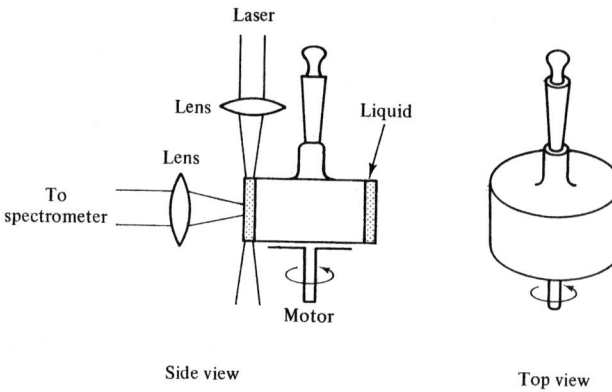

Laser

Lens

Lens

To spectrometer

Liquid

Motor

Side view

Top view

**FIGURE 16-9** Rotating cell for Raman spectroscopy of highly colored solutions. By spinning the sample in the laser beam, intense, localized overheating of the sample is avoided. Thus spectra can be obtained on highly absorbing solutions.

taining Raman spectra of highly absorbing materials. Such materials can be readily destroyed by intense, localized overheating caused by the laser. These cells have not only been applied to liquids, but also to gases, solids, and semimicro samples. A divided rotating cell and a gated detection system have been used to obtain difference spectra.

Raman spectra are often obtained with solution samples. Because overtone and combination bands are quite weak in Raman spectra, a wide variety of solvents can be employed. Water is a weak Raman scatterer. Hence inorganic materials and biological samples can be readily examined. Solvents such as $CCl_4$, $CS_2$, and $CHCl_3$ are also frequently used.

Gas-phase samples require higher-powered lasers and a more complicated sampling system than liquid samples. Nevertheless, it is relatively easy to obtain Raman spectra on gaseous samples. Special cells for handling gases are commercially available, and many others have been described in the literature.

Solid samples are also relatively easy to use. Powders can be tamped into an open-end capillary. Potassium bromide pellets mounted at 45° to the laser beam are quite frequently used. Films, fibers, coatings, and several other types of solid samples can be examined directly without any sample preparation.

*Wavelength Selection Devices.* A high-quality wavelength selection device is needed in Raman spectrometry. The wavelength selector must achieve the desired resolution (typically, $<5$ cm$^{-1}$) and separate the relatively weak Raman lines from the intense Rayleigh-scattered radiation. Double or even triple monochromators can be employed to achieve the required rejection of stray radiation.

The replacement of ruled gratings with holo-

graphic gratings in modern spectrometers has led to significant improvements in stray radiation rejection. Holographic gratings allow Raman signals to be obtained with good signal-to-noise ratios in the low-frequency region near the exciting line. Concave holographic gratings have also been used in Raman spectrometers. With these gratings no additional optical elements are needed other than entrance and exit slits. Stray radiation is again reduced with this approach, and throughput is improved. In fact, one commercial monochromator uses a concave grating and achieves stray radiation levels sufficiently low to permit Raman spectra to be obtained with a single monochromator. Although stray radiation levels are higher than with a double or triple monochromator, throughput is significantly better with the single monochromator.

*Transducer and Readout Systems.* Photomultiplier tubes (PMTs) are most often used as transducers in Raman spectrometry because of the weak signals obtained. Tubes with Ga-As or multialkali photocathodes (see Section 4-4) are quite popular. These have high quantum efficiencies and nearly constant response over the visible region (see Figure 4-23); both of these features are desirable for Raman measurements.

The signals obtained in Raman spectrometry vary over a wide dynamic range. The early Raman spectrometers employed dc signal processing techniques. These have been replaced by photon counting systems in modern Raman instruments. The weak Raman signals and the relatively low background make photon counting and Raman spectrometry nearly ideal partners. In fact Raman spectrometry is one of the few techniques that is able to utilize the signal-to-noise ratio ($S/N$) advantage of the photon counting method. Recent photon counting systems can also produce accurate results at very high count rates, which allows the measurement of widely varying Raman intensities. Some instruments employ automatic switching between photon counting and dc signal processing as radiation levels change to increase the dynamic range even further. Since photon counting is an inherently digital technique, instruments employing it are readily interfaced to computers.

Recently, intensified diode arrays have been shown to be useful for obtaining complete Raman spectra in a short time period. Although intensified array detectors do not have quite the detectability of PMTs, with the powerful lasers available adequate Raman signals can be obtained. As the new pulsed and CW lasers become more common, the combination of a powerful laser and an intensified array detector is certain to become widely used. Such techniques have made it feasible to obtain time-resolved Raman spectra for studies of the dynamics of molecular processes.

Microcomputers and microprocessors are playing increasingly important roles in Raman instruments. Their major use is in signal averaging. Here, an automated spectrometer can use long integration times and photon counting for *S/N* enhancement. With time-resolved Raman spectrometry of repetitive processes, computer averaging of multiple spectra can be used. In addition, recent monochromators have been operated under computer control. Often these use the simple sine bar, linear wavelength drive (see Figure 3-46) rather than the more complicated cosecant drive; the computer then converts wavelength to the Raman shift in wavenumbers. Computers have also been used to perform many other functions, including digital filtering, smoothing, calculation of depolarization ratios, line-shape analysis, spectral matching, and wavelength calibration.

*Signal-to-Noise Ratio Considerations.* The signal-to-noise ratio (*S/N*) expressions in Raman spectrometry are quite similar to those in molecular luminescence spectrometry (see Table 15-2). The major differences are that the scattering signal is the analytical signal in Raman spectrometry, and fluorescence from any source is a background signal.

Since photon counting is widely used in Raman spectrometry, we will write the *S/N* expression in terms of the number of Raman counts $n_R$ observed in time $t$.

With photon counting we will assume that amplifier-readout and excess dark current noise are negligible. We can thus write an expression similar to equation 5-61 for the *S/N*.

$$\frac{S}{N} = \frac{n_R}{[(n_R + n_d + n_F + n_{sc}) + \xi_1^2(n_R + n_F + n_{sc})^2]^{1/2}}$$

(16-13)

where $n_F$ is the number of fluorescence counts from the analyte, concomitants and the cell in time $t$, $n_{sc}$ is the number of counts due to elastic scattering, $n_d$ is the number of dark counts, and $\xi_1$ is the source flicker factor.

Equation 16-13 shows that for weak Raman signals, the *S/N* can be improved by increasing $n_R$, or by reducing $n_d$, $n_F$, or $n_{sc}$. The Raman signal can be increased by using a higher source irradiance or a higher excitation frequency and by improving the collection efficiency. The dark signal can be reduced by cooling the PMT. The elastic scattering signal can be reduced by using a monochromator with better stray radiation rejection. Reducing the fluorescence background is often one of the more difficult tasks in Raman spectrometry. Sometimes changing to a different excitation frequency will reduce fluorescence. It may also be necessary to clean up the sample or to purify the solvent if the fluorescence is from concomitants and not sufficiently re-

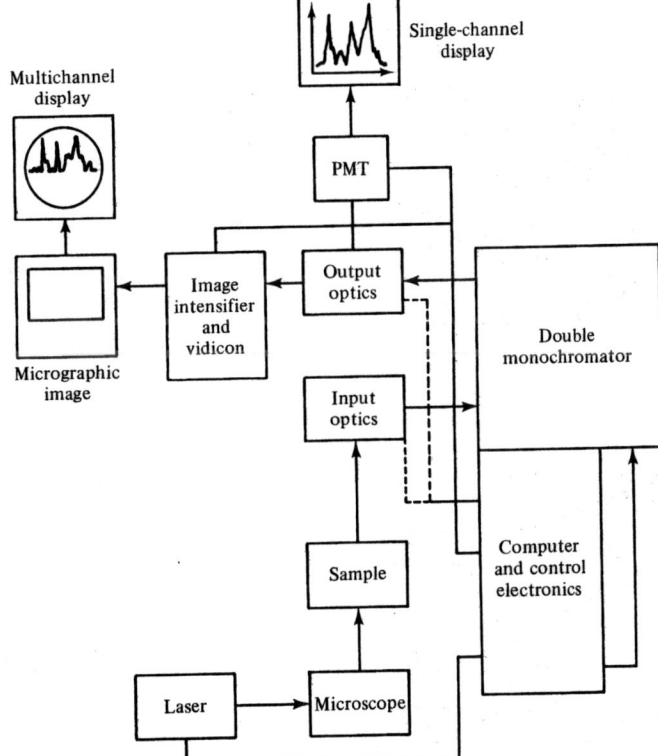

**FIGURE 16-10** Block diagram of Raman microprobe. A laser beam of approximately 1 μm² area can be scanned over the sample surface. In the point mode of operation, the optical microscope is used to select the area of interest on the surface. The laser beam is then directed to this point and the spectrum is obtained. Either single channel detection or multichannel detection can be employed. In the global mode, the spectrometer is set to a particular Raman shift characteristic of the species of interest and the beam is scanned over the surface. With a vidicon tube readout, the spatial distribution of Raman scattering at the selected Raman shift is obtained.

duced by switching laser lines. Analyte fluorescence can be reduced by using another laser line or by time gating the detection system as described below. Some nonlinear Raman techniques (see later) are useful in fluorescence discrimination.

*The Raman Microprobe.* The Raman microprobe is a instrument that couples a Raman spectrometer with an optical microscope. This allows the examination of surfaces by a technique that is sensitive to chemical structure. A block diagram of a commercial Raman microprobe, called the MOLE (Molecular Optics Laser Examiner), is shown in Figure 16-10. An example that illustrates the various modes of operation of the Raman microprobe is shown in Figure 16-11.

The Raman microprobe is a significant development for examining surfaces. A major application has been to the study of inclusions in minerals. In addition, it can be applied to polymers, films, organic and in-

organic materials, solid-state reactions, semiconductors, and biological specimens such as single cells.

*Fourier Transform Raman Spectrometry.* There has long been interest in applying the Fourier transform techniques that are so successful in the IR region to Raman spectrometry. However, there are many reasons to suspect that the combination of interferometry and Raman spectrometry would not be advantageous. First, Raman measurements are normally signal shot noise limited, so that the multiplex advantage of Fourier transform techniques does not apply. In addition, the shot noise of the Rayleigh-scattered line is redistributed over the weak Raman lines. Also, the throughput advantage of interferometry is not important because of the high intensity of the Rayleigh-scattered radiation compared to the Raman lines.

One way to solve most of the problems noted above is to filter the scattered laser radiation before it

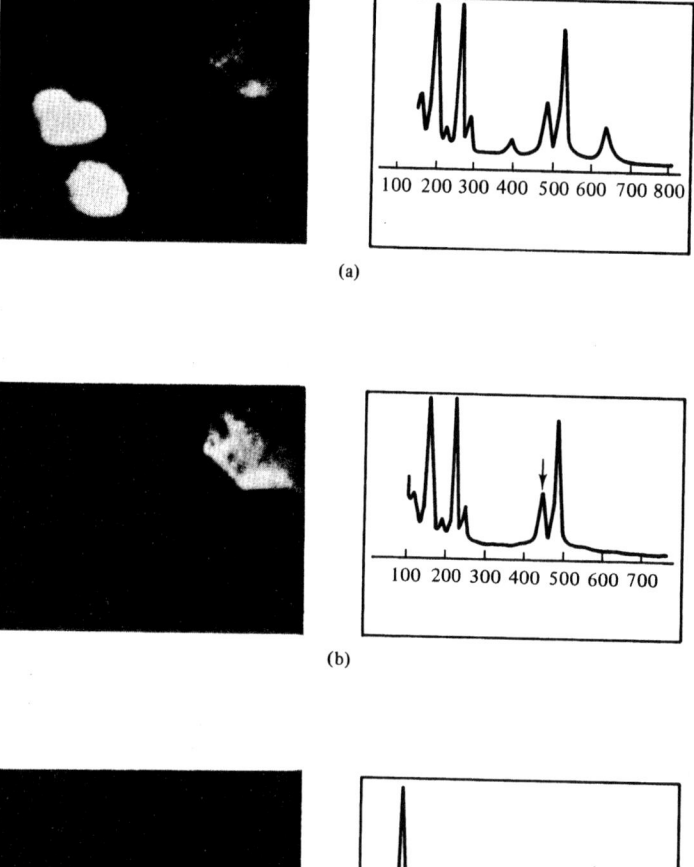

(a)

(b)

(c)

**FIGURE 16-11** Example of an application of the Raman microprobe. In (a) a sample is placed on the microscope slide and photographed (left). On the right is the total Raman spectrum taken from the entire field of view. This spectrum shows that the particles are $TiO_2$ and $SrSO_4$. In (b) the laser is directed to the top right particle, and the Raman spectrum obtained (right). This shows that this particle is $SrSO_4$. The image on the left in (b) was obtained by setting the spectrometer at the Raman shift indicated by the arrow on the spectrum and scanning the entire specimen. In (c) the Raman spectrum from the two left particles is shown (right). These are identified as $TiO_2$ particles. The image on the left results when the Raman spectrometer is set to the shift indicated by the arrow on the spectrum. (Reprinted with permission from J. G. Grasselli, M. K. Snavely, and B. J. Bulkin, *Chemical Applications of Raman Spectroscopy*, Wiley, New York, 1981. Copyright © John Wiley & Sons, Inc., 1981.)

enters the interferometer. Multistage interference filters and very high rejection ratio filters called Chevron filters have been used to achieve rejection ratios in the range $10^{-9}$ to $10^{-10}$. In addition, it has been proposed to carry out FT-Raman measurements with excitation in the near-IR region. In this region, the fluorescence stimulated is minimal and photodecomposition of the sample can be largely avoided. Although spectra are weaker because of the wavelength dependence of scattering and detectors are much less sensitive, very high average–power lasers can be used. With solid-state detectors rather than photomultipliers, the multiplex advantage comes back into play, and it has been calculated that these detectors should outperform the PMT at moderate signal levels. In any case, theoretical analysis has suggested and experimental results have demonstrated that FT-Raman measurements in the near-IR region provide $S/N$ values in the range $10^2$ to $10^4$, which is in the same range as ordinary Raman measurements.

In addition to the freedom from fluorescence and photochemistry, FT-Raman benefits from using the same interferometer and computer system that is used for FTIR measurements. We should see further improvements in instrumentation and many new applications of this method.

## Comparison of Raman and IR Spectrometry

Although the spectra obtained by Raman scattering and IR absorption spectrometry have much in common, there are also many important differences between the two techniques. This is especially true when one considers the instrumentation and techniques for sample handling. Some of the advantages of Raman in comparison to IR spectrometry include:

1. Water is a useful solvent in Raman spectrometry, whereas it is generally a poor solvent for IR studies.

2. The optics and cell materials for Raman spectrometry are made from glass or quartz instead of the salts used in IR measurements. This greatly simplifies sample handling for Raman methods.

3. The properties of the laser sources used in Raman spectrometry make it relatively easy to probe micro-samples, surfaces, films, powders, solutions, gases, and many other sample types.

4. Transducers in Raman instruments are standard UV-visible devices (PMTs, diode arrays, etc.) instead of the thermal detectors employed in IR spectrometry. Since Raman detectors respond very rapidly, Raman spectrometry can be used to study short-lived or transient species and to follow the kinetics of rapid reactions.

5. A single Raman spectrometer can cover the entire range of vibrational frequencies, whereas even with FTIR systems, changes in detectors or beam splitters must be made to cover this range. With conventional IR spectrophotometers, two or more instruments must be used over this range.

6. Raman spectra are usually much simpler than IR spectra because overtone and combination bands are not very intense. Overlapping bands are thus much less common.

7. Totally symmetric vibrations can be observed with Raman spectrometry, whereas they are not with IR spectrometry.

8. Polarization measurements add an extra dimension to the information obtained by Raman spectrometry. This aids in band assignments and structure determinations.

9. Raman intensities are directly proportional to concentration and to the laser power.

On the other hand, IR spectrophotometry is still widely used for many applications. Among the advantages of IR over Raman spectrometry are:

1. Because of the intensity of overtone and combination bands, IR spectrophotometry is more sensitive to small structural differences. Hence it is more useful in qualitative analysis and complementary to Raman spectrometry for structural elucidation studies. Extensive libraries of infrared spectra have been compiled.

2. Infrared instruments are generally less expensive than Raman instruments. Raman measurements are susceptible to spectral artifacts from grating imperfections (e.g., ghosts) and other sources. The monochromators used in Raman spectrometers must be of higher quality than those used in IR spectrometry. Alignment is often simpler with IR spectrophotometers.

3. Because Raman spectra depend highly on laser power, cell geometry, and instrument characteristics, it is difficult to compare Raman intensities from instrument to instrument. With IR spectrophotometers, the nature of absorption measurements (e.g., ratio measurement) makes this comparison easier.

4. Detection limits for IR spectrometry are often superior to those obtained with Raman spectrometry unless resonance enhancement (see later in this section) is utilized. Neither technique is considered particularly good for trace analysis.

5. The efficiency of the Raman process is quite low. Even in favorable situations a very small fraction (e.g., $10^{-8}$) of the incident photons are converted to scattered photons. As a result, broadband fluorescence emission can completely obscure the Raman signals.

Raman and IR spectrometry should be viewed as complementary techniques, particularly for structural studies. The differences in the selection rules lead to different spectra for some types of molecules. In any case the combination of IR and Raman spectrometry is very powerful for elucidating chemical structures. For organic functional group identification, IR spectrophotometry is usually used. Raman spectrometry is often the method of choice for inorganic and biological substances whose spectra are most conveniently obtained in aqueous solutions.

## Resonance Raman Spectrometry

Raman spectra obtained when the frequency of the exciting radiation coincides with or is in the region of an electronic absorption band, are called **resonance Raman spectra**. They often show a significant enhancement of some of the Raman lines. Resonance Raman lines can be $10^2$ to $10^6$ times more intense than in ordinary Raman spectrometry. The enhancement is restricted to vibrational modes that couple with the electronic transition. That is, those vibrations that exhibit a large change in equilibrium geometry upon electronic excitation are enhanced. This usually means that enhancements are observed for totally symmetric vibrations and vibrations that vibronically couple the two electronic states. As a result of enhancement, resonance Raman spectrometry provides low detection limits ($10^{-6}$ to $10^{-8}$ M for some species) and is suitable for trace analysis. Resonance Raman spectra are often quite simple, since only the bands related to the chromophoric group in a molecule are enhanced.

In the resonance Raman effect, the incident photon is absorbed, promoting an electron into an excited vibronic state, as depicted in Figure 16-12a. Immediate relaxation to a vibrational level of the ground state results in resonance Raman emission. The resonance Raman effect is closely related to fluorescence emission, as can be seen in Figure 16-12b. As might be

expected, with fluorescent molecules, fluorescence emission can be a major interference in resonance Raman spectrometry. The fluorescence appears as a broadband background signal, which can obscure the resonance Raman signals. The resonance Raman signals can be enhanced relative to fluorescence signals by time-resolution techniques. The resonance Raman process is nearly instantaneous compared to the much slower fluorescence emission. Hence measurements made very early in time during pulsed laser excitation will enhance the Raman signal over the fluorescence signal, which takes time to develop and decay. In practice the enhancement is only 10- to 100-fold.

The intense $Ar^+$ and $Kr^+$ lasers that became available in the late 1960s made the resonance Raman technique practical. Today, tunable dye lasers are frequently used. With a tunable laser the **Raman excitation spectrum** can be obtained by scanning the laser and observing the Raman scattering at a fixed value of the Raman shift. Raman spectra obtained with the excitation source tuned to a frequency just slightly lower than that necessary to cause an electronic transition are also slightly enhanced (often less than 10-fold). These spectra are called **preresonance Raman spectra**.

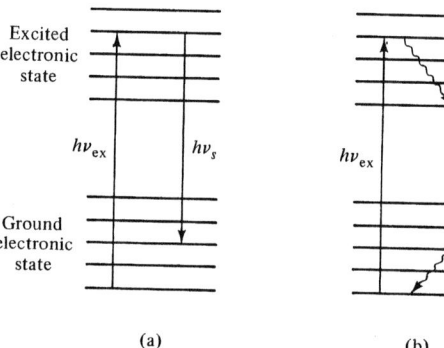

**FIGURE 16-12**   Energy-level diagrams illustrating resonance Raman scattering for an overtone transition (a) and fluorescence emission (b). In the resonance Raman case the excited electron immediately relaxes into a vibrational level of the ground electronic state giving up a Stokes photon $v_s$. In fluorescence there is relaxation from the excited vibronic state to the lowest vibrational level of the excited electronic state prior to emission. The resonance Raman process is nearly instantaneous, which results in narrow bands in the spectrum. Fluorescence spectra in solutions are typically broad because of the many close vibrational states. When a pulsed laser is used, the fluorescence emission builds up and then decays with time (usually on the nanosecond time scale), while the resonance Raman signal appears instantaneously (actually femtoseconds).

The resonance Raman effect can substantially improve the selectivity of Raman measurements on complex molecules. Because an electronic transition is somewhat localized in one part of a complex molecule, bands of the chromophore can be observed with little interference from other Raman bands of the molecule.

## Applications

Raman spectrometry is widely used in the chemical sciences. We consider here only a few typical applications of the Raman technique in qualitative and quantitative analyses.

*Organic Applications.* Raman spectrometry has been applied to organic systems since the 1930s. Because of its complementary nature, Raman spectrometry is often used in conjunction with IR to give qualitative and structural information.

Group frequencies regions are used to identify functional groups just as they are in IR spectrophotometry. Raman scattering is particularly useful for such groups as —C—S—, —S—S—, —C—C—, —N=N—, and —C=C—; IR spectrophotometry is well suited to identify such groups as —O—H, C=O, P=O, —NO$_2$, and S=O. Correlation charts and spectral compilations are now available for functional group and compound identification.

Raman and IR spectra are both used for structural studies of organic compounds. Figure 16-13 compares the IR and Raman spectra of cystine in the crystalline

state. Note the simplicity and the sharpness of the Raman spectrum. The combination of Raman and IR spectrometry is extremely powerful for structure elucidation.

Raman spectrometry is also applied in measurements of physical properties, such as hydrogen-bond strengths, acid dissociation constants, and phase transitions. In addition, Raman spectrometry is becoming quite useful in organic chemistry for the study of short-lived species, such as reaction intermediates or excited states.

*Inorganic Applications.* Raman spectrometry is an essential technique in inorganic structural analysis. It is widely used in conjunction with IR data for the vibrational analysis of inorganic solids. Such studies allow the assignment of internal modes of the ions and crystal lattice modes.

With transition metal complexes, the vibrational energies of many metal–ligand bonds are in the region 100 to 700 cm$^{-1}$, a region that is difficult to study by IR spectrophotometry. Such metal–ligand vibrations are frequently Raman active and readily observed by Raman spectrometry.

The ease of obtaining Raman spectra in aqueous solutions has also contributed to the popularity of Raman measurements on inorganic compounds. Highly colored samples can be studied with the rotating cell technique described earlier. Resonance enhancement is becoming increasingly useful in inorganic applications. Figure 16-14 shows the resonance Raman spec-

**FIGURE 16-13** Infrared (a) and Raman (b) spectra of crystalline cystine. Note that the NH$_3^+$ stretching vibration dominates the 3000-cm$^{-1}$ region in the IR spectrum, whereas the Raman spectrum shows sharp bands due to CH and CH$_2$ stretching modes. The NH$_3^+$ deformation and the carboxylate antisymmetric vibrations near 1600 cm$^{-1}$ are much stronger in the IR spectrum. The strong Raman band at 410 cm$^{-1}$ is due to the —S—S— stretch; this band is somewhat obscured and weak in the IR spectrum. [With permission of VCH Verlagsgesselschaft from B. Schrader, *Angew. Chem. Int. Ed. Engl.*, *12*, 884 (1973).]

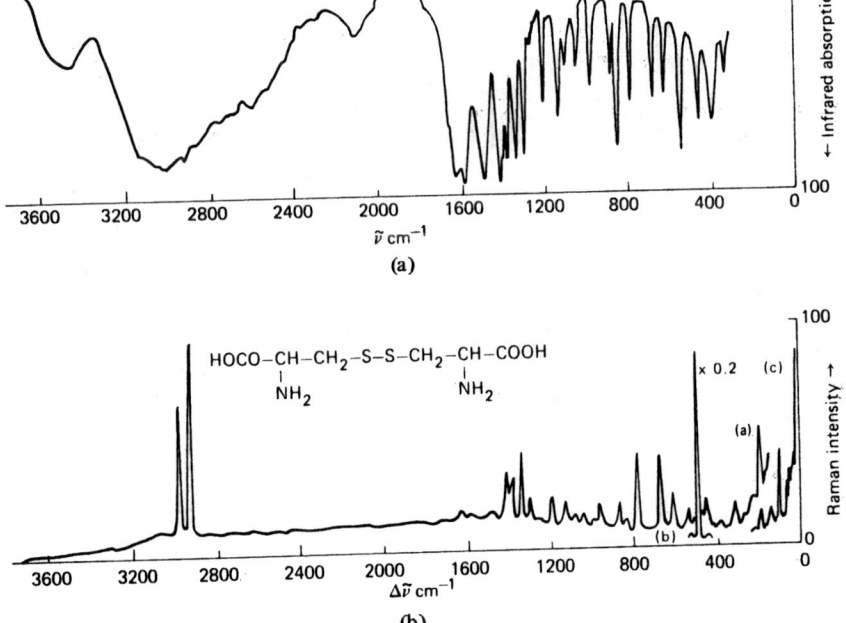

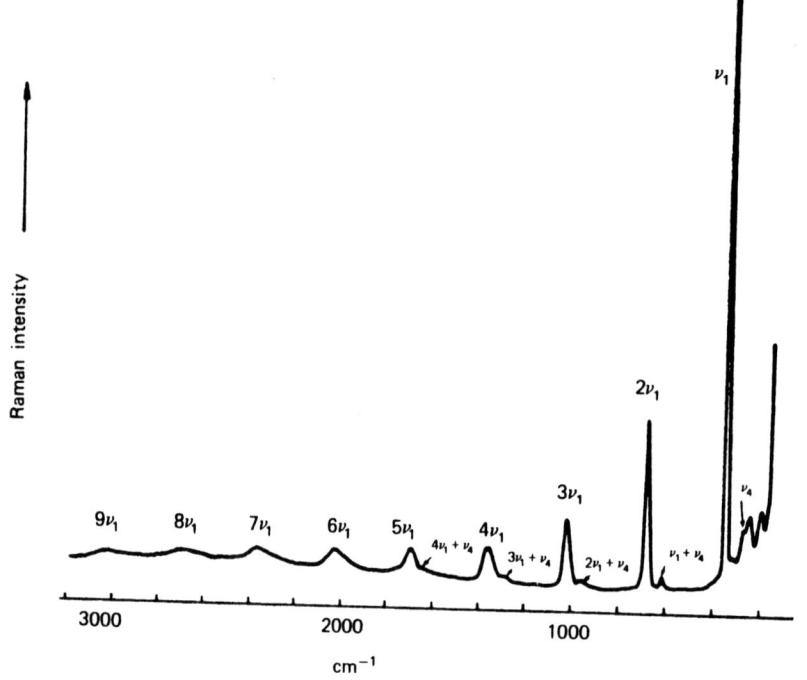

**FIGURE 16-14** Resonance Raman spectrum of $Cs_4Mo_2Cl_8$. The exciting frequency was brought into coincidence with the electronic transition involving the δ electron of the Mo–Mo bond. This results in resonance enhancement of the totally symmetric Mo–Mo vibration. Note the overtone progression of the Mo–Mo fundamental. [Reprinted with permission from R. J. H. Clark and M. L. Franks, *J. Chem. Soc. Chem. Commun., 9,* 316 (1974).]

trum of $Cs_4Mo_2Cl_8$. By bringing the exciting line into coincidence with the lowest allowed electronic transition, the overtone progression of the totally symmetric Mo–Mo vibration is observed.

*Biological Applications.* Raman spectrometry has recently become an important technique for obtaining structural information on biological systems. The ability to use small sample volumes and aqueous solutions is a major factor. In addition, the ease with which information can be obtained on samples such as lipids and membranes has also contributed to the increasing use of Raman techniques. Raman measurements are also quite sensitive to conformational changes, which are of great importance in biological systems. Since many molecules of biological interest are also fluorescent, a major problem has been to separate the Raman scattering from the fluorescence background.

Conventional Raman spectrometry has been used to monitor conformational changes of peptides and proteins in aqueous solutions. Nucleic acids and polynucleotides have also been studied. For example, the ionization equilibria of the bases and of the phosphate groups in nucleic acids have been investigated with the aid of Raman spectrometry. The secondary structure of RNA has been studied, and the binding of metal ions to various sites on nucleic acids has been monitored. Raman spectrometry has been used to measure the rates of hydrolysis of several nucleic acids.

Because of its poor detection limits, conventional Raman spectrometry is of limited utility in studying

actual biological processes. Normally, concentrations on the order of 0.1 M are required. Resonance Raman spectrometry, however, can readily detect molecules at the $10^{-5}$ M concentration level. With resonance Raman, only that part of the molecule closely associated with the electronic transition is enhanced. Thus interactions involving a chromophoric group may be studied in large molecules such as proteins. For example in hemoglobin, it is possible to selectively enhance vibrations associated with the heme group (chromophore) by appropriate choice of excitation frequency.

*Surface-Enhanced Raman Scattering.* Conventional Raman scattering and resonance Raman techniques have been applied to studies of adsorption on surfaces. Surface-enhanced Raman scattering (SERS) is a recent technique that allows very small amounts of adsorbates to be investigated. Originally, an enhancement of $\approx 10^6$ was found in the Raman spectrum of pyridine on a silver electrode, after an electrochemical redox cycle had occurred. Enhancements were later observed for a variety of adsorbates on several metallic surfaces.

Many experimental details of the SERS phenomenon have been elucidated and several attempts have been made to develop a theoretical model to explain the enhancement. One model involves the coupling of a molecular dipole with its image in the metal, which leads to an increased polarizability. Other models have involved a resonance Raman effect of some kind. Still others have invoked the roughness of the surface. At present the theoretical explanation for the SERS effect is incomplete. However, SERS is being applied to several problems involving surface adsorption and catalysis.

## Nonlinear Raman Scattering

Ordinary or spontaneous Raman scattering is a linear process that originates from induced sample polarization $P$ that is dependent on the first power of the electric field strength of the incident beam. Nonlinear Raman effects can be observed at high laser irradiances where the electric field strength exceeds about $10^9$ V m$^{-1}$. These effects depend on the polarization induced by second- and higher-order terms in the electric field strength. Nonlinear Raman effects are usually treated theoretically by introducing the phenomenological nonlinear susceptibilities as was done in Section 4-3 when treating nonlinear optics. When higher-order terms are included, the polarization can be described by equation 4-20. Although nonlinear effects were observed in the early days of laser-Raman spectrometry, the introduction of the dye laser was necessary to make their meas-

urement practical. The nonlinear Raman techniques described below have begun to open up many new chemical applications including several in analytical chemistry.

Energy-level diagrams illustrating the nonlinear Raman processes are shown in Figure 16-15. The ordinary or spontaneous Raman process was illustrated in Figure 16-5. The techniques considered here are stimulated Raman scattering (SRS), the hyper-Raman effect, inverse Raman scattering (IRS), stimulated Raman gain (SRG), coherent anti-Stokes Raman spectroscopy (CARS), and coherent Stokes Raman spectroscopy (CSRS, pronounced *scissors*).

*Stimulated Raman Scattering.* Stimulated Raman scattering is a nonlinear technique in which the spontaneous Raman emission is amplified by factors as large as $10^{13}$. The energy-level diagram for SRS is identical to that for ordinary Raman scattering, as shown in Figure 16-15a. In SRS, however, the laser irradiance is such that a high population is created in the virtual state. Interaction with the laser field causes stimulated Raman scattering. The scattered beam is coherent and coincident with the direction of the incident beam.

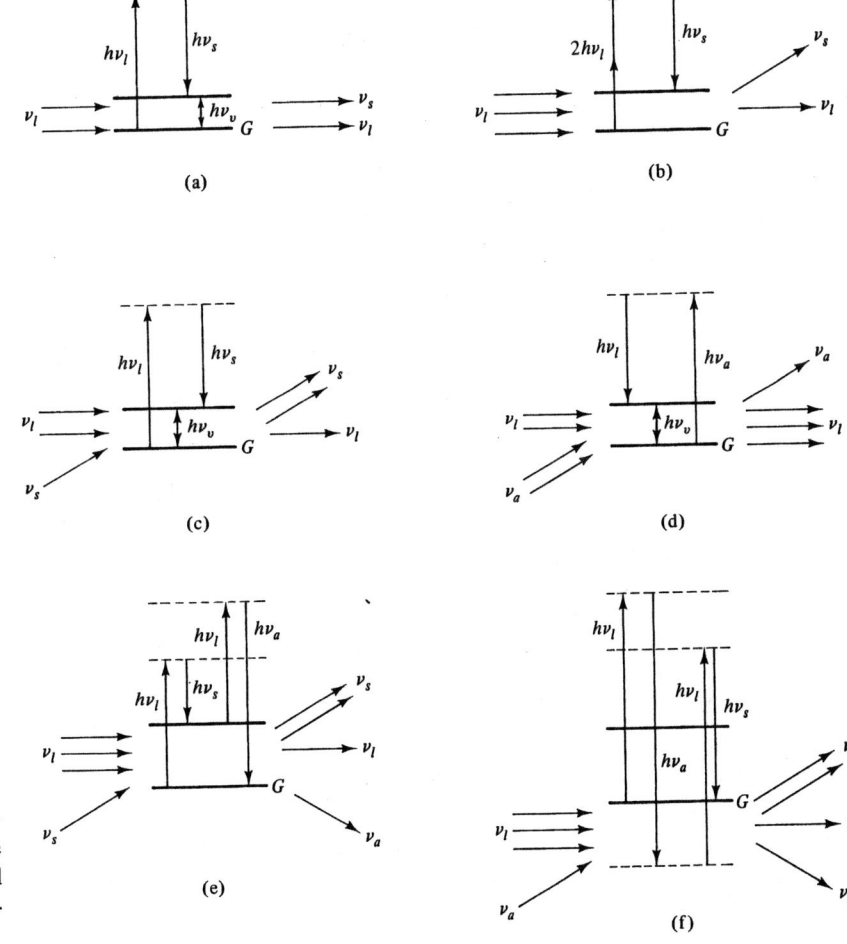

**FIGURE 16-15** Nonlinear Raman processes. The ground vibrational level in each case is labeled $G$. Dashed lines indicate virtual levels while solid lines indicate real vibrational levels. The arrows on the left of each diagram are incident photons, while those on the right are photons leaving the sample. Note that the number of photons in each process is conserved. $\nu_l$ = laser frequency, $\nu_s$ = Stokes frequency, $\nu_a$ = anti-Stokes frequency, $\nu_v$ = vibrational frequency. (a) stimulated Raman scattering (SRS); (b) the hyper-Raman effect; (c) stimulated Raman gain; (d) inverse Raman spectroscopy; (e) CARS; (f) CSRS.

Stimulated Raman scattering has not found much analytical utility because only certain lines emit in the stimulated mode. When the laser power exceeds the threshold needed for stimulated emission of one mode, raising the laser power further results only in increasing the intensity of that mode. **Raman shifters** use the SRS technique for increasing the tuning ranges of some lasers. In such a Raman shifter, laser radiation is focused into a cell containing hydrogen or deuterium gas. By the SRS process an intense beam is created that is shifted from the laser frequency by the vibrational frequency of the $H_2$ or $D_2$ molecules.

*The Hyper-Raman Effect.* The hyper-Raman effect is illustrated in Figure 16-15b. It can be considered as Raman scattering resulting from two incident laser photons. The hyper-Raman effect is a second-order process arising from the quadratic electric field term. The effect is extremely weak and experimentally difficult to measure. Unlike normal Raman, which depends on polarizability, the hyper-Raman effect depends on the hyperpolarizability. The symmetry factors controlling the latter are different from those controlling polarizability; hence selection rules are not the same as for linear Raman scattering. In particular, some vibrations which are both infrared and Raman inactive are hyper-Raman active, which enables them to be studied spectroscopically.

*Stimulated Raman Gain and Inverse Raman Spectroscopy.* The stimulated Raman gain technique illustrated in Figure 16-15c is closely related to inverse Raman spectroscopy shown in Figure 16-15d. In fact, the latter would be more appropriately called stimulated Raman loss spectroscopy. The SRG process may be viewed as stimulated emission at the Stokes frequency ($v_s$). Two laser beams are directed into the sample: a pump laser of frequency $v_l$ and a tunable probe laser. The intensity of either beam is monitored as the probe laser is scanned. When the probe laser is scanned through the Stokes frequency $v_s$, the molecules vibrate at the frequency $v_v$, where $v_v = v_l - v_s$. As a result, there is a net loss of photons in the pump beam and a net gain in the probe beam. We can describe the SRG process in terms of a "reaction":

$$v_l = v_s + v_v$$

Here the net result is that a photon of frequency $v_l$ is annihilated, while a photon of frequency $v_s$ is created.

In IRS, the probe beam is scanned through the anti-Stokes frequency $v_a$, while the pump beam is set at $v_l$ (see Figure 16-15d). Now the pump beam experiences a gain, while the probe beam undergoes a loss

in intensity. This stimulated Raman loss occurs at the anti-Stokes frequency. The process corresponds to

$$v_a = v_l + v_v$$

Here a photon of frequency $v_a$ is converted into a photon of frequency $v_l$.

Both SRG and IRS are coherent processes which result in the Raman output being coherent and highly directional. The coherence and directionality allow the Raman output beam to be readily separated from fluorescence emission, which is incoherent. Because of the directionality, for example, the solid angle of collection in these techniques can be as much as five orders of magnitude smaller than that used in ordinary Raman methods. Also, large signals are produced in these techniques. Typically, with a 10-kW pulsed laser the stimulated Raman signal can be three or four orders of magnitude larger than ordinary Raman signals. In addition, IRS has a spectral advantage in discriminating against fluorescence in that the signal beam (probe beam) is at a higher frequency than the incident pump beam.

The "reactions" given above show that SRG and IRS should be linearly dependent on the intensity at $v_l$ or $v_a$ since only one photon at these frequencies is involved. Since only a single step is involved in both processes, a linear dependence on concentration should result.

*Coherent Anti-Stokes and Coherent Stokes Raman Spectroscopy.* The CARS technique is the most widely used nonlinear Raman method. As illustrated in Figure 16-15e, the CARS technique uses two laser sources, a pump beam at $v_l$ and a Stokes shifted probe beam at $v_s$. If these two beams are focused into the sample at sufficient power, a third beam of frequency $v_3 = 2v_l - v_s$ is generated. The probe beam, normally from a tunable laser, is scanned to generate the third frequency continuously. The power generated at this third frequency is normally very weak until the pump and probe frequencies differ by the frequency of a vibrational transition, $v_v = v_l - v_s$. When this resonance condition is met, $v_3 = 2v_l - v_l + v_v = v_l + v_v = v_a$.

Let us consider what happens when the frequency difference $v_l - v_s$ matches the vibrational frequency $v_v$. In the first step we can consider that a photon at $v_l$ scatters as a photon at $v_s$ and a vibration at $v_v$. Note that this process is stimulated since the lasers have frequencies $v_l$ and $v_s$. The vibrational excitation at $v_v$ can now mix with the photons at $v_l$ or those at $v_s$ to form new photons. We will consider here only the mixing with photons of frequency $v_l$ (CARS output). This mixing produces photons having sum ($v_l + v_v = v_a$) and difference ($v_l - v_v = v_s$) frequencies. As before, we can write the steps involved as "reactions":

$$\nu_l = \nu_s + \nu_v \qquad \text{incident pump photon}$$

$$\nu_s = \nu_l - \nu_v \qquad \text{incident probe photon}$$

$$\nu_l + \nu_v = \nu_a \qquad \text{sum frequency}$$

$$\nu_l - \nu_v = \nu_s \qquad \text{difference frequency}$$

Note that the overall reaction is $2\nu_l = \nu_s + \nu_a$ and the new CARS photon appears at $\nu_a = 2\nu_l - \nu_s$.

We are now prepared to understand some of the features of CARS. Since two photons at $\nu_l$ and one photon at $\nu_s$ are required to produce the CARS photon at $\nu_a$, the CARS signal should vary with the square of the pump laser power and linearly with the probe laser power. The CARS signal should also vary with the square of the sample concentration. This is because the number of vibrations created in the first step will depend linearly on concentration. Similarly, the number of vibrations mixing with the pump laser will depend linearly on concentration, giving the overall process a square dependence. Even in the absence of the resonance condition, photons are generated at $\nu_a$. Thus there is a continuous background level associated with CARS; the signal is greatly enhanced when resonance occurs.

The CSRS technique involves a process similar to CARS, as can be seen in Figure 16-15f. Here Stokes radiation is generated by the mixing of a pump frequency with an anti-Stokes-shifted probe frequency. The CSRS output occurs at $\nu_s = 2\nu_l - \nu_a$. CARS is by far the most popular of the two techniques primarily, because the frequency generated is in the anti-Stokes region, where it can be spectrally distinguished from fluorescence.

Like IRS, CARS is a coherent process. The coherence, directionality, and spectral nature of CARS allow the CARS output to be readily separated from incoherent background such as fluorescence emission. The major limitation to CARS is the nonresonant background mentioned above. In conventional CARS this background has limited the limits of detection to the 1% level in solution. Resonance enhancement has been used to improve the detection limits (DLs) in CARS. When the pump frequency approaches the frequency of an allowed electronic transition, a greatly enhanced CARS output is obtained. With resonance enhancement, DLs of $10^{-7}$ M have been reported.

At present CARS is still a new and relatively expensive technique. It has been shown to be useful in combustion diagnostics, in studies of strongly fluorescent biological compounds, and in HPLC detection. As less expensive and more routinely useful lasers become available, the nonlinear Raman techniques, particularly CARS, are certain to become more widely used.

## 16-3 TURBIDIMETRY AND NEPHELOMETRY

**Turbidimetry** and **nephelometry** are methods that measure the concentrations of particulate matter in a suspension. Both are based on the elastic scattering of radiation. Turbidimetric methods measure the decrease that occurs in the transmitted radiation as a result of particle scattering, while nephelometric methods measure the radiant power of the scattered radiation itself.

Turbidimetric methods are preferred when the concentration of suspended particles is high. Here it is simple to detect changes in the transmitted radiant power, whereas when the amount of scattering is high, nephelometry suffers from interferences and nonlinearities. On the other hand, when the concentration of scatterers is low and the radiant power of scattering is small, nephelometric methods are preferred.

The scattering of radiation by suspended particles is often called the **Tyndall effect**. In the UV and visible regions the scattering particles are usually of colloidal size, from 1 nm to 1 μm in diameter. The scattering can be Rayleigh, Debye, or Mie, depending on the size of the particles.

We consider here briefly the instrumentation, theory, and a few applications of turbidimetry and nephelometry.

### Instrumentation and Theory

The instrumentation used in turbidimetry and nephelometry is normally quite simple, as illustrated in Figure 16-16. In both methods an incident beam passes through the sample cell and strikes a radiation transducer. An interference filter is often used, either before or after the sample cell, so that the detected beam is in the blue region of the spectrum, where scattering intensity is high. In turbidimetry, the transducer is placed at an angle of 0° with respect to the incident beam, whereas in nephelometry a 90° geometry is used.

Turbidimetric measurements are often made with simple filter photometers or modifications of these. Nephelometric measurements, on the other hand, can be made with fluorescence instruments. In some cases more elegant light-scattering photometers, such as those described in Section 16-4, are employed. These can detect the scattered radiation as a function of angle. The sample cells are often rectangular absorption (turbidimetry) or fluorescence (nephelometry) cells.

When an incident beam of radiation is scattered by suspended particles, the radiant power appearing at

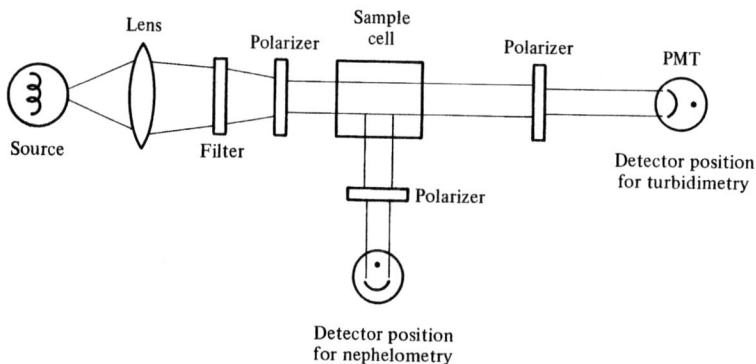

**FIGURE 16-16** Instrumentation for turbidimetry and nephelometry. In turbidimetry the detector is placed in line with the source, while in nephelometry a 90° configuration is employed. In some instruments, polarizers can be inserted in the incident, and/or scattered beams for increased selectivity. This results because scattering changes the plane of polarization of polarized incident radiation. In turbidimetry, for example, crossed polarizers in the incident and transmitted beams can discriminate against absorption by the particles or solution. In nephelometry, unpolarized incident radiation should produce linearly polarized radiation at 90° if the particles are isotropic.

any angle is a function of the concentration of particles, their size and shape, their refractive indices relative to the medium, and the wavelength of the incident radiation. Although the comprehensive Mie theory of radiation scattering is applicable to turbidimetry and nephelometry, it is seldom applied. Instead, these techniques are usually based on empirical relationships between the scattered or transmitted radiant power and the concentrations of the particles.

Let us consider a parallel, monochromatic, incident beam of radiant power $\Phi_0$ passing through a solution containing suspended particles. The attenuation by scattering in a dilute suspension follows the relationship

$$\Phi = \Phi_0 e^{-\tau b} \qquad (16\text{-}14)$$

where $\Phi$ is the transmitted radiant power, $b$ is the pathlength of the medium and $\tau$ is called the **turbidity coefficient** or simply the **turbidity**, with the units of cm$^{-1}$. For small values of $\tau$, a linear relationship is usually obtained between the turbidity and the concentration of suspended particles. Thus a relationship that closely resembles Beer's law results:

$$-\log\frac{\Phi}{\Phi_0} = kbc \qquad (16\text{-}15)$$

Here the proportionality constant $k$ is given by $k = 0.434\tau/c$.

In turbidimetric measurements, equation 16-15 is applied just as Beer's law is applied in spectrophotometry. Standards of known concentration are used to prepare a calibration curve of $-\log(\Phi_s/\Phi_r)$ vs. $c$, where $\Phi_s$ is the radiant power transmitted with the sample (standard) in the cell and $\Phi_r$ is that with the reference (usually the solvent) in the cell. The resulting working curve is then used to obtain the concentrations of unknowns. There is, of course, a mutual interference between ordinary absorption and turbidimetry. Scattering from turbid solutions causes an interference in absorption measurements, while absorbing species interfere in turbidimetric measurements.

In nephelometry, the scattered radiant power $\Phi_{sc}$ is empirically related to the concentration of suspended particles by the equation

$$\Phi_{sc} = \Phi_0 k_{sc} c \qquad (16\text{-}16)$$

where $k_{sc}$ is an experimentally obtained constant and $\Phi_0$ is the incident radiant power. A working curve of $\Phi_{sc}$ vs. $c$ is prepared under carefully controlled conditions. Usually, the value of $\Phi_{sc}$ is obtained by subtracting the scattering signal of an appropriate blank solution from the scattering signal of the sample. Since measurements are made on fluorescence instruments, fluorescence should be absent in nephelometric procedures unless some discrimination method is used. Scattering has been detected in the presence of fluorescence by the same types of time discrimination methods used to distinguish Raman scattering from fluorescence.

In both turbidimetry and nephelometry, the size and shape of the particles formed have a large effect on the radiant power of scattering. In most analytical applications suspensions are produced by the addition of various reagents. Hence factors such as pH, temperature, concentration of reagents, order of mixing, ionic strength, and time before measurement must be carefully reproduced to obtain precise and accurate results.

## Applications

Turbidimetric or nephelometric measurements are often used to estimate the amount of suspended matter in liquids. The determination of the clarity of water is one example. These methods are also applied to determine the effectiveness of water treatment processes.

The concentrations of various species in solution can also be determined by turbidimetry or nephelometry. Here precipitating agents are added to the sample to form the suspended particles. Often a surface-active agent, such as gelatin, is added to prevent coagulation beyond a certain size.

A few of the species that have been determined by turbidimetry or nephelometry are given in Table 16-3. The most widely used method is that employed for the determination of sulfate. With nephelometry, sulfate concentrations as low as a few $\mu g\ mL^{-1}$ have been obtained with relative standard deviations on the order of 5%.

Turbidimetric methods have also been applied to determine endpoints in precipitation titrations. With titrations, the instrument employed can be as simple as a white light source and a photocell placed on opposite sides of the titration vessel with an appropriate readout device. Turbidimetric detection has been employed for titrations of fluoride with calcium, of bromide with silver, and of sulfate with barium.

## 16-4 LASER SCATTERING METHODS

Radiation scattering methods have been widely used for many years in the determination of molecular weights, particles sizes, and particles shapes. Several commercial

instruments, called light-scattering photometers, are available for such measurements. Many of these are similar in design to the photometer shown in Figure 16-17.

In recent years laser sources have begun to replace conventional sources in such photometers. There are several reasons why CW lasers are nearly ideal for radiation scattering measurements. First, the spatial coherence, directionality, and low divergence of the laser beam reduce some of the optical constraints placed on instruments with conventional sources. This permits scattering to be measured at very low angles. Second, the laser allows a high output power to be transmitted in a beam of low cross-sectional area; this permits measurements to be made on small sample volumes. The outputs of many laser beams are polarized, which is used to advantage in many radiation scattering measurements.

In this section we first discuss the principles of molecular weight determinations by scattering methods. Then, low-angle laser light-scattering techniques are described. These methods are not only applied to determine the molecular weights of isolated materials, but they are also used as molecular weight–sensitive detectors for size-exclusion chromatography. Information regarding molecular geometry is also provided. Laser sources are also used in Brillouin scattering measurements and in quasi-elastic (dynamic) light-scattering methods. The information obtainable from these techniques is briefly discussed in this section.

## Molecular Weight Determinations

To see how radiation scattering can be applied to molecular weight determinations, it is useful to consider again the equation that relates the intensity of Rayleigh scattered radiation to polarizability and scattering angle (equation 16-1). Let us consider the sample to be a collection of isotropic polarizable particles (e.g., molecules). The polarizability $\alpha'$ is directly proportional to

**TABLE 16-3**
Turbidimetric and nephelometric methods

| Species | Particles formed | Reagent | Method[a] |
|---------|------------------|---------|-----------|
| Ag | AgCl | NaCl | T, N |
| As | As | $KH_2PO_2$ | T |
| Au | Au | $SnCl_2$ | T |
| Ca | $CaC_2O_4$ | $H_2C_2O_4$ | T, N |
| $Cl^-$ | AgCl | $AgNO_3$ | T |
| K | $K_2NaCo(NO_2)_6$ | $Na_3Co(NO_2)_6$ | T, N |
| $SO_4^{2-}$ | $BaSO_4$ | $BaCl_2$ | T, N |
| Se | Se | $SnCl_2$ | T |
| Te | Te | $NaH_2PO_2$ | T |

[a] T, Turbidimetry; N, nephelometry.

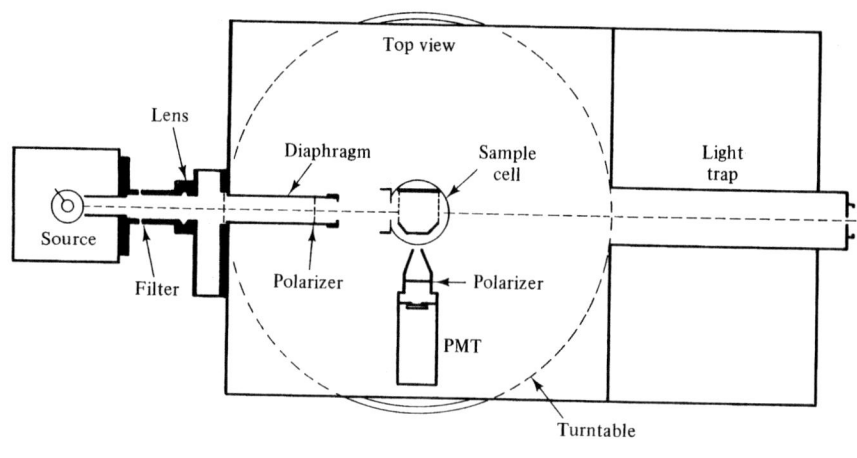

Top view

Lens

Diaphragm

Sample cell

Light trap

Source

Filter

Polarizer

Polarizer

PMT

Turntable

(a)

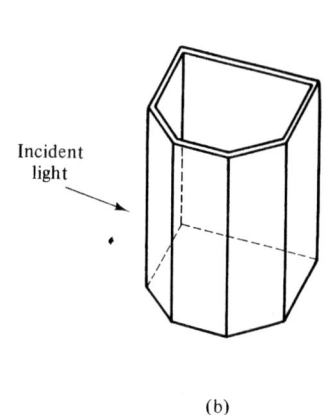

Incident light

(b)

**FIGURE 16-17** Light-scattering photometer. In (a), collimated radiation from a mercury source and interference filter is polarized and passes through the sample cell. The scattered radiation passes through a second polarizer and is detected by a photomultiplier tube (PMT). The PMT is mounted on a turntable, which can be positioned to receive scattered radiation at any angle from 0° to 180°. The semioctagonal cell (b) has faces that allow measurements at 45°, 90°, and 135° to the incident beam. Often the walls not used by the incident or detected beams are blackened to absorb radiation. The polarizers allow the use of polarized light or the measurement of depolarization ratios, as discussed in the text.

molecular weight $[\alpha' \approx M\eta_0(d\eta/dc)/(2\pi N)]$, where $M$ is the molecular weight, $N$ is Avogadro's number, $\eta_0$ is the refractive index of the pure solvent, and $d\eta/dc$ is the refractive index change of the solution with concentration. Hence the intensity of Rayleigh scattering from a single particle is proportional to the square of the molecular weight.

Experimentally, one does not determine the scattering from a single particle, of course, but rather the scattering per unit volume. For a sample of concentration $c$ (g cm$^{-3}$) with $Nc/M$ particles per unit volume, the scattered radiation intensity per unit volume is obtained from equation 16-1 by substituting the value of $\alpha'$ given in the previous paragraph and multiplying by the number of particles per unit volume, $Nc/M$. For unpolarized incident radiation, the scattered irradiance per unit volume, $(E_{sc})'_\theta$ can be written

$$(E_{sc})'_\theta = \frac{2\pi^2\eta_0^2(d\eta/dc)^2\,(1\,+\,\cos^2\theta)E_0cM}{N\lambda^4d^2}$$

(16-17)

Note again that $\theta$ is taken as 0° for forward scattering and 180° for scattering back toward the source.

The **Rayleigh ratio** $R_\theta$ is defined by the equation

$$R_\theta = \frac{(E_{sc})'_\theta d^2}{(1\,+\,\cos^2\theta)E_0}$$

$$= \frac{2\pi^2\eta_0^2(d\eta/dc)^2cM}{N\lambda^4}$$

(16-18)

This is often written simply as

$$R_\theta = KcM$$

(16-19)

where $K = 2\pi^2\eta_0^2(d\eta/dc)^2/(N\lambda^4)$ with the units cm$^2$ g$^{-2}$ mol. Actually, it is something of a misnomer to call $R_\theta$ a ratio since it has units of cm$^{-1}$, but this is standard practice.

In theory, then, measurements of the Rayleigh ratio should allow determination of the molecular weight $M$. In practice, however, the equations above hold only for pure Rayleigh scattering (i.e., small particles) and for ideal solutions. For nonideal solutions, local fluctuations in concentration must be taken into account. This is usually accomplished by a virial expansion involving virial coefficients $A_2$ and $A_3$. In dilute solutions only the second virial coefficient $A_2$ is used. Adding this extra concentration dependence and rearranging equation 16-19 yields

$$\frac{Kc}{R_\theta} = \frac{1}{M} + 2A_2c \qquad (16\text{-}20)$$

A plot of the left-hand side of equation 16-20 versus concentration allows extrapolation to $c = 0$ and determination of the molecular weight $M$.

For large particles, an additional correction for destructive interference must be made. The upper limit of the relative particle size $(d_s/\lambda)$ for pure Rayleigh scattering is $\approx \frac{1}{20}$. For irradiation by the green mercury line (546 nm), this upper size limit is on the order of 200 Å. Debye scattering treats the case where the major dimension of the particle is large, and destructive interference occurs. In Debye scattering, we define a particle scattering factor $P(\theta)$, which is a function of both the observation angle $\theta$ and the relative particle size $(d_s/\lambda)$. The scattering factor is obtained by measuring the Rayleigh ratio at an arbitrary angle $\theta$ and dividing by the ratio $R_0$ obtained for forward scattering (0°) where no interference should occur. In practice, $R_0$ is obtained by extrapolation to 0° or by measurements at very low angles. The scattering factor is given by

$$P(\theta) = \frac{R_\theta}{R_0} = \frac{\text{scattered irradiance (large particle)}}{\text{scattered irradiance (no interference)}}$$

$$(16\text{-}21)$$

The scattering factor serves as an additional correction in equation 16-20, which becomes

$$\frac{Kc}{R_\theta} = \frac{1}{MP(\theta)} + 2A_2c \qquad (16\text{-}22)$$

Extrapolation of the left-hand side of equation 16-22 to zero concentration and zero scattering angle gives $M^{-1}$.

A general procedure, known as a Zimm plot, is to plot the left-hand side of equation 16-22 against $\sin^2[(\theta/2) + kc]$, where $k$ is an arbitrary constant chosen to space the data regularly. An example of such a plot is shown in Figure 16-18.

The Zimm plot can provide a great deal of information about macromolecules. First, the molecular weight can be obtained as we have seen. Second, the initial slope of the line resulting from the extrapolation to zero scattering angle can give the second virial coefficient. The initial slope of the line resulting from the extrapolation to zero concentration can give a parameter known as the radius of gyration, which is related to the size and shape of the molecule. Finally, the Zimm plot data can give the dependence of the scattering

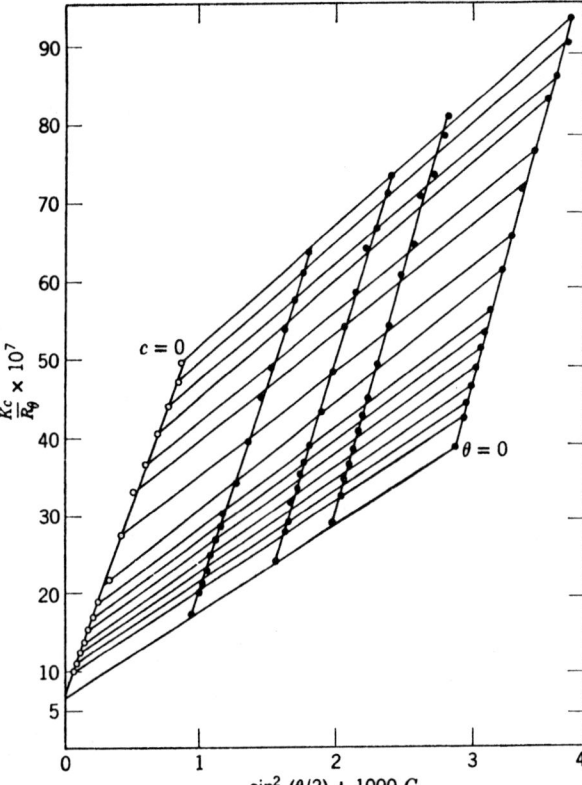

**FIGURE 16-18** Zimm plot for a solution of cellulose nitrate in acetone. The data points on each line with a nearly vertical slope represent values obtained at one concentration at different angles. Note that the data fall on a grid of parallel lines since the concentration and angular dependencies are independent of each other. The values obtained on extrapolating to zero concentration are at the left-hand edge of the grid; the values obtained on extrapolating to zero angle are on the lower edge of the plot. (Reprinted with permission from G. Oster, "Light Scattering," in *Techniques of Chemistry*, vol. 1, pt. IIIA, A. Weissberger and B. W. Rossiter, eds., Wiley-Interscience, New York, 1972. Copyright © John Wiley & Sons, Inc., 1972.)

factor $P(\theta)$ on concentration and angle, which yields molecular shape information.

Thus far we have only considered isotropic scattering particles. For such particles the radiation scattered at 90° is linearly polarized perpendicular to the propagation axis. When anisotropic particles are present, the radiation scattered at 90° is not completely linearly polarized; a weak parallel component is also present. The ratio of parallel to perpendicular components is the depolarization ratio which will be greater than zero for anisotropic particles. Hence measurements of the polarization of the scattered radiation can give information on the anisotropy of molecules. For this reason most light-scattering photometers (see Figure 16-17) have provision for measuring depolarization ratios.

Although conventional light-scattering measurements can provide much information, they are time consuming and tedious. Also, traditional instruments, such as that shown in Figure 16-17, cannot obtain scattering intensities at very low angles (e.g., less than about 10°) because of background scattering from dust and the need to avoid observing the transmitted beam. Serious errors can result from extrapolations to zero scattering angle if data are obtained only at fairly large angles.

### Low-Angle Scattering

Soon after the laser was developed, it was widely recognized to be a useful source in light-scattering investigations. However, in early instruments the conventional mercury source was replaced with a He-Ne laser and the unique characteristics of the laser radiation were not employed. In the mid-1970s laser light-scattering photometers were developed with optical systems designed around the laser.

The first low-angle laser light-scattering (LALLS) photometer, developed by Chromatix, Inc., was designed to take advantage of the excellent spatial coherence of laser radiation and to permit scattering intensities to be measured at angles below 2°. As shown in Figure 16-19, a low output power (3 mW) He-Ne laser is used as the source. The small scattering volume reduces the probability of a foreign particle such as a dust particle being present within the irradiated region and is advantageous when the amount of available material is limited. Irradiation with the He-Ne laser (633

nm) reduces the effect of sample absorption and fluorescence. Although scattering intensities vary with the inverse fourth power of wavelength, the high intensity of the laser leads to excellent signal-to-noise ratios.

A major advantage of the LALLS method is that the scattering angles are so near 0° that extrapolation to zero angle is often not required. Hence equation 16-22 reduces to equation 16-20. All that is required to calculate molecular weights is to obtain the Rayleigh ratio as a function of concentration and to extrapolate these results to infinite dilution. In practice the Rayleigh ratio is obtained from the equation

$$R_\theta = \frac{S_\theta}{S_0} \frac{T}{\Omega b} \qquad (16\text{-}23)$$

where $S_\theta$ is the photomultiplier readout signal for the scattered radiation, $S_0$ is the PMT signal for the incident radiation, $T$ is the transmittance of the incident attenuator, $\Omega$ is the solid angle through which the scattered radiation is collected, and $b$ the length of the scattering volume parallel to the incident beam. Note that $\Omega$ and $b$ are geometrical constants that can be independently obtained. Also, $T$ can be measured optically. Hence absolute values of $R_\theta$ are obtained without standards.

Although the LALLS method greatly simplifies the determination of molecular weights, size and shape information are lost. These are only obtainable through measurements of scattering versus angle and concentration.

The LALLS method has also been applied to size-exclusion chromatography. A commercial LALLS detector is available from Chromatix, Inc. The detector is usually used in series with a conventional, concentration-sensitive detector, such as a differential refrac-

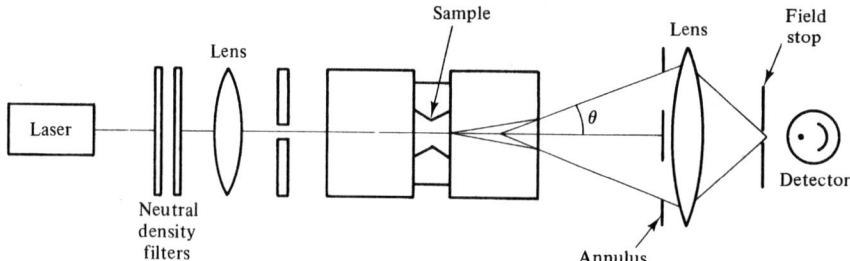

**FIGURE 16-19**  Schematic diagram of a commercial low-angle laser light-scattering photometer. Radiation from a low power (3 mW) He-Ne laser is focused to a small diameter (e.g., 100 μm) to achieve a small scattering volume (as low as $2 \times 10^{-5}$ cm³). Sample volumes as small as 10 μL can be studied. The annulus shown defines the scattering angle that is measured. After the scattering signal is obtained, the annulus is removed, and the incident beam is permitted to pass through the sample to the PMT. Because the incident irradiance is on the order of $10^9$ times the irradiance of the scattered beam, a series of neutral density filters is inserted during the measurement of the signal $S_0$ (see equation 16-23).

tive index detector. The concentration detector response is used to evaluate the concentration dependence of the LALLS signal. A typical application of this system is to determine the molecular weight distribution of a polymer sample.

### Quasi-Elastic Light Scattering

The determination of particle sizes has become increasingly important in recent years. One of the promising new techniques for particle-size analysis in solution is **quasi-elastic light scattering** (QLS), also known as *dynamic light scattering* and *photon correlation spectrometry*. The QLS technique can obtain particle-size information in a few minutes with particles ranging from less than 50 Å to about 2 μm in diameter.

The QLS technique involves the measurement of the Doppler broadening of the Rayleigh-scattered light as a result of Brownian motion (translational diffusion) of the particle. The name *quasi-elastic* distinguishes the technique from the elastic light-scattering techniques discussed previously and from inelastic light-scattering methods, such as Raman spectrometry.

In quasi-elastic light scattering the thermal motion of the particles gives rise to time fluctuations in the scattering intensity and a broadening of the Rayleigh line. The Rayleigh line has a Lorentzian shape. In macromolecular solutions, concentration fluctuations are usually dominant. Under these conditions the width of the central line is directly proportional to the translational diffusion coefficient $D_T$. Conventional spectrometers and even interferometers cannot be used to measure the small line widths (1 Hz to 1 MHz).

To measure the Doppler widths, the QLS photometer uses optical mixing or light-beating techniques. These techniques translate the optical frequencies involved ($\approx 5 \times 10^{14}$ Hz with a He-Ne laser) to frequencies near 0 Hz that can be readily measured. In most QLS photometers, a photomultiplier tube (PMT) is used as a nonlinear mixer since its output is proportional to the square of the electric field falling on its photosensitive surface.

Let us consider the QLS photometer illustrated in Figure 16-20. The sample is irradiated with laser radiation. The scattered radiation at some angle θ is incident on a PMT. The PMT output can be processed by photon counting techniques as shown or as a photocurrent. Photon counting is preferred because of its digital processing compatibility. Particle-size information is obtained from an autocorrelation analysis of the processed signal as described below.

In order to see how the PMT acts as a mixer, let us assume that the scattered radiation contains two frequencies $\omega_1$ and $\omega_2$. The electric field **E** can be written as

$$\mathbf{E} = \mathbf{E}_1 \sin \omega_1 t + \mathbf{E}_2 \sin \omega_2 t \qquad (16\text{-}24)$$

The PMT output signal $S(\omega)$ is proportional to the square of the electric field and can be written

$$S(\omega) = A\{\mathbf{E}_1^2 \sin^2 \omega_1 t + \mathbf{E}_2^2 \sin^2 \omega_2 t$$
$$+ \mathbf{E}_1 \mathbf{E}_2 \left[\cos (\omega_2 - \omega_1)t - \cos (\omega_2 + \omega_1)t\right]\}$$

$$(16\text{-}25)$$

where $A$ is a proportionality constant. The PMT cannot respond directly to frequencies $\omega_1$, $\omega_2$, or the sum term since these are greater than $10^{14}$ Hz for visible laser sources. The PMT can respond, however, to the difference frequency term ($\omega_2 - \omega_1$) which can be as small

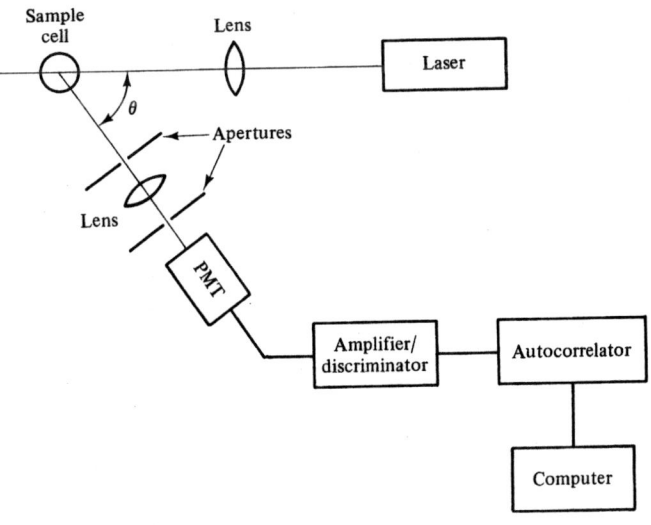

**FIGURE 16-20** Block diagram of a quasi-elastic light-scattering photometer. A laser source is incident on the sample. The scattered radiation containing the Doppler-broadening information is incident on a PMT. Photon counting signal processing is used. The autocorrelation function of the scattering signal is used to obtain the translational diffusion coefficient $D_T$, which is related to particle size.

as a few Hz. When multiple frequencies are present, a difference spectrum is generated which is centered at 0 Hz. The time dependence of the intensity fluctuations is then used to obtain the particle-size information. Optical mixing can be accomplished by beating the scattered light against a small portion of the source beam (heterodyne detection) or by beating the scattered light against itself (self-beating), as illustrated in Figure 16-20. Figure 16-21a shows the PMT output signal (proportional to the scattered radiation intensity) vs. time.

The next step is to take the autocorrelation function of the signal. With autocorrelation, the signal is multiplied by a delayed version of itself, and the product is time-averaged. The time-averaged product is ob-

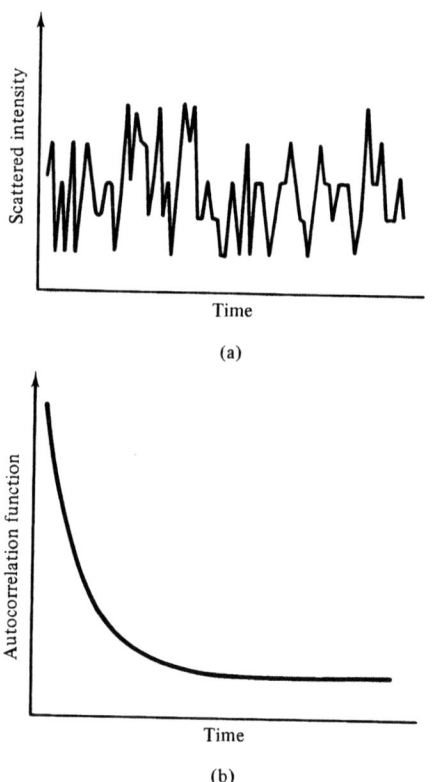

**FIGURE 16-21** Scattered-radiation intensity fluctuations from an aqueous solution of 2.02 μm diameter polystyrene spheres (a) and autocorrelation function of intensity fluctuations (b). The autocorrelation function at small delay times is usually ignored, because in that region shot noise is highly correlated. (Adapted from B. J. Berne and R. Pecora, *Dynamic Light Scattering with Applications to Chemistry, Biology and Physics*, Wiley-Interscience, New York, 1976; copyright © John Wiley & Sons, Inc., 1976, reprinted with permission.)

tained at various delay times and plotted vs. the delay time. The autocorrelation function is the Fourier transform of the power spectrum. Since the scattered radiation has a Lorentzian line shape, its Fourier transform should be an exponential decay as illustrated in Figure 16-21b. According to the theory of dynamic light scattering, the time constant of the exponential is directly related to the translational diffusion coefficient, $D_T$. The particle size is obtained from $D_T$ and particle-shape information. If the particle is spherical, the Stokes–Einstein relationship is used to calculate the particle diameter $d$:

$$d = \frac{kT}{3\pi\eta D_T} \tag{16-26}$$

where $k$ is the Boltzmann constant, $T$ is the temperature, and $\eta$ is the solvent viscosity. Similar equations are available for nonspherical particles.

The QLS method has been used to determine the sizes of polymer lattices and resins and to monitor the growth of particles during such processes as emulsification and polymerization. It is most applicable where it is not desirable to desolvate the particles. For example, micelles and microemulsions exist only in solution; these can be readily studied by the QLS technique. Quasi-elastic light scattering is also widely applicable to the investigation of biopolymers and biocolloids. It has been used to study natural and synthetic polypeptides, nucleic acids, ribosomes, vesicles, viruses, and muscle fibers.

## Brillouin Scattering

Like quasi-elastic light scattering, Brillouin scattering also results from thermal fluctuations. In this case the fluctuations of density in a liquid can be regarded as arising from the presence of sound waves associated with the thermal energy of the medium. In the early 1920s, Brillouin predicted the presence of a doublet in the frequency distribution of the scattered light due to such thermal sound waves. This scattering can be regarded as the reflection of light waves by sound waves which arise thermally. Brillouin further demonstrated that this reflection should be accompanied by frequency shifts that are related to the frequency of the sound waves effective along the direction of observation.

For compressed gases and liquids, the frequency shifts can be expressed by the equation

$$\Delta\nu = \pm\frac{(2\nu_0\nu_s\eta)}{c}\frac{\sin\theta}{2} \tag{16-27}$$

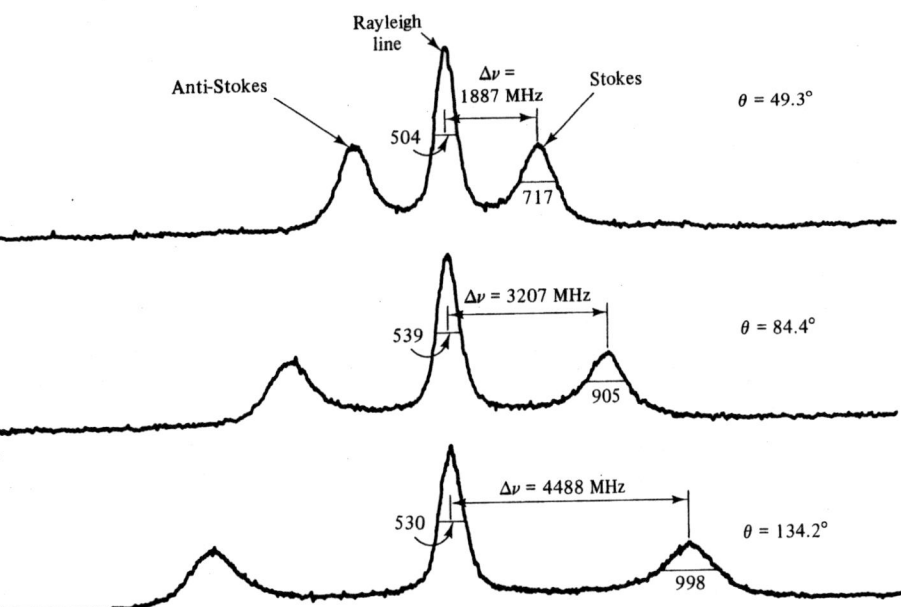

**FIGURE 16-22** Brillouin spectra of $CCl_4$ (20°C) at three different scattering angles. An He-Ne laser was used as the incident source. The shifts and widths of the components are given in MHz. [With permission of Marcel Dekker, Inc., from R. S. Krishnan, "Brillouin Scattering," in *The Raman Effect*, vol. 1, Marcel Dekker, New York, 1971; adapted from W. S. Gornall, G. I. A. Stegeman, B. P. Stoicheff, R. H. Stolen, and V. Volterra, *Phys. Rev. Lett.*, **17**, 297 (1966).]

where $\nu_0$ is the frequency of the incident radiation, $v_s$ is the velocity of sound in the medium, $c$ is the speed of light, $\eta$ is the refractive index of the medium, and $\theta$ is the scattering angle. Equation 16-27 predicts that there should be Brillouin lines appearing on either side of the unshifted Rayleigh line. The frequency shifts are generally on the order of 0.05 Å. Because of the small frequency shifts and low intensities, Brillouin scattering was not observed experimentally until the 1930s. Figure 16-22 shows Brillouin spectra of $CCl_4$ taken at various scattering angles.

The frequency shifts observed in Brillouin scattering require high-resolution spectrometers to measure. High-dispersion grating spectrometers were once used, but in recent years the scanning Fabry–Perot interferometer has become the wavelength-selection device of choice. A typical setup for observing Brillouin scattering is shown in Figure 16-23.

The frequency shifts of the Brillouin lines have been used to obtain hypersonic velocities [the frequencies of sound waves are often classified as *hypersonic* (>500 MHz), *ultrasonic* (20 kHz to 500 MHz), *audible* (20 Hz to 20 kHz), and *infrasonic* (<20 Hz)] in many liquids. These have been compared with acoustic velocities obtained with ultrasonic waves. Some liquids, such as benzene, carbon tetrachloride, glycerine, octane, and quinoline, show considerable dispersion of sound velocity with the hypersonic velocity being higher than the ultrasonic velocity.

Measurement of the widths of the Brillouin lines allows the determination of the hypersonic absorption coefficient and relaxation time. Often measurements are made as a function of temperature. As the temperature of the liquid increases, the Brillouin components become broader, and their position shifts toward the central line. Brillouin scattering is useful for studying the behavior of liquids near the critical temperature. When the temperature of the liquid approaches the critical temperature, the intensity of the central line increases rapidly, while the Brillouin components get much weaker and disappear in the background. The width of the central component is used to measure the thermal conductivity near the critical temperature. Brillouin scattering has also been noted in viscous liquids and has provided experimental evidence for the solid-like behavior of such liquids toward hypersonic waves. In addition, hypersonic velocities have been obtained

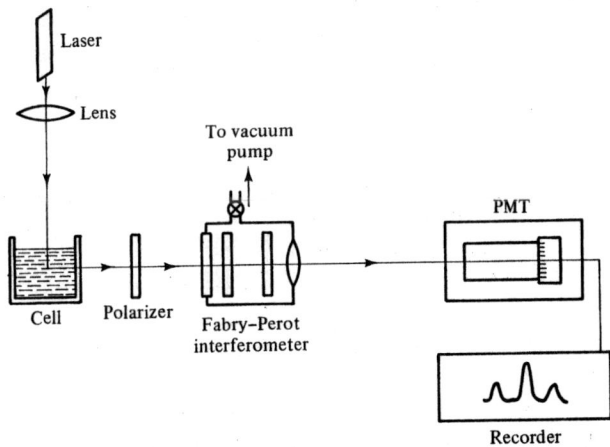

**FIGURE 16-23** Apparatus for recording Brillouin spectra. A laser source is used, and the scattering is measured at 90°. A pressure-scanned Fabry–Perot interferometer is the wavelength selection device and a PMT is the transducer.

on amorphous and crystalline solids with Brillouin scattering.

## 16-5  REMOTE SENSING WITH LASERS

Laser-based techniques are being rapidly developed for remote-sensing applications. Lasers have been used to determine the optical characteristics of the atmosphere, to probe the upper atmosphere, and to monitor atmospheric pollutants. In addition, laser sources have been mounted in aircraft and pointed downward to study the hydrosphere. Many of the laser-based methods used in such studies are based on radiation scattering. In this section we consider briefly some of the techniques that have proven most useful in remote sensing. Many of these are called **laser radar** or **lidar** methods. The principles of lidar systems are first introduced and then a few of the most important applications are discussed.

### Laser Radar

Following the introduction of lasers in the 1960s, it was soon realized that these sources allow optical measurements to be made at sites remote from the probed area. Indeed, studies of the atmosphere with laser sources date from 1963 when the first laser radar system was developed to study aerosol distributions. In the years that followed, enormous advances have been made in the development of lidar systems (*lidar* stands for *li*ght *de*tection *a*nd *r*anging).

The term "lidar" is actually a generic term that does not imply a particular detection process. For example, lidar systems can use Rayleigh scattering, Mie scattering, Raman scattering, fluorescence, and differential absorption. Typically, such systems are constructed inside astronomical domes or placed in shelters on the tops of buildings. They have also been placed in mobile vans and in aircraft. It is anticipated that the Space Shuttle or a space station will be widely used for global monitoring with lidar in the future.

A typical large-scale lidar system uses a pulsed laser source and a telescope receiver at the same location. The laser transmitter and the telescope receiver can be collinear or offset. In the offset geometry the laser beam enters the field of view of the telescope optics only beyond some preestablished distance. The offset geometry avoids detector saturation from the near-field back-scattered radiation. With collinear geometries near-field back-scatter is overcome by either gating the PMT detector or by using a fast optical shutter.

The various scattering methods and differential absorption are most widely used for ground-based remote sensing. Although fluorescence methods have low detection limits (DLs), they suffer from quenching effects with the more abundant atmospheric species. The atomic fluorescence (AF) method is best suited for monitoring trace constituents, such as sodium, in the upper atmosphere. Molecular fluorescence is also highly useful. It finds major applications for detecting oil spills and monitoring water quality from the air.

Because the cross section for Mie scattering is much larger than that for Rayleigh scattering (due to the much larger particles detected in Mie scattering), Mie scattering often dominates in the atmosphere. Thus quite low concentrations of atmospheric particles can be detected with Mie scattering. Raman scattering is also employed in atmospheric studies when molecular specificity is desired. Raman back-scattering techniques are highly useful in air pollution studies, because most of the molecules of interest are Raman active. Although the cross sections are small for Raman scattering, the resonance Raman technique can be employed to improve detection limits. Raman scattering is highly useful for remote monitoring of effluent plumes from industrial operations. Here the concentrations of the pollutants of interest can be in the tens to hundreds of $\mu g \ mL^{-1}$.

Differential absorption lidar or DIAL can achieve low limits of detection and excellent spatial resolution. In the DIAL technique, two laser beams are used, and the differential attenuation is obtained from the back-scattered signals. One laser is tuned to a molecular transition while the other is slightly detuned. The ratio of the two signals gives the differential absorption. The large Mie scattering cross section is used to advantage in this technique to provide both spatial resolution and a strong return beam. The DIAL technique is most advantageous for long-range monitoring of specific molecular constituents.

### Applications

Although the use of lasers in remote sensing is still in its infancy, major advances in our knowledge have already been made. For example, it is now common to obtain information on the particulate and aerosol content of the atmosphere using lidar. Particle sizes can readily be determined from Mie scattering. Multiwavelength lidar has been employed to obtain aerosol size distributions.

Atomic fluorescence measurements have shown that there is a "sodium layer" in the upper atmosphere. Several groups have investigated the spatial and temporal behavior of this layer during meteor showers. For example, a fourfold increase in the sodium concentration was observed during one shower. It was concluded that meteor ablation was probably the source of sodium.

Raman back-scattering measurements have been used to measure the oxygen/nitrogen balance in the atmosphere. Similarly, measurements of the $CO_2/N_2$ ratio and long-terms drifts in this and the $O_2/N_2$ ratio are being made. Spatial profiles of nitrogen in the atmosphere have been obtained with Raman back-scattering and the temperature profile calculated from the molecular concentration profile and the pressure. The DIAL technique, with the multiwavelength $CO_2$ laser, has been used to measure the vertical profiles of atmospheric ozone and atmospheric water vapor.

Differential absorption and Raman back-scattering are being increasingly used in pollution studies. For example, the DIAL technique was used to obtain the $NO_2$ distribution over a chemical factory; the Raman technique was used to measure $SO_2$ in the plume of a power plant.

Laser-induced fluorescence has been used for a variety of high-altitude environmental studies. Fluorescent dyes have been dissolved in water and used to study water movement. Oil pipelines have been monitored for leakage and oil spills have been detected and classified. In some remote studies of water quality, the Raman band of water has been used as a built-in reference for fluorescence measurements to compensate for laser fluctuations and for changes in some of the variables inherent in quantitative fluorescence measurements.

From this brief discussion it is apparent that remote sensing is a rapidly growing field. Improvements in laser and detector technology, particularly with infrared detectors, are making atmospheric monitoring studies routine. Already laser remote sensors have helped make major contributions to our understanding of the atmosphere and our immediate environment.

## PROBLEMS

**16-1.** A molecule has a Raman shift of 859 cm$^{-1}$ when excited by an argon ion laser at 514.5 nm. What are the wavelengths in nm of the Stokes and anti-Stokes Raman lines? Which of these is more intense, and why?

**16-2.** A tunable dye laser has approximately the same irradiance at 400 nm as at 500 nm. It is used in a Rayleigh scattering experiment. Compare the irradiance of Rayleigh scattering at 90° at the two laser wavelengths.

**16-3.** A typical Raman spectrum of an organic compound covers Raman shifts ranging from 0 to 3500 cm$^{-1}$. For excitation by an argon ion laser at 488.0 nm, what is the equivalent wavelength range? Recalculate for the 514.5-nm argon ion laser line. What types of photomultiplier tubes would be most suitable for use with these two sources to acquire Raman spectra (see Figure 4-23).

**16-4.** Assume that the ratio of anti-Stokes to Stokes Raman intensities is governed entirely by the Boltzmann distribution. For temperatures of 25°C and 35°C, find the ratio of intensities for the $CCl_4$ lines at 218 cm$^{-1}$ and 790 cm$^{-1}$.

**16-5.** From the Stokes Brillouin component of $CCl_4$ with $\Delta\nu = 1887$ MHz (see Figure 16-22), find the hypersonic velocity $v_s$ if $\eta = 1.46$.

**16-6.** Calculate the percentage decrease in radiant power when a 546.1-nm radiation beam passes through a 3-cm-thick layer of a polymer solution with a turbidity $\tau$ of 0.01 cm$^{-1}$. What would the percentage decrease in radiant power be for the same solution if the wavelength were changed to 365.0 nm? Assume purely Rayleigh scattering.

**16-7.** For isotropic scattering the Raleigh ratio and the turbidity are related by $R_\theta = 3\tau(1 + \cos^2\theta)/16\pi$. For the solution described in problem, 16-6, find the Rayleigh ratio at a 60° angle of observation.

**16-8.** The absorbance of a solution is measured and found to be 0.18. Upon removing the 1.00-cm cuvette from the spectrophotometer, the solution was found to be slightly turbid. The turbidity was then measured and found to be 0.014. What was the true absorbance of the solution?

**16-9.** Contrast normal IR spectrometry and Raman spectrometry with respect to applications, factors limiting the $S/N$, and instrumentation.

**16-10.** A Raman shifter with $H_2$ gas was used to produce new wavelengths from a fixed wavelength KrF laser. The output wavelength of the excimer laser is 249 nm and the fundamental vibrational frequency of $H_2$ is 4400 cm$^{-1}$. Find the wavelengths of the five Stokes and the two anti-Stokes lines produced at the output of the Raman shifter.

**16-11.** In a CARS experiment, a laser pump beam at 500 nm was incident on a cell containing $CO_2$ gas. The symmetric stretch of $CO_2$ is Raman active with a vibrational frequency of 1340 cm$^{-1}$. What should be the approximate wavelength of the probe beam? At what wavelength will the CARS output appear?

**16-12.** In a low-angle light-scattering experiment to determine the molecular weight of a polymer, values of the Rayleigh ratio were obtained at various concentrations as shown below. The value of $K$ at the 632.8 nm He-Ne laser line was $2.25 \times 10^{-8}$ cm$^2$ g$^{-2}$ mol.

| $R_\theta/(10^{-6}$ cm$^{-1})$ | 1.6 | 1.8 | 2.1 | 2.6 | 3.0 |
|---|---|---|---|---|---|
| $c/(10^{-3}$ g mL$^{-1})$ | 1.0 | 1.2 | 1.5 | 2.0 | 2.5 |

Determine the molecular weight $M$ of the polymer and the second virial coefficient $A_2$.

**16-13.** The translational diffusion coefficient for spherical particles in an aqueous solution as determined by quasi-elastic light scattering is $2.0 \times 10^{-8}$ cm$^2$ s$^{-1}$ at 25°C. If the viscosity of the water solvent at this temperature is $0.890 \times 10^{-2}$ poise, find the diameter $d$ of the particles.

# REFERENCES

The references below are general references on light scattering, including quasielastic and low-angle laser light scattering.

1. M. Kerker, *The Scattering of Light and Other Electromagnetic Radiation*, Academic Press, New York, 1969. A classic book on light scattering.

2. B. J. Berne and R. Pecora, *Dynamic Light Scattering with Applications to Chemistry, Biology and Physics*, Wiley-Interscience, New York, 1976. An outstanding book on dynamic light scattering.

3. G. Oster, "Light Scattering," in *Techniques of Chemistry*, vol. 1, pt. IIIA, A. Weissberger and B. W. Rossiter, eds., Wiley-Interscience, New York, 1972.

4. B. Chu, *Laser Light Scattering*, Academic Press, New York, 1974. Highly mathematical, but a very useful book. Contains a wealth of information.

5. N. C. Ford, "Theory and Practice of Correlation Spectroscopy," in *Measurements of Suspended Particles by Quasi-elastic Light Scattering*, B. E. Dahneke, ed., Wiley, New York, 1983.

6. M. L. McConnell, "Particle Size Determination by Quasielastic Light Scattering," *Anal. Chem.*, 53, 1007A (1981).

7. A. T. Young, "Rayleigh Scattering," *Phys. Today*, 42 (Jan. 1982).

8. W. Kaye, "Low-Angle Laser Light Scattering," *Anal. Chem.*, 45, 221A (1973).

The following are references that deal specifically with Raman spectrometry, including in many cases resonance Raman and nonlinear Raman methods.

9. D. A. Long, *Raman Spectroscopy*, McGraw-Hill, New York, 1977. A modern and useful text on Raman spectroscopy. Includes theory, instrumentation, and applications.

10. J. C. Grasselli, M. K. Snavely, and B. J. Bulkin, *Chemical Applications of Raman Spectroscopy*, Wiley, New York, 1981. A small and quite readable book on applications of Raman spectrometry to inorganic, organic, and biological systems.

11. N. B. Colthup, L. H. Daly, and S. E. Wiberley, *Introduction to Infrared and Raman Spectroscopy*, 2nd ed., Academic Press, New York, 1974. Includes survey of theory, instrumentation, and applications.

12. M. C. Tobin, *Laser Raman Spectroscopy*, Wiley-Interscience, New York, 1971.

13. M. D. Levenson, *Introduction to Nonlinear Laser Spectroscopy*, Academic Press, New York, 1982. Includes an excellent chapter on nonlinear Raman techniques.

14. G. L. Eesely, *Coherent Raman Spectroscopy*, Pergamon Press, Elmsford, N.Y., 1981. A very good discussion of nonlinear Raman methods.

15. B. S. Hudson, "Coherent Anti-Stokes Raman Scattering Spectroscopy," in *New Applications of Lasers to Chemistry*, G. M. Hieftje, ed., American Chemical Society, Washington, D.C., 1978.

16. J. J. Valentini, "Coherent Anti-Stokes Raman Spectroscopy," in *Spectrometric Techniques*, vol. IV, G. A. Vanasse, ed., Academic Press, New York, 1985.

17. E. S. Yeung, "Spectroscopy by Inverse Raman Scattering," in *New Applications of Lasers to Chemistry*, G. M. Hieftje, ed., American Chemical Society, Washington, D.C., 1978.

18. W. H. Woodruff and S. Farquharson, "Time-Resolved Resonance Raman Spectroscopy (TR$^3$) and Related Vidicon Raman Spectrography: Vibration Spectra in Nanoseconds," in *New Applications of Lasers to Chemistry*, G. M. Hieftje, ed., American Chemical Society, Washington, D.C., 1978.

19. J. C. Wright, "Nonlinear Optics," in *Lasers in Chemical Analysis*, G. M. Hieftje, J. C. Travis, and F. E. Lytle, eds., Humana Press, Clifton, N.J., 1981. Includes a good discussion of nonlinear Raman methods.

20. A. Anderson, ed., *The Raman Effect*, vol. 1., Marcel Dekker, New York, 1971. Includes a chapter on Brillouin scattering and one on the stimulated Raman effect.

21. M. D. Morris and D. J. Wallan, "Resonance Raman Spectroscopy—Current Applications and Prospects," *Anal. Chem.*, 51, 182A (1979).

22. S. A. Borman, "Nonlinear Raman Spectroscopy," *Anal. Chem.*, 54, 1021A (1982). A very good introductory review of the various nonlinear techniques.

23. B. S. Hudson, "Resonance Raman in the Far-Ultraviolet Region," *Spectroscopy*, 1(1) (1986).

24. T. Hirschfield and B. Chase, "FT-Raman Spectroscopy: Development and Justification," *Appl. Spectrosc.*, 40, 133 (1986).

25. B. Wopenka and J. D. Pasteris, "Limitations to Quantitative Analysis of Fluid Inclusions in Geological Samples by Laser Raman Microprobe Spectroscopy," *Appl. Spectrosc.*, 40, 144 (1986).

# CHAPTER 17

# Spectrochemical Techniques on the Horizon

There are a number of spectrochemical techniques that do not readily fit into the usual emission, absorption, photoluminescence, and scattering categories. Among these are photoacoustic spectrometry, thermal lensing spectrometry, laser-enhanced ionization spectrometry and several new laser-based methods. In this, the final chapter, these newer spectrochemical methods are described. Many of these have great potential in trace analysis.

We have chosen to title this chapter "Spectrochemical Techniques on the Horizon" because the methods discussed are relatively new. It can be argued that photoacoustic spectrometry is now a well-established method. However, it can also be argued that we have only scratched the surface in applying photoacoustic spectrometry to analytical problems. Because of this and also because commercial developments in the photoacoustic area are very recent, we have chosen to include photoacoustic methods in this chapter.

The remaining methods discussed here have not been applied routinely in analytical chemistry. Many are techniques that depend on laser sources. These have the potential to become standard tools of the analytical chemist in the future, but they await further experimental, theoretical, and economic developments before they can become routine tools. We begin by discussing photoacoustic and thermal lensing spectrometry which are related to each other. Then the atomic laser ionization methods are presented. In the final section several of the new laser-based methods are described.

## 17-1 PHOTOACOUSTIC SPECTROMETRY

In spectrophotometry, one measures the source radiant power transmitted by the sample. The absorbed source radiation produces excited states which deactivate by radiative processes. A signal proportional to the fraction of the absorbed radiant power that is emitted as photons is monitored in photoluminescence spectrometry. In **photoacoustic spectrometry** (PAS), the signal detected is related to the other fraction, the fraction of the absorbed radiation that is converted eventually to heat. If the excitation source is modulated, the periodic heating and cooling of the sample results in pressure waves at the source modulation frequency; these are detected with an acoustic transducer such as a microphone or piezoelectric device.

The photoacoustic (PA) effect, also termed the *optoacoustic* or *acousto-optic effect*, was first reported by Alexander Graham Bell in 1880. Until the mid-1970s, there was little interest in PAS, but in the last decade there have been rapid developments in the theory, instrumentation, and applications.

## Instrumentation

A general block diagram of a photoacoustic spectrometer is shown in Figure 17-1. In commercial instruments for the UV-visible-NIR region, source radiation over a selected wavelength band is usually provided by a 300 to 1000-W Xe arc lamp and a monochromator. The excitation radiation is modulated with a mechanical rotating chopper. Laser sources have also been employed in place of the conventional source-monochromator combination. A modulator is still required with a CW laser, whereas with pulsed lasers, no external modulator is necessary. The photoacoustic sample compartment is a sealed chamber with appropriate windows for transmission of the excitation radiation. It contains the sample and also houses the acoustic transducer and preamplifier. For gas samples, the sample fills the chamber and is in direct contact with the transducer. For condensed samples, the gas-coupling method is commonly employed. The heat generated by absorption of radiation in the sample is transferred to the surface of the sample to create pressure waves that are carried to the transducer by a coupling gas. Direct coupling techniques have also been employed, particularly for liquids. In this case, the transducer (usually piezoelectric) is inserted into the sample without the intervention of a gas medium. Typically, the PA signal is further processed with a lock-in amplifier. The phase of the PA signal lags behind the reference signal that tracks the modulated incident radiation. Thus the phase of the lock-in amplifier is adjusted to maximize the PA signal.

To obtain a PA spectrum, it is necessary to compensate or correct for the variation of the incident radiant power with wavelength and time. A sample of carbon black is often used as a reference because it absorbs essentially all the incident radiation. The reference signal derived is proportional to the incident spectral radiant power and defines the maximum PA signal obtainable. In single-beam instruments, the sample and reference spectra are measured at separate times and stored. After data acquisition, a normalized spectrum is produced by ratioing the sample PA signal at each wavelength to the corresponding reference PA signal. In double-beam instruments, the excitation beam is split into two beams which are simultaneously incident on the test and reference samples. The resulting sample and reference PA signals are separated, detected, and processed and then ratioed to provide online normalization during a wavelength scan. In some instruments, a small fraction of the excitation beam is diverted to a reference thermal detector to provide a reference signal for normalization. In all cases, the normalized PA spectrum can be related to the absorption spectrum of the sample in much the same way as the excitation spectrum is in photoluminescence spectrometry.

The design and materials of construction of the photoacoustic cell and the characteristics of the acoustic transducer are critical. The detection limit of photoacoustic measurements is determined by the ratio of the blank noise to the response obtained for a given amount of radiation absorbed by the sample. The blank noise is that observed in the absence of an absorber or with a nonabsorbing sample and includes additive noise components independent of the source and dependent on the source intensity. Source-independent noise includes electrical noise from the microphone-amplifier combi-

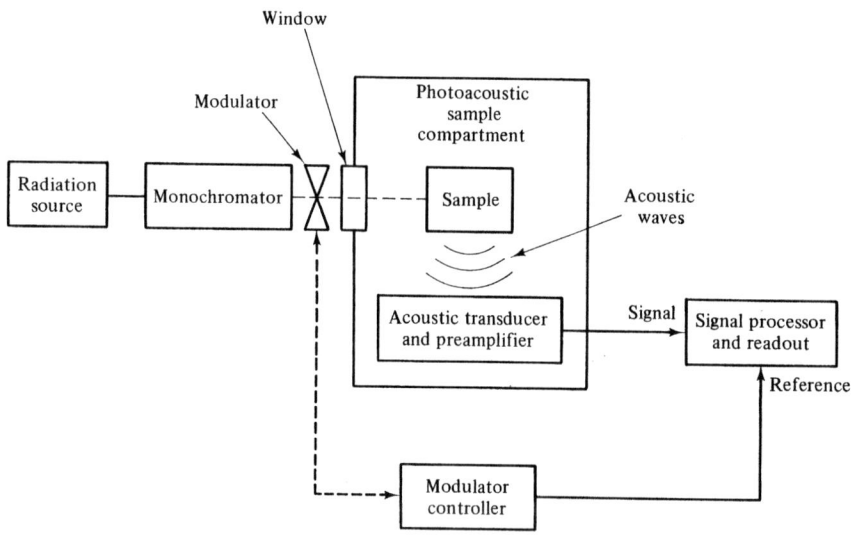

**FIGURE 17-1** Block diagram of a photoacoustic spectrometer. Radiation from the source is focused on the sample after selection of the desired wavelength band with a monochromator and modulation of the source beam. The sample is sealed in the photoacoustic sample compartment which also contains an acoustic transducer and preamplifier. The radiation absorbed by the sample is converted to heat and finally to pressure oscillations that are detected by the acoustic transducer. The signal processor (a lock-in amplifier) provides phase sensitive detection of the photoacoustic signal relative to a reference signal related to the modulated incident beam.

nation, noise due to Brownian motion of molecules, and noise generated by mechanical vibrations. The last type of noise is minimized by insulating the cell from external acoustic waves and incorporating acoustic baffles within the sample compartment. Source-dependent background signals and noise arise from the acoustic waves due to radiation absorbed by the cell windows and walls. Because the detector does not directly respond to scattered or reflected radiation, these optical signals are much less of a problem in PAS than in photoluminescence spectrometry. However, scattered or reflected radiation can eventually strike the cell walls or windows, be absorbed, and generate a PA signal.

Under many conditions, the PA signal is inversely proportional to the distance of the sample from the transducer. Thus cell volumes are commonly less than 1 cm$^3$. If the cell volume is too small, the acoustic waves produced by the sample suffer significant dissipation to the cell walls before they reach the transducer.

The PA signal is proportional to $1/f^n$, where $f$ is the modulation frequency and $n$ is typically in the range 0.5 to 1.5. In practice, modulation frequencies from 10 to 1000 Hz are used. Although very low frequencies produce the largest magnitude acoustic wave, the $S/N$ is better at moderate frequencies because of the $1/f$ character of the source-independent noise. The frequency response of most microphones is also quite poor at low frequencies.

## Principles

The theory for the generation of photoacoustical signals is quite complex and highly dependent on the nature of the sample. We will summarize the results of mathematical treatments in a semiquantitative fashion.

For gas samples and liquid samples with direct coupling, a fraction of the radiation absorbed is converted into an acoustic energy that can be detected. Thus the radiant power absorbed ($\Phi_A$) and the PA signal ($E_{PA}$) are proportional to $\Phi_0(1 - e^{-k(\lambda)b})$, where $\Phi_0$ is the incident radiant power, $k(\lambda)$ is the absorption coefficient in cm$^{-1}$, and $b$ is the sample pathlength. Usually, PAS is used with weakly absorbing gases or liquids in which case the PA signal is proportional to $\Phi_0 k(\lambda)b$. If one species is responsible for the absorption, $k(\lambda)$ and $E_{PA}$ are proportional to the analyte concentration.

For gas samples, multiple-pass or resonance cells are sometimes used to improve detectability. In the latter case, one or more cell dimensions are adjusted to be equal to one-half the wavelength of the acoustic wave to create a standing wave and amplification for that signal [$\lambda/2 = v_s/(2f)$, where $v_s$ is the velocity of

sound (344 m s$^{-1}$)]. For modulation frequencies around 50 Hz, the appropriate dimension is several meters.

A gas-piston model is often evoked to describe the generation of the PA signal from solids. Radiation is absorbed by species in the upper layers of a solid. Nonradiative deactivation produces heat. The heat diffuses to the surface and is transferred to the gas molecules near the surface. With source modulation, the gas molecules next to the surface act like a piston as they are periodically heated and cooled by the thermal waves generated in the sample. Except for very weakly absorbing samples, the contribution of vibrations due to expansion and contraction of the solid to the total PA signal is minor in contrast to gas and liquid samples.

A solid sample can be characterized by its thickness ($b$), its optical absorption length ($b_{op}$), and its thermal diffusion length ($b_{th}$). The **optical absorption length** is usually defined as $1/k(\lambda)$, where $k(\lambda)$ is the spectral absorption coefficient in cm$^{-1}$ at the wavelength of the incident radiation. As the radiation passes through the sample, its intensity decreases exponentially with penetration depth according to Beer's law. Thus the optical absorption pathlength is the penetration depth where the intensity has decreased to $1/e$ of its initial value. Most of the radiation is absorbed in the first few optical absorption lengths. The amplitude of the thermal wave generated by the radiation absorbed in a particular sublayer of the solid decays exponentially as it propagates to the surface with a thermal absorption coefficient ($a_{th}$) in cm$^{-1}$. The **thermal diffusion length** is defined as $1/a_{th}$ and thus represents the distance over which the amplitude of the thermal wave decays to $1/e$ of its initial value. Most of the detected acoustic signal arises from absorption occurring in the first few thermal diffusion lengths.

For a thermally thick ($b >> b_{th}$) and optically thick ($b >> b_{op}$) sample, the magnitude of the PA wave generated in the coupling gas, and thus the measured PA signal, is related to the optical and thermal diffusion lengths as follows:

$$E_{PA} \propto \frac{b_{th}}{b_{op}} \left[ \frac{2}{(b_{th}/b_{op} + 1)^2 + 1} \right]^{1/2} \quad (17\text{-}1)$$

The PA signal is also proportional to the incident source radiant power, the ambient pressure of the coupling gas, and the square root of the thermal diffusivity of the coupling gas.

For a sample that is more thermally thick than optically thick ($b_{th} << b_{op}$), equation 17-1 reduces to

$$E_{PA} \propto \frac{b_{th}}{b_{op}} \propto k(\lambda)b_{th} \quad (17\text{-}2)$$

Thus the attenuation of radiation in the thermal diffusion length is small and the amount of heat generated in this region and $E_{PA}$ are proportional to the optical absorption coefficient. If one species is responsible for the absorption, the PA signal is directly proportional to its concentration. This is somewhat analogous to photoluminescence measurements in which a linear relationship is realized only when the fraction of radiation absorbed over the viewed pathlength is small.

As the optical absorption coefficient of the material increases, the amount of radiation absorbed in the thermal diffusion region increases and the optical absorption pathlength becomes smaller. Eventually, the relationship between the PA signal and the absorption coefficient exhibits a negative deviation. In the limit of a very opaque material for which the optical pathlength is much smaller than the thermal diffusion length ($b_{op} \ll b_{th}$), all the radiation is absorbed within the thermal diffusion region. From equation 17-1, $E_{PA}$ becomes independent of the absorption coefficient. This condition is termed **photoacoustic saturation.**

The phase of the PA signal relative to the incident radiation also varies with the optical and thermal diffusion lengths. The range of linearity can be extended about an order of magnitude by measuring the dependence of the phase angle on the absorption coefficient. It is also possible to combine magnitude and phase information and extend the range of linearity even more.

Theories have also been developed to account for scattering in particulate solids. Thermally thin conditions ($b \ll b_{th}$) have also been investigated. This applies to situations in which a thin transparent coating is on top of a thick substrate. Here the thermal and optical properties of both layers must be considered. By adjusting the phase angle it is sometimes possible to selectively probe the absorption characteristics of only one of the layers.

The thermal diffusion length is given by

$$b_{th} = \left(\frac{D_{th}}{\pi f}\right)^{1/2} \tag{17-3}$$

where $D_{th}$ is the thermal diffusivity of the sample in $cm^2\ s^{-1}$. Thus the thickness of sample probed can be adjusted by varying the modulation frequency. Higher frequencies reduce the thermal diffusion length, which is useful for highly absorbing samples where photoacoustic saturation is a problem. The difference in PA spectra taken with two modulation frequencies can yield absorption information about an underlying layer. The relatively new technique of Fourier transform PAS can give depth profile information.

## Applications

Photoacoustic spectrometry has been used to study the absorption properties of gas, liquid, and solid samples in the UV, visible, and IR regions. With gas samples it is possible to measure absorption coefficient as low as $10^{-10}\ cm^{-1}$ in a 10-cm-pathlength cell. Most analytical applications have involved determination of trace gaseous pollutants. Detection limits in the range 0.1 to 10 ppb (v/v) have been achieved for such species as methane, $NH_3$, $NO$, $NO_2$, and $SO_2$ with laser sources.

For liquid samples, absorbances as low as $10^{-6}$ to $10^{-7}$ have been measured. This is one to two orders of magnitude smaller than achieved with conventional spectrophotometry. This ability to detect very small absorbances cannot be utilized in standard absorption measurements, as it would be difficult to prepare a blank that would compensate for absorption of all concomitants at this level. However, photoacoustic HPLC detectors have been developed that provide detection limits over an order of magnitude better than conventional HPLC absorption detectors. In this application, the high irradiance of pulsed lasers has proved useful. It is possible to discriminate the PA signal of the sample from that of cell wall and window absorption by the difference in the arrival times of the different acoustic waves at the acoustic transducer.

The versatility of PAS has been well demonstrated with solid samples. It provides a nondestructive technique that can be used to measure the absorption properties of a diverse set of samples such as semiconductors, polished oxides, polymers, and biological materials (e.g., leaves and blood smears). Sample preparation is simple since dissolution of samples is not required. This is particularly useful for materials which cannot be readily dissolved or pressed into KBr disks. The absorption spectra of adsorbed and chemisorbed species on surfaces can be measured. Moreover, the phase and modulation frequency can be varied to conduct depth profiling.

Photoacoustic spectrometry has been used to measure the absorption properties of both weakly and strongly absorbing solids. In the former case, the absorption coefficients of nearly transparent materials can be measured down to $10^{-6}\ cm^{-1}$ without having to account for the reflection and scattering properties of the sample as must be done in optical transmission measurements. The absorption spectra of highly absorbing solids can be obtained because the pathlength probed is small (e.g., 0.01 to 0.001 cm) if the thermal diffusion length is small; there is no need to prepare very thin samples.

For UV-visible measurements, PAS provides the same type of information as diffuse reflection spectrometry. In general, PAS is more useful for weakly reflecting samples. In the IR region, PAS provides an alternative to attenuated total reflectance (ATR) and Fourier transform infrared (FTIR) spectroscopy. Because of the low radiance of conventional IR sources, Fourier transform instruments have been developed in which a photoacoustic cell is used as the detector; a *S/N* advantage is realized as with conventional thermal detectors. Recently, there has been interest in a new coherent Raman technique called **photoacoustic Raman spectroscopy** (PARS). In PARS, a given energy state is selectively populated by the stimulated Raman effect (see Section 16-2) induced by two incident laser beams. The frequency difference of the two beams is adjusted to equal the frequency of a Raman-active transition. Relaxation of the excited molecules generates an acoustic wave which is detected by a microphone. At present PARS has only been applied to obtaining vibration-rotation spectra of gases and gaseous mixtures. The PARS signal has been shown to vary linearly with gas concentrations, and detection limits for gases have been estimated to be in the ppm range.

The use of PAS for qualitative analysis of solids in now established. Its use as a quantitative tool requires further assessment. Accuracy is limited by absorption of radiation by concomitants in the sample and differences among the bulk properties of different types of samples. Even if the analyte is the only absorbing species at the analysis wavelength, the PA signal for a given amount of analyte is affected by the matrix-dependent thermal diffusivity, thermal diffusion pathlength, and scattering properties.

## 17-2 THERMAL LENSING SPECTROMETRY

The **thermal lens effect,** also known as **thermal blooming,** was accidentally discovered in the mid-1960s when a Raman scattering cell filled with an organic liquid was placed inside the cavity of a He-Ne laser in order to increase the scattering signal. When the laser was turned on, large power transients were noted in the laser output that would last for seconds. These transients were observed to buildup and decay with time and were very sensitive to cell positioning. Upon placing liquid samples in the cavity, the laser beam spot size was observed to increase as if a diverging lens were present in the cavity. It was later shown that the thermal lens effect could be observed with the liquid sample placed outside the laser cavity, and this is the configuration most often used today.

The thermal lensing effect is relatively easy to understand qualitatively. The fundamental (TEM$_{00}$) mode of a laser gives a beam with a spatial intensity-

profile described by a Gaussian function. When a weakly absorbing liquid is placed in the laser beam, the liquid is heated as a result of optical absorption. A temperature profile is established in the liquid that matches the intensity profile of the laser. As a result of heating, the liquid expands, which leads to a refractive index gradient in the sample. Most liquids have a positive coefficient of thermal expansion and a negative temperature coefficient of the index of refraction, leading to a diverging lens. In certain cases, however, the temperature coefficient of the index of refraction is positive, resulting in a converging lens.

Although the effect was originally thought to be a nuisance, thermal lensing has since been shown to be a valuable spectroscopic tool. For example, the thermal lensing effect can be used to measure extremely small absorptions in ostensibly "transparent" liquids. Thermal lensing has also been used to increase our understanding of nonradiative processes in luminescence spectrometry, of vibrational-translational relaxation, of thermal diffusion in gases, and of photolytic reactions.

In this section we first describe the instrumentation for observing the thermal lens effect and discuss a model of thermal lens formation. Analytical applications of thermal lensing spectrometry are then presented, and several closely related techniques are considered briefly.

### Instrumentation

A simple single-beam thermal lensing spectrometer is illustrated in Figure 17-2. A shutter defines the start of a thermal lensing experiment. A computer data acquisition system is shown. With a computer system the beginning of each experiment is defined by a start pulse generated by the computer. This pulse causes the shutter to open, which allows the laser beam to strike the sample and lens formation to begin. The photodiode current, proportional to the beam center intensity, is converted to voltage prior to analog-to-digital conversion. Typical thermal lensing transients are shown in Figure 17-3 as a function of the number of transients averaged.

Several different instrumental arrangements have been used for thermal lensing spectrometry. It is common to use a chopper or modulator instead of the shutter shown in Figure 17-2. Here lock-in detection is used to monitor the result of lens formation. A two-laser system has been used in which an intense dye laser beam causes lens formation and a weak fixed-wavelength probe beam allows the lens to be monitored. This arrangement in which one laser acts as a stimulus and a second probes the response is often called a *pump/probe geometry*. In addition, a differential system has been used to cancel

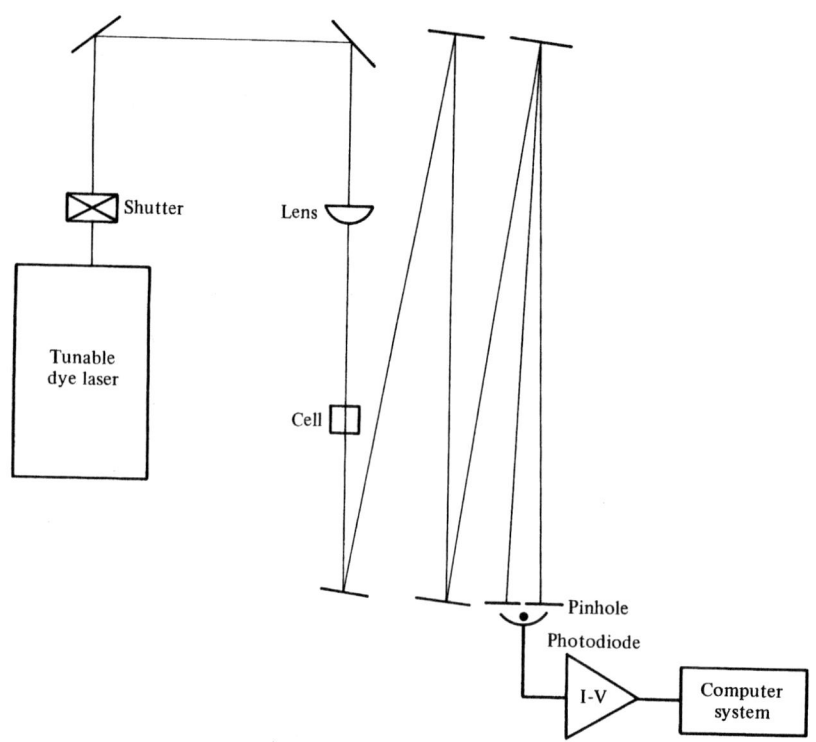

**FIGURE 17-2**   Simple thermal lensing spectrometer. A CW tunable dye laser is used both to form the thermal lens and to monitor its buildup and decay. A shutter controls the start of an experiment. The beam from a dye laser passes through a lens to define the beam waist and into the sample cell. Mirrors placed after the cell fold the beam prior to its detection by a photodiode. The folded beam configuration provides a large distance between the cell and detector and allows measurements of beam size in the far field. A pinhole serves as a limiting aperture to allow only the center of the beam to strike the photodiode.

any lens formed by the solvent itself. Here two cells are used and they are symmetrically positioned around the focus of the lens used to define the beam. Solvent alone is placed in one cell and the sample solution in the other. A diverging lens, placed after the beam focus, causes an increase in beam divergence. If the same type of lens is placed before the beam focus, it causes a decreased convergence before the focus and a decreased divergence after the focus. Thus if the two cells

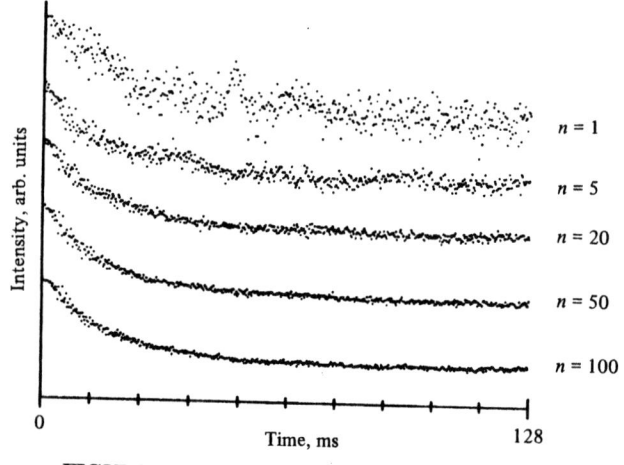

**FIGURE 17-3**   Thermal lensing signals. The laser beam intensity is monitored in the far field. The number of transients averaged (*n*) is shown. Note the improved *S/N* upon averaging. The traces are displaced from each other for clarity. (With permission from C. M. Phillips, Ph.D. dissertation, Michigan State University, 1986.)

are positioned correctly, the lens formed by the solvent, if it is not too strong, is effectively canceled.

## Models of Thermal Lens Formation

To model the formation of the thermal lens, several assumptions must be made. We will assume that the electric field of the laser beam is described by a Gaussian function ($TEM_{00}$ mode) and that the beam propagates along the *z* axis. As the Gaussian beam passes through an absorbing medium, a time-dependent temperature increase develops in the medium. For a temperature rise $\Delta T$, the refractive index distribution as a function of radial distance *r* and time *t* is given by

$$\eta(r,t) = \eta_0 + \frac{d\eta}{dT} \Delta T(r,t) \qquad (17\text{-}4)$$

where $\eta_0$ is the bulk refractive index of the medium and $d\eta/dT$ is the temperature coefficient of refractive index.

To solve for the temperature increase $\Delta T(r,t)$, the heat convection equation is used in which the heat gain of the system is equated to the difference in the heat input by absorption and the heat loss due to thermal diffusion. The heat input is proportional to the laser beam energy absorbed per unit time, which is related to the attenuation of the beam per unit length of the beam as it passes through the medium. The laser beam attenuation is of course related to the absorbance of the sample. By assuming small attenuations and a parabolic function for $\Delta T$, the time-dependent focal length *f(t)* of the lens can be obtained:

$$f(t) = \frac{\pi\kappa(1 + \tau_l/2t)\omega^2}{0.24(d\eta/dT)bk(\lambda)\Phi} \quad (17\text{-}5)$$

where $\kappa$ is the thermal conductivity of the medium (W cm$^{-1}$ K$^{-1}$), $\tau_l$ is the characteristic time constant of the lens, $\omega$ is the beam waist (cm), $b$ is the pathlength (cm), $k(\lambda)$ is the absorption coefficient (cm$^{-1}$), and $\Phi$ is the laser beam incident radiant power ($W$). Note from equation 17-5 that at $t = 0$ (at the beginning of the experiment), $d\eta/dT = 0$ and there is no lens [$f(0) = \infty$]. The lens strength builds up over time (usually milliseconds to seconds depending on $\kappa$ and $\omega$) to its steady-state focal length $f(\infty)$ given by

$$f(\infty) = \frac{\pi\kappa\omega^2}{0.24(d\eta/dT)bk(\lambda)\Phi} \quad (17\text{-}6)$$

When steady state has been reached, the rate of heat input supplied by the laser beam equals the rate of heat loss through diffusion.

The change in focal length of the lens can be related to the change in beam irradiance in the far field. Usually, the relative change between $t = 0$ and $t = \infty$ (steady state) is measured. If the cell is positioned at the confocal distance $z_c = \pi\omega_0^2/\lambda$ ($\lambda$ is the laser wavelength and $\omega_0$ is the minimum beam waist) and the amount of absorption is small, the relative change in irradiance is given by

$$\frac{\Delta E}{E} = 2.303A\left[\frac{-0.24\Phi(d\eta/dT)}{\lambda\kappa}\right] \quad (17\text{-}7)$$

$$= 2.303KA$$

where $\Delta E/E = [E(0) - E(\infty)]/E(\infty)$, $A$ is the solution absorbance, and $K$ is the thermal lensing enhancement factor encompassing all the terms in brackets.

Note from above that $K$ represents the factor by which the thermal lensing signal is enhanced over the normal absorption signal. In normal absorption spectrometry the source radiant power does not affect the magnitude of the signal obtained since a ratio measurement is made. However, in thermal lensing, the magnitude of the signal is directly proportional to the source radiant power; the enhancement can be quite large when lasers are used as sources. Because of the dependence of $K$ on thermal conductivity and refractive index gradient, different solvents give different enhancement factors. Most nonpolar solvents have larger enhancement factors than polar solvents because the long-range dipolar interactions in polar solvents allows efficient heat conduction away from the heat source.

Enhancement factors per milliwatt of laser power are 8.93 and 7.02 for $CCl_4$ and benzene, respectively, while methanol (3.06) and water (0.21) show much lower enhancements.

The parabolic model above gives good qualitative agreement to experimental data. This model has been modified, however, to take into account lens aberrations. This more extensive aberrant lens model better fits the experimental data in the case of a relatively strong lens and better predicts the time dependence of lens formation. In the aberrant lens model, the irradiance as a function of time is given by

$$E(t) = E(0)\left[1 - 0.577\theta\left(1 + \frac{t_l}{t}\right)^{-1}\right] \quad (17\text{-}8)$$

where $\theta$ is the so-called thermal lens parameter [$\theta = 0.24\Phi bk(\lambda)(d\eta/dT)/(\lambda\kappa)$].

### Analytical Applications

Because of the enhancement of the thermal lensing signal over ordinary absorption, the effect should be very useful in trace analysis. Absorption coefficients as low as $10^{-6}$ cm$^{-1}$ should be measurable. For compounds that have reasonably high molar absorptivities (e.g., $\varepsilon \geq 10^5$ L mol$^{-1}$ cm$^{-1}$), concentrations as low as $10^{-11}$ M should be detectable. Indeed, Cu(II)-EDTA has been determined at ultra-trace levels with a 4-mW He-Ne laser. It was predicted that $10^{-10}$ M would be detectable with current 1-mW lasers.

Thermal lensing has also been applied to several practical analytical problems. For example, phosphate was determined in seawater as the reduced heteropolymolybdate (see Section 13-5) by thermal lensing spectrometry. Detection limits in the pg mL$^{-1}$ range were claimed. Many of the standard colorimetric procedures appear to be applicable to thermal lensing methods.

Matrix interferences have not often been reported with the thermal lensing technique. Much of the initial work was, understandably, carried out with ultrapure solvents containing the analyte at ultratrace levels. As the method is applied to practical situations, interferences will undoubtedly appear. It was recently found, for example, that the presence of NaCl strongly affects the thermal lensing signal of the reduced antimonyl molybdophosphate complex. A 28% enhancement in the thermal lensing signal was found in a matrix equivalent to that of seawater over an aqueous dilute solution. Solvent extractions, followed by thermal lensing measurements, did not solve the matrix problem because of ionic strength effects on the extraction efficiency. A good deal of future analytical work is needed before the technique will be routinely useful with "dirty" samples.

Thermal lensing should also be applicable to HPLC detection because of its low detection limits and the enhancements found in many organic solvents. Indeed, some work has been done to apply thermal lensing spectrometry in this manner. Thermal lensing would seem most suitable for use with microbore-column HPLC, where extremely low detection limits are needed.

### Related Thermo-Optical Methods

There are several other related photothermal methods. Photoacoustic spectrometry is, of course, also based on nonradiational energy dissipation in the sample. Several other methods are based on changes in refractive index caused by heating the sample. Interferometric detection of the small phase shifts that accompany refractive index changes has been used to measure very weak absorbers. Similarly, the deflection of a probe laser beam by a surface heated by laser absorption forms the basis of *photothermal deflection* spectrometry. Other closely related methods that have not yet received as much attention as thermal lensing are *photothermal diffraction* and *photothermal refraction* methods. All the photothermal methods are complementary to molecular fluorescence spectrometry, in that they achieve low detection limits when fluorescence quantum efficiencies are low.

## 17-3 LASER IONIZATION OF ATOMS

Among the promising new atomic spectrometric methods are those based on the detection of ions produced as a result of irradiating an atomic vapor with the beam from a pulsed laser. Laser ionization techniques can offer extremely low detection limits for certain elements. Indeed, single atom detection has been achieved in special cases. The increased ionization that occurs when the beam from a pulsed laser irradiates the sample was originally called the **optogalvanic effect**. This name has largely been supplanted by more descriptive terms as discussed below.

Ions can be produced from atoms by several different schemes as illustrated in Figure 17-4. Usually, a flame is employed to produce the atomic vapor. In schemes 1–3, ionization occurs by collisions from an excited state of the atom that is populated by absorption of laser radiation. As shown, excitation of the atom can occur by absorption of a single photon, absorption of two photons of different wavelengths, or absorption of two identical photons. Techniques based on collisional

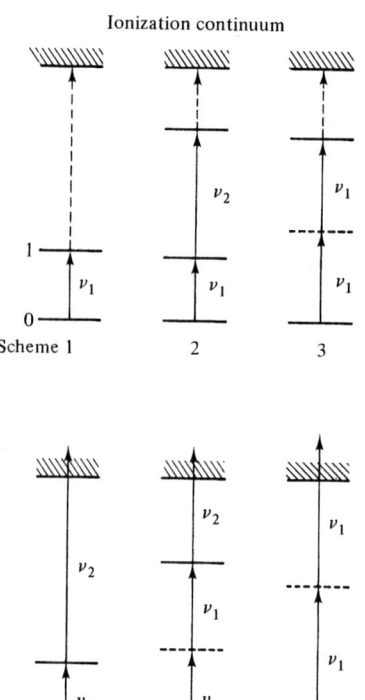

**FIGURE 17-4** Laser-assisted ionization schemes. The solid lines indicate laser-induced transitions, while the dashed lines indicate collision-induced steps. In scheme 1 a laser is tuned to a resonance transition of the analyte (energy $h\nu_1$), while ionization from the excited level occurs by collisions. Scheme 2 is similar to scheme 1 except that stepwise excitation occurs with two different laser photons (energies $h\nu_1$ and $h\nu_2$). Scheme 3 shows excitation via a two-photon process followed by collisional ionization. Schemes 4 to 6 show photoionization pathways. Scheme 4 represents two-step photoionization, while scheme 5 illustrates two-photon excitation followed by photoionization. Scheme 6 shows a single laser two-photon photoionization pathway.

ionization are often called **laser-enhanced ionization (LEI) methods**. In schemes 4–6, similar absorption processes produce excited state atoms, but ionization occurs by photon absorption (photoionization). Techniques based on these schemes are called **resonance ionization** or **dual laser ionization (DLI) methods**. In all laser ionization methods the ionization signal is proportional to the population of the excited atomic level. Hence, if possible, it is desirable to saturate this level.

The ions produced by the schemes shown in Figure 17-4 can be detected in several ways. Since flames are the most popular atomic vapor cells for analytical applications based on laser ionization, ion detection is usually accomplished by measuring the current in a circuit containing a pair of biased collection electrodes as shown in Figure 17-5. Detection of the ions produced has several advantages over detection of the photons emitted by fluorescence processes. These advantages are responsible for the extremely low detection limits achieved for certain elements. First, biased electrodes can collect essentially all the ions produced. By contrast, only a fraction (i.e., the fractional solid angle $\Omega/4\pi$) of the emitted photons are collected in fluorescence techniques. By detecting ions, the quantum efficiency losses in photon detectors are avoided and there are no spectral interferences in the detection process; there can still be spectral interferences in the excitation process as well as nonspectral interferences (e.g., vaporization interferences). Unlike resonance AF, scattered laser light is not detected with ionization methods. Ion detection also does not suffer from self-absorption losses.

The laser ionization methods do have certain disadvantages, however, when compared to techniques that use optical detection. First, the continuous background ionization from the analyte, the atomizer, or from concomitants must be discriminated against. This is usually done by using a high-pass filter and a gated integrator to process the ionization signals, as shown in Figure 17-5. The integrator is turned on only when the laser pulse irradiates the flame. Since the pulsed lasers used have relatively low duty cycles, excellent discrimination

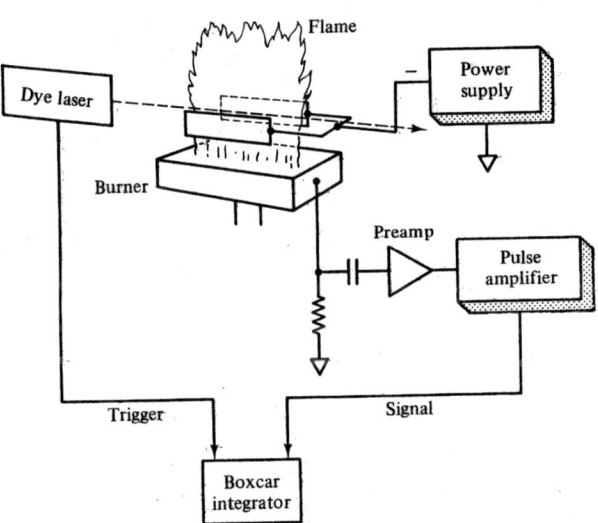

**FIGURE 17-5** Apparatus for laser-enhanced ionization. Radiation from a dye laser is incident on a flame where atomization, excitation and ionization of analyte atoms occurs. Collection electrodes biased at a high negative voltage (e.g., −1000 V) collect ions while electrons move to the grounded burner top. The ionization signal is a fast current pulse that is processed by a preamplifier, a pulse amplifier, and a boxcar integrator.

can be achieved. However, the noise in the background ionization signal can influence the $S/N$. Laser ionization methods are also susceptible to certain interferences that are not present in optical detection methods. Concomitants in the sample that affect the laser-induced ionization rate cause an interference. Continuous ionization in the flame can produce charge sheaths at the electrodes. Concomitants that influence these charge sheaths can also cause interferences. In addition, laser ionization methods have not been very successful with elements of high ionization potential. The best detection limits have been achieved with elements of low-to-moderate ionization potential.

## Laser-Enhanced Ionization Methods

Enhancement of the ionization of metal vapors in flames irradiated by intense laser beams was first observed in 1976 by scientists at the National Bureau of Standards and by van Dijk and Alkemade in the Netherlands. These initial studies prompted many other workers to begin research in the area, and an international conference on laser ionization methods was held in 1986. Research studies have been aimed at understanding the fundamentals of the ionization process, improving the ion collection and detection systems, extending the range of elements detected, and applying LEI in trace analysis.

*Theory of LEI.* The complete quantum mechanical theory of LEI has not been developed. However, many of the features of LEI are revealed by a simple rate-equation argument. We will assume here that a dye laser excites a resonance transition of the analyte (scheme 1 of Figure 17-4) in a flame and that ionization occurs by collisions with some collision partner $x$. We will also neglect the time-dependent profile of the laser pulse and the transport of the ions through the flame to the electrodes.

If the collection electrodes are biased so as to collect all the ions produced by LEI, the current density obtained is directly proportional to the rate of ionization per unit volume. From the kinetic theory of gases, the rate constant $(k_j)^{ci}$ ($s^{-1}$) for collisional ionization from an excited level $j$ by collision partner $x$ can be expressed as the product of the number density of collision partners $n_x$ ($cm^{-3}$), the average relative velocity of collision partners $(8kT/\mu\pi)^{1/2}$ ($cm\ s^{-1}$), the collision cross section $\sigma_{xj}^{ci}$ for ionization of the analyte in level $j$ ($cm^2$), and the fraction of collisions with sufficient energy to cause ionization. Thus

$$(k_j)^{ci} = n_x \left(\frac{8kT}{\mu\pi}\right)^{1/2} \sigma_{xj}^{ci} \left[e^{-(E_{ion} - E_j)/kT}\right] \quad (17-9)$$

where $k$ is Boltzmann's constant, $T$ is the temperature, $E_{ion}$ is the ionization energy, $(E_{ion} - E_j)$ is the energy difference between the ionization continuum and the excited level $j$, and $\mu$ is the reduced mass of the collision pair. If the laser is tuned to a resonance transition of the analyte which populates level $j$ only (as shown in scheme 1 of Figure 17-4), the total rate of ionization $dn_{ion}/dt$ is given by

$$\frac{dn_{ion}}{dt} = (k_j)^{ci}n_j \qquad (17\text{-}10)$$

where $n_j$ is the number density of the laser-excited level. The latter quantity is proportional to the laser irradiance unless the transition is saturated.

Equation 17-9 explains the success of LEI in providing large enhancements of the ionization rate. In flames of 2500 K, the thermal energy $kT$ is approximately 0.22 eV. The ionization energy for most elements is greater than 4 eV; for Na, $E_{ion}$ is 5.14 eV. Each electron volt of optical excitation decreases the energy that must be supplied by collisions and increases the ionization rate constant by a factor of $e^{(1/0.22)} \approx 100$. Tunable dye lasers can provide excitation energies in the range 2 to 5 eV, which implies enhancements of 4 to 10 orders of magnitude in the ionization rates. Of course, there are practical limits to the enhancement imposed by optical and electrical saturation effects. Nonetheless, enhancements of several orders of magnitude have been achieved.

*Performance Characteristics.* Several different flames have been used to produce atomic vapors for LEI. Air–acetylene and nitrous oxide–acetylene flames are most commonly used for analytical applications. Dye lasers pumped by flashlamps, $N_2$ lasers, or Nd:YAG lasers have been most successful, although LEI has been observed with CW dye lasers. Collection electrodes have been rods made of tungsten or molybdenum, flat plates made of similar materials, or probes made of tungsten, irridium, or even nichrome wire for cool flames. The electrodes have been placed in the flame or outside the flame; in the latter case electrical contact with the flame is necessary (e.g., through the grounded burner). Usually, the LEI signal is a fast current pulse. This is processed by a boxcar integrator, as shown in Figure 17-5.

Ionization signals have been observed for many of the elements, although easily ionized elements often give the best detection limits. For many elements a linear dynamic range of four to five orders of magnitude can be achieved. Detection limits for several elements are shown in Table 17-1. By comparing these values with DLs for AF (Table 11-5) and AA (Table 10-3), it

**TABLE 17-1**
Detection limits for LEI

| Element | Detection limit (ng mL$^{-1}$) |
|---------|-------------------------------|
| Ag | 1 |
| Ba | 0.2 |
| Bi | 2 |
| Ca | 0.1 |
| Cr | 2 |
| Cu | 100 |
| Fe | 2 |
| Ga | 0.07 |
| In | 0.008 |
| K | 1 |
| Li | 0.001 |
| Mg | 0.1 |
| Mn | 0.3 |
| Na | 0.05 |
| Ni | 8 |
| Pb | 0.6 |
| Tl | 0.09 |

can be seen that with flame atomization, LEI achieves the lowest DLs for Ba, Ga, In, K, Li, Mn, Na, Pb, and Tl. In some cases (K, Mn, Na, and Pb), electrothermal AA provides even lower DLs than LEI.

The LEI method can be used with nonresonance transitions as shown in schemes 2 and 3 of Figure 17-4. Stepwise excitation or two-photon excitation to levels close to the ionization potential has been shown to provide excellent DLs for several elements. Stepwise excitation can also provide additional selectivity through the use of two optical transitions.

The major problems remaining to be solved with LEI are those involving high ionization potential elements and the interferences caused by easily ionized matrix constituents. Stepwise excitation methods or the use of photoionization techniques (see later in this chapter) may extend ionization methods to elements of high ionization energy. Our present models of the ion collection process are rather crude and need to be improved. Better models along with empirical tests of various electrode shapes and positions should lead to new electrode configurations that improve ion collection and reduce interferences. The LEI method has not been very successfully coupled with atomizers other than flames. Electrothermal atomization with stepwise excitation schemes may prove useful. Attempts to couple normal LEI with ICP atomization have not been successful due to the extensive ionization in ICPs, and the RF noise generated by ICPs. However, LEI has been detected optically in plasmas by measuring the decrease in AF that accompanies the enhancement of ionization. This has been termed the **ionization dip**, or more appropriately, the **fluorescence dip**. The LEI method is so young that many new developments should occur with further research.

## Resonance Ionization Methods

Resonance ionization differs from LEI in that laser radiation is used to photoionize the atom (or molecule) in the former instead of relying on collisions from an excited state. Resonance ionization spectroscopy (RIS) originated in the mid-1970s at the Oak Ridge National Laboratory. Today there are several different forms of RIS. The original RIS method uses laser photoionization and a proportional counter for electron detection. A RIS ionization source has been combined with a mass spectrometer to provide excellent detectability with extremely high selectivity. The RIS process has also been used in conjunction with flame atomizers where it is more commonly called **dual laser** or **direct laser ionization** (DLI).

A proportional counter RIS detector is illustrated in Figure 17-6. Proportional counters are capable of counting single electrons. They are usually filled with 90% Ar and 10% $CH_4$. If charged particles (ions and electrons) are formed in the counter, they are accelerated by the nonlinear electric field of a small-diameter wire held at a high voltage. As the charged particle moves it collides with atoms of the counting gas producing more charges and thus amplification. In the experiment shown in Figure 17-6, a low concentration of

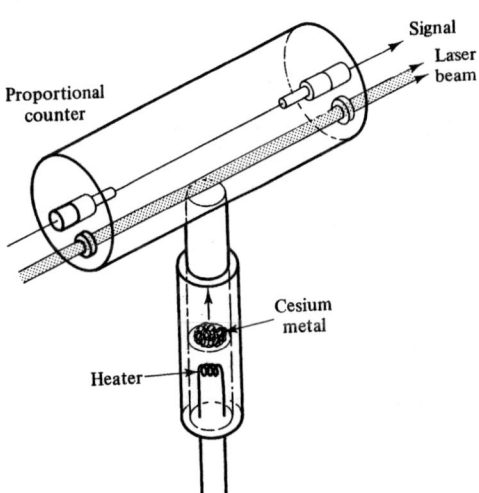

**FIGURE 17-6** Proportional counter detection system for RIS. A small amount of Cs is introduced into the counter with the counting gas (Ar and $CH_4$). A laser beam causes photoionization of the Cs. Acceleration of the charged particles by the field of a small wire produces additional charges and thus amplification. The signal is processed by counting electronics (not shown).

Cs is introduced into the counting gas. A pulsed dye laser tuned to 455 nm excites the $6^2S_{1/2}$–$7^2P_{3/2}$ transition. The 455-nm photons are also capable of photoionization of atoms in the $7p$ level. The proportional counter signal is linearly related to the number of ions produced. By this experiment, Oak Ridge scientists were able to conclude that at the lowest signal level measured, a single Cs atom in $10^{19}$ atoms of the counting gas was ionized.

Because of its youth, the RIS technique has only been applied at present to a few problems. In chemical physics, it has been used to study the diffusion of selected atoms among many other atoms and molecules. In chemical kinetics, RIS has been applied to the study of reactions of highly reactive atoms such as alkali halide atoms. Here, the ability to detect extremely low concentrations is a distinct advantage because corrosion of the apparatus and the possibility of side reactions can be minimized.

In analytical chemistry, RIS has been receiving increased attention because of its single-atom detection capabilities. At the National Bureau of Standards (NBS) an apparatus has been constructed for laser ablation and RIS with a proportional counter as detector. Here one laser is used to ablate atoms from a solid, while a second laser is used to photoionize. With this system, NBS scientists were able to detect $5 \times 10^{11}$ Na atoms per $cm^3$ of electronics-purity Si. The detection limit was such that less than 1 Na atom per integrated-circuit device could be detected in principle. The combination of RIS and mass spectrometry is also of considerable interest. One possible use of a RIS/MS instrument is for isotopic selectivity. Here the total selectivity is the product of the selectivities of photionization and mass separation. An alternative approach is laser-induced fluorescence of the ions selected by a conventional mass spectrometer.

The DLI method has also been receiving attention for flame-based determinations. The DLI technique is closely related experimentally to LEI in that the detection system employs a biased probe or a pair of biased probes. The major difference is that in DLI either a second laser is used to photoionize excited atoms by schemes 4 and 5 in Figure 17-4 or direct photoionization by scheme 6 is employed. Thus the ionization scheme is closely related to that of RIS. Schemes 4 and 5 are most commonly employed. Here the first laser is used to produce a large population of excited atoms (saturation is desirable), while the second laser promotes those excited atoms into the ionization continuum. In DLI the rate constant $(k_j)^{pi}$ for photoionization from excited level $j$ is given by

$$(k_j)^{pi} = \sigma_j^{pi} E_0 \qquad (17\text{-}11)$$

where $\sigma_j^{pi}$ is the cross section for photoionization from level $j$ and $E_0$ is the incident photon irradiance of the ionizing beam ($s^{-1}$ $cm^{-2}$). In any DLI experiment in a flame, the collisional ionization given by equation 17-10 cannot be eliminated, so that the total rate of ionization in DLI, $dn_{ion}/dt$, is given by

$$\frac{dn_{ion}}{dt} = [(k_j)^{pi} + (k_j)^{ci}]n_j \qquad (17\text{-}12)$$

We are assuming here that any ground-state thermal ionization is negligible or has been discriminated against by a high-pass filter and gated electronics.

Although photoionization and collisional ionization compete in DLI, conditions can be chosen in some cases so that photoionization will dominate. For this to happen, the overshoot energy of the ionizing laser beam (energy in excess of that needed to reach the ionization continuum) should be small. Photoionization cross sections decrease with increasing overshoot energy. In addition, the energy defect ($E_{ion} - E_j$) should be large enough that photoionization is favored over collisional ionization. Thus, optimizing DLI involves carefully selecting the excitation/ionization scheme. When the energetics are favorable, DLI can provide an enhancement of 100 to 1000 times over LEI.

Several experimental configurations can be used for DLI. In the original approach an $N_2$ pumped dye laser was used to provide excitation. A portion of the pumping laser beam was intercepted and directed collinearly into the flame with the dye laser beam to provide ionization. The two beams must spatially and temporally overlap for excitation/ionization to occur. A Nd:YAG pumped dye laser has also been employed. Here one of the harmonics of the pumping laser provides the ionizing beam. The most versatile approach would probably involve pumping two dye lasers with a Nd:YAG laser and frequency doubling one of the dye lasers for the ionizing beam.

The DLI method has not been applied to many analytical problems at present. For selected elements (Na, Li) it has achieved excellent detection limits. The method has been applied, however, to flame diagnostic measurements. In one application, flame temperatures were determined by DLI. A time-resolved DLI technique was used to determine the mobility coefficient and the diffusion coefficient of a selected ion. These coefficients can be related to flame temperature by an equation originally developed by Einstein. Because the ions are formed in a very small area and detected by probes immersed in the flame, excellent spatial resolution can be achieved. This should make it possible to measure spatial profiles of flame temperatures.

## Summary

In summary, the laser ionization methods are relatively new analytical tools. As such they have not yet been applied to a variety of practical problems as have such established techniques as AA and AE. The laser ionization process competes with atomic fluorescence. Ionization and fluorescence are in fact somewhat exclusive processes. Much research remains to be done to develop the quantum mechanical basis of the methods and to establish various performance characteristics. Since they all depend on laser sources, the future of these techniques is linked to future developments in pulsed lasers. However, already the ionization methods can achieve outstanding detection limits for certain elements. They have great potential as future routine analytical tools.

## 17-4 MISCELLANEOUS LASER-BASED TECHNIQUES

In this section we consider briefly a few additional methods that have been made feasible by laser radiation. Only those techniques that appear to have promise in analytical applications are considered. Most of these methods involve nonlinear phenomena, where the signal at the detector varies nonlinearly with the irradiance of the source.

### Intracavity Absorption

One method that has been successfully used to enhance the detection of weakly absorbing samples is to place the sample inside the laser cavity. There can be several reasons for the enhancements observed. First, there is an increase in the effective pathlength for absorption when the sample is inside the laser resonator. Inside the cavity the laser photons are reflected back and forth many times by the cavity mirrors. If one of the cavity mirrors is totally reflecting and the other has a reflectance $\rho$ of less than unity, the average number of passes $n$ for each laser photon is $n = 1/(1 - \rho)$, and the effective pathlength is $nb$, where $b$ is the geometric pathlength. The enhancement factor is then $n$, the number of passes. If $\rho = 0.99$, an enhancement factor of 100 is obtained. Of course, the same enhancement could be achieved outside the cavity by using a multipass cell of equivalent effective pathlength.

If the laser is operated close to its lasing threshold, an additional enhancement can occur. The laser will oscillate only when the total gain overcomes the total losses in the cavity. If the gain of the active medium only slightly exceeds that necessary for oscillation, small changes in the absorption losses in the cavity can cause very large changes in the laser output. Unfortunately, there is also an increase in laser instability when the gain is near threshold, which makes this method of enhancement difficult to use in practice.

With multimode dye lasers an additional enhancement effect can be observed. If selective modes are attenuated by absorption by molecules inside the cavity, the gain for the other modes increases. By using a frequency-selective detector, the power loss in the attenuated modes can be measured and used to obtain the amount of sample absorption.

Intracavity absorption has been used to make quantitative measurements of several molecular species. By placing a gas cell containing $I_2$ inside the cavity of a CW dye laser an enhancement of $10^5$ over a single-pass external cell was observed. An $I_2$ vapor pressure of $1.5 \times 10^{-6}$ torr could be detected. Free radicals, such as OH, $NH_2$, and HCO, have also been quantitatively determined by intracavity absorption.

## Molecular Multiphoton Ionization

Laser ionization methods for atoms were described earlier in this chapter. Lasers can also be used to ionize molecules via multiple photon processes. **Multiphoton ionization** (MPI) was first reported in the mid-1970s and has since been widely used in molecular spectroscopy.

The MPI process involves a multiphoton excitation to an intermediate vibronic state followed by a multiphoton excitation to the ionization continuum through a dense manifold of autoionizing states. Early experiments were carried out in the gas phase in cells containing collection electrodes. Here only the total ion current (electrons, positive ions, and negative ions) could be obtained without any specific means to identify the detected species. In more recent work a mass spectrometer has been used to identify the ionic fragments obtained. In addition, the MPI technique has been extended to solutions with simple collection electrodes used to monitor the total charge produced by the incident laser beam.

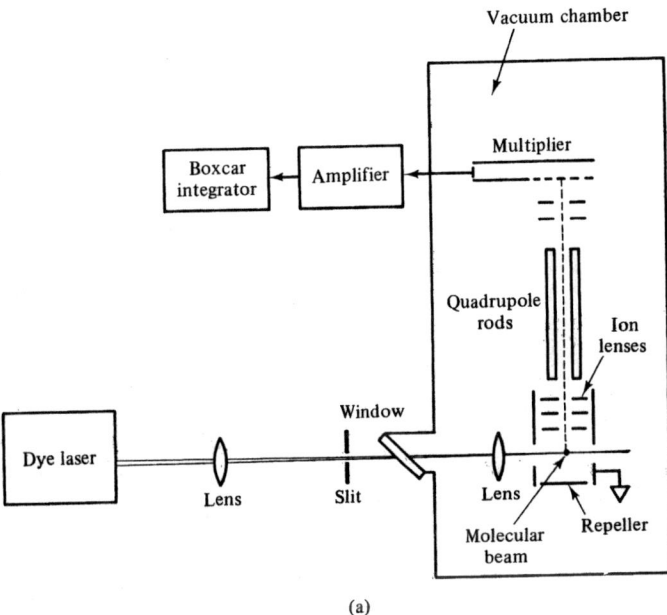

**FIGURE 17-7** Diagram of laser molecular ionization mass spectrometer (a) and a molecular beam apparatus (b). In (a) an $N_2$ pumped tunable dye laser is focused into a molecular beam at right angles to the laser beam. Ions produced are mass separated by the quadrupole mass spectrometer. The ion signal produced is processed by a boxcar integrator synchronized to the firing of the laser. The collimated molecular beam in (b) produces a pencil-shaped beam of atoms or molecules. The sample, a metal or a salt, is placed in an oven and heated to the desired temperature. A small hole in the side of the oven allows the beam of molecules to effuse into a high-vacuum region. The pinhole provides collimation of the beam. Often a velocity selector is used to allow only molecules in a particular velocity range to pass. This can be a pair of slits displaced relative to each other by a certain angle and separated by a certain distance. Only molecules with a particular velocity pass both slits.

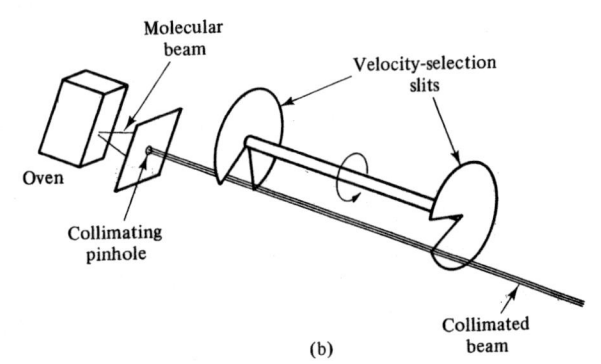

A block diagram of a laser multiphoton ionization molecular beam mass spectrometer is shown in Figure 17-7a. An apparatus for producing a collimated molecular beam is illustrated in Figure 17-7b. With the MPI mass spectrometer system either the total ion current or the ion current corresponding to a selected fragment can be monitored. Figure 17-8a shows the MPI spectrum of benzene in the total ion current mode, while Figure 17-8b shows the spectrum obtained while monitoring $C^+$, the most abundant ion. By tuning the dye laser to various wavelengths and recording the mass spectrum of the positive ions, a two-dimensional vibronic/mass spectrum is obtained. That is, mass spectra can be measured corresponding to different vibrational excitation bands.

Although this technique is quite new, it has obvious potential in analytical chemistry. Trace analysis of atmospheric pollutants is one potential application. The selectivity is such that analyses of isomeric mixtures may be feasible without GC separations. Recent variations in technique have included using the laser ioni-

zation source with a time-of-flight mass spectrometer with both pulsed and continuous molecular beam introduction systems.

Multiphoton ionization has also been actively studied in solution. Here various pulsed lasers (e.g., an $N_2$ laser, an $N_2$ pumped dye laser or an excimer laser) have been used to provide the excitation/ionization energy. The current produced as a result of irradiation by the pulsed laser beam is measured with two electrodes; a boxcar integrator is used to acquire the photoionization signal. Multiphoton ionization signals have been obtained for polycyclic aromatic hydrocarbons and several drugs of analytical importance. The multiphoton ionization method shows considerable promise as an HPLC detector. Indeed, a windowless flow cell has been designed that allows simultaneous detection by molecular fluorescence, the photoacoustic effect and photoionization methods. These techniques were chosen because they monitor all the major deactivation pathways for excited molecules.

## Doppler-Free Absorption Spectroscopy

In principle, electronic spectra of molecules in the gas phase are capable of providing a wealth of information about rotational and vibrational levels of electronic states. In many cases, however, the Doppler widths of the lines making up an electronic absorption or fluorescence band mask a good deal of the structural information. (Refer to Section 7-4 for information about Doppler broadening.) With the advent of laser sources, techniques have been developed to resolve the various rotational and vibrational lines and to investigate in greater detail the structure of excited states. These methods are called **Doppler-free methods** and they are limited only by the natural linewidths which are usually two to three orders of magnitude smaller than Doppler widths. We discuss briefly here a few of the methods that have been used to obtain high-resolution Doppler-free spectra.

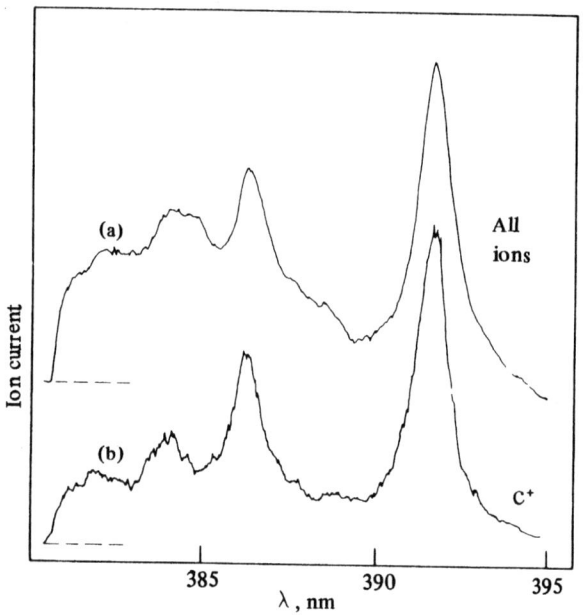

**FIGURE 17-8** Multiphoton ionization spectra. Curve a shows the MPI spectrum of benzene. This was obtained while scanning the laser wavelength and monitoring the total positive ion current. For curve b the most abundant ion $C^+$ was selected with the mass spectrometer and monitored while the laser was scanned. (With permission from D. A. Lichtin, L. Zandee, and R. B. Bernstein, "Potential Analytical Aspects of Laser Multiphoton Ionization Mass Spectrometry," in *Lasers in Chemical Analysis*, G. M. Hieftje, J. C. Travis, and F. E. Lytle, eds., Humana Press, Clifton, N.J., 1981.)

*Collimated Molecular Beam Spectroscopy.* One approach for reducing the Doppler width is to introduce the sample as a well-collimated molecular beam (see Figure 17-7b). In such a beam the velocity component perpendicular to the beam axis can be a factor of 200 or more lower than that in the beam direction. To obtain absorption spectra, a monochromatic laser beam is crossed perpendicular to the molecular beam and tuned over the absorption profile. Since the velocity component is small in the direction of the laser beam, a reduction in the Doppler width is achieved, in many cases to the point where the natural linewidth becomes the limiting factor.

To take full advantage of the high resolution afforded by reducing the Doppler width, a very narrow laser bandwidth must be used. With dye lasers, this usually involves using one or more Fabry–Perot etalons in the laser cavity, employing bandwidth narrowing outside the cavity or switching to a single-mode laser. Because of very small absorption pathlengths and relatively low molecular densities, the absorption spectrum is often obtained by measuring the total fluorescence from the laser-excited levels (fluorescence excitation spectrum). As an example of the improved resolution that can be obtained, Figure 17-9 shows excitation spectra of $NO_2$ taken in an ordinary gas cell and in a collimated molecular beam.

*Saturation Spectroscopy.* Saturation phenomena have been studied since the early days of powerful laser sources. We have previously discussed saturation in Sections 11-3 and 15-2 in conjunction with atomic and molecular fluorescence. Selective saturation of molecular transitions can be used to produce Doppler-free spectra.

Consider a gas-phase collection of free atoms that can absorb different frequencies of laser radiation because of the Doppler effect. Atoms moving away from the source absorb lower frequencies than those moving toward the source. The absorption lines of such atoms are said to be **inhomogeneously broadened** because different atoms are responsible for different parts of the absorption line. If any frequency within the absorption line could excite every atom with equal probability, the line would be **homogeneously broadened.** Atomic and molecular beams crossed with a laser beam can produce homogeneously broadened lines because the Doppler width is so small, as discussed previously.

Let us now consider a two-laser experiment on a collection of gas-phase atoms. We will use an intense monochromatic pump laser beam to saturate the atomic population and a weak tunable probe beam to monitor the saturation effect. If the pump beam is of very narrow bandwidth, only those atoms that have the appropriate Doppler velocities will absorb the laser radiation. Hence we can cause saturation of only those atoms within a

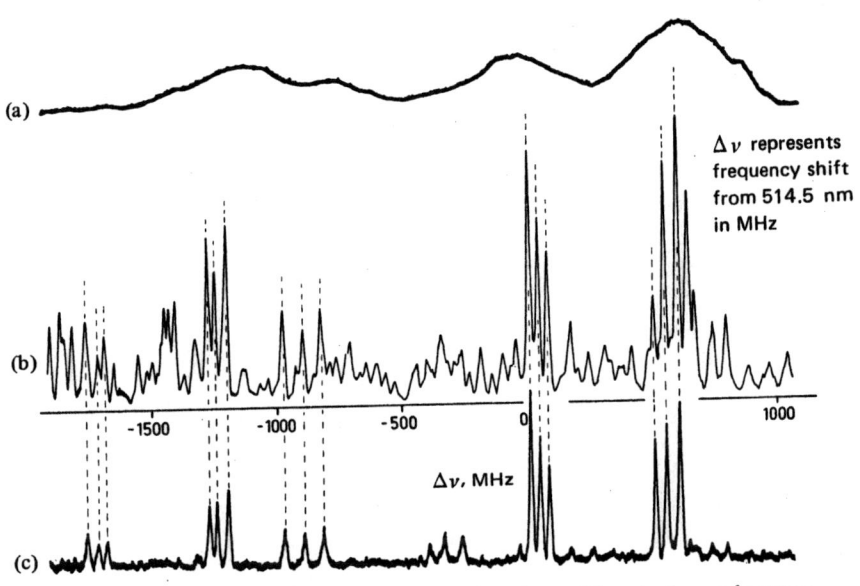

**FIGURE 17-9** Excitation spectra of $NO_2$ taken with a single-mode argon laser tunable over a small wavelength range near 514.5 nm. In (a) the $NO_2$ was in an ordinary gas cell at $10^{-3}$ torr. In (b) the $NO_2$ was in a collimated molecular beam and the total fluorescence was observed. The great number of resolved lines in the Doppler-free spectrum (b) correspond to transitions between rotational levels in the ground state and in an electronically excited state. Note that these are totally hidden in (a) by the Doppler broadening. In (c) a fluorescence monochromator was used so that a selected fluorescence transition could be studied as a function of excitation wavelength. The selected fluorescence transition terminated at a specific vibrational level of the ground electronic state. (Adapted with permission from W. Demtroder, "Investigations of Small Molecules by Modern Spectroscopic Techniques," in *Case Studies in Atomic Physics*, M. R. C. McDowell and E. W. McDaniel, eds., vol. 6, North-Holland, Amsterdam, 1976.)

particular Doppler velocity subset. If a tunable probe laser is now scanned across the Doppler-broadened absorption line of the atoms, there will be a dip in the absorption coefficient at the pump laser wavelength as shown in Figure 17-10a. Such a line is said to have a *hole burned in it*. The hole is caused by the depletion of the ground-state population with the appropriate Doppler shift. By modulating the pump laser or by using polarization techniques, one can separate the hole from the remainder of the line and obtain the narrow profile shown in Figure 17-10b. The width of the resulting line is related to the homogeneous line width.

Figure 17-11 compares the saturated absorption spectrum of the Balmer α line of atomic deuterium to the ordinary emission spectrum. The increased resolution by suppressing the Doppler broadening is apparent. Experiments such as these in high-resolution spectroscopy have lead to improved values of fundamental constants and may yield improved definitions of the meter, the second, and the speed of light. Saturation spectroscopy allows the determination of term values with increased accuracy and the measurement of fine and hyperfine spectral features.

Another method to suppress the Doppler effect is by means of **Lamb dip spectroscopy**. In this technique, two laser waves of the same frequency are sent through the sample in opposite directions. This can be accom-

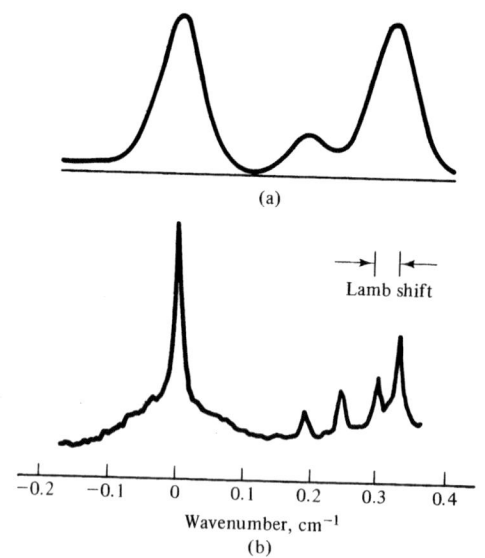

(a)

Lamb shift

| | | | | | | |
|---|---|---|---|---|---|---|
| −0.2 | −0.1 | 0 | 0.1 | 0.2 | 0.3 | 0.4 |

Wavenumber, cm⁻¹

(b)

**FIGURE 17-11** Spectra of the Balmer α lines of atomic deuterium taken by ordinary emission spectrometry (a) and saturated absorption spectrometry (b). The extra splitting in (b) is due to the Lamb shift, which is predicted by quantum electrodynamics.

plished by putting the sample in a resonant cavity since the standing waves in such a cavity are equivalent to traveling waves propagating in opposite directions. If the laser radiation is sufficiently intense, it can burn a hole in the absorption profile of the gas by saturating the population within a certain Doppler velocity subset. The radiation traveling in one direction will burn a hole at $\nu_L = \nu_m + \delta\nu$ (see Figure 17-10), while the beam traveling in the opposite direction will burn a hole at $\nu_L = \nu_m - \delta\nu$. Under the special case that both beams are tuned to the absorption maximum ($\nu_L = \nu_m$), the same set of atoms having zero velocity along the optical axis will interact with both waves simultaneously. In this case the absorption coefficient will show a dip, called the **Lamb dip**, in the center of the Doppler-broadened profile. The dip can be easily monitored if the absorbing sample is inside the laser cavity. Here the laser output increases sharply when the wavelength is tuned to that of the line center because the cavity losses are at a minimum. The Lamb dip can also be measured by monitoring the total fluorescence as the laser is tuned across the center of the line.

Although we have illustrated saturation spectroscopy with examples from atomic systems, it can also by used for studies of molecules. For obtaining Doppler-free spectra, saturation spectroscopy has some advantages over molecular beam spectroscopy. First, the apparatus is much simpler. Also collisional broadening effects can be studied with saturation spectroscopy by measuring the Doppler-free linewidth as a function of

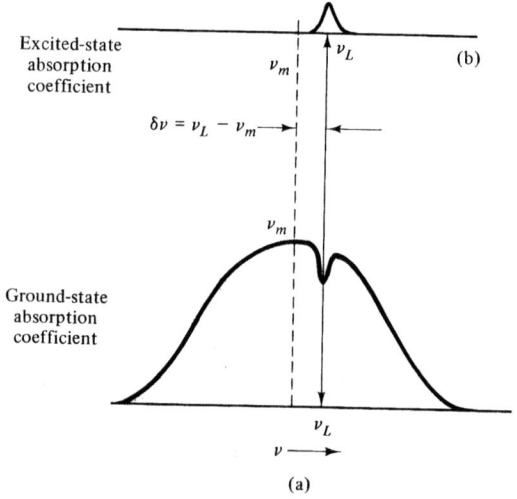

Excited-state absorption coefficient

$\nu_m$   $\nu_L$   (b)

$\delta\nu = \nu_L - \nu_m$

$\nu_m$

Ground-state absorption coefficient

$\nu_L$

$\nu \longrightarrow$

(a)

**FIGURE 17-10** Hole burning in an inhomogeneously Doppler-broadened transition. A laser of frequency $\nu_L$ depopulates the part of the ground state population that is in resonance (a) and creates an anomalous peak in the excited state distribution (b). The ground-state absorption coefficient shows a dip, while the excited-state absorption coefficient shows a peak. The hole is shifted from the absorption maximum $\nu_m$ by a frequency shift $\delta\nu$ given by $\delta\nu = \nu_L - \nu_m$.

collision gas pressure. This is, of course, not possible with a molecular beam.

*Optical Double-Resonance Methods.* Doppler-free spectra can be obtained by using two lasers in a double-resonance experiment as shown in Figure 17-12. This method has been used in both the visible and in the infrared region of the spectrum. Infrared-microwave double-resonance experiments have been successfully performed on a variety of molecules.

*Multiphoton Absorption.* Multiphoton absorption can be used to produce atoms and molecules in highly excited electronic states. With infrared lasers molecules in highly excited vibrational levels can be created. Since the selection rules for multiple-photon processes differ from those of single photon processes, new transitions and states can be studied.

In addition geometries exist in which no momentum is transferred from the radiation field to the absorbing species. In such cases, Doppler broadening from the thermal motion of the absorbers is suppressed. For example, consider a two-photon process where the photons come from two laser beams and are of different wavelengths. If the two photons are simultaneously absorbed by the sample and supplied by counter-propagating laser waves, Doppler-free spectra are obtained.

In Doppler-free two-photon spectroscopy we can consider one photon to be Doppler-shifted relative to the atoms or molecules by a positive $\delta\nu$ while the other

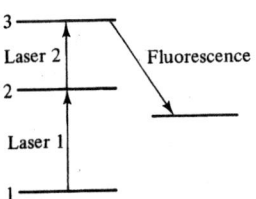

**FIGURE 17-12** Two-laser optical double-resonance method. Laser 1 is tuned to excite the molecules from level 1 to level 2. If the laser wavelength corresponds to the center wavelength of the absorption profile, only molecules with the appropriate velocity components are excited. If a second laser is now tuned to excite the molecules from level 2 to level 3, only those molecules with a narrow velocity distribution will reach level 3. The double resonance can be detected by monitoring the fluorescence from level 3 which will be essentially Doppler free if the two waves travel parallel or antiparallel through the absorbing sample.

photon (from the laser) is Doppler-shifted by $\delta\nu$. The total energy of the two-photon absorption is independent of the velocity of the atoms or molecules absorbing. In Doppler-free saturation spectroscopy only a small fraction of all the molecules have the appropriate velocity to absorb the laser radiation. By contrast, in Doppler-free multiphoton absorption, all molecules have an equal probability of absorption. This increase can in some cases overcome the low cross sections that are typical of multiphoton processes.

# REFERENCES

The references below are general references on recently developed laser spectrometric methods.

1. M. D. Levenson, *Introduction to Nonlinear Laser Spectroscopy*, Academic Press, New York, 1982. Includes chapters on saturation spectroscopy and multiphoton absorption and a useful catalog of nonlinear phenomena.

2. D. S. Kliger, ed., *Ultrasensitive Laser Spectroscopy*, Academic Press, New York, 1983. Covers several topics relevant here, including thermal lensing, laser ionization, photoacoustic spectrometry, two-photon spectroscopy, and laser intercavity-enhanced spectrometry.

3. N. Omenetto, *Analytical Laser Spectroscopy*, Wiley, New York, 1979. An excellent book dealing with laser principles and the applications of lasers in analytical chemistry.

4. G. M. Hieftje, ed., *New Applications of Lasers to Chemistry*, American Chemical Society, Washington, D.C., 1978. Contains excellent chapters on laser applications written by experts in the field.

5. G. M. Hieftje, "Approaching the Limit in Atomic Spectrochemical Analysis," *J. Chem. Educ.*, **59**, 900 (1982). Part of seven articles taken from a symposium "Approaching the Limits of Chemical Analysis" at the Spring 1982 National American Chemical Society Meeting. Covers a variety of laser techniques for single-atom detection.

6. C. Th. J. Alkemade, "Single Atom Detection," *Appl. Spectrosc.*, **35**, 1 (1981). Covers theory of single-atom detection and possible techniques to achieve it.

7. E. H. Piepmeier, ed., *Analytical Applications of Lasers*, Wiley, New York, 1986.

The following references are devoted wholly or in part to photoacoustic spectrometry.

8. J. F. McClelland, "Photoacoustic Spectroscopy," *Anal. Chem.*, **55**, 89A (1983).

9. D. Betteridge and P. J. Meylor, "Analytical Aspects of Photoacoustic Spectroscopy," *CRC Crit. Rev. Anal. Chem.*, **14**, 267 (1984).

10. A. Rosencwaig, "Photoacoustic Spectroscopy: A New Tool for Investigation of Solids," *Anal. Chem.*, *47*, 592A (1975).

11. A. Rosencwaig, *Photoacoustics and Photoacoustic Spectroscopy*, Wiley, New York, 1980.

The following references discuss photothermal methods, such as thermal lensing spectrometry.

12. R. L. Swofford, "Analytical Aspects of Thermal Lensing Spectroscopy," in *Lasers in Chemical Analysis*, G. M. Hieftje, J. C. Travis, and F. E. Lytle, eds., Humana Press, Clifton, N.J., 1981.

13. C. M. Phillips, S. R. Crouch, and G. E. Leroi, "Matrix Effects in Thermal Lensing Spectrometry: Determination of Phosphate in Saline Solutions," *Anal. Chem.*, *58*, 1710 (1986).

14. M. D. Morris and K. Peck, "Photothermal Effects in Chemical Analysis," *Anal. Chem.*, *58*, 811A (1986).

15. N. J. Dovichi and J. M. Harris, "Laser-Induced Thermal Lens Effect for Calorimetric Trace Analysis," *Anal. Chem.*, *51*, 728 (1979).

The references below deal with laser ionization methods.

16. J. C. Travis and J. R. DeVoe, "The Optogalvanic Effect," in *Lasers in Chemical Analysis*, G. M. Hieftje, J. C. Travis, and F. E. Lytle, eds., Humana Press, Clifton, N. J., 1981.

17. S. L. Chin and P. Lambropoulos, *Multiphoton Ionization of Atoms*, Academic Press, Toronto, 1984.

18. J. C. Travis, G. C. Turk, and R. B. Green, "Laser-Enhanced Ionization for Trace Metal Analysis," in *New Applications of Lasers to Chemistry*, G. M. Hieftje, ed., American Chemical Society, Washington, D.C., 1978.

19. F. M. Curran, K. C. Lin, P. M. Hunt, G. E. Leroi, and S. R. Crouch, "Energy Consideration in Dual Laser Ionization Processes in Flames," *Anal. Chem.*, *55*, 2382 (1983).

20. K. C. Lin, P. M. Hunt, and S. R. Crouch, "Flame Temperature Determination by Dual Laser Ionization," *Chem. Phys. Lett.*, *90*, 111 (1982).

21. G. S. Hurst, "Resonance Ionization Spectroscopy," *Anal. Chem.*, *53*, 1448A (1981).

22. J. C. Travis, G. C. Turk, and R. B. Green, "Laser-Enhanced Ionization Spectrometry, *Anal. Chem.*, *54*, 1007A (1982).

23. M. H. Nayfeh, "Laser Detection of Single Atoms," *Am. Sci.*, *67*, 204, March-April, (1979).

24. G. S. Hurst, M. H. Nayfeh, and J. P. Young, "One-Atom Detection Using Resonance Ionization Spectroscopy," *Phys. Rev. A*, *15*, 2283 (1977).

25. J. C. Travis, "Limits to Sensitivity in Laser Enhanced Ionization," *J. Chem. Educ.*, *59*, 909 (1982). Part of the Spring 1982 ACS symposium on "Approaching the Limits of Chemical Analysis."

The following references deal with specialized methods of interest in this chapter.

26. D. A. Lichtin, L. Zandee, and R. B. Bernstein, "Potential Analytical Aspects of Laser Multiphoton Ionization Mass Spectrometry," in *Lasers in Chemical Analysis*, G. M. Hieftje, J. C. Travis, and F. E. Lytle, eds., Humana Press, Clifton, N. J., 1981.

25. R. E. Smalley, "Mass-Selective Laser Photoionization," *J. Chem. Educ.*, *59*, 934 (1982). Also based on a presentation at the Spring 1982 ACS meeting.

26. J. M. Hayes and G. J. Small, "Supersonic Jets, Rotational Cooling and Analytical Chemistry," *Anal. Chem.*, *55*, 565A (1983).

# APPENDIX A

# Statistical Concepts

In most analytical determinations, practical considerations limit the number of measurements of a given sample and the number of samples in a total population that we can analyze. Because of this and the uncertainties associated with measuring the magnitude of a quantity of a given sample, it is important to report some estimate of the accuracy and precision of results. The **accuracy** indicates how close the measured magnitude of a quantity is to the true magnitude of the quantity in the population, whereas the **precision** indicates the reproducibility in the magnitude of the measured quantity or the variability of that magnitude of the quantity among samples in the population. Accuracy is affected by both systematic and random errors. **Systematic errors**, also known as **determinate errors**, result from factors which cause the magnitude of the measured quantity to be biased too high or too low by either a fixed fraction or an absolute amount. They are caused by factors affecting the measurement which are consistently the same in or for a given population. For example, a concomitant present in all samples in a population may cause the measured analyte concentration to be too high. Theoretically, a systematic error can be eliminated, estimated, or corrected if the cause or factor responsible for the error is known.

**Random error** is caused by uncontrolled fluctuations in variables or conditions that affect the magnitude of the measured quantity. The precision is a measure of the magnitude of the random error. Random error causes measurement results to vary in magnitude and sign. Random error can be reduced in some cases by better control of the variables that cause the fluctuation. For example, random fluctuations in temperature from sample to sample may cause the signal measured for a given analyte concentration to vary. Better control of the temperature will reduce the random error due to temperature fluctuations. Also the effect of random error can be reduced by measuring many samples from a population. This is not true for systematic errors.

Statistics provide us with a formalized strategy to evaluate the uncertainty in measurements due to random errors and to determine if systematic errors are also present. In this appendix, statistical concepts for evaluating errors are presented and their application is demonstrated. The references for this appendix should be consulted for a more detailed discussion.

## A-1 STATISTICAL QUANTITIES

Typically, we measure a given quantity or variable in a given population $n$ times. This could be $n$ measurements on a given sample or one measurement for each of $n$ samples. The magnitude of the $i$th measurement of the quantity is denoted $X_i$. If $n$ is large, we can make a plot of the relative frequency with which a given magnitude of $X$ is obtained. This frequency distribution is an estimate of the probability distribution for the random variable that would be obtained if all samples in the population were measured. If the variable can only have discrete magnitudes or values, the probability distribution indicates the probability associated with each value of the random variable. If the variable is continuous or we can assume an infinite number of values, the probability distribution can be used to estimate the probability that the measured magnitude of the variable is within a finite interval.

### Mean and Standard Deviation

The **arithmetic experimental mean** $\overline{X}$ and **experimental standard deviation** (SD) $s$ are calculated from equations A-1 and A-2, respectively, for $n$ measurements of $X_i$.

$$\overline{X} = \frac{\sum_{i=1}^{n} X_i}{n} \tag{A-1}$$

$$s = \sqrt{\frac{\sum_{i=1}^{n} (X_i - \overline{X})^2}{n - 1}} \tag{A-2}$$

$$= \sqrt{\frac{1}{n-1}\left[\left(\sum_{i=1}^{n} X_i\right)^2 - \frac{\sum_{i=1}^{n} (X_i)^2}{n}\right]}$$

The second form of equation A-2 is useful for calculation purposes and is the equation preprogrammed into scientific calculators. The experimental mean is a measure of the central tendency of the measurements or probability distribution and is an estimate of the population or true mean, $\mu$, the mean for an infinite number of measurements or for all members in the population.

The absolute error is $|\mu - \overline{X}|$ and the relative error is $|\mu - \overline{X}|/\mu$, which is often reported as a *percentage*. The absolute or relative error can be due to both random and systematic error.

The experimental SD is a measure of the scatter or reproducibility of measurements and hence the breadth or dispersion of the probability distribution; it is an estimate of the population or true standard deviation $\sigma$, and accounts for only the random errors in the measurement. The variance is the square of the SD ($\sigma^2$ or $s^2$). The experimental and population relative standard deviations (RSDs) are $s/\overline{X}$ and $\sigma/\mu$, respectively, or the SD divided by the mean. Often the precision of a technique is reported as the percent RSD.

### Probability Distributions

The shape of the probability distribution depends on the particular population. Many times a **Gaussian** or **normal distribution** applies when fluctuations in many independent factors cause the random error in the measured quantity or variable. The standardized normal distribution $P(z)$ is given by

$$P(z) = (\sigma\sqrt{2\pi})^{-1} \exp\left(\frac{-z^2}{2}\right) \tag{A-3}$$

and is plotted v. $z$ in Figure A-1. The variable $z$ is defined as

$$z = \frac{X - \mu}{\sigma} \tag{A-4}$$

and is called the $z$ statistic or the standard normal deviate. Note that the mean has been normalized to zero and the SD is normalized to 1.

If we are concerned with the distribution of the means rather than of individual measurements (e.g., the distribution of $\overline{X}$ calculated from $n$ individual measurements of $X$), then $X$ is replaced by $\overline{X}$ in equation

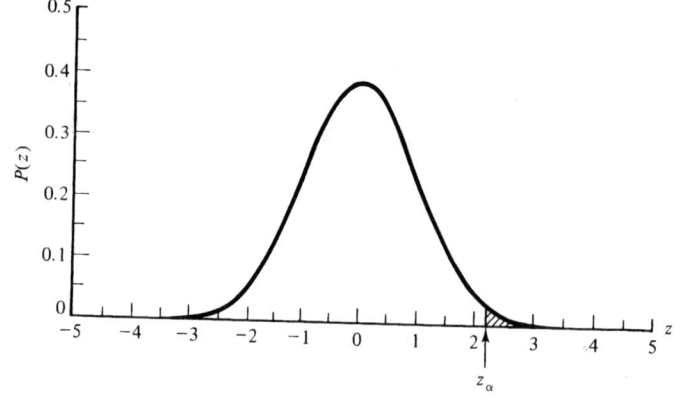

**FIGURE A-1**   Standardized normal probability distribution. The shaded area represents the probability that $z > z_\alpha$. $P$ is the probability.

A-4 so that

$$z = \frac{\overline{X} - \mu}{\sigma/\sqrt{n}} \quad (A-5)$$

Clearly, $z$ decreases with increasing $n$ because $\overline{X}$ is a better estimate of $\mu$.

If the probability distribution is known, then the probability $P$ that the random variable $X$ or $z$ is within a given interval can be calculated. The probability $\alpha$ that $z$ is equal to or greater than a specified value denoted $z_\alpha$, which is always greater than or equal to zero, is given by

$$\alpha = P(z > z_\alpha) \quad (A-6)$$

$$= (\sigma\sqrt{2\pi})^{-1} \int_{z_\alpha}^{\infty} \exp\left(\frac{-z^2}{2}\right) dz$$

Table A-1 indicates how $\alpha$ depends upon $z_\alpha$ and $\alpha$ is

represented by the shaded area in Figure A-1. From $\alpha$ we can calculate the probability that $z$ is less than or equal to $z_\alpha$ $[P(z < z_\alpha)]$, which is just $(1 - \alpha)$. Similarly, $2\alpha$ is the probability that $|z| > z_\alpha$, and $(1 - 2\alpha)$ is the probability that $|z| < z_\alpha$. Since the normal distribution is symmetrical, $(0.5 - \alpha)$ is the probability that $0 < z < z_\alpha$. We will see momentarily how the $z$ statistic and the normal distribution table can be used in practice.

For the $z$ statistic, it is assumed that the population SD ($\sigma$) is known or in practical terms a good estimate of $\sigma$ is available. If the number of measurements used to calculate the SD is less than 30, we cannot assume that $s = \sigma$ for a calculation of the $z$ statistic. In this case we must use the $t$ statistic as defined in the equation

$$t = \frac{\overline{X} - \mu}{s/\sqrt{n}} \quad (A-7)$$

**TABLE A-1**
Probability table for the normal distribution

| z | α | z | α | z | α | z | α | z | α | z | α | z | α | z | α | z | α |
|---|---|---|---|---|---|---|---|---|---|---|---|---|---|---|---|---|---|
| 0.00 | .5000 | 0.35 | .3632 | 0.70 | .2420 | 1.05 | .1469 | 1.40 | .0808 | 1.75 | .0401 | 2.10 | .0179 | 2.45 | .0071 | 2.80 | .0026 |
| 0.01 | .4960 | 0.36 | .3594 | 0.71 | .2389 | 1.06 | .1446 | 1.41 | .0793 | 1.76 | .0392 | 2.11 | .0174 | 2.46 | .0069 | 2.81 | .0025 |
| 0.02 | .4920 | 0.37 | .3557 | 0.72 | .2358 | 1.07 | .1423 | 1.42 | .0778 | 1.77 | .0384 | 2.12 | .0170 | 2.47 | .0068 | 2.82 | .0024 |
| 0.03 | .4880 | 0.38 | .3520 | 0.73 | .2327 | 1.08 | .1401 | 1.43 | .0764 | 1.78 | .0375 | 2.13 | .0166 | 2.48 | .0066 | 2.83 | .0023 |
| 0.04 | .4840 | 0.39 | .3483 | 0.74 | .2296 | 1.09 | .1379 | 1.44 | .0749 | 1.79 | .0367 | 2.14 | .0162 | 2.49 | .0064 | 2.84 | .0023 |
| 0.05 | .4801 | 0.40 | .3446 | 0.75 | .2266 | 1.10 | .1357 | 1.45 | .0735 | 1.80 | .0359 | 2.15 | .0158 | 2.50 | .0062 | 2.85 | .0022 |
| 0.06 | .4761 | 0.41 | .3409 | 0.76 | .2236 | 1.11 | .1335 | 1.46 | .0721 | 1.81 | .0351 | 2.16 | .0154 | 2.51 | .0060 | 2.86 | .0021 |
| 0.07 | .4721 | 0.42 | .3372 | 0.77 | .2206 | 1.12 | .1314 | 1.47 | .0708 | 1.82 | .0344 | 2.17 | .0150 | 2.52 | .0059 | 2.87 | .0021 |
| 0.08 | .4681 | 0.43 | .3336 | 0.78 | .2177 | 1.13 | .1292 | 1.48 | .0694 | 1.83 | .0336 | 2.18 | .0146 | 2.53 | .0057 | 2.88 | .0020 |
| 0.09 | .4641 | 0.44 | .3300 | 0.79 | .2148 | 1.14 | .1271 | 1.49 | .0681 | 1.84 | .0329 | 2.19 | .0143 | 2.54 | .0055 | 2.89 | .0019 |
| 0.10 | .4602 | 0.45 | .3264 | 0.80 | .2119 | 1.15 | .1251 | 1.50 | .0668 | 1.85 | .0322 | 2.20 | .0139 | 2.55 | .0054 | 2.90 | .0019 |
| 0.11 | .4562 | 0.46 | .3228 | 0.81 | .2090 | 1.16 | .1230 | 1.51 | .0655 | 1.86 | .0314 | 2.21 | .0136 | 2.56 | .0052 | 2.91 | .0018 |
| 0.12 | .4522 | 0.47 | .3192 | 0.82 | .2061 | 1.17 | .1210 | 1.52 | .0643 | 1.87 | .0307 | 2.22 | .0132 | 2.57 | .0051 | 2.92 | .0018 |
| 0.13 | .4483 | 0.48 | .3156 | 0.83 | .2033 | 1.18 | .1190 | 1.53 | .0630 | 1.88 | .0301 | 2.23 | .0129 | 2.58 | .0049 | 2.93 | .0017 |
| 0.14 | .4443 | 0.49 | .3121 | 0.84 | .2005 | 1.19 | .1170 | 1.54 | .0618 | 1.89 | .0294 | 2.24 | .0125 | 2.59 | .0048 | 2.94 | .0016 |
| 0.15 | .4404 | 0.50 | .3085 | 0.85 | .1977 | 1.20 | .1151 | 1.55 | .0606 | 1.90 | .0287 | 2.25 | .0122 | 2.60 | .0047 | 2.95 | .0016 |
| 0.16 | .4364 | 0.51 | .3050 | 0.86 | .1949 | 1.21 | .1131 | 1.56 | .0594 | 1.91 | .0281 | 2.26 | .0119 | 2.61 | .0045 | 2.96 | .0015 |
| 0.17 | .4325 | 0.52 | .3015 | 0.87 | .1922 | 1.22 | .1112 | 1.57 | .0582 | 1.92 | .0274 | 2.27 | .0116 | 2.62 | .0044 | 2.97 | .0015 |
| 0.18 | .4286 | 0.53 | .2981 | 0.88 | .1894 | 1.23 | .1093 | 1.58 | .0571 | 1.93 | .0268 | 2.28 | .0113 | 2.63 | .0043 | 2.98 | .0014 |
| 0.19 | .4247 | 0.54 | .2946 | 0.89 | .1867 | 1.24 | .1075 | 1.59 | .0559 | 1.94 | .0262 | 2.29 | .0110 | 2.64 | .0041 | 2.99 | .0014 |
| 0.20 | .4207 | 0.55 | .2912 | 0.90 | .1841 | 1.25 | .1056 | 1.60 | .0548 | 1.95 | .0256 | 2.30 | .0107 | 2.65 | .0040 | 3.00 | .0013 |
| 0.21 | .4168 | 0.56 | .2877 | 0.91 | .1814 | 1.26 | .1038 | 1.61 | .0537 | 1.96 | .0250 | 2.31 | .0104 | 2.66 | .0039 | 3.01 | .0013 |
| 0.22 | .4129 | 0.57 | .2843 | 0.92 | .1788 | 1.27 | .1020 | 1.62 | .0526 | 1.97 | .0244 | 2.32 | .0102 | 2.67 | .0038 | 3.02 | .0013 |
| 0.23 | .4090 | 0.58 | .2810 | 0.93 | .1762 | 1.28 | .1003 | 1.63 | .0516 | 1.98 | .0239 | 2.33 | .0099 | 2.68 | .0037 | 3.03 | .0012 |
| 0.24 | .4052 | 0.59 | .2776 | 0.94 | .1736 | 1.29 | .0985 | 1.64 | .0505 | 1.99 | .0233 | 2.34 | .0096 | 2.69 | .0036 | 3.04 | .0012 |
| 0.25 | .4013 | 0.60 | .2743 | 0.95 | .1711 | 1.30 | .0963 | 1.65 | .0495 | 2.00 | .0228 | 2.35 | .0094 | 2.70 | .0035 | 3.05 | .0011 |
| 0.26 | .3974 | 0.61 | .2709 | 0.96 | .1685 | 1.31 | .0951 | 1.66 | .0485 | 2.01 | .0222 | 2.36 | .0091 | 2.71 | .0034 | 3.06 | .0011 |
| 0.27 | .3936 | 0.62 | .2676 | 0.97 | .1660 | 1.32 | .0934 | 1.67 | .0475 | 2.02 | .0217 | 2.37 | .0089 | 2.72 | .0033 | 3.07 | .0011 |
| 0.28 | .3897 | 0.63 | .2643 | 0.98 | .1635 | 1.33 | .0918 | 1.68 | .0465 | 2.03 | .0212 | 2.38 | .0087 | 2.73 | .0032 | 3.08 | .0010 |
| 0.29 | .3859 | 0.64 | .2611 | 0.99 | .1611 | 1.34 | .0901 | 1.69 | .0455 | 2.04 | .0207 | 2.39 | .0084 | 2.74 | .0031 | 3.09 | .0010 |
| 0.30 | .3821 | 0.65 | .2578 | 1.00 | .1587 | 1.35 | .0885 | 1.70 | .0446 | 2.05 | .0202 | 2.40 | .0082 | 2.75 | .0030 | 3.10 | .0010 |
| 0.31 | .3783 | 0.66 | .2546 | 1.01 | .1562 | 1.36 | .0869 | 1.71 | .0436 | 2.06 | .0197 | 2.41 | .0080 | 2.76 | .0029 | 3.11 | .0009 |
| 0.32 | .3745 | 0.67 | .2514 | 1.02 | .1539 | 1.37 | .0853 | 1.72 | .0427 | 2.07 | .0192 | 2.42 | .0078 | 2.77 | .0028 | 3.12 | .0009 |
| 0.33 | .3707 | 0.68 | .2483 | 1.03 | .1515 | 1.38 | .0838 | 1.73 | .0418 | 2.08 | .0188 | 2.43 | .0075 | 2.78 | .0027 | 3.13 | .0009 |
| 0.34 | .3669 | 0.69 | .2451 | 1.04 | .1492 | 1.39 | .0823 | 1.74 | .0409 | 2.09 | .0183 | 2.44 | .0073 | 2.79 | .0026 | 3.14 | .0008 |

If this is compared to the $z$ statistic, we note that the only difference is that $\sigma$ has been replaced by the experimental standard deviation, $s$. As for the $z$ statistic, the difference between the experimental and population mean has been normalized to the SD. The probability distribution for the $t$ statistic is similar to that for the $z$ statistic, but it is broader and the breadth of the distribution varies with $n$. Probability tables (see Table A-2) indicate the probability $P(t > t_\alpha) = \alpha$ that $t$ is equal to or greater than a specified value of $t$ denoted $t_\alpha$, which is always positive. Note in the table that the value of $t_\alpha$ varies with the number of measurements $n$ used to calculate $s$ or the number of degrees of freedom $\nu$ defined as $n - 1$. As for the $z$ statistic, $P(t < t_\alpha)$ is $(1 - \alpha)$, $P(|t| < t_\alpha)$ is $(1 - 2\alpha)$, and here $P(|t| > t_\alpha)$ is $2\alpha$, and $P(0 < t < t_\alpha)$ is $(0.5 - \alpha)$.

For a given $\alpha$, $t_\alpha$ is greater than $z_\alpha$ because $t$ is a more conservative statistic which takes into account the uncertainty in determining $s$. Hence to include the chance that the calculated magnitude of $s$ is greater than the actual $\sigma$ due to random error, $t$ must be greater than $z$ (or the experimental mean must be further from the population mean) to obtain the same level of confidence $\alpha$ that the difference between the means is due only to random error. Note in Table A-2, that $t_\alpha$ is quite large for small $n$ (e.g., 2 or 3) due to the high probability of error in the estimate of $s$, but as $n$ increases for a given value of $\alpha$, $t_\alpha$ decreases and for $n = \infty$, $t_\alpha = z_\alpha$.

Usually, the mean and SD are estimated from the same $n$ measurements. However, if they are not, it is important to realize that $n$ in the definitions of the $z$ and $t$ statistics (equations A-5 and A-7) is the number of measurements used to obtain the mean, while for the $t$ statistic, $\nu$ is determined by the number of measurements used to obtain the SD. Often, $\sigma/n^{1/2}$ or $s/n^{1/2}$, which are denoted $\hat{\sigma}$ and $\hat{s}$, is defined as the SD of the mean, while $\sigma$ or $s$ is the SD of an individual result. The SD of the mean represents the SD of a probability distribution of the means rather than of individual measurements.

**TABLE A-2**
Critical values of $t$

| $n$ | $t_{.100}$ | $t_{.050}$ | $t_{.025}$ | $t_{.010}$ | $t_{.001}$ | $\nu$ |
|---|---|---|---|---|---|---|
| 2 | 3.078 | 6.314 | 12.706 | 31.821 | 63.657 | 1 |
| 3 | 1.886 | 2.920 | 4.303 | 6.965 | 9.925 | 2 |
| 4 | 1.638 | 2.353 | 3.182 | 4.541 | 5.841 | 3 |
| 5 | 1.533 | 2.132 | 2.776 | 3.747 | 4.604 | 4 |
| 6 | 1.476 | 2.015 | 2.571 | 3.365 | 4.032 | 5 |
| 7 | 1.440 | 1.943 | 2.447 | 3.143 | 3.707 | 6 |
| 8 | 1.415 | 1.895 | 2.365 | 2.998 | 3.499 | 7 |
| 9 | 1.397 | 1.860 | 2.306 | 2.896 | 3.355 | 8 |
| 10 | 1.383 | 1.833 | 2.262 | 2.821 | 3.250 | 9 |
| 11 | 1.372 | 1.812 | 2.228 | 2.764 | 3.169 | 10 |
| 12 | 1.363 | 1.796 | 2.201 | 2.718 | 3.106 | 11 |
| 13 | 1.356 | 1.782 | 2.179 | 2.681 | 3.055 | 12 |
| 14 | 1.350 | 1.771 | 2.160 | 2.650 | 3.012 | 13 |
| 15 | 1.345 | 1.761 | 2.145 | 2.624 | 2.977 | 14 |
| 16 | 1.341 | 1.753 | 2.131 | 2.602 | 2.947 | 15 |
| 17 | 1.337 | 1.746 | 2.120 | 2.583 | 2.921 | 16 |
| 18 | 1.333 | 1.740 | 2.110 | 2.567 | 2.898 | 17 |
| 19 | 1.330 | 1.734 | 2.101 | 2.552 | 2.878 | 18 |
| 20 | 1.328 | 1.729 | 2.093 | 2.539 | 2.861 | 19 |
| 21 | 1.325 | 1.725 | 2.086 | 2.528 | 2.845 | 20 |
| 22 | 1.323 | 1.721 | 2.080 | 2.518 | 2.831 | 21 |
| 23 | 1.321 | 1.717 | 2.074 | 2.508 | 2.819 | 22 |
| 24 | 1.319 | 1.714 | 2.069 | 2.500 | 2.807 | 23 |
| 25 | 1.318 | 1.711 | 2.064 | 2.492 | 2.797 | 24 |
| 26 | 1.316 | 1.708 | 2.060 | 2.485 | 2.787 | 25 |
| 27 | 1.315 | 1.706 | 2.056 | 2.479 | 2.779 | 26 |
| 28 | 1.314 | 1.703 | 2.052 | 2.473 | 2.771 | 27 |
| 29 | 1.313 | 1.701 | 2.048 | 2.467 | 2.763 | 28 |
| 30 | 1.311 | 1.699 | 2.045 | 2.462 | 2.756 | 29 |
| inf. | 1.282 | 1.645 | 1.960 | 2.326 | 2.576 | inf. |

## Hypothesis Testing

The $z$ and $t$ statistics are used to make statistical statements about data, to indicate the significance of the difference between the true and experimental means, or to indicate the range in which the true mean will lie with a given level of confidence. If the true mean is known or there is an expected value, the test statistics can be used to calculate the probability that the difference between the true mean and experimental mean is significant or real (i.e., due to a systematic error) and not totally ascribable to random error. To do this, the test statistic is calculated, and this experimental value of $z$ or $t$ (if $n < 30$) is compared to the value of $z_\alpha$ or $t_\alpha$ which is close to but smaller than the experimental value. The $\alpha$ value corresponding to this value of $t_\alpha$ or $z_\alpha$ is noted. From this $\alpha$ value we can say there is less than a probability $\alpha$ that the difference between the experimental and expected value is totally due to random error or a with a confidence level of $(1 - \alpha)$ that some of the difference is due to systematic error. Note that we are using a one-tailed test, because we are only concerned with $\overline{X}$ being more positive or more negative than $\mu$; the experimental value of $t$ or $z$ is calculated so that the test statistic is positive. For example, if $\overline{X} = 2.00$ and $\mu = 2.05$, the experimental value of $z$ equals $|2.00 - 2.05|/(\sigma/n^{1/2})$ or $0.05/(\sigma/n^{1/2})$.

In the procedure described above, we are actually setting up a null hypothesis that the experimental and theoretical mean are really the same and only apparently different due to random error. An error of the first kind is rejecting the null hypothesis when it is true. To make this error small, we chose $t_\alpha$ or $z_\alpha$ to be large enough that $\alpha$ is small. If $\alpha$ is small, there is a high degree of confidence that the means are significantly different, but we cannot say what part of the difference is due to random error and what part is due to systematic error; there is still a probability $\alpha$ that all the difference is due to random error. We also can say that with a confidence level of $(1 - \alpha)$ that the means are different although the difference may only be partially due to systematic error.

If the experimental value of $t$ or $z$ is less than the value of $t_\alpha$ or $z_\alpha$ for a given value of $\alpha$, we cannot accept the null hypothesis that there is no significant difference between the means or say with a $(1 - \alpha)$ level of confidence that the difference is only due to random error even though there is a possibility that the difference is totally due to systematic error. We can only say that we cannot reject the null hypothesis. An error of the second kind is accepting the null hypothesis when it is false. The probability $\beta$ of doing this can be calculated, but is beyond the scope of this book. In U.S. law, the null hypothesis is that the defendant is innocent. We wish to minimize an error of the first kind by making $\alpha$ small so that an innocent defendant is not found guilty. An error of the second kind is to find a guilty defendant innocent.

## Confidence Intervals

We can also use the $t$ and $z$ statistics to predict the probability that an interval around the experimental mean encloses the true mean according to the equations

$$\mu = \overline{X} \pm \frac{z\sigma}{\sqrt{n}} \qquad \text{(A-8)}$$

$$\mu = \overline{X} \pm \frac{ts}{\sqrt{n}} \qquad \text{(A-9)}$$

Equation A-8 is used if $n \geq 30$, while equation A-9 is used if $n < 30$. The confidence level depends on the value of $z$ or $t$ chosen and the number of measurements used to obtain $s$ in equation A-9. Here a two-tailed test is used since $\mu$ can be greater or less than $\overline{X}$. Thus the value of $\alpha$ is chosen so that $(1 - 2\alpha)$ equals the level of confidence desired. For equation A-8, choosing $z$ equal to 1, 2, or 3 corresponds to confidence levels of 68.2%, 95.6%, and 99.8%, respectively. Values of $z$ of 1.96 and 2.58 correspond to 95% and 99% confidence levels since $\alpha$ equals 0.025 and 0.005, respectively. Note that the error interval estimated by the equations A-8 and A-9 only accounts for random error. It can be reduced, for a given level of confidence and SD, by increasing the number of measurements $n$.

To illustrate the discussion above, consider the following example. Five measurements of quantity $X$ are made from which we calculate that $\overline{X} = 1.01$ and $s = 0.010$. We expected a value of 1.00 ($\mu = 1.00$). The $t$ statistic is $(1.01 - 1.00)/(0.01/5^{1/2}) = 2.24$ and $\nu = 5 - 1 = 4$. From Table A-2, we can see that $2.24 > t_{0.05} = 2.13$, but $2.24 < t_{0.025} = 2.776$. Thus we can say that 1.01 is significantly different than 1.00 or the difference of 0.01 is not totally due to randomness at the 95% confidence level. However, at the 97.5% confidence level, we cannot say that the difference is significant since at this level of confidence the difference can be totally attributed to random error. This in no way implies that the total difference is due to random error since the observed difference is normally due to random and systematic error. We can also say at the 95% confidence level that the range $1.01 \pm 2.776 \times 0.01/5^{1/2} = 1.01 \pm 0.012$ encloses the true mean since for $\nu = 4$, $t_\alpha = 2.776$ for $\alpha = 0.025$.

Note that we cannot use the $z$ statistic for the data above because only five measurements are used to calculate the SD. For the $z$ statistic (see Table A-1)

$z < z_\alpha$ for $\alpha = 0.0125$. This predicts a significant difference at the 98.8% confidence level which is too liberal a confidence level. Similarly, for the two-tailed test, the interval for 95% confidence would be smaller by a factor of $(1.96/2.776) = 0.71$ since $z_\alpha = 1.96$ for $\alpha = 0.025$. Again this is not warranted due to the uncertainty in the estimate of the SD.

The $t$ and $z$ statistics can be used only if the normal distribution applies. In this case, test statistics are also available to compare the experimental variance to the population variance (chi-square test) and to compare two experimental standard deviations ($F$-test). Other test statistics are available for different distributions and, in addition, we can test the normalcy of data and evaluate if other test statistics must be employed. Sometimes the data can be made to follow the normal distribution by mathematical transformation. The log normal distribution is an example where the logarithm of values follows the normal distribution. This distribution applies sometimes to the concentration distribution of trace elements in environmental samples where negative concentrations are not possible.

## A-2  PROPAGATION OF UNCERTAINTIES

If a result $R$ is computed from or is dependent on a number of measurements or variables ($A, B, C, \ldots$), we can estimate both the systematic error and the SD in $R$ from the systematic errors and standard deviations in $A, B, C \ldots$ with **propagation of uncertainty** mathematics. Thus, if

$$R = f(A, B, C, \ldots) \qquad \text{(A-10)}$$

partial differentiation yields

$$dR = \frac{\partial f}{\partial A}\, dA + \frac{\partial f}{\partial B}\, dB + \frac{\partial f}{\partial C}\, dC + \cdots \qquad \text{(A-11)}$$

where the variables $A$, $B$, and $C$ are assumed to be independent.

If the infinitesimal differentials are replaced by small finite errors, the latter equation becomes

$$\Delta R = \frac{\partial R}{\partial A}\, \Delta A + \frac{\partial R}{\partial B}\, \Delta B + \frac{\partial R}{\partial C}\, \Delta C + \cdots \qquad \text{(A-12)}$$

for systematic errors where $\Delta R$, $\Delta A$, $\Delta B$, and $\Delta C$ are estimates of the systematic errors in $R$, $A$, $B$, and $C$,

**TABLE A-3**
Propagation of random uncertainties for common functions

1.  $R = A \pm B$
    $\sigma_R = (\sigma_A^2 + \sigma_B^2)^{1/2}$

2.  $R = A \times B$
    $\sigma_R = (B^2\sigma_A^2 + A^2\sigma_B^2)^{1/2}$
    $R = \dfrac{A}{B}$
    $\sigma_R = \left(\dfrac{\sigma_A^2}{B^2} + \dfrac{A^2\sigma_B^2}{B^4}\right)^{1/2}$
    $\dfrac{\sigma_R}{R} = \left[\left(\dfrac{\sigma_A}{A}\right)^2 + \left(\dfrac{\sigma_B}{B}\right)^2\right]^{1/2}$  for $R = A \times B$  or $\dfrac{A}{B}$

3.  $R = \log_{10} A = (0.4343)\ln A$
    $\sigma_R = \dfrac{(0.4343)\sigma_A}{A}$

4.  $R = e^A$
    $\sigma_R = e^A\sigma_A = R\sigma_A$

5.  $R = \dfrac{A - C}{B - C}$
    $\dfrac{\sigma_R}{R} = \dfrac{1}{B - C}\left(\dfrac{\sigma_A^2}{R^2} + \sigma_B^2 + \left(\dfrac{1 - R}{R}\right)^2 \sigma_R^2\right)^{1/2}$

respectively. Usually, the absolute values of each product in the sum are added together unless the sign of the systematic error is known.

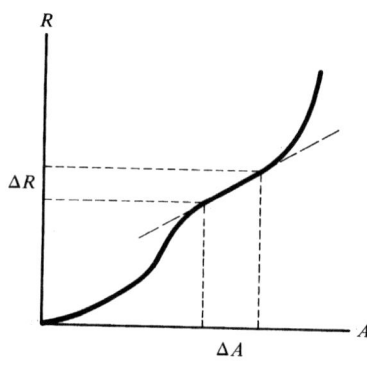

**FIGURE A-2**  Graphical interpretation of propagation of uncertainty mathematics. The partial derivative of the function $R$ with respect to a variable $A$ is the slope of the function with respect to $A$ at a particular value of $A$. This indicates how a change in $A$ causes a change in $R$. Thus if we reflect the uncertainty in $A$, $\Delta A$, through this slope, we can determine the uncertainty in $R$, $\Delta R$, caused by $\Delta A$. To obtain an accurate value of $\Delta R$, it is necessary that the derivative be constant over $\Delta A$.

For random errors, equation A-11 becomes

$$s_R = \sqrt{\left(\frac{\partial R}{\partial A}\right)^2 s_A^2 + \left(\frac{\partial R}{\partial B}\right)^2 s_B^2 + \left(\frac{\partial R}{\partial C}\right)^2 s_C^2 + \cdots}$$

(A-13)

where $s_R$, $s_A$, $s_B$, and $s_C$ are the standard deviations in $R$, $A$, $B$, and $C$, respectively. In equations A-12 and A-13 it is assumed that errors or standard deviations are much smaller than mean values (e.g., $\Delta A \ll A$, $s_A \ll A$). If the variables are nonindependent or correlated, then cross products which represent covariances must be included.

Formulae for random errors are shown in Table A-3 for several common functions. The graphical interpretation of propagation of uncertainty mathematics is presented in Figure A-2.

## REFERENCES

1. W. Mendenhall, *Introduction to Probability and Statistics*, 5th ed., Duxbury Press, North Scituate, Mass., 1979.

2. *Precision Measurement and Calibration-Statistical Concepts and Procedures*, NBS Special Publication 300, vol. 1, National Bureau of Standards, Washington, D.C., February 1969.

3. M. G. Natrella, *Experimental Statistics*, NBS Handbook 91, National Bureau of Standards, Washington, D.C., August 1963.

4. P. R. Bevington, *Data Reduction and Error Analysis for the Physical Sciences*, McGraw-Hill, New York, 1969.

5. P. Moritz, "The Application of Mathematical Statistics in Analytical Chemistry," in *Comprehensive Analytical Chemistry*, G. Svehla, ed., vol. 11, chap. 1, Elsevier, Amsterdam, 1981.

6. H. D. Young, *Statistical Treatment of Experimental Data*, McGraw-Hill, New York, 1962.

7. J. H. Zar, *Biostatistical Analysis*, 2nd ed., Prentice-Hall, Englewood Cliffs, N.J., 1984.

8. A. Savitzky and M. J. E. Golay, "Smoothing and Differentiation of Data by Simplified Least Squares Procedures," *Anal. Chem.*, *36*, 1627 (1964).

9. C. G. Enke and T. A. Nieman, "Signal-to-Noise Ratio Enhancement by Least-Squares Polynominal Smoothing," *Anal. Chem.*, *48*, 705A (1976).

10. R. J. Harris, *A Primer of Multivariate Statistics*, 2nd ed., Academic Press, Orlando, Fla., 1985.

11. D. L. Massart, A. Dijkstra, and L. Kaufman, *Evaluation and Optimization of Laboratory Methods and Analytical Procedures*, Elsevier, Amsterdam, 1978.

12. J. C. Miller and J. N. Miller, *Statistics for Analytical Chemistry*, Halsted Press, New York, 1984.

13. R. Caulcutt and R. Boddy, *Statistics for Analytical Chemists*, Chapman & Hall, New York, 1983.

14. W. J. Dixon and F. J. Massey, Jr., *Introduction to Statistical Analysis*, 3rd ed., McGraw-Hill, New York, 1969.

15. O. L. Davies, ed., *Statistical Methods in Research and Production*, 4th ed., Hafner Press, New York, 1972.

16. E. B. Wilson, Jr., *An Introduction to Scientific Research*, McGraw-Hill, New York, 1952.

17. C. A. Bennett and N. L. Franklin, *Statistical Analysis in Chemistry and the Chemical Industry*, Wiley, New York, 1954.

18. H. A. Laitinen and W. E. Harris, *Chemical Analysis*, 2nd ed., chaps. 26 and 27, McGraw-Hill, New York, 1975.

19. W. J. Youden, *Statistical Methods for Chemists*, Wiley, New York, 1951.

20. G. E. P. Box, W. G. Hunter, and J. S. Hunter, *Statistics for Experimenters*, Wiley, New York, 1978.

21. N. R. Draper and H. Smith, *Applied Regression Analysis*, 2nd ed., Wiley, New York, 1981.

22. S. L. Meyer, *Data Analysis for Scientists and Engineers*, Wiley, New York, 1975.

# APPENDIX B

# Properties
# of Optical Materials

Many different materials and coatings are used in optical components and systems. In this appendix a few of the most important properties of materials are presented. Additional information can be found in *Handbook of Optics*, W. G. Driscoll and W. Vaughan, eds, (McGraw-Hill, New York, 1978), and in catalogs and guides available from optics houses such as Melles Griot, Oriel, Ealing, Newport, and many others.

## B-1 REFRACTIVE INDICES

The refractive index of optical materials that are used to construct lenses, windows, filters, or dispersive devices is a crucial property. The refractive index is a function of wavelength and temperature. Table B-1 lists the refractive indices of several important optical materials, usually measured at 589 nm. If a different wavelength was used, it is indicated in parentheses. Where applicable, the ordinary refractive index ($\eta_o$) and the extraordinary refractive index ($\eta_e$) are given. These were also obtained at 589 nm except where noted. The value of $\eta_e$ varies with the propagation direction, and the minimum or maximum value is reported.

A refractive index difference across an interface separating two media causes a reflection loss at the interface according to the Fresnel equations 3-8 and 3-9. Table B-2 gives the spectral reflectances for several common interfaces at 589 nm for radiation incident normal to the interface. It is assumed in this table that $\eta_{glass} = 1.500$.

Since a glass lens has two glass–air interfaces, reflection losses are approximately $1 - (0.96)^2 \approx 0.08$ or 8% for each lens unless they are antireflection-loss coated. For a system with multiple lenses, reflection losses can rapidly accumulate and drastically reduce the throughput. Magnesium fluoride with a thin-film refractive index of $\approx 1.38$ at 550 nm is commonly used as a coating. When a thin coating of $MgF_2$ is used, instead of a single reflection at each air–glass interface, there are two, a reflection at the air–$MgF_2$ interface followed by one at the $MgF_2$–glass interface. These two reflections destructively interfere giving rise to a total reflectance at normal incidence at one lens surface of less than 1.5% for a coating of one-quarter wavelength thickness. For a complete lens, the reflection loss is $\approx 3\%$ for a single-layer coating at normal incidence. The reflectance can be minimized for various angles of incidence and wavelengths by applying coatings of ap-

**TABLE B-1**
Refractive indices of optical materials at 20°C

| Material | $\eta$ at 589 nm | | Material | $\eta$ at 589 nm | |
|---|---|---|---|---|---|
| Air | 1.003 | | AgCl | 2.000 | (3.9 $\mu$m) |
| $H_2O$ | 1.333 | | Sapphire | 1.769 | (579 nm) |
| Fused silica | 1.458 | | NaCl | 1.544 | |
| | 1.513 | (240 nm) | | 1.522 | (4.0 $\mu$m) |
| Borosilicate | 1.517 | | ZnS | 2.35 | |
| Crown glass | 1.548 | (313 nm) | Cryolite | 1.34 | |
| KCl | 1.490 | | ADP | 1.525 | ($\eta_o$, 579 nm) |
| | 1.471 | (4.71 $\mu$m) | | 1.479 | ($\eta_e$, 579 nm) |
| KI | 1.666 | | KDP | 1.510 | ($\eta_o$, 579 nm) |
| | 1.627 | (4.13 $\mu$m) | | 1.469 | ($\eta_e$, 579 nm) |
| $MgF_2$ | 1.378 | ($\eta_o$) | Calcite | 1.658 | ($\eta_o$) |
| | 1.390 | ($\eta_e$) | | 1.486 | ($\eta_e$) |
| Crystal quartz | 1.544 | ($\eta_o$) | MgO | 1.773 | (361 nm) |
| | 1.553 | ($\eta_e$) | | 1.723 | (1.01 $\mu$m) |
| KBr | 1.560 | | Benzene | 1.500 | |
| | 1.535 | (4.26 $\mu$m) | | | |
| 1 M KCl (aq) | 1.342 | | | | |

propriate thicknesses. The $MgF_2$ coating also protects the glass from chemical deterioration. Multilayer broadband coatings can give reflectances of less than 0.6% over much of the visible region. Multilayer coatings are also available to minimize the reflectance at a particular wavelength; these can reduce the reflective loss to less than 0.25% at the chosen wavelength.

## B-2 TRANSMISSION CHARACTERISTICS

Absorption of the incident radiation by the optical material can be a significant problem in spectrochemical analysis. The absorption characteristics often dictate the type of material suitable for use in lenses, windows, cuvettes, and other components. The transmission characteristics of several typical UV and IR transmitting glasses are shown in Figure B-1.

Often, absorption coefficient data are given for optical materials. The transmittance $T$ of the glass (neglecting surface reflection losses) is given by $T = e^{-kb}$, where $k$ is the absorption coefficient (cm$^{-1}$) and $b$ is the thickness of the glass (cm). Figure B-2 shows $k$ vs. $\lambda$ for several materials.

Note that glass, quartz, and fused silica all absorb in the IR region at wavelengths longer than 2 $\mu$m. Hence for IR spectrometry, halide salts (NaCl, NaBr, AgCl, KBr) are used for optical components. The IR transmitting regions for several IR optical materials are listed in Table B-3. The water solubility of these materials is also given.

## B-3 REFLECTANCE OF MIRROR MATERIALS

The reflectance properties of typical mirror coatings are illustrated in Figure B-3. Mirrors are usually front-sur-

**TABLE B-2**
Reflectances for several interfaces

| Interface | Spectral reflectance | Interface | Spectral reflectance |
|---|---|---|---|
| Glass–air | 0.0403 | NaCl–air | 0.045 |
| Glass–$H_2O$ | 0.0035 | KCl–air | 0.038 |
| Glass–1 M KCl | 0.0031 | Sapphire–air | 0.076 |
| Glass–benzene | $1.1 \times 10^{-7}$ | Sapphire–$H_2O$ | 0.0198 |

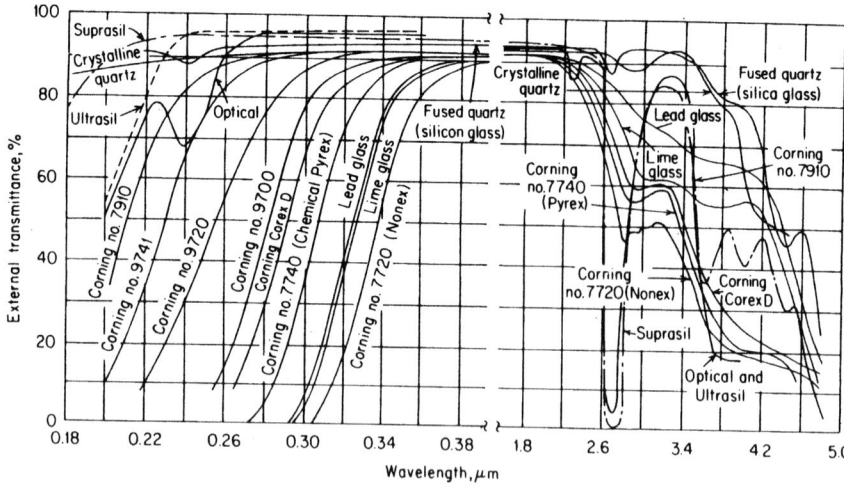

**FIGURE B-1** Transmittance vs. wavelength for several UV and IR transmitting glasses. Ultrasil is a type of fused quartz and suprasil is fused silica. (Reprinted with permission from *Handbook of Optics*, W. Driscoll and W. Vaughan, eds., McGraw-Hill, New York, 1978.)

employed in the UV-visible region. The $MgF_2$ prevents oxidation of the aluminum and thus a deterioration of its reflectance.

In addition to metallic reflecting coatings, thin dielectric films are used for extremely high reflectance mirrors. These are typically multiple layers of quarter-wavelength films with alternating layers of high and low refractive indexes. The reflectances of such thin-film coatings often exceed 99%.

The substrate on which the reflecting surface is coated is normally crown glass, Pyrex, fused silica, or glass–ceramic. The latter is suitable for use in situations demanding extremely high thermal stability. Crown glass is suitable when high thermal stability is not required, Pyrex when moderate thermal stability is needed, and fused silica for all but the most demanding situations.

## B-4 OTHER PROPERTIES

In addition to the properties which control the transmittance and reflectance of radiation, there are several other important properties to consider when selecting optical materials. The thermal characteristics of the material are quite important, as mentioned above, for mirror substrates, lenses, and cell materials. The specific heat, thermal conductivity, and coefficient of linear expansion determine how a material responds to changes in temperature. The elasticity and rigidity of a material may also be important to consider in specific applications. For use as cell windows for liquid samples, the solubility of the material in the various solvents that are commonly used is an important consideration. (See, for example, Table B-3 for the water solubility of various IR window materials.)

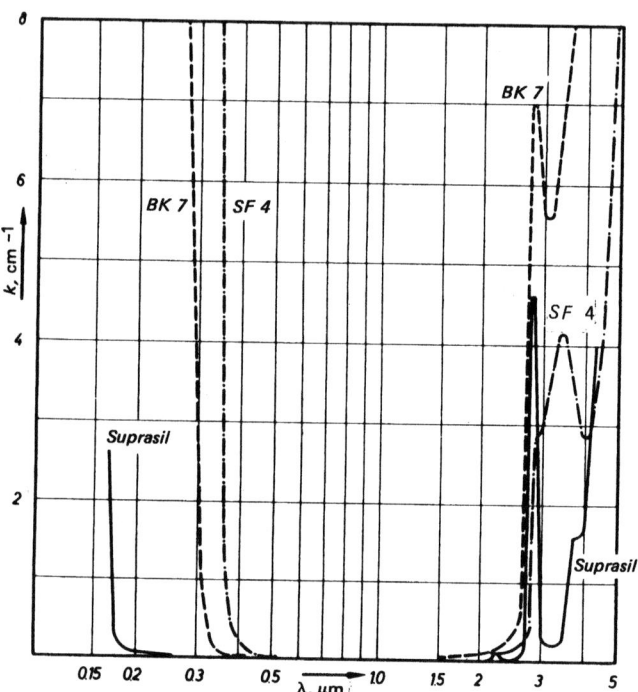

**FIGURE B-2** Absorption coefficient vs. wavelength for several glasses. The BK7 is a crown glass and the SF4 is a heavy flint glass. (With permission of Marcel Dekker, Inc., from H. W. Faust, "Prism Systems, Spectrographs and Spectrometers," chap. 3 in *Analytical Emission Spectroscopy*, vol. I, pt. I, E. L. Grove, ed., Marcel Dekker, New York, 1971.)

face coated so that radiation does not pass through glass, which would cause additional reflection losses and aberrations. Aluminum coated with $MgF_2$ is commonly

**TABLE B-3**
Properties of infrared window materials

| Optical material | Upper wavelength limit ($\mu$m)[a] | Water solubility (g/100 g) |
|---|---|---|
| MgF$_2$ (Irtran 1) | 6.5 | Insoluble |
| MgO (Irtran 5) | 8.5 | Insoluble |
| LiF | 9 | 0.27 |
| CaF$_2$ (Irtran 3) | 12 | 0.0017 |
| ZnS (Irtran 2) | 14.5 | Insoluble |
| BaF$_2$ | 15 | 0.17 |
| NaF | 15 | 4.22 |
| ZnSe (Irtran 4) | 24 | Insoluble |
| NaCl | 26 | 35.7 |
| AgCl | 28 | Insoluble |
| KCl | 30 | 34.7 |
| CdTe (Irtran 6) | 30 | Insoluble |
| KRS-6[b] | 35 | 0.32 |
| KRS-5[b] | 40 | 0.05 |
| KBr | 40 | 53.5 |
| KI | 45 | 127.5 |
| CsBr | 55 | 124.3 |
| CsI | 80 | 44 |
| Diamond | 80 | Insoluble |

[a]The lower wavelength limit of all materials is < 1.0 $\mu$m.
[b]The KRS materials are mixed crystal ThI$_2$–ThBr$_2$ eutectics.

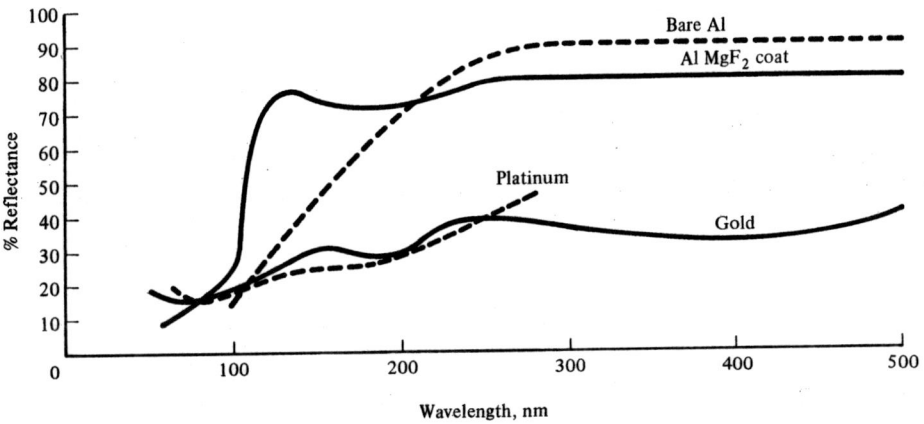

**FIGURE B-3** Reflectance of mirror materials vs. wavelength.

# REFERENCES

1. W. L. Wolfe, "Properties of Optical Materials," in *Handbook of Optics*, W. G. Driscoll and W. Vaughan, eds., McGraw-Hill, New York, 1978.

2. J. A. Dobrowolski, "Coatings and Filters," in *Handbook of Optics*, W. G. Driscoll and W. Vaughan, eds., McGraw-Hill, New York, 1978.

3. S. Musikant, *Optical Materials: An Introduction to Selection and Application*, Marcel Dekker, New York, 1984.

4. *Electro-Optics Handbook*, Technical Series EOH-11, RCA Corp., Lancaster, Pa., 1974.

5. J. Wilson and J. F. B. Hawkes, *Optoelectronics: An Introduction*, Prentice-Hall, Englewood Cliffs, N.J., 1983.

In addition, the catalogs from many of the optics companies (Oriel, Melles Griot, Corion, Ealing, Rolyn, Newport, Physitec, etc.) often contain a wealth of information about optical materials.

# APPENDIX C

# Characteristics of Optical Filters

This appendix gives the transmittance characteristics of various optical filters. Figure C-1 shows the transmittance curves for 12 different colored-glass bandpass filters for the UV-visible region. Such filters are commonly employed for rejection of stray radiation, for order sorting, and for many other applications that do not require the narrow bandpass characteristics of interference filters. The transmission curves for a set of

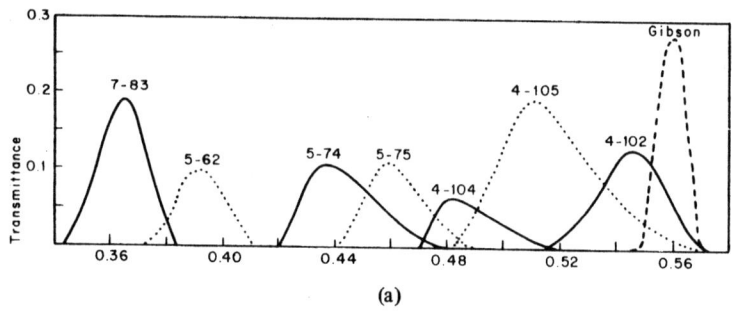

(a)

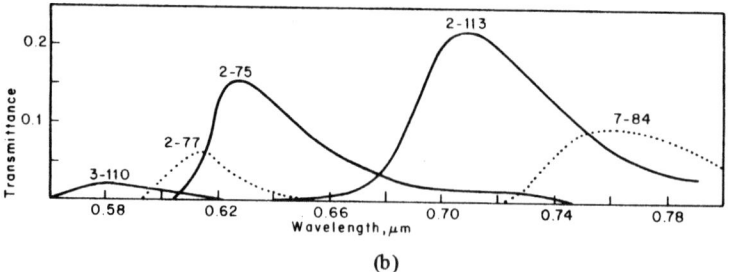

(b)

**FIGURE C-1** Transmittances of Corning glass filters. The numerical designations are the product numbers for specific filters. [With permission from K. S. Gibson, *J. Opt. Soc. Am., 25,* 131 (1935).]

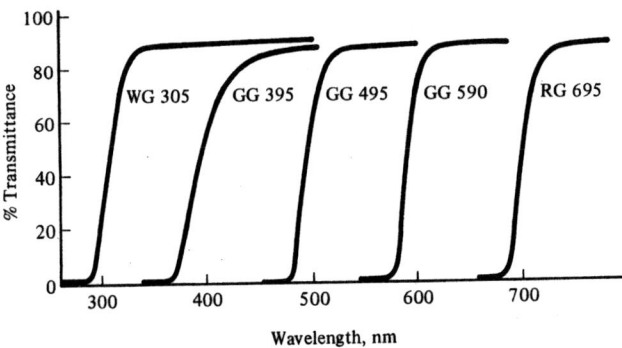

**FIGURE C-2** Transmittance of Schott colored-glass cutoff filters. Specific Schott glass types are indicated by the code (WG 305, GG 395, etc.). (Courtesy of Melles Griot.)

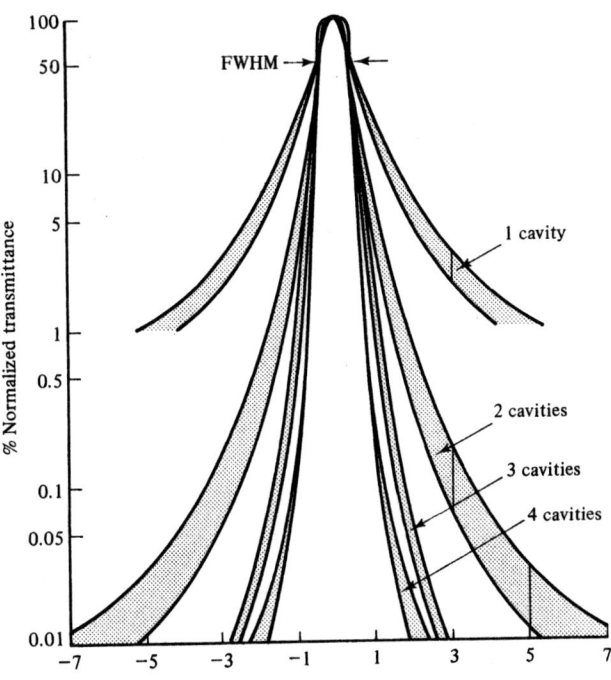

**FIGURE C-3** Transmittance curves for bandpass interference filters with one to four cavities. The shaded regions indicate nominal tolerance limits for each filter type. (Courtesy of Melles Griot.)

colored-glass cutoff filters are shown in Figure C-2. Both short- and long-wavelength cutoff filters for the UV-visible region are readily available.

Neutral density filters are filters with absorbances (the obsolete term *optical density* is often used) that are constant over a broad wavelength range. They are widely used to attenuate optical beams without influencing the chromaticity of the beam. Neutral density filters are made by vacuum depositing thin films of metallic alloys on glass or fused silica substrates. The compositions are chosen to give flat spectral response characteristics. Such filters are available in absorbances from 0 to 6.0. They can be used in the wavelength range between 350 and 700 nm for glass and 200 to 700 nm for fused silica. Filter absorbances are additive. Thus filters can be combined to give values not otherwise available. Circular linear-wedge neutral density filters are also made. These have absorbances that are linear functions of the angle of rotation of the wedge. Some manufacturers produce calibrated circular disk neutral density filters. The absorbance of these varies with rotation. They typically provide absorbances from 0.1 to 2.0 with a rotation of 285°. Another type of variable neutral density filter consists of three thin-film polarizers. Two polarizers are aligned with their polarization axes parallel. The third polarizer is rotated to give an absorbance that is variable from 0 to 3.0 absorbance units. The relative absorbance (after accounting for an insertion loss) is independent of wavelength (450 to 750 nm), polarization state, and incidence angle. This type of filter is highly useful for calibrating photometric instruments and checking the linearity of detectors.

Interference filters are typically of the Fabry–Perot type or the multiple cavity type (see Section 3-5). Multiple-cavity filters have steeper rejection slopes and nearly square passband tops instead of Gaussian or Lorentzian shapes. Alternating layers of high (ZnS) and low (cryolite, $Na_3AlF_6$) refractive index materials are deposited

on top of each other to form an interference cavity. Cavities are separated by a layer with one-half wavelength (or a multiple) thickness, which itself makes no contribution (absentee layer). Filters with 2-4 cavities are typical. Figure C-3 shows typical transmittance curves for multiple-cavity filters as a function of the number of cavities. The transmittance curves for a set of visible interference filters are shown in Figure C-4. Filters can be made with any desired central wavelength from 200 nm to 25 μm. Interference filter sets with values spaced every 10 nm are available as stock items. In addition, filters for isolating or rejecting various laser lines (Ar ion, He-Ne, Nd:YAG, etc.) and for isolating various

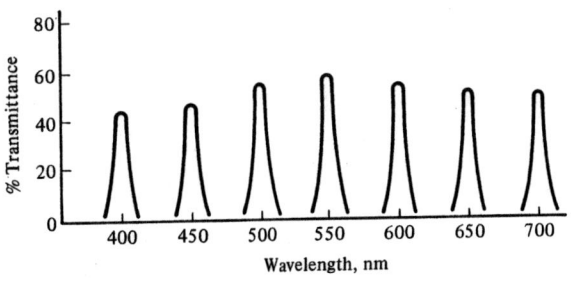

**FIGURE C-4** Transmittance curves for a set of multiple-cavity interference filters. (Courtesy of Melles Griot.)

atomic lines are often stock items. Continuously variable interference filters are also available in which the central wavelength changes with position along the length of the filter. One commercial filter (Oriel Corp.) isolates wavelengths from 400 to 700 nm along a 60-mm length. Circular wedge interference filters are also available.

Interference filters can be angle tuned over a small wavelength range as described in Section 3-5. Filters with 5- to 10-nm bandpass values can be tuned over about 10 nm (<30° rotation) without drastic changes in the bandpass shape. Filters with narrower bandpass have less tuning range.

Although filters for the UV-visible region are most common, IR interference filters are also readily available. One manufacturer (Oriel Corp.) produces IR interference filters in the range 1.8 to 5.5 μm with central wavelengths every 0.1 μm. The bandwidths (FWHM values) normally vary from 0.1 to 0.15 μm. Special filters with central wavelengths as long as 20 μm can be purchased. Short- and long-wavelength cutoff filters are also available for the IR region.

## REFERENCES

Many of the general optics references for Chapter 3 are also relevant here.

1. *Optics and Filters*, Oriel Corporation, Stamford, Conn., vol. III of Oriel catalog.
2. *Optical Filters and Coatings*, Corion, Corp., Holliston, Mass.
3. *Optics Guide 3*, Melles Griot, Irvine, Calif. More than a catalog—almost an optics textbook.
4. E. Hecht and A. Zajac, *Optics*, Addison-Wesley, Reading, Mass., 1974. Good discussions on the principles and construction of optical filters.
5. J. A. Dobrowolski, "Coatings and Filters," in *Handbook of Optics*, W. G. Driscoll and W. Vaughan, eds., McGraw-Hill, New York, 1978.

# APPENDIX D

# Photomultiplier Tube Specifications

Manufacturers of photomultiplier tubes (PMTs) such as RCA, Hamamatsu, and EMI provide detailed specification sheets for these devices which indicate their characteristics and operating considerations. As an example, some of the figures from the specification sheet for an RCA 1P28 PMT are reproduced here (Figures D-1 and D-2). This tube or its equivalent is used in many spectrometers.

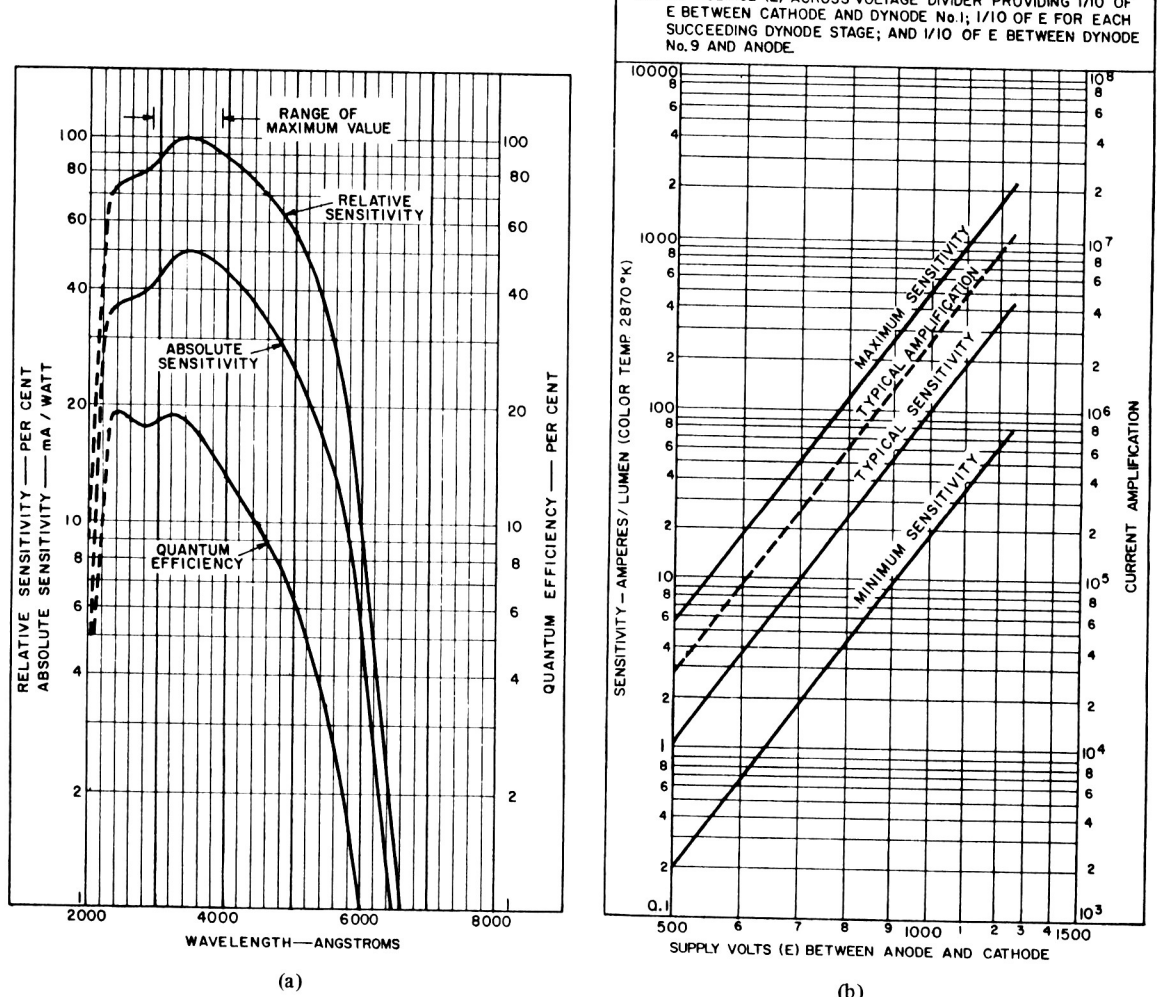

(a)                                                    (b)

**FIGURE D-1**    Spectral response, sensitivity, and current amplification characteristics. In (a) plots of quantum efficiency $[K(\lambda)]$ and absolute sensitivity [the cathodic responsivity, $R(\lambda)$] vs. wavelength are shown. From these data, the photocathodic current or pulse rate at a given wavelength for a given incident radiant power can be calculated. A plot of relative sensitivity is the cathodic responsivity normalized to 100% for the wavelength of maximum response [i.e., $R(\lambda)/(R(\lambda_m)]$. In (b) the dependence of the typical current amplification (the gain $m$) on the PMT bias voltage $(E_b)$ is illustrated. Note the linear log-log relationship between $m$ and $E_b$. The sensitivity [anodic responsivity, $mR(\lambda)$] is also shown as a function of $E_b$. The maximum, typical, and minimum plots illustrate that the anodic responsivity can vary about two orders of magnitude among tubes of the same type. This behavior is due primarily to a variation in gain among tubes. Thus the figure can be used to estimate $m$ within an order of magnitude. Note that photometric units (e.g., lumens) are often used instead of more convenient radiometric units (e.g., watts). (Courtesy of RCA New Products Division, Lancaster, Penn.)

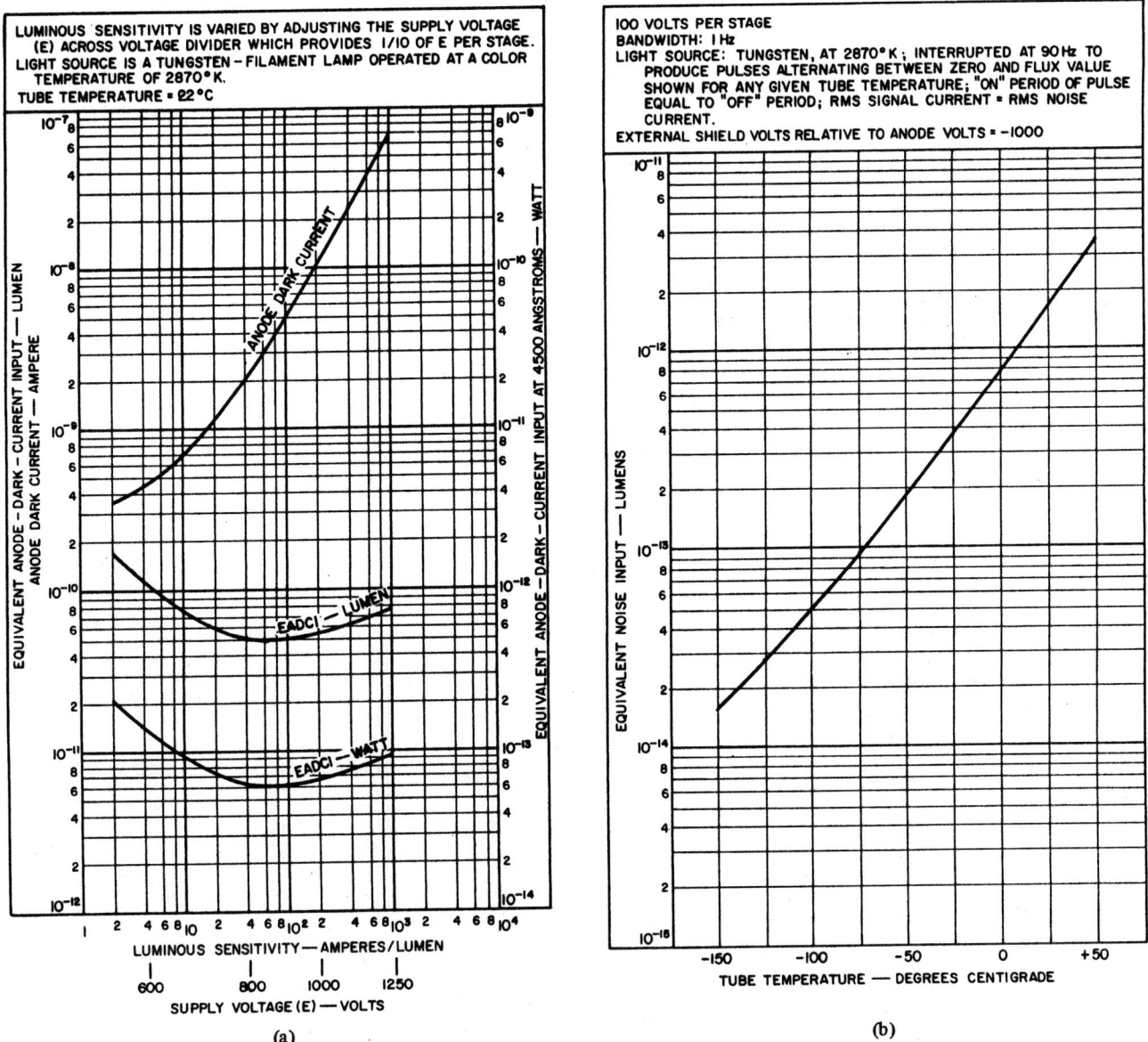

**FIGURE D-2** Dark current, EADCI, and ENI characteristics. In (a) the dependence of the anodic dark current on the bias voltage ($E_b$) is shown. At higher bias voltages, the log-log plot is reasonably linear because the dark current is primarily due to cathodic thermal emission and proportional to the gain ($i_{ad} = mi_{cd}$). Also shown are plots of the equivalent anode dark current input (EADCI), the incident radiant power necessary to produce a photoanodic current equal to the anodic dark current. At about 900 V, the ratio of gain to dark current is optimal. The effect of temperature on the equivalent noise input (ENI) is illustrated in (b). The ENI is the incident radiant power that produces a photoanodic current equal to the anodic dark current rms noise (i.e., an $S/N$ of unity if measurements are dark-current noise limited). As the temperature is reduced, the dark current and dark current noise decrease exponentially. (Courtesy of RCA New Products Division, Lancaster, Penn.)

# APPENDIX E

# Sample Preparation Methods

Analytical methods usually require some sample treatment prior to the final, instrumental measurement. For the majority of the spectrochemical methods discussed in this book, the analytical sample must be in the form of a solution. Some samples are obtained initially as solutions (e.g., blood serum, other biological fluids, samples from lakes, rivers, oceans, and wastewaters). Solid samples, however, must normally be converted into solution samples in order to mix, aliquot, dilute, separate, or otherwise treat and present to the spectrometer. Sample cleanup methods are used, if necessary, for separation of the analyte, removal of primary interferences, or preconcentration of the analyte. This appendix briefly reviews the most common sample dissolution and separation methods employed with spectrochemical methods. The exact preparation method to be used depends highly on the nature of the initial sample, the likelihood of potential interferences being present, and the characteristics of the specific spectrochemical methods to be employed. The references for this appendix should be consulted for specific details. In addition, a good deal of information can be found in the "cookbooks" published by instrument manufacturers and in various applications-oriented monographs.

## E-1 SAMPLE DISSOLUTION METHODS

Inorganic samples are brought into solution by **mineral acid dissolution** or **decomposition with fluxes**. Metallic samples, some alloys, and metallic oxides, as well as pure solids to prepare standards, are commonly dissolved in mineral acids such as HCl, $HNO_3$, $H_3PO_4$, HF, $HClO_4$, and $H_2SO_4$. Heating is often required and the type of acid or combination of acids used depends on the composition of the sample (e.g., HF is primarily used for decomposing silicate rocks and minerals). In any acid dissolution method, potential contamination of the sample or standard by impurities in the acid must be considered, particularly for trace determinations.

Many mineral samples and materials of geological interest are solubilized by forming a soluble fused salt by heating a mixture of an appropriate flux material and the sample at high temperatures in special inert crucibles (e.g., carbon or platinum). Common flux materials are $Na_2CO_3$ (m.p. 851°C), $Na_2O_2$ (decomposes at high temperature), $K_2S_2O_7$ (m.p. 300°C), and $LiBO_2$ (m.p. 840°C). The use of fusion techniques for dissolution is subject to contamination from the high concentration of the flux material or from the crucible and volatilization losses at the high temperatures employed.

Analysis of organic materials, including biological solids, for trace metals usually requires oxidative treatment to convert the carbon and hydrogen to $CO_2$ and $H_2$. Oxidative decompositions are **wet-ashing procedures**, also called **digestions**, and **dry-ashing methods**. In wet-ashing methods, the sample is treated with oxidative mineral acids such as $HNO_3$, $HClO_4$, and $H_2SO_4$ or mixtures of these acids at elevated temperatures. Precipitation and volatilization of analytes, the potential of explosions, and contamination by impurities in the acids must be considered. Dry ashing is a very simple procedure, but prone to losses of some trace elements (e.g., Hg) by volatilization. In dry ashing the organic matter is heated in steps up to $\approx 500°C$ in a muffle furnace in the presence of atmospheric oxygen until carbonaceous material has been oxidized to $CO_2$, after which the sample can be dissolved in dilute acid. The use of microwave ovens for dissolution and ashing procedures is rapidly increasing.

## E-2 SEPARATION AND PRECONCENTRATION

A wide variety of separation techniques, including solvent extraction, ion exchange, and precipitation, have been applied prior to spectrochemical determinations for preconcentration purposes and for gross separations from major matrix constituents. Solvent extraction techniques are commonly used to extract the metals of interest into an organic solvent with the use of a complexing agent. If the volume of the organic solvent is less than that of the aqueous sample and the extraction efficiency is near unity, preconcentration results.

In flame atomic spectrometry, the organic phase is often directly introduced into the flame and can provide significant signal enhancements particularly with spray chamber nebulizers. Such enhancements can result from alteration of the physical properties of the solution (i.e., lower surface tension and increased vapor pressure), which improves the nebulization and desolvation efficiencies and, in some cases, increases in flame temperature or excitation efficiency due to chemiluminescence. Ammonium pyrrolidine dithiocarbamate (APDC) is widely used to extract many transition metals simultaneously in conjunction with methyl isobutyl ketone (MIBK) as the solvent.

Cation, anion, and chelating (e.g., Chelex 100) ion-exchange resins are often used in separating and preconcentrating trace elements. A large volume of the sample solution is passed through the resin column and certain ions are selectively retained depending on the type of resin. For a cation or chelating exchange resins, a small volume of a strong acid is then used to elute the retained metals. Large concentration factors can be achieved because of the high capacity of many resins.

In certain situations undesired major constituents of a sample can be separated by precipitation methods. For example, proteins are often removed from biological fluids prior to spectrochemical analysis by precipitation with trichloroacetic acid or occasionally sodium tungstate and centrifugation. Deproteinization can prevent clogging and fouling of nebulizers in plasma and flame spectroscopy; the protein-free solution more nearly matches the viscosity of aqueous standards.

Precipitation reactions can also be used to separate the species of interest by forming an insoluble compound. The precipitate is then redissolved and analyzed. For example, calcium and magnesium can be determined by precipitation with 8-hydroxyquinoline followed by determination with atomic spectrometry on the redissolved precipitate. In some cases, trace metals are coprecipitated in the precipitate of a metal added in excess.

Many other separation and preconcentration methods have been employed with spectrochemical methods. Simple evaporation is an effective but slow method for preconcentration of aqueous solutions of low salt content. Dialysis has been used for protein removal particularly in continuous flow determinations of various analytes in biological fluids.

# REFERENCES

1. R. Bock, *A Handbook of Decomposition Methods in Analytical Chemistry*, translated and revised by I. L. Marr, International Textbook, London, 1979.
2. T. T. Gorsuch, *The Destruction of Organic Matter*, Pergamon Press, Oxford, 1970.
3. J. Dolezal, P. Povondra, and Z. Sulcek, *Decomposition Techniques in Inorganic Analysis*, Iliffe Books, London, 1968.
4. T. C. O'Haver, "Chemical Aspects of Elemental Analysis," chap. III in *Trace Analysis: Spectroscopic Methods for Elements*, J. D. Winefordner, ed., Wiley-Interscience, New York, 1976.

5. I. M. Korennan, *Analytical Chemistry of Low Concentrations*, translated by J. Schmorak, Israel Program for Scientific Translations, Jerusalem, 1968.

6. G. Tolg, *Ultramicro Elemental Analysis*, translated by C. G. Thalmayer, Wiley, New York, 1970.

7. R. E. Thiers, "Contamination in Trace Element Analysis and its Control," in *Methods of Biochemical Analysis*, D. Glick, ed., vol. V, Wiley-Interscience, New York, 1957, p. 274.

8. P. J. Elving and I. M. Kolthoff, eds., *Treatise on Analytical Chemistry*, 2nd ed., pt. 1, vol. 5, Wiley-Interscience, New York, 1982. This volume contains chapters on decomposition and dissolution of inorganic and organic samples and on techniques for separation of interferents.

9. A. Mizuikc, *Enrichment Techniques in Inorganic Trace Analysis*, Springer-Verlag, Berlin, Heidelberg, 1983.

# APPENDIX F

# Atomic and Molecular Transitions

The particles (electrons, nuclei) in a molecule or atom can be described by the time-independent Schrödinger equation

$$H\psi = E\psi \qquad \text{(F-1)}$$

where $\psi$ is the wavefunction, $H$ is the Hamiltonian operator, and $E$ is the energy of a state characterized by $\psi$. The product of $\psi$ and its complex conjugate $\psi^*$ represents a probability distribution for the position of a particle in an atom or molecule in the state $\psi$. The Hamiltonian operator is composed of potential energy and kinetic energy operators which can be determined for any atom or molecule. For atoms or molecules with only one electron, equation F-1 can be solved to obtain analytical functions for the wavefunctions and energy levels. For most atoms and molecules, such an exact solution is not possible. In this case, approximation methods (e.g., linear variation methods, perturbation theory) and numerical methods are used to provide estimates of the wavefunctions and the energy-level structure.

The wavefunction can also be used to calculate the **transition moment** $R$ as shown in the equation

$$R = \int \psi_i^* \mu \psi_j \, d\tau \qquad \text{(F-2)}$$

for a transition between states $i$ and $j$, where $\mu$ is the electric dipole moment operator ($\mu = er$), $e$ is the charge, $r$ is the distance between the charges, and $d\tau$ indicates the integration over all space. The square of the transition moment or $R^2$ in J cm$^3$ is called the **transition probability**.

We will concern ourselves primarily with electric dipole transitions involving the first-order interaction of the oscillating electric vector of the radiation field with the electric dipole moment of an atom or molecule. Radiative transitions can also arise due to the interaction of the radiation field with electric quadrupole moments or magnetic dipole moments. These second-order effects are extremely weak compared to first-order electric dipole transitions.

Only transitions between certain energy levels are probable as expressed by selection rules. Equation F-2 is used to establish selection rules. Often, the symmetry of wavefunctions can be used to show that $R$ is zero for certain transitions called **forbidden transitions**. If $R$ is finite, the transition is an **allowed transition**. The magnitude of $R^2$ indicates the probability of the transition and hence the intensity of the transition. Transitions that are dipole forbidden may occur because they are electric quadrupole or magnetic dipole allowed; these are usually weak in intensity.

## F-1 EINSTEIN COEFFICIENTS

There are many ways to express the intensity of transitions. It is important to be able to interconvert between these quantities. The different Einstein coefficients express the probability in terms of the rate constants for the transitions. The transition probability is related to the Einstein coefficient for absorption $(B_{ij})$ in $cm^3 J^{-1} s^{-1}$ Hz by

$$B_{ij} = \frac{8\pi^3 R^2}{3h^2 g_i} \qquad (F\text{-}3)$$

where $g_i$ is the degeneracy or statistical weight of the lower state. When degeneracies are involved, $R^2$ represents the sum of the squares of all of the transition moments for all possible sublevels in the upper and lower states. The Einstein coefficients for stimulated emission $(B_{ji})$ and spontaneous emission $(A_{ji})$ are related to $B_{ij}$ (see Section 4-1) as follows:

$$B_{ji} = B_{ij} \frac{g_i}{g_j} \qquad (F\text{-}4)$$

$$A_{ji} = \frac{8\pi h v_m^3 B_{ij} g_i}{g_j c^3} = \frac{8\pi h B_{ij} g_i}{g_j \lambda_m^3} \qquad (F\text{-}5)$$

where $g_j$ is the statistical weight of the upper state and $v_m$ and $\lambda_m$ are the frequency and wavelength of maximum absorption or emission. Combining equations F-3 and F-5 yields the following relationship between the Einstein coefficient for spontaneous emission and the transition probability:

$$A_{ji} = \frac{64\pi^4 v_m^3 R^2}{3hc^3 g_j} = \frac{64\pi^4 R^2}{3h \lambda_m^3 g_j} \qquad (F\text{-}6)$$

It is common to use the classical three-dimensional harmonic oscillator as a model for transitions; it is assumed that the electron responsible for absorption is attracted to the center of a molecule or an atom by a Hooke's law type of force. With this model the wavefunctions are known. The classical transition probability can be calculated from equation F-2 and is given by

$$R^2(\text{classical}) = \frac{3e^2 h g_i}{32\pi^3 \varepsilon_0 m_e v_m} = \frac{3e^2 h \lambda_m g_i}{32\pi^3 \varepsilon_0 m_e c} \qquad (F\text{-}7)$$

where $m_e$ is the mass of an electron, $e$ is the charge in coulombs, and the factor $g_i$ is added to account for the degeneracy of the lower state. From this equation and equations F-3 and F-6, the classical Einstein coefficients

can be shown to be

$$B_{ij}(\text{classical}) = \frac{e^2}{4\varepsilon_0 h v_m m_e} = \frac{e^2 \lambda_m}{4\varepsilon_0 hc m_e} \qquad (F\text{-}8)$$

$$A_{ij}(\text{classical}) = \frac{2\pi e^2 v_m^2 g_i}{\varepsilon_0 m_e c^3 g_j} = \frac{2\pi e^2 g_i}{\varepsilon_0 m_e c \lambda_m^2 g_j} \qquad (F\text{-}9)$$

## F-2 OSCILLATOR STRENGTHS

The classical expressions above are used to define another measure of transition strengths, termed the **oscillator strength** $f$ which describes the number of electrons per atom or molecule undergoing a radiative transition. This dimensionless quantity is the ratio of a theoretically or experimentally derived measure of the transition strength to the respective value predicted by the classical harmonic oscillator model [e.g., $R^2/R^2(\text{classical})$ or $B_{ij}/B_{ij}(\text{classical})$]. In particular, the absorption oscillator strength $(f_{ij})$ is related to the transition probability and Einstein coefficients by

$$f_{ij} = \frac{32R^2\pi^3 \varepsilon_0 m_e v_m}{3e^2 h g_i} = \frac{32R^2\pi^3 \varepsilon_0 m_e c}{3e^2 h \lambda_m g_i} \qquad (F\text{-}10)$$

$$f_{ij} = \frac{4\varepsilon_0 h v_m m_e B_{ij}}{e^2} = \frac{4\varepsilon_0 hc m_e B_{ij}}{e^2 \lambda_m}$$

$$= \frac{2.50 \times 10^{-34} B_{ij}}{\lambda_m} \qquad (F\text{-}11)$$

$$f_{ij} = \frac{A_{ji} \varepsilon_0 m_e c^3 g_j}{2g_i \pi e^2 v_m^2} = \frac{A_{ji} \lambda_m^2 \varepsilon_0 m_e c g_j}{2g_i \pi e^2}$$

$$= \frac{1.50 \times 10^4 A_{ji} \lambda_m^2 g_j}{g_i} \qquad (F\text{-}12)$$

$$A_{ji} = \frac{2\pi e^2 g_i f_{ij}}{\varepsilon_0 m_e c g_j \lambda_m^2} = \frac{6.67 \times 10^{-5} g_i f_{ij}}{g_j \lambda_m^2} \qquad (F\text{-}13)$$

The evaluated constants in the last forms of equations F-11 to F-13 apply if $\lambda_m$ is expressed in meters. The relationship shown in equation F-13 is used extensively in Chapter 7.

The emission oscillator strength $(f_{ji})$ is related to the absorption oscillator strength by

$$f_{ji} = f_{ij} \frac{g_i}{g_j} \qquad (F\text{-}14)$$

In the literature, the oscillator strength is often tabulated as a weighted value, the "*gf*" factor. It is related to the oscillator strengths for emission and absorption by

$$gf = g_i f_{ji} = g_j f_{ij} \qquad (F-15)$$

Oscillator strength is a convenient way to express transition strength because it is dimensionless and has a nominal range from zero to unity (or a small integer if there is degeneracy). Typically, $f_{ij}$, for an electronic transition from the ground state to the first excited state ($f_{01}$), is on the order of unity for a strong spectral line or band. Experimental values of the oscillator strength can be greater than unity although this is not predicted by the simple model used.

## F-3 RELATIONSHIPS BETWEEN EXPERIMENTAL AND FUNDAMENTAL QUANTITIES

Absorption coefficients [$k(v)$], absorptivities [$a(v)$ or $\varepsilon(v)$], and absorption cross sections [$\sigma(v)$] can be directly calculated from the measured absorbance with the equations (see Section 3-1)

$$A = 0.434k(v)b = 0.434\sigma(v)bn = a(v)bc = \varepsilon(v)bc \qquad (F-16)$$

where $c$ is the concentration of the absorber. Thus these experimentally derived measures of absorption strength are related by

$$k(v) = \sigma(v)n = 2.303a(v)c = 2.303\varepsilon(v)c \qquad (F-17)$$

Because the absorbance is usually measured over a wavelength interval much smaller than the width of the absorption line or band, these measures of absorption strength are spectral quantities that vary over the profile of the absorption line or band. By contrast, fundamental quantities such as transition probabilities, Einstein coefficients, or oscillator strengths are related to transition strength for a whole line or band. We now show how fundamental quantities are related to the experimentally derived quantities.

Consider radiation of spectral energy density $U_v$ (J cm$^{-3}$ Hz$^{-1}$) incident on a sample containing $n_i$ molecules or atoms in lower state $i$ per cm$^3$. From equation 4-8, the number of transitions per cm$^3$ per second produced by the incident radiation across the whole absorption line or band is $B_{ij}U_v n_i$, where $U_v$ is assumed constant over the absorption profile.

The radiant power absorbed from the incident light beam of 1 cm$^2$ cross section in a layer of thickness $db$

at a given frequency $v$ or the decrease in spectral irradiance $dE_v$ (in W cm$^{-2}$ Hz$^{-1}$) is given by

$$dE_v = -B_{ij}U_v n_i hv S_v \, db \qquad (F-18)$$

where $S_v$ is the profile of the line or band in Hz$^{-1}$ normalized so that $\int S_v \, dv = 1$. With the relationship $E_v = U_v c$, equation F-18 becomes

$$dE_v = -B_{ij}\frac{E_v}{c} n_i hv S_v \, db \qquad (F-19)$$

If equation 3-10 is written in terms of spectral irradiance, the decrease in spectral irradiance over pathlength $db$ is given by

$$dE_v = -k(v)E_v \, db \qquad (F-20)$$

where $k(v)$ is the frequency-dependent absorption coefficient in cm$^{-1}$. If we set the right sides of equations F-19 and F-20 equal to each other, we note that

$$k(v) = \frac{B_{ij}n_i hv S_v}{c} \qquad (F-21)$$

Equation F-21 can also be written in terms of wavelength

$$k(\lambda) = \frac{B_{ij}n_i h\lambda S_\lambda}{c} \qquad (F-22)$$

where $S_\lambda$ is the line profile in m$^{-1}$ ($S_\lambda = S_v c/\lambda^2$).

By using the relationship between $B_{ij}$ and the transition probability $R^2$ (equation F-3), equations F-21 and F-22 become

$$k(v) = \frac{8\pi^3 n_i v R^2 S_v}{3hcg_i} \qquad (F-23)$$

$$k(\lambda) = \frac{8\pi^3 n_i \lambda R^2 S_\lambda}{3hcg_i} \qquad (F-24)$$

With equation F-5, the absorption coefficient can also be expressed in terms of the Einstein coefficient for spontaneous emission as follows:

$$k(v) = \frac{A_{ji}c^2 g_j S_v n_i}{8\pi v^2 g_i} \qquad (F-25)$$

$$k(\lambda) = \frac{A_{ji}\lambda^4 g_j S_\lambda n_i}{8\pi c g_i} \qquad (F-26)$$

Finally, we express absorption coefficients in terms of

the absorption oscillator strength with equation F-13:     case

$$k(\nu) = \frac{f_{ij} S_\nu n_i e^2}{4\varepsilon_0 m_e c} \tag{F-27}$$

$$k(\lambda) = \frac{f_{ij} S_\lambda \lambda^2 n_i e^2}{4\varepsilon_0 m_e c^2} = 8.817 \times 10^{-15} f_{ij} S_\lambda \lambda^2 n_i \tag{F-28}$$

where the evaluated constant has units of meters.

It is common to calculate the integrated absorption coefficient ($\bar{k}$) [the integral of $k(\nu)$ over the absorption line or band profile ($p$) in cm$^{-1}$ Hz], in which

$$\bar{k} = \int_p k(\nu)\, d\nu = \frac{B_{ij} n_i h}{c} \int_p \nu S_\nu\, d\nu$$

$$= \frac{8\pi^3 n_i R^2}{3hcg_i} \int_p \nu S_\nu\, d\nu \tag{F-29}$$

This equation provides a way to calculate transition probabilities or Einstein coefficients from experimental measurements. If the absorption band or line is narrow,

---

**TABLE F-1**

Relationships between quantities indicating absorption strength

| | | |
|---|---|---|
| Absorption cross section (cm$^2$)[a] | $\sigma(\nu) = \dfrac{B_{ij} h \nu S_\nu}{c}$ | (F-33) |
| | $\sigma(\nu) = \dfrac{8\pi^3 R^2 \nu S_\nu}{3hcg_i}$ | (F-34) |
| Molar absorptivity (L mol$^{-1}$ cm$^{-1}$)[b] | $\varepsilon(\nu) = \dfrac{B_{ij} h \nu N S_\nu}{2303c}$ | (F-35) |
| | $\varepsilon(\nu) = \dfrac{8\pi^3 N R^2 \nu S_\nu}{6909hcg_i}$ | (F-36) |
| Integrated molar absorptivity (L mol$^{-1}$ cm$^{-1}$ Hz)[c] | $\bar{\varepsilon} = 1.065\varepsilon_m\, \Delta\nu = \dfrac{1.065\varepsilon_m c\, \Delta\lambda}{\lambda_m^2}$ | (F-37) |
| | $\bar{\varepsilon} \approx \dfrac{B_{ij} h \nu_m N}{2303c} = \dfrac{B_{ij} h N}{2303\lambda_m}$ | (F-38) |
| | $\bar{\varepsilon} \approx \dfrac{8\pi^3 N R^2 \nu_m}{6906hcg_i} = \dfrac{8\pi^3 N R^2}{6909h\lambda_m g_i}$ | (F-39) |
| | $\bar{\varepsilon}(\text{classical}) = \dfrac{e^2 N}{9212\varepsilon_0 cm_e}$ | (F-40) |
| Oscillator strength[d] | $f_{ij} = \dfrac{9212\varepsilon_0 m_e c\bar{\varepsilon}(\exp)}{e^2 N} = 1.44 \times 10^{-19}\, \bar{\varepsilon}(\exp)$ | (F-41) |
| | $f_{ij} = 1.53 \times 10^{-19}\varepsilon_m\, \Delta\nu = 4.59 \times 10^{-11}\, \dfrac{\varepsilon_m\, \Delta\lambda}{\lambda_m^2}$ | (F-42) |

[a] From equations F-17, F-21, and F-23.

[b] From equations F-17, F-21, and F-23 and $n = Nc_M/1000$, where $c_M$ is the molar concentration and $N$ is Avogadro's number.

[c] $\bar{\varepsilon}$ is the integral of $\varepsilon(\nu)$ over the absorption band. In equation F-37, a Gaussian band shape is assumed so that the area is 1.065 times the product of the band height and half-width or the maximum molar absorptivity ($\varepsilon_m$) and band half-width ($\Delta\nu$ in Hz or $\Delta\lambda$ in m) in this case. Equation F-40 is derived from equations F-39 and F-7.

[d] The experimental value of $\bar{\varepsilon}$ divided by the classical value (equation F-40).

$\nu$ can be considered constant over the absorption band and equation F-29 reduces to

$$\bar{k} = \frac{B_{ij}n_ih\nu_m}{c} = \frac{8\pi^3 n_i R^2 \nu_m}{3hcg_i} \qquad \text{(F-30)}$$

where $\nu_m$ is the frequency of maximum absorption. This approximation is only good for very narrow lines such as observed in electronic atomic spectra or some pure rotational spectra, but is often quoted in the literature for broad molecular electronic or vibrational absorption bands.

By substituting the expression for the classical transition probability (equation F-7) into equation F-30, we obtain an expression for the classical absorption coefficient:

$$\bar{k}(\text{classical}) = \frac{e^2 n_i}{4\varepsilon_0 m_e c} \qquad \text{(F-31)}$$

This now allows us to calculate the oscillator strength from the experimentally determined integrated absorption coefficient [$f_{ij} = \bar{k}(\text{exp})/\bar{k}(\text{classical})$] or

$$f_{ij} = \frac{4\varepsilon_0 \bar{k}(\text{exp})m_e c}{e^2 n_i} \qquad \text{(F-32)}$$

In molecular spectrometry, it is more common to deal with the absorption cross section and particularly, the molar absorptivity. Using the relationship between the absorption coefficient and the absorption cross section or molar absorptivity (equation F-17) and the last few equations, we can derive similar expressions that relate $\varepsilon(\nu)$, $\sigma(\nu)$, and the integrated molar absorptivity [$\bar{\varepsilon} = \int \varepsilon(\nu)\, d\nu$] to the transition probability, Einstein coefficients, and the oscillator strength. These are summarized in Table F-1.

## F-4  UNIT CONSIDERATIONS

The units in the equations presented in the appendix are complex and can be confusing because they vary among literature sources. This is particularly true for expressions involving the Einstein $B$ coefficients. It must be remembered that the product of radiation field intensity and the $B$ coefficient must have units of $s^{-1}$. Throughout this appendix, we have assumed that the radiation field intensity is expressed in terms of a spectral energy density $U_\nu$ in J cm$^{-3}$ Hz$^{-1}$ so that $B$ has units of J$^{-1}$ cm$^3$ Hz s$^{-1}$. This is true even in expressions that contain $\lambda$ rather than $\nu$ such as the second form of equation F-5 ($c/\lambda$ substituted for $\nu$). If the spectral energy density is expressed in J cm$^{-3}$ nm$^{-1}$, the $B$ coefficient has units of J$^{-1}$ cm$^3$ nm s$^{-1}$. To convert the expressions given into the form appropriate for $B$ coefficients in nm, we use the relationship $d\lambda = d\nu \, \lambda^2/c$ and substitute $B_\lambda c/\lambda^2$ for $B_\nu$, where the subscript indicates the type of units. If the spectral energy density is expressed in photons cm$^{-3}$ Hz$^{-1}$, the $B$ coefficient has units of cm$^3$ Hz s$^{-1}$. In this case, the conversion involves substituting $B_p/h\nu$ for $B_e$. In some literature, the radiation field intensity is given in terms of spectral irradiance in W cm$^{-2}$ Hz$^{-1}$, in which case $B$ has the units W$^{-1}$ cm$^2$ Hz s$^{-1}$. To convert to expressions with $B$ coefficients in these units, $B_E c$ is substituted for $B_U$, where the subscript $E$ denotes irradiance units and $U$ denotes energy density units.

Another potential point of confusion has to do with the units of the charge of an electron ($e$). Throughout this book we express $e$ in SI units of coulombs (i.e., $e = 1.6 \times 10^{-19}$ C). Some authors express $e$ in terms of electrostatic units (esu), in which case $e = 4.80 \times 10^{-10}$ esu. Expressions containing $e^2$ in SI units can be converted to those in esu (or Gaussian) units by substituting $4\pi\varepsilon_0 e^2(\text{esu})$ for $e^2(\text{C})$. Note that $4\pi\varepsilon_0 = 1.113 \times 10^{-10}$ C$^2$ N$^{-1}$ m$^{-2}$ or $1.113 \times 10^{-19}$ C$^2$ dyne$^{-1}$ cm$^{-2}$ and that $e^2(\text{C})/4\pi\varepsilon_0 = 2.56 \times 10^{-38}$ kg m$^3$ s$^{-1}$ or $2.56 \times 10^{-29}$ g cm$^3$ s$^{-1}$.

# Units, Constants, Conversion Factors, Abbreviations, and Quantum Numbers

## 1. COMMONLY USED UNITS, THEIR ABBREVIATIONS, AND PREFIXES

| ABBREVIATION* | UNIT |
|---|---|
| atm | atmosphere |
| A | ampere |
| Å | angstrom |
| cal | calorie |
| cd | candela |
| C | coulomb |
| °C | degrees Celsius |
| dyn | dyne |
| erg | erg |
| eV | electron volt |
| F | farad |
| g | gram |
| G | gauss |
| H | henry |
| Hz | hertz (cycles per second) |
| J | joule |
| K | degrees Kelvin |
| lm | lumen |
| L | liter |
| m | meter |
| mol | mole |
| M | moles per liter |
| N | newton |
| P | poise (for viscosity) |
| rad | radian (plane angle) |
| s | second |
| sr | steradian (solid angle) |
| V | volt |
| W | watt |
| Ω | ohm |

*Note that unit abbreviations are uppercase when the unit is named after a person. The symbol for liter is capitalized to distinguish it from the number one.

### PREFIXES

| | | |
|---|---|---|
| g | giga | $10^9$ |
| M | mega | $10^6$ |
| k | kilo | $10^3$ |
| c | centi | $10^{-2}$ |
| m | milli | $10^{-3}$ |
| μ | micro | $10^{-6}$ |
| n | nano | $10^{-9}$ |
| p | pico | $10^{-12}$ |
| f | femto | $10^{-15}$ |
| a | atto | $10^{-18}$ |

## 2. PHYSICAL CONSTANTS

| | | |
|---|---|---|
| $c$ | speed of light in vacuum | $2.99792458 \times 10^{10}$ cm s$^{-1}$ |
| $e$ | elementary charge | $1.60210 \times 10^{-19}$ C |
| $h$ | Planck's constant | $6.6260755 \times 10^{-34}$ J s |
| | | $6.6260755 \times 10^{-27}$ erg s |
| $k$ | Boltzmann's constant | $1.38054 \times 10^{-23}$ J K$^{-1}$ |
| | | $1.38054 \times 10^{-16}$ erg K$^{-1}$ |
| $m_e$ | rest mass of electron | $9.1093897 \times 10^{-28}$ g |
| $N$ | Avogadro's number | $6.0221367 \times 10^{23}$ mol$^{-1}$ |
| $R$ | gas constant | $8.31441$ J mol$^{-1}$ K$^{-1}$ |
| | | $1.98719$ cal mol$^{-1}$ K$^{-1}$ |
| $v_{air}$ | speed of light in air | $2.997056 \times 10^{10}$ cm s$^{-1}$ |
| $\varepsilon_0$ | permittivity of free space | $8.854 \times 10^{-12}$ C$^2$ N$^{-1}$ m$^{-2}$ |
| $\mu_0$ | permeability of free space | $4\pi \times 10^{-7}$ N s$^2$ C$^{-2}$ |
| $\sigma$ | Stefan–Boltzmann constant | $5.6697 \times 10^{-12}$ W cm$^{-2}$ K$^{-4}$ |

## 3. EVALUATED CONSTANTS

$$2hc^2 = 1.190 \times 10^{16} \text{ W nm}^4 \text{ cm}^{-2} \text{ sr}^{-1}$$

$$\frac{hc}{k} = 1.438 \times 10^7 \text{ nm K}$$

$$hc = 1.986 \times 10^{-16} \text{ J nm}$$

$$\frac{h}{k} = 4.799 \times 10^{-11} \text{ s K}$$

$$4\pi\varepsilon_0 = 1.113 \times 10^{-10} \text{ C}^2 \text{ N}^{-1} \text{ m}^{-2}$$

$$= 1.113 \times 10^{-19} \text{ C}^2 \text{ dyne}^{-1} \text{ cm}^{-2}$$

$$\frac{e^2}{4\pi\varepsilon_0} = 2.56 \times 10^{-38} \text{ kg m}^3 \text{ s}^{-1}$$

$$= 2.56 \times 10^{-29} \text{ g cm}^3 \text{ s}^{-1}$$

$$\frac{4\varepsilon_0 hcm_e}{e^2} = 2.50 \times 10^{-34} \text{ J s cm}^{-2} \text{ Hz}^{-1}$$

$$\frac{e^2}{4c^2\varepsilon_0 m_e} = 8.82 \times 10^{-15} \text{ m}$$

$$\frac{2\pi e^2}{\varepsilon_0 m_e c} = 6.67 \times 10^{-5} \text{ m}^2 \text{ s}^{-1}$$

$$\frac{\varepsilon_0 m_e c}{2\pi e^2} = 1.50 \times 10^4 \text{ m}^{-2} \text{ s}$$

## 4. CONVERSION FACTORS AND IDENTITIES

calorie = 4.186 J

dyne = g cm s$^{-2}$

electron volt = $1.6 \times 10^{-19}$ J

erg = $10^{-7}$ J = $9.8687 \times 10^{-10}$ L atm

erg = dyne cm = g cm$^2$ s$^{-2}$

joule = volt coulomb = N m = kg m$^2$ s$^{-2}$

nanometer = 10 Å = $10^{-3}$ $\mu$m = $\dfrac{c}{v} \times 10^7$

watt = J s$^{-1}$

wavenumber = $\dfrac{10^7}{\lambda \text{ (in nm)}} = \dfrac{v}{c}$

## 5. COMMON ABBREVIATIONS

| | |
|---|---|
| ac | alternating current |
| AA | atomic absorption |
| AAC | continuum source atomic absorption |
| AAL | line source atomic absorption |
| AAS | atomic absorption spectrometry |
| ADP | ammonium dihydrogen phosphate or adenosine diphosphate |
| AE | atomic emission |
| AES | atomic emission spectrometry |
| AF | atomic fluorescence |
| AFC | continuum source atomic fluorescence |
| AFL | line atomic fluorescence |
| AFS | atomic fluorescence spectrometry |
| AMP | adenosine monophosphate |
| ATP | adenosine triphosphate |
| ATR | attenuated total reflectance |
| A.U. | absorbance units |
| BEC | background equivalent concentration |
| BL | bioluminescence |
| CARS | coherent anti-stokes Raman spectroscopy |
| CCD | charge-coupled device |
| CID | charge injection device |
| CL | chemiluminescence |
| CMC | critical micelle concentration |
| CSRS | coherent Stokes Raman spectroscopy |
| CW | continuous wave |
| dB | decibel |
| dc | direct current |
| DB | double beam |
| DCP | direct-coupled plasma |
| DIAL | differential absorption lidar |
| DL | detection limit |
| DLI | dual or direct laser ionization |
| DW | dual-wavelength |

| | |
|---|---|
| EADCI | equivalent anode dark current input |
| EDL | electrodeless discharge lamp |
| ENI | equivalent noise input |
| FAD | flavin dinucleotide |
| FIA | flow injection analysis |
| FIR | far infrared |
| FL | fluorescence |
| FMN | flavin mononucleotide |
| F/n | F-number |
| FTIR | Fourier transform infrared |
| FWHM | full width at half maximum |
| GC | gas chromatography |
| HCL | hollow cathode lamp |
| HPLC | high-performance liquid chromatography |
| ICP | inductively coupled plasma |
| IDT | image dissector tube |
| IR | infrared |
| IRS | inverse Raman spectroscopy |
| ISIT | intensified silicon intensified target |
| KDP | potassium dihydrogen phosphate |
| lidar | light detection and ranging |
| LALLS | low-angle laser light scattering |
| LEI | laser-enhanced ionization |
| LIBS | laser-induced breakdown spectroscopy |
| LTE | local thermodynamic equilibrium |
| LTP | low-temperature phosphorescence |
| MIP | microwave-induced plasma |
| MOS | metal-oxide semiconductor |
| MPI | multiphoton ionization |
| NA | numerical aperture |
| NAD$^+$ | nicotinamide adenine dinucleotide |
| NADP$^+$ | nicotinamide adenine dinucleotide phosphate |
| NEP | noise equivalent power |
| NIR | near infrared |
| NPS | noise power spectrum |
| op amp | operational amplifier |
| OPL | optical pathlength |
| PA | photoacoustic |
| PAS | photoacoustic spectroscopy |
| PD | photodiode |
| PMT | photomultiplier tube |
| PT | vacuum phototube |
| Q | quencher |
| QLS | quasi-elastic light scattering |
| rms | root mean square |
| RIA | radioimmunoassay |
| RIS | resonance ionization spectroscopy |
| RSD | relative standard deviation |
| RTP | room-temperature phosphorescence |

| | |
|---|---|
| S | singlet state |
| SB | single beam |
| *S/B* | signal-to-background ratio |
| SD | standard deviation |
| SERS | surface-enhanced Raman spectroscopy |
| SIT | silicon-intensified target |
| *S/N* | signal-to-noise ratio |
| SRG | stimulated Raman gain |
| SRS | stimulated Raman scattering |
| T | triplet state |
| TEM | transverse electric and magnetic |
| TGL | thermal gradient lamp |
| TGS | triglycine sulfate |
| UV | ultraviolet |
| YAG | yittrium-aluminum-garnet |
| ZAA | Zeeman atomic absorption |

## 6. QUANTUM NUMBERS

### ATOMIC

$j$ = total angular momentum quantum number
$J$ = resultant or total angular momentum or inner quantum number
$l$ = orbital or azimuthal angular momentum quantum number
$L$ = resultant or total orbital angular momentum quantum number
$m_l$ = orbital magnetic quantum number
$m_s$ = spin magnetic quantum number
$M_J$ = resultant total magnetic quantum number
$M_L$ = resultant orbital magnetic quantum number
$M_s$ = resultant spin magnetic quantum number
$n$ = principal quantum number
$s$ = electron spin quantum number
$S$ = resultant or total spin quantum number

### MOLECULAR

$J$ = (rotational) angular momentum quantum number
$S$ = resultant or total spin quantum number
$v$ = vibrational quantum number
$\lambda$ = quantum number for axial component of the orbital angular momentum
$\Lambda$ = total axial electronic angular momentum quantum number
$\Omega$ = total axial angular momentum quantum number

# Symbols

The symbols listed below are used in two or more chapters or several times in at least one chapter. IUPAC recommendations are used where possible. By necessity some symbols have more than one meaning which can be determined from the context. Subscripts and superscripts are used with many symbols to designate specific values. Their common meanings and conventions for their use are presented after the lists of Roman and Greek symbols.

## ROMAN SYMBOLS

| | | |
|---|---|---|
| $a$ | = | absorptivity, L g$^{-1}$ cm$^{-1}$ |
| $a$ | = | damping constant or $a$-parameter for Voigt profile, dimensionless |
| $a_{th}$ | = | thermal diffusion coefficient, cm$^{-1}$ |
| $A$ | = | absorbance, dimensionless |
| $A$ | = | area, cm$^2$ |
| $A(\lambda)$ | = | spectral absorbance, dimensionless |
| $A_t$ | = | integral or total absorption, m or Hz |
| $A_{ji}$ | = | Einstein coefficient for spontaneous emission between states $j$ and $i$, s$^{-1}$ |
| $b$ | = | pathlength of sample cell or length of base of prism, cm |
| $B$ | = | radiance or line radiance, W cm$^{-2}$ sr$^{-1}$ |
| $B$ | = | background signal |
| $B$ or $\overline{B}$ | = | rotational constant B in J or $\overline{B}$ in cm$^{-1}$ |
| $B_{ij}$ and $B_{ji}$ | = | Einstein coefficient for stimulated absorption between states $i$ and $j$ or stimulated emission between states $j$ and $i$, cm$^3$ J$^{-1}$ s$^{-1}$ Hz |
| $B_\lambda$ | = | spectral radiance, W cm$^{-2}$ sr$^{-1}$ nm$^{-1}$ |
| $B(\lambda)$ | = | cumulative radiance, W cm$^{-2}$ sr$^{-1}$ |
| $B_\lambda^b$ | = | spectral radiance for blackbody radiator, W cm$^{-2}$ sr$^{-1}$ nm$^{-1}$ |
| $c$ | = | speed of light, m s$^{-1}$ |
| $c$ | = | concentration (usually of analyte or the absorbing species) |
| $C$ | = | capacitance, F |
| $C$ | = | correlation coefficient, dimensionless |
| $C_F$ | = | coefficient of finesse, dimensionless |
| $C(\tau)$ | = | correlation function |
| $d$ | = | distance (between a source or scatterer and a detector), spacing between grating grooves, or thickness of dielectric spacing in an interference filter, cm |
| $d_m$ | = | modulation depth, dimensionless |
| $d_s$ | = | major dimension of a scatterer, cm |
| $D$ | = | detectivity, W$^{-1}$ |

| | | | | |
|---|---|---|---|---|
| $D$ | = density (photographic), dimensionless | | $h$ | = height (usually of beam), cm |
| $D$ | = diameter of lens, mirror or beam, cm | | $H$ | = radiant exposure, J cm$^{-2}$ |
| | | | $H$ | = Hamiltonian operator |
| $D$ | = diffusion coefficient, cm$^2$ s$^{-1}$ | | $H$ | = slit height, cm |
| $D^*$ | = normalized detectivity, W$^{-1}$ cm Hz$^{1/2}$ | | $|H(f)|$ | = amplitude transfer function, dimensionless |
| $\overline{D}$ | = centrifugal distortion constant, cm$^{-1}$ | | $H(j\omega)$ | = transfer function, dimensionless |
| $D_a$ | = angular dispersion, rad nm$^{-1}$ | | $H_\lambda$ | = spectral radiant exposure, J cm$^{-2}$ nm$^{-1}$ |
| $\overline{D}_e$ and $\overline{D}_o$ | = dissociation energy relative to zero-point, vibrational energy ($\overline{D}_e$) or bottom of potential well ($\overline{D}_o$), cm$^{-1}$ | | $i$ | = current, cathodic current, anodic current, A |
| $D_l$ | = linear dispersion, cm nm$^{-1}$ (mm nm$^{-1}$) | | $I$ | = moment of inertia, kg m$^2$ |
| | | | $I$ | = radiant intensity, W sr$^{-1}$ |
| $D_{th}$ | = thermal diffusivity, cm$^2$ s$^{-1}$ | | $I_\lambda$ | = spectral radiant intensity, W sr$^{-1}$ nm$^{-1}$ |
| $e$ | = instantaneous voltage, V | | $J$ | = radiant emissivity, W cm$^{-3}$ sr$^{-1}$ |
| $e_f$ | = gas expansion factor in an atomizer, dimensionless | | $J_\lambda$ | = spectral radiant emissivity, W cm$^{-3}$ sr$^{-1}$ nm$^{-1}$ |
| $E$ | = irradiance, W cm$^{-2}$ | | $k$ | = free-space propagation number ($k = 2\pi/\lambda$), rad nm$^{-1}$ |
| $E$ | = voltage or readout voltage, V | | | |
| $E$ | = energy, J | | $k$ | = $k(\lambda) = k(\nu)$ = absorption coefficient, cm$^{-1}$ |
| $\mathbf{E}$ | = instantaneous value of electric field, V m$^{-1}$ | | $k$ | = force constant, N cm$^{-1}$ |
| $E_\lambda$ | = spectral irradiance, W cm$^{-2}$ nm$^{-1}$ | | $k$ | = Boltzmann's constant, J K$^{-1}$ |
| $f$ | = fraction of stray radiation, dimensionless | | $k$ | = confidence factor for detection limit calculation, dimensionless |
| $f$ | = frequency (audio range), Hz | | $k$ | = number of dynodes in PMT, dimensionless |
| $f$ | = focal length of lens or mirror or monochromator, cm | | $k$ | = overall nebulization-atomization efficiency, cm$^{-3}$ (concentration units)$^{-1}$ |
| $f_a$ | = fraction of excitation radiation absorbed by the analyte, dimensionless | | | |
| | | | $k$ | = rate constant, s$^{-1}$ |
| $f(c)$ | = calibration function | | $k_D^0$ | = peak absorption coefficient for pure Doppler broadening, cm$^{-1}$ |
| $f_d$ | = discriminator level in photon counting, dimensionless | | | |
| $f_{ij}$ | = absorption oscillator strength between states $i$ and $j$, dimensionless | | $K$ | = equilibrium constant, units variable |
| | | | $K$ | = bandwidth constant, A |
| $f_p$ | = primary absorption (correction) factor, dimensionless | | $K_D$ | = dielectric constant, dimensionless |
| | | | $K_m$ | = relative permeability, dimensionless |
| $f_{pr}$ | = prefilter factor, dimensionless | | $K_q$ | = Stern-Volmer quenching constant, L mol$^{-1}$ |
| $f_s$ | = secondary absorption (correction) factor, dimensionless | | $l$ | = pathlength in an atomizer or length, cm |
| $\Delta f$ | = noise equivalent bandpass, Hz | | | |
| $F$ | = solution transport rate, mL s$^{-1}$ | | $L$ | = pathlength or width in an atomizer, cm |
| $F$ | = finesse, dimensionless | | | |
| $g_i$ | = statistical weight of the $i$th state, dimensionless | | $L$ | = inductance, H |
| | | | $m$ | = slope or calibration sensitivity, signal or absorbance units (concentration units)$^{-1}$ |
| $g(S)$ | = analytical function | | | |
| $G$ | = gain of signal modifier, V A$^{-1}$, or gain of active medium in laser cavity, dimensionless | | $m$ | = current gain of photomultiplier, dimensionless |
| | | | $m$ | = order (for grating, etalon, or interference filter), dimensionless |
| $h$ | = Planck's constant, J s | | $m$ | = mass, g |

| | | |
|---|---|---|
| $m$ | = | magnification, dimensionless |
| $m$ | = | demodulation factor, dimensionless |
| $m_A$ | = | characteristic concentration or atomic absorption sensitivity, mass or concentration units |
| $M$ | = | radiant emittance, W cm$^{-2}$ |
| $M$ | = | molecular or atomic weight, g mol$^{-1}$ |
| $M_\lambda$ | = | spectral radiant emittance, W cm$^{-2}$ nm$^{-1}$ |
| $n$ | = | number density, cm$^{-3}$ |
| $n$ | = | mean number of events counted in time $t$, dimensionless |
| $n$ | = | number of measurements made to determine the mean or standard deviation, dimensionless |
| $n_r$ | = | mean number of readout counts in a time $t$, dimensionless |
| $n(298)$ or $n(T)$ | = | moles of gas before and after combustion, respectively, dimensionless |
| $N$ | = | Avogadro's number, mol$^{-1}$ |
| OPL | = | optical pathlength, m |
| $p$ | = | partial pressure, atm |
| $p$ | = | degree of polarization, dimensionless |
| $P$ | = | polarization, C cm$^{-2}$ |
| $P$ | = | angular momentum, erg s |
| $P(e)$ | = | noise power, V$^2$ Hz$^{-1}$ |
| $P(n)$ | = | probability of counting $n$ random events, dimensionless |
| $P(\theta)$ | = | particle scattering factor, dimensionless |
| $\Delta P$ | = | pressure differential across capillary tube |
| $q$ | = | displacement of internuclear distance from equilibrium value, m |
| $Q$ | = | quality factor for laser cavity or bandpass filter, dimensionless |
| $Q$ | = | flow rate of gases into burner or plasma, L s$^{-1}$ |
| $Q$ | = | radiant energy, J |
| $Q_\lambda$ | = | spectral radiant energy, J nm$^{-1}$ |
| $Q(\lambda)$ | = | transducer sensitivity, A W$^{-1}$ |
| $r$ | = | rate or pulse rate, s$^{-1}$ |
| $r$ | = | anisotropy, dimensionless |
| $r$ | = | radius, or distance, or internuclear distance, cm |
| $r$ | = | fraction of reference readout signal due to stray radiation, dimensionless |
| $R$ | = | transition moment, J$^{1/2}$ cm$^{3/2}$ |
| $R$ | = | resistance, $\Omega$ |
| $R$ | = | gas constant, J mol$^{-1}$ K$^{-1}$ |
| $R$ | = | radius of curvature for a lens or mirror, cm |

| | | |
|---|---|---|
| $R$ | = | resolving power, dimensionless |
| $R^2$ | = | transition probability, J cm$^3$ |
| $R_d$ | = | reciprocal linear dispersion, nm cm$^{-1}$ (Å mm$^{-1}$) |
| $R_\theta$ | = | Rayleigh ratio, cm$^{-1}$ |
| $R(\lambda)$ | = | detector responsivity, A W$^{-1}$ |
| $s$ | = | standard deviation |
| $s$ | = | spectral bandpass, nm |
| $S$ | = | analytical signal |
| $S_{bk}$ | = | blank signal |
| $S_1$ and $S_2$ | = | object and image distance from a lens or mirror, respectively, cm |
| $S_\nu$ or $S_\lambda$ | = | spectral line or band profile, Hz$^{-1}$ or nm$^{-1}$ |
| $t$ | = | time or integration time, s |
| $t_p$ | = | pulse width, s |
| $t$ | = | Student $t$ statistic, dimensionless |
| $t(\lambda)$ | = | slit function, dimensionless |
| $T$ | = | transmittance, dimensionless |
| $T$ | = | transmission factor or optical efficiency, dimensionless |
| $T$ | = | period of waveform, s |
| $T$ | = | temperature, K |
| $T(\lambda)$ | = | spectral transmittance, dimensionless |
| $U$ | = | energy density, J cm$^{-3}$ |
| $U_\lambda$ | = | spectral energy density J cm$^{-3}$ nm$^{-1}$ |
| $U_\lambda^b$ | = | spectral energy density for blackbody radiator, J cm$^{-3}$ nm$^{-1}$ |
| $v$ | = | velocity of light or a particle, cm s$^{-1}$ |
| $\bar{v}_r$ | = | average relative velocity of collision partners, cm s$^{-1}$ |
| $V$ | = | volume or volume element viewed, L or cm$^3$ |
| $V$ | = | potential energy, J |
| $V(\lambda)$ | = | spectral luminous efficiency, dimensionless |
| $w$ | = | width (usually of beam), cm |
| $W$ | = | slit width, cm |
| $W_D$ or $W_D'$ | = | width of dispersion element or beam emerging from a dispersion element, respectively, cm |
| $x$ | = | thickness of layer or distance, cm |
| $y$ | = | distance, m |
| $Y$ | = | luminescence or fluorescence power yield, dimensionless |
| $Y_\lambda$ or $Y_\nu$ | = | luminescence or fluorescence spectral power yield, nm$^{-1}$ or Hz |
| $z$ | = | $z$ statistic, dimensionless |
| $z$ | = | distance, m |
| $Z$ | = | atomic number, dimensionless |
| $Z(T)$ | = | internal electronic partition function, dimensionless |

## GREEK SYMBOLS

| | |
|---|---|
| $\alpha$ | = absorptance or absorption factor, dimensionless |
| $\alpha$ | = probability that a specified value is exceeded, dimensionless |
| $\alpha$ | = angle of incidence with respect to normal for a grating or apex angle of prism, dimensionless |
| $\alpha$ | = secondary emission factor for PMT, dimensionless |
| $\alpha$ | = polarizability, $C^2 J^{-1} m^2$ ($C^2 N^{-1} m$) |
| $\alpha_i$ | = degree of ionization, dimensionless |
| $\alpha(\lambda)$ | = spectral absorptance (usually analyte), dimensionless |
| $\beta$ | = angle of diffraction with respect to normal grating, dimensionless |
| $\beta$ | = molecular constant for Morse potential function, $cm^{-1}$ |
| $\beta_a$ | = free atom fraction, dimensionless |
| $\beta_s$ | = local desolvation efficiency or fraction desolvated in atomizer, dimensionless |
| $\beta_v$ | = local volatilization efficiency or fraction vaporized in atomizer, dimensionless |
| $\gamma$ | = blaze angle for grating |
| $\gamma$ | = analytical sensitivity, (conc. units)$^{-1}$ |
| $\delta$ | = phase difference, rad |
| $\delta$ | = current gain per dynode stage, dimensionless |
| $\delta(a, v_r)$ | = Voigt integral, dimensionless |
| $\varepsilon$ | = permittivity, $C^2 N^{-1} m^{-2}$ |
| $\varepsilon$, $\varepsilon(\lambda)$, or $\varepsilon(v)$ | = molar absorptivity, $L \ mol^{-1} \ cm^{-1}$ |
| $\bar{\varepsilon}$ | = integrated molar absorptivity, $L \ mol^{-1} \ cm^{-1} \ Hz$ |
| $\varepsilon_a$ | = overall atomization efficiency, dimensionless |
| $\varepsilon_n$ | = nebulization efficiency, dimensionless |
| $\varepsilon(\lambda)$ | = spectral emissivity, dimensionless |
| $\eta$ | = refractive index, dimensionless |
| $\eta$ | = viscosity of solution, poise |
| $\eta$ | = collection efficiency of anode in PT or first dynode in PMT, dimensionless |
| $\theta$ | = angle, rad or degrees (°) |
| $\kappa$ | = thermal conductivity, $J \ cm^{-1} \ s^{-1} \ K^{-1}$ |
| $\lambda$ | = wavelength, nm, μm, or Å |
| $\Delta\lambda$ | = resolution or wavelength difference between two wavelengths, nm |
| $\Delta\lambda$ | = half-width of line or band, nm |

| | |
|---|---|
| $\Delta\lambda_{eff}$ | = effective width of an absorption profile, nm |
| $\Delta\lambda_f$ | = free spectral range, nm |
| $\mu$ | = mean |
| $\mu$ | = permeability, $N \ s^2 \ C^{-2}$ |
| $\mu$ | = dipole moment, C m |
| $\mu$ | = reduced mass, g |
| $v$ | = frequency (in optical range), Hz or $s^{-1}$ |
| $v$ | = degrees of freedom, dimensionless |
| $\bar{v}$ | = wavenumber, $cm^{-1}$ |
| $\Delta v$ | = half-width or frequency difference, Hz |
| $\Delta\bar{v}$ | = Raman wavenumber shift, $cm^{-1}$ |
| $\xi$ | = signal flicker factor, dimensionless |
| $\rho$ | = reflectance or fraction of radiant power reflected at in interface, dimensionless |
| $\rho$ | = depolarization ratio, dimensionless |
| $\rho$ | = density, $g \ mL^{-1}$ |
| $\rho(\lambda)$ | = spectral reflectance, dimensionless |
| $\sigma$ | = population standard deviation or rms noise |
| $\sigma$, $\sigma(v)$, or $\sigma(\lambda)$ | = absorption cross section, $cm^2$ |
| $\sigma$ | = collisional cross section, $cm^2$ |
| $\sigma$ | = Stefan–Boltzmann constant, $W \ cm^{-2} \ K^{-4}$ |
| $\tau$ | = time constant, s |
| $\tau$ | = lifetime of decay, s |
| $\tau$ | = turbidity coefficient, $cm^{-1}$ |
| $\tau_c$ | = coherence time, s |
| $Y$ | = throughput factor of wavelength selection device, $cm^2 \ sr$ |
| $\phi$ | = luminescence or fluorescence quantum efficiency, dimensionless |
| $\phi$ | = phase angle, rad |
| $\phi_r$ | = rotational correlation time, s |
| $\phi_v$ or $\phi_\lambda$ | = spectral luminescence or fluorescence quantum efficiency, $Hz^{-1}$ or $nm^{-1}$, respectively |
| $\Phi$ | = radiant power, W |
| $\Phi_n$ | = noise equivalent power, W |
| $\Phi_\lambda$ | = spectral radiant power, $W \ nm^{-1}$ |
| $\chi$ | = background signal flicker factor, dimensionless |
| $\chi_e$ | = anharmonicity constant, dimensionless |
| $\chi_n$ | = $n$th-order electric susceptibility, dimensionless |
| $\psi$ | = wavefunction |
| $\omega$ | = angular frequency ($2\pi v$), $rad \ s^{-1}$ |
| $\omega$ | = laser beam waist, cm |
| $\Omega$ | = solid angle, sr |

## SUBSCRIPTS AND SUPERSCRIPTS

### 1. General

   a.   A prime is used to designate or distinguish a measured or observed quantity from the ideal value of that quantity. A prime is also used in photoluminescence expressions to distinguish the emission wavelength ($\lambda$) or quantities related to the emission wavelength from the excitation wavelength or quantities dependent on the excitation wavelength. In addition, a prime is used to distinguish common units (e.g., $m^3$) of polarizability from SI units. Finally, a prime is used to distinguish quantities related to the triplet state from quantities related to the singlet state.

   b.   A bar is used to denote the average or mean value of a quantity or the integrated value of that quantity (e.g., the quantity integrated over the width of a line or band).

   c.   A subscript $m$ is used, often in conjunction with other subscripts, to denote the maximum value of that quantity (e.g., the value at the maximum of a line or band).

   d.   An asterisk denotes an excited state or quantities related to the volume element excited and viewed in photoluminescence measurements.

   e.   A preceding $\Delta$ indicates a change or difference.

   f.   Subscripts $i$ and $j$ indicates the $i$th or $j$th value of a quantity (e.g., the $i$th measurement or the $i$th state). The subscripts $ij$ and $ji$ denote the value of the quantity for a transition between states $i$ and $j$ or $j$ and $i$, respectively.

   g.   Subscript 0 indicates the initial value (e.g., at time zero). It also is used to designate the central wavelength or frequency of a band passed by a wavelength selector or an electronic filter, the excitation radiation incident on the sample or the radiation incident on a detector, or the value of the quantity in vacuum.

### 2. Radiometric quantities [e.g., radiance ($B$)]

   a.   Subscripts $\lambda$, $\nu$, and $\bar{\nu}$ designate the spectral quantity in $nm^{-1}$, $Hz^{-1}$, or $cm^{-1}$, respectively.

   b.   Subscript $\Delta\lambda$ designates the value of quantity over the wavelength interval $\Delta\lambda$.

   c.   Superscript $b$ indicates the value of that quantity for a blackbody radiator.

   d.   Subscript $v$ denotes the quantity in photometric units (lumens instead of watts).

   e.   Subscript $p$ denotes the quantity in photons per second instead of watts.

### 3. Specific subscripts used by themselves or in conjunction with other subscripts:

| | |
|---|---|
| $a$ | absorption, anodic (current), amplifier (noise), analyte, aberration-limited, anti-Stokes, or adiabatic |
| $aB$ | anodic background (current) |
| $ac$ | alternating current |
| $ad$ | anodic dark (current) |
| $ap$ | anodic photo (current) |
| $ar$ | amplifier-readout (noise) |
| $aS$ | anodic signal (current) |
| $A$ | analyte absorbance or due to absorbed radiation |
| $b$ | bias (voltage), nonanalyte, or blaze |
| $bc$ | background corrected |
| $bE$ | background emission |
| $bF$ | background fluorescence |
| $bk$ | blank |
| $bL$ | background luminescence |
| $bP$ | background phosphorescence |
| $B$ | background |
| $c$ | cathodic (current), concentration, cell wall, corrected, critical, corner, cutoff, correlation, or cycle |
| $cB$ | cathodic background (current); the subscript $c$ is often dropped and understood for this and the following five subscripts |
| $cd$ | cathodic dark (current) |
| $cp$ | cathodic photo (current) |
| $cr$ | cathodic reference (current) |
| $cs$ | cathodic signal (current) |
| $ct$ | total cathodic (current) |
| $ci$ | collisional ionization |
| $C$ | continuum |
| $CL$ | chemiluminescence |
| $d$ | dark, dissociation, diffraction-limited, delay, or diabatic |
| $dc$ | direct current |
| $dt$ | total dark (current) |
| $D$ | Doppler or dispersion element |
| $e$ | electronic, extraordinary, equilibrium, electron, or excitation |
| $ec$ | external conversion |
| $em$ | emission |
| $ex$ | excess (noise) or excitation |
| $exp$ | experimental |
| $E$ | voltage signal or analyte emission |
| $f$ | final, flicker (noise), fuel, filter, or feedback |
| $fs$ | full scale |
| $F$ | analyte fluorescence |
| $FC$ | analyte fluorescence with continuum source excitation |

| | |
|---|---|
| *FL* | analyte fluorescence with line source excitation |
| *g* | geometric |
| *i* | current, image, or imaging |
| *ic* | internal conversion |
| in | input or induced |
| ion | ion or ionization |
| *isc* | intersystem crossing |
| *l* | lifetime or laser |
| *lr* | logarithmic reference |
| *ls* | logarithmic sample |
| *L* | analyte luminescence, load, Lorentzian or line source |
| *m* | modulation |
| min | minimum |
| *M* | in mol $L^{-1}$ |
| *n* | counts, nuclear, or noise |
| *nr* | nonradiative |
| *N* | natural |
| *o* | output, oxidant, observation, or ordinary |
| op | optical |
| 0*t* | 0% *T* |
| *p* | photo (current), photoionization or projected (area or diameter) |
| *pi* | photoionization |
| p-p | peak-to-peak |
| *P* | analyte phosphorescence |
| *PA* | photoacoustic |
| *q* | quantization, quantum (noise), or quenching |
| *Q* | quencher |
| *r* | readout, reference species, due source radiation passed by reference solution, |
| | rotational, or radiative |
| *rt* | total reference |
| *R* | Raman |
| *s* | shot (noise), due to source radiation passed by sample or standard, due to standard or sample, spin, support gas, slit-width limited, Stokes', signal, of sound, of source, or scatterer |
| *sc* | scattering |
| *sp* | spontaneous |
| *st* | total sample |
| *S* | analytical signal or singlet (state) |
| *SR* | stray radiation |
| *t* | total, at time *t*, or threshold |
| *tCL* | total chemiluminescence |
| *tE* | total emission |
| *tF* | total fluorescence |
| th | thermal or theoretical |
| *tL* | total luminescence |
| *tP* | total phosphorescence |
| *T* | temperature, transmittance, translational, or triplet |
| *v* | vibrational |
| *V* | Voigt |
| *w* | window |
| *x* | unknown value in sample, collisional partner, or along *x* axis |
| *x* + *s* | value after standard addition |
| $\theta$ | at angle theta |
| $\parallel$ | parallel polarization component |
| $\perp$ | perpendicular polarization component |
| 1/2 | half (life) |

# Index

*Note:* Italicized page numbers indicate
  definitions.